CHEMISTRY
PRINCIPLES AND REACTIONS

Fourth Edition

A Core Text

CHEMISTRY
PRINCIPLES AND REACTIONS

Fourth Edition

A Core Text

WILLIAM L. MASTERTON
University of Connecticut

CECILE N. HURLEY
University of Connecticut

HARCOURT COLLEGE PUBLISHERS

FORT WORTH PHILADELPHIA SAN DIEGO NEW YORK ORLANDO AUSTIN
SAN ANTONIO TORONTO MONTREAL LONDON SYDNEY TOKYO

About the cover: The new Guggenheim Museum (designed by Frank O. Gehry) recently opened in Bilbao, Spain. Half-millimeter-thick "fish-scale" panels of titanium (a fourth-period transition metal) cover most of the building and are guaranteed to last one hundred years. The museum is a fine example of how chemical principles can contribute to a monumental work of art. *(Photo by Erika Barahona Ede © FMGB Guggenheim Bilbao Museoa)*

Publisher: John Vondeling
Publisher: Emily Barrosse
Marketing Strategist: Pauline Mula
Developmental Editor: Ed Dodd
Project Editor: Bonnie Boehme
Production Manager: Charlene Catlett Squibb
Art Director: Caroline McGowan
Text Designer: Ruth A. Hoover
Frontispiece Credit: Charles D. Winters

Chemistry: Principles and Reactions, Fourth Edition
ISBN: 0-03-026036-1
Library of Congress Catalog Card Number: 99-069839

Address for domestic orders:
Harcourt College Publishers, 6277 Sea Harbor Drive, Orlando, FL 32887-6777
1-800-782-4479
e-mail   collegesales@harcourt.com

Address for international orders:
International Customer Service, Harcourt, Inc.
6277 Sea Harbor Drive, Orlando, FL 32887-6777
(407) 345-3800
Fax (407) 345-4060
e-mail   hbintl@harcourt.com

Address for editorial correspondence:
Harcourt College Publishers, Public Ledger Building, Suite 1250, 150 S. Independence Mall West, Philadelphia, PA 19106-3412

Web Site Address
http://www.harcourtcollege.com

Printed in the United States of America

0123456789   048   10 9 8 7 6 5 4 3 2

To

Loris and **Jim**

For increasing reaction rates and lowering boiling points

William L. Masterton received his Ph.D. in physical chemistry from the University of Illinois in 1953. Two years later, he arrived at the University of Connecticut, where he taught general chemistry and a graduate course in chemical thermodynamics. He received numerous teaching awards; the one of which he is most proud came from the Student Senate at UConn. Bill wrote, with co-author Emil Slowinski, the all-time best-selling general chemistry textbook, *Chemical Principles,* which sold well over one and a half million copies. Bill has also written a definitive account of the Lizzie Borden case entitled *Lizzie Didn't Do It,* published almost simultaneously with this book. Bill's field of research, solution thermodynamics, prepared him well for making maple syrup each March at the family farmhouse in New Hampshire.

Cecile Nespral Hurley received her M.S. at the University of California, Los Angeles. Since 1979, she has served as Lecturer and Coordinator of Freshman Chemistry at the University of Connecticut, where she directed a groundbreaking National Science Foundation–supported project on cooperative learning in general chemistry. She is one of a prestigious group of University Teaching Fellows, who are selected by their fellow faculty members as models of teaching excellence and dedication. In addition, she coordinates the High-School Cooperative Program in Chemistry, through which superior Connecticut high-school students take the University's general chemistry course at their schools. In her spare time, she roots for the UConn Women's Basketball Huskies and roots out weeds from her country garden, which she likes to imagine rivals Monet's at Giverny.

CONTENTS OVERVIEW

1   Matter and Measurement   1

2   Atoms, Molecules, and Ions   28

3   Mass Relations in Chemistry; Stoichiometry   55

4   Reactions in Aqueous Solution   83

5   Gases   114

6   Electronic Structure and the Periodic Table   143

7   Covalent Bonding   179

8   Thermochemistry   215

9   Liquids and Solids   246

10   Solutions   280

11   Rate of Reaction   309

12   Gaseous Chemical Equilibrium   350

13   Acids and Bases   382

14   Equilibria in Acid-Base Solutions   412

15   Complex Ions   437

16   Precipitation Equilibria   460

17   Spontaneity of Reaction   482

18   Electrochemistry   513

19   Nuclear Chemistry   547

20   Chemistry of the Metals   573

21   Chemistry of the Nonmetals   595

22   Organic Chemistry   625

Appendix 1:   Units, Constants, and Reference Data   A.1

Appendix 2:   Properties of the Elements   A.7

Appendix 3:   Exponents and Logarithms   A.9

Appendix 4:   Nomenclature   A.14

Appendix 5:   Molecular Orbitals   A.17

Appendix 6:   Answers to Even-Numbered and Challenge Questions & Problems   A.24

Index/Glossary   I.1

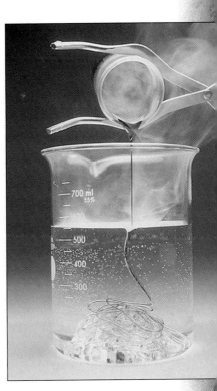

PREFACE

*C*hemistry: Principles and Reactions differs in mass and volume from other general chemistry texts. The typical text today runs anywhere from 1000 to 1300 pages and weighs 4 to 6 pounds. This book is several hundred pages shorter and at least a couple of pounds lighter.

An obvious advantage of a shorter text is the lower cost to the student (this book sells for much less than the $90 to $110 for the average general chemistry text). A further advantage is that this book can be covered in its entirety in a year course. With most of the general chemistry texts on the market, the instructor has to decide which chapters to omit. Frequently some of the most interesting topics are skipped, such as complex ions (Chapter 15 in this book) and nuclear chemistry (Chapter 19). Another casualty due to time restraints is often descriptive inorganic chemistry (Chapters 20 and 21 in this book).

You may be curious as to how we achieved this reduction in size. It was *not* done by lowering the level. Our criterion for including material was its importance and relevance to the student, not its difficulty. Beyond that, we have developed chemical principles slowly and carefully, devoting as much space to a topic as experience has shown to be necessary for student understanding.

How then did we arrive at a text at least 300 pages shorter than most general chemistry texts? For one thing, we have eliminated repetition and duplication wherever possible. For example, the book contains—

■ one and only one method of balancing redox equations, the half-equation approach introduced in Chapter 4.
■ one and only one way of working gas law problems, using the ideal gas law in all cases (Chapter 5).
■ one and only one way of calculating ΔH (Chapter 8), using enthalpies of formation.
■ one and only one equilibrium constant for gas phase reactions (Chapter 12), the thermodynamic constant K, often referred to as K_p. This simplifies not only the treatment of gaseous equilibrium but also the discussion of reaction spontaneity (Chapter 17) and electrochemistry (Chapter 18).

Certain topics ordinarily covered in texts of 1,000 or more pages have been deleted, abbreviated, or relegated to an appendix. Items in this category are—

■ *biochemistry,* traditionally covered in the last chapter of general chemistry texts. Interesting as this material is, it requires a background in organic chemistry that first-year college students do not have. Our last chapter (Chapter 22) is devoted to the concepts of organic chemistry, including isomerism.
■ *molecular orbital theory* (Appendix 5). Our experience has been that, important as this approach to chemical bonding is, it doesn't go over well with most general chemistry students.
■ *nomenclature of complex ions and organic compounds* (Appendix 4). We believe this material is of questionable value in a beginning course.

■ *qualitative analysis,* summarized in a few pages in Chapter 16. An extended discussion of the qual scheme and the chemistry behind it belongs in a lab manual, not a textbook.

Through three editions, this text has earned a reputation as being student-oriented. We have tried to continue and expand that tradition this time around. In evaluating reviewers' comments, we always ask the question, "Will this (recommended) change enhance the student's understanding or appreciation of chemistry?" All the changes we made in this fourth edition had that objective in mind.

Along these lines, it may be helpful to point out several unique features of this text. (These are described in more detail under "To the Student," page xvii)

■ ***CD-ROM References*** appear in almost every chapter. These marginal callouts direct students to specific topic screens on the *Saunders Interactive General Chemistry CD-ROM* (Version 2.5). Additionally, for both the students' and professors' reference, a list of videos and animations from the CD-ROM that are appropriate to given sections of the text appears after the Table of Contents.

■ In-text ***Examples*** start off with a **Strategy** section, which outlines the reasoning to be followed in the **Solution.** This helps to distinguish the "why" from the "how" of problem solving. About half of the examples end with a **Reality Check,** designed to convince the student that solving a problem begins the learning process rather than ending it. It encourages students to make sure that the answer is reasonable and/or relates directly to a chemical principle they have learned.

■ ***Chapter Highlights,*** at the conclusion of each chapter, list the **Key Concepts, Key Equations,** and **Key Terms** covered. Included among the key terms are items, marked with a ■, introduced in earlier chapters. This section ends with a **Summary Problem,** which serves to tie together all of the concepts covered in a chapter. This helps the student to see the forest as well as the individual trees.

■ End-of-chapter ***Questions & Problems*** are divided into four categories. Most are classified under special headings; problems of this type are arranged in matched pairs, only one of which is answered in Appendix 6. **Unclassified** questions and problems typically illustrate more than one concept. Then there are several **Conceptual Questions** of a qualitative nature designed to test how well the student understands chemical concepts. Finally, there are a few **Challenge Problems,** which we hope will pique the interest of student and instructor alike. As an aid for students, selected, fully worked-out solutions for selected problems from each chapter, identified by a **WEB**, are posted at:

http://www.harcourtcollege.com/chem/general/masterton4/student/

The function of a chemistry teacher and hence of a textbook goes well beyond explaining chemical principles. Students must be convinced that chemistry is so relevant to their lives that it is worth devoting the time and effort required to master these principles. Real-life applications are emphasized throughout the exercises and problems in this text. They also appear in a special feature, **Chemistry: Beyond the Classroom,** found near the end of every chapter. New to this edition are discussions of amines and alkaloids (Chapter 4), automobile airbags (Chapter 5), polymers (Chapter 9), organic acids (Chapter 13), and cholesterol (Chapter 22).

Throughout the text we have presented short biographies of some of the pioneers in chemistry, ranging from Antoine Lavoisier (1748–1794) to Glenn Seaborg

(1912–1999). These sketches are referred to as **Chemistry: The Human Side;** they emphasize the personalities as well as the accomplishments of these individuals.

ANCILLARY MATERIAL

A large number of auxiliary materials have been developed for use with this text. These include:

For the Instructor

Instructor's Manual by William L. Masterton. Included are lecture outlines and lists of demonstrations for each chapter. Worked-out solutions are provided for all of the end-of-chapter problems that do not have answers in the appendix.

   Printed Test Bank by David Treichel (Nebraska Wesleyan University). Features over 1,000 multiple-choice, five-part questions.

   ExaMaster™ Computerized Test Bank is the software version of the printed Test Bank. Instructors can create thousands of questions in a multiple-choice format. A command reformats the multiple-choice question into a short-answer question. Adding or modifying existing problems, as well as incorporating graphics, can be done. ExaMaster™ also has gradebook capabilities for recording and graphing students' grades.

   Overhead Transparencies set includes 142 full-color acetates with sizable labels for viewing in large lecture halls. The illustrations and tables chosen are those most often used in the classroom, and are marked with the icon **OHT** for easy identification.

   Cooperative Learning Workbook by Cecile N. Hurley. A collection of worksheets (about three per chapter) that students work on in groups. Designed to stimulate group activity and discussion, the questions provided on each worksheet are equally conceptually oriented and quantitatively oriented. The booklet includes instructions for use, how to guide student discussion, and supporting data on the success of cooperative learning at University of Connecticut.

   Chemical Principles in the Laboratory, **Seventh Edition,** by Emil Slowinski and Wayne Wolsey (both of Macalester College) and William Masterton provides detailed directions and advanced study assignments. This manual contains 43 experiments that have been selected with regard to cost and safety. All the experiments have been thoroughly class-tested. Alternatively, the ***Chemical Principles in the Laboratory with Qualitative Analysis,*** **Sixth Edition, Alternate Version,** is available, with eight experiments covering qualitative analysis. An **Instructor's Manual** is available for each version of *Chemical Principles in the Laboratory,* and each Instructor's Manual provides lists of equipment and chemicals needed for each experiment.

For the Student

Study Guide/Workbook by Cecile N. Hurley. Worked examples and problem-solving techniques help the student understand the principles of general chemistry. Each chapter is outlined for the student with fill-in-the-blanks, and exercises and self-tests allow the students to gauge their mastery of the chapter.

Student Solutions Manual by Cassandra T. Eagle and David G. Farrar (both of Appalachian State University). Complete solutions to all the problems answered in the text, including the Challenge Problems. References to the main text's sections and tables are provided as a guide for problem-solving techniques employed by the authors. Selected solutions from each chapter, identified by a **WEB**, are posted at:

http://www.harcourtcollege.com/chem/general/masterton4/student/

Lecture Outline by Ronald O. Ragsdale (University of Utah). Organized to follow class lectures to free students from extensive note taking during lectures.

Chemistry Internet Resource Guide by Susan M. Young provides students with a tool to help understand the ways in which the Internet can assist their education. In addition to general information on chemistry Web sites, the Guide will help them to understand the learning aids offered on ChemSource, the Harcourt general chemistry Web site.

Multimedia Ancillary Materials

Saunders Interactive General Chemistry CD-ROM **(Version 2.5) with ActivChemistry,** by John Kotz and William Vining and produced by Archipelago Productions. Considered the best general chemistry CD-ROM available by allowing the students to interact with the information presented. Divided into chapters, the CD-ROM allows one to watch a reaction in progress, change a variable in an experiment and experience the result, follow stepwise solutions to problems, explore the periodic table, and listen to tips and suggestions on problem solving and understanding concepts. The CD-ROM includes original graphics, over 100 video clips of chemical experiments, which are enhanced by sound and narration, and several hundred molecular models and animations. With ActivChemistry, students can also perform simulations of laboratory experiments.

The CD-ROM also includes molecular modeling software from Oxford Molecular Group that can be used to view hundreds of models, rotate the models for a fuller understanding of their structures, and measure bond lengths and bond angles.

The CD-ROM has been used by thousands of students worldwide since its introduction in 1996. It can be purchased as a package with the textbook or as a stand-alone product, and will run on either Windows™ or Macintosh® platforms.

The **2001 Chemistry Instructor's Resource CD-ROM** is a dynamic lecture tool containing imagery from all of the 2001 Harcourt chemistry titles. It can be used in conjunction with commercial presentation software such as PowerPoint™, Persuasion™, and Podium™. The CD-ROM is for both Macintosh® and Windows™ platforms.

ChemSource World Wide Web Site at

http://www.harcourtcollege.com/chem/general/masterton4/

PowerPoint™ files, selected movie clips and animations, quizzing and testing, real-world applications, molecular models, teaching tips, and more features give the instructor many tools to enhance the lecture presentation and to adjust the curriculum.

Lecture Outlines created with **PowerPoint**™ are also available for the professor. Relating to the text on a per chapter basis, Powerpoint™ users can edit the content with their own material or import material from our 2001 Chemistry Instructor's Resource CD-ROM.

CalTech Chemistry Animation Project (CAP) is a set of six video units of unmatched quality and clarity that cover the chemical topics of Atomic Orbitals, Valence Shell Electron Pair Repulsion Theory, Crystals and Unit Cells, Molecular Orbitals in Diatomic Molecules, Periodic Trends, and Hybridization and Resonance. Available to qualified adopters.

Periodic Table Videodisc: Reactions of the Elements by Alton Banks, North Carolina State University, features still and live footage of the elements, their uses, and their reactions with air, water, acids, and bases. Available to qualified adopters. Also available in CD-ROM format through JCE:Software, Chemistry Department, University of Wisconsin, Madison, WI 53706, (800) 991-5534.

Shakhashiri Demonstration Videotapes feature well-known instructor Bassam Shakhashiri of the University of Wisconsin-Madison performing 50 three- to five-minute chemical demonstrations. An accompanying manual describes each demonstration and includes discussion questions.

Harcourt College Publishers may provide complimentary instructional aids and supplements or supplemental packages to those adopters qualified under our adoption policy. Please contact your sales representative for more information. If, as an adopter or potential user, you receive supplements you do not need, please return them to your sales representative or send them to:

Attn: Returns Department
Troy Warehouse
465 South Lincoln Drive
Troy, MO 63379

Through the services of the Harcourt Custom Publishing Group, portions of *Chemistry: Principles and Reactions,* Fourth Edition, can be packaged according to individual needs. Instructors who wish to augment *Chemistry: Principles and Reactions,* Fourth Edition, with their own material, make selected chapters available in courses with a different focus than that of the textbook as a whole, or package *Chemistry: Principles and Reactions,* Fourth Edition, with select chapters from other Harcourt College Publishers textbooks should contact their local sales representative.

ACKNOWLEDGMENTS

We are indebted to a great many people who have used this book, instructors and students alike, for suggestions as to how we might improve the fourth edition. Reviewers who have helped us include:

Steven Albrecht, *Oregon State University*
Charles W. Armbruster, *University of Missouri, St. Louis*
William H. Brown, *Beloit College*
John DeKorte, *Glendale Community College (Arizona)*
Harry A. Frank, *University of Connecticut, Storrs*
Dean F. Keeley, *The University of Southwestern Louisiana*
Don Kleinfelter, *University of Tennessee*
Tammy J. Melton, *Middle Tennessee State University*
Don Mencer, *Pennsylvania State University, Hazelton*

Kathy Nabona, *Austin Community College*
S. F. Pavkovic, *Loyola University of Chicago*
Richard Prigodich, *Trinity College*
Ronald Ragsdale, *University of Utah*
Thomas G. Richmond, *University of Utah*
Larry A. Scheich, *Saint Norbert College*
David Treichel, *Nebraska Wesleyan University*
Deborah Warnaar, *James Madison University*

We are particularly grateful to two members of the chemistry staff at the University of Connecticut. *Harry Frank* convinced us to revise the treatment of the solubility rules (Chapter 4) and kinetic theory (Chapter 5). *Bob Bohn* helped clarify the treatment of electronic structure (Chapter 6) and chemical bonding (Chapter 7). A former co-author and UConn chemistry professor, *Emil Slowinski* has given us valuable advice on this and previous editions. Although he retired from textbook writing more than ten years ago, Slow's philosophy and humor still permeate this book.

Reviewers of the first three editions include:

Linda Atwood, *California Polytechnic State University, San Luis Obispo*
Peter Baine, *California State University, Long Beach*
John E. Bauman, *University of Missouri, Columbia*
Jesse Binford, *University of South Florida*
Janice Bradley, *Lake City Community College*
James Carroll, *University of Nebraska, Omaha*
James D. Cherry, *Enrico Fermi High School*
Coran Cluff, *Brigham Young University*
Robert Conley, *New Jersey Institute of Technology*
Marsha C. Davies, *Creighton University*
M. Davis, *University of Texas, El Paso*
Cassandra Eagle, *Appalachian State University*
Gordon Eggleton, *Southeastern Oklahoma State University*
William B. Euler, *University of Rhode Island*
Elizabeth S. Friedman, *Los Angeles Valley College*
Steven D. Gammon, *University of Idaho*
Frederick A. Grimm, *The University of Tennessee, Knoxville*
Wyman K. Grindstaff, *Southwest Missouri State University*
Anthony W. Harmon, *The University of Tennessee at Martin*
Sammye Sue Harrill, *Mount St. Mary Academy*
Daniel T. Haworth, *Marquette University*
Douglas W. Hensley, *Louisiana Tech University*
David Hilderbrand, *South Dakota State University*
Grant N. Holder, *Appalachian State University*
Barbara Hopkins, *Xavier University*
Pushkar Kaul, *Boston College*
David L. Keeling, *California Polytechnic State University*
Paul B. Kelter, *The University of Wisconsin, Oshkosh*
Michael Kenney, *Marquette University*
Leslie Kinsland, *University of Southwestern Louisiana*
Donald Kleinfelter, *The University of Tennessee*

James Long, *University of Oregon*
Carol Martinez, *Albuquerque Technical, Vocational Institute*
James McClure, *Southern Illinois University at Edwardsville*
Gregory Neyhart, *North Carolina State University*
Deborah M. Nycz, *Broward Community College*
William E. Ohnesorge, *Lehigh University*
George Patterson, *Suffolk University*
Virgil L. Payne, *Florida Atlantic University*
Paul S. Poskozim, *Northeastern Illinois University*
Lawrence Potts, *Gustavus Adolphus College*
Ronald Ragsdale, *University of Utah*
Henry D. Schreiber, *Virginia Military Institute*
Al Shina, *San Diego City College*
Richard L. Snow, *Brigham Young University*
Steven Socol, *Southern Utah University*
Thomas W. Sottery, *University of Southern Maine*
Robert S. Sprague, *Lehigh University*
Joseph Stenson, *Delaware Valley College*
Donald Titus, *Temple University*
Richard Treptow, *Chicago State University*
William Van Doorne, *Calvin College*
Paul Walter, *Skidmore College*
Charles A. Wilkie, *Marquette University*
Robert A. Wilkins, *Andrews University*
Stanley M. Williamson, *University of California, Santa Cruz*
Sidney H. Young, *University of South Alabama*

John Vondeling, of Harcourt College Publishers, has contributed a great deal to this and previous editions, first as an editor and more recently as "publisher" (whatever that means). Our developmental editor, *Ed Dodd,* is new to this edition; he and *Mary Castellion* have devoted a great deal of time and effort to it. *Bonnie Boehme,* our project editor, has been a joy to work with. She combines unusual competence with a sparkling personality that has cheered us on numerous occasions.

William L. Masterton
Cecile N. Hurley
University of Connecticut
Storrs
March, 2000

TO THE STUDENT

Over the next several months, you will probably receive a lot of advice from your instructor, teaching assistant, and fellow students about how to study chemistry. We hesitate to add our advice; experience as teachers and parents has taught us that students tend to do surprisingly well without it. We would, however, like to acquaint you with some of the learning tools in this text. They are described in the pages that follow.

EXAMPLES

In a typical chapter, you will find ten or more examples, each designed to illustrate a particular principle. These have answers, screened in color. More important, they contain a strategy statement, which describes the reasoning behind the solution. You will find it helpful to get into the habit of working all problems this way. First, spend a few moments deciding how the problem should be solved. Then, and only then, set up the arithmetic to solve it.

Many of the examples end with a **Reality Check,** which encourages you to check whether the answer makes sense. We hope you will get into the habit of doing this when you work problems on your own on quizzes and examinations.

CD-ROM REFERENCES

These marginal callouts (identified by a ⊙ icon) refer to specific topic screens in the *Saunders Interactive General Chemistry CD-ROM* (Version 2.5) (available for purchase with this text). In addition, a list of animations and videos from the CD-ROM that are appropriate to given sections of the text appears after the Table of Contents.

MARGINAL NOTES

Sprinkled throughout the text are a number of short notes that have been placed in the margin. Many of these are of the "now, hear this" variety; a few bring you up to date on current research in chemistry, in progress when the text was written. Some, probably fewer than we think, are humorous.

CHAPTER HIGHLIGHTS

At the end of each chapter, you will find a brief review of the material covered in that chapter. The "Chapter Highlights" include—

- the **Key Concepts** introduced in the chapter. These are indexed to the corresponding examples and end-of-chapter problems. If you have trouble working a particular problem, it may help to go back and re-read the example that covers the same concept.

■ the **Key Equations** and **Key Terms** in the chapter. If a particular term is unfamiliar to you, refer to the index at the back of the book. You will find the term defined in the glossary that is incorporated into the index.

■ a **Summary Problem,** covering all or nearly all of the key concepts in the chapter. You can test your understanding of the chapter by working this problem; you may wish to do this as part of your preparation for examinations. A major advantage of a summary problem is that it ties together many different ideas, showing how they correlate with one another.

QUESTIONS & PROBLEMS

At the end of each chapter is a set of **Questions & Problems.** Most of these are classified, that is, grouped by type under a particular heading. The classified problems are in matched pairs. The second member of each pair illustrates the same principle as the first; it is numbered in color and answered in Appendix 6. Selected solutions for problems from each chapter, identified by a **WEB** icon, are posted at:

http://www.harcourtcollege.com/chem/general/masterton4/student/

Your instructor may assign unanswered problems as homework. After these problems have been discussed, you should work the corresponding answered problems to make sure you know what's going on.

Each chapter also contains a smaller number of **Unclassified, Conceptual,** and **Challenge Problems.** All of the challenge problems are answered in Appendix 6.

MATHEMATICS REVIEW

Appendix 3 touches on just about all the mathematical techniques you will use in general chemistry. Exponential notation and logarithms (natural and base 10) are emphasized.

CONTENTS

1 MATTER AND MEASUREMENTS   1

1.1 Types of Matter   3
1.2 Measurements   7
1.3 Properties of Substances   15
   Chemistry: The Human Side: *Antoine Lavoisier*   16
   Chemistry Beyond the Classroom: *Carbon Allotropes*   20
Chapter Highlights   22
Summary Problem   22
Questions & Problems   23

2 ATOMS, MOLECULES, AND IONS   28

2.1 Atoms and the Atomic Theory   29
   Chemistry: The Human Side: *John Dalton*   30
2.2 Components of the Atom   31
2.3 Introduction to the Periodic Table   35
2.4 Molecules and Ions   38
2.5 Formulas of Ionic Compounds   42
2.6 Names of Compounds   44
   Chemistry Beyond the Classroom: *Ethyl Alcohol and the Law*   48
Chapter Highlights   49
Summary Problem   50
Questions & Problems   51

3 MASS RELATIONS IN CHEMISTRY; STOICHIOMETRY   55

3.1 Atomic Masses   56
3.2 The Mole   60
3.3 Mass Relations in Chemical Formulas   62
3.4 Mass Relations in Reactions   67
   Chemistry Beyond the Classroom: *Hydrates*   74
Chapter Highlights   75
Summary Problem   76
Questions & Problems   76

4 REACTIONS IN AQUEOUS SOLUTION   83

4.1 Solute Concentrations; Molarity   84
4.2 Precipitation Reactions   88

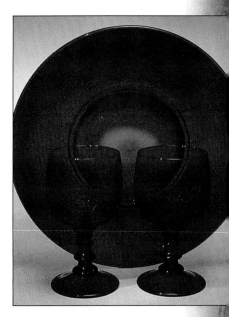

4.3   Acid-Base Reactions   91
Chemistry: The Human Side: Svante August Arrhenius   97
4.4   Oxidation-Reduction Reactions   98
Chemistry: Beyond the Classroom: Amines   105
Chapter Highlights   106
Summary Problem   107
Questions & Problems   107

5 GASES   114

5.1   Measurements on Gases   115
5.2   The Ideal Gas Law   118
5.3   Gas Law Calculations   120
5.4   Stoichiometry of Gaseous Reactions   123
Chemistry: The Human Side: Amadeo Avogadro   125
5.5   Gas Mixtures: Partial Pressures and Mole Fractions   125
5.6   Kinetic Theory of Gases   128
5.7   Real Gases   133
Chemistry: Beyond the Classroom: Airbags   135
Chapter Highlights   136
Summary Problem   137
Questions & Problems   137

6 ELECTRONIC STRUCTURE AND THE PERIODIC TABLE   143

6.1   Light, Photon Energies, and Atomic Spectra   144
6.2   The Hydrogen Atom   148
6.3   Quantum Numbers, Energy Levels, and Orbitals   151
6.4   Electron Configurations in Atoms   156
Chemistry: The Human Side: Glenn Seaborg   160
6.5   Orbital Diagrams of Atoms   162
6.6   Electron Arrangements in Monatomic Ions   164
6.7   Periodic Trends in the Properties of Atoms   166
Chemistry: Beyond the Classroom: The Aurora   171
Chapter Highlights   173
Summary Problem   174
Questions & Problems   174

7 COVALENT BONDING   179

7.1   Lewis Structures; The Octet Rule   181
Chemistry: The Human Side: G. N. Lewis   191
7.2   Molecular Geometry   192
7.3   Polarity of Molecules   200
7.4   Atomic Orbitals; Hybridization   203
Chemistry: Beyond the Classroom: The Noble Gases   208

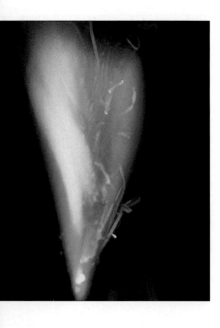

Chapter Highlights   209
Summary Problem   210
Questions & Problems   211

8 THERMOCHEMISTRY   215

8.1   Principles of Heat Flow   216
8.2   Measurement of Heat Flow; Calorimetry   219
8.3   Enthalpy   222
8.4   Thermochemical Equations   223
8.5   Enthalpies of Formation   228
8.6   Bond Enthalpy   232
8.7   The First Law of Thermodynamics   235
    Chemistry: Beyond the Classroom: *Energy Balance in the Human Body*   238
Chapter Highlights   240
Summary Problem   240
Questions & Problems   241

9 LIQUIDS AND SOLIDS   246

9.1   Liquid-Vapor Equilibrium   248
9.2   Phase Diagrams   254
9.3   Molecular Substances; Intermolecular Forces   256
9.4   Network Covalent, Ionic, and Metallic Solids   262
9.5   Crystal Structures   267
    Chemistry: The Human Side: *Dorothy Hodgkin*   271
    Chemistry: Beyond the Classroom: *Synthetic Organic Polymers*   272
Chapter Highlights   273
Summary Problem   274
Questions & Problems   275

10 SOLUTIONS   280

10.1   Concentration Units   281
10.2   Principles of Solubility   286
10.3   Colligative Properties   291
    Chemistry: Beyond the Classroom: *Maple Syrup*   301
Chapter Highlights   302
Summary Problem   302
Questions & Problems   303

11 RATE OF REACTION   309

11.1   Meaning of Reaction Rate   310
11.2   Reaction Rate and Concentration   313
11.3   Reactant Concentration and Time   317
11.4   Models for Reaction Rate   323

Chemistry: The Human Side: *Henry Eyring*   327
11.5 Reaction Rate and Temperature   328
11.6 Catalysis   331
11.7 Reaction Mechanisms   333
Chemistry: Beyond the Classroom: *The Ozone Story*   338
Chapter Highlights   339
Summary Problem   340
Questions & Problems   341

12 GASEOUS CHEMICAL EQUILIBRIUM   350

12.1 The $N_2O_4-NO_2$ Equilibrium System   352
12.2 The Equilibrium Constant Expression   355
12.3 Determination of K 359
12.4 Applications of the Equilibrium Constant   361
12.5 Effect of Changes in Conditions on an Equilibrium System   366
Chemistry: Beyond the Classroom: *An Industrial Application of Gaseous Equilibrium*   372
Chapter Highlights   374
Summary Problem   374
Questions & Problems   375

13 ACIDS AND BASES   382

13.1 Brønsted-Lowry Acid-Base Model   383
13.2 The Ion Product of Water   384
13.3 pH and pOH   385
13.4 Weak Acids and Their Equilibrium Constants   389
13.5 Weak Bases and Their Equilibrium Constants   398
13.6 Acid-Base Properties of Salt Solutions   402
Chemistry: Beyond the Classroom: *Organic Acids*   404
Chapter Highlights   406
Summary Problem   407
Questions & Problems   407

14 EQUILIBRIA IN ACID-BASE SOLUTIONS   412

14.1 Buffers   413
14.2 Acid-Base Indicators   420
14.3 Acid-Base Titrations   422
Chemistry: Beyond the Classroom: *Acid Rain*   429
Chapter Highlights   431
Summary Problem   431
Questions & Problems   432

15 COMPLEX IONS   437

15.1 Composition of Complex Ions   438
15.2 Geometry of Complex Ions   442

*Chemistry: **The Human Side**: Alfred Werner*   446
15.3   Electronic Structure of Complex Ions   447
15.4   Formation Constants of Complex Ions   452
*Chemistry: **Beyond the Classroom**: Chelates: Natural and Synthetic*   454
Chapter Highlights   456
Summary Problem   456
Questions & Problems   457

16 Precipitation Equilibria   460

16.1   Precipitate Formation; Solubility Product Constant (K_{sp})   461
16.2   Dissolving Precipitates   468
16.3   Qualitative Analysis   472
*Chemistry: **Beyond the Classroom**: Fluoridation of Drinking Water*   475
Chapter Highlights   476
Summary Problem   476
Questions & Problems   477

17 Spontaneity of Reaction   482

17.1   Spontaneous Processes   483
17.2   Entropy, S   486
17.3   Free Energy, G   490
17.4   Standard Free Energy Change, $\Delta G°$ 492
*Chemistry: **The Human Side**: J. Willard Gibbs*   493
17.5   Effect of Temperature, Pressure, and Concentration on Reaction Spontaneity   497
17.6   The Free Energy Change and the Equilibrium Constant   500
17.7   Additivity of Free Energy Changes; Coupled Reactions   502
*Chemistry: **Beyond the Classroom**: Global Warming*   504
Chapter Highlights   505
Summary Problem   506
Questions & Problems   507

18 Electrochemistry   513

18.1   Voltaic Cells   515
18.2   Standard Voltages   518
18.3   Relations Between $E°$, $\Delta G°$, and K   525
18.4   Effect of Concentration on Voltage   528
18.5   Electrolytic Cells   531
*Chemistry: **The Human Side**: Michael Faraday*   534
18.6   Commercial Cells   535
*Chemistry: **Beyond the Classroom**: Corrosion of Metals*   539
Chapter Highlights   540
Summary Problem   541
Questions & Problems   542

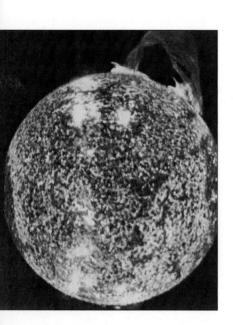

19 NUCLEAR REACTIONS   547

19.1   Radioactivity   549
  Chemistry: The Human Side: *The Curies*   553
19.2   Rate of Radioactive Decay   554
19.3   Mass-Energy Relations   557
19.4   Nuclear Fission   561
19.5   Nuclear Fusion   564
  Chemistry: Beyond the Classroom: *Biological Effects of Radiation*   566
Chapter Highlights   567
Summary Problem   568
Questions & Problems   568

20 CHEMISTRY OF THE METALS   573

20.1   Metallurgy   574
20.2   Reactions of the Alkali and Alkaline Earth Metals   581
20.3   Redox Chemistry of the Transition Metals   584
  Chemistry: Beyond the Classroom: *Essential Metals in Nutrition*   590
Chapter Highlights   591
Summary Problem   592
Questions & Problems   593

21 CHEMISTRY OF THE NONMETALS   595

21.1   The Elements and Their Preparation   597
21.2   Hydrogen Compounds of Nonmetals   603
21.3   Oxygen Compounds of Nonmetals   608
21.4   Oxoacids and Oxoanions   611
  Chemistry: Beyond the Classroom: *Arsenic and Selenium*   619
Chapter Highlights   620
Summary Problem   620
Questions & Problems   621

22 ORGANIC CHEMISTRY   625

22.1   Saturated Hydrocarbons: Alkanes   627
22.2   Unsaturated Hydrocarbons: Alkenes and Alkynes   631
22.3   Aromatic Hydrocarbons and Their Derivatives   633
22.4   Oxygen Compounds; Functional Groups   635
22.5   Isomerism in Organic Compounds   640
  Chemistry: Beyond the Classroom: *Cholesterol*   645
Chapter Highlights   646
Summary Problem   647
Questions & Problems   648

Appendix 1: Units, Constants, and Reference Data   A . 1

Appendix 2: Properties of the Elements   A . 7

Appendix 3: Exponents and Logarithms   A . 9

Appendix 4: Nomenclature   A . 1 4

Appendix 5: Molecular Orbitals   A . 1 7

Appendix 6: Answers to Even-Numbered and Challenge Questions &
            Problems   A . 2 4

Index/Glossary   I . 1

LIST OF VIDEOS AND ANIMATIONS

On the following pages is a list of videos and animations that appear on the *Saunders Interactive General Chemistry CD-ROM* **(Version 2.5)** and illustrate discussions within the text. In the list below, the page numbers refer to the page in the text where the CD-ROM correlation appears.

Chapter 1: MATTER AND MEASUREMENTS

Page 5, Screen 1.6, Compounds and Molecules.
Video: Frank DiSalvo Discusses the Value of the Periodic Table of Elements

Page 6, Screen 1.14, Separation of Mixtures.
Video: Separation of a Mixture of Sand and Potassium Chromate
Video: Kitchen Chemistry: Separation of Dyes

Page 9, Screen 1.10, Temperature.
Video: Transfer of Molecular Momentum

Chapter 2: ATOMS, MOLECULES, AND IONS

Page 30, Screen 2.5, The Dalton Atomic Theory.
Video: Magnesium Burns in Air, Producing Magnesium Oxide

Page 31, Screen 2.2, Introduction to Atoms.
Video: The Big Bang
Video: A Closer Look: Fuel for the Space Shuttle

Page 31, Screen 2.8, Electrons.
Animation: Deflection of Cathode Rays

Page 31, Screen 2.10, Protons.
Animation: Canal-Ray Experiment

Page 32, Screen 2.11, The Nucleus of the Atom.
Animation: Rutherford's α Particle Experiment

Page 34, Screen 2.7, Evidence of Subatomic Particles.
Animation: Separation of Radiation by an Electric Field

Page 37, Screen 2.17, Chemical Periodicity.
Video: Lithium in Water
Video: Sodium in Water
Video: Potassium in Water

Page 40, Screen 3.7, Ions—Cations and Anions.
Animation: Cation (Mg, 12 Protons, 12 Electrons)
Animation: Anion (F, 9 Protons, 9 Electrons)

Page 42, Screen 3.10, Ionic Compounds.
Video & Animation: The Reaction of Sodium and Chlorine (Yields Sodium Chloride)
Video & Animation: A Closer Look: The Ionic Crystal Lattice (KBr)

Chapter 3: MASS RELATIONS IN CHEMISTRY; STOICHIOMETRY

Page 64, Screen 3.18, Determining Empirical Formulas.
Video: $Zn(s) + S(s) \rightarrow ZnS(s)$

Page 65, Screen 5.8, Using Stoichiometry (2).
Animation: Carbon-Hydrogen Analysis Unit

Page 67, Screens 4.2, Chemical Equations, & 4.4, Balancing Chemical Equations.
Screen 4.2 Video: Bromine, $Br_2(l)$ and Aluminum, $Al(s)$ [$2Al(s) + 3Br(l) \rightarrow Al_2Br_6(s)$]
Screen 4.4 Video: Reactions That Form Oxides (Phosphorus, P_4, Reacts Vigorously with Oxygen to Give Tetraphosphorus Decaoxide)
Screen 4.4 Animation: Combustion Reactions (Reaction of Propane and Oxygen)

Page 69, Screens 5.2, Weight Relations in Chemical Reactions, & 5.3, Calculations in Stoichiometry.
Screen 5.2 Video & Animation: $P_4(s) + 6Cl_2(g) \rightarrow 4PCl_3(l)$
Screen 5.3 Video: Weight Relations in Chemical Reactions [$2Mg(s) + O_2(g) \rightarrow 2MgO(s)$]

Page 70, Screen 5.4, Reactions Controlled by the Supply of One Reactant.
Video: Limiting Reactant [$2CH_3OH(l) + 3O_2(g) \rightarrow 2CO_2(g) + 4H_2O(g)$]

Page 71, Screen 5.5, Limiting Reactants.
Video: Limiting Reactant: The Details (Zinc and Hydrochloric Acid Reacting to Form Zinc Chloride and Hydrogen Gas)

Page 74, Screen 3.14, Hydrated Compounds.
Video: Dehydration of Copper(II) Sulfate Pentahydrate

Chapter 4: REACTIONS IN AQUEOUS SOLUTION

Page 84, Screen 5.9, Solutions.
Video & Animation: $H_2O(l) + KMnO_4(s) \rightarrow KMnO_4(aq)$

Page 84, Screen 5.10, Solution Concentration.
Animation: A Closer Look: Ion Concentrations in Solution ($CaCl_2$ Dissolving in Water)

Page 85, Screen 5.11, Preparing Solutions of Known Concentrations (1).
Video: Direct Addition

Page 86, Screen 4.5, Compounds in Aqueous Solution.
Video: $Mg(s) + 2HCl(aq) \rightarrow MgCl_2(aq) + H_2(g)$
Animation: Strong Electrolytes (HCl)
Animation: Weak Electrolytes (Acetic Acid)
Animation: Nonelectrolytes (Ethanol)

Page 88, Screens 4.11, Types of Reactions in Aqueous Solution, & 4.12, Precipitation Reactions.
Screen 4.11 Video: Precipitation Reactions [$Pb(NO_3)_2(aq) + 2KI(aq) \rightarrow PbI_2(s) + 2KNO_3(aq)$]
Screen 4.12 Video: $Ag^+(aq) + Cl^-(aq) \rightarrow AgCl(s)$
Screen 4.12 Video: $Fe^{3+}(aq) + 3OH^-(aq) \rightarrow Fe(OH)_3(s)$

Page 89, Screen 4.10, Equations for Reactions in Aqueous Solution—Net Ionic Equations.

Video: Net Ionic Equations: Reaction of Potassium Chromate and Lead Nitrate

Page 90, Screen 5.13, Stoichiometry of Reactions in Solution.
Video: Redox Reaction Between Two Ions in Solution $[5Fe^{2+}(aq) + MnO_4^-(aq) + 8H_3O^+(aq) \rightarrow 5Fe^{3+}(aq) + Mn^{2+}(aq) + 12H_2O(l)]$

Page 91, Screens 4.11, Types of Reactions in Aqueous Solution, & 4.13, Acid-Base Reactions.
Screen 4.11 Video: Acid-Base Reactions $[HCl(aq) + NH_3(aq) \rightarrow NH_4^+(aq) + Cl^-(aq)]$
Screen 4.13 Video: A Closer Look: Detecting Reactions Having No Visible Change

Page 92, Screens 4.7, Acids, & 4.8, Bases.
Screen 4.7 Animation: Strong Acids (HCl)
Screen 4.7 Animation: Weak Acids (HF)
Screen 4.7 Animation: A Closer Look: H^+ Ions in Water
Screen 4.8 Animation: Strong Base (NaOH)
Screen 4.8 Animation: Weak Base (NH_3)

Page 95, Screens 5.14, Titrations, & 5.15, Titration Simulation.
Screen 5.14 Video: Titration (Adding a Base Solution of NaOH to an Impure Sample of Oxalic Acid ($H_2C_2O_4$) to Determine Its Purity)
Screen 5.15 Animation: Titration Simulation

Page 98, Screens 4.11, Types of Reactions in Aqueous Solution, & 4.15, Oxidation-Reduction Reactions.
Screen 4.11 Video: Oxidation-Reduction Reactions (Reaction of Copper and Nitric Acid) $[Cu(s) + 2NO_3^-(aq) + 4H^+(aq) \rightarrow 2NO_2(g) + Cu^{2+}(aq) + 2H_2O(l)]$
Screen 4.15 Video: Magnesium Burns in Air $[2Mg(s) + O_2(g) \rightarrow 2MgO(s)]$

Page 98, Screen 4.16, Redox Reactions and Electron Transfer.
Video: Copper Wire in a Silver Nitrate Solution

Page 99, Screen 4.17, Oxidation Numbers.
Video: Reaction of NO and O_2 $[2NO(g) + O_2(g) \rightarrow 2NO_2(g)]$
Video: A Closer Look: Disproportion Reaction of H_2O_2

Page 100, Screen 4.18, Recognizing Oxidation-Reduction Reactions.
Video: $2Fe(s) + 3Cl_2(g) \rightarrow 2[Fe^{3+}, 3Cl^-](s)$
Video: $2Fe(s) + O_2(g) \rightarrow 2[Fe^{2+}, O^{2-}](s)$

Page 100, Screen 21.3, Balancing Equations for Redox Reactions.
Video: Complex Reaction Occurring in Acidic Solution (Iron(II) Ion Reacts with Permanganate Ion in Solution to Yield Mn^{2+} and Fe^{3+})

Chapter 5: GASES

Page 116, Screen 12.2, Properties of Gases.
Video: Gas Pressure (A Bicycle Pump Forces Gas Molecules into a Tire)

Page 118, Screen 12.3, Gas Laws.
Animation: Boyle's Law
Animation: Charles's Law
Animation: Avogadro's Law

Page 125, Screen 12.5, Gas Density.
Video: Bromine Vapor

Page 125, Screen 12.8, Gas Mixtures and Partial Pressures.
Video: Ammonium Nitrate Explodes

Page 128, Screens 12.9, The Kinetic Molecular Theory of Gases, & 12.10, Gas Laws and the Kinetic Molecular Theory.

Screen 12.9 Animation: Gases on the Molecular Scale at High and Low Temperatures
Screen 12.10 Animation: Three Gas Laws: Avogadro's (Pressure and Number of Moles, $P \propto n$)
Screen 12.10 Animation: Three Gas Laws: P and T (Pressure and Absolute Temperature, $P \propto T$)
Screen 12.10 Animation: Three Gas Laws: Boyle's (Pressure and Volume, $P \propto 1/V$)

Page 131, Screen 12.12, Application of the Kinetic Molecular Theory.
Animation: Gaseous Diffusion of $NH_3(g)$ and $HCl(g)$

Page 132, Screen 12.11, Distribution of Molecular Speeds.
Animation: Boltzmann Distributions

Chapter 6: ELECTRONIC STRUCTURE AND THE PERIODIC TABLE

Page 146, Screen 7.5, Planck's Equation.
Video: When a Metal Bar Is Heated, It Emits Electromagnetic Radiation

Page 147, Screen 7.6, Atomic Line Spectra.
Animation: Atomic Line Spectrum of Excited H Atoms

Page 148, Screen 7.7, Bohr's Model of the Hydrogen Atom.
Animation: Bohr's Model of the Hydrogen Atom

Page 150, Screen 7.8, Wave Properties of the Electron.
Animation: Electron Diffraction

Page 156, Screen 8.5, Atomic Subshell Energies.
Animation: Relationship Between Subshell Energies in Single Electron Atoms and Order of Filling in Many-Electron Atoms

Page 163, Screen 8.3, Spinning Electrons and Magnetism.
Animation: Spinning Electrons and Magnetic Polarity
Video: Diamagnetic (Liquid Nitrogen) and Paramagnetic (Liquid Oxygen) Substances

Page 167, Screen 8.10, Atomic Properties and Periodic Trends: Size.
Animation: Atomic Size

Page 168, Screen 8.6, Effective Nuclear Charge, Z^*.
Animation: Effective Nuclear Charge, Z^*

Page 169, Screen 8.12, Atomic Properties and Periodic Trends: Ionization Energy
Animation: Ionization Energies Required for Magnesium

Chapter 7: COVALENT BONDING

Page 180, Screen 9.3, Chemical Bond Formation.
Animation: Interatomic Interactions
Animation: Biography: Linus Pauling and the Protein Molecules' Helical Structure

Page 188, Screen 9.7, Electron-Deficient Compounds.
Animation: Exceptions to the Octet Rule (BF_3)

Page 188, Screen 9.8, Free Radicals.
Video & Animation: NO_2 Readily Dimerizes to Form N_2O_4 at Low Temperatures
Video: Current Issues in Chemistry: Nitric Oxide

Page 193, Screen 9.15, Ideal Electron Repulsion Shapes.
Animation: Ideal Electron Repulsion Shapes

Page 203, Screens 10.3, Valence Bond Theory, & 10.4, Hybrid Orbitals.

Screen 10.3 Animation: Bond Formation: Potential Energy vs. Distance for H_2
Screen 10.4 Animation: Atomic Orbitals for One Carbon Atom → Hybrid Orbitals for Carbon Atom in a Methane Molecule

Page 207, Screens 10.5, Sigma Bonding, & 10.7, Multiple Bonding.
Screen 10.5 Animation: sp^3 Hybrid Orbital
Screen 10.7 Animation: σ Bonds Only
Screen 10.7 Animation: π Bonds Only
Screen 10.7 Animation: π and σ Bonds

Chapter 8: THERMOCHEMISTRY

Page 218, Screen 6.12, Energy Changes in Chemical Processes.
Animation: Endothermic and Exothermic Systems
Video: Chemistry and You: Hand Warmers and Cold Packs

Page 222, Screen 6.14, Enthalpy Change and ΔH
Animation: Enthalpy Change and ΔH

Page 227, Screen 6.16, Hess's Law.
Animation: Three-Step Formation of Sulfuric Acid Represented Using an Enthalpy Diagram

Page 235, Screen 6.13, The First Law of Thermodynamics.
Video & Animation: The First Law of Thermodynamics (Using $CO_2(s)$ to illustrate)

Chapter 9: LIQUIDS AND SOLIDS

Page 247, Screen 13.2, Intermolecular Forces (1).
Animation: The Kinetic Molecular Theory

Page 247, Screen 13.11, Properties of Liquids.
Video: Surface Tension (Paper Clip and Water)
Video: Capillary Action (Paper Placed in Water)
Video: Viscosity (Glycerol and Ethanol)

Page 248, Screen 13.9, Properties of Liquids (Vapor Pressure).
Animation: System in a State of Dynamic Equilibrium
Animation: Chemistry and You: Butane Lighters and Propane Tanks

Page 250, Screen 13.8, Properties of Liquids (Enthalpy of Vaporization).
Animation: Enthalpy of Vaporization

Page 251, Screen 13.10, Properties of Liquids (Boiling Point).
Video: Boiling Point

Page 256, Screen 13.14, Solid Structures (Molecular Solids).
Animation: C_{60} Molecule

Page 257, Screen 13.5, Intermolecular Forces (3).
Animation: Dipole–Induced Dipole Forces
Animation: Induced Dipole–Induced Dipole Forces
Video: Sublimation and Condensation of Iodine Crystals

Page 258, Screen 13.4, Intermolecular Forces (2).
Animation: Ion-Dipole Forces: $Na^+(g) + 6H_2O(l) \rightarrow$ $[Na(H_2O)_6]^+(aq)$; $\Delta H_{rxn} = -405$ kJ
Animation: Dipole-Dipole Forces

Page 261, Screen 13.7, The Weird Properties of Water.
Animation: Boiling Points of Simple Hydrogen-Containing Compounds
Animation: A Consequence of Hydrogen Bonding (Ice)

Page 263, Screen 13.15, Solid Structures (Network Solids).
Animation: Diamond and Silicon

Page 264, Screen 3.11, Properties of Ionic Compounds.

Video: Potassium Bromide Crystals Cleave Along Perpendicular Planes

Page 268, Screen 13.12, Solid Structures (Crystalline and Amorphous Solids).
Animation: Crystalline Solids (Simple Cubic Unit Cell)

Chapter 10: SOLUTIONS

Page 281, Screens 5.10, Solution Concentration, & 14.2, Solubility.
Screen 5.10 Animation: A Closer Look: Ion Concentrations in Solution ($CaCl_2$ Dissolving in Water)
Screen 14.2 Video: Saturated Solution $[Pb(NO_3)_2(aq) + 2KI(aq) \rightarrow PbI_2(s) + 2KI(aq)]$
Screen 14.2 Video: Unsaturated Solution $[NiCl_2 + 6H_2O(s) \rightarrow NiCl_2(aq)]$
Screen 14.2 Video: Supersaturated Solution $[NaCH_3CO_2(aq) \rightarrow NaCH_3CO_2(s)]$

Page 282, Screen 5.12, Preparing Solutions of Known Concentration (2).
Video: Solution by Dilution

Page 286, Screen 14.3, The Solution Process.
Animation: Intermolecular Forces

Page 288, Screen 14.4, Energetics of Solution Formation.
Animation: Dissolving Ionic Compounds $[KF(s) \rightarrow K^+(aq) + F^-(aq)]$
Video: Chemistry and You: Why Don't Oil and Water Mix?

Page 290, Screen 14.5, Factors Affecting Solubility (1).
Video: Chemistry and You: Why Soft Drinks Go Flat

Page 292, Screen 14.8, Colligative Properties (2).
Animation: Boiling Point (Water and Ethylene Glycol)
Animation: Freezing Point (Pure Water Without Solute and Ethylene Glycol Solution)

Page 295, Screen 14.9, Colligative Properties (3).
Animation: Osmosis
Animation: Semipermeable Membrane
Video: A Semipermeable Membrane (an Egg in Acetic Acid, Water, and Corn Syrup)

Chapter 11: RATE OF REACTION

Page 310, Screen 15.2, Rates of Chemical Reactions.
Video & Animation: Initial Rate
Video & Animation: Instantaneous Rate
Video & Animation: Average Reaction Rate

Page 313, Screen 15.4, Control of Reaction Rates (Concentration Dependence).
Video: Magnesium Metal Added to Two Different Molar Solutions of $HCl(aq)$

Page 323, Screen 15.9, Microscopic View of Reactions (1).
Animation: Collision Theory: Contact/Collision
Animation: Collision Theory: Sufficient Energy
Animation: Collision Theory: Proper Orientation

Page 326, Screen 15.10, Microscopic View of Reactions (2).
Animation: Transition State
Animation: A Closer Look: Reaction Coordinate Diagrams (Single-Step Reaction)
Animation: A Closer Look: Reaction Coordinate Diagrams (Two-Step Reaction)

Page 328, Screen 15.11, Control of Reaction Rates (Temperature Dependence).
Video: Bleaching Food Dye

Page 331, Screen 15.14, Catalysis and Reaction Rate.
Video: Catalysis of H_2O_2 Decomposition
Video: Interview: James Cusumano: What Is a Catalyst?/Catalysts and Society/Catalysis and the Environment

Page 333, Screen 15.12, Reaction Mechanisms.
Animation: The Decomposition of Ozone
Animation: The Formation of Hydrazine

Page 334, Screen 15.13, Reaction Mechanisms and Rate Equations.
Animation: Proposing Mechanisms [for the reaction $NO_2(g) + CO(g) \rightarrow NO(g) + CO_2(g)$]
Animation: A Closer Look: Isotopic Labeling Studies

Chapter 12: GASEOUS CHEMICAL EQUILIBRIUM

Page 355, Screen 16.5, The Meaning of the Equilibrium Constant
Video: $2NO(g) + O_2(g) \rightleftharpoons 2NO_2(g)$; $K = 2.26 \times 10^{12}$
Video: $PbI_2(s) \rightleftharpoons Pb^{2+}(aq) + 2I^-(aq)$; $K = 8.7 \times 10^{-9}$

Page 366, Screen 16.11, Disturbing an Equilibrium (Le Châtelier's Principle).
Animation: Demonstration of Le Châtelier's Principle

Page 366, Screen 16.13, Disturbing an Equilibrium (Addition or Removal of a Reagent).
Animation: Isobutane/Butane Equilibrium

Page 368, Screen 16.14, Disturbing an Equilibrium (Volume Changes).
Animation: Disturbing an Equilibrium (Volume Decrease)

Page 370, Screen 16.12, Disturbing an Equilibrium (Temperature Changes).
Animation & Video: Disturbing an Equilibrium (Temperature Increase)

Chapter 13: ACIDS AND BASES

Page 383, Screen 17.2, Brønsted Acids and Bases.
Animation: Brønsted Acid
Animation: Brønsted Base
Animation: Acid-Base Reactions and Conjugate Pairs

Page 384, Screen 17.3, The Acid-Base Properties of Water.
Animation: Autoionization [$2H_2O(l) \rightleftharpoons H_3O^+(aq) + OH^-(aq)$]

Page 386, Screen 17.4, The pH Scale.
Animation: A Closer Look: The pH Meter

Page 387, Screen 17.5, Strong Acids and Bases.
Video: Water Added to Calcium Hydride

Chapter 14: EQUILIBRIA IN ACID-BASE SOLUTIONS

Page 413, Screen 18.8, Buffer Solutions.
Video: A Buffered Solution
Video: An Unbuffered Solution

Page 416, Screen 18.10, Preparing Buffer Solutions.
Video: Preparing Buffer Solutions

Page 417, Screen 18.11, Adding Reagents to Buffer Solutions.
Video: Measuring pH of a Buffer Solution

Page 423, Screen 18.3, Acid-Base Reactions (Strong Acids + Strong Bases).
Video: Acid-Base Reactions (Strong Acids + Strong Bases)
Animation: Mixing HCl and NaOH

Page 425, Screen 18.5, Acid-Base Reactions (Weak Acids + Strong Bases).
Animation: Mixing $H_2PO_4^-$ and OH^-

Chapter 15: COMPLEX IONS

Page 439, Screens 17.11, Lewis Acids and Bases, and 17.12, Cationic Lewis Acids.
Screen 17.11 Animation: Lewis Acid-Base Reaction [$BF_3 + NH_3 \rightarrow F_3B—NH_3$]
Screen 17.12 Video & Animation: Copper Sulfate and Ammonia
Screen 17.11 Animation: Chemistry and You: Hemoglobin, Oxygen, and Carbon Monoxide Poisoning

Chapter 16: PRECIPITATION EQUILIBRIA

Page 461, Screen 19.2, Precipitation Reactions.
Video: $Pb(NO_3)_2(aq) + 2KI(aq) \rightarrow PbI_2(s) + KNO_3(aq)$

Page 461, Screen 19.4, Solubility Product Constant.
Video: Lead Chloride Put in Water

Page 462, Screen 19.5, Determining K_{sp}.
Video: Experimentally Determined Measurements

Page 464, Screen 19.7, Can a Precipitation Reaction Occur?
Video: Predicting with Q, the Reaction Quotient ($Q < K_{sp}$)
Video: Predicting with Q, the Reaction Quotient ($Q > K_{sp}$)
Video: Predicting with Q, the Reaction Quotient ($Q = K_{sp}$)

Page 467, Screen 19.8, The Common Ion Effect.
Video: $PbCl_2(s) \rightleftharpoons Pb^{2+}(aq) + 2Cl^-(aq)$

Page 468, Screen 19.11, Solubility and pH.
Video: Dissolving a Precipitate

Page 470, Screen 19.12, Complex Ion Formation and Solubility.
Video & Animation: Formation of $Ag(NH_3)_2^+$
Video: Chemistry and You: Floor Wax and Complex Ions

Page 472, Screen 19.9, Using Solubility.
Video: Ion Identification and Separation (Using a Solution with Ag^+, Pb^{2+}, and Cu^{2+})

Chapter 17: SPONTANEITY OF REACTION

Page 483, Screen 20.2, Reaction Spontaneity.
Video: What Controls Whether a Reaction Occurs?

Page 486, Screen 20.4, Entropy.
Animation: $S°$ (gases) $\gg$ $S°$ (liquids) $>$ $S°$ (solids)
Animation: The Entropy of a Substance Increases with Temperature
Animation: The More Complex the Molecule, the Greater the Value of $S°$
Animation: Entropies of Ionic Solids Depend on Coulombic Attractions
Animation: Entropy Usually Increases When a Pure Liquid or Solid Dissolves in a Solvent
Video & Animation: A Closer Look: Entropy Changes for Phase Changes

Page 490, Screen 20.6, The Second Law of Thermodynamics.

Animation: Demonstration of the Second Law of Thermodynamics

Page 490, Screen 20.7, Gibbs Free Energy.
Video & Animation: Putting ΔH and ΔS together

Page 501, Screen 20.9, Thermodynamics and the Equilibrium Constant.
Animation: Product-Favored
Animation: Reaction-Favored

Chapter 18: ELECTROCHEMISTRY

Page 514, Screen 21.2, Redox Reactions.
Video: Electron Transfer: Copper Metal Reacts Spontaneously with a Solution of Silver Nitrate

Page 515, Screen 21.4, Electrochemical Cells.
Animation: An Electrochemical Cell

Page 528, Screen 21.7, Electrochemical Cells at Nonstandard Conditions.
Video: Change in Cell Potential with Concentration

Page 531, Screen 21.10, Electrolysis.
Video: Chemical Change from Electrical Energy

Animation: Chemistry and You: Commercial Production of Aluminum

Page 536, Screen 21.8, Batteries.
Animation: Common Dry Cell
Animation: Alkaline Battery
Animation: Mercury Battery
Animation: Lead Storage Battery
Animation: Ni-Cad Battery

Page 539, Screen 21.9, Corrosion.
Animation: Redox Reactions in the Environment (Formation of Rust)
$[2Fe(s) + 2H_2O(l) + O_2(g) \rightarrow 2Fe(OH)_2(s)$
Animation: A Closer Look: Why Doesn't Aluminum Rust?

Chapter 22: ORGANIC CHEMISTRY

Page 632, Screen 11.4, Hydrocarbons and Addition Reactions.
Animation: Bromination
Animation: Hydrogenation

Page 646, Screen 11.7, Fats and Oils.
Animation: Stearic Acid, a Saturated Fatty Acid
Animation: Linolenic Acid, an Unsaturated Fatty Acid

MATTER AND MEASUREMENTS

A heterogeneous mixture of the minerals yellow orpiment and red realgar in a rock sample. *(Dave Davidson/Tom Stack and Associates)*

CHAPTER OUTLINE

1.1   TYPES OF MATTER   1.3   PROPERTIES OF SUBSTANCES

1.2   MEASUREMENTS

Almost certainly, this is your first college course in chemistry; perhaps it is your first exposure to chemistry at any level. Unless you are a chemistry major, you may wonder why you are taking this course and what you can expect to gain from it. To address that question, it will be helpful to look at some of the ways in which chemistry contributes to other disciplines.

If you're planning to be an engineer, you can be sure that many of the materials you will work with have been synthesized by chemists. Some of these materials are organic (carbon-containing). They could be familiar plastics like polyethylene (Chapter 22) or the more esoteric plastics used in unbreakable windows and nonflammable clothing. Other materials, including metals (Chapter 20) and semiconductors, are inorganic in nature.

Perhaps you are a health science major, looking forward to a career in medicine or pharmacy. If so, you will want to become familiar with the properties of aqueous solutions (Chapters 4, 10, 14, and 16), which include blood and other body fluids. Chemists today are involved in the synthesis of a variety of life-saving products. These range from drugs used in chemotherapy (Chapter 15) to new antibiotics used against resistant microorganisms.

Beyond career preparation, an objective of a college education is to stimulate your curiosity about the world around you or, to abuse a cliché, to make you a "better informed citizen." In this text, we will look at some of the chemistry-related topics referred to on the evening news—

- essential versus toxic elements in your diet (Chapters 2, 20).

- the effect of ethyl alcohol and other drugs on your body (Chapters 2, 22).

- the effect of acid rain on the environment (Chapter 14).

- depletion of the ozone layer (Chapter 11).

- global warming (Chapter 17).

- pros and cons of nuclear power (Chapter 19).

Many of these topics are controversial even within the scientific community. Consider global warming, for example. Everyone agrees that average global temperatures are rising; 1998 was the warmest year on record. Debate centers on the cause of this phenomenon. The consensus of the scientific community is that global warming is due in large part to human activities and that we must take steps *now* to control it. Yet there are experts in the field who disagree about the cause and the urgency of the problem. One such person went so far as to state, "Another 20 years of doing nothing is unlikely to mess things up."

As a voter, perhaps as a juror, and certainly as a participant in the twenty-first century, you will have to decide for yourself what to do about global warming and many other chemistry-related topics listed above. To do that intelligently, you need a basic knowledge of chemistry as well as the other natural sciences. That's really why you're taking a course in general chemistry.

This chapter begins the study of chemistry by—

- considering the different types of matter: pure substances versus mixtures, elements versus compounds (Section 1.1).
- looking at the kinds of measurements on which chemistry is founded, the uncertainties associated with these measurements, and a method used to convert measured quantities from one unit to another (Section 1.2).
- focusing on certain physical properties including color, density, and water solubility, which can be used to identify substances (Section 1.3).

⌐ Chemistry deals with the properties and reactions of substances.

1.1   TYPES OF MATTER

Matter is anything that has mass and occupies space. It exists in three phases: solid, liquid, and gas. A solid has a rigid shape and a fixed volume. A liquid has a fixed volume but is not rigid in shape; it takes on the shape of the container. A gas has neither a fixed volume nor a rigid shape; it takes on both the volume and the shape of the container.

⊙ See the *Saunders Interactive General Chemistry CD-ROM*, Screen 1.3, States of Matter.

Matter can be classified into two categories—

- pure substances, each of which has a fixed composition and a unique set of properties.
- mixtures, composed of two or more substances.

Pure substances are either elements or compounds (Figure 1.1), whereas mixtures can be either homogeneous or heterogeneous.

⌐ Most materials you encounter are mixtures.

Elements

An **element** is a type of matter that cannot be broken down into two or more pure substances. There are 112 known elements,* of which 91 occur naturally.

⊙ See Screen 1.5, Elements and Atoms.

*In 1999, the discovery of three new elements was reported; if confirmed, that would raise the total to 115.

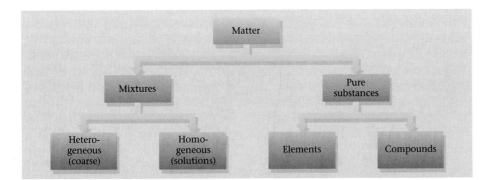

FIGURE 1.1
Classification of matter. **OHT**

TABLE 1.1 **Some Familiar Elements with Their Percentage Abundances**

Element	Symbol	Percentage Abundance	Element	Symbol	Percentage Abundance
Aluminum	A1	7.5	Magnesium	Mg	1.9
Bromine	Br	0.00025	Manganese	Mn	0.09
Calcium	Ca	3.4	Mercury	Hg	0.00005
Carbon	C	0.08	Nitrogen	N	0.03
Chlorine	Cl	0.2	Oxygen	O	49.4
Chromium	Cr	0.018	Phosphorus	P	0.12
Copper	Cu	0.007	Potassium	K	2.4
Gold	Au	0.0000005	Silicon	Si	25.8
Hydrogen	H	0.9	Silver	Ag	0.00001
Iodine	I	0.00003	Sodium	Na	2.6
Iron	Fe	4.7	Sulfur	S	0.06
Lead	Pb	0.0016	Zinc	Zn	0.008

Many elements are familiar to all of us. The charcoal used in outdoor grills is nearly pure carbon. Electrical wiring, jewelry, and water pipes are often made from copper, a metallic element. Another such element, aluminum, is used in many household utensils.

Some elements come in and out of fashion, so to speak. Fifty years ago, elemental silicon was a chemical curiosity. Today, ultrapure silicon has become the basis for the multibillion-dollar semiconductor industry. Lead, on the other hand, is an element moving in the other direction. A generation ago it was widely used to make paint pigments, plumbing connections, and gasoline additives. Today, because of the toxicity of lead compounds, all of these applications have been banned in the United States.

In chemistry, an element is identified by its symbol. This consists of one or two letters, usually derived from the name of the element. Thus the symbol for carbon is C; that for aluminum is Al. Sometimes the symbol comes from the Latin name of the element or one of its compounds. The two elements copper and mercury, which were known in ancient times, have the symbols Cu *(cuprum)* and Hg *(hydrargyrum)*.

Table 1.1 lists the names and symbols of several elements that are probably familiar to you. In either free or combined form, they are commonly found in the laboratory or in commercial products. The abundances listed measure the relative amount of each element in the Earth's crust, the atmosphere, and the oceans.

Curiously, several of the most familiar elements are really quite rare. An example is mercury, which has been known since at least 500 B.C., even though its abundance is only 0.00005%. It can easily be prepared by heating the red mineral cinnabar (Figure 1.2). Beyond that, its unique properties make it extremely useful. You are perhaps most familiar with mercury as the liquid used in thermometers and barometers; ancient peoples used it to extract gold and silver from their ores.

In contrast, aluminum (abundance = 7.5%), despite its usefulness, was little more than a chemical curiosity until about a century ago. It occurs in combined form in clays and rocks, from which it cannot be extracted. In 1886 two young chemists, Charles Hall in the United States and Paul Herroult in France, independently worked out a process for extracting aluminum from a relatively rare ore,

(a)

(b)

FIGURE 1.2
Cinnabar and mercury. (a) The mineral cinnabar, from which mercury is obtained. (b) Mercury, an element. *(Charles D. Winters)*

bauxite. That process is still used today to produce the element. By an odd coincidence, Hall and Herroult were born in the same year (1863) and died in the same year (1914).

Compounds

A **compound** is a pure substance that contains more than one element. Water is a compound of hydrogen and oxygen. The compounds methane, acetylene, and naphthalene all contain the elements carbon and hydrogen, in different proportions.

Compounds have fixed compositions. That is, a given compound always contains the same elements in the same percentages by mass. A sample of pure water contains precisely 11.19% hydrogen and 88.81% oxygen. In contrast, mixtures can vary in composition. For example, a mixture of hydrogen and oxygen might contain 5, 10, 25, or 60% hydrogen, along with 95, 90, 75, or 40% oxygen.

The properties of compounds are very different from those of the elements they contain. Ordinary table salt, sodium chloride, is a white, unreactive solid. As you can guess from its name, it contains the two elements sodium and chlorine. Sodium (Na) is a shiny, extremely reactive metal. Chlorine (Cl) is a poisonous, greenish-yellow gas. Clearly, when these two elements combine to form sodium chloride, a profound change takes place (Figure 1.3).

Many different methods can be used to resolve compounds into their elements. Sometimes, but not often, heat alone is sufficient. Mercury(II) oxide, a

See Screen 1.6, Compounds and Molecules.

FIGURE 1.3
Sodium, chlorine, and sodium chloride. (a) Sodium, a metallic element that is soft enough to be cut with a knife. (b) Chlorine, a nonmetallic element that is a gas. (c) Sodium chloride, the crystalline chemical compound formed when sodium combines with chlorine. *(Charles D. Winters)*

(a)   (b)   (c)

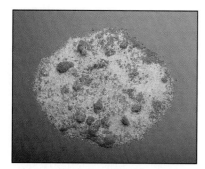

FIGURE 1.4
A heterogeneous mixture of copper sulfate crystals (blue) and sand. *(Charles D. Winters)*

All gaseous mixtures, including air, are solutions.

 See Screen 1.14, Separation of Mixtures.

FIGURE 1.5
Two mixtures. On the left is brass, a *homogeneous* mixture of copper and zinc. On the right is a piece of granite, a *heterogeneous* mixture that contains discrete regions of different minerals (feldspar, mica, and quartz). *(Charles D. Winters)*

compound of mercury and oxygen, decomposes to its elements when heated to 600°C. Joseph Priestley, an English chemist, discovered oxygen more than 200 years ago when he carried out this reaction by exposing a sample of mercury(II) oxide to an intense beam of sunlight focused through a powerful lens. The mercury vapor formed is a deadly poison. Sir Isaac Newton, who distilled large quantities of mercury in his laboratory, suffered the effects in his later years.

Another method of resolving compounds into elements is electrolysis, which involves passing an electric current through a compound, usually in the liquid state. By electrolysis it is possible to separate water into the gaseous elements hydrogen and oxygen. Thirty years ago it was proposed to use the hydrogen produced by electrolysis to raise the *Titanic* from its watery grave off the coast of Newfoundland. It didn't work.

Mixtures

A **mixture** contains two or more substances combined in such a way that each substance retains its chemical identity. When you shake copper sulfate with sand (Figure 1.4), the two substances do not react with one another. In contrast, when sodium is exposed to chlorine gas, a new compound, sodium chloride, is formed.

There are two types of mixtures:

1. Homogeneous or uniform mixtures are ones in which the composition is the same throughout. Another name for a homogeneous mixture is a **solution,** which is made up of a solvent, usually taken to be the substance present in largest amount, and one or more solutes. Most commonly, the solvent is a liquid, whereas solutes may be solids, liquids, or gases. Soda water is a solution of carbon dioxide (solute) in water (solvent). Seawater is a more complex solution in which there are several solid solutes, including sodium chloride; the solvent is water. It is also possible to have solutions in the solid state. Brass (Figure 1.5) is a solid solution containing the two metals copper (67%–90%) and zinc (10%–33%).

2. Heterogeneous or nonuniform mixtures are those in which the composition varies throughout. Most rocks fall into this category. In a piece of granite (Figure 1.5), several components can be distinguished, differing from one another in color.

Many different methods can be used to separate the components of a mixture from one another. A couple of methods that you may have carried out in the laboratory are—

- *filtration,* used to separate a heterogeneous solid-liquid mixture. The mixture is passed through a barrier with fine pores, such as filter paper. Copper sulfate, which is water-soluble, can be separated from sand by shaking with water. On filtration the sand remains on the paper and the copper sulfate solution passes through it.
- *distillation,* used to resolve a homogeneous solid-liquid mixture. The liquid vaporizes, leaving a residue of the solid in the distilling flask. The liquid is obtained by condensing the vapor. Distillation can be used to separate the components of a water solution of copper sulfate (Figure 1.6).

A more complex but more versatile separation method is *chromatography,* a technique widely used in teaching, research, and industrial laboratories to separate all kinds of mixtures. This method takes advantage of differences in solubility and/or extent of adsorption on a solid surface. In *gas-liquid chromatography,* a mixture of volatile liquids and gases is introduced into one end of a heated glass tube.

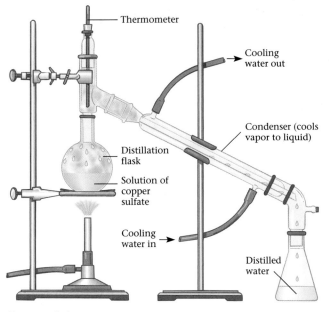

FIGURE 1.6
Apparatus for a simple distillation. The two components of a water solution of copper sulfate are being separated. Water is vaporized by heating, the vapor is cooled as it passes through the condenser, and liquid water collects in the flask at the lower right. The copper sulfate, a blue solid, does not vaporize and remains in the distillation flask. **OHT**

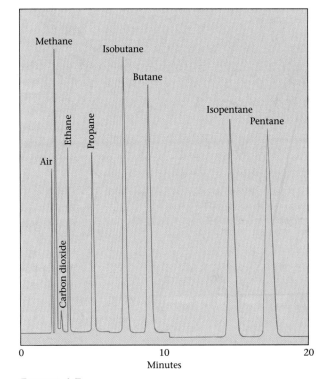

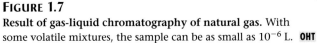

FIGURE 1.7
Result of gas-liquid chromatography of natural gas. With some volatile mixtures, the sample can be as small as 10^{-6} L. **OHT**

As little as one microliter (10^{-6} L) of sample may be used. The tube is packed with an inert solid whose surface is coated with a viscous liquid. An unreactive "carrier gas," often helium, is passed through the tube. The components of the sample gradually separate as they vaporize into the helium or condense into the viscous liquid. Usually the more volatile fractions move faster and emerge first; successive fractions activate a detector and recorder. The end result is a plot such as that shown in Figure 1.7.

Gas-liquid chromatography (GLC) finds many applications outside the chemistry laboratory. If you've ever had an emissions test on the exhaust system of your car, GLC was almost certainly the analytical method used. Pollutants such as carbon monoxide and unburned hydrocarbons appear as peaks on a graph such as that shown in Figure 1.7. A computer determines the areas under these peaks, which are proportional to the concentrations of pollutants, and prints out a series of numbers that tells the inspector whether your car passed or failed the test. Many of the techniques used to test people for drugs (marijuana, cocaine, and others) or alcohol also make use of gas-liquid chromatography.

1.2 MEASUREMENTS

Chemistry is a quantitative science. The experiments that you carry out in the laboratory and the calculations that you perform almost always involve measured quantities with specified numerical values. Consider, for example, the following

TABLE **1.2** Metric Prefixes					
Factor	Prefix	Abbreviation	Factor	Prefix	Abbreviation
10^6	mega	M	10^{-3}	**milli**	m
10^3	**kilo**	k	10^{-6}	micro	μ
10^{-1}	deci	d	10^{-9}	**nano**	n
10^{-2}	**centi**	c	10^{-12}	pico	p

set of directions for the preparation of aspirin (measured quantities are shown in italics).

> Add *2.0 g* of salicylic acid, *5.0 mL* of acetic anhydride, and *5 drops* of 85% H_3PO_4 to a 50-mL Erlenmeyer flask. Heat in a water bath at *75°C* for *15 minutes*. Add cautiously *20 mL* of water and transfer to an ice bath at *0°C*. Scratch the inside of the flask with a stirring rod to initiate crystallization. Separate aspirin from the solid-liquid mixture by filtering through a Buchner funnel *10 cm* in diameter.

See Screen 1.16, The Metric System.

Scientific measurements are expressed in the **metric system.** As you know, this is a decimal-based system in which all of the units of a particular quantity are related to one another by factors of 10. The more common prefixes used to express these factors are listed in Table 1.2.

In this section we will look at four familiar quantities that you will almost certainly measure in the laboratory: length, volume, mass, and temperature. Other quantities will be introduced in later chapters as they are needed.

Instruments and Units

See Screen 1.15, Units of Measurement.

The standard unit of *length* in the metric system is the meter, which is a little larger than a yard. The meter was originally intended to be 1/40,000,000 of the Earth's meridian that passes through Paris. It is now defined as the distance light travels in 1/299,792,458 of a second.

Other units of length are expressed in terms of the meter, using the prefixes listed in Table 1.2. You are familiar with the centimeter, the millimeter, and the kilometer:

$$1 \text{ cm} = 10^{-2} \text{ m} \qquad 1 \text{ mm} = 10^{-3} \text{ m} \qquad 1 \text{ km} = 10^3 \text{ m}$$

The dimensions of very tiny particles are often expressed in nanometers:

$$1 \text{ nm} = 10^{-9} \text{ m}$$

Volume is most commonly expressed in one of three units—

- cubic centimeters $1 \text{ cm}^3 = (10^{-2} \text{ m})^3 = 10^{-6} \text{ m}^3$
- liters (L) $1 \text{ L} = 10^{-3} \text{ m}^3 = 10^3 \text{ cm}^3$
- milliliters (mL) $1 \text{ mL} = 10^{-3} \text{ L} = 10^{-6} \text{ m}^3$

Notice that a milliliter is equal to one cubic centimeter:

$$1 \text{ mL} = 1 \text{ cm}^3$$

The device most commonly used to measure volumes in general chemistry is the graduated cylinder. A pipet or buret (Figure 1.8) is used when greater accuracy is required. A pipet is calibrated to deliver a fixed volume of liquid—for example,

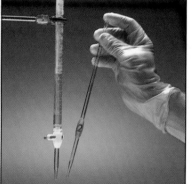

FIGURE 1.8

Measuring volume. A buret *(left)* delivers an accurately measured variable volume of liquid. A pipet *(right)* delivers a fixed volume (e.g., 25.00 mL) of liquid. *(Charles D. Winters)*

FIGURE 1.9
Weighing a solid. The solid sample plus the paper on which it rests weigh 144.998 g. The pictured balance is a single-pan analytical balance. *(Charles D. Winters)*

25.00 mL—when filled to the mark and allowed to drain. Variable volumes can be delivered accurately by a buret, perhaps to ± 0.01 mL.

In the metric system, **mass** is most commonly expressed in grams, kilograms, or milligrams:

$$1 \text{ g} = 10^{-3} \text{ kg} \qquad 1 \text{ mg} = 10^{-3} \text{ g}$$

This book weighs about 1.5 kg. The megagram, more frequently called the *metric ton,* is

$$1 \text{ Mg} = 10^6 \text{ g} = 10^3 \text{ kg}$$

> Writing "m" in upper case or lower case makes a big difference.

Properly speaking, there is a distinction between mass and weight. *Mass* is a measure of the amount of matter in an object; *weight* is a measure of the gravitational force acting on the object. Chemists often use these terms interchangeably; we determine the mass of an object by "weighing" it on a balance (Figure 1.9).

Temperature is the factor that determines the direction of heat flow. When two objects at different temperatures are placed in contact with one another, heat flows from the one at the higher temperature to the one at the lower temperature.

> See Screen 1.10, Temperature.

Temperature is measured indirectly, by observing its effect on the properties of a substance. A mercury-in-glass thermometer takes advantage of the fact that mercury, like other substances, expands as temperature increases. When the temperature rises, the mercury in the thermometer expands up a narrow tube. The total volume of the tube is only about 2% of that of the bulb at the base. In this way, a rather small change in volume is made readily visible.

Thermometers used in chemistry are marked in degrees *Celsius* (referred to as degrees centigrade until 1948). On this scale, named after the Swedish astronomer Anders Celsius (1701–1744), the freezing point of water is taken to be 0°C. The normal boiling point of water is 100°C. Household thermometers in the United States are commonly marked in *Fahrenheit* degrees. Daniel Fahrenheit (1686–1736) was a German instrument maker who was the first to use the mercury-in-glass thermometer. On this scale, the normal freezing and boiling points of water are taken to be 32° and 212°, respectively (Figure 1.10, page 10). It follows that (212°F–32°F) = 180°F covers the same temperature interval as (100°C–0°C) = 100°C. This leads to the general relation between the two scales:

> Many countries still use degrees centigrade.

$$t_{°F} = 1.8 \, t_{°C} + 32°$$

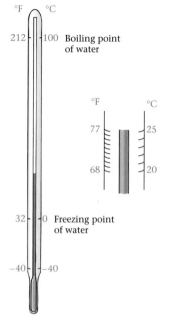

FIGURE 1.10
The Fahrenheit and Celsius temperature scales. The distance between the freezing and boiling points of water is 180° on the Fahrenheit scale and 100° on the Celsius scale. Thus the Celsius degree is 9/5 as large as the Fahrenheit degree, as is evident from the magnified section of the thermometer at the right. **OHT**

Notice (Figure 1.10) that the two scales coincide at $-40°$, as can readily be seen from the equation just written:

$$\text{At } -40°C: \qquad t_{°F} = 1.8(-40°) + 32° = -72° + 32° = -40°$$

For many purposes in chemistry, the most convenient unit of temperature is the **kelvin (K)**; note the absence of the degree sign. The kelvin is defined to be 1/273.16 of the difference between the lowest attainable temperature (0 K) and the triple point of water* (0.01°C). The relationship between temperature in K and in °C is

$$T_K = t_{°C} + 273.15$$

This scale is named after Lord Kelvin, a British scientist who showed in 1848, at the age of 24, that it is impossible to reach a temperature lower than 0 K.

EXAMPLE 1.1   Express normal body temperature, 98.60°F, in °C and K.

Strategy   Use the relations $t_{°F} = 1.8t_{°C} + 32°$ and $T_K = t_{°C} + 273.15$. Solve algebraically for the desired quantity, $t_{°C}$ in the first case, T_K in the second.

Solution

$$\text{To find } t_{°C}: \qquad 98.60° = 1.8t_{°C} + 32°$$

$$\text{Solving:} \qquad t_{°C} = \frac{98.60° - 32°}{1.8} = \boxed{37.00°C}$$

$$\text{To find } T_K: \qquad T_K = 37.00 + 273.15 = \boxed{310.15K}$$

As you can see from this discussion, a wide number of different units can be used to express measured quantities in the metric system. This proliferation of units has long been of concern to scientists. In 1960 a self-consistent set of metric units was proposed. This so-called International System of Units (SI) is discussed in Appendix 1. The SI units for the four quantities discussed are

Length: meter   (m)   *Mass:*   kilogram (kg)
Volume: cubic meter (m³)   *Temperature:* kelvin   (K)

Uncertainties in Measurements: Significant Figures

Every measurement carries with it a degree of uncertainty whose magnitude depends on the nature of the measuring device and the skill with which it is used. Suppose, for example, you measure out 8 mL of liquid using the 100-mL graduated cylinder shown in Figure 1.11, page 11. Here the volume is uncertain to perhaps ± 1 mL. With such a crude measuring device, you would be lucky to obtain a volume between 7 and 9 mL. To obtain greater precision, you could use a narrow 10-mL cylinder, which has divisions in small increments. You might now measure a

---

*The triple point of water (Chapter 9) is the one unique temperature at which ice, liquid water, and water vapor can coexist in contact with one another.

volume within 0.1 mL of the desired value, in the range of 7.9 to 8.1 mL. Using a buret, the uncertainty could be reduced to ± 0.01 mL.

Anyone making a measurement has a responsibility to indicate the uncertainty associated with it. Such information is vital to anyone who wants to repeat the experiment or judge its precision. The three volume measurements referred to earlier could be reported as

$$8 \pm 1 \text{ mL} \qquad \text{(large graduated cylinder)}$$
$$8.0 \pm 0.1 \text{ mL} \qquad \text{(small graduated cylinder)}$$
$$8.00 \pm 0.01 \text{ mL} \qquad \text{(buret)}$$

In this text, we will drop the $\pm$ notation and simply write

$$8 \text{ mL} \qquad 8.0 \text{ mL} \qquad 8.00 \text{ mL}$$

When we do this, it is understood that there is an ***uncertainty of at least one unit in the last digit***—that is, 1 mL, 0.1 mL, 0.01 mL, respectively. This method of citing the degree of confidence in a measurement is often described in terms of **significant figures,** the meaningful digits obtained in a measurement. In 8.00 mL there are three significant figures; each of the three digits has experimental meaning. Similarly, there are two significant figures in 8.0 mL and one significant figure in 8 mL.

Frequently we need to know the number of significant figures in a measurement reported by someone else (Example 1.2).

> There's a big difference between 8 mL and 8.00 mL, perhaps as much as half a milliliter.

EXAMPLE 1.2   Three different students weigh the same object, using different balances. They report the following masses:
(a) 15.02 g   (b) 15.0 g   (c) 0.01502 kg
How many significant figures are there in each value?

Strategy   Assume each student reported the mass in such a way as to indicate the uncertainty associated with the measurement. Then follow a commonsense approach.

Solution

(a) 4
(b) 3 The zero after the decimal point is significant. It indicates that the object was weighed to the nearest 0.1 g.
(c) 4 The zeros at the left are not significant. They are there only because the mass was expressed in kilograms rather than grams. Note that 15.02 g and 0.01502 kg represent the same mass.

Reality Check   If you express these masses in exponential notation as 1.502×10^1 g, 1.50×10^1 g, and 1.502×10^{-2} kg, the number of significant figures becomes obvious: 4 in (a), 3 in (b), and 4 in (c).

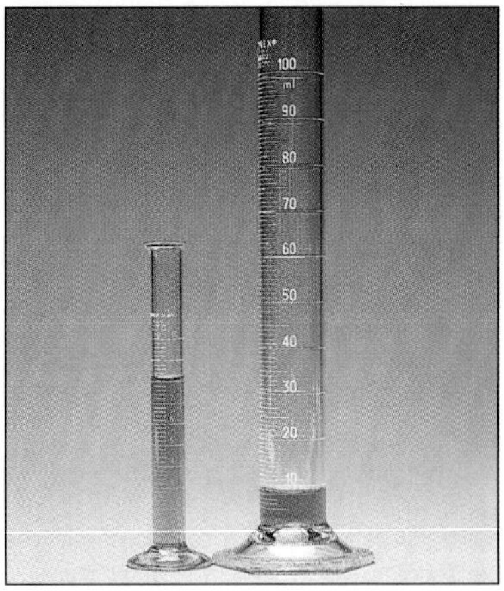

FIGURE 1.11
Uncertainty in measuring volume. The uncertainty depends on the nature of the measuring device. Eight milliliters of liquid can be measured with less uncertainty in the 10-mL graduated cylinder than in the 100-mL graduated cylinder. *(Marna G. Clarke)*

Sometimes the number of significant figures in a reported measurement is ambiguous. Suppose that a piece of metal is reported to weigh 500 g. You cannot be sure how many of these digits are meaningful. Perhaps the metal was weighed to the nearest gram (500 ± 1 g). If so, the 5 and the two zeros are significant; there are three significant figures. Then again, the metal might have been weighed only to the nearest 10 g (500 ± 10 g). In this case, only the 5 and one zero are known accurately; there are two significant figures. About all you can do in such cases is to wish the person who carried out the weighing had used exponential notation. The mass should have been reported as

$$5.00 \times 10^2 \text{ g} \quad \text{(3 significant figures)}$$

or

$$5.0 \times 10^2 \text{ g} \quad \text{(2 significant figures)}$$

or

$$5 \times 10^2 \text{ g} \quad \text{(1 significant figure)}$$

In general, *any ambiguity concerning the number of significant figures in a measurement can be resolved by using exponential notation* (often referred to as "scientific notation"), discussed in Appendix 3.

Most measured quantities are not end results in themselves. Instead, they are used to calculate other quantities, often by multiplication or division. The precision of any such derived result is limited by that of the measurements on which it is based. **When measured quantities are multiplied or divided, the number of significant figures in the result is the same as that in the quantity with the smallest number of significant figures.**

EXAMPLE 1.3 An overseas flight leaves New York in the late afternoon and arrives in London 8.50 hours later. The airline distance from New York to London is about 5.6×10^3 km, depending to some extent on the flight path followed. What is the average speed of the plane, in kilometers per hour?

Strategy Calculate the average speed by taking the quotient

$$\text{speed} = \frac{\text{distance traveled}}{\text{time elapsed}}$$

Count the number of significant figures in the numerator and in the denominator; the smaller of these two numbers is the number of significant figures in the quotient.

Solution The average speed will appear on your calculator as

$$\frac{5.6 \times 10^3 \text{ km}}{8.50 \text{ h}} = 658.8235294 \text{ km/h}$$

There are three significant figures in the denominator and two in the numerator. The answer should have two significant figures; round off the average speed to 6.6×10^2 km/h.

Expanding on the calculation in Example 1.3, we can see the reason behind the rule for the number of significant figures retained in multiplication or division. When we say that the distance is 5.6×10^3 km, we mean that it lies between 5.5×10^3 and 5.7×10^3 km. Similarly, if the time is quoted to ± 0.01 h, its true value should lie between 8.49 and 8.51 h. We see then that the average speed might be as large as

$$\frac{5.7 \times 10^3 \text{ km}}{8.49 \text{ h}} = 6.7 \times 10^2 \text{ km/h}$$

On the other hand, the average speed could be as little as

$$\frac{5.5 \times 10^3 \text{ km}}{8.51 \text{ h}} = 6.5 \times 10^2 \text{ km/h}$$

Looking at the results of these calculations, we see that it is entirely reasonable to report the average speed to be 6.6×10^2 km/h, implying an uncertainty of $\pm 0.1 \times 10^2$ km/h.

When measured quantities are added or subtracted, the uncertainty in the result is found in a quite different way than in multiplication and division. It is determined by counting the number of decimal places, that is, the number of digits to the right of the decimal point for each measured quantity. *When measured quantities are added or subtracted, the number of decimal places in the result is the same as that in the quantity with the greatest uncertainty and hence the smallest number of decimal places.*

To illustrate this rule, suppose you want to find the total mass of a solution made up of 10.21 g of instant coffee, 0.2 g of sugar, and 256 g of water.

	Mass	Uncertainty	
Instant coffee	10.21 g	± 0.01 g	2 decimal places
Sugar	0.2 g	± 0.1 g	1 decimal place
Water	256 g	± 1 g	0 decimal places
Total mass	266 g		

Because there are no digits after the decimal point in the mass of water, there are none in the total mass. Looking at it another way, we can say that the total mass, 266 g, has an uncertainty of ± 1 g, as does the mass of water, the quantity with the greatest uncertainty.

In applying the rules governing the use of significant figures, you should keep in mind that certain numbers involved in calculations are exact rather than approximate. To illustrate this situation, consider the equation relating Fahrenheit and Celsius temperatures:

$$t_{\,^\circ \mathrm{F}} = 1.8 t_{\,^\circ \mathrm{C}} + 32^\circ$$

The numbers 1.8 and 32 are exact. Hence they do not limit the number of significant figures in a temperature conversion; that limit is determined only by the precision of the thermometer used to measure temperature.

A different type of exact number arises in certain calculations. Suppose you are asked to determine the amount of heat evolved when *one kilogram* of coal burns. The implication is that because "one" is spelled out, *exactly* one kilogram of coal burns. The uncertainty in the answer should be independent of the amount of coal.

A number that is spelled out (one, two, . . .) does not affect the number of significant figures

Conversion of Units

It is often necessary to convert a measurement expressed in one unit (e.g., cubic centimeters) to another unit (liters). To do this we follow what is known as a **conversion factor** approach. For example, to convert a volume of 536 cm^3 to liters, the relation

$$1 \text{ L} = 1000 \text{ cm}^3$$

is used. Dividing both sides of this equation by 1000 cm^3 gives a quotient equal to 1:

$$\frac{1 \text{ L}}{1000 \text{ cm}^3} = \frac{1000 \text{ cm}^3}{1000 \text{ cm}^3} = 1$$

See Screen 1.17, Using Numerical Information.

The quotient 1 L/1000 cm³, which is called a *conversion factor,* is multiplied by 536 cm³. Because the conversion factor equals 1, this does not change the actual volume. However, it does accomplish the desired conversion of units. The cm³ in the numerator and denominator cancel to give the desired unit: liters.

$$536 \text{ cm}^3 \times \frac{1 \text{ L}}{1000 \text{ cm}^3} = 0.536 \text{ L}$$

The relation 1 L = 1000 cm³ can be used equally well to convert a volume in liters, say, 1.28 L, to cubic centimeters. In this case, the necessary conversion factor is obtained by dividing both sides of the equation by 1 L:

$$\frac{1000 \text{ cm}^3}{1 \text{ L}} = \frac{1 \text{ L}}{1 \text{ L}} = 1$$

Volume and exact conversions. The conversion factors are exact within the English or metric systems (*left,* 32 fluid oz/1 qt); the liter (*right*) is exactly equal to 100 mL. (*Charles D. Winters*)

Multiplying 1.28 L by the quotient 1000 cm³/1 L converts the volume from liters to cubic centimeters:

$$1.28 \text{ L} \times \frac{1000 \text{ cm}^3}{1 \text{ L}} = 1280 \text{ cm}^3 = 1.28 \times 10^3 \text{ cm}^3$$

Notice that a single relation (1 L = 1000 cm³) gives two conversion factors:

$$\frac{1 \text{ L}}{1000 \text{ cm}^3} \quad \text{and} \quad \frac{1000 \text{ cm}^3}{1 \text{ L}}$$

To go from cubic centimeters to liters, use the ratio 1 L/1000 cm³; to go from liters to cubic centimeters, use the ratio 1000 cm³/1 L. In general, when you make a conversion, **choose the factor that cancels out the initial unit.**

$$\text{initial quantity} \times \text{conversion factor(s)} = \text{desired quantity}$$

Conversions between English and metric units can be made using Table 1.3.

TABLE 1.3	**Relations Between Length, Volume, and Mass Units** OHT				
Metric		**English**		**Metric-English**	
Length					
1 km	= 10^3 m	1 ft	= 12 in	1 in	= 2.54 cm*
1 cm	= 10^{-2} m	1 yd	= 3 ft	1 m	= 39.37 in
1 mm	= 10^{-3} m	1 mi	= 5280 ft	1 mi	= 1.609 km
1 nm	= 10^{-9} m = 10Å				
Volume					
1 m³	= 10^6 cm³ = 10^3 L	1 gal	= 4 qt = 8 pt	1 ft³	= 28.32 L
1 cm³	= 1 mL = 10^{-3} L	1 qt (U.S. liq)	= 57.75 in³	1 L	= 1.057 qt (U.S. liq)
Mass					
1 kg	= 10^3 g	1 lb	= 16 oz	1 lb	= 453.6 g
1 mg	= 10^{-3} g	1 short ton	= 2000 lb	1 g	= 0.03527 oz
1 metric ton	= 10^3 kg			1 metric ton	= 1.102 short ton

*This conversion factor is exact; the inch is defined to be exactly 2.54 cm. The other factors listed in this column are approximate, quoted to four significant figures. Additional digits are available if needed for very accurate calculations. For example, the pound is defined to be 453.59237 g.

EXAMPLE 1.4   According to a highway sign, the distance from St. Louis to Chicago is 295 miles. Express this distance in kilometers.

Strategy   Use Table 1.3 to find a relation between miles and kilometers. Write the conversion factor in such a way that miles cancel out and are replaced by kilometers.

Solution   From Table 1.3, the required relation is

$$1 \text{ mile} = 1.609 \text{ km}$$

Because the initial unit, miles, is in the numerator, the conversion factor must have miles in the denominator, 1.609 km/1 mile:

$$295 \text{ miles} \times \frac{1.609 \text{ km}}{1 \text{ mile}} = \boxed{475 \text{ km}}$$

Reality Check   Because a kilometer is shorter than a mile, it makes sense that the number of kilometers (475) should be larger than the number of miles (295).

A highway sign in Missouri.
(Beverly March)

We wouldn't need to do calculations like this if the United States joined the rest of the world in using metric units.

Frequently it is necessary to carry out more than one conversion to work a problem. This can be done by setting up successive conversion factors (Example 1.5).

EXAMPLE 1.5   A certain U.S. car has a fuel efficiency rating of 36.2 mi/gal. Convert this to kilometers per liter.

Strategy   Use Table 1.3 to find a relation between kilometers and miles and between gallons and liters. Sometimes there is no direct relation shown in the table; in that case, use two relations to accomplish the desired conversion. Set up the arithmetic in a single expression, writing conversion factors in such way that, after cancellation, only the desired unit remains.

Solution   The relations to be used are

$$1 \text{ mile} = 1.609 \text{ km}$$
$$1 \text{ gallon} = 4 \text{ quarts} \qquad \text{(This is an exact relation.)}$$
$$1 \text{ L} = 1.057 \text{ quart}$$

Using these relations as conversion factors gives

$$36.2 \, \frac{\text{mi}}{\text{gal}} \times \frac{1.609 \text{ km}}{1 \text{ mi}} \times \frac{1 \text{ gal}}{4 \text{ qt}} \times \frac{1.057 \text{ qt}}{1 \text{ L}} = \boxed{15.4 \text{ km/L}}$$

Notice that three conversion factors are required. First miles are converted to kilometers to obtain the fuel efficiency in kilometers per gallon. Then gallons are converted to quarts, and, finally, quarts to liters.

1.3 PROPERTIES OF SUBSTANCES

Every pure substance has its own unique set of properties that serve to distinguish it from all other substances. A chemist most often identifies an unknown substance by measuring its properties and comparing them with the properties recorded in the chemical literature for known substances.

CHEMISTRY ▪ The Human Side

ANTOINE LAVOISIER
(1743–1794)

⌐Lavoisier was executed because he was a tax collector; chemistry had nothing to do with it.

The discussion in Section 1.2 emphasizes the importance of making precise numerical measurements. Chemistry was not always so quantitative. The following recipe for finding the philosopher's stone was recorded more than 300 years ago.

Take all the mineral salts there are, also all salts of animal and vegetable origin. Add all the metals and minerals, omitting none. Take two parts of the salts and grate in one part of the metals and minerals. Melt this in a crucible, forming a mass that reflects the essence of the world in all its colors. Pulverize this and pour vinegar over it. Pour off the red liquid into English wine bottles, filling them half-full. Seal them with the bladder of an ox (*not* that of a pig). Punch a hole in the top with a coarse needle. Put the bottles in hot sand for three months. Vapor will escape through the hole in the top, leaving a red powder. . . .

One man more than any other transformed chemistry from an art to a science. Antoine Lavoisier was born in Paris; he died on the guillotine during the French Revolution. Above all else, Lavoisier understood the importance of carefully controlled, quantitative experiments. These were described in his book *Elements of Chemistry.* Published in 1789, it is illustrated with diagrams by his wife.

The results of one of Lavoisier's quantitative experiments are shown in Table A; the data are taken directly from Lavoisier. If you add up the masses of reactants and products (expressed in arbitrary units), you find them to be the same, 510. As Lavoisier put it, "In all of the operations of men and nature, nothing is created. An equal quantity of matter exists before and after the experiment."

TABLE A **Quantitative Experiment on the Fermentation of Wine (Lavoisier)**

Reactants	Mass (Relative)	Products	Mass (Relative)
Water	400	Carbon dioxide	35
Sugar	100	Alcohol	58
Yeast	10	Acetic acid	3
		Water	409
		Sugar (unreacted)	4
		Yeast (unreacted)	1

This was the first clear statement of the law of conservation of mass (Chapter 2), which was the cornerstone for the growth of chemistry in the nineteenth century. Again, to quote Lavoisier, "it is on this principle that the whole art of making experiments is founded."

(Photo credit: Northwind Picture Archive)

The properties used to identify a substance must be **intensive;** that is, they must be independent of amount. The fact that a sample weighs 4.02 g or has a volume of 229 mL tells us nothing about its identity; mass and volume are **extensive** properties; that is, they depend on amount. Beyond that, substances may be identified on the basis of their—

- **chemical properties,** observed when the substance takes part in a **chemical reaction,** a change that converts it to a new substance. For example, the fact that mercury(II) oxide decomposes to mercury and oxygen on heating to 600°C can be used to identify it. Again, the chemical inertness of helium helps to distinguish it from other, more reactive gases, such as hydrogen and oxygen.
- **physical properties,** observed without changing the chemical identity of a substance. Two such properties that are particularly useful for identifying a substance are
  - *melting point,* the temperature at which a substance changes from the solid to the liquid state.
  - *boiling point,* the temperature at which bubbles filled with vapor form within a liquid. If a substance melts at 0°C and boils at 100°C, we are inclined to suspect that it might just be water.

In the remainder of this section we will consider a few other physical properties that can be measured without changing the identity of a substance.

Density

The **density** of a substance is the ratio of mass to volume:

$$\text{density} = \frac{\text{mass}}{\text{volume}} \qquad d = \frac{m}{V}$$

Note that even though mass and volume are extensive properties, the ratio of mass to volume is intensive. Samples of copper weighing 1.00 g, 10.5 g, 264 g, . . . all have the same density, 8.94 g/mL at 25°C.

Density can be found in a straightforward way by measuring, independently, the mass and volume of a sample (Example 1.6).

EXAMPLE 1.6   To determine the density of ethyl alcohol, a student pipets a 5.00-mL sample into an empty flask weighing 15.246 g. He finds that the mass of the flask + ethyl alcohol = 19.171 g. Calculate the density of ethyl alcohol.

Strategy   Determine the mass of the alcohol by subtracting the mass of the empty flask from the mass of the flask and the alcohol. The volume is given. Take the quotient of mass/volume as the density.

Solution

mass of ethyl alcohol = 19.171 g − 15.246 g = 3.925 g

volume of ethyl alcohol = 5.00 mL

density = 3.925 g/5.00 mL = 0.785 g/mL

Note that the density is expressed to three significant figures because the volume (5.00 mL) contains only three significant figures.

Taste is a physical property, but it's never measured in the lab.

Properties of gold. The color of gold is an intensive property. The quantity of gold in a sample is an extensive property. The fact that gold can be stored in the air without undergoing any chemical reaction with oxygen in the air is a chemical property. The temperature at which gold melts (1063°C) is a physical property.
(Charles D. Winters)

(a)

(b)

Density. (a) The pieces of styrofoam and rock are similar in size but very different in density. (b) The block has a lower density than water and floats, and the ring has a higher density than water and sinks. *(Charles D. Winters)*

⌐ The "100 g of water" in this expression is an exact quantity.

⌐ When the temperature changes, the amount of solute in solution changes but the mass of water stays the same.

In a practical sense, density can be treated as a conversion factor to relate mass and volume. Knowing that mercury has a density of 13.6 g/mL, we can calculate the mass of 2.6 mL of mercury:

$$2.6 \text{ mL} \times 13.6 \frac{\text{g}}{\text{mL}} = 35 \text{ g}$$

or the volume occupied by one kilogram of mercury:

$$1.000 \text{ kg} \times \frac{10^3 \text{ g}}{1 \text{ kg}} \times \frac{1 \text{ mL}}{13.6 \text{ g}} = 73.5 \text{ mL}$$

Solubility

The process by which a solute dissolves in a solvent is ordinarily a physical rather than a chemical change. The extent to which it dissolves can be expressed in various ways. A common method is to state the number of grams of the substance that dissolves in 100 g of solvent at a given temperature. At 20°C, about 32 g of potassium nitrate dissolves in 100 g of water. At 100°C, the solubility of this solid is considerably greater, about 246 g/100 g of water.

EXAMPLE 1.7   Taking the solubility of potassium nitrate, KNO_3, to be 246 g/100 g of water at 100°C and 32 g/100 g of water at 20°C, calculate
(a) the mass of water required to dissolve one hundred grams of KNO_3 at 100°C.
(b) The amount of KNO_3 that remains in solution when the mixture in (a) is cooled to 20°C.

Strategy   The solubility at a particular temperature gives you a relationship between grams of solute (KNO_3) and grams of solvent (water). This in turn leads to the conversion factor required to calculate the mass of water in (a) or that of KNO_3 in (b). Note that the temperature is 100°C in (a), 20°C in (b).

Solution

(a) Mass of water required = $100 \text{ g } KNO_3 \times \dfrac{100 \text{ g water}}{246 \text{ g } KNO_3}$ = **40.7 g water**

(b) Because the solution contains 40.7 g of water,

mass of KNO_3 in solution = $40.7 \text{ g water} \times \dfrac{32 \text{ g } KNO_3}{100 \text{ g water}}$ = **13 g KNO_3**

Reality Check   The remaining 87 g of potassium nitrate crystallizes out of solution on cooling. Most solids are less soluble at low temperatures.

Color; Absorption Spectrum

Some of the substances you work with in general chemistry can be identified at least tentatively by their color. Gaseous nitrogen dioxide has a brown color; vapors of bromine and iodine are red and violet, respectively. A water solution of copper sulfate is blue, and a solution of potassium permanganate is purple (Figure 1.12).

The colors of gases and liquids are due to the selective absorption of certain components of visible light. Bromine, for example, absorbs in the violet and blue regions of the spectrum (Table 1.4). The subtraction of these components from vis-

TABLE 1.4   **Relation Between Color and Wavelength**

Wavelength (nanometers)	Color Absorbed	Color Transmitted
< 400 nm	Ultraviolet	Colorless
400–450 nm	Violet	} Red, Orange, Yellow
450–500 nm	Blue	
500–550 nm	Green	} Purple
550–580 nm	Yellow	
580–650 nm	Orange	} Blue, Green
650–700 nm	Red	
> 700 nm	Infrared	Colorless

FIGURE 1.12

Drops of potassium permanganate falling into water. The purple color of the solution results from absorption at approximately 550 nm.
(Charles D. Winters)

ible light accounts for the red color of bromine liquid or vapor. The purple (blue-red) color of a potassium permanganate solution results from absorption in the green region.

The wavelengths of visible light range from 400 to 700 nm. For a substance to be colored, it must absorb somewhere within this region. However, many colorless substances absorb in the ultraviolet (< 400 nm) or infrared (> 700 nm) regions. Ozone in the upper atmosphere absorbs harmful, high-energy ultraviolet radiation from the Sun. Carbon dioxide absorbs infrared radiation given off by the Earth's surface, preventing it from escaping into outer space, and thereby contributing to global warming.

Organic compounds, all of which contain (at a minimum) the elements carbon and hydrogen, absorb at well-characterized wavelengths in the infrared. By exposing an organic compound such as ethyl alcohol to infrared radiation covering a range of wavelengths, an *absorption spectrum* of the type shown in Figure 1.13 can be obtained. This spectrum serves as a fingerprint to distinguish ethyl alcohol from all other substances.

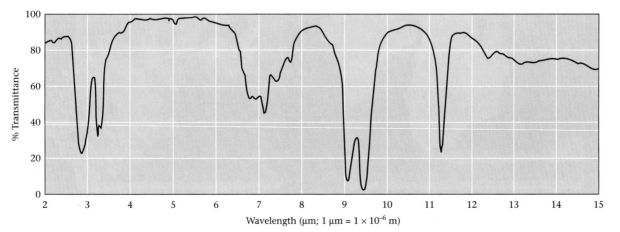

FIGURE 1.13

Infrared spectrum of ethyl alcohol. The percent transmittance is the percent of the infrared radiation that passes through the sample without being absorbed. The absorption varies with the wavelength and each downward spike is the absorption at that wavelength.

CHEMISTRY
Beyond the Classroom

Carbon Allotropes

Several elements exist in more than one form in the same physical state; these forms are called allotropes. All of us are familiar with two allotropic forms of solid carbon: diamond and graphite. As you can see from Figure A, these allotropes have quite different structures. Diamond is a rigid, three-dimensional network; graphite has a two-dimensional layer structure. This difference is reflected in their properties. Diamond is one of the hardest substances known; graphite is soft enough to be used as a lubricant. A "lead" pencil really contains a graphite rod, thin layers of which rub off on the paper as you write.

At room temperature and atmospheric pressure, graphite is the stable form of carbon. Diamond, in principle, should slowly transform to graphite under ordinary conditions. Fortunately, for the owners of diamond rings, this transition occurs at zero rate unless the diamond is heated to about 1500°C, at which temperature the conversion occurs rapidly. For understandable reasons, no one has ever become very excited over the commercial possibilities of this process. The more difficult task of converting graphite to diamond has aroused much greater enthusiasm.

Diamond has a higher density than graphite (3.51 versus 2.26 g/cm^3); as the more dense phase, it should be favored by high pressures. Theoretically, at 25°C and 15,000 atm graphite should turn to diamond. However, under those conditions the reaction has a negligible rate. At higher temperatures it goes faster, but the required pressure goes up too; at 2000°C, a pressure of about 100,000 atm is needed. In 1954 scientists at the General Electric laboratories were able to achieve these high temperatures and pressures and converted graphite carbon to diamond. At present, all industrial diamonds are synthetic.

(a)

(b)

FIGURE A
Two allotropes of carbon. The black spheres represent the arrangements of carbon atoms. (a) Graphite is made of two-dimensional layers of carbon atoms. (b) Diamond is made of a rigid three-dimensional network of carbon atoms. *(Photos, Charles D. Winters; models, S. M. Young).*

In 1985 a third allotropic form of carbon was discovered by a group of chemists including Richard E. Smalley and Robert F. Curl at Rice University and Harold W. Kroto at the University of Sussex, England. Eleven years later they were awarded a Nobel Prize for their research. The structure of this allotrope is shown at the top of Figure B. Sixty carbon atoms are arranged in a nearly spherical cage. More exactly, the geometry of this molecule is that of a polygon with 32 faces; 12 of these are pentagons, 20 are hexagons.

If you're a sports fan, you've almost certainly seen this structure before: It is that of a soccer ball. (The next time you have nothing to do, try counting the number of pentagons and hexagons on a soccer ball.) Smalley and his colleagues considered names such as "carbosoccer" or "soccerene" for this allotropic form of carbon. In the end, though, they called it "buckminsterfullerene" in honor of the controversial architect R. Buckminster Fuller, who was fascinated by geodesic domes that at least vaguely resembled truncated soccer balls. Less formally, buckminsterfullerene is called "buckyball."

Macroscopic quantities of buckyball did not become available until 1990, when a group of physicists worked out a process for its synthesis. They vaporized graphite electrodes in a helium atmosphere, producing a black soot containing up to 20% of this allotrope of carbon. The process is slow and relatively expensive; the price of buckyball per gram is many times that of gold.

In 1991 Japanese scientists discovered another carbon structure related to those of both graphite and buckminsterfullerene. This is a hollow cylindrical tube called a "nanotube" (Figure C) because it has a diameter of about one nanometer (10^{-9} m). Typically, nanotubes have a length of at least one micrometer (10^{-6} m). The carbon atoms that form the framework of the tube have an open hexagonal pattern like graphite (or rolled-up chicken wire). The tubes are closed at both ends by carbon atoms forming a hemisphere in the buckyball (or geodesic dome) pattern. Nanotubes have a tensile strength far greater than that of ordinary carbon fibers, which suggests that they might be used in an infinitely flexible, unbreakable "graphite" fly rod (quite possibly selling for less than $100,000).

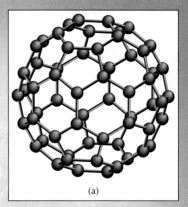

(a)

(b)

Figure B

Buckminsterfullerene (buckyball). (a) The arrangement of the 60 carbon atoms in a buckyball. (b) The identical arrangement of the hexagons and pentagons on the surface of a soccer ball.

(b, George Semple)

Figure C

Carbon nanotube model. The tube can be imagined as half buckyballs (C_{60}) held together by C_6 rings. Each tube has a diameter of about 1.4 nm, hence the name nanotube.

(Lawrence Berkeley Laboratory)

CHAPTER HIGHLIGHTS

Key Concepts

1. Convert between °F, °C, and K.
   (**Example 1.1; Problems 15–18, 66**)
2. Determine the number of significant figures in a measured quantity.
   (**Example 1.2; Problems 19, 20**)
3. Determine the number of significant figures in a calculated quantity.
   (**Example 1.3; Problems 21–24**)
4. Use conversion factors to change the units of a measured quantity.
   (**Examples 1.4, 1.5, 1.7; Problems 31–42, 51, 52, 55, 58, 67, 69**)
5. Relate density to mass and volume.
   (**Example 1.6; Problems 43–50, 56–60, 68**)

Key Equations

Fahrenheit temperature	$t_{°F} = 1.8\, t_{°C} + 32°$
Kelvin temperature	$T_K = t_{°C} + 273.15$
Density	$d = m/V$

Key Terms

centi	kilo-	property
chemical reaction	metric system	—chemical
compound	milli-	—extensive
conversion factor	mixture	—intensive
density	—homogeneous	—physical
element	—heterogeneous	significant figures
kelvin (K)	nano-	solution

Summary Problem

Potassium permanganate is a purple compound containing three elements: potassium, manganese, and oxygen. It has a density of 2.703 g/cm^3; its melting point is 2.40×10^2 °C. At 20°C, its solubility is 6.38 g/100 g water; at 75°C, the solubility is 32.3 g/100 g water.

(a) What are the symbols of the three elements in potassium permanganate?
(b) List the physical properties of potassium permanganate given above.
(c) What is the volume of 246 g of potassium permanganate?
(d) Express the density in pounds per cubic foot.
(e) Express the melting point of potassium permanganate in °F and K.
(f) How much potassium permanganate can be dissolved in 63.5 g of water at 20°C?
(g) How much water is required to dissolve 28.9 g of potassium permanganate at 75°C?

(h) Can one dissolve 12.0 g of potassium permanganate in 45.0 g of water at 75°C? If so, how much potassium permanganate remains dissolved when the solution (12.0 g potassium permanganate/45.0 g water) is cooled from 75°C to 20°C? How much, if any, crystallizes out?
(i) Why is potassium permanganate purple? What colors of visible light does it absorb?

Express all your answers to the correct number of significant figures; use the conversion factor approach throughout.

Answers

(a) K, Mn, O

(b) density, melting point, solubility, color

(c) 91.0 cm^3

(d) 168.7 lb/ft^3

(e) 464°F; 513 K

(f) 4.05 g

(g) 89.5 g

(h) yes; 2.87 g will dissolve; 9.1 g will crystallize out

(i) See Figure 1.12. It absorbs green and yellow.

Questions & Problems

Problem numbers in blue indicate that the answer is available in Appendix 6 at the back of the book.
WEB indicates that the solution is posted at **http://www.harcourtcollege.com/chem/general/masterton4/student/**

The questions and problems listed here are typical of those at the end of each chapter. Some are conceptual. Most require calculations, writing equations, or other quantitative work. The headings identify the primary topic of each set of questions or problems, such as "Symbols and Formulas" or "Significant Figures." Those in the "Unclassified" category may involve more than one concept, including, perhaps, topics from a pre-

ceding chapter. "Challenge Problems," listed at the end of the set, require extra skill and/or effort. The "Classified" questions and problems (Problems 1–52 in this set) occur in matched pairs, one below the other, and illustrate the same concept. For example, Questions 1 and 2 are nearly identical in nature; the same is true of Questions 3 and 4, and so on.

Types of Matter

1. Classify each of the following as element, compound, or mixture.
    (a) silver
    (b) ethyl alcohol
    (c) milk
    (d) aluminum
2. Classify each of the following as element, compound, or mixture.
    (a) platinum
    (b) table salt
    (c) soy sauce
    (d) sugar
3. Classify the following as solution or heterogeneous mixture.
    (a) maple syrup
    (b) seawater
    (c) melted rocky road ice cream
4. Classify the following as solution or heterogeneous mixture.
    (a) wine
    (b) gasoline
    (c) batter for chocolate chip cookies

5. Suggest a method to separate the different components in
    (a) a seawater-sand mixture.
    (b) a solution of ethyl alcohol in grape juice.
6. Suggest a method to separate the different components in
    (a) a car's gaseous emission.
    (b) a mixture of carbon and water.
7. Write the symbol for the following elements.
    (a) chlorine   **(b)** phosphorus
    (c) potassium   **(d)** mercury
8. Write the symbol for the following elements.
    (a) sodium   **(b)** nitrogen   **(c)** nickel   **(d)** lead
9. Write the name of the element represented by the following symbols:
    (a) Au   **(b)** Al   **(c)** Ag   **(d)** Mg
10. Write the name of the element represented by the following symbols:
    (a) Si   **(b)** S   **(c)** Fe   **(d)** Zn

Measurements

11. What quantity and unit is given for
    (a) a bag of potatoes?
    (b) the freezing point of water?
    (c) a bottle of vinegar?

12. What quantity and unit is given for
    (a) an aspirin tablet?
    (b) a bottle of fruit juice?
    (c) the thermostat setting for a furnace?
13. Write the appropriate symbol in the blank ($>$, $<$, or $=$).
    (a) 37.12 g _____ 0.3712 kg
    (b) 28 m³ _____ 28×10^2 cm³
    (c) 525 mm _____ 525×10^6 nm
14. Write the appropriate symbol in the blank ($>$, $<$, or $=$).
    (a) 303 m _____ 303×10^3 km
    (b) 500 g _____ 0.500 kg
    (c) 1.50 cm³ _____ 1.50×10^3 nm³
15. A glass of lukewarm milk is suggested for people who cannot sleep. Milk at 52°C can be characterized as lukewarm. What is the temperature of lukewarm milk in °F? In K?
WEB 16. A recipe for apple pie calls for a preheated 350°F (three significant figures) oven. Express this temperature setting in °C and in K.
17. Dry ice is prepared by freezing carbon dioxide at -56.5°C and 5 atm. What is the freezing point of carbon dioxide in Kelvin?
18. Computers are not supposed to be in very warm rooms. The highest temperature tolerated for maximum performance is 308 K. What is this temperature in °C? In °F?

Significant Figures

19. How many significant figures are there in each of the following?
    (a) 0.136 m   (b) 0.0001050 g
    (c) 2.700×10^3 nm   (d) 6×10^{-4} L   (e) 56003 cm³
20. How many significant figures are there in each of the following?
    (a) 12.7040 g   (b) 200.0 cm   (c) 276.2 tons
    (d) 4.00×10^3 mL   (e) 100°C
21. The volume of a sphere is $\frac{4}{3}\pi r^3$ where r is the radius. One student measured the radius to be 4.30 cm. Another measured the radius to be 4.33 cm. What is the difference in volume between the two measurements?
22. A graduated cylinder has a circular cross section with a radius of 2.500 cm. What is the volume of water in the graduated cylinder with a measured height of 1.20 cm? (The volume of a cylinder is $\pi r^2 h$, where r is the radius and h is the height.)
23. How many significant figures are there in the values of x obtained from
    (a) $x = \dfrac{34.0300 \text{ g}}{12.09 \text{ cm}^3}$
    (b) $x = (0.00630 \text{ cm})(2.003 \text{ cm})(200.0 \text{ cm})$
    (c) $x = 32.647 \text{ in} - 32.327 \text{ in}$
    (d) $x = \dfrac{236.45 \text{ g} - 1.3 \text{ g}}{(3.4561 \text{ cm})(32.567 \text{ cm}^2)}$

24. Calculate the following to the correct number of significant figures.
    (a) $x = \dfrac{2.63 \text{ g}}{4.982 \text{ cm}^3}$
    (b) $x = \dfrac{13.54 \text{ mi}}{5.00 \text{ hr}}$
    (c) $x = 13.2 \text{ g} + 1468 \text{ g} + 0.04 \text{ g}$
    (d) $x = \dfrac{2 \text{ g} + 0.127 \text{ g} + 459 \text{ g}}{6.2 \text{ cm}^3 - 0.567 \text{ cm}^3}$
25. Round off the following quantities to the indicated number of significant figures.
    (a) 132.505 g (four significant figures)
    (b) 298.693 cm (five significant figures)
    (c) 13.452 lb (two significant figures)
    (d) 345 oz (two significant figures)
26. Round off the following quantities to the indicated number of significant figures.
    (a) 7.4855 g (three significant figures)
    (b) 298.693 cm (five significant figures)
    (c) 11.698 lb (one significant figure)
    (d) 12.05 oz (three significant figures)
27. Express the following measurements in scientific notation:
    (a) 4633.2 mg   (b) 0.000473 L   (c) 127,000.0 cm³
28. Express the following measurements in scientific notation:
    (a) 4020.6 mL   (b) 1.006 g   (c) 100.1°C
29. Which of the following statements use exact numbers?
    (a) There are 12 dozen apples in a gross of apples.
    (b) The tomatoes that you are buying weigh 4.5 lb.
    (c) Your temperature is 99.2°F.
30. Which of the following statements use exact numbers?
    (a) One foot equals 12 in.
    (b) We will deliver 13 potted geraniums to your house tomorrow.
    (c) This car has a gasoline mileage of 22.6 mi/gal.

Conversion Factors

31. Convert 6743 nm to
    (a) Å   (b) inches   (c) miles
WEB 32. Convert 22.3 mL to
    (a) liters   (b) in³   (c) quarts
33. The height of a horse is usually measured in hands. One hand is exactly 1/3 ft.
    (a) How tall (in feet) is a horse of 19.2 hands?
    (b) How tall (in meters) is a horse of 17.8 hands?
    (c) A horse of 20.5 hands is to be transported in a trailer. The roof of the trailer needs to provide 3.0 ft of vertical clearance. What is the minimum height of the trailer in feet?

34. At sea, distances are measured in nautical miles and speeds are expressed in knots.

$$1 \text{ nautical mile} = 6076.12 \text{ ft}$$

$$1 \text{ knot} = 1 \text{ nautical mi/hr (exactly)}$$

(a) How many miles are there in one nautical mile?

(b) How many meters are there in one nautical mile?

(c) A ship is traveling at a rate of 22 knots. Express the ship's speed in miles per hour.

35. A light-year is defined to be the distance traveled by light in one year. If the speed of light is 3.0×10^{10} cm/s, how many miles are there in one light-year?

36. The unit of land measure in the metric system is the hectare. The English system uses the acre. A hectare is equal to the area of a square of exactly one hundred meters on a side. If one hectare is equal to 2.47 acres, how many square feet are there in one acre?

37. Cholesterol in blood is measured in milligrams of cholesterol per deciliter of blood. If the unit of measurement were changed to grams of cholesterol per milliliter of blood, what would a cholesterol reading of 185 mg/dL translate to?

38. An average adult has 6.0 L of blood. The Red Cross usually takes one pint of blood from each donor at a donation. What percentage (by volume) of a person's blood does a blood donor give in one donation?

39. Some states have reduced the legal limit for alcohol sobriety from 0.10% to 0.080% alcohol by volume in blood plasma.

(a) How many milliliters of alcohol are there in 3.0 qt of blood plasma at the lower legal limit?

(b) How many milliliters of alcohol are there in 3.0 qt of blood plasma at the higher legal limit?

(c) How much less alcohol is there in 3.0 qt of blood plasma with the reduced sobriety level?

40. The area of the 48 contiguous states is 3.02×10^6 mi^2. Assume that these states are completely flat (no mountains and no valleys). What volume of water, in liters, would cover these states with a rainfall of two inches?

41. When the *Exxon Valdez* ran aground off the coast of Alaska, 2.5×10^5 barrels of crude oil were spilled. There are exactly 42 gal to a barrel. If the oil was allowed to flow and fill a two-car garage with dimensions $8 \times 25 \times 25$ ft, how many of these garages would be filled by the oil spilled from the tanker?

42. Silver dollars must contain 90.0% silver. A silver dollar has a mass of 27.0 g. In February 1999 the price of silver was $4.32 an ounce. In February 1999 did the silver dollar have more value as currency or as a source for silver?

Physical and Chemical Properties

43. The volume occupied by egg whites from four "large" eggs is 112 mL. The mass of the egg whites from these four eggs is 1.20×10^2 g. What is the average density of the egg white from one "large" egg?

44. The cup is a measure of volume widely used in cookbooks. One cup is equivalent to 225 mL. What is the density of clover honey (in grams per milliliter) if three quarters of a cup has a mass of 252 g?

45. A metal slug weighing 25.17 g is added to a flask with a volume of 59.7 mL. It is found that 43.7 g of methanol ($d = 0.791$ g/mL) must be added to the metal to fill the flask. What is the density of the metal?

46. A solid with an irregular shape and a mass of 11.33 g is added to a graduated cylinder filled with water ($d = 1.00$ g/mL) to the 35.0-mL mark. After the solid sinks to the bottom, the water level is read to be at the 42.3-mL mark. What is the density of the solid?

47. A cube of ice ($d = 0.917$ g/mL) is 2.50 in on a side. How many milliliters of water ($d = 1.00$ g/mL) are obtained when the ice melts?

48. A sheet of aluminum foil that is 11 in wide and 12 in long weighs 8.9 g. If aluminum has a density of 2.70 g/cm^3, what is the thickness, in millimeters, of the foil? (Assume that the foil is pure aluminum.)

49. Air is 21% oxygen by volume. Oxygen has a density of 1.31 g/L. What is the volume, in liters, of a room that holds enough air to contain 55 kg of oxygen?

WEB **50.** Vinegar contains 5.00% acetic acid by mass and has a density of 1.01 g/mL. What mass (in grams) of acetic acid is present in 5.00 L of vinegar?

51. The solubility of ammonium bromide in water at 20°C is 75.5 g/100 g water. Its solubility at 50°C is 99.2 g/100 g water. Calculate

(a) the mass of ammonium bromide that dissolves in 62.0 g of water at 20°C.

(b) the mass of water required to dissolve 125 g of ammonium bromide at 50°C.

(c) the mass of ammonium bromide that would not remain in solution if a solution made up of 29.0 g of ammonium bromide in 35.0 g of water at 50°C is cooled to 20°C.

52. The solubility of potassium chloride is 37.0 g/100 g water at 30°C. Its solubility at 70°C is 48.3 g/100 g water.

(a) Calculate the mass of potassium chloride that dissolves in 48.6 g of water at 30°C.

(b) Calculate the mass of water required to dissolve 52.0 g of potassium chloride at 70°C.

(c) If 30.0 g of KCl were added to 75.0 g of water at 30°C, would it all disappear? If the temperature were increased to 70°C, would it then all dissolve?

Unclassified

53. The following data refer to the compound water. Classify each as a chemical or a physical property.

(a) It is a colorless liquid at 25°C and 1 atm.
(b) It reacts with sodium to form hydrogen gas as one of the products.
(c) Its melting point is 0°C.
(d) It is insoluble in carbon tetrachloride.

54. The following data refer to the element phosphorus. Classify each as a physical or a chemical property.
(a) It exists in several forms, for example, white, black, and red phosphorus.
(b) It is a solid at 25°C and 1 atm.
(c) It is insoluble in water.
(d) It burns in chlorine to form phosphorus trichloride.

55. A cup of brewed coffee is made with about 9.0 g of ground coffee beans. If a student brews three cups of gourmet coffee a day, how much does the student spend on a year's supply of gourmet coffee that sells at $8.99/lb?

56. Lead has a density of 11.34 g/cm³ and oxygen has a density of 1.31×10^{-3} g/cm³ at room temperature. How many cm³ are occupied by one gram of lead? By one gram of oxygen? Comment on the difference in volume for the two elements.

57. A roll of aluminum foil sold in the supermarket is $66\frac{2}{3}$ yards by 12 inches. It has a mass of 0.83 kg. If the density of aluminum foil is 2.70 g/cm³, what is the thickness of the foil in inches?

58. The Kohinoor Diamond ($d = 3.51$ g/cm³) is 108 carats. If one carat has a mass of 2.00×10^2 mg, what is the mass of the Kohinoor Diamond in pounds? What is the volume of the diamond in cubic inches?

59. A pycnometer is a device used to measure density. It weighs 20.455 g empty and 31.486 g when filled with water ($d = 1.00$ g/cm³). Pieces of an alloy are put into the empty, dry pycnometer. The mass of the alloy and pycnometer is 28.695 g. Water is added to the alloy to exactly fill the pycnometer. The mass of the pycnometer, water, and alloy is 38.689 g. What is the density of the alloy?

60. Titanium is used in airplane bodies because it is strong and light. It has a density of 4.55 g/cm³. If a cylinder of titanium is 7.75 cm long and has a mass of 153.2 g, calculate the diameter of the cylinder. ($V = \pi r^2 h$, where V is the volume of the cylinder, r is its radius, and h is the height.)

Conceptual Questions

61. How do you distinguish
(a) density from solubility?
(b) an element from a compound?
(c) a solution from a heterogeneous mixture?

62. How do you distinguish
(a) chemical properties from physical properties?
(b) distillation from filtration?
(c) a solute from a solution?

63. Why is the density of a regular soft drink larger than that of a diet soft drink?

64. Mercury, ethyl alcohol, and lead are poured into a cylinder. Three distinct layers are formed. The densities of the three substances are

$$\text{mercury} = 13.55 \text{ g/cm}^3$$
$$\text{ethyl alcohol} = 0.78 \text{ g/cm}^3$$
$$\text{lead} = 11.4 \text{ g/cm}^3$$

Sketch the cylinder with the three layers. Identify the substance in each layer.

65. Given the following solubility curves, answer the following questions:
(a) In which of the two compounds can more solute be dissolved in the same amount of water when the temperature is decreased?
(b) At what temperature is the solubility of both compounds the same?
(c) Will an increase in temperature always increase solubility of a compound? Explain.

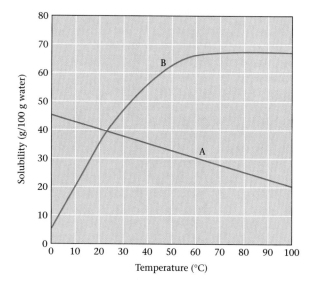

Challenge Problems

66. At what point is the temperature in °F exactly twice that in °C?

67. Oil spreads on water to form a film about 100 nm thick (two significant figures). How many square kilometers of ocean will be covered by the slick formed when one barrel of oil is spilled (1 barrel = 31.5 U.S. gal)?

68. A laboratory experiment requires twelve grams of aluminum wire ($d = 2.70$ g/cm^3). The diameter of the wire is 0.200 in. Determine the length of the wire, in centimeters, to be used for this experiment. The volume of a cylinder is $\pi r^2 \ell$, where r = radius and ℓ = length.

69. An average adult breathes about 8.50×10^3 L of air per day. The concentration of lead in highly polluted urban air is 7.0×10^{-6} g of lead per one m^3 of air. Assume that 75% of the lead is present as particles less than 1.0×10^{-6} m in diameter, and that 50% of the particles below that size are retained in the lungs. Calculate the mass of lead absorbed in this manner in one year by an average adult living in this environment.

2 ATOMS, MOLECULES, AND IONS

The element sulfur has the symbol S. *(Charles D. Winters)*

CHAPTER OUTLINE

2.1 ATOMS AND THE ATOMIC THEORY

2.2 COMPONENTS OF THE ATOM

2.3 INTRODUCTION TO THE PERIODIC TABLE

2.4 MOLECULES AND IONS

2.5 FORMULAS OF IONIC COMPOUNDS

2.6 NAMES OF COMPOUNDS

See the *Saunders Interactive General Chemistry CD-ROM*, Screen 2.3, **Origins of Atomic Theory.**

To learn chemistry, you must become familiar with the building blocks that chemists use to describe the structure of matter. These include—

- *atoms* (Section 2.1), which in turn are composed of electrons, protons, and neutrons (Section 2.2).

- *molecules,* the building blocks of several elements and a great many compounds. Molecular substances can be identified by their formulas (Section 2.4) or their names (Section 2.6).

- *ions,* species of opposite charge found in all ionic compounds. Using relatively simple principles, it is possible to derive the formulas (Section 2.5) and names (Section 2.6) of ionic compounds.

Early in this chapter (Section 2.3) we will introduce a classification system for elements known as the *periodic table*. It will prove useful in this chapter and throughout the remainder of this text.

2.1 ATOMS AND THE ATOMIC THEORY

In 1808, an English scientist and schoolteacher, John Dalton, developed the atomic model of matter that underlies modern chemistry. Three of the main postulates of modern atomic theory, all of which Dalton suggested, are stated below and illustrated in Figure 2.1.

1. *An element is composed of tiny particles called atoms.* All atoms of a given element show the same chemical properties. Atoms of different elements show different properties.
2. *In an ordinary chemical reaction, atoms move from one substance to another, but no attom of any element disappears or is changed into an atom of another element.*
3. *Compounds are formed when atoms of two or more elements combine.* In a given compound, the relative numbers of atoms of each kind are definite and constant. In general, these relative numbers can be expressed as integers or simple fractions.

On the basis of Dalton's theory, the **atom** can be defined as the smallest particle of an element that can enter into a chemical reaction.

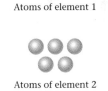

Atoms of element 1

Atoms of element 2

Compound 1

Compound 2

Different combinations produce different compounds

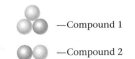

Atoms of different elements have different masses

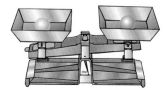

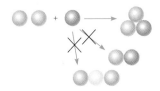

No atom disappears or is changed in a chemical reaction

FIGURE 2.1

Some features of Dalton's atomic theory. OHT

CHEMISTRY ▪ The Human Side

JOHN DALTON
(1766—1844)

⊙ See Screen 2.5, The Dalton Atomic Theory.

Dalton was a quiet, unassuming man and a devout Quaker. When presented to King William IV of England, Dalton refused to wear the colorful court robes because of his religion. His friends persuaded him to wear the scarlet robes of Oxford University, from which he had a doctor's degree. Dalton was color-blind, so he saw himself clothed in gray.

Dalton was a prolific scientist who made contributions to biology and physics as well as chemistry. At a college in Manchester, England, he did research and spent as many as 20 hours a week lecturing in mathematics and the physical sciences. Dalton never married; he said once, "My head is too full of triangles, chemical properties, and electrical experiments to think much of marriage."

Dalton's atomic theory explained three of the basic laws of chemistry:

The **law of conservation of mass:** This states that *there is no detectable change in mass in an ordinary chemical reaction.* If atoms are conserved in a reaction (postulate 2 of the atomic theory), mass will also be conserved.

The **law of constant composition:** This tells us that *a compound always contains the same elements in the same proportions by mass.* If the atom ratio of the elements in a compound is fixed (postulate 3), their proportions by mass must also be fixed.

The **law of multiple proportions:** This law, formulated by Dalton himself, was crucial to establishing atomic theory. It applies to situations in which two elements form more than one compound. The law states that in these compounds, *the masses of one element that combine with a fixed mass of the second element are in a ratio of small whole numbers.*

The validity of this law depends on the fact that atoms combine in simple, whole-number ratios (postulate 3). Its relation to atomic theory is further illustrated in Figure A.

(Photo Credit: The E.F. Smith Memorial Collection in the History of Chemistry, Department of Special Collections, Van Pelt–Dietrich Library, University of Pennsylvania)

FIGURE A
Chromium-oxygen compounds and the law of multiple proportions. Chromium forms two different compounds with oxygen, as shown by their different colors. In the green compound on the left, there are two chromium atoms for every three oxygen atoms (2Cr:3O) and 2.167 g of chromium per gram of oxygen. In the red compound on the right there is one chromium atom for every three oxygen atoms (1Cr:3O) and 1.083 g of chromium per gram of oxygen. The ratio of the chromium masses, 2.167:1.083, is that of two small whole numbers, 2.167:1.083 = 2:1, an illustration of the law of multiple proportions. *(Charles D. Winters)*

2.2   COMPONENTS OF THE ATOM

Like any useful scientific theory, the atomic theory raised more questions than it answered. Scientists wondered whether atoms, tiny as they are, could be broken down into still smaller particles. Nearly 100 years passed before the existence of subatomic particles was confirmed by experiment. Two future Nobel laureates did pioneer work in this area. J. J. Thomson was an English physicist working at the Cavendish Laboratory at Cambridge. Ernest Rutherford, at one time a student of Thomson's (Figure 2.2), was a native of New Zealand. Rutherford carried out his research at McGill University in Montreal and at Manchester and Cambridge in England. He was clearly the greatest experimental physicist of his time, and one of the greatest of all time.

FIGURE 2.2
**J. J. Thomson and Ernest Rutherford
(right).** They are talking, perhaps about nuclear physics, but more likely about yesterday's cricket match. *(AIP Niels Bohr Library, Bainbridge Collection)*

Electrons

The first evidence for the existence of subatomic particles came from studies of the conduction of electricity through gases at low pressures. When the glass tube shown in Figure 2.3 is partially evacuated and connected to a spark coil, an electric current flows through it. Associated with this flow are colored rays of light called *cathode rays,* which are bent by both electric and magnetic fields. From a careful study of this deflection, J. J. Thomson showed in 1897 that the rays consist of a stream of negatively charged particles, which he called **electrons.** We now know that electrons are common to all atoms, carry a unit negative charge (-1), and have a very small mass, roughly 1/2000 that of the lightest atom.

⊙ See Screen 2.2, Introduction to Atoms, & 2.4, The Discovery of Atomic Structure: Milestones.

⊙ See Screen 2.8, Electrons.

Every atom contains a definite number of electrons. This number, which runs from 1 to more than 100, is characteristic of a neutral atom of a particular element. All atoms of hydrogen contain one electron; all atoms of the element uranium contain 92 electrons. We will have more to say in Chapter 6 about how these electrons are arranged relative to one another. Right now, you need only know that they are found in the outer regions of the atom, where they form what amounts to a cloud of negative charge.

Protons and Neutrons; the Atomic Nucleus

A series of experiments carried out under the direction of Ernest Rutherford in 1911 shaped our ideas about the nature of the atom. He and his students bombarded a piece of thin gold foil (Figure 2.4, page 32) with α-particles (helium atoms minus their electrons). With a fluorescent screen, they observed the extent

⊙ See Screen 2.10, Protons.

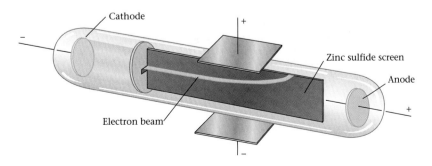

FIGURE 2.3
A cathode ray passing through an electric field. The ray consists of fast-moving electrons. Within the electric field, the ray is deflected toward the positive pole, showing that it is negatively charged. **OHT**

See Screen 2.11, The Nucleus of the Atom.

See Screen 2.14, Summary of Atomic Composition.

FIGURE 2.4

Rutherford's α-particle scattering experiment. The zinc sulfide screen fluoresces when struck by an α-particle. Most of the α-particles are deflected very little, but a few are scattered at large angles. The large deflections occur when the α-particles collide with heavy, positively charged nuclei. **OHT**

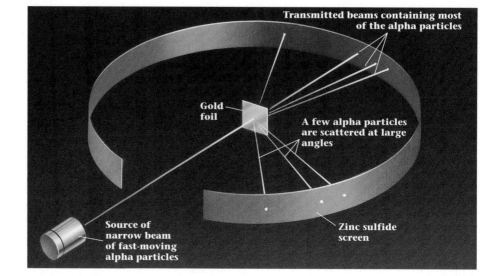

Before this experiment, it was believed that positive and negative particles were more or less uniformly distributed through the atom.

The diameter of an atom is 10,000 times that of its nucleus.

to which the α-particles were scattered. Most of the particles went through the foil unchanged in direction; a few, however, were reflected back at acute angles. This was a totally unexpected result, inconsistent with the model of the atom in vogue at that time. In Rutherford's words, "It was as though you had fired a 15-inch shell at a piece of tissue paper and it had bounced back and hit you." By a mathematical analysis of the forces involved, Rutherford showed that the scattering was caused by a small, positively charged **nucleus** at the center of the gold atom. Most of the atom is empty space, which explains why most of the bombarding particles passed through the gold foil undeflected.

Since Rutherford's time scientists have learned a great deal about the properties of atomic nuclei. For our purposes in chemistry, the nucleus of an atom can be considered to consist of two different types of particles (Table 2.1):

1. The **proton,** which has a mass nearly equal to that of an ordinary hydrogen atom. The proton carries a unit positive charge (+ 1), equal in magnitude to that of the electron (− 1).
2. The **neutron,** an uncharged particle with a mass slightly greater than that of a proton.

Because protons and neutrons are much heavier than electrons, most of the mass of an atom (>99.9%) is concentrated in the nucleus, even though the volume of the nucleus is much smaller than that of the atom.

TABLE 2.1	Properties of Subatomic Particles		
Particle	Location	Relative Charge	Relative Mass*
Proton	Nucleus	+ 1	1.00728
Neutron	Nucleus	0	1.00867
Electron	Outside nucleus	− 1	0.00055

*These are expressed in atomic mass units (Chapter 3).

Atomic Number

All the atoms of a particular element have the same number of protons in the nucleus. This number is a basic property of an element, called its **atomic number** and given the symbol Z:

$$Z = \text{number of protons}$$

In a neutral atom, the number of protons in the nucleus is exactly equal to the number of electrons outside the nucleus. Consider, for example, the elements hydrogen ($Z = 1$) and uranium ($Z = 92$). All hydrogen atoms have one proton in the nucleus; all uranium atoms have 92. In a neutral hydrogen atom there is one electron outside the nucleus; in a uranium atom there are 92.

H atom:	1 proton, 1 electron	$Z = 1$
U atom:	92 protons, 92 electrons	$Z = 92$

Mass Numbers; Isotopes

The **mass number** of an atom, given the symbol A, is found by adding up the number of protons and neutrons in the nucleus:

$$A = \text{number of protons} + \text{number of neutrons}$$

All atoms of a given element have the same number of protons, hence the same atomic number. They may, however, differ from one another in mass and therefore in mass number. This can happen because, although the number of protons in an atom of an element is fixed, the number of neutrons is not. It may vary and often does. Consider the element hydrogen ($Z = 1$). There are three different kinds of hydrogen atoms. They all have one proton in the nucleus. A light hydrogen atom (the most common type) has no neutrons in the nucleus ($A = 1$). Another type of hydrogen atom (deuterium) has one neutron ($A = 2$). Still a third type (tritium) has two neutrons ($A = 3$).

Atoms that contain the same number of protons but a different number of neutrons are called **isotopes.** The three kinds of hydrogen atoms just described are isotopes of that element. They have masses that are very nearly in the ratio $1:2:3$. Among the isotopes of the element uranium are the following:

Isotope	Z	A	Number of Protons	Number of Neutrons
uranium-235	92	235	92	143
uranium-238	92	238	92	146

The composition of a nucleus is shown by its **nuclear symbol.** Here, the atomic number appears as a subscript at the lower left of the symbol of the element. The mass number is written as a superscript at the upper left.

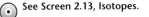
Mass number $\longrightarrow A$
Atomic number $\longrightarrow Z$ X $\longleftarrow$ element symbol

The nuclear symbols for the isotopes of hydrogen and uranium referred to above are

$$^1_1H,\ ^2_1H,\ ^3_1H \qquad ^{235}_{92}U,\ ^{238}_{92}U$$

> Number of neutrons = $A - Z$.

⊙ See Screen 2.13, Isotopes.

$^2_1H_2O$ ice and $^1_1H_2O$ ice. Solid deuterium oxide *(left)* is more dense than liquid water and sinks, whereas the ordinary water ice *(right)* is less dense than liquid water and floats. *(Charles D. Winters)*

Quite often, isotopes of an element are distinguished from one another by writing the mass number after the symbol of the element. The isotopes of uranium are often referred to as U-235 and U-238.

EXAMPLE 2.1

(a) An isotope of cobalt (Co) is used in radiation therapy for certain types of cancer. Write nuclear symbols for three isotopes of cobalt ($Z = 27$) in which there are 29, 31, and 33 neutrons, respectively.

(b) One of the most harmful components of nuclear waste is a radioactive isotope of strontium, $^{90}_{38}Sr$; it can be deposited in your bones, where it replaces calcium. How many protons are there in the nucleus of Sr-90? How many neutrons?

Strategy Remember the definitions of atomic number and mass number and where they appear in the nuclear symbol.

Solution

(a) The mass numbers are

$$27 + 29 = 56 \qquad 27 + 31 = 58 \qquad 27 + 33 = 60$$

Thus the nuclear symbols are $^{56}_{27}Co$, $^{58}_{27}Co$, $^{60}_{27}Co$.

(Cobalt-60 is the isotope used in cancer treatment; it is radioactive.)

(b) The number of protons is given by the atomic number (left subscript) and is **38.** The mass number (left superscript) is 90. The number of neutrons is $90 - 38 =$ **52.**

Nuclear Stability; Radioactivity

There are eight known isotopes of carbon:

$$^{9}_{6}C \qquad ^{10}_{6}C \qquad ^{11}_{6}C \qquad ^{12}_{6}C \qquad ^{13}_{6}C \qquad ^{14}_{6}C \qquad ^{15}_{6}C \qquad ^{16}_{6}C$$

Of these, two ($^{12}_{6}C$, $^{13}_{6}C$) are stable in the sense that they do not decompose over time. In contrast, the three lighter nuclei ($^{9}_{6}C$, $^{10}_{6}C$, $^{11}_{6}C$) and the three heavier nuclei ($^{14}_{6}C$, $^{15}_{6}C$, $^{16}_{6}C$) are unstable; as time passes they decompose to other nuclei.

Whether or not a given nucleus will be stable depends on its neutron-to-proton ratio. The 264 known stable nuclei, shown as red dots in Figure 2.5, fall within a relatively narrow "belt of stability"; $^{12}_{6}C$ and $^{13}_{6}C$ are in that category. Isotopes falling outside that belt because they have too few neutrons ($^{9}_{6}C$, $^{10}_{6}C$, $^{11}_{6}C$) or too many neutrons ($^{14}_{6}C$, $^{15}_{6}C$, $^{16}_{6}C$) are unstable.

As you can see from Figure 2.5, the neutron-to-proton ratio required for stability varies with atomic number. For light elements ($Z < 20$), this ratio is close to 1. For example, the isotopes $^{12}_{6}C$, $^{14}_{7}N$, and $^{16}_{8}O$ are stable. As atomic number increases, the ratio increases; the "belt of stability" shifts to higher numbers of neutrons. With very heavy isotopes such as $^{206}_{82}Pb$, the stable neutron-to-proton ratio is about 1.5:

$$(206 - 82)/82 = 124/82 = 1.51$$

Unstable isotopes decompose (*decay*) by a process referred to as **radioactivity.** Ordinarily the result is the *transmutation* of elements; the atomic number of the product nucleus differs from that of the reactant. For example, radioactive de-

See Screen 2.7, Evidence of Subatomic Particles.

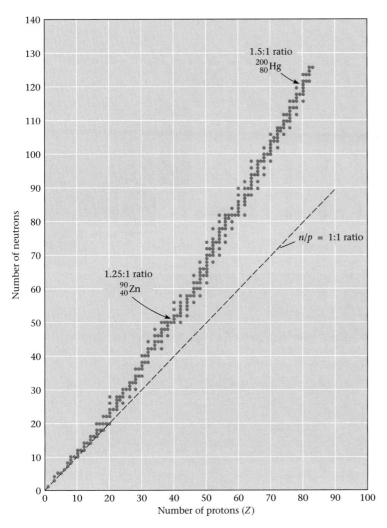

FIGURE 2.5
Neutron-to-proton ratios of stable isotopes. The ratios of stable isotopes (blue dots) fall within a narrow range, referred to as the "belt of stability." For light isotopes of small atomic number the stable ratio is 1:1. For heavier isotopes the ratio gradually increases to about 1.5:1. Isotopes outside the band of stability are unstable and radioactive. There are no stable isotopes for elements of atomic number greater than 83 (Bi). **OHT**

cay of $^{14}_{6}C$ produces a stable isotope of nitrogen, $^{14}_{7}N$. The radiation given off (Figure 2.6) may be in the form of—

- *beta* (β) particles identical in their properties to electrons.
- *alpha* (α) particles, which are $^{4}_{2}He$ nuclei, carrying a +2 charge because the two extranuclear electrons of the helium atom are missing.
- *gamma* (γ) rays, which consist of high-energy radiation.

2.3   INTRODUCTION TO THE PERIODIC TABLE

From a microscopic point of view, an element is a substance all of whose atoms have the same number of protons, that is, the same atomic number. The chemical properties of elements depend upon their atomic numbers, which can be read from the periodic table. A complete **periodic table** that lists symbols, atomic numbers, and atomic masses is given on the inside front cover of this text. For our purposes in this chapter, the abbreviated table shown in Figure 2.7, page 36, will suffice.

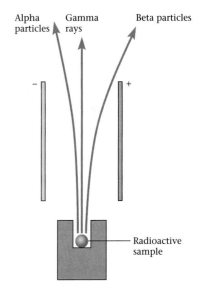

FIGURE 2.6
Nuclear radiation. In the presence of an electric field, alpha particles are deflected toward the negative pole, showing that they are positively charged, and beta particles are deflected toward the positive pole, showing that they are negatively charged. Because they are lighter, beta particles are deflected more than alpha particles. Gamma rays pass straight through the field and so must have no charge. **OHT**

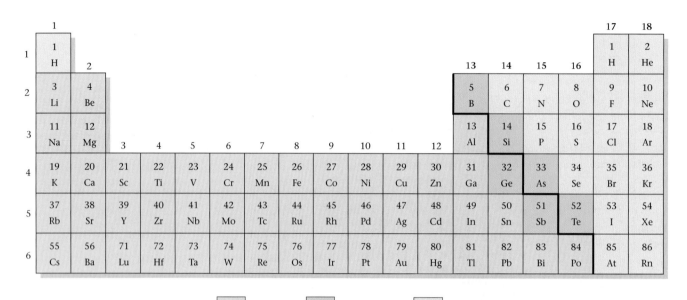

	Metals		Metalloids		Nonmetals

FIGURE 2.7

Periodic table. The group numbers stand above the columns. The numbers at the left of the rows are the period numbers. (A complete periodic table is given inside the front cover.) **OHT**

See Screen 2.16, The Periodic Table.

Other periodic tables label the groups differently, but the elements have the same position.

You'll have to wait until Chapter 7 to learn why some group numbers are in bold type.

Periods and Groups

The horizontal rows in the table are referred to as **periods.** The first period consists of the two elements hydrogen (H) and helium (He). The second period starts with lithium (Li) and ends with neon (Ne).

The vertical columns are known as **groups** or **families.** Historically, many different systems have been used to designate the different groups. Both Arabic and Roman numerals have been used in combination with the letters A and B. The system used in this text is the one recommended by the International Union of Pure and Applied Chemistry (IUPAC) in 1985. The groups are numbered from 1 to **18,** starting at the left.

Elements falling in Groups 1, 2, **13**, **14**, **15**, **16**, **17**, and **18*** are referred to as **main-group elements.** The ten elements in the center of periods 4 through 6 are called **transition metals;** they fall in Groups 3 through 12. The first transition series (period 4) starts with Sc (Group 3) and ends with Zn (Group 12).

The metals in groups **13**, **14**, and **15**, which lie to the right of the transition metals (Ga, In, Tl, Sn, Pb, Bi), are often referred to as *post-transition metals.*

Certain main groups are given special names. The elements in Group 1, at the far left of the periodic table, are called *alkali metals;* those in Group 2 are referred to as *alkaline earth metals.* Moving to the right, the elements in Group **17** are called *halogens;* at the far right, the *noble* (unreactive) *gases* constitute Group **18.**

Elements in the same main group show very similar chemical properties. For example—

*Prior to 1985, Groups **13** to **18** were commonly numbered 3 to 8 or 3A to 8A in the United States.

- lithium (Li), sodium (Na), and potassium (K) in Group 1 all react vigorously with water to produce hydrogen gas.
- helium (He), neon (Ne), and argon (Ar) in Group **18** do not react with any other substances.

On the basis of observations such as these, we can say that *the periodic table is an arrangement of elements, in order of increasing atomic number, in horizontal rows of such a length that elements with similar chemical properties fall directly beneath one another in vertical groups.*

⊙ See Screen 2.17, Chemical Periodicity.

The person whose name is most closely associated with the periodic table is Dmitri Mendeleev (1836–1907), a Russian chemist. In writing a textbook of general chemistry, Mendeleev devoted separate chapters to families of elements with similar properties, including the alkali metals, the alkaline earth metals, and the halogens. Reflecting on the properties of these and other elements, he proposed in 1869 a primitive version of today's periodic table. Mendeleev shrewdly left empty spaces in his table for new elements yet to be discovered. Indeed, he predicted detailed properties for three such elements (scandium, gallium, and germanium). By 1886 all of these elements had been discovered and found to have properties very similar to those he had predicted.

Metals and Nonmetals

The diagonal line or stairway that starts to the left of boron in the periodic table (Figure 2.7) separates metals from nonmetals. The more than 80 elements to the left and below that line, shown in blue in the table, have the properties of **metals;** in particular, they have high electrical conductivities. Elements above and to the right of the stairway are **nonmetals** (yellow); about 18 elements fit in that category.

Along the diagonal line in the periodic table are several elements that are difficult to classify exclusively as metals or nonmetals. They have properties between those of elements in the two classes. In particular, their electrical conductivities are intermediate between those of metals and nonmetals. The six elements

B	Si	Ge	As	Sb	Te
boron	silicon	germanium	arsenic	antimony	tellurium

are often called **metalloids.**

⌐ These elements, particularly Si, are used in semiconductors.

Sulfur   Selenium   Silver

(a)   (b)   (c)

Three elements. Sulfur (Group **16**) is a nonmetal. Selenium (Group **16**) is a metalloid. Silver (Group **11**) is a metal. *(Charles D. Winters)*

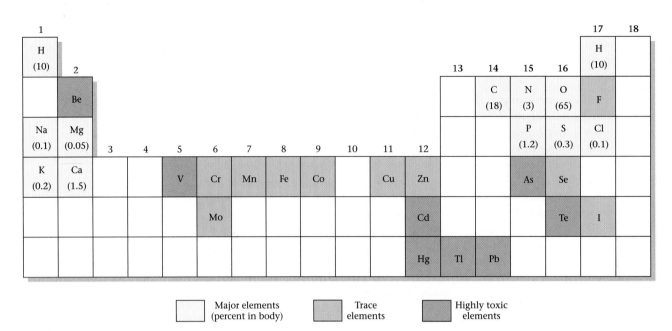

FIGURE 2.8
Biologically important elements and highly toxic elements. For the major elements in the body, their percent abundance in the body is given below the symbols. **OHT**

Figure 2.8 shows the biologically important elements. The "good guys," essential to life, include the major elements (yellow), which account for 99.9% of total body mass, and the trace elements (green) required in very small quantities. In general, the abundances of elements in the body parallel those in the world around us, but there are some important exceptions. Aluminum and silicon, although widespread in nature, are missing in the human body. The reverse is true of carbon, which makes up only 0.08% of the Earth's crust but 18% of the body, where it occurs in a variety of organic compounds including proteins, carbohydrates, and fats.

The "bad guys," shown in red in Figure 2.8, are toxic, often lethal, even in relatively small quantities. Several of the essential trace elements *become* toxic if their concentrations in the body increase. Selenium is a case in point. You need about 0.00005 g/day to maintain good health but 0.001 g/day can be deadly. That's a good thing to keep in mind if you're taking selenium supplements.

2.4 MOLECULES AND IONS

Isolated atoms rarely occur in nature; only the noble gases (He, Ne, Ar, . . .) consist of individual, nonreactive atoms. Atoms tend to combine with one another in various ways to form more complex structural units. Two such units, which serve as building blocks for a great many elements and compounds, are molecules and ions.

Molecules

Two or more atoms may combine with one another to form an uncharged **molecule.** The atoms involved are usually those of nonmetallic elements. Within the molecule, atoms are held to one another by strong forces called *covalent bonds,*

Super Glue is weak compared with covalent bonds.

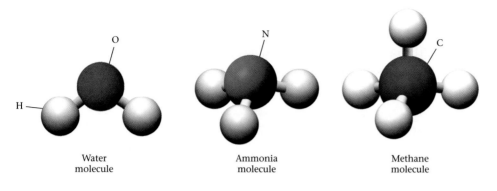

FIGURE 2.9
Ball-and-stick models of water (H₂O), ammonia (NH₃), and methane (CH₄). The "sticks" represent covalent bonds between H atoms and O, N, or C atoms. The models illustrate the geometry of the molecules (discussed in Chapter 7). **OHT**

which consist of shared pairs of electrons (Chapter 7). Forces between neighboring molecules, in contrast, are quite weak.

Molecular substances most often are represented by **molecular formulas,** in which the number of atoms of each element is indicated by a subscript written after the symbol of the element. Thus we interpret the molecular formulas for water (H_2O), ammonia (NH_3), and methane (CH_4) to mean that in—

⬛ the water molecule there are two hydrogen atoms and one oxygen atom.
⬛ the ammonia molecule, there is one nitrogen atom and three hydrogen atoms.
⬛ the methane molecule, there is one carbon atom and four hydrogen atoms.

The structures of molecules are sometimes represented by **structural formulas,** which show the bonding pattern within the molecule. The structural formulas of water, ammonia, and methane are

$$H—O—H \qquad H—N—H \atop \qquad\qquad\quad | \atop \qquad\qquad\quad H \qquad \begin{matrix} H \\ | \\ H—C—H \\ | \\ H \end{matrix}$$

The dashes represent covalent bonds. The geometries of these molecules are shown in Figure 2.9.

Sometimes we represent a molecular substance with a formula intermediate between a structural formula and a molecular formula. A **condensed structural formula** suggests the bonding pattern in the molecule and highlights the presence of a reactive group of atoms within the molecule. Consider, for example, the organic compounds commonly known as methyl alcohol and methylamine. Their structural formulas are

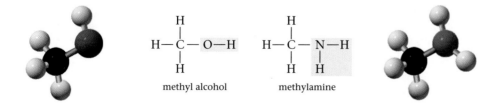

methyl alcohol   methylamine

The condensed structural formulas of these compounds are written as

$$CH_3\,OH \qquad CH_3\,NH_2$$

See Screen 3.4, Representing Compounds.

Structural formulas help us predict chemical properties.

		Group 17
Group 15	Group 16	H_2 (g)
N_2 (g)	O_2 (g)	F_2 (g)
P_4 (s)	S_8 (s)	Cl_2 (g)
		Br_2 (l)
		I_2 (s)

FIGURE 2.10
Molecular elements and their physical states at room temperature: gaseous (*g*), liquid (*l*), or solid (*s*). **OHT**

⊙ See the Screen 3.2, Elements that Exist as Molecules.

⊙ See Screen 3.7, Ions—Cations and Anions.

⌐ Metals form cations, nonmetals form anions; C, P, and the metalloids do not form monatomic ions.

These formulas take up considerably less space and emphasize the presence in the molecule of—

■ the OH group found in all *alcohols*, including ethyl alcohol, CH_3CH_2OH, the alcohol found in intoxicating beverages such as beer and wine.
■ the NH_2 group found in certain *amines*, including ethylamine, $CH_3CH_2NH_2$.

EXAMPLE 2.2   Give the molecular formulas of (a) ethyl alcohol and (b) ethylamine.

Strategy   To find the molecular formulas, simply add up the atoms of each type and use the sums as subscripts in the formulas.

Solution

(a) C_2H_6O (b) C_2H_7N

Reality Check   Note that although molecular formulas give the composition of the molecule, they reveal nothing about the way the atoms fit together. In that sense they are less useful than structural formulas.

Elements as well as compounds can exist as discrete molecules. In hydrogen gas, the basic building block is a molecule consisting of two hydrogen atoms joined by a covalent bond:

$$H—H$$

Other molecular elements are shown in Figure 2.10.

Ions

When an atom loses or gains electrons, charged particles called **ions** are formed. Metal atoms typically tend to lose electrons to form positively charged ions called **cations** (pronounced CAT-i-ons). Examples include the Na^+ and Ca^{2+} ions, formed from atoms of the metals sodium and calcium:

$$Na\ atom \longrightarrow Na^+\ ion + e^-$$
$$(11p^+, 11e^-)\quad (11p^+, 10e^-)$$

$$Ca\ atom \longrightarrow Ca^{2+}\ ion + 2e^-$$
$$(20p^+, 20e^-)\quad (20p^+, 18e^-)$$

(The arrows separate *reactants*, Na and Ca atoms, from *products*, cations and electrons.)

Nonmetal atoms form negative ions (**anions**—pronounced AN-i-ons) by gaining electrons. Consider, for example, what happens when atoms of the nonmetals chlorine and oxygen acquire electrons:

$$Cl\ atom + e^- \longrightarrow Cl^-\ ion$$
$$(17p^+, 17e^-)\quad\quad (17p^+, 18e^-)$$

$$O\ atom + 2e^- \longrightarrow O^{2-}\ ion$$
$$(8p^+, 8e^-)\quad\quad (8p^+, 10e^-)$$

Notice that when an ion is formed, the number of protons in the nucleus is unchanged. It is the number of electrons that increases or decreases.

EXAMPLE 2.3   Give the number of protons, neutrons, and electrons in $^{27}_{13}Al^{3+}$, a cation found in rubies and sapphires.

Strategy   The atomic number directly gives the number of protons. The +3 charge of the ion tells us that there are three more protons than electrons. The number of neutrons is deduced in the usual way by subtracting the atomic number from the mass number.

Solution   no. protons = $\boxed{13}$ no. electrons = 13 − 3 = $\boxed{10}$
no. neutrons = 27 − 13 = $\boxed{14}$

Reality Check   In any cation, such as Al^{3+}, there are more protons than electrons; in an anion, such as Cl^-, there are more electrons than protons.

The ions dealt with to this point (e.g., Na^+, Cl^-) are **monatomic;** that is, they are derived from a single atom by the loss or gain of electrons. Many of the most important ions in chemistry are **polyatomic,** containing more than one atom. Examples include the hydroxide ion (OH^-) and the ammonium ion (NH_4^+). In these and other polyatomic ions, the atoms are held together by covalent bonds, for example,

$$(O-H)^- \qquad \left(\begin{array}{c} H \\ | \\ H-N-H \\ | \\ H \end{array}\right)^+$$

In a very real sense, you can think of a polyatomic ion as a "charged molecule."

Because a bulk sample of matter is electrically neutral, ionic compounds always contain both cations (positively charged particles) and anions (negatively charged particles). Ordinary table salt, sodium chloride, is made up of an equal number of Na^+ and Cl^- ions. The structure of sodium chloride is shown in Figure 2.11. Notice that—

- there are two kinds of structural units in NaCl, the Na^+ and Cl^- ions.
- there are no discrete molecules; Na^+ and Cl^- ions are bonded together in a continuous network.

Ionic compounds are held together by strong electrical forces between oppositely charged ions (e.g., Na^+, Cl^-). These forces are referred to as **ionic bonds.**

We write +3 when describing the charge but 3+ when using it as a superscript in the formula of an ion.

Aluminum oxide. Corundum *(left)* is an Al_2O_3-containing mineral. Ruby *(top right)* is Al_2O_3 with Al^{3+} ions replaced by Cr^{3+} ions, and sapphire *(bottom right)* is Al_2O_3 with some Al^{3+} ions replaced by Fe^{3+} or Ti^{4+} ions. *(Charles D. Winters)*

You can't buy a bottle of Na^+ ions.

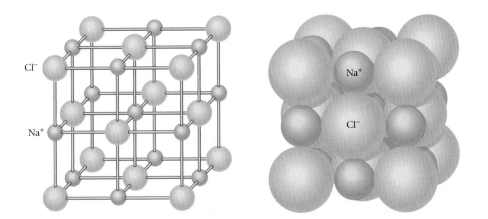

FIGURE 2.11
Sodium chloride structure. In these two ways of showing the structure, the small spheres represent Na^+ ions and the large spheres, Cl^- ions. Note that in any sample of sodium chloride there are equal numbers of Na^+ and Cl^- ions, but no NaCl molecules.

Typically, ionic compounds are solids at room temperature and have relatively high melting points (mp $NaCl = 801°C$, $CaCl_2 = 772°C$). To melt an ionic compound requires that oppositely charged ions be separated from one another, thereby breaking ionic bonds.

2.5 FORMULAS OF IONIC COMPOUNDS

See Screen 3.10, Ionic Compounds.

When a metal such as sodium (Na) or calcium (Ca) reacts with a nonmetal such as chlorine (Cl_2), the product is ordinarily an ionic compound. The formula of that compound (e.g., $NaCl$, $CaCl_2$) shows the simplest ratio between cation and anion (one Na^+ ion for one Cl^- ion; one Ca^{2+} ion for two Cl^- ions). In that sense, the formulas of ionic compounds are simplest formulas. Notice that the symbol of the metal (Na, Ca) always appears first in the formula, followed by that of the nonmetal.

To predict the formula of an ionic compound, you need to know the charges of the two ions involved. Then you can apply the principle of electrical neutrality, which requires that *the total positive charge of the cations in the formula must equal the total negative charge of the anions.* Consider, for example, the ionic compound calcium chloride. The ions present are Ca^{2+} and Cl^-. For the compound to be electrically neutral, there must be two Cl^- ions for every Ca^{2+} ion. The formula of calcium chloride must be $CaCl_2$, indicating that the simplest ratio of Cl^- to Ca^{2+} ions is 2:1.

Cations and Anions with Noble-Gas Structures

The charges of ions formed by atoms of the main-group elements can be predicted by applying a simple principle:

Atoms that are close to a noble gas (Group **18**) *in the periodic table form ions that contain the same number of electrons as the neighboring noble gas atom.*

This is reasonable; noble gas atoms must have an extremely stable electronic structure, because they are so unreactive. Other atoms might be expected to acquire noble gas electronic structures by losing or gaining electrons.

Applying this principle, you can deduce the charges of ions formed by main-group atoms:

Group	No. of Electrons in Atom	Charge of Ion Formed	Examples
1	1 more than noble gas atom	+1	Na^+, K^+
2	2 more than noble gas atom	+2	Mg^{2+}, Ca^{2+}
16	2 less than noble gas atom	−2	O^{2-}, S^{2-}
17	1 less than noble gas atom	−1	F^-, Cl^-

Two other ions that have noble-gas structures are

Al^{3+} (Al has three more e^- than the preceding noble gas, Ne)

N^{3-} (N has three fewer e^- than the following noble gas, Ne)

Cations of the Transition and Post-Transition Metals

Several metals that are farther removed from the noble gases in the periodic table form positive ions. These include the transition metals in Groups 3 to 12 and the post-transition metals in Groups **13** to **15**. The cations formed by these metals typically have charges of $+1$, $+2$, or $+3$ and ordinarily do not have noble-gas structures. We will postpone to Chapter 4 a general discussion of the specific charges of cations formed by these metals.

Many of the transition and post-transition metals form more than one cation. Consider, for example, iron in Group 8. This metal forms two different series of compounds with nonmetals. In one series, iron is present as a $+2$ cation

$$Fe^{2+}: \quad FeCl_2, FeBr_2, \ldots$$

In the other series, iron exists as a $+3$ cation

$$Fe^{3+}: \quad FeCl_3, FeBr_3, \ldots$$

Ionic compounds containing poly-atomic ions. Potassium dichromate ($K_2Cr_2O_7$, orange), potassium permanganate ($KMnO_4$, dark purple), and potassium nitrate (KNO_3, white).
(Charles D. Winters)

EXAMPLE 2.4 Predict the formulas of the ionic compounds

(a) formed by barium with iodine.   (b) formed by aluminum with oxygen.
(c) containing Cu^{2+} and oxide ions.   (d) containing Cu^+ and oxide ions.

Strategy First, identify the charges of the cation and anion. Then balance positive with negative charges to arrive at the formula.

Solution

(a) BaI_2 : one Ba^{2+} ion requires two I^- ions.
(b) Al_2O_3 : two Al^{3+} ions (total charge $= +6$) require three O^{2-} ions (total charge $= -6$).
(c) CuO : one Cu^{2+} ion balances one O^{2-} ion.
(d) Cu_2O : two Cu^+ ions balance one O^{2-} ion.

Polyatomic Ions

Table 2.2 lists some of the polyatomic ions that you will need to know, along with their names and charges. Notice that—

- there are only two common polyatomic cations, NH_4^+ and Hg_2^{2+}. *All other cations considered in this text are derived from individual metal atoms* (e.g., Na^+ from Na, Ca^{2+} from Ca, . . .).
- most of the polyatomic anions contain one or more oxygen atoms; collectively these species are called **oxoanions.**

See Screen 3.8, Polyatomic Ions.

Nearly all cations are monatomic; the majority of anions are poly-atomic.

TABLE 2.2 Some Common Polyatomic Ions OHT			
+1	**−1**	**−2**	**−3**
NH_4^+ (ammonium)	OH^- (hydroxide)	CO_3^{2-} (carbonate)	PO_4^{3-} (phosphate)
Hg_2^{2+} (mercury I)	NO_3^- (nitrate)	SO_4^{2-} (sulfate)	
	ClO_3^- (chlorate)	CrO_4^{2-} (chromate)	
	ClO_4^- (perchlorate)	$Cr_2O_7^{2-}$ (dichromate)	
	CN^- (cyanide)	HPO_4^{2-} (hydrogen phosphate)	
	$C_2H_3O_2^-$ (acetate)		
	MnO_4^- (permanganate)		
	HCO_3^- (hydrogen carbonate)		
	$H_2PO_4^-$ (dihydrogen phosphate)		

Sodium carbonate crystals. *(Charles D. Winters)*

EXAMPLE 2.5   Using Table 2.2, predict the formulas of (a) strontium hydroxide, (b) sodium carbonate, and (c) ammonium phosphate.

Strategy   The reasoning is entirely similar to that in Example 2.4. The only difference is that you must know the formulas and charges of the polyatomic ions in Table 2.2.

Solution

(a) One Sr^{2+} ion requires two OH^- ions. The formula is $Sr(OH)_2$. Parentheses are used to indicate that there are two polyatomic OH^- ions for every Sr^{2+}.

(b) Two Na^+ ions require one CO_3^{2-} ion. The formula is Na_2CO_3.

(c) Three NH_4^+ ions are required for one PO_4^{3-} ion. The formula is $(NH_4)_3PO_4$.

2.6   NAMES OF COMPOUNDS

A compound can be identified either by its formula (e.g., NaCl) or by its name (sodium chloride). In this section, you will learn the rules used to name ionic and simple molecular compounds. To start with, it will be helpful to show how individual ions within ionic compounds are named.

Ions

Monatomic cations take the name of the metal from which they are derived. Examples include

$$Na^+ \text{ sodium} \qquad K^+ \text{ potassium}$$

There is one complication. As mentioned earlier, certain metals in the transition and post-transition series form more than one cation, for example, Fe^{2+} and Fe^{3+}. To distinguish between these cations, the charge must be indicated in the name. This is done by putting the charge as a Roman numeral in parentheses after the name of the metal:

$$Fe^{2+} \text{ iron(II)} \qquad Fe^{3+} \text{ iron(III)}$$

(An older system used the suffixes $-ic$ for the ion of higher charge and *-ous* for the ion of lower charge. These were added to the stem of the Latin name of the metal, so that the Fe^{3+} ion was referred to as ferric and the Fe^{2+} ion as ferrous.)

Monatomic anions are named by adding the suffix *-ide* to the stem of the name of the nonmetal from which they are derived.

			H^-	hydride
N^{3-}	nitride	O^{2-} oxide	F^-	fluoride
		S^{2-} sulfide	Cl^-	chloride
		Se^{2-} selenide	Br^-	bromide
		Te^{2-} telluride	I^-	iodide

Polyatomic ions, as you have seen (Table 2.2), are given special names. Certain nonmetals in Groups 15 to 17 of the periodic table form more than one polyatomic ion containing oxygen **(oxoanions).** The names of several such oxoanions are shown in Table 2.3. From the entries in the table, you should be able to deduce the following rules.

TABLE 2.3 Oxoanions of Nitrogen, Sulfur, and Chlorine OHT		
Nitrogen	**Sulfur**	**Chlorine**
NO_3^- nit*rate* NO_2^- nit*rite*	SO_4^{2-} sulf*ate* SO_3^{2-} sulf*ite*	ClO_4^- *per*chlor*ate* ClO_3^- chlor*ate* ClO_2^- chlor*ite* ClO^- *hypo*chlor*ite*

1. When a nonmetal forms two oxoanions, the suffix *-ate* is used for the anion with the larger number of oxygen atoms. The suffix *-ite* is used for the anion containing fewer oxygen atoms.
2. When a nonmetal forms more than two oxoanions, the prefixes *per-* (largest number of oxygen atoms) and *hypo-* (fewest oxygen atoms) are used as well.

⌐ Cl, Br, and I form more than two oxoanions.

Ionic Compounds

The name of an ionic compound consists of two words. The first word names the cation and the second names the anion. This is, of course, the same order in which the ions appear in the formula.

⊙ See Screen 3.13, Naming Ionic Compounds.

EXAMPLE 2.6   Name the following ionic compounds:
(a) CaS   (b) $Al(NO_3)_3$ (c) $FeCl_2$

Strategy   To name an ionic compound, you must know the rules for naming individual ions, as discussed previously.

Solution

(a) calcium sulfide   (b) aluminum nitrate   (c) iron(II) chloride

---

In naming the compounds of transition or post-transition metals, we ordinarily indicate the charge of the metal cation by a Roman numeral:

$Cr(NO_3)_3$ chromium(III) nitrate $SnCl_2$ tin(II) chloride

In contrast, we never use Roman numerals with compounds of the Group 1 or Group 2 metals; they always form cations with charges of $+1$ or $+2$, respectively.

Binary Molecular Compounds

When two nonmetals combine with each other, the product is most often a binary molecular compound. There is no simple way to deduce the formulas of such compounds. There is, however, a systematic way of naming molecular compounds that differs considerably from that used with ionic compounds.

The systematic name of a binary molecular compound, which contains two different nonmetals, consists of two words.

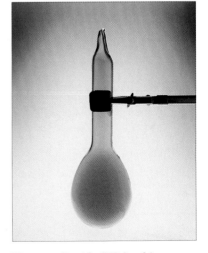

Nitrogen dioxide (NO_2), a binary molecular compound. It is a reddish-brown gas at 25°C and 1 atm. *(Charles D. Winters)*

TABLE 2.4	Greek Prefixes Used in Nomenclature				
Number*	Prefix	Number	Prefix	Number	Prefix
2	di	5	penta	8	octo
3	tri	6	hexa	9	nona
4	tetra	7	hepta	10	deca

*The prefix mono (1) is seldom used.

1. The first word gives the name of the element that appears first in the formula; a Greek prefix (Table 2.4) is used to show the number of atoms of that element in the formula.
2. The second word consists of—

  ■ the appropriate Greek prefix designating the number of atoms of the second element.
  ■ the stem of the name of the second element.
  ■ the suffix -*ide*.

To illustrate these rules, consider the names of the several oxides of nitrogen:

See Screen 3.5, Binary Compounds of the Nonmetals.

N_2O_5	dinitrogen *penta*oxide	N_2O_3	dinitrogen *tri*oxide
N_2O_4	dinitrogen *tetra*oxide	NO	nitrogen ox*ide*
NO_2	nitrogen *di*oxide	N_2O	dinitrogen ox*ide*

When the prefixes tetr*a*, pent*a*, hex*a*, . . . are followed by the letter "o," the *a* is sometimes dropped. For example, N_2O_5 is often referred to as *dinitrogen pentoxide*.

EXAMPLE 2.7 Give the names of
(a) SO_2 (b) SO_3 (c) PCl_3 (d) Cl_2O_7

Strategy Start with the prefix denoting the number of atoms (if there is more than one) of the first element followed by the name of that element. Repeat for the second element, ending with the suffix -*ide*.

Solution

(a) sulfur dioxide   (b) sulfur trioxide
(c) phosphorus trichloride   (d) dichlorine heptaoxide

Many of the best-known binary compounds of the nonmetals have acquired common names. These are widely—and in some cases exclusively—used. Examples include

H_2O	water	PH_3	phosphine
H_2O_2	hydrogen peroxide	AsH_3	arsine
NH_3	ammonia	NO	nitric oxide
N_2H_4	hydrazine	N_2O	nitrous oxide
C_2H_2	acetylene	CH_4	methane

Water (H_2O), a binary molecular compound. It is never called "dihydrogen oxide." *(Charles D. Winters)*

Acids

A few binary molecular compounds containing H atoms ionize in water to form H^+ ions. These are called **acids.** One such compound is hydrogen chloride, HCl; in water solution it exists as aqueous H^+ and Cl^- ions. The water solution of hydrogen chloride is given a special name: It is referred to as *hydrochloric acid*. A similar situation applies with HBr and HI:

Pure Substance		Water Solution	
HCl(g)	Hydrogen chloride	$H^+(aq)$, $Cl^-(aq)$	Hydrochloric acid
HBr(g)	Hydrogen bromide	$H^+(aq)$, $Br^-(aq)$	Hydrobromic acid
HI(g)	Hydrogen iodide	$H^+(aq)$, $1^-(aq)$	Hydriodic acid

Most acids contain oxygen in addition to hydrogen atoms. Such species are referred to as **oxoacids.** Two oxoacids that you are likely to encounter in the general chemistry laboratory are

$$HNO_3 \quad \text{nitric acid} \qquad H_2SO_4 \quad \text{sulfuric acid}$$

The names of oxoacids are simply related to those of the corresponding oxoanions. The *-ate* suffix of the anion is replaced by *-ic* in the acid. In a similar way, the suffix *-ite* is replaced by the suffix *-ous*. The prefixes *per-* and *hypo-* found in the name of the anion are retained in the name of the acid.

ClO_4^-	*perchlorate* ion	$HClO_4$	*perchloric* acid
ClO_3^-	*chlorate* ion	$HClO_3$	*chloric* acid
ClO_2^-	*chlorite* ion	$HClO_2$	*chlorous* acid
ClO^-	*hypochlorite* ion	$HClO$	*hypochlorous* acid

EXAMPLE 2.8   Give the names of
(a) HNO_2 (b) H_2SO_3 (c) HIO

Strategy   In (a) and (b), refer back to Table 2.3 for the name of the oxoanion. In (c), reason by analogy with chlorine oxoacids.

Solution

(a) nitrous acid   (b) sulfurous acid   (c) hypoiodous acid

CHEMISTRY
Beyond the Classroom

Ethyl Alcohol and the Law

There is a strong correlation between a driver's blood alcohol concentration (BAC) and the likelihood that he or she will be involved in an accident (Figure A). At a BAC of 0.08% his or her chance of colliding with another car is four times greater than that of a sober driver.

Blood alcohol concentration can be determined directly by gas chromatography (Chapter 1). However, this approach is impractical for testing a driver on the highway. It requires that the suspect be transported to a hospital, where trained medical personnel can take a blood sample, then preserve and analyze it.

It is simpler and quicker to measure a suspect's breath alcohol concentration (BrAc). This can be converted to BAC by multiplying by 2100; one volume of blood contains about 2100 times as much alcohol as the same volume of breath. In practice, this calculation is done automatically in the instrument used to measure BrAc; it reads directly the BAC of 0.05%, 0.08%, or whatever.

The first instruments used by police to determine BrAc were developed in the 1930s. Until about 1980, the standard method involved adding $K_2Cr_2O_7$, which reacts chemically with ethyl alcohol. Potassium dichromate has a bright orange-red color whose intensity fades as reaction occurs. The extent of the color change is a measure of the amount of alcohol present.

The standard instrument used in police stations today (Intoxilizer 5000) measures infrared absorption (Chapter 1) at three wavelengths (3390, 3480, 3800 nm) where ethyl alcohol absorbs. At the scene of an accident, a police officer may use a less accurate hand-held instrument about the size of a deck of cards, which estimates BAC by an electrochemical process (Chapter 18).

The greatest uncertainty in instruments such as the Intoxilizer lies in the conversion from BrAC to BAC values. The factor of 2100 referred to previously can vary considerably depending on the circumstances under which the sample is taken. Simultaneous measurements of BrAc and BAC values suggest that the factor can be anywhere between 1800 and 2400. This means that a calculated BAC value of 0.100% could be as low as 0.086% or as high as 0.114%.

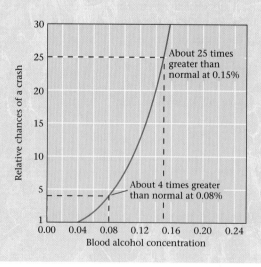

FIGURE A
Blood alcohol concentration and risk of a crash. With increasing alcohol concentration in the blood, the risk of an automobile crash rises rapidly to 25 times the normal risk of a crash (that is, the risk with no alcohol consumption).
(Source: U.S. Dept of Transportation, Washington D.C.)

After drinking an alcoholic beverage, a person's BAC rises to a maximum, typically in 30 to 90 minutes, and then drops steadily at the rate of about 0.02% per hour. The maximum BAC depends on the amount of alcohol consumed and the person's body weight. The data cited in Table A are for male subjects; females show maximum BACs about 20% higher. Thus a man weighing 60 kg who takes three drinks may be expected to reach a maximum BAC of 0.09%; a woman of the same weight consuming the same amount of alcohol may show a BAC of

$$0.09\% \times 1.2 = 0.11\%$$

An instrument used prior to 1980 to determine BAC. *(Charles D. Winters)*

TABLE A Maximum Blood Alcohol Concentrations (Percentages)*						
	No. Drinks (1 drink = 1 oz 100-proof liquor or a 12-oz beer)					
Body Weight	**1**	**2**	**3**	**4**	**5**	**6**
50 kg = 110 lb	0.04	0.07	**0.10**	**0.14**	**0.17**	**0.21**
60 kg = 132 lb	0.03	0.06	0.09	**0.12**	**0.14**	**0.17**
70 kg = 154 lb	0.02	0.05	0.07	**0.10**	**0.12**	**0.15**
80 kg = 176 lb	0.02	0.04	0.06	0.09	**0.11**	**0.13**
90 kg = 198 lb	0.02	0.04	0.06	0.08	**0.10**	**0.11**

*Values are for men; values for women are about 20% higher; BAC values of 0.10% or higher are shown in boldface type.

CHAPTER HIGHLIGHTS

Key Concepts

1. Relate a nuclear symbol to the number of protons and neutrons in the nucleus.
   (Examples 2.1, 2.3: Problems 7–16)
2. Relate structural, condensed structural, and molecular formulas.
   (Example 2.2; Problems 27, 28, 45)
3. Predict formulas of ionic compounds from ionic charges.
   (Examples 2.4, 2.5; Problems 33, 34)
4. Relate names to formulas for—
   ▪ ionic compounds
   (Example 2.6; Problems 35–42)
   ▪ binary molecular compounds
   (Example 2.7; Problems 29–32, 41, 42)
   ▪ oxoacids
   (Example 2.8; Problems 39–41)

Key Terms

acid
atom
atomic number
electron
formula
—condensed structural
—molecular
—structural
ion
—anion
—cation
—monatomic
—polyatomic

ionic bonds
isotope
main-group element
mass number
metal
metalloid
molecule
neutron
nonmetal
nuclear symbol
nucleus
oxoacid
oxoanion

periodic table
—group
—period
proton
radioactivity
transition metal

Summary Problem

Iodine is a dark purple solid. It was once popular as an antiseptic.

(a) What is its molecular formula?

(b) How many protons and electrons are there in a molecule of iodine? In an iodide ion?

(c) What is the atomic number for iodine?

(d) Write the nuclear symbol for the iodine atom with 53 protons and 78 neutrons.

(e) In what group and period does iodine belong in the periodic table? Is it a metal, nonmetal, or metalloid?

(f) I-123 is a radioactive isotope used as a diagnostic imaging tool. How many neutrons are there in I-123?

(g) A compound is made up of one atom of iodine and three atoms of chlorine. What is the molecular formula of this compound? What is the name of this compound?

(h) When iodine and barium combine, an ionic compound is formed. Give its name and formula.

(i) Iodine and oxygen combine to form oxoanions entirely analogous to the oxoanions of chlorine. Give the name and the formula of the oxoanion of iodine made up of one atom of oxygen and one atom of iodine. Give the name and formula of the corresponding oxoacid.

Answers

(a) I_2

(b) 106 protons, 106 electrons; 53 protons, 54 electrons

(c) 53

(d) $^{131}_{53}I$

(e) period 5, group **17**, nonmetal

(f) 70

(g) ICl_3; iodine trichloride

(h) BaI_2; barium iodide

(i) IO^-, hypoiodite; HIO, hypoiodous acid

Problem numbers in blue indicate that the answer is available in Appendix 6 at the back of the book.
WEB indicates that the solution is posted at **http://www.harcourtcollege.com/chem/general/masterton4/student/**

Atomic Theory and Laws

1. State in your own words the law of constant composition.

2. State in your own words the law of conservation of mass. State the law in its modern form.

3. Two basic laws of chemistry are the law of conservation of mass and the law of constant composition. Which of these laws (if any) do the following statements illustrate?

(a) A sealed bag of popcorn has the same mass before and after it is put in a microwave oven. (Assume no breaks develop in the bag.)

(b) Hydrogen has three isotopes. One has a mass number A equal to its atomic number (Z), in another $A = 2Z$, and in a third, $A = 3Z$.

(c) Analysis of water in the rain collected in the Amazon rainforest and that formed in a test tube by combining hydrogen and oxygen gives the same value for the percentage of oxygen.

4. Which of the laws described in Question 3 do the following statements illustrate?

(a) When copper(II) oxide decomposes, the total mass of the copper and oxygen formed equals the mass of copper oxide decomposed.

(b) A teaching assistant writes "highly improbable" in a student's report that states that her unknown is $S_{1.2}O_{2.7}$.

(c) Analysis of marble from Carara, Italy, and Burlington, Vermont, gives the same value for percentage of calcium.

Nuclear Symbols and Isotopes

5. Who discovered the electron? Describe the experiment that led to the deduction that electrons are negatively charged particles.

6. Who discovered the nucleus? Describe the experiment that led to this discovery.

7. Radon is a radioactive gas that can cause lung cancer. It has been detected in the basements of some homes. How many protons are there in a Rn-220 atom? How many neutrons?

8. Selenium is widely sold as a dietary supplement. It is advertised to "protect" women from breast cancer. Write the nuclear symbol for naturally occurring selenium. It has 34 protons and 46 neutrons.

9. How do the two isotopes N-14 and N-15 differ from each other? Write nuclear symbols for both.

10. Consider the isotopes of carbon, C-12 and C-13.

(a) Write the nuclear symbol for both isotopes.

(b) Do C-12 and C-13 have the same number of protons? Neutrons? Electrons?

11. Nuclei with the same mass number but different atomic numbers are called isobars. Consider Na-21 and write the nuclear symbol for

(a) an isotope of Na-21 that has one more neutron than Na-21.

(b) an isobar of Na-21 with atomic number 10.

(c) a nucleus with 11 protons and 12 neutrons. Is this nucleus an isotope of Na-21?

12. Review the definition for isobars given in Question 11. Consider Ca-40, Ca-41, K-41, and Ar-41.

(a) Which of these are isobars? Which are isotopes?

(b) What do Ca-40 and Ca-41 have in common?

(c) Correct the statement (if it is incorrect): Atoms of Ca-41, K-41, and Ar-41 have the same number of neutrons.

13. Uranium-235 is the isotope of uranium commonly used in nuclear power plants. How many

(a) protons are in its nucleus?

(b) neutrons are in its nucleus?

(c) electrons are in a uranium atom?

(d) neutrons, protons, and electrons are in the U^{2+} ion formed from this isotope?

WEB 14. Selenium-75 is used in the diagnosis of disorders related to the pancreas. How many

(a) protons are in its nucleus?

(b) neutrons are in its nucleus?

(c) electrons are in a selenium atom?

(d) neutrons, protons, and electrons are in the Se^{2-} ion formed from this isotope?

15. Complete the table below. If necessary, use the periodic table.

Nuclear Symbol	Charge	Number of Protons	Number of Neutrons	Number of Electrons
^7Li	————	————	————	————
————	————	26	30	23
	−3	7	7	————
$^{25}_{12}Mg^{2+}$	————	————	————	————

16. Complete the table below. Use the periodic table if necessary.

Nuclear Symbol	Charge	Number of Protons	Number of Neutrons	Number of Electrons
$^{64}_{30}Zn$	————	————	————	————
————	+4	14	14	————
————	−2	16	16	————

17. Give the number of protons and electrons in

(a) a C_{60} molecule.   (c) a CO_2 molecule.

(b) a CN^- ion.   (d) an N^{3-} ion.

18. Give the number of protons and electrons in
(a) an N_2 molecule (identified in 1772).
(b) a N_3^- unit (synthesized in 1890).
(c) an N_5^+ unit (synthesized in 1999).
(d) an N_5N_5 salt (a U.S. Air Force research team's synthesis project).

Elements and the Periodic Table

19. Give the symbols for
(a) calcium (b) gold (c) antimony
(d) krypton (e) potassium
20. Name the elements whose symbols are
(a) S (b) Sc (c) Se (d) Si (e) Sr
21. Classify the elements in Question 19 as metals, nonmetals, or metalloids.
22. Classify the elements in Question 20 as metals, nonmetals, or metalloids.
23. How many metals are there in the following groups?
(a) Group 1 (b) Group 13 (c) Group 17
WEB 24. How many nonmetals are in the following periods?
(a) period 2 (b) period 4 (c) period 6
25. Which period of the periodic table
(a) has no metals?
(b) has no nonmetals?
(c) has one post-transition metal and two metalloids?
26. Which group in the periodic table
(a) has one metalloid and no nonmetals?
(b) has no nonmetals or transition metals?
(c) has no metals or metalloids?

Names and Formulas of Ionic and Molecular Compounds

27. Write the condensed structural formulas and molecular formulas for the following molecules. The reactive groups are shown in red.

(a)
```
    H
    |
H — C — C = O        (acetic acid)
    |   |
    H   O — H
```

(b)
```
    H
    |
H — C — Cl        (methyl chloride)
    |
    H
```

28. Given the following condensed formulas, write the molecular formulas for the following molecules.
(a) dimethylamine $(CH_3)_2NH$
(b) propyl alcohol $CH_3(CH_2)_2OH$
29. Write the formulas for the following molecules.
(a) methane (b) carbon tetraiodide
(c) hydrogen peroxide (d) nitrogen oxide
(e) silicon dioxide

30. Write the formulas for the following molecules.
(a) water (b) ammonia
(c) hydrazine (d) sulfur hexafluoride
(e) phosphorus pentachloride
31. Write the names of the following molecules.
(a) CO_2 (b) C_2H_2 (c) XeF_4
(d) S_4N_4 (e) CCl_4
32. Write the names of the following molecules.
(a) Se_2Cl_2 (b) N_2O_4 (c) PH_3 (d) IF_7 (e) SiC
33. Give the formulas of compounds in which
(a) the cation is Ba^{2+}, the anion is I^- or N^{3-}.
(b) the anion is O^{2-}, the cation is Fe^{2+} or Fe^{3+}.
34. Give the formulas of all the compounds containing no ions other than K^+, Ca^{2+}, Cl^-, and S^{2-}.
35. Write the formulas of the following ionic compounds.
(a) iron(III) carbonate (b) sodium azide (N_3^-)
(c) calcium sulfate (d) copper(I) sulfide
(e) lead(IV) oxide
WEB 36. Write formulas for the following ionic compounds.
(a) cobalt(II) acetate (b) barium oxide
(c) aluminum sulfide (d) potassium permanganate
(e) sodium hydrogen carbonate
37. Write the names of the following ionic compounds.
(a) $FeCr_2O_7$ (b) $Al_2(SO_4)_3$ (c) $NaC_2H_3O_2$
(d) Ba_3N_2 (e) Rb_2S
38. Write the names of the following ionic compounds
(a) KNO_3 (b) $(NH_4)_2SO_4$ (c) $Mg_3(PO_4)_2$
(d) $FeCl_3$ (e) Cr_2O_7
39. Write formulas for the following ionic compounds.
(a) hydrochloric acid (b) sodium nitrite
(c) chromium(III) sulfite (d) potassium chlorate
(e) iron(III) perbromate
40. Write the names of the following ionic compounds.
(a) $HCl(aq)$ (b) $HClO_3(aq)$ (c) $Fe_2(SO_3)_3$
(d) $Ba(NO_2)_2$ (e) $NaClO$
41. Complete the following table.

Name	Formula
nitrous acid	
	$Ni(IO_3)_2$
gold(III) sulfide	
	$H_2SO_3(aq)$
nitrogen trifluoride	

42. Complete the following table.

Name	Formula
sodium dichromate	
	BrI_3
copper(II) hypochlorite	
	S_2Cl_2
potassium nitride	

Unclassified

43. Identify each of the following elements.

(a) A halogen with 53 protons in its nucleus.

(b) A member of the same period as tin whose cation has a + 2 charge and has 36 electrons.

(c) A post-transition metal in period 4.

(d) A metalloid in Group 13.

(e) A transition metal whose cations have either 27 or 28 electrons.

44. Write the formulas and names of the following:

(a) A molecule made up of two atoms of an element with seven protons and three atoms of an element in Group 16, period 2.

(b) The major inorganic component of bones, made up of calcium and an anion containing four oxygen atoms and one phosphorus atom.

(c) The zinc salt of the anion derived from the sulfate ion by the loss of one oxygen atom.

45. A molecule of ethylamine is made up of two carbon atoms, seven hydrogen atoms, and one nitrogen atom.

(a) Write its molecular formula.

(b) The reactive group in ethylamine is NH_2. Write its condensed structural formula.

46. Hydrogen-1 can take the form of a molecule, an anion (H^-), or a cation (H^+).

(a) How many protons, electrons, and neutrons are there in each possible species?

(b) Write the name and formula for the compound formed between hydrogen and a metal in Group 2 with 12 protons.

(c) What is the general name of the aqueous compounds in which hydrogen is a cation?

47. Criticize each of the following statements.

(a) In an ionic compound, the number of cations is always the same as the number of anions.

(b) The molecular formula for strontium bromide is $SrBr_2$.

(c) The mass number is always equal to the atomic number.

(d) For any ion, the number of electrons is always more than the number of protons.

48. Which of the following statements is/are always true? Never true? Usually true?

(a) Compounds containing carbon atoms are molecular.

(b) A molecule is made up of nonmetal atoms.

(c) An ionic compound has at least one metal atom.

Conceptual Problems

49. Use the law of conservation of mass to determine which numbered box(es) represent(s) the product mixture after the substances in the box at the top of the next column undergo a reaction.

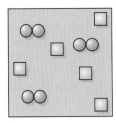

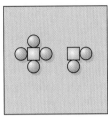
(1)

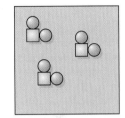
(2)

(3)   (4)

50. Using the laws of constant composition and the conservation of mass, complete the molecular picture of hydrogen molecules ($\bigcirc\!\!-\!\!\bigcirc$) reacting with chlorine molecules ($\square\!\!-\!\!\square$) to give hydrogen chloride ($\square\!\!-\!\!\bigcirc$) molecules.

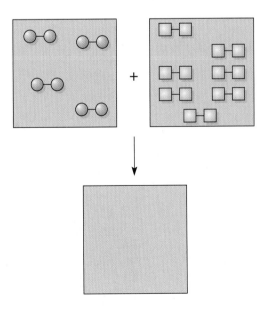

51. If squares represent carbon and spheres represent chlorine, make a representation of liquid CCl_4.

52. If squares represent Cl atoms and spheres represent K atoms, make a representation of a KCl crystal.

53. Scientists are trying to synthesize elements with more than 114 protons. State the expected atomic number of

(a) the newest inert gas.

(b) the new element with properties similar to those of the alkaline earth metals.

(c) the new element that will behave like the halogens.

(d) the new (nontransition) metal whose ion will have a $+2$ charge.

(e) the new element that will start period 8.

54. Write the nuclear symbol for the element whose mass number is 234 and has 60% more neutrons than protons.

55. Mercury(II) oxide, a red powder, can be decomposed by heating to produce liquid mercury and oxygen gas. When a sample of this compound is decomposed, 3.87 g of oxygen and 48.43 g of mercury are produced. In a second experiment, 15.68 g of mercury is allowed to react with an excess of oxygen and 16.93 g of red mercury(II) oxide is produced. Show that these results are consistent with the law of constant composition.

Challenge Problems

56. Three compounds containing only carbon and hydrogen are analyzed. The results for the analysis of the first two compounds are given below:

Compound	Mass of Carbon (g)	Mass of Hydrogen (g)
A	28.5	2.39
B	34.7	11.6
C	16.2	—

Which, if any, of the following results for the mass of hydrogen in compound C follows the law of multiple proportions?

(a) 5.84 g   (b) 3.47 g   (c) 2.72 g

57. Ethane and ethylene are two gases containing only hydrogen and carbon atoms. In a certain sample of ethane, 4.53 g of hydrogen is combined with 18.0 g of carbon. In a sample of ethylene, 7.25 g of hydrogen is combined with 43.20 g of carbon.

(a) Show how the data illustrate the law of multiple proportions.

(b) Suggest reasonable formulas for the two compounds.

58. Calculate the average density of a single Al-27 atom by assuming that it is a sphere with a radius of 0.143 nm. The masses of a proton, electron, and neutron are 1.6726×10^{-24} g, 9.1094×10^{-28} g, and 1.6749×10^{-24} g, respectively. The volume of a sphere is $4\pi r^3/3$, where r is its radius. Express the answer in grams per cubic centimeter. The density of aluminum is found experimentally to be 2.70 g/cm^3. What does that suggest about the packing of aluminum atoms in the metal?

59. The mass of a beryllium atom is 1.4965×10^{-23} g. Using that fact and other information in this chapter, find the mass of a Be^{2+} ion.

60. Each time you inhale, you take in about 500 mL (two significant figures) of air, each milliliter of which contains 2.5×10^{19} molecules. In delivering the Gettysburg Address, Abraham Lincoln is estimated to have inhaled about 200 times.

(a) How many molecules did Lincoln take in?

(b) In the entire atmosphere, there are about 1.1×10^{44} molecules. What fraction of the molecules in the Earth's atmosphere was inhaled by Lincoln at Gettysburg?

(c) In the next breath that you take, how many molecules were inhaled by Lincoln at Gettysburg?

MASS RELATIONS IN CHEMISTRY; STOICHIOMETRY

3

Antimony iodide can be produced by the reaction between antimony (Sb) and iodine (I_2).
(Charles D. Winters)

CHAPTER OUTLINE

3.1   ATOMIC MASSES   3.4   MASS RELATIONS IN
                                            REACTIONS
3.2   THE MOLE

3.3   MASS RELATIONS IN
        CHEMICAL FORMULAS

To this point, our study of chemistry has been largely qualitative, involving very few calculations. However, chemistry is a quantitative science. Atoms of elements differ from one another not only in composition (number of protons, electrons, neutrons), but also in mass. Chemical formulas of compounds tell us not only the atom ratios in which elements are present but also the mass ratios.

The general topic of this chapter is stoichiometry (stoy-key-OM-e-tree), the study of mass relations in chemistry. Whether dealing with atomic masses (Section 3.1), molar masses (Section 3.2), chemical formulas (Section 3.3), or chemical reactions (Section 3.4), you will be answering some very practical questions that ask "how much—" or "how many—." For example—

■ how many atoms are there in a gram of an element? (Section 3.1)

■ how much iron can be obtained from a ton of iron ore? (Section 3.3)

■ how much nitrogen gas is required to form a kilogram of ammonia? (Section 3.4)

3.1   ATOMIC MASSES

Individual atoms are far too small to be weighed on a balance. However, as you will soon see, it is possible to determine quite accurately the relative masses of different atoms and molecules. Indeed, it is possible to go a step further and calculate the actual masses of these tiny building blocks of matter.

Atomic Masses; the Carbon-12 Scale

Relative masses of atoms of different elements are expressed in terms of their **atomic masses** (often referred to as atomic weights). The atomic mass of an element indicates how heavy, on the average, one atom of that element is compared with an atom of another element.

To set up a scale of atomic masses, it is necessary to establish a standard value for one particular species. The modern atomic mass scale is based on the most common isotope of carbon, $^{12}_{6}C$. This isotope is assigned a mass of exactly 12 **atomic mass units** (amu):

1 amu = 1/12 mass C atom ≈ mass H atom.

$$\text{mass of C-12 atom} = 12 \text{ amu (exactly)}$$

It follows that an atom half as heavy as a C-12 atom would weigh 6 amu, an atom twice as heavy as C-12 would have a mass of 24 amu, and so on.

In the periodic table, atomic masses are listed directly below the symbol of the element. In the table on the inside front cover of this text, atomic masses are cited to four significant figures. That ordinarily will be sufficient for our purposes, although more precise values are available (see the alphabetical list of elements opposite the periodic table).

Notice that hydrogen has an atomic mass of 1.008 amu; helium has an atomic mass of 4.003 amu. This means that, on the average, a helium atom has a mass that is about one-third that of a C-12 atom:

$$\frac{4.003 \text{ amu}}{12.00 \text{ amu}} = 0.3336$$

or about four times that of a hydrogen atom:

$$\frac{4.003 \text{ amu}}{1.008 \text{ amu}} = 3.971$$

In general, for two elements X and Y:

$$\frac{\text{atomic mass X}}{\text{atomic mass Y}} = \frac{\text{mass of atom of X}}{\text{mass of atom of Y}}$$

Atomic Masses and Isotopic Abundances

Relative masses of individual atoms can be determined using a mass spectrometer (Figure 3.1). Gaseous atoms or molecules at very low pressures are ionized by removing one or more electrons. The cations formed are accelerated by a potential of 500 to 2000 V toward a magnetic field, which deflects the ions from their straight-line path. The extent of deflection is inversely related to the mass of the ion. By measuring the voltages required to bring two ions of different mass to the same point on the detector, it is possible to determine their relative masses. For example, using a mass spectrometer, it is found that a $^{19}_{9}\text{F}$ atom is 1.583 times as heavy as a $^{12}_{6}\text{C}$ atom and so has a mass of

$$1.583 \times 12.00 \text{ amu} = 19.00 \text{ amu}$$

As it happens, naturally occurring fluorine consists of a single isotope, $^{19}_{9}\text{F}$. It follows that the atomic mass of the element fluorine must be the same as that of F-19, 19.00 amu. The situation with most elements is more complex, because they

> Atomic masses give relative masses of different atoms.

> See the *Saunders Interactive General Chemistry CD-ROM*, Screen 2.15, **Atomic Mass.**

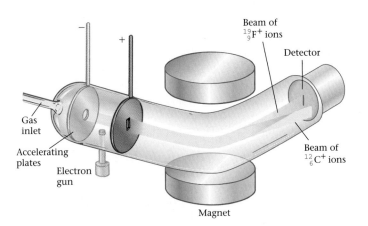

FIGURE 3.1

The mass spectrometer. A beam of gaseous ions is deflected in the magnetic field toward the detector. Light ions are deflected more than heavy ones. By comparing the accelerating voltages required to bring ions to the same point on the detector, it is possible to determine the relative masses of the ions. **OHT**

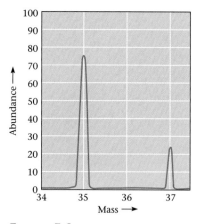

FIGURE 3.2

Mass spectrum of chlorine.
Elemental chlorine (Cl_2) contains only
two isotopes: 34.97 amu (75.53%) and
36.97 amu (24.47%). **OHT**

occur in nature as a mixture of two or more isotopes. To determine the atomic mass of such an element, it is necessary to know not only the masses of the individual isotopes but also their atom percents **(isotopic abundances)** in nature.

Fortunately, isotopic abundances as well as isotopic masses can be determined by mass spectroscopy. The situation with chlorine, which has two stable isotopes, Cl-35 and Cl-37, is shown in Figure 3.2. The atomic masses of the two isotopes are determined in the usual way. The relative abundances of these isotopes are proportional to the heights of the recorder peaks or, more accurately, to the areas under these peaks. For chlorine, the data obtained from the mass spectrometer are

	Atomic Mass	Abundance
Cl-35	34.97 amu	75.53%
Cl-37	36.97 amu	24.47%

We interpret this to mean that, in elemental chlorine, 75.53% of the atoms have a mass of 34.97 amu, and the remaining atoms, 24.47% of the total, have a mass of 36.97 amu. With this information we can readily calculate the atomic mass of chlorine using the general equation

$$\text{atomic mass Y} = (\text{atomic mass Y}_1) \times \frac{\% \text{Y}_1}{100\%} + (\text{atomic mass Y}_2) \times \frac{\% \text{Y}_2}{100\%} + \cdots$$

where $Y_1, Y_2, \ldots$ are isotopes of element Y.

$$\text{atomic mass Cl} = 34.97 \text{ amu} \times \frac{75.53}{100.0} + 36.97 \text{ amu} \times \frac{24.47}{100.0} = 35.46 \text{ amu}$$

Atomic masses calculated in this manner, using data obtained with a mass spectrometer, can in principle be precise to seven or eight significant figures. The accuracy of tabulated atomic masses is limited mostly by variations in natural abundances. Sulfur is an interesting case in point. It consists largely of two isotopes, $^{32}_{16}S$ and $^{34}_{16}S$. The abundance of sulfur-34 varies from about 4.18% in sulfur deposits in Texas and Louisiana to 4.34% in volcanic sulfur from Italy. This leads to an uncertainty of 0.006 amu in the atomic mass of sulfur.

If the atomic mass of an element is known *and* if it has only two stable isotopes, their abundances can be calculated from the general equation cited above.

EXAMPLE 3.1   Carbon (atomic mass = 12.01 amu) consists of two isotopes: C-12 (12.00 amu) and C-13 (13.00 amu). What is the abundance of the heavier isotope?

Strategy   The key to solving this problem is to realize that *the abundances of the isotopes have to add to 100%*. If we let x = abundance of C-13, then the abundance of C-12 must be $(100 - x)$.

Solution   Substituting in the general equation cited above:

$$12.01 \text{ amu} = 12.00 \text{ amu} \frac{(100 - x)}{100} + 13.00 \text{ amu} \frac{(x)}{100}$$

$$= 12.00 \text{ amu} + 1.00 \text{ amu} \frac{(x)}{100}$$

Solving: $\dfrac{x}{100} = \dfrac{12.01 \text{ amu} - 12.00 \text{ amu}}{1.00 \text{ amu}} = 0.01$ $x = 100(0.01) = 1$

We conclude that elemental carbon contains 1% C-13 ; the remaining 99% is the more abundant isotope, C-12.

Reality Check   It makes sense that carbon consists mostly of C-12, because the atomic mass of the element is very close to 12.

Masses of Individual Atoms; Avogadro's Number

For most purposes in chemistry, it is sufficient to know the relative masses of different atoms. Sometimes, however, it is necessary to go one step further and calculate the mass in grams of individual atoms. Let us consider how this can be done.

To start with, consider the elements helium and hydrogen. A helium atom is about four times as heavy as a hydrogen atom (He = 4.003 amu, H = 1.008 amu). It follows that a sample containing 100 helium atoms weighs four times as much as a sample containing 100 hydrogen atoms. Again, comparing samples of the two elements containing a million atoms each, the masses will be in a 4 (helium) to 1 (hydrogen) ratio. Turning this argument around, it follows that a sample of helium weighing four grams must contain the same number of atoms as a sample of hydrogen weighing one gram. More exactly:

No. of He atoms in 4.003 g helium = no. of H atoms in 1.008 g hydrogen

This reasoning is readily extended to other elements. A sample of any element with a mass in grams equal to its atomic mass contains the same number of atoms, N_A, regardless of the identity of the element.

The question now arises as to the numerical value of N_A; that is, how many atoms are there in 4.003 g of helium, 1.008 g of hydrogen, 32.07 g of sulfur, and so on? As it happens, this problem is one that has been studied for at least a century. Several ingenious experiments have been designed to determine this number, known as **Avogadro's number** and given the symbol N_A (see Problem 86, end of chapter). As you can imagine, it is huge. (Remember that atoms are tiny. There must be a lot of them in 4.003 g of He, 1.008 g of H, and so on.) To four significant figures,

$$N_A = 6.022 \times 10^{23}$$

To get some idea of how large this number is, suppose the entire population of the world were assigned to counting the atoms in 4.003 g of helium. If each person counted one atom per second and worked a 48-hour week, the task would take more than ten million years.

The importance of Avogadro's number in chemistry should be clear. *It represents the number of atoms of an element in a sample whose mass in grams is numerically equal to the atomic mass of the element.* Thus there are

6.022×10^{23} H atoms in 1.008 g H	atomic mass H = 1.008 amu
6.022×10^{23} He atoms in 4.003 g He	atomic mass He = 4.003 amu
6.022×10^{23} S atoms in 32.07 g S	atomic mass S = 32.07 amu

Knowing Avogadro's number and the atomic mass of an element, it is possible to calculate the mass of an individual atom (Example 3.2a). You can also determine the number of atoms in a weighed sample of any element (Example 3.2b).

If a nickel weighs twice as much as a dime, there are equal numbers of coins in 1000 g of nickels and 500 g of dimes.

Most people have better things to do.

See Screen 2.19, Moles and Molar Masses of the Elements.

Glass is made red by adding a compound of selenium to sand. *(Charles D. Winters)*

EXAMPLE 3.2   When selenium (Se) is added to glass, it gives the glass a brilliant red color. Taking Avogadro's number to be 6.022×10^{23}, calculate

(a) the mass of a selenium atom.
(b) the number of selenium atoms in a 1.000-g sample of the element.

Strategy   The atomic mass of Se, from the periodic table, is 78.96 amu. It follows that

$$6.022 \times 10^{23} \text{ Se atoms} = 78.96 \text{ g Se}$$

This relation yields the required conversion factors.

Solution

(a) mass of Se atom $= 1$ Se atom $\times \dfrac{78.96 \text{ g Se}}{6.022 \times 10^{23} \text{ Se atoms}} = \boxed{1.311 \times 10^{-22} \text{ g}}$

(b) no. of Se atoms $= 1.000$ g $\times \dfrac{6.022 \times 10^{23} \text{ Se atoms}}{78.96 \text{ g}}$

$\qquad\qquad\qquad = \boxed{7.627 \times 10^{21} \text{ Se atoms}}$

Reality Check   Because atoms are so tiny, we expect their mass to be very small; 1.311×10^{-22} g sounds reasonable. Conversely it takes a lot of atoms, in this case 7.627×10^{21}, to weigh one gram.

3.2   THE MOLE

The quantity represented by Avogadro's number is so important that it is given a special name, the **mole.** A mole represents 6.022×10^{23} items, whatever they may be (Figure 3.3).

$$1 \text{ mol H atoms} = 6.022 \times 10^{23} \text{ H atoms}$$

$$1 \text{ mol O atoms} = 6.022 \times 10^{23} \text{ O atoms}$$

$$1 \text{ mol H}_2 \text{ molecules} = 6.022 \times 10^{23} \text{ H}_2 \text{ molecules}$$

$$1 \text{ mol H}_2\text{O molecules} = 6.022 \times 10^{23} \text{ H}_2\text{O molecules}$$

$$1 \text{ mol electrons} = 6.022 \times 10^{23} \text{ electrons}$$

$$1 \text{ mol pennies} = 6.022 \times 10^{23} \text{ pennies}$$

(One mole of pennies is a lot of money. It's enough to pay all the expenses of the United States for the next billion years or so.)

A mole represents not only a specific number of particles but also a definite mass of a substance as represented by its formula (O, O_2, H_2O, NaCl, . . .). ***The molar mass, $\mathcal{M}$, in grams per mole, is numerically equal to the sum of the masses (in amu) of the atoms in the formula.***

> The size of the mole was chosen to make this relation true.

Formula	Sum of Atomic Masses	Molar Mass, $\mathcal{M}$
O	16.00 amu	16.00 g/mol
O_2	$2(16.00 \text{ amu}) = 32.00 \text{ amu}$	32.00 g/mol
H_2O	$2(1.008 \text{ amu}) + 16.00 \text{ amu} = 18.02 \text{ amu}$	18.02 g/mol
NaCl	$22.99 \text{ amu} + 35.45 \text{ amu} = 58.44 \text{ amu}$	58.44 g/mol

Notice that the formula of a substance must be known to find its molar mass. It would be ambiguous, to say the least, to refer to the "molar mass of hydrogen." One mole of hydrogen atoms, represented by the symbol H, weighs 1.008 g; the molar mass of H is 1.008 g/mol. One mole of hydrogen molecules, represented by the formula H_2, weighs 2.016 g; the molar mass of H_2 is 2.016 g/mol.

Mole-Gram Conversions

As you will see later in this chapter, it is often necessary to convert from moles of a substance to mass in grams or vice versa. Such conversions are readily made by using the general relation

$$m = \mathcal{M} \times n$$

where m is the mass in grams, $\mathcal{M}$ is the molar mass (g/mol), and n is the amount in moles.

FIGURE 3.3
One-mole amounts of sugar, baking soda, and copper nails. *(Charles D. Winters)*

EXAMPLE 3.3   Calcium carbonate is the principal ingredient of the chalk used in most classrooms. Determine the number of moles of calcium carbonate in a stick of chalk containing 14.8 g of calcium carbonate.

Strategy   Find the molar mass of calcium carbonate and use it to convert 14.8 g to moles. Before calculating the molar mass, you must come up with the formula of calcium carbonate (Chapter 2).

Solution   The formula is $CaCO_3$, so the molar mass is

$$\mathcal{M} = [40.08 + 12.01 + 3(16.00)] \text{ g/mol} = 100.09 \text{ g/mol}$$

$$n = 14.8 \text{ g } CaCO_3 \times \frac{1 \text{ mol } CaCO_3}{100.09 \text{ g } CaCO_3} = \boxed{0.148 \text{ mol } CaCO_3}$$

EXAMPLE 3.4   Acetylsalicylic acid, $C_9H_8O_4$, is the active ingredient of aspirin. What is the mass in grams of 0.287 mol of acetylsalicylic acid?

Strategy   Find the molar mass of $C_9H_8O_4$ and use it to obtain the conversion factor required to convert 0.287 mol to mass in grams.

Solution   The molar mass of $C_9H_8O_4$ is

$$\mathcal{M} = [9(12.01) + 8(1.008) + 4(16.00)] \text{ g/mol} = 180.15 \text{ g/mol}$$

Hence

$$\text{mass } C_9H_8O_4 = 0.287 \text{ mol } C_9H_8O_4 \times \frac{180.15 \text{ g } C_9H_8O_4}{1 \text{ mol } C_9H_8O_4} = \boxed{51.7 \text{ g } C_9H_8O_4}$$

Reality Check   Because every known substance has a molar mass greater than 1 g/mol, it follows that the mass of a sample in grams (51.7 g) will be larger than the number of moles (0.287 mol).

Calcium carbonate. Calcium carbonate can also be found as the main constituent of the white cliffs of Dover. *(Andrea Pistolesi/The Image Bank)*

⊙ **See Screen 3.16, Using Molar Mass.**

Conversions of the type we have just carried out come up over and over again in chemistry. They will be required in nearly every chapter of this text. Clearly, you must know what is meant by a mole. Remember, a mole always represents a certain number of items, 6.022×10^{23}. Its mass, however, differs with the substance involved: A mole of H_2O, 18.02 g, weighs considerably more than a mole of H_2, 2.016 g, even though they both contain the same number of molecules. In the same way, a dozen bowling balls weigh a lot more than a dozen eggs, even though each involves the same number of items.

3.3 MASS RELATIONS IN CHEMICAL FORMULAS

As you will see shortly, the formula of a compound can be used to determine the mass percents of the elements present. Conversely, if the percentages of the elements are known, the simplest formula can be determined. Knowing the molar mass of a molecular compound, it is possible to go one step further and find the molecular formula. In this section we will consider how these three types of calculations are carried out.

Percent Composition from Formula

See Screen 3.17, Percent Composition.

The **percent composition** of a compound is specified by citing the mass percents of the elements present. For example, in a 100-g sample of water there are 11.19 g of hydrogen and 88.81 g of oxygen. Hence the percentages of the two elements are

$$\frac{11.19 \text{ g H}}{100.00 \text{ g}} \times 100\% = 11.19\% \text{ H} \qquad \frac{88.81 \text{ g O}}{100.00 \text{ g}} \times 100\% = 88.81\% \text{ O}$$

We would say that the percent composition of water is 11.19% H, 88.81% O.

Knowing the formula of a compound, you can readily calculate the mass percents of its constituent elements. It is convenient to start with one mole of the compound (Example 3.5).

Baking soda ($NaHCO_3$). By whatever name it is known and whatever the sample size, the percentage composition is the same. (Charles D. Winters)

EXAMPLE 3.5   Sodium hydrogen carbonate, commonly called bicarbonate of soda, is used in many commercial products to relieve an upset stomach. It has the formula $NaHCO_3$. What are the mass percents of Na, H, C, and O in sodium hydrogen carbonate?

Strategy   Find the mass in grams of each element in one mole of $NaHCO_3$. Then find

$$\% \text{ element} = \frac{\text{mass element}}{\text{total mass compound}} \times 100\%$$

Solution   It is convenient to set up a table to determine the mass of each element in one mole of $NaHCO_3$.

	n	$\times$	$\mathcal{M}$	$=$	m
Na	1 mol	$\times$	22.99 g/mol	$=$	22.99 g
H	1 mol	$\times$	1.008 g/mol	$=$	1.008 g
C	1 mol	$\times$	12.01 g/mol	$=$	12.01 g
O	3 mol	$\times$	16.00 g/mol	$=$	48.00 g
					84.01 g $NaHCO_3$

Because 84.01 g of $NaHCO_3$ contains 22.99 g of Na, 1.008 g of H, 12.01 g of C, and 48.00 g of O,

$$\text{mass \% Na} = \frac{22.99 \text{ g}}{84.01 \text{ g}} \times 100\% = \boxed{27.36\%} \qquad \text{mass \% H} = \frac{1.008 \text{ g}}{84.01 \text{ g}} \times 100\%$$

$$= \boxed{1.200\%}$$

$$\text{mass \% C} = \frac{12.01 \text{ g}}{84.01 \text{ g}} \times 100\% = \boxed{14.30\%} \qquad \text{mass \% O} = \frac{48.00 \text{ g}}{84.01 \text{ g}} \times 100\%$$

$$= \boxed{57.14\%}$$

Reality Check   The mass percents add up to 100, as they should:
27.36% + 1.200% + 14.30% + 57.14% = 100.00%

The calculations in Example 3.5 illustrate an important characteristic of formulas. In one mole of $NaHCO_3$, there is 1 mol of Na (22.99 g), 1 mol of H (1.008 g), 1 mol of C (12.01 g), and 3 mol of O (48.00 g). In other words, the mole ratio is 1 mol Na : 1 mol H : 1 mol C : 3 mol O. This is the same as the atom ratio in $NaHCO_3$, 1 atom Na : 1 atom H : 1 atom C : 3 atoms O. In general, *the subscripts in a formula represent not only the atom ratio in which the different elements are combined but also the mole ratio.* For example,

Formula	Atom Ratio	Mole Ratio
H_2O	2 atoms H : 1 atom O	2 mol H : 1 mol O
KNO_3	1 atom K : 1 atom N : 3 atoms O	1 mol K : 1 mol N : 3 mol O
$C_{12}H_{22}O_{11}$	12 atoms C : 22 atoms H : 11 atoms O	12 mol C : 22 mol H : 11 mol O

The formula of a compound can also be used in a straightforward way to find the mass of an element in a weighed sample of the compound (Example 3.6).

Can you explain why the mole ratio must equal the atom ratio?

EXAMPLE 3.6   An iron-containing mineral responsible for the red color of soils in many parts of the country is limonite, which has the formula $Fe_2O_3 \cdot \frac{3}{2}H_2O$. (The dot means that there is 3/2 mol of H_2O for every mole of Fe_2O_3.) What mass of iron in grams can be obtained from a metric ton (10^3 kg = 10^6 g) of limonite?

$\frac{1}{3}Fe_2O_3 \cdot H_2O$

Strategy   First find the mass percent of iron. Then, using the mass percent, determine how much iron there is in one metric ton of the ore.

Solution   The molar mass of limonite is

$$[111.7 + 3(16.00) + \tfrac{3}{2}(18.02)]\text{g/mol} = 186.7 \text{ g/mol}$$

In one mole of limonite there is

$$2 \text{ mol Fe} \times \frac{55.85 \text{ g Fe}}{1 \text{ mol Fe}} = 111.7 \text{ g Fe}$$

Thus

$$\% \text{ Fe} = \frac{111.7 \text{ g}}{186.7 \text{ g}} \times 100\% = 59.83\%$$

In one metric ton of limonite

$$\text{mass Fe} = 1.000 \times 10^6 \text{ g limonite} \times \frac{59.83 \text{ g Fe}}{100.0 \text{ g limonite}} = \boxed{5.983 \times 10^5 \text{ g Fe}}$$

Red soil in the southeastern United States. The color shows the high iron content of this soil. *(Adam Jones/ Dembinsky Photo Associates)*

See Screen 3.18, Determining Empirical Formulas.

Simplest Formula from Chemical Analysis

A major task of chemical analysis is to determine the formulas of compounds. The formula found by the approach described here is the **simplest formula,** which gives the simplest whole-number ratio of the atoms present. For an ionic compound, the simplest formula is ordinarily the only one that can be written (e.g., $CaCl_2$, Cr_2O_3). For a molecular compound, the molecular formula is a whole-number multiple of the simplest formula, where that number may be 1, 2, ...

Compound	Simplest Formula	Molecular Formula	Multiple
Water	H_2O	H_2O	1
Hydrogen peroxide	HO	H_2O_2	2
Propylene	CH_2	C_3H_6	3

The analytical data leading to the simplest formula may be expressed in various ways. You may know—

- the masses of the elements in a weighed sample of the compound.
- the mass percents of the elements in the compound.
- the masses of products obtained by reaction of a weighed sample of the compound.

The strategy used to calculate the simplest formula depends to some extent on which of these types of information is given. The basic objective in each case is to find the number of moles of each element, then the simplest mole ratio, and finally the simplest formula.

Mass data → mole ratio = atom ratio → simplest formula.

EXAMPLE 3.7 A 25.00-g sample of an orange compound contains 6.64 g of potassium, 8.84 g of chromium, and 9.52 g of oxygen. Find the simplest formula.

Strategy First (1), convert the masses of the three elements to moles. Knowing the number of moles (n) of K, Cr, and O, you can then (2) calculate the mole ratios. Finally (3), equate the mole ratio to the atom ratio, which gives you the simplest formula.

Solution

(1) $n_K = 6.64 \text{ g K} \times \dfrac{1 \text{ mol K}}{39.10 \text{ g K}} = 0.170 \text{ mol K}$

$n_{Cr} = 8.84 \text{ g Cr} \times \dfrac{1 \text{ mol Cr}}{52.00 \text{ g Cr}} = 0.170 \text{ mol Cr}$

$n_O = 9.52 \text{ g O} \times \dfrac{1 \text{ mol O}}{16.00 \text{ g O}} = 0.595 \text{ mol O}$

(2) To find the mole ratios, divide by the smallest number, 0.170 mol K:

$$\frac{0.170 \text{ mol Cr}}{0.170 \text{ mol K}} = \frac{1.00 \text{ mol Cr}}{1 \text{ mol K}} \qquad \frac{0.595 \text{ mol O}}{0.170 \text{ mol K}} = \frac{3.50 \text{ mol O}}{1 \text{ mol K}}$$

The mole ratio is 1 mol of K : 1 mol of Cr : 3.50 mol of O.

(3) As pointed out earlier, the mole ratio is the same as the atom ratio. To find the simplest whole-number atom ratio, multiply throughout by 2:

$$2 \text{ K} : 2 \text{ Cr} : 7 \text{ O}$$

The simplest formula of the orange compound is $K_2Cr_2O_7$.

Reality Check   A mole ratio of 1.00 A:1.00 B:3.33 C would imply a formula $A_3B_3C_{10}$; if the mole ratio were 1.00 A:2.50 B:5.50 C, the formula would be $A_2B_5C_{11}$. In general, multiply through by the smallest whole number that will give integers for all the subscripts.

Sometimes you will be given the mass percents of the elements in a compound. If that is the case, one extra step is involved. ***Assume a 100-g sample and calculate the mass of each element in that sample.***

Suppose, for example, you are told that the percentages of K, Cr, and O in a compound are 26.6%, 35.4%, and 38.0%, respectively. It follows that in a 100.0-g sample there are

26.6 g K, 35.4 g Cr, 38.0 g O

Working with these masses, you can go through the same procedure followed in Example 3.7 to arrive at the same answer. (Try it!)

The most complex problem of this type requires you to determine the simplest formula of a compound given only the raw data obtained from its analysis. Here, an additional step is involved; you have to determine the masses of the elements present in a fixed mass of the compound (Example 3.8).

⊙ **See Screen 5.8, Using Stoichiometry (2).**

EXAMPLE 3.8   The compound that gives fermented grape juice, malt liquor, and vodka their intoxicating properties is ethyl alcohol, which contains the elements carbon, hydrogen, and oxygen. When a sample of ethyl alcohol is burned in air, it is found that

$$5.00 \text{ g ethyl alcohol} \longrightarrow 9.55 \text{ g } CO_2 + 5.87 \text{ g } H_2O$$

What is the simplest formula of ethyl alcohol?

Strategy   The first step (1) is to calculate the masses of C, H, and O in the 5.00-g sample. To do this note that all of the carbon has been converted to carbon dioxide. One mole of C (12.01 g) forms one mole of CO_2 (44.01 g). Hence to find the mass of carbon in 9.55 g of CO_2, you use the conversion factor:

$$12.01 \text{ g C}/44.01 \text{ g } CO_2$$

By the same token, all the hydrogen ends up as water. Because there are two moles of H (2.016 g) in one mole of H_2O (18.02 g), the conversion factor is

$$2.016 \text{ g H}/18.02 \text{ g } H_2O$$

The mass of oxygen in the sample cannot be found in a similar way because some of the oxygen in the products comes from the air required to burn the reactants. Instead we find the mass of oxygen by difference:

$$\text{mass O} = \text{mass of sample} - (\text{mass of C} + \text{mass of H})$$

Once the masses of the elements are obtained, it's all downhill; follow the same path as in Example 3.7.

Solution

(1) Find the mass of each element in the sample.

$$\text{mass C} = 9.55 \text{ g } CO_2 \times \frac{12.01 \text{ g C}}{44.01 \text{ g } CO_2} = 2.61 \text{ g C}$$

An ethyl alcohol–powered bus in France. *(Sebart/Jerrican/Photo Researchers, Inc.)*

$$\text{mass H} = 5.87 \text{ g H}_2\text{O} \times \frac{2.016 \text{ g H}}{18.02 \text{ g H}_2\text{O}} = 0.657 \text{ g H}$$

$$\text{mass O} = 5.00 \text{ g sample} - (2.61 \text{ g C} + 0.657 \text{ g H}) = 1.73 \text{ g O}$$

(2) Find the number of moles of each element.

$$n_C = 2.61 \text{ g C} \times \frac{1 \text{ mol C}}{12.01 \text{ g C}} = 0.217 \text{ mol C}$$

$$n_H = 0.657 \text{ g H} \times \frac{1 \text{ mol H}}{1.008 \text{ g H}} = 0.652 \text{ mol H}$$

$$n_O = 1.73 \text{ g O} \times \frac{1 \text{ mol O}}{16.00 \text{ g O}} = 0.108 \text{ mol O}$$

(3) Find the mole ratios and then the simplest formula.

Sometimes it's not so easy to convert the atom ratio to simplest formula (cf. Problem 39).

$$\frac{0.217 \text{ mol C}}{0.108 \text{ mol O}} = \frac{2.01 \text{ mol C}}{1 \text{ mol O}} \qquad \frac{0.652 \text{ mol H}}{0.108 \text{ mol O}} = \frac{6.04 \text{ mol H}}{1 \text{ mol O}}$$

Rounding off to whole numbers, the mole ratios are

$$2 \text{ mol C} : 6 \text{ mol H} : 1 \text{ mol O}$$

The simplest formula of ethyl alcohol is C_2H_6O.

Molecular Formula from Simplest Formula

See Screen 3.19, Determining Molecular Formulas.

Chemical analysis always leads to the simplest formula of a compound because it gives only the simplest atom ratio of the elements. As pointed out earlier, the molecular formula is a whole-number multiple of the simplest formula. That multiple may be 1 as in H_2O, 2 as in H_2O_2, 3 as in C_3H_6, or some other integer. To find the multiple, one more piece of data is needed: the molar mass.

EXAMPLE 3.9 Vitamin C may (or may not) help prevent the common cold. Its simplest formula is found by analysis to be $C_3H_4O_3$. From another experiment, the molar mass is found to be about 180 g/mol. What is the molecular formula of vitamin C?

Strategy Calculate the molar mass corresponding to the simplest formula, that is, $\mathcal{M}_{C_3H_4O_3}$. Then find the multiple by dividing the actual molar mass, 180 g/mol, by $\mathcal{M}_{C_3H_4O_3}$.

Solution

$$\mathcal{M}_{C_3H_4O_3} = 3(12.01 \text{ g/mol}) + 4(1.008 \text{ g/mol}) + 3(16.00 \text{ g/mol}) = 88.06 \text{ g/mol}$$

The ratio of the actual molar mass to that of $C_3H_4O_3$ is

$$\frac{180 \text{ g/mol}}{88.06 \text{ g/mol}} = 2.04$$

The multiple is 2; the molecular formula is $C_6H_8O_6$.

3.4   MASS RELATIONS IN REACTIONS

A chemist who carries out a reaction in the laboratory needs to know how much **product** can be obtained from a given amount of starting materials **(reactants).** To do this, he or she starts by writing a balanced chemical equation.

Writing and Balancing Chemical Equations

Chemical reactions are represented by chemical equations, which identify reactants and products. Formulas of reactants appear on the left side of the equation; those of products are written on the right. In a balanced chemical equation, there are the same number of atoms of a given element on both sides. The same situation holds for a chemical reaction that you carry out in the laboratory; atoms are conserved. For that reason, *any calculation involving a reaction must be based on the balanced equation for that reaction.*

Beginning students are sometimes led to believe that writing a chemical equation is a simple, mechanical process. Nothing could be further from the truth. One point that seems obvious is often overlooked. *You cannot write an equation unless you know what happens in the reaction that it represents.* All the reactants and all the products must be identified. Moreover, you must know their formulas and physical states.

To illustrate how a relatively simple equation can be written and balanced, consider a reaction used in the Titan rocket motor to launch astronauts into space (Figure 3.4). The reactants were two liquids, hydrazine and dinitrogen tetraoxide, whose molecular formulas are N_2H_4 and N_2O_4, respectively. The products of the reaction are gaseous nitrogen, N_2, and water vapor. To write a balanced equation for this reaction, proceed as follows:

1. Write a "skeleton" equation in which the formulas of the reactants appear on the left and those of the products on the right. In this case,

$$N_2H_4 + N_2O_4 \longrightarrow N_2 + H_2O$$

2. Indicate the physical state of each reactant and product, after the formula, by writing

- (*g*)   for a gaseous substance
- (*l*)   for a pure liquid
- (*s*)   for a solid
- (*aq*)   for an ion or molecule in water (aqueous) solution

In this case

$$N_2H_4(l) + N_2O_4(l) \longrightarrow N_2(g) + H_2O(g)$$

3. Balance the equation. To accomplish this, start by writing a coefficient of 4 for H_2O, thus obtaining 4 oxygen atoms on both sides:

$$N_2H_4(l) + N_2O_4(l) \longrightarrow N_2(g) + 4H_2O(g)$$

Now consider the hydrogen atoms. There are $4 \times 2 = 8$ H atoms on the right. To obtain 8 H atoms on the left, write a coefficient of 2 for N_2H_4:

$$2N_2H_4(l) + N_2O_4(l) \longrightarrow N_2(g) + 4H_2O(g)$$

FIGURE 3.4
Titan rocket. *(NASA)*

See Screens 4.2, Chemical Equations, & 4.4, Balancing Chemical Equations.

Using the letters (*s*), (*l*), (*g*), (*aq*) lends a sense of physical reality to the equation.

Finally, consider nitrogen. There are a total of $(2 \times 2) + 2 = 6$ nitrogen atoms on the left. To balance nitrogen, write a coefficient of 3 for N_2:

$$2N_2H_4(l) + N_2O_4(l) \longrightarrow 3N_2(g) + 4H_2O(g)$$

This is the final balanced equation for the reaction of hydrazine with dinitrogen tetraoxide.

Three points concerning the balancing process are worth noting.

1. Equations are balanced by adjusting coefficients in front of formulas, never by changing subscripts within formulas. On paper, the equation discussed above could have been balanced by writing N_6 on the right, but that would have been absurd. Elemental nitrogen exists as diatomic molecules, N_2; there is no such thing as an N_6 molecule.

2. In balancing an equation, it is best to start with an element that appears in only one species on each side of the equation. In this case, either oxygen or hydrogen is a good starting point. Nitrogen would have been a poor choice, however, because there are nitrogen atoms in both reactant molecules, N_2H_4 and N_2O_4.

3. In principle, there are an infinite number of balanced equations that can be written for any reaction. The equations

$$4N_2H_4(l) + 2N_2O_4(l) \longrightarrow 6N_2(g) + 8H_2O(g)$$

$$N_2H_4(l) + \tfrac{1}{2}N_2O_4(l) \longrightarrow \tfrac{3}{2}N_2(g) + 2H_2O(g)$$

are balanced in that there are the same number of atoms of each element on both sides. Ordinarily, the equation with the simplest whole-number coefficients

$$2N_2H_4(l) + N_2O_4(l) \longrightarrow 3N_2(g) + 4H_2O(g)$$

is preferred.

Frequently, when you are asked to balance an equation, the formulas of products and reactants are given. Sometimes, though, you will have to derive the formulas, given only the names (Examples 3.10).

6N is no good either.

It's time to review Table 2.2 (page 43)!

EXAMPLE 3.10   Crystals of sodium hydroxide (lye) react with carbon dioxide of the air to form a white powder, sodium carbonate, and a colorless liquid, water. Write a balanced equation for this chemical reaction.

Strategy   To translate names into formulas, recall the discussion in Section 2.6, Chapter 2. The physical states are given or implied. To balance the equation, you could start with either sodium or hydrogen.

Solution   The skeleton equation is

$$NaOH(s) + CO_2(g) \longrightarrow Na_2CO_3(s) + H_2O(l)$$

Because there are two Na atoms on the right, a coefficient of 2 is written in front of NaOH:

$$2NaOH(s) + CO_2(g) \longrightarrow Na_2CO_3(s) + H_2O(l)$$

Careful inspection shows that all other atoms are now balanced.

Mass Relations from Equations

The principal reason for writing balanced equations is to make it possible to relate the masses of reactants and products. Calculations of this sort are based on a very important principle:

See Screens 5.2, Weight Relation in Chemical Reactions, & 5.3, Calculations in Stoichiometry.

> *The coefficients of a balanced equation represent numbers of moles of reactants and products.*

To show that this statement is valid, recall the equation

$$2\,N_2H_4(l) + N_2O_4(l) \longrightarrow 3\,N_2(g) + 4\,H_2O(g)$$

The coefficients in this equation represent numbers of molecules, that is,

2 molecules N_2H_4 + 1 molecule $N_2O_4 \longrightarrow$ 3 molecules N_2 + 4 molecules H_2O

A balanced equation remains valid if each coefficient is multiplied by the same number, including Avogadro's number, N_A:

$2N_A$ molecules N_2H_4 + $1N_A$ molecule $N_2O_4 \longrightarrow$
$3N_A$ molecules N_2 + $4N_A$ molecules H_2O

Recall from Section 3.2 that a mole represents Avogadro's number of items, N_A. Thus the equation above can be written

2 mol N_2H_4 + 1 mol $N_2O_4 \longrightarrow$ 3 mol N_2 + 4 mol H_2O

The quantities 2 mol of N_2H_4, 1 mol of N_2O_4, 3 mol of N_2, and 4 mol of H_2O are chemically equivalent to each other in this reaction. Hence they can be used in conversion factors such as

$$\frac{2\ \text{mol}\ N_2H_4}{3\ \text{mol}\ N_2}, \frac{3\ \text{mol}\ N_2}{1\ \text{mol}\ N_2O_4}, \text{or a variety of other combinations}$$

If you wanted to know how many moles of hydrazine were required to form 1.80 mol of elemental nitrogen, the conversion would be

$$n_{N_2H_4} = 1.80\ \text{mol}\ N_2 \times \frac{2\ \text{mol}\ N_2H_4}{3\ \text{mol}\ N_2} = 1.20\ \text{mol}\ N_2H_4$$

To find the number of moles of nitrogen produced from 2.60 mol of N_2O_4,

$$n_{N_2} = 2.60\ \text{mol}\ N_2O_4 \times \frac{3\ \text{mol}\ N_2}{1\ \text{mol}\ N_2O_4} = 7.80\ \text{mol}\ N_2$$

Simple mole relationships of the type just discussed are readily extended to relate—

- moles of one substance to grams of another (Example 3.11a).
- grams of one substance to grams of another (Examples 3.11b and c).

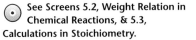

N_2 + $3H_2$

$\downarrow$

$2NH_3$

EXAMPLE 3.11 Ammonia used to make fertilizers for lawns and gardens is made by reacting nitrogen with hydrogen. The balanced equation for the reaction is

$$N_2(g) + 3H_2(g) \longrightarrow 2NH_3(g)$$

Determine

(a) the mass in grams of ammonia, NH_3, formed when 1.34 mol of N_2 reacts.
(b) the mass in grams of N_2 required to form 1.00 kg of NH_3.
(c) the mass in grams of H_2 required to react with 6.00 g of N_2.

Strategy   In each case you use the mole ratios given by the coefficients of the balanced equation to relate moles of one substance to moles of another. Beyond that, use molar mass to relate moles of one substance to mass in grams of that substance. Before starting, decide on the path you will follow to go from the quantity given to that required, that is,

To convert between moles of A and grams of A, use molar mass. To convert between moles of A and moles of B, use the balanced equation.

(a) $n_{N_2} \longrightarrow n_{NH_3} \longrightarrow$ mass of NH_3
(b) mass of $NH_3 \longrightarrow n_{NH_3} \longrightarrow n_{N_2} \longrightarrow$ mass of N_2
(c) mass of $N_2 \longrightarrow n_{N_2} \longrightarrow n_{H_2} \longrightarrow$ mass of H_2

Conversions indicated by colored arrows involve mole ratios given by the coefficients of the balanced equation. The other conversions require only molar masses and are essentially identical to those carried out in Examples 3.3 and 3.4.

Solution

(a) mass of NH_3 = 1.34 mol $N_2 \times \dfrac{2 \text{ mol } NH_3}{1 \text{ mol } N_2} \times \dfrac{17.03 \text{ g } NH_3}{1 \text{ mol } NH_3}$ = $\boxed{45.6 \text{ g } NH_3}$

(b) mass of N_2 = 1.00×10^3 g $NH_3 \times \dfrac{1 \text{ mol } NH_3}{17.03 \text{ g } NH_3} \times \dfrac{1 \text{ mol } N_2}{2 \text{ mol } NH_3} \times \dfrac{28.02 \text{ g } N_2}{1 \text{ mol } N_2}$

= $\boxed{823 \text{ g } N_2}$

(c) mass of H_2 = 6.00 g $N_2 \times \dfrac{1 \text{ mol } N_2}{28.02 \text{ g } N_2} \times \dfrac{3 \text{ mol } H_2}{1 \text{ mol } N_2} \times \dfrac{2.016 \text{ g } H_2}{1 \text{ mol } H_2}$ = $\boxed{1.30 \text{ g } H_2}$

Reality Check   Another approach in (c) would be to note from the balanced equation that 28.02 g (1 mol) of N_2 reacts with 6.048 g (3 mol) of H_2, so,

$$\text{mass of } H_2 = 6.00 \text{ g } N_2 \times \dfrac{6.048 \text{ g } H_2}{28.02 \text{ g } N_2} = 1.30 \text{ g } H_2$$

Limiting Reactant and Theoretical Yield

See Screen 5.4, Reactions Controlled by the Supply of One Reactant.

When the two elements antimony and iodine are heated in contact with one another (Figure 3.5), they react to form antimony(III) iodide.

$$2Sb(s) + 3I_2(s) \longrightarrow 2SbI_3(s)$$

The coefficients in this equation show that two moles of Sb (243.6 g) reacts with exactly three moles of I_2 (761.4 g) to form two moles of SbI_3 (1005.0 g). Put another way, the maximum quantity of SbI_3 that can be obtained under these conditions, assuming the reaction goes to completion and no product is lost, is 1005.0 g. This quantity is referred to as the **theoretical yield** of SbI_3.

Ordinarily, in the laboratory, reactants are not mixed in exactly the ratio required for reaction. Instead, an excess of one reactant, usually the cheaper one, is used. For example, 3.00 mol of Sb could be mixed with 3.00 mol of I_2. In that case, after the reaction is over, 1.00 mol of Sb remains unreacted.

excess Sb = 3.00 mol Sb originally − 2.00 mol Sb consumed

= 1.00 mol Sb

The Reactants   The Reaction   The Product

FIGURE 3.5
Reaction of antimony with iodine. When antimony (gray powder) comes in contact with iodine (violet vapor), a vigorous reaction takes place *(middle)*. The product is SbI_3, a red solid *(right)*. *(Charles D. Winters)*

The 3.00 mol of I_2 should be completely consumed in forming the 2.00 mol of SbI_3:

$$n_{SbI_3} \text{ formed} = 3.00 \text{ mol } I_2 \times \frac{2 \text{ mol } SbI_3}{3 \text{ mol } I_2} = 2.00 \text{ mol } SbI_3$$

After the reaction is over, the solid obtained would be a mixture of product, 2.00 mol of SbI_3, with 1.00 mol of unreacted Sb.

In situations such as this, a distinction is made between the *excess reactant* (Sb) and the **limiting reactant,** I_2. The amount of product formed is determined (limited) by the amount of limiting reactant. With 3.00 mol of I_2, only 2.00 mol of SbI_3 is obtained, regardless of how large an excess of Sb is used.

Under these conditions, the theoretical yield of product is the amount produced if the limiting reactant is completely consumed. In the case just cited, the theoretical yield of SbI_3 is 2.00 mol, the amount formed from the limiting reactant, I_2.

Often you will be given the amounts of two different reactants and asked to determine which is the limiting reactant and to calculate the theoretical yield of product. To do this, it helps to follow a systematic, three-step procedure.

1. *Calculate the amount of product that would be formed if the first reactant were completely consumed.*
2. *Repeat this calculation for the second reactant; that is, calculate how much product would be formed if all of that reactant were consumed.*
3. *Choose the smaller of the two amounts calculated in (1) and (2). This is the theoretical yield of product; the reactant that produces the smaller amount is the limiting reactant.* The other reactant is in excess; only part of it is consumed.

To illustrate how this approach works, suppose you want to make grilled cheese sandwiches from 6 slices of cheese and 18 pieces of bread. The cheese available would make 6 grilled cheese sandwiches; the bread would make 9. Clearly, the cheese is the limiting reactant; there is an excess of bread. The theoretical yield is 6 sandwiches.

See Screen 5.5, Limiting Reactants.

You have 62 buns, 95 hamburgers, and 51 slices of cheese. What is the limiting reactant for making double cheeseburgers?

EXAMPLE 3.12 Consider the reaction

$$2Sb(s) + 3I_2(s) \longrightarrow 2SbI_3(s)$$

Determine the limiting reactant and the theoretical yield when

(a) 1.20 mol of Sb and 2.40 mol of I_2 are mixed.
(b) 1.20 g of Sb and 2.40 g of I_2 are mixed. What mass of excess reactant is left when the reaction is complete?

Strategy Follow the three steps outlined above. In steps (1) and (2), follow the strategy described in Example 3.11 to calculate the amount of product formed. In (a), a simple one-step conversion is required; in (b), the path is longer because you have to go from mass of reactant to mass of product. To find the mass of reactant left over in (b), calculate how much is required to give the theoretical yield of product. Subtract that from the starting amount to find the amount left.

Solution

(a) (1) n_{SbI_3} from Sb = $1.20 \text{ mol Sb} \times \dfrac{2 \text{ mol SbI}_3}{2 \text{ mol Sb}} = 1.20 \text{ mol SbI}_3$

(2) n_{SbI_3} from I_2 = $2.40 \text{ mol I}_2 \times \dfrac{2 \text{ mol SbI}_3}{3 \text{ mol I}_2} = 1.60 \text{ mol SbI}_3$

(3) Because 1.20 mol is the smaller amount of product, that is the theoretical yield of SbI_3. This amount of SbI_3 is produced by the antimony, so Sb is the limiting reactant.

(b) (1) mass of SbI_3 from Sb = $1.20 \text{ g Sb} \times \dfrac{1 \text{ mol Sb}}{121.8 \text{ g Sb}} \times \dfrac{2 \text{ mol SbI}_3}{2 \text{ mol Sb}} \times \dfrac{502.5 \text{ g SbI}_3}{1 \text{ mol SbI}_3}$

$= 4.95 \text{ g SbI}_3$

(2) mass of SbI_3 from I_2 = $2.40 \text{ g I}_2 \times \dfrac{1 \text{ mol I}_2}{253.8 \text{ g I}_2} \times \dfrac{2 \text{ mol SbI}_3}{3 \text{ mol I}_2} \times \dfrac{502.5 \text{ g SbI}_3}{1 \text{ mol SbI}_3}$

$= 3.17 \text{ g SbI}_3$

(3) The reactant that yields the smaller amount (3.17 g) of SbI_3 is I_2. Hence I_2 is the limiting reactant. The smaller amount, 3.17 g of SbI_3, is the theoretical yield.

(4) mass of Sb required = $3.17 \text{ g SbI}_3 \times \dfrac{121.8 \text{ g Sb}}{502.5 \text{ g SbI}_3} = 0.768 \text{ g Sb}$

mass of Sb left = 1.20 g − 0.768 g = 0.43 g

Remember that in deciding on the theoretical yield of product, you *choose the smaller of the two calculated amounts.* To see why this must be the case, refer back to Example 3.12b. There 1.20 g of Sb was mixed with 2.40 g of I_2. Calculations show that the theoretical yield of SbI_3 is 3.17 g, and 0.43 g of Sb is left over. Thus

$$1.20 \text{ g Sb} + 2.40 \text{ g I}_2 \longrightarrow 3.17 \text{ g SbI}_3 + 0.43 \text{ g Sb}$$

This makes sense: 3.60 g of reactants yields a total of 3.60 g of products, including the unreacted antimony. Suppose, however, that 4.95 g of SbI_3 was chosen as the theoretical yield. The following nonsensical situation would arise.

$$1.20 \text{ g Sb} + 2.40 \text{ g I}_2 \longrightarrow 4.95 \text{ g SbI}_3$$

This violates the law of conservation of mass; 3.60 g of reactants cannot form 4.95 g of product.

Actual yield. The quantity of product isolated (the pink solid) is usually less than the theoretical yield. *(Charles D. Winters)*

Experimental Yield; Percent Yield

The theoretical yield is the maximum amount of product that can be obtained. In calculating the theoretical yield, it is assumed that the limiting reactant is 100% converted to product. In the real world, that is unlikely to happen. Some of the limiting reactant may be consumed in competing reactions. Some of the product may be lost in separating it from the reaction mixture. For these and other reasons, the experimental yield is ordinarily less than the theoretical yield. Put another way, the **percent yield** is expected to be less than 100%:

$$\text{percent yield} = \frac{\text{experimental yield}}{\text{theoretical yield}} \times 100\%$$

⊙ **See Screen 5.6, Percent Yield.**

⌐ You never do quite as well as you hoped, so you never get 100%.

EXAMPLE 3.13   When methylamine, CH_3NH_2, is treated with acid, the following reaction occurs:

$$CH_3NH_2(aq) + H^+(aq) \longrightarrow CH_3NH_3^+(aq)$$

When 3.00 g of CH_3NH_2 reacts with 0.100 mol of H^+, the product contains 2.60 g of $CH_3NH_3^+$. What is the theoretical yield? The percent yield?

Strategy   Start by determining the limiting reactant. The theoretical yield is the amount of $CH_3NH_3^+$ that would be formed if all of the limiting reactant were completely consumed. Then use the definition above to find the percent yield.

Solution

Mass of $CH_3NH_3^+$ from CH_3NH_2:

$$3.00 \text{ g } CH_3NH_2 \times \frac{1 \text{ mol } CH_3NH_2}{31.06 \text{ g } CH_3NH_2} \times \frac{1 \text{ mol } CH_3NH_3^+}{1 \text{ mol } CH_3NH_2} \times \frac{32.07 \text{ g } CH_3NH_3^+}{1 \text{ mol } CH_3NH_3^+} = 3.10 \text{ g}$$

Mass of $CH_3NH_3^+$ from 0.100 mol H^+:

$$0.100 \text{ mol of } H^+ \times \frac{1 \text{ mol } CH_3NH_3^+}{1 \text{ mol } H^+} \times \frac{32.07 \text{ g } CH_3NH_3^+}{1 \text{ mol } CH_3NH_3^+} = 3.21 \text{ g}$$

The limiting reactant is CH_3NH_2; the H^+ ion is (slightly) in excess. The theoretical yield is 3.10 g.

$$\text{percent yield} = \frac{2.60 \text{ g}}{3.10 \text{ g}} \times 100\% = 83.9\%$$

Hydrates

Ionic compounds often separate from water solution with molecules of water incorporated into the solid. Such compounds are referred to as **hydrates.** An example is hydrated copper sulfate, which contains five moles of H_2O for every mole of $CuSO_4$. Its formula is $CuSO_4 \cdot 5H_2O$; a dot is used to separate the formulas of the two compounds $CuSO_4$ and H_2O. A Greek prefix is used to show the number of moles of water; the systematic name of $CuSO_4 \cdot 5H_2O$ is copper(II) sulfate pentahydrate.

As you can perhaps guess from Table A, hydrates have been around for a long time. The compound $Na_2SO_4 \cdot 10H_2O$ was first obtained from Hungarian spring waters by the alchemist Johann Glauber in the seventeenth century. At about the same time, $MgSO_4 \cdot 7H_2O$ was isolated from mineral springs at Epsom in England. Both Glauber's salt and Epsom salts were once used as mild laxatives.

See Screen 3.14, Hydrated Compounds.

TABLE A	Some Familiar Hydrates	
Composition	**Systematic Name**	**Common Name**
$Na_2SO_4 \cdot 10H_2O$	Sodium sulfate decahydrate	Glauber's salt
$Na_2CO_3 \cdot 10H_2O$	Sodium carbonate decahydrate	Washing soda
$CaSO_4 \cdot 2H_2O$	Calcium sulfate dihydrate	Gypsum
$CaSO_4 \cdot \frac{1}{2}H_2O$	Calcium sulfate hemihydrate	Plaster of Paris
$MgSO_4 \cdot 7H_2O$	Magnesium sulfate heptahydrate	Epsom salts
$FeSO_4 \cdot 6H_2O$	Iron(II) sulfate hexahydrate	Green vitriol
$CuSO_4 \cdot 5H_2O$	Copper(II) sulfate pentahydrate	Blue vitriol
$CoSO_4 \cdot 6H_2O$	Cobalt(II) sulfate hexahydrate	Red vitriol

Certain hydrates, notably $Na_2CO_3 \cdot 10H_2O$ and $FeSO_4 \cdot 7H_2O$, lose all or part of their water of hydration when exposed to dry air. This process is referred to as *efflorescence;* the glassy (vitreous) hydrate crystals crumble to a powder. Frequently, dehydration is accompanied by a color change (Figure A). When $CoCl_2 \cdot 6H_2O$ is exposed to dry air or is heated, it loses water and changes color from red to purple or blue. Crystals of this compound are used as humidity indicators and as an ingredient of invisible ink. Writing becomes visible only when the paper is heated, driving off water and leaving a blue residue.

Hydrates often have quite different crystal shapes from those of the corresponding anhydrous compounds. Sodium urate monohydrate, $NaC_5H_3N_4O_3 \cdot H_2O$, deposits in human joints (particularly the big toe) as sharp, needle-like crystals (Figure B). These are responsible for the excruciating pain associated with the illness called gout. In contrast, anhydrous sodium urate crystals have a nearly spherical shape.

Molecular as well as ionic substances can form hydrates, but of an entirely different nature. In these crystals, sometimes referred to as *clathrates,* a molecule (such as CH_4, $CHCl_3$) is quite literally trapped in an ice-like cage of water molecules. Perhaps the best-known molecular hydrate is that of chlorine, which has the approximate composition $Cl_2 \cdot 7.3H_2O$. This compound was discovered by the great English physicist and electrochemist Michael Faraday in 1823. You can make it by bubbling chlorine gas through calcium chloride solution at 0°C; the hydrate comes down as feathery white crystals. In the winter of 1914, the German army used chlorine in chemical warfare on the Russian front against the soldiers of the Tsar. They were puzzled by its ineffectiveness; not until spring was deadly chlorine gas liberated from the hydrate, which is stable at cold temperatures.

Recently, geologists have discovered huge deposits of gas hydrates at the bottom of the ocean, where the pressure is high and the temperature slightly above 0°C. Trapped

FIGURE A

Copper and cobalt hydrates.
$CuSO_4 \cdot 5H_2O$ *(upper left)* is blue; the anhydrous solid—that is, copper sulfate without the water—is white *(lower left)*. $CoCl_2 \cdot 5H_2O$ *(upper right)* is pink. Lower hydrates of cobalt such as $CoCl_2 \cdot 4H_2O$ are purple *(lower right)* or blue. *(Marna G. Clarke)*

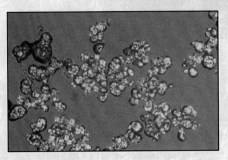
(a)

(b)

FIGURE B
Sodium urate crystals.
(a) Spherical anhydrous crystals.
(b) Needle-shaped monohydrate crystals that cause gout. *(Courtesy of P. A. Dieppe, P. A. Bacon, A. N. Bamji and I. Watt and the Gower Medical Publishing Co. Ltd., London.)*

within these solids are methane, CH_4, and other combustible hydrocarbons found in natural gas. One hydrate field in a deep region of the Atlantic Ocean is the size of the state of Vermont and is estimated to contain enough methane to produce 350 times as much energy as is annually consumed in the United States. Whether this remarkable discovery will prove practical as a source of energy is another question. Getting the hydrate crystals to the surface and using the methane they contain poses a formidable engineering problem.

CHAPTER HIGHLIGHTS

Key Concepts

1. Relate the atomic mass of an element to isotopic masses and abundances.
   (Example 3.1; Problems 3–10)

2. Use Avogadro's number to calculate the mass of an atom or molecule.
   (Example 3.2; Problems 13–18, 25, 26, 73, 74)

3. Use molar mass to relate—
   ▪ moles to mass of a substance.
   (Examples 3.3, 3.4; Problems 21–26)
   ▪ molecular formula to simplest formula.
   (Example 3.9; Problems 43–46)

4. Use the formula of a compound to find percent composition or mass of an element in a sample.
   (Examples 3.5, 3.6; Problems 27–32)

5. Find the simplest formula of a compound from chemical analysis.
   (Examples 3.7, 3.8; Problems 35–46)

6. Balance chemical equations by inspection.
   (Example 3.10; Problems 47–52, 63–67)

7. Use a balanced equation to—
   ▪ relate masses of reactants and products.
   (Example 3.11; Problems 53–62, 76)
   ▪ find the limiting reactant, theoretical yield, and percent yield.
   (Examples 3.12, 3.13; Problems 63–70)

Key Terms

Terms marked ■ were introduced in an earlier chapter but are essential to an understanding of the material covered in this chapter.

atomic mass	molar mass	reactant
atomic mass unit	mole	simplest formula
Avogadro's number	■ molecular formula	yield
hydrate	percent composition	—percent
isotopic abundance	product	—theoretical
limiting reactant		

Summary Problem

Consider silicon, an element widely used in semiconductors.

(a) It consists of three isotopes with atomic masses 27.977 amu, 28.977 amu, and 29.974 amu. Their abundances are 92.23%, 4.67%, and 3.10%, respectively. What is the atomic mass of silicon?

(b) What is the average mass, in grams, of a silicon atom?

(c) How many moles are there in 38.00 g of silicon?

(d) When silicon reacts with oxygen, silicon dioxide is formed. Write a balanced equation for this reaction.

(e) What is the molar mass of silicon dioxide?

(f) How many grams of silicon will react with 0.150 mol of oxygen?

(g) If 35.00 g of silicon reacts with an excess of oxygen, how many grams of silicon dioxide are formed?

(h) If 22.0 g of silicon reacts with 22.0 g of oxygen, how many grams of silicon dioxide are formed?

(i) If 37.20 g of silicon dioxide is produced from the reaction in (h), what is the percent yield?

(j) Talc is a common mineral that contains silicon. It is made up of 19.2% Mg, 50.6% O, 0.529% H, and 29.6% Si. What is the simplest formula for talc?

Answers

(a) 28.09 amu
(b) 4.665×10^{-23}
(c) 1.353
(d) $Si(s) + O_2(g) \rightarrow SiO_2(s)$
(e) 60.09 g

(f) 4.21
(g) 74.87
(h) 41.3
(i) 90.1%
(j) $Mg_3O_{12}H_2Si_4$

Questions & Problems

Problem numbers in blue indicate that the answer is available in Appendix 6 at the back of the book.
WEB indicates that the solution is posted at **http://www.harcourtcollege.com/chem/general/masterton4/student/**

Atomic Masses

1. Arrange the following in order of increasing mass.
  (a) a sodium ion   **(b)** a selenium atom
  (c) a sulfur (S_8) molecule   **(d)** a scandium atom
2. Calculate the mass ratio of a bromine atom to an atom of
  (a) neon   **(b)** calcium   **(c)** helium
3. Strontium has four isotopes with the following masses:

83.9134 amu (0.56%), 85.9094 amu (9.86%), 86.9089 amu (7.00%), and 87.9056 (82.58%). Calculate the atomic mass of strontium.

4. Oxygen consists of three isotopes with atomic masses 16.00, 17.00, and 18.00 amu. Their abundances are 99.76%, 0.04%, and 0.20%, respectively. What is the atomic mass of oxygen?

5. Naturally occurring europium (Eu) consists of two isotopes.

One of the isotopes has a mass of 150.960 amu and 48.03% abundance. What is the atomic mass of the other isotope?

WEB 6. Naturally occurring boron has only two isotopes. B-10 has an atomic mass of 10.0129 amu and an abundance of 19.9%. What is the atomic mass of the second isotope, B-11?

7. Oxygen (atomic mass = 15.9994 amu) has three isotopes with masses 15.9949 amu, 16.9993 amu, and 17.9992 amu. The abundance of the last isotope is 0.204%. Estimate the abundance of the other two isotopes.

8. Chromium (atomic mass = 51.9961 amu) has four isotopes. Their masses are 49.94605 amu, 51.94051 amu, 52.94065 amu, and 53.93888 amu. The first two isotopes have a total abundance of 87.87%, and the last isotope has an abundance of 2.365%. What is the abundance of the third isotope? Estimate the abundances of the first two isotopes.

9. Neon consists of three isotopes, Ne-20, Ne-21, and Ne-22. Their abundances are 90.48%, 0.27%, and 9.22%, respectively. Sketch the mass spectrum for neon.

10. Chlorine has two isotopes, Cl-35 and Cl-37. Their abundances are 75.53% and 24.47%, respectively. Assume that the only hydrogen isotope present is H-1.
(a) How many different HCl molecules are possible?
(b) What is the sum of the mass numbers of the two atoms in each molecule?
(c) Sketch the mass spectrum for HCl if all the positive ions are obtained by removing a single electron from an HCl molecule.

Avogadro's Number and the Mole

11. The meat from one hazelnut has a mass of 0.985 g.
(a) What is the mass of a millionth of a mole (10^{-6}) of hazelnut meats? (A millionth of a mole is also called a micromole.)
(b) How many moles are there in a pound of hazelnut meats?

12. One chocolate chip used in making chocolate chip cookies has a mass of 0.324 g.
(a) How many chocolate chips are there in one mole of chocolate chips?
(b) If a cookie needs 15 chocolate chips, how many cookies can one make with a billionth (1×10^{-9}) of a mole of chocolate chips? (A billionth of a mole is scientifically known as a *nanomole*.)

13. Nickel has atomic mass 58.69 amu. Calculate
(a) the mass of three trillion (trillion means 10^{12}) nickel atoms.
(b) the number of atoms in 0.020 oz of nickel.

WEB 14. The atomic mass of tungsten, W, is 183.9 amu. Calculate
(a) the mass, in grams, of a tungsten atom.
(b) the number of atoms in a picogram of tungsten (*pico* means 10^{-12}).

15. Determine
(a) the mass of 0.357 mol of gold.
(b) the number of atoms in 0.357 g of gold.
(c) the number of moles of electrons in 0.357 g of gold.

16. The isotope Si-28 has a mass of 27.977 amu. For ten grams of Si-28, calculate
(a) the number of moles.
(b) the number of atoms.
(c) the total number of protons, neutrons, and electrons.

17. How many electrons are there in
(a) a silver atom?
(b) one mole of silver atoms?
(c) a metric ton (1000 kg) of silver?

18. Consider the isotope Cl-37 (atomic mass = 37.00 amu). In this isotope, how many neutrons are there in
(a) ten atoms?
(b) ten chloride ions?
(c) one mole of chloride ions?
(d) one gram of Cl-37?

Molar Mass and Mole-Gram Conversions

19. Calculate the molar masses (in grams per mole) of
(a) gallium, Ga, a metal that literally melts in your hands.
(b) plaster of Paris, $CaSO_4 \cdot \frac{1}{2}H_2O$, used in making casts for broken bones.
(c) benzoyl peroxide, $C_{14}H_{10}O_4$, the active ingredient in many lotions used to treat acne.

20. Calculate the molar masses (in grams per mole) of
(a) cane sugar, $C_{12}H_{22}O_{11}$.
(b) laughing gas, N_2O.
(c) vitamin A, $C_{20}H_{30}O$.

21. Convert the following to moles.
(a) 4.00×10^3 g of hydrazine, a rocket propellant
(b) 12.5 g of tin(II) fluoride, the active ingredient in fluoride toothpaste
(c) 13 g of caffeine, $C_4H_5N_2O$

22. Convert the following to moles.
(a) 35.00 g of CF_2Cl_2, a chlorofluorocarbon that destroys the ozone layer in the atmosphere
(b) 100.0 mg of iron(II) sulfate, an iron supplement prescribed for anemia
(c) 2.00 g of valium ($C_{15}H_{13}ClN_2O$—diazepam)

23. Calculate the mass in grams of 7.58 mol of
(a) nitrogen dioxide, the brown gas responsible for the color of smog.
(b) aspartame, $C_{14}H_{18}N_2O_5$, the artificial sweetener.
(c) urea, CH_4N_2O, commonly used as a fertilizer.

WEB 24. Calculate the mass in grams of 17.5 mol of
(a) nitrogen atoms   (b) nitrogen gas
(c) ammonia gas

25. Complete the following table for citric acid, $C_6H_8O_7$, the acid found in many citrus fruits.

	Number of Grams	Number of Moles	Number of Molecules	Number of O Atoms
(a)	0.1364			
(b)		1.248		
(c)			4.32×10^{22}	
(d)				5.55×10^{19}

26. Complete the following table for TNT (trinitrotoluene), $C_7H_5(NO_2)_3$.

	Number of Grams	Number of Moles	Number of Molecules	Number of N Atoms
(a)	127.2	_____	_____	_____
(b)	_____	0.9254	_____	_____
(c)	_____	_____	1.24×10^{28}	_____
(d)	_____	_____	_____	7.5×10^{22}

Percent Composition from Formula

27. Turquoise has the following chemical formula: $CuAl_6(PO_4)_4(OH)_8 \cdot 4H_2O$. Calculate the mass percent of each element in turquoise.

28. The active ingredient of some antiperspirants is $Al_2(OH)_5Cl$. Calculate the mass percent of each element in this ingredient.

29. One of the most common rocks on Earth is feldspar. One type of feldspar has the formula $CaAl_2Si_2O_8$. How many grams of aluminum can be obtained from 1276 kg of feldspar?

30. Cisplatin, an antitumor ingredient, has the formula $Pt(NH_3)_2Cl_2$. How many milligrams of platinum are present in 250.0 mg of cisplatin?

31. The active ingredient in Pepto-Bismol (an over-the-counter remedy for upset stomach) is bismuth subsalicylate, $C_7H_5BiO_4$. Analysis of a 1.500-g sample of Pepto-Bismol yields 346 mg of bismuth. What percent by mass is bismuth subsalicylate? (Assume that there is no other bismuth-containing compound in Pepto-Bismol.)

32. A tablet of Tylenol has a mass of 0.611 g. It contains 251 mg of its active ingredient, acetaminophen, $C_8H_9NO_2$.

    (a) What is the mass percent of acetaminophen in a tablet of Tylenol?

    (b) Assume that all the nitrogen in the tablet is in the acetaminophen. How many grams of nitrogen are present in a tablet of Tylenol?

33. Combustion analysis of 1.00 g of the male sex hormone, testosterone, yields 2.90 g of CO_2 and 0.875 g H_2O. What are the mass percents of carbon, hydrogen, and oxygen in testosterone?

34. Hexachlorophene, a compound made up of atoms of carbon, hydrogen, chlorine, and oxygen, is an ingredient in germicidal soaps. Combustion of a 1.000-g sample yields 1.407 g of carbon dioxide, 0.134 g of water, and 0.523 g of chlorine gas. What are the mass percents of carbon, hydrogen, oxygen, and chlorine in hexachlorophene?

Simplest Formula from Analysis

35. If 7.35 g of chromium reacts directly with oxygen to form 10.74 g of a metal oxide, what is the simplest formula of the metal oxide? Name the oxide.

WEB **36.** Arsenic reacts with chlorine to form a chloride. If 1.587 g of arsenic reacts with 3.755 g of chlorine, what is the simplest formula of the chloride? Name the chloride.

37. Determine the simplest formulas of the following compounds:

    (a) saccharin, the artificial sweetener, which has the composition 45.90% C, 2.75% H, 26.20% O, 17.50% S, and 7.65% N.

    (b) allicin, the compound that gives garlic its characteristic odor, which has the composition 6.21% H, 44.4% C, 9.86% O, and 39.51% S.

    (c) sodium thiosulfate, the fixer used in developing photographic film, which has the composition 30.36% O, 29.08% Na, 40.56% S.

38. Determine the simplest formulas of the following compounds:

    (a) the food enhancer monosodium glutamate (MSG), which has the composition 35.51% C, 4.77% H, 37.85% O, 8.29% N, and 13.60% Na.

    (b) zircon, a diamond-like mineral, which has the composition 34.91% O, 15.32% Si, and 49.76% Zr.

    (c) nicotine, which has the composition 74.0% C, 8.65% H, and 17.4% N.

39. Ibuprofen, the active ingredient in Advil, is made up of carbon, hydrogen, and oxygen atoms. When a sample of ibuprofen, weighing 5.000 g, burns in oxygen, 13.86 g of CO_2 and 3.926 g of water are obtained. What is the simplest formula of ibuprofen?

40. Methyl salicylate is a common "active ingredient" in liniments such as Ben-Gay. It is also known as oil of wintergreen. It is made up of carbon, hydrogen, and oxygen atoms. When a sample of methyl salicylate weighing 5.287 g is burned in excess oxygen, 12.24 g of carbon dioxide and 2.505 g of water are formed. What is the simplest formula for oil of wintergreen?

41. Halothane is a commonly used gaseous anesthetic. It is made up of carbon, hydrogen, bromine, fluorine, and chlorine atoms. A ten-gram sample of halothane when burned in oxygen gas yields 4.459 g of CO_2 and 0.4559 g of H_2O. A second ten-gram sample is burned in hydrogen and yields 1.847 g of HCl and 4.099 g of HBr. What is the simplest formula for halothane?

42. Sevin is a common plant insecticide, particularly effective against Japanese beetles. It is made up of carbon, hydrogen, nitrogen, and oxygen atoms. A 10.94-g sample of Sevin burned in oxygen yields 28.73 g of CO_2 and 5.386 g of H_2O. Another experiment determines that the insecticide contains 6.963% nitrogen by mass. What is the simplest formula for Sevin?

43. Dimethylhydrazine, the fuel used in the Apollo lunar descent module, has a molar mass of 60.10 g/mol. It is made up of carbon, hydrogen, and nitrogen atoms. The combustion of 2.859 g of the fuel in excess oxygen yields 4.190 g of carbon dioxide and 3.428 g of water. What are the simplest and molecular formulas for dimethylhydrazine?

44. Hexamethylenediamine ($\mathcal{M} = 116.2$ g/mol), a compound made up of carbon, hydrogen, and nitrogen atoms, is used in the production of nylon. When 6.315 g of hexamethylenediamine is burned in oxygen, 14.36 g of carbon dioxide and 7.832 g of water are obtained. What are the simplest and molecular formulas of this compound?

45. A certain hydrate of potassium aluminum sulfate (alum) has the formula $KAl(SO_4)_2 \cdot xH_2O$. When a hydrate sample weighing 5.459 g is heated to remove all the water, 2.583 g of $KAl(SO_4)_2$ remains. What is the mass percent of water in the hydrate? What is x?

46. Epsom salts are hydrates of magnesium sulfate. The formula for Epsom salts is $MgSO_4 \cdot 7H_2O$. A 7.834-g sample is heated until a constant mass is obtained indicating that all the water has been evaporated off. What is the mass of the anhydrous magnesium sulfate? What percentage of the hydrate is water?

Balancing Equations

47. Balance the following equations:
(a) $Cr(s) + O_2(g) \rightarrow Cr_2O_3(s)$
(b) $NaNO_3(s) + H_2SO_4(l) \rightarrow Na_2SO_4(s) + HNO_3(g)$
(c) $BF_3(g) + H_2O(g) \rightarrow B_2O_3(s) + HF(g)$

48. Balance the following equations:
(a) $TiO_2(s) + Cl_2(g) + C(s) \rightarrow TiCl_4(l) + CO(g)$
(b) $Br_2(l) + I_2(s) \rightarrow IBr_3(g)$
(c) $C_2H_8N_2(s) + N_2O_4(g) \rightarrow N_2(g) + CO_2(g) + H_2O(g)$

49. Write balanced equations for the reaction of aluminum metal with the following nonmetals:
(a) sulfur   (b) bromine   (c) nitrogen
(d) oxygen (forming O^{2-} ions)
(e) oxygen (forming O_2^{2-}, or peroxide ions)

50. Write balanced equations for the reaction of sulfur with the following metals to form solids that you can take to be ionic when the anion is S^{2-}.
(a) potassium   (b) magnesium   (c) aluminum
(d) calcium   (e) iron (forming Fe^{2+} ions)

51. Write a balanced equation for
(a) the combustion (reaction with oxygen gas) of glucose, $C_6H_{12}O_6$, to give carbon dioxide and water.
(b) the reaction between xenon tetrafluoride gas and water to give xenon, oxygen, and hydrogen fluoride gases.
(c) the reaction between aluminum and iron(III) oxide to give aluminum oxide and iron.
(d) the formation of ammonia gas from its elements.
(e) the reaction between sodium chloride, sulfur dioxide gas, steam, and oxygen to give sodium sulfate and hydrogen chloride gas.

52. Write a balanced equation for
(a) the reaction between fluorine gas and water to give oxygen difluoride and hydrogen fluoride gases.
(b) the reaction between oxygen and ammonia gases to give nitrogen dioxide gas and water.
(c) the burning of gold(III) sulfide in hydrogen to give gold metal and dihydrogen sulfide gas.
(d) the decomposition of sodium hydrogen carbonate to sodium carbonate, water, and carbon dioxide gas.
(e) the reaction between sulfur dioxide gas and liquid hydrogen fluoride to give sulfur tetrafluoride gas and water.

Mole–Mass Relations in Reactions

53. One way to remove nitrogen oxide (NO) from smoke stack emissions is to react it with ammonia:

$$4NH_3(g) + 6NO(g) \longrightarrow 5N_2(g) + 6H_2O$$

Fill in the blanks below:
(a) 15.4 mol of NH_3 reacts with _____ mol of NO.
(b) 2.441 mol of NO yields _____ mol of water.
(c) 0.1431 mol of nitrogen requires _____ mol of NH_3.
(d) 21.2 mol of NH_3 yields _____ mol of nitrogen.

54. Phosphine gas reacts with oxygen according to the following equation:

$$4PH_3(g) + 8O_2(g) \longrightarrow P_4O_{10}(s) + 6H_2O(g)$$

Fill in the blanks below:
(a) 12.7 mol of O_2 reacts with _____ mol of PH_3.
(b) 5.43 mol of PH_3 yields _____ mol of P_4O_{10}.
(c) 1.003 mol of H_2O requires yields _____ mol of O_2.
(d) 1.76 mol of oxygen produces _____ mol of P_4O_{10}.

55. Use the equation given in Problem 53 to calculate
(a) the mass of water produced from 0.839 mol of ammonia.
(b) the mass of NO required to react with 3.402 mol of ammonia.
(c) the mass of ammonia required to produce 12.0 g of nitrogen gas.
(d) the mass of ammonia required to react with 115 g of NO.

56. Use the equation given in Problem 54 to calculate
(a) the mass of tetraphosphorus decaoxide produced from 12.43 mol of phosphine.
(b) the mass of PH_3 required to form 0.739 mol of water.
(c) the mass of oxygen gas that yields 1.000 g of water.
(d) the mass of oxygen required to react with 20.50 g of phosphine.

57. The combustion of liquid chloroethylene, C_2H_3Cl, yields carbon dioxide, steam, and hydrogen chloride gas.
(a) Write a balanced equation for the reaction.
(b) How many moles of oxygen are required to react with 35.00 g of chloroethylene?
(c) If 25.00 g of chloroethylene reacts with an excess of oxygen, how many grams of each product are formed?

WEB 58. Sand consists mainly of silicon dioxide. When sand is heated with an excess of coke (carbon), pure silicon and carbon monoxide are produced.
(a) Write a balanced equation for the reaction.
(b) How many moles of silicon dioxide are required to form 12.72 g of silicon?
(c) How many grams of carbon monoxide are formed when 44.99 g of silicon is produced?

59. When corn is allowed to ferment, ethyl alcohol is produced from the glucose in corn according to the reaction

$$C_6H_{12}O_6(s) \longrightarrow 2C_2H_5OH(l) + 2CO_2(g)$$

(a) What volume of ethyl alcohol ($d = 0.789$ g/cm^3) is produced from 100.0 lb of glucose?

(b) Gasohol is a mixture of 10 cm^3 of ethyl alcohol per 90 cm^3 of gasoline. How many grams of glucose are required to produce the ethyl alcohol in ten liters of gasohol?

60. A wine cooler contains about 4.5% ethyl alcohol, C_2H_5OH, by mass. Assume that the alcohol in 10.00 kg of wine cooler is produced by the fermentation of glucose in grapes according to the reaction in Problem 59.

(a) How many grams of glucose are needed to produce the ethyl alcohol in the wine cooler?

(b) What volume of carbon dioxide gas ($d = 1.80$ g/L) is produced at the same time under the same conditions of temperature and pressure?

61. Oxygen masks for producing O_2 in emergency situations contain potassium superoxide, KO_2. It reacts with CO_2 and H_2O in exhaled air to produce oxygen:

$$4KO_2(s) + 2H_2O(g) + 4CO_2(g) \longrightarrow 4KHCO_3(s) + 3O_2(g)$$

If a person wearing such a mask exhales 0.702 g CO_2/min, how many grams of KO_2 are consumed in 25 minutes?

62. A crude oil burned in electrical generating plants contains about 1.2% sulfur by mass. When the oil burns, the sulfur forms sulfur dioxide gas:

$$S(s) + O_2(g) \longrightarrow SO_2(g)$$

How many liters of SO_2 ($d = 2.60$ g/L) are produced when 1.00×10^4 kg of oil burns at the same temperature and pressure?

Limiting Reactant; Theoretical Yield

63. A gaseous mixture containing 4.15 mol of hydrogen gas and 7.13 mol of oxygen gas reacts to form steam.

(a) Write a balanced equation for the reaction.

(b) What is the limiting reactant?

(c) What is the theoretical yield of steam in moles?

(d) How many moles of the excess reactant remain unreacted?

64. Aluminum reacts with sulfur gas to form aluminum sulfide. Initially, 1.18 mol of aluminum and 2.25 mol of sulfur are combined.

(a) Write a balanced equation for the reaction.

(b) What is the limiting reactant?

(c) What is the theoretical yield of aluminum sulfide in moles?

(d) How many moles of excess reactant remain unreacted?

65. Titanium(IV) oxide (TiO_2) is the substance used as the white pigment in paint. It is prepared by the combustion of titanium(IV) chloride. Chlorine gas is also produced.

(a) Write a balanced equation for the reaction.

(b) How much titanium(IV) chloride must react with ex-

cess oxygen to prepare 175 g of the pigment? The reaction is found to be 85% efficient.

66. Silicon nitride (Si_3N_4) is a ceramic. It is made by reacting silicon and nitrogen at high temperatures.

(a) Write a balanced equation for the reaction.

(b) How much nitrogen must react with excess silicon to prepare 2.00×10^2 kg of silicon nitride if the reaction yield is 72%?

67. The space shuttle uses aluminum metal and ammonium perchlorate in its reusable booster rockets. The products of the reaction are aluminum oxide, aluminum chloride, nitrogen oxide gas, and steam. The reaction mixture contains 7.00 g of aluminum and 9.32 g of ammonium perchlorate.

(a) Write a balanced equation for the reaction.

(b) What is the theoretical yield of aluminum oxide?

(c) If 1.56 g of aluminum oxide is formed, what is the percent yield?

(d) How many grams of excess reactant are unused?

68. Oxyacetylene torches used for welding reach temperatures near 2000°C. The reaction involved in the combustion of acetylene is

$$2C_2H_2(g) + 5O_2(g) \longrightarrow 4CO_2(g) + 2H_2O(g)$$

(a) Starting with 175 g of both acetylene and oxygen, what is the theoretical yield, in grams, of carbon dioxide?

(b) If 68.5 L ($d = 1.85$ g/L) of carbon dioxide is produced, what is the percent yield at the same conditions of temperature and pressure?

(c) How much of the reactant in excess is unused?

69. Aspirin, $C_9H_8O_4$, is prepared by reacting salicylic acid, $C_7H_6O_3$, with acetic anhydride, $C_4H_6O_3$, in the reaction

$$C_7H_6O_3(s) + C_4H_6O_3(l) \longrightarrow C_9H_8O_4(s) + C_2H_4O_2(l)$$

A student is told to prepare 45.0 g of aspirin. She is also told to use a 55.0% excess of acetic anhydride and to expect to get an 85.0% yield in the reaction. How many grams of each reactant should she use?

70. A student prepares phosphorous acid, H_3PO_3, by reacting solid phosphorus triiodide with water.

$$PI_3(s) + 3H_2O(l) \longrightarrow H_3PO_3(s) + 3HI(g)$$

The student needs to obtain 0.250 L of H_3PO_3 ($d = 1.651$ g/cm^3). The procedure calls for a 45.0% excess of water and a yield of 75.0%. How much phosphorus triiodide should be weighed out? What volume of water ($d = 1.00$ g/cm^3) should be used?

Unclassified

71. Iron reacts with oxygen. Different masses of iron are burned in a constant amount of oxygen. The product, an oxide of iron, is weighed. The graph on page 81 is obtained when the mass of product used is plotted against the mass of iron used.

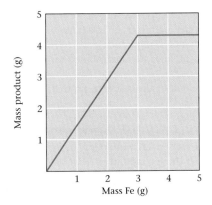

(a) How many grams of product are obtained when 0.50 g of iron is used?
(b) What is the limiting reactant when 2.00 g of iron is used?
(c) What is the limiting reactant when 5.00 g of iron is used?
(d) How many grams of iron react exactly with the amount of oxygen supplied?
(e) What is the simplest formula of the product?

72. Magnesium ribbon reacts with acid to produce hydrogen gas and magnesium ions. Different masses of magnesium ribbon are added to 10 mL of the acid. The volume of the hydrogen gas obtained is a measure of the number of moles of hydrogen produced by the reaction. Various measurements are given in the table below.

Experiment	Mass of Mg Ribbon (g)	Volume of Acid Used (mL)	Volume of H₂ Gas (mL)
1	0.020	10.0	21
2	0.040	10.0	42
3	0.080	10.0	82
4	0.120	10.0	122
5	0.160	10.0	122
6	0.200	10.0	122

(a) Draw a graph of the results by plotting the mass of Mg versus the volume of the hydrogen gas.
(b) What is the limiting reactant in experiment 1?
(c) What is the limiting reactant in experiment 3?
(d) What is the limiting reactant in experiment 6?
(e) Which experiment uses stoichiometric amounts of each reactant?
(f) What volume of gas would be obtained if 0.300 g of Mg ribbon were used? If 0.010 g were used?

73. Carbon tetrachloride, CCl_4, was a popular dry-cleaning agent until it was shown to be carcinogenic. It has a density of 1.589 g/cm³. What volume of carbon tetrachloride will contain a total of 6.00×10^{25} molecules of CCl_4?

74. Some brands of salami contain 0.090% sodium benzoate ($NaC_7H_5O_2$) by mass as a preservative. If you eat 6.00 oz of this salami, how many atoms of sodium will you consume, assuming salami contains no other source of that element?

75. All the fertilizers listed below contribute nitrogen to the soil. If all these fertilizers are sold for the same price per gram of nitrogen, which will cost the least per 50-lb bag?

> urea $(NH_2)_2CO$
> ammonia NH_3
> ammonium nitrate NH_4NO_3
> guanidine $HNC(NH_2)_2$

76. Most wine is prepared by the fermentation of the glucose in grape juice by yeast:

$$C_6H_{12}O_6(aq) \longrightarrow 2C_2H_5OH(aq) + 2CO_2(g)$$

How many grams of glucose should there be in grape juice to produce 725 mL of wine that is 11.0% ethyl alcohol, C_2H_5OH ($d = 0.789$ g/cm³), by volume?

77. Perhaps the simplest way to calculate Avogadro's number is to compare the charge on the electron, first determined by Robert Millikan in 1909, with the charge on a mole of electrons, determined electrochemically (Chapter 18). These charges, in coulombs (C), are given in Appendix 1. Use them to calculate Avogadro's number to five significant figures.

Conceptual Problems

78. Represent the following equation pictorially (see Question 79, to see a pictorial drawing), using squares to represent A and circles to represent B.

$$A_3 + B_2 \longrightarrow AB_4$$

After you have "drawn" the equation, use your drawing as a guide to balance it.

79. The reaction between compounds made up of M (squares) and X (circles) is shown pictorially below. Write a balanced equation to represent the picture shown using smallest whole-number coefficients.

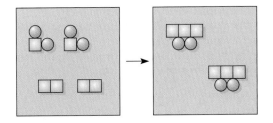

80. Consider the following diagram where atom X is represented by a square and atom Y is represented by a circle.

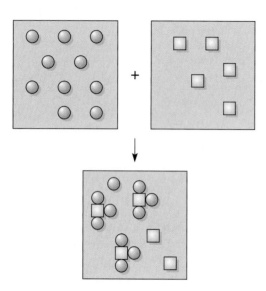

(a) Write the equation for the reaction represented by the diagram.
(b) If each circle stands for a mole of Y and each square a mole of X, how many moles of X did one start with? How many moles of Y?
(c) Using the same representation described in part (b), how many moles of product are formed? How many moles of X and Y are left unreacted?

81. Nitrogen reacts with hydrogen to form ammonia. Represent each nitrogen atom by a square and each hydrogen atom with a circle. Starting with five molecules of both hydrogen and nitrogen, show pictorially what you have after the reaction is complete.

82. When 4.0 mol of CCl_4 reacts with an excess of HF, 3.0 mol of CCl_2F_2 (Freon) is obtained. The equation for the reaction is

$$CCl_4(l) + 2HF(g) \longrightarrow CCl_2F_2(l) + 2HCl(g)$$

State which of the statements are true about the reaction and make the false statements true.
(a) The theoretical yield for CCl_2F_2 is 3.0 mol.
(b) The theoretical yield for HCl is 71 g.
(c) The percent yield for the reaction is 75%.
(d) The theoretical yield cannot be determined unless the exact amount of HF is given.
(e) From just the information given above, it is impossible to calculate how much HF is unreacted.

(f) For this reaction, as well as for any other reaction, the total number of moles of reactants is equal to the total number of moles of product.
(g) Half a mole of HF is consumed for every mole of CCl_4 used.
(h) At the end of the reaction, no CCl_4 is theoretically left unreacted.

83. Suppose that the atomic mass of C-12 is taken to be 5.000 amu and that a mole is defined as the number of atoms in 5.000 kg of carbon-12. How many atoms would there be in one mole under these conditions? (*Hint:* There are 6.022×10^{23} C atoms in 12.00 g of C-12.)

84. Suppose that Si-28 ($^{28}_{14}Si$) is taken as the standard for expressing atomic masses and assigned an atomic mass of 10.00 amu. Estimate the molar mass of lithium nitride.

Challenge Problems

85. Chlorophyll, the substance responsible for the green color of leaves, has one magnesium atom per chlorophyll molecule and contains 2.72% magnesium by mass. What is the molar mass of chlorophyll?

86. By x-ray diffraction it is possible to determine the geometric pattern in which atoms are arranged in a crystal and the distances between atoms. In a crystal of silver, four atoms effectively occupy the volume of a cube 0.409 nm on an edge. Taking the density of silver to be 10.5 g/cm^3, calculate the number of atoms in one mole of silver.

87. A 5.025-g sample of calcium is burned in air to produce a mixture of two ionic compounds, calcium oxide and calcium nitride. Water is added to this mixture. It reacts with calcium oxide to form 4.832 g of calcium hydroxide. How many grams of calcium oxide are formed? How many grams of calcium nitride?

88. A mixture of potassium chloride and potassium bromide weighing 3.595 g is heated with chlorine, which converts the mixture completely to potassium chloride. The total mass of potassium chloride after the reaction is 3.129 g. What percentage of the original mixture was potassium bromide?

89. A sample of an oxide of vanadium weighing 4.589 g was heated with hydrogen gas to form water and another oxide of vanadium weighing 3.782 g. The second oxide was treated further with hydrogen until only 2.573 g of vanadium metal remained.
(a) What are the simplest formulas of the two oxides?
(b) What is the total mass of water formed in the successive reactions?

90. A sample of cocaine, $C_{17}H_{21}O_4N$, is diluted with sugar, $C_{12}H_{22}O_{11}$. When a 1.00-mg sample of this mixture is burned, 1.00 mL of carbon dioxide ($d = 1.80$ g/L) is formed. What is the percentage of cocaine in this mixture?

> He takes up the waters of the sea in his
> hand, leaving the salt;
> He disperses it in mist through the skies;
> He recollects and sprinkles it like grain in
> six-rayed snowy stars over the earth,
> There to lie till he dissolves the bonds again.
>
> —Henry David Thoreau
> *Journal* (January 5, 1856)

REACTIONS IN AQUEOUS SOLUTION

4

Rust results when metallic iron
(Fe) is oxidized to an iron ion
(Fe^{3+}). *(Charles D. Winters)*

CHAPTER OUTLINE

**4.1   SOLUTE CONCENTRATIONS;
         MOLARITY**

4.2   PRECIPITATION REACTIONS

4.3   ACID–BASE REACTIONS

**4.4   OXIDATION–REDUCTION
         REACTIONS**

See the *Saunders Interactive General
Chemistry CD-ROM,* Screen 5.9,
Solutions.

Most of the reactions considered in Chapter 3 involved pure substances reacting with each other. However, most of the reactions you will carry out in the laboratory or hear about in lecture take place in water (aqueous) solution. Beyond that, most of the reactions that occur in the world around you involve ions or molecules dissolved in water. For these reasons, among others, you need to become familiar with some of the more important types of aqueous reactions. These include—

■ precipitation reactions (Section 4.2).

■ acid-base reactions (Section 4.3).

■ oxidation-reduction reactions (Section 4.4).

The emphasis is on writing and balancing chemical equations for these reactions. All of these reactions involve ions in solution. The corresponding equations are given a special name: net ionic equations. They can be used to carry out stoichiometric calculations similar to those discussed in Chapter 3.

To carry out these calculations for solution reactions, you need to be familiar with a concentration unit called molarity, which tells you how many moles of a species there are in a given volume of solution.

4.1   SOLUTE CONCENTRATIONS; MOLARITY

See Screen 5.10, Solution
Concentration.

The concentration of a solute in solution can be expressed in terms of its **molarity:**

$$\text{molarity } (M) = \frac{\text{moles of solute}}{\text{liters of solution}}$$

[NH₃] means "concentration of ammonia in moles per liter."

The symbol [ ] is commonly used to represent the molarity of a species in solution. For a solution containing 1.20 mol of substance A in 2.50 L of solution,

$$[A] = \frac{1.20 \text{ mol}}{2.50 \text{ L}} = 0.480 \text{ mol/L} = 0.480 \, M$$

One liter of such a solution would contain 0.480 mol of A; 100 mL of solution would contain 0.0480 mol of A, and so on.

The molarity of a soluble solute can vary over a wide range. With sodium hydroxide, for example, we can prepare a 6 *M* solution, a 1 *M* solution, a 0.1 *M* solution, and so on. The words "concentrated" and "dilute" are often used in a qualita-

(a)   (b)   (c)

FIGURE 4.1

Preparing one liter of 0.100 *M* potassium chromate. (a) The weighed K_2CrO_4 (19.4 g) is transferred to a 1000-mL volumetric flask. (b) Enough water is added to fully dissolve all of the solid by swirling. (c) More water is then added to bring the water level up to the one-liter mark on the neck of the flask. Finally, the flask must be shaken repeatedly until a homogeneous solution is formed. *(Marna G. Clarke)*

tive way to describe these solutions. We would describe a 6 *M* solution of NaOH as concentrated; it contains a relatively large amount of solute per liter. A 0.1 *M* NaOH solution is dilute, at least in comparison to 1 *M* or 6 *M*.

To prepare a solution to a desired molarity, you first calculate the amount of solute required. This is then dissolved in enough solvent to form the required volume of solution. Suppose, for example, you want to make one liter of 0.100 *M* K_2CrO_4 solution (Figure 4.1). You first weigh out 19.4 g (0.100 mol) of K_2CrO_4 ($\mathcal{M}$ = 194.20 g/mol). Then stir with enough water to form one liter (1000 mL) of solution.

The molarity of a solution can be used to calculate—

- the number of moles of solute in a given volume of solution.
- the volume of solution containing a given number of moles of solute.

Here, as in so many other cases, a conversion factor approach is used (Example 4.1).

See Screen 5.11, Preparing Solutions of Known Concentrations (1).

EXAMPLE 4.1   The bottle labeled "concentrated hydrochloric acid" in the lab contains 12.0 mol of HCl per liter of solution. That is, [HCl] = 12.0 *M*.

(a) How many moles of HCl are there in 25.0 mL of this solution?
(b) What volume (*V*) of concentrated hydrochloric acid must be taken to contain 1.00 mol of HCl?

Strategy   The required conversion factors are

$$\frac{12.0 \text{ mol HCl}}{1 \text{ L}} \quad \text{or} \quad \frac{1 \text{ L}}{12.0 \text{ mol HCl}}$$

Solution

(a) $n_{HCl} = 25.0 \text{ mL} \times \dfrac{1 \text{ L}}{1000 \text{ mL}} \times \dfrac{12.0 \text{ mol HCl}}{1 \text{ L}} = $ 0.300 mol HCl

(b) $V = 1.00 \text{ mol HCl} \times \dfrac{1 \text{ L}}{12.0 \text{ mol HCl}} = $ 0.0833 L (83.3 mL)

⊙ See Screen 4.5, Compounds in Aqueous Solution.

When an ionic solid dissolves in water, the cations and anions separate from one another. This process can be represented by a chemical equation in which the reactant is the solid and the products are the positive and negative ions in water solution. For the dissolving of sodium chloride, we write

$$NaCl(s) \longrightarrow Na^+(aq) + Cl^-(aq)$$

Similarly, for sodium carbonate and magnesium chloride

$$Na_2CO_3(s) \longrightarrow 2Na^+(aq) + CO_3^{2-}(aq)$$

$$MgCl_2(s) \longrightarrow Mg^{2+}(aq) + 2Cl^-(aq)$$

It is important to realize that compounds such as these are completely ionized in water, as they are in the solid state; there are no discrete molecules of NaCl, Na_2CO_3, or $MgCl_2$. Ionic solids are commonly described as **strong electrolytes.** The ions they contain are good conductors of electricity in aqueous solution (Figure 4.2).

Let us return for a moment to the equation just written for the dissolving of magnesium chloride. It tells us that *one* mole of $MgCl_2$ yields *one* mole of Mg^{2+}

(a) (b) (c)

FIGURE 4.2

Electrical conductivity test. For electrical current to flow and light the bulb, the solution in which the electrodes are immersed must contain ions, which carry electrical charge. The solutions or liquids shown are (a) pure water, (b) a solution of sucrose (table sugar) and pure water, and (c) a solution of sodium chloride (NaCl) and pure water. Only the NaCl solution contains ions. (*Marna G. Clarke*)

3	4	5	6	7	8	9	10	11	12
			Cr^{3+}	Mn^{2+}	Fe^{2+} Fe^{3+}	Co^{2+} Co^{3+}	Ni^{2+}	Cu^{2+}	Zn^{2+}
								Ag^+	Cd^{2+}
									Hg^{2+}

FIGURE 4.3
Charges of transition metal cations commonly found in aqueous solution.

ions and *two* moles of Cl^- ions in solution. It follows that in any solution of magnesium chloride:

$$\text{molarity } Mg^{2+} = \text{molarity } MgCl_2 \qquad \text{molarity } Cl^- = 2 \times \text{molarity } MgCl_2$$

Similar relationships hold for ionic solids containing polyatomic ions (Table 2.2, page 43) or transition metal cations (Figure 4.3).

$$(NH_4)_3PO_4(s) \longrightarrow 3NH_4^+(aq) + PO_4^{3-}(aq)$$

$$\text{molarity } NH_4^+ = 3 \times \text{molarity } (NH_4)_3PO_4 \qquad \text{molarity } PO_4^{3-} = \text{molarity } (NH_4)_3PO_4$$

$$Cr_2(SO_4)_3(s) \longrightarrow 2Cr^{3+}(aq) + 3SO_4^{2-}(aq)$$

$$\text{molarity } Cr^{3+} = 2 \times \text{molarity } Cr_2(SO_4)_3 \qquad \text{molarity } SO_4^{2-} = 3 \times \text{molarity } Cr_2(SO_4)_3$$

EXAMPLE 4.2 Give the concentration, in moles per liter, of each ion in

(a) 0.080 M K_2SO_4 (b) 0.40 M $FeCl_3$

Strategy To go from concentration of solute to concentration of an individual ion, you must know the conversion factor relating moles of ions to moles of solute. To find this conversion factor, it is helpful to write the equation for the solution process.

Solution

(a) $K_2SO_4(s) \rightarrow 2K^+(aq) + SO_4^{2-}(aq)$

The conversion factors are 2 mol K^+/1 mol K_2SO_4 and 1 mol SO_4^{2-}/1 mol K_2SO_4.

$$[K^+] = \frac{0.080 \text{ mol } K_2SO_4}{1 \text{ L}} \times \frac{2 \text{ mol } K^+}{1 \text{ mol } K_2SO_4} = \boxed{0.16 \, M \, K^+}$$

$$[SO_4^{2-}] = \frac{0.080 \text{ mol } K_2SO_4}{1 \text{ L}} \times \frac{1 \text{ mol } SO_4^{2-}}{1 \text{ mol } K_2SO_4} = \boxed{0.080 \, M \, SO_4^{2-}}$$

(b) $FeCl_3(s) \rightarrow Fe^{3+}(aq) + 3Cl^-(aq)$

$$[Fe^{3+}] = \frac{0.40 \text{ mol } FeCl_3}{1 \text{ L}} \times \frac{1 \text{ mol } Fe^{3+}}{1 \text{ mol } FeCl_3} = \boxed{0.40 \, M \, Fe^{3+}}$$

$$[Cl^-] = \frac{0.40 \text{ mol } FeCl_3}{1 \text{ L}} \times \frac{3 \text{ mol } Cl^-}{1 \text{ mol } FeCl_3} = \boxed{1.2 \, M \, Cl^-}$$

⌐ For aluminum sulfate: $[Al^{3+}] = 2[Al_2(SO_4)_3]$, $[SO_4^{2-}] = 3[Al_2(SO_4)_3]$.

Reality Check For K_2SO_4, the concentration of K^+ should be twice that for SO_4^{2-}; it is! Similarly, for $FeCl_3$, $[Cl^-] = 3 \times [Fe^{3+}]$.

FIGURE 4.4

Precipitation diagram. Choose the cation row and read across to the anion column. If the block is blank, no precipitate will form. If the block is colored, a precipitate will form from dilute solution. Where a formula is given, that is the only cation-anion combination in that block that will precipitate.

	NO_3^-	Cl^-	SO_4^{2-}	OH^-	CO_3^{2-}	PO_4^{3-}
Group 1 cations (Na^+, K^+) and NH_4^+						
Group 2 cations (Mg^{2+}, Ca^{2+}, Ba^{2+})			$BaSO_4$	$Mg(OH)_2$		
Transition metal cations (Figure 4.3)		$AgCl$				

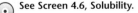 See Screens 4.11, Types of Reactions in Aqueous Solution, & 4.12, Precipitation Reactions.

See Screen 4.6, Solubility.

4.2 PRECIPITATION REACTIONS

Sometimes when water solutions of two different ionic compounds are mixed, an insoluble solid separates out of solution. The **precipitate** that forms is itself ionic; the cation comes from one solution, the anion from the other. To predict the occurrence of reactions of this type, you must know which ionic substances are insoluble in water.

The precipitation diagram shown in Figure 4.4 enables you to determine whether or not a precipitate will form when dilute solutions of two ionic solutes are mixed. If a cation in solution 1 mixes with an anion in solution 2 to form an insoluble compound (colored squares), that compound will precipitate. Cation-anion combinations that lead to the formation of a soluble compound (white squares) will not give a precipitate. For example, if solutions of $NiCl_2$ (Ni^{2+}, Cl^- ions) and $NaOH$ (Na^+, OH^-) ions are mixed (Figure 4.5)—

- a precipitate of $Ni(OH)_2$, an insoluble compound, will form.
- $NaCl$, a soluble compound, will not precipitate.

FIGURE 4.5

Precipitation of nickel hydroxide ($Ni(OH)_2$). The precipitate forms when solutions of nickel chloride ($NiCl_2$) and sodium hydroxide ($NaOH$) are mixed. *(Charles D. Winters)*

EXAMPLE 4.3 Using the precipitation diagram (Figure 4.4), predict what will happen when the following pairs of aqueous solutions are mixed.

(a) $CuSO_4$ and $NaNO_3$ (b) Na_2CO_3 and $CaCl_2$

Strategy First decide what cation and anion are present in each solution. Then write the formulas of the two possible precipitates, combining a cation of one solution with the anion of the other solution. Check Figure 4.4 to see if one or both of these compounds are insoluble. If so, a precipitation reaction occurs.

Solution

(a) Ions present in first solution: Cu^{2+}, SO_4^{2-}; second solution: Na^+, NO_3^-
Possible precipitates: $Cu(NO_3)_2$, Na_2SO_4
From Figure 4.4, both of these compounds must be soluble, so no precipitate forms.
(b) Ions present: Na^+, CO_3^{2-}; Ca^{2+}, Cl^-
Possible precipitates: $NaCl$, $CaCO_3$
Sodium chloride is soluble, but calcium carbonate is not. When these two solutions are mixed, $CaCO_3$ precipitates (Figure 4.6).

Net Ionic Equations

The precipitation reaction that occurs when solutions of Na_2CO_3 and $CaCl_2$ are mixed can be represented by a simple equation. The product of the reaction is solid $CaCO_3$, formed by the reaction between Ca^{2+} and CO_3^{2-} ions in aqueous solution. The equation for the reaction is

$$Ca^{2+}(aq) + CO_3^{2-}(aq) \longrightarrow CaCO_3(s)$$

Notice that this equation includes only those ions that participate in the reaction. To be specific, Na^+ and Cl^- ions do not appear. They are "spectator ions," which are present in solution before and after the precipitation of calcium carbonate.

Equations such as this, which involve ions and exclude any species that do not take part in the reaction, are referred to as **net ionic equations.** We will use net ionic equations throughout this chapter and indeed the entire text to represent a wide variety of reactions in aqueous solution. Like all equations, net ionic equations must show—

- *atom balance.* There must be the same number of atoms of each element on both sides. In the preceding equation, there is one Ca atom, one C atom, and three oxygen atoms on both sides.
- *charge balance.* There must be the same total charge on both sides. In this equation, the total charge is zero on both sides.

$$Ca^{2+}(aq) + CO_3^{2-}(aq) \longrightarrow CaCO_3(s)$$

FIGURE 4.6
Precipitation of calcium carbonate ($CaCO_3$). The precipitate forms when solutions of sodium carbonate ($NaCO_3$) and calcium chloride ($CaCl_2$) are mixed. *(Charles D. Winters)*

⊙ See Screen 4.10, Equations for Reactions in Aqueous Solution—Net Ionic Equations.

EXAMPLE 4.4   Write a net ionic equation for any precipitation reaction that occurs when solutions of the following ionic compounds are mixed.

(a) NaOH and $Cu(NO_3)_2$ (b) $BaCl_2$ and Ag_2SO_4 (c) $(NH_4)_3PO_4$ and K_2CO_3

Strategy   Follow the procedure of Example 4.3 to decide whether a precipitate will form. If it does, write its formula, followed by (s), on the right side of the equation. On the left (reactant) side, write the formulas of the ions (aq) required to produce the precipitate. Finally, balance the equation.

Solution

(a) Ions present: Na^+, OH^-; Cu^{2+}, NO_3^-
Possible precipitates: $NaNO_3$, $Cu(OH)_2$
$NaNO_3$ is soluble, but $Cu(OH)_2$ is not.

Equation: $Cu^{2+}(aq) + 2OH^-(aq) \longrightarrow Cu(OH)_2(s)$

(b) Ions present: Ba^{2+}, Cl^-; Ag^+, SO_4^{2-}
Possible precipitates: $BaSO_4$, AgCl
Both compounds are insoluble, so two reactions occur.

Equations: $Ba^{2+}(aq) + SO_4^{2-}(aq) \longrightarrow BaSO_4(s)$
$Ag^+(aq) + Cl^-(aq) \longrightarrow AgCl(s)$

(c) Ions present: NH_4^+, PO_4^{3-}, K^+, CO_3^{2-}

Both possible products are soluble, so there is no precipitation reaction and no equation.

⌐ To write a net ionic equation, you first have to identify the ions.

Although we have introduced net ionic equations to represent precipitation reactions, they have a much wider application. Indeed, we will use them for all kinds of reactions in water solution. In particular, *all of the chemical equations written throughout this chapter are net ionic equations.*

Stoichiometry

⊙ See Screen 5.13, Stoichiometry of Reactions in Solution.

The approach followed in Chapter 3 to calculate mole-mass relations in reactions is readily applied to solution reactions represented by net ionic equations.

EXAMPLE 4.5   Consider the net ionic equation derived in Example 4.4 for the reaction that occurs when solutions of $Cu(NO_3)_2$ and NaOH are mixed:

$$Cu^{2+}(aq) + 2OH^-(aq) \longrightarrow Cu(OH)_2(s)$$

What volume of 0.106 M $Cu(NO_3)_2$ solution is required to form 6.52 g of solid $Cu(OH)_2$?

Strategy   The first three steps in the path are

$$\text{mass } Cu(OH)_2 \longrightarrow \text{moles } Cu(OH)_2 \longrightarrow \text{moles } Cu^{2+}$$

But because one mole of $Cu(NO_3)_2$ produces one mole of Cu^{2+}, it follows that

$$\text{moles } Cu(NO_3)_2 = \text{moles } Cu^{2+}$$

Finally, knowing the number of moles of $Cu(NO_3)_2$, you can use molarity as a conversion factor to find the volume of solution required.

Solution   Setting up the calculation as a continuous series of conversions:

$$V_{Cu(NO_3)_2} = 6.52 \text{ g } Cu(OH)_2 \times \frac{1 \text{ mol } Cu(OH)_2}{97.57 \text{ g } Cu(OH)_2} \times \frac{1 \text{ mol } Cu^{2+}}{1 \text{ mol } Cu(OH)_2} \times \frac{1 \text{ mol } Cu(NO_3)_2}{1 \text{ mol } Cu^{2+}}$$

$$\times \frac{1.00 \text{ L } Cu(NO_3)_2}{0.106 \text{ mol } Cu(NO_3)_2} = \boxed{0.630 \text{ L}} \ Cu(NO_3)_2 \text{ solution}$$

Limiting reactant problems involving net ionic equations are solved in the ordinary way.

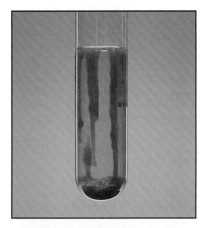

Precipitation of iron(III) hydroxide (Fe(OH)₃). The red, gelatinous precipitate forms when aqueous solutions of sodium hydroxide (NaOH) and iron(III) nitrate (Fe(NO₃)₃) are mixed.
(Charles D. Winters)

EXAMPLE 4.6   When aqueous solutions of sodium hydroxide and iron(III) nitrate are mixed, a red gelatinous precipitate forms. Calculate the mass of precipitate formed when 50.00 mL of 0.200 M NaOH and 30.00 mL of 0.125 M $Fe(NO_3)_3$ are mixed.

Strategy   Data are given for two reactants, so this is a limiting-reactant problem, similar to those worked in Chapter 3. There are, however, a couple of preliminary steps. First (1), decide on the net ionic equation for the precipitation reaction. Second (2), calculate the number of moles of each reactant. Then apply the three-step procedure (3–5) described on page 71.

Solution

(1) Following the procedure described earlier in this section, you should arrive at the net ionic equation:

$$Fe^{3+}(aq) + 3OH^-(aq) \longrightarrow Fe(OH)_3(s)$$

(2) $n_{Fe^{3+}} = 0.03000 \text{ L Fe(NO}_3)_3 \times \dfrac{0.125 \text{ mol Fe(NO}_3)_3}{1 \text{ L Fe(NO}_3)_3} \times \dfrac{1 \text{ mol Fe}^{3+}}{1 \text{ mol Fe(NO}_3)_3}$

   $= 3.75 \times 10^{-3} \text{ mol Fe}^{3+}$

Note from the formula of iron(III) nitrate that there is one mole of Fe^{3+} per mole of $Fe(NO_3)_3$.

$n_{OH^-} = 0.05000 \text{ L NaOH} \times \dfrac{0.200 \text{ mol NaOH}}{1 \text{ L NaOH}} \times \dfrac{1 \text{ mol OH}^-}{1 \text{ mol NaOH}} = 1.00 \times 10^{-2} \text{ mol OH}^-$

(3) If Fe^{3+} is limiting,

   $n_{Fe(OH)_3} = 3.75 \times 10^{-3} \text{ mol Fe}^{3+} \times \dfrac{1 \text{ mol Fe(OH)}_3}{1 \text{ mol Fe}^{3+}} = 3.75 \times 10^{-3} \text{ mol Fe(OH)}_3$

(4) If OH^- is limiting,

   $n_{Fe(OH)_3} = 1.00 \times 10^{-2} \text{ mol OH}^- \times \dfrac{1 \text{ mol Fe(OH)}_3}{3 \text{ mol OH}^-} = 3.33 \times 10^{-3} \text{ mol Fe(OH)}_3$

(5) Because 3.33×10^{-3} is less than 3.75×10^{-3}, OH^- is the limiting reactant. The theoretical yield of $Fe(OH)_3$ is 3.33×10^{-3} mol. The molar mass of $Fe(OH)_3$ is 106.87 g/mol, so

   $\text{mass Fe(OH)}_3 = 3.33 \times 10^{-3} \text{ mol Fe(OH)}_3 \times \dfrac{106.87 \text{ g Fe(OH)}_3}{1 \text{ mol Fe(OH)}_3} = \boxed{0.356 \text{ g Fe(OH)}_3}$

Reality Check Another way to decide that OH^- is limiting is to realize that you need three moles of OH^- for every mole of Fe^{3+} and you don't have that much:

$$3 \times n_{Fe^{3+}} = 3 \times (3.75 \times 10^{-3}) = 1.13 \times 10^{-2} > n_{OH^-}$$

4.3 ACID–BASE REACTIONS

You are probably familiar with a variety of aqueous solutions that are either acidic or basic (Figure 4.7). Acidic solutions have a sour taste and affect the color of certain organic dyes known as acid-base indicators. For example, litmus turns from

See Screens 4.11, Types of Reactions in Aqueous Solution, & 4.13, Acid-Base Reactions.

(a)   (b)

FIGURE 4.7

Acidic (a) and basic (b) household solutions.
Many common household items, including vinegar, lemon juice, and cola drinks, are acidic. Common basic household items include ammonia and most detergents and cleaning agents. *(Marna G. Clarke)*

blue to red in acidic solution. Basic solutions have a slippery feeling and change the colors of indicators (e.g., red to blue for litmus).

The species that give these solutions their characteristic properties are called acids and bases. In this chapter, we use the definitions first proposed by Svante Arrhenius more than a century ago.

An acid *is a species that produces H^+ ions in water solution.*

A base *is a species that produces OH^- ions in water solution.*

We will consider more general definitions of acids and bases in Chapter 13 (Brønsted-Lowry) and Chapter 15 (Lewis).

Strong and Weak Acids and Bases

See Screens 4.7, Acids, & 4.8, Bases.

There are two types of acids, strong and weak, which differ in the extent of their ionization in water. **Strong acids** ionize completely, forming H^+ ions and anions. A typical strong acid is HCl. It undergoes the following reaction on addition to water:

$$HCl(aq) \longrightarrow H^+(aq) + Cl^-(aq)$$

In a solution prepared by adding 0.1 mol of HCl to water, there is 0.1 mol of H^+ ions, 0.1 mol of Cl^- ions, and no HCl molecules. There are six common strong acids, whose names and formulas are listed in Table 4.1.

All acids other than those listed in Table 4.1 can be taken to be weak. A weak acid is only partially ionized to H^+ ions in water. All of the **weak acids** considered in this chapter are molecules containing an ionizable hydrogen atom. Their general formula can be represented as HB; the general ionization reaction in water is

$$HB(aq) \rightleftharpoons H^+(aq) + B^-(aq)$$

The double arrow implies that this reaction does not go to completion. Instead, a mixture is formed containing significant amounts of both products and reactants. With the weak acid hydrogen fluoride

$$HF(aq) \rightleftharpoons H^+(aq) + F^-(aq)$$

a solution prepared by adding 0.1 mol of HF to a liter of water contains about 0.01 mol of H^+ ions, 0.01 mol of F^- ions, and 0.09 mol of HF molecules.

Bases, like acids, are classified as strong or weak. A **strong base** in water solution is completely ionized to OH^- ions and cations. As you can see from Table 4.1, the strong bases are the hydroxides of the Group 1 and Group 2 metals. These are

Hydrochloric acid as known in hardware stores. This strong acid is used to clean metal and stone surfaces. (*Charles D. Winters*)

You need to know the strong acids and bases to work with acid-base reactions.

Sulfuric acid ionization:
$H_2SO_4(aq) \rightarrow H^+(aq) + HSO_4^-(aq)$.

TABLE 4.1	**Common Strong Acids and Bases** OHT		
Acid	**Name of Acid**	**Base**	**Name of Base**
HCl	Hydrochloric acid	LiOH	Lithium hydroxide
HBr	Hydrobromic acid	NaOH	Sodium hydroxide
HI	Hydriodic acid	KOH	Potassium hydroxide
HNO_3	Nitric acid	$Ca(OH)_2$	Calcium hydroxide
$HClO_4$	Perchloric acid	$Sr(OH)_2$	Strontium hydroxide
H_2SO_4	Sulfuric acid	$Ba(OH)_2$	Barium hydroxide

typical ionic solids, completely ionized both in the solid state and in water solution. The equations written to represent the processes by which NaOH and $Ca(OH)_2$ dissolve in water are

$$NaOH(s) \longrightarrow Na^+(aq) + OH^-(aq)$$

$$Ca(OH)_2(s) \longrightarrow Ca^{2+}(aq) + 2OH^-(aq)$$

In a solution prepared by adding 0.1 mol of NaOH to water, there is 0.1 mol of Na^+ ions, 0.1 mol of OH^- ions, and no NaOH molecules.

Weak bases produce OH^- ions in a quite different manner. They react with H_2O molecules, acquiring H^+ ions and leaving OH^- ions behind. The reaction of ammonia, NH_3, is typical:

$$NH_3(aq) + H_2O \rightleftharpoons NH_4^+(aq) + OH^-(aq)$$

As with all weak bases, this reaction does not go to completion. In a solution prepared by adding 0.1 mol of ammonia to a liter of water, there is about 0.001 mol of NH_4^+, 0.001 mol of OH^-, and nearly 0.099 mol of NH_3.

A common class of weak bases consists of the organic molecules known as *amines*. An amine can be considered to be a derivative of ammonia in which one or more hydrogen atoms have been replaced by hydrocarbon groups.

In the simplest case, methylamine, a hydrogen atom is replaced by a $—CH_3$ group to give the CH_3NH_2 molecule, which reacts with water in a manner very similar to NH_3:

$$CH_3NH_2(aq) + H_2O \rightleftharpoons CH_3NH_3^+(aq) + OH^-(aq)$$

As we have pointed out, strong acids and bases are completely ionized in water. As a result, compounds such as HCl and NaOH are strong electrolytes like NaCl. In contrast, molecular weak acids and weak bases are poor conductors because their water solutions contain relatively few ions. Hydrofluoric acid and ammonia are commonly described as **weak electrolytes.**

Equations for Acid-Base Reactions

When an acidic water solution is mixed with a basic water solution, an acid-base reaction takes place. The nature of the reaction and hence the equation written for it depend on whether the acid and base involved are strong or weak.

1. *Strong acid–strong base.* Consider what happens when a solution of a strong acid such as HNO_3 is added to a solution of a strong base such as NaOH. Because HNO_3 is a strong acid, it is completely converted to H^+ and NO_3^- ions in solution. Similarly, with the strong base NaOH, the solution species are the Na^+ and OH^- ions. When the solutions are mixed, the H^+ and OH^- ions react with each other to form H_2O molecules. This reaction, referred to as **neutralization,** is represented by the net ionic equation

$$H^+(aq) + OH^-(aq) \longrightarrow H_2O$$

The Na^+ and NO_3^- ions take no part in the reaction and so do not appear in the equation.

There is considerable evidence to indicate that the neutralization reaction occurs when any strong base reacts with any strong acid in water solution. It follows that the neutralization equation written above applies to any strong acid–strong base reaction.

A hydrocarbon group has a string of C and H atoms.

H^+ and OH^- are the reactive species; Na^+ and NO_3^- are spectator ions.

When $HClO_4$ reacts with $Ca(OH)_2$, the equation is $H^+(aq) + OH^-(aq) \rightarrow H_2O$.

2. Weak acid–strong base. When a strong base such as NaOH is added to a solution of a weak acid, HB, a two-step reaction occurs. The first step is the ionization of the HB molecule to H^+ and B^- ions; the second is the neutralization of the H^+ ions produced in the first step by the OH^- ions of the NaOH solution.

$$(1)\ HB(aq) \rightleftharpoons H^+(aq) + B^-(aq)$$

$$(2)\ H^+(aq) + OH^-(aq) \longrightarrow H_2O$$

The equation for the overall reaction is obtained by adding the two equations just written and canceling H^+ ions:

$$HB(aq) + OH^-(aq) \longrightarrow B^-(aq) + H_2O$$

For the reaction between solutions of sodium hydroxide and hydrogen fluoride, the net ionic equation is

$$HF(aq) + OH^-(aq) \longrightarrow F^-(aq) + H_2O$$

When an acid is weak, like HF, its formula appears in the equation.

Here, as always, spectator ions such as Na^+ are not included in the net ionic equation.

3. Strong acid–weak base. As an example of a reaction of a strong acid with a weak base, consider what happens when an aqueous solution of a strong acid like HCl is added to an aqueous solution of ammonia, NH_3. Again, we consider the reaction to take place in two steps. The first step is the reaction of NH_3 with H_2O to form NH_4^+ and OH^- ions. Then, in the second step, the H^+ ions of the strong acid neutralize the OH^- ions formed in the first step.

$$(1)\ NH_3(aq) + H_2O \rightleftharpoons NH_4^+(aq) + OH^-(aq)$$

$$(2)\ H^+(aq) + OH^-(aq) \longrightarrow H_2O$$

The overall equation is obtained by summing those for the individual steps. Canceling species (OH^-, H_2O) that appear on both sides, we obtain the net ionic equation

$$H^+(aq) + NH_3(aq) \longrightarrow NH_4^+(aq)$$

In another case, for the reaction of a strong acid such as HNO_3 with methylamine, CH_3NH_2, the net ionic equation is

$$H^+(aq) + CH_3NH_2(aq) \longrightarrow CH_3NH_3^+(aq)$$

Table 4.2 summarizes the equations written for the three types of acid-base reactions just discussed. You should find it helpful in writing the equations called for in Example 4.7.

TABLE 4.2   Types of Acid-Base Reactions [OHT]

Reactants	Reacting Species	Net Ionic Equation
Strong acid–strong base	H^+ OH^-	$H^+(aq) + OH^-(aq) \longrightarrow H_2O$
Weak acid–strong base	HB OH^-	$HB(aq) + OH^-(aq) \longrightarrow H_2O + B^-(aq)$
Strong acid–weak base	H^+ B	$H^+(aq) + B(aq) \longrightarrow BH^+(aq)$

EXAMPLE 4.7   Write a net ionic equation for each of the following reactions in dilute water solution.

(a) Nitrous acid (HNO_2) with sodium hydroxide (NaOH)
(b) Ethylamine ($CH_3CH_2NH_2$) with perchloric acid ($HClO_4$)
(c) Hydrobromic acid (HBr) with potassium hydroxide (KOH)

Strategy   Decide whether the acid and base are strong or weak. Then decide which of the three types of acid-base reactions is involved. Finally, use Table 4.2 to derive the proper equation.

Solution

(a) HNO_2 is weak; NaOH is strong. This is a weak acid–strong base reaction.

$$HNO_2(aq) + OH^-(aq) \longrightarrow H_2O + NO_2^-(aq)$$

(b) $HClO_4$ is strong (Table 4.1); $CH_3CH_2NH_2$ is weak. This is a strong acid–weak base reaction.

$$H^+(aq) + CH_3CH_2NH_2(aq) \longrightarrow CH_3CH_2NH_3^+(aq)$$

(c) HBr is a strong acid; KOH is a strong base.

$$H^+(aq) + OH^-(aq) \longrightarrow H_2O$$

Acid–Base Titrations

Acid-base reactions in water solution are commonly used to determine the concentration of a dissolved species or its percentage in a solid mixture. This is done by carrying out a **titration,** measuring the volume of a *standard solution* (a solution of known concentration) required to react with a measured amount of sample.

The experimental setup for a titration is shown in Figure 4.8. The flask contains vinegar, a water solution of a weak organic acid called acetic acid. A solution of sodium hydroxide of known concentration is added from a buret. The net ionic equation for the acid-base reaction that occurs is

$$HC_2H_3O_2(aq) + OH^-(aq) \longrightarrow C_2H_3O_2^-(aq) + H_2O$$

⊙ See Screens 5.14, Titrations, & 5.15, Titration Simulation.

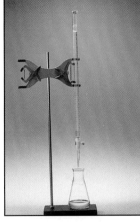
(a)

(b)

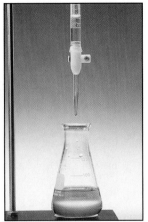
(c)

FIGURE 4.8
Titration of vinegar with sodium hydroxide (NaOH). (a) The flask contains vinegar (a dilute solution of acetic acid) and an acid-base indicator (phenolphthalein) that is colorless in acid solution and pink in basic solution. The buret contains a sodium hydroxide solution of known concentration. (b) The sodium hydroxide solution is slowly added. It reacts with the acetic acid in the vinegar solution. (c) The equivalence point is reached and the volume of NaOH used is noted. The volume of NaOH together with its concentration is used in determining the concentration of acetic acid in the vinegar. *(Charles D. Winters)*

The objective of the titration is to determine the point at which reaction is complete, called the **equivalence point.** This is reached when the number of moles of OH^- added is exactly equal to the number of moles of acetic acid, $HC_2H_3O_2$, originally present. To determine this point, an **acid-base indicator** such as phenolphthalein is used. It should change color (colorless to pink) at the equivalence point.

EXAMPLE 4.8   In a titration it is found that 25.0 mL of 0.500 M NaOH is required to react with a 15.0-mL sample of vinegar. What is the molarity of acetic acid in the sample?

Strategy   As pointed out above, the equation is

$$HC_2H_3O_2(aq) + OH^-(aq) \longrightarrow C_2H_3O_2^-(aq) + H_2O$$

The number of moles of OH^- is readily calculated, knowing the volume and molarity of NaOH. From the equation it is clear that the number of moles of acetic acid must equal the number of moles of OH^-. Finally, the molarity of $HC_2H_3O_2$ can be calculated because the number of moles and the volume are known.

Solution

$$n_{HC_2H_3O_2} = n_{OH^-} = (25.0 \times 10^{-3}\,L)(0.500\,mol/L) = 1.25 \times 10^{-2}\,mol$$

$$[HC_2H_3O_2] = \frac{1.25 \times 10^{-2}\,mol}{15.0 \times 10^{-3}\,L} = \boxed{0.833\,M}$$

Reality Check   Because the volume of acidic solution (15.0 mL) is less than that of base (25.0 mL), and the acid and base react in a 1:1 mole ratio, the concentration of acetic acid must be greater than that of NaOH. It is; 0.833 M > 0.500 M.

An antacid tablet in water. The fizz is due to CO_2 being released. *(Charles D. Winters)*

EXAMPLE 4.9   The principal ingredient of certain commercial antacids is calcium carbonate, $CaCO_3$. A student titrates an antacid tablet weighing 0.542 g with hydrochloric acid; the reaction is

$$CaCO_3(s) + 2H^+(aq) \longrightarrow Ca^{2+}(aq) + CO_2(g) + H_2O$$

If 38.5 mL of 0.200 M HCl is required for complete reaction, what is the percentage of $CaCO_3$ in the antacid tablet?

Strategy   From the titration data, the number of moles of H^+ is readily calculated. Then follow the path

$$n_{H^+} \longrightarrow n_{CaCO_3} \longrightarrow \text{mass of } CaCO_3 \longrightarrow \%\,CaCO_3$$

Solution

(1) $n_{H^+} = 0.0385\,L \times \dfrac{0.200\,mol\,HCl}{1\,L} \times \dfrac{1\,mol\,H^+}{1\,mol\,HCl} = 7.70 \times 10^{-3}\,mol\,H^+$

(2) $n_{CaCO_3} = 7.70 \times 10^{-3}\,mol\,H^+ \times \dfrac{1\,mol\,CaCO_3}{2\,mol\,H^+} = 3.85 \times 10^{-3}\,mol\,CaCO_3$

(3) mass of $CaCO_3 = 3.85 \times 10^{-3}\,mol\,CaCO_3 \times \dfrac{100.09\,g\,CaCO_3}{1\,mol\,CaCO_3} = 0.385\,g\,CaCO_3$

(4) $\%\,CaCO_3 = \dfrac{0.385\,g}{0.542\,g} \times 100\% = \boxed{71.0\%}$

CHEMISTRY ▪ The Human Side

For reasons that are by no means obvious, Sweden produced a disproportionate number of outstanding chemists in the eighteenth and nineteenth centuries. Jons Jakob Berzelius (1779–1848) determined with amazing accuracy the atomic masses of virtually all the elements known in his time. In his spare time, he invented such modern laboratory tools as the beaker, the flask, the pipet, and the ringstand.

Svante Arrhenius, like Berzelius, was born in Sweden and spent his entire professional career there. According to Arrhenius, the concept of strong and weak acids and bases came to him on May 13, 1883, when he was 24 years old. He added, "I could not sleep that night until I had worked through the entire problem."

Almost exactly one year later, Arrhenius submitted his Ph.D. thesis at the University of Uppsala. He proposed that salts, strong acids, and strong bases are completely ionized in dilute water solution. Today, it seems quite reasonable that solutions of NaCl, HCl, and NaOH contain, respectively, Na^+ and Cl^- ions, H^+ and Cl^- ions, and Na^+ and OH^- ions. It did not seem nearly so obvious to the chemistry faculty at Uppsala in 1884. Arrhenius's dissertation received the lowest passing grade "approved not without praise."

Arrhenius sent copies of his Ph.D. thesis to several well-known chemists in Europe and America. Most ignored his ideas; a few were openly hostile. A pair of young chemists gave positive responses: Jacobus Van't Hoff (age 32) at Amsterdam (Holland) and Wilhelm Ostwald (also 32) at Riga (Latvia). For some years, these three young men were referred to, somewhat disparagingly, as "ionists" or "ionians." As time passed, the situation changed. The first Nobel Prize in chemistry was awarded to Van't Hoff in 1901. Two years later, in 1903, Arrhenius became a Nobel laureate; Ostwald followed in 1909.

Among other contributions of Arrhenius, the most important were probably in chemical kinetics (Chapter 11). In 1889 he derived the relation for the temperature dependence of reaction rate. In quite a different area in 1896 Arrhenius published an article, "On the Influence of Carbon Dioxide in the Air on the Temperature of the Ground." He presented the basic idea of the greenhouse effect, discussed in Chapter 17.

In his later years, Arrhenius turned his attention to popularizing chemistry. He wrote several different textbooks that were well received. In 1925, under pressure from his publisher to submit a manuscript, Arrhenius started getting up at 4 A.M. to write. As might be expected, rising at such an early hour had an adverse effect on his health. Arrhenius suffered a physical breakdown in 1925, from which he never really recovered, dying two years later.

(Photo credit: The E.F. Smith Memorial Collection in the History of Chemistry, Dept. of Special Collections, Van Pelt–Dietrich Library, University of Pennsylvania)

Svante August Arrhenius
(1859–1927)

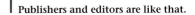
Publishers and editors are like that.

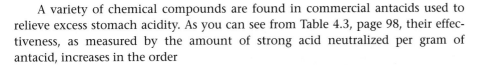

A variety of chemical compounds are found in commercial antacids used to relieve excess stomach acidity. As you can see from Table 4.3, page 98, their effectiveness, as measured by the amount of strong acid neutralized per gram of antacid, increases in the order

$$NaHCO_3 < CaCO_3 < MgCO_3 < Mg(OH)_2 < Al(OH)_3$$

TABLE 4.3   Commercial Antacids

Compound	(g/mol)	Product*	$\dfrac{n_{H^+}}{n_{antacid}}$	$\dfrac{n_{H^+}}{g_{antacid}}$
NaHCO$_3$	84.01	Alka-Seltzer	1	0.0119
CaCO$_3$	100.09	Rolaids, Tums	2	0.0200
MgCO$_3$	84.31	Gaviscon	2	0.0237
Mg(OH)$_2$	58.32	Mylanta, Maalox	2	0.0343
Al(OH)$_3$	78.00	Maalox	3	0.0385

* Many commercial products contain a mixture of antacids.

4.4   OXIDATION–REDUCTION REACTIONS

See Screens 4.11, Types of Reactions in Aqueous Solution, & 4.15, Oxidation-Reduction Reactions.

See Screen 4.16, Redox Reactions and Electron Transfer.

Another common type of reaction in aqueous solution involves a transfer of electrons between two species. Such a reaction is called an **oxidation-reduction** or **redox reaction.** Many familiar reactions fit into this category, including the reaction of metals with acid.

In a redox reaction, one species *loses* (i.e., donates) electrons and is said to be *oxidized*. The other species, which *gains* (or receives) electrons, is *reduced*. To illustrate, consider the redox reaction that takes place when zinc pellets are added to hydrochloric acid (Figure 4.9). The net ionic equation for the reaction is

$$Zn(s) + 2H^+(aq) \longrightarrow Zn^{2+}(aq) + H_2(g)$$

This equation can be split into two half-equations, one of oxidation and the other of reduction. Zinc atoms are oxidized to Zn^{2+} ions by losing electrons. The oxidation half-equation is

$$\text{oxidation:} \qquad Zn(s) \longrightarrow Zn^{2+}(aq) + 2e^-$$

At the same time, H^+ ions are reduced to H_2 molecules by gaining electrons; the reduction half-equation is

$$\text{reduction:} \qquad 2H^+(aq) + 2e^- \longrightarrow H_2(g)$$

From this example, it should be clear that—

FIGURE 4.9

Redox reaction of zinc with a strong acid. The zinc atoms are oxidized to Zn^{2+} ions in solution; the H^+ ions are reduced to H_2 molecules. *(Charles D. Winters)*

- *oxidation and reduction occur together,* in the same reaction; you can't have one without the other.
- *there is no net change in the number of electrons in a redox reaction.* Those given off in the oxidation half-reaction are taken on by another species in the reduction half-reaction.

The two species that exchange electrons in a redox reaction are given special names. The ion or molecule that accepts electrons is called the **oxidizing agent;** by accepting electrons it brings about the oxidation of another species. Conversely, the species that donates electrons is called the **reducing agent;** when reaction occurs it reduces the other species.

To illustrate these concepts consider the reaction

$$Zn(s) + 2H^+(aq) \longrightarrow Zn^{2+}(aq) + H_2(g)$$

The H^+ ion is the oxidizing agent; it brings about the oxidation of zinc. By the same token, zinc acts as a reducing agent; it furnishes the electrons required to reduce H^+ ions.

In earlier sections of this chapter, we showed how to write and balance equations for precipitation reactions (Section 4.2) and acid-base reactions (Section 4.3). In this section we will concentrate on balancing redox equations, given the identity of reactants and products. To do that, it is convenient to introduce a new concept, oxidation number.

> The oxidizing agent is reduced; the reducing agent is oxidized.

Oxidation Number

The concept of **oxidation number** is used to simplify the electron bookkeeping in redox reactions. For a monatomic ion (e.g., Na^+, S^{2-}), the oxidation number is, quite simply, the charge of the ion ($+1$, -2). In a molecule or polyatomic ion, the oxidation number of an element is a "pseudo-charge" obtained in a rather arbitrary way, assigning bonding electrons to the atom with the greater attraction for electrons.

In practice, oxidation numbers in all kinds of species are assigned according to a set of four arbitrary rules.

1. *The oxidation number of an element in an elementary substance is 0.* For example, the oxidation number of chlorine in Cl_2 or of phosphorus in P_4 is 0.

2. *The oxidation number of an element in a monatomic ion is equal to the charge of that ion.* In the ionic compound NaCl, sodium has an oxidation number of $+1$, chlorine an oxidation number of -1. The oxidation numbers of aluminum and oxygen in Al_2O_3 (Al^{3+}, O^{2-} ions) are $+3$ and -2, respectively.

3. *Certain elements have the same oxidation number in all or almost all their compounds.* The Group 1 metals always exist as $+1$ ions in their compounds and hence are assigned an oxidation number of $+1$. By the same token, Group 2 elements always have oxidation numbers of $+2$ in their compounds. Fluorine always has an oxidation number of -1.

Oxygen is ordinarily assigned an oxidation number of -2 in its compounds. (An exception arises in compounds containing the peroxide ion, O_2^{2-}, where the oxidation number of oxygen is -1.)

Hydrogen in its compounds ordinarily has an oxidation number of $+1$. (The major exception is in metal hydrides such as NaH and CaH_2, where hydrogen is present as the H^- ion and hence is assigned an oxidation number of -1.)

> Oxidation numbers are calculated, not determined experimentally.

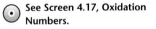 See Screen 4.17, Oxidation Numbers.

Rusting and oxidation number. As iron rusts, its oxidation number changes from 0 in Fe(s) to $+3$ in Fe^{3+}.
(Charles D. Winters)

4. *The sum of the oxidation numbers in a neutral species is 0; in a polyatomic ion, it is equal to the charge of that ion.* The application of this very useful principle is illustrated in Example 4.10.

EXAMPLE 4.10   What is the oxidation number of sulfur in Na_2SO_4? Of manganese in MnO_4^-?

Strategy   First look for elements whose oxidation number is always or almost always the same (Rule 3). Then solve for the oxidation number of the remaining element by applying Rule 4.

Solution   In Na_2SO_4, the oxidation numbers of sodium and oxygen are $+1$ and -2, respectively. Sodium sulfate, like all compounds, is neutral, so the sum of the oxidation numbers is zero. Letting x be the oxidation number of sulfur we have

$$0 = 2(+1) + x + 4(-2) \qquad x = \text{oxid. no. S} = \boxed{+6}$$

In the MnO_4^- ion, oxygen has an oxidation number of -2. The ion has a charge of -1, so the sum of the oxidation numbers must be -1 (Rule 4). Letting y be the oxidation number of manganese

$$-1 = y + 4(-2) \qquad y = \text{oxid. no. Mn} = \boxed{+7}$$

In the SO_4^{2-} ion and the SO_3 molecule, the oxidation number of S is also $+6$.

The concept of oxidation number leads directly to a working definition of the terms oxidation and reduction. **Oxidation** is defined as *an increase in oxidation number* and **reduction** as *a decrease in oxidation number.* Consider once again the reaction of zinc with a strong acid:

$$Zn(s) + 2H^+(aq) \longrightarrow Zn^{2+}(aq) + H_2(g)$$

Zn is oxidized (oxid. no.: $0 \longrightarrow +2$)

H^+ is reduced (oxid. no.: $+1 \longrightarrow 0$)

These definitions are of course compatible with the interpretation of oxidation and reduction in terms of loss and gain of electrons. An element that loses electrons must increase in oxidation number. The gain of electrons always results in a decrease in oxidation number.

See Screen 4.18, Recognizing Oxidation-Reduction Reactions.

An easy way to recognize a redox equation is to note changes in oxidation number of two different elements. The net ionic equation

$$2Al(s) + 3Cu^{2+}(aq) \longrightarrow 2Al^{3+}(aq) + 3Cu(s)$$

must represent a redox reaction because aluminum increases in oxidation number, from 0 to $+3$, and copper decreases from $+2$ to 0. In contrast, the reaction

$$CO_3^{2-}(aq) + 2H^+(aq) \longrightarrow CO_2(g) + H_2O$$

is not of the redox type because each element has the same oxidation number in both reactants and products: $O = -2$, $H = +1$, $C = +4$.

Balancing Half-Equations (Oxidation or Reduction)

Before you can balance an overall redox equation, you have to be able to balance two **half-equations,** one for oxidation (electron loss) and one for reduction (electron gain). Sometimes that's easy. Given the oxidation half-equation

See Screen 21.3, Balancing Equations for Redox Reactions.

$$Fe^{2+}(aq) \longrightarrow Fe^{3+}(aq) \qquad (\text{oxid. no. Fe: } +2 \longrightarrow +3)$$

it is clear that mass and charge balance can be achieved by adding an electron to the right:

$$Fe^{2+}(aq) \longrightarrow Fe^{3+}(aq) + e^-$$

In another case, this time a reduction half-equation,

$$Cl_2(g) \longrightarrow Cl^-(aq) \qquad \text{(oxid. no. Cl: } 0 \longrightarrow -1)$$

mass balance is obtained by writing a coefficient of 2 for Cl^-; charge is then balanced by adding two electrons to the left. The balanced half-equation is

$$Cl_2(g) + 2e^- \longrightarrow 2Cl^-(aq)$$

(Throughout this discussion we will show balanced oxidation half-equations in yellow, balanced reduction half-equations in green.)

Sometimes, though, it is by no means obvious how a given half-equation is to be balanced. This commonly happens when elements other than those being oxidized or reduced take part in the reaction. Most often, these elements are oxygen (oxid. no. $= -2$) and hydrogen (oxid. no. $= +1$). Consider, for example, the half-equation for the reduction of the permanganate ion,

$$MnO_4^-(aq) \longrightarrow Mn^{2+}(aq) \qquad \text{(oxid. no. Mn: } +7 \longrightarrow +2)$$

or the oxidation of chromium(III) hydroxide,

$$Cr(OH)_3(s) \longrightarrow CrO_4^{2-}(aq) \qquad \text{(oxid. no. Cr: } +3 \longrightarrow +6)$$

To balance half-equations such as these, proceed as follows:

(a) *Balance the atoms of the element being oxidized or reduced.*
(b) *Balance oxidation number by adding electrons.* For a reduction half-equation, add the appropriate number of electrons to the left; for an oxidation half-equation, add electrons to the right.
(c) *Balance charge by adding H^+ ions in acidic solution, OH^- ions in basic solution.*
(d) *Balance hydrogen by adding H_2O molecules.*
(e) *Check to make sure that oxygen is balanced.* If it is, the half-equation is almost certainly balanced correctly with respect to both mass and charge.

Example 4.11 shows how these rules are applied to balance half-equations.

> It's important to carry out these steps in the order indicated.

> You have to assign oxidation numbers to decide how many electrons to use.

EXAMPLE 4.11   Balance the following half-equations:

(1) $MnO_4^-(aq) \rightarrow Mn^{2+}(aq)$ (acidic solution)
(2) $Cr(OH)_3(s) \rightarrow CrO_4^{2-}(aq)$ (basic solution)

Strategy   Follow the rules outlined above, step by step in the proper order, and see what happens.

Solution

(1) (a) Because there is one atom of Mn on both sides, no adjustment is required.

$$MnO_4^-(aq) \longrightarrow Mn^{2+}(aq)$$

(b) Because manganese is reduced from an oxidation number of $+7$ to $+2$, five electrons must be added to the left.

$$MnO_4^-(aq) + 5e^- \longrightarrow Mn^{2+}(aq)$$

Manganese in four different oxidation states. From left to right: Mn^{2+} in solution, solid MnO_2, MnO_4^{2-} (manganate ion) in solution, MnO_4^- (permanganate ion) in solution.
(Charles D. Winters)

(c) There is a total charge of -6 on the left versus $+2$ on the right. To balance, add eight H^+ to the left to give a charge of $+2$ on both sides.

$$MnO_4^-(aq) + 8H^+(aq) + 5e^- \longrightarrow Mn^{2+}(aq)$$

(d) To balance the eight H^+ ions on the left, add four H_2O molecules to the right.

$$MnO_4^-(aq) + 8H^+(aq) + 5e^- \longrightarrow Mn^{2+}(aq) + 4H_2O$$

(e) Note that there are the same number of oxygen atoms, four, on both sides, as there should be. The equation shown in green is the correctly balanced reduction half-equation.

(2) (a) Again, there is one chromium atom on both sides.

$$Cr(OH)_3(s) \longrightarrow CrO_4^{2-}(aq)$$

(b) Because the oxidation number of chromium increases from $+3$ to $+6$, add three electrons to the right.

$$Cr(OH)_3(s) \longrightarrow CrO_4^{2-}(aq) + 3e^-$$

(c) There is a charge of zero on the left, -5 on the right. To balance charge, add five OH^- ions to the left.

$$Cr(OH)_3(s) + 5OH^-(aq) \longrightarrow CrO_4^{2-}(aq) + 3e^-$$

(d) There are eight hydrogens on the left, none on the right. Add four H_2O molecules to the right.

$$Cr(OH)_3(s) + 5OH^-(aq) \longrightarrow CrO_4^{2-}(aq) + 4H_2O + 3e^-$$

(e) There are eight oxygen atoms on both sides; the oxidation half-equation is properly balanced.

Balancing Redox Equations

The process used to balance an overall redox equation is relatively straightforward, provided you know how to balance half-equations. Follow a systematic, four-step procedure.

(1) *Split the equation into two half-equations,* one for reduction, the other for oxidation.
(2) *Balance one of the half-equations* with respect to both atoms and charge as described above (steps a through e).
(3) *Balance the other half-equation.*
(4) *Combine the two half-equations in such a way as to eliminate electrons.*

Suppose, for example, the two balanced half-equations are

$$A(s) \longrightarrow A^{2+}(aq) + 2e^-$$
$$B^{3+}(aq) + 3e^- \longrightarrow B(s)$$

Multiplying the first half-equation by 3, the second by 2, and adding gives

$$3A(s) + 2B^{3+}(aq) \longrightarrow 3A^{2+}(aq) + 2B(s)$$

The electrons, six on both sides, cancel.

EXAMPLE 4.12 Balance the following redox equations.

(a) $Fe^{2+}(aq) + MnO_4^-(aq) \rightarrow Fe^{3+}(aq) + Mn^{2+}(aq)$ (acidic solution)
(b) $Cl_2(g) + Cr(OH)_3(s) \rightarrow Cl^-(aq) + CrO_4^{2-}(aq)$ (basic solution)

Strategy Follow the four-step procedure described above. Actually, if you look carefully at the text preceding this example, you'll find that all the half-equations have already been balanced! The color coding should help you find them.

Solution

(a) (1) oxidation: $Fe^{2+}(aq) \rightarrow Fe^{3+}(aq)$
reduction: $MnO_4^-(aq) \rightarrow Mn^{2+}(aq)$
(2), (3) The balanced half-equations, as obtained previously, are

$$Fe^{2+}(aq) \longrightarrow Fe^{3+}(aq) + e^-$$

$$MnO_4^-(aq) + 8H^+(aq) + 5e^- \longrightarrow Mn^{2+}(aq) + 4H_2O$$

(4) To eliminate electrons, multiply the oxidation half-equation by 5 and add to the reduction half-equation.

$$5[Fe^{2+}(aq) \longrightarrow Fe^{3+}(aq) + e^-]$$

$$MnO_4^-(aq) + 8H^+(aq) + 5e^- \longrightarrow Mn^{2+}(aq) + 4H_2O$$

$$\overline{5Fe^{2+}(aq) + MnO_4^-(aq) + 8H^+(aq) \longrightarrow 5Fe^{3+}(aq) + Mn^{2+}(aq) + 4H_2O}$$

> Multiply the half-equations by coefficients (1, 2, . . .) so that there are equal numbers of electrons on both sides.

(b) (1) reduction: $Cl_2(g) \rightarrow Cl^-(aq)$
oxidation: $Cr(OH)_3(s) \rightarrow CrO_4^{2-}(aq)$
(2), (3) Earlier, the balanced half-equations were found to be

$$Cl_2(g) + 2e^- \longrightarrow 2Cl^-(aq)$$

$$Cr(OH)_3(s) + 5OH^-(aq) \longrightarrow CrO_4^{2-}(aq) + 4H_2O + 3e^-$$

(4) Multiply the reduction half-equation by 3, the oxidation half-equation by 2, then add. This will produce $6e^-$ on both sides, so they will cancel.

$$3[Cl_2(g) + 2e^- \longrightarrow 2Cl^-(aq)]$$

$$2[Cr(OH)_3(s) + 5OH^-(aq) \longrightarrow CrO_4^{2-}(aq) + 4H_2O + 3e^-]$$

$$\overline{3Cl_2(g) + 2Cr(OH)_3(s) + 10OH^-(aq) \longrightarrow 6Cl^-(aq) + 2CrO_4^{2-}(aq) + 8H_2O}$$

Reality Check It's a good idea to check mass and charge balance in the final equation. In part (b), for example,

	Cl Atoms	Cr Atoms	O Atoms	H Atoms	Charge
Left	6	2	16	16	− 10
Right	6	2	16	16	− 10

Stoichiometric calculations for redox reactions in water solution are carried out in much the same way as those for precipitation reactions (Example 4.5) or acid-base reactions (Examples 4.8, 4.9).

EXAMPLE 4.13   As you found in Example 4.12, the balanced equation for the reaction between MnO_4^- and Fe^{2+} in acidic solution is

$$MnO_4^-(aq) + 8H^+(aq) + 5Fe^{2+}(aq) \longrightarrow 5Fe^{3+}(aq) + Mn^{2+}(aq) + 4H_2O$$

What volume of 0.684 M $KMnO_4$ solution is required to react completely with 27.50 mL of 0.250 M $Fe(NO_3)_2$ (Figure 4.10)?

Strategy   (1) Start by calculating the number of moles of Fe^{2+}. Then (2) use the coefficients of the balanced equation to find the number of moles of MnO_4^-. Finally, (3), use molarity as a conversion factor to find the volume of $KMnO_4$ solution.

Solution

(1) $n_{Fe^{2+}} = 0.02750 \text{ L} \times \dfrac{0.250 \text{ mol } Fe(NO_3)_2}{1 \text{ L}} \times \dfrac{1 \text{ mol } Fe^{2+}}{1 \text{ mol } Fe(NO_3)_2}$

$$= 6.88 \times 10^{-3} \text{ mol } Fe^{2+}$$

(2) $n_{MnO_4^-} = 0.00688 \text{ mol } Fe^{2+} \times \dfrac{1 \text{ mol } MnO_4^-}{5 \text{ mol } Fe^{2+}} = 0.00138 \text{ mol } MnO_4^-$

(3) $n_{KMnO_4} = n_{MnO_4^-} = 1.38 \times 10^{-3} \text{ mol } KMnO_4$

$$V = 1.38 \times 10^{-3} \text{ mol } KMnO_4 \times \dfrac{1 \text{ L}}{0.684 \text{ mol } KMnO_4} = 2.02 \times 10^{-3} \text{ L} \quad (2.02 \text{ mL})$$

(a)   (b)   (c)

FIGURE 4.10
A redox titration. (a) A solution of Fe^{2+} in an acidic solution ready to be titrated with a solution of potassium permanganate ($KMnO_4$) (in the buret). (b) When the potassium permanganate is added, a redox reaction occurs. The equation for the reaction is derived and balanced in the text. As the reaction takes place, the purple color characteristic of MnO_4^- fades; the Fe^{2+} formed is pale yellow. (c) Just past the equivalence point a small excess of MnO_4^- gives a light purple color to the solution. *(Charles D. Winters)*

Amines

The organic compounds called amines (page 93) can be classified according to the number of hydrocarbon groups bonded to nitrogen (Table A). Most amines of low molar mass are volatile compounds with distinctly unpleasant odors. For example, $(CH_3)_3N$ is a gas at room temperature (bp = 3°C) with an odor somewhere between that of ammonia and spoiled fish.

TABLE A	Types of Amines	
Type	**General Formula**	**Example**
Primary	RNH_2	$CH_3{-}\overset{\displaystyle |}{\underset{\displaystyle |}{N}}{-}H$
		H
Secondary	R_2NH	$CH_3{-}\overset{\displaystyle |}{\underset{\displaystyle |}{N}}{-}CH_3$
		H
Tertiary	R_3N	$CH_3{-}\overset{\displaystyle |}{\underset{\displaystyle |}{N}}{-}CH_3$
		CH_3
	where R = CH_3, C_2H_5, . . .	

The reaction of amines with H^+ ions (page 94) has an interesting practical application. Amines of high molar mass, frequently used as drugs, have very low water solubilities. They can be converted to a water-soluble form by treatment with strong acid. For example,

$$C_9H_{10}NO_2{-}\overset{\displaystyle |}{\underset{\displaystyle |}{N}}{-}C_2H_5(s) + HCl(aq) \longrightarrow [C_9H_{10}NO_2{-}\overset{\displaystyle H}{\overset{\displaystyle |}{\underset{\displaystyle |}{N}}}{-}C_2H_5]^+(aq) + Cl^-(aq)$$

$$\underset{C_2H_5}{} \qquad\qquad \underset{C_2H_5}{}$$

novocaine   novocaine hydrochloride

Novocaine hydrochloride is about 200 times as soluble as novocaine itself. When your dentist injects "Novocaine," the liquid in the syringe is a water solution of novocaine hydrochloride.

Alkaloids such as caffeine, coniine, and morphine (Figure A) are amines that are extracted from plants.

FIGURE A
Flowers and unripe seed capsules of the opium poppy. Within the capsule is a milky, gummy substance that contains the alkaloid morphine. *(Scott Camazine/Photo Researchers, Inc.)*

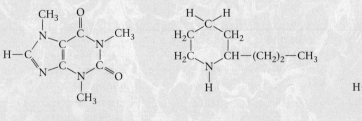

caffeine   coniine

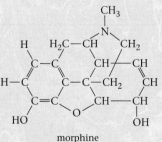

morphine

continued

Caffeine occurs in tea leaves, coffee beans, and cola nuts. Morphine is obtained from unripe opium poppy seed pods. Conline, extracted from hemlock, is the alkaloid that killed Socrates. He was sentenced to death because of unconventional teaching methods; teacher evaluations had teeth in them in ancient Greece.

During the Civil War (1860–1865) morphine was widely used as a pain-killer injected via the hypodermic syringe developed during the 1850s. Soldiers (and desperate physicians at prisoner-of-war camps such as Andersonville, Georgia) discovered that morphine had the side effect of relieving dysentery, a disease as deadly as combat. When the war ended, more than a hundred thousand veterans were addicted to morphine, and the United States faced (and largely ignored) its first drug crisis.

CHAPTER HIGHLIGHTS

Key Concepts

1. Relate molarity of a solute to—
   - number of moles and volume
   (Example 4.1; Problems 1–4)
   - molarities of ions
   (Example 4.2; Problems 5, 6)

2. Apply the precipitation diagram (Figure 4.4) to predict precipitation reactions and write net ionic equations for them.
   (Examples 4.3, 4.4; Problems 7–16)

3. With the aid of Tables 4.1 and 4.2, write net ionic equations for acid-base reactions.
   (Example 4.7; Problems 21–30)

4. Determine oxidation numbers of elements in compounds and polyatomic ions.
   (Example 4.10; Problems 41–44)

5. Balance redox half-equations and overall equations.
   (Examples 4.11, 4.12; Problems 45–64, 71)

6. Use balanced net ionic equations to carry out stoichiometric calculations for precipitation, acid-base, and redox reactions.
   (Examples 4.5, 4.6, 4.8, 4.9, 4.13; Problems 17–20, 31–40, 61–72, 81–84)

Key Terms

acid	—weak	oxidation number
—strong	equivalence point	oxidizing agent
—weak	half-equation	precipitate
acid-base indicator	■ limiting reactant	redox reaction
base	molarity	reducing agent
—strong	■ mole	reduction
—weak	net ionic equation	■ theoretical yield
electrolyte	neutralization	titration
—strong	oxidation	

Nitric acid and barium hydroxide are a strong acid and a strong base, respectively. They participate in a variety of acid-base reactions with other species. In addition, nitric acid is often involved in redox reactions because the nitrate ion is readily reduced. Finally, barium hydroxide can form precipitates involving either the barium ion or the hydroxide ion.

(a) Write net ionic equations for the reaction between aqueous solutions of
  (1) barium hydroxide and iron(III) nitrate.
  (2) nitric acid and aqueous ammonia.
  (3) barium hydroxide and nitric acid.
  (4) barium hydroxide and acetic acid ($HC_2H_3O_2$).
  (5) nitric acid and tin(II) ions forming nitrogen oxide gas and tin(IV) ions.

(b) How would you prepare 750.0 mL of 0.0325 M barium hydroxide solution from solid barium hydroxide? What is the concentration of each ion in the prepared solution?

(c) How many milliliters of a concentrated (12.0 M) nitric acid solution are needed for a reaction that requires 4.50 mol of H^+ ions?

(d) When 30.00 mL of 0.09250 M barium hydroxide solution reacts with 45.00 mL of 0.3375 M nickel(II) sulfate, two precipitates form. Calculate the mass of each precipitate and the concentration of all ions after the reaction, assuming that the final volume is the sum of the initial volumes.

(e) In the analysis of an alloy, copper is oxidized to copper(II) ions and nitrate ions are reduced to nitrogen oxide gas. It is found that 32.55 mL of a solution that is 0.4250 M in nitrate is required to react exactly with a 2.000-g sample of the alloy. Calculate the mass percent of copper in the alloy.

(f) To determine the molarity of a solution of hydrochloric acid, it is titrated with 0.1000 M barium hydroxide. It is determined that 22.50 mL of barium hydroxide is required to react with 25.00 mL of hydrochloric acid. What is the molarity of the hydrochloric acid?

Answers

(a) **(1)** $Fe^{3+}(aq) + 3OH^-(aq) \rightarrow Fe(OH)_3(s)$
  (2) $H^+(aq) + NH_3(aq) \rightarrow NH_4^+(aq)$
  (3) $H^+(aq) + OH^-(aq) \rightarrow H_2O$
  (4) $HC_2H_3O_2(aq) + OH^-(aq) \rightarrow C_2H_3O_2^-(aq) + H_2O$
  (5) $8H^+(aq) + 2NO_3^-(aq) + 3Sn^{2+}(aq) \rightarrow 2NO(g) + 4H_2O + 3Sn^{4+}(aq)$

(b) Dissolve 4.18 g of $Ba(OH)_2$ in enough water to form 750.0 mL of solution; 0.0325 M Ba^{2+}; 0.0650 M OH^-

(c) 375 mL

(d) 0.6476 g $BaSO_4$; 0.2573 g $Ni(OH)_2$; 0.1655 M Ni^{2+}; 0.1655 M SO_4^{2-}; $[Ba^{2+}] = [OH^-] \approx 0$

(e) 65.94%

(f) 0.1800 M

Problem numbers in blue indicate that the answer is available in Appendix 6 at the back of the book.
WEB indicates that the solution is posted at **http://www.harcourtcollege.com/chem/general/masterton4/student/**

Molarity

1. Starting with the solid and adding water, how would you prepare 2.00 L of 0.685 M
  (a) $Ni(NO_3)_2$ **(b)** $CuCl_2$ **(c)** $C_6H_8O_6$ (vitamin C)
2. How would you prepare from the solid and pure water
  (a) 0.400 L of 0.155 M $Sr(OH)_2$

  (b) 1.75 L of 0.333 M $(NH_4)_2CO_3$
3. You are asked to prepare a 0.8500 M solution of aluminum nitrate. You find that you have only 50.00 g of the solid.
  (a) What is the maximum volume of solution that you can prepare?

(b) How many milliliters of this prepared solution are required to furnish 0.5000 mol of aluminum nitrate to a reaction?

(c) If 2.500 L of the prepared solution is required, how much more aluminum nitrate would you need?

(d) Fifty milliliters of a 0.450 M solution of aluminum nitrate is needed. How would you prepare the required solution from the solution prepared in (a)?

WEB **4.** A reagent bottle is labeled 0.450 M K_2CO_3.

(a) How many moles of K_2CO_3 are present in 45.6 mL of this solution?

(b) How many milliliters of this solution are required to furnish 0.800 mol of K_2CO_3?

(c) Assuming no volume change, how many grams of K_2CO_3 do you need to add to 2.00 L of this solution to obtain a 1.000 M solution of K_2CO_3?

(d) If 50.0 mL of this solution is added to enough water to make 125 mL of solution, what is the molarity of the diluted solution?

5. How many moles of ions are present in water solutions prepared by dissolving 0.2375 mol of

(a) iron(II) chloride   **(b)** aluminum sulfate

(c) nitric acid   **(d)** sodium dichromate

6. How many moles of ions are present in water solutions prepared by dissolving 0.4132 mol of

(a) magnesium chloride

(b) manganese(II) nitride

(c) cobalt(III) nitrate

(d) iron(II) permanganate

Precipitation Reactions

7. Write the formulas of the following compounds and decide which are soluble in water.

(a) barium chloride

(b) magnesium hydroxide

(c) chromium(III) carbonate

(d) potassium phosphate

8. Follow the instructions for Question 7 for the following compounds:

(a) sodium sulfate   **(b)** iron(III) nitrate

(c) silver chloride   **(d)** chromium(III) hydroxide

9. Describe how you would prepare

(a) cadmium(II) carbonate from a solution of cadmium(II) nitrate.

(b) copper(II) hydroxide from a solution of sodium hydroxide.

(c) magnesium carbonate from a solution of magnesium chloride.

10. Name the reagent, if any, that you would add to a solution of cobalt(III) chloride to precipitate

(a) cobalt(III) phosphate   **(b)** cobalt(III) carbonate

(c) cobalt(III) hydroxide

11. Write net ionic equations for the formation of

(a) a green precipitate when solutions of nickel(II) nitrate and sodium hydroxide are mixed.

(b) two different precipitates when solutions of cadmium(II) sulfate and barium hydroxide are mixed.

WEB **12.** Write net ionic equations to explain the formation of

(a) a red precipitate when solutions of iron(III) chloride and sodium hydroxide are mixed.

(b) two different precipitates, one blue and the other white, when solutions of silver hydroxide and copper(II) chloride are mixed.

13. Decide whether a precipitate will form when the following solutions are mixed. If a precipitate forms, write a net ionic equation for the reaction.

(a) copper(II) sulfate and sodium chloride

(b) manganese(II) nitrate and sodium hydroxide

(c) silver nitrate and hydrochloric acid

(d) cobalt(II) sulfate and barium hydroxide

(e) ammonium carbonate and potassium hydroxide

14. Follow the directions of Question 13 for solutions of the following.

(a) potassium nitrate and magnesium sulfate

(b) silver nitrate and potassium carbonate

(c) ammonium carbonate and cobalt(III) chloride

(d) sodium phosphate and barium hydroxide

(e) barium nitrate and potassium hydroxide

15. Write a net ionic equation for any precipitation reaction that occurs when 0.1 M solutions of the following are mixed.

(a) zinc nitrate and nickel(II) chloride

(b) potassium phosphate and calcium nitrate

(c) sodium hydroxide and zinc nitrate

(d) iron(III) nitrate and barium hydroxide

16. Follow the directions for Question 15 for the following pairs of solutions.

(a) sodium phosphate and barium chloride

(b) zinc sulfate and potassium hydroxide

(c) ammonium sulfate and sodium chloride

(d) cobalt(III) nitrate and sodium phosphate

17. What volume of 0.500 M nickel(II) sulfate is required to react completely with

(a) 25.6 mL of 0.250 M barium hydroxide?

(b) 46.8 mL of 1.050 M sodium phosphate?

(c) 15.0 mL of 0.896 M ammonium carbonate?

18. What volume of 0.773 M barium nitrate is required to react completely with

(a) 20.0 mL of 0.0937 M cobalt(III) sulfate?

(b) 43.3 mL of 0.396 M potassium carbonate?

(c) 29.5 mL of 0.631 M ammonium phosphate?

19. Aluminum ions react with carbonate ions to form an insoluble compound, aluminum carbonate.

(a) Write the net ionic equation for this reaction.

(b) What is the molarity of a solution of aluminum chloride if 30.0 mL is required to react with 35.5 mL of 0.137 M sodium carbonate?

(c) How many grams of aluminum carbonate are formed in (b)?

20. A 50.00-mL sample of 0.0250 M silver nitrate is mixed with 0.0400 M chromium(III) chloride.

(a) What is the minimum volume of chromium(III) chloride required to completely precipitate silver chloride?

(b) How many grams of silver chloride are produced from (a)?

Acid-Base Reactions

21. For an acid-base reaction, what is the reacting species, that is, the ion or molecule that appears in the chemical equation, in the following acids?

(a) perchloric acid   (b) hydriodic acid
(c) nitrous acid   (d) nitric acid
(e) lactic acid ($HC_3H_5O_3$)

22. Follow the directions of Problem 21 for the following acids.

(a) sulfurous acid   (b) chlorous acid
(c) perchloric acid   (d) sulfuric acid
(e) formic acid ($HCHO_2$)

23. For an acid-base reaction, what is the reacting species, that is, the ion or molecule that appears in the chemical equation, in the following bases?

(a) aqueous ammonia   (b) strontium hydroxide
(c) sodium hydroxide   (d) pyridine, C_5H_5N

24. Follow the directions of Problem 23 for the following bases.

(a) potassium hydroxide
(b) aniline, $C_6H_5NH_2$
(c) dimethylamine, $(CH_3)_2NH$
(d) barium hydroxide

25. Classify the following compounds as acids or bases, weak or strong.

(a) sulfurous acid   (b) ammonia
(c) barium hydroxide   (d) hydriodic acid

26. Follow the directions of Question 25 for

(a) perchloric acid   (b) cesium hydroxide
(c) carbonic acid, H_2CO_3 (d) ethylamine, $C_2H_5NH_2$

27. Write a balanced net ionic equation for each of the following acid-base reactions in water.

(a) nitrous acid and barium hydroxide
(b) potassium hydroxide and hydrofluoric acid
(c) aniline ($C_6H_5NH_2$) and perchloric acid

28. Write a balanced net ionic equation for each of the following acid-base reactions in water.

(a) acetic acid ($HC_2H_3O_2$) with strontium hydroxide
(b) diethylamine, $(C_2H_5)_2NH$, with sulfuric acid
(c) aqueous hydrogen cyanide (HCN) with sodium hydroxide

29. Consider the equation

$$H^+(aq) + OH^-(aq) \longrightarrow H_2O$$

For which of the following pairs would this be the correct equation for the acid-base reaction in solution? If it is not correct, write the proper equation for the acid-base reaction between the pair.

(a) hydrochloric acid and calcium hydroxide
(b) HBr and CH_3NH_2
(c) nitric acid and ammonia

(d) H_2SO_4 and KOH
(e) HF and barium hydroxide

30. Follow the directions of Question 29 for the following pairs.

(a) nitric acid and $C_2H_5NH_2$
(b) perchloric acid and cesium hydroxide
(c) $HC_2H_3O_2$ and LiOH
(d) sulfuric acid and calcium hydroxide
(e) barium hydroxide and hydriodic acid

31. What volume of 0.285 M strontium hydroxide is required to neutralize 25.00 mL of 0.275 M hydrofluoric acid (HF)?

WEB **32.** What is the molarity of a solution of nitric acid if 0.216 g of barium hydroxide is required to neutralize 20.00 mL of nitric acid?

33. What is the volume of 1.222 M sodium hydroxide required to react with

(a) 32.5 mL of 0.569 M sulfurous acid? (One mole of sulfurous acid reacts with two moles of hydroxide ion.)
(b) 5.00 g of oxalic acid ($H_2C_2O_4$)? (One mole of oxalic acid reacts with two moles of hydroxide ion.)
(c) 15.0 g of concentrated acetic acid ($HC_2H_3O_2$) that is 88% by mass pure?

34. What is the volume of 0.885 M hydrochloric acid required to react with

(a) 25.00 mL of 0.288 M aqueous ammonia?
(b) 10.00 g of sodium hydroxide?
(c) 25.0 mL of a solution ($d = 0.928$ g/cm^3) containing 10.0% by mass of methylamine (CH_3NH_2)?

35. Boric acid (H_3BO_3) can be used to neutralize bases. The equation for the reaction is

$$H_3BO_3(s) + 2OH^-(aq) \longrightarrow 2H_2O + HBO_3^{2-}(aq)$$

What volume of 0.216 M barium hydroxide can be neutralized by 10.00 g of H_3BO_3?

36. Some hydrochloric acid is spilled on the lab floor. Sodium hydrogen carbonate is sprinkled on the spill to neutralize the acid. The balanced equation for the reaction that takes place is

$$NaHCO_3(s) + H^+(aq) \longrightarrow Na^+(aq) + CO_2(g) + H_2O$$

If 75 mL of 3.00 M HCl were spilled, what is the minimum amount (in grams) of $NaHCO_3$ that must be sprinkled to neutralize all the acid in the spill?

37. A capsule of vitamin C, a weak acid, is analyzed by titrating it with 0.425 M sodium hydroxide. It is found that 6.20 mL of base is required to react with a capsule weighing 0.628 g. What is the percentage of vitamin C ($C_6H_8O_6$) in the capsule? (One mole of vitamin C reacts with one mole of hydroxide ion.)

38. The percentage of sodium hydrogen carbonate, $NaHCO_3$, in a powder for stomach upsets is found by titrating with 0.275 M hydrochloric acid. If 15.5 mL of hydrochloric acid is required to react with 0.500 g of the sample, what is the percentage of sodium hydrogen carbonate in the sample? The reaction is given in Problem 36.

39. An artificial fruit beverage contains 12.0 g of tartaric acid, $H_2C_4H_4O_6$, to achieve tartness. It is titrated with a basic solution that has a density of 1.045 g/cm^3 and contains 5.00 mass percent KOH. What volume of the basic solution is required? (One mole of tartaric acid reacts with two moles of hydroxide ion.)

40. Lactic acid, $C_3H_6O_3$, is the acid present in sour milk. A 0.100-g sample of pure lactic acid requires 12.95 mL of 0.0857 M sodium hydroxide for complete reaction. How many moles of hydroxide ion are required to neutralize one mole of lactic acid?

Redox Reactions

41. Assign oxidation numbers to each element in
  (a) nitrite ion   (b) sulfite ion
  (c) potassium oxide   (d) oxalate ion ($C_2O_4^{2-}$)

42. Assign oxidation numbers to each element in
  (a) dinitrogen trioxide   (b) sulfur dioxide
  (c) the dichromate ion   (d) the hypochlorite ion

43. Assign oxidation numbers to each element in
  (a) HIO_3 (b) $NaMnO_4$ (c) SnO_2
  (d) NOF (e) NaO_2

44. Assign oxidation numbers to each element in
  (a) P_2O_5 (b) NH_3 (c) CO_3^{2-}
  (d) $S_2O_3^{2-}$ (e) N_2H_4

45. Classify each of the following half-reactions as oxidation or reduction.
  (a) $O_2(g) \rightarrow O^{2-}(aq)$
  (b) $MnO_4^-(aq) \rightarrow MnO_2(s)$
  (c) $Cr_2O_7^{2-}(aq) \rightarrow Cr^{3+}(aq)$
  (d) $Cl^-(aq) \rightarrow Cl_2(g)$

46. Classify each of the following half-reactions as oxidation or reduction.
  (a) $TiO_2(s) \rightarrow Ti^{3+}(aq)$ (b) $Zn^{2+}(aq) \rightarrow Zn(s)$
  (c) $NH_4^+(aq) \rightarrow N_2(g)$ (d) $CH_3OH(aq) \rightarrow CH_2O(aq)$

47. Balance the half-equations in Question 45. Balance (a) and (b) in acidic medium, (c) and (d) in basic medium.

48. Balance the half-equations in Question 46. Balance (a) and (b) in basic medium, (c) and (d) in acidic medium.

49. Classify each of the following half-equations as oxidation or reduction and balance.
  (a) (basic) $ClO^-(aq) \rightarrow Cl^-(aq)$
  (b) (acidic) $NO_3^-(aq) \rightarrow NO(g)$
  (c) (basic) $Ni^{2+}(aq) \rightarrow Ni_2O_3(s)$
  (d) (acidic) $Mn^{2+}(aq) \rightarrow MnO_2(s)$

50. Classify each of the following half-equations as oxidation or reduction and balance.
  (a) (acidic) $Mn^{2+}(aq) \rightarrow MnO_4^-(aq)$
  (b) (basic) $CrO_4^{2-}(aq) \rightarrow Cr^{3+}(aq)$
  (c) (basic) $PbO_2(s) \rightarrow Pb^{2+}(aq)$
  (d) (acidic) $ClO_2^-(aq) \rightarrow ClO^-(aq)$

51. For each unbalanced equation given below—
  ■ write unbalanced half-reactions.
  ■ identify the species oxidized and the species reduced.
  ■ identify the oxidizing and reducing agents.
  (a) $Ag(s) + NO_3^-(aq) \rightarrow Ag^+(aq) + NO(g)$
  (b) $CO_2(g) + H_2O \rightarrow C_2H_4(g) + O_2(g)$

52. Follow the directions of Question 51 for the following unbalanced equations.
  (a) $H_2O_2(aq) + Ni^{2+}(aq) \rightarrow Ni^{3+}(aq) + H_2O$
  (b) $Cr_2O_7^{2-}(aq) + Sn^{2+}(aq) \rightarrow Cr^{3+}(aq) + Sn^{4+}(aq)$

53. Balance the equations in Question 51 in acid.

54. Balance the equations in Question 52 in acid.

55. Write balanced equations for the following reactions in acid solution.
  (a) $P_4(s) + Cl^-(aq) \rightarrow PH_3(g) + Cl_2(g)$
  (b) $MnO_4^-(aq) + NO_2^-(aq) \rightarrow Mn^{2+}(aq) + NO_3^-(aq)$
  (c) $HBrO_3(aq) + Bi(s) \rightarrow HBrO_2(aq) + Bi_2O_3(s)$
  (d) $CrO_4^{2-}(aq) + SO_3^{2-}(aq) \rightarrow Cr^{3+}(aq) + SO_4^{2-}(aq)$

56. Write balanced equations for the following reactions in acid solution.
  (a) $Ni^{2+}(aq) + IO_4^-(aq) \rightarrow Ni^{3+}(aq) + I^-(aq)$
  (b) $O_2(g) + Br^-(aq) \rightarrow H_2O + Br_2(l)$
  (c) $Ca(s) + Cr_2O_7^{2-}(aq) \rightarrow Ca^{2+}(aq) + Cr^{3+}(aq)$
  (d) $IO_3^-(aq) + Mn^{2+}(aq) \rightarrow I^-(aq) + MnO_2(s)$

57. Write balanced equations for the following reactions in basic solution.
  (a) $SO_2(g) + I_2(aq) \rightarrow SO_3(g) + I^-(aq)$
  (b) $Zn(s) + NO_3^-(aq) \rightarrow NH_3(aq) + Zn^{2+}(aq)$
  (c) $ClO^-(aq) + CrO_2^-(aq) \rightarrow Cl^-(aq) + CrO_4^{2-}(aq)$
  (d) $K(s) + H_2O \rightarrow K^+(aq) + H_2(g)$

58. Write balanced equations for the following reactions in basic solution.
  (a) $Ni(OH)_2(s) + N_2H_4(aq) \rightarrow Ni(s) + N_2(g)$
  (b) $Fe(OH)_3(s) + Cr^{3+}(aq) \rightarrow Fe(OH)_2(s) + CrO_4^{2-}(aq)$
  (c) $MnO_4^-(aq) + BrO_3^-(aq) \rightarrow MnO_2(s) + BrO_4^-(aq)$
  (d) $H_2O_2(aq) + IO_4^-(aq) \rightarrow IO_2^-(aq) + O_2(g)$

59. Write balanced net ionic equations for the following reactions in acid solution.
  (a) Solid phosphorus (P_4) reacts with hypochlorous acid, $HClO$, to form phosphoric acid, H_3PO_4, and chloride ion.
  (b) Copper is oxidized by nitrate ion to form copper(II) ion and nitrogen dioxide gas.
  (c) An aqueous solution of chlorine is reduced to chloride ion; at the same time iodide ions are oxidized to iodate ions, IO_3^-.

WEB 60. Write balanced net ionic equations for the following reactions in acid solution.
  (a) Zinc is dissolved in hydrochloric acid, forming aqueous zinc chloride and hydrogen gas.
  (b) Solid copper(II) sulfide is dissolved in nitric acid, forming copper(II) nitrate (aq), sulfur, and nitrogen oxide gas.
  (c) Antimony(III) ion reacts with periodate ion, IO_4^-, yielding antimony(V) ion and iodide ion.

61. Iodine reacts with thiosulfate ion, $S_2O_3^{2-}$, to give iodide ion and the tetrathionate ion, $S_4O_6^{2-}$.
  (a) Write a balanced net ionic equation for the reaction.
  (b) If 25.0 g of iodine is dissolved in enough water to make 1.50 L of solution, what volume of 0.244 M sodium thiosulfate will be needed for complete reaction?

62. A solution of potassium permanganate reacts with oxalic acid, $H_2C_2O_4$, to form carbon dioxide and solid manganese(IV) oxide (MnO_2).

(a) Write a balanced net ionic equation for the reaction.
(b) If 20.0 mL of 0.300 M potassium permanganate is required to react with 13.7 mL of oxalic acid, what is the molarity of the oxalic acid?
(c) What is the mass of manganese(IV) oxide formed?

63. Hydrogen gas is bubbled into a solution of barium hydroxide that has sulfur in it. The unbalanced equation for the reaction that takes place is

$$H_2(g) + S(s) + OH^-(aq) \longrightarrow S^{2-}(aq) + H_2O$$

(a) Balance the equation.
(b) What volume of 0.349 M Ba(OH)$_2$ is required to react completely with 3.00 g of sulfur?

64. Consider the reaction between silver and nitric acid for which the unbalanced equation is

$$Ag(s) + H^+(aq) + NO_3^-(aq) \longrightarrow Ag^+(aq) + NO_2(g) + H_2O$$

(a) Balance the equation.
(b) If 42.50 mL of 12.0 M nitric acid furnishes enough H^+ to react with silver, how many grams of silver react?

65. A wire weighing 1.232 g and containing 87.1% Fe is dissolved in HCl. The iron is completely oxidized to Fe^{3+} by bromine water. The solution is then treated with tin(II) chloride to bring about the reaction

$$Sn^{2+}(aq) + 2Fe^{3+}(aq) \longrightarrow Sn^{4+}(aq) + 2Fe^{2+}(aq)$$

If 49.3 mL of tin(II) chloride solution is required for complete reaction, what is its molarity?

66. Hair-bleaching solutions contain hydrogen peroxide, H_2O_2. The hydrogen peroxide content can be determined by reacting H_2O_2 with a potassium dichromate acidic solution. The unbalanced equation for the reaction is

$$H_2O_2(aq) + Cr_2O_7^{2-}(aq) + H^+(aq) \longrightarrow O_2(g) + Cr^{3+}(aq) + H_2O$$

A 30.0-g bleach solution of H_2O_2 needed 54.6 mL of 0.715 M $K_2Cr_2O_7$ to completely react with the H_2O_2 in the hair bleach. What is the mass percent of H_2O_2 in the bleach?

67. Laws passed in some states define a drunk driver as one who drives with a blood alcohol level of 0.10% by mass or higher. The level of alcohol can be determined by titrating blood plasma with potassium dichromate according to the unbalanced equation:

$$H^+(aq) + Cr_2O_7^{2-}(aq) + C_2H_5OH(aq) \longrightarrow$$
$$Cr^{3+}(aq) + CO_2(g) + H_2O$$

Assuming that the only substance that reacts with dichromate in blood plasma is alcohol, is a person legally drunk if 38.94 mL of 0.0723 M potassium dichromate is required to titrate a 50.0-g sample of blood plasma?

68. Laundry bleach is a solution of sodium hypochlorite (NaClO). To determine the hypochlorite (ClO^-) content of bleach (which is responsible for its bleaching action), sulfide ion is added in basic solution. The balanced equation for the reaction is

$$ClO^-(aq) + S^{2-}(aq) + H_2O \longrightarrow Cl^-(aq) + S(s) + 2OH^-(aq)$$

The chloride ion resulting from the reduction of HClO is precipitated as AgCl. When 50.0 mL of laundry bleach ($d = 1.02$ g/cm^3) is treated as described above, 4.95 g of AgCl is obtained. What is the mass percent of NaClO in the bleach?

Unclassified

69. A sample of limestone weighing 1.005 g is dissolved in 75.00 mL of 0.2500 M hydrochloric acid. The following reaction occurs:

$$CaCO_3(s) + 2H^+(aq) \longrightarrow Ca^{2+}(aq) + CO_2(g) + H_2O$$

It is found that 19.26 mL of 0.150 M NaOH is required to titrate the excess HCl left after reaction with the limestone. What is the mass percent of $CaCO_3$ in the limestone?

70. The iron content of hemoglobin is determined by destroying the hemoglobin molecule and producing small water-soluble ions and molecules. The iron in the aqueous solution is reduced to iron(II) ion and then titrated against potassium permanganate. In the titration, iron(II) is oxidized to iron(III) and permanganate is reduced to manganese(II) ion. A 5.00-g sample of hemoglobin requires 32.3 mL of a 0.002100 M solution of potassium permanganate. What is the mass percent of iron in hemoglobin?

71. Gold metal will dissolve only in *aqua regia*, a mixture of concentrated hydrochloric acid and concentrated nitric acid in 3:1 volume ratio. The products of the reaction between gold and the concentrated acids are $AuCl_4^-(aq)$, NO(g), and H_2O.

(a) Write a balanced net ionic equation for the redox reaction, treating HCl and HNO$_3$ as strong acids.
(b) What stoichiometric ratio of hydrochloric acid to nitric acid should be used?
(c) What volumes of 12 M HCl and 16 M HNO$_3$ are required to furnish the Cl^- and NO_3^- ions to react with 25.0 g of gold?

72. The standard set by OSHA for the maximum amount of ammonia permitted in the workplace is 5.00×10^{-3}% by mass. To determine a factory's compliance, 100.0 mL of air ($d = 1.19$ g/L) is bubbled into 100.0 mL of 0.02500 M HCl at the same temperature and pressure. Ammonia in the air bubbled in reacts with H^+ as follows:

$$NH_3(aq) + H^+(aq) \longrightarrow NH_4^+(aq)$$

The unreacted hydrogen ions required 57.00 mL of 0.03500 M NaOH for complete neutralization. Is the factory compliant with the OSHA standards for ammonia in the workplace?

Conceptual Questions

73. Classify each of the following as a precipitation, acid-base, or redox reaction.

(a) the reaction between solutions of sulfuric acid and barium nitrate

(b) the reaction between solutions of sulfuric acid and calcium hydroxide

(c) the reaction of HCl with Al to evolve H_2

(d) the reaction of a solution of tin(II) chloride with air to form SnO_2

74. Using circles to represent cations and squares to represent anions, represent the reactions that occur between aqueous solutions of

(a) Na^+ and Cl^- **(b)** Ag^+ and Cl^-

75. Assuming that the circles represent cations and squares represent anions, match the incomplete net ionic equations to their pictorial representations

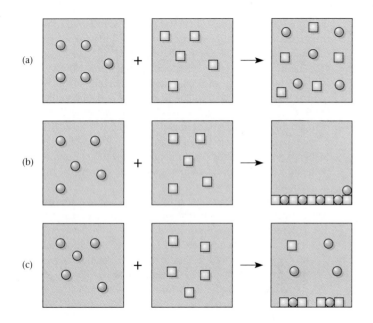

(1) $2Na^+ + SO_4^{2-} \longrightarrow$ _____
(2) $Mg^{2+} + 2OH^- \longrightarrow$ _____
(3) $Ba^{2+} + CO_3^{2-} \longrightarrow$ _____

76. Using squares to represent atoms of one element (or cations) and circles to represent the atoms of the other element (or anions), represent the principal species in the following pictorially. (You may represent the hydroxide anion as a single circle.)

(a) a solution of HCl   **(b)** a solution of HF

(c) a solution of KOH   **(d)** a solution of HNO_2

77. The following figures represent species before and after they are dissolved in water. Classify each species as weak electrolyte, strong electrolyte, or nonelectrolyte. You may assume that species that dissociate during solution break up as ions.

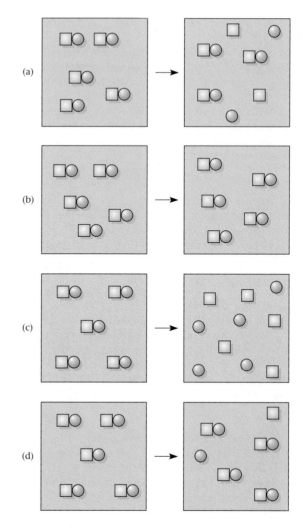

78. A student is asked to identify the metal nitrate present in an aqueous solution. The cation in the solution can either be Na^+, Ba^{2+}, Ag^+, or Ni^{2+}. Results of solubility experiments are as follows:

> unknown + chloride ions—no precipitate
> unknown + carbonate ions—precipitate
> unknown + sulfate ions—precipitate

What is the cation in the solution?

79. Use the following reactions to arrange the cations W^+, X^+, Y^+, and Z^+ in order of their increasing ability as oxidizing agents.

$$W^+ + Z \longrightarrow \text{no reaction}$$

$$X^+ + W \longrightarrow \text{no reaction}$$

$$Y^+ + X \longrightarrow X^+ + Y$$

$$Y^+ + Z \longrightarrow Z^+ + Y$$

80. Based on the relative oxidizing strengths of the cations obtained in Question 79, which of the following reactions would you expect to occur?

    (a) $Y^+ + W \rightarrow$ _____

    (b) $Z^+ + X \rightarrow$ _____

    (c) $W^+ + X \rightarrow$ _____

Challenge Problems

81. Calcium in blood or urine can be determined by precipitation as calcium oxalate, CaC_2O_4. The precipitate is dissolved in strong acid and titrated with potassium permanganate. The products of the reaction are carbon dioxide and manganese(II) ion. A 24-hour urine sample is collected from an adult patient, reduced to a small volume, and titrated with 26.2 mL of 0.0946 M $KMnO_4$. How many grams of calcium oxalate are in the sample? Normal range for Ca^{2+} output for an adult is 100 to 300 mg per 24 hours. Is the sample within the normal range?

82. Stomach acid is approximately 0.020 M HCl. What volume of this acid is neutralized by an antacid tablet that weighs 330 mg and contains 41.0% $Mg(OH)_2$, 36.2% $NaHCO_3$, and 22.8% NaCl? The reactions involved are

$$Mg(OH)_2(s) + 2H^+(aq) \longrightarrow Mg^{2+}(aq) + 2H_2O$$

$$HCO_3^-(aq) + H^+(aq) \longrightarrow CO_2(g) + H_2O$$

83. Copper metal can reduce silver ions to metallic silver. The copper is oxidized to copper ions according to the reaction

$$2Ag^+(aq) + Cu(s) \longrightarrow Cu^{2+}(aq) + 2Ag(s)$$

A copper strip with a mass of 2.00 g is dipped into a solution of $AgNO_3$. After some time has elapsed, the copper strip is coated with silver. The strip is removed from the solution, dried, and weighed. The coated strip has a mass of 4.18 g. What are the masses of copper and silver metals in the strip? (*Hint:* Remember that the copper metal is being used up as silver metal forms.)

84. A solution contains both iron(II) and iron(III) ions. A 50.00-mL sample of the solution is titrated with 35.0 mL of 0.0280 M $KMnO_4$, which oxidizes Fe^{2+} to Fe^{3+}. The permanganate ion is reduced to manganese(II) ion. Another 50.00-mL sample of the solution is treated with zinc, which reduces all the Fe^{3+} to Fe^{2+}. The resulting solution is again titrated with 0.0280 M $KMnO_4$; this time 48.0 mL is required. What are the concentrations of Fe^{2+} and Fe^{3+} in the solution?

R eligious faith is a most filling vapor.
  It swirls occluded in us under tight
    Compression to uplift us out of weight —
As in those buoyant bird bones thin as paper,
To give them still more buoyancy in flight.
Some gas like helium must be innate.

—Robert Frost
Innate Helium

5 GASES

A green plant oxidizes water to oxygen gas. *(Charles D. Winters)*

CHAPTER OUTLINE

5.1 MEASUREMENTS ON GASES

5.2 THE IDEAL GAS LAW

5.3 GAS LAW CALCULATIONS

5.4 STOICHIOMETRY OF GASEOUS REACTIONS

5.5 GAS MIXTURES: PARTIAL PRESSURES AND MOLE FRACTIONS

5.6 KINETIC THEORY OF GASES

5.7 REAL GASES

By far the most familiar gas to all of us is the air we breathe. The Greeks considered air to be one of the four fundamental elements of nature, along with earth, water, and fire. Late in the eighteenth century, Cavendish, Priestley, and Lavoisier studied the composition of air, which is primarily a mixture of nitrogen and oxygen with smaller amounts of argon, carbon dioxide, and water vapor. Today it appears that the concentrations of some of the minor components of the atmosphere may be changing, with adverse effects on the environment. The depletion of the ozone layer and increases in the amounts of "greenhouse" gases are topics for the evening news.

Ozone is a vital but minor component of the upper atmosphere (see Chapter 11).

All gases resemble one another closely in their physical behavior. Their volumes respond in almost exactly the same way to changes in pressure, temperature, or amount of gas. In fact, it is possible to write a simple equation relating these four variables that is valid for all gases. This equation, known as the ideal gas law, is the central theme of this chapter; it is introduced in Section 5.2. The law is applied to—

- pure gases in Section 5.3.

- gases in chemical reactions in Section 5.4.

- gas mixtures in Section 5.5.

Section 5.6 considers the kinetic theory of gases, the molecular model on which the ideal gas law is based. Finally, in Section 5.7 we describe the extent to which real gases deviate from the law.

5.1 MEASUREMENTS ON GASES

To completely describe the state of a gaseous substance, its volume, amount, temperature, and pressure are specified. The first three of these quantities were discussed in earlier chapters and will be reviewed briefly in this section. Pressure, a somewhat more abstract quantity, will be examined in more detail.

Volume, Amount, and Temperature

A gas expands uniformly to fill any container in which it is placed. This means that the volume of a gas is the volume of its container. Volumes of gases can be expressed in liters, cubic centimeters, or cubic meters:

$$1 \text{ L} = 10^3 \text{ cm}^3 = 10^{-3} \text{ m}^3$$

Most commonly, the amount of matter in a gaseous sample is expressed in terms of the number of moles (n). In some cases, the mass m in grams is given instead. These two quantities are related through the molar mass, $\mathcal{M}$, in grams per mole.

$$m = \mathcal{M} \times n$$

The temperature of a gas is ordinarily measured using a thermometer marked in degrees Celsius. However, as we will see in Section 5.2, *in any calculation involving the physical behavior of gases, temperatures must be expressed on the Kelvin scale.* To convert between °C and K, use the relation introduced in Chapter 1:

$$T_K = t_{°C} + 273.15$$

Typically, in gas law calculations, temperatures are expressed only to the nearest degree. In that case, the Kelvin temperature can be found by simply adding 273 to the Celsius temperature.

Pressure

Pressure is defined as force per unit area. You are probably familiar with the English unit "pound per square inch," often abbreviated psi. When we say that a gas exerts a pressure of 15 psi, we mean that the pressure on the walls of the gas container is 15 pounds (of force) per square inch of wall area.

A device commonly used to measure atmospheric pressure is the mercury *barometer* (Figure 5.1), first constructed by Evangelista Torricelli in the seventeenth century. This consists of a closed gas tube filled with mercury inverted over a pool of mercury. The pressure exerted by the mercury column exactly equals that of the atmosphere. Hence the height of the column is a measure of the atmospheric pressure. At or near sea level, it typically varies from 740 to 760 mm, depending on weather conditions.

The pressure of a confined gas can be measured by a *manometer* of the type shown in Figure 5.2. You may have seen such a device in the doctor's office, where it is used for measuring blood pressure. Here again the fluid used is mercury. If the level in the inner tube (A) is lower than that in the outer tube (B), the pressure of the gas is greater than that of the atmosphere. If the reverse is true, the gas pressure is less than atmospheric.

Because of the way in which gas pressure is measured, it is often expressed in **millimeters of mercury (mm Hg).*** Thus we might say that the atmospheric pressure on a certain day is 752 mm Hg. This means that the pressure of the air is equal to that exerted by a column of mercury 752 mm high.

Another unit commonly used to express gas pressure is the standard atmosphere, or simply **atmosphere (atm).** This is the pressure exerted by a column of mercury 760 mm high with the mercury at 0°C. If we say that a gas has a pressure of 0.98 atm, we mean that the pressure is 98% of that exerted by a mercury column 760 mm high.

⊙ See the *Saunders Interactive General Chemistry CD-ROM*, Screen 12.2, Properties of Gases.

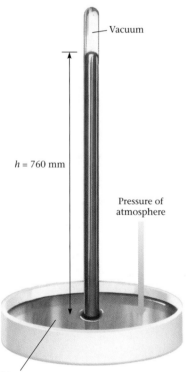

FIGURE 5.1

A mercury barometer. This is the type of barometer first constructed by Torricelli. The pressure of the atmosphere pushes the mercury in the dish to rise into the glass tube. The height of the column of mercury is a measure of the atmospheric pressure. **OHT**

---

*The pressure exerted by a column of mercury depends on its density, which varies slightly with temperature. To get around this ambiguity, the *torr* was defined to be the pressure exerted by 1 mm of mercury at certain specified conditions, notably 0°C. Over time, the unit torr has become a synonym for millimeter of mercury. Throughout this text, we will use millimeter of mercury rather than torr because the former has a clearer physical meaning.

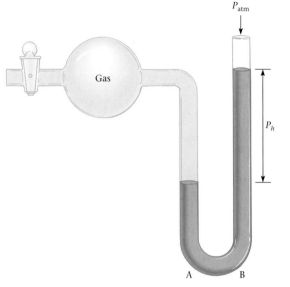

FIGURE 5.2
A manometer open to the atmosphere, used to measure gas pressure in a closed system. The pressure of the gas is given by $P_{gas} = P_{atm} + P_h$. In the figure, the gas pressure is greater than the atmospheric pressure.   **OHT**

In the International System (Appendix 1), the standard unit of pressure is the *pascal* (Pa). A pascal is a very small unit; it is approximately the pressure exerted by a film of water 0.1 mm high on the surface beneath it. A related unit is the **bar** (10^5 Pa). A bar is nearly, but not quite, equal to an atmosphere:

$$1.013 \text{ bar} = 1 \text{ atm} = 760 \text{ mm Hg}$$

EXAMPLE 5.1   A balloon with a volume (V) of 2.36×10^4 m^3 contains 4.68×10^6 g of helium at 18°C and 1.20 bar. Express the volume of the balloon in liters, the amount in moles (n), the temperature (T) in K, and the pressure (P) in both atmospheres and millimeters of mercury.

Strategy   Use the following conversion factors:

$$\frac{1 \text{ L}}{10^{-3} \text{ m}^3} \qquad \frac{1 \text{ mol He}}{4.003 \text{ g He}} \qquad \frac{760 \text{ mm Hg}}{1.013 \text{ bar}} \qquad \frac{1 \text{ atm}}{1.013 \text{ bar}}$$

For the temperature conversion, use the relation: $T_K = t_{°C} + 273$

> In these relations, "760 mm Hg" and "1 atm" are exact; "1.013 bar" has four significant figures.

Solution

$$V = 2.36 \times 10^4 \text{ m}^3 \times \frac{1 \text{ L}}{10^{-3} \text{ m}^3} = \boxed{2.36 \times 10^7 \text{ L}}$$

$$n_{He} = 4.68 \times 10^6 \text{ g He} \times \frac{1 \text{ mol He}}{4.003 \text{ g He}} = \boxed{1.17 \times 10^6 \text{ mol He}}$$

$$T = 18 + 273 = \boxed{291 \text{ K}}$$

$$P = 1.20 \text{ bar} \times \frac{1 \text{ atm}}{1.013 \text{ bar}} = \boxed{1.18 \text{ atm}}$$

$$P = 1.20 \text{ bar} \times \frac{760 \text{ mm Hg}}{1.013 \text{ bar}} = \boxed{9.00 \times 10^2 \text{ mm Hg}}$$

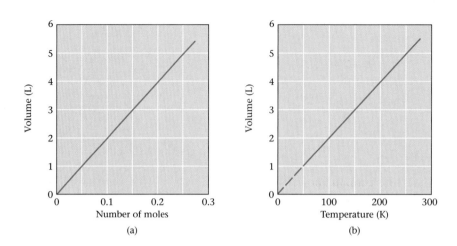

FIGURE 5.3

Relation of gas volume (V) to number of moles (n) and temperature (T) at constant pressure (P). The volume of a gas at constant pressure is directly proportional to (a) the number of moles of gas and (b) the absolute temperature. The volume-temperature plot must be extrapolated to reach zero because most gases liquefy at low temperatures well above 0 K. **OHT**

5.2 THE IDEAL GAS LAW

All gases closely resemble each other in the dependence of volume on amount, temperature, and pressure.

1. Volume is directly proportional to amount. Figure 5.3a shows a typical plot of volume (V) versus number of moles (n) for a gas. Notice that the graph is a straight line passing through the origin. The general equation for such a plot is

$$V = k_1 n \qquad (\text{constant } T, P)$$

where k_1 is a constant; that is, it is independent of individual values of V and n and of the nature of the gas. This is the equation of a direct proportionality.

2. Volume is directly proportional to absolute temperature. The dependence of volume (V) on the Kelvin temperature (T) is shown in Figure 5.3b. The graph is a straight line through the origin. The equation of the line is

$$V = k_2 T \qquad (\text{constant } n, P)$$

where k_2 is a constant. This relationship was first suggested, in a different form, by two French scientists, Jacques Charles (1746–1823) and Joseph Gay-Lussac (1778–1850), both of whom were balloonists. Hence it is often referred to as the law of Charles and Gay-Lussac or, simply, as *Charles's law.*

3. Volume is inversely proportional to pressure. Figure 5.4 shows a typical plot of volume (V) versus pressure (P). Notice that V decreases as P increases. The graph is a hyperbola. The general relation between the two variables is

$$V = k_3/P \qquad (\text{constant } n, T)$$

The quantity k_3, like k_1 and k_2, is a constant. This is the equation of an inverse proportionality. The fact that volume is inversely proportional to pressure was first established in 1660 by Robert Boyle, an Irish experimental scientist. The equation above is one form of *Boyle's law.*

The three equations relating the volume, pressure, temperature, and amount of a gas can be combined into a single equation. Because V is directly proportional to both n and T,

$$V = k_1 n \qquad V = k_2 T$$

⊙ See Screens 12.3, Gas Laws, & 12.4, The Ideal Gas Law.

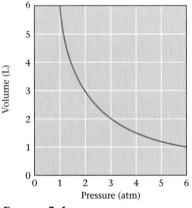

FIGURE 5.4

Relation of gas volume (V) to pressure (P) at constant temperature (T). The volume of a fixed quantity of gas at constant temperature is inversely proportional to the pressure. In this case, the volume decreases from 6 L to 1 L when the pressure increases from 1 atm to 6 atm. **OHT**

⌐ Gay-Lussac isolated boron and was the first to prepare HF.

and inversely proportional to P,

$$V = \frac{k_3}{P}$$

it follows that

$$V = \text{constant} \times \frac{n \times T}{P}$$

We can evaluate the constant ($k_1 k_2 k_3$) in this equation by taking advantage of *Avogadro's law,* which states that equal volumes of all gases at the same temperature and pressure contain the same number of moles. For this law to hold, the constant must be the same for all gases. Ordinarily it is represented by the symbol **R**. Both sides of the equation are multiplied by P to give the **ideal gas law**

$$\boxed{PV = nRT}$$

where P is the pressure, V the volume, n the number of moles, and T the Kelvin temperature. Experimentally, it is found that the ideal gas law predicts remarkably well the experimental behavior of real gases (e.g., H_2, N_2, O_2 . . .) at ordinary temperatures and pressures.

The value of the gas constant R can be calculated from experimental values of P, V, n, and T. Consider, for example, the situation that applies at 0°C and 1 atm. These conditions are often referred to as **standard temperature and pressure (STP)** for a gas. At STP, one mole of any gas occupies a volume of 22.4 L. Solving the ideal gas law for R:

$$R = \frac{PV}{nT}$$

Substituting $P = 1.00$ atm, $V = 22.4$ L, $n = 1.00$ mol, and $T = 0 + 273 = 273$ K,

$$R = \frac{1.00 \text{ atm} \times 22.4 \text{ L}}{1.00 \text{ mol} \times 273 \text{ K}} = 0.0821 \text{ L} \cdot \text{atm}/(\text{mol} \cdot \text{K})$$

Notice that R has the units of atmospheres, liters, moles, and K. These units must be used for pressure, volume, amount, and temperature in any problem in which this value of R is used.

Throughout most of this chapter, we will use 0.0821 L·atm/(mol·K) as the value of R. For certain purposes, however, R must be expressed in different units (Table 5.1).

Molar volume. The cube has a volume of 22.4 L, which is the volume of one mole at STP. *(Charles D. Winters)*

R is a constant, independent of P, V, n, and T, but its numerical value depends on the units used.

	TABLE 5.1	**Values of R in Different Units**	
Value		**Where Used**	**How Obtained**
$0.0821 \dfrac{\text{L} \cdot \text{atm}}{\text{mol} \cdot \text{K}}$		Gas law problems with V in liters, P in atm	From known values of P, V, T, n
$8.31 \dfrac{\text{J}}{\text{mol} \cdot \text{K}}$		Equations involving energy in joules	$1 \text{ L} \cdot \text{atm} = 101.3 \text{ J}$
$8.31 \times 10^3 \dfrac{\text{g} \cdot \text{m}^2}{\text{s}^2 \cdot \text{mol} \cdot \text{K}}$		Calculation of molecular speed (page 130)	$1 \text{ J} = 10^3 \dfrac{\text{g} \cdot \text{m}^2}{\text{s}^2}$

5.3 GAS LAW CALCULATIONS

The ideal gas law can be used to solve a variety of problems. We will show how you can use it to find—

- the final state of a gas, knowing its initial state and the changes in P, V, n, or T that occur.
- one of the four variables, P, V, n, or T, given the values of the other three.
- the molar mass or density of a gas.

Final and Initial State Problems

A gas commonly undergoes a change from an initial to a final state. Typically, you are asked to determine the effect on V, P, n, or T of a change in one or more of these variables. For example, starting with a sample of gas at 25°C and 1.00 atm, you might be asked to calculate the pressure developed when the sample is heated to 95°C at constant volume.

The ideal gas law is readily applied to problems of this type. A relationship between the variables involved is derived from this law. In this case, pressure and temperature change, while n and V remain constant.

$$\text{initial state:} \qquad P_1 V = nRT_1$$

$$\text{final state:} \qquad P_2 V = nRT_2$$

> To obtain a two-point equation, write the gas law twice and divide to eliminate constants.

Dividing the second equation by the first cancels V, n, and R, leaving the relation

$$P_2/P_1 = T_2/T_1 \qquad \text{(constant } n, V)$$

Applying this general relation to the problem just described,

$$P_2 = P_1 \times \frac{T_2}{T_1} = 1.00 \text{ atm} \times \frac{368 \text{ K}}{298 \text{ K}} = 1.23 \text{ atm}$$

Similar "two-point" equations can be derived from the ideal gas law to solve any problem of this type.

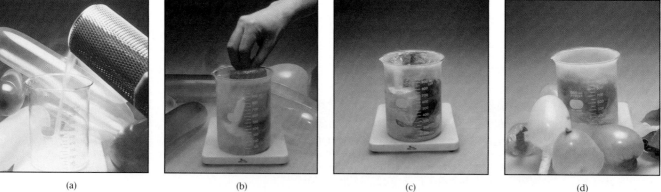

(a)   (b)   (c)   (d)

An illustration of Charles's law. (a) Liquid nitrogen is poured into a beaker. (b) Inflated balloons are pushed into the very cold liquid ($T = 77$ K). (c) The volume of the air in the balloons, as predicted by Charles's law, decreases; the balloons all fit into the beaker. (d) When removed from the liquid nitrogen, the volume of the air in the balloons expands as the air warms up. *(Charles D. Winters)*

EXAMPLE 5.2   A 250-mL flask, open to the atmosphere, contains 0.0110 mol of air at 0°C. On heating, part of the air escapes; how much remains in the flask at 100°C? (3 sig. fig.)

Strategy   The key to working this problem is to realize that *pressure and volume remain constant*. The pressure is that of the atmosphere; the volume is that of the flask, 250 mL. Look for a two-point relation between n and T at constant P and V. Then substitute for n_1, T_1, and T_2; solve for n_2.

Solution

$$\text{initial state:} \quad n_1 T_1 = PV/R$$

$$\text{final state:} \quad n_2 T_2 = PV/R$$

It follows that

$$n_2 T_2 = n_1 T_1 \qquad n_2 = n_1 \times \frac{T_1}{T_2}$$

Because $n_1 = 0.0110$ mol; $T_1 = 0 + 273 = 273$ K; and $T_2 = 100 + 273 = 373$ K,

$$n_2 = 0.0110 \text{ mol} \times \frac{273 \text{ K}}{373 \text{ K}} = \boxed{0.00805 \text{ mol}}$$

Reality Check   We conclude that 0.00805/0.0110, or about 73%, of the air remains in the flask at 100°C. About 27% of the air escapes when the temperature is raised.

Calculation of P, V, n, or T

Frequently, values are known for three of these quantities (perhaps V, n, and T); the other one (P) must be calculated. This is readily done by direct substitution into the ideal gas law.

EXAMPLE 5.3   The ozone-friendly compound now used as a refrigerant in car air conditioners has the molecular formula $C_2F_4H_2$. If 2.50 g of this compound is introduced into an evacuated 500.0-mL container at 10°C (2 sig. fig.), what pressure in atmospheres is developed?

Freon, CF_2Cl_2, formerly was used.

Strategy   Substitute directly into the ideal gas law and solve for P. Note that V, n, and T have to be in units consistent with $R = 0.0821$ L·atm/(mol·K).

Solution   Converting to the appropriate units,

$$V = 500.0 \text{ mL} \times \frac{1 \text{ L}}{1000 \text{ mL}} = 0.5000 \text{ L}$$

$$T = 10 + 273 = 283 \text{ K}$$

The molar mass of $C_2F_4H_2$ is 102.04 g/mol. Hence

$$n = 2.50 \text{ g } C_2F_4H_2 \times \frac{1 \text{ mol } C_2F_4H_2}{102.04 \text{ g } C_2F_4H_2} = 0.0245 \text{ mol}$$

Substituting into the ideal gas law:

$$P = \frac{nRT}{V} = \frac{0.0245 \text{ mol} \times 0.0821 \text{ L·atm/(mol·K)} \times 283 \text{ K}}{0.5000 \text{ L}} = 1.14 \text{ atm}$$

Molar Mass and Density

⊙ **See Screen 12.6, Using Gas Laws:**
Determining Molar Mass.

The ideal gas law offers a simple approach to the experimental determination of the molar mass of a gas. Indeed, this approach can be applied to volatile liquids like acetone (Example 5.4). All you need to know is the mass of a sample confined to a container of fixed volume at a particular temperature and pressure.

EXAMPLE 5.4   Acetone is widely used as a nail polish remover. A sample of liquid acetone is placed in a 3.00-L flask and vaporized by heating to 95°C at 1.02 atm. The vapor filling the flask at this temperature and pressure weighs 5.87 g. Calculate the molar mass of acetone.

Strategy   Perhaps the simplest approach is to use the ideal gas law to solve for n (V, T, and P are known). When the number of moles and the mass in grams, m, are known, the molar mass $\mathcal{M}$ is readily calculated from the relation

$$m = \mathcal{M} \times n$$

Solution

$$n = \frac{PV}{RT} = \frac{(1.02\text{ atm})(3.00\text{ L})}{(0.0821\text{ L}\cdot\text{atm/mol}\cdot\text{K})(368\text{ K})} = 0.101\text{ mol}$$

Substituting in the equation $m = \mathcal{M} \times n$: $5.87\text{ g} = \mathcal{M} \times 0.101\text{ mol}$
Solving for $\mathcal{M}$: $\mathcal{M} = 5.87\text{ g}/0.101\text{ mol} = \boxed{58.1\text{ g/mol}}$

Reality Check   Acetone has the structural formula $(CH_3)_2CO$; its molar mass is $[2(15.0) + 12.0 + 16.0]\text{g/mol} = 58.0\text{ g/mol}$.

Densities of gases, ordinarily expressed in grams per liter, can be calculated by an approach very similar to that followed in Example 5.4.

EXAMPLE 5.5   Taking the molar mass of dry air to be 29.0 g/mol, calculate the density of air at 27°C and 1.00 atm.

Strategy   Use the ideal gas law to calculate the number of moles in *one liter* of air at 27°C and 1.00 atm. Then use the relation

$$m = \mathcal{M} \times n$$

to find the mass of one liter of air and hence its density.

⌐ We might call 0.0406 mol/L the
molar density.

Solution

$$n = \frac{PV}{RT} = \frac{(1.00\text{ atm})(1.00\text{ L})}{(0.0821\text{ L}\cdot\text{atm/mol}\cdot\text{K})(300\text{ K})} = 0.0406\text{ mol}$$

$$m = (29.0\text{ g/mol})(0.0406\text{ mol}) = 1.18\text{ g}$$

$$\text{density} = \frac{1.18\text{ g}}{1.00\text{ L}} = \boxed{1.18\text{ g/L}}$$

Reality Check   Gases at room temperature and atmospheric pressure typically have densities of 0.1 to 10 g/L, so this answer is reasonable.

(a)   (b)

Gas density. (a) The blue balloon contains hydrogen, which is less dense than air because it has a lower molar mass than any gas in the air. The red balloon contains argon, which is more dense than air because it has a higher molar mass than air. (b) Gas density decreases with rising temperature, allowing the balloons to rise, as shown in this time exposure taken at the Great Reno Balloon Race. *(a, Charles D. Winters; b, Ken M. Jones/Photo Researchers, Inc.)*

Another approach to calculating gas densities is to use the ideal gas law to derive the general relation

$$\text{density} = \frac{\mathcal{M}P}{RT}$$

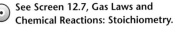

See Screen 12.5, Gas Density.

From this equation we see that the density of a gas is dependent on—

- *pressure.* Compressing a gas increases its density.
- *temperature.* Hot air rises because a gas becomes less dense when its temperature is increased. Hot air balloons are filled with air at a temperature higher than that of the atmosphere, making the air inside less dense than that outside. First used in France in the eighteenth century, they are now seen in balloon races and other sporting events. Heat is supplied on demand by a propane burner.
- *molar mass.* Hydrogen ($\mathcal{M} = 2.016$ g/mol) has the lowest molar mass and the lowest density (at a given P and T) of all gases. Hence it has the greatest lifting power in "lighter-than-air" balloons. However, hydrogen has not been used in balloons carrying passengers since 1937, when the *Hindenburg*, a hydrogen-filled airship, exploded and burned. Helium ($\mathcal{M} = 4.003$ g/mol) is slightly less effective than hydrogen but a lot safer to work with because it is nonflammable. It is used in a variety of balloons, ranging from the small ones used at parties to meteorological balloons with volumes of a billion liters.

5.4 STOICHIOMETRY OF GASEOUS REACTIONS

As pointed out in Chapter 3, a balanced equation can be used to relate moles or grams of substances taking part in a reaction. Where gases are involved, these relations can be extended to include volumes. To do this, we use the ideal gas law and the conversion factor approach described in Chapter 3.

See Screen 12.7, Gas Laws and Chemical Reactions: Stoichiometry.

This is the most important application of the ideal gas law to chemistry.

EXAMPLE 5.6

Hydrogen peroxide is the active ingredient in commercial preparations for bleaching hair. What mass of hydrogen peroxide must be used to produce 1.00 L of oxygen gas at 25°C and 1.00 atm? The equation for the reaction is

$$2H_2O_2(aq) \longrightarrow O_2(g) + 2H_2O$$

Strategy First convert volume of $O_2(g)$ to moles using the ideal gas law. Then find the mass of H_2O_2 by the conversion factor approach described in Chapter 3; the path is $n_{O_2} \rightarrow n_{H_2O_2} \rightarrow m_{H_2O_2}$.

Solution

(1) $n_{O_2} = \dfrac{PV}{RT} = \dfrac{1.00 \text{ atm} \times 1.00 \text{ L}}{0.0821 \text{ L} \cdot \text{atm/(mol} \cdot \text{K)} \times 298 \text{ K}} = 0.0409 \text{ mol } O_2$

(2) mass of $H_2O_2 = 0.0409 \text{ mol } O_2 \times \dfrac{2 \text{ mol } H_2O_2}{1 \text{ mol } O_2} \times \dfrac{34.02 \text{ g } H_2O_2}{1 \text{ mol } H_2O_2} = \boxed{2.78 \text{ g } H_2O_2}$

Reality Check A gas at room temperature and atmospheric pressure has a molar volume of about 25 L, so we have about $1/25 = 0.04$ mol of O_2. According to the chemical equation, we need about 0.08 mol of H_2O_2, which is a little less than 3 g. Check!

In going a mile, a typical engine takes in 1000 L of air.

EXAMPLE 5.7

Octane, C_8H_{18}, is one of the hydrocarbons in gasoline. On combustion (burning in oxygen), octane produces carbon dioxide and water. How many liters of oxygen, measured at 0.974 atm and 24°C, are required to burn 1.00 g of octane?

Strategy First write a balanced equation for the reaction. Then calculate the number of moles of oxygen required to burn 1.00 g of C_8H_{18} ($\mathcal{M} = 114.22$ g/mol). Finally, use the ideal gas law to find the volume of oxygen.

Solution

(1) The balanced equation for the reaction is

$$2C_8H_{18}(l) + 25 \, O_2(g) \longrightarrow 16CO_2(g) + 18H_2O(l)$$

(2) $n_{O_2} = 1.00 \text{ g } C_8H_{18} \times \dfrac{1 \text{ mol } C_8H_{18}}{114.22 \text{ g } C_8H_{18}} \times \dfrac{25 \text{ mol } O_2}{2 \text{ mol } C_8H_{18}} = 0.109 \text{ mol } O_2$

(3) $V_{O_2} = \dfrac{nRT}{P} = \dfrac{0.109 \text{ mol} \times 0.0821 \text{ L} \cdot \text{atm/(mol} \cdot \text{K)} \times 297 \text{ K}}{0.974 \text{ atm}} = \boxed{2.73 \text{ L}}$

FIGURE 5.5
Electrolysis of water (H_2O). The volume of hydrogen (H_2) formed in the tube at the left is twice the volume of oxygen (O_2) formed in the tube at the right, in accordance with the equation, $2H_2O(l) \rightarrow 2H_2(g) + O_2(g)$.
(Charles D. Winters)

Perhaps the first stoichiometric relationship to be discovered was the *law of combining volumes*, proposed by Gay-Lussac in 1808: **The volume ratio of any two gases in a reaction at constant temperature and pressure is the same as the reacting mole ratio.**

To illustrate the law, consider the reaction

$$2H_2O(l) \longrightarrow 2H_2(g) + O_2(g)$$

As you can see from Figure 5.5, the volume of hydrogen produced is twice that of the other gaseous product, oxygen.

The law of combining volumes, like so many relationships involving gases, is readily explained by the ideal gas law. At constant temperature and pressure, volume is directly proportional to number of moles ($V = k_1 n$). It follows that for gaseous species involved in reactions, the volume ratio must be the same as the mole ratio given by the coefficients of the balanced equation.

CHEMISTRY · The Human Side

Avogadro's law, referred to on page 119, was proposed in 1811 by an Italian physicist at the University of Turin with the improbable name of Lorenzo Romano Amadeo Carlo Avogadro di Quarequa e di Cerreto (1776–1856).

Avogadro suggested this relationship to explain the law of combining volumes. Today it seems obvious. For example, in the reaction

$$2H_2O(l) \longrightarrow 2H_2(g) + O_2(g)$$

the volume of hydrogen, like the number of moles, is twice that of oxygen. Hence, equal volumes of these gases must contain the same number of moles or molecules. This was by no means obvious to Avogadro's contemporaries. Berzelius, among others, dismissed Avogadro's ideas because he did not believe diatomic molecules composed of identical atoms (H_2, O_2) could exist. Dalton went a step further; he refused to accept the law of combining volumes because he thought it implied splitting atoms.

As a result of arguments like these, Avogadro's ideas lay dormant for nearly half a century. They were revived by another Italian scientist, Stanislao Cannizzaro, professor of chemistry at the University of Genoa. At a conference held in Karlsruhe in 1860, he persuaded the chemistry community of the validity of Avogadro's law and showed how it could be used to determine molar and atomic masses.

The quantity now called "Avogadro's number" (6.02×10^{23}/mol) was first estimated in 1865, nine years after Avogadro died. Not until well into the twentieth century did it acquire its present name. It seems appropriate to honor Avogadro in this way for the contributions he made to chemical theory.

AMADEO AVOGADRO
(1776–1856)

(Photo credit: Courtesy of E. F. Smith Memorial Collection, Van Pelt Library, University of Pennsylvania)

5.5   GAS MIXTURES: PARTIAL PRESSURES AND MOLE FRACTIONS

Because the ideal gas law applies to all gases, you might expect it to apply to gas mixtures. Indeed it does. For a mixture of two gases A and B, the total pressure is given by the expression

$$P_{tot} = n_{tot} \frac{RT}{V} = (n_A + n_B) \frac{RT}{V}$$

Separating the two terms on the right,

$$P_{tot} = n_A \frac{RT}{V} + n_B \frac{RT}{V}$$

The terms $n_A RT/V$ and $n_B RT/V$ are, according to the ideal gas law, the pressures that gases A and B would exert if they were alone. These quantities are referred to as **partial pressures,** P_A and P_B.

$$P_A = \text{partial pressure A} = n_A RT/V$$

$$P_B = \text{partial pressure B} = n_B RT/V$$

See Screen 12.8, Gas Mixtures and Partial Pressures.

Partial pressure is the pressure a gas would exert if it occupied the entire volume by itself.

Collecting a gas by water displacement. Hydrogen gas is being generated in the flask at the left by an acid solution dripping onto a metal. When a gas is collected by displacing water, it becomes saturated with water vapor. The partial pressure of $H_2O(g)$ in the collecting flask is equal to the vapor pressure of liquid water at the temperature of the system. *(Marna G. Clarke)*

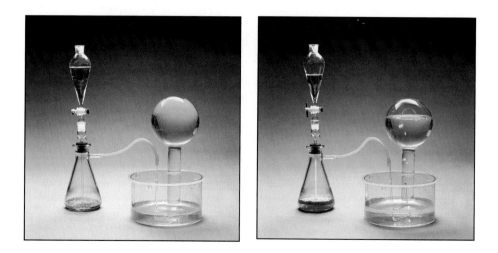

Substituting P_A and P_B for $n_A RT/V$ and $n_B RT/V$ in the equation for P_{tot},

$$P_{tot} = P_A + P_B$$

The relation just derived was first proposed by John Dalton in 1801; it is often referred to as **Dalton's law** of partial pressures:

The total pressure of a gas mixture is the sum of the partial pressures of the components of the mixture.

To illustrate Dalton's law, consider a gaseous mixture of hydrogen and helium in which

$$P_{H_2} = 2.46 \text{ atm} \qquad P_{He} = 3.69 \text{ atm}$$

It follows from Dalton's law that

$$P_{tot} = 2.46 \text{ atm} + 3.69 \text{ atm} = 6.15 \text{ atm}$$

Wet Gases; Partial Pressure of Water

When a gas such as hydrogen is collected by bubbling through water (Figure 5.6), it picks up water vapor; molecules of H_2O escape from the liquid and enter the gas phase. Dalton's law can be applied to the resulting gas mixture:

$$P_{tot} = P_{H_2O} + P_{H_2}$$

In this case, P_{tot} is the measured pressure. The partial pressure of water vapor, P_{H_2O}, is equal to the **vapor pressure** of liquid water. It has a fixed value at a given temperature (see Appendix 1). The partial pressure of hydrogen, P_{H_2}, can be calculated by subtraction. The number of moles of hydrogen in the wet gas, n_{H_2}, can then be determined using the ideal gas law.

EXAMPLE 5.8 A student prepares a sample of hydrogen gas by electrolyzing water at 25°C. She collects 152 mL of H_2 at a total pressure of 758 mm Hg. Using Appendix 1 to find the vapor pressure of water, calculate

(a) the partial pressure of hydrogen.

(b) the number of moles of hydrogen collected.

Strategy

(a) Use Dalton's law to find the partial pressure of hydrogen, P_{H_2}.

(b) Use the ideal gas law to calculate n_{H_2}, with P_{H_2} as the pressure.

Solution

(a) From Appendix 1, P_{H_2O} = 23.76 mm Hg at 25°C. The total pressure, P_{tot}, is 758 mm Hg.

$$P_{H_2} = P_{tot} - P_{H_2O} = 758 \text{ mm Hg} - 23.76 \text{ mm Hg} = \boxed{734 \text{ mm Hg}}$$

(b) $n_{H_2} = \dfrac{(P_{H_2})V}{RT} = \dfrac{(734/760 \text{ atm})(0.152 \text{ L})}{(0.0821 \text{ L·atm/mol·K})(298 \text{ K})} = \boxed{0.00600 \text{ mol } H_2}$

Vapor pressure, like density and solubility, is an intensive physical property that is characteristic of a particular substance. The vapor pressure of water at 25°C is 23.76 mm Hg, independent of volume or the presence of another gas. Like density and solubility, vapor pressure varies with temperature; for water it is 55.3 mm Hg at 40°C, 233.7 mm Hg at 70°C, and 760.0 mm Hg at 100°C. We will have more to say in Chapter 9 about the temperature dependence of vapor pressure.

Partial Pressure and Mole Fraction

As pointed out earlier, the following relationship applies to a mixture containing gas A (and gas B):

$$P_A = \frac{n_A RT}{V} \qquad P_{tot} = \frac{n_{tot} RT}{V}$$

Dividing P_A by P_{tot} gives

$$\frac{P_A}{P_{tot}} = \frac{n_A}{n_{tot}}$$

The fraction n_A/n_{tot} is referred to as the **mole fraction** of A in the mixture. It is the fraction of the total number of moles that is accounted for by gas A. Using X_A to represent the mole fraction of A (i.e., $X_A = n_A/n_{tot}$),

$$\boxed{P_A = X_A P_{tot}}$$

In other words, *the partial pressure of a gas in a mixture is equal to its mole fraction multiplied by the total pressure.* This relation is commonly used to calculate partial pressures of gases in a mixture when the total pressure and the composition of the mixture are known (Example 5.9).

If a mixture contains equal numbers of A and B molecules, $X_A = X_B = 0.50$ and $P_A = P_B = \frac{1}{2}P_{tot}$.

EXAMPLE 5.9

Chemical analysis of dry air shows that the mole fractions of nitrogen, oxygen, and argon are 0.781, 0.210, and 0.009, respectively. Calculate the partial pressure of each of these gases on a day when the barometric pressure is 747 mm Hg.

Strategy   The barometric pressure is the total pressure; use the equation $P_A = X_A P_{tot}$ to find the partial pressures.

Solution

$$P_{N_2} = 0.781 \times 747 \text{ mm Hg} = \boxed{583 \text{ mm Hg}}$$

$$P_{O_2} = 0.210 \times 747 \text{ mm Hg} = \boxed{157 \text{ mm Hg}}$$

$$P_{Ar} = 0.009 \times 747 \text{ mm Hg} = \boxed{7 \text{ mm Hg}}$$

Reality Check The partial pressures (583 mm Hg, 157 mm Hg, 7 mm Hg) add up to the total pressure, 747 mm Hg, as they should according to Dalton's law. They are also in the same ratio as the mole fractions (0.781, 0.210, 0.009).

5.6 KINETIC THEORY OF GASES

See Screens 12.9, The Kinetic Molecular Theory of Gases, & 12.10, Gas Laws and The Kinetic Molecular Theory.

The fact that the ideal gas law applies to all gases indicates that the gaseous state is a relatively simple one from a molecular standpoint. Gases must have certain common properties that cause them to follow the same natural law. Between about 1850 and 1880, James Maxwell, Rudolf Clausius, Ludwig Boltzmann, and others developed the **kinetic theory** of gases. They based it on the idea that all gases behave similarly as far as particle motion is concerned.

Molecular Model

By the use of the kinetic theory, it is possible to derive or explain the experimental behavior of gases. To do this we start with a simple molecular model which assumes that—

- *gases are mostly empty space.* The total volume of the molecules is negligibly small compared with that of the container to which they are confined.
- *gas molecules are in constant, chaotic motion.* They collide frequently with one another and with the container walls. As a result, their velocities are constantly changing.
- *collisions are elastic.* There are no attractive forces that would tend to make molecules "stick" to one another or to the container walls.
- *gas pressure is caused by collisions of molecules with the walls of the container* (Figure 5.7). As a result, pressure increases with the energy and frequency of these collisions.

In air, a molecule undergoes about 10 billion collisions per second.

Expression for Pressure, P

Applying the laws of physics to this simple model, it can be shown that the pressure (P) exerted by a gas in a container of volume V is

$$P = \frac{Nmu^2}{3V}$$

where N is the number of molecules, m is the mass of a molecule, and u is the average speed.* We will not attempt to derive this equation. However, it makes sense, at least qualitatively. In particular—

- the ratio N/V expresses the concentration of gas molecules in the container. The

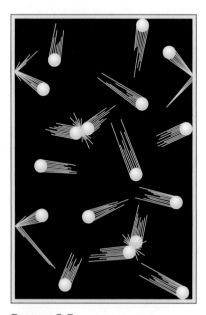

FIGURE 5.7
The kinetic molecular model of a gas. Gas molecules are in constant motion and their collisions are elastic. Collisions with the walls cause gas pressure. **OHT**

*More rigorously, u^2 is the average of the squares of the speeds of all molecules.

more molecules there are in a given volume, the greater the collision frequency so the greater the pressure.

■ the product mu^2 is a measure of the energy of collision. (When a Cadillac traveling at 100 mph collides with a brick wall, the energy transferred is much greater than would be obtained with a bicycle at 5 mph.) Hence, as this equation predicts, pressure is directly related to mu^2.

Average Kinetic Energy of Translational Motion, E_t

The kinetic energy, E_t, of a gas molecule of mass m moving at speed u is

$$E_t = \frac{mu^2}{2}$$

From the equation written above for P, we see that $mu^2 = 3PV/N$. Hence

$$E_t = \frac{3PV}{2N}$$

But the ideal gas law tells us that $PV = nRT$, so

$$E_t = \frac{3nRT}{2N}$$

This equation can be simplified by noting that the number of molecules, N, in a gas sample is equal to the number of moles, n, multiplied by Avogadro's number, N_A, that is, $N = n \times N_A$. Making this substitution in the above equation and simplifying, we obtain the final expression for the **average translational kinetic energy** of a gas molecule

$$E_t = \frac{3RT}{2N_A}$$

This equation contains three constants (3/2, R, N_A) and only one variable, the temperature T. It follows that—

■ *at a given temperature, molecules of different gases (e.g., H_2, O_2, . . .) must all have the same average kinetic energy of translational motion.*

■ *the average translational kinetic energy of a gas molecule is directly proportional to the Kelvin temperature, T.*

⌐ A baseball in motion has translational energy.

Average Speed, u

By equating the first and last expressions written above for E_t, we see that

$$\frac{mu^2}{2} = \frac{3RT}{2N_A}$$

Solving this equation for u^2:

$$u^2 = \frac{3RT}{mN_A}$$

⌐ E_t depends only upon T; u depends on both T and $\mathcal{M}$.

But the product m (the mass of a molecule) times N_A (the number of molecules in a mole) is simply the molar mass $\mathcal{M}$, so we can write

$$u^2 = \frac{3RT}{\mathcal{M}}$$

FIGURE 5.8
Relation of molecular speed to molar mass. When ammonia gas, which is injected into the left arm of the tube, comes in contact with hydrogen chloride, which is injected into the right arm of the tube, they react to form solid ammonium chloride: $NH_3(g) + HCl(g) \rightarrow NH_4Cl(s)$. Because NH_3 (M = 17 g/mol) moves faster than HCl (M = 36.5 g/mol), the ammonium chloride forms closer to the HCl end of the tube. *(Marna G. Clarke)*

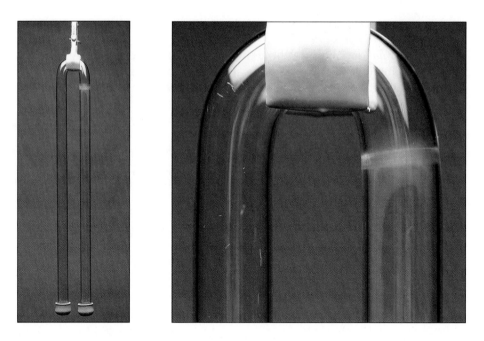

Taking the square root of both sides of this equation, we arrive at the final expression for the **average speed, u,**

$$u = \left(\frac{3RT}{M} \right)^{1/2}$$

From this relation you can see that the average speed u is—

■ *directly proportional to the square root of the absolute temperature.* For a given gas at two different temperatures, T_2 and T_1, the quantity M is constant, and we can write

$$\frac{u_2}{u_1} = \left(\frac{T_2}{T_1} \right)^{1/2}$$

■ *inversely proportional to the square root of molar mass* (M) (Figure 5.8). For two different gases A and B at the same temperature (T constant):

$$\frac{u_B}{u_A} = \left(\frac{M_A}{M_B} \right)^{1/2}$$

EXAMPLE 5.10   Calculate the average speed, u, of an N_2 molecule at 25°C.

Strategy   Use the equation $u = (3RT/M)^{1/2}$; remember to use the proper value of $R = 8.31 \times 10^3$ g·m²/(s²·mol·K). Be careful about units!

Solution

$$u = \left(\frac{3RT}{M} \right)^{1/2} = \left[\frac{3 \times 8.31 \times 10^3 \dfrac{\text{g·m}^2}{\text{s}^2 \cdot \text{mol} \cdot \text{K}} \times 298 \text{ K}}{28.02 \dfrac{\text{g}}{\text{mol}}} \right]^{1/2} = \boxed{515 \text{ m/s}}$$

Reality Check   Notice that the average speed is very high. In miles per hour it is

$$515 \frac{\text{m}}{\text{s}} \times \frac{1 \text{ mi}}{1.609 \times 10^3 \text{ m}} \times 3.600 \times 10^3 \frac{\text{s}}{\text{hr}} = 1.15 \times 10^3 \text{ mi/hr}$$

Effusion of Gases; Graham's Law

One way to check the validity of calculations made from kinetic theory is to study the process of **effusion,** the flow of gas particles through tiny pores or pinholes. The relative rates of effusion of different gases depend on two factors: the pressures of the gases and the relative speeds of their particles. If two different gases A and B are compared at the same pressure, only their speeds are of concern, and

$$\frac{\text{rate of effusion B}}{\text{rate of effusion A}} = \frac{u_B}{u_A}$$

where u_A and u_B are average speeds. As pointed out earlier, at a given temperature,

$$\frac{u_B}{u_A} = \left(\frac{\mathcal{M}_A}{\mathcal{M}_B}\right)^{1/2}$$

It follows that, at constant pressure and temperature,

$$\boxed{\frac{\text{rate of effusion B}}{\text{rate of effusion A}} = \left(\frac{\mathcal{M}_A}{\mathcal{M}_B}\right)^{1/2}}$$

This relation in a somewhat different form was discovered experimentally by the Scottish chemist Thomas Graham in 1829. Graham was interested in a wide variety of chemical and physical problems, among them the separation of the components of air. **Graham's law** can be stated as

At a given temperature and pressure, the rate of effusion of a gas, in moles per unit time, is inversely proportional to the square root of its molar mass.

Graham's law tells us qualitatively that light molecules effuse more rapidly than heavier ones (Figure 5.8). In quantitative form, it allows us to determine molar masses of gases (Example 5.11).

EXAMPLE 5.11   In an effusion experiment, argon gas is allowed to expand through a tiny opening into an evacuated flask of volume 120 mL for 32.0 s, at which point the pressure in the flask is found to be 12.5 mm Hg. This experiment is repeated with a gas X of unknown molar mass at the same T and P. It is found that the pressure in the flask builds up to 12.5 mm Hg after 48.0 s. Calculate the molar mass of X.

Strategy   The key to solving this problem is to realize that because P, T, and V are the same in the two experiments, *the number of moles of Ar and X effusing into the flask is the same:*

$$n_{Ar} = n_X = n$$

Solution   Comparing the two rates in moles per second:

$$\frac{\text{rate Ar}}{\text{rate X}} = \frac{n/32.0\ s}{n/48.0\ s} = \frac{48.0}{32.0} = 1.50$$

Applying Graham's law

$$1.50 = \left(\frac{\mathcal{M}_X}{\mathcal{M}_{Ar}}\right)^{1/2}$$

Solving by squaring both sides and substituting $\mathcal{M}_{Ar} = 39.95$ g/mol

$$\mathcal{M}_X = (39.95\ \text{g/mol}) \times (1.50)^2 = \boxed{89.9\ \text{g/mol}}$$

Reality Check   Because it takes longer for gas X to effuse, it must have a larger molar mass than argon. It does!

 See Screen 12.12, Application of the Kinetic Molecular Theory.

High molar mass   Low molar mass

Vacuum chamber

Effusion of gases. A gas with a higher molar mass (yellow spheres) effuses into a vacuum more slowly than a gas with a lower molar mass (green spheres).

A practical application of Graham's law arose during World War II, when scientists were studying the fission of uranium atoms as a source of energy. It became necessary to separate $^{235}_{92}U$, which is fissionable, from the more abundant isotope of uranium, $^{238}_{92}U$, which is not fissionable. Because the two isotopes have almost identical chemical properties, chemical separation was not feasible. Instead, an effusion process was worked out using uranium hexafluoride, UF_6. This compound is a gas at room temperature and low pressures. Preliminary experiments indicated that $^{235}_{92}UF_6$ could indeed be separated from $^{238}_{92}UF_6$ by effusion. The separation factor is very small, because the rates of effusion of these two species are nearly equal:

$$\frac{\text{rate of effusion of } ^{235}_{92}UF_6}{\text{rate of effusion of } ^{238}_{92}UF_6} = \left(\frac{352.0}{349.0}\right)^{1/2} = 1.004$$

so a great many repetitive separations are necessary. An enormous plant was built for this purpose in Oak Ridge, Tennessee. In this process, UF_6 effuses many thousands of times through porous barriers. The lighter fractions move on to the next stage while heavier fractions are recycled through earlier stages. Eventually, a nearly complete separation of the two isotopes is achieved.

Distribution of Molecular Speeds

 See Screen 12.11, Distribution of Molecular Speeds.

As shown in Example 5.10, the average speed of an N_2 molecule at 25°C is 515 m/s; that of H_2 is even higher, 1920 m/s. However, not all molecules in these gases have these speeds. The motion of particles in a gas is utterly chaotic. In the course of a second, a particle undergoes millions of collisions with other particles. As a result, the speed and direction of motion of a particle are constantly changing. Over a period of time, the speed will vary from almost zero to some very high value, considerably above the average.

In 1860 James Clerk Maxwell, a Scottish physicist and one of the greatest theoreticians the world has ever known, showed that different possible speeds are distributed among particles in a definite way. Indeed, he developed a mathematical expression for this distribution. His results are shown graphically in Figure 5.9 for O_2 at 25°C and 1000°C. On the graph, the relative number of molecules having a certain speed is plotted against that speed. At 25°C, this number increases rapidly with the speed, up to a maximum of about 400 m/s. This is the most probable

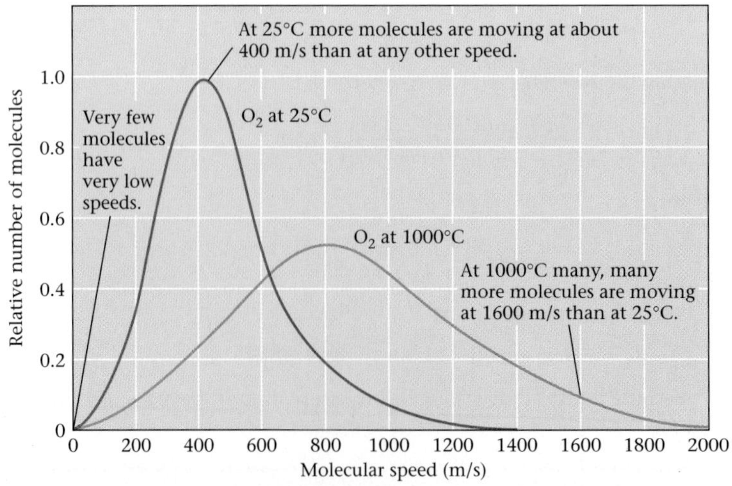

> In a gas sample at any instant, gas molecules are moving at a variety of speeds.

FIGURE 5.9

Distribution of molecular speeds of oxygen molecules at 25°C and 1000°C. At the higher temperature the fraction of molecules moving at very high speeds is greater. **OHT**

speed of an oxygen molecule at 25°C. Above about 400 m/s, the number of molecules moving at any particular speed decreases. For speeds in excess of about 1600 m/s, the fraction of molecules drops off to nearly zero. In general, most molecules have speeds rather close to the average value.

As temperature increases, the speed of the molecules increases. The distribution curve for molecular speeds (Figure 5.9) shifts to the right and becomes broader. The chance of a molecule having a very high speed is much greater at 1000°C than at 25°C. Note, for example, that a large number of molecules have speeds greater than 1600 m/s at 1000°C; almost none have that speed at 25°C. The general principle here is one that we will find very useful when we look at the effect of temperature on reaction rate in Chapter 11.

Sort of like people—most of them go along with the crowd.

5.7 REAL GASES

In this chapter, the ideal gas law has been used in all calculations, with the assumption that it applies exactly. Under ordinary conditions, this assumption is a good one; however, all real gases deviate at least slightly from the ideal gas law. Table 5.2 shows the extent to which two gases, O_2 and CO_2, deviate from ideality at different temperatures and pressures. The data compare the experimentally observed molar volume, V_m

$$\text{molar volume} = V_m = V/n$$

The molar volume is V when $n = 1$.

with the molar volume calculated from the ideal gas law V_m°:

$$V_m^\circ = RT/P$$

It should be obvious from Table 5.2 that deviations from ideality become larger at *high pressures and low temperatures*. Moreover, the deviations are larger for CO_2 than for O_2. All of these effects can be correlated in terms of a simple, common-sense observation:

In general, the closer a gas is to the liquid state, the more it will deviate from the ideal gas law.

A gas is liquefied by going to low temperatures and/or high pressures. Moreover, as you can see from Table 5.2, carbon dioxide is much easier to liquefy than oxygen.

TABLE 5.2 Real Versus Ideal Gases; Percent Deviation* in Molar Volume

P(atm)	O_2			CO_2		
	50°C	0°C	−50°C	50°C	0°C	−50°C
1	−0.0%	−0.1%	−0.2%	−0.4%	−0.7%	−1.4%
10	−0.4%	−1.0%	−2.1%	−4.0%	−7.1%	
40	−1.4%	−3.7%	−8.5%	−17.9%		
70	−2.2%	−6.0%	−14.4%	−34.2%	Condenses to	
100	−2.8%	−7.7%	−19.1%	−59.0%	liquid	

*Percent deviation $= \dfrac{(V_m - V_m^\circ)}{V_m^\circ} \times 100\%$

From a molecular standpoint, deviations from the ideal gas law arise because it neglects two factors:

1. attractive forces between gas particles.
2. the finite volume of gas particles.

We will now consider in turn the effect of these two factors on the molar volumes of real gases.

Attractive Forces

Notice that in Table 5.2 all the deviations are negative; the observed molar volume is less than that predicted by the ideal gas law. This effect can be attributed to attractive forces between gas particles. These forces tend to pull the particles toward one another, reducing the space between them. As a result, the particles are crowded into a smaller volume, just as if an additional external pressure were applied. The observed molar volume, V_m, becomes less than V_m°, and the deviation from ideality is *negative*:

> Attractive forces make the molar volume smaller than expected.

$$\frac{V_m - V_m^\circ}{V_m^\circ} < 0$$

The magnitude of this effect depends on the strength of the attractive forces and hence on the nature of the gas. Intermolecular attractive forces are stronger in CO_2 than they are in O_2, which explains why the deviation from ideality of V_m is greater with carbon dioxide and why carbon dioxide is more readily condensed to a liquid than is oxygen.

Particle Volume

Figure 5.10 shows a plot of V_m/V_m° versus pressure for methane at 25°C. Up to about 150 atm, methane shows a steadily increasing negative deviation from ideality, as might be expected on the basis of attractive forces. At 150 atm, V_m is only about 70% of V_m°.

At very high pressures, methane behaves quite differently. Above 150 atm, the ratio V_m/V_m° *increases,* becoming 1 at about 350 atm. Above that pressure, methane shows a *positive* deviation from the ideal gas law:

> Particle volume makes the molar volume larger than expected.

$$\frac{V_m - V_m^\circ}{V_m^\circ} > 0$$

FIGURE 5.10
Deviation of methane gas from ideal gas behavior. Below about 350 atm, attractive forces between methane (CH_4) molecules cause the observed molar volume at 25°C to be less than that calculated from the ideal gas law. At 350 atm, the effect of the attractive forces is just balanced by that of the finite volume of CH_4 molecules, and the gas appears to behave ideally. Above 350 atm, the effect of finite molecular volume predominates and $V_m > V_m^\circ$. **OHT**

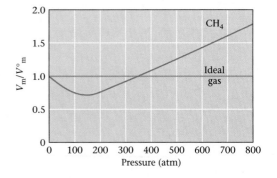

This effect is by no means unique to methane; it is observed with all gases. If the data in Table 5.2 are extended to very high pressures, oxygen and carbon dioxide behave like methane; V_m becomes larger than V_m°.

An increase in molar volume above that predicted by the ideal gas law is related to the finite volume of gas particles. These particles contribute to the observed volume, making V_m greater than V_m°. Ordinarily, this effect becomes evident only at high pressures, where the particles are quite close to one another.

Airbags

S ince driver's-side airbags were made mandatory in 1984, it is estimated that they have saved 5000 lives (Figure A). The way they work is relatively simple. A mechanical sensor in the front of the vehicle is set off by any sudden impact equivalent to hitting a brick wall at 10 mph. The sensor sends an electrical signal to a gas generator attached to the airbag. The bag fills with gas in less than a tenth of a second, preventing the head or chest of the driver from striking the steering column.

Most airbags today are filled with nitrogen gas generated by a three-step reaction involving sodium azide (NaN_3), potassium nitrate (KNO_3), and silicon dioxide (SiO_2). In the first step sodium azide decomposes to form nitrogen gas:

$$(1) \quad 2NaN_3(s) \longrightarrow 2Na(s) + 3N_2(g)$$

The sodium metal formed as a by-product would react violently with any moisture present. The function of the potassium nitrate is to prevent this by reacting with the sodium:

$$(2) \quad 2KNO_3(s) + 10Na(s) \longrightarrow K_2O(s) + 5Na_2O(s) + N_2(g)$$

Finally, the strongly basic metal oxides react with silicon dioxide to form a glassy product:

$$(3) \quad K_2O(s) + Na_2O(s) + SiO_2(s) \longrightarrow glass$$

By applying the principles discussed in Section 5.4, you can calculate that for every liter of nitrogen (at 50°C and 1.25 atm) generated by Reaction 1, about 2.0 g of sodium azide and 0.63 g of potassium nitrate are required. This means that to inflate a 15-L airbag, the gas generator should contain about 40 g of NaN_3 and 20 g of KNO_3.

Unfortunately, there are now more than 100 documented cases in which airbags have caused fatalities rather than preventing them. Many of the victims were infants or small children; it is now recommended that no child under the age of 12 should sit in the front seat of a vehicle equipped with a passenger-side airbag. Equally disturbing, adult drivers have been seriously injured or killed because they were sitting too close to the steering wheel when the airbag deployed; the minimum distance should be 25 cm (10 in.). That's easy to accomplish for a 6-ft 6-in. basketball player with long legs. It's not so easy for a short person who has trouble reaching the gas or brake pedal, let alone seeing over the steering wheel.

The solution to these problems is to develop "smart" airbag systems that can sense the severity of a collision, the size and weight of the driver, and how far his or her head is from the steering wheel. If desirable, the airbag can fill more slowly, perhaps in stages. Airbags of this type became available in a few 2000 model cars, notably the Ford *Taurus*.

CHEMISTRY

Beyond the Classroom

FIGURE A
A historic collision. The crash in 1990 of two Chrysler LeBaron cars, each equipped with an airbag, resulted in the destruction shown in this photo. Both airbags deployed, and both drivers walked away with minor injuries. The crash made history as the first collision between two airbag-equipped cars.

CHAPTER HIGHLIGHTS

Key Concepts

1. Convert between different units of pressure, volume, temperature, and amount of gas.
   (Example 5.1; Problems 1–4)
2. Use the ideal gas law to—
   ■ solve initial and final state problems.
   (Example 5.2; Problems 5–14)
   ■ calculate P, V, T, or n.
   (Example 5.3; Problems 15–20)
   ■ calculate density or molar mass.
   (Examples 5.4, 5.5; Problems 21–30)
   ■ relate amounts and volumes of gases in reactions.
   (Examples 5.6, 5.7; Problems 31–36, 56, 59, 60, 74)
3. Use Dalton's law to relate partial pressures to total pressures and mole fractions.
   (Examples 5.8, 5.9; Problems 37–44)
4. Calculate the speeds of gas molecules.
   (Example 5.10; Problems 48–50, 71)
5. Use Graham's law to relate rate of effusion to molar mass.
   (Example 5.11; Problems 45–48, 61)

Key Equations

Ideal gas law	$PV = nRT$
Dalton's law	$P_{tot} = P_A + P_B \qquad P_A = X_A P_{tot}$
Average translational kinetic energy	$E_t = 3RT/2N_A$
Average speed	$u = (3RT/\mathcal{M})^{1/2}$
Graham's law	$\dfrac{\text{rate}_B}{\text{rate}_A} = (\mathcal{M}_A/\mathcal{M}_B)^{1/2}$

Key Terms

atmosphere (atm)
bar
■ density
effusion
■ Kelvin scale

kinetic theory
millimeter of mercury (mm Hg)
■ molar mass
■ mole
mole fraction

partial pressure
R (gas constant)
STP
vapor pressure

When calcium carbonate, the main component of marble and chalk, is added to a solution of hydrochloric acid, the following reaction takes place:

$$CaCO_3(s) + 2H^+(aq) \longrightarrow CO_2(g) + H_2O + Ca^{2+}(aq)$$

(a) If 639 mL of CO_2 is produced at 21°C and 758 mm Hg, what is the yield of CO_2 in moles? In grams?

(b) The gas collected in (a) is transferred without loss to a flask at 21°C with twice the volume in (a). What is the pressure in the flask? At what temperature would the pressure in the larger flask return to the original pressure in (a)?

(c) What is the density of CO_2 at 21°C and 758 mm Hg?

(d) Suppose the carbon dioxide is collected at 21°C in a flask that contains nitrogen gas with a pressure of 135 mm Hg. After CO_2 is collected, the total pressure is 805 mm Hg. What is the partial pressure of CO_2? What mass of CO_2 is collected in the 639-mL flask?

(e) What is the mole fraction of nitrogen in the mixture in (d)?

(f) What volume of 0.150 M HCl is required to generate 0.350 L of CO_2 at 25°C and 1.15 atm?

(g) What volume of CO_2 can be generated at 25°C and 1.15 atm if 5.00 g of $CaCO_3$ is added to 75.00 mL of 0.888 M HCl?

(h) Compare the rate of effusion of CO_2 to N_2 at the same temperature and pressure. Compare the time required for equal numbers of moles of CO_2 and N_2 to effuse.

(i) The average speed of a carbon dioxide molecule at 25°C is 411 m/s. What is the average speed of a nitrogen molecule at the same temperature?

Answers

(a) 0.0264 mol; 1.16 g   **(d)** 6.70×10^2 mm Hg; 1.03 g   **(g)** 708 mL
(b) 379 mm Hg; 315°C   **(e)** 0.168   **(h)** 0.798; 1.25
(c) 1.82 g/L   **(f)** 0.219 L   **(i)** 515 m/s

Problem numbers in blue indicate that the answer is available in Appendix 6 at the back of the book.
WEB indicates that the solution is posted at **http://www.harcourtcollege.com/chem/general/masterton4/student/**

Measurements on Gases

1. A 6.00-ft cylinder has a radius of 26 in. It contains 189 lb of helium at 25°C. Express the volume of the cylinder ($V = \pi r^2 h$) in liters, the amount of helium in moles, and the temperature in Kelvin.

2. A ten-gallon methane tank contains 1.243 mol of methane (CH_4) at 74°F. Express the volume of the tank in liters, the amount of methane in the tank in grams, and the temperature of the tank in Kelvin.

3. Complete the following table of pressure conversions.

mm Hg	atm	kPa	bar
396			
	1.15		
		97.1	
			1.00

4. Complete the following table of pressure conversions.

mm Hg	atm	kPa	bar
1215			
	0.714		
		143	
			0.904

Initial and Final States

5. A 5.00-L balloon is filled with air at 812 mm Hg and 25°C. What is the pressure in another 5.00-L balloon filled with the same amount of air at −15°C?

6. A five-liter cylinder is filled with butane at 27°C until the pressure in the tank is 1.00 atm. What is the pressure in the tank

(a) on a winter day when the temperature is $-10°C$? (2 sig. fig.)

(b) on a warm day when the temperature is $40°C$? (2 sig. fig.)

7. A sample of air is originally at $32°C$. If P and n are kept constant, to what temperature must the air be cooled to

(a) decrease its volume by 25%?

(b) decrease its volume to 25% of its original volume?

8. A sample of CO_2 gas at $22°C$ and 1.00 atm has a volume of 2.00 L. Determine the ratio of the original volume to the final volume when

(a) the pressure and amount of gas remain unchanged and the Celsius temperature is doubled.

(b) the pressure and amount of gas remain unchanged and the Kelvin temperature is doubled.

9. A basketball is inflated in a garage at $25°C$ to a gauge pressure of 8.0 psi. Gauge pressure is the pressure above atmospheric pressure, which is 14.7 psi. The ball is used on the driveway at a temperature of $-7°C$ and feels flat. What is the actual pressure of the air in the ball? What is the gauge pressure?

WEB **10.** A tire is inflated to a gauge pressure of 28.0 psi at $71°F$. Gauge pressure is the pressure above atmospheric pressure, which is 14.7 psi. After several hours of driving, the air in the tire has a temperature of $115°F$. What is the gauge pressure of the air in the tire? What is the actual pressure of the air in the tire? Assume that the tire volume changes are negligible.

11. A glass tube is filled with neon gas at $32°C$ and 1.31 atm pressure. The tube is sealed, then put in a freezer at $-12°C$. What is the pressure in the tube when it is in the freezer?

12. On a cold winter day, a person takes in a breath of 0.50 L of air at 751 mm Hg and $-6°C$. What is the volume of this air in the lungs at $37°C$ and 1.00 atm pressure?

13. A 25.00-L flask has 1.35 mol of hydrogen gas at $25°C$ and a pressure of 1.05 atm. Nitrogen gas is added to the flask at the same temperature until the pressure rises to 1.64 atm. How many moles of nitrogen gas are added?

14. A balloon filled with helium has a volume of 1.28×10^3 L at sea level where the pressure is 0.998 atm and the temperature is $31°C$. The balloon is taken to the top of a mountain where the pressure is 0.753 atm and the temperature is $-25°C$. What is the volume of the balloon at the top of the mountain?

Ideal Gas Law; Calculation of One Variable

15. A two-liter plastic soft drink bottle can withstand a pressure of 5 atm. Half a cup (approximately 120 mL) of ethyl alcohol, C_2H_5OH, $(d = 0.789$ g/mL$)$ is poured into a soft drink bottle at room temperature. The bottle is then heated to $100°C$ (3 sig. fig.), changing the liquid alcohol to a gas. Will the soft drink bottle withstand the pressure, or will it explode?

16. Arterial blood contains about 0.25 mg of oxygen per milliliter. What is the pressure exerted by the oxygen in one liter of arterial blood at normal body temperature of $37°C$?

17. Compressed-air tanks used by scuba divers have a volume of 8.0 L and are filled with air to a pressure of 135 atm at $20°C$

(2 sig. fig.). How many grams of helium are required to fill a tank under these conditions?

18. An uninflated ball has a mass of 732.2 g. Nitrogen is pumped into the ball at $25°C$ and 1.25 atm. The inflated ball has a mass of 764.9 g. What is the pressure at $25°C$ in an identical ball filled with the same mass of helium?

19. Use the ideal gas law to complete the following table for butane (C_4H_{10}) gas.

Pressure	Volume	Temperature	Moles	Grams
_____	1.75 L	19°C	1.66	
0.895 atm	_____	6°C	_____	14.0
433 mm Hg	92.4 mL	_____	0.395	_____
1.74 bar	8.66 L	98°F	_____	_____

WEB **20.** Complete the following table for dinitrogen tetroxide gas.

Pressure	Volume	Temperature	Moles	Grams
1.77 atm	4.98 L	43.1°C	_____	_____
673 mm Hg	488 mL	_____	0.783	_____
0.899 bar	_____	912°C	_____	6.25
_____	1.15 L	39°F	0.166	_____

Ideal Gas Law; Density and Molar Mass

21. Calculate the densities (in grams per liter) of the following gases at $75°F$ and 1.33 bar.

(a) argon **(b)** ammonia **(c)** acetylene (C_2H_2)

22. Calculate the densities (in grams per liter) of the following gases at $97°C$ and 755 mm Hg.

(a) hydrogen chloride **(b)** sulfur dioxide

(c) butane (C_4H_{10})

23. Acetone (C_3H_6O) is widely used as a solvent. Assuming that acetone vapor behaves ideally at $56°C$ and 1.00 atm, what is the ratio of the density of acetone vapor to liquid acetone under these conditions? (The density of liquid acetone at $56°C$ is 0.791 g/cm^3.)

24. Recent measurements show that the atmosphere on Venus consists mostly of carbon dioxide. The temperature at the surface of Venus is $460°C$; the pressure is 75 atm. Compare the density of carbon dioxide on Venus's surface with that on the Earth's surface at $25°C$ and one atmosphere.

25. Phosgene is a highly toxic gas made up of carbon, oxygen, and chlorine atoms. Its density at 1.05 atm and $25°C$ is 4.24 g/L.

(a) What is the molar mass of phosgene?

(b) Phosgene is made up of 12.1% C, 16.2% O, and 71.7% Cl. What is the molecular formula of phosgene?

26. Cyclopropane mixed in the proper ratio with oxygen can be used as an anesthetic. At 755 mm Hg and $25°C$, it has a density of 1.71 g/L.

(a) What is the molar mass of cyclopropane?

(b) Cyclopropane is made up of 85.7% C and 14.3% H. What is the molecular formula of cyclopropane?

27. The gas in the discharge cell of a laser contains (in mole percent) 11% CO_2, 5.3% N_2, and 84% He.
  (a) What is the molar mass of this mixture?
  (b) Calculate the density of this gas mixture at 32°C and 758 mm Hg.
  (c) What is the ratio of the density of this gas to that of air ($\mathcal{M} = 29.0$ g/mol) at the same conditions?

28. Exhaled air contains 74.5% N_2, 15.7% O_2, 3.6% CO_2, and 6.2% H_2O (mole percent).
  (a) Calculate the molar mass of exhaled air.
  (b) Calculate the density of exhaled air at 37°C and 757 mm Hg, and compare the value obtained with that for ordinary air ($\mathcal{M} = 29.0$ g/mol).

29. A 2.00-g sample of $SX_6(g)$ has a volume of 329.5 cm^3 at 1.00 atm and 20°C. Identify the element X. Name the compound.

30. A 1.58-g sample of $C_2H_3X_3(g)$ has a volume of 297 mL at 769 mm Hg and 35°C. Identify the element X.

Stoichiometry of Gaseous Reactions

31. Ammonia gas is used widely in the fertilizer industry. It is produced by the reaction

$$N_2(g) + 3H_2(g) \longrightarrow 2NH_3(g)$$

How many liters of ammonia can be produced from 7.54 L of hydrogen? Assume 100% yield and that all the gases are measured at the same temperature and pressure.

32. Dichlorine oxide is used as a bacteriocide to purify water. It is produced by the reaction

$$SO_2(g) + 2Cl_2(g) \longrightarrow SOCl_2(l) + Cl_2O(g)$$

How many liters of chlorine gas would be required to react with 21.7 L of sulfur dioxide? Assume 100% yield and constant temperature and pressure.

33. Dinitrogen oxide, commonly called nitrous oxide, is used as a propellant gas for whipped-cream dispensers. It is prepared by heating ammonium nitrate to 250°C. Water vapor is also formed.
  (a) Write a balanced equation for the decomposition of ammonium nitrate into nitrous oxide and steam.
  (b) What volume of dinitrogen oxide gas is formed at 250°C and 1.0 atm when 5.00 g of NH_4NO_3 is heated? Assume 100% yield.

34. Nitric acid can be prepared by bubbling dinitrogen pentoxide into water.

$$N_2O_5(g) + H_2O \longrightarrow 2H^+(aq) + 2NO_3^-(aq)$$

  (a) How many moles of H^+ are obtained when 1.50 L of N_2O_5 at 25°C and 1.00 atm pressure is bubbled into water?
  (b) The solution obtained in (a) after reaction is complete has a volume of 437 mL. What is the molarity of the nitric acid obtained?

35. Nitroglycerine is an explosive used by the mining industry. It detonates according to the following equation:

$$4C_3H_5N_3O_9(l) \longrightarrow 12CO_2(g) + 6N_2(g) + 10H_2O(g) + O_2(g)$$

What volume is occupied by the gases produced when 10.00 g of nitroglycerine explodes? The total pressure is 1.45 atm at 523°C.

36. Ammonium nitrate can be used as an effective explosive because it decomposes into a large number of gaseous products. At a sufficiently high temperature, ammonium nitrate decomposes into nitrogen, oxygen, and steam.
  (a) Write a balanced equation for the decomposition of ammonium nitrate.
  (b) If 1.00 kg of ammonium nitrate is sealed into a 50.0-L steel drum and heated to 787°C, what is the pressure in the drum, assuming 100% decomposition?

Dalton's Law

37. To prevent a condition called the "bends," deep-sea divers breathe a mixture containing, in mole percent, 10.0% O_2, 10.0% N_2, and 80.0% He. What is the partial pressure of each gas if the total pressure is 825 mm Hg?

38. A certain laser uses a gas mixture made up of 9.00 g HCl, 2.00 g H_2, and 165.0 g Ne. What pressure is exerted by the mixture in a 75.0-L tank at 22°C? Which gas has the smallest partial pressure?

39. A sample of oxygen gas is collected over water at 25°C (vp $H_2O(l) = 23.8$ mm Hg). The wet gas occupies a volume of 7.28 L at a total pressure of 1.25 bar. If all the water is removed, what volume will the dry oxygen occupy at a pressure of 1.07 atm and a temperature of 37°C?

WEB **40.** A sample of gas collected over water at 42°C occupies a volume of one liter. The wet gas has a pressure of 0.986 atm. The gas is dried and the dry gas occupies 1.04 L with a pressure of 1.00 atm at 90°C. Using this information, calculate the vapor pressure of water at 42°C.

41. Consider two bulbs separated by a valve. Both bulbs are maintained at the same temperature. Assume that when the valve between the two bulbs is closed, the gases are sealed in their respective bulbs. When the valve is closed, the following data apply:

	Bulb A	Bulb B
Gas	**Ne**	**CO**
V	2.50 L	2.00 L
P	1.09 atm	0.773 atm

Assuming no temperature change, determine the final pressure inside the system after the valve connecting the two bulbs is opened. Ignore the volume of the tube connecting the two bulbs.

42. Follow the instructions of Problem 41 for the following set-up:

Gas	Bulb A Ar	Bulb B Cl$_2$
V	4.00 L	1.00 L
P	2.50 atm	1.00 atm

43. When methane (CH$_4$) is burned in oxygen, carbon dioxide and water vapor are formed. When 8.00 L of methane at 125°C and 1.00 atm is burned in an excess of oxygen, the products are collected into a separate 10.0-L flask.
  (a) Write a balanced equation for the reaction.
  (b) What is the total pressure of the products in the flask at 125°C?
  (c) What is the partial pressure of each of the products in the flask?

44. When ethylene chloride (C$_2$H$_3$Cl) is burned in oxygen, carbon dioxide, water vapor, and hydrogen chloride gas are formed. Twenty-five grams of ethylene chloride is burned in an excess of oxygen. The products are collected at 75°C into a separate 5.00-L flask.
  (a) Write a balanced equation for the reaction.
  (b) What is the total pressure of the products in the flask?
  (c) What is the partial pressure of each of the products in the flask?

Kinetic Theory

45. What is the ratio of the rate of effusion of the most abundant gas, nitrogen, to the lightest gas, hydrogen?

46. A gas effuses 1.55 times faster than propane (C$_3$H$_8$) at the same temperature and pressure.
  (a) Is the gas heavier or lighter than propane?
  (b) What is the molar mass of the gas?

47. Rank the following gases

$$NO \quad Ar \quad N_2 \quad N_2O_5$$

in order of
  (a) increasing speed of effusion through a tiny opening.
  (b) increasing time of effusion.

48. A balloon filled with nitrogen gas has a small leak. Another balloon filled with hydrogen gas has an identical leak. How much faster will the hydrogen balloon deflate?

49. At what temperature will a molecule of uranium hexafluoride, the densest gas known, have the same average speed as a molecule of the lightest gas, hydrogen, at 37°C?

50. What is the average speed of
  (a) a bromine molecule at 28°C?
  (b) a krypton atom at −28°C?

Real Gases

51. The normal boiling points of CO and SO$_2$ are −192°C and −10°C, respectively.
  (a) At 25°C and 1 atm, which gas would you expect to have a molar volume closest to the ideal value?
  (b) If you wanted to reduce the deviation from ideal gas behavior, in what direction would you change the temperature? The pressure?

52. A sample of methane gas (CH$_4$) is at 50°C and 20 atm. Would you expect it to behave more or less ideally if
  (a) the pressure were reduced to 1 atm?
  (b) the temperature were reduced to −50°C?

53. Using Figure 5.10
  (a) estimate the density of methane gas at 200 atm.
  (b) compare the value obtained in (a) to that calculated from the ideal gas law.

WEB **54.** Using Figure 5.10
  (a) estimate the density of methane gas at 100 atm.
  (b) compare the value obtained in (a) to that calculated from the ideal gas law.
  (c) determine the pressure at which the calculated density and the actual density of methane will be the same.

Unclassified

55. The pressure exerted by a column of liquid is directly proportional to its density and height. If you wanted to fill a barometer with ethyl alcohol (d = 0.78 g/cm^3), how long a tube should you use? (d Hg = 13.6 g/cm^3)

56. Assume that an automobile burns octane C$_8$H$_{18}$ (d = 0.692 g/cm^3).
  (a) Write the equation for the combustion of octane to carbon dioxide and water.
  (b) A car has a fuel efficiency of 22 mi/gal of octane. What volume of carbon dioxide at 25°C and 1.00 atm is generated when that car goes on a 75-mi trip?

57. Given that 1.00 mol of neon and 1.00 mol of hydrogen chloride gas are in separate containers at the same temperature and pressure, calculate each of the following ratios.
  (a) volume Ne/volume HCl
  (b) density Ne/density HCl
  (c) average translational energy Ne/average translational energy HCl
  (d) number of Ne atoms/number of HCl molecules

58. A mixture of 3.5 mol of Kr and 3.9 mol of He occupies a 10.00-L container at 300 K. Which gas has the larger
  (a) average translational energy?
  (b) partial pressure?
  (c) mole fraction?
  (d) effusion rate?

59. An intermediate reaction used in the production of nitrogen-containing fertilizers is that between ammonia and oxygen:

$$4NH_3(g) + 5O_2(g) \longrightarrow 4NO(g) + 6H_2O(g)$$

A 150.0-L reaction chamber is charged with reactants to the following partial pressures at 500°C: $P_{NH_3} = 1.3$ atm, $P_{O_2} = 1.5$ atm. What is the limiting reactant?

60. A Porsche 928 S4 engine has a cylinder volume of 618 cm³. The cylinder is full of air at 75°C and 1.00 atm.

    (a) How many moles of oxygen are in the cylinder? (Mole percent of oxygen in air = 21.0.)

    (b) Assume that the hydrocarbons in gasoline have an average molar mass of 1.0×10^2 g/mol and react with oxygen in a 1:12 mole ratio. How many grams of gasoline should be injected into the cylinder to react with the oxygen?

61. At 25°C and 380 mm Hg, the density of sulfur dioxide is 1.31 g/L. The rate of effusion of sulfur dioxide through an orifice is 4.48 mL/s. What is the density of a sample of gas that effuses through an identical orifice at the rate of 6.78 mL/s under the same conditions? What is the molar mass of the gas?

62. Glycine is an amino acid made up of carbon, hydrogen, oxygen, and nitrogen atoms. Combustion of a 0.2036-g sample gives 132.9 mL of CO_2 at 25°C and 1.00 atm and 0.122 g of water. What are the percentages of carbon and hydrogen in glycine? Another sample of glycine weighing 0.2500 g is treated in such a way that all the nitrogen atoms are converted to $N_2(g)$. This gas has a volume of 40.8 mL at 25°C and 1.00 atm. What is the percentage of nitrogen in glycine? What is the percentage of oxygen? What is the empirical formula of glycine?

Conceptual Questions

63. Sketch a cylinder with ten molecules of helium (He) gas. The cylinder has a movable piston. Label this sketch *before*. Make an *after* sketch to represent

    (a) a decrease in temperature at constant pressure.

    (b) a decrease in pressure from 1000 mm Hg to 500 mm Hg at constant temperature.

    (c) five molecules of H_2 gas added at constant temperature and pressure.

64. A rigid sealed cylinder has seven molecules of neon (Ne).

    (a) Make a sketch of the cylinder with the neon molecules at 25°C. Make a similar sketch of the same seven molecules in the same cylinder at −80°C.

    (b) Make the same sketches asked for in part (a), but this time attach a pressure gauge to the cylinder.

65. Tank A has ammonia at 300 K. Tank B has nitrogen gas at 150 K. Tanks A and B have the same volume. Compare the pressures in tanks A and B if

    (a) tank B has twice as many moles of nitrogen as tank A has of ammonia.

    (b) tank A has the same number of moles of ammonia as tank B has of nitrogen. (Try to do this without a calculator!)

66. Two tanks have the same volume and are kept at the same temperature. Compare the pressure in both tanks if

    (a) tank A has 2.00 mol of carbon dioxide and tank B has 2.00 mol of helium.

    (b) tank A has 2.00 g of carbon dioxide and tank B has 2.00 g of helium. (Try to do this without a calculator!)

67. Consider the sketch below. Each square in bulb A represents a mole of atoms X. Each circle in bulb B represents a mole of atoms Y. The bulbs have the same volume, and the temperature is kept constant. When the valve is opened, atoms of X react with atoms of Y according to the following equation:

$$2X(g) + Y(g) \longrightarrow X_2Y(g)$$

The gaseous product is represented as □O□ and each □O□ represents one mole of product.

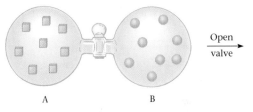

A   B

    (a) If $P_A = 2.0$ atm, what is P_B before the valve is opened and the reaction is allowed to occur? What is $P_A + P_B$?

    (b) Redraw the sketch above to represent what happens after the valve is opened.

    (c) What is P_A? What is P_B? What is $P_A + P_B$? Compare your answer with the answer in part (a).

68. The following figure shows three 1.00-L bulbs connected by valves. Each bulb contains neon gas with amounts proportional to the number of atoms pictorially represented in each chamber. All three bulbs are maintained at the same temperature. Unless stated otherwise, assume that the valves connecting chambers are closed and seal the gases in their respective chambers. Assume also that the volume between chambers is negligible.

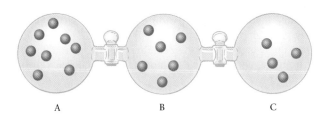

A   B   C

    (a) Which bulb has the lowest pressure?

    (b) If the pressure in bulb A is 2.00 atm, what is the pressure in bulb C?

    (c) If the pressure in bulb A is 2.00 atm, what is the sum of $P_A + P_B + P_C$?

    (d) If the pressure in bulb A is 2.00 atm and the valve between bulbs A and B is opened, redraw the above figure to accurately represent the gas atoms in all the bulbs. What is $P_A + P_B$? What is $P_A + P_B + P_C$? Compare your answer with your answer in part (c).

    (e) Follow the same instructions as in (d) if all the valves are open.

69. Consider three sealed steel tanks, labeled X, Y, and Z. Each tank has the same volume and the same temperature. In each tank, one mole of CH_4 is represented by a circle, one mole of oxygen by a square, and one mole of SO_2 by a triangle. Assume that no reaction takes place between these molecules.

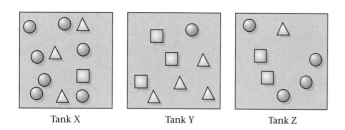

Tank X   Tank Y   Tank Z

    (a) In which tank is the total pressure highest?
    (b) In which tank is the partial pressure of SO_2 highest?
    (c) In which tank is the mass of all three gases the same?
    (d) Which tank has the heaviest contents?

Challenge Problems

70. Consider an ideal gas that exerts a pressure of 23.76 mm Hg at 25°C. Assuming n and V are held constant, what would its pressure be at 40°C? 70°C? 100°C? Compare the numbers you have just calculated with the vapor pressures of water at these temperatures. Can you suggest a reason why the two sets of numbers are so different?

71. The escape velocity required for gas molecules to overcome the Earth's gravity and go off to outer space is 1.12×10^3 m/s at 15°C. Calculate the molar mass of a species with that velocity. Would you expect to find He and H_2 molecules in the Earth's atmosphere? How about argon atoms?

72. A tube 5.0 ft long is evacuated. Samples of NH_3 and HCl, at the same temperature and pressure, are introduced simultaneously through tiny openings at opposite ends of the tube. When the two gases meet, a white ring of $NH_4Cl(s)$ forms. How far from the end at which ammonia was introduced will the ring form?

73. The Rankine temperature scale resembles the Kelvin scale in that 0° is taken to be the lowest attainable temperature (0°R = 0 K). However, the Rankine degree is the same size as the Fahrenheit degree, whereas the Kelvin degree is the same size as the Celsius degree. What is the value of the gas constant in L-atm/(mol-°R)?

74. A 0.2500-g sample of an Al-Zn alloy reacts with HCl to form hydrogen gas:

$$Al(s) + 3H^+(aq) \longrightarrow Al^{3+}(aq) + \tfrac{3}{2}H_2(g)$$

$$Zn(s) + 2H^+(aq) \longrightarrow Zn^{2+}(aq) + H_2(g)$$

The hydrogen produced has a volume of 0.147 L at 25°C and 755 mm Hg. What is the percentage of zinc in the alloy?

75. The buoyant force on a balloon is equal to the mass of air it displaces. The gravitational force on the balloon is equal to the sum of the masses of the balloon, the gas it contains, and the balloonist. If the balloon and the balloonist together weigh 168 kg, what would the diameter of a spherical hydrogen-filled balloon have to be in meters if the rig is to get off the ground at 22°C and 758 mm Hg? (Take $\mathcal{M}_{air}$ = 29.0 g/mol.)

76. A mixture in which the mole ratio of hydrogen to oxygen is 2:1 is used to prepare water by the reaction

$$2H_2(g) + O_2(g) \longrightarrow 2H_2O(g)$$

The total pressure in the container is 0.950 atm at 25°C before the reaction. What is the final pressure in the container at 125°C after the reaction, assuming an 88.0% yield and no volume change?

77. The volume fraction of a gas A in a mixture is defined by the equation

$$\text{volume fraction A} = \frac{V_A}{V}$$

where V is the total volume and V_A is the volume that gas A would occupy alone at the same temperature and pressure. Assuming ideal gas behavior, show that the volume fraction is the same as the mole fraction. Explain why the volume fraction differs from the mass fraction.

ELECTRONIC STRUCTURE AND THE PERIODIC TABLE

6

Metallic copper in a flame emits energy with a wavelength that produces a green flame. *(Charles D. Winters)*

CHAPTER OUTLINE

6.1 LIGHT, PHOTON ENERGIES, AND ATOMIC SPECTRA

6.2 THE HYDROGEN ATOM

6.3 QUANTUM NUMBERS, ENERGY LEVELS, AND ORBITALS

6.4 ELECTRON CONFIGURATIONS IN ATOMS

6.5 ORBITAL DIAGRAMS OF ATOMS

6.6 ELECTRON ARRANGEMENTS IN MONATOMIC IONS

6.7 PERIODIC TRENDS IN THE PROPERTIES OF ATOMS

Chemical properties of atoms and molecules depend on their electronic structures.

In Chapter 2 we briefly considered the structure of the atom. You will recall that every atom has a tiny, positively charged nucleus, made up of protons and neutrons. The nucleus is surrounded by negatively charged electrons. The number of protons in the nucleus is characteristic of the atoms of a particular element and is referred to as the atomic number. In a neutral atom, the number of electrons is equal to the number of protons and hence to the atomic number.

In this chapter, we focus on electron arrangements in atoms, paying particular attention to the relative energies of different electrons *(energy levels)* and their spatial locations *(orbitals)*. Specifically, we consider the nature of the energy levels and orbitals available to—

- the single electron in the hydrogen atom (Section 6.2).

- the several electrons in more complex atoms (Section 6.3).

With this background, we show how electron arrangements in multielectron atoms and the monatomic ions derived from them can be described in terms of—

- *electron configurations,* which show the number of electrons in each energy level (Sections 6.4, 6.6).

- *orbital diagrams,* which show the arrangement of electrons within orbitals (Section 6.5, 6.6).

The electron configuration or orbital diagram of an atom of an element can be deduced from its position in the periodic table. Beyond that, position in the table can be used to predict (Section 6.7) the relative sizes of atoms and ions *(atomic radius, ionic radius)* and the relative tendencies of atoms to give up or acquire electrons *(ionization energy, electronegativity)*.

Before dealing with electronic structures as such, it will be helpful to examine briefly the experimental evidence on which such structures are based (Section 6.1). In particular, we need to look at the phenomenon of *atomic spectra*.

Fireworks. The different colors are created by the atomic spectra of different elements. *(Eugen Gebhardt/FPG International)*

6.1 LIGHT, PHOTON ENERGIES, AND ATOMIC SPECTRA

Fireworks displays are fascinating to watch. Neon lights and sodium vapor lamps can transform the skyline of a city with their brilliant colors. The eerie phenomenon of the aurora borealis is an unforgettable experience when you see it for the first time. All of these events relate to the generation of light and its transmission through space.

The Wave Nature of Light: Wavelength and Frequency

Light travels through space as a wave, which is made up of successive crests, which rise above the midline, and troughs, which sink below it. Waves have three primary characteristics (Figure 6.1), two of which are of particular interest at this point:

1. **Wavelength** (λ), the distance between two consecutive crests or troughs, most often measured in meters or nanometers ($1 \text{ nm} = 10^{-9} \text{ m}$).
2. **Frequency** (ν), the number of wave cycles (successive crests or troughs) that pass a given point in unit time. If 10^8 cycles pass a particular point in one second,

$$\nu = 10^8/s = 10^8 \text{ Hz}$$

The frequency unit *hertz* (Hz) represents one cycle per second.

The speed at which a wave moves through space can be found by multiplying the length of a wave cycle (λ) by the number of cycles passing a point in unit time (ν). For light,

$$\lambda\nu = c$$

where c, the speed of light, is a constant, 2.998×10^8 m/s. To use this equation with this value of c—

- λ should be expressed in meters.
- ν should be expressed in reciprocal seconds (hertz).

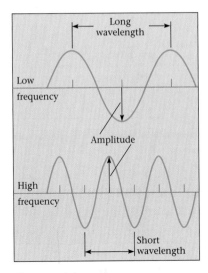

FIGURE 6.1
Characteristics of waves. The *amplitude* (ψ) is the height of a crest or the depth of a trough. The *wavelength* (λ) is the distance between successive crests or troughs. The *frequency* (ν) is the number of wave cycles (successive crests or troughs) that pass a given point in a given time.   **OHT**

EXAMPLE 6.1   The green light associated with the aurora borealis is emitted by excited (high-energy) oxygen atoms at 557.7 nm. What is the frequency of this light?

Strategy   Use the equation: $\lambda\nu = c$, taking $c = 2.998 \times 10^8$ m/s. Note that λ must be expressed in meters.

Solution

$$\lambda = 557.7 \text{ nm} \times \frac{1 \text{ m}}{10^9 \text{ nm}} = 557.7 \times 10^{-9} \text{ m}$$

$$\nu = \frac{2.998 \times 10^8 \text{ m/s}}{557.7 \times 10^{-9} \text{ m}} = 5.376 \times 10^{14}/s = \boxed{5.376 \times 10^{14} \text{ Hz}}$$

λ is the Greek letter lambda; ν is the Greek letter nu.

Light visible to the eye is a tiny portion of the entire *electromagnetic spectrum* (Figure 6.2), covering only the narrow wavelength region 400 to 700 nm. You can see the red glow given off by charcoal on a barbecue grill, but the heat given off is largely in the infrared (IR) region, above 700 nm. Microwave ovens produce radiation at even longer wavelengths. At the other end of the spectrum, below 400 nm, are ultraviolet (UV) radiation, responsible for sunburn, and x-rays, used for diagnostic purposes in medicine.

See the *Saunders Interactive General Chemistry CD-Rom*, Screen 7.4, The Electromagnetic Spectrum.

The Particle Nature of Light; Photon Energies

A hundred years ago it was generally supposed that all the properties of light could be explained in terms of its wave nature. A series of investigations carried out be-

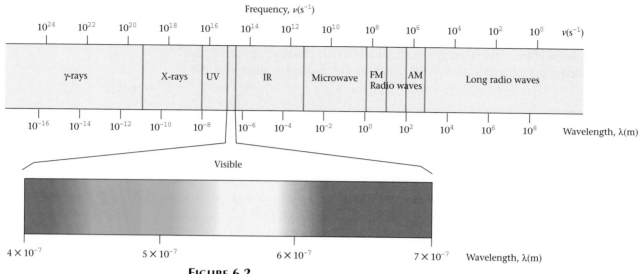

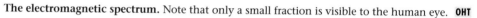

Visible

FIGURE 6.2

The electromagnetic spectrum. Note that only a small fraction is visible to the human eye. **OHT**

tween 1900 and 1910 by Max Planck (blackbody radiation) and Albert Einstein (photoelectric effect) discredited that notion. Today we consider light to be generated as a stream of particles called **photons,** whose energy E is given by the equation

⊙ **See Screen 7.5, Planck's Equation.**

$$E = h\nu = hc/\lambda$$

Throughout this text, we will use the SI unit *joule* (J) to express energy. A joule is a rather small quantity. One joule of electrical energy would keep a 10-W lightbulb burning for only a tenth of a second. For that reason, we will often express energies in *kilojoules* (1 kJ = 10^3 J). The quantity h appearing in Planck's equation is referred to as Planck's constant.

$$h = 6.626 \times 10^{-34}\,\text{J}\cdot\text{s}$$

Notice from this equation that energy is *inversely* related to wavelength. This explains why you put on a sunscreen to protect yourself from UV solar radiation (< 400 nm) and a "lead apron" when dental x-rays are being taken. Conversely, IR (> 700 nm) and microwave radiation are of relatively low energy (but don't try walking on hot coals).

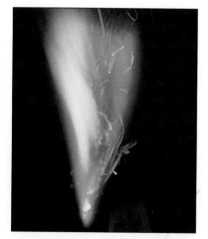

Energy and wavelength. A copper wire held in a flame colors the flame green. The energy of the photons of this light can be calculated from its wavelength. *(Charles D. Winters)*

EXAMPLE 6.2   Referring back to Example 6.1, calculate
(a) the energy, in joules, of a photon emitted by an excited oxygen atom.
(b) the energy, in kilojoules, of a mole of such photons.

Strategy   Use the equation $E = hc/\lambda$. In part (b), remember that 1 mol = 6.022 $\times$ 10^{23} particles.

Solution

(a) $E = \dfrac{hc}{\lambda} = \dfrac{(6.626 \times 10^{-34}\,\text{J}\cdot\text{s})(2.998 \times 10^8\,\text{m/s})}{557.7 \times 10^{-9}\,\text{m}} = \boxed{3.562 \times 10^{-19}\,\text{J}}$

This seems like a tiny amount of energy, but remember that it's coming from a single oxygen atom.

(b) $E = 3.562 \times 10^{-19}\,\text{J} \times \dfrac{1\,\text{kJ}}{10^3\,\text{J}} \times \dfrac{6.022 \times 10^{23}}{1\,\text{mol}} = \boxed{2.145 \times 10^2\,\text{kJ/mol}}$

Reality Check   This is roughly comparable to energy effects in chemical reactions; about 240 kJ of heat is evolved when a mole of hydrogen burns.

Atomic Spectra

In the seventeenth century, Sir Isaac Newton showed that visible (white) light from the Sun can be broken down into its various color components by a prism. The **spectrum** obtained is continuous; it contains essentially all wavelengths between 400 and 700 nm.

The situation with high-energy atoms of gaseous elements is quite different (Figure 6.3). Here the spectrum consists of discrete lines given off at specific wavelengths. Each element has a characteristic spectrum that can be used to identify it. In the case of sodium, there are two strong lines in the yellow region at 589.0 nm and 589.6 nm. These lines account for the yellow color of sodium vapor lamps used to illuminate highways.

The fact that the photons making up atomic spectra have only certain discrete wavelengths implies that they can only have certain discrete energies, because

$$E = h\nu = hc/\lambda$$

Because these photons are produced when an electron moves from one energy level to another, the electronic energy levels in an atom must be *quantized,* that is, limited to particular values. Moreover, it would seem that by measuring the spectrum of an element it should be possible to unravel its electronic energy levels. This is indeed possible, but it isn't easy. Gaseous atoms typically give off hundreds, even thousands, of spectral lines.

One of the simplest of atomic spectra, and the most important from a theoretical standpoint, is that of hydrogen. When energized by a high-voltage discharge, gaseous hydrogen atoms emit radiation at wavelengths that can be grouped into

See Screen 7.6, Atomic Line Spectra.

Atomic spectroscopy can identify metals at concentrations as low as 10^{-7} mol/L.

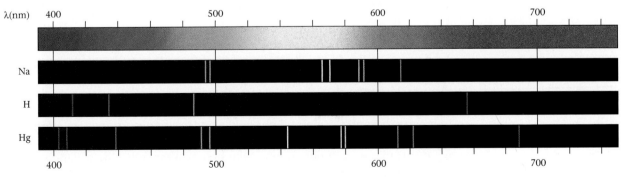

FIGURE 6.3
Continuous and line emission spectra. From the top down: the continuous visible spectrum; the line emission spectra for sodium (Na), hydrogen (H), and mercury (Hg). **OHT**

TABLE 6.1 Wavelengths (nm) of Lines in the Atomic Spectrum of Hydrogen		
Ultraviolet (Lyman Series)	Visible (Balmer Series)	Infrared (Paschen Series)
121.53	656.28	1875.09
102.54	486.13	1281.80
97.23	434.05	1093.80
94.95	410.18	1004.93
93.75	397.01	
93.05		

Balmer was a Swiss high-school teacher.

several different series (Table 6.1). The first of these to be discovered, the Balmer series, lies in the visible region. It consists of a strong line at 656.28 nm followed by successively weaker lines, closer and closer together, at lower wavelengths.

6.2 THE HYDROGEN ATOM

The hydrogen atom, containing a single electron, has played a major role in the development of models of electronic structure. In 1913 Niels Bohr, a Danish physicist, offered a theoretical explanation of the atomic spectrum of hydrogen. His model was based largely on classical mechanics. In 1922 this model earned him the Nobel Prize in physics. By that time, Bohr had become director of the Institute of Theoretical Physics at Copenhagen. There he helped develop the new discipline of quantum mechanics, used by other scientists to construct a more sophisticated model for the hydrogen atom.

Bohr, like all the other individuals mentioned in this chapter, was not a chemist. His only real contact with chemistry came as an undergraduate at the University of Copenhagen. His chemistry teacher, Niels Bjerrum, who later became his close friend and sailing companion, recalled that Bohr set a record for broken glassware that lasted half a century.

Bohr Model

 See Screen 7.7, Bohr's Model of the Hydrogen Atom.

Bohr assumed that a hydrogen atom consists of a central proton about which an electron moves in a circular orbit. He related the electrostatic force of attraction of the proton for the electron to the centrifugal force due to the circular motion of the electron. In this way, Bohr was able to express the energy of the atom in terms of the radius of the electron's orbit. To this point, his analysis was purely classical, based on Coulomb's law of electrostatic attraction and Newton's laws of motion. To progress beyond this point, Bohr boldly and arbitrarily assumed, in effect, that the electron in the hydrogen atom can have only certain definite energies. Using arguments that we will not go into, Bohr obtained the following equation for the energy of the hydrogen electron:

$E_n = -2.180 \times 10^{-18}\ \text{J}/n^2$

$$E_n = -R_H/n^2$$

where E_n is the energy of the electron, R_H is a quantity called the Rydberg constant (modern value = 2.180×10^{-18} J), and **n** is an integer called the principal quantum number. Depending on the state of the electron, **n** can have any positive, integral value, that is,

$$\mathbf{n} = 1, 2, 3, \ldots$$

Before proceeding with the Bohr model, let us make three points:

1. In setting up his model, Bohr designated zero energy as the point at which the proton and electron are completely separated. Energy has to be absorbed to reach that point. This means that the electron, in all its allowed energy states within the atom, must have an energy below zero; that is, it must be negative, hence the minus sign in the equation:

$$E_\mathbf{n} = -R_H/\mathbf{n}^2$$

2. Ordinarily the hydrogen electron is in its lowest-energy state, referred to as the **ground state** or ground level, for which **n** = 1. When an electron absorbs enough energy, it moves to a higher, **excited state.** In a hydrogen atom, the first excited state has **n** = 2, the second **n** = 3, and so on.

3. When an excited electron gives off energy as a photon of light, it drops back to a lower energy state. The electron can return to the ground state (from **n** = 2 to **n** = 1, for example) or to a lower excited state (from **n** = 3 to **n** = 2). In every case, the energy of the photon ($h\nu$) evolved is equal to the difference in energy between the two states:

$$\Delta E = h\nu = E_{hi} - E_{lo}$$

where E_{hi} and E_{lo} are the energies of the higher and lower states, respectively.

Using this expression for ΔE and the equation $E_\mathbf{n} = -R_H/\mathbf{n}^2$, it is possible to relate the frequency of the light emitted to the quantum numbers, $\mathbf{n}_{hi}$ and $\mathbf{n}_{lo}$, of the two states:

$$h\nu = -R_H\left[\frac{1}{(\mathbf{n}_{hi})^2} - \frac{1}{(\mathbf{n}_{lo})^2}\right]$$

$$\nu = \frac{R_H}{h}\left[\frac{1}{(\mathbf{n}_{lo})^2} - \frac{1}{(\mathbf{n}_{hi})^2}\right]$$

The last equation written is the one Bohr derived in applying his model to the hydrogen atom. Given

$$R_H = 2.180 \times 10^{-18}\,\text{J} \qquad h = 6.626 \times 10^{-34}\,\text{J}\cdot\text{s}$$

you can use the equation to find the frequency or wavelength of any of the lines in the hydrogen spectrum.

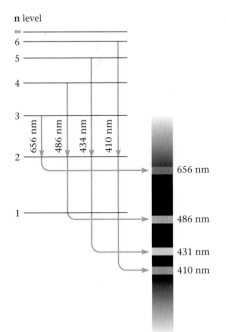

Some Balmer series lines for hydrogen. The lines in the visible region result from transitions from levels with values of **n** greater than 2 to the **n** = 2 level. **OHT**

EXAMPLE 6.3 Calculate the wavelength in nanometers of the line in the Balmer series that results from the transition **n** = 3 to **n** = 2.

Strategy Use the above equation to find the frequency; $\mathbf{n}_{lo}$ = 2, $\mathbf{n}_{hi}$ = 3. Then use the equation $\lambda = c/\nu$ to find the wavelength.

Solution

$$\nu = \frac{2.180 \times 10^{-18}\,\text{J}}{6.626 \times 10^{-34}\,\text{J}\cdot\text{s}}\left[\frac{1}{4} - \frac{1}{9}\right] = 4.570 \times 10^{14}/\text{s}$$

$$\lambda = \frac{c}{\nu} = \frac{2.998 \times 10^{8}\,\text{m/s}}{4.570 \times 10^{14}/\text{s}} \times \frac{10^{9}\,\text{nm}}{1\,\text{m}} = 656.0\,\text{nm}$$

Reality Check Compare this value with that listed in Table 6.1 for the first line in the Balmer series.

All of the lines in the Balmer series (Table 6.1) come from transitions to the level **n** = 2 from higher levels (**n** = 3, 4, 5, . . .). Similarly, lines in the Lyman series arise when electrons fall to the **n** = 1 level from higher levels (**n** = 2, 3, 4, . . .). For the Paschen series, which lies in the infrared, the lower level is always **n** = 3.

Quantum Mechanical Model

Bohr's theory for the structure of the hydrogen atom was highly successful. Scientists of the day must have thought they were on the verge of being able to predict the allowed energy levels of all atoms. However, the extension of Bohr's ideas to atoms with two or more electrons gave, at best, only qualitative agreement with experiment. Consider, for example, what happens when Bohr's theory is applied to the helium atom. For helium, the errors in calculated energies and wavelengths are of the order of 5% instead of the 0.1% error with hydrogen. There appeared to be no way the theory could be modified to make it work well with helium or other atoms. Indeed, it soon became apparent that there was a fundamental problem with the Bohr model. The idea of an electron moving about the nucleus in a well-defined orbit at a fixed distance from the nucleus had to be abandoned.

Scientists in the 1920s, speculating on this problem, became convinced that an entirely new approach was required to treat electrons in atoms and molecules. In 1924 a young French scientist, Louis de Broglie (1892–1987), in his doctoral thesis at the Sorbonne made a revolutionary suggestion. He reasoned that if light could show the behavior of particles (photons) as well as waves, then perhaps an electron, which Bohr had treated as a particle, could behave like a wave. In a few years, de Broglie's postulate was confirmed experimentally. This led to the development of a whole new discipline, first called wave mechanics, more commonly known today as *quantum mechanics*.

The quantum mechanical atom differs from the Bohr model in several ways. In particular, according to quantum mechanics—

■ *the kinetic energy of an electron is inversely related to the volume of the region to which it is confined.* This phenomenon has no analog in classical mechanics, but it helps to explain the stability of the hydrogen atom. Consider what happens when an electron moves closer and closer to the nucleus. The electrostatic energy decreases; that is, it becomes more negative. If this were the only factor, the electron should radiate energy and "fall" into the nucleus. However, the kinetic energy is increasing at the same time, because the electron is moving within a smaller and smaller volume. The two effects oppose each other; at some point a balance is reached and the atom is stable.

See Screen 7.8, Wave Properties of the Electron.

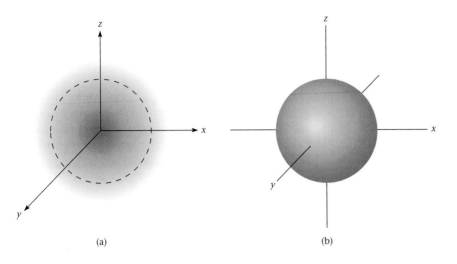

(a)　　　　　　　　　　　　(b)

FIGURE 6.4

Two different ways of showing the electron distribution in the ground state of the hydrogen atom. In (a) the depth of color is proportional to the probability of finding the electron at a given point. The dotted line encloses the volume within which there is a 90% probability of finding the electron. In (b), the orbital encloses that volume; the electron density variation within the orbital is not shown. **OHT**

■ *it is impossible to specify the precise position of an electron in an atom at a given instant.* Neither can we describe in detail the path that an electron takes about the nucleus. (After all, if we can't say where the electron is, we certainly don't know how it got there.) The best we can do is to estimate the *probability* of finding the electron within a particular region.

In 1926 Erwin Schrödinger (1887–1961), an Austrian physicist, made a major contribution to quantum mechanics. He wrote down a rather complex differential equation to express the wave properties of an electron in an atom. This equation can be solved, at least in principle, to find the amplitude (height) ψ of the electron wave at various points in space. The quantity ψ (psi) is known as the *wave function*. Although we will not use the Schrödinger wave equation in any calculations, you should realize that much of our discussion of electronic structure is based on solutions to that equation for the electron in the hydrogen atom.

For the hydrogen electron, the square of the wave function, ψ^2, is directly proportional to the probability of finding the electron at a particular point. If ψ^2 at point A is twice as large as at point B, then we are twice as likely to find the electron at A as at B. Putting it another way, over time the electron will turn up at A twice as often as at B.

Figure 6.4a, an *electron cloud* diagram, shows how ψ^2 for the hydrogen electron in its ground state (**n** = 1) varies moving out from the nucleus. The depth of the color is supposed to be directly proportional to ψ^2 and hence to the probability of finding the electron at a point. As you can see, the color fades moving out from the nucleus in any direction; the value of ψ^2 drops accordingly.

Another, more common way of showing the electron distribution in the ground state of the hydrogen atom is to draw the *orbital* (Figure 6.4b) within which there is a 90% chance of finding the electron. Notice that the orbital is spherical, which means that the probability is independent of direction; the electron is equally likely to be found north, south, east, or west of the nucleus.

⊙ See Screen 7.10, Schrödinger's Equation and Wave Functions.

In case you're curious, the equation is:
$$\frac{d^2\psi}{dx^2} + \frac{8\pi^2 m(E - V)}{h^2}\psi = 0$$
(and that's just in one dimension).

6.3 QUANTUM NUMBERS, ENERGY LEVELS, AND ORBITALS

Although the Schrödinger equation is a rather complex one, it can be solved exactly for the hydrogen atom. Indeed, it turns out that there are many solutions for ψ, each associated with a set of numbers called **quantum numbers.** There are

⊙ See Screen 7.12, Quantum Numbers and Orbitals.

three such numbers, given the symbols **n**, ℓ, and $\mathbf{m}_\ell$. A wave function corresponding to a particular set of three quantum numbers (e.g., **n** = 2, ℓ = 1, $\mathbf{m}_\ell$ = 0) is referred to as an **atomic orbital.** Orbitals differ from one another in their energy and in the shape and spatial orientation of their electron cloud.

For reasons we will discuss later, a fourth quantum number is required to completely describe a specific electron in a multielectron atom. The fourth quantum number is given the symbol $\mathbf{m}_s$. Each electron in an atom has a set of four quantum numbers, **n**, ℓ, $\mathbf{m}_\ell$, and $\mathbf{m}_s$. We will now discuss the quantum numbers of electrons as they are used in atoms beyond hydrogen.

First Quantum Number, *n*; Principal Energy Levels

The first quantum number, **n**, designates the **principal energy level.** As **n** increases, the energy of the electron increases, and, on the average, it is found farther out from the nucleus. The quantum number **n** can take on only integral values, starting with 1:

$$\boxed{\mathbf{n} = 1, 2, 3, 4, \; \ldots}$$

An electron for which **n** = 1 is said to be in the first principal level. If **n** = 2, we are dealing with the second principal level, and so on.

Second Quantum Number, ℓ; Sublevels (s, p, d, f)

Each principal level includes one or more **sublevels.** The sublevels are denoted by the second quantum number, ℓ. The quantum numbers **n** and ℓ are related; ℓ can take on any integral value starting with 0 and going up to a maximum of (**n** − 1). That is,

$$\boxed{\ell = 0, 1, 2, \; \ldots \;, (\mathbf{n} - 1)}$$

If **n** = 1, there is only one possible value of ℓ—namely 0. This means that, in the first principal level, there is only one sublevel, for which ℓ = 0. If **n** = 2, two values of ℓ are possible, 0 and 1. In other words, there are two sublevels (ℓ = 0 and ℓ = 1) within the second principal energy level. In the same way,

if **n** = 3: $\quad \ell$ = 0, 1, or 2 $\qquad$ (three sublevels)

if **n** = 4: $\quad \ell$ = 0, 1, 2, or 3 $\qquad$ (four sublevels)

In general, ***in the nth principal level, there are* n *different sublevels.***

Another method is commonly used to designate sublevels. Instead of giving the quantum number ℓ, the letters s, p, d, or f* indicate the sublevels ℓ = 0, 1, 2, or 3, respectively. That is,

quantum number, ℓ	0	1	2	3
type of sublevel		s	p	d	f

Usually, in designating a sublevel, a number is included (Table 6.2) to indicate the principal level as well. Thus reference is made to a 1s sublevel (**n** = 1, ℓ = 0), a 2s sublevel (**n** = 2, ℓ = 0), a 2p sublevel (**n** = 2, ℓ = 1), and so on.

---

*These letters come from the adjectives used by spectroscopists to describe spectral lines: *sharp, principal, diffuse,* and *fundamental.*

This relation between **n** and ℓ comes from the Schrödinger equation.

n	1	2		3			4			
ℓ	0	0	1	0	1	2	0	1	2	3
Sublevel	1s	2s	2p	3s	3p	3d	4s	4p	4d	4f

TABLE 6.2   Sublevel Designations for the First Four Principal Levels OHT

For the hydrogen atom, the energy of an electron is independent of the value of ℓ. (This is implied by the equation $E = R_H/\mathbf{n}^2$, which involves only one quantum number.) For multielectron atoms, the situation is quite different; the energy is dependent on ℓ as well as on $\mathbf{n}$. Within a given principal level (same value of $\mathbf{n}$), sublevels increase in energy in the order

$$\mathbf{n}s < \mathbf{n}p < \mathbf{n}d < \mathbf{n}f$$

Thus a 2p sublevel has a slightly higher energy than a 2s sublevel. By the same token, when $\mathbf{n} = 3$, the 3s sublevel has the lowest energy, the 3p is intermediate, and the 3d has the highest energy.

> Which would have the higher energy, 4p or 3s?

Third Quantum Number, m_ℓ; Orbitals

Each sublevel contains one or more orbitals, which differ from one another in the value assigned to the third quantum number, $\mathbf{m}_\ell$. This quantum number determines the direction in space of the electron cloud surrounding the nucleus. The value of $\mathbf{m}_\ell$ is related to that of ℓ. For a given value of ℓ, $\mathbf{m}_\ell$ can have any integral value, including 0, between ℓ and $-\ell$, that is

$$\mathbf{m}_\ell = \ell, \ldots, +1, 0, -1, \ldots, -\ell$$

To illustrate how this rule works, consider an s sublevel ($\ell = 0$). Here $\mathbf{m}_\ell$ can have only one value, 0. This means that an s sublevel contains only one orbital, referred to as an s orbital. The s orbital shown in Figure 6.5a is identical to that

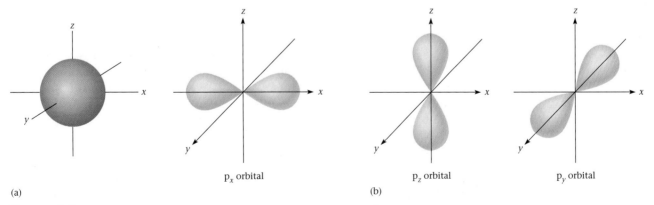

(a)

(b)

p_x orbital

p_z orbital

p_y orbital

FIGURE 6.5

The s and p orbitals. (a) All s orbitals are spherically symmetrical; the density of the electron cloud is independent of the direction from the nucleus (x, y, or z). (b) The electron density of the three p orbitals is directed along the x, y, or z axes. The three p orbitals are located at 90° angles to each other. OHT

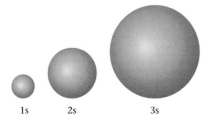

s Orbitals. The relative sizes of the 90% contours (see Figure 6.5a) are shown for the 1s, 2s, and 3s orbitals.

shown in Figure 6.4b for a very good reason; the hydrogen electron in its ground state occupies the 1s orbital. All s orbitals are spherical; they differ from one another only in size. As **n** increases, the radius of the orbital becomes larger. This means that an electron in a 2s orbital is more likely to be found far out from the nucleus than one in a 1s orbital.

For a p sublevel ($\ell = 1$), $\mathbf{m}_\ell = 1$, 0, or -1. Within a given p sublevel, there are three different orbitals described by the quantum numbers $\mathbf{m}_\ell = 1$, 0, and -1. Commonly, p orbitals are referred to as p_x, p_y, and p_z orbitals (Figure 6.5b).

For the d and f sublevels:

d sublevel:	$\ell = 2$	$\mathbf{m}_\ell = 2, 1, 0, -1, -2$	5 orbitals
f sublevel:	$\ell = 3$	$\mathbf{m}_\ell = 3, 2, 1, 0, -1, -2, -3$	7 orbitals

(The shapes and orientations of d orbitals are shown in Chapter 15.)

In general, for a sublevel of quantum number ℓ, there are a total of $2\ell + 1$ orbitals. All of the orbitals in a given sublevel (e.g., $2p_x$, $2p_y$, $2p_z$) have essentially the same energy.

Fourth Quantum Number, $\mathbf{m}_s$; Electron Spin

 See Screen 8.2, Spinning Electrons and Magnetism.

FIGURE 6.6
Electron spin. The spins can be represented as clockwise and counterclockwise, with the different values of $\mathbf{m}_s$ of $+\frac{1}{2}$ and $-\frac{1}{2}$. **OHT**

The fourth quantum number, $\mathbf{m}_s$, is associated with **electron spin.** An electron has magnetic properties that correspond to those of a charged particle spinning on its axis. Either of two spins is possible, clockwise or counterclockwise (Figure 6.6).

The quantum number $\mathbf{m}_s$ was introduced to make theory consistent with experiment. In that sense, it differs from the first three quantum numbers, which came from the solution to the Schrödinger wave equation for the hydrogen atom. This quantum number is not related to **n**, ℓ, or $\mathbf{m}_\ell$. It can have either of two possible values:

$$\mathbf{m}_s = +1/2 \qquad \text{or} \qquad -1/2$$

Electrons that have the same value of $\mathbf{m}_s$ (i.e., both $+\frac{1}{2}$ or both $-\frac{1}{2}$) are said to have *parallel* spins. Electrons that have different $\mathbf{m}_s$ values (i.e., one $+\frac{1}{2}$ and the other $-\frac{1}{2}$) are said to have *opposed* spins.

Pauli Exclusion Principle

 See Screen 8.4, The Pauli Exclusion Principle.

Our model for electronic structure is a pragmatic blend of theory and experiment.

The four quantum numbers that characterize an electron in an atom have now been considered. There is an important rule, called the **Pauli exclusion principle,** that relates to these numbers. It requires that *no two electrons in an atom can have the same set of four quantum numbers.* This principle was first stated in 1925 by Wolfgang Pauli, a colleague of Bohr, again to make theory consistent with the properties of atoms.

The Pauli exclusion principle has an implication that is not obvious at first glance. It requires that no more than two electrons can fit into an orbital. Moreover, if two electrons occupy the same orbital they must have opposed spins. To see that this is the case, consider the 2s orbital. Any electron in this orbital must have

$$\mathbf{n} = 2 \qquad \ell = 0 \qquad \mathbf{m}_\ell = 0$$

To satisfy the Pauli exclusion principle, the electrons in this orbital must have different $\mathbf{m}_s$ values. But there are only two possible values of $\mathbf{m}_s$. Hence only two

TABLE 6.3 **Allowed Sets of Quantum Numbers for Electrons in Atoms** OHT

Level **n**	1	2				3								
Sublevel ℓ	0	0	1			0	1			2				
Orbital $\mathbf{m}_\ell$	0	0	1	0	−1	0	1	0	−1	2	1	0	−1	−2
Spin $\mathbf{m}_s$ $\begin{array}{l}\uparrow = +\frac{1}{2} \\ \downarrow = -\frac{1}{2}\end{array}$	↑↓	↑↓	↑↓	↑↓	↑↓	↑↓	↑↓	↑↓	↑↓	↑↓	↑↓	↑↓	↑↓	↑↓

electrons can enter the orbital. If the orbital is filled, one electron must have $\mathbf{m}_s = +\frac{1}{2}$ and the other $\mathbf{m}_s = -\frac{1}{2}$. In other words, the two electrons must have opposed spins.

Capacities of Principal Levels, Sublevels, and Orbitals

The rules that we have given for quantum numbers fix the capacities of principal levels, sublevels, and orbitals. In summary,

See Screen 7.11, Shells, Subshells, and Orbitals.

1. Each principal level of quantum number **n** contains a total of **n** sublevels.
2. Each sublevel of quantum number ℓ contains a total of $2\ell + 1$ orbitals; that is,

> an s sublevel ($\ell = 0$) contains 1 orbital
>
> a p sublevel ($\ell = 1$) contains 3 orbitals
>
> a d sublevel ($\ell = 2$) contains 5 orbitals
>
> an f sublevel ($\ell = 3$) contains 7 orbitals

These three rules summarize everything we've said in this section.

3. Each orbital can hold two electrons, which must have opposed spins.

Applying these rules to the first three principal energy levels, we obtain Table 6.3. At the bottom of the table, each arrow indicates an electron. In each orbital, there are two electrons with opposed spins. The number of electrons in a sublevel is found by adding up the electrons in the orbitals within that sublevel. For example, a p sublevel ($\ell = 1$) can hold six electrons, two in each of three orbitals. To find the total number of electrons that can fit into a principal level, we add the number of electrons required to fill the sublevels within that principal level.

EXAMPLE 6.4

(a) What is the capacity for electrons of the 3d sublevel?
(b) How many electrons can fit into the principal level for which **n** = 4?

Strategy Apply the three rules just cited. Notice particularly that the capacity of a sublevel is found by multiplying the number of orbitals in that sublevel by 2.

Solution

(a) Each d sublevel contains five orbitals, so its capacity is

$$5 \times 2e^- = \boxed{10e^-}$$

(b) By rule (1), there must be four sublevels. These are 4s, 4p, 4d, 4f. Remember that an s sublevel has one orbital, a p sublevel three, a d sublevel five, and an f sublevel seven:

$$1(2e^-) + 3(2e^-) + 5(2e^-) + 7(2e^-) = \boxed{32e^-}$$

TABLE 6.4   Capacities of Electron Levels and Sublevels in Atoms

Level n	Total Number of Electrons in Level ($2n^2$)	Maximum Number of Electrons in Sublevels $[2(2\ell + 1)]$			
		s	p	d	f
1	2	2	—	—	—
2	8	2	6	—	—
3	18	2	6	10	—
4	32	2	6	10	14

Using the approach in Example 6.4 you can readily obtain the electron capacity of any principal level or sublevel. In Table 6.4, these capacities are listed through $n = 4$. Notice that—

- *a sublevel of quantum number ℓ has a capacity of* $2(2\ell + 1)$. For example, a 3d sublevel ($\ell = 2$) has a capacity of $2(2 \times 2 + 1) = 10e^-$.
- *a principal level of quantum number n has a capacity of* $2n^2$. For example, the total capacity of the fourth principal energy level is $2(4)^2 = 32e^-$.

6.4   ELECTRON CONFIGURATIONS IN ATOMS

See Screen 8.7, Atomic Electron Configuration.

Given the rules referred to in Section 6.3, it is possible to assign quantum numbers to each electron in an atom. Beyond that, electrons can be assigned to specific principal levels, sublevels, and orbitals. There are several ways to do this. Perhaps the simplest way to describe the arrangement of electrons in an atom is to give its **electron configuration,** which shows the number of electrons, indicated by a superscript, in each sublevel. For example, a species with the electron configuration

$$1s^2 2s^2 2p^5$$

has two electrons in the 1s sublevel, two electrons in the 2s sublevel, and five electrons in the 2p sublevel.

In this section, you will learn how to predict the electron configurations of atoms of elements. There are a couple of different ways of doing this, which we consider in turn. It should be emphasized that, throughout this discussion, we refer to *isolated gaseous atoms in the ground state.*

Electron Configuration from Sublevel Energies

See Screen 8.5, Atomic Subshell Energies.

Electron configurations are readily obtained if the order of filling sublevels is known. Electrons enter the available sublevels in order of increasing sublevel energy. Ordinarily, a sublevel is filled to capacity before the next one starts to fill. The relative energies of different sublevels can be obtained from experiment. Figure 6.7 is a plot of these energies for atoms through the $n = 4$ principal level.

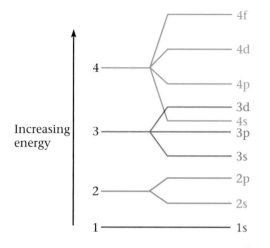

FIGURE 6.7
Electron energy sublevels in the order of increasing energy. The order shown is the order of sublevel filling as atomic number increases, starting at the bottom with 1s. **OHT**

From Figure 6.7 it is possible to predict the electron configurations of atoms of elements with atomic numbers 1 through 36. Because an s sublevel can hold only two electrons, the 1s is filled at helium ($1s^2$). With lithium ($Z = 3$), the third electron has to enter a new sublevel: This is the 2s, the lowest sublevel of the second principal energy level. Lithium has one electron in this sublevel ($1s^2 2s^1$). With beryllium ($Z = 4$), the 2s sublevel is filled ($1s^2 2s^2$). The next six elements fill the 2p sublevel. Their electron configurations are

$_5$B	$1s^2 2s^2 2p^1$	$_8$O	$1s^2 2s^2 2p^4$
$_6$C	$1s^2 2s^2 2p^2$	$_9$F	$1s^2 2s^2 2p^5$
$_7$N	$1s^2 2s^2 2p^3$	$_{10}$Ne	$1s^2 2s^2 2p^6$

These are ground-state configurations; $1s^2 2s^1 2p^2$ would be an excited state for boron.

Beyond neon, electrons enter the third principal level. The 3s sublevel is filled at magnesium:

$$_{12}\text{Mg} \qquad 1s^2 2s^2 2p^6 3s^2$$

Six more electrons are required to fill the 3p sublevel with argon:

$$_{18}\text{Ar} \qquad 1s^2 2s^2 2p^6 3s^2 3p^6$$

After argon, an "overlap" of principal energy levels occurs. The next electron enters the *lowest* sublevel of the fourth principal level (4s) instead of the *highest* sublevel of the third principal level (3d). Potassium ($Z = 19$) has one electron in the 4s sublevel; calcium ($Z = 20$) fills it with two electrons:

$$_{20}\text{Ca} \qquad 1s^2 2s^2 2p^6 3s^2 3p^6 4s^2$$

Now the 3d sublevel starts to fill with scandium ($Z = 21$). Recall that a d sublevel has a capacity of ten electrons. Hence the 3d sublevel becomes filled at zinc ($Z = 30$):

$$_{30}\text{Zn} \qquad 1s^2 2s^2 2p^6 3s^2 3p^6 4s^2 3d^{10}$$

The next sublevel, 4p, is filled at krypton ($Z = 36$):

$$_{36}\text{Kr} \qquad 1s^2 2s^2 2p^6 3s^2 3p^6 4s^2 3d^{10} 4p^6$$

The order of filling, through $Z = 36$, is 1s, 2s, 2p, 3s, 3p, 4s, 3d, 4p.

EXAMPLE 6.5 Find the electron configuration of the sulfur atom and the nickel atom.

Strategy Determine the number of electrons in the atom from its atomic number. Fill the appropriate sublevels using the energy diagram in Figure 6.7.

Solution The sulfur atom has atomic number 16, so there are $16e^-$. Two electrons go into the 1s sublevel, two into the 2s sublevel, and six into the 2p sublevel. Two more go into the 3s sublevel, and the remaining four go into the 3p sublevel. The electron configuration of the S atom is

$$1s^2 2s^2 2p^6 3s^2 3p^4$$

The nickel atom has 28 electrons. Proceeding as above, the electron configuration for nickel is

$$1s^2 2s^2 2p^6 3s^2 3p^6 4s^2 3d^8$$

The 4s sublevel fills before the 3d sublevel.

Reality Check You can use the periodic table to check your answers. See the discussion on pages 159–162.

Often, to save space, electron configurations are shortened; the **abbreviated electron configuration** starts with the preceding noble gas. For the elements sulfur and nickel,

	Electron Configuration	Abbreviated Electron Configuration
$_{16}$S	$1s^2 2s^2 2p^6 3s^2 3p^4$	[Ne] $3s^2 3p^4$
$_{28}$Ni	$1s^2 2s^2 2p^6 3s^2 3p^6 4s^2 3d^8$	[Ar]$4s^2 3d^8$

The symbol [Ne] indicates that the first 10 electrons in the sulfur atom have the neon configuration $1s^2 2s^2 2p^6$; similarly, [Ar] represents the first 18 electrons in the nickel atom.

Filling of Sublevels and the Periodic Table

In principle, a diagram such as Figure 6.7 can be extended to include all sublevels occupied by electrons in any element. As a matter of fact, that is a relatively simple thing to do; such a diagram is in effect incorporated into the periodic table introduced in Chapter 2.

TABLE 6.5 Abbreviated Electron Configurations of Group 1 and 2 Elements

Group 1		Group 2	
$_3$Li	[He] **2s^1**	$_4$Be	[He] **2s^2**
$_{11}$Na	[Ne] **3s^1**	$_{12}$Mg	[Ne] **3s^2**
$_{19}$K	[Ar] **4s^1**	$_{20}$Ca	[Ar] **4s^2**
$_{37}$Rb	[Kr] **5s^1**	$_{38}$Sr	[Kr] **5s^2**
$_{55}$Cs	[Xe] **6s^1**	$_{56}$Ba	[Xe] **6s^2**

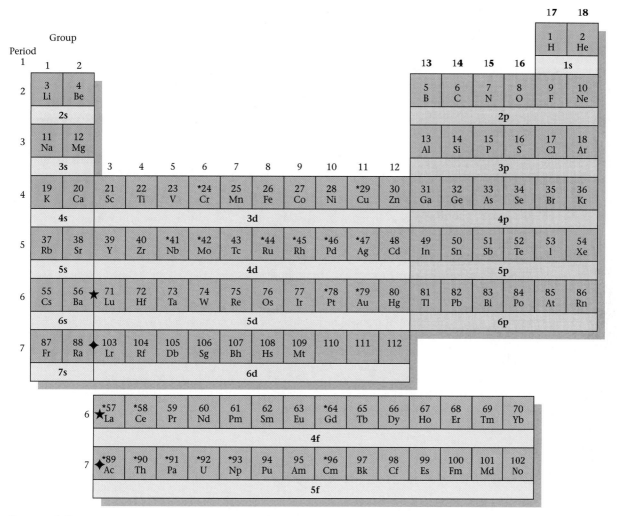

FIGURE 6.8

The periodic table and electron configurations. The periodic table can be used to deduce the electron configurations of atoms. The color code in the figure shows the energy sublevels being filled across each period. Elements marked with asterisks have electron configurations slightly different from those predicted by the table. **OHT**

To understand how position in the periodic table relates to the filling of sublevels, consider the metals in the first two groups. Atoms of the Group 1 elements all have one s electron in the outermost principal energy level (Table 6.5). In each Group 2 atom, there are two s electrons in the outermost level. A similar relationship applies to the elements in any group:

The atoms of elements in a group of the periodic table have the same distribution of electrons in the outermost principal energy level.

This means that the order in which electron sublevels are filled is determined by position in the periodic table. Figure 6.8 shows how this works. Notice the following points:

The periodic table works because an element's chemical properties depend on the number of the outer electrons.

CHEMISTRY ▪ The Human Side

GLENN THEODORE SEABORG
(1912–1999)

One name, more than any other, is associated with the actinide elements: Glenn Seaborg (1912–1999). Between 1940 and 1957, Seaborg and his team at the University of California, Berkeley, prepared nine of these elements (at. no. 94–102) for the first time. Moreover, in 1945 Seaborg made the revolutionary suggestion that the actinides, like the lanthanides, were filling an f sublevel. For these accomplishments, he received the 1951 Nobel Prize in chemistry.

Glenn Seaborg was born in a small town in middle America, Ishpeming, Michigan. After obtaining a bachelor's degree in chemistry from UCLA, he spent the rest of his scientific career at Berkeley, first as a Ph.D. student and then as a faculty member. During World War II, he worked on the Manhattan Project. Along with other scientists, Seaborg recommended that the devastating power of the atomic bomb be demonstrated by dropping it on a barren island before United Nations observers. Later he headed the U.S. Atomic Energy Commission under Presidents Kennedy, Johnson, and Nixon.

Sometimes referred to as the "gentle giant" (he was 6 ft 4 in. tall), Seaborg had a charming, self-deprecating sense of humor. He recalled that friends advised him *not* to publish his theory about the position of the actinides in the periodic table, lest it ruin his scientific reputation. Seaborg went on to say that, "I had a great advantage. I didn't have any scientific reputation, so I went ahead and published it." Late in his life there was considerable controversy as to whether a transuranium element should be named for him; that honor had always been bestowed posthumously. Seaborg commented wryly that, "They don't want to do it because I'm still alive and they can prove it." Element 106 was named seaborgium (Sg) in 1997; he considered this his greatest honor, even above the Nobel Prize.

(Photo credit: Lawrence Berkeley Laboratory)

> The periodic table was created before electrons were discovered or atomic number defined.

1. *The elements in Groups 1 and 2 are filling an s sublevel.* Thus Li and Be in the second period fill the 2s sublevel. Na and Mg in the third period fill the 3s sublevel, and so on.

2. *The elements in Groups 13 through 18 (six elements in each period) fill p sublevels,* which have a capacity of six electrons. In the second period, the 2p sublevel starts to fill with B ($Z = 5$) and is completed with Ne ($Z = 10$). In the third period, the elements Al ($Z = 13$) through Ar ($Z = 18$) fill the 3p sublevel.

3. *The transition metals, in the center of the periodic table, fill d sublevels.* Remember that a d sublevel can hold ten electrons. In the fourth period, the ten elements Sc ($Z = 21$) through Zn ($Z = 30$) fill the 3d sublevel. In the fifth period, the 4d sublevel is filled by the elements Y ($Z = 39$) through Cd ($Z = 48$). The ten transition metals in the sixth period fill the 5d sublevel. Elements 103 to 112 in the seventh, incomplete period are believed to be filling the 6d sublevel.

4. *The two sets of 14 elements listed separately at the bottom of the table are filling f sublevels with a principal quantum number two less than the period number.* That is—

■ 14 elements in the sixth period ($Z = 57$ to 70) are filling the 4f sublevel. These elements are sometimes called rare earths or, more commonly, **lanthanides,** after the name of the first element in the series, lanthanum (La). Modern separation techniques, notably chromatography, have greatly increased the availability of compounds of these elements. A brilliant red phosphor used in color TV receivers contains a small amount of europium oxide, Eu_2O_3. This is added to yttrium oxide, Y_2O_3, or gadolinium oxide, Gd_2O_3. Cerium(IV) oxide is used to coat interior surfaces of "self-cleaning" ovens, where it prevents the build-up of tar deposits.

■ 14 elements in the seventh period ($Z = 89$ to 102) are filling the 5f sublevel. The first element in this series is actinium (Ac); collectively, these elements are referred to as **actinides.** All these elements are radioactive; only thorium and uranium occur in nature. The other actinides have been synthesized in the laboratory by nuclear reactions. Their stability decreases rapidly with increasing atomic number. The longest-lived isotope of nobelium ($_{102}$No) has a half-life of about 3 minutes; that is, in 3 minutes half of the sample decomposes. Nobelium and the preceding element, mendeleevium ($_{101}$Md), were identified in samples containing one to three atoms of No or Md.

Color TV. Several of the lanthanide elements are essential in creating the color. *(Charles D. Winters)*

Electron Configuration from the Periodic Table

Figure 6.8 (or any periodic table) can be used to deduce the electron configuration of any element. It is particularly useful for heavier elements such as iodine (Example 6.6).

EXAMPLE 6.6   For the iodine atom, write
(a) the electron configuration.   (b) the abbreviated electron configuration.

Strategy   To obtain the electron configuration, use Figure 6.8. Go across each period in succession, noting the sublevels occupied, until you get to iodine. To find the abbreviated configuration, start with the preceding noble gas, krypton.

Solution

(a) Period 1: the 1s sublevel fills ($1s^2$)
    Period 2: the 2s and 2p sublevels fill ($2s^2 2p^6$)
    Period 3: the 3s and 3p sublevels fill ($3s^2 3p^6$)
    Period 4: the 4s, 3d, and 4p sublevels fill ($4s^2 3d^{10} 4p^6$)
    Period 5: the 5s and 4d sublevels fill; five electrons enter the 5p ($5s^2 4d^{10} 5p^5$)
Putting it all together, the electron configuration of iodine must be

$$1s^2 2s^2 2p^6 3s^2 3p^6 4s^2 3d^{10} 4p^6 5s^2 4d^{10} 5p^5$$

(b) [Kr] accounts for the first 36e: [Kr] $5s^2 4d^{10} 5p^5$

To obtain electron configurations from the periodic table, consider what sublevels are filled going across each period.

As you can see from Figure 6.8, the electron configurations of several elements (marked*) differ slightly from those predicted. In every case, the difference involves a shift of one or, at the most, two electrons from one sublevel to another of very similar energy. For example, in the first transition series, two elements, chromium and copper, have an extra electron in the 3d as compared with the 4s orbital.

	Predicted	Observed
$_{24}$Cr	[Ar] $4s^2 3d^4$	[Ar] $4s^1 3d^5$
$_{29}$Cu	[Ar] $4s^2 3d^9$	[Ar] $4s^1 3d^{10}$

These anomalies reflect the fact that the 3d and 4s orbitals have very similar energies. Beyond that, it has been suggested that there is a slight increase in stability with a half-filled (Cr) or completely filled (Cu) 3d sublevel.

6.5 ORBITAL DIAGRAMS OF ATOMS

For many purposes, electron configurations are sufficient to describe the arrangements of electrons in atoms. Sometimes, however, it is useful to go a step further and show how electrons are distributed among orbitals. In such cases, **orbital diagrams** are used. Each orbital is represented by parentheses ( ), and electrons are shown by arrows written ↑ or ↓, depending on spin.

To show how orbital diagrams are obtained from electron configurations, consider the boron atom ($Z = 5$). Its electron configuration is $1s^2 2s^2 2p^1$. The pair of electrons in the 1s orbital must have opposed spins ($+\frac{1}{2}, -\frac{1}{2}$, or ↑↓). The same is true of the two electrons in the 2s orbital. There are three orbitals in the 2p sublevel. The single 2p electron in boron could be in any one of these orbitals. Its spin could be either "up" or "down". The orbital diagram is ordinarily written

$$\begin{array}{cccc} & 1s & 2s & 2p \\ _5B & (\uparrow\downarrow) & (\uparrow\downarrow) & (\uparrow)(\ )(\ ) \end{array}$$

with the first electron in an orbital arbitrarily designated by an up arrow, ↑.

With the next element, carbon, a complication arises. In which orbital should the sixth electron go? It could go in the same orbital as the other 2p electron, in which case it would have to have the opposite spin, ↓. It could go into one of the other two orbitals, either with a parallel spin, ↑, or an opposed spin, ↓. Experiment shows that there is an energy difference among these arrangements. The most stable is the one in which the two electrons are in different orbitals with parallel spins. The orbital diagram of the carbon atom is

$$\begin{array}{cccc} & 1s & 2s & 2p \\ _6C & (\uparrow\downarrow) & (\uparrow\downarrow) & (\uparrow)(\uparrow)(\ ) \end{array}$$

Similar situations arise frequently. There is a general principle that applies in all such cases; **Hund's rule** (Friedrich Hund 1896–1997) predicts that, ordinarily,

When several orbitals of equal energy are available, as in a given sublevel, electrons enter singly with parallel spins.

Only after all the orbitals are half-filled do electrons pair up in orbitals.

Following this principle, the orbital diagrams for the elements boron through neon are shown in Figure 6.9. Notice that—

■ *in all filled orbitals, the two electrons have opposed spins.* Such electrons are often referred to as being *paired*. There are four paired electrons in the B, C, and N atoms, six in the oxygen atom, eight in the fluorine atom, and ten in the neon atom.

■ *in accordance with Hund's rule, within a given sublevel there are as many half-filled orbitals as possible.* Electrons in such orbitals are said to be *unpaired*. There is one unpaired electron in atoms of B and F, two unpaired electrons in C and O atoms, and three unpaired electrons in the N atom. When there are two or more unpaired electrons, as in C, N, and O, those electrons have parallel spins.

Atom	Orbital diagram					Electron configuration
B	$(\uparrow\downarrow)$	$(\uparrow\downarrow)$	$(\uparrow)$	$(\ )$	$(\ )$	$1s^22s^22p^1$
C	$(\uparrow\downarrow)$	$(\uparrow\downarrow)$	$(\uparrow)$	$(\uparrow)$	$(\ )$	$1s^22s^22p^2$
N	$(\uparrow\downarrow)$	$(\uparrow\downarrow)$	$(\uparrow)$	$(\uparrow)$	$(\uparrow)$	$1s^22s^22p^3$
O	$(\uparrow\downarrow)$	$(\uparrow\downarrow)$	$(\uparrow\downarrow)$	$(\uparrow)$	$(\uparrow)$	$1s^22s^22p^4$
F	$(\uparrow\downarrow)$	$(\uparrow\downarrow)$	$(\uparrow\downarrow)$	$(\uparrow\downarrow)$	$(\uparrow)$	$1s^22s^22p^5$
Ne	$(\uparrow\downarrow)$	$(\uparrow\downarrow)$	$(\uparrow\downarrow)$	$(\uparrow\downarrow)$	$(\uparrow\downarrow)$	$1s^22s^22p^6$
	1s	2s		2p		

FIGURE 6.9
Orbital diagrams for atoms with five to ten electrons. Orbitals of equal energy are all occupied by unpaired electrons before pairing begins. **OHT**

Hund's rule, like the Pauli exclusion principle, is based on experiment. It is possible to determine the number of unpaired electrons in an atom. With solids, this is done by studying their behavior in a magnetic field. If there are unpaired electrons present, the solid will be attracted into the field. Such a substance is said to be *paramagnetic*. If the atoms in the solid contain only paired electrons, it is slightly repelled by the field. Substances of this type are called *diamagnetic*. With gaseous atoms, the atomic spectrum can also be used to establish the presence and number of unpaired electrons.

See Screen 8.3, Spinning Electrons and Magnetism.

EXAMPLE 6.7   Construct orbital diagrams for atoms of sulfur and iron.

Strategy   Start with the electron configuration, obtained as in Section 6.4. Then write the orbital diagram, recalling the number of orbitals per sublevel, putting two electrons of opposed spin in each orbital within a completed sublevel, and applying Hund's rule where sublevels are partially filled.

Solution   Recall from Example 6.5 that the electron configuration of sulfur is $1s^22s^22p^63s^23p^4$. Its orbital diagram is

	1s	2s	2p	3s	3p
$_{16}$S	$(\uparrow\downarrow)$	$(\uparrow\downarrow)$	$(\uparrow\downarrow)(\uparrow\downarrow)(\uparrow\downarrow)$	$(\uparrow\downarrow)$	$(\uparrow\downarrow)(\uparrow)(\uparrow)$

The atomic number of iron is 26; its electron configuration is $1s^22s^22p^63s^23p^64s^23d^6$. All the orbitals are filled except those in the 3d sublevel, which is populated according to Hund's rule to give four unpaired electrons.

	1s	2s	2p	3s	3p	4s	3d
$_{26}$Fe	$(\uparrow\downarrow)$	$(\uparrow\downarrow)$	$(\uparrow\downarrow)(\uparrow\downarrow)(\uparrow\downarrow)$	$(\uparrow\downarrow)$	$(\uparrow\downarrow)(\uparrow\downarrow)(\uparrow\downarrow)$	$(\uparrow\downarrow)$	$(\uparrow\downarrow)(\uparrow)(\uparrow)(\uparrow)(\uparrow)$

Reality Check   To construct an orbital diagram, start with the electron configuration and apply Hund's rule.

6.6 ELECTRON ARRANGEMENTS IN MONATOMIC IONS

See Screen 8.8, Electron Configuration of Ions.

The discussion so far in this chapter has focused on electron configurations and orbital diagrams of neutral atoms. It is also possible to assign electronic structures to monatomic ions, formed from atoms by gaining or losing electrons. In general, when a monatomic ion is formed from an atom, *electrons are added to or removed from sublevels in the highest principal energy level.*

Ions with Noble Gas Structures

Important ideas are worth repeating.

As pointed out in Chapter 2, elements close to a noble gas in the periodic table form ions that have the same number of electrons as the noble gas atom. This means that these ions have noble gas electron configurations. Thus the three elements preceding Ne (N, O, and F) and the three elements following neon (Na, Mg, and Al) all form ions with the neon configuration, $1s^2 2s^2 2p^6$. The three nonmetal atoms achieve this structure by gaining electrons to form anions:

$$_7N\ (1s^2 2s^2 2p^3) + 3e^- \longrightarrow {}_7N^{3-}\ (1s^2 2s^2 2p^6)$$

$$_8O\ (1s^2 2s^2 2p^4) + 2e^- \longrightarrow {}_8O^{2-}\ (1s^2 2s^2 2p^6)$$

$$_9F\ (1s^2 2s^2 2p^5) + e^- \longrightarrow {}_9F^-\ (1s^2 2s^2 2p^6)$$

The three metal atoms acquire the neon structure by losing electrons to form cations:

$$_{11}Na\ (1s^2 2s^2 2p^6 3s^1) \longrightarrow {}_{11}Na^+\ (1s^2 2s^2 2p^6) + e^-$$

$$_{12}Mg\ (1s^2 2s^2 2p^6 3s^2) \longrightarrow {}_{12}Mg^{2+}\ (1s^2 2s^2 2p^6) + 2e^-$$

$$_{13}Al\ (1s^2 2s^2 2p^6 3s^2 3p^1) \longrightarrow {}_{13}Al^{3+}\ (1s^2 2s^2 2p^6) + 3e^-$$

The species N^{3-}, O^{2-}, F^-, Ne, Na^+, Mg^{2+}, and Al^{3+} are said to be *isoelectronic;* that is, they have the same electron configuration.

There are a great many monatomic ions that have noble gas configurations; Figure 6.10 shows 24 ions of this type. Note, once again, that ions in a given main group have the same charge (+1 for Group 1, +2 for Group 2, −2 for Group 16, −1 for Group 17). This explains, in part, the chemical similarity among elements

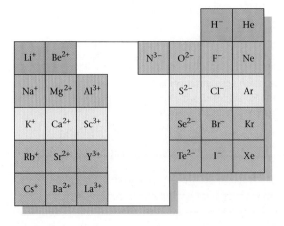

FIGURE 6.10
Cations, anions, and atoms with ground state noble gas electron configurations. Atoms and ions shown in the same color are *isoelectronic;* that is, they have the same electron configurations. **OHT**

in the same main group. In particular, ionic compounds formed by such elements have similar chemical formulas. For example,

- halides of the alkali metals have the general formula MX, where M = Li, Na, K, . . . and X = F, Cl, Br, ...
- halides of the alkaline earth metals have the general formula MX_2, where M = Mg, Ca, Sr, . . . and X = F, Cl, Br, ...
- oxides of the alkaline earth metals have the general formula MO, where M = Mg, Ca, Sr, ...

Transition Metal Cations

The transition metals to the right of the scandium subgroup do not form ions with noble gas configurations. To do so, they would have to lose four or more electrons. The energy requirement is too high for that to happen. However, as pointed out in Chapter 2, these metals do form cations with charges of + 1, + 2, or + 3. Applying the principle that, in forming cations, electrons are removed from the sublevel of highest **n**, you can predict correctly that *when transition metal atoms form positive ions, the outer s electrons are lost first.* Consider, for example, the formation of the Mn^{2+} ion from the Mn atom:

$$_{25}Mn \quad [Ar]\, 4s^2 3d^5 \qquad _{25}Mn^{2+} \quad [Ar]\, 3d^5$$

Notice that it is the 4s electrons that are lost rather than the 3d electrons. This is known to be the case because the Mn^{2+} ion has been shown to have five unpaired electrons (the five 3d electrons). If two 3d electrons had been lost, the Mn^{2+} ion would have had only three unpaired electrons.

All the transition metals form cations by a similar process, that is, loss of outer s electrons. Only after those electrons are lost are electrons removed from the inner d sublevel. Consider, for example, what happens with iron, which you will recall, forms two different cations. First the 4s electrons are lost to give the Fe^{2+} ion:

$$_{26}Fe\,(Ar\, 4s^2 3d^6) \longrightarrow {}_{26}Fe^{2+}\,(Ar\, 3d^6) + 2e^-$$

Then an electron is removed from the 3d level to form the Fe^{3+} ion:

$$_{26}Fe^{2+}\,(Ar\, 3d^6) \longrightarrow {}_{26}Fe^{3+}\,(Ar\, 3d^5) + e^-$$

See Screen 8.14, Properties of Transition Metals.

The s electrons are "first in" with the atoms and "first out" with the cations.

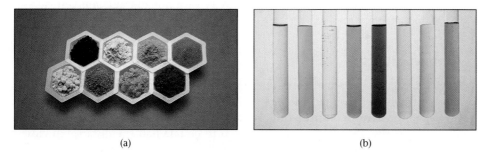

(a)   (b)

Transition metal ions. Transition metal ions impart color to many of their compounds and solutions. (a) Bottom row *(left to right):* iron(III) chloride, copper(II) sulfate, manganese(II) chloride, cobalt(II) chloride. Top row *(left to right):* chromium(III) nitrate, iron(II) sulfate, nickel(II) sulfate, potassium dichromate. (b) Solutions of the compounds shown in (a) in the order listed above. *(Charles D. Winters)*

In the Fe^{2+} and Fe^{3+} ions, as in all transition metal ions, there are no outer s electrons.

You will recall that for fourth period *atoms,* the 4s sublevel fills before the 3d. In the corresponding *ions,* the electrons come out of the 4s sublevel before the 3d. This is sometimes referred to as the "first in, first out" rule.

EXAMPLE 6.8   Give the electron configuration of

(a) Zn^{2+} (b) Se^{2-}

Strategy   First obtain the electron configuration of the corresponding atom, as in Section 6.4. Then add or remove electrons from sublevels of highest **n**.

Solution

(a) The electron configuration of Zn ($Z = 30$) is $1s^2 2s^2 2p^6 3s^2 3p^6 4s^2 3d^{10}$. Two electrons are removed from the 4s sublevel to form Zn^{2+}:

$$Zn^{2+} \qquad 1s^2 2s^2 2p^6 3s^2 3p^6 3d^{10}$$

(b) The electron configuration of Se ($Z = 34$) is $1s^2 2s^2 2p^6 3s^2 3p^6 4s^2 3d^{10} 4p^4$. Two electrons are added to the 4p sublevel to form Se^{2-}:

$$Se^{2-} \qquad 1s^2 2s^2 2p^6 3s^2 3p^6 4s^2 3d^{10} 4p^6$$

This is the electron configuration of the noble gas krypton ($Z = 36$).

6.7   PERIODIC TRENDS IN THE PROPERTIES OF ATOMS

One of the most fundamental principles of chemistry is the periodic law, which states that

The chemical and physical properties of elements are a periodic function of atomic number.

This is, of course, the principle behind the structure of the periodic table. Elements within a given vertical group resemble one another chemically because chemical properties repeat themselves at regular intervals of 2, 8, 18, or 32 elements.

In this section we will consider how the periodic table can be used to correlate properties on an atomic scale. In particular, we will see how atomic radius, ionic radius, ionization energy, and electronegativity vary horizontally and vertically in the periodic table.

Atomic Radius

Strictly speaking, the "size" of an atom is a rather nebulous concept. The electron cloud surrounding the nucleus does not have a sharp boundary. However, a quantity called the **atomic radius** can be defined and measured, assuming a spherical atom. Ordinarily, the atomic radius is taken to be one half the distance of closest approach between atoms in an elemental substance (Figure 6.11).

The atomic radii of the main-group elements are shown at the top of Figure 6.12. Notice that, in general, atomic radii—

■ decrease across a period from left to right in the periodic table.
■ increase down a group in the periodic table.

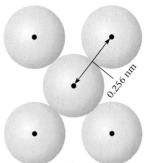

Cu

(a) Atomic radius = $\dfrac{0.256 \text{ nm}}{2}$ = 0.128 nm

Cl_2

(b) Atomic radius = $\dfrac{0.198 \text{ nm}}{2}$ = 0.099 nm

FIGURE 6.11
Atomic radii. The radii are determined by assuming that atoms in closest contact in an element touch one another. The atomic radius is taken to be one half of the closest internuclear distance. (a) Arrangement of copper atoms in metallic copper, giving an atomic radius of 0.128 nm for copper. (b) Chlorine atoms in a chlorine (Cl_2) molecule, giving an atomic radius of 0.099 nm for chlorine.

Atomic radii (nm)

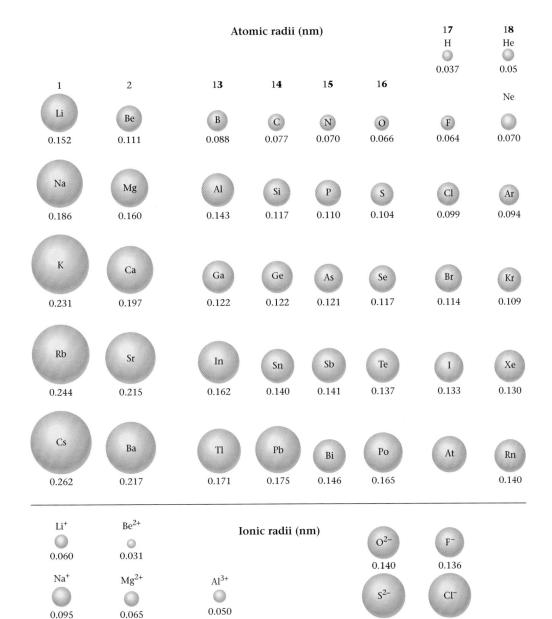

1	2	13	14	15	16	17	18
						H 0.037	He 0.05
Li 0.152	Be 0.111	B 0.088	C 0.077	N 0.070	O 0.066	F 0.064	Ne 0.070
Na 0.186	Mg 0.160	Al 0.143	Si 0.117	P 0.110	S 0.104	Cl 0.099	Ar 0.094
K 0.231	Ca 0.197	Ga 0.122	Ge 0.122	As 0.121	Se 0.117	Br 0.114	Kr 0.109
Rb 0.244	Sr 0.215	In 0.162	Sn 0.140	Sb 0.141	Te 0.137	I 0.133	Xe 0.130
Cs 0.262	Ba 0.217	Tl 0.171	Pb 0.175	Bi 0.146	Po 0.165	At	Rn 0.140

Group 13 elements. The periodic table can be used to predict the relative atomic radii of these three elements. The atomic radius of boron *(left)* is smaller than that of aluminum *(center)*, which is smaller than that of indium *(right)*. *(Charles D. Winters)*

⊙ **See Screen 8.10, Atomic Properties and Periodic Trends: Size.**

Ionic radii (nm)

Li⁺ 0.060	Be²⁺ 0.031		O²⁻ 0.140	F⁻ 0.136
Na⁺ 0.095	Mg²⁺ 0.065	Al³⁺ 0.050	S²⁻ 0.184	Cl⁻ 0.181
K⁺ 0.133	Ca²⁺ 0.099	Ga³⁺ 0.062	Se²⁻ 0.198	Br⁻ 0.195
Rb⁺ 0.148	Sr²⁺ 0.113	In³⁺ 0.081	Te²⁻ 0.221	I⁻ 0.216
Cs⁺ 0.169	Ba²⁺ 0.135	Tl³⁺ 0.095		

The values above are:
Li⁺ 0.060, Be²⁺ 0.031; Na⁺ 0.095, Mg²⁺ 0.065, Al³⁺ 0.050; K⁺ 0.133, Ca²⁺ 0.099, Ga³⁺ 0.062; Rb⁺ 0.148, Sr²⁺ 0.113, In³⁺ 0.081; Cs⁺ 0.169, Ba²⁺ 0.135, Tl³⁺ 0.095; O²⁻ 0.140, F⁻ 0.136; S²⁻ 0.184, Cl⁻ 0.181; Se²⁻ 0.198, Br⁻ 0.195; Te²⁻ 0.221, I⁻ 0.216.

FIGURE 6.12
Atomic and ionic radii of the main-group elements. Negative ions are always larger than atoms of the same element, whereas positive ions are always smaller than atoms of the same element. **OHT**

It is possible to explain these trends in terms of the electron configurations of the corresponding atoms. Consider first the increase in radius observed as we move down the table, let us say among the alkali metals (Group 1). All these elements have a single s electron outside a filled level or filled p sublevel. Electrons in these inner levels are much closer to the nucleus than the outer s electron and hence effectively shield it from the positive charge of the nucleus. To a first approximation, each inner electron cancels the charge of one proton in the nucleus, so the outer s electron is attracted by a net positive charge of +1. In this sense, it has the properties of an electron in the hydrogen atom. Because the average distance of the electron from the hydrogen nucleus increases with the principal quantum number, $\mathbf{n}$, the radius increases moving from Li (2s electron) to Na (3s electron) and so on down the group.

The decrease in atomic radius moving across the periodic table can be explained in a similar manner. Consider, for example, the third period, where electrons are being added to the third principal energy level. The added electrons should be relatively poor shields for each other because they are all at about the same distance from the nucleus. Only the ten electrons in inner, filled levels ($\mathbf{n} = 1$, $\mathbf{n} = 2$) are expected to shield the outer electrons from the nucleus. This means that the charge felt by an outer electron, called the *effective nuclear charge,* should increase steadily with atomic number as we move across the period. As effective nuclear charge increases, the outermost electrons are pulled in more tightly, and atomic radius decreases.

See Screen 8.6, Effective Nuclear Charge, Z^*.

Ionic Radius

See Screen 8.11, Atomic Properties and Periodic Trends: Ion Sizes.

The radii of cations and anions derived from atoms of the main-group elements are shown at the bottom of Figure 6.12. The trends referred to previously for atomic radii are clearly visible with ionic radius as well. Notice, for example, that **ionic radius** increases moving down a group in the periodic table. Moreover, the radii of both cations (left) and anions (right) decrease from left to right across a period.

Comparing the radii of cations and anions with those of the atoms from which they are derived—

- *positive ions are smaller than the metal atoms from which they are formed.* The Na^+ ion has a radius, 0.095 nm, only a little more than half that of the Na atom, 0.186 nm.
- *negative ions are larger than the nonmetal atoms from which they are formed.* The radius of the Cl^- ion, 0.181 nm, is nearly twice that of the Cl atom, 0.099 nm.

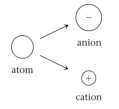

As a result of these effects, anions in general are larger than cations. Compare, for example, the Cl^- ion (radius = 0.181 nm) with the Na^+ ion (radius = 0.095 nm). This means that in sodium chloride, and indeed in the vast majority of all ionic compounds, most of the space in the crystal lattice is taken up by anions.

The differences in radii between atoms and ions can be explained quite simply. A cation is smaller than the corresponding metal atom because the excess of protons in the ion draws the outer electrons in closer to the nucleus. In contrast, an extra electron in an anion adds to the repulsion between outer electrons, making a negative ion larger than the corresponding nonmetal atom.

EXAMPLE 6.9   Using only the periodic table, arrange each of the following sets of atoms and ions in order of increasing size.

(a) Mg, Al, Ca   (b) S, Cl, S^{2-} (c) Fe, Fe^{2+}, Fe^{3+}

Strategy   Recall that radius decreases across a period and increases moving down a group. An atom is larger than the corresponding cation but smaller than the corresponding anion.

Solution

(a) Compare the other atoms with Mg. Al, to the right, is smaller than Mg. Ca, below Mg, is larger. The predicted order is $Al < Mg < Ca$.
(b) Compare with the S atom. Cl, to the right, is smaller. The S^{2-} anion is larger than the S atom. The predicted order is $Cl < S < S^{2-}$.
(c) Compare with the Fe^{2+} ion. The Fe atom, with no charge, is larger. The Fe^{3+} ion, with a +3 charge, is smaller. The predicted order is $Fe^{3+} < Fe^{2+} < Fe$ (Figure 6.13).

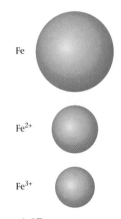

FIGURE 6.13
Relative sizes of the iron atom and its ions.

Ionization Energy

Ionization energy is a measure of how difficult it is to remove an electron from a gaseous atom. Energy must always be *absorbed* to bring about ionization, so ionization energies are always *positive* quantities.

The (first) ionization energy is the energy change for the removal of the outermost electron from a gaseous atom to form a +1 ion:

$$M(g) \longrightarrow M^+(g) + e^-   \Delta E_1 = \text{first ionization energy}$$

The more difficult it is to remove electrons, the larger the ionization energy.

Ionization energies of the main-group elements are listed in Figure 6.14. Notice that ionization energy—

- increases across the periodic table from left to right.
- decreases moving down the periodic table.

Comparing Figures 6.12 and 6.14 shows an inverse correlation between ionization energy and atomic radius. The smaller the atom, the more tightly its elec-

See Screen 8.12, Atomic Properties and Periodic Trends: Ionization Energy.

1	2		13	14	15	16	17	18
							H 1312	He 2372
Li 520	Be 900		B 801	C 1086	N 1402	O 1314	F 1681	Ne 2081
Na 496	Mg 738		Al 578	Si 786	P 1012	S 1000	Cl 1251	Ar 1520
K 419	Ca 590		Ga 579	Ge 762	As 944	Se 941	Br 1140	Kr 1351
Rb 403	Sr 550		In 558	Sn 709	Sb 832	Te 869	I 1009	Xe 1170
Cs 376	Ba 503		Tl 589	Pb 716	Bi 703	Po 812	At	Rn 1037

FIGURE 6.14
First ionization energies of the main-group elements, in kilojoules per mole. In general, ionization energy decreases moving down the periodic table groups and across the periods, although there are several exceptions. **OHT**

trons are held to the positively charged nucleus and the more difficult they are to remove. Conversely, in a large atom such as that of a Group 1 metal, the electron is relatively far from the nucleus, so less energy has to be supplied to remove it from the atom.

EXAMPLE 6.10   Consider the three elements B, C, and Al. Using only the periodic table, predict which of the three elements has

(a) the largest atomic radius; the smallest atomic radius.
(b) the largest ionization energy; the smallest ionization energy.

Strategy   Because these three elements form a block,

$$B \quad C$$
$$Al$$

in the periodic table, it is convenient to compare both carbon and aluminum with boron. Recall the trends for atomic radius and ionization energy.

Solution

(a) B is larger than C but smaller than Al, so   Al must be the largest atom and C the smallest.
(b) B has a smaller ionization energy than C, but a larger ionization energy than Al. Hence Al has the smallest ionization energy and C the largest.

If you look carefully at Figure 6.14, you will note a few exceptions to the general trends referred to above and illustrated in Example 6.10. For example, the ionization energy of B (801 kJ/mol) is *less* than that of Be (900 kJ/mol). This happens because the electron removed from the boron atom comes from the 2p as opposed to the 2s sublevel for beryllium. Because 2p is higher in energy than 2s, it is not too surprising that less energy is required to remove an electron from that sublevel.

Electronegativity

The ionization energy of an atom is a measure of its tendency to lose electrons; the larger the ionization energy, the more difficult it is to remove an electron. There are several different ways of comparing the tendencies of different atoms to gain electrons. The most useful of these for our purposes is the **electronegativity,** which measures the ability of an atom to attract to itself the electron pair forming a covalent bond.

The greater the electronegativity of an atom, the greater its attraction for electrons. Table 6.6 shows a scale of electronegativities first proposed by Linus Pauling. Each element is assigned a number, ranging from 4.0 for the most electronegative element, fluorine, to 0.8 for cesium, the least electronegative. Among the main-group elements, electronegativity increases moving from left to right in the periodic table. Ordinarily, it decreases moving down a group. You will find Table 6.6 very helpful when we discuss covalent bonding in Chapter 7.

⌐ Electronegativity is a positive
quantity.

TABLE 6.6 Electronegativity Values OHT

H 2.2							—*
Li 1.0	Be 1.6	B 2.0	C 2.5	N 3.0	O 3.5	F 4.0	—*
Na 0.9	Mg 1.3	Al 1.6	Si 1.9	P 2.2	S 2.6	Cl 3.2	—*
K 0.8	Ca 1.0	Sc 1.4	Ge 2.0	As 2.2	Se 2.5	Br 3.0	Kr 3.3
Rb 0.8	Sr 0.9	Y 1.2	Sn 1.9	Sb 2.0	Te 2.1	I 2.7	Xe 3.0
Cs 0.8	Ba 0.9						

*The noble gases He, Ne, and Ar are not listed because they form no stable compounds.

CHEMISTRY
Beyond the Classroom

The Aurora

What awesome sights are these northern lights, whatever be their meaning
As they dance and prance their ritual dance, shifting, fading, gleaming, . . .

—Robert H. Eather in "Majestic Lights,"
1980, American Geophysical Union

Far and away the most spectacular atmospheric phenomenon is the light display called the aurora (*aurora borealis* or "northern lights" in the northern hemisphere; *aurora australis* in the southern hemisphere). Still pictures (Figure A) can never capture their full glory; the rays of light pulsate constantly, alternately waxing and waning. Usually the aurora is brightly colored, most often red, green, or blue. The lights can appear as high as 1000 km in the night sky; their greatest intensity is typically found 100 to 150 km above the Earth's surface.

The aurora borealis is concentrated in an oval-shaped area near the Earth's north magnetic pole. Occasionally it becomes visible much farther south, which explains why it was studied extensively by Benjamin Franklin in the United States and John Dalton in England. If you've never seen an aurora (most people haven't), you could travel to

FIGURE A
The aurora borealis swirling over northwestern Canada.
(George Lepp/Tony Stone Images)

(continued)

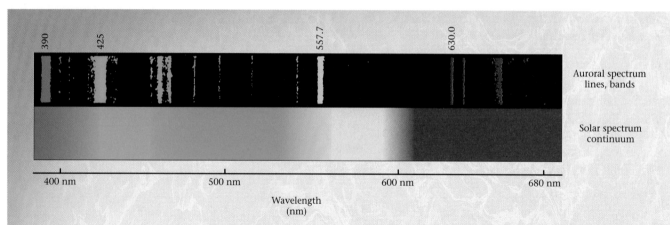

Auroral spectrum
lines, bands

Solar spectrum
continuum

Wavelength
(nm)

FIGURE B

Colors in the aurora. The red and green colors characteristic of the aurora are due to lines in
the spectrum of atomic oxygen at 630.0 nm and 557.7 nm. *(From "Majestic Lights" by Robert H.*
Eather, American Geophysical Union)

Barrow, Alaska, where there is a display every night of the year. (Take a sweater along, because the average nighttime temperature in Barrow is somewhere below −20°C). If you
live in New York City, you have about 1 chance in 20 of seeing the aurora on a clear,
moonless night when there is a power outage. Rural New England (e.g., Storrs, CT) where
the bright lights are dimmer would be a better choice (for seeing the aurora, that is). To
further improve your chances, note that northern lights are seen most often—

- around the equinoxes in March and September, between 9 P.M. and midnight.
- a couple of days after a major solar disturbance ("Sun spots").

Hundreds of years ago, native Americans interpreted the aurora to be the spirits of
the dead fighting one another in the sky; the red glow was supposed to reflect the blood
spilled in these conflicts. Today, scientists have a more mundane explanation. We at-
tribute the colors of the aurora to the visible spectrum produced when electrons and protons from the Sun interact with and energize atoms, molecules, and ions in the upper atmosphere. Two types of spectra are observed (Figure B)—

- *line spectra,* very similar to those discussed in this chapter. These result from electron
  transitions in atoms, mostly O and N, or to a lesser extent, in monatomic ions such as
  O^+ or N^+. The two brightest lines in the auroral spectrum, at 557.7 nm (green) and
  630.0 nm (red), are both due to atomic oxygen.
- *band spectra,* where the emission of light is spread over a range of perhaps 10 to
  100 nm in a particular region of the spectrum. These arise from electron transitions in
  molecules (O_2, N_2) or molecular ions (O_2^+, N_2^+). The blue color sometimes seen in au-
  roras is associated with spectral bands around 400 nm in N_2^+ and N_2.

Curiously, neither of the two lines referred to above, at 557.7 nm and 630.0 nm, is
observed in the normal spectrum of oxygen, as obtained with laboratory instruments at
the Earth's surface. There, excited oxygen atoms lose their energy, not by emission of
light but by collision with other atoms. In the upper atmosphere, atoms are very far apart
(the total pressure at 100 km is only 10^{-8} atm). As a result, they seldom collide with one
another, so excited oxygen atoms stay around long enough to emit visible light.

CHAPTER HIGHLIGHTS

Key Concepts

1. Relate wavelength, frequency, and energy.
   (**Examples 6.1, 6.2: Problems 1–8, 62, 63, 65**)
2. Use the Bohr model to identify lines in the hydrogen spectrum.
   (**Example 6.3; Problems 9–16**)
3. Identify the quantum numbers and capacities corresponding to energy levels, sublevels, and orbitals.
   (**Example 6.4; Problems 17–28, 37, 38, 68, 73, 78, 79**)
4. Write electron configurations, full or abbreviated, for atoms and ions.
   (**Examples 6.5, 6.6, 6.8; Problems 29–36, 49–52**)
5. Write orbital diagrams for atoms and ions.
   (**Example 6.7; Problems 39–48**)
6. Identify periodic trends in radii, ionization energy, and electronegativity.
   (**Examples 6.9, 6.10; Problems 53–60**)

Key Equations

Frequency-wavelength $\lambda \nu = c = 2.998 \times 10^8 \text{ m/s}$

Energy-frequency $E = h\nu = hc/\lambda$ $h = 6.626 \times 10^{-34} \text{ J} \cdot \text{s}$

Bohr model $E_{\mathbf{n}} = -R_H/\mathbf{n}^2$ $R_H = 2.180 \times 10^{-18} \text{ J}$

Key Terms

actinide
atomic orbital (s, p, d, f)
atomic radius
electron configuration
—abbreviated
electron spin
electronegativity
excited state
frequency

ground state
ionic radius
Hund's rule
ionization energy
lanthanide
■ main-group element
■ noble gas
orbital diagram
Pauli exclusion principle

photon
principal energy level
quantum number
spectrum
sublevel (s, p, d, f)
■ transition metal
wavelength

Summary Problem

Consider the element nickel ($Z = 28$).

(a) There is a line in the nickel spectrum at 212.59 nm. In what region of the spectrum (ultraviolet, visible, or infrared) is this line found? What is the frequency of the line? What is the energy difference between the two levels responsible for this line in kJ/mol?

(b) The ionization energy of nickel from the ground state is 757 kJ/mol. Assume that the transition in (a) is from the ground state to an excited state. If that is the case, calculate the ionization energy from the excited state.

(c) Give the electron configuration of the Ni atom; the Ni^{2+} ion.

(d) Give the orbital diagram (beyond argon) for Ni and Ni^{2+}.

(e) How many unpaired electrons are there in the Ni atom? In the Ni^{2+} ion?

(f) How many s electrons are there in the Ni atom? In the Ni^{2+} ion?

(g) Rank Ni, Ni^{2+}, and Ni^{3+} in order of increasing size.

Answers

(a) ultraviolet; 1.410×10^{15} s^{-1}; 562.7 kJ/mol

(b) 194 kJ

(c) $1s^2\ 2s^2\ 2p^6\ 3s^2\ 3p^6\ 4s^2\ 3d^8$; $1s^2\ 2s^2\ 2p^6\ 3s^2\ 3p^6\ 3d^8$

(d)

	4s	3d
Ni:	(↑↓)	(↑↓)(↑↓)(↑↓)(↑)(↑)
Ni^{2+}:	( )	(↑↓)(↑↓)(↑↓)(↑)(↑)

(e) 2; 2

(f) 8; 6

(g) $Ni^{3+} < Ni^{2+} < Ni$

Questions & Problems

Problem numbers in blue indicate that the answer is available in Appendix 6 at the back of the book.
WEB indicates that the solution is posted at **http://www.harcourtcollege.com/chem/general/masterton4/student/**

Energy, Wavelength, and Frequency

1. Magnetic resonance imaging (MRI) is a powerful diagnostic tool used in medicine. The imagers used in hospitals operate at a frequency of 4.00×10^2 MHz (1 MHz = 10^6 Hz). Calculate

  (a) the wavelength.

  (b) the energy in joules per photon.

  (c) the energy in kilojoules per mole.

2. A photon of violet light has a wavelength of 423 nm. Calculate

  (a) the frequency.

  (b) the energy in joules per photon.

  (c) the energy in kilojoules per mole.

3. A line in the spectrum of neon has a wavelength of 837.8 nm.

  (a) In what spectral range does the absorption occur?

  (b) Calculate the frequency of this absorption.

  (c) What is the energy in kilojoules per mole?

4. Carbon dioxide absorbs energy at a wavelength of 1498 nm.

  (a) In what spectral range does the absorption occur?

  (b) Calculate the frequency of this absorption.

  (c) What is the energy absorbed by one photon?

5. The ionization energy of potassium is 419 kJ/mol. Do x-rays with a wavelength of 78 nm have sufficient energy to ionize rubidium?

6. A laser with a wavelength of 643 nm is turned on for two minutes. During that time it emits 59 J of energy. How many photons are emitted by the laser while it is on?

7. Energy from radiation can cause chemical bonds to break. To break the nitrogen-nitrogen bond in N_2 gas, 941 kJ/mol is required.

  (a) Calculate the wavelength of the radiation that could break the bond.

  (b) In what spectral range does this radiation occur?

WEB 8. Microwave ovens heat food by the energy given off by microwaves. These microwaves have a wavelength of 5.00×10^6 nm.

  (a) How much energy in kilojoules per mole is given off by a microwave oven?

  (b) Compare the energy obtained in (a) with that given off by the ultraviolet rays ($\lambda \approx 100$ nm) of the Sun that you absorb when you try to get a tan.

Bohr Model of the Hydrogen Atom

9. Consider the transition from the energy levels **n** = 1 to **n** = 3.

  (a) What is the wavelength associated with this transition?

  (b) In what spectral region does this transition occur?

  (c) Is energy absorbed?

10. Consider the transition from the energy levels $n = 4$ to $n = 2$.

(a) What is the frequency associated with this transition?

(b) In what spectral region does this transition occur?

(c) Is energy absorbed?

11. According to the Bohr model, the radius of a circular orbit is given by the equation

$$r(\text{in nm}) = 0.0529\, n^2$$

Draw successive orbits for the hydrogen electron at $n = 1, 2, 3$, and 4. Indicate by arrows transitions between orbits that lead to lines in the

(a) Lyman series ($n_{lo} = 1$).

(b) Balmer series ($n_{lo} = 2$).

12. Calculate E_n for $n = 1, 2, 3$, and 4 ($R_H = 2.180 \times 10^{-18}$ J). Make a one-dimensional graph showing energy, at different values of n, increasing vertically. On this graph, indicate by vertical arrows transitions in the

(a) Lyman series ($n_{lo} = 1$).

(b) Balmer series ($n_{lo} = 2$).

13. The Brackett series lines in the atomic spectrum of hydrogen result from transitions from $n > 4$ to $n = 4$.

(a) Calculate the wavelength, in nanometers, of a line in this series resulting from the $n = 6$ to $n = 4$ transition.

(b) In what region of the spectrum are these lines formed?

14. For the Pfund series, $n_{lo} = 5$.

(a) Calculate the wavelength in nanometers of a transition from $n = 6$ to $n = 5$.

(b) In what region of the spectrum are these lines formed?

15. In the Paschen series, $n_{lo} = 3$. Calculate the longest wavelength possible for a transition in this series.

16. A line in the Lyman series ($n_{lo} = 1$) occurs at 97.23 nm. Calculate n_{hi} for the transition associated with this line.

Energy Levels, Sublevels, and Quantum Numbers

17. What are the possible values for m_ℓ for

(a) the d sublevel?

(b) the s sublevel?

(c) all sublevels where $n = 2$?

WEB **18.** What are the possible values for m_ℓ for

(a) the p sublevel?

(b) the f sublevel?

(c) all sublevels where $n = 3$?

19. For the following pairs of orbitals, indicate which is higher in energy in a many-electron atom.

(a) 3s or 3p   (b) 4s or 4d

(c) 2s or 3d   (d) 5s or 4s

20. For the following pairs of orbitals, indicate which is lower in energy in a many-electron atom.

(a) 3p or 4p   (b) 4p or 4d

(c) 1s or 2s   (d) 5f or 4d

21. What type of electron orbital (i.e., s, p, d, or f) is designated by

(a) $n = 2$, $\ell = 1$, $m_\ell = -1$?

(b) $n = 1$, $\ell = 0$, $m_\ell = 0$?

(c) $n = 5$, $\ell = 2$, $m_\ell = 2$?

22. What type of electron orbital (i.e., s, p, d, or f) is designated by

(a) $n = 3$, $\ell = 2$, $m_\ell = -1$?

(b) $n = 6$, $\ell = 3$, $m_\ell = 2$?

(c) $n = 4$, $\ell = 3$, $m_\ell = 3$?

23. State the total capacity for electrons in

(a) $n = 4$.

(b) a 3s sublevel.

(c) a d sublevel.

(d) a p orbital.

24. Give the number of orbitals in

(a) n = 3.

(b) a 4p sublevel.

(c) an f sublevel.

(d) a d sublevel.

25. What is the

(a) minimum n value for ℓ designated as **d**?

(b) letter used to designate the sublevel with $\ell = 2$?

(c) number of orbitals in a sublevel where ℓ is designated as p?

(d) number of different sublevels when $n = 3$?

26. What is the

(a) minimum n value for ℓ designated as s?

(b) letter used to designate the sublevel with $\ell = 4$?

(c) number of orbitals in a sublevel where ℓ is designated as d?

(d) number of different sublevels when $n = 1$?

27. Given the following sets of electron quantum numbers, indicate those that could not occur and explain your answer.

(a) $1, 0, 0, -\frac{1}{2}$ (b) $1, 1, 0, +\frac{1}{2}$ (c) $3, 2, -2, +\frac{1}{2}$

(d) $2, 1, 2, +\frac{1}{2}$ (e) $4, 0, 2, +\frac{1}{2}$

28. Given the following sets of electron quantum numbers, indicate those that could not occur and explain your answer.

(a) $3, 0, 0, -\frac{1}{2}$ (b) $2, 2, 1, -\frac{1}{2}$ (c) $3, 2, 1, +\frac{1}{2}$

(d) $3, 1, 1, +\frac{1}{2}$ (e) $4, 2, -2, 0$

29. Write the ground state electron configuration for

(a) N   (b) Na   (c) Ne   (d) Ni   (e) Si

30. Write the ground state electron configuration for

(a) B   (b) Ba   (c) Be   (d) Bi   (e) Br

31. Write the abbreviated ground state electron configuration for

(a) S   (b) Se   (c) Sb   (d) Sc   (e) Si

32. Write the abbreviated ground state electron configuration for

(a) Hg   (b) Al   (c) As   (d) W   (e) At

33. Give the symbol of the element of lowest atomic number whose ground state has

(a) a completed f subshell.

(b 20 p electrons.

(c) two 4d electrons.

(d) five 5p electrons.

34. Give the symbol of the element of lowest atomic number whose ground state has

(a) a p electron.
(b) four f electrons.
(c) a completed d subshell.
(d) six s electrons.

35. What fraction of the total number of electrons is in d sublevels in
(a) C   (b) Cl   (c) Co

36. What fraction of the total number of electrons is in p sublevels in
(a) Mg   (b) Mn   (c) Mo

37. Which of the following electron configurations are for atoms in the ground state? In the excited state? Which are impossible?
(a) $1s^2\,2s^2\,3s^2$ (b) $1s^2\,2p^3$
(c) $1s^2\,2s^3\,2p^5$ (d) $1s^1\,2s^2\,2p^7$
(e) $2s^2\,2p^6\,3s^1$ (f) $1s^2\,2s^2\,2p^6\,3s^2\,3d^1$

WEB **38.** Which of the following electron configurations are for atoms in the ground state? In the excited state? Which are impossible?
(a) $1s^2 2s^1 3s^1$ (b) $2s^2 2p^2$
(c) $1s^2 2s^2 2p^3$ (d) $1s^2 1p^1 2s^2$
(e) $1s^2 2s^2 2p^6 3s^2 3p^6 3d^1$ (f) $1s^2 2s^2 2p^6 3s^2 3p^6 3d^1 3f^1$

Orbital Diagrams; Hund's Rule

39. Give the orbital diagram for an atom of
(a) Na   (b) O   (c) Co   (d) Cl

40. Give the orbital diagram of
(a) Li   (b) P   (c) F   (d) Fe

41. Give the symbol of the atom with the following orbital diagram beyond argon.

	4s	3d	4p
(a)	(↑↓)	(↑↓)(↑↓)(↑↓)(↑)(↑)	( )( )( )
(b)	(↑↓)	(↑↓)(↑↓)(↑)(↑)(↑)	( )( )( )
(c)	(↑↓)	(↑↓)(↑↓)(↑↓)(↑↓)(↑↓)	(↑)(↑)( )

42. Give the symbol of the atom with the following orbital diagram

	1s	2s	2p	3s	3p
(a)	(↑↓)	(↑↓)	(↑↓)(↑↓)(↑↓)	(↑↓)	( )( )( )
(b)	(↑↓)	(↑↓)	(↑↓)(↑↓)(↑↓)	(↑↓)	(↑)(↑)(↑)
(c)	(↑↓)	(↑↓)	(↑↓)(↑)(↑)	( )	( )( )( )

43. Give the symbols of
(a) all the elements in which all the 5d orbitals are half-full.
(b) all the nonmetals in period 3 that have two or more unpaired electrons.
(c) all the elements in Group 1 where the 5p sublevel is full.
(d) all the metalloids that have paired 4p electrons.

44. Give the symbols of
(a) all the elements in Group **17** that have filled 4p sublevels.
(b) all the nonmetals in the fourth period that have no unpaired electrons.

(c) all the metalloids that have filled 3s sublevels.
(d) all the elements in the second period that have one unpaired electron.

45. Give the number of unpaired electrons in an atom of
(a) mercury   (b) manganese   (c) magnesium

46. Give the number of unpaired electrons in an atom of
(a) phosphorus   (b) potassium   (c) plutonium (Pu)

47. In what main group(s) of the periodic table do element(s) have the following number of filled p orbitals in the outermost principal level?
(a) 0   (b) 1   (c) 2   (d) 3

48. Give the symbol of the main-group metals in period 4 with the following number of unpaired electrons per atom. (Transition metals are not included.)
(a) 0   (b) 1   (c) 2   (d) 3

Electron Arrangement in Ions

49. Write the ground state electron configuration for
(a) Li, Li^+ (b) O, O^{2-}
(c) Sc, Sc^{3+} (d) Co^{2+}, Co^{3+}

50. Write the ground state electron configuration for
(a) P, P^{3-} (b) Ca, Ca^{2+}
(c) Ti, Ti^{4+} (d) Mn^{2+}, Mn^{4+}

51. How many unpaired electrons are there in the following ions?
(a) Al^{3+} (b) Cl^- (c) Sr^{2+} (d) Ag^+

52. How many unpaired electrons are there in the following ions?
(a) Hg^{2+} (b) F^- (c) Sb^{3+} (d) Fe^{3+}

Trends in the Periodic Table

53. Arrange the elements Rb, Te, and I in order of
(a) increasing atomic radius.
(b) increasing first ionization energy.
(c) increasing electronegativity.

54. Arrange the elements Mg, S, and Cl in order of
(a) increasing atomic radius.
(b) increasing first ionization energy.
(c) increasing electronegativity.

55. Which of the four atoms Na, P, Cl, or K
(a) has the largest atomic radius?
(b) has the highest ionization energy?
(c) is the most electronegative?

56. Which of the four atoms Rb, Sr, Sb, or Cs
(a) has the smallest atomic radius?
(b) has the lowest ionization energy?
(c) is the least electronegative?

57. Select the larger member of each pair.
(a) K and K^+ (b) O and O^{2-}
(c) Tl and Tl^{3+} (d) Cu^+ and Cu^{2+}

58. Select the smaller member of each pair.
(a) N and N^{3-} (b) Ba and Ba^{2+}
(c) Se and Se^{2-} (d) Co^{2+} and Co^{3+}

59. List the following species in order of decreasing radius.
(a) K, Ca, Ca^{2+}, Rb   (b) S, Te^{2-}, Se, Te

WEB 60. List the following species in order of increasing radius.
(a) Co, Co^{2+}, Co^{3+} (b) Cl, Cl^-, Br^-

Unclassified

61. An argon-ion laser is used in some laser light shows. The argon ion has strong emissions at 485 nm and 512 nm.
(a) What is the color of these emissions?
(b) What is the energy associated with these emissions in kilojoules per mole?
(c) Write the ground state electron configuration and orbital diagram of Ar^+.

62. A lightbulb radiates 8.5% of the energy supplied to it as visible light. If the wavelength of the visible light is assumed to be 565 nm, how many photons per second are emitted by a 75-W lightbulb? (1 W = 1 J/s)

63. A carbon dioxide laser produces radiation of wavelength 10.6 micrometers (1 micrometer = 10^{-6} meter). If the laser produces about one joule of energy per pulse, how many photons are produced per pulse?

64. Name and give the symbol for the element with the characteristic given below:
(a) Its electron configuration is $1s^2 2s^2 2p^6 3s^2 3p^5$.
(b) Lowest ionization energy in Group **14**.
(c) Its 2^+ ion has the configuration $[_{18}Ar]\, 3d^5$.
(d) It is the alkali metal with the smallest atomic radius.
(e) Largest ionization energy in the fourth period.

Conceptual Questions

65. Compare the energies and wavelengths of two photons, one with a low frequency, the other with a high frequency.

66. Consider the following transitions
1. $n = 3$ to $n = 1$
2. $n = 2$ to $n = 3$
3. $n = 4$ to $n = 3$
4. $n = 3$ to $n = 5$
(a) For which of the transitions is energy absorbed?
(b) For which of the transitions is energy emitted?
(c) Which transitions involve the ground state?
(d) Which transition absorbs the most energy?
(e) Which transition emits the most energy?

67. Explain in your own words the physical meaning of the electron distribution in Figure 6.4.

68. State the relationship, if any, between the total number of orbitals and
(a) n   (b) ℓ (c) m_s

69. Explain the difference between
(a) the Bohr model of the atom and the quantum mechanical model.
(b) wavelength and frequency.
(c) the geometries of the three different p orbitals.

70. Explain in your own words what is meant by
(a) the Pauli exclusion principle.
(b) Hund's rule.
(c) a line in an atomic spectrum.
(d) the principal quantum number.

71. Indicate whether each of the following statements is true or false. If false, correct the statement.
(a) An electron transition from $n = 3$ to $n = 1$ gives off energy.
(b) Light emitted by an $n = 4$ to $n = 2$ transition will have a longer wavelength than that from an $n = 5$ to $n = 2$ transition.
(c) A sublevel of $\ell = 3$ has a capacity of ten electrons.
(d) An atom of Group **13** has three unpaired electrons.

72. Criticize or comment on the following statements:
(a) The energy of a photon is inversely proportional to its wavelength.
(b) The energy of the hydrogen electron is inversely proportional to the quantum number ℓ.
(c) Electrons start to enter the fifth principal level as soon as the fourth is full.

73. No currently known elements contain electrons in g ($\ell = 4$) orbitals in the ground state. If an element is discovered that has electrons in the g orbital, what is the lowest value for n in which these g orbitals could exist? What are the possible values of m_ℓ? How many electrons could a set of g orbitals hold?

74. Explain why
(a) negative ions are larger than their corresponding atoms.
(b) scandium, a transition metal, forms an ion with a noble-gas structure.
(c) electronegativity decreases down a group in the periodic table.

75. Indicate whether each of the following is true or false.
(a) Effective nuclear charge stays about the same when one goes down a group.
(b) Group **17** elements have seven electrons in their outer level.
(c) Energy is given off when an electron is removed from an atom.

Challenge Problems

76. The energy of any one-electron species in its nth state (n = principal quantum number) is given by $E = -BZ^2/n^2$, where Z is the charge on the nucleus and B is 2.180×10^{-18} J. Find the ionization energy of the Li^{2+} ion in its first excited state in kilojoules per mole.

77. In 1885, Johann Balmer, a mathematician, derived the following relation for the wavelength of lines in the visible spectrum of hydrogen

$$\lambda = \frac{364.5\, n^2}{(n^2 - 4)}$$

where λ is in nanometers and n is an integer that can be 3, 4, 5, . . . Show that this relation follows from the Bohr equation and the equation using the Rydberg constant. Note that in the Balmer series, the electron is returning to the $n = 2$ level.

78. Suppose the rules for assigning quantum numbers were as follows

$n = 1, 2, 3, \ldots$
$\ell = 0, 1, 2, \ldots, n$

$m_\ell = 0, 1, 2, \ldots, \ell + 1$

$m_s = +\frac{1}{2}$ or $-\frac{1}{2}$

Prepare a table similar to Table 6.2 based on these rules for **n** = 1 and **n** = 2. Give the electron configuration for an atom with eight electrons.

79. Suppose that the spin quantum number could have the values $\frac{1}{2}$, 0, and $-\frac{1}{2}$. Assuming that the rules governing the values of the other quantum numbers and the order of filling sublevels were unchanged,

(a) what would be the electron capacity of an s sublevel? A p sublevel? A d sublevel?

(b) how many electrons could fit in the **n** = 3 level?

(c) what would be the electron configuration of the element with atomic number 8? 17?

80. In the photoelectric effect, electrons are ejected from a metal surface when light strikes it. A certain minimum energy, E_{min}, is required to eject an electron. Any energy absorbed beyond that minimum gives kinetic energy to the electron. It is found that when light at a wavelength of 540 nm falls on a cesium surface, an electron is ejected with a kinetic energy of 2.60×10^{-20} J. When the wavelength is 400 nm, the kinetic energy is 1.54×10^{-19} J.

(a) Calculate E_{min} for cesium in joules.

(b) Calculate the longest wavelength, in nanometers, that will eject electrons from cesium.

COVALENT BONDING

7

Maximum repulsion for the four balloons is achieved with a tetrahedral geometry. *(Charles D. Winters)*

CHAPTER OUTLINE

7.1 LEWIS STRUCTURES; THE OCTET RULE

7.2 MOLECULAR GEOMETRY

7.3 POLARITY OF MOLECULES

7.4 ATOMIC ORBITALS; HYBRIDIZATION

See the *Saunders Interactive General Chemistry CD-ROM*, Screen 9.3, **Chemical Bond Formation.**

E arlier we referred to the forces that hold nonmetal atoms to one another, covalent bonds. These bonds consist of an electron pair shared between two atoms. To represent the covalent bond in the H_2 molecule, two structures can be written:

$$H:H \quad \text{or} \quad H\text{—}H$$

These structures can be misleading if they are taken to mean that the two electrons are fixed in position between the two nuclei. A more accurate picture of the electron density in H_2 is shown in Figure 7.1. At a given instant, the two electrons may be located at any of various points about the two nuclei. However, they are more likely to be found between the nuclei than at the far ends of the molecule.

To understand the stability of the electron-pair bond, consider the graph shown in Figure 7.2, where we plot the energy of interaction between two hydrogen atoms as a function of distance. At large distances of separation (far right) the system consists of two isolated H atoms that do not interact with each other. As the atoms come closer together (moving to the left in Figure 7.2), they experience an attraction that leads gradually to an energy minimum. At an internuclear distance of 0.074 nm and an attractive energy of 436 kJ, the system is in its most stable state; we refer to that state as the H_2 molecule. If the atoms are brought closer together, forces of repulsion become increasingly important and the energy curve rises steeply.

The existence of the energy minimum shown in Figure 7.2 is directly responsible for the stability of the H_2 molecule. The attractive forces that bring about this minimum result from two factors:

1. Locating two electrons between the two protons of the H_2 molecule lowers the electrostatic energy of the system. Attractive energies between oppositely charged particles (electron-proton) slightly exceed the repulsive energies between particles of like charge (electron-electron, proton-proton).

2. When two hydrogen atoms come together to form a molecule, the electrons are spread over the entire volume of the molecule instead of being confined to a particular atom. As pointed out in Chapter 6, quantum mechanics tells us that increasing the volume available to an electron decreases its kinetic energy. We often describe this situation by saying that the two 1s orbitals of the hydrogen atom "overlap" to form a new bonding orbital. At any rate, calculations suggest that this is the principal factor accounting for the stability of the H_2 molecule.

This chapter is devoted to the covalent bond as it exists in molecules and polyatomic ions. We consider—

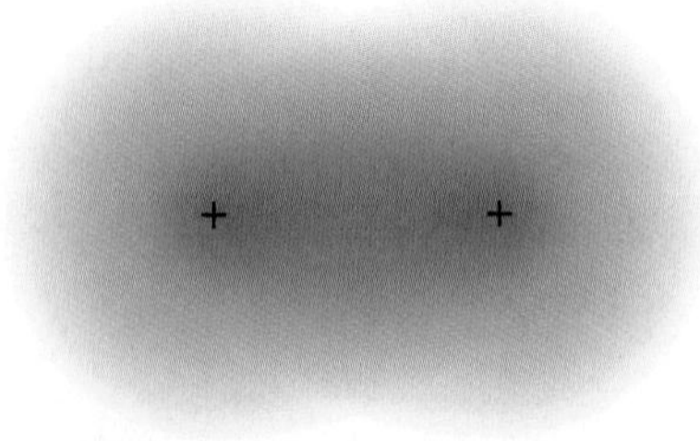

FIGURE 7.1

Electron density in H_2. The depth of color is proportional to the probability of finding an electron in a particular region. **OHT**

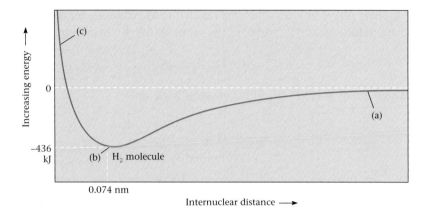

FIGURE 7.2
Energy of two hydrogen atoms as a function of the distance between their nuclei. (a) At zero energy, the H atoms are separated. (b) The minimum in the curve, which occurs at the observed internuclear distance of 0.074 nm, corresponds to the most stable state of the H_2 molecule. (c) At internuclear distances less than 0.074 nm, the energy of interaction rises rapidly because of repulsion between the hydrogen nuclei. **OHT**

- the distribution of outer level *(valence)* electrons in species in which atoms are joined by covalent bonds. These distributions are most simply described by *Lewis structures* (Section 7.1).

- molecular geometries. The so-called *VSEPR model* can be used to predict the angles between covalent bonds formed by a central atom (Section 7.2).

- the polarity of covalent bonds and the molecules they form (Section 7.3). Most bonds and many molecules are polar in the sense that they have a positive and a negative pole.

- the distribution of valence electrons among *atomic orbitals,* using the valence bond approach (Section 7.4).

⊙ **See Screen 9.2, Valence Electrons.**

7.1   LEWIS STRUCTURES; THE OCTET RULE

The idea of the covalent bond was first suggested by the American physical chemist G. N. Lewis in 1916. He pointed out that the electron configuration of the noble gases appears to be a particularly stable one. Noble-gas atoms are themselves extremely unreactive. Moreover, as pointed out in Chapter 6, a great many monatomic ions have noble-gas structures. Lewis suggested that **atoms, by sharing electrons to form an electron-pair bond, can acquire a stable noble-gas structure.** Consider, for example, two hydrogen atoms, each with one electron. The process by which they combine to form an H_2 molecule can be shown as

Noble-gas structures are stable in molecules, as they are in atoms and ions.

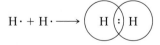

⊙ **See Screen 9.4, Lewis Electron Dot Structures.**

using dots to represent electrons; the circles emphasize that the pair of electrons in the covalent bond can be considered to occupy the 1s orbital of either hydrogen atom. In that sense, each atom in the H_2 molecule has the electronic structure of the noble gas helium, with the electron configuration $1s^2$.

This idea is readily extended to simple molecules of compounds formed by nonmetal atoms. An example is the HF molecule. You will recall that a fluorine atom has the electron configuration $1s^2 2s^2 2p^5$. It has seven electrons in its outermost principal energy level (**n** = 2). These are referred to as **valence electrons.** If these are shown as dots about the symbol of the element, the fluorine atom can be represented as

$$:\ddot{\text{F}}\cdot$$

The combination of a hydrogen with a fluorine atom leads to

$$\text{H}\cdot + \cdot\ddot{\text{F}}: \longrightarrow$$

| Shared electrons are counted for both atoms.

As you can see, the fluorine atom "owns" six valence electrons outright and shares two others. Putting it another way, the F atom is surrounded by eight valence electrons; its electron configuration has become $1s^2 2s^2 2p^6$, which is that of the noble gas neon. This, according to Lewis, explains why the HF molecule is stable in contrast to species such as H_2F, H_3F, . . . none of which exist.

These structures (without the circles) are referred to as **Lewis structures.** In writing Lewis structures, only the valence electrons written above are shown, because they are the ones that participate in covalent bonding. For the main-group elements, the only ones dealt with here, the number of valence electrons is equal to the last digit of the group number in the periodic table (Table 7.1). Notice that elements in a given main group all have the same number of valence electrons. This explains why such elements behave similarly when they react to form covalently bonded species.

In the Lewis structure of a molecule or polyatomic ion, valence electrons ordinarily occur in pairs. There are two kinds of electron pairs.

1. A pair of electrons shared between two atoms is a **covalent bond,** ordinarily shown as a straight line between bonded atoms.
2. An **unshared pair** of electrons, owned entirely by one atom, is shown as a pair of dots on that atom. (An unshared pair is often referred to, more picturesquely, as a *lone pair.*)

The Lewis structures for the species OH^-, H_2O, NH_3, and NH_4^+ are

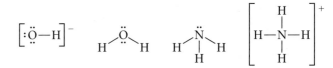

| This is why we put the last digit of the group number in bold type.

TABLE 7.1 Lewis Structures of Atoms Commonly Forming Covalent Bonds								
Group:	1	2	13	14	15	16	17	18
No. of valence e^-:	1	2	3	4	5	6	7	8
	H·							
		·Be·	·Ḃ·	·Ċ·	·N̈·	·Ö·	:F̈·	
				·S̈i·	·P̈·	·S̈·	:C̈l·	
				·Ġe·	·Äs·	·S̈e·	:B̈r·	:K̈r:
					·S̈b·	·T̈e·	:Ï·	:Ẍe:

Notice that in each case the oxygen or nitrogen atom is surrounded by eight valence electrons. In each species, a single electron pair is shared between two bonded atoms. These bonds are called **single bonds.** There is one single bond in the OH^- ion, two in the H_2O molecule, three in NH_3, and four in NH_4^+. There are three unshared pairs in the hydroxide ion, two in the water molecule, one in the ammonia molecule, and none in the ammonium ion.

Bonded atoms can share more than one electron pair. A **double bond** occurs when bonded atoms share two electron pairs; in a **triple bond,** three pairs of electrons are shared. In ethylene (C_2H_4) and acetylene (C_2H_2), the carbon atoms are linked by a double bond and triple bond, respectively. Using two parallel lines to represent a double bond and three for a triple bond, we write the structures of these molecules as

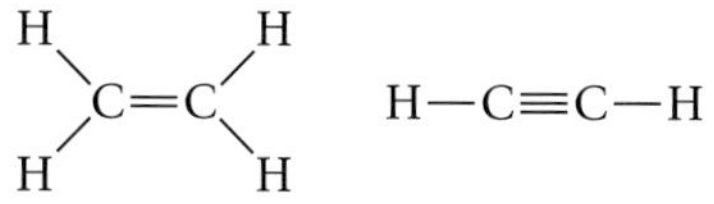

ethylene, C_2H_4 acetylene, C_2H_2

Note that each carbon is surrounded by eight valence electrons and each hydrogen by two.

These examples illustrate the principle that atoms in covalently bonded species tend to have noble-gas electronic structures. This generalization is often referred to as the **octet rule.** Nonmetals, except for hydrogen, achieve a noble-gas structure by sharing in an *octet* of electrons (eight). Hydrogen atoms, in molecules or polyatomic ions, are surrounded by a *duet* of electrons (two).

Writing Lewis Structures

For very simple species, Lewis structures can often be written by inspection. Usually, though, you will save time by following these steps:

1. Count the number of valence electrons. For a molecule, simply sum up the valence electrons of the atoms present. For a polyatomic anion, one electron is added for each unit of negative charge. For a polyatomic cation, a number of electrons equal to the positive charge must be subtracted.

2. Draw a skeleton structure for the species, joining atoms by single bonds. In some cases, only one arrangement of atoms is possible; in others, there are two or more alternative structures. Most of the molecules and polyatomic ions dealt with in this chapter consist of a *central atom* bonded to two or more *terminal atoms,* located at the outer edges of the molecule or ion. For such species (e.g., NH_4^+, SO_2, CCl_4), it is relatively easy to derive the skeleton structure. *The central atom is usually the one written first in the formula* (N in NH_4^+, S in SO_2, C in CCl_4); *put this in the center of the molecule or ion. Terminal atoms are most often hydrogen, oxygen, or a halogen; bond these atoms to the central atom.*

3. Determine the number of valence electrons still available for distribution. To do this, deduct two valence electrons for each single bond written in Step 2.

4. Determine the number of valence electrons required to fill out an octet for each atom (except H) **in the skeleton.** Remember that shared electrons are counted for both atoms.

(a) If the number of electrons available (Step 3) is equal to the number required (Step 4), distribute the available electrons as unshared pairs, satisfying the octet rule for each atom.

Forming a multiple bond "saves" electrons because bonding pairs are counted for both atoms.

Most molecules follow the octet rule.

See Screen 9.5, Drawing Lewis Electron Dot Structures.

It would be nice if all shortages could be solved so simply.

(b) If the number of electrons available (Step 3) is less than the number required (Step 4), the skeleton structure must be modified by changing single to multiple bonds. If you are two electrons short, convert a single bond to a double bond; if there is a deficiency of four electrons, convert a single bond to a triple bond (or two single bonds to double bonds). Multiple bond formation is pretty much limited to the four atoms: C, N, O, and S. Hydrogen and halogen atoms never form double bonds.

The application of these steps and some further guiding principles are shown in Example 7.1. Later we will consider how to draw Lewis structures of species that do not follow the octet rule.

Hypochlorite ions in action. The $HClO^-$ ion is the active bleaching agent in Clorox™. *(Charles D. Winters)*

EXAMPLE 7.1   Draw Lewis structures of

(a) the hypochlorite ion, OCl^- (b) methyl alcohol, CH_4O

Strategy   Follow the steps listed above. For the OCl^- ion, only one skeleton is possible; for methyl alcohol, keep in mind that the carbon atom ordinarily forms four bonds. Hydrogen must be a terminal atom because it forms only one bond.

Solution

(a)

(1) The number of valence electrons is 6 (from O) + 7 (from Cl) + 1 (from the −1 charge) = 14.
(2) The skeleton structure is $[O{-}Cl]^-$.
(3) The number of electrons available for distribution is 14 (originally) − 2 (used in skeleton) = 12.
(4) The number of electrons required to give each atom an octet is 6 (for O) + 6 (for Cl) = 12.

The number of available electrons is the same as the number required. This skeleton structure is correct; there are no multiple bonds. The Lewis structure is

$$\left[:\ddot{O}{-}\ddot{Cl}:\right]^-$$

(b)

(1) The number of valence electrons is 4 (from C) + 4 (1 from each H) + 6 (from O) = 14.
(2) Because carbon forms four bonds and H must always be a terminal atom, the only reasonable skeleton is

$$\begin{array}{c} H \\ | \\ H{-}C{-}O{-}H \\ | \\ H \end{array}$$

(3) The skeleton structure contains five single bonds, using up $5 \times 2e^- = 10e^-$. The number of electrons available for distribution is $14 - 10 = 4$.
(4) The number of electrons needed to satisfy the octet rule is 4 (for O). Again, the number of electrons available is equal to the number required. The Lewis structure is

Reality Check   Whenever you write a Lewis structure, check to see if it obeys the octet rule. The structures written for OCl^- and CH_4O do just that; each atom except H is surrounded by eight electrons.

EXAMPLE 7.2 Draw Lewis structures for

(a) SO_2 (b) N_2

Strategy Follow the four steps for writing Lewis structures.

Solution

(a)
(1) There are 18 valence electrons. Both sulfur (two atoms) and oxygen (one atom) are in Group **16**; $3 \times 6 = 18$.
(2) The skeleton structure, with sulfur as the central atom, is

$$O—S—O$$

The two single bonds account for four electrons.
(3) There are $18 - 4 = 14$ electrons available for distribution.
(4) Sixteen electrons are required to give each atom an octet (four for S and six for each O). There is a deficiency of two electrons. This means that a single bond in the skeleton must be converted to a double bond. The Lewis structure of SO_2 is

(b)
(1) There are ten valence electrons; nitrogen is in Group **15**.
(2) The skeleton structure is

$$N—N$$

(3) There are $10 - 2 = 8$ electrons available for distribution.
(4) Each N needs six electrons for an octet, so 12 electrons are needed. This means that there is a deficiency of $12 - 8 = 4$ electrons. Convert the single bond between the two N atoms to a triple bond. The Lewis structure is

$$:N≡N:$$

Resonance Forms

In certain cases, the Lewis structure does not adequately describe the properties of the ion or molecule that it represents. Consider, for example, the SO_2 structure derived in Example 7.2. This structure implies that there are two different kinds of sulfur-to-oxygen bonds in SO_2. One of these appears to be a single bond, the other a double bond. Yet experiment shows that there is only one kind of bond in the molecule.

One way to explain this situation is to assume that each of the bonds in SO_2 is intermediate between a single and a double bond. To express this concept, two structures are written:

$$\ddot{O}=\overset{..}{S}-\ddot{O}: \longleftrightarrow :\ddot{O}-\overset{..}{S}=\ddot{O}$$

with the understanding that the true structure is intermediate between them. These are referred to as *resonance forms*. The concept of **resonance** is invoked whenever a single Lewis structure does not adequately reflect the properties of a substance.

⊙ See Screen 9.6, Resonance Structures.

⌐ The double-headed arrow is used to separate resonance structures.

Another species for which it is necessary to invoke the idea of resonance is the nitrate ion. Here three equivalent structures can be written

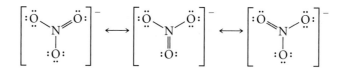

to explain the experimental observation that the three nitrogen-to-oxygen bonds in the NO_3^- ion are identical in all respects.

Resonance can also occur with many organic molecules, including benzene, C_6H_6, which is known to have a hexagonal ring structure. Benzene can be considered to be a resonance hybrid of the two forms:

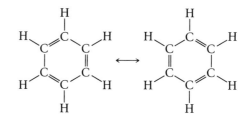

These structures are commonly abbreviated as

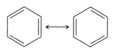

with the understanding that, at each corner of the hexagon, there is a carbon attached to a hydrogen atom.

We will encounter other examples of molecules and ions whose properties can be interpreted in terms of resonance. In all such species,

1. Resonance forms do not imply different kinds of molecules with electrons shifting eternally between them. There is only one type of SO_2 molecule; its structure is intermediate between those of the two resonance forms drawn for sulfur dioxide.

2. Resonance can be anticipated when it is possible to write two or more Lewis structures that are about equally plausible. In the case of the nitrate ion, the three structures we have written are equivalent. One could, in principle, write many other structures, but none of them would put eight electrons around each atom.

3. Resonance forms differ only in the distribution of electrons, not in the arrangement of atoms. The molecule

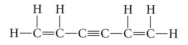

is not a resonance structure of benzene, even though it has the same molecular formula, C_6H_6. Indeed, it is an entirely different substance with different chemical and physical properties. Species such as these with the same molecular formula but different arrangements of atoms are referred to as *isomers*.

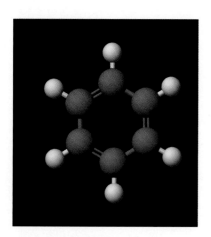

Benzene ball-and-stick model, showing double bonds.

EXAMPLE 7.3   Write three resonance forms for SO_3.

Strategy   Write a Lewis structure for SO_3 following the four steps for writing Lewis structures. Then write all the resonance forms by changing the position of the multiple bond. Do *not* change the skeleton structure.

Solution   The Lewis structure and all the resonance forms for SO_3 are

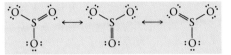

Reality Check   The three sulfur-to-oxygen bonds in SO_3 are identical in all respects. Writing resonance forms is one way of showing that.

Formal Charge

Often it is possible to write two different Lewis structures for a molecule differing in the arrangement of atoms, that is,

See Screen 9.13, Formal Charge.

$$A—A—B \quad \text{or} \quad A—B—A$$

Sometimes both structures represent real compounds that are isomers of each other. More often, only one structure exists in nature. For example, methyl alcohol (CH_4O) has the structure

$$H—\overset{\displaystyle H}{\underset{\displaystyle H}{\overset{|}{\underset{|}{C}}}}—\ddot{O}—H$$

In contrast, the structure

$$H—\ddot{C}\overset{|}{\underset{\displaystyle H}{}}—\ddot{O}\overset{|}{\underset{\displaystyle H}{}}—H$$

does not correspond to any real compound even though it obeys the octet rule.

There are several ways to choose the more plausible of two structures differing in their arrangement of atoms. As pointed out in Example 7.1, the fact that carbon almost always forms four bonds leads to the correct structure for methyl alcohol. Another approach involves a concept called **formal charge**, which can be applied to any atom within a Lewis structure. The formal charge is the difference between the number of valence electrons in the free atom and the number assigned to that atom in the Lewis structure. The assigned electrons include—

Formal charge is the charge an atom would have if valence electrons in bonds were distributed evenly.

- all the unshared electrons owned by that atom.
- one half of the bonding electrons shared by that atom.

This definition of formal charge, C_f, leads to the following equation

$$C_f = X - (Y + Z/2)$$

where X = the number of valence e^- in the free atom, which is equal to the last digit of the group number in the periodic table;
Y = the number of unshared e^- owned by the atom in the Lewis structure;
Z = the number of bonding e^- shared by the atom in the Lewis structure.

To show how this works, let's calculate the formal charges of carbon and oxygen in the two structures written above for methyl alcohol:

(1) $H-\overset{\overset{\displaystyle H}{|}}{\underset{\underset{\displaystyle H}{|}}{C}}-\overset{..}{\underset{..}{O}}-H$

(2) $H-\overset{\overset{\displaystyle H}{|}}{\underset{\underset{\displaystyle H}{|}}{C}}-\overset{..}{\underset{..}{O}}-H$

For C: $X = 4, Y = 0, Z = 8$
       $C_f = 4 - (0 + 8/2) = 0$
For O: $X = 6, Y = 4, Z = 4$
       $C_f = 6 - (4 + 4/2) = 0$

For C: $X = 4, Y = 2, Z = 6$
       $C_f = 4 - (2 + 6/2) = -1$
For O: $X = 6, Y = 2, Z = 6$
       $C_f = 6 - (2 + 6/2) = +1$

Ordinarily, the more likely Lewis structure is the one in which—

■ the formal charges are as close to zero as possible.
■ any negative formal charge is located on the most strongly electronegative atom.

Applying these rules, we can see that structure (1) for methyl alcohol is preferred over structure (2). In (1), both carbon and oxygen have formal charges of zero. In (2), a negative charge is assigned to carbon, which is actually less electronegative than oxygen (2.5 versus 3.5).

Formal charge is not an infallible guide to predicting Lewis structures.

The concept of formal charge has a much wider applicability than this short discussion might imply. In particular, it can be used to predict situations in which conventional Lewis structures, written in accordance with the octet rule, may be incorrect (Table 7.2).

Exceptions to the Octet Rule: Electron-Deficient Molecules

See Screen 9.7, Electron-Deficient Compounds.

Although most of the molecules and polyatomic ions referred to in general chemistry follow the octet rule, there are some familiar species that do not. Among these are molecules containing an odd number of valence electrons. Nitric oxide, NO, and nitrogen dioxide, NO_2, fall in this category:

NO   no. of valence electrons = 5 + 6 = 11

NO_2 no. of valence electrons = 5 + 6(2) = 17

See Screen 9.8, Free Radicals.

For such *odd electron* species (sometimes called free radicals) it is impossible to write Lewis structures in which each atom obeys the octet rule. In the NO mole-

TABLE 7.2 Possible Structures for BeF_2 and BF_5			
Structure I	C_f	**Structure II**	C_f
$:\overset{..}{F}=Be=\overset{..}{F}:$	Be = −2 F = +1	$:\overset{..}{\underset{..}{F}}-Be-\overset{..}{\underset{..}{F}}:$	Be = 0 F = 0
$\overset{:\overset{..}{F}}{\diagdown}\underset{\overset{\displaystyle\|}{:F:}}{B}\overset{\overset{..}{F}:}{\diagup}$	B = −1 F = +1,0,0	$\overset{:\overset{..}{F}}{\diagdown}\underset{\overset{\displaystyle\|}{:\overset{..}{\underset{..}{F}}:}}{B}\overset{\overset{..}{F}}{\diagup}$	B = 0 F = 0

cule, the unpaired electron is put on the nitrogen atom, giving both atoms a formal charge of zero:

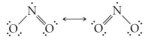

In NO_2, the best structure one can write again puts the unpaired electron on the nitrogen atom:

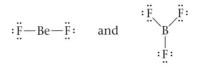

Elementary oxygen, like NO and NO_2, is paramagnetic (Figure 7.3). Experimental evidence suggests that the O_2 molecule contains two unpaired electrons *and* a double bond. It is impossible to write a conventional Lewis structure for O_2 that has these two characteristics. A more sophisticated model of bonding, using molecular orbitals (Appendix 5), is required to explain the properties of oxygen.

There are a few species in which the central atom violates the octet rule in the sense that it is surrounded by two or three electron pairs rather than four. Examples include the fluorides of beryllium and boron, BeF_2 and BF_3. Although one could write multiple bonded structures for these molecules in accordance with the octet rule (Table 7.2), experimental evidence suggests the structures

in which the central atom is surrounded by four and six valence electrons, respectively, rather than eight. Another familiar substance in which boron is surrounded by only three pairs of electrons rather than four is boric acid, H_3BO_3, used as an insecticide and fungicide.

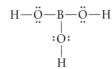

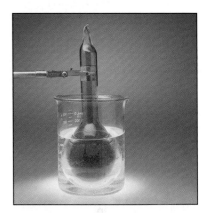

Two nitrogen oxides. NO_2 (*top*), a red-brown gas, has an unpaired electron on the N atom. When cooled, NO_2 forms N_2O_4 (*bottom*) by combination of the unpaired electrons on two NO_2 molecules to form an N—N covalent bond that converts the original two molecules into one. (*Charles D. Winters*)

FIGURE 7.3
Oxygen (O_2) in a magnetic field. The liquid oxygen, which is blue, is attracted into a magnetic field between the poles of an electromagnet. Both the paramagnetism and the blue color are due to the unpaired electrons in the O_2 molecule. (*S. Ruven Smith*)

Exceptions to the Octet Rule: Expanded Octets

In most molecules, the central atom is surrounded by 8 electrons. Rarely, it is surrounded by 4 (BeF_2) or 6 (BF_3). Occasionally, the number is 10 (PCl_5) or 12 (SF_6).

The largest class of molecules to violate the octet rule consists of species in which the central atom is surrounded by more than four pairs of valence electrons. Typical molecules of this type are phosphorus pentachloride, PCl_5, and sulfur hexafluoride, SF_6. The Lewis structures of these molecules are

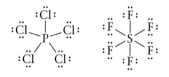

As you can see, the central atoms in these molecules have **expanded octets.** In PCl_5, the phosphorus atom is surrounded by 10 valence electrons (5 shared pairs); in SF_6, there are 12 valence electrons (6 shared pairs) around the sulfur atom.

In molecules of this type, the terminal atoms are most often halogens (F, Cl, Br, I); in a few molecules, oxygen is a terminal atom. The central atom is a nonmetal in the third, fourth, or fifth period of the periodic table. Most frequently, it is one of the following elements:

	Group 15	Group 16	Group 17	Group 18
3rd period	P	S	Cl	
4th period	As	Se	Br	Kr
5th period	Sb	Te	I	Xe

All these atoms have d orbitals available for bonding (3d, 4d, 5d). These are the orbitals in which the extra pairs of electrons are located in such species as PCl_5 and SF_6. Because there is no 2d sublevel, C, N, and O never form expanded octets.

Sometimes, as with PCl_5 and SF_6, it is clear from the formula that the central atom has an expanded octet. Often, however, it is by no means obvious that this is the case. At first glance, formulas such as ClF_3 or XeF_4 look completely straightforward. However, when you try to draw the Lewis structure it becomes clear that an expanded octet is involved. The number of electrons available after the skeleton is drawn is *greater* than the number required to give each atom an octet. When that happens, ***distribute the extra electrons (two or four) around the central atom as unshared pairs.***

EXAMPLE 7.4 Draw the Lewis structure of XeF_4.

Strategy Follow the usual four-step sequence. If there is an electron surplus, add the extra electrons to the central xenon atom as unshared pairs.

Solution

(1) The number of valence electrons is 8 (from Xe) + 28 (from four F atoms) = 36.
(2) The skeleton is

$$
\begin{array}{c}
\text{F} \\
| \\
\text{F}\!-\!\text{Xe}\!-\!\text{F} \\
| \\
\text{F}
\end{array}
$$

(3) The number of electrons available for distribution is 36 − 8 = 28.
(4) Each fluorine atom needs six electrons to complete its octet; the number of electrons required = 4(6) = 24.

CHEMISTRY · The Human Side

Born in Massachusetts, G. N. Lewis grew up in Nebraska, then came back east to obtain his B.S. (1896) and Ph.D. (1899) at Harvard. Although he stayed on for a few years as an instructor, Lewis seems never to have been happy at Harvard. A precocious student and an intellectual rebel, he was repelled by the highly traditional atmosphere that prevailed in the chemistry department there in his time. Many years later, he refused an honorary degree from his alma mater.

After leaving Harvard, Lewis made his reputation at MIT, where he was promoted to full professor in only four years. In 1912, he moved across the country to the University of California, Berkeley, as dean of the College of Chemistry and department head. He remained there for the rest of his life. Under his guidance, the chemistry department at Berkeley became perhaps the most prestigious in the country. Among the faculty and graduate students that he attracted were five future Nobel Prize winners: Harold Urey in 1934, William Giauque in 1949, Glenn Seaborg in 1951, Willard Libby in 1960, and Melvin Calvin in 1961.

In administering the chemistry department at Berkeley, Lewis demanded excellence in both research and teaching. Virtually the entire staff was involved in the general chemistry program; at one time eight full professors carried freshman sections.

Lewis's interest in chemical bonding and structure dated from 1902. In attempting to explain "valence" to a class at Harvard, he devised an atomic model to rationalize the octet rule. His model was deficient in many respects; for one thing, Lewis visualized cubic atoms with electrons located at the corners. Perhaps this explains why his ideas of atomic structure were not published until 1916. In that year, Lewis conceived of the electron-pair bond, perhaps his greatest single contribution to chemistry. At that time, it was widely believed that all bonds were ionic; Lewis's ideas were rejected by many well-known organic chemists.

In 1923, Lewis published a classic book (later reprinted by Dover Publications) titled *Valence and the Structure of Atoms and Molecules*. Here, in Lewis's characteristically lucid style, we find many of the basic principles of covalent bonding discussed in this chapter. Included are electron-dot structures, the octet rule, and the concept of electronegativity. Here too is the Lewis definition of acids and bases (Chapter 15). That same year, Lewis published with Merle Randall a text called *Thermodynamics and the Free Energy of Chemical Substances*. Today, a revised edition of that text is still widely used in graduate courses in chemistry.

The years from 1923 to 1938 were relatively unproductive for G. N. Lewis insofar as his own research was concerned. The applications of the electron-pair bond came largely in the areas of organic and quantum chemistry; in neither of these fields did Lewis feel at home. In the early 1930s, he published a series of relatively minor papers dealing with the properties of deuterium. Then in 1939 he began to publish in the field of photochemistry. Of approximately 20 papers in this area, several were of fundamental importance, comparable in quality to the best work of his early years. Retired officially in 1945, Lewis died a year later while carrying out an experiment on fluorescence.

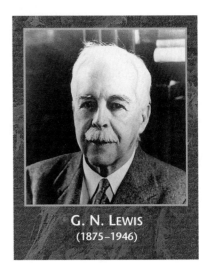

G. N. LEWIS
(1875–1946)

Lewis certainly deserved a Nobel Prize, but he never received one.

(Photo credit: Dr. Glenn Seaborg, University of California, Lawrence Berkeley Laboratory)

There is a surplus of electrons; 28 are available and only 24 are required to fill out octets. The four extra electrons are added to the central Xe atom. The Lewis structure for XeF_4 is

$$:\ddot{F}: \\ | \\ :\ddot{F} - \ddot{Xe} - \ddot{F}: \\ | \\ :\ddot{F}:$$

The extra electrons go to the central atom.

In this molecule there are six pairs of electrons around the xenon atom.

7.2   MOLECULAR GEOMETRY

The geometry of a diatomic molecule such as Cl_2 or HCl can be described very simply. Because two points define a straight line, the molecule must be linear.

$$Cl—Cl \qquad H—Cl$$

With molecules containing three or more atoms, the geometry is not so obvious. Here, the angles between bonds, called **bond angles,** must be considered. For example, a molecule of the type YX_2, where Y represents the central atom and X an atom bonded to it, can be—

- *linear,* with a bond angle of 180°:   X—Y—X

- *bent,* with a bond angle less than 180°: $X{\diagup}^{Y}{\diagdown}_{X}$

⊙ **See Screen 9.14, Molecular Shape.**

The major features of **molecular geometry** can be predicted on the basis of a quite simple principle—electron-pair repulsion. This principle is the essence of the *valence-shell electron-pair repulsion (VSEPR) model,* first suggested by N. V. Sidgwick and H. M. Powell in 1940. It was developed and expanded later by R. J. Gillespie and R. S. Nyholm. According to the **VSEPR model,** *the valence electron pairs surrounding an atom repel one another. Consequently, the orbitals containing those electron pairs are oriented to be as far apart as possible.*

⌐ This is a very useful rule; to apply it you start with the Lewis structure.

In this section we apply this model to predict the geometry of some rather simple molecules and polyatomic ions. In all these species, a central atom is surrounded by from two to six pairs of electrons.

Ideal Geometries with Two to Six Electron Pairs on the Central Atom

We begin by considering species in which a central atom, A, is surrounded by from two to six electron pairs, all of which are used to form single bonds with terminal atoms, X. These species have the general formulas AX_2, AX_3, . . . , AX_6. It is understood that there are no unshared pairs around atom A.

To see how the bonding orbitals surrounding the central atom are oriented with respect to one another, consider Figure 7.4, which shows the positions taken

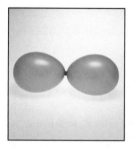

FIGURE 7.4
VSEPR electron pair geometries. The balloons, by staying as far apart as possible, illustrate the geometries (*left to right*) for two to six electron pairs. (*Charles D. Winters*)

naturally by two to six balloons tied together at the center. The balloons, like the orbitals they represent, arrange themselves to be as far from one another as possible.

Figure 7.5 (page 194) shows the geometries predicted by the VSEPR model for molecules of the types AX_2 to AX_6. The geometries for two and three electron pairs are those associated with species in which the central atom has less than an octet of electrons. Molecules of this type include BeF_2 and BF_3, which have the Lewis structures

Two electron pairs are as far apart as possible when they are directed at 180° to one another. This gives BeF_2 a **linear** structure. The three electron pairs around the boron atom in BF_3 are directed toward the corners of an equilateral triangle; the bond angles are 120°. We describe this geometry as **triangular planar.**

In species that follow the octet rule, the central atom is surrounded by four electron pairs. If each of these pairs forms a single bond with a terminal atom, a molecule of the type AX_4 results. The four bonds are directed toward the corners of a regular **tetrahedron.** All the bond angles are 109.5°, the tetrahedral angle. This geometry is found in many polyatomic ions such as NH_4^+ and SO_4^{2-} and in a wide variety of organic molecules, the simplest of which is methane, CH_4.

> This puts the four electron pairs as far apart as possible.

Molecules of the type AX_5 and AX_6 require that the central atom have an expanded octet. The geometries of these molecules are shown at the bottom of Figure 7.5. In PF_5, the five bonding pairs are directed toward the corners of a **triangular bipyramid,** a figure formed when two triangular pyramids are fused together, base to base. Three of the fluorine atoms are located at the corners of an equilateral triangle with the phosphorus atom at the center; the other two fluorine atoms are directly above and below the P atom. In SF_6, the six bonds are directed toward the corners of a regular **octahedron,** a figure with eight sides but *six* vertices. An octahedron can be formed by fusing two square pyramids base to base. Four of the fluorine atoms in SF_6 are located at the corners of a square with the S atom at the center; one fluorine atom is directly above the S atom, another directly below it.

See Screen 9.15, Ideal Electron Repulsion Shapes.

Effect of Unshared Pairs on Molecular Geometry

In many molecules and polyatomic ions, one or more of the electron pairs around the central atom is unshared. The VSEPR model is readily extended to predict the geometries of these species. In general,

1. The **electron-pair geometry** is approximately the same as that observed when only single bonds are involved. The bond angles are ordinarily a little smaller than the ideal values listed in Figure 7.5.

2. The **molecular geometry** is quite different when one or more unshared pairs are present. In describing molecular geometry, we refer only to the positions of the bonded atoms. These positions can be determined experimentally; positions of unshared pairs cannot be established by experiment. Hence, the locations of unshared pairs are not specified in describing molecular geometry.

See Screen 9.16, Determining Molecular Shape.

FIGURE 7.5

Molecular geometries for molecules with two to six electron-pair bonds around a central atom (A). **OHT**

Species type	Orientation of electron pairs	Predicted bond angles	Example	Ball and stick model
AX_2	Linear	180°	BeF_2	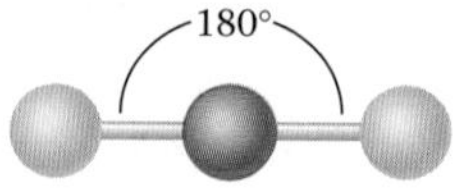
AX_3	Triangular planar	120°	BF_3	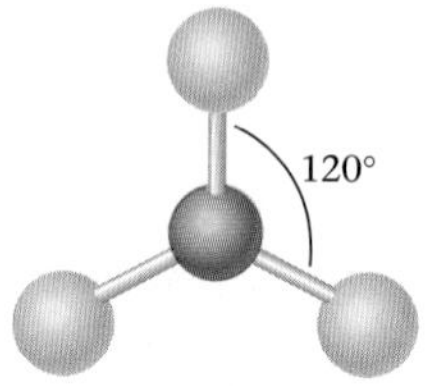
AX_4	Tetrahedron	109.5°	CH_4	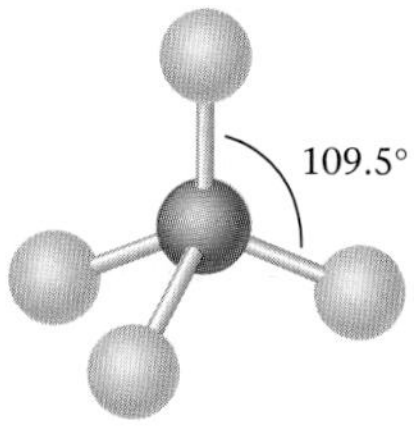
AX_5	Triangular bipyramid	90° 120° 180°	PF_5	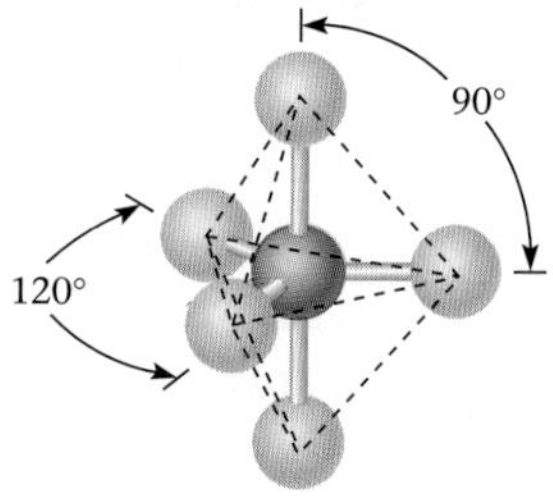
AX_6	Octahedron	90° 180°	SF_6	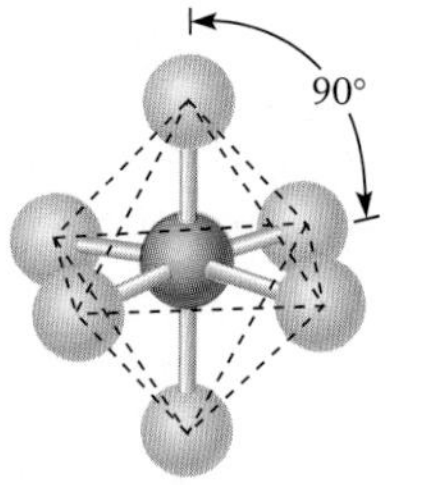

With these principles in mind, consider the NH_3 molecule:

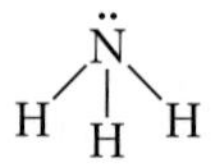

The apparent orientation of the four electron pairs around the N atom in NH_3 is shown at the left of Figure 7.6. Notice that, as in CH_4, the four pairs are directed toward the corners of a regular tetrahedron. The diagram at the right of Figure 7.6 shows the positions of the atoms in NH_3. The nitrogen atom is located above the center of an equilateral triangle formed by the three hydrogen atoms. The molecular geometry of the NH_3 molecule is described as a **triangular pyramid.** The ni-

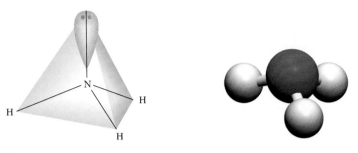

FIGURE 7.6
Two ways of showing the geometry of the NH₃ molecule. The orientation of the electron pairs, including the unshared pair (blue lobe), is shown at the left. The orientation of the atoms is shown at the right. The nitrogen atom is located directly above the center of the equilateral triangle formed by the three hydrogen atoms. The NH₃ molecule is described as a triangular pyramid.

trogen atom is at the apex of the pyramid, and the three hydrogen atoms form its triangular base. The molecule is three-dimensional, as the word "pyramid" implies.

The development we have just gone through for NH₃ is readily extended to the water molecule, H₂O. Here the Lewis structure shows that the central oxygen atom is surrounded by two single bonds and two unshared pairs:

$$H \overset{\overset{\displaystyle ..}{O}}{}_{..} H$$

The diagram at the left of Figure 7.7 emphasizes that the four electron pairs are oriented tetrahedrally. At the right, the positions of the atoms are shown. Clearly they are not in a straight line; the H₂O molecule is **bent.**

Experiments show that the bond angles in NH₃ and H₂O are slightly less than the ideal value of 109.5°. In NH₃ (three single bonds, one unshared pair around N), the bond angle is 107°. In H₂O (two single bonds, two unshared pairs around O), the bond angle is about 105°.

These effects can be explained in a rather simple way. An unshared pair is attracted by one nucleus, that of the atom to which it belongs. In contrast, a bonding pair is attracted by two nuclei, those of the two atoms it joins. Hence the electron cloud of an unshared pair is expected to spread out over a larger volume than that of a bonding pair. In NH₃, this tends to force the bonding pairs closer to one

Unshared pairs reduce bond angles below ideal values.

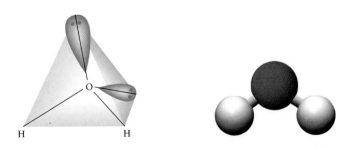

FIGURE 7.7
Two ways of showing the geometry of the H₂O molecule. At the left, the two unshared pairs are shown. As you can see from the drawing at the right, H₂O is a bent molecule. The bond angle, 105°, is a little smaller than the tetrahedral angle, 109.5°. The unshared pairs spread out over a larger volume than that occupied by the bonding pairs.

| TABLE 7.3   Geometries with Two, Three, or Four Electron Pairs Around a Central Atom OHT | | | | |
No. of Terminal Atoms (X) + Unshared Pairs (E)	Species Type	Ideal Bond Angles*	Molecular Geometry	Examples
2	AX_2	180°	Linear	BeF_2, CO_2
3	AX_3	120°	Triangular planar	BF_3, SO_3
	AX_2E	120°*	Bent	GeF_2, SO_2
4	AX_4	109.5°	Tetrahedron	CH_4
	AX_3E	109.5°*	Triangular pyramid	NH_3
	AX_2E_2	109.5°*	Bent	H_2O

*In these species, the observed bond angle is ordinarily somewhat less than the ideal value.

another, thereby reducing the bond angle. Where there are two unshared pairs, as in H_2O, this effect is more pronounced. In general, the VSEPR model predicts that unshared electron pairs will occupy slightly more space than bonding pairs.

Table 7.3 summarizes the molecular geometries of species in which a central atom is surrounded by two, three, or four electron pairs. The table is organized in terms of the number of terminal atoms, X, and unshared pairs, E, surrounding the central atom, A.

EXAMPLE 7.5   Predict the geometries of the following molecules:

(a) BeH_2 (b) OF_2 (c) PF_3

Strategy   Draw the Lewis structure as a first step. Then decide what type (AX_2, AX_3, etc.) the molecule is, focusing on the central atom. Remember, X represents a terminal atom, E an unshared pair of electrons.

Solution

(a) The Lewis structure for BeH_2 is

$$H—Be—H$$

The central atom has no unshared electron pairs and two bonded atoms. The molecule is of type AX_2. It is linear with a bond angle of 180°.

(b) The Lewis structure of OF_2 is

The skeleton does not neccessarily imply geometry.

The central oxygen atom has two unshared pairs and two bonds. The molecule is of type AX_2E_2 (like H_2O); it should be bent with a bond angle somewhat less than the ideal value of 109.5°. The observed value is 103°.

(c) The Lewis structure of PF_3 is

The central phosphorus atom has one unshared pair and three bonded atoms. The molecule is of type AX_3E; it should be a triangular pyramid (like NH_3) with a bond angle somewhat less than 109.5° (actually, 104°).

Reality Check BeH_2 is like BeF_2, OF_2 like H_2O, PF_3 like NH_3.

In many expanded-octet molecules, one or more of the electron pairs around the central atom is unshared. Recall, for example, the Lewis structure of xenon tetrafluoride, XeF_4 (Example 7.4).

There are six electron pairs around the xenon atom; four of these are covalent bonds to fluorine and the other two pairs are unshared. This molecule is classified as AX_4E_2.

Geometries of molecules such as these can be predicted by the VSEPR model. The results are shown in Figure 7.8 (page 198). The structures listed include those of all types of molecules having five or six electron pairs around the central atom, one or more of which may be unshared. Note that—

- in molecules of the type AX_4E_2, the two lone pairs occupy opposite rather than adjacent vertices of the octahedron.
- in the molecules AX_4E, AX_3E_2, and AX_2E_3 the lone pairs occupy successive positions in the equilateral triangle at the center of the triangular bipyramid.

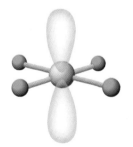

Xenon tetrafluoride, an AX_4E_2 molecule. *(Charles D. Winters/Susan M. Young)*

Multiple Bonds

The VSEPR model is readily extended to species in which double or triple bonds are present. A simple principle applies: ***Insofar as molecular geometry is concerned, a multiple bond behaves like a single bond.*** This makes sense. The four electrons in a double bond, or the six electrons in a triple bond, must be located between the two atoms, as are the two electrons in a single bond. This means that the electron pairs in a multiple bond must occupy the same region of space as those in a single bond. Hence the "extra" electron pairs in a multiple bond have no effect on geometry.

To illustrate this principle, consider the CO_2 molecule. Its Lewis structure is

$$:\ddot{O}{=}C{=}\ddot{O}:$$

The central atom, carbon, has two double bonds and no unshared pairs. For purposes of determining molecular geometry, we pretend that the double bonds are single bonds, ignoring the extra bonding pairs. The bonds are directed to be as far apart as possible, giving a 180° O—C—O bond angle. The CO_2 molecule, like BeF_2, is linear:

$$\underbrace{F{-}Be{-}F}_{180°} \qquad \underbrace{O{=}C{=}O}_{180°}$$

FIGURE 7.8

Molecular geometries for molecules with expanded octets and unshared electron pairs. The green spheres represent terminal atoms (X) and the open ellipses represent unshared electron pairs (E). For example, AX_4E represents a molecule in which the central atom is surrounded by four covalent bonds and one unshared electron pair. **OHT**

5 ELECTRON PAIRS

Species type	Structure	Description	Example	Bond angles
AX_5	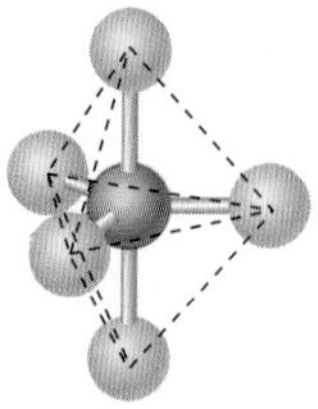	Triangular bipyramidal	PF_5	90°, 120°, 180°
AX_4E	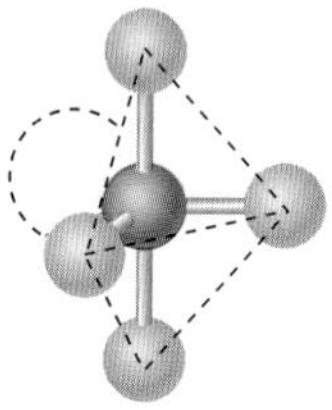	See-saw	SF_4	90°, 120°, 180°
AX_3E_2	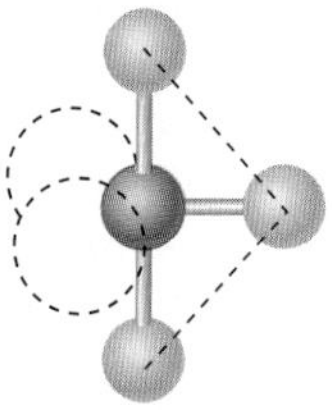	T-shaped	ClF_3	90°, 180°
AX_2E_3	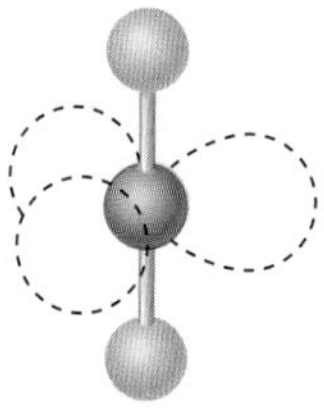	Linear	XeF_2	180°

6 ELECTRON PAIRS

Species type	Structure	Description	Example	Bond angles
AX_6	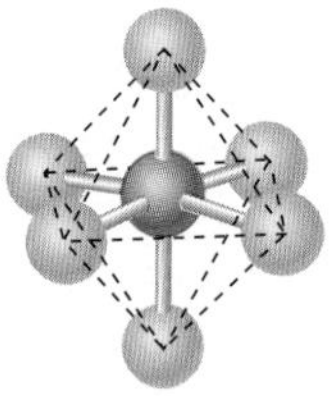	Octahedral	SF_6	90°, 180°
AX_5E	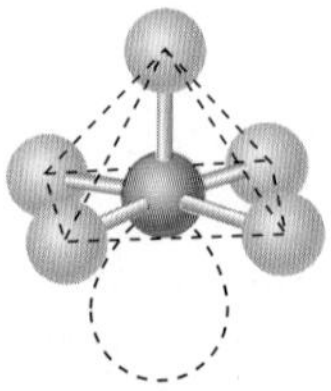	Square pyramidal	ClF_5	90°, 180°
AX_4E_2	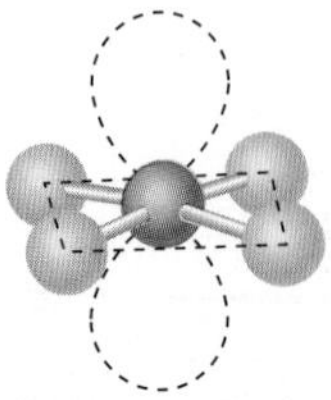	Square planar	XeF_4	90°, 180°

This principle can be restated in a somewhat different way for molecules in which there is a single central atom. The geometry of such a molecule depends only on—

■ *the number of terminal atoms, X, bonded to the central atom, irrespective of whether the bonds are single, double, or triple.*
■ *the number of unshared pairs, E, around the central atom.*

This means that Table 7.3 can be used in the usual way to predict the geometry of a species containing multiple bonds.

EXAMPLE 7.6   Predict the geometries of the ClO_3^- ion, the NO_3^- ion, and the N_2O molecule, which have the Lewis structures

(a) $\left[:\ddot{O}-\ddot{Cl}-\ddot{O}: \atop \quad\ :\ddot{O}: \right]^-$ (b) $\left[:\ddot{O}-N-\ddot{O}: \atop \quad\ \|\atop\quad :O: \right]^-$ (c) $:\ddot{N}=N=\ddot{O}:$

Strategy   Classify each species as AX_mE_n and use Table 7.3.

Solution

(a) The central atom, chlorine, is bonded to three oxygen atoms; it has one unshared pair. The ClO_3^- ion is of the type AX_3E. It is a triangular pyramid; the ideal bond angle is 109.5°.
(b) The central atom, nitrogen, is bonded to three oxygen atoms; it has no unshared pairs. The NO_3^- ion is of the type AX_3. It has the geometry of an equilateral triangle, with the nitrogen atom at the center; the bond angle is 120°. The ion is triangular planar.
(c) The central nitrogen atom is bonded to two other atoms with no unshared pairs. The molecule, type AX_2, is linear, with a bond angle of 180°.

> AX_3E_2 means that the central atom A is bonded to three X atoms and has two unshared electron pairs.

The VSEPR model applies equally well to molecules in which there is no single central atom. Consider the acetylene molecule, C_2H_2. Recall that here the two carbon atoms are joined by a triple bond:

$$H-C\equiv C-H$$

Each carbon atom behaves as if it were surrounded by two electron pairs. Both of the bond angles (H—C≡C and C≡C—H) are 180°. The molecule is linear; the four atoms are in a straight line. The two extra electron pairs in the triple bond do not affect the geometry of the molecule.

In ethylene, C_2H_4, there is a double bond between the two carbon atoms. The molecule has the geometry to be expected if each carbon atom had only three pairs of electrons around it.

$$\begin{matrix} H & & H \\ & \diagdown\, C{=}C\,\diagup & \\ H & & H \end{matrix}$$

The six atoms are located in a plane with bond angles of approximately 120°. Actually, the double bond between the carbon atoms occupies slightly more space than a single bond joining carbon to hydrogen. As a result, the H—C=C angles are slightly larger than 120°, and the H—C—C angles are slightly less.

$$\text{H} \quad 122° \quad \text{H}$$
$$116° \quad \text{C}=\text{C} \quad 116°$$
$$\text{H} \quad 122° \quad \text{H}$$

See Screens 9.11, Bond Polarity and Electronegativity, & 9.17, Molecular Polarity.

7.3 POLARITY OF MOLECULES

Covalent bonds and molecules held together by such bonds may be—

- *polar.* As a result of an unsymmetrical distribution of electrons, the bond or molecule contains a positive and a negative pole and is therefore a *dipole.*
- *nonpolar.* A symmetrical distribution of electrons leads to a bond or molecule with no positive or negative poles.

Polar and Nonpolar Covalent Bonds

The two electrons in the H_2 molecule are shared equally by the two nuclei. Stated another way, a bonding electron is as likely to be found in the vicinity of one nucleus as another. Bonds of this type are described as nonpolar. **Nonpolar bonds** are formed whenever the two atoms joined are identical, as in H_2 and F_2.

In the HF molecule, the distribution of the bonding electrons is somewhat different from that found in H_2 or F_2. Here the density of the electron cloud is greater about the fluorine atom. The bonding electrons, on the average, are shifted toward fluorine and away from the hydrogen (atom Y in Figure 7.9). Bonds in which the electron density is unsymmetrical are referred to as **polar bonds.**

Atoms of two different elements always differ at least slightly in their electronegativity (recall Table 6.6, page 171). Hence covalent bonds between unlike atoms are always polar. Consider, for example, the H—F bond. Because fluorine has a higher electronegativity (4.0) than does hydrogen (2.2), bonding electrons are displaced toward the fluorine atom. The H—F bond is polar, with a partial negative charge at the fluorine atom and a partial positive charge at the hydrogen atom.

The extent of polarity of a covalent bond is related to the difference in electronegativities of the bonded atoms. If this difference is large, as in HF ($\Delta EN = 1.8$), the bond is strongly polar. Where the difference is small, as in H—C($\Delta EN = 0.3$), the bond is only slightly polar.

> All molecules, except those of elements, have polar bonds.

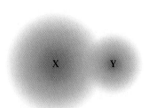

FIGURE 7.9
Bond polarity. If atom X is more electronegative than atom Y, the electron cloud of the bonding electrons will be concentrated around atom X. Thus the bond is polar.

Polar and Nonpolar Molecules

A **polar molecule** is one that contains positive and negative poles. There is a partial positive charge (positive pole) at one point in the molecule and a partial negative charge (negative pole) at a different point. As shown in Figure 7.10, polar molecules orient themselves in the presence of an electric field. The positive pole in the molecule tends to align with the external negative charge, and the negative pole with the external positive charge. In contrast, there are no positive and nega-

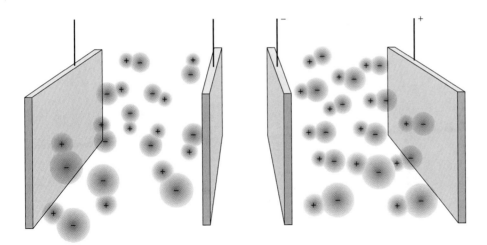

FIGURE 7.10
Orientation of polar molecules in an electric field. With the field off, polar molecules are randomly oriented. With the field on, polar molecules such as HF align their positive and negative ends toward the negative and positive poles of the field. Nonpolar molecules such as H_2 do not line up. **OHT**

Field off   Field on

tive poles in a **nonpolar molecule.** In an electric field, nonpolar molecules, such as H_2, show no preferred orientation.

The extent to which molecules tend to orient themselves in an electrical field is a measure of their *dipole moment.* A polar molecule such as HF has a dipole moment; a nonpolar molecule such as H_2 or F_2 has a dipole moment of zero.

If a molecule is diatomic, it is easy to decide whether it is polar or nonpolar. A diatomic molecule has only one kind of bond; hence the polarity of the molecule is the same as the polarity of the bond. Hydrogen and fluorine (H_2, F_2) are nonpolar because the bonded atoms are identical and the bond is nonpolar. Hydrogen fluoride, HF, on the other hand, has a polar bond, so the molecule is polar. The bonding electrons spend more time near the fluorine atom so that there is a negative pole at that end and a positive pole at the hydrogen end. This is sometimes indicated by writing

$$H \!\!+\!\!\longrightarrow F$$

The arrow points toward the negative end of the polar bond (F atom); the plus sign is at the positive end (H atom). The HF molecule is called a **dipole;** it contains positive and negative poles.

If a molecule contains more than two atoms it is not so easy to decide whether it is polar or nonpolar. In this case, not only bond polarity but also molecular geometry determines the polarity of the molecule. To illustrate what is involved, consider the molecules shown in Figure 7.11 (page 202).

1. In BeF_2 there are two polar Be—F bonds; in both bonds, the electron density is concentrated around the more electronegative fluorine atom. However, because the BeF_2 molecule is linear, the two $Be\!\!+\!\!\longrightarrow F$ dipoles are in opposite directions and cancel one another. The molecule has no net dipole and hence is nonpolar. From a slightly different point of view, in BeF_2 the centers of positive and negative charge coincide with each other at the Be atom. There is no way that a BeF_2 molecule can line up in an electric field.

2. Because oxygen is more electronegative than hydrogen (3.5 versus 2.2) an O—H bond is polar, with the electron density higher around the oxygen atom. In the bent H_2O molecule, the two $H \!\!+\!\!\longrightarrow O$ dipoles do not cancel each other.

The following types of molecules are nonpolar: A_2, AX_2 (linear), AX_3 (triangular planar), AX_4 (tetrahedral).

FIGURE 7.11

Polarity of molecules. All bonds in
these molecules are polar, as shown by
the $\longmapsto$ symbol, in which the arrow
points to the more negative end of
the bond and the + indicates the
more positive end. In BeF_2 and CCl_4
the bond dipoles cancel and the molecules are nonpolar. In H_2O and $CHCl_3$
the molecules are polar, with net
dipoles shown by the broad arrows
pointing towards the negative poles. **OHT**

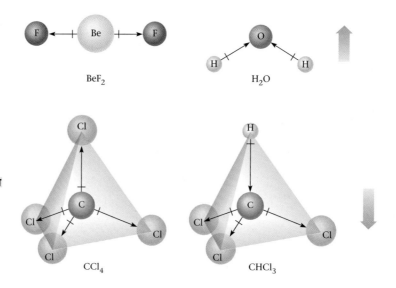

Instead, they add to give the H_2O molecule a net dipole. The center of negative
charge is located at the O atom; this is the negative pole of the molecule. The center of positive charge is located midway between the two H atoms; the positive
pole of the molecule is at that point. The H_2O molecule is polar. It tends to line up
in an electric field with the oxygen atom oriented toward the positive electrode.

3. Carbon tetrachloride, CCl_4, is another molecule that, like BeF_2, is nonpolar despite the presence of polar bonds. Each of its four bonds is a dipole,
$C\longmapsto Cl$. However, because the four bonds are arranged symmetrically around the
carbon atom, they cancel. As a result, the molecule has no net dipole; it is nonpolar. If one of the Cl atoms in CCl_4 is replaced by hydrogen, the situation changes.
In the $CHCl_3$ molecule, the $H\longmapsto C$ dipole does not cancel with the three
$C\longmapsto Cl$ dipoles. Hence $CHCl_3$ is polar.

There are two criteria for determining the polarity of a molecule; bond polarity and molecular geometry. *If the polar A—X bonds in a molecule AX_mE_n are
arranged symmetrically around the central atom A, the molecule is nonpolar.*

EXAMPLE 7.7   Determine whether each of the following is polar or nonpolar:

(a) SO_2 (b) BF_3 (c) CO_2

Strategy   Write the Lewis structure of the molecule and classify it as AX_mE_n. Using
Table 7.3 or Figure 7.5, decide on the geometry of the molecule. Then decide whether the
dipoles cancel, in which case it is nonpolar.

Solution

(a) The Lewis structure of SO_2 is shown on page 185; it is of the type AX_2E. The molecule is
bent, so the dipoles do not cancel. The SO_2 molecule is polar.
(b) The Lewis structure of BF_3 is shown on page 189; it is of the type AX_3. The molecule is
an equilateral triangle with the boron atom at the center. The three polar bonds cancel one
another; BF_3 is nonpolar.
(c) The Lewis structure of CO_2 is shown on page 197; it is of the type AX_2. The molecule is
linear, so the two $C\longmapsto O$ dipoles cancel each other; CO_2 is nonpolar.

7.4 ATOMIC ORBITALS; HYBRIDIZATION

See Screens 10.3, Valence Bond Theory, & 10.4, Hybrid Orbitals.

In the 1930s a theoretical treatment of the covalent bond was developed by, among others, Linus Pauling, then at the California Institute of Technology. The *atomic orbital* or **valence bond model** won him the Nobel Prize in chemistry in 1954. Eight years later, Pauling won the Nobel Peace Prize for his efforts to stop nuclear testing.

Pauling, a versatile chemist with contributions in many areas, died in 1994.

According to this model, a covalent bond consists of a pair of electrons of opposed spin within an orbital. For example, a hydrogen atom forms a covalent bond by accepting an electron from another atom to complete its 1s orbital. Using orbital diagrams, we could write

	1s
isolated H atom	(↑)
H atom in a stable molecule	(↑↓)

The second electron, shown in color, is contributed by another atom. This could be another H atom in H_2, an F atom in HF, a C atom in CH_4, and so on.

This simple model is readily extended to other atoms. The fluorine atom (electron configuration $1s^2 2s^2 2p^5$) has a half-filled p orbital:

	1s	2s	2p
isolated F atom	(↑↓)	(↑↓)	(↑↓)(↑↓)(↑)

By accepting an electron from another atom, F can complete this 2p orbital:

	1s	2s	2p
F atom in HF, F_2, . . .	(↑↓)	(↑↓)	(↑↓)(↑↓)(↑↓)

According to this model, it would seem that for an atom to form a covalent bond, it must have an unpaired electron. Indeed, the number of bonds formed by an atom should be determined by its number of unpaired electrons. Because hydrogen has an unpaired electron, an H atom should form one covalent bond, as indeed it does. The same holds for the F atom, which forms only one bond. The noble-gas atoms He and Ne, which have no unpaired electrons, should not form bonds at all; they don't.

When this simple idea is extended beyond hydrogen, the halogens, and the noble gases, problems arise. Consider, for example, the three atoms Be ($Z = 4$), B ($Z = 5$), and C ($Z = 6$):

	1s	2s	2p
Be atom	(↑↓)	(↑↓)	( )( )( )
B atom	(↑↓)	(↑↓)	(↑)( )( )
C atom	(↑↓)	(↑↓)	(↑)(↑)( )

Notice that the beryllium atom has no unpaired electrons, the boron atom has one, and the carbon atom two. Simple valence bond theory would predict that Be, like He, should not form covalent bonds. A boron atom should form one bond, carbon two. Experience tells us that these predictions are wrong. Beryllium forms two bonds in BeF_2; B forms three bonds in BF_3. Carbon ordinarily forms four bonds, not two.

To explain these and other discrepancies, simple valence bond theory must be modified. It is necessary to invoke a new kind of orbital, called a **hybrid orbital.**

Hybrid Orbitals: sp, sp², sp³, sp³d, sp³d²

The number of orbitals is shown by the superscript.

The formation of the BeF_2 molecule can be explained by assuming that, as two fluorine atoms approach Be, the atomic orbitals of the beryllium atom undergo a significant change. Specifically, the 2s orbital is mixed or *hybridized* with a 2p orbital to form two new *sp hybrid orbitals:*

one s atomic orbital + one p atomic orbital ⟶ *two* sp hybrid orbitals

Notice (Figure 7.12) that **the number of hybrid orbitals formed is equal to the number of atomic orbitals mixed.** This is always true in hybridization of orbitals. Moreover, the energies of the hybrid orbitals are intermediate between those of the atomic orbitals from which they are derived.

In the BeF_2 molecule, there are two electron-pair bonds. These electron pairs are located in the two sp hybrid orbitals just described. In each orbital—

- one electron is a valence electron contributed by beryllium; remember that Be has two such electrons (·Be·).
- one electron is a valence electron contributed by a fluorine atom.

The orbital diagram for the beryllium atom in the BeF_2 molecule can be written as

	1s	2s	2p
Be in BeF_2	(↑↓)	(↑↓) (↑↓)( )( )	

(The blue arrows indicate the two electrons supplied by the fluorine atoms. The horizontal lines enclose the hybrid orbitals involved in bond formation.)

A similar argument can be used to explain why boron forms three bonds and carbon four.

In the case of boron:

one s atomic orbital + two p atomic orbitals ⟶ *three sp² hybrid orbitals*

With carbon:

one s atomic orbital + three p atomic orbitals ⟶ *four sp³ hybrid orbitals*

Thus we have

	1s	2s	2p
B in BF_3	(↑↓)	(↑↓) (↑↓)(↑↓)( )	
C in CH_4	(↑↓)	(↑↓) (↑↓)(↑↓)(↑↓)	

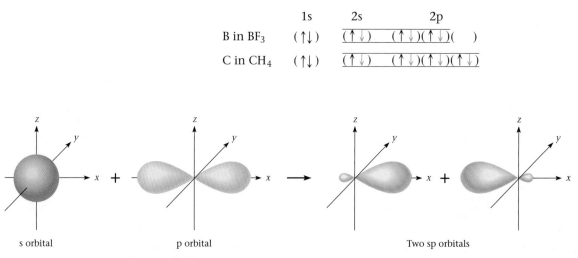

s orbital   p orbital   Two sp orbitals

FIGURE 7.12
Formation of sp hybrid orbitals. The mixing of one s orbital and one p orbital gives two sp hybrid orbitals. **OHT**

TABLE 7.4 Hybrid Orbitals and Their Geometries

Number of Electron Pairs	Atomic Orbitals	Hybrid Orbitals	Orientation	Examples
2	s, p	sp	Linear	BeF_2, CO_2
3	s, two p	sp^2	Triangular planar	BF_3, SO_3
4	s, three p	sp^3	Tetrahedron	CH_4, NH_3, H_2O
5	s, three p, d	sp^3d	Triangular bipyramid	PCl_5, SF_4, ClF_3
6	s, three p, two d	sp^3d^2	Octahedron	SF_6, ClF_5, XeF_4

(Remember that boron has three valence electrons and carbon has four; each orbital contains one of these electrons along with another, shown in blue, contributed by an F or H atom.)

You will recall that the bond angles in NH_3 and H_2O are very close to that in CH_4. This suggests that the four electron pairs surrounding the central atom in NH_3 and H_2O, like those in CH_4, occupy sp^3 hybrid orbitals. In NH_3, three of these orbitals are filled by bonding electrons, the other by the unshared pair on the nitrogen atom. In H_2O, two of the sp^3 orbitals of the oxygen atom contain bonding electron pairs; the other two contain unshared pairs. The situation in NH_3 and H_2O is not unique. In general, we find that **unshared as well as shared electron pairs can be located in hybrid orbitals.**

> Hybridization: sp (AX_2), sp^2 (AX_3, AX_2E), sp^3 (AX_4, AX_3E, AX_2E_2).

The extra electron pairs in an expanded octet are accommodated by using d orbitals. The phosphorus atom (five valence electrons) in PCl_5 and the sulfur atom (six valence electrons) in SF_6 make use of 3d as well as 3s and 3p orbitals:

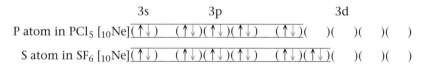

The orbitals used by the five pairs of bonding electrons surrounding the phosphorus atom in PCl_5 are **sp^3d hybrid orbitals.**

one s orbital + three p orbitals + one d orbital ⟶ *five sp^3d hybrid orbitals*

Similarly, in SF_6, the six pairs of bonding electrons are located in **sp^3d^2 hybrid orbitals:**

one s orbital + three p orbitals + two d orbitals ⟶ *six sp^3d^2 hybrid orbitals*

Table 7.4 gives the orientation in space of hybrid orbitals, as found mathematically by quantum mechanics. Notice that *the geometries are exactly as predicted by VSEPR theory.* In each case, the several hybrid orbitals are directed to be as far apart as possible; recall Figure 7.5 (page 194).

> ⊙ **See Screen 10.6, Determining Hybrid Orbitals.**

EXAMPLE 7.8   Give the hybridization of

(a) carbon in CF_4 (b) phosphorus in PF_3 (c) sulfur in SF_4

Strategy   Draw the Lewis structure for the molecule and determine the number of electron pairs (single bonds or unshared pairs) around the central atom. The possible hybridizations are sp (two pairs), sp^2 (three pairs), sp^3 (four pairs), sp^3d (five pairs), and sp^3d^2 (six pairs).

Solution

(a) For CF$_4$, the Lewis structure is

There are four bonds and no unshared electron pairs around C, the central atom. The hybridization is sp^3.

(b) The Lewis structure of PF$_3$ is

There are three bonds and one unshared pair, making a total of four pairs. The hybridization is sp^3.

(c) Sulfur tetrafluoride has the Lewis structure

There are five electron pairs around sulfur: four bonds and one unshared pair. The hybridization is sp^{3}d.

Reality Check If you're asked to predict the hybridization, as you are here, the geometry as in Example 7.6, or the polarity as in Example 7.7, you have to start with Lewis structures; that's why they are so important.

Multiple Bonds

In Section 7.2, we saw that insofar as geometry is concerned, a multiple bond acts as if it were a single bond. In other words, the extra electron pairs in a double or triple bond have no effect on the geometry of the molecule. This behavior is related to hybridization. ***The extra electron pairs in a multiple bond (one pair in a double bond, two pairs in a triple bond) are not located in hybrid orbitals.***

To illustrate this rule, consider the ethylene (C$_2$H$_4$) and acetylene (C$_2$H$_2$) molecules. You will recall that the bond angles in these molecules are 120° for ethylene and 180° for acetylene. This implies sp^2 hybridization in C$_2$H$_4$ and sp hybridization in C$_2$H$_2$ (see Table 7.4). Using blue lines to represent hybridized electron pairs,

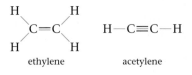

ethylene   acetylene

In both cases, only one of the electron pairs in the multiple bond occupies a hybrid orbital.

EXAMPLE 7.9   What is the hybridization of the nitrogen atom in

(a) N$_2$ (b) NO$_3^-$

Strategy The first step is to draw Lewis structures. To find the hybridization of the nitrogen, place in hybrid orbitals—

- unshared electron pairs.
- electron pairs forming single bonds.
- one and only one of the electron pairs in a multiple bond.

Solution

(a) $:N\equiv N:$

For each N atom, hybridize one unshared pair and one of the electron pairs in the triple bond; the hybridization is sp.

(b) $\left[\begin{array}{c} :\ddot{O}-N-\ddot{O}: \\ \parallel \\ :O: \end{array} \right]$

Three of the four electron pairs around the nitrogen atom occupy hybrid orbitals; the hybridization is sp^2.

Sigma and Pi Bonds

We have noted that the extra electron pairs in a multiple bond are not hybridized and have no effect on molecular geometry. At this point, you may well wonder what happened to those electrons. Where are they in molecules like C_2H_4 and C_2H_2?

To answer this question, it is necessary to consider the shape or spatial distribution of the orbitals filled by bonding electrons in molecules. From this point of view, we can distinguish between two types of bonding orbitals. The first of these, and by far the more common, is called a *sigma* bonding orbital. It consists of a single lobe:

A   B

in which the electron density is concentrated in the region directly between the two bonded atoms, A and B. A **sigma** (σ) bond consists of an electron pair occupying a sigma bonding orbital.

The unhybridized electron pairs associated with multiple bonds occupy orbitals of a quite different shape, called *pi* bonding orbitals. Such an orbital consists of two lobes, one above the bond axis, the other below it.

A ———— B

Along the bond axis itself, the electron density is zero. The electron pair of a **pi** (π) bond occupies a pi bonding orbital. There is one π bond in the C_2H_4 molecule, two in C_2H_2.

The geometries of the bonding orbitals in ethylene and acetylene are shown in Figure 7.13 (page 208). Sigma orbitals are shown in blue, pi orbitals in red. Notice that—

- the nature of the π bond in C_2H_4 prohibits free rotation about the carbon-carbon bond. For that to happen the electron density of the π orbital would have to be high in areas outside the plane of the paper. In effect, the pi bond is fixed in position, so the C_2H_4 molecule is planar.
- the two π bonding orbitals in C_2H_2, both of which consist of two lobes, are oriented at right angles to one another. This is difficult to show in a two-

See Screens 10.5, Sigma Bonding, & 10.7, Multiple Bonding.

Sigma and pi orbitals, like s and p, differ in shape; each orbital can hold $2e^-$.

FIGURE 7.13

Bonding orbitals in ethylene (CH_2=CH_2) and acetylene (CH≡CH). The sigma bond backbones are shown in blue. The pi bonds (one in ethylene and two in acetylene) are shown in red. Note that a pi bonding orbital consists of two lobes.

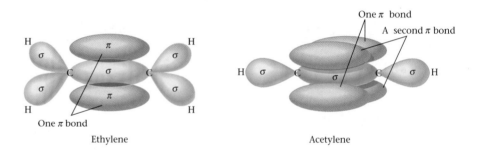

Ethylene   Acetylene

dimensional figure. In effect, the four lobes of the two π bonds (red) are wrapped around the central σ bond (blue) like a bun around a hot dog.

In general, to find the number of σ and π bonds in a species, remember that—

- all single bonds are sigma bonds.
- one of the electron pairs in a multiple bond is a sigma bond; the others are pi bonds.

EXAMPLE 7.10   Give the number of pi and sigma bonds in

(a) N_2 (b) NO_3^-

Strategy   Refer back to Example 7.9 for the Lewis structures. Then apply the two rules just cited.

Solution

(a) In N_2 there is a triple bond consisting of one σ bond and two π bonds.
(b) In NO_3^-, there is one double bond and two single bonds; this translates to three σ bonds and one π bond.

Reality Check   There are two pi bonds in a triple bond, one in a double bond, and none in a single bond.

CHEMISTRY
Beyond the Classroom

The Noble Gases

The modern periodic table contains six relatively unreactive gases that were unknown to Mendeleev: the noble gases that make up Group **18** at the far right of the table. The first of these elements to be isolated was argon, which makes up about 0.9% of air. The physicist Lord Rayleigh found that the density of atmospheric nitrogen, obtained by removing O_2, CO_2, and H_2O from air, was slightly greater than that of chemically pure N_2 ($\mathcal{M} = 28.02$ g/mol). Following up on that observation, Sir William Ramsey separated argon ($\mathcal{M} = 39.95$ g/mol) from air. Over a three-year period between 1895 and 1898, this remarkable Scotsman, who never took a formal course in chemistry, isolated three more noble gases: Ne, Kr, and Xe. In effect, Ramsey added a whole new group to the periodic table.

Helium, the first member of the group, was detected in the spectrum of the Sun in 1868. Because of its low density ($\frac{1}{7}$ that of air), helium is used in all kinds of balloons and in synthetic atmospheres to make breathing easier for people suffering from emphysema.

45. There are thre

Which of these mc

46. There are two

F

N

Is either molecule

Hybridization

47. Give the hybr
in Question 29.

48. Give the hybr
in Question 30.

49. Give the hybr
in Question 31.

50. Give the hybr
in Question 32.

51. Give the hybr
in Question 33.

WEB **52.** Give the hybr
in Question 34.

53. In each of the
has an expanded c
around the central
  (a) ClF_2^-

54. Follow the dir
atomic ions.
  (a) XeF_3^-

55. Give the hybr
vent dimethylsul
shown.)

56. Give the hybri

57. What is the hy
  (a)
  (c) $\ddot{N}\equiv\ddot{N}$

Research laboratories use helium as a liquid coolant to achieve very low temperatures (bp He = $-269°C$). Argon and, more recently, krypton are used to provide an inert atmosphere in lightbulbs, thereby extending the life of the tungsten filament. In neon signs, a high voltage is passed through a glass tube containing neon at very low pressures. The red glow emitted corresponds to an intense line at 640 nm in the neon spectrum.

Until about 40 years ago, these elements were referred to as "inert gases"; they were believed to be entirely unreactive toward other substances. In 1962 Neil Bartlett, a 29-year-old chemist at the University of British Columbia, shook up the world of chemistry by preparing the first noble-gas compound. In the course of his research on platinum-fluorine compounds, he isolated a reddish solid that he showed to be $O_2^+(PtF_6^-)$. Bartlett realized that the ionization energy of Xe (1170 kJ/mol) is virtually identical to that of the O_2 molecule (1165 kJ/mol). This encouraged him to attempt to make the analogous compound $XePtF_6$. His success opened up a new era in noble-gas chemistry.

The most stable binary compounds of xenon are the three fluorides, XeF_2, XeF_4, and XeF_6. Xenon difluoride can be prepared quite simply by exposing a 1:1 mol mixture of xenon and fluorine to ultraviolet light; colorless crystals of XeF_2 (mp = 129°C) form slowly.

$$Xe(g) + F_2(g) \longrightarrow XeF_2(s)$$

The higher fluorides are prepared using excess fluorine (Figure A). All these compounds are stable in dry air at room temperature. However, they react with water to form compounds in which one or more of the fluorine atoms has been replaced by oxygen. Thus xenon hexafluoride reacts rapidly with water to give the trioxide

$$XeF_6(s) + 3H_2O(l) \longrightarrow XeO_3(s) + 6HF(g)$$

Xenon trioxide is highly unstable; it detonates if warmed above room temperature.

The chemistry of xenon is much more extensive than that of any other noble gas. Only one binary compound of krypton, KrF_2, has been prepared. It is a colorless solid that decomposes at room temperature. The chemistry of radon is difficult to study because all its isotopes are radioactive. Indeed, the radiation given off is so intense that it decomposes any reagent added to radon in an attempt to bring about a reaction.

FIGURE A
Crystals of xenon tetrafluoride
(XeF_4). *(Argonne National Laboratory)*

CHAPTER HIGHLIGHTS

Key Concepts

1. Draw Lewis structures for molecules and polyatomic ions.
  (Examples 7.1, 7.2, 7.4; Problems 1–20, 67, 68)

2. Write resonance forms.
  (Example 7.3; Problems 21–28)

3. Use Table 7.3 and Figure 7.8, applying the VSEPR model, to predict molecular geometry.
  (Examples 7.5, 7.6; Problems 29–40, 73–75)

4. Having derived the geometry of a species, predict whether it will be polar.
  (Example 7.7; Problems 41–46)

5. State the hybridization of an atom in a bonded species.
  (Examples 7.8, 7.9; Problems 47–60)

6. State the number of pi and sigma bonds in a species.
  (Example 7.10; Problems 61–64)

25. The skeleton stru

Draw the resonar
26. Borazine, $B_3N_3H_6$

Draw the resonar
27. What is the form
the following species
(a) oxygen in H(
(b) nitrogen in N
(c) sulfur in (S=
28. Follow the direct
(a) the central n
(b) xenon in Xel
(c) bromine in B

Molecular Geom

29. Predict the geom
(a) SCO   (b)
30. Predict the geom
(a) SO_2 (b) F
31. Predict the geom
(a) KrF_2 (b)
32. Predict the geom
(a) NNO   (b)
33. Predict the geom
(a) ClO_2^- (b
WEB **34.** Predict the geom
(a) ClO_4^- (b
35. Give all the idea
following molecules
geometry.)
(a) O=C=O
(b) H—B—H
     |
     H

(c) H—O—N

     H
     |
(d) H—C—C=
     | |
     H H

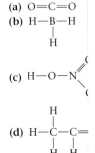

(a) How many sigma and pi bonds are there in aspirin?
(b) What are the approximate values of the angles marked (in blue) A, B, and C?
(c) What is the hybridization of each atom marked (in red) 1, 2, and 3?

Species	Atoms Around Central Atom A	Unshared Pairs Around A	Geometry	Hybridization	Polarity (Assume all X atoms the same)
AX_2E_2	_____	_____	_____	_____	_____
_____	3	0	_____	_____	_____
AX_4E_2	_____	_____	_____	_____	_____
_____ _____	_____	triangular bipyramid	_____	_____

Conceptual Problems

71. Based on the concept of formal charge, what is the central atom in
(a) HCN (do not include H as a possibility)?
(b) NOCl (Cl is always a terminal atom)?
72. Given the following electronegativities

$$C = 2.5 \quad N = 3.0 \quad S = 2.6$$

what is the central atom in CNS^-?
73. Describe the geometry of the species in which there are, around the central atom
(a) four single bonds, two unshared pairs of electrons.
(b) five single bonds.
(c) two single bonds, one unshared pair of electrons.
(d) three single bonds, two unshared pairs of electrons.
(e) two single bonds, two unshared pairs of electrons.
(f) five single bonds, one unshared pair of electrons.
74. Consider the following molecules: SiH_4, PH_3, H_2S. In each case, a central atom is surrounded by four electron pairs. In which of these molecules would you expect the bond angle to be less than 109.5°? Explain your reasoning.
75. In each of the following molecules, a central atom is surrounded by a total of three atoms or unshared electron pairs: $SnCl_2$, BCl_3, SO_2. In which of these molecules would you expect the bond angle to be less than 120°? Explain your reasoning.
76. Give the formula of an ion or molecule in which an atom of
(a) N forms three bonds using sp^3 hybrid orbitals.
(b) N forms two pi bonds and one sigma bond.
(c) O forms one sigma and one pi bond.
(d) C forms four bonds in three of which it uses sp^2 hybrid orbitals.
(e) Xe forms two bonds using sp^3d^2 hybrid orbitals.
77. Explain the meaning of the following terms.
(a) expanded octet
(b) resonance
(c) unshared electron pair
(d) odd-electron species

70. Complete the table below.

Challenge Problems

78. A compound of chlorine and fluorine, ClF_x, reacts at about 75°C with uranium to produce uranium hexafluoride and chlorine fluoride, ClF. A certain amount of uranium produced 5.63 g of uranium hexafluoride and 457 mL of chlorine fluoride at 75°C and 3.00 atm. What is x? Describe the geometry, polarity, and bond angles of the compound and the hybridization of chlorine. How many sigma and pi bonds are there?
79. Draw the Lewis structure and describe the geometry of the hydrazine molecule, N_2H_4. Would you expect this molecule to be polar?
80. Consider the polyatomic ion IO_6^{5-}. How many pairs of electrons are there around the central iodine atom? What is its hybridization? Describe the geometry of the ion.
81. It is possible to write a simple Lewis structure for the SO_4^{2-} ion, involving only single bonds, which follows the octet rule. However, Linus Pauling and others have suggested an alternative structure, involving double bonds, in which the sulfur atom is surrounded by six electron pairs.
(a) Draw the two Lewis structures.
(b) What geometries are predicted for the two structures?
(c) What is the hybridization of sulfur in each case?
(d) What are the formal charges of the atoms in the two structures?
82. Phosphoryl chloride, $POCl_3$, has the skeleton structure

Write
(a) a Lewis structure for $POCl_3$ following the octet rule. Calculate the formal charges in this structure.
(b) a Lewis structure where all the formal charges are zero. (The octet rule need not be followed.)

*S*ome say the world will end in fire,
Some say in ice.
From what I've tasted of desire
I hold with those who favor fire.

—Robert Frost
Fire and Ice

THERMOCHEMISTRY

8

Magnesium ribbon reacts with oxygen to give MgO, a white solid, and 602 kJ of heat per mole of magnesium. *(Charles D. Winters)*

CHAPTER OUTLINE

8.1   PRINCIPLES OF HEAT FLOW

8.2   MEASUREMENT OF HEAT FLOW; CALORIMETRY

8.3   ENTHALPY

8.4   THERMOCHEMICAL EQUATIONS

8.5   ENTHALPIES OF FORMATION

8.6   BOND ENTHALPY

8.7   THE FIRST LAW OF THERMODYNAMICS

This chapter deals with energy and heat, two terms used widely by both the general public and scientists. Energy, in the vernacular, is equated with pep and vitality. Heat conjures images of blast furnaces and sweltering summer days. Scientifically, these terms have quite different meanings. *Energy* can be defined as the capacity to do work. *Heat* is a particular form of energy that is transferred from a body at a high temperature to one at a lower temperature when they are brought into contact with each other. Two centuries ago, heat was believed to be a material fluid (caloric); we still use the phrase "heat flow" to refer to heat transfer or to heat effects in general.

Thermochemistry refers to the study of the heat flow that accompanies chemical reactions. Our discussion of this subject will focus on—

- the basic principles of heat flow (Section 8.1).

- the experimental measurement of the magnitude and direction of heat flow, known as *calorimetry* (Section 8.2).

- the concept of enthalpy (heat content) and *enthalpy change, ΔH* (Section 8.3).

- the calculation of ΔH for reactions, using *thermochemical equations* (Section 8.4) and *enthalpies of formation* (Section 8.5).

- heat effects in the breaking and formation of covalent bonds (Section 8.6).

- the relation between heat and other forms of energy, as expressed by the first law of thermodynamics (Section 8.7).

8.1 PRINCIPLES OF HEAT FLOW

In any discussion of heat flow, it is important to distinguish between system and surroundings. The **system** is that part of the universe on which attention is focused. In a very simple case (Figure 8.1), it might be a sample of water in contact with a hot plate. The **surroundings,** which exchange energy with the system, make up in principle the rest of the universe. For all practical purposes, however, they include only those materials in close contact with the system. In Figure 8.1, the surroundings would consist of the hot plate, the beaker holding the water sample, and the air around it.

When a chemical reaction takes place, we consider the substances involved, reactants and products, to be the system. The surroundings include the vessel in which the reaction takes place (test tube, beaker, and so on) and the air or other material in thermal contact with the reaction system.

The universe is a big place.

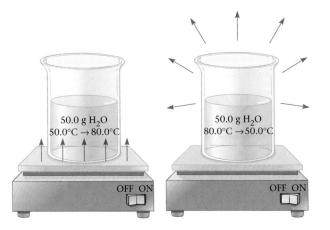

State Properties

The *state* of a system is described by giving its composition, temperature, and pressure. The system at the left of Figure 8.1 consists of

50.0 g of $H_2O(l)$ at 50.0°C and 1 atm

When this system is heated, its state changes, perhaps to one described as

50.0 g of $H_2O(l)$ at 80.0°C and 1 atm

   Certain quantities, called **state properties,** depend only on the state of the system, not on the way the system reached that state. Putting it another way, if X is a state property, then

$$\Delta X = X_{final} - X_{initial}$$

That is, the change in X is the difference between its values in final and initial states. Most of the quantities that you are familiar with are state properties; volume is a common example. You may be surprised to learn, however, that heat flow is *not* a state property; its magnitude depends on how a process is carried out (Section 8.7).

The distance between two cities depends on path so it isn't a state property.

Direction and Sign of Heat Flow

Consider again the setup in Figure 8.1. If the hot plate is turned on, there is a flow of heat from the surroundings into the system, 50.0 g of water. This situation is described by stating that the heat flow, q, for the system is a positive quantity.

   q is + when heat flows into the system from the surroundings

Usually, when heat flows into a system, its temperature rises. In this case, the temperature of the 50.0-g water sample might increase from 50.0 to 80.0°C. When the hot plate in Figure 8.1 is shut off, the hot water gives off heat to the surrounding air. In this case, q for the system is a negative quantity.

   q is − when heat flows out of the system into the surroundings

As is usually the case, the temperature of the system drops when heat flows out of it into the surroundings. The 50.0-g water sample might cool from 80.0°C back to 50.0°C.

This same reasoning can be applied to a reaction in which the system consists of the reaction mixture (products and reactants). We can distinguish between—

■ *an* **endothermic** *process (q < 0), in which heat flows from the surroundings into the reaction system.* An example is the melting of ice:

$$H_2O(s) \longrightarrow H_2O(l) \qquad q > 0$$

The melting of ice *absorbs* heat from the surroundings, which might be the water in a glass of iced tea. The temperature of the surroundings drops, perhaps from 25 to 0°C, as they give up heat to the system.

■ *an* **exothermic** *process (q < 0), in which heat flows from the reaction system into the surroundings.* A familiar example is the combustion of methane, the major component of natural gas.

$$CH_4(g) + 2O_2(g) \longrightarrow CO_2(g) + 2H_2O(l) \qquad q < 0$$

This reaction *evolves* heat to the surroundings, which might be the air around a Bunsen burner in the laboratory or a potato being baked in a gas oven. In either case, the effect of the heat transfer is to raise the temperature of the surroundings.

Magnitude of Heat Flow

In any process, we are interested not only in the direction of heat flow but also in its magnitude. We will express q in the units introduced in Chapter 6, joules and kilojoules. The joule is named for James Joule (1818–1889), who carried out very precise thermometric measurements that established the first law of thermodynamics (Section 8.7).

In the past, chemists used the calorie* as an energy unit. This is the amount of heat required to raise the temperature of one gram of water one degree Celsius. The calorie is a somewhat larger energy unit than the joule:

$$1 \text{ cal} = 4.184 \text{ J} \qquad 1 \text{ kcal} = 4.184 \text{ kJ}$$

Most of the remainder of this chapter is devoted to a discussion of the magnitude of the heat flow in chemical reactions or phase changes. However, we will focus on a simpler process in which the only effect of the heat flow is to change the temperature of a system. In general, the relationship between the magnitude of the **heat flow,** q, and the temperature change, Δt, is given by the equation

$$q = C \times \Delta t \qquad (\Delta t = t_{final} - t_{initial})$$

The quantity C appearing in this equation is known as the **heat capacity** of the system. It represents the amount of heat required to raise the temperature of the system 1°C and has the units J/°C.

For a pure substance of mass m, the expression for q can be written as

$$\boxed{q = m \times c \times \Delta t}$$

The quantity c is known as the **specific heat;** it is defined as the amount of heat required to raise the temperature of one gram of a substance 1°C.

Specific heat, like density or melting point, is an intensive property that can be used to identify a substance. Water has an unusually large specific heat,

See the *Saunders Interactive General Chemistry CD-ROM*, Screen 6.12, **Energy Changes in Chemical Processes.**

endo = into system; exo = out of system.

An endothermic process. The icicle melts as heat is absorbed by the ice. *(Charles D. Winters)*

See Screens 6.7, Heat Capacity, & 6.8, Heat Capacity of Pure Substances.

*The "calorie" referred to by nutritionists is actually a kilocalorie (1 kcal = 10^3 cal). On a 2000-calorie per day diet, you eat food capable of producing 2000 kcal = 2×10^3 kcal = 2×10^6 cal of energy.

TABLE 8.1   Specific Heats of a Few Common Substances

	c (J/g·°C)		c (J/g·°C)
$Br_2(l)$	0.474	$Cu(s)$	0.382
$Cl_2(g)$	0.478	$Fe(s)$	0.446
$C_2H_5OH(l)$	2.43	$H_2O(g)$	1.87
$C_6H_6(l)$	1.72	$H_2O(l)$	4.18
$CO_2(g)$	0.843	$NaCl(s)$	0.866

Shoreline in Australia. Because of its high specific heat, the water warms up and cools down more slowly than the adjacent beach and buildings. *(David Jacobs/Tony Stone Images)*

4.18 J/g·°C. This explains why swimming is not a popular pastime in northern Minnesota in May. Even if the air temperature rises to 90°F, the water temperature will remain below 60°F. Metals have a relatively low specific heat (Table 8.1). When you warm water in a stainless steel saucepan, for example, nearly all of the heat is absorbed by the water, very little by the steel.

EXAMPLE 8.1   How much heat is given off by a 50.0-g sample of copper when it cools from 80.0 to 50.0°C?

Strategy   Calculate q by first determining Δt and then substituting into the equation $q = m \times c \times \Delta t$. The specific heat of copper is given in Table 8.1.

Solution   The temperature change is

$$\Delta t = t_{final} - t_{initial} = 50.0°C - 80.0° = -30.0°C$$

$$q = 50.0 \text{ g} \times 0.382 \frac{J}{g \cdot °C} \times (-30.0°C) = -573 \text{ J}$$

Reality Check   The negative sign indicates that heat flows from copper to the surroundings. If the sample were heated from 50.0 to 80.0°C, heat would flow in the opposite direction and q for the copper would be $+573$ J.

8.2   MEASUREMENT OF HEAT FLOW; CALORIMETRY

To measure the heat flow in a reaction, it is carried out in a device known as a **calorimeter.** The apparatus contains water and/or other materials of known heat capacity. The walls of the calorimeter are insulated so that there is no exchange of heat with the air outside the calorimeter. It follows that the only heat flow is between the reaction system and the calorimeter. The heat flow for the reaction system is equal in magnitude but opposite in sign to that for the calorimeter:

$$q_{reaction} = -q_{calorimeter}$$

Notice that if the reaction is exothermic ($q_{reaction} < 0$), $q_{calorimeter}$ must be positive; that is, heat flows from the reaction mixture into the calorimeter. Conversely, if the reaction is endothermic, the calorimeter gives up heat to the reaction mixture.

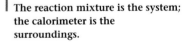 **See Screen 6.18, Calorimetry.**

The reaction mixture is the system; the calorimeter is the surroundings.

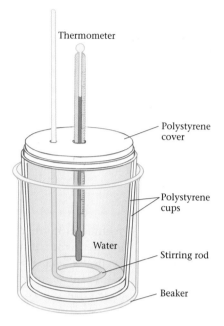

Thermometer

Polystyrene cover

Polystyrene cups

Water

Stirring rod

Beaker

FIGURE 8.2

Coffee-cup calorimeter. The heat given off by a reaction is absorbed by the water. If you know the mass of the water, its specific heat (4.18 J/g·°C), and the temperature change as read on the thermometer, you can calculate the heat flow, q, for the reaction. **OHT**

● **See Screen 6.9, Calculating Heat Transfer.**

The equation just written is basic to calorimetric measurements. It allows you to calculate the amount of heat absorbed or evolved in a reaction if you know the heat capacity, C_{cal}, and the temperature change, Δt, of the calorimeter.

$$q_{reaction} = -C_{cal} \times \Delta t$$

Coffee-Cup Calorimeter

Figure 8.2 shows a simple **coffee-cup calorimeter** used in the general chemistry laboratory. It consists of a polystyrene foam cup partially filled with water. The cup has a tightly fitting cover through which an accurate thermometer is inserted. Because polystyrene foam is a good insulator, there is very little heat flow through the walls of the cup. Essentially all the heat evolved by a reaction taking place within the calorimeter is absorbed by the water. This means that, to a good degree of approximation, the heat capacity of the coffee-cup calorimeter is that of the water:

$$C_{cal} = m_{water} \times c_{water} = m_{water} \times 4.18 \frac{J}{g \cdot °C}$$

and hence

$$q_{reaction} = -m_{water} \times 4.18 \frac{J}{g \cdot °C} \times \Delta t$$

EXAMPLE 8.2 When 1.00 g of ammonium nitrate, NH_4NO_3, is added to 50.0 g of water in a coffee-cup calorimeter, it dissolves:

$$NH_4NO_3(s) \longrightarrow NH_4^+(aq) + NO_3^-(aq)$$

and the temperature of the water drops from 25.00 to 23.32°C. Assuming that all the heat absorbed by the reaction comes from the water, calculate q for the reaction system.

Strategy   Apply the relation

$$q_{reaction} = -m_{water} \times 4.18 \, \frac{J}{g \cdot °C} \times \Delta t$$

Solution

$$q_{reaction} = -50.0 \, g \times 4.18 \, \frac{J}{g \cdot °C} \times (23.32°C - 25.00°C) = \boxed{+351 \, J}$$

Reality Check   Notice that q is positive; this is an endothermic process.

Bomb Calorimeter

A coffee-cup calorimeter is suitable for measuring heat flow for reactions in solution. However, it cannot be used for reactions involving gases, which would escape from the cup. Neither would it be appropriate for reactions in which the products reach high temperatures. The **bomb calorimeter,** shown in Figure 8.3, is more versatile. To use it, a weighed sample of the reactant(s) is added to the heavy-walled metal vessel called a "bomb." This is then sealed and lowered into the insulated outer container. An amount of water sufficient to cover the bomb is added, and the entire apparatus is closed. The initial temperature is measured precisely. The reaction is then started, perhaps by electrical ignition. In an exothermic reaction, the hot products give off heat to the walls of the bomb and to the water. The final temperature is taken to be the highest value read on the thermometer.

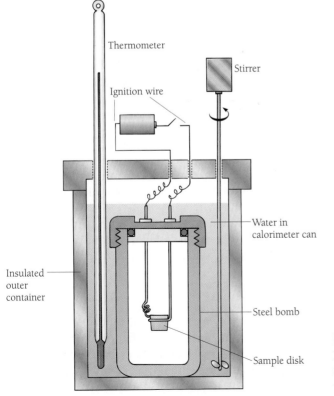

Thermometer

Stirrer

Ignition wire

Water in calorimeter can

Insulated outer container

Steel bomb

Sample disk

FIGURE 8.3
Bomb calorimeter. The heat flow, q, for the reaction is calculated from the temperature change multiplied by the heat capacity of the calorimeter, which is determined in a preliminary experiment. **OHT**

All of the heat given off by the reaction is absorbed by the calorimeter, which consists of the metal bomb and the water that surrounds it. In other words,

$$q_{\text{reaction}} = -q_{\text{calorimeter}}$$

The heat absorbed by the calorimeter is equal to the product of its heat capacity, C_{cal}, and the temperature change, Δt. Hence

$$q_{\text{reaction}} = -C_{\text{cal}} \times \Delta t$$

To find the heat capacity, C_{cal}, the experiment is repeated using the same bomb and the same amount of water. This time, though, we carry out a reaction for which the amount of heat evolved is known; the temperature increase is again measured carefully. Suppose the reaction is known to evolve 93.3 kJ of heat and the temperature rises from 20.00 to 30.00°C. It follows that

$$C_{\text{cal}} = \frac{93.3 \text{ kJ}}{10.00°\text{C}} = 9.33 \text{ kJ/°C}$$

Knowing the heat capacity of the calorimeter, the heat flow for any reaction taking place within the calorimeter can be calculated (Example 8.3).

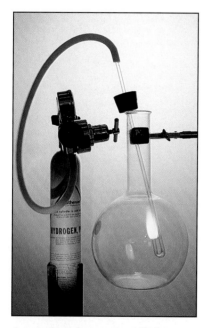

An exothermic reaction. Hydrogen, a colorless gas, reacts with chlorine, a pale yellow gas, to form colorless hydrogen chloride gas. *(Charles D. Winters)*

EXAMPLE 8.3   The reaction between hydrogen and chlorine, $H_2(g) + Cl_2(g) \rightarrow 2HCl(g)$, can be studied in a bomb calorimeter. It is found that when a 1.00-g sample of H_2 reacts completely, the temperature rises from 20.00 to 29.82°C. Taking the heat capacity of the calorimeter to be 9.33 kJ/°C, calculate the amount of heat evolved in the reaction.

Strategy   Use the equation:

$$q_{\text{reaction}} = -C_{\text{cal}} \times \Delta t$$

Solution

$$q_{\text{reaction}} = -9.33 \frac{\text{kJ}}{°\text{C}} \times (29.82°\text{C} - 20.00°\text{C}) = \boxed{-91.6 \text{ kJ}}$$

⊙ See Screens 6.14, Enthalpy Change and ΔH, & 6.15, Enthalpy Changes for Chemical Reactions.

8.3   ENTHALPY

We have referred several times to the "heat flow for the reaction system," symbolized as q_{reaction}. At this point, you may well find this concept a bit nebulous and wonder if it could be made more concrete by relating q_{reaction} to some property of reactants and products. This can indeed be done; the situation is particularly simple for reactions taking place at constant pressure. Under that condition, the heat flow for the reaction system is equal to the difference in **enthalpy** (H) between products and reactants. That is,

$$q_{\text{reaction}} \text{ at constant pressure} = \Delta H = H_{\text{products}} - H_{\text{reactants}}$$

Enthalpy is a type of chemical energy, sometimes referred to as "heat content." Reactions that occur in the laboratory in an open container or in the world around us take place at a constant pressure, that of the atmosphere. For such reactions, the equation just written is valid, making enthalpy a very useful quantity.

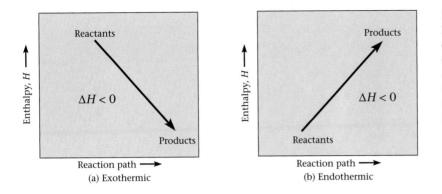

(a) Exothermic

(b) Endothermic

FIGURE 8.4

Energy diagram showing ΔH for a reaction. In an exothermic reaction (a), the products have a lower enthalpy than the reactants; thus ΔH is negative, and heat is given off to the surroundings. In an endothermic reaction (b), the products have a higher enthalpy than the reactants so ΔH is positive, and heat is absorbed from the surroundings. **OHT**

Figure 8.4a shows the enthalpy relationship between reactants and products for an exothermic reaction such as

$$CH_4(g) + 2O_2(g) \longrightarrow CO_2(g) + 2H_2O(l) \qquad \Delta H < 0$$

Here, the products, 1 mol of $CO_2(g)$ and 2 mol of $H_2O(l)$, have a lower enthalpy than the reactants, 1 mol of $CH_4(g)$ and 2 mol of $O_2(g)$. The decrease in enthalpy is the source of the heat evolved to the surroundings. Figure 8.4b shows the situation for an endothermic process such as

$$H_2O(s) \longrightarrow H_2O(l) \qquad \Delta H > 0$$

Liquid water has a higher enthalpy than ice, so heat must be transferred from the surroundings to melt the ice.

In general, the following relations apply for reactions taking place at constant pressure.

> When wood burns, ΔH is negative; when ice melts, ΔH is positive.

> ⊙ See Screen 6.11, Heat Associated with Phase Changes.

exothermic reaction:	$q = \Delta H < 0$	$H_{products} < H_{reactants}$
endothermic reaction:	$q = \Delta H > 0$	$H_{products} > H_{reactants}$

The enthalpy of a substance, like its volume, is a state property. A sample of one gram of liquid water at 25.00°C and 1 atm has a fixed enthalpy, H. In practice, no attempt is made to determine absolute values of enthalpy. Instead, scientists deal with changes in enthalpy, which are readily determined. For the process

$$1.00 \text{ g } H_2O \ (l, 25.00°C, 1 \text{ atm}) \longrightarrow 1.00 \text{ g } H_2O \ (l, 26.00°C, 1 \text{ atm})$$

ΔH is 4.18 J because the specific heat of water is 4.18 J/g·°C.

8.4   THERMOCHEMICAL EQUATIONS

A chemical equation that shows the enthalpy relation between products and reactants is called a **thermochemical equation.** This type of equation contains, at the right of the balanced chemical equation, the appropriate value and sign for ΔH.

To see where a thermochemical equation comes from, consider again the process by which ammonium nitrate dissolves in water:

$$NH_4NO_3(s) \longrightarrow NH_4^+(aq) + NO_3^-(aq)$$

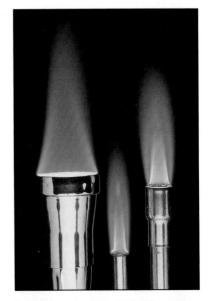

Laboratory burners fueled by natural gas, which is mostly methane.

(Charles D. Winters)

FIGURE 8.5

Endothermic reaction in a cold pack. When the seal separating the compartments containing ammonium nitrate and water is broken the following endothermic reaction occurs:

$$NH_4NO_3(s) \longrightarrow NH_4^+(aq) + NO_3^-(aq) \qquad \Delta H = +28.1 \text{ kJ}$$

As a result the temperature, as read on the thermometer, drops. *(Marna G. Clarke)*

Recall (Example 8.2) that a simple experiment with a coffee-cup calorimeter shows that when one gram of NH_4NO_3 dissolves, $q_{reaction} = 351$ J. The calorimeter is open to the atmosphere, the pressure is constant, and

$$\Delta H \text{ for dissolving } 1.00 \text{ g of } NH_4NO_3 = 351 \text{ J} = 0.351 \text{ kJ}$$

When one mole (80.05 g) of NH_4NO_3 dissolves, ΔH should be 80 times as great:

$$\Delta H \text{ for dissolving } 1.00 \text{ mol of } NH_4NO_3 = 0.351 \frac{\text{kJ}}{\text{g}} \times 80.05 \text{ g} = 28.1 \text{ kJ}$$

The thermochemical equation for this reaction (Figure 8.5) must then be

$$NH_4NO_3(s) \longrightarrow NH_4^+(aq) + NO_3^-(aq) \qquad \Delta H = +28.1 \text{ kJ}$$

By an entirely analogous procedure, the thermochemical equation for the formation of HCl from the elements (Example 8.3) is found to be

$$H_2(g) + Cl_2(g) \longrightarrow 2HCl(g) \qquad \Delta H = -185 \text{ kJ}$$

In other words, 185 kJ of heat is evolved when two moles of HCl are formed from H_2 and Cl_2.

These thermochemical equations are typical of those used throughout this text. It is important to realize that—

- the sign of ΔH indicates whether the reaction, when carried out at constant pressure, is endothermic (positive ΔH) or exothermic (negative ΔH).
- in interpreting a thermochemical equation, the coefficients represent numbers of moles (ΔH is −185 kJ when **1 mol** H_2 + **1 mol** Cl_2 → **2 mol** HCl).
- the phases (physical states) of all species must be specified, using the symbols (s), (l), (g), or (aq). The enthalpy of one mole of $H_2O(g)$ at 25°C is 44 kJ larger than that of one mole of $H_2O(l)$; the difference, which represents the heat of vaporization of water, is clearly significant.
- the value quoted for ΔH applies when products and reactants are at the same temperature, ordinarily taken to be 25°C unless specified otherwise.

This explains why burns from steam are more painful than those from boiling water.

Rules of Thermochemistry

To make effective use of thermochemical equations, three basic rules of thermochemistry are applied.

1. The magnitude of ΔH is directly proportional to the amount of reactant or product. This is a common-sense rule, consistent with experience. The amount of heat that must be absorbed to boil a sample of water is directly proportional to its mass. In another case, the more gasoline you burn in your car's engine, the more energy you produce.

This rule allows you to find ΔH corresponding to any desired amount of reactant or product. To do this, you follow the conversion-factor approach used in Chapter 3 with ordinary chemical equations. Consider, for example,

$$H_2(g) + Cl_2(g) \longrightarrow 2HCl(g) \qquad \Delta H = -185 \text{ kJ}$$

The thermochemical equation allows us to relate the enthalpy change to amounts of reactants and products, leading to conversion factors such as

$$\frac{-185 \text{ kJ}}{1 \text{ mol Cl}_2} \qquad \frac{1 \text{ mol H}_2}{-185 \text{ kJ}} \qquad \frac{-185 \text{ kJ}}{2 \text{ mol HCl}}$$

EXAMPLE 8.4   Consider the thermochemical equation for the formation of two moles of HCl from the elements. Calculate ΔH when

(a) 1.00 mol of HCl is formed.   (b) 1.00 g of Cl_2 reacts.

Strategy   Use the conversion factor approach. In (a), only one conversion is required, from moles of HCl to ΔH in kilojoules. In (b), you first have to convert to moles of Cl_2 and then to ΔH in kilojoules.

Solution

(a) $\Delta H = 1.00 \text{ mol HCl} \times \dfrac{-185 \text{ kJ}}{2 \text{ mol HCl}} = \boxed{-92.5 \text{ kJ}}$

This means that the thermochemical equation for the formation of one mole of HCl would be

$$\tfrac{1}{2}H_2(g) + \tfrac{1}{2}Cl_2(g) \longrightarrow HCl(g) \qquad \Delta H = -92.5 \text{ kJ}$$

(b) $\Delta H = 1.00 \text{ g Cl}_2 \times \dfrac{1 \text{ mol Cl}_2}{70.90 \text{ g Cl}_2} \times \dfrac{-185 \text{ kJ}}{1 \text{ mol Cl}_2} = \boxed{-2.61 \text{ kJ}}$

This relation between ΔH and amounts of substances is equally useful in dealing with chemical reactions or phase changes.

EXAMPLE 8.5   Using a coffee-cup calorimeter, it is found that when an ice cube weighing 24.6 g melts, it absorbs 8.19 kJ of heat. Calculate ΔH for the phase change represented by the thermochemical equation $H_2O(s) \rightarrow H_2O(l)$ $\qquad \Delta H = ?$

Strategy   The data given lead directly to the conversion factor

$$\frac{8.19 \text{ kJ}}{24.6 \text{ g ice}}$$

You need to know ΔH when one mole (18.02 g) of ice melts; two conversions are required.

Solution

$$\Delta H = 1 \text{ mol } H_2O \times \frac{18.02 \text{ g } H_2O}{1 \text{ mol } H_2O} \times \frac{8.19 \text{ kJ}}{24.6 \text{ g } H_2O} = \boxed{6.00 \text{ kJ}}$$

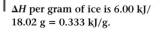 ΔH per gram of ice is 6.00 kJ/ 18.02 g = 0.333 kJ/g.

This calculation shows that 6.00 kJ of heat must be absorbed to melt one mole of ice:

$$H_2O(s) \longrightarrow H_2O(l) \qquad \Delta H = +6.00 \text{ kJ}$$

Reality Check   Because one mole of H_2O (18.0 g) is about $\frac{3}{4}$ of 24.6 g, ΔH should be about $\frac{3}{4}$ of 8.19 kJ; it is.

The heat absorbed when a solid melts ($s \rightarrow l$) is referred to as the **heat of fusion;** that absorbed when a liquid vaporizes ($l \rightarrow g$) is called the **heat of vaporization.** Heats of fusion (ΔH_{fus}) and vaporization (ΔH_{vap}) are most often expressed in kilojoules per mole (kJ/mol). Values for several different substances are given in Table 8.2.

2. ΔH for a reaction is equal in magnitude but opposite in sign to ΔH for the reverse reaction. Another way to state this rule is to say that the amount of heat evolved in a reaction is exactly equal to the amount of heat absorbed in the reverse reaction. This again is a common-sense rule. If 6.00 kJ of heat is absorbed when a mole of ice melts,

$$H_2O(s) \longrightarrow H_2O(l) \qquad \Delta H = +6.00 \text{ kJ}$$

then 6.00 kJ of heat should be evolved when a mole of liquid water freezes.

$$H_2O(l) \longrightarrow H_2O(s) \qquad \Delta H = -6.00 \text{ kJ}$$

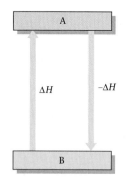

Thermochemistry Rule 2.

EXAMPLE 8.6   Given

$$H_2(g) + \tfrac{1}{2}O_2(g) \longrightarrow H_2O(l) \qquad \Delta H = -285.8 \text{ kJ}$$

calculate ΔH for the equation

$$2H_2O(l) \longrightarrow 2H_2(g) + O_2(g)$$

Strategy   Note that for the required equation the coefficients are twice as great as in the given equation; the equation is also reversed. Apply Rules 1 and 2 in succession.

Solution   Applying Rule 1:

$$2H_2(g) + O_2(g) \longrightarrow 2H_2O(l) \qquad \Delta H = 2(-285. \text{ kJ}) = -571.6 \text{ kJ}$$

Applying Rule 2:

$$2H_2O(l) \longrightarrow 2H_2(g) + O_2(g) \qquad \Delta H = \boxed{+571.6 \text{ kJ}}$$

---

3. The value of ΔH for a reaction is the same whether it occurs in one step or in a series of steps (Figure 8.6). If a thermochemical equation can be expressed as the sum of two or more equations,

$$\text{equation} = \text{equation (1)} + \text{equation (2)} + \ldots$$

TABLE 8.2 ΔH(kJ/mol) for Phase Changes

Substance		mp (°C)	ΔH_{fus}*	bp (°C)	ΔH_{vap}*
Benzene	C_6H_6	5	9.84	80	30.8
Bromine	Br_2	−7	10.8	59	29.6
Mercury	Hg	−39	2.33	357	59.4
Naphthalene	$C_{10}H_8$	80	19.3	218	43.3
Water	H_2O	0	6.00	100	40.7

*Values of ΔH_{fus} are given at the melting point, values of ΔH_{vap} at the boiling point. The heat of vaporization of water decreases from 44.9 kJ/mol at 0°C to 44.0 kJ/mol at 25°C to 40.7 kJ/mol at 100°C.

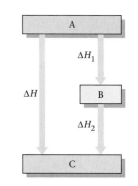

FIGURE 8.6
Thermochemistry Rule 3, Hess's law. ΔH is independent of path because H is a state property. **OHT**

then ΔH for the overall equation is the sum of the ΔH's for the individual equations:

$$\Delta H = \Delta H_1 + \Delta H_2 + \ldots$$

This relationship is referred to as **Hess's law,** after Germaine Hess, professor of chemistry at the University of St. Petersburg, who deduced it in 1840. Hess's law is a direct consequence of the fact that enthalpy is a state property, dependent only on initial and final states. This means that, in Figure 8.6, ΔH *must* equal the sum of ΔH_1 and ΔH_2, because the final and initial states are the same for the two processes.

Hess's law is very convenient for obtaining values of ΔH for reactions that are difficult to carry out in a calorimeter. Consider, for example, the formation of the toxic gas carbon monoxide from the elements

$$C(s) + \tfrac{1}{2}O_2(g) \longrightarrow CO(g)$$

It is difficult, essentially impossible, to measure ΔH for this reaction because when carbon burns the major product is always carbon dioxide, CO_2. It is possible, however, to calculate ΔH using thermochemical data for two other reactions that are readily carried out in the laboratory (Example 8.7).

⊙ See Screen 6.16, Hess's Law.

EXAMPLE 8.7   Given

$$(1) \quad C(s) + O_2(g) \longrightarrow CO_2(g) \qquad \Delta H = -393.5 \text{ kJ}$$

$$(2) \; 2CO(g) + O_2(g) \longrightarrow 2CO_2(g) \qquad \Delta H = -566.0 \text{ kJ}$$

calculate ΔH for the reaction

$$C(s) + \tfrac{1}{2}O_2(g) \longrightarrow CO(g) \qquad \Delta H = ?$$

Strategy   The "trick" here is to work with the given information until you arrive at two equations that will add to give the equation you want ($C + \tfrac{1}{2}O_2 \rightarrow CO$). To do this, focus on CO, which, unlike CO_2 and O_2, appears in only one thermochemical equation. Notice that you want *one* mole (not two) of CO on the *right* side (not the left side) of the equation.

Solution   To get one mole of CO on the right side, reverse Equation (2) and divide the coefficients by 2. Applying Rule 1 and Rule 2 in succession,

$$CO_2(g) \longrightarrow CO(g) + \tfrac{1}{2}O_2(g) \qquad \Delta H = +566.0 \text{ kJ}/2 = +283.0 \text{ kJ}$$

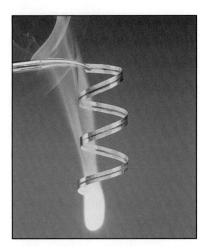

Enthalpy of formation. Magnesium ribbon reacts with oxygen to give MgO, a white solid, and 601.7 kJ of heat per mole of MgO formed. Hence, $\Delta H_f^{\circ} \text{ MgO}(s) = -601.7$ kJ/mol. *(Charles D. Winters)*

 See Screen 6.17, Standard Enthalpy of Formation.

┌ Often called "heat of formation."

Now, add Equation (1) and simplify:

$$CO_2(g) \longrightarrow CO(g) + \tfrac{1}{2}O_2(g) \qquad \Delta H = +283.0 \text{ kJ}$$

(1) $\underline{\qquad C(s) + O_2(g) \longrightarrow CO_2(g) \qquad\qquad \Delta H = -393.5 \text{ kJ}}$

$$C(s) + \tfrac{1}{2}O_2(g) \longrightarrow CO(g) \qquad\qquad \Delta H = -110.5 \text{ kJ}$$

Reality Check Notice that thermochemical equations can be added in exactly the same manner as algebraic equations; in this case $1CO_2$ and $\tfrac{1}{2}O_2$ canceled when the equations were added.

Summarizing the rules of thermochemistry:

1. ΔH is directly proportional to the amount of reactant or product.
2. ΔH changes sign when a reaction is reversed.
3. ΔH for a reaction has the same value regardless of the number of steps.

8.5 ENTHALPIES OF FORMATION

We have now written several thermochemical equations. In each case, we have cited the corresponding value of ΔH. Literally thousands of such equations would be needed to list the ΔH values for all the reactions that have been studied. Clearly, there has to be some more concise way of recording data of this sort. These data should be in a form that can easily be used to calculate ΔH for any reaction. It turns out that there is a simple way to do this, using quantities known as enthalpies of formation.

Meaning of ΔH_f°

The standard molar **enthalpy of formation** of a compound, ΔH_f°, is equal to the enthalpy change when one mole of the compound is formed at a constant pressure of 1 atm and a fixed temperature, ordinarily 25°C, from the elements in their stable states at that pressure and temperature. From the equations

$$Ag(s, 25°C) + \tfrac{1}{2}Cl_2(g, 25°C, 1 \text{ atm}) \longrightarrow AgCl(s, 25°C) \qquad \Delta H = -127.1 \text{ kJ}$$

$$\tfrac{1}{2}N_2(g, 1 \text{ atm}, 25°C) + O_2(g, 1 \text{ atm}, 25°C) \longrightarrow NO_2(g, 1 \text{ atm}, 25°C) \quad \Delta H = +33.2 \text{ kJ}$$

it follows that

$$\Delta H_f^{\circ} \text{ AgCl}(s) = -127.1 \text{ kJ/mol} \qquad \Delta H_f^{\circ} \text{ NO}_2(g) = +33.2 \text{ kJ/mol}$$

Enthalpies of formation are listed for a variety of compounds in Table 8.3. Notice that, with a few exceptions, enthalpies of formation are negative quantities. This means that the formation of a compound from the elements is ordinarily exothermic. Conversely, when a compound decomposes to the elements, heat usually must be absorbed.

You will note from Table 8.3 that there are no entries for elemental species such as $Br_2(l)$ and $O_2(g)$. This is a consequence of the way in which enthalpies of formation are defined. In effect, the enthalpy of formation of an element in its stable state at 25°C and 1 atm is taken to be zero. That is,

$$\Delta H_f^{\circ} \text{ Br}_2(l) = \Delta H_f^{\circ} \text{ O}_2(g) = 0$$

TABLE 8.3 **Standard Enthalpies of Formation at 25°C (kJ/mol) of Compounds at 1 atm, Aqueous Ions at 1 M** OHT

Compounds

AgBr(s)	−100.4	CaCl$_2$(s)	−795.8	H$_2$O(g)	−241.8	NH$_4$NO$_3$(s)	−365.6
AgCl(s)	−127.1	CaCO$_3$(s)	−1206.9	H$_2$O(l)	−285.8	NO(g)	+90.2
AgI(s)	−61.8	CaO(s)	−635.1	H$_2$O$_2$(l)	−187.8	NO$_2$(g)	+33.2
AgNO$_3$(s)	−124.4	Ca(OH)$_2$(s)	−986.1	H$_2$S(g)	−20.6	N$_2$O$_4$(g)	+9.2
Ag$_2$O(s)	−31.0	CaSO$_4$(s)	−1434.1	H$_2$SO$_4$(l)	−814.0	NaCl(s)	−411.2
Al$_2$O$_3$(s)	−1675.7	CdCl$_2$(s)	−391.5	HgO(s)	−90.8	NaF(s)	−573.6
BaCl$_2$(s)	−858.6	CdO(s)	−258.2	KBr(s)	−393.8	NaOH(s)	−425.6
BaCO$_3$(s)	−1216.3	Cr$_2$O$_3$(s)	−1139.7	KCl(s)	−436.7	NiO(s)	−239.7
BaO(s)	−553.5	CuO(s)	−157.3	KClO$_3$(s)	−397.7	PbBr$_2$(s)	−278.7
BaSO$_4$(s)	−1473.2	Cu$_2$O(s)	−168.6	KClO$_4$(s)	−432.8	PbCl$_2$(s)	−359.4
CCl$_4$(l)	−135.4	CuS(s)	−53.1	KNO$_3$(s)	−494.6	PbO(s)	−219.0
CHCl$_3$(l)	−134.5	Cu$_2$S(s)	−79.5	MgCl$_2$(s)	−641.3	PbO$_2$(s)	−277.4
CH$_4$(g)	−74.8	CuSO$_4$(s)	−771.4	MgCO$_3$(s)	−1095.8	PCl$_3$(g)	−287.0
C$_2$H$_2$(g)	+226.7	Fe(OH)$_3$(s)	−823.0	MgO(s)	−601.7	PCl$_5$(g)	−374.9
C$_2$H$_4$(g)	+52.3	Fe$_2$O$_3$(s)	−824.2	Mg(OH)$_2$(s)	−924.5	SiO$_2$(s)	−910.9
C$_2$H$_6$(s)	−84.7	Fe$_3$O$_4$(s)	−1118.4	MgSO$_4$(s)	−1284.9	SnO$_2$(s)	−580.7
C$_3$H$_8$(g)	−103.8	HBr(g)	−36.4	MnO(s)	−385.2	SO$_2$(g)	−296.8
CH$_3$OH(l)	−238.7	HCl(g)	−92.3	MnO$_2$(s)	−520.0	SO$_3$(g)	−395.7
C$_2$H$_5$OH(l)	−277.7	HF(g)	−271.1	NH$_3$(g)	−46.1	ZnI$_2$(s)	−208.0
CO(g)	−110.5	HI(g)	+26.5	N$_2$H$_4$(l)	+50.6	ZnO(s)	−348.3
CO$_2$(g)	−393.5	HNO$_3$(l)	−174.1	NH$_4$Cl(s)	−314.4	ZnS(s)	−206.0

Cations				Anions			
Ag$^+$(aq)	+105.6	Hg^{2+}(aq)	+171.1	Br$^-$(aq)	−121.6	HPO$_4^{2-}$(aq)	−1292.1
Al^{3+}(aq)	−531.0	K$^+$(aq)	−252.4	CO$_3^{2-}$(aq)	−677.1	HSO$_4^-$(aq)	−887.3
Ba^{2+}(aq)	−537.6	Mg^{2+}(aq)	−466.8	Cl$^-$(aq)	−167.2	I$^-$(aq)	−55.2
Ca^{2+}(aq)	−542.8	Mn^{2+}(aq)	−220.8	ClO$_3^-$(aq)	−104.0	MnO$_4^-$(aq)	−541.4
Cd^{2+}(aq)	−75.9	Na$^+$(aq)	−240.1	ClO$_4^-$(aq)	−129.3	NO$_2^-$(aq)	−104.6
Cu$^+$(aq)	+71.7	NH$_4^+$(aq)	−132.5	CrO$_4^{2-}$(aq)	−881.2	NO$_3^-$(aq)	−205.0
Cu^{2+}(aq)	+64.8	Ni^{2+}(aq)	−54.0	Cr$_2$O$_7^{2-}$(aq)	−1490.3	OH$^-$(aq)	−230.0
Fe^{2+}(aq)	−89.1	Pb^{2+}(aq)	−1.7	F$^-$(aq)	−332.6	PO$_4^{3-}$(aq)	−1277.4
Fe^{3+}(aq)	−48.5	Sn^{2+}(aq)	−8.8	HCO$_3^-$(aq)	−692.0	S^{2-}(aq)	+33.1
H$^+$(aq)	0.0	Zn^{2+}(aq)	−153.9	H$_2$PO$_4^-$(aq)	−1296.3	SO$_4^{2-}$(aq)	−909.3

The standard enthalpies of formation of ions in aqueous solution listed at the bottom of Table 8.3 are relative values, established by taking

$$\Delta H_f^\circ \, H^+(aq) = 0$$

From the values listed we see that the Cu^{2+} ion has a heat of formation *greater* than that of H$^+$ by about 65 kJ/mol, and Cd^{2+} has a heat of formation about 76 kJ/mol *less* than that of H$^+$.

Calculation of $\Delta H°$

Enthalpies of formation can be used to calculate $\Delta H°$ for a reaction. To do this, apply this general rule:

(a)   (b)   (c)

FIGURE 8.7

The thermite reaction. (a) A piece of burning magnesium acts as a fuse when inserted into a pot containing a finely divided mixture of aluminum powder and iron(III) oxide. (b) The burning magnesium initiates the exothermic reaction:

$$2Al(s) + Fe_2O_3(s) \longrightarrow 2Fe(s) + Al_2O_3(s) \qquad \Delta H° = -851.5 \text{ kJ}$$

(c) Enough heat is generated to produce molten iron, which can be seen flowing out of the broken pot onto the protective mat and the bottom of the stand. *(Charles D. Winters)*

*The **standard enthalpy change**, $\Delta H°$, for a given thermochemical equation is equal to the sum of the standard enthalpies of formation of the product compounds minus the sum of the standard enthalpies of formation of the reactant compounds.*

Using the symbol Σ to represent "the sum of,"

> This is a very useful equation, one we will use again and again.

$$\boxed{\Delta H° = \Sigma\, \Delta H_f°\, \text{products} - \Sigma\, \Delta H_f°\, \text{reactants}}$$

In applying this equation—

■ *elements in their standard states can be omitted,* because their heats of formation are zero. To illustrate, consider the thermite reaction once used to weld rails (Figure 8.7).

$$2Al(s) + Fe_2O_3(s) \longrightarrow 2Fe(s) + Al_2O_3(s)$$

We can write, quite simply,

$$\Delta H° = \Sigma\, \Delta H_f°\, \text{products} - \Sigma\, \Delta H_f°\, \text{reactants} = \Delta H_f°\, Al_2O_3(s) - \Delta H_f°\, Fe_2O_3(s)$$

■ *the coefficients of products and reactants in the thermochemical equation must be taken into account.* For example, consider the reaction:

$$2Al(s) + 3Cu^{2+}(aq) \longrightarrow 2Al^{3+}(aq) + 3Cu(s)$$

for which:

$$\Delta H° = 2\Delta H_f° Al^{3+}(aq) - 3\Delta H_f° Cu^{2+}(aq)$$

Strictly speaking, $\Delta H°$ calculated from enthalpies of formation listed in Table 8.3 represents the enthalpy change at 25°C and 1 atm. Actually, ΔH is independent of pressure and varies relatively little with temperature, changing by perhaps 1 to 10 kJ per 100°C.

EXAMPLE 8.8   Calculate $\Delta H°$ for the combustion of one mole of propane, C_3H_8, according to the equation

$$C_3H_8(g) + 5O_2(g) \longrightarrow 3CO_2(g) + 4H_2O(l)$$

Strategy   Use the equation: $\Delta H° = \Sigma \Delta H_f°$ products $- \Sigma \Delta H_f°$ reactants. Take enthalpies of formation from Table 8.3. Remember that the values in the table are for one mole of compound; if there are n moles in the equation, multiply the molar enthalpy of formation by n.

Solution   Expressing $\Delta H°$ in terms of enthalpies of formation,

$$\Delta H° = 3\Delta H_f° CO_2(g) + 4\Delta H_f° H_2O(l) - [\Delta H_f° C_3H_8(g) + 5\Delta H_f° O_2(g)]$$

The unit for $\Delta H_f°$ is kJ/mol; that for $\Delta H°$ is kJ.

Taking the enthalpy of formation of $O_2(g)$ to be zero and substituting values for the other substances from Table 8.3,

$$\Delta H° = 3 \text{ mol}\left(-393.5\ \frac{kJ}{mol}\right) + 4 \text{ mol}\left(-285.8\ \frac{kJ}{mol}\right) - 1 \text{ mol}\left(-103.8\ \frac{kJ}{mol}\right)$$

$$= -2219.9 \text{ kJ}$$

The relation between $\Delta H°$ and enthalpies of formation is perhaps used more often than any other in thermochemistry. Its validity depends on the fact that enthalpy is a state property. For any reaction, $\Delta H°$ can be obtained by imagining that the reaction takes place in two steps. First, the reactants (compounds or ions) are converted to the elements:

$$\text{reactants} \longrightarrow \text{elements} \qquad \Delta H_1° = -\Sigma \Delta H_f° \text{ reactants}$$

Then the elements are converted to products:

$$\text{elements} \longrightarrow \text{products} \qquad \Delta H_2° = \Sigma \Delta H_f° \text{ products}$$

Applying Hess's law,

$$\Delta H° = \Delta H_1° + \Delta H_2° = \Sigma \Delta H_f° \text{ products} - \Sigma \Delta H_f° \text{ reactants}$$

The relation between $\Delta H°$ and $\Delta H_f°$ can also be used to calculate the heat of formation of a substance from an experimentally determined heat of reaction. Example 8.9 illustrates the calculations involved.

EXAMPLE 8.9   The thermochemical equation for the combustion of benzene, C_6H_6, is

$$C_6H_6(l) + \tfrac{15}{2}O_2(g) \longrightarrow 6CO_2(g) + 3H_2O(l) \qquad \Delta H° = -3267.4 \text{ kJ}$$

Using Table 8.3 to obtain heats of formation for CO_2 and H_2O, calculate the heat of formation of benzene.

Strategy Work with the general relation $\Delta H° = \Sigma \Delta H_f°$ products $- \Sigma \Delta H_f°$ reactants. All the heats of formation are known except that of benzene; $\Delta H°$ is also known. With only one unknown, the equation is readily solved for $\Delta H_f° C_6H_6$.

Solution Taking the heat of formation of $O_2(g)$ to be zero, the relation for $\Delta H°$ is

$$\Delta H° = 6\Delta H_f° CO_2(g) + 3\Delta H_f° H_2O(l) - \Delta H_f° C_6H_6(l)$$

Substituting values for $\Delta H°$, $\Delta H_f° CO_2$, and $\Delta H_f° H_2O(l)$, we have

$$-3267.4 \text{ kJ} = 6 \text{ mol}\left(-393.5 \frac{\text{kJ}}{\text{mol}}\right) + 3 \text{ mol}\left(-285.8 \frac{\text{kJ}}{\text{mol}}\right) - 1 \text{ mol } [\Delta H_f° C_6H_6(l)]$$

Solving,

$$\boxed{\Delta H_f° C_6H_6(l) = +49.0 \text{ kJ/mol}}$$

Reality Check The heats of formation of all the organic compounds listed in Table 8.3, including acetylene, C_2H_2, and ethyl alcohol, C_2H_5OH, were determined this way. Organic compounds cannot be made by the direct reaction of the elements.

Enthalpy changes for reactions in solution can be determined using standard enthalpies of formation of aqueous ions, applying the general relation

$$\Delta H° = \Sigma \Delta H_f° \text{ products} - \Sigma \Delta H_f° \text{ reactants}$$

and taking account of the fact that $\Delta H_f° H^+(aq) = 0$ (Example 8.10).

FIGURE 8.8

Reaction of an acid with a carbonate. Hydrochloric acid added to sodium carbonate solution produces CO_2 gas. *(Charles D. Winters)*

EXAMPLE 8.10 When hydrochloric acid is added to a solution of sodium carbonate, carbon dioxide gas is formed (Figure 8.8). The equation for the reaction is

$$2H^+(aq) + CO_3^{2-}(aq) \longrightarrow CO_2(g) + H_2O(l)$$

Calculate $\Delta H°$ for this thermochemical equation.

Strategy Apply the general relation between $\Delta H°$ and enthalpies of formation. Obtain enthalpies of formation from Table 8.3. Remember that $\Delta H_f° H^+(aq) = 0$.

Solution

$$\Delta H° = \Delta H_f° CO_2(g) + \Delta H_f° H_2O(l) - [2\Delta H_f° H^+(aq) + \Delta H_f° CO_3^{2-}(aq)]$$

Taking $\Delta H_f° H^+(aq)$ to be zero and obtaining the other enthalpies of formation from Table 8.3,

$$\Delta H° = 1 \text{ mol}\left(-393.5 \frac{\text{kJ}}{\text{mol}}\right) + 1 \text{ mol}\left(-285.8 \frac{\text{kJ}}{\text{mol}}\right) - 1 \text{ mol}\left(-677.1 \frac{\text{kJ}}{\text{mol}}\right)$$

$$= \boxed{-2.2 \text{ kJ}}$$

Reality Check Notice that the reaction is exothermic, but only slightly so. If you carry out the reaction in the laboratory, you find that the solution temperature increases slightly.

8.6 BOND ENTHALPY

For many reactions, ΔH is a large negative number; the reaction gives off a lot of heat. In other cases, ΔH is positive; heat must be absorbed for the reaction to occur. You may well wonder why the enthalpy change should vary so widely from

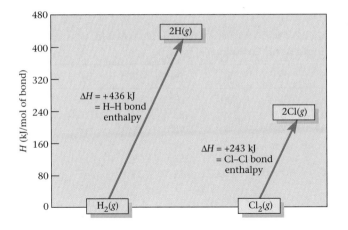

FIGURE 8.9
Bond enthalpies for H₂ and Cl₂. The H—H bond enthalpy is greater than the Cl—Cl bond enthalpy (+436 kJ/mol versus +243 kJ/mol). This means that the bond in H_2 is stronger than that in Cl_2. **OHT**

one reaction to another. Is there some basic property of the molecules involved in the reaction that determines the sign and magnitude of ΔH?

These questions can be answered on a molecular level in terms of a quantity known as bond enthalpy. (More commonly but less properly it is called bond energy). The **bond enthalpy** is defined as ΔH when one mole of bonds is broken in the gaseous state. From the equations (see also Figure 8.9)

$$H_2(g) \longrightarrow 2H(g) \qquad \Delta H = +436 \text{ kJ}$$

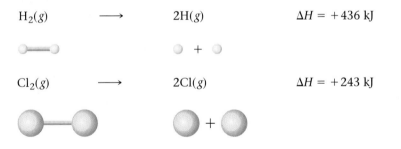

$$Cl_2(g) \longrightarrow 2Cl(g) \qquad \Delta H = +243 \text{ kJ}$$

| What is ΔH_f° for H(g)? Cl(g)?

it follows that the H—H bond enthalpy is +436 kJ/mol, and that for Cl—Cl is 243 kJ/mol. In both reactions, one mole of bonds (H—H and Cl—Cl) is broken.

Bond enthalpies for a variety of single and multiple bonds are listed in Table 8.4 on page 234. Note that bond enthalpy is always a positive quantity; heat is always absorbed when chemical bonds are broken. Conversely, heat is given off when bonds are formed from gaseous atoms. Thus

$$H(g) + Cl(g) \longrightarrow HCl(g) \qquad \Delta H = -431 \text{ kJ}$$

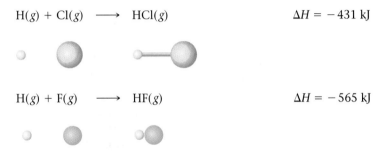

$$H(g) + F(g) \longrightarrow HF(g) \qquad \Delta H = -565 \text{ kJ}$$

Using bond enthalpies, it is possible to explain why certain gas phase reactions are endothermic and others are exothermic. In general, a reaction is expected to be endothermic (i.e., heat must be absorbed) if—

TABLE 8.4   Bond Enthalpies

Single Bond Enthalpy (kJ/mol)

	H	C	N	O	S	F	Cl	Br	I
H	436	414	389	464	339	565	431	368	297
C		347	293	351	259	485	331	276	218
N			159	222	—	272	201	243	—
O				138	—	184	205	201	201
S					226	285	255	213	—
F						153	255	255	277
Cl							243	218	209
Br								193	180
I									151

Muliple Bond Enthalpy (kJ/mol)

C=C	612	N=N	418	C≡C	820
C=N	615	N=O	607	C≡N	890
C=O	715	O=O	498	C≡O	1075
C=S	477	S=O	498	N≡N	941

The higher the bond enthalpy, the stronger the bond.

- *the bonds in the reactants are stronger than those in the products.* Consider, for example, the decomposition of hydrogen fluoride to the elements, which can be shown to be endothermic from the data in Table 8.3.

$$2HF(g) \longrightarrow H_2(g) + F_2(g) \qquad \Delta H = -2\Delta H_f^\circ \, HF = +542.2 \text{ kJ}$$

The bonds in HF (565 kJ/mol) are stronger than the average of those in H_2 and F_2: (436 kJ/mol + 153 kJ/mol)/2 = 295 kJ/mol.

- *there are more bonds in the reactants than in the products.* For the reaction

$$2H_2O(g) \longrightarrow 2H_2(g) + O_2(g) \qquad \Delta H = -2\Delta H_f^\circ \, H_2O(g) = +483.6 \text{ kJ}$$

there are four moles of O—H bonds in two moles of H_2O as compared with only three moles of bonds in the products (two mol H—H bonds, one mole O=O bonds).

You will note from Table 8.4 that the bond enthalpy is larger for a multiple bond than for a single bond between the same two atoms. Thus

But, the double bond enthalpy is less than twice that for a single bond.

$$\text{bond enthalpy C—C} = 347 \text{ kJ/mol}$$
$$\text{bond enthalpy C=C} = 612 \text{ kJ/mol}$$
$$\text{bond enthalpy C≡C} = 820 \text{ kJ/mol}$$

This effect is a reasonable one; the greater the number of bonding electrons, the more difficult it should be to break the bond between two atoms.

We should point out a serious limitation of the bond enthalpies listed in Table 8.4. *Whenever the bond involves two different atoms* (e.g., O—H) *the value listed is approximate rather than exact,* because it represents an average taken over two or more different species. Consider, for example, the O—H bond where we find

$$\text{H—O—H}(g) \longrightarrow H(g) + OH(g) \qquad \Delta H = +499 \text{ kJ}$$
$$\text{H—O}(g) \longrightarrow H(g) + O(g) \qquad \Delta H = +428 \text{ kJ}$$

Both of these reactions involve breaking a mole of O—H bonds, yet the experimental values of ΔH are quite different. The bond enthalpy listed in Table 8.3, 464 kJ/mol, is an average of these two values.

This limitation explains why, whenever possible, we use enthalpies of formation (ΔH_f°) rather than bond enthalpies to calculate the value of ΔH for a reaction. Calculations involving enthalpies of formation are expected to be accurate within ± 0.1 kJ; the use of bond enthalpies can result in an error of 10 kJ or more.

8.7   THE FIRST LAW OF THERMODYNAMICS

So far in this chapter our discussion has focused on thermochemistry, the study of the heat effects in chemical reactions. Thermochemistry is a branch of *thermodynamics,* which deals with all kinds of energy effects in all kinds of processes. Thermodynamics distinguishes between two types of energy. One of these is heat (q); the other is **work,** represented by the symbol w. The thermodynamic definition of work is quite different from its colloquial meaning. Quite simply, *work includes all forms of energy except heat.*

The law of conservation of energy states that energy (E) can neither be created nor destroyed; it can only be transferred between system and surroundings. That is,

$$\Delta E_{system} = -\Delta E_{surroundings}$$

The first law of thermodynamics goes a step further. Taking account of the fact that there are two kinds of energy, heat and work, the first law states:

*In any process, the total change in energy of a system, **ΔE**, is equal to the sum of the heat, q, and the work, w, transferred between the system and the surroundings.*

$$\Delta E = q + w$$

In applying the first law, note (Figure 8.10) that q and w are positive when heat or work enters the system from the surroundings. If the transfer is in the opposite direction, from system to surroundings, q and w are negative.

See Screen 6.13, The First Law of Thermodynamics.

If work is done on the system or if heat is added to it, its energy increases.

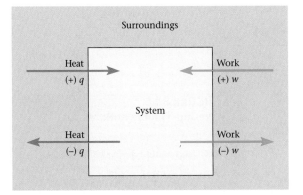

FIGURE 8.10

Heat and work. These quantities are positive when they enter a system and negative when they leave. **OHT**

(a)

(b)

An automobile that runs on natural gas. (a) The vehicle pulled up to a CNG (compressed natural gas) pump. The gas is 90% methane, according to a label on the pump. (b) The car's gas tank (which occupies the dark space inside the back of the trunk) is being filled. *(Dennis Bradbury Photos)*

EXAMPLE 8.11   Calculate ΔE of a gas for a process in which the gas

(a) absorbs 20 J of heat and does 12 J of work by expanding.
(b) evolves 30 J of heat and has 52 J of work done on it as it contracts.

Strategy   First decide on the signs (+ or −) of q and w. Then substitute into the equation for ΔE.

Solution

(a) $q = +20$ J; $w = -12$ J, because the gas does work on the surroundings.

$$\Delta E = +20 \text{ J} - 12 \text{ J} = \boxed{+8 \text{ J}}$$

(b) $q = -30$ J; $w = +52$ J, because work is done on the gas by the surroundings.

$$\Delta E = -30 \text{ J} + 52 \text{ J} = \boxed{+22 \text{ J}}$$

Ordinarily, when a chemical reaction is carried out in the laboratory, any energy evolved is in the form of heat. Consider, for example, the reaction with oxygen of methane, the principal constituent of natural gas.

$$CH_4(g) + 2O_2(g) \longrightarrow CO_2(g) + 2H_2O(l) \qquad \Delta E = -885 \text{ kJ}$$

When you ignite methane in a Bunsen burner, the amount of heat evolved is very close to 885 kJ/mol. There is a small work effect, due to the decrease in volume that occurs when the reaction takes place (Figure 8.11), but this amounts to less than 1% of the energy change.

The situation changes if methane is used as a substitute for gasoline in an internal combustion engine. Here a significant fraction of the energy evolved in combustion is converted to useful work, propelling your car uphill, overcoming friction, charging the battery, or whatever. Depending on the efficiency of the engine, as much as 25% of the available energy might be converted to work; the amount of heat evolved through the tail pipe or radiator drops accordingly.

Finally, the energy available from the above reaction might be used to operate a fuel cell such as those involved in the space program. In that case, as much as 818 kJ/mol of useful electrical work could be obtained; relatively little heat is evolved. Summarizing this discussion in terms of an energy balance (per mole of methane reacting):

	ΔE	q	w
Bunsen burner	−885 kJ	−890 kJ	+5 kJ
Automobile engine	−885 kJ	−665 kJ	−220 kJ
Fuel cell	−885 kJ	−67 kJ	−818 kJ

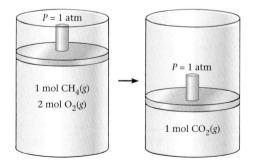

FIGURE 8.11

Pressure-volume work. When the reaction

$$CH_4(g) + 2O_2(g) \longrightarrow CO_2(g) + 2H_2O(l)$$

is carried out in a cylinder fitted with a piston that exerts a pressure of 1 atm, a contraction occurs (3 mol gas $\longrightarrow$ 1 mol gas). The piston falls and a small amount of work is done on the reaction system. **OHT**

Notice that ΔE, like ΔH, is a *state property*; it has the same value regardless of how or where or why the reaction is carried out. In contrast, q and w are path-dependent; their values vary depending on whether the reaction is carried out in the atmosphere, an engine, or an electrical cell.

ΔH Versus ΔE

As noted earlier, for a reaction at constant pressure, such as that taking place in an open coffee-cup calorimeter, the heat flow is equal to the change in enthalpy. If a reaction is carried out at constant volume (as is the case in a sealed bomb calorimeter) and there is no mechanical or electrical work involved, no work is done. Under these conditions, with $w = 0$, the heat flow is equal to the change in energy, ΔE. Hence we have

$$\Delta H = q_p \qquad \Delta E = q_v$$

(q_p = heat flow at constant pressure, q_v = heat flow at constant V).

Almost always, when you carry out a reaction in the laboratory, you do so at constant pressure, that of the atmosphere. That is why we devoted 90% of the space in this chapter to ΔH rather than ΔE.

We can obtain a relation between ΔH and ΔE by starting with the fundamental thermodynamic relation between the enthalpy of a substance, H, and its energy, E:

$$H = E + PV \qquad (P = \text{pressure}, V = \text{volume})$$

where V is the volume of a substance at pressure P. Ordinarily, the PV product in this equation is very small; for one mole of a gas at 25°C, using the ideal gas equation,

$$PV = (1.00 \text{ mol})\left(0.0821 \frac{\text{L} \cdot \text{atm}}{\text{mol} \cdot \text{K}}\right)(298 \text{ K}) = 24.5 \text{ L} \cdot \text{atm}$$

Using the conversion factor: 1 L·atm = 0.1013 kJ we see that the PV product for a gas at 25°C is only about 2.5 kJ

> The liter-atmosphere is an energy unit, about 1/10 as large as a kilo-joule.

$$24.5 \text{ L} \cdot \text{atm} \times \frac{0.1013 \text{ kJ}}{1 \text{ L} \cdot \text{atm}} \approx 2.5 \text{ kJ}$$

For liquids and solids, which ordinarily have volumes much smaller than those of gases, the PV product is so small as to be completely negligible.

This means that the enthalpy of a substance is very nearly (but not quite) equal to its energy. More important to us, it means that the difference between ΔH and ΔE in a chemical reaction is small. In general, we can say that

$$\Delta H = \Delta E + \Delta(PV)$$

$$= \Delta E + (PV)_{\text{products}} - (PV)_{\text{reactants}}$$

For the reaction:

$$CH_4(g) + 2O_2(g) \longrightarrow CO_2(g) + 2H_2O(l)$$

we see from the equation just written and the preceding discussion that

$$\Delta H = \Delta E + PV \, CO_2(g) + 0 - [PV \, CH_4(g) + 2PV \, O_2(g)]$$

$$= \Delta E - 2 \,(PV) = \Delta E - 5.0 \text{ kJ}$$

Recall that ΔE for this reaction is -885 kJ; it follows that ΔH is -890 kJ. The difference between ΔE and ΔH is less than 1%, which is typical for most reactions.

Energy Balance in the Human Body

All of us require energy to maintain life processes and do the muscular work associated with such activities as studying, writing, walking, or jogging. The fuel used to produce that energy is the food we eat. In this discussion, we focus on energy input or energy output and the balance between them.

Energy Input

Energy values of food can be estimated on the basis of the content of carbohydrate, protein, and fat:

carbohydrate:	17 kJ (4.0 kcal) of energy per gram*
protein:	17 kJ (4.0 kcal) of energy per gram
fat:	38 kJ (9.0 kcal) of energy per gram

Alcohol, which doesn't fit into any of these categories, furnishes about 29 kJ (7.0 kcal) per gram.

Using these guidelines, it is possible to come up with the data given in Table A, which lists approximate energy values for standard portions of different types of foods. With a little practice, you can use the table to estimate your energy input within ± 10%.

Fats and oils. Gram for gram, they provide more than twice as many calories as carbohydrates or proteins. *(Charles D. Winters)*

TABLE A Energy Values of Food Portions*

	kJ	kcal
8-oz glass (250 g) skim milk	380	90
whole milk	630	150
beer	420	100
1 portion vegetable	100	25
1 portion fruit	250	60
1 portion bread or starchy vegetable	330	80
1 oz (30 g) lean meat	230	55
medium-fat meat	310	75
high-fat meat	420	100
1 teaspoon (5 g) fat or oil	190	45
1 teaspoon (5 g) sugar	80	20

*Adapted from Lori A. Smolin and Mary B. Grosvenor: *Nutrition,* 3rd ed. Saunders College Publishing, Philadelphia, 2000.

Energy Output

Energy output can be divided, more or less arbitrarily, into two categories.

1. Metabolic energy. This is the largest item in most people's energy budget; it comprises the energy necessary to maintain life. Metabolic rates are ordinarily estimated based on a person's sex, weight, and age. For a young adult female, the metabolic rate is about 100 kJ per day per kilogram of body weight. For a 20-year-old woman weighing 120 lb:

*In most nutrition texts, energy is expressed in kilocalories rather than kilojoules, although that situation is changing. Remember that 1 kcal = 4.18 kJ; 1 kg = 2.20 lb.

$$\text{metabolic rate} = 120 \text{ lb} \times \frac{1 \text{ kg}}{2.20 \text{ lb}} \times 100 \frac{\text{kJ}}{\text{d} \cdot \text{kg}}$$

$$= 5500 \text{ kJ/d} \ (1300 \text{ kcal/day})$$

For a young adult male of the same weight, the figure is about 15% higher.

2. Muscular activity. Energy is consumed whenever your muscles contract, whether in writing, walking, playing tennis, or even sitting in class. Generally speaking, energy expenditure for muscular activity ranges from 50% to 100% of metabolic energy, depending on lifestyle. If the 20-year-old woman referred to earlier is a student who does nothing more strenuous than lift textbooks (and that not too often), the energy spent on muscular activity might be

$$0.50 \times 5500 \text{ kJ} = 2800 \frac{\text{kJ}}{\text{d}}$$

Her total energy output per day would be 5500 kJ + 2800 kJ = 8300 kJ (2000 kcal).

Energy Balance

To maintain constant weight, your daily energy input, as calculated from the foods you eat, should be about 700 kJ (170 kcal) greater than output. The difference allows for the fact that about 40 g of protein is required to maintain body tissues and fluids. If the excess of input over output is greater than 700 kJ/day, the unused food (carbohydrate, protein, or fat) is converted to fatty tissue and stored as such in the body.

Fatty tissue consists of about 85% fat and 15% water; its energy value is

$$0.85 \times 9.0 \frac{\text{kcal}}{\text{g}} \times \frac{454 \text{ g}}{1 \text{ lb}} = 3500 \text{ kcal/lb}$$

This means that to lose one pound of fat, energy input from foods must be decreased by about 3500 kcal. To lose weight at a sensible rate of one pound per week, it is necessary to cut down by 500 kcal (2100 kJ) per day.

It's also possible, of course, to lose weight by increasing muscular activity. Table B shows the amount of energy consumed per hour with various types of exercise. In principle, you can lose a pound a week by climbing mountains for an hour each day, provided you're not already doing that (most people aren't). Alternatively, you could spend $1\frac{1}{2}$ hours a day ice skating or water skiing, depending on the weather.

TABLE B   Energy Consumed by Various Types of Exercise*

kJ/hour	kcal/hour	
1000–1250	240–300	Walking (3 mph), bowling, golf (pulling cart)
1250–1500	300–360	Volleyball, calisthenics, golf (carrying clubs)
1500–1750	360–420	Ice skating, roller skating
1750–2000	420–480	Tennis (singles), water skiing
2000–2500	480–600	Jogging (5 mph), downhill skiing, mountain climbing
>2500	>600	Running (6 mph), basketball, soccer

*Adapted from Jane Brody: *Jane Brody's Nutrition Book: A Lifetime Guide to Good Eating for Better Health and Weight Control by the Personal Health Columnist for the New York Times*. Norton, New York, 1981.

CHAPTER HIGHLIGHTS

Key Concepts

1. Relate heat flow to specific heat, m, and Δt.
   (Example 8.1; Problems 1–4, 62, 71)
2. Calculate q for a reaction from calorimetric data.
   (Examples 8.2, 8.3; Problems 5–14)
3. Use a thermochemical equation to relate ΔH to mass of reactant or product and/or to ΔH of the reverse reaction.
   (Examples 8.4–8.6; Problems 15–24)
4. Use Hess's law to calculate ΔH.
   (Example 8.7; Problems 25–30, 57)
5. Relate ΔH to enthalpies of formation.
   (Examples 8.8–8.10; Problems 31–46)
6. Use the first law of thermodynamics to relate ΔE to q and w.
   (Example 8.11; Problems 49, 50)
7. Relate ΔH and ΔE.
   (Problems 51–54, 61)

Key Equations

Heat flow	$q = C \times \Delta t = m \times c \times \Delta t$
Calorimetry	$q_{reaction} = -C_{cal} \times \Delta t$
Standard enthalpy change	$\Delta H° = \Sigma \, \Delta H°_f \text{ products} - \Sigma \, \Delta H°_f \text{ reactants}$

Key Terms

bond enthalpy	heat	specific heat
calorimeter	—capacity	state property
—bomb	—of fusion	surroundings
—coffee-cup	—of vaporization	system
endothermic	Hess's law	thermochemical equation
enthalpy	■ joule	work
enthalpy of formation	■ kilojoule	
exothermic	■ mole	

Summary Problem

Dimethylhydrazine ($N_2H_2(CH_3)_2$) is used as a rocket fuel. When it reacts with oxygen, the thermochemical equation for the reaction is

$$\tfrac{1}{2}N_2H_2(CH_3)_2(l) + 2O_2(g) \longrightarrow CO_2(g) + 2H_2O(g) + \tfrac{1}{2}N_2(g) \qquad \Delta H = -901.6 \text{ kJ}$$

(a) Calculate ΔH for the following thermochemical equations:

(1) $N_2H_2(CH_3)_2(l) + 4O_2(g) \longrightarrow 2CO_2(g) + 4H_2O(g) + N_2(g)$

(2) $CO_2(g) + 2H_2O(g) + \frac{1}{2}N_2(g) \longrightarrow \frac{1}{2}N_2H_2(CH_3)_2(l) + 2O_2(g)$

(b) The heat of vaporization of $H_2O(l)$ is 44.0 kJ/mol. Calculate ΔH for the equation

$$\frac{1}{2}N_2H_2(CH_3)_2(l) + 2O_2(g) \longrightarrow CO_2(g) + 2H_2O(l) + \frac{1}{2}N_2(g)$$

(c) The heats of formation of CO_2 and $H_2O(g)$ are -393.5 and -241.8 kJ/mol, respectively. Calculate ΔH_f° of $N_2H_2(CH_3)_2(l)$.

(d) How much heat is evolved when 10.00 g of $N_2H_2(CH_3)_2(l)$ is burned in an open container?

(e) The temperature of a bomb calorimeter increases by 1.78°C when it absorbs 8.55 kJ of heat. Calculate $C_{calorimeter}$.

(f) In the calorimeter described in (e), dimethylhydrazine, $N_2H_2(CH_3)_2(l)$, is burned at 25.00°C. The temperature in the calorimeter increases to 29.55°C. How much dimethylhydrazine was burned?

(g) For the reaction, $\Delta H = q_p = -901.6$ kJ. Calculate $\Delta E = q_v$ at 25°C for the reaction.

Answers

(a) -1803.2 kJ; $+901.6$ kJ

(b) -989.6 kJ

(c) $+49.0$ kJ/mol

(d) 300.0 kJ

(e) 4.80 kJ/°C

(f) 0.728 g

(g) -905.4 kJ

Questions & Problems

Problem numbers in blue indicate that the answer is available in Appendix 6 at the back of the book.

WEB indicates that the solution is posted at **http://www.harcourtcollege.com/chem/general/masterton4/student/**

Specific Heat and Calorimetry

1. Gold has a specific heat of 0.129 J/g·°C. When a 5.00-g piece of gold absorbs 1.33 J of heat, what is the change in temperature?

2. Titanium is a metal used in jet engines. Its specific heat is 0.523 J/g·°C. If 5.88 g of titanium absorbs 4.78 J, what is the change in temperature?

3. Stainless steel accessories in cars are usually plated with chromium to give them a shiny surface and to prevent rusting. When 5.00 g of chromium at 23.00°C absorbs 62.5 J of heat, the temperature increases to 50.8°C. What is the specific heat of chromium?

4. The specific heat of aluminum is 0.902 J/g·°C. How much heat is absorbed by an aluminum pie tin with a mass of 473 g to raise its temperature from room temperature (23.00°C) to oven temperature (375°F)?

5. Sodium chloride is added in cooking to enhance the flavor of food. When 1.00 mol of NaCl dissolves, 3.91 kJ of heat is absorbed.

(a) Write a balanced equation for the solution process.

(b) Is the process exothermic?

(c) What is q_{H_2O}?

(d) What is the final temperature of the solution if 1.00 mol of NaCl is dissolved in 175 g of water at 22°C? (Specific heat of water = 4.18 J/g·°C)

(e) How much heat is absorbed when 5.73 g of NaCl is dissolved?

WEB 6. When 1.34 g of potassium bromide dissolves in 74.0 g of water in a coffee-cup calorimeter, the temperature drops from 18.000° to 17.279°C. Assume all the heat absorbed in the solution process comes from the water. (Specific heat of water = 4.18 J/g·°C.)

(a) Write a balanced equation for the solution process.

(b) Is the process exothermic or endothermic?

(c) What is q when 1.34 g of potassium bromide dissolves?

(d) What is q when one mole of potassium bromide dissolves?

7. In earlier times, ethyl ether was commonly used as an anesthetic. It is, however, highly flammable. When five milliliters of ethyl ether, $C_4H_{10}O(l)$, ($d = 0.714$ g/mL) is burned in a bomb calorimeter, the temperature rises from 23.5°C to 39.7°C. The calorimeter heat capacity is 10.34 kJ/°C.

(a) What is q for the calorimeter?

(b) What is q when 5.00 mL of ethyl ether is burned?

(c) What is q for the combustion of one mole of ethyl ether?

8. Fructose is a sugar commonly found in fruit. A sample of fructose, $C_6H_{12}O_6$, weighing 4.50 g is burned in a bomb calorimeter. The heat capacity of the calorimeter is 2.115 ×

10^4 J/°C. The temperature in the calorimeter rises from 23.49°C to 27.71°C.

    (a) What is q for the calorimeter?

    (b) What is q when 4.50 g of fructose is burned?

    (c) What is q for the combustion of one mole of fructose?

9. Isooctane is a primary component of gasoline and gives gasoline its octane rating. Burning 1.00 mL of isooctane ($d = 0.688$ g/mL) releases 33.0 kJ of heat. When 10.00 mL of isooctane is burned in a bomb calorimeter, the temperature in the bomb rises from 23.2°C to 66.5°C. What is the heat capacity of the bomb calorimeter?

10. When one mole of caffeine ($C_8H_{10}N_4O_2$) is burned in air, 4.96×10^3 kJ of heat is evolved. Five grams of caffeine is burned in a bomb calorimeter. The temperature is observed to increase by 11.37°C. What is the heat capacity of the calorimeter in J/°C?

11. Urea, $(NH_2)_2CO$, is a commonly used fertilizer. When 237.1 mg of urea is burned, 2.495 kJ is given off. If 5.000 mg of urea is burned in a bomb calorimeter (heat capacity = 5326 J/°C) initially at 23.15°C, what is the calorimeter temperature when combustion is complete?

12. Methanol, CH_3OH, is used as a fuel in some engines. When one mole of methanol burns, 1453 kJ of heat is evolved. Two milliliters of methanol ($d = 0.791$ g/mL) is burned in a bomb calorimeter, heat capacity = 1.231 kJ/°C. The temperature of the bomb when combustion is completed is 80.1°C. What was the initial temperature of the bomb?

13. Naphthalene, $C_{10}H_8$, is the compound present in moth balls. When one mole of naphthalene is burned, 5.15×10^3 kJ of heat is evolved. A sample of naphthalene burned in a bomb calorimeter (heat capacity = 9832 J/°C) increases the temperature in the calorimeter from 25.1°C to 28.4°C. How many milligrams of naphthalene were burned?

14. Salicylic acid, $C_7H_6O_3$, is one of the starting materials in the manufacture of aspirin. When 1.00 g of salicylic acid burns in a bomb calorimeter, the temperature rises from 23.11°C to 28.91°C. The temperature in the bomb calorimeter increases by 2.48°C when the calorimeter absorbs 9.37 kJ. How much heat is given off when one mole of salicylic acid is burned?

Rules of Thermochemistry

15. Nitrogen oxide (NO) has been found to be a key component in many biological processes. It also can react with oxygen to give the brown gas NO_2. When one mole of NO reacts with oxygen, 57.0 kJ of heat is evolved.

    (a) Write the thermochemical equation for the reaction between one mole of nitrogen oxide and oxygen.

    (b) Is the reaction exothermic or endothermic?

    (c) Draw an energy diagram showing the path of this reaction. (Figure 8.4 is an example of such an energy diagram.)

    (d) What is ΔH when 5.00 g of nitrogen oxide reacts?

    (e) How many grams of nitrogen oxide must react with an excess of oxygen to liberate ten kilojoules of heat?

WEB **16.** Calcium carbide, CaC_2, is the raw material for the production of acetylene (used in welding torches). Calcium carbide is produced by reacting calcium oxide with carbon, producing carbon monoxide as a by-product. When one mole of calcium carbide is formed, 464.8 kJ is absorbed.

    (a) Write a thermochemical equation for this reaction.

    (b) Is the reaction exothermic or endothermic?

    (c) Draw an energy diagram showing the path of this reaction. (Figure 8.4 is an example of such an energy diagram.)

    (d) What is ΔH when 1.00 g of $CaC_2(g)$ is formed?

    (e) How many grams of carbon are used up when 20.00 kJ of heat is absorbed?

17. Calcium chloride is a compound frequently found in first-aid hot packs. It gives off heat when dissolved in water. The following reaction takes place.

$$CaCl_2(s) \longrightarrow Ca^{2+}(aq) + 2Cl^-(aq) \qquad \Delta H = -81.4 \text{ kJ}$$

    (a) Calculate ΔH when one mole of calcium chloride precipitates from solution.

    (b) What is ΔH when 10.00 g of calcium chloride precipitates?

18. Consider the dissociation of water into ions.

$$H_2O(l) \longrightarrow H^+(aq) + OH^-(aq) \qquad \Delta H = 55.8 \text{ kJ}$$

    (a) Calculate ΔH when one mole of water is formed from the ions.

    (b) What is ΔH when 1.00 g of water is formed?

19. Nitroglycerine, $C_3H_5(NO_3)_3(l)$, is an explosive most often used in mine or quarry blasting. It is a powerful explosive because four gases (N_2, O_2, CO_2, and steam) are formed when nitroglycerine is detonated. In addition, 6.26 kJ of heat is given off per gram of nitroglycerine detonated.

    (a) Write a balanced thermochemical equation for the reaction.

    (b) What is ΔH when 4.65 mol of products is formed?

20. Strontium metal is responsible for the red color in fireworks. Fireworks manufacturers use strontium carbonate, which can be produced by combining strontium metal, graphite (C), and oxygen gas. The formation of one mole of $SrCO_3$ releases 1.220×10^3 kJ of energy.

    (a) Write a balanced thermochemical equation for the reaction.

    (b) What is ΔH when 10.00 L of oxygen at 25°C and 1.00 atm are used by the reaction?

21. A typical fat in the body is glyceryl trioleate $C_{57}H_{104}O_6$. When it is metabolized in the body, it combines with oxygen to produce carbon dioxide, water, and 3.022×10^4 kJ of heat per mole of fat.

    (a) Write a balanced thermochemical equation for the metabolism of fat.

    (b) How many kilojoules of energy must be evolved in the form of heat if you want to get rid of five pounds of this fat by combustion?

    (c) How many nutritional calories is this? (1 nutritional calorie = 1×10^3 calories)?

22. Using the same metabolism of fat equation from Question

21, how many grams of fat would have to be burned to heat 100.0 mL of water ($d = 1.00$ g/mL) from 22.00°C to 25.00°C? The specific heat of water is 4.18 J/g·°C.

23. Which requires the absorption of a greater amount of heat: vaporizing 100.0 g of naphthalene or boiling 100.0 g of water? (Use Table 8.2.)

24. Which evolves more heat—freezing 100.0 g of bromine or condensing 100.0 g of water vapor?

Hess's Law

25. A student is asked to calculate the amount of heat involved in changing 10.0 g of liquid bromine at room temperature (22.5°C) to vapor at 59.0 C. To do this, one must use Tables 8.1 and 8.2 for information on the specific heat, boiling point, and heat of vaporization of bromine. In addition, the following step-wise process must be followed.

(a) Calculate ΔH for: $Br_2(l, 22.5°C) \longrightarrow Br_2(l, 59.0°C)$
(b) Calculate ΔH for: $Br_2(l, 59.0°C) \longrightarrow Br_2(g, 59.0°C)$
(c) Using Hess's law, calculate ΔH for: $Br_2(l, 22.5°C) \longrightarrow Br_2(g, 59.0°C)$

WEB 26. Follow the step-wise process outlined in Problem 25 to calculate the amount of heat involved in condensing 100.00 g of benzene gas (C_6H_6) at 80.00°C to liquid benzene at 25.00°C. Use Tables 8.1 and 8.2 for the specific heat, boiling point, and heat of vaporization of benzene.

27. To produce silicon, used in semiconductors, from sand (SiO_2), a reaction is used that can be broken down into three steps:

$$SiO_2(s) + 2C(s) \longrightarrow Si(s) + 2CO(g) \qquad \Delta H = 689.9 \text{ kJ}$$

$$Si(s) + 2Cl_2(g) \longrightarrow SiCl_4(g) \qquad \Delta H = -657.0 \text{ kJ}$$

$$SiCl_4(g) + 2Mg(s) \longrightarrow 2MgCl_2(s) + Si(s) \qquad \Delta H = -625.6 \text{ kJ}$$

(a) Write the thermochemical equation for the overall reaction for the formation of silicon from silicon dioxide; CO and $MgCl_2$ are by-products.
(b) What is ΔH for the formation of one mole of silicon?
(c) Is the overall reaction exothermic?

28. A lead ore, galena, consisting mainly of lead(II) sulfide, is the principal source of lead. To obtain the lead, the ore is first heated in the air to form lead oxide.

$$PbS(s) + \tfrac{3}{2}O_2(g) \longrightarrow PbO(s) + SO_2(g) \qquad \Delta H = -415.4 \text{ kJ}$$

The oxide is then reduced to metal with carbon.

$$PbO(s) + C(s) \longrightarrow Pb(s) + CO(g) \qquad \Delta H = +108.5 \text{ kJ}$$

Calculate ΔH for the reaction of one mole of lead(II) sulfide with oxygen and carbon, forming lead, sulfur dioxide, and carbon monoxide.

29. Given the following thermochemical equations,

$$C_2H_2(g) + \tfrac{5}{2}O_2(g) \longrightarrow 2CO_2(g) + H_2O(l) \qquad \Delta H = -1299.5 \text{ kJ}$$

$$C(s) + O_2(g) \longrightarrow CO_2(g) \qquad \Delta H = -393.5 \text{ kJ}$$

$$H_2(g) + \tfrac{1}{2}O_2(g) \longrightarrow H_2O(l) \qquad \Delta H = -285.8 \text{ kJ}$$

calculate ΔH for the decomposition of one mole of acetylene, $C_2H_2(g)$, to its elements in their stable state at 25°C and 1 atm.

30. Given the following thermochemical equations

$$2H_2(g) + O_2(g) \longrightarrow 2H_2O(l) \qquad \Delta H = -571.6 \text{ kJ}$$

$$N_2O_5(g) + H_2O(l) \longrightarrow 2HNO_3(l) \qquad \Delta H = -73.7 \text{ kJ}$$

$$\tfrac{1}{2}N_2(g) + \tfrac{3}{2}O_2(g) + \tfrac{1}{2}H_2(g) \longrightarrow HNO_3(l) \qquad \Delta H = -174.1 \text{ kJ}$$

calculate ΔH for the formation of one mole of dinitrogen pentoxide from its elements in their stable state at 25°C and 1 atm.

$\Delta H°$ and Heats of Formation

31. Write thermochemical equations for the decomposition of one mole of the following compounds into the elements in their stable states at 25°C and 1 atm.
(a) potassium chlorate (s)   (b) carbon tetrachloride (l)
(c) hydrazine (l)   (d) magnesium hydroxide (s)

32. Write thermochemical equations for the formation of one mole of the following compounds from the elements in their stable states at 25°C and 1 atm.
(a) dinitrogen tetraoxide (g)   (b) calcium sulfate (s)
(c) silver chloride (s)   (d) hydrogen iodide (g)

33. Given

$$2CuO(s) \longrightarrow 2Cu(s) + O_2(g) \qquad \Delta H° = 314.6 \text{ kJ}$$

(a) Determine the heat of formation of CuO.
(b) Calculate $\Delta H°$ for the formation of 13.58 g of CuO.

WEB 34. Given

$$2Al_2O_3(s) \longrightarrow 4Al(s) + 3O_2(g) \qquad \Delta H° = 3351.4 \text{ kJ}$$

(a) What is the heat of formation of aluminum oxide?
(b) What is $\Delta H°$ for the formation of 12.50 g of aluminum oxide?

35. Limestone, $CaCO_3$, when subjected to a temperature of 900°C in a kiln, decomposes to calcium oxide and carbon dioxide. How much heat is evolved or absorbed when one gram of limestone decomposes? (Use Table 8.3.)

36. When ammonia reacts with oxygen, nitrogen oxide gas and steam are formed. How much heat is evolved or absorbed when one liter of ammonia gas at 23.0°C and at 765 mm Hg reacts?

37. Use Table 8.3 to obtain $\Delta H°$ for the following thermochemical equations:
(a) $Cl_2(g) + 2I^-(aq) \rightarrow 2Cl^-(aq) + I_2(s)$
(b) $2Al^{3+}(aq) + 3Zn(s) \rightarrow 3Zn^{2+}(aq) + 2Al(s)$
(c) $Cd^{2+}(aq) + SO_2(g) + 2H_2O(l) \rightarrow$
$$SO_4^{2-}(aq) + 4H^+(aq) + Cd(s)$$

38. Use Table 8.3 to obtain $\Delta H°$ for the following thermochemical equations:
(a) $Cl_2(g) + H_2(g) + 2OH^-(aq) \rightarrow 2Cl^-(aq) + 2H_2O(l)$

(b) $5NO(g) + 3MnO_4^-(aq) + 4H^+(aq) \rightarrow$
$$5NO_3^-(aq) + 3Mn^{2+}(aq) + 2H_2O(l)$$
(c) $6Fe^{2+}(aq) + CrO_4^{2-}(aq) + 8H^+(aq) \rightarrow$
$$Cr(s) + 6Fe^{3+}(aq) + 4H_2O(l)$$

39. Use the appropriate table to calculate $\Delta H°$ for
(a) the reaction between calcium hydroxide and carbon dioxide to form calcium carbonate and steam.
(b) the decomposition of one mole of liquid sulfuric acid to steam, oxygen, and sulfur dioxide gas.

40. Use the appropriate table to calculate $\Delta H°$ for
(a) the reaction between copper(II) oxide and carbon monoxide to give copper metal and carbon dioxide.
(b) the decomposition of one mole of methyl alcohol (CH_3OH) to methane and oxygen gases.

41. Butane, C_4H_{10}, is widely used as a fuel for disposable lighters. When one mole of butane is burned in oxygen, carbon dioxide and steam are formed and 2658.3 kJ of heat is evolved.
(a) Write a thermochemical equation for the reaction.
(b) Using Table 8.3, calculate the standard heat of formation of butane.

42. When one mole of calcium carbonate reacts with ammonia, solid calcium cyanamide, $CaCN_2$, and liquid water are formed. The reaction absorbs 90.1 kJ of heat.
(a) Write a balanced thermochemical equation for the reaction.
(b) Using Table 8.3, calculate $\Delta H_f°$ for calcium cyanamide.

43. When ethylene gas, C_2H_4, reacts with fluorine gas, hydrogen fluoride and carbon tetrafluoride gases are formed and 2496.7 kJ of heat is given off. What is $\Delta H_f°$ for CF_4?

44. Use the appropriate tables to determine the heat of formation of $Cr_2O_3(s)$ given $\Delta H° = -1854.9$ kJ for the reaction

$8H^+(aq) + Cr_2O_7^{2-}(aq) + 2Al(s) \longrightarrow$
$$2Al^{3+}(aq) + Cr_2O_3(s) + 4H_2O(l)$$

45. When ammonia reacts with dinitrogen oxide gas ($\Delta H_f° = 82.05$ kJ/mol), liquid water and nitrogen gas are formed. How much heat is liberated or absorbed by the reaction that produces 345 mL of nitrogen gas at 25°C and 717 mg Hg?

46. Glucose, $C_6H_{12}O_6(s)$, ($\Delta H_f° = -1275.2$ kJ/mol) is converted to ethyl alcohol, $C_2H_5OH(l)$, and carbon dioxide in the fermentation of grape juice. What quantity of heat is liberated when 750.0 mL of wine containing 12.0% ethyl alcohol by volume ($d = 0.789$ g/cm³) is produced by the fermentation of grape juice?

First Law of Thermodynamics

47. How many kJ are equal to 3.27 L·atm of work?
48. How many L·atm are equal to 12.2 kJ of work?
49. Calculate
(a) q when a system does 54 J of work and its energy decreases by 72 J.
(b) ΔE for a gas that releases 38 J of heat and has 102 J of work done on it.

50. Find
(a) ΔE when a gas absorbs 18 J of heat and has 13 J of work done on it.
(b) q when 72 J of work is done on a system and its energy is increased by 61 J.

51. Determine the difference between ΔH and ΔE at 25°C for
$$2CO(g) + O_2(g) \longrightarrow 2CO_2(g)$$

52. For the vaporization of one mole of water at 100°C determine
(a) ΔH (Table 8.3)   **(b)** ΔPV (in kilojoules)   **(c)** ΔE

53. Consider the combustion of one mole of acetylene, C_2H_2, which yields carbon dioxide and liquid water.
(a) Using Table 8.3, calculate ΔH.
(b) Calculate ΔE at 25°C.

54. Consider the combustion of propane, C_3H_8, the fuel that is commonly used in portable gas barbeque grills. The products of combustion are carbon dioxide and liquid water.
(a) Write a thermochemical equation for the combustion of one mole of propane.
(b) Calculate ΔE for the combustion of propane at 25°C.

Unclassified

55. The BTU (British thermal unit) is the unit of energy most commonly used in the United States. One BTU is the amount of heat required to raise the temperature of one pound of water 1°F. What is the specific heat of water in BTU/g·°C?

56. A 75-W halogen lightbulb and a 20-W (2 significant figures) fluorescent bulb both give off about 1200 lumens of light. Suppose both bulbs are used for three hours in a room that is 13 × 15 × 8.0 ft. Compare
(a) the amounts of heat generated (a joule is defined to be a watt-second).
(b) the temperature changes observed if all the heat were absorbed by the air in the room. The specific heat of air is 1.007 J/g·°C; take its density to be 1.20 g/L.

57. Given the following reactions,

$N_2H_4(l) + O_2(g) \longrightarrow N_2(g) + 2H_2O(g)$ $\Delta H° = -534.2$ kJ
$H_2(g) + \frac{1}{2}O_2(g) \longrightarrow H_2O(g)$ $\Delta H° = -241.8$ kJ

Calculate the heat of formation of hydrazine.

58. A 12-oz can of most colas has about 120 nutritional calories (1 nutritional calorie = 1 kilocalorie). Approximately how many minutes of walking are required to burn up as energy the calories taken in after drinking a can of cola? (Walking uses up about 250 kcal/h.)

59. In World War II, the Germans made use of otherwise unusable airplane parts by grinding them up into powdered aluminum. This was made to react with ammonium nitrate to produce powerful bombs. The products of this reaction were nitrogen gas, steam, and aluminum oxide. If 10.00 kg of ammonium nitrate is mixed with 10.00 kg of powdered aluminum, how much heat is generated?

60. One way to lose weight is to exercise. Playing tennis for half an hour consumes about 225 kcal of energy. How long would you have to play tennis to lose one pound of body fat? (One gram of body fat is equivalent to 32 kJ of energy.)

61. Consider the reaction of methane with oxygen. Suppose that the reaction is carried out in a furnace used to heat a house. If $q = -890$ kJ and $w = +5$ kJ, what is ΔE? ΔH at 25°C?

62. Brass has a density of 8.25 g/cm^3 and a specific heat of 0.362 J/g·°C. A cube of brass 22.00 mm on an edge is heated in a Bunsen burner flame to a temperature of 95.0°C. It is then immersed into 20.0 mL of water ($d = 1.00$ g/mL, $c = 4.18$ J/g·°C) at 22.0°C in an insulated container. Assuming no heat loss, what is the final temperature of the water?

63. Some solutes have large heats of solution, and care should be taken in preparing solutions of these substances. The heat evolved when sodium hydroxide dissolves is 44.5 kJ/mol. What is the final temperature of the water, originally at 20.0°C, used to prepare 1.25 L of 6.00 M NaOH solution? Assume that all the heat is absorbed by 1.25 L of water, specific heat = 4.18 J/g·°C.

64. Microwave ovens convert radiation to energy. A microwave oven uses radiation with a wavelength of 12.5 cm. Assuming that all the energy from the radiation is converted to heat without loss, how many moles of photons are required to raise the temperature of a cup of water (350.0 g, specific heat = 4.18 J/g·°C) from 23.0°C to 99.0°C?

65. On complete combustion at constant pressure, a 1.00-L sample of a gaseous mixture at 0°C and 1.00 atm (STP) evolves 75.65 kJ of heat. If the gas is a mixture of ethane (C_2H_6) and propane (C_3H_8), what is the mole fraction of ethane in the mixture?

Conceptual Questions

66. Using Table 8.2, write thermochemical equations for the following.
- (a) mercury thawing
- (b) bromine vaporizing
- (c) benzene freezing
- (d) mercury condensing
- (e) naphthalene subliming

67. Draw a cylinder with a movable piston containing six molecules of a liquid. A pressure of 1 atm is exerted on the piston. Next draw the same cylinder after the liquid has been vaporized. A pressure of one atmosphere is still exerted on the piston. Is work done *on* the system or *by* the system?

68. Redraw the cylinder in Question 67 after work has been done on the system.

69. Equal masses of liquid A, initially at 100°C, and liquid B, initially at 50°C, are combined in an insulated container. The final temperature of the mixture is 80°C. All the heat flow occurs between the two liquids. The two liquids do not react with each

other. Is the specific heat of liquid A larger than, equal to, or smaller than the specific heat of liquid B?

70. Which statement(s) is/are true about bond enthalpy?
- (a) Energy is required to break a bond.
- (b) ΔH for the formation of a bond is always a negative number.
- (c) Bond enthalpy is defined only for bonds broken or formed in the gaseous state.
- (d) Because the presence of π bonds does not influence the geometry of a molecule, the presence of π bonds does not affect the value of the bond enthalpy between two atoms either.
- (e) The bond enthalpy for a double bond between atoms A and B is twice that for a single bond between atoms A and B.

Challenge Problems

71. On a hot day, you take a six-pack of soda on a picnic, cooling it with ice. Each empty (aluminum) can weighs 12.5 g and contains 12.0 oz of soda. The specific heat of aluminum is 0.902 J/g·°C; take that of soda to be 4.10 J/g·°C.
- (a) How much heat must be absorbed from the six-pack to lower the temperature from 25.0° to 5.0°C?
- (b) How much ice must be melted to absorb this amount of heat? (ΔH_{fus} of ice is given in Table 8.2.)

72. A cafeteria sets out glasses of tea at room temperature. The customer adds ice. Assuming that the customer wants to have some ice left when the tea cools to 0°C, what fraction of the total volume of the glass should be left empty for adding ice? Make any reasonable assumptions needed to work this problem.

73. The thermite reaction was once used to weld rails:

$$2Al(s) + Fe_2O_3(s) \longrightarrow Al_2O_3(s) + 2Fe(s)$$

- (a) Using heat of formation data, calculate ΔH for this reaction.
- (b) Take the specific heats of Al_2O_3 and Fe to be 0.77 and 0.45 J/g·°C, respectively. Calculate the temperature to which the products of this reaction will be raised starting at room temperature, by the heat given off in the reaction.
- (c) Will the reaction produce molten iron (mp Fe = 1535°C, $\Delta H_{fus} = 270$ J/g)?

74. A sample of sucrose, $C_{12}H_{22}O_{11}$, is contaminated by sodium chloride. When the contaminated sample is burned in a bomb calorimeter, sodium chloride does not burn. What is the percentage of sucrose in the sample if a temperature increase of 1.67°C is observed when 3.000 g of the sample is burned in the calorimeter? Sucrose gives off 5.64×10^3 kJ/mol when burned. The heat capacity of the calorimeter is 22.51 kJ/°C.

9 LIQUIDS AND SOLIDS

Carbon tetrachloride and octane (bottom and top layer, respectively) are nonpolar molecules immiscible in a water solution of copper sulfate. Water is a polar molecule. *(Charles D. Winters)*

CHAPTER OUTLINE

9.1 LIQUID-VAPOR EQUILIBRIUM

9.2 PHASE DIAGRAMS

9.3 MOLECULAR SUBSTANCES; INTERMOLECULAR FORCES

9.4 NETWORK COVALENT, IONIC, AND METALLIC SOLIDS

9.5 CRYSTAL STRUCTURES

In Chapter 5 we pointed out that at ordinary temperatures and pressures all gases follow the ideal gas law. Unfortunately, there is no simple "equation of state" that can be written to correlate the properties of liquids or solids. There are two reasons for this difference in behavior.

1. *Molecules are much closer to one another in liquids and solids.* In the gas state, particles are typically separated by ten molecular diameters or more; in liquids or solids, they touch one another. This explains why liquids and solids have densities so much larger than those of gases. At 100°C and 1 atm, $H_2O(l)$ has a density of 0.95 g/mL; that of $H_2O(g)$ under the same conditions is only 0.00059 g/mL. For the same reason, liquids and solids are much less compressible than gases. When the pressure on liquid water is increased from 1 to 2 atm, the volume decreases by 0.0045%; the same change in pressure reduces the volume of an ideal gas by 50%.

2. *Intermolecular forces, which are essentially negligible with gases, play a much more important role in liquids and solids.* Among the effects of these forces is the phenomenon of *surface tension,* shown by liquids. Molecules at the surface are attracted inward by unbalanced intermolecular forces; as a result, liquids tend to form spherical drops, for which the ratio of surface area to volume is as small as possible (Figure 9.1). Water, in which there are relatively strong intermolecular forces, has a high surface tension. That of most organic liquids is lower, which explains why they tend to "wet" or spread out on solid surfaces more readily than water. The wetting ability of water can be increased by adding a soap or detergent, which drastically lowers the surface tension.

Our discussion of liquids and solids will focus on—

■ phase equilibria. Section 9.1 considers several different phenomena related to gas-liquid equilibria, including *vapor pressure, boiling point behavior,* and *critical properties.* Section 9.2 deals with phase diagrams, which describe all three types of phase equilibria (gas-liquid, gas-solid, and liquid-solid).

■ the relationships among particle structure, interparticle forces, and physical properties. Section 9.3 explores these relationships for molecular substances; Section 9.4 extends the discussion to nonmolecular solids (network covalent, ionic, and metallic). Section 9.5 is devoted to the crystal structure of ionic and metallic solids.

See the *Saunders Interactive General Chemistry CD-ROM,* Screen 13.2, **Intermolecular Forces (1).**

See Screen 13.11, **Properties of Liquids.**

FIGURE 9.1
Surface tension. The water droplets are almost spherical and cling to the edge of the leaf because of the high surface tension of water. *(Charles D. Winters)*

Liquid-vapor equilibrium. Under the conditions shown, the bromine liquid and vapor have established a dynamic equilibrium. *(Charles D. Winters)*

See Screen 13.9, Properties of Liquids (Vapor Pressure).

9.1 LIQUID-VAPOR EQUILIBRIUM

All of us are familiar with the process of vaporization, in which a liquid is converted to a gas, commonly referred to as a *vapor*. In an open container, *evaporation* continues until all the liquid is gone. If the container is closed, the situation is quite different. At first, the movement of molecules is primarily in one direction, from liquid to vapor. Here, however, the vapor molecules cannot escape from the container. Some of them collide with the surface and re-enter the liquid. As time passes and the concentration of molecules in the vapor increases, so does the rate of condensation. When the rate of condensation becomes equal to the rate of vaporization, the liquid and vapor are in a state of **dynamic equilibrium:**

$$\text{liquid} \rightleftharpoons \text{vapor}$$

The double arrow implies that the forward and reverse processes are occurring at the same rate, which is characteristic of a dynamic equilibrium. The vapor, like any gas, can be assumed to obey the ideal gas law.

Vapor Pressure

Once equilibrium between liquid and vapor is reached, the number of molecules per unit volume in the vapor does not change with time. This means that *the pressure exerted by the vapor over the liquid remains constant.* The pressure of vapor in equilibrium with a liquid is called the vapor pressure. This quantity is a characteristic property of a given liquid at a particular temperature. It varies from one liquid to another, depending on the strength of the intermolecular forces. At 25°C, the vapor pressure of water is 24 mm Hg; that of ether, in which intermolecular forces are weaker, is 537 mm Hg.

It is important to realize that *so long as both liquid and vapor are present, the pressure exerted by the vapor is independent of the volume of the container.* If a small amount of liquid is introduced into a closed container, some of it will vaporize, establishing its equilibrium vapor pressure. The greater the volume of the container, the greater will be the amount of liquid that vaporizes to establish that pressure. The ratio n/V stays constant, so $P = nRT/V$ does not change. Only if all the liquid vaporizes will the pressure drop below the equilibrium value.

EXAMPLE 9.1   Consider a sample of 1.00 g of water, which has an equilibrium vapor pressure of 24 mm Hg at 25°C.

(a) If this sample is introduced into a 10.0-L flask, how much liquid water will remain when equilibrium is established?
(b) What would be the minimum volume of a flask in which the 1.00-g sample would be completely vaporized?
(c) Suppose the 1.00-g sample is introduced into a 60.0-L flask. What will be the pressure exerted by the vapor?

Strategy   In each case, apply the ideal gas law to the vapor, *not* to the liquid. In (a), calculate the mass of $H_2O(g)$ required to establish a vapor pressure of 24 mm Hg in a 10.0-L flask at 25°C; the mass of $H_2O(l)$ left can then be found by subtraction. In (b), calculate the volume occupied by 1.00 g of $H_2O(g)$ at 25°C and 24 mm Hg. In (c), calculate the pressure exerted by 1.00 g of $H_2O(g)$ at 25°C in a 60.0-L flask.

Solution

(a) $n_{H_2O(g)} = \dfrac{PV}{RT} = \dfrac{(24/760 \text{ atm})(10.0 \text{ L})}{(0.0821 \text{ L} \cdot \text{atm/mol} \cdot \text{K})(298 \text{ K})} = 0.013 \text{ mol}$

$m_{H_2O(g)} = 0.013 \text{ mol} \times \dfrac{18.02 \text{ g}}{1 \text{ mol}} = 0.23 \text{ g}$

$m_{H_2O(l)} = 1.00 \text{ g} - 0.23 \text{ g} = \boxed{0.77 \text{ g}}$

(b) $V = \dfrac{nRT}{P} = \dfrac{(1.00/18.02 \text{ mol})(0.0821 \text{ L} \cdot \text{atm/mol} \cdot \text{K})(298 \text{ K})}{(24/760 \text{ atm})} = \boxed{43 \text{ L}}$

(c) $P = \dfrac{nRT}{V} = \dfrac{(1.00/18.02 \text{ mol})(0.0821 \text{ L} \cdot \text{atm/mol} \cdot \text{K})(298 \text{ K})}{60.0 \text{ L}} = 0.0226 \text{ atm}$

$= 0.0226 \text{ atm} \times \dfrac{760 \text{ mm Hg}}{1 \text{ atm}} = \boxed{17.2 \text{ mm Hg}}$

Reality Check   A pressure of 17.2 mm Hg is less than the equilibrium vapor pressure of water, which makes sense. Because 1.00 g of $H_2O(l)$ vaporizes completely in a 43-L flask, there is none left to establish equilibrium in a 60.0-L flask, and the pressure drops below its equilibrium value.

Vapor Pressure Versus Temperature

The vapor pressure of a liquid always increases as temperature rises. Water evaporates more readily on a hot, dry day. Stoppers in bottles of volatile liquids such as ether or gasoline pop out when the temperature rises.

The vapor pressure of water, which is 24 mm Hg at 25°C, becomes 92 mm Hg at 50°C and 1 atm (760 mm Hg) at 100°C. The data for water are plotted at the top of Figure 9.2. As you can see, the graph of vapor pressure versus temperature is not a straight line, as it would be if pressure were plotted versus temperature for an ideal gas. Instead, the slope increases steadily as temperature rises, reflecting the fact that more molecules vaporize at higher temperatures. At 100°C, the concentration of H_2O molecules in the vapor in equilibrium with liquid is 25 times as great as at 25°C.

In working with the relationship between two variables, such as vapor pressure and temperature, scientists prefer to deal with linear (straight-line) functions. Straight-line graphs are easier to construct and to interpret. In this case, it is possible to obtain a linear function by making a simple shift in variables. Instead of plotting vapor pressure (P) versus temperature (T), we plot the *natural logarithm** of the vapor pressure ($\ln P$) versus the reciprocal of the absolute temperature ($1/T$). Such a plot for water is shown at the bottom of Figure 9.2. As with all other liquids, a plot of $\ln P$ versus $1/T$ is a straight line.

The general equation of a straight line is

$$y = mx + b \qquad (m = \text{slope}, \ b = y\text{-intercept})$$

The y-coordinate in Figure 9.2b is $\ln P$, and the x-coordinate is $1/T$. The slope, which is a negative quantity, turns out to be $-\Delta H_{vap}/R$, where ΔH_{vap} is the molar

How would these answers change if 2.00 g of water were used?
Answer: (a) 1.77 g; (b) 86 L; (c) 24 mm Hg.

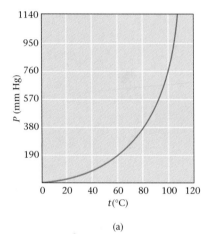

(a)

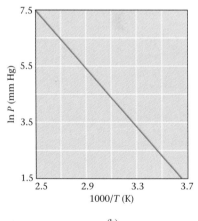

(b)

FIGURE 9.2

Vapor pressure versus temperature. (a) A plot of vapor pressure (P) versus temperature t (°C) for water (or any other liquid) is a curve with a steadily increasing slope. (b) A plot of the logarithm of the vapor pressure versus the reciprocal of the absolute temperature, $1/T$ (K), is a straight line. (To provide larger numbers, we have plotted $1000 \times 1/T$.) **OHT**

---

*Many natural laws are most simply expressed in terms of natural logarithms, which are based on the number $e = 2.71828. \ldots$ If $y = e^x$, then the natural logarithm of y, $\ln y$, is equal to x.

heat of vaporization and R is the gas constant, in the proper units. Hence the equation of the straight line in Figure 9.2b is

$$\ln P = \frac{-\Delta H_{vap}}{RT} + b$$

See Screen 13.8, Properties of Liquids (Enthalpy of Vaporization).

For many purposes, it is convenient to have a two-point relation between the vapor pressures (P_2, P_1) at two different temperatures (T_2, T_1). Such a relation is obtained by first writing separate equations at each temperature:

$$\text{at } T_2: \quad \ln P_2 = b - \frac{\Delta H_{vap}}{RT_2}$$

$$\text{at } T_1: \quad \ln P_1 = b - \frac{\Delta H_{vap}}{RT_1}$$

On subtraction, the constant b is eliminated, and we obtain

$$\ln P_2 - \ln P_1 = \ln \frac{P_2}{P_1} = -\frac{\Delta H_{vap}}{R}\left[\frac{1}{T_2} - \frac{1}{T_1}\right] = \frac{\Delta H_{vap}}{R}\left[\frac{1}{T_1} - \frac{1}{T_2}\right]$$

This equation is known as the *Clausius-Clapeyron equation*. Rudolph Clausius was a prestigious nineteenth-century German scientist. B. P. E. Clapeyron, a French engineer, first proposed a modified version of the equation in 1834.

In using the Clausius-Clapeyron equation, the units of ΔH_{vap} and R must be consistent. If ΔH is expressed in joules, then R must be expressed in joules per mole per kelvin. Recall (Table 5.1, page 119) that

$$R = 8.31 \text{ J/mol} \cdot \text{K}$$

This equation can be used to calculate any one of the five variables (P_2, P_1, T_2, T_1, and ΔH_{vap}), knowing the values of the other four. For example, we can use it to find the vapor pressure (P_1) at temperature T_1, knowing P_2 at T_2 and the value of the heat of vaporization (Example 9.2).

EXAMPLE 9.2

A certain organic compound has a vapor pressure of 91 mm Hg at 40°C. Calculate its vapor pressure at 25°C, taking the heat of vaporization to be 5.31×10^4 J/mol.

Strategy It is convenient to use the subscript 2 for the higher temperature and pressure. Substitute into the Clausius-Clapeyron equation, solving for P_1. Remember to express temperature in K and take $R = 8.31$ J/mol·K.

We'll use this value of R frequently in future chapters.

Solution

$$\ln \frac{P_2}{P_1} = \ln \frac{91 \text{ mm Hg}}{P_1} = \frac{5.31 \times 10^4 \text{ J/mol}}{8.31 \text{ J/mol} \cdot \text{K}}\left[\frac{1}{298} - \frac{1}{313}\right] = 1.03$$

Taking inverse logs,

$$\frac{91 \text{ mm Hg}}{P_1} = e^{1.03} = 2.8$$

(e is the base of natural logarithms; on many calculators, to find the number whose natural logarithm is 1.03, you first enter 1.03, then press in succession the INV and ln x keys).

Solving for P_1,

$$P_1 = \frac{91 \text{ mm Hg}}{2.8} = \boxed{32 \text{ mm Hg}}$$

Reality Check   This value is reasonable in the sense that lowering the temperature should reduce the vapor pressure.

Boiling Point

When a liquid is heated in an open container, bubbles form, usually at the bottom, where heat is applied. The first small bubbles are air, driven out of solution by the increase in temperature. Eventually, at a certain temperature, large vapor bubbles form throughout the liquid. These vapor bubbles rise to the surface, where they break. When this happens, the liquid is said to be boiling. For a pure liquid, the temperature remains constant throughout the boiling process.

The temperature at which a liquid boils depends on the pressure above it. To understand why this is the case, consider Figure 9.3. This shows vapor bubbles rising in a boiling liquid. For a vapor bubble to form, the pressure within it, P_1, must be at least equal to the pressure above it, P_2. Because P_1 is simply the vapor pressure of the liquid, it follows that **_a liquid boils at a temperature at which its vapor pressure is equal to the pressure above its surface._** If this pressure is 1 atm (760 mm Hg), the temperature is referred to as the **normal boiling point.** (When the term "boiling point" is used without qualification, normal boiling point is implied.) The normal boiling point of water is 100°C; its vapor pressure is 760 mm Hg at that temperature.

As you might expect, the boiling point of a liquid can be reduced by lowering the pressure above it. Water can be made to boil at 25°C by evacuating the space

See Screen 13.10, Properties of Liquids (Boiling Point).

The vapor pressure of a substance at its normal bp is 760 mm Hg.

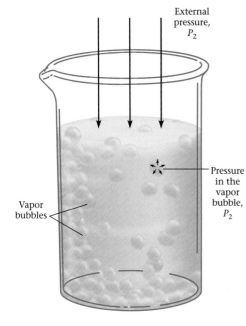

External pressure, P_2

Pressure in the vapor bubble, P_2

Vapor bubbles

FIGURE 9.3
Boiling and vapor pressure. A liquid boils when it reaches the temperature at which the vapor pressure in its vapor bubbles (P_1) exceeds the pressure above the liquid (P_2).

above it. When a pressure of 24 mm Hg, the equilibrium vapor pressure at 25°C, is reached, the water starts to boil. Chemists often take advantage of this effect in purifying a high-boiling compound that might decompose or oxidize at its normal boiling point. They distill it at a reduced temperature under vacuum and condense the vapor.

Or in the Presidential Range of New Hampshire (WLM).

If you have been fortunate enough to camp in the high Sierras or the Rockies, you may have noticed that it takes longer at high altitudes to cook foods in boiling water. The reduced pressure lowers the temperature at which water boils in an open container and thus slows down the physical and chemical changes that take place when foods like potatoes or eggs are cooked. In principle, this problem can be solved by using a pressure cooker. In that device, the pressure that develops is high enough to raise the boiling point of water above 100°C. Pressure cookers are indeed used in places like Salt Lake City, Utah (elevation 1340 m, bp H_2O)(l) = 95°C), but not by mountain climbers, who have to carry all their equipment on their backs.

Critical Temperature and Pressure

Consider an experiment in which liquid carbon dioxide is introduced into an otherwise evacuated glass tube, which is then sealed (Figure 9.4). At 0°C, the pressure above the liquid is 34 atm, the equilibrium vapor pressure of CO_2(l) at that temperature. As the tube is heated, some of the liquid is converted to vapor, and the pressure rises, to 44 atm at 10°C and 56 atm at 20°C. Nothing spectacular happens (unless there happens to be a weak spot in the tube) until 31°C is reached, where

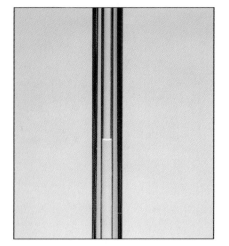

(a)

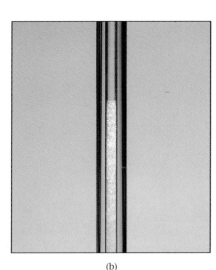

(b)

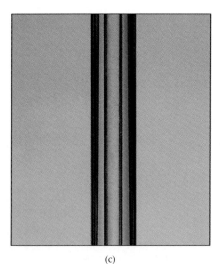
(c)

FIGURE 9.4

Critical temperature. (a) Liquid carbon dioxide under pressure is sealed in an evacuated glass tube; some vaporizes. (b) When the tube is heated, vapor bubbles form in the liquid. (c) Suddenly at 31°C, which is the critical temperature of CO_2, the meniscus (the surface of the liquid in the tube) disappears. Above this temperature, only one phase is present, no matter what the applied pressure is. *(Marna G. Clarke)*

TABLE 9.1 Critical Temperatures (°C)					
Permanent Gases		**Condensable Gases**		**Liquids**	
Helium	−268	Carbon dioxide	31	Ethyl ether	194
Hydrogen	−240	Ethane	32	Ethyl alcohol	243
Nitrogen	−147	Propane	97	Benzene	289
Argon	−122	Ammonia	132	Bromine	311
Oxygen	−119	Chlorine	144	Water	374
Methane	−82	Sulfur dioxide	158		

the vapor pressure is 73 atm. Suddenly, as the temperature goes above 31°C, the meniscus between the liquid and vapor disappears! The tube now contains only one phase.

It is impossible to have liquid carbon dioxide at temperatures above 31°C, no matter how much pressure is applied. Even at pressures as high as 1000 atm, carbon dioxide gas does not liquefy at 35 or 40°C. This behavior is typical of all substances. There is a temperature, called the **critical temperature,** above which the liquid phase of a pure substance cannot exist. The pressure that must be applied to cause condensation at that temperature is called the **critical pressure.** Quite simply, the critical pressure is the vapor pressure of the liquid at the critical temperature.

Table 9.1 lists the critical temperatures of several common substances. The species in the column at the left all have critical temperatures below 25°C. They are often referred to as "permanent gases." Applying pressure at room temperature will not condense a permanent gas. It must be cooled as well. When you see a truck labeled "liquid nitrogen" on the highway, you can be sure that the cargo trailer is refrigerated to at least −147°C, the critical temperature of N_2.

"Permanent gases" are most often stored and sold in steel cylinders under high pressures, often 150 atm or greater. When the valve on a cylinder of N_2 or O_2 is opened, gas escapes, and the pressure drops accordingly. The substances listed in the center column of Table 9.1, all of which have critical temperatures above 25°C, are handled quite differently. They are available commercially as liquids in high-pressure cylinders. When the valve on a cylinder of propane is opened, the gas that escapes is replaced by vaporization of liquid. The pressure quickly returns to its original value. Only when the liquid is completely vaporized does the pressure drop as gas is withdrawn. This indicates that almost all of the propane is gone, and it's time to recharge the tank.

Above the critical temperature and pressure, a substance is referred to as a *supercritical fluid*. Such fluids have unusual solvent properties that have led to many practical applications. Supercritical carbon dioxide is used most commonly because it is cheap, nontoxic, and relatively easy to liquefy (critical $T = 31°C$, $P = 73$ atm). It was first used more than 20 years ago to extract caffeine from coffee; dichloromethane, CH_2Cl_2, long used for this purpose, is both a narcotic and a potential carcinogen. Today more than 10^5 metric tons of decaf coffee are made using supercritical CO_2. It is also used on a large scale to extract nicotine from tobacco and various objectionable impurities from the hops used to make beer.

A pressure gauge on a propane tank doesn't indicate how much gas you have left.

9.2 PHASE DIAGRAMS

⊙ See Screen 13.17, Phase Diagrams.

In the preceding section, we discussed several features of the equilibrium between a liquid and its vapor. For a pure substance, at least two other types of phase equilibria need to be considered. One is the equilibrium between a solid and its vapor, the other between solid and liquid at the melting (freezing) point. Many of the important relations in all these equilibria can be shown in a **phase diagram.** A phase diagram is a graph that shows the pressures and temperatures at which different phases are in equilibrium with each other. The phase diagram of water is shown in Figure 9.5. This figure, which covers a wide range of temperatures and pressures, is not drawn to scale.

To understand what a phase diagram implies, consider first the three lines AB, AC, and AD in Figure 9.5. Each of these lines shows the pressures and temperatures at which two adjacent phases are at equilibrium.

1. Line AB is a portion of the vapor pressure–temperature curve of liquid water. At any temperature and pressure along this line, liquid water is in equilibrium with water vapor. At point A on the curve, these two phases are in equilibrium at 0°C and about 5 mm Hg (more exactly, 0.01°C and 4.56 mm Hg). At B, corresponding to 100°C, the pressure exerted by the vapor in equilibrium with liquid water is 1 atm; this is the normal boiling point of water. The extension of line AB beyond point B gives the equilibrium vapor pressure of the liquid above the normal boiling point. The line ends at 374°C, the critical temperature of water, where the pressure is 218 atm.

2. Line AC represents the vapor pressure curve of ice. At any point along this line, such as point A (0°C, 5 mm Hg) or point C, which might represent −3°C and 3 mm Hg, ice and vapor are in equilibrium with each other.

3. Line AD gives the temperatures and applied pressures at which liquid water is in equilibrium with ice.

┌ When *T* and *P* fall on a line, two phases are in equilibrium.

Point A on a phase diagram is the only one at which all three phases, liquid, solid, and vapor, are in equilibrium with each other. It is called the **triple point.** For water, the triple-point temperature is 0.01°C. At this temperature, liquid water and ice have the same vapor pressure, 4.56 mm Hg.

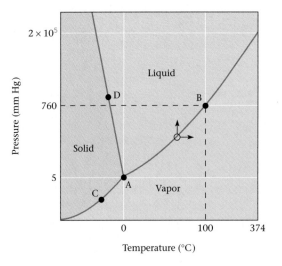

FIGURE 9.5
Phase diagram of water (not to scale). The lines represent the temperatures and pressures at which phases are in equilibrium. The triple point is at 0.01°C, 4.56 mm Hg; the critical point is at 374°C, 1.66×10^{-5} mm Hg (218 atm). **OHT**

In the three areas of the phase diagram labeled solid, liquid, and vapor, only one phase is present. To understand this, consider what happens to an equilibrium mixture of two phases when the pressure or temperature is changed. Suppose we start at the point on AB indicated by an open circle. Here liquid water and vapor are in equilibrium with each other, let us say at 70°C and 234 mm Hg. If the pressure on this mixture is increased, condensation occurs. The phase diagram confirms this; increasing the pressure at 70°C *(vertical arrow)* puts us in the liquid region. In another experiment, the temperature might be increased at a constant pressure. This should cause the liquid to vaporize. The phase diagram shows that this is indeed what happens. An increase in temperature *(horizontal arrow)* shifts us to the vapor region.

Freeze-dried dinner. This meal (of the type carried by campers) contains freeze-dried pasta and vegetables. *(Charles D. Winters)*

EXAMPLE 9.3   Consider a sample of H_2O at point D in Figure 9.5.

(a) What phase(s) is (are) present?
(b) If the temperature of the sample were reduced at constant pressure, what would happen?
(c) How would you convert the sample to vapor without changing the temperature?

Strategy   Use the phase diagram in Figure 9.5. Note that P increases moving up vertically; T increases moving to the right.

Solution

(a) Point D is on the solid-liquid equilibrium line. Ice and liquid water are present.
(b) Move to the left to reduce T. This penetrates the solid area, which implies that the solid freezes completely.
(c) Reduce the pressure to below the triple-point value, perhaps to 4 mm Hg.

Sublimation

The process by which a solid changes directly to vapor without passing through the liquid phase is called **sublimation.** A solid can sublime only at temperatures below the triple point; above that temperature it will melt to liquid (Figure 9.5). At temperatures below the triple point, a solid can be made to sublime by reducing the pressure of the vapor above it to less than the equilibrium value. To illustrate what this means, consider the conditions under which ice sublimes. This happens on a cold, dry, winter day when the temperature is below 0°C and the pressure of water vapor in the air is less than the equilibrium value (4.5 mm Hg at 0°C). The rate of sublimation can be increased by evacuating the space above the ice. This is how foods are freeze-dried. The food is frozen, put into a vacuum chamber, and evacuated to a pressure of 1 mm Hg or less. The ice crystals formed on freezing sublime, which leaves a product whose mass is only a fraction of that of the original food.

Iodine sublimes more readily than ice because its triple-point pressure, 90 mm Hg, is much higher. Sublimation occurs on heating (Figure 9.6) below the triple-point temperature, 114°C. If the triple point is exceeded, the solid melts. Solid carbon dioxide (Dry Ice) has a triple-point pressure above 1 atm (5.2 atm at −57°C). Liquid carbon dioxide cannot be made by heating Dry Ice in an open container. Solid CO_2 always passes directly to vapor at 1 atm pressure.

FIGURE 9.6
Sublimation. Solid iodine passes directly to the vapor at any temperature below the triple point, 114°C. The vapor condenses back to a solid on a cold surface like that of the upper flask, which is filled with ice. *(Charles D. Winters)*

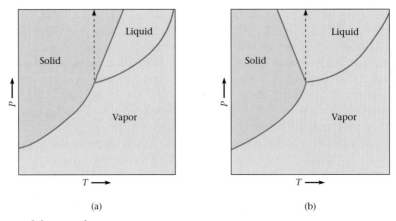

(a)   (b)

Melting Point

For a pure substance, the **melting point** is identical to the **freezing point.** It represents the temperature at which solid and liquid phases are in equilibrium. Melting points are usually measured in an open container, that is, at atmospheric pressure. For most substances, the melting point at 1 atm (the "normal" melting point) is virtually identical with the triple-point temperature. For water, the difference is only 0.01°C.

Although the effect of pressure on melting point is very small, its direction is still important. To decide whether the melting point will be increased or decreased by compression, a simple principle is applied. ***An increase in pressure favors the formation of the more dense phase.***

Two types of behavior are shown in Figure 9.7.

1. *The solid is the more dense phase* (Figure 9.7a). The solid-liquid equilibrium line is inclined to the right, shifting away from the *y*-axis as it rises. At higher pressures, the solid becomes stable at temperatures above the normal melting point. In other words, the melting point is raised by an increase in pressure. This behavior is shown by most substances.

2. *The liquid is the more dense phase* (Figure 9.7b). The liquid-solid line is inclined to the left, toward the *y*-axis. An increase in pressure favors the formation of liquid; that is, the melting point is decreased by raising the pressure. Water is one of the few substances that behaves this way; ice is less dense than liquid water. The effect is exaggerated for emphasis in Figure 9.7b. Actually, an increase in pressure of 134 atm is required to lower the melting point of ice by 1°C.

9.3   MOLECULAR SUBSTANCES; INTERMOLECULAR FORCES

See Screen 13.14, Solid Structures
(Molecular Solids).

Molecules are the characteristic structural units of gases, most liquids, and many solids. As a class, molecular substances tend to have the following characteristics. They are:

1. *Nonconductors of electricity when pure.* Molecules are uncharged, so they cannot carry an electric current. In most cases (e.g., iodine, I_2, and ethyl alcohol, C_2H_5OH), water solutions of molecular substances are also nonconductors. A few polar molecules, including HCl, react with water to form ions:

$$HCl(g) \longrightarrow H^+(aq) + Cl^-(aq)$$

and hence produce a conducting water solution.

2. *Insoluble in water but soluble in nonpolar solvents such as CCl₄ or benzene.* Iodine is typical of most molecular substances; it is only slightly soluble in water (0.0013 mol/L at 25°C), much more soluble in benzene (0.48 mol/L). A few molecular substances, including ethyl alcohol, are very soluble in water. As you will see later in this section, such substances have intermolecular forces similar to those in water.

3. *Low melting and boiling.* Many molecular substances are gases at 25°C and 1 atm (e.g., N_2, O_2, and CO_2), which means that they have boiling points below 25°C. Others (such as H_2O and CCl_4) are liquids with melting (freezing) points below room temperature. Of the molecular substances that are solids at ordinary temperatures, most are low melting. For example, iodine melts at 114°C; the melting point of naphthalene, used in mothballs, is 80°C. The upper limit for melting and boiling points of most molecular substances is about 300°C.

The generally low melting and boiling points of molecular substances reflect the fact that the forces between molecules **(intermolecular forces)** are weak. To melt or boil a molecular substance, the molecules must be set free from one another. This requires only that enough energy be supplied to overcome the weak attractive forces between molecules. The strong covalent bonds within molecules remain intact when a molecular substance melts or boils.

The boiling points of different molecular substances are directly related to the strength of the intermolecular forces involved. *The stronger the intermolecular forces, the higher the boiling point of the substance.* In the remainder of this section, we examine the nature of the three different types of intermolecular forces: *dispersion forces*, *dipole forces*, and *hydrogen bonds*.

Dispersion (London) Forces

The most common type of intermolecular force, found in all molecular substances, is referred to as a **dispersion force.** It is basically electrical in nature, involving an attraction between temporary or *induced* dipoles in adjacent molecules. To understand the origin of dispersion forces, consider Figure 9.8.

On the average, electrons in a nonpolar molecule, such as H_2, are as close to one nucleus as to the other. However, at a given instant, the electron cloud may be concentrated at one end of the molecule (position 1A in Figure 9.8). This momentary concentration of the electron cloud on one side of the molecule creates a temporary dipole in H_2. One side of the molecule, shown in deeper color in Figure 9.8, acquires a partial negative charge; the other side has a partial positive charge of equal magnitude.

Molecular liquids. The bottom layer, carbon tetrachloride (CCl_4), and the top layer, octane (C_8H_{18}), are nonpolar molecular liquids that are not soluble in water. The middle layer is a water solution of blue copper sulfate. *(Charles D. Winters)*

It's much easier to separate two molecules than to separate two atoms within a molecule.

See Screen 13.5, Intermolecular Forces (3).

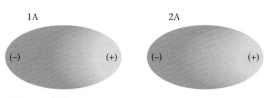

FIGURE 9.8

Dispersion force. Temporary dipoles in adjacent molecules line up to create an electrical attraction force known as the dispersion force. Deeply shaded areas indicate regions where the electron cloud is momentarily concentrated and creates partial charges, indicated by (+) and (−).

TABLE 9.2 Effect of Molar Mass on Boiling Points of Molecular Substances OHT								
Noble Gases*			**Halogens**			**Hydrocarbons**		
	$\mathcal{M}$ (g/mol)	bp (°C)		$\mathcal{M}$ (g/mol)	bp (°C)		$\mathcal{M}$ (g/mol)	bp (°C)
He	4	−269	F_2	38	−188	CH_4	16	−161
Ne	20	−246	Cl_2	71	−34	C_2H_6	30	−88
Ar	40	−186	Br_2	160	59	C_3H_8	44	−42
Kr	84	−152	I_2	254	184	$n\text{-}C_4H_{10}$	58	0

*Strictly speaking, the noble gases are "atomic" rather than molecular. However, like molecules, the noble-gas atoms are attracted to one another by dispersion forces.

This temporary dipole induces a similar dipole (an induced dipole) in an adjacent molecule. When the electron cloud in the first molecule is at 1A, the electrons in the second molecule are attracted to 2A. These temporary dipoles, both in the same direction, lead to an attractive force between the molecules. This is the dispersion force.

All molecules have dispersion forces. The strength of these forces depends on two factors–

- the number of electrons in the atoms that make up the molecule.
- the ease with which electrons are *dispersed* to form temporary dipoles.

Both these factors increase with increasing molecular size. Large molecules are made up of more and/or larger atoms. The outer electrons in larger atoms are relatively far from the nucleus and are easier to disperse than the electrons in small atoms. In general, molecular size and molar mass parallel one another. Thus within a given class of substances (Table 9.2) we can say that *as molar mass increases, dispersion forces become stronger and the boiling point of nonpolar molecular substances increases.*

Dispersion forces are weak in H_2, much stronger in I_2.

See Screen 13.4, Intermolecular Forces (2).

Dipole Forces

Polar molecules, like nonpolar molecules, are attracted to one another by dispersion forces. In addition, they experience **dipole forces** as illustrated in Figure 9.9, which shows the orientation of polar molecules, such as ICl, in a crystal. Adjacent molecules line up so that the negative pole of one molecule (small Cl atom) is as close as possible to the positive pole (large I atom) of its neighbor. Under these conditions, there is an electrical attractive force, referred to as a dipole force, between adjacent polar molecules.

When iodine chloride is heated to 27°C, the weak intermolecular forces are unable to keep the molecules rigidly aligned, and the solid melts. Dipole forces are still important in the liquid state, because the polar molecules remain close to one another. Only in the gas, where the molecules are far apart, do the effects of dipole forces become negligible. Hence boiling points as well as melting points of polar compounds such as ICl are somewhat higher than those of nonpolar substances of comparable molar mass. This effect is shown in Table 9.3.

TABLE 9.3   Boiling Points of Nonpolar Versus Polar Substances

	Nonpolar			Polar	
Formula	$\mathcal{M}$ (g/mol)	bp (°C)	Formula	$\mathcal{M}$ (g/mol)	bp (°C)
N_2	28	−196	CO	28	−192
SiH_4	32	−112	PH_3	34	−88
GeH_4	77	−90	AsH_3	78	−62
Br_2	160	59	ICl	162	97

EXAMPLE 9.4   Explain, in terms of intermolecular forces, why

(a) the boiling point of O_2 (−183°C) is higher than that of N_2 (−196°C).

(b) the boiling point of NO (−151°C) is higher than that of either O_2 or N_2.

Strategy   Determine whether the molecule is polar or nonpolar and identify the inter-molecular forces present. Remember, dispersion forces are always present and increase with molar mass.

Solution

(a) Only dispersion forces are involved with these nonpolar molecules. The molar mass of O_2 is greater (32.0 g/mol versus 28.0 g/mol for N_2), so its dispersion forces are stronger, making its boiling point higher.

(b) Dispersion forces in NO (30 g/mol) are comparable in strength to those in O_2 and N_2. The polar NO molecule shows an additional type of intermolecular force not present in N_2 or O_2: the dipole force. As a result, its boiling point is the highest of the three substances.

> In most molecules, dispersion forces are stronger than dipole forces.

Hydrogen Bonds

Ordinarily, polarity has a relatively small effect on boiling point. In the series HCl → HBr → HI, boiling point increases steadily with molar mass, even though polarity decreases moving from HCl to HI. However, when hydrogen is bonded to

See Screen 13.6, Hydrogen Bonding.

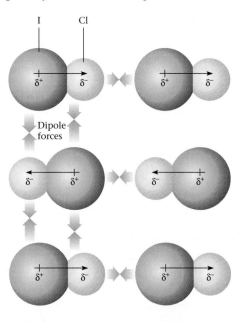

FIGURE 9.9

Dipole forces in the ICl crystal. The δ+ and δ− indicate partial charges on the I and Cl atoms in the polar molecules. The existence of these partial charges causes the molecules to line up in the pattern shown. Adjacent molecules are attracted to each other by the dipole forces between the δ+ of one molecule and the δ− of another molecule. **OHT**

TABLE 9.4 **Effect of Hydrogen Bonding on Boiling Point** OHT

	bp(°C)		bp(°C)		bp(°C)
NH_3	−33	H_2O	100	HF	19
PH_3	−88	H_2S	−60	HCl	−85
AsH_3	−63	H_2Se	−42	HBr	−67
SbH_3	−18	H_2Te	−2	HI	−35

Note: Molecules in color show hydrogen bonding.

a small, highly electronegative atom (N, O, F), polarity has a much greater effect on boiling point. Hydrogen fluoride, HF, despite its low molar mass (20 g/mol), has the highest boiling point of all the hydrogen halides. Water ($\mathcal{M}$ = 18 g/mol) and ammonia ($\mathcal{M}$ = 17 g/mol) also have abnormally high boiling points (Table 9.4). In these cases, the effect of polarity reverses the normal trend expected from molar mass alone.

The unusually high boiling points of HF, H_2O, and NH_3 result from an unusually strong type of dipole force called a **hydrogen bond.** The hydrogen bond is a force exerted between an H atom bonded to an F, O, or N atom in one molecule and an unshared pair on the F, O, or N atom of a neighboring molecule:

$$:X\!-\!H\text{ - - - - }:X\!-\!H \qquad X = N,\ O,\ or\ F$$

$$\uparrow$$

$$\text{hydrogen bond}$$

There are two reasons why hydrogen bonds are stronger than ordinary dipole forces:

1. The difference in electronegativity between hydrogen (2.2) and fluorine (4.0), oxygen (3.5), or nitrogen (3.0) is quite large. It causes the bonding electrons in molecules such as HF, H_2O, and NH_3 to be primarily associated with the more electronegative atom (F, O, or N). So the hydrogen atom, insofar as its interaction with a neighboring molecule is concerned, behaves almost like a bare proton.

2. The small size of the hydrogen atom allows the unshared pair of an F, O, or N atom of one molecule to approach the H atom in another very closely. It is significant that hydrogen bonding occurs only with these three nonmetals, all of which have small atomic radii.

Hydrogen bonds can exist in many molecules other than HF, H_2O, and NH_3. The basic requirement is simply that hydrogen be bonded to a fluorine, oxygen, or nitrogen atom with at least one unshared pair. Consider, for example, the two compounds whose condensed structural formulas are

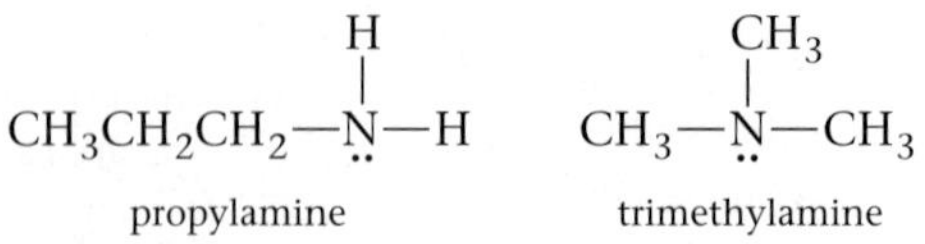

$$\begin{array}{cc} \overset{\displaystyle H}{\underset{\displaystyle \ddot{N}}{|}} & \overset{\displaystyle CH_3}{\underset{\displaystyle \ddot{N}}{|}} \\ CH_3CH_2CH_2\!-\!\overset{|}{\underset{..}{N}}\!-\!H & CH_3\!-\!\overset{|}{\underset{..}{N}}\!-\!CH_3 \\ \text{propylamine} & \text{trimethylamine} \end{array}$$

Propylamine, in which there are hydrogen atoms bonded to nitrogen, can show hydrogen bonding; it is a liquid with a normal boiling point of 49°C. Trimethylamine, which like propylamine has the molecular formula C_3H_9N, cannot hydrogen bond to itself; it boils at 3°C.

EXAMPLE 9.5   Would you expect to find hydrogen bonds in

(a) ethyl alcohol?

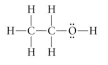

(b) dimethyl ether?

(c) hydrazine, N_2H_4?

Strategy   For hydrogen bonding to occur, hydrogen must be bonded to F, O, or N. In (c), draw the Lewis structure first.

Solution

(a) There should be hydrogen bonds in ethyl alcohol, because an H atom is bonded to oxygen.

(b) All the hydrogen atoms are bonded to carbon in dimethyl ether, so there should be no hydrogen bonds.

(c) The Lewis structure of hydrazine is

$$H - \ddot{N} - \ddot{N} - H$$
$$\mid \quad \mid$$
$$H \quad H$$

Because hydrogen is bonded to nitrogen, hydrogen bonding can occur between neighboring N_2H_4 molecules. The boiling point of hydrazine (114°C) is much higher than that of molecular O_2 (−183°C), which has the same molar mass.

Reality Check   In ethyl alcohol, the H atom bonded to oxygen in one molecule forms a hydrogen bond with an oxygen in an adjacent molecule. The same situation applies in hydrazine if you substitute nitrogen for oxygen.

Water has many unusual properties in addition to its high boiling point. As pointed out in Chapter 8, it has a very high specific heat, 4.18 J/g·°C. Its heat of vaporization per gram, 2.26 kJ/g, is the highest of all molecular substances. Both of these properties reflect the hydrogen-bonded structure of the liquid. Many of these bonds have to be broken when the liquid is heated; all of them disappear on boiling.

In contrast to most substances, water expands on freezing. Ice is one of the very few solids that has a density less than that of the liquid from which it is formed (*d* ice at 0°C = 0.917 g/cm³; *d* water at 0°C = 1.000 g/cm³). This behavior is an indirect result of hydrogen bonding. When water freezes to ice, an open hexagonal pattern of molecules results (Figure 9.10, page 262). Each oxygen atom in an ice crystal is bonded to four hydrogens. Two of these are attached by ordinary covalent bonds at a distance of 0.099 nm. The other two form hydrogen bonds 0.177 nm in length. The large proportion of empty space in the ice structure explains why ice is less dense than water.

⊙ **See Screen 13.7, The Weird Properties of Water.**

If ice were more dense than water, skating would be a lot less popular.

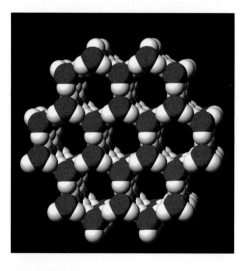

FIGURE 9.10

The structure of ice. In ice, the water molecules are arranged in an open pattern that gives ice its low density. Each oxygen atom (red) is bonded covalently to two hydrogen atoms (off-white) and forms hydrogen bonds with two other hydrogen atoms. *(Model by S. M. Young)*

EXAMPLE 9.6   What types of intermolecular forces are present in H_2? CCl_4? SCO? NH_3?

Strategy   Determine whether the molecules are polar or nonpolar; only polar molecules show dipole forces. Check the Lewis structures for H atoms bonded to F, O, or N. All molecules have dispersion forces.

Solution   The H_2 and CCl_4 molecules are nonpolar (Chapter 7). Thus only dispersion forces are present among neighboring molecules. Both SCO and NH_3 are polar molecules, because they are unsymmetrical:

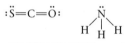

There are dipole forces as well as dispersion forces with SCO molecules. In NH_3, there are hydrogen bonds and dispersion forces.

We have now discussed three types of intermolecular forces: dispersion forces, dipole forces, and hydrogen bonds. You should bear in mind that all these forces are relatively weak compared with ordinary covalent bonds. Consider, for example, the situation in H_2O. The total intermolecular attractive energy in ice is about 50 kJ/mol; the O—H bond enthalpy is an order of magnitude greater, about 464 kJ/mol. This explains why it is a lot easier to melt ice or boil water than it is to decompose water into the elements. Even at a temperature of 1000°C and 1 atm, only about one H_2O molecule in a billion decomposes to hydrogen and oxygen atoms.

9.4 NETWORK COVALENT, IONIC, AND METALLIC SOLIDS

Virtually all substances that are gases or liquids at 25°C and 1 atm are molecular. In contrast, there are three types of nonmolecular solids (Figure 9.11). These are—

- **network covalent solids,** in which atoms are joined by a continuous network of covalent bonds. The entire crystal, in effect, consists of one huge molecule.
- *ionic solids,* held together by strong electrical forces (ionic bonds) between oppositely charged ions adjacent to one another
- *metallic solids,* in which the structural units are electrons (e^-) and cations, which may have charges of +1, +2, or +3.

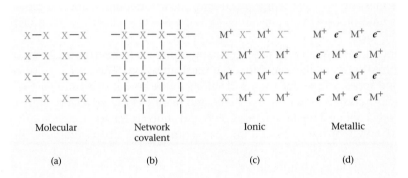

FIGURE 9.11

Diagrams of four types of substances (see text discussion). X represents a nonmetal atom, — represents a covalent bond, M^+ a cation, X^- an anion, and e^- an electron.

Network Covalent Solids

As a class, network covalent solids are—

■ *high melting, often with melting points about 1000°C.* To melt the solid, covalent bonds between atoms must be broken. In this respect, solids of this type differ markedly from molecular solids, which have much lower melting points.

■ *insoluble in all common solvents.* For solution to occur, covalent bonds throughout the solid would have to be broken.

■ *poor electrical conductors.* In most network covalent substances (graphite is an exception), there are no mobile electrons to carry a current.

Substances that form network covalent solids include both elements and compounds. As pointed out in Chapter 1 (page 20) graphite and diamond, two allotropic forms of carbon, have structures in which carbon atoms are bonded to one another throughout the crystal. In graphite the C—C bonds extend continuously through two dimensions, leading to a layer structure. In contrast, diamond has a three-dimensional, tetrahedral structure. Both diamond and graphite have melting points above 3500°C.

Perhaps the simplest compound with a network covalent structure is quartz, the most common form of SiO_2 and the major component of sand. In quartz, each silicon atom bonds tetrahedrally to four oxygen atoms. Each oxygen atom bonds to two silicons, thus linking adjacent tetrahedra to one another (Figure 9.12). Notice that the network of covalent bonds extends throughout the entire crystal. Unlike most pure solids, quartz does not melt sharply to a liquid. Instead, it turns

⊙ See Screen 13.15, Solid Structures (Network Solids).

⊙ See Screen 13.12, Solid Structures (Crystalline and Amorphous Solids).

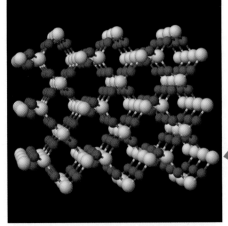
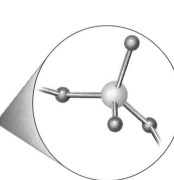

FIGURE 9.12

Crystal structure of quartz. The Si (off-white) and O (red) atoms form six-membered rings. Each Si atom is bonded tetrahedrally to four O atoms.

(Model by S. M. Young)

FIGURE 9.13
Samples of natural quartz and
amethyst. *(Charles D. Winters)*

 See Screen 13.16, Silicate Minerals.

This type of water softener
shouldn't be used if you're trying
to reduce sodium intake.

See Screen 3.11, Properties of Ionic
Compounds.

to a viscous mass over a wide temperature range, first softening at about 1400°C.
The viscous fluid probably contains long —Si—O—Si—O— chains, with enough
bonds broken to allow flow.

Crystalline quartz occurs in nature in many different forms (Figure 9.13), of
which amethyst is generally considered the most attractive. The purple color of
this gemstone appears to be due to imperfections in the crystal structure. The perfect, pure quartz crystals used in wristwatches and mantle clocks are made synthet-
ically by slowly cooling a melt containing SiO_2 and aqueous NaOH.

More than 90% of the rocks and minerals found in the Earth's crust are silicates, which are essentially ionic. Typically the anion has a network covalent
structure in which SiO_4^{4-} tetrahedra are bonded to one another in one, two, or
three dimensions. The structure shown at the left of Figure 9.14, where the anion
is a one-dimensional infinite chain, is typical of fibrous minerals such as diopside,
$CaSiO_3 \cdot MgSiO_3$. Asbestos has a related structure in which two chains are linked
together to form a double strand.

The structure shown at the right of Figure 9.14 is typical of layer minerals
such as talc, $Mg_3(OH)_2Si_4O_{10}$. Here SiO_4^{4-} tetrahedra are linked together to form
an infinite sheet. The layers are held loosely together by weak dispersion forces, so
they easily slide past one another. As a result, talcum powder, like graphite, has a
slippery feeling.

Among the three-dimensional silicates are the *zeolites*, which contain cavities
or tunnels in which Na^+ or Ca^{2+} ions may be trapped. Synthetic zeolites with
made-to-order holes are used in home water softeners. When hard water containing Ca^{2+} ions flows through a zeolite column, an exchange reaction occurs. If
we represent the formula of the zeolite as NaZ, where Z^- represents a complex,
three-dimensional anion, the water-softening reaction can be represented by the
equation

$$Ca^{2+}(aq) + 2NaZ(s) \longrightarrow CaZ_2(s) + 2Na^+(aq)$$

Sodium ions migrate out of the cavities; Ca^{2+} ions from the hard water move in to
replace them.

Ionic Solids

An ionic solid consists of cations and anions (e.g., Na^+, Cl^-). No simple, discrete
molecules are present in NaCl or other ionic compounds; rather, the ions are held

FIGURE 9.14
Two infinite silicate lattices. The Si
atoms are shown as black dots and
there is an O atom at each corner of
the tetrahedra. Asbestos has a double
anion chain like that shown on the
left. Talc is composed of infinite
sheets. A portion of the talc structure
is shown on the right.

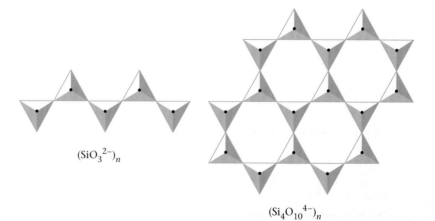

$(SiO_3^{2-})_n$

$(Si_4O_{10}^{4-})_n$

in a regular, repeating arrangement by strong ionic bonds, electrostatic interactions between oppositely charged ions. Because of this structure shown in Figure 2.11 on page 41, ionic solids have the following properties:

1. *Ionic solids are nonvolatile and high-melting* (typically at 600°C to 2000°C). Ionic bonds must be broken to melt the solid, separating oppositely charged ions from each other. Only at high temperatures do the ions acquire enough kinetic energy for this to happen.

2. *Ionic solids do not conduct electricity because the charged ions are fixed in position*. They become good conductors, however, when melted or dissolved in water. In both cases, in the melt or solution, the ions (such as Na^+ and Cl^-) are free to move through the liquid and thus can conduct an electric current.

3. *Many, but not all, ionic compounds* (e.g., NaCl but not $CaCO_3$) ***are soluble in water, a polar solvent.*** In contrast, ionic compounds are insoluble in nonpolar solvents such as benzene (C_6H_6) or carbon tetrachloride (CCl_4).

The relative strengths of different ionic bonds can be estimated from Coulomb's law, which gives the electrical energy of interaction between a cation and anion in contact with one another:

$$E = \frac{k \times Q_1 \times Q_2}{d}$$

Here, Q_1 and Q_2 are the charges of anion and cation, and d, the distance between the centers of the two ions, is the sum of the ionic radii (Appendix 2):

$$d = r_{cation} + r_{anion}$$

The quantity k is a constant whose magnitude need not concern us. Because the cation and anion have opposite charges, E is a negative quantity. This makes sense; energy is evolved when two oppositely charged ions, originally far apart with $E = 0$, approach one another closely. Conversely, energy has to be absorbed to separate the ions from each other.

From Coulomb's law, the strength of the ionic bond should depend on two factors:

1. *The charges of the ions.* The bond in CaO (+2, −2 ions) is considerably stronger than that in NaCl (+1, −1 ions). This explains why the melting point of calcium oxide (2927°C) is so much higher than that of sodium chloride (801°C).

2. *The size of the ions.* The ionic bond in NaCl (mp = 801°C) is somewhat stronger than that in KBr (mp = 734°C) because the internuclear distance is smaller in NaCl:

$$d_{NaCl} = r_{Na^+} + r_{Cl^-} = 0.095 \text{ nm} + 0.181 \text{ nm} = 0.276 \text{ nm}$$

$$d_{KBr} = r_{K^+} + r_{Br^-} = 0.133 \text{ nm} + 0.195 \text{ nm} = 0.328 \text{ nm}$$

Metals

Figure 9.11d illustrates a simple model of bonding in metals known as the **electron-sea model.** The metallic crystal is pictured as an array of positive ions, for example, Na^+, Mg^{2+}. These are anchored in position, like buoys in a mobile "sea" of electrons. These electrons are not attached to any particular positive ion but rather can wander through the crystal. The electron-sea model explains many of the characteristic properties of metals:

Charged particles must move to carry a current.

Asbestos in its natural state has a fibrous structure. *(Paul Silverman/ Fundamental Photographs)*

The band theory of metals is described in Appendix 5.

1. *High electrical conductivity.* The presence of large numbers of relatively mobile electrons explains why metals have electrical conductivities several hundred times greater than those of typical nonmetals. Silver is the best electrical conductor but is too expensive for general use. Copper, with a conductivity close to that of silver, is the metal most commonly used for electrical wiring. Although a much poorer conductor than copper, mercury is used in many electrical devices, such as silent light switches, where a liquid conductor is required.

2. *High thermal conductivity.* Heat is carried through metals by collisions between electrons, which occur frequently. Saucepans used for cooking commonly contain aluminum, copper, or stainless steel; their handles are made of a nonmetallic material that is a good thermal insulator.

3. *Ductility and malleability.* Most metals are ductile (capable of being drawn out into a wire) and malleable (capable of being hammered into thin sheets; Figure 9.15). In a metal, the electrons act like a flexible glue holding the atomic nuclei together. As a result, metal crystals can be deformed without shattering.

4. *Luster.* Polished metal surfaces reflect light. Most metals have a silvery white metallic color because they reflect light of all wavelengths. Because electrons are not restricted to a particular bond, they can absorb and re-emit light over a wide wavelength range. Gold and copper absorb some light in the blue region of the visible spectrum and so appear yellow (gold) or red (copper).

5. *Insolubility in water and other common solvents.* No metals dissolve in water; electrons cannot go into solution, and cations cannot dissolve by themselves. The only liquid metal, mercury, dissolves many metals, forming solutions called *amalgams*. An Ag-Sn-Hg amalgam is used in filling teeth.

FIGURE 9.15
Metal malleability. Blacksmiths rely on the malleability of metal horseshoes. *(A.J. Copley/Visuals Unlimited)*

TABLE 9.5 **Structures and Properties of Types of Substances** [OHT]

Type	Structural Particles	Forces Within Particles	Forces Between Particles	Properties	Examples
Molecular	Molecules				
(a) nonpolar		Covalent bond	Dispersion	Low mp, bp; often gas or liquid at 25°C; nonconductors; insoluble in water, soluble in organic solvents	H_2 CCl_4
(b) polar		Covalent bond	Dispersion, dipole, H bond	Similar to nonpolar but generally higher mp and bp, more likely to be water-soluble	HCl NH_3
Network covalent	Atoms	—	Covalent bond	Hard, very high-melting solids; nonconductors; insoluble in common solvents	C SiO_2
Ionic	Ions	—	Ionic bond	High mp; conductors in molten state or water solution; often soluble in water, insoluble in organic solvents	NaCl MgO $CaCO_3$
Metallic	Cations, mobile electrons	—	Metallic bond	Variable mp; good conductors in solid; insoluble in common solvents	Na Fe

(a)

(b)

(c)

In general, the melting points of metals cover a wide range, from −39°C for mercury to 3410°C for tungsten. This variation in melting point corresponds to a similar variation in the strength of the metallic bond. Generally speaking, the lowest-melting metals are those that form +1 cations, like sodium (mp = 98°C) and potassium (mp = 64°C).

Much of what has been said about the four structural types of solids in Sections 9.3 and 9.4 is summarized in Table 9.5 and Example 9.7.

Solids with different structures.
(a) Diamond, a network covalent solid. (b) Potassium dichromate, $K_2Cr_2O_7$, an ionic solid. (c) Manganese, a metallic solid. *(a, c, Charles D. Winters; b, Marna Clarke)*

EXAMPLE 9.7   A certain substance is a liquid at room temperature and is insoluble in water. Suggest which of the four types of basic structural units is present in the substance and list additional experiments that could be carried out to confirm your prediction.

Strategy   Use Table 9.5, which shows the general properties of the four different types of substances. Note that a given property (e.g., high electrical conductivity) may be shown by more than one type of substance.

Solution   The fact that the substance is a liquid suggests that it is probably molecular. The fact that it is insoluble in water agrees with this classification. To confirm that it is molecular, measure the conductivity, which should be zero. This test eliminates mercury, a liquid metal.

9.5   CRYSTAL STRUCTURES

Solids tend to crystallize in definite geometric forms that often can be seen by the naked eye. In ordinary table salt, cubic crystals of NaCl are clearly visible. Large, beautifully formed crystals of such minerals as fluorite, CaF_2, are found in nature. It is possible to observe distinct crystal forms of many metals under a microscope.

Crystals have definite geometric forms because the atoms or ions present are arranged in a definite, three-dimensional pattern. The nature of this pattern can be deduced by a technique known as x-ray diffraction. The basic information that comes out of such studies has to do with the dimensions and geometric form of the **unit cell,** the smallest structural unit that, repeated over and over again in three dimensions, generates the crystal. In all, there are 14 different kinds of unit cells. Our discussion will be limited to a few of the simpler unit cells found in metals and ionic solids.

Metals

Three of the simpler unit cells found in metals, shown in Figure 9.16 (page 268), are the following:

FIGURE 9.16

Three types of unit cells. In each case, there is an atom at each of the eight corners of the cube. In the body-centered cubic unit cell, there is an additional atom in the center of the cube. In the face-centered cubic unit cell, there is an atom in the center of each of the six faces. **OHT**

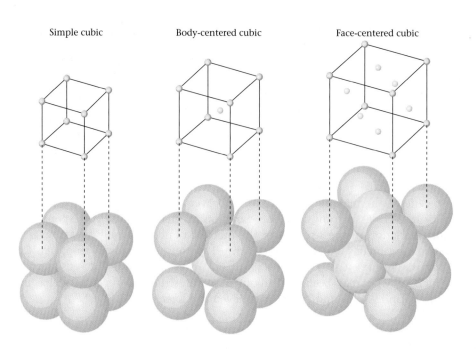

Simple cubic   Body-centered cubic   Face-centered cubic

See Screen 13.12, Solid Structures (Crystalline and Amorphous Solids).

1. Simple cubic cell (SC). This is a cube that consists of eight atoms whose centers are located at the corners of the cell. Atoms at adjacent corners of the cube touch one another.

2. Face-centered cubic cell (FCC). Here there is an atom at each corner of the cube and one in the center of each of the six faces of the cube. In this structure, atoms at the corners of the cube do not touch one another; they are forced slightly apart. Instead, contact occurs along a face diagonal. The atom at the center of each face touches atoms at opposite corners of the face.

3. Body-centered cubic cell (BCC). This is a cube with atoms at each corner and one in the center of the cube. Here again, corner atoms do not touch each other. Instead, contact occurs along the body diagonal; the atom at the center of the cube touches atoms at opposite corners.

Table 9.6 lists three other ways in which these types of cubic cells differ from one another.

1. *Number of atoms per unit cell.* Keep in mind that a huge number of unit cells are in contact with each other, interlocking to form a three-dimensional crystal. This means that several of the atoms in a unit cell do not belong exclusively to that cell. Specifically—

TABLE 9.6 Properties of Cubic Unit Cells			
	Simple	**BCC**	**FCC**
Number of atoms per unit cell	1	2	4
Relation between side of cell, s, and atomic radius, r	$2r = s$	$4r = s\sqrt{3}$	$4r = s\sqrt{2}$
% of empty space	47.6	32.0	26.0

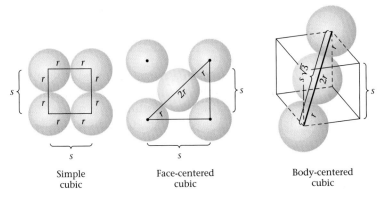

FIGURE 9.17
Relation between atomic radius (r) and length of edge (s) for cubic cells. In the simple cubic cell, $2r = s$. In the face-centered cubic cell, the face diagonal is equal to $s\sqrt{2}$ (hypotenuse of a right triangle) and to $4r$. Thus, $s\sqrt{2} = 4r$. In the body-centered cubic cell, the body diagonal is equal to $s\sqrt{3}$ (diagonal of a cube) and to $4r$. Thus $s\sqrt{3} = 4r$. **OHT**

Simple cubic   Face-centered cubic   Body-centered cubic

- an atom at the corner of a cube forms a part of eight different cubes that touch at that point. (To convince yourself of this, look back at Figure 2.11 (page 41); focus on the small sphere in the center.) In this sense, only $\frac{1}{8}$ of a corner atom belongs to a particular cell.
- an atom at the center of the face of a cube is shared by another cube that touches that face. In effect, only $\frac{1}{2}$ of that atom can be assigned to a given cell. This means, for example, that the number of atoms per FCC unit cell is

$$8 \text{ corner atoms} \times \tfrac{1}{8} + 6 \text{ face atoms} \times \tfrac{1}{2} = 4 \text{ atoms per cube}$$

2. Relation between side of cell (s) and atomic radius (r). To see how these two quantities are related, consider an FCC cell in which atoms touch along a face diagonal. As you can see from Figure 9.17—

- the distance along the face diagonal, d, is equal to four atomic radii

$$d = 4r$$

- d can be related to the length of a side of the cell by the Pythagorean theorem, $d^2 = s^2 + s^2 = 2s^2$. Taking the square root of both sides,

$$d = s\sqrt{2}$$

Equating the two expressions for d, we have

$$4r = s\sqrt{2}$$

This relation offers an experimental way of determining the atomic radius of a metal, if the nature and dimensions of the unit cell are known.

EXAMPLE 9.8   Silver crystallizes with a face-centered cubic (FCC) unit cell 0.407 nm on an edge. Calculate the atomic radius of silver.

Strategy   Find the appropriate relation between the radius and the length of a side of an FCC cell. Substitute 0.407 nm for s.

Solution   The equation is

$$4r = s\sqrt{2}$$

Substituting $s = 0.407$ nm and solving for r:

$$r = \frac{0.407 \text{ nm} \times \sqrt{2}}{4} = \frac{0.407 \text{ nm} \times 1.41}{4} = \boxed{0.144 \text{ nm}}$$

Reality Check This is the way the atomic radius of silver (listed in Appendix 2) was determined.

Golf balls and oranges pack naturally in an FCC structure.

3. *Percentage of empty space.* Metal atoms in a crystal, like marbles in a box, tend to pack closely together. As you can see from Table 9.6, nearly half of a simple cubic unit cell is empty space. This makes the SC structure very unstable. A body-centered cubic structure has less waste space; about 20 metals have this type of unit cell. A still more efficient way of packing spheres of the same size is the face-centered cubic structure, where the fraction of empty space is only 0.26. About 40 different metals have a structure based on either the face-centered cubic cell or a close relative in which the packing is equally efficient (hexagonal closest-packed structure).

Ionic Crystals

The geometry of ionic crystals, in which there are two different kinds of ions, is more difficult to describe than that of metals. However, in many cases the packing can be visualized in terms of the unit cells described above. Lithium chloride, LiCl, is a case in point. Here, the larger Cl^- ions form a face-centered cubic lattice (Figure 9.18). The smaller Li^+ ions fit into "holes" between the Cl^- ions. This puts a Li^+ ion at the center of each edge of the cube.

In the sodium chloride crystal, the Na^+ ion is slightly too large to fit into holes in a face-centered lattice of Cl^- ions (Figure 9.18). As a result, the Cl^- ions are pushed slightly apart so that they are no longer touching, and only Na^+ ions are in contact with Cl^- ions. However, the relative positions of positive and negative ions remain the same as in LiCl: Each anion is surrounded by six cations and each cation by six anions.

NaCl is FCC in both Na^+ and Cl^- ions; CsCl is BCC in both Cs^+ and Cl^- ions.

The structures of LiCl and NaCl are typical of all the alkali halides (Group 1 cation, Group **17** anion) except those of cesium. Because of the large size of the Cs^+ ion, CsCl crystallizes in a quite different structure. Here, each Cs^+ ion is located at the center of a simple cube outlined by Cl^- ions. The Cs^+ ion at the center touches all the Cl^- ions at the corners; the Cl^- ions do not touch each other. As you can see, each Cs^+ ion is surrounded by eight Cl^- ions, and each Cl^- ion is surrounded by eight Cs^+ ions.

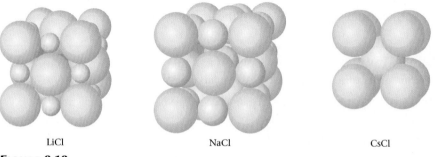

LiCl   NaCl   CsCl

FIGURE 9.18

Three types of lattices in ionic crystals. In LiCl, the Cl^- ions are in contact with each other, forming a face-centered cubic lattice. In NaCl, the Cl^- ions are forced slightly apart by the larger Na^+ ions. In CsCl, the large Cs^+ ion at the center touches the Cl^- ions at each corner of the cube.

CHEMISTRY ▪ *The Human Side*

T he crystal structures discussed in this section were determined by a powerful technique known as x-ray diffraction (Figure A). By studying the pattern produced when the scattered rays strike a target, it is possible to deduce the geometry of the unit cell. With molecular crystals, one can go a step further, identifying the geometry and composition of the molecule. A father and son team of two English physicists, William H. and William L. Bragg, won the Nobel Prize in 1915 for pioneer work in this area.

The Braggs and their successors strongly sought out talented women scientists to develop the new field of x-ray crystallography. Foremost among these women was Dorothy Crowfoot Hodgkin, who spent almost all of her professional career at Oxford University in England. As the years passed she unraveled the structures of successively more complex natural products. These included penicillin, which she studied between 1942 and 1949, vitamin B-12 (1948–1957), and her greatest triumph, insulin, which she worked on for more than 30 years. Among her students at Oxford was Margaret Thatcher, the future prime minister. In 1964, Dorothy Hodgkin became the third (and, so far, the last) woman to win the Nobel Prize in chemistry).*

Hodgkin's accomplishments were all the more remarkable when you consider some of the obstacles she had to overcome. At the age of 24, she developed rheumatoid arthritis, a crippling disease of the immune system. Gradually she lost the use of her hands; ultimately she was confined to a wheelchair. Dorothy Hodgkin succeeded because she combined a first-rate intellect with an almost infinite capacity for hard work. Beyond that, she was one of those rare individuals who inspire loyalty by taking genuine pleasure in the successes of other people. A colleague referred to her as "the gentle genius."

Dorothy's husband, Thomas Hodgkin, was a scholar in his own right with an interest in the history of Africa. Apparently realizing that hers was the greater talent, he acted as a "house-husband" for their three children so she would have more time to devote to research. One wonders how Dorothy and Thomas Hodgkin reacted to the 1964 headline in a London tabloid, "British Wife Wins Nobel Prize."

*To learn more about Dorothy Hodgkin and other women Nobel laureates, we recommend the book *Nobel Prize Women in Science* by Sharon Bertsch McGrayne, published in 1993.

DOROTHY HODGKIN
(1910–1994)

(Photo Credit: UPI/Bettman Archive)

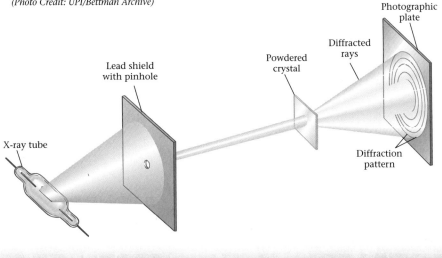
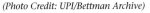

FIGURE A
X-ray diffraction. Knowing the angles and intensities at which x-rays are diffracted by a crystal, it is possible to calculate the distances between layers of atoms.

Synthetic Organic Polymers

I n this chapter we saw that covalent bonding can lead to two quite different structures: small, discrete molecules and network structures in which all the atoms are linked into a huge unit. In a sense, polymers are intermediate between these two extremes; they consist of very large molecules with molar masses ranging from a few thousand to a million grams per mole. Most polymers are organic; that is, they contain carbon, hydrogen, and often other nonmetal atoms, particularly oxygen and nitrogen. Some polymers occur naturally; included in this category are such vital substances as cellulose, starch, proteins, and DNA.

Since about 1930, chemists have synthesized a wide variety of polymers by combining small molecular units *(monomers)*. Some idea of the magnitude of the polymer industry may be gained when you realize that the total mass of synthetic organic polymers made annually in the United States exceeds the total mass of its population.

We distinguish between two different kinds of synthetic polymers—

- *addition polymers,* in which monomer units add directly to one another; typically, only one kind of monomer is involved.
- *condensation polymers,* in which monomer units combine by splitting out (condensing) a simple molecule such as H_2O. Typically, two different monomers react with one another to form a condensation polymer.

The simplest and most common addition polymer is polyethylene, a solid derived from the monomer ethylene. The polymerization process can be represented as

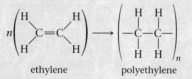

ethylene   polyethylene

when *n* is a very large number, of the order of 2000.

Depending on the conditions of polymerization, the product may be—

- *branched* polyethylene, which has a structure of the type

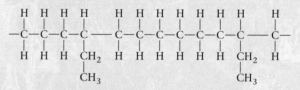

Here, neighboring chains are arranged in a somewhat random fashion, producing a soft, flexible solid (Figure A). The plastic bags at the vegetable counters of supermarkets are made of this material.

- *linear* polyethylene, which consists almost entirely of unbranched chains:

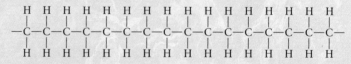

Neighboring chains in linear polyethylene line up nearly parallel to each other. This gives a polymer that approaches a crystalline material. It is used for bottles, toys, and other semirigid objects.

FIGURE A
Branched and linear polyethylene. The polyethylene bottle at the left is made of pliable, branched polyethylene. The one at the right is made of semirigid, linear polyethylene. *(Marna G. Clarke)*

A model of linear polyethylene.

The two forms of polyethylene differ slightly in density. Linear polyethylene is referred to in the recycling business as *high-density* polyethylene, represented by the symbol "HDPE #2" on the bottom of a plastic bottle. The corresponding symbol for branched polyethylene is "LDPE #4," indicating low-density polyethylene. (The smaller the number, the easier it is to recycle.)

To make a condensation polymer, each monomer must have reactive groups at both ends of the molecule. Consider, for example, the polymer known as polyethylene terephthalate (recycling symbol PETE #1). It is made from the two monomers, ethylene glycol and terephthalic acid; the first step in the reaction is

$$HO-CH_2-CH_2-OH + HO-\underset{O}{\overset{}{C}}-\bigcirc-\underset{O}{\overset{}{C}}-OH \longrightarrow$$

ethylene glycol   terephthalic acid

$$HO-CH_2-CH_2-O-\underset{O}{\overset{}{C}}-\bigcirc-\underset{O}{\overset{}{C}}-OH + H_2O$$

Because there are reactive —OH groups at each end of the dimer formed, reaction can continue to form the polymer:

$$(-O-CH_2-CH_2-O-\underset{O}{\overset{}{C}}-\bigcirc-\underset{O}{\overset{}{C}}-O-)_n$$

This polymer is widely used as a plastic in soft-drink bottles, as a fabric (Dacron), and as a very fine film (Mylar) used in audio- and videotapes.

CHAPTER HIGHLIGHTS

Key Concepts

1. Use the ideal gas law to determine whether a liquid will completely vaporize.
   (Example 9.1; Problems 1–6, 58, 66)
2. Use the Clausius-Clapeyron equation to relate the vapor pressure of a liquid to temperature.
   (Example 9.2; Problems 7–12)
3. Use a phase diagram to determine the phase(s) present at a given T and P.
   (Example 9.3; Problems 13–20)
4. Identify intermolecular forces in different substances and evaluate their effects on physical properties.
   (Examples 9.4–9.6; Problems 21–34)
5. Classify substances as molecular, network covalent, ionic, or metallic on the basis of their properties.
   (Example 9.7; Problems 35–44, 56, 63)
6. Use the information in Table 9.6 to compare different types of unit cells.
   (Example 9.8; Problems 45–48)

Key Equations

Vapor pressure–temperature (Claysius-Clapeyron equation)

$$\ln \frac{P_2}{P_1} = \frac{\Delta H}{R}\left[\frac{1}{T_1} - \frac{1}{T_2}\right]$$

Unit cell

$$2r = s \text{ (SC)} \qquad 4r = s\sqrt{2} \text{ (FCC)} \qquad 4r = s\sqrt{3} \text{ (BCC)}$$

Key Terms

boiling point
—normal
critical conditions
—pressure
—temperature
cubic cell
—body-centered (BCC)
—face-centered (FCC)

—simple (SC)
dipole force
dispersion force
dynamic equilibrium
electron-sea model
freezing point
hydrogen bond
intermolecular forces

melting point
network covalent solid
phase diagram
sublimation
triple point
unit cell
■ vapor pressure

Summary Problem

Consider propylamine, an organic compound of structural formula

$$H_3C—CH_2—CH_2—NH_2$$

(a) What kind of intermolecular forces would you expect to find in propylamine?

(b) How would you expect the normal boiling point of propylamine to compare to that of nitrogen? To methylamine (CH_3NH_2)? To ammonium chloride? To silicon dioxide (SiO_2)?

(c) Would you expect propylamine to conduct electricity? How would its solubility in water compare to that of nitrogen gas? To silicon dioxide?

(d) Give the symbol of a solid nonmetal that is a better conductor than propylamine. Give the formula of a solid compound that, unlike propylamine, has a network structure.

(e) The normal boiling point of propylamine is 48.7°C, its normal freezing point is −83°C, and its critical temperature is 234°C. Draw a rough sketch of the phase diagram of propylamine, assuming that the solid is more dense than the liquid.

(f) If you want to purify propylamine by sublimation, at what approximate temperature would you operate?

(g) A sample of propylamine vapor at 250°C and 1 atm is cooled at constant pressure. At what temperature does the liquid first appear?

(h) Taking the normal boiling point of propylamine to be 48.7°C and its heat of vaporization to be 30.7 kJ/mol, estimate its normal vapor pressure at 25.0°C. Estimate the temperature at which its vapor pressure is 0.500 atm.

(i) A sample of 1.00 mL of propylamine ($d = 0.719$ g/mL) is placed in a one-liter flask at 22°C, where its vapor pressure is 269 mm Hg. Show by calculation whether there is any liquid left in the flask when equilibrium is established.

Answers

(a) dispersion, dipole, H bonds

(b) higher than N_2 and CH_3NH_2; lower than NH_4Cl and SiO_2

(c) No; higher than N_2 and SiO_2

(d) C(Graphite) or Si; SiO_2, NaCl

(e) See Figure 9.7a for the general configuration. Put in the appropriate temperatures and pressures.

(f) below about $-83°C$

(g) $48.7°C$

(h) 305 mm Hg; $30.5°C$

(i) no

Problem numbers in blue indicate that the answer is available in Appendix 6 at the back of the book.

WEB indicates that the solution is posted at **http://www.harcourtcollege.com/chem/general/masterton4/student/**

Liquid-Vapor Equilibrium

1. Benzene, a known carcinogen, was once widely used as a solvent. A sample of benzene vapor in a flask of constant volume exerts a pressure of 325 mm Hg at 80°C. The flask is slowly cooled.

(a) Assuming no condensation, use the ideal gas law to calculate the pressure of the vapor at 50°C; at 60°C.

(b) Compare your answers in (a) to the equilibrium vapor pressures of benzene: 269 mm Hg at 50°C, 389 mm Hg at 60°C.

(c) On the basis of your answers to (a) and (b), predict the pressure exerted by the benzene at 50°C; at 60°C.

2. Methyl alcohol can be used as a fuel instead of, or combined with, gasoline. A sample of methyl alcohol, CH_3OH, in a flask of constant volume exerts a pressure of 254 mm Hg at 57°C. The flask is slowly cooled.

(a) Assuming no condensation, use the ideal gas law to calculate the pressure of the vapor at 35°C; at 45°C.

(b) Compare your answers in (a) to the equilibrium vapor pressures of methyl alcohol: 203 mm Hg at 35°C; 325 mm Hg at 45°C.

(c) On the basis of your answers to (a) and (b), predict the pressure exerted by the methyl alcohol in the flask at 35°C; at 45°C.

(d) What physical states of methyl alcohol are present in the flask at 35°C? at 45°C?

3. Naphthalene, $C_{10}H_8$, is the substance present in some moth balls. Its vapor pressure at 25°C is 0.300 mm Hg.

(a) How many milligrams of naphthalene will sublime into an evacuated 1.000-L flask?

(b) If 0.700 mg of naphthalene is used, what will the final pressure be? What physical state(s) of naphthalene is (are) in the flask?

(c) If 4.00 mg of naphthalene is used, what will the final pressure be? What physical state(s) of naphthalene is (are) in the flask?

4. The vapor pressure of $I_2(s)$ at 30°C is 0.466 mm Hg.

(a) How many milligrams of iodine will sublime into an evacuated 1.00-L flask?

(b) If 2.0 mg of I_2 is used, what will the final pressure be?

(c) If 10.0 mg of I_2 is used, what will the final pressure be?

5. Ethyl ether, $C_4H_{10}O$, was a widely used anesthetic in the early days of surgery. It has a vapor pressure of 537 mm Hg at room temperature, 25°C. A ten-milliliter sample ($d = 0.708$ g/mL) is placed in a sealed 0.250-L flask.

(a) Will there be any liquid left in the flask at equilibrium at 25°C?

(b) What is the maximum volume the flask can have if equilibrium is to be established between liquid and vapor?

(c) If the flask has a volume of 5.00 L, what is the pressure of ethyl ether in the flask?

6. Isooctane, C_8H_{18}, is a major component of gasoline. It has a vapor pressure of 45.2 mm Hg at 25°C. Ten milliliters ($d = 0.692$ g/mL) of isooctane is sealed in a car's 15.00-L gas tank.

(a) Will there be any liquid in the tank at equilibrium at 25°C?

(b) What is the maximum volume the gas tank can have if equilibrium is to be established between liquid and vapor?

(c) If the gas tank had a volume of 35.0 L, what would be the pressure of isooctane in the gas tank?

7. Methyl alcohol, CH_3OH, has a normal boiling point of 64.7°C and has a vapor pressure of 203 mm Hg at 35°C. Estimate

(a) its heat of vaporization (ΔH_{vap}).

(b) its vapor pressure at 40.0°C.

8. Chloroform, $CHCl_3$, was once used as an anesthetic. In spy movies it is the liquid put in handkerchiefs to render victims unconscious. Its vapor pressure is 197 mm Hg at 23°C and 448 mm Hg at 45°C. Estimate its

(a) heat of vaporization.

(b) normal boiling point.

9. Mt. McKinley in Alaska has an altitude of 20,320 ft. Water ($\Delta H_{vap} = 40.7$ kJ/mol) boils at 77°C atop Mt. McKinley. What is the normal atmospheric pressure at the summit?

WEB **10.** Glacier National Park in Montana is a favorite vacation spot for backpackers. It is about 4100 ft above sea level with an atmospheric pressure of 681 mm Hg. At what temperature does water ($\Delta H_{vap} = 40.7$ kJ/mol) boil at Glacier National Park?

11. Mexico City has an elevation of 7400 ft above sea level. In Mexico City, water (ΔH_{vap} = 40.7 kJ/mol) boils at about 93°C. In a pressure cooker, boiling water has a vapor pressure of 1.500×10^3 mm Hg.

(a) At what temperature does water boil in a pressure cooker?

(b) How much higher is the boiling point in a pressure cooker in Mexico City?

12. Using the data in Question 11, calculate the difference in the normal boiling point and the boiling point in a pressure cooker in a city at sea level.

Phase Diagrams

13. Referring to Figure 9.5, state what phase(s) is (are) present at

(a) 1 atm, 100°C   (b) 0.5 atm, 100°C

(c) 0.5 atm, 0°C

14. Referring to Figure 9.5, state what phase(s) is (are) present at

(a) −30°C, 5 mm Hg   (b) 25°C, 1 atm

(c) 70°C, 20 mm Hg

15. Argon gas has its triple point at −189.3°C and 516 mm Hg. It has a critical point at −122°C and 48 atm. The density of the solid is 1.65 g/cm³, whereas that of the liquid is 1.40 g/cm³. Sketch the phase diagram for argon and use it to fill in the blanks below with the words "boils," "melts," "sublimes," or "condenses."

(a) Solid argon at 500 mm Hg _____ when the temperature is increased.

(b) Solid argon at 2 atm _____ when the temperature is increased.

(c) Argon gas at −150°C _____ when the pressure is increased.

(d) Argon gas at −165°C _____ when the pressure is increased.

16. Iodine has a triple point at 114°C, 90 mm Hg. Its critical temperature is 535°C. The density of the solid is 4.93 g/cm³, and that of the liquid is 4.00 g/cm³. Sketch the phase diagram for iodine and use it to fill in the blanks using either "liquid" or "solid."

(a) Iodine vapor at 80 mm Hg condenses to the _____ when cooled sufficiently.

(b) Iodine vapor at 125°C condenses to the _____ when enough pressure is applied.

(c) Iodine vapor at 700 mm Hg condenses to the _____ when cooled above the triple point temperature.

17. Given the following data about benzene:

normal boiling point = 80°C

$d_{solid} > d_{liquid}$

triple point = 6°C at 36 mm Hg

critical point = 289°C at 48 atm

(a) Construct an appropriate phase diagram for benzene.

(b) Estimate the vapor pressure of benzene at 50°C.

WEB **18.** Given the following data about oxygen:

normal boiling point = −183°C

normal melting point = −218°C

triple point = −219°C at 2 mm Hg

critical point = −119°C at 50 atm

(a) Construct an appropriate phase diagram for oxygen.

(b) Is solid oxygen denser than liquid oxygen?

(c) Estimate the vapor pressure of oxygen at −200°C.

19. A pure substance A has a liquid vapor pressure of 320 mm Hg at 125°C, 800 mm Hg at 150°C, and 60 mm Hg at the triple point, 85°C. The melting point of A decreases slightly as pressure increases.

(a) Sketch a phase diagram for A.

(b) From the phase diagram, estimate the boiling point.

(c) What changes occur when, at a constant pressure of 320 mm Hg, the temperature drops from 150°C to 100°C?

20. A pure substance X has the following properties: mp = 90°C, increasing slightly as pressure increases; normal bp = 120°C; liquid vp = 65 mm Hg at 100°C, 20 mm Hg at triple point.

(a) Draw a phase diagram for X.

(b) Label solid, liquid, and vapor regions of the diagram.

(c) What changes occur if, at a constant pressure of 100 mm Hg, the temperature is raised from 100°C to 150°C?

Intermolecular Forces

21. Arrange the following in order of decreasing boiling point.

(a) I_2 (b) F_2 (c) Cl_2 (d) Br_2

22. Arrange the following in order of increasing boiling point.

(a) Ar   (b) He   (c) Ne   (d) Xe

23. Which of the following would you expect to show dispersion forces? Dipole forces?

(a) SCl_2 (b) $SiCl_4$ (c) CO_2 (d) $HCl(g)$

24. Which of the following would you expect to show dispersion forces? Dipole forces?

(a) $CHCl_3$ (b) Cl_2 (c) CCl_4 (d) H_2O

25. Which of the following would show hydrogen bonding?

(a) CH_3F (b) HO—OH

(c) NH_3 (d) H_3C—O—CH_3

26. Which of the following would show hydrogen bonding?

(a) CH_3OH (b) CH_3—N—CH_3 with CH_3 below N

(c) CH_3NH_2 (d) HF

27. Explain in terms of forces between structural units why

(a) HI has a higher boiling point than HBr.

(b) GeH_4 has a higher boiling point than SiH_4.

(c) H_2O_2 has a higher melting point than C_3H_8.

(d) NaCl has a higher boiling point than CH_3OH.

28. Explain in terms of forces between structural units why

(a) Br_2 has a lower melting point than NaBr.

(b) C_2H_5OH has a higher boiling point than butane, C_4H_{10}.

(c) H_2O has a higher boiling point than H_2Te.

(d) Acetic acid CH_3—C—OH has a lower boiling point
                         $\parallel$
                         O

than benzoic acid, C_6H_5—C—OH
                          $\parallel$
                          O

29. In which of the following processes is it necessary to break covalent bonds as opposed to simply overcoming intermolecular forces?

(a) subliming Dry Ice
(b) vaporizing chloroform ($CHCl_3$)
(c) decomposing water into H_2 and O_2
(d) changing chlorine molecules into chlorine atoms

30. In which of the following processes is it necessary to break covalent bonds as opposed to simply overcoming intermolecular forces?

(a) melting mothballs made of naphthalene
(b) dissolving HBr gas in water to form hydrobromic acid
(c) vaporizing ethyl alcohol, C_2H_5OH
(d) changing ozone, O_3, to oxygen gas, O_2

31. For each of the following pairs, choose the member with the lower boiling point. Explain your reason in each case.

(a) CO_2 or SO_2 (b) HBr or HCl
(c) H_2 or O_2 (d) F_2 or Cl_2

WEB 32. Follow the directions for Question 31 for the following substances.

(a) AsH_3 or PH_3 (b) C_6H_6 or $C_{10}H_8$
(c) NH_3 or PH_3 (d) LiCl or C_3H_8

33. What are the strongest attractive forces that must be overcome to

(a) boil silicon hydride SiH_4?
(b) vaporize calcium chloride?
(c) dissolve Cl_2 in carbon tetrachloride, CCl_4?
(d) melt iodine?

34. What are the strongest attractive forces that must be overcome to

(a) melt ice?
(b) sublime bromine?
(c) boil chloroform ($CHCl_3$)?
(d) vaporize benzene (C_6H_6)?

Types of Substances

35. Classify each of the following solids as metallic, network covalent, ionic, or molecular.

(a) It melts below 100°C and is insoluble in water.
(b) It conducts electricity only when melted.
(c) It is insoluble in water and conducts electricity.

36. Classify each of the following solids as metallic, network covalent, ionic, or molecular.

(a) It is insoluble in water, melts above 500°C, and does not conduct electricity either as a solid, dissolved in water, or molten.
(b) It dissolves in water, but does not conduct electricity as an aqueous solution, as a solid, or when molten.

(c) It dissolves in water, melts above 100°C, and conducts electricity when present in an aqueous solution.

37. Of the four general types of solids, which one(s)

(a) are generally low-boiling?
(b) are ductile and malleable?
(c) are generally soluble in nonpolar solvents?

38. Of the four general types of solids, which one(s)

(a) are generally insoluble in water?
(b) have very high melting points?
(c) conduct electricity as solids?

39. Classify each of the following species as molecular, network covalent, ionic, or metallic.

(a) K   (b) $CaCO_3$ (c) C_8H_{18}
(d) C (diamond)   (e) HCl(g)

40. Classify each of the following species as molecular, network covalent, ionic, or metallic.

(a) mercury   (b) IBr(s)   (c) sand
(d) $Cr_2(SO_4)_3$ (e) $Cl_2(g)$

41. Give the formula of a solid containing oxygen that is

(a) a polar molecule   (b) ionic
(c) network covalent   (d) a nonpolar molecule

42. Give the formula of a solid containing carbon that is

(a) molecular (b) ionic
(c) network covalent   (d) metallic

43. Describe the structural units in

(a) NaI   (b) N_2 (c) KO_2 (d) Au

44. Describe the structural units in

(a) C (graphite)   (b) SiC   (c) $FeCl_2$ (d) C_2H_2

Crystal Structure

45. Barium has an atomic radius of 0.217 nm. The volume of its cubic unit cell is 0.127 nm³. What is the geometry of the barium unit cell?

46. Sodium has an atomic radius of 0.186 nm. The edge of its cubic unit cell is 0.430 nm. What is the geometry of the sodium unit cell?

47. Aluminum (atomic radius = 0.143 nm) crystallizes with a face-centered cubic unit cell. What is the length of a side of the cell?

48. Xenon crystallizes with a face-centered cubic unit cell. The edge of the unit cell is 0.620 nm. What is the atomic radius of xenon?

49. Potassium iodide has a unit cell similar to that of sodium chloride (Figure 9.18). The ionic radii of K^+ and I^- are 0.133 nm and 0.216 nm, respectively. How long is

(a) one side of the cube?
(b) the face diagonal of the cube?

WEB 50. In the LiCl structure shown in Figure 9.18, the chloride ions form a face-centered cubic unit cell 0.513 nm on an edge. The ionic radius of Cl^- is 0.181 nm.

(a) Along a cell edge, how much space is there between the Cl^- ions?
(b) Would an Na^+ ion ($r = 0.095$ nm) fit into this space? A K^+ ion ($r = 0.133$ nm)?

51. For a cell of the CsCl type (Figure 9.18), how is the length of one side of the cell, s, related to the sum of the radii of the ions, $r_{cation} + r_{anion}$?

52. Consider the CsCl cell (Figure 9.18). The ionic radii of Cs^+ and Cl^- are 0.169 and 0.181 nm, respectively. What is the length of
   (a) the body diagonal?
   (b) the side of the cell?

53. Consider the CsCl unit shown in Figure 9.18. How many Cs^+ ions are there per unit cell? How many Cl^- ions? (Note that each Cl^- ion is shared by eight cubes.)

54. Consider the sodium chloride unit cell shown in Figure 9.18. Looking only at the front face (five large Cl^- ions, four small Na^+ ions),
   (a) how many cubes share each of the Na^+ ions in this face?
   (b) how many cubes share each of the Cl^- ions in this face?

Unclassified

55. The density of liquid mercury at 20°C is 13.6 g/cm³, its vapor pressure is 1.2×10^{-3} mm Hg.
   (a) What volume (in cm³) is occupied by one mole of $Hg(l)$ at 20°C?
   (b) What volume (in cm³) is occupied by one mole of $Hg(g)$ at 20°C and the equilibrium vapor pressure?
   (c) The atomic radius of Hg is 0.155 nm. Calculate the volume (in cm³) of one mole of Hg atoms ($V = 4\pi r^3/3$).
   (d) From your answers to (a), (b), and (c), calculate the percentage of the total volume occupied by the atoms in $Hg(l)$ and $Hg(g)$ at 20°C and 1.2×10^{-3} mm Hg.

56. Elemental boron is almost as hard as diamond. It is insoluble in water, does not conduct electricity at room temperature, and melts at 2300°C. What type of solid is boron?

57. An experiment is performed to determine the vapor pressure of formic acid. A 30.0-L volume of helium gas at 20.0°C is passed through 10.00 g of liquid formic acid (HCOOH) at 20.0°C. After the experiment, 7.50 g of liquid formic acid remains. Assume that the helium gas becomes saturated with formic acid vapor and the total gas volume and temperature remain constant. What is the vapor pressure of formic acid at 20.0°C?

58. Mercury is an extremely toxic substance. Inhalation of the vapor is just as dangerous as swallowing the liquid. How many milliliters of mercury will saturate a room that is $15 \times 12 \times 8.0$ ft with mercury vapor at 25°C? The vapor pressure of Hg at 25°C is 0.00163 mm Hg and its density is 13 g/mL.

Conceptual Problems

59. Represent pictorially using ten molecules
   (a) water freezing.
   (b) water vaporizing.
   (c) water being electrolyzed into hydrogen and oxygen.

60. Which of the following statements are true?

   (a) The critical temperature must be reached to change liquid to gas.
   (b) To melt a solid at constant pressure, the temperature must be above the triple point.
   (c) CHF_3 can be expected to have a higher boiling point than $CHCl_3$ because CHF_3 has hydrogen bonding.
   (d) One metal crystallizes in a body-centered cubic cell and another in a face-centered cubic cell of the same volume. The two atomic radii are related by the factor $\sqrt{1.5}$.

61. Criticize or comment on each of the following statements.
   (a) Vapor pressure remains constant regardless of volume.
   (b) The only forces that affect boiling point are dispersion forces.
   (c) The strength of the covalent bonds within a molecule has no effect on the melting point of the molecular substance.
   (d) A compound at its critical temperature is always a gas regardless of pressure.

62. Differentiate between
   (a) a covalent bond and a hydrogen bond.
   (b) normal boiling point and a boiling point.
   (c) the triple point and the critical point.
   (d) a phase diagram and a vapor pressure curve.
   (e) volume effect and temperature effect on vapor pressure.

63. Four shiny solids are labeled A, B, C, and D. Given the following information about the solids, deduce the identity of A, B, C, and D.
   (1) The solids are a graphite rod, a silver bar, a lump of "fool's gold" (iron sulfide), and iodine crystals.
   (2) B, C, and D are insoluble in water. A is slightly soluble.
   (3) Only C can be hammered into a sheet.
   (4) C and D conduct electricity as solids, B conducts when melted, A does not conduct as a solid, melted, or dissolved in water.

Challenge Problems

64. Iron crystallizes in a body-centered unit cell. Its atomic radius is 0.124 nm. Its density is 7.86 g/cm³. Using this information, estimate Avogadro's number.

65. A flask with a volume of 10.0 L contains 0.400 g of hydrogen gas and 3.20 g of oxygen gas. The mixture is ignited and the reaction

$$2H_2(g) + O_2(g) \longrightarrow 2H_2O$$

goes to completion. The mixture is cooled to 27°C. Assuming 100% yield,
   (a) What physical state(s) of water is (are) present in the flask?
   (b) What is the final pressure in the flask?
   (c) What is the pressure in the flask if 3.2 g of each gas is used?

66. Trichloroethane, $C_2H_3Cl_3$, is the active ingredient in aerosols that claim to stain proof men's ties. Trichloroethane has a vapor pressure of 100.0 mm Hg at 20.0°C and boils at 74.1°C. An uncovered cup ($\frac{1}{2}$ pint) of trichloroethane ($d = 1.325$ g/mL) is kept in an 18-ft³ refrigerator at 39°F. What percentage (by mass) of the trichloroethane is left as a liquid when equilibrium is established?

67. It has been suggested that the pressure exerted on a skate blade is sufficient to melt the ice beneath it and form a thin film of water, which makes it easier for the blade to slide over the ice. Assume that a skater weighs 120 lb and the blade has an area of 0.10 in.². Calculate the pressure exerted on the blade (1 atm = 15 lb/in.²). From information in the text, calculate the decrease in melting point at this pressure. Comment on the plausibility of this explanation and suggest another mechanism by which the water film might be formed.

68. As shown in Figure 9.18, Li^+ ions fit into a closely packed array of Cl^- ions, but Na^+ ions do not. What is the value of the r_{cation}/r_{anion} ratio at which a cation just fits into a structure of this type?

69. When the temperature drops from 20°C to 10°C, the pressure of a cylinder of compressed N_2 drops by 3.4%. The same temperature change decreases the pressure of a propane (C_3H_8) cylinder by 42%. Explain the difference in behavior.

10 SOLUTIONS

Sodium acetate crystals form quickly in a supersaturated solution when a small speck of solute is added. *(Charles D. Winters)*

CHAPTER OUTLINE

10.1   CONCENTRATION UNITS   **10.3   COLLIGATIVE PROPERTIES**

10.2   PRINCIPLES OF SOLUBILITY

In the course of a day, you use or make solutions many times. Your morning cup of coffee is a solution of solids (sugar and coffee) in a liquid (water). The gasoline you fill your gas tank with is a solution of several different liquid hydrocarbons. The soda you drink at a study break is a solution containing a gas (carbon dioxide) in a liquid (water).

A solution is a homogeneous mixture of a **solute** (substance being dissolved) distributed through a **solvent** (substance doing the dissolving). Solutions exist in any of the three physical states: gas, liquid, or solid. Air, the most common gaseous solution, is a mixture of nitrogen, oxygen, and lesser amounts of other gases. Many metal alloys are solid solutions. An example is the U.S. "nickel" coin (25% Ni, 75% Cu). The most familiar solutions are those in the liquid state, especially ones in which water is the solvent. Aqueous solutions are most important for our purposes in chemistry and will be emphasized in this chapter.

This chapter covers several of the physical aspects of solutions, including—

■ methods of expressing solution concentrations by specifying the relative amounts of solute and solvent (Section 10.1).

■ factors affecting solubility, including the nature of the solute and the solvent, the temperature, and the pressure (Section 10.2).

■ the effect of solutes on such solvent properties as vapor pressure, freezing point, and boiling point (Section 10.3).

10.1   CONCENTRATION UNITS

Several different methods are used to express relative amounts of solute and solvent in a solution. Two concentration units, *molarity* and *mole fraction*, were referred to in previous chapters. Two others, *mass percent* and *molality*, are considered for the first time.

Molarity (*M*)

In Chapter 4, molarity was the concentration unit of choice in dealing with solution stoichiometry. You will recall that molarity is defined as

$$\text{molarity } (M) = \frac{\text{moles solute}}{\text{liters solution}}$$

See the *Saunders Interactive General Chemistry CD-ROM*, Screens 5.10, Solution Concentration, & 14.2, Solubility.

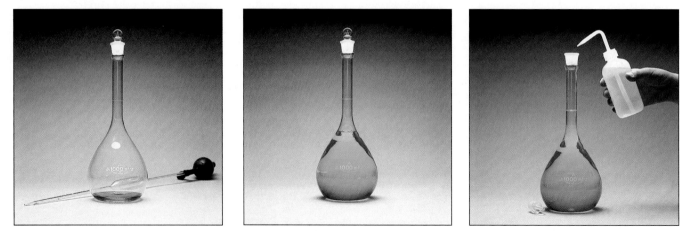

(a)   (b)   (c)

FIGURE 10.1

Preparation of one liter of 0.100 M CuSO₄ by dilution. (a) A volumetric flask containing 50.0 mL of 2.00 M CuSO₄ that was accurately measured using the pipet. (b) Water has been added to just below the narrow neck of the flask. (c) Finally, water is carefully added to bring the level exactly to the mark on the neck of the flask corresponding to 1.00 L. *(Marna G. Clarke)*

A solution can be prepared to a specified molarity by weighing out the calculated mass of solute and dissolving in enough solvent to form the desired volume of solution. Alternatively, you can start with a more concentrated solution and dilute with water to give a solution of the desired molarity (Figure 10.1). The calculations are straightforward if you keep a simple point in mind: Adding solvent cannot change the number of moles of solute, B, that is, n_B.

$$n_B \text{ in concentrated solution} = n_B \text{ in dilute solution}$$

See Screen 5.12, Preparing Solutions of Known Concentration (2).

In both solutions, the number of moles of solute can be found by multiplying the molarity, [B], by the volume in liters, V. Hence,

$$[B]_c V_c = [B]_d V_d$$

where the subscripts c and d stand for concentrated and dilute solutions, respectively.

EXAMPLE 10.1   How would you prepare 1.00 L of 0.100 M CuSO₄ starting with 2.00 M CuSO₄?

Strategy   This question might be restated: What volume of 2.00 M CuSO₄ should be diluted with water to give 1.00 L of 0.100 M CuSO₄? To do this, you use the equation $[CuSO_4]_c V_c = [CuSO_4]_d V_d$ to calculate V_c, the volume of the more concentrated solution.

Solution   Solving for V_c

$$V_c = \frac{[CuSO_4]_d \times V_d}{[CuSO_4]_c} = \frac{0.100 \; M \times 1.00 \; L}{2.00 \; M} = 0.0500 \; L = 50.0 \; mL$$

Measure out 50.0 mL of 2.00 M CuSO₄ and dilute with enough water (about 950 mL) to form 1.00 L of 0.100 M solution.

Reality Check   The molarity of CuSO₄ you *have* (2.00 M) is 20 times what you *need* (0.100 M), so it's reasonable to use only 1/20 of the volume you want (50.0 mL → 1.00 L).

The advantage of preparing solutions by the method illustrated in Example 10.1 and Figure 10.1 is that only volume measurements are involved. If you wander into the general chemistry storeroom, you're likely to find concentrated "stock" solutions of various chemicals. Storeroom personnel prepare the more dilute solutions that you use in the laboratory on the basis of calculations like those shown in the example.

It's easier to dilute a concentrated solution than to start from "scratch."

Mole Fraction (X)

Recall from Chapter 5 the defining equation for mole fraction (X) of a component A:

$$X_A = \frac{\text{moles A}}{\text{total moles}} = \frac{n_A}{n_{tot}}$$

The mole fractions of all components of a solution (A, B, . . .) must add to unity:

$$X_A + X_B + \cdots = 1$$

EXAMPLE 10.2   What are the mole fractions of CH_3OH and H_2O in a solution prepared by dissolving 1.20 g of methyl alcohol in 16.8 g of water?

Strategy   First (1), convert grams to moles for both components. Then (2), using the defining equation, calculate the mole fraction of CH_3OH. Finally (3), obtain the mole fraction of H_2O, most simply by using the relation: $X_{CH_3OH} + X_{H_2O} = 1$.

Solution

(1) $n_{CH_3OH} = 1.20 \text{ g CH}_3OH \times \dfrac{1 \text{ mol CH}_3OH}{32.04 \text{ g CH}_3OH} = 0.0375 \text{ mol CH}_3OH$

$n_{H_2O} = 16.8 \text{ g H}_2O \times \dfrac{1 \text{ mol H}_2O}{18.02 \text{ g H}_2O} = 0.932 \text{ mol H}_2O$

(2) $X_{CH_3OH} = \dfrac{n_{CH_3OH}}{n_{CH_3OH} + n_{H_2O}} = \dfrac{0.0375}{0.0375 + 0.932} = \boxed{0.0387}$

(3) $X_{H_2O} = 1 - X_{CH_3OH} = 1 - 0.0387 = \boxed{0.9613}$

Reality Check   It's reasonable that the mole fraction of water should be much greater than that of methyl alcohol. For one thing, the mass of water is larger (16.8 g versus 1.20 g); for another, the molar mass of water is smaller (18.02 g/mol versus 32.04 g/mol).

96% of the molecules in this solution are H_2O; the rest are CH_3OH.

Mass Percent; Parts per Million

The **mass percent** of solute in solution is expressed quite simply:

$$\text{mass percent of solute} = \frac{\text{mass solute}}{\text{total mass solution}} \times 100\%$$

In a solution prepared by dissolving 24 g of NaCl in 152 g of water,

$$\text{mass percent of NaCl} = \frac{24 \text{ g}}{24 \text{ g} + 152 \text{ g}} \times 100\% = \frac{24}{176} \times 100\% = 14\%$$

When the amount of solute is very small, as with trace impurities in water, concentration is often expressed in **parts per million (ppm):**

Parts per billion is defined similarly.

$$\text{ppm solute} = \frac{\text{mass solute}}{\text{total mass solution}} \times 10^6$$

Comparing the defining equations for mass percent and parts per million, it should be clear that

$$\text{ppm} = \text{mass percent} \times 10^4$$

In the United States and Canada, drinking water cannot by law contain more than 5×10^{-7} g of mercury per gram of sample. In parts per million that would be

$$\text{ppm Hg} = \frac{5 \times 10^{-7} \text{ g Hg}}{\text{g sample}} \times 10^6 = 0.5$$

Molality (*m*)

The concentration unit **molality,** given the symbol m, is defined as the number of moles of solute per kilogram (1000 g) of *solvent.*

Molality and molarity are concentration units; morality is something else.

$$\text{molality } (m) = \frac{\text{moles solute}}{\text{kilograms solvent}}$$

The molality of a solution is readily calculated if the masses of solute and solvent are known (Example 10.3).

EXAMPLE 10.3   A solution used for intravenous feeding contains 4.80 g of glucose, $C_6H_{12}O_6$, in 90.0 g of water. What is the molality of glucose?

Strategy   First (1), calculate the number of moles of glucose ($\mathcal{M} = 180.16$ g/mol), then (2), the number of kilograms of water. Finally (3), use the defining equation to calculate molality.

Solution

(1) $n_{\text{glucose}} = 4.80 \text{ g glucose} \times \dfrac{1 \text{ mol glucose}}{180.16 \text{ g glucose}} = 0.0266 \text{ mol glucose}$

(2) $\text{kg water} = 90.0 \text{ g water} \times \dfrac{1 \text{ kg water}}{1000 \text{ g water}} = 0.0900 \text{ kg water}$

(3) $\text{molality} = \dfrac{0.0266 \text{ mol glucose}}{0.0900 \text{ kg water}} = \boxed{0.296 \text{ } m}$

Conversions Between Concentration Units

It is frequently necessary to convert from one concentration unit to another. This problem arises, for example, in making up solutions of hydrochloric acid. Typically, the analysis or assay that appears on the label (Figure 10.2) does not give the molarity or molality of the acid. Instead, it lists the mass percent of solute and the density of the solution.

Conversions between concentration units are relatively straightforward provided you *first decide on a fixed amount of solution.* The amount chosen depends on

```
ACTUAL ANALYSIS. LOT  320037       MEETS A.C.S. SPECIFICATIONS

*  Assay (HCl)(by acidimetry) . . . . . . . . . . .        37.7      %
   Appearance . . . . . . .                             Passes Test
   Color (APHA) . . . . . .                           <  5
   Specific Gravity at 60°/60°F . . . .                   1.1906
   Residue after Ignition . . . . . . . .                 0.00005   %
   Free Chlorine (Cl) . . . . . . . .                   Passes Test
*  Bromide (Br) . . . . . . .                          <  0.005     %
                      Trace Impurities (in ppm):
   Ammonium (NH₄) . . . . . . .                        <  3
   Sulfate (SO₄) . . . . . . .                            0.25
   Sulfite (SO₃) . . . .                                  0.2
   Arsenic (As) . . . . . .                            <  0.004
   Copper (Cu) . . . . . .                                0.0004
   Iron (Fe) . . . . . .                                  0.002
   Heavy Metals (as Pb) . . . . . .                    <  0.05
   Nickel (Ni) . . . . . . .                              0.0004

         *Assay value tends to be less than reported due to vapor loss.
                    especially when opening container.
```

FIGURE 10.2
The label on a bottle of concentrated hydrochloric acid. The label gives the mass percent of HCl in the solution (known as the *assay*) and the density (or *specific gravity*) of the solution. The molality, molarity, and mole fraction of HCl in the solution can be calculated from this information. *(Marna G. Clarke)*

the unit in which concentration is originally expressed. Some suggested starting quantities are listed below.

When the Original Concentration Is	Start With
Mass percent	100 g solution
Molarity (*M*)	1.00 L solution
Molality (*m*)	1000 g solvent
Mole fraction (*X*)	1 mol (solute + solvent)

> Complex problems can usually be solved if you know where to start.

To illustrate the approach used, let us consider how the molarity of HCl can be calculated from the data shown on the label (Figure 10.2).

EXAMPLE 10.4 Calculate the molarity of a concentrated solution of hydrochloric acid that is 37.7% by mass HCl; the solution has a density of 1.19 g/mL.

Strategy   Start with 100.0 g of solution. First (1), knowing the percent by mass of HCl (37.7%) and its molar mass (36.46 g/mol), you can readily calculate the number of moles of HCl in 100.0 g of solution. Then (2), knowing the density of the solution, you should be able to calculate the volume occupied by 100.0 g of solution. Finally (3), knowing both the number of moles and the volume in liters, use the defining equation to calculate the molarity of HCl.

Solution

(1) $n_{HCl} = 100.0 \text{ g soln.} \times \dfrac{37.7 \text{ g HCl}}{100.0 \text{ g soln.}} \times \dfrac{1 \text{ mol HCl}}{36.46 \text{ g HCl}} = 1.03 \text{ mol HCl}$

(2) $V = 100.0 \text{ g} \times \dfrac{1 \text{ mL}}{1.19 \text{ g}} \times \dfrac{1 \text{ L}}{1000 \text{ mL}} = 0.0840 \text{ L}$

(3) $[HCl] = 1.03 \text{ mol}/0.0840 \text{ L} = \boxed{12.3 \text{ M}}$

This same approach can be used to convert between molarity and molality (Example 10.5). Again, the density must be known.

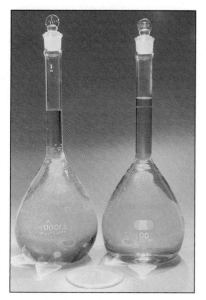

Molarity and molality. On the left is a 0.10 M solution of potassium chromate. On the right is a 0.10 m solution that contains the same amount of potassium chromate (19.4 g, in dish). The 0.10 M solution was made by placing the solid in the flask and adding water to give 1 L of solution. The 0.10 m solution was prepared by placing the solid in the flask and adding 1000 g of water. You can see that the 0.10 m solution has a slightly larger volume. *(Charles D. Winters)*

See Screen 14.3, The Solution Process.

EXAMPLE 10.5 A 1.13 M solution of KOH has a density of 1.05 g/mL. Calculate its molality.

Strategy Start with one liter of solution, which contains 1.13 mol of KOH. Then find, in order, (1) the total mass of solution, knowing the density; (2) the mass of KOH, knowing the molarity; and (3) the mass of water. Finally (4), use the defining equation to calculate molality.

Solution

(1) total mass of 1 L of soln. $= 1.00 \text{ L} \times \dfrac{1000 \text{ mL}}{1 \text{ L}} \times \dfrac{1.05 \text{ g}}{1 \text{ mL}} = 1050 \text{ g}$

(2) mass of KOH in 1 L of soln. $= 1.13 \text{ mol KOH} \times \dfrac{56.11 \text{ g KOH}}{1 \text{ mol KOH}} = 63.4 \text{ g KOH}$

(3) mass of water in 1 L of soln. $= 1050 \text{ g} - 63.4 \text{ g} = 987 \text{ g}$

(4) molality $= \dfrac{1.13 \text{ mol KOH}}{0.987 \text{ kg water}} = \boxed{1.14 \ m}$

Reality Check To convert between a concentration unit based on volume (molarity) and one based on mass (mass percent, mole fraction, and molality) you must know the density of the solution.

You will notice from Example 10.5 that the molarity and molality of the KOH solutions are very close to one another (1.13 M, 1.14 m). This is generally true of *dilute water solutions;* one liter of a dilute aqueous solution contains approximately one kilogram of water. For concentrated or nonaqueous solutions, molarity and molality ordinarily differ considerably from each other.

10.2 PRINCIPLES OF SOLUBILITY

The extent to which a solute dissolves in a particular solvent depends on several factors. The most important of these are—

- the nature of solvent and solute particles and the interactions between them.
- the temperature at which the solution is formed.
- the pressure of a gaseous solute.

In this section we consider in turn the effect of each of these factors on solubility.

Solute–Solvent Interactions

In discussing solubility, it is sometimes stated that "like dissolves like." A more meaningful way to express this idea is to say that two substances with intermolecular forces of about the same type and magnitude are likely to be very soluble in one another. To illustrate, consider the hydrocarbons pentane, C_5H_{12}, and hexane, C_6H_{14}, which are completely soluble in each other. Molecules of these nonpolar substances are held together by dispersion forces of about the same magnitude. A pentane molecule experiences little or no change in intermolecular forces when it goes into solution in hexane.

Most nonpolar substances have very small water solubilities. Petroleum, a mixture of hydrocarbons, spreads out in a thin film on the surface of a body of water rather than dissolving. The mole fraction of pentane, C_5H_{12}, in a saturated water solution is only 0.0001. These low solubilities are readily understood in terms of the structure of liquid water, which you will recall (Chapter 9) is strongly hydrogen-bonded. Dissimilar intermolecular forces between C_5H_{12} (dispersion) and H_2O (H bonds) lead to low solubility.

Of the relatively few organic compounds that dissolve readily in water, many contain —OH groups. Three familiar examples are methyl alcohol, ethyl alcohol, and ethylene glycol, all of which are infinitely soluble in water.

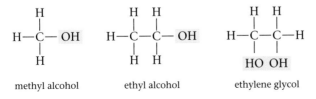

methyl alcohol   ethyl alcohol   ethylene glycol

In these compounds, as in water, the principal intermolecular forces are hydrogen bonds. When a substance like methyl alcohol dissolves in water, it forms hydrogen bonds with H_2O molecules. These hydrogen bonds, joining a CH_3OH molecule to an H_2O molecule, are about as strong as those in the pure substances.

Not all organic compounds that contain —OH groups are soluble in water (Table 10.1). As molar mass increases, the polar —OH group represents an increasingly smaller portion of the molecule. At the same time, the nonpolar hydrocarbon portion becomes larger. As a result, solubility decreases with molar mass. Butanol, $CH_3CH_2CH_2CH_2$ OH, is much less soluble in water than methyl alcohol, CH_3 OH. The hydrocarbon portion, shaded in green, is much larger in butanol.

The solubility (or insolubility) of different vitamins is of concern in nutrition. Molecules of vitamins B and C contain several —OH groups that can form hydrogen bonds with water (Figure 10.3, page 288). As a result, they are water-soluble, readily excreted by the body, and must be consumed daily. In contrast, vitamins A, D, E, and K, whose molecules are relatively nonpolar, are water-insoluble. These vitamins are not so readily excreted; they tend to stay behind in fatty tissues. This means that the body can draw on its reservoir of vitamins A, D, E, and K to deal with sporadic deficiencies. Conversely, megadoses of these vitamins can lead to very high, possibly toxic, concentrations in the body.

As we noted in Chapter 4, the solubility of ionic compounds in water varies tremendously from one solid to another. The extent to which solution occurs depends on a balance between two forces, both electrical in nature:

Solubility and intermolecular forces. Motor oil (*left*) is made up of nonpolar molecules. It does not mix well with water because water is polar. Ethylene glycol (*right*), commonly used as antifreeze, mixes with water in all proportions because both form hydrogen bonds. (*Charles D. Winters*)

You can't remove the "hot" taste of chili peppers by drinking water because the compound responsible is nonpolar and water-insoluble.

TABLE 10.1	**Solubilities of Alcohols in Water**	
Substance	**Formula**	**Solubility (g solute/L H_2O)**
Methyl alcohol	CH_3OH	Completely soluble
Ethyl alcohol	CH_3CH_2OH	Completely soluble
Propanol	$CH_3CH_2CH_2OH$	Completely soluble
Butanol	$CH_3CH_2CH_2CH_2OH$	74
Pentanol	$CH_3CH_2CH_2CH_2CH_2OH$	27
Hexanol	$CH_3CH_2CH_2CH_2CH_2CH_2OH$	6.0
Heptanol	$CH_3CH_2CH_2CH_2CH_2CH_2CH_2OH$	1.7

FIGURE 10.3

Molecular structures of vitamin D_2 and vitamin B_6. Polar groups are shown in color. Vitamin D_2 is water-insoluble and vitamin B_6 is water-soluble.

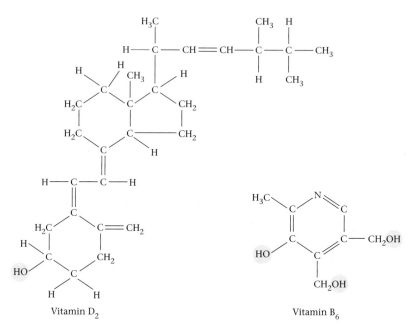

Vitamin D_2

Vitamin B_6

See Screen 14.4, Energetics of Solution Formation.

1. The force of attraction between H_2O molecules and the ions, which tends to bring the solid into solution. If this factor predominates, the compound is very soluble in water, as is the case with NaCl, NaOH, and many other ionic solids.

2. The force of attraction between oppositely charged ions, which tends to keep them in the solid state. If this is the major factor, the water solubility is very low. The fact that $CaCO_3$ and $BaSO_4$ are almost insoluble in water implies that interionic attractive forces predominate with these ionic solids.

Effect of Temperature on Solubility

When an excess of a solid such as sucrose ($C_{12}H_{22}O_{11}$, commonly called sugar) is shaken with water, it forms a **saturated solution.** An equilibrium is established between molecules in the solid state and in solution

$$C_{12}H_{22}O_{11}(s) \rightleftharpoons C_{12}H_{22}O_{11}(aq)$$

At 20°C, the saturated solution contains 204 g of sucrose per 100 g of water.

A similar type of equilibrium is established when a gas such as oxygen is bubbled through water.

$$O_2(g) \rightleftharpoons O_2(aq)$$

At 20°C and 1 atm, 0.00138 mol of O_2 dissolves per liter of water.

The effect of a temperature change on solubility equilibria such as these can be predicted by applying a simple principle. *An increase in temperature always shifts the position of an equilibrium to favor an endothermic process.* This means that if the solution process absorbs heat ($\Delta H_{soln.} > 0$), an increase in temperature increases the solubility. Conversely, if the solution process is exothermic ($\Delta H < 0$), an increase in temperature decreases the solubility.

Dissolving a solid in a liquid is usually an endothermic process; heat must be absorbed to break down the crystal lattice.

$$\text{solid} + \text{liquid} \rightleftharpoons \text{solution} \qquad \Delta H_{soln.} > 0$$

Ions in solution. The positive and negative ions are surrounded by water molecules attracted by the oppositely charged ends of their dipoles. **OHT**

Consistent with this effect, the solubilities of solids usually increase as the temperature rises. More sugar dissolves in hot coffee than in cold coffee (Figure 10.4).

Cooling a saturated solution usually causes a solid to crystallize out of solution, because the solubility is smaller at the lower temperature. Sometimes, however, this doesn't happen; the excess solute stays in solution on cooling, giving what is known as a **supersaturated solution** (Figure 10.5). This happens when a saturated solution of sucrose at 80°C (362 g of $C_{12}H_{22}O_{11}$)/100 g of water) is cooled carefully to 20°C, where the solubility is 204 g/100 g of water. The excess sucrose stays in solution until a small seed crystal of sucrose is added, whereupon crystallization quickly takes place. The excess solute (362 g − 204 g = 158 g of $C_{12}H_{22}O_{11}$/100 g of water) comes out of solution, establishing equilibrium between the saturated solution and the sucrose crystals.

The crystallization of excess solute is a common problem in the preparation of candies and in the storage of jam and honey. From these supersaturated solutions, sugar separates either as tiny crystals, causing the "graininess" in fudge, or as large crystals, which often appear in honey kept for a long time (Figure 10.6, page 290).

When a gas condenses to a liquid, heat is always evolved. By the same token, heat is usually evolved when a gas dissolves in a liquid:

$$\text{gas} + \text{liquid} \rightleftharpoons \text{solution} \qquad \Delta H_{\text{soln.}} < 0$$

This means that the reverse process (gas coming out of solution) is endothermic. Hence, it is favored by an increase in temperature; typically, gases become less soluble as the temperature rises. This rule is followed by all gases in water. You have probably noticed this effect when heating water in an open pan or beaker. Bubbles of air are driven out of the water by an increase in temperature. The reduced solubility of oxygen in water at high temperatures (Figure 10.7a, page 290) may explain why trout congregate at the bottom of deep pools on hot summer days when the surface water is depleted of dissolved oxygen.

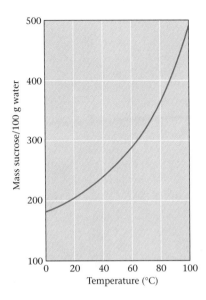

FIGURE 10.4
Solubility of table sugar (sucrose). The solubility of sugar, $C_{12}H_{22}O_{11}$, in water increases exponentially with temperature.

(a) (b) (c)

FIGURE 10.5
Crystallization in a supersaturated solution of sodium acetate ($NaC_2H_3O_2$). The transition from (a) to (b) to (c) is rapid once crystallization is initiated by dropping a tiny seed crystal of sodium acetate into a supersaturated solution of this ionic compound. *(Charles D. Winters)*

FIGURE 10.6
Rock candy. The candy is formed by the crystallization of sugar from a saturated solution that is cooled slowly.
(Marna G. Clarke)

See Screen 14.5, Factors Affecting Solubility (1).

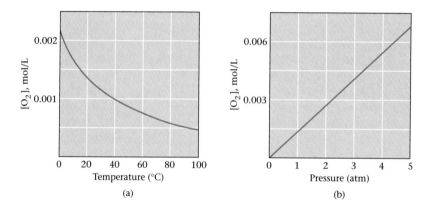

(a)   (b)

FIGURE 10.7
Oxygen solubility. The solubility of $O_2(g)$ in water decreases as temperature rises (a) and increases as pressure increases (b). In (a), the pressure is held constant at 1 atm; in (b), the temperature is held constant at 25°C. **OHT**

Effect of Pressure on Solubility

Pressure has a major effect on solubility only for gas-liquid systems. At a given temperature, raising the pressure increases the solubility of a gas. Indeed, at low to moderate pressures, gas solubility is directly proportional to pressure (Figure 10.7b).

$$C_g = kP_g$$

where P_g is the partial pressure of the gas over the solution, C_g is its concentration in the solution, and k is a constant characteristic of the particular gas-liquid system. This relation is called **Henry's law** after its discoverer, William Henry (1775–1836), a friend of John Dalton.

Henry's law arises because increasing the pressure raises the concentration of molecules in the gas phase. To balance this change and maintain equilibrium, more gas molecules enter the solution, increasing their concentration in the liquid phase.

Henry's law. As soon as the pressure is released, carbon dioxide begins to bubble out of solution in a carbonated beverage. *(Charles D. Winters)*

EXAMPLE 10.6 The solubility of pure nitrogen in blood at body temperature, 37°C, and one atmosphere is 6.2×10^{-4} M. If a diver breathes air ($X\,N_2 = 0.78$) at a depth where the total pressure is 2.5 atm, calculate the concentration of nitrogen in his blood.

Strategy Use Henry's law in any problem involving gas solubility and pressure. Perhaps the simplest approach is to use the data at one atmosphere to calculate k. Then apply Henry's law to calculate C_g at the higher pressure. Note that it is the *partial* pressure of N_2 that is required; use the relation $P_{N_2} = X_{N_2} \times P_{tot}$ to find it.

Solution At 1 atm:

$$k = \frac{\text{concentration of } N_2}{\text{pressure of } N_2} = \frac{6.2 \times 10^{-4}\,M}{1.00\,\text{atm}} = 6.2 \times 10^{-4}\,M/\text{atm}$$

At the higher pressure:

$$P_{N_2} = X_{N_2} \times P_{tot} = 0.78 \times 2.5\,\text{atm} = 2.0\,\text{atm}$$

$$[N_2] = kP_{N_2} = 6.2 \times 10^{-4}\,\frac{M}{\text{atm}} \times 2.0\,\text{atm} = \boxed{1.2 \times 10^{-3}\,M}$$

The influence of partial pressure on gas solubility is used in making carbonated beverages such as beer, sparkling wines, and many soft drinks. These beverages are bottled under pressures of CO_2 as high as 4 atm. When the bottle or can is opened, the pressure above the liquid drops to 1 atm, and the carbon dioxide rapidly bubbles out of solution. Pressurized containers for shaving cream, whipped cream, and cheese spreads work on a similar principle. Pressing a valve reduces the pressure on the dissolved gas, causing it to rush from solution, carrying liquid with it as a foam.

Another consequence of the effect of pressure on gas solubility is the painful, sometimes fatal, affliction known as the "bends." This occurs when a person goes rapidly from deep water (high pressure) to the surface (lower pressure), where gases are less soluble. The rapid decompression causes air, dissolved in blood and other body fluids, to bubble out of solution. These bubbles impair blood circulation and affect nerve impulses. To minimize these effects, deep-sea divers and aquanauts breathe a helium-oxygen mixture rather than compressed air (nitrogen-oxygen). Helium is only about one-third as soluble as nitrogen, and hence much less gas comes out of solution on decompression.

Scuba divers have to worry about this.

10.3 COLLIGATIVE PROPERTIES

The properties of a solution differ considerably from those of the pure solvent. Those solution properties that depend primarily on the *concentration of solute particles* rather than their nature are called **colligative properties.** Such properties include vapor pressure lowering, osmotic pressure, boiling point elevation, and freezing point depression. This section considers the relations between colligative properties and solute concentration, starting with **nonelectrolytes,** which exist in solution as molecules, and then going on (briefly) to electrolytes, which form ions in solution.

The relationships among colligative properties and solute concentration are best regarded as limiting laws. They are approached more closely as the solution becomes more dilute. In practice, the relationships discussed in this section are valid, for nonelectrolytes, to within a few percent at concentration as high as 1 *M*. At higher concentrations, solute-solute interactions lead to larger deviations.

Vapor Pressure Lowering (Nonelectrolytes)

You may have noticed that concentrated aqueous solutions evaporate more slowly than does pure water. This reflects the fact that the vapor pressure of water over the solution is less than that of pure water (Figure 10.8).

Vapor pressure lowering is a true colligative property; that is, it is independent of the nature of the solute but directly proportional to its concentration. For example, the vapor pressure of water above a 0.10 *M* solution of either glucose or sucrose at 0°C is the same, about 0.008 mm Hg less than that of pure water. In 0.30 *M* solution, the vapor pressure lowering is almost exactly three times as great, 0.025 mm Hg.

The relationship between solvent vapor pressure and concentration is ordinarily expressed as

$$\boxed{P_1 = X_1 P_1^\circ}$$

In this equation, P_1 is the vapor pressure of solvent over the solution, P_1° is the vapor pressure of the pure solvent at the same temperature, and X_1 is the mole frac-

Solvent molecules
Solute molecules

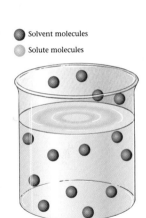
$X_{solvent} = 1.0$

$X_{solvent} = 0.5$

FIGURE 10.8
Raoult's law. Adding a solute lowers the concentration of solvent molecules in the liquid phase. To maintain equilibrium, the concentration of solvent molecules in the gas phase must decrease, thereby lowering the solvent vapor pressure. **OHT**

tion of solvent. Note that because X_1 in a solution must be less than 1, so P_1 must be less than P_1°. This relationship is called **Raoult's law;** François Raoult (1830–1901) carried out a large number of careful experiments on vapor pressures and freezing point lowering.

To obtain a direct expression for vapor pressure lowering, note that $X_1 = 1 - X_2$, where X_2 is the mole fraction of solute. Substituting $1 - X_2$ for X_1 in Raoult's law,

$$P_1 = (1 - X_2)P_1^\circ$$

Rearranging,

$$P_1^\circ - P_1 = X_2 P_1^\circ$$

Vapor-pressure lowering is directly proportional to solute mole fraction.

The quantity $(P_1^\circ - P_1)$ is the vapor pressure lowering (ΔP). It is the difference between the solvent vapor pressure in the pure solvent and in solution.

$$\Delta P = X_2 P_1^\circ$$

EXAMPLE 10.7

A solution contains 102 g of sugar, $C_{12}H_{22}O_{11}$, in 375 g of water. Calculate the vapor pressure lowering at 25°C (vp of pure water = 23.76 mm Hg).

Strategy   First (1), calculate the number of moles of sugar ($\mathcal{M}$ = 342.30 g/mol) and water ($\mathcal{M}$ = 18.02 g/mol). That information allows you to calculate (2) the mole fraction of sugar. Finally (3), use Raoult's law to find the vapor pressure lowering.

Solution

(1) $n_{sugar} = 102 \text{ g sugar} \times \dfrac{1 \text{ mol sugar}}{342.30 \text{ g sugar}} = 0.298 \text{ mol sugar}$

$n_{water} = 375 \text{ g water} \times \dfrac{1 \text{ mol water}}{18.02 \text{ g water}} = 20.8 \text{ mol water}$

(2) $X_{sugar} = \dfrac{0.298}{0.298 + 20.8} = \dfrac{0.298}{21.1} = 0.0141$

(3) $\Delta P = X_{sugar} \times P_{H_2O}^\circ = 0.0141 \times 23.76 \text{ mm Hg} = \boxed{0.335 \text{ mm Hg}}$

Reality Check   You expect the vapor pressure lowering to be small (0.335 mm Hg) because the mole fraction of solute is low (0.0141).

Boiling Point Elevation and Freezing Point Lowering (Nonelectrolytes)

See Screen 14.8, Colligative Properties (2).

When a solution of a nonvolatile solute is heated, it does not begin to boil until the temperature exceeds the boiling point of the solvent. The difference in temperature is called the **boiling point elevation,** ΔT_b.

$$\Delta T_b = T_b - T_b^\circ$$

where T_b and T_b° are the boiling points of the solution and the pure solvent, respectively. As boiling continues, pure solvent distills off, the concentration of solute increases, and the boiling point continues to rise (Figure 10.9).

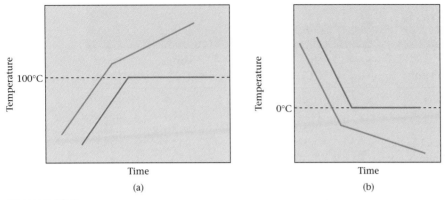

FIGURE 10.9

Heating (a) and cooling (b) curves for pure water (red) and an aqueous solution (green).
For pure water, the temperature remains constant during boiling or freezing. For the solution, the temperature changes steadily during the phase change because water is being removed, increasing the concentration of solute.

When a solution is cooled, it does not begin to freeze until a temperature below the freezing point of the pure solvent is reached. The **freezing point lowering,** ΔT_f, is defined to be a positive quantity:

$$\Delta T_f = T_f^\circ - T_f$$

where T_f°, the freezing point of the solvent, lies above T_f, the freezing point of the solution. As freezing takes place, pure solvent freezes out, the concentration of solute increases, and the freezing point continues to drop. This is what happens with "ice beer." When beer is cooled below 0°C, pure ice separates and the percentage of ethyl alcohol increases.

Boiling point elevation is a direct result of vapor pressure lowering. At any given temperature, a solution of a nonvolatile solute* has a vapor pressure *lower* than that of the pure solvent. Hence a *higher* temperature must be reached before the solution boils, that is, before its vapor pressure becomes equal to the external pressure. Figure 10.10 illustrates this reasoning graphically.

The freezing point lowering, like the boiling point elevation, is a direct result of the lowering of the solvent vapor pressure by the solute. Notice from Figure 10.10 that the freezing point of the solution is the temperature at which the solvent in solution has the same vapor pressure as the pure solid solvent. This implies that it is pure solvent (e.g., ice) that separates when the solution freezes.

Boiling point elevation and freezing point lowering, like vapor pressure lowering, are colligative properties. They are directly proportional to solute concentration, generally expressed as molality, m. The relevant equations are

$$\Delta T_b = k_b \text{(molality)}$$
$$\Delta T_f = k_f \text{(molality)}$$

The proportionality constants in these equations, k_b and k_f, are called the *molal boiling point constant* and the *molal freezing point constant,* respectively. Their mag-

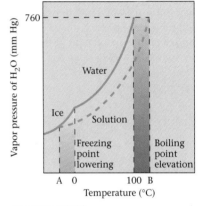

FIGURE 10.10

Vapor pressure lowering. Because a nonvolatile solute lowers the vapor pressure of a solvent, the boiling point of a solution will be higher and the freezing point lower than the corresponding values for the pure solvent. Water solutions freeze *below* 0°C at point A and boil *above* 100°C at point B. **OHT**

Solutes raise the boiling point and lower the freezing point.

*Volatile solutes ordinarily lower the boiling point because they contribute to the total vapor pressure of the solution.

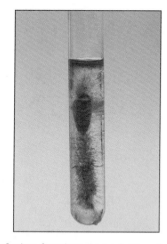

Solution freezing. As a solution of water and purple dye freezes, pure ice crystals form and the liquid solution phase becomes more and more concentrated. In the photo, ice crystals have formed around the inside of the test tube and the concentrated solution of purple dye remains in the center. *(Charles D. Winters)*

	fp (°C)	k_f (°C/m)	bp (°C)	k_b (°C/m)
TABLE 10.2 Molal Freezing Point and Boiling Point Constants				
Solvent				
Water	0.00	1.86	100.00	0.52
Acetic acid	16.66	3.90	117.90	2.53
Benzene	5.50	5.10	80.10	2.53
Cyclohexane	6.50	20.2	80.72	2.75
Camphor	178.40	40.0	207.42	5.61
p-Dichlorobenzene	53.1	7.1	174.1	6.2
Naphthalene	80.29	6.94	217.96	5.80

nitudes depend on the nature of the solvent (Table 10.2). Note that when the solvent is water,

$$k_b = 0.52°C/m \qquad k_f = 1.86°C/m$$

EXAMPLE 10.8 An antifreeze solution is prepared containing 50.0 cm³ of ethylene glycol, $C_2H_6O_2$ (d = 1.12 g/cm³), in 50.0 g of water. Calculate the freezing point of this 50-50 mixture.

Strategy First (1) calculate the number of moles of $C_2H_6O_2$ ($\mathcal{M}$ = 62.07 g/mol). Then (2) apply the defining equation to calculate the molality. Finally (3), use the equation ΔT_f = (1.86°C/m) × molality to find the freezing point lowering.

Solution

(1) $n_{C_2H_6O_2}$ = 50.0 cm³ × $\dfrac{1.12\ g}{1\ cm³}$ × $\dfrac{1\ mol}{62.04\ g}$ = 0.903 mol

(2) molality = $\dfrac{mol\ of\ C_2H_6O_2}{kg\ of\ water}$ = $\dfrac{0.903\ mol}{0.0500\ kg}$ = 18.1 m

(3) $\Delta T_f = k_f$ (molality) = 1.86°C/m × 18.1 m = 33.7°C

The freezing point of the solution is 33.7°C below that of pure water (0°C). Hence the solution should freeze at −33.7°C.

Reality Check Actually, the freezing point is somewhat lower, about −37°C (−35°F), which reminds us that the equation used, $\Delta T_f = k_f m$, is a limiting law, strictly valid only in very dilute solution.

Propylene glycol, HO—(CH₂)₃—OH, is much less toxic.

You take advantage of freezing point lowering when you add antifreeze to your automobile radiator in winter. Ethylene glycol is the solute commonly used. It has a high boiling point (197°C), is virtually nonvolatile at 100°C, and raises the boiling point of water. Hence antifreeze that contains ethylene glycol does not boil away in summer driving.

Osmotic Pressure (Nonelectrolytes)

One interesting effect of vapor-pressure lowering is shown at the left of Figure 10.11. We start with two beakers, one containing pure water and the other con-

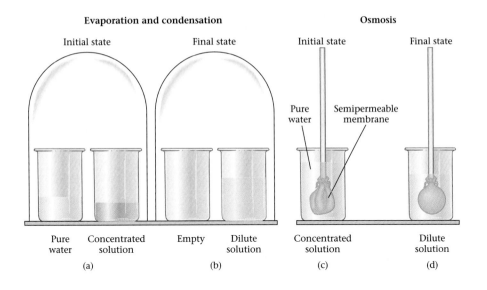

Evaporation and condensation

| Initial state | Final state |

| Pure water | Concentrated solution | Empty | Dilute solution |
| (a) | | (b) | |

Osmosis

| Initial state | Final state |

Pure water   Semipermeable membrane

| Concentrated solution | Dilute solution |
| (c) | (d) |

FIGURE 10.11

Evaporation and condensation (a and b); osmosis (c and d). Water tends to move spontaneously from a region where its vapor pressure is high to a region where it is low. In a → b, movement of water molecules occurs through the air trapped under the bell jar. In c → d, water molecules move by osmosis through a semipermeable membrane. The driving force is the same in the two cases, although the mechanism differs. **OHT**

taining a sugar solution. These are placed next to each other under a bell jar (Figure 10.11a). As time passes, the liquid level in the beaker containing the solution rises. The level of pure water in the other beaker falls. Eventually, by evaporation and condensation, all the water is transferred to the solution (Figure 10.11b). At the end of the experiment, the beaker that contained pure water is empty. The driving force behind this process is the difference in vapor pressure of water in the two beakers. *Water moves from a region where its vapor pressure or mole fraction is high ($X_1 = 1$ in pure water) to one in which its vapor pressure or mole fraction is lower ($X_1 < 1$ in sugar solution).*

The apparatus shown in Figure 10.11c and d can be used to achieve a result similar to that found in the bell jar experiment. In this case, a sugar solution is separated from water by a semipermeable membrane. This may be an animal bladder, a slice of vegetable tissue, or a piece of parchment. The membrane, by a mechanism that is not well understood, allows water molecules to pass through it, but not sugar molecules. As before, water moves from a region where its mole fraction is high (pure water) to a region where it is lower (sugar solution). This process, taking place through a membrane permeable only to the solvent, is called **osmosis.** As a result of osmosis, the water level rises in the tube and drops in the beaker (Figure 10.11d).

The osmotic pressure, π, is equal to the external pressure, P, just sufficient to prevent osmosis (Figure 10.12, page 296). If P is less than π, osmosis takes place in the normal way, and water moves through the membrane into the solution (Figure 10.12a). By making the external pressure large enough, it is possible to reverse this process (Figure 10.12b). When $P > \pi$, water molecules move through the membrane from the solution to pure water. This process, called *reverse osmosis,* is used to obtain fresh water from seawater in arid regions of the world, including Saudi Arabia.

Osmotic pressure, like vapor pressure lowering, is a colligative property. For any nonelectrolyte B, π is directly proportional to molarity, [B]. The equation relating these two quantities is very similar to the ideal gas law:

$$\pi = \frac{n_B RT}{V} = [B]RT$$

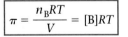 See Screen 14.9, Colligative Properties (3).

Reverse osmosis absorbs energy, but there's plenty of that in Saudi Arabia.

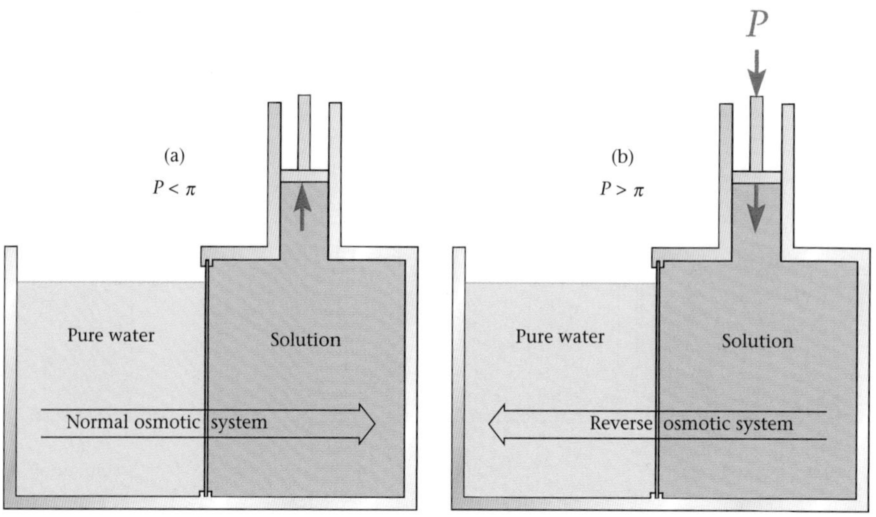

FIGURE 10.12

Reverse osmosis. (a) When the external pressure (P) is less than the osmotic pressure (π) ($P < \pi$), normal osmosis occurs. (b) When the external pressure exceeds the osmotic pressure, water flows in the opposite direction, producing reverse osmosis. Reverse osmosis can be used to obtain fresh water from seawater. **OHT**

(a)

$P < \pi$

Pure water

Solution

Normal osmotic system

(b)

$P > \pi$

Pure water

Solution

Reverse osmotic system

where R is the gas law constant, 0.0821 L·atm/mol·K, and T is the Kelvin temperature. Even in dilute solution, the osmotic pressure is quite large. Consider, for example, a 0.10 M solution at 25°C:

$$\pi = (0.10 \text{ mol/L}) \left(0.0821 \frac{\text{L·atm}}{\text{mol·K}} \right) (298 \text{ K}) = 2.4 \text{ atm}$$

A pressure of 2.4 atm is equivalent to that of a column of water 25 m (more than 80 ft) high.

If a cucumber is placed in a concentrated brine solution, it shrinks and assumes the wrinkled skin of a pickle. The skin of the cucumber acts as a semipermeable membrane. The water solution inside the cucumber is more dilute than the solution surrounding it. As a result, water flows out of the cucumber into the brine (Figure 10.13).

When a dried prune is placed in water, the skin also acts as a semipermeable membrane. This time the solution inside the prune is more concentrated than the water, so that water flows into the prune, making the prune less wrinkled.

Nutrient solutions used in intravenous feeding must be *isotonic* with blood; that is, they must have the same osmotic pressure as blood. If the solution is too dilute, its osmotic pressure will be less than that of the fluids inside blood cells; in

Dishwashers' hands get wrinkled, too.

FIGURE 10.13

Effect of osmosis on cucumbers and prunes. When a cucumber is pickled, water moves out of the cucumber by osmosis into the concentrated brine solution. A prune placed in pure water swells as water moves into the prune, again by osmosis. *(Marna G. Clarke)*

that case, water will flow into the cell until it bursts. Conversely, if the nutrient solution has too high a concentration of solutes, water will flow out of the cell until
it shrivels and dies.

Determination of Molar Masses of Nonelectrolytes from Colligative Properties

Colligative properties, particularly freezing point depression, can be used to determine molar masses of a wide variety of nonelectrolytes. The approach used is illustrated in Example 10.9.

EXAMPLE 10.9   A student dissolves 1.50 g of a newly prepared compound in 75.0 g
of cyclohexane. She measures the freezing point of the solution to be 2.70°C; that of pure
cyclohexane is 6.50°C. Cyclohexane has a k_f of 20.2°C/m. Using these data, calculate the
molar mass of the compound.

Strategy   First (1), knowing the freezing point of the solution and that of the pure solvent, calculate ΔT_f. Then (2) use the equation $\Delta T_f = k_f$ (molality) to find the molality.
Finally (3), calculate the number of moles, n, and then the molar mass, $\mathcal{M}$.

Solution

(1) $\Delta T_f = T_f^\circ - T_f = 6.50°C - 2.70°C = 3.80°C$

(2) $\Delta T_f = k_f$(molality)

$$\text{molality} = \frac{\Delta T_f}{k_f} = \frac{3.80°C}{20.2°C/m} = 0.188 \; m$$

(3) $n = 75.0 \text{ g solvent} \times \dfrac{1 \text{ kg}}{10^3 \text{ g}} \times \dfrac{0.188 \text{ mol}}{\text{kg solvent}} = 0.0141 \text{ mol}$

$$\mathcal{M} = \frac{\text{mass in grams of solute}}{n} = \frac{1.50 \text{ g}}{0.0141 \text{ mol}} = \boxed{106 \text{ g/mol}}$$

In carrying out a molar mass determination by freezing point depression, we
must choose a solvent in which the solute is readily soluble. Usually, several such
solvents are available. Of these, we tend to pick one that has the largest k_f. This
makes ΔT_f large and thus reduces the percent error in the freezing point measurement. From this point of view, cyclohexane or other organic solvents are better
choices than water, because their k_f values are larger.

Molar masses can also be determined using other colligative properties.
Osmotic pressure measurements are often used, particularly for solutes of high
molar mass, where the concentration is likely to be quite low. The advantage of using osmotic pressure is that the effect is relatively large. Consider, for example, a
0.0010 M aqueous solution, for which

$$\pi \text{ at } 25°C = 0.024 \text{ atm} = 18 \text{ mm Hg}$$
$$\Delta T_f \approx 1.86 \times 10^{-3}°C$$
$$\Delta T_b \approx 5.2 \times 10^{-4}°C$$

A pressure of 18 mm Hg can be measured relatively accurately; temperature differences of the order of 0.001°C are essentially impossible to measure accurately.

EXAMPLE 10.10 A solution contains 1.0 g of hemoglobin dissolved in enough water to form 0.100 L of solution. The osmotic pressure at 20°C is found to be 2.75 mm Hg. Calculate

(a) the molarity of hemoglobin.
(b) the molar mass of hemoglobin.

Strategy The molarity of hemoglobin [Hem] is readily calculated from the relation $\pi = $ [Hem]RT. Then the molar mass can be obtained from the defining equation for molarity.

Solution

> Molar masses of polymers like polyethylene can be measured this way.

(a) $\pi = (2.75/760)$ atm $R = 0.0821$ L·atm/mol·K $T = 293$ K

$$[\text{Hem}] = \pi/RT$$

$$[\text{Hem}] = \frac{(2.75/760) \text{ atm}}{(0.0821 \text{ L·atm/mol·K})(293 \text{ K})} = 1.50 \times 10^{-4} \text{ mol/L}$$

(b) From the defining equation for molarity,

$$\text{molarity} = \frac{\text{moles solute}}{\text{liters solution}} = \frac{\text{grams solute}/\mathcal{M}}{\text{liters solution}}$$

Solving for the molar mass,

$$\mathcal{M} = \frac{\text{grams solute}}{(\text{molarity})(\text{liters solution})}$$

Recall that there is 1.0 g of hemoglobin in 0.100 L of solution. Hence

$$\mathcal{M} = \frac{1.0 \text{ g}}{(1.50 \times 10^{-4} \text{ mol/L})(0.100 \text{ L})} = 6.7 \times 10^{4} \text{ g/mol}$$

Reality Check This seems like a very high molar mass, but hemoglobin is a very large molecule.

Colligative Properties of Electrolytes

As noted earlier, colligative properties of solutions are directly proportional to the concentration of solute *particles*. On this basis, it is reasonable to suppose that, at a given concentration, an electrolyte should have a greater effect on these properties than does a nonelectrolyte. When one mole of a nonelectrolyte such as glucose dissolves in water, one mole of solute molecules is obtained. On the other hand, one mole of the electrolyte NaCl yields two moles of ions (1 mol of Na$^+$, 1 mol of Cl$^-$). With $CaCl_2$, three moles of ions are produced per mole of solute (1 mol of Ca^{2+}, 2 mol of Cl$^-$).

This reasoning is confirmed experimentally. Compare, for example, the vapor pressure lowerings for 1.0 M solutions of glucose, sodium chloride, and calcium chloride at 25°C.

	Glucose	NaCl	CaCl$_2$
ΔP	0.42 mm Hg	0.77 mm Hg	1.3 mm Hg

With many electrolytes, ΔP is so large that the solid, when exposed to moist air, picks up water *(deliquesces)*. This occurs with calcium chloride, whose saturated solution has a vapor pressure only 30% that of pure water. If dry CaCl$_2$ is exposed to air in which the relative humidity is greater than 30%, it absorbs water and forms a saturated solution. Deliquescence continues until the vapor pressure of the solution becomes equal to that of the water in the air.

The freezing points of electrolyte solutions, like their vapor pressures, are lower than those of nonelectrolytes at the same concentration. Sodium chloride and calcium chloride are used to lower the melting point of ice on highways; their aqueous solutions can have freezing points as low as −21 and −55°C, respectively. The equation for the freezing point lowering of an electrolyte is similar to that for nonelectrolytes, except for the introduction of a multiplier, i, called the *Van't Hoff factor*. For aqueous solutions of electrolytes,

$$\Delta T_f = 1.86°C/m \times \text{molality} \times i$$

Similar equations apply for other colligative properties.

$$\Delta T_b = 0.52°C/m \times \text{molality} \times i$$

$$\pi = \text{molarity} \times R \times T \times i$$

If we assume that the ions of an electrolyte behave independently, i should be equal to **the number of moles of ions per mole of electrolyte.** Thus i should be 2 for NaCl and $MgSO_4$, 3 for $CaCl_2$ and Na_2SO_4, and so on.

Potassium chloride is sold for home use because it's kinder to the environment.

Truck applying salt, NaCl, on a snow-packed road. *(Charles D. Winters)*

EXAMPLE 10.11   Estimate the freezing points of 0.20 m solutions of

(a) KNO$_3$ (b) Cr(NO$_3$)$_3$

Assume that i is the number of moles of ions formed per mole of electrolyte.

Strategy   Find i and apply the equation $\Delta T_f = 1.86°C/m \times \text{molality} \times i$.

Solution

(a) One mole of KNO_3 forms two moles of ions:

$$KNO_3(s) \longrightarrow K^+(aq) + NO_3^-(aq)$$

Hence i should be 2, and we have

$$\Delta T_f = (1.86°C)(0.20)(2) = 0.74°C \qquad T_f = -0.74°C$$

(b) For Cr(NO$_3$)$_3$, $i = 4$:

$$Cr(NO_3)_3(s) \longrightarrow Cr^{3+}(aq) + 3NO_3^-(aq)$$
$$\Delta T_f = (1.86°C)(0.20)(4) = 1.5°C \qquad T_f = -1.5°C$$

In effect, $i = 1$ for a nonelectrolyte like glucose.

The data in Table 10.3 on page 300 suggest that the situation is not as simple as this discussion implies. The observed freezing point lowerings of NaCl and $MgSO_4$ are smaller than would be predicted with $i = 2$. For example, 0.50 m solutions of NaCl and $MgSO_4$ freeze at −1.68 and −0.995°C, respectively; the predicted freezing point is −1.86°C. Only in very dilute solution does the multiplier i approach the predicted value of 2.

For a nonelectrolyte, $i = 1$.

This behavior is generally typical of electrolytes. Their colligative properties deviate considerably from ideal values, even at concentrations below 1 m. There are at least a couple of reasons for this effect.

1. Because of electrostatic attraction, an ion in solution tends to surround itself with more ions of opposite than of like charge (Figure 10.14). The existence of

TABLE 10.3 Freezing Point Lowerings of Solutions				
	ΔT_f Observed (°C)		i (Calc from ΔT_f)	
Molality	NaCl	MgSO$_4$	NaCl	MgSO$_4$
0.00500	0.0182	0.0160	1.96	1.72
0.0100	0.0360	0.0285	1.94	1.53
0.0200	0.0714	0.0534	1.92	1.44
0.0500	0.176	0.121	1.89	1.30
0.100	0.348	0.225	1.87	1.21
0.200	0.685	0.418	1.84	1.12
0.500	1.68	0.995	1.81	1.07

this *ionic atmosphere,* first proposed by Peter Debye, a Dutch physical chemist, in 1923, prevents ions from acting as completely independent solute particles. The result is to make an ion somewhat less effective than a nonelectrolyte molecule in its influence on colligative properties.

2. Oppositely charged ions may interact strongly enough to form a discrete species called an *ion pair.* This effect is essentially nonexistent with electrolytes such as NaCl, in which the ions have low charges (+1, −1). However, with MgSO$_4$ (+2, −2 ions), ion pairing plays a major role. Even at concentrations as low as 0.1 m, there are more MgSO$_4$ ion pairs than free Mg^{2+} and SO$_4^{2+}$ ions.

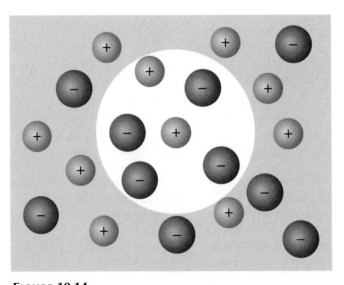

FIGURE 10.14
Ion atmosphere. An ion, on the average, is surrounded by more ions of opposite charge than of like charge.

Maple Syrup

The collection of maple sap and its conversion to syrup or sugar illustrate many of the principles covered in this chapter. Moreover, in northern New England, making maple syrup is an interesting way to spend the month of March (Figure A), which separates midwinter from "mud season."

The driving force behind the flow of maple sap is by no means obvious. The calculated osmotic pressure of sap, a 2% solution of sucrose ($\mathcal{M}$ = 342 g/mol), is

$$\pi = \frac{20/342 \text{ mol}}{1 \text{ L}} \times 0.0821 \frac{\text{L·atm}}{\text{mol·K}} \times 280 \text{ } K = 1.3 \text{ atm}$$

This is sufficient to push water to a height of about 45 ft; many maple trees are taller than that. Besides, a maple tree continues to bleed sap for several days after it has been cut down. An alternative theory suggests that sap is forced out of the tree by bubbles of $CO_2(g)$, produced by respiration. When the temperature drops at night, the carbon dioxide goes into solution, and the flow of sap ceases. This would explain the high sensitivity of sap flow to temperature; the aqueous solubility of carbon dioxide doubles when the temperature falls by 15°C.

If you want to make your own maple syrup on a small scale, perhaps 10 to 20 L per season, there are a few principles to keep in mind.

1. Make sure the trees you tap are *maples;* hemlocks would be a particularly poor choice. Identify the trees to be tapped in the fall, before the leaves fall. If possible, select sugar maples, which produce about one liter of maple syrup per tree per season. Other types of maples are less productive.

2. It takes 20 to 40 L of sap to yield one liter of maple syrup. To remove the water, you could freeze the sap, as the colonists did 200 years ago. The ice that forms is pure water; by discarding it, you increase the concentration of sugar in the remaining solution. Large-scale operators today use reverse osmosis to remove about half of the water. The remainder must be boiled off. The characteristic flavor of maple syrup is caused by compounds formed on heating, such as

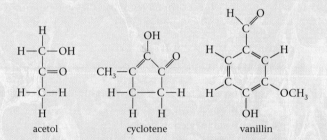

acetol   cyclotene   vanillin

It's best to boil off the water outdoors. If you do this in the kitchen, you may not be able to open the doors and windows for a couple of weeks. Wood tends to swell when it absorbs a few hundred liters of water.

3. When the concentration of sugar reaches 66%, you are at the maple syrup stage. The calculated boiling point elevation (660 g of sugar, 340 g of water)

$$\Delta T_b = 0.52°\text{C} \times \frac{660/342}{0.340} = 3.0°\text{C}$$

is somewhat less than the observed value, about 4°C. The temperature rises very rapidly around 104°C (Figure B), which makes thermometry the method of choice for detecting the end point. Shortly before that point, add a drop or two of vegetable oil to prevent foaming; perhaps this is the source of the phrase "spreading oil on troubled waters."

FIGURE A
WLM collecting sap from red maple trees.

WLM makes delicious maple syrup. —CNH

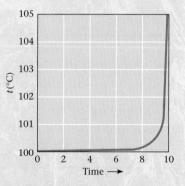

FIGURE B
Boiling maple sap. It seems to take forever to boil down maple sap until you reach 101°C. Then the temperature rises rapidly to the end point, 104°C. Further *careful* heating produces maple sugar.

CHAPTER HIGHLIGHTS

Key Concepts

1. Make dilution calculations.
   (Example 10.1; Problems 11–15)
2. Calculate a concentration (molarity, mole fraction, mass percent, molality).
   (Examples 10.2, 10.3; Problems 1–8)
3. Convert from one concentration unit to another.
   (Examples 10.4, 10.5; Problems 9, 10, 16–18)
4. Apply Henry's law to relate gas solubility to partial pressure.
   (Example 10.6; Problems 23–26, 57)
5. Apply Raoult's law to calculate the vapor pressure of a solution.
   (Example 10.7; Problems 27–30)
6. Relate freezing point, boiling point, or osmotic pressure of a solution to solute concentration.
   (Examples 10.8, 10.11; Problems 31, 32, 35–38, 49–51)
7. Use colligative properties to determine molar mass of a solute.
   (Examples 10.9, 10.10; Problems 33, 34, 39, 41, 42, 44, 47, 48)

Key Equations

Henry's law	$C_g = kP_g$
Raoult's law	$P_1 = X_1 P_1^\circ$
Osmotic pressure	$\pi = MRT$
Boiling point, freezing point	$\Delta T_b = k_b \times m \times i \qquad \Delta T_f = k_f \times m \times i$
	(i = no. of moles of particles per mole of solute)

Key Terms

- boiling point
  —elevation
  colligative property
- electrolyte
- freezing point
  —lowering
  mass percent
  molality

- molar mass
- molarity
- mole fraction
  nonelectrolyte
  osmosis
  osmotic pressure
- partial pressure
  parts per million (ppm)

  saturated solution
- solute
- solvent
  supersaturated solution
- vapor pressure
  —lowering

Summary Problem

Consider naphthalene, $C_{10}H_8$ ($\mathcal{M}$ = 128.2 g/mol), the active ingredient of some moth balls. A solution of naphthalene is prepared by mixing 35.0 g of naphthalene with 1.20 L of carbon disulfide, CS_2 (d = 1.263 g/mL). Assume the volume remains 1.20 L when the solution is prepared.

(a) What is the mass percent of naphthalene in the solution?
(b) What is the concentration of naphthalene in parts per million?
(c) What is the density of the solution?
(d) What is the molarity of the solution?
(e) What is the molality of the solution?
(f) The vapor pressure of pure CS_2 at 25°C is 358 mm Hg. Assume that the vapor pressure exerted by naphthalene at 25°C is negligible. (It is 1 mm Hg at 50°C.) What is the vapor pressure of the solution at this temperature?
(g) What is the osmotic pressure of the solution at 25°C?
(h) The normal boiling point of CS_2 is 46.56°C (k_b = 2.34 °C/m). What is the normal boiling point of the solution?
(i) Vasopressin is a hormone secreted by the pituitary gland to regulate the rate at which water is reabsorbed by the kidneys. To determine its molar mass, 1.00 g of vasopressin is dissolved in 5.00 g of naphthalene (k_f = 6.94°C/m). The freezing point of the mixture is determined to be 79.01°C; that of pure naphthalene is 80.29°C. What is the molar mass of vasopressin?

Answers

(a) 2.26%
(b) 2.26×10^4 ppm
(c) 1.29 g/mL
(d) 0.228 *M*
(e) 0.180 *m*

(f) 353 mm Hg
(g) 5.58 atm
(h) 46.98°C
(i) 1.09×10^3 g/mol

Questions & Problems

Problem numbers in blue indicate that the answer is available in Appendix 6 at the back of the book.
WEB indicates that the solution is posted at **http://www.harcourtcollege.com/chem/general/masterton4/student/**

Concentrations of Solutions

1. A solution is prepared by dissolving 5.721 g of copper(II) sulfate in 125 mL of water (*d* = 1.00 g/mL). Calculate
    (a) the mass percent of copper(II) sulfate in the solution.
    (b) the mole fraction of copper ions in the solution.

2. Ethyl alcohol, C_2H_6O, is the alcohol present in liquor. A solution is made up by adding 25.0 mL of ethyl alcohol (*d* = 0.789 g/mL) to 45.0 mL of isopropyl alcohol, C_3H_8O (*d* = 0.7855 g/mL). Assuming volumes are additive, calculate
    (a) the mass percent of ethyl alcohol in the solution.
    (b) the volume percent of isopropyl alcohol in the solution.
    (c) the mole fraction of ethyl alcohol in the solution.

3. Solutions introduced directly into the bloodstream have to be "isotonic" with blood; that is, they must have the same osmatic pressure as blood. An aqueous NaCl solution has to be 0.90% by mass to be isotonic with blood. What is the molarity of sodium ions in solution? Take the density of the solution to be 1.00 g/mL.

4. For a solution of acetic acid (CH_3COOH) to be called "vinegar," it must contain 5.00% acetic acid by mass. If a vinegar is made up only of acetic acid and water, what is the molarity of acetic acid in the vinegar? The density of vinegar is 1.006 g/mL.

5. Silver ions can be found in some of the city water piped into homes. The average concentration of silver ions in city water is 0.028 ppm.
    (a) How many milligrams of silver ions would you ingest daily if you drank eight glasses (eight oz/glass) of city water daily?
    (b) How many liters of city water are required to recover 1.00 g of silver chemically?

6. Lead is a poisonous metal that especially affects children because they retain a larger fraction of lead than adults do. Lead levels of 0.250 ppm in a child cause delayed cognitive development. How many moles of lead present in 1.00 g of a child's blood would 0.250 ppm represent?

7. Complete the following table for aqueous solutions of potassium permanganate.

	Mass of Solute	Volume of Solution	Molarity
(a)	15.00 g	325 mL	_____
(b)	_____	435 mL	1.76 *M*
(c)	2.99 g	_____	0.532 *M*

8. Complete the following table for aqueous solutions of sodium chromate.

	Mass of Solute	Volume of Solution	Molarity
(a)	_____	750.0 mL	2.757 M
(b)	1.500 g	_____	0.1155 M
(c)	12.00 g	1.50 L	_____

9. Complete the following table for aqueous solutions of urea, $CO(NH_2)_2$.

	Molality	Mass Percent Solvent	Ppm Solute	Mole Fraction Solvent
(a)	2.577	_____	_____	_____
(b)	_____	45.0	_____	_____
(c)	_____	_____	4768	_____
(d)	_____	_____	_____	0.815

WEB 10. Complete the following table for aqueous solutions of caffeine, $C_8H_{10}O_2N_4$.

	Molality	Mass Percent Solvent	Ppm Solute	Mole Fraction Solvent
(a)	_____	_____	_____	0.900
(b)	_____	_____	1269	_____
(c)	_____	85.5	_____	_____
(d)	0.2560	_____	_____	_____

11. Describe how you would prepare 465 mL of 0.3550 M potassium dichromate solution starting with
    (a) solid potassium dichromate.
    (b) 0.750 M potassium dichromate solution.
12. Describe how you would prepare 1.00 L of 0.750 M barium hydroxide solution starting with
    (a) solid barium hydroxide.
    (b) 6.00 M barium hydroxide solution.
13. A solution is prepared by diluting 125 mL of 0.1500 M iron(III) sulfate solution with water to a final volume of 0.900 L. Calculate
    (a) the molarity of iron(III) sulfate, iron(III) ions, and sulfate ions.
    (b) the number of moles of iron(III) ions present in the original solution.
14. A solution is prepared by diluting 0.450 L of 0.773 M cobalt(II) chloride solution with water to a final volume of 1.25 L.
    (a) What are the molarities of cobalt(II) chloride, cobalt(II) ion, and chloride ion in the diluted solution?
    (b) How many grams of cobalt(II) chloride were dissolved to give the original solution?
15. A bottle of commercial concentrated aqueous ammonia is labeled "29.89% NH_3 by mass; density = 0.8960 g/mL."
    (a) What is the molarity of the ammonia solution?

(b) If 250.0 mL of the commercial ammonia is diluted with water to make 3.00 L of solution, what is the molarity of the diluted solution?
16. A bottle of phosphoric acid is labeled "85.0% H_3PO_4 by mass; density = 1.689 g/cm³." Calculate the molarity, molality, and mole fraction of the phosphoric acid in solution.
17. Complete the following table for aqueous solutions of potassium hydroxide.

	Density (g/mL)	Molarity	Molality	Mass Percent of Solute
(a)	1.05	1.13	_____	_____
(b)	1.29	_____	_____	30.0
(c)	1.43	_____	14.2	_____

18. Complete the following table for aqueous solutions of ammonium sulfate.

	Density (g/mL)	Molarity	Molality	Mass Percent of Solute
(a)	1.06	0.886	_____	_____
(b)	1.15	_____	_____	26.0
(c)	1.23	_____	3.11	_____

Solubilities

19. Choose the member of each set that you would expect to be more soluble in water. Explain your answer.
    (a) chloromethane, CH_3Cl, or methanol, CH_3OH
    (b) nitrogen triiodide or potassium iodide
    (c) lithium chloride or ethyl chloride, C_2H_5Cl
    (d) ammonia or methane
WEB 20. Choose the member of each set that you would expect to be more soluble in water. Explain your answer.
    (a) naphthalene, $C_{10}H_8$, or hydrogen peroxide, H—O—O—H
    (b) silicon dioxide or sodium hydroxide
    (c) chloroform, $CHCl_3$, or hydrogen chloride
    (d) methyl alcohol, CH_3OH, or methyl ether, H_3C—O—CH_3
21. Consider the process by which ammonium chloride dissolves in water:

$$NH_4Cl(s) \longrightarrow NH_4{}^+(aq) + Cl^-(aq)$$

    (a) Using data from Table 8.3, calculate ΔH for this reaction.
    (b) Would you expect the solubility of NH_4Cl to increase if the temperature is increased?
22. Consider the process by which calcium carbonate dissolves in water:

$$CaCO_3(s) \longrightarrow Ca^{2+}(aq) + CO_3{}^{2-}(aq)$$

    (a) Using data from Table 8.3, calculate ΔH for this reaction.
    (b) Would you expect the solubility of $CaCO_3$ to increase if the temperature is increased?

23. The Henry's law constant for the solubility of hydrogen gas in water is 8.13×10^{-4} M/atm at 25°C.

(a) Express the solubility of hydrogen gas in M/mm Hg.

(b) If the partial pressure of H_2 at 25°C is 425 mm Hg, what is the concentration of dissolved H_2 in mol/L at 25°C?

(c) How many grams of hydrogen gas can be dissolved in 1.00 mL of water at 425 mm Hg and 25°C?

24. The solubility of nitrogen gas in water is 6.40×10^{-4} M/atm at 25°C. The vapor pressure of water at 25°C is 23.8 mm Hg.

(a) Express the solubility of nitrogen gas in M/mm Hg.

(b) How many grams of nitrogen gas can be dissolved in 1.00 L of water at 1.00 atm total pressure and 25°C?

(c) How many grams of nitrogen gas can be dissolved in 525 mL of water at 0.332 atm and 25°C?

25. A carbonated beverage is made by saturating water with carbon dioxide at 0°C and a pressure of 3.0 atm. The bottle is then opened at room temperature (25°C), and comes to equilibrium with air in the room containing CO_2 ($P_{CO_2} = 3.4 \times 10^{-4}$ atm).

(a) What is the concentration of carbon dioxide in the bottle before it is opened?

(b) What is the concentration of carbon dioxide in the bottle after it has been opened and come to equilibrium with the air?

The Henry's law constant for the solubility of CO_2 in water is 0.0769 M/atm at 0°C and 0.0313 M/atm at 25°C.

26. The Henry's law constant for the solubility of oxygen in water is 3.30×10^{-4} M/atm at 12°C and 2.85×10^{-4} M/atm at 22°C. Air is 21 mol% oxygen.

(a) How many grams of oxygen can be dissolved in one liter of a trout stream at 12°C (54°F) at an air pressure of 1.00 atm?

(b) How many grams of oxygen can be dissolved per liter in the same trout stream at 22°C (72°F) at the same pressure as in (a)?

(c) A nuclear power plant is responsible for the stream's increase in temperature. What percentage of dissolved oxygen is lost by this increase in the stream's temperature?

Colligative Properties

27. Calculate the vapor pressure of water over each of the following ethylene glycol $C_2H_6O_2$ solutions at 22°C (vp pure water = 19.83 mm Hg). Ethylene glycol can be assumed to be nonvolatile.

(a) $X_{\text{ethylene glycol}} = 0.288$

(b) % ethylene glycol by mass = 39.0%

(c) 2.42 m ethylene glycol

28. Repeat the calculations called for in Problem 27 at 96°C (vp pure water = 657.6 mm Hg).

29. The vapor pressure of pure CCl_4 at 65°C is 504 mm Hg. How many grams of naphthalene ($C_{10}H_8$) must be added to

25.00 g of CCl_4 so that the vapor pressure of CCl_4 over the solution is 483 mm Hg? Assume the vapor pressure of naphthalene at 65°C is negligible.

WEB 30. How would you prepare 1.00 L of an aqueous oxalic acid ($H_2C_2O_4$) solution ($d = 1.05$ g/mL) with a vapor pressure of 21.97 mm Hg at 24°C (vp pure water = 22.38 mm Hg)?

31. Calculate the osmotic pressure of the following solutions of glucose, $C_6H_{12}O_6$, at 27°C.

(a) 0.439 M glucose

(b) 12.0 g of glucose dissolved in enough water to make 725 mL of solution

(c) 22.5% glucose by mass (density of the solution = 1.08 g/mL)

32. Adrenaline is the hormone that is responsible for supplying cells with extra glucose in times of stress. Its molar mass is 182 g/mol. What is the osmotic pressure at 25°C of a 75.0-mL solution of adrenaline prepared by dissolving 25.0 mg of adrenaline in water?

33. Lysozyme, extracted from egg whites, is an enzyme that cleaves bacterial cell walls. A 20.0-mg sample of this enzyme is dissolved in enough water to make 225 mL of solution. At 23°C the solution has an osmotic pressure of 0.118 mm Hg. Estimate the molar mass of lysozyme.

34. Insulin is a hormone responsible for the regulation of glucose levels in the blood. An aqueous solution of insulin has an osmotic pressure of 2.5 mm Hg at 25°C. It is prepared by dissolving 0.100 g of insulin in enough water to make 125 mL of solution. What is the molar mass of insulin?

35. Calculate the freezing point and normal boiling point of each of the following solutions:

(a) 25.0% by mass glycerine $C_3H_8O_3$ in water.

(b) 28.0 g of propylene glycol, $C_3H_8O_2$, in 325 mL of water ($d = 1.00$ g/cm^3)

(c) 25.0 mL of ethanol, C_2H_5OH ($d = 0.780$ g/mL), in 735 g of water ($d = 1.00$ g/cm^3)

36. How many grams of the following nonelectrolytes would have to be dissolved in 100.0 g of cyclohexane (see Table 10.2) to increase the boiling point by 2.0°C? To decrease the freezing point by 1.0°C?

(a) citric acid, $C_6H_8O_7$

(b) caffeine, $C_8H_{10}N_4O_2$

37. What is the freezing point and normal boiling point of a solution made by adding 50.0 mL of isopropyl alcohol, C_3H_7OH, to 138.0 mL of water? The densities of isopropyl alcohol and water are 0.785 g/cm^3 and 1.00 g/cm^3, respectively.

38. An automobile radiator is filled with an antifreeze solution prepared by mixing four volumes of ethylene glycol, $C_2H_6O_2$ ($d = 1.12$ g/cm^3), to six volumes of water ($d = 1.00$ g/cm^3). Estimate the freezing point of the mixture. Will this mixture protect automobile engines in Connecticut if the lowest temperature expected is -20°F?

39. The Rast method uses camphor ($C_{10}H_{16}O$) as a solvent for determining the molar mass of a compound. When 2.50 g of cortisone acetate is dissolved in 50.00 g of camphor ($k_f = 40.0$°C/m), the freezing point of the mixture is determined to be

173.44°C; that of pure camphor is 178.40°C. What is the molar mass of cortisone acetate?

WEB **40.** When 13.66 g of lactic acid, $C_3H_6O_3$, is mixed with 115 g of stearic acid, the mixture freezes at 62.7°C. The freezing point of pure stearic acid is 69.4°C. What is the freezing point constant of stearic acid?

41. Lauryl alcohol is obtained from the coconut and is an ingredient in many hair shampoos. Its empirical formula is $C_{12}H_{26}O$. A solution of 5.00 g of lauryl alcohol in 100.0 g of benzene boils at 80.78°C. Using Table 10.2, find the molecular formula of lauryl alcohol.

42. A compound contains 42.9% C, 2.4% H, 16.6% N, and 38.1% O. The addition of 3.16 g of this compound to 75.0 mL of cyclohexane ($d = 0.779$ g/cm³) gives a solution with a freezing point at 0.0°C. Using Table 10.2, determine the molecular formula of the compound.

43. Pure phenol has a boiling point of 182.0°C. Its boiling point constant, k_b, is 3.56°C/m. A sample of phenol is contaminated by benzoic acid, $C_7H_6O_2$. The boiling point of the contaminated sample is 182.6°C. How pure is the sample? (Express your answer as mass percent of phenol.)

44. It has been shown that antioxidants play a role in protecting organisms against certain cancers. Analysis of β-carotene, an antioxidant and source of vitamin A, shows that it is made up of carbon and hydrogen atoms only. It is 89.5% C by mass. When 50.0 mg of β-carotene is dissolved in 5.00 g of cyclohexane ($k_f = 20.2$°C/m), the freezing point depression of the solution is 0.376°C. What are the simplest and molecular formulas of β-carotene?

45. Aqueous solutions introduced into the bloodstream by injection must have the same osmotic pressure as blood; that is, they must be "isotonic" with blood. At 25°C, the average osmotic pressure of blood is 7.7 atm. What is the molarity of an isotonic saline solution (NaCl in H_2O)? Recall that NaCl is an electrolyte; assume complete conversion to Na^+ and Cl^- ions.

46. Refer to Problem 45 and determine the concentration of a glucose solution isotonic with blood.

47. A biochemist isolates a new protein and determines its molar mass by osmotic pressure measurements. A 50.0-mL solution is prepared by dissolving 225 mg of the protein in water. The solution has an osmotic pressure of 4.18 mm Hg at 25°C. What is the molar mass of the new protein?

48. The molar mass of a type of hemoglobin was determined by osmotic pressure measurement. A student measured an osmotic pressure of 4.60 mm Hg for a solution at 20°C containing 3.27 g of hemoglobin in 0.200 L of solution. What is the molar mass of hemoglobin?

49. Estimate the freezing and boiling points (1 atm) of 0.10 m solutions of
  (a) KCl   (b) $FeCl_3$ (c) $(NH_4)_2SO_4$

50. Arrange 0.10 m solutions of the following solutes in order of decreasing freezing point and boiling point.
  (a) $Ni(NO_3)_2$ (b) CH_3OH
  (c) $Al_2(SO_4)_3$ (d) $KMnO_4$

51. What is the freezing point of maple syrup (66% sucrose)? Sucrose is $C_{12}H_{22}O_{11}$.

52. Assume that 30 L of maple sap yields one kilogram of maple syrup (66% sucrose, $C_{12}H_{22}O_{11}$). What is the molality of the sucrose solution after one fourth of the water content of the sap has been removed?

Unclassified

53. A sucrose ($C_{12}H_{22}O_{11}$) solution that is 45.0% sucrose by mass has a density of 1.203 g/mL at 25°C. Calculate its
  (a) molarity.
  (b) molality.
  (c) vapor pressure (vp H_2O at 25°C = 23.76 mm Hg).
  (d) normal boiling point.

54. An aqueous solution made up of 32.47 g of iron(III) chloride in 100.0 mL of solution has a density of 1.249 g/mL at 25°C. Calculate its
  (a) molarity.
  (b) molality.
  (c) osmotic pressure at 25°C (assume $i = 4$).
  (d) freezing point.

55. Copper(II) sulfate is often added to water in swimming pools to prevent algal growth. Starting with solid copper(II) sulfate, how would you prepare 10.0 L of a solution that is 16% by mass with a density of 1.160 g/mL? What is the molarity of the resulting solution?

56. Ethyl iodide (C_2H_5I) boils at 72.5°C and has a density of 1.933 g/mL.
  (a) A solution prepared by dissolving 0.300 mol of a nonelectrolyte in 750.0 mL of ethyl iodide boils at 73.5°C. What is the boiling point constant (k_b) for ethyl iodide?
  (b) Another solution is prepared by dissolving 12.5 g of an unknown nonelectrolyte in 100.0 mL of ethyl iodide. The resulting solution boils at 74.9°C. What is the molar mass of the nonelectrolyte?

57. The Henry's law constant for the solubility of radon in water at 30°C is 9.57×10^{-6} M/mm Hg. Radon is present with other gases in a sample taken from an aquifer at 30°C. Radon has a mole fraction of 2.7×10^{-6} in the gaseous mixture. The gaseous mixture is shaken with water at a total pressure of 28 atm. Calculate the concentration of radon in the water. Express your answers using the following concentration units.
  (a) molarity
  (b) ppm (Assume that the water sample has a density of 1.00 g/mL.)

58. Twenty-five milliliters of a solution ($d = 1.107$ g/mL) containing 15.25% by mass of sulfuric acid is added to 50.0 mL of 2.45 M barium chloride.
  (a) What is the expected precipitate?
  (b) How many grams of precipitate are obtained?
  (c) What is the chloride concentration after precipitation is complete?

Conceptual Problems

59. One mole of $CaCl_2$ is represented as □⊖ where □ represents Ca and ○ represents Cl. Complete the picture showing only the calcium and chloride ions. The water molecules need not be shown.

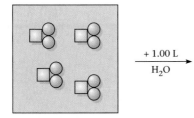

What is the molarity of Ca^{2+}? Of Cl^-?

60. One mole of Na_2S is represented as ⊟○ where □ represents Na and ○ represents S. Complete the picture showing only the sodium and sulfide ions. The water molecules need not be shown.

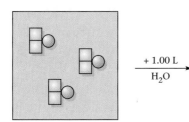

What is the molarity of Na^+? of S^{2-}?

61. A single-celled animal lives in a fresh-water lake. The cell is transferred into ocean water. Does it stay the same, shrink, or burst? Explain why.

62. Use Figure 10.4 to answer the following questions.
  (a) How many grams of sucrose can be dissolved in 80 g of water at 40°C to make a saturated solution?
  (b) A supersaturated solution is prepared by dissolving 750 g of sucrose in 125 g of boiling water (100°C). The solution is then slowly cooled to 25°C. One milligram of sucrose is added. How many grams of sucrose crystallize out?

63. Describe the following solutions as saturated, unsaturated, or supersaturated. Use Figure 10.4.
  (a) 90 g of sucrose in 50 g of water at 20°C
  (b) 58 g of sucrose in 20 g of water at 20°C
  (c) 12 g of sucrose in 35 g of water at 80°C

64. A certain gaseous solute dissolves in water, evolving 12.0 kJ of heat. Its solubility at 25°C and 4.00 atm is 0.0200 *M*. Would you expect the solubility to be greater or less than 0.0200 *M* at
  (a) 5°C and 6 atm?   **(b)** 50°C and 2 atm?
  (c) 20°C and 4 atm?   **(d)** 25°C and 1 atm?

65. The freezing point of 0.20 *m* HF is −0.38°C. Is HF primarily nonionized in this solution (HF molecules), or is it dissociated to H^+ and F^- ions?

66. The freezing point of 0.10 *M* $KHSO_3$ is −0.38°C. Which of the following equations best represents what happens when $KHSO_3$ dissolves in water?
  (a) $KHSO_3(s) \rightarrow KHSO_3(aq)$
  (b) $KHSO_3(s) \rightarrow K^+(aq) + HSO_3^-(aq)$
  (c) $KHSO_3(s) \rightarrow K^+(aq) + SO_3^{2-}(aq) + H^+(aq)$

67. Explain why
  (a) the freezing point of 0.10 *m* $CaCl_2$ is lower than the freezing point of 0.10 *m* $CaSO_4$.
  (b) the solubility of solids in water usually increases as the temperature increases.
  (c) pressure must be applied to cause reverse osmosis to occur.
  (d) 0.10 *M* $BaCl_2$ has a higher osmotic pressure than 0.10 *M* glucose.
  (e) *molarity* and *molality* are nearly the same in dilute solutions.

68. In your own words, explain
  (a) why seawater has a lower freezing point than fresh water.
  (b) why one often obtains a "grainy" product when making fudge (a supersaturated sugar solution).
  (c) why the concentrations of solutions used for intravenous feeding must be controlled carefully.
  (d) why fish in a lake (and fishermen) seek deep, shaded places during summer afternoons.
  (e) why champagne "fizzes" in a glass.

69. Criticize the following statements.
  (a) A saturated solution is always a concentrated solution.
  (b) The water solubility of a solid always decreases with a drop in temperature.
  (c) For all aqueous solutions, molarity and molality are equal.
  (d) The freezing point depression of a 0.10 *m* $CaCl_2$ solution is twice that of a 0.10 *m* KCl solution.
  (e) A 0.10 *M* sucrose solution and a 0.10 *M* NaCl solution have the same osmotic pressure.

70. Explain, in your own words
  (a) how to determine experimentally whether a pure substance is an electrolyte or a nonelectrolyte.
  (b) why a cold glass of beer goes "flat" upon warming.
  (c) why the molality of a solute is ordinarily larger than its mole fraction.
  (d) why the boiling point is raised by the presence of a solute.

71. Beaker A has 1.00 mol of chloroform, $CHCl_3$, at 27°C. Beaker B has 1.00 mol of carbon tetrachloride, CCl_4, also at 27°C. Equal masses of a nonvolatile, nonreactive solute are added to both beakers. In answering the questions on page 308, the following data may be helpful.

	CHCl$_3$ (A)	CCl$_4$ (B)
Vapor pressure at 27°C	0.276 atm	0.164 atm
Boiling point	61.26°C	76.5°C
k_b (°C/m)	3.63	5.03

Write $<$, $>$, $=$, or *more information needed* in the blanks provided.

(a) Vapor pressure of solvent over beaker B _____ vapor pressure of solvent over beaker A.

(b) Boiling point of solution in beaker A _____ boiling point of solution in beaker B.

(c) Vapor pressure of pure CHCl$_3$ _____ vapor pressure of solvent over beaker A.

(d) Vapor pressure lowering of solvent in beaker A _____ vapor pressure lowering of solvent in beaker B.

(e) Mole fraction of solute in beaker A _____ mole fraction of solute in beaker B.

Challenge Problems

72. What is the density of an aqueous solution of potassium nitrate that has a normal boiling point of 103.0°C and an osmotic pressure of 122 atm at 25°C?

73. A solution contains 158.2 g of KOH per liter; its density is 1.13 g/mL. A lab technician wants to prepare 0.250 m KOH, starting with 100.0 mL of this solution. How much water or solid KOH should be added to the 100.0-mL portion?

74. Show that the following relation is generally valid for all solutions:

$$\text{molality} = \frac{\text{molarity}}{d - \dfrac{\mathcal{M}(\text{molarity})}{1000}}$$

where d is solution density (g/cm^3) and $\mathcal{M}$ is the molar mass of the solute. Using this equation, explain why molality approaches molarity in dilute solution when water is the solvent, but not with other solvents.

75. The water-soluble nonelectrolyte X has a molar mass of 410 g/mol. A 0.100-g mixture containing this substance and sugar ($\mathcal{M}$ = 342 g/mol) is added to 1.00 g of water to give a solution whose freezing point is $-0.500°C$. Estimate the mass percent of X in the mixture.

76. A martini, weighing about 5.0 oz (142 g), contains 30.0% by mass of alcohol. About 15% of the alcohol in the martini passes directly into the bloodstream (7.0 L for an adult). Estimate the concentration of alcohol in the blood (g/cm^3) of a person who drinks two martinis before dinner. (A concentration of 0.0010 g/cm^3 or more is frequently considered indicative of intoxication in a "normal" adult.)

77. When water is added to a mixture of aluminum metal and sodium hydroxide, hydrogen gas is produced. This is the reaction used in commercial drain cleaners:

$$2Al(s) + 6H_2O(l) + 2OH^-(aq) \longrightarrow 2Al(OH)_4{}^-(aq) + 3H_2(g)$$

A sufficient amount of water is added to 49.92 g of NaOH to make 0.600 L of solution; 41.28 g of Al is added to this solution and hydrogen gas is formed.

(a) Calculate the molarity of the initial NaOH solution.

(b) How many moles of hydrogen were formed?

(c) The hydrogen was collected over water at 25°C and 758.6 mm Hg. The vapor pressure of water at this temperature is 23.8 mm Hg. What volume of hydrogen was generated?

78. It is found experimentally that the volume of a gas that dissolves in a given amount of water is independent of the pressure of the gas; that is, if 5 cm^3 of a gas dissolves in 100 g of water at 1 atm pressure, 5 cm^3 will dissolve at a pressure of 2 atm, 5 atm, 10 atm, . . . Show that this relationship follows logically from Henry's law and the ideal gas law.

Not every collision,
not every punctilious trajectory
by which billiard-ball complexes
arrive at their calculable meeting
places leads to reaction

Men (and women) are not
as different from molecules
as they think

—Roald Hoffmann
Men and Molecules

RATE OF REACTION

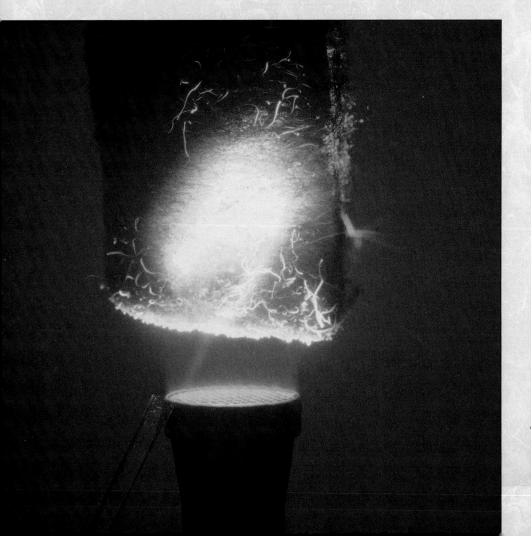

Steel wool burns more rapidly in
pure oxygen. *(Charles D. Winters)*

CHAPTER OUTLINE

11.1   MEANING OF REACTION RATE

11.2   REACTION RATE AND
CONCENTRATION

11.3   REACTANT CONCENTRATION
AND TIME

11.4   MODELS FOR REACTION RATE

11.5   REACTION RATE AND
TEMPERATURE

11.6   CATALYSIS

11.7   REACTION MECHANISMS

For a chemical reaction to be feasible, it must occur at a reasonable rate. Consequently, it is important to be able to control the rate of reaction. Most often, this means making it occur more rapidly. When you carry out a reaction in the general chemistry laboratory, you want it to take place quickly. A research chemist trying to synthesize a new drug has the same objective. Sometimes, though, it is desirable to reduce the rate of reaction. The aging process, a complex series of biological oxidations, believed to involve "free radicals" with unpaired electrons such as

$$\cdot \ddot{O}\text{—H} \quad \text{and} \quad \cdot \ddot{O}\text{—}\ddot{O}\!:^{-}$$

At least all of us who are over the age of 10.

is one we would all like to slow down.

This chapter sets forth the principles of *chemical kinetics*, the study of reaction rates. The main emphasis is on those factors that influence rate. These include—

■ the concentrations of reactants (Sections 11.2, 11.3).

■ the process by which the reaction takes place (Section 11.4).

■ the temperature (Section 11.5).

■ the presence of a catalyst (Section 11.6).

■ the reaction mechanism (Section 11.7).

11.1   MEANING OF REACTION RATE

See the *Saunders Interactive General Chemistry CD-ROM,* **Screen 15.2, Rates of Chemical Reactions.**

To discuss **reaction rate** meaningfully, it must be defined precisely. ***The rate of reaction is a positive quantity that expresses how the concentration of a reactant or product changes with time.*** To illustrate what this means, consider the reaction

$$N_2O_5(g) \longrightarrow 2NO_2(g) + \tfrac{1}{2}O_2(g)$$

As you can see from Figure 11.1, the concentration of N_2O_5 decreases with time; the concentrations of NO_2 and O_2 increase. Because these species have different coefficients in the balanced equation, their concentrations do not change at the same rate. When *one* mole of N_2O_5 decomposes, *two* moles of NO_2 and *one-half* mole of O_2 are formed. This means that

$$-\Delta[N_2O_5] = \frac{\Delta[NO_2]}{2} = \frac{\Delta[O_2]}{\tfrac{1}{2}}$$

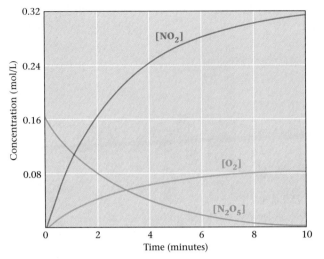

FIGURE 11.1

Changes in reactant and product concentrations with time. For the reaction $N_2O_5(g) \rightarrow 2NO_2(g) + \frac{1}{2}O_2(g)$, the concentrations of NO_2 and O_2 increase with time, whereas that of N_2O_5 decreases. The reaction rate is defined as $-\Delta[N_2O_5]/\Delta t = \Delta[NO_2]/2\,\Delta t = \Delta[O_2]/\frac{1}{2}\,\Delta t$ **OHT**

where $\Delta[\ ]$ refers to the change in concentration in moles per liter. The minus sign in front of the N_2O_5 term takes account of the fact that $[N_2O_5]$ decreases as the reaction takes place; the numbers in the denominator of the terms on the right $(2, \frac{1}{2})$ are the coefficients of these species in the balanced equation. The rate of reaction can now be defined by dividing by the change in time, Δt:

$$\text{rate} = \frac{-\Delta[N_2O_5]}{\Delta t} = \frac{\Delta[NO_2]}{2\Delta t} = \frac{\Delta[O_2]}{\frac{1}{2}\,\Delta t}$$

More generally, for the reaction

$$a\text{A} + b\text{B} \longrightarrow c\text{C} + d\text{D}$$

where A, B, C, and D represent substances in the gas phase (*g*) or in aqueous solution (*aq*), and *a, b, c, d* are their coefficients in the balanced equation,

$$\text{rate} = \frac{-\Delta[A]}{a\,\Delta t} = \frac{-\Delta[B]}{b\,\Delta t} = \frac{\Delta[C]}{c\,\Delta t} = \frac{\Delta[D]}{d\,\Delta t}$$

⌐ This is the defining equation for rate.

To illustrate the use of this expression, suppose that for the formation of ammonia:

$$N_2(g) + 3\,H_2(g) \longrightarrow 2\,NH_3(g)$$

molecular nitrogen is disappearing at the rate of 0.10 mol/L per minute, that is, $\Delta[N_2]/\Delta t = -0.10$ mol/L·min. From the coefficients of the balanced equation, we see that the concentration of H_2 must be decreasing three times as fast: $\Delta[H_2]/\Delta t = -0.30$ mol/L·min. By the same token, the concentration of NH_3 must be increasing at the rate of 2×0.10 mol/L·min: $\Delta[NH_3]/\Delta t = 0.20$ mol/L·min. It follows that

$$\text{rate} = \frac{-\Delta[N_2]}{\Delta t} = \frac{-\Delta[H_2]}{3\,\Delta t} = \frac{\Delta[NH_3]}{2\,\Delta t} = \frac{0.10\ \text{mol}}{\text{L}\cdot\text{min}}$$

By defining rate this way, it is independent of which species we focus on, N_2, H_2, or NH_3.

Notice that reaction rate has the units of concentration divided by time. We will always express concentration in moles per liter. Time, on the other hand, can

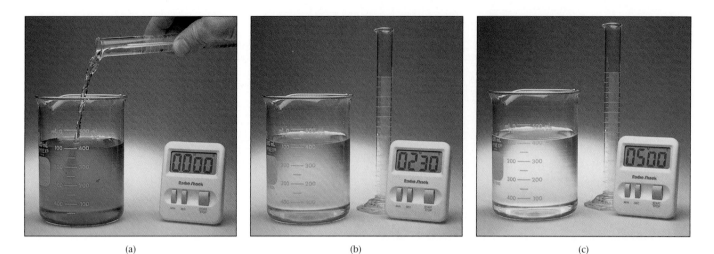

(a)   (b)   (c)

Measurement of reaction rate by observing color change. (a) Bleach is added to a solution of a blue dye to give an initial solution with known concentrations of bleach and dye. The timer is started. (b) and (c) With time the dye fades as it reacts with the bleach. The changing concentration of the dye could be measured by following the color change with a spectrophotometer. *(Charles D. Winters)*

be expressed in seconds, minutes, hours, ... A rate of 0.10 mol/L·min corresponds to

$$0.10 \frac{\text{mol}}{\text{L} \cdot \text{min}} \times \frac{1 \text{ min}}{60 \text{ s}} = 1.7 \times 10^{-3} \frac{\text{mol}}{\text{L} \cdot \text{s}}$$

or

$$0.10 \frac{\text{mol}}{\text{L} \cdot \text{min}} \times \frac{60 \text{ min}}{1 \text{ h}} = 6.0 \frac{\text{mol}}{\text{L} \cdot \text{h}}$$

Measurement of Rate

For the reaction

$$N_2O_5(g) \longrightarrow 2NO_2(g) + \tfrac{1}{2}O_2(g)$$

the rate could be determined by measuring—

- the absorption of visible light by the NO_2 formed; this species has a reddish-brown color, whereas N_2O_5 and O_2 are colorless.
- the change in pressure that results from the increase in the number of moles of gas (1 mol reactant → $2\tfrac{1}{2}$ mol product).

The graphs shown in Figure 11.1 were plotted from data obtained by measurements of this type.

To find the rate of decomposition of N_2O_5, it is convenient to use Figure 11.2, which is a magnified version of a portion of Figure 11.1. If a tangent is drawn to the curve of concentration versus time, its slope at that point must equal $\Delta[N_2O_5]/\Delta t$. But because the reaction rate is $-\Delta[N_2O_5]/\Delta t$, it follows that

$$\text{rate} = -\text{slope of tangent}$$

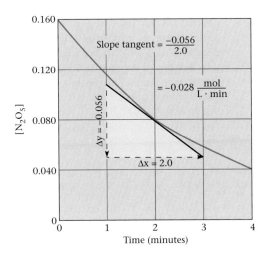

FIGURE 11.2

Determination of the instantaneous rate at a particular concentration. To determine the rate of reaction, plot concentration versus time and take the tangent to the curve at the desired point. For the reaction $N_2O_5(g) \rightarrow 2NO_2(g) + \frac{1}{2}O_2(g)$, it appears that the reaction rate at $[N_2O_5] = 0.080\ M$ is 0.028 mol/L·min.   **OHT**

From Figure 11.2 it appears that the slope of the tangent at $t = 2$ min is -0.028 mol/L·min. Hence

$$\text{rate at 2 min} = -(-0.028\ \text{mol/L·min}) = 0.028\ \text{mol/L·min}$$

11.2   REACTION RATE AND CONCENTRATION

Ordinarily, reaction rate is directly related to reactant concentration. The higher the concentration of starting materials, the more rapidly a reaction takes place. Pure hydrogen peroxide, in which the concentration of H_2O_2 molecules is about 40 mol/L, is an extremely dangerous substance. In the presence of trace impurities, it decomposes explosively

$$H_2O_2(l) \longrightarrow H_2O(g) + \frac{1}{2}O_2(g)$$

at a rate too rapid to measure. The hydrogen peroxide you buy in a drugstore is a dilute aqueous solution in which $[H_2O_2] \approx 1\ M$. At this relatively low concentration, decomposition is so slow that the solution is stable for several months.

The dependence of reaction rate on concentration is readily explained. Ordinarily, *reactions occur as the result of collisions between reactant molecules.* The higher the concentration of molecules, the greater the number of collisions in unit time and hence the faster the reaction. As reactants are consumed, their concentrations drop, collisions occur less frequently, and reaction rate decreases. This explains the common observation that reaction rate drops off with time, eventually going to zero when the limiting reactant is consumed.

Rate Expression and Rate Constant

The dependence of reaction rate on concentration is readily determined for the decomposition of N_2O_5. Figure 11.3 (page 314) shows what happens when reaction rate is plotted versus $[N_2O_5]$. As you would expect, rate increases as concentration increases, going from zero when $[N_2O_5] = 0$ to about 0.06 mol/L·min when $[N_2O_5] = 0.16\ M$. Moreover, as you can see from the figure, the plot of rate versus

See Screen 15.4, Control of Reaction Rates (Concentration Dependence).

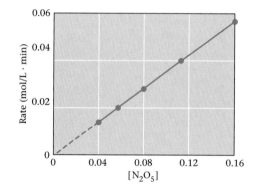

FIGURE 11.3

A plot of rate vs. concentration for the decomposition of N_2O_5 is a straight line. The line, if extrapolated, passes through the origin. This means that rate is directly proportional to concentration; that is, rate = $k[N_2O_5]$.

concentration is a straight line through the origin, which means that rate must be directly proportional to the concentration.

$$\text{rate} = k[N_2O_5]$$

This equation is referred to as the **rate expression** for the decomposition of N_2O_5. It tells how the rate of the reaction

$$N_2O_5(g) \longrightarrow 2NO_2(g) + \tfrac{1}{2}O_2(g)$$

depends on the concentration of reactant. The proportionality constant k is called a **rate constant.** It is independent of the other quantities in the equation.

As we will see shortly, the rate expression can take various forms, depending on the nature of the reaction. It can be quite simple, as in the N_2O_5 decomposition, or exceedingly complex.

| Rate depends on concentration, but rate constant does not.

Order of Reaction Involving a Single Reactant

Rate expressions have been determined by experiment for a large number of reactions. For the process

$$A \longrightarrow \text{products}$$

the rate expression has the general form

$$\text{rate} = k[A]^m$$

| Remember, [A] means the concentration of A in moles per liter.

⊙ **See Screen 15.6, Concentration-Time Relationships.**

The power to which the concentration of reactant A is raised in the rate expression is called the **order of the reaction,** m. If m is 0, the reaction is said to be "zero-order." If $m = 1$, the reaction is "first-order"; if $m = 2$, it is "second-order" and so on. Ordinarily the reaction order is integral (0, 1, 2, . . .), but fractional orders such as $\tfrac{3}{2}$ are possible.

The order of a reaction must be determined experimentally; *it cannot be deduced from the coefficients in the balanced equation.* This must be true because there is only one reaction order, but there are many different ways in which the equation for the reaction can be balanced. For example, although we wrote

$$N_2O_5(g) \longrightarrow 2NO_2(g) + \tfrac{1}{2}O_2(g)$$

to describe the decomposition of N_2O_5, it could have been written

$$2N_2O_5(g) \longrightarrow 4NO_2(g) + O_2(g)$$

The reaction is still first-order no matter how the equation is written.

One way to find the order of a reaction is to measure the **initial rate** (i.e., the rate at $t = 0$) as a function of the concentration of reactant. Suppose, for example, that we make up two different reaction mixtures differing only in the concentration of reactant A. We now measure the rates at the beginning of reaction, before the concentration of A has decreased appreciably. This gives two different initial rates ($rate_1$, $rate_2$) corresponding to two different starting concentrations of A, $[A]_1$ and $[A]_2$. From the rate expression,

$$rate_2 = k[A]_2{}^m \qquad rate_1 = k[A]_1{}^m$$

Dividing the second rate by the first,

$$\frac{rate_2}{rate_1} = \frac{[A]_2{}^m}{[A]_1{}^m} = \left(\frac{[A]_2}{[A]_1}\right)^m$$

Because all the quantities in this equation are known except m, the reaction order can be calculated (Example 11.1).

See Screen 15.5, Determination of Rate Equation (Method of Initial Rates).

Another approach is to measure concentration as a function of time (Section 11.3).

EXAMPLE 11.1   The initial rate of decomposition of acetaldehyde, CH_3CHO, at 600°C

$$CH_3CHO(g) \longrightarrow CH_4(g) + CO(g)$$

was measured at a series of concentrations with the following results:

$[CH_3CHO]$	0.10 M	0.20 M	0.30 M	0.40 M
Rate (mol/L·s)	0.085	0.34	0.76	1.4

Using these data, determine the reaction order; that is, determine the value of m in the equation: $rate = k[CH_3CHO]^m$.

Strategy   Choose the first two concentrations, 0.10 M and 0.20 M. Calculate the ratio of the rates, the ratio of the concentrations, and finally the order of reaction, using the general relation derived above.

Solution

$$\frac{rate_2}{rate_1} = \frac{0.34 \text{ mol/L·s}}{0.085 \text{ mol/L·s}} = 4.0$$

$$\frac{[CH_3CHO]_2}{[CH_3CHO]_1} = \frac{0.20 \text{ M}}{0.10 \text{ M}} = 2.0$$

Hence the general relation becomes $4.0 = (2.0)^m$. Clearly, $m = 2$; the reaction is second-order.

Reality Check   You would get the same result ($m = 2$) if you used any two points. Try it!

Once the order of the reaction is known, the rate constant is readily calculated. Consider, for example, the decomposition of acetaldehyde, where we have shown that the rate expression is

$$rate = k[CH_3CHO]^2$$

There are two variables in this equation, rate and concentration, and two constants, k and reaction order.

The data in Example 11.1 show that the rate at 600°C is 0.085 mol/L·s when the concentration is 0.10 mol/L. It follows that

$$k = \frac{\text{rate}}{[CH_3CHO]^2} = \frac{0.085 \text{ mol/L·s}}{(0.10 \text{ mol/L})^2} = 8.5 \text{ L/mol·s}$$

The same value of k would be obtained, within experimental error, using any other data pair.

Having established the value of k and the reaction order, the rate is readily calculated at any concentration. Again, using the decomposition of acetaldehyde as an example, we have established that

$$\text{rate} = 8.5 \frac{L}{\text{mol·s}} [CH_3CHO]^2$$

If the concentration of acetaldehyde were 0.50 M,

$$\text{rate} = 8.5 \frac{L}{\text{mol·s}} (0.50 \text{ mol/L})^2 = 2.1 \text{ mol/L·s}$$

Order of Reaction with More Than One Reactant

Many reactions (indeed, most reactions) involve more than one reactant. For a reaction between two species A and B,

$$aA + bB \longrightarrow \text{products}$$

the general form of the rate expression is

$$\text{rate} = k[A]^m \times [B]^n$$

In this equation m is referred to as "the order of the reaction with respect to A." Similarly, n is "the order of the reaction with respect to B." The **overall order** of the reaction is the sum of the exponents, $m + n$. If $m = 1$, $n = 2$, then the reaction is first-order in A, second-order in B, and third-order overall.

When more than one reactant is involved, the order can be determined by holding the initial concentration of one reactant constant while varying that of the other reactant. From rates measured under these conditions, it is possible to deduce the order of the reaction with respect to the reactant whose initial concentration is varied.

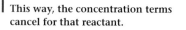

This way, the concentration terms cancel for that reactant.

Reaction rate and concentration. On the left, solid zinc is reacting with a dilute solution of sulfuric acid ($Zn(s) + 2H^+ \rightarrow Zn^{2+}(aq) + H_2(g)$). On the right, the same reaction in more concentrated sulfuric acid has a much faster rate. *(Charles Steele)*

To see how this is done, consider the reaction between A and B referred to above. Suppose we run two different experiments in which the initial concentrations of A differ ($[A]_1$, $[A]_2$) but that of B is held constant at $[B]$. Then

$$\text{rate}_1 = k[A]_1{}^m \times [B]^n \qquad \text{rate}_2 = k[A]_2{}^m \times [B]^n$$

Dividing the second equation by the first:

$$\frac{\text{rate}_2}{\text{rate}_1} = \frac{k[A]_2{}^m \times [B]^n}{k[A]_1{}^m \times [B]^n} = \frac{[A]_2{}^m}{[A]_1{}^m} = \left(\frac{[A]_2}{[A]_1}\right)^m$$

Knowing the two rates and the ratio of the two concentrations, we can readily find the value of m.

EXAMPLE 11.2   Consider the reaction at 55°C:

$$(CH_3)_3CBr(aq) + OH^-(aq) \longrightarrow (CH_3)_3COH(aq) + Br^-(aq)$$

A series of experiments is carried out with the following results:

	Expt. 1	Expt. 2	Expt. 3	Expt. 4	Expt. 5
$[(CH_3)_3CBr]$	0.50	1.0	1.5	1.0	1.0
$[OH^-]$	0.050	0.050	0.050	0.10	0.20
Rate (mol/L·s)	0.0050	0.010	0.015	0.010	0.010

These are initial rates.

Find the order of the reaction with respect to both $(CH_3)_3CBr$ and OH^-. Write the rate expression for the reaction.

Strategy   To find the order of the reaction with respect to $(CH_3)_3CBr$, choose two experiments, perhaps 1 and 3, where $[OH^-]$ is constant. A similar approach can be used to find n; compare experiments 2 and 5, where $[(CH_3)_3CBr]$ is constant. To write the rate expression, use the calculated reaction orders.

Solution

(1) $\dfrac{\text{rate}_3}{\text{rate}_1} = \left(\dfrac{[(CH_3)_3CBr]_3}{[(CH_3)_3CBr]_1}\right)^m \qquad \dfrac{0.015}{0.0050} = \left(\dfrac{1.5}{0.50}\right)^m \qquad 3 = 3^m$

Clearly, $m = 1$.

(2) $\dfrac{\text{rate}_5}{\text{rate}_2} = \left(\dfrac{[OH^-]_5}{[OH^-]_2}\right)^n \qquad \dfrac{0.010}{0.010} = \left(\dfrac{0.20}{0.050}\right)^n \qquad 1 = 4^n$

In this case, $n = 0$; the rate is independent of the concentration of OH^-. The rate expression is

$$\text{rate} = k[(CH_3)_3CBr]$$

Any number raised to the power zero equals 1.

11.3   REACTANT CONCENTRATION AND TIME

The rate expression

$$\text{rate} = k[N_2O_5]$$

shows how the rate of decomposition of N_2O_5 changes with concentration. From a practical standpoint, however, it is more important to know the relation between concentration and *time* rather than between concentration and rate. Suppose, for example, you are studying the decomposition of dinitrogen pentaoxide. Most

See Screen 15.6, Concentration-Time Relationships.

There are no speedometers for reactions, but every lab has a clock.

FIGURE 11.4

A first-order reaction plot. The rate constant for a first-order reaction can be determined from the slope of a plot of ln[A] versus time. The reaction illustrated, at 67°C, is $N_2O_5(g) \rightarrow 2NO_2(g) + \frac{1}{2} O_2(g)$. Using two points on the line, one can find the slope of the line, which is the rate constant. In this case, k is found to be about 0.35/min.

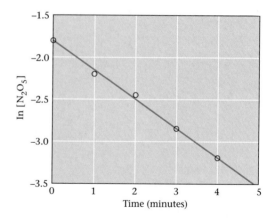

likely, you would want to know how much N_2O_5 is left after 5 min, 1 h, or several days. An equation relating *rate* to concentration does not answer that purpose.

Using calculus, it is possible to develop **integrated rate equations** relating reactant concentration to time. We now examine several such equations, starting with first-order reactions.

First-Order Reactions

For the decomposition of N_2O_5 and other **first-order reactions** of the type

$$A \longrightarrow \text{products} \qquad \text{rate} = k[A]$$

it can be shown by using calculus that the relationship between concentration and time is*

$$\boxed{\ln \frac{[A]_o}{[A]} = kt}$$

where $[A]_o$ is the original concentration of reactant, $[A]$ is its concentration at time t, k is the first-order rate constant, and the abbreviation "ln" refers to the natural logarithm.

Because $\ln a/b = \ln a - \ln b$, the first-order equation can be written in the form

$$\ln [A]_o - \ln [A] = kt$$

Solving for ln [A],

$$\ln [A] = \ln [A]_o - kt$$

Comparing this equation to the general equation of a straight line,

$$y = b + mx \qquad (b = y\text{-intercept}, m = \text{slope})$$

it is clear that a plot of ln [A] versus t should be a straight line with a y-intercept of $\ln [A]_o$ and a slope of $-k$. This is indeed the case, as you can see from Figure 11.4 where we have plotted $\ln [N_2O_5]$ versus t for the decomposition of N_2O_5. Drawing

*Throughout Section 11.3, balanced chemical equations are written in such a way that the coefficient of the reactant is 1. In general, if the coefficient of the reactant is a, where a may be 2 or 3 or . . . , then k in each integrated rate equation must be replaced by the product ak. (See Problem 81.)

the best straight line through the points and taking the slope based on two points on the y axis ($y = -1.8$, $x = 0$) and the x axis ($y = -3.5$, $x = 4.9$),

$$\text{slope} = \frac{-1.8 - (-3.5)}{0 - (4.9)} = \frac{1.7}{-4.9} = -0.35$$

It follows that the rate constant is 0.35/min; the integrated first-order equation for the decomposition of N_2O_5 is

$$\ln \frac{[N_2O_5]_o}{[N_2O_5]} = \frac{0.35}{\text{min}} t \qquad \text{(at 67°C)}$$

> The slope is negative but k is always positive.

EXAMPLE 11.3

For the decomposition of N_2O_5 at 67°C, where $k = 0.35$/min, calculate

(a) the concentration after 4.0 min, starting at 0.160 M.
(b) the time required for the concentration to drop from 0.160 to 0.100 M.
(c) the time required for half a sample of N_2O_5 to decompose.

Strategy   In each case, the equation

$$\ln [N_2O_5]_o/[N_2O_5] = (0.35/\text{min})t$$

is used. In (a) and (b), two of the three variables, $[N_2O_5]_o$, $[N_2O_5]$, and t, are known; the other is readily calculated. In (c), you should be able to find the ratio $[N_2O_5]_o/[N_2O_5]$; knowing that ratio, the time can be calculated.

Solution

(a) Substituting in the integrated first-order equation,

$$\ln \frac{0.160\ M}{[N_2O_5]} = \frac{0.35}{\text{min}} (4.0\ \text{min}) = 1.4$$

Taking inverse logarithms,

$$\frac{0.160\ M}{[N_2O_5]} = e^{1.4} = 4.0 \qquad [N_2O_5] = \frac{0.160\ M}{4.0} = 0.040\ M$$

> If $\ln x = y$, then $x = e^y$.

(b) Solving the concentration-time relation for t,

$$t = \frac{1}{k} \ln \frac{[A]_o}{[A]} = \frac{1}{0.35/\text{min}} \ln \frac{0.160\ M}{0.100\ M} = \frac{\ln 1.60}{0.35/\text{min}} = 0.47\ \text{min}/0.35 = 1.3\ \text{min}$$

(c) When half of the sample has decomposed,

$$[N_2O_5] = [N_2O_5]_o/2 \qquad [N_2O_5]_o = 2[N_2O_5] \qquad [N_2O_5]_o/[N_2O_5] = 2$$

Using the equation in (b),

$$t = \frac{1}{k} \ln 2 = \frac{0.693}{k}$$

With $k = 0.35$/min, we have

$$t = \frac{0.693}{0.35} = 2.0\ \text{min}$$

The analysis of Example 11.3c reveals an important feature of a first-order reaction: *The time required for one half of a reactant to decompose via a first-*

TABLE 11.1	**Decomposition of N_2O_5 at 67°C ($t_{1/2} = 2.0$ min)**				
t(min)	0.0	2.0	4.0	6.0	8.0
$[N_2O_5]$	0.160	0.080	0.040	0.020	0.010
Fraction of N_2O_5 decomposed	0	$\frac{1}{2}$	$\frac{3}{4}$	$\frac{7}{8}$	$\frac{15}{16}$
Fraction of N_2O_5 left	1	$\frac{1}{2}$	$\frac{1}{4}$	$\frac{1}{8}$	$\frac{1}{16}$
Number of half-lives	0	1	2	3	4

order reaction has a fixed value, independent of concentration. This quantity, called the **half-life,** is given by the expression

See Screen 15.8, Half-Life: First-Order Reactions.

$$t_{1/2} = \frac{\ln 2}{k} = \frac{0.693}{k} \qquad \text{first-order reaction}$$

where k is the rate constant. For the decomposition of N_2O_5, where $k = 0.35$/min, $t_{1/2} = 2.0$ min. Thus every 2 minutes, one half of a sample of N_2O_5 decomposes (Table 11.1).

Notice that for a first-order reaction the rate constant has the units of reciprocal time, for example, min^{-1}. This suggests a simple physical interpretation of k (at least where k is small); it is the fraction of reactant decomposing in unit time. For a first-order reaction in which

$$k = 0.010/\text{min} = 0.010 \ \text{min}^{-1}$$

we can say that 0.010 (i.e., 1.0%) of the sample decomposes per minute.

Perhaps the most important first-order reaction is that of radioactive decay, in which an unstable nucleus decomposes (Chapter 2). Letting X be the amount of a radioactive isotope present at time t:

$$\text{rate} = kX$$

and

$$\ln \frac{X_0}{X} = kt$$

where X_0 is the initial amount. The amount of a radioactive isotope can be expressed in terms of moles, grams, or number of atoms.

EXAMPLE 11.4 Plutonium-240, produced in nuclear reactors, has a half-life of 6.58×10^3 years. Calculate

(a) the first-order rate constant for the decay of plutonium-240.
(b) the fraction of a sample that will remain after one hundred years.

Strategy This is a standard type of first-order rate calculation. Use the equation $k = 0.693/t_{1/2}$ to calculate k and then the equation $\ln X_0/X = kt$ to find the fraction left, X/X_0.

Solution

(a) $k = 0.693/(6.58 \times 10^3 \ \text{yr}) = \boxed{1.05 \times 10^{-4} \ \text{yr}}$

(b) $\ln \dfrac{X_0}{X} = 1.05 \times 10^{-4} \ \text{yr}^{-1} \times 1.00 \times 10^2 \ \text{yr} = 1.05 \times 10^{-2}$

Taking the inverse ln, $X_0/X = 1.01$.
Hence $X/X_0 = 1/1.01 = \boxed{0.99}$. That is, 99% remains.

Reality Check This problem illustrates the difficulty inherent in storing or disposing of a long-lived radioactive species. Such isotopes cannot be released into the environment in the hope that they will decompose rapidly. In the case of plutonium-240, its radiation level would be virtually unchanged a century from now.

Half-lives can be interpreted in terms of the level of radiation of the corresponding isotopes. Uranium has a very long half-life (4.5×10^9 yr), so it gives off radiation very slowly. At the opposite extreme is fermium-258, which decays with a half-life of 3.8×10^{-4} s. You would expect the rate of decay to be quite high. Within a second virtually all the radiation from fermium-258 is gone. Species such as this produce very high radiation during their brief existence.

Zero- and Second-Order Reactions

For a **zero-order reaction,**

$$A \longrightarrow \text{products} \qquad \text{rate} = k[A]^0 = k$$

because any quantity raised to the zero power, including [A], is equal to 1. In other words, the rate of a zero-order reaction is constant, independent of concentration. As you might expect from our earlier discussion, zero-order reactions are relatively rare. Most of them take place at solid surfaces, where the rate is independent of concentration in the gas phase. A typical example is the thermal decomposition of hydrogen iodide on gold:

$$HI(g) \xrightarrow{\text{Au}} \tfrac{1}{2}H_2(g) + \tfrac{1}{2}I_2(g)$$

When the gold surface is completely covered with HI molecules, increasing the concentration of HI(g) has no effect on reaction rate.

It can readily be shown (Problem 70) that the concentration-time relation for a zero-order reaction is

$$\boxed{[A] = [A]_o - kt}$$

Comparing this equation with that for a straight line,

$$y = b + mx \qquad (b = y\text{-intercept}, \ m = \text{slope})$$

it should be clear that a plot of [A] versus t should be a straight line with a slope of $-k$. Putting it another way, if a plot of concentration versus time is linear, the reaction must be zero-order; the rate constant k is numerically equal to the slope of that line but has the opposite sign.

For a **second-order reaction** involving a single reactant, such as acetaldehyde (recall Example 11.1),

$$A \longrightarrow \text{products} \qquad \text{rate} = k[A]^2$$

it is again necessary to resort to calculus* to obtain the concentration-time relationship

$$\boxed{\dfrac{1}{[A]} - \dfrac{1}{[A]_o} = kt}$$

> In a zero-order reaction, all the reactant is consumed in a finite time.

*If you are taking a course in calculus, you may be surprised to learn how useful it can be in the real world (e.g., chemistry). The general rate expressions for zero-, first-, and second-order reactions are

$$-d[A]/dt = k \qquad -d[A]/dt = k[A] \qquad -d[A]dt = k[A]^2$$

Integrating these questions from 0 to t and from $[A]_o$ to [A], you should be able to derive the equations for zero-, first-, and second-order reactions.

TABLE 11.2 Characteristics of Zero-, First-, and Second-Order Reactions of the Form A (g) → products; [A], [A]$_o$ = conc. A at t and t = 0, respectively OHT

Order	Rate Expression	Conc.-Time Relation	Half-Life	Linear Plot
0	rate = k	$[A]_o - [A] = kt$	$[A]_o/2k$	[A] vs. t
1	rate = k[A]	$\ln \dfrac{[A]_o}{[A]} = kt$	$0.693/k$	ln [A] vs. t
2	rate = k[A]^2	$\dfrac{1}{[A]} - \dfrac{1}{[A]_o} = kt$	$1/k[A]_o$	$\dfrac{1}{[A]}$ vs. t

where the symbols [A], [A]$_o$, t, and k have their usual meanings. For a second-order reaction, a plot of 1/[A] versus t should be linear.

The characteristics of zero-, first-, and second-order reactions are summarized in Table 11.2. To determine reaction order, the properties in either of the last two columns at the right of the table can be used (Example 11.5).

See Screen 15.7, Determination of Rate Equation (Graphical Methods).

EXAMPLE 11.5 The following data were obtained for the gas-phase decomposition of hydrogen iodide:

Time (h)	0	2	4	6
[HI]	1.00	0.50	0.33	0.25

Is this reaction zero-, first-, or second-order in HI?

Strategy Note from the last column of Table 11.2 that—

- if the reaction is zero-order, a plot of [A] versus t should be linear.
- if the reaction is first-order, a plot of ln [A] versus t should be linear.
- if the reaction is second-order, a plot of 1/[A] versus t should be linear.

Using these data, prepare these plots and determine which one is a straight line.

Solution It is useful to prepare a table listing [HI], ln [HI], and 1/[HI] as a function of time.

t(h)	[HI]	ln [HI]	1/[HI]
0	1.00	0	1.0
2	0.50	−0.69	2.0
4	0.33	−1.10	3.0
6	0.25	−1.39	4.0

If it is not obvious from this table that the only linear plot is that of 1/[HI] versus time, that point should be clear from Figure 11.5 (page 323). This is a second-order reaction.

Reality Check Notice that the concentration drops from 1.00 M to 0.50 M in 2 h. If the reaction were zero-order, it would be all over after four hours. If it were first-order, the concentration would be 0.25 M after four hours.

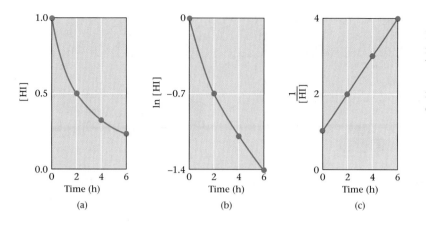

FIGURE 11.5
Determination of reaction order. One way to determine reaction order is to search for a linear relation between some function of concentration and time (Table 11.2). Because a plot of 1/[HI] versus t (plot (c)) is linear, the decomposition of HI must be second-order. **OHT**

11.4 MODELS FOR REACTION RATE

So far in this chapter we have approached reaction rate from an experimental point of view, describing what happens in the laboratory or the world around us. Now we change emphasis and try to explain why certain reactions occur rapidly while others take place slowly. To do this, we look at a couple of models that chemists have developed to predict rate constants for reactions.

Collision Model; Activation Energy

⊙ See Screen 15.9, Microscopic View of Reactions (1).

Consider the reaction

$$CO(g) + NO_2(g) \longrightarrow CO_2(g) + NO(g)$$

Above 600 K, this reaction takes place as a direct result of collisions between CO and NO_2 molecules. When the concentration of CO doubles (Figure 11.6), the number of these collisions in a given time increases by a factor of 2; doubling the concentration of NO_2 has the same effect. Assuming that reaction rate is directly proportional to the collision rate, the following relation should hold:

$$\text{reaction rate} = k[CO] \times [NO_2]$$

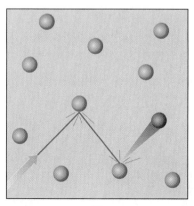

(a)

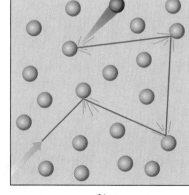

(b)

FIGURE 11.6
Concentration and molecular collisions. A red molecule must collide with a blue molecule for a reaction to take place. (a) A single red molecule moves among 10 blue molecules and collides with two of them per second. (b) With double the number of blue molecules, the number of collisions doubles to four per second. **OHT**

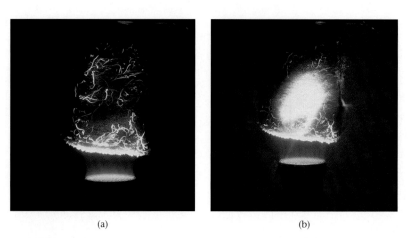

(a)   (b)

Combustion and oxygen concentration. (a) Steel wool in the flame reacts with atmospheric oxygen. (b) In a stream of pure oxygen, the steel wool burns much faster. *(Charles D. Winters)*

Experimentally, this prediction is confirmed; the reaction is first-order in both carbon monoxide and nitrogen dioxide.

There is a restriction on this simple model for the CO-NO$_2$ reaction. According to the kinetic theory of gases, for a reaction mixture at 700 K and concentrations of 0.10 M, every CO molecule should collide with about 10^9 NO$_2$ molecules in one second. If every collision were effective, the reaction should be over in a fraction of a second. In reality, this does not happen; under these conditions, the half-life is about 10 s. This implies that not every CO-NO$_2$ collision leads to reaction.

There are a couple of reasons why collision between reactant molecules does not always lead to reaction. For one thing, the molecules have to be properly oriented with respect to one another when they collide. Suppose, for example, that the carbon atom of a CO molecule strikes the nitrogen atom of an NO$_2$ molecule (Figure 11.7). This is very unlikely to result in the transfer of an oxygen atom from NO$_2$ to CO, which is required for reaction to occur.

A more important factor that reduces the number of effective collisions has to do with the kinetic energy of reactant molecules. These molecules are held to-

⌐ **There would be an explosion.**

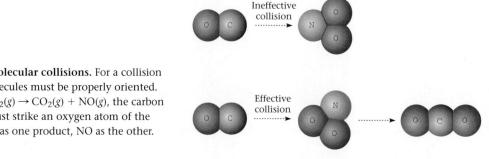

FIGURE 11.7

Effective and ineffective molecular collisions. For a collision to result in reaction, the molecules must be properly oriented. For the reaction CO(g) + NO$_2$(g) → CO$_2$(g) + NO(g), the carbon atom of the CO molecule must strike an oxygen atom of the NO$_2$ molecule, forming CO$_2$ as one product, NO as the other.

OHT

gether by strong chemical bonds. Only if the colliding molecules are moving very rapidly will the kinetic energy be large enough to supply the energy required to break these bonds. Molecules with small kinetic energies bounce off one another without reacting. As a result, only a small fraction of collisions are effective.

For every reaction, there is a certain minimum energy that molecules must possess for collision to be effective. This is referred to as the **activation energy.** It has the symbol E_a and is expressed in kilojoules. For the reaction between one mole of CO and one mole of NO_2, E_a is 134 kJ. The colliding molecules (CO and NO_2) must have a total kinetic energy of at least 134 kJ/mol if they are to react. The activation energy for a reaction is a positive quantity ($E_a > 0$) whose value depends on the nature of the reaction.

The collision model of reaction rates just developed can be made quantitative. We can say that the rate constant for a reaction, k, is a product of three factors:

$$k = p \times Z \times f$$

where—

- **p**, called a *steric factor,* takes into account the fact that only certain orientations of colliding molecules are likely to lead to reaction (recall Figure 11.7). Unfortunately, the collision model cannot predict the value of p; about all we can say is that it is likely to be less than 1, sometimes much less.
- **Z**, the *collision frequency,* which gives the number of molecular collisions occurring in unit time at unit concentrations of reactants. This quantity can be calculated quite accurately from the kinetic theory of gases, but we will not describe that calculation.
- **f**, the *fraction of collisions in which the energy of the colliding molecules is equal to or greater than E_a*. It can be shown that

$$f = e^{-E_a/RT}$$

where e is the base of natural logarithms, R is the gas constant, and T is the absolute temperature in K.

Substituting this expression for f into the equation written above for k, we obtain the basic equation for the collision model:

$$k = p \times Z \times e^{-E_a/RT}$$

This equation tells us, among other things, that *the larger the value of E_a, the smaller the rate constant.* Thus

$$\text{if } E_a = 0,\ e^{-E_a/RT} = e^0 = 1$$

$$E_a = RT,\ e^{-E_a/RT} = e^{-1} = 0.37$$

$$E_a = 2RT,\ e^{-E_a/RT} = e^{-2} = 0.14$$

and so on. This makes sense; the larger the activation energy, the smaller the fraction of molecules having enough energy to react on collision, so the slower the rate of reaction.

Table 11.3, page 326, compares observed rate constants for several reactions with those predicted by collision theory, arbitrarily taking $p = 1$. As you might expect, the calculated k's are too high, suggesting that the steric factor is indeed less than 1.

High-energy molecules get the job done.

		Collision	Transition-State
TABLE 11.3 Observed and Calculated Rate Constants for Second-Order Gas Phase Reactions			
Reaction	k Observed	Model	Model
$NO + O_3 \rightarrow NO_2 + O_2$	6.3×10^7	4.0×10^9	3.2×10^7
$NO + Cl_2 \rightarrow NOCl + Cl$	5.2	130	1.6
$NO_2 + CO \rightarrow NO + CO_2$	1.2×10^{-4}	6.4×10^{-4}	1.0×10^{-4}
$2NO_2 \rightarrow 2NO + O_2$	5.0×10^{-3}	1.0×10^{-1}	1.2×10^{-2}

Transition-State Model; Activation Energy Diagrams

See Screen 15.10, Microscopic View of Reactions (2).

Figure 11.8 is an energy diagram for the CO-NO_2 reaction. Reactants CO and NO_2 are shown at the left. Products CO_2 and NO are at the right; they have an energy 226 kJ less than that of the reactants. The enthalpy change, ΔH, for the reaction is −226 kJ. In the center of the figure is an intermediate called an *activated complex*. This is an unstable, high-energy species that must be formed before the reaction can occur. It has an energy 134 kJ greater than that of the reactants and 360 kJ greater than that of the products. The state of the system at this point is often referred to as a *transition state,* intermediate between reactants and products.

The exact nature of the activated complex is difficult to determine. For this reaction, the activated complex might be a "pseudomolecule" made up of CO and NO_2 molecules in close contact. The path of the reaction might be more or less as follows:

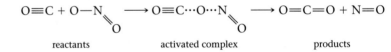

| reactants | activated complex | products |

In the activated complex, electrons are often in excited states.

The dotted lines stand for "partial bonds" in the activated complex. The N—O bond in the NO_2 molecule has been partially broken. A new bond between carbon and oxygen has started to form.

The idea of the activated complex was developed by, among others, Henry Eyring at Princeton in the 1930s. It forms the basis of the *transition-state model* for reaction rate, which assumes that the activated complex—

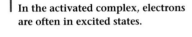

FIGURE 11.8

Reaction energy diagram. During the reaction initiation, 134 kJ—the activation energy E_a—must be furnished to the reactants for every mole of CO that reacts. This energy activates each CO—NO_2 complex to the point at which reaction can occur. **OHT**

CHEMISTRY ▪ *The Human Side*

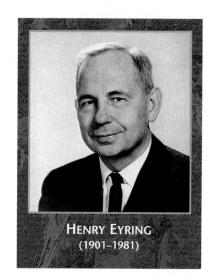

HENRY EYRING
(1901–1981)

Chemical kinetics, the subject of this chapter, evolved from an art into a science in the first half of the 20th century. One person more than any other was responsible for that change: Henry Eyring. Born in a Mormon settlement in Mexico, Henry emigrated to Arizona with his parents in 1912, when the Mexican revolution broke out. Many years later, he became a U.S. citizen, appearing before a judge who had just naturalized Albert Einstein.

Eyring's first exposure to kinetics came as a young chemistry instructor at the University of Wisconsin in the 1920s. There he worked on the thermal decomposition of N_2O_5 (Section 11.2). In 1931 Eyring came to Princeton, where he developed the transition-state model of reaction rates. Using this model and making reasonable guesses about the nature of the activated complex, Eyring calculated rate constants that agreed well with experiment. After 15 productive years at Princeton, Eyring returned to his Mormon roots at the University of Utah in Salt Lake City. While serving as dean of the graduate school, he found time to do ground-breaking research on the kinetics of life processes, including the mechanism of aging.

Henry Eyring's approach to research is best described in his own words:

> I perceive myself as rather uninhibited, with a certain mathematical facility and more interest in the broad aspects of a problem than the delicate nuances. I am more interested in discovering what is over the next rise than in assiduously cultivating the beautiful garden close at hand.

Henry's enthusiasm for chemistry spilled over into just about every other aspect of his life. Each year he challenged his students to a 50-yard dash held at the stadium of the University of Utah. He continued these races until he was in his mid-70s. Henry never won, but on one occasion he did make the CBS evening news.

(Photo credit: AIP Emilio Segre Visual Archives)

- is in equilibrium, at low concentrations, with the reactants.
- may either decompose to products by "climbing" over the energy barrier or, alternatively, revert back to reactants.

With these assumptions, it is possible to develop an expression for the rate constant k. Although the expression is too complex to discuss here, we show some of the results in Table 11.3.

The transition-state model is generally somewhat more accurate than the collision model (at least with $p = 1$). Another advantage is that it explains why the activation energy is ordinarily much smaller than the bond enthalpies in the reactant molecules. Consider, for example, the reaction

$$CO(g) + NO_2(g) \longrightarrow CO_2(g) + NO(g)$$

where $E_a = 134$ kJ. This is considerably smaller than the amount of energy required to break any of the bonds in the reactants.

$$C\equiv O \quad 1075 \text{ kJ} \qquad N=O \quad 607 \text{ kJ} \qquad N-O \quad 222 \text{ kJ}$$

Forming the activated complex shown in Figure 11.8 requires the absorption of relatively little energy, because it requires only the weakening of reactant bonds rather than their rupture.

11.5 REACTION RATE AND TEMPERATURE

See Screen 15.11, Control of Reaction Rates (Temperature Dependence).

Hibernating animals lower their body temperature, slowing down life processes.

The rates of most reactions increase as the temperature rises. A person in a hurry to prepare dinner applies this principle in using a pressure cooker to raise the temperature for cooking potatoes, apples, or a pot roast (not all at the same time, we trust). By storing the leftovers in a refrigerator, the chemical reactions responsible for food spoilage are slowed down. As a general and very approximate rule, it is often stated that an increase in temperature of 10°C doubles the reaction rate. If this rule holds, foods should cook twice as fast in a pressure cooker at 110°C as in an open saucepan and deteriorate four times as rapidly at room temperature (25°C) as they do in a refrigerator at 5°C.

The effect of temperature on reaction rate can be explained in terms of kinetic theory. Recall from Chapter 5 that raising the temperature greatly increases the fraction of molecules having very high speeds and hence high kinetic energies (kinetic energy = $mu^2/2$). These are the molecules that are most likely to react when they collide. The higher the temperature, the larger the fraction of molecules that can provide the activation energy required for reaction. This effect is apparent from Figure 11.9, where the distribution of kinetic energies is shown at two different temperatures. Notice that the fraction of molecules having a kinetic energy equal to or greater than the activation energy E_a (shaded area) is considerably larger at the higher temperature. Hence the fraction of effective collisions increases; this is the major factor causing reaction rate to increase with temperature.

The Arrhenius Equation

You will recall from Section 11.4 that the collision model yields the following expression for the rate constant:

$$k = p \times Z \times e^{-E_a/RT}$$

The steric factor, p, is presumably temperature-independent. The collision number,

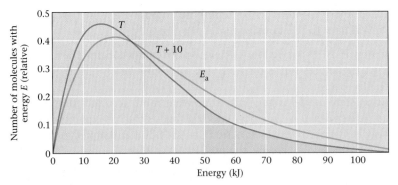

FIGURE 11.9

Temperature and activation energy. When T is increased to $T + 10$, the number of molecules with high energies increases. Hence many more molecules possess sufficient energy to react and the reaction occurs more rapidly. If E_a is 50 kJ/mol, any molecule with this or a higher energy can react. The number of molecules that react at T is represented by the green area under the T curve, and the number that react at $T + 10$ is represented by the green area *plus* the yellow area under the $T + 10$ curve. **OHT**

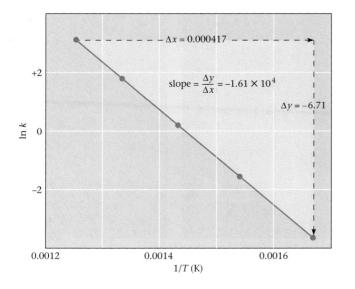

FIGURE 11.10

Arrhenius plot. A plot of ln k versus $1/T$ (k = rate constant, T = Kelvin temperature) is a straight line. From the slope of this line, the activation energy can be determined: $E_a = -R \times$ slope. For the CO—NO_2 reaction shown here:

$$E_a = -(8.31 \text{ J/mol} \cdot \text{K})(-1.61 \times 10^4 \text{ K})$$
$$= 1.34 \times 10^5 \text{ J/mol} = 134 \text{ kJ/mol. } \text{ OHT}$$

Z, is relatively insensitive to temperature. For example, when the temperature increases from 500 to 600 K, Z increases by less than 10%. Hence to a good degree of approximation we can write, as far as temperature dependence is concerned,

$$k = Ae^{-E_a/RT}$$

where A is a constant. Taking the natural logarithm of both sides of this equation,

$$\ln k = \ln A - E_a/RT$$

This equation was first shown to be valid by Svante Arrhenius (Chapter 4) in 1889. It is commonly referred to as the **Arrhenius equation.**

Comparing the Arrhenius equation to the general equation for a straight line,

$$y = b + mx$$

it should be clear that a plot of ln k ("y") versus $1/T$ ("x") should be linear. The slope of the straight line, m, is equal to $-E_a/R$. Figure 11.10 shows such a plot for the CO-NO_2 reaction. The slope appears to be about -1.61×10^4 K. Hence

$$\frac{-E_a}{R} = -1.61 \times 10^4 \text{ K}$$

$$E_a = 8.31 \frac{\text{J}}{\text{mol} \cdot \text{K}} (1.61 \times 10^4 \text{ K}) = 1.34 \times 10^5 \text{ J/mol} = 134 \text{ kJ/mol}$$

Two-Point Equation Relating k and T

The Arrhenius equation can be expressed in a different form by following the procedure used with the Clausius-Clapeyron equation in Chapter 9. At two different temperatures, T_2 and T_1,

$$\ln k_2 = \ln A - \frac{E_a}{RT_2}$$

$$\ln k_1 = \ln A - \frac{E_a}{RT_1}$$

Temperature and reaction rate. A chemical reaction in the Cyalume® light sticks produces light as it takes place. (a) The sticks outside the beaker at room temperature are brighter than the two sticks cooled to 0.2°C because the reaction rate is larger at the higher temperature. (b) With the sticks in the beaker heated well above room temperature, to 59°C, these sticks are now much brighter than those at room temperature. *(Charles D. Winters)*

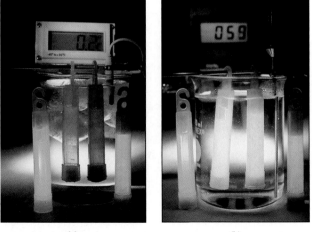

(a)   (b)

Subtracting the second equation from the first eliminates ln A, and we obtain

$$\ln k_2 - \ln k_1 = \frac{-E_a}{R}\left[\frac{1}{T_2} - \frac{1}{T_1}\right]$$

$$\boxed{\ln \frac{k_2}{k_1} = \frac{E_a}{R}\left[\frac{1}{T_1} - \frac{1}{T_2}\right]}$$

Taking $R = 8.31$ J/mol·K, the activation energy is expressed in joules per mole.

EXAMPLE 11.6   For a certain reaction, the rate constant doubles when the temperature increases from 15 to 25°C. Calculate

(a) the activation energy, E_a.
(b) the rate constant at 100°C, taking k at 25°C to be 1.2×10^{-2} L/mol·s.

Strategy   Use the two-point form of the Arrhenius equation. In (a), the two temperatures and the ratio k_2/k_1 are known, so E_a is readily calculated. The value of E_a is used in (b) to calculate k_2, knowing k_1, T_2, and T_1. Remember to express temperatures in K.

Solution

(a) $k_2/k_1 = 2.00$ ln $k_2/k_1 =$ ln $2.00 = 0.693$ $T_2 = 298$ K $T_1 = 288$ K
Substituting into the two-point form of the Arrhenius equation,

> With $R = 8.31$ J/mol·K, E_a must be expressed in J/mol.

$$0.693 = \frac{E_a}{8.31 \text{ J/mol·K}}\left[\frac{1}{288 \text{ K}} - \frac{1}{298 \text{ K}}\right]$$

Solving $E_a = 4.9 \times 10^4$ J/mol = $\boxed{49 \text{ kJ/mol}}$

(b) $\ln \dfrac{k_2}{k_1} = \dfrac{49,000 \text{ J/mol}}{8.31 \text{ J/mol·K}}\left[\dfrac{1}{298 \text{ K}} - \dfrac{1}{373 \text{ K}}\right] = 3.98$ $k_2/k_1 = e^{3.98} = 53$

Taking $k_1 = 1.2 \times 10^{-2}$ L/mol·s

$$k_2 = 53(1.2 \times 10^{-2} \text{ L/mol·s}) = \boxed{0.64 \text{ L/mol·s}}$$

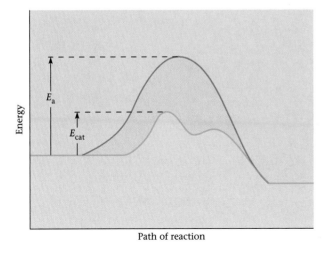

FIGURE 11.11

Catalysis and activation energy. By changing the path by which a reaction occurs, a catalyst can lower the activation energy that is required and so speed up a reaction.

11.6 CATALYSIS

A **catalyst** *is a substance that increases the rate of a reaction without being consumed by it.* It does this by changing the reaction path to one with a lower activation energy. Frequently the catalyzed path consists of two or more steps. In this case, the activation energy for the uncatalyzed reaction exceeds that for any of the steps in the catalyzed reaction (Figure 11.11).

> See Screen 15.14, Catalysis and Reaction Rate.

Heterogeneous Catalysis

A *heterogeneous catalyst* is one that is in a different phase from the reaction mixture. Most commonly, the catalyst is a solid that increases the rate of a gas-phase or liquid-phase reaction. An example is the decomposition of N_2O on gold:

$$N_2O(g) \xrightarrow{\text{Au}} N_2(g) + \tfrac{1}{2}O_2(g)$$

In the catalyzed decomposition, N_2O is chemically adsorbed on the surface of the solid. A chemical bond is formed between the oxygen atom of an N_2O molecule and a gold atom on the surface. This weakens the bond joining nitrogen to oxygen, making it easier for the N_2O molecule to break apart. Symbolically, this process can be shown as

$$N\equiv N{-}O(g) + Au(s) \longrightarrow N\equiv N\,\text{---}\,O\,\text{---}\,Au(s) \longrightarrow N\equiv N(g) + O(g) + Au(s)$$

where the broken lines represent weak covalent bonds.

Perhaps the most familiar example of heterogeneous catalysis is the series of reactions that occur in the catalytic converter of an automobile (Figure 11.12, page 332). Typically this device contains 1 to 3 g of platinum metal mixed with rhodium. The platinum catalyzes the oxidation of carbon monoxide and unburned hydrocarbons such as benzene, C_6H_6:

$$2CO(g) + O_2(g) \xrightarrow{\text{Pt}} 2CO_2(g)$$

$$C_6H_6(l) + \tfrac{15}{2}O_2(g) \xrightarrow{\text{Pt}} 6CO_2(g) + 3H_2O(l)$$

The rhodium acts as a catalyst to convert oxides of nitrogen to the free element:

$$CO(g) + NO(g) \xrightarrow{\text{Rh}} CO_2(g) + \tfrac{1}{2}N_2(g)$$

> ⌐ Many industrial processes use heterogeneous catalysts.

FIGURE 11.12

Automobile catalytic converter.
Catalytic converters contain a "three-way" catalyst designed to convert CO to CO_2, unburned hydrocarbons to CO_2 and H_2O, and NO to N_2. The active components of the catalysts are the precious metals platinum and rhodium; palladium is sometimes used as well.

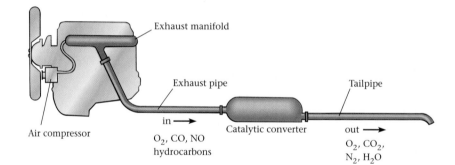

One problem with heterogeneous catalysis is that the solid catalyst is easily "poisoned." Foreign materials deposited on the catalytic surface during the reaction reduce or even destroy its effectiveness. A major reason for using unleaded gasoline is that lead metal poisons the Pt-Rh mixture in the catalytic converter.

Homogeneous Catalysis

A *homogeneous catalyst* is one that is present in the same phase as the reactants. It speeds up the reaction by forming a reactive intermediate that decomposes to give products. In this way, the catalyst provides an alternative process of lower activation energy.

An example of a reaction that is subject to homogeneous catalysis is the decomposition of hydrogen peroxide in aqueous solution:

$$2H_2O_2(aq) \longrightarrow 2H_2O + O_2(g)$$

Under ordinary conditions, this reaction occurs very slowly. However, if a solution of sodium iodide, NaI, is added, reaction occurs almost immediately; you can see the bubbles of oxygen forming.

The catalyzed decomposition of hydrogen peroxide is believed to take place by a two-step path:

$$\text{Step 1: } H_2O_2(aq) + I^-(aq) \longrightarrow H_2O + IO^-(aq)$$

$$\underline{\text{Step 2: } H_2O_2(aq) + IO^-(aq) \longrightarrow H_2O + O_2(g) + I^-(aq)}$$

$$2H_2O_2(aq) \longrightarrow 2H_2O + O_2(g)$$

Notice that the end result is the same as in the direct reaction.

The I^- ions are not consumed in the reaction. For every I^- ion used up in the first step, one is produced in the second step. The activation energy for this two-step process is much smaller than for the uncatalyzed reaction.

Enzymes

Many reactions that take place slowly under ordinary conditions occur readily in living organisms in the presence of catalysts called *enzymes*. Enzymes are protein molecules of high molar mass. An example of an enzyme-catalyzed reaction is the decomposition of hydrogen peroxide:

$$2H_2O_2(aq) \longrightarrow 2H_2O + O_2(g)$$

FIGURE 11.13
Enzyme catalysis. An enzyme in the potato is catalyzing the decomposition of a hydrogen peroxide solution, as shown by the bubbles of oxygen. *(Charles D. Winters)*

In blood or tissues, this reaction is catalyzed by an enzyme called catalase (Figure 11.13). When 3% hydrogen peroxide is used to treat a fresh cut or wound, oxygen gas is given off rapidly. The function of catalase in the body is to prevent the build-up of hydrogen peroxide, a powerful oxidizing agent.

Many enzymes are extremely specific. For example, the enzyme maltase catalyzes the hydrolysis of maltose:

$$C_{12}H_{22}O_{11}(aq) + H_2O \xrightarrow{\text{maltase}} 2C_6H_{12}O_6(aq)$$

maltose   glucose

This is the only function of maltase, but it is one that no other enzyme can perform. There are many such digestive enzymes required for the metabolism of carbohydrates, proteins, and fats. It has been estimated that without enzymes, it would take upwards of 50 years to digest a meal.

Enzymes, like all other catalysts, lower the activation energy for reaction. They can be enormously effective; it is not uncommon for the rate constant to increase by a factor of 10^{12} or more. However, from a commercial standpoint, enzymes have some drawbacks. A particular enzyme operates best over a narrow range of temperature. An increase in temperature frequently deactivates an enzyme by causing the molecule to "unfold," changing its characteristic shape. Recently, chemists have discovered that this effect can be prevented if the enzyme is immobilized by bonding to a solid support. Among the solids that have been used are synthetic polymers, porous glass, and even stainless steel. The development of immobilized enzymes has led to a host of new products. The sweetener aspartame, used in many diet soft drinks, is made using the enzyme aspartase in immobilized form.

┌ That would make dieting a lot easier.

11.7 REACTION MECHANISMS

A **reaction mechanism** is a description of a path, or a sequence of steps, by which a reaction occurs at the molecular level. In the simplest case, only a single step is involved. This is a collision between two reactant molecules. This is the mechanism for the reaction of CO with NO_2 at high temperatures, above about 600 K:

$$CO(g) + NO_2(g) \longrightarrow NO(g) + CO_2(g)$$

See Screen 15.12, Reaction Mechanisms.

Reaction mechanisms frequently change with *T*, sometimes with *P*.

At lower temperatures the reaction between carbon monoxide and nitrogen dioxide takes place by a quite different mechanism. Two steps are involved:

$$NO_2(g) + NO_2(g) \longrightarrow NO_3(g) + NO(g)$$

$$\underline{CO(g) + NO_3(g) \longrightarrow CO_2(g) + NO_2(g)}$$

$$CO(g) + NO_2(g) \longrightarrow NO(g) + CO_2(g)$$

Notice that the overall reaction, obtained by summing the individual steps, is identical with that for the one-step process. The rate expressions are quite different, however:

high temperatures: $\quad$ rate = $k[CO] \times [NO_2]$

low temperatures: $\quad$ rate = $k[NO_2]^2$

See Screen 15.13, Reaction Mechanisms and Rate Equations.

In general, the nature of the rate expression and hence *the reaction order depend on the mechanism by which the reaction takes place.*

Elementary Steps

The individual steps that constitute a reaction mechanism are referred to as **elementary steps.** These steps may be *unimolecular*

$$A \longrightarrow B + C \qquad rate = k[A]$$

bimolecular

$$A + B \longrightarrow C + D \qquad rate = k[A] \times [B]$$

or, in rare cases, *termolecular*

$$A + B + C \longrightarrow D + E \qquad rate = k[A] \times [B] \times [C]$$

(A color screen is used to distinguish elementary steps from overall reactions.)

Notice from the rate expressions just written that *the rate of an elementary step is equal to a rate constant k multiplied by the concentration of each reactant molecule.* This rule is readily explained. Consider, for example, a step in which two molecules, A and B, collide effectively with each other to form C and D. As pointed out earlier, the rate of collision and hence the rate of reaction will be directly proportional to the concentration of each reactant.

Slow Steps

Often, one step in a mechanism is much slower than any other. If this is the case, *the slow step is* **rate-determining.** That is, the rate of the overall reaction can be taken to be that of the slow step. Consider, for example, a three-step reaction:

Step 1: $A \longrightarrow B \qquad$ (fast)

Step 2: $B \longrightarrow C \qquad$ (slow)

Step 3: $\underline{C \longrightarrow D} \qquad$ (fast)

$$A \longrightarrow D$$

The rate at which A is converted to D (the overall reaction) is approximately equal to the rate of conversion of B to C (the slow step).

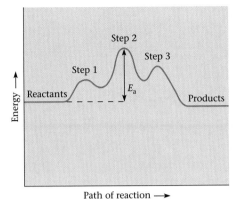

FIGURE 11.14
Reaction plot for a reaction with a three-step mechanism. The second of the three steps is rate-determining.

To understand the rationale behind this rule, and its limitations, consider an analogous situation. Suppose three people (A, B, and C) are assigned to grade general chemistry examinations that contain three questions. On the average, A spends 10 s grading Question 1 and B spends 15 s grading Question 2. In contrast, C, the ultimate procrastinator, takes 5 min to grade Question 3. The rate at which exams are graded is

$$\frac{1 \text{ exam}}{10 \text{ s} + 15 \text{ s} + 300 \text{ s}} = \frac{1 \text{ exam}}{325 \text{ s}} = 0.00308 \text{ exam/s}$$

This is approximately equal to the rate of the slower grader:

$$\frac{1 \text{ exam}}{300 \text{ s}} = 0.00333 \text{ exam/s}$$

Extrapolating from general chemistry exams to chemical reactions, we can say that—

- the overall rate of the reaction cannot exceed that of the slowest step.
- if that step is by far the slowest, its rate will be approximately equal to that of the overall reaction.
- the slowest step in a mechanism is ordinarily the one with the highest activation energy (Figure 11.14).

Deducing a Rate Expression from a Proposed Mechanism

As we have seen, rate expressions for reactions must be determined experimentally. Once this has been done, it is possible to derive a plausible mechanism compatible with the observed rate expression. This, however, is a rather complex process and we will not attempt it here. Instead we will consider the reverse process, which is much more straightforward. *Given a mechanism for a several-step reaction, how can you deduce the rate expression corresponding to that mechanism?*

In principle, at least, all you have to do is to apply the rules just cited.

1. *Find the slowest step and equate the rate of the overall reaction to the rate of that step.*
2. *Find the rate expression for the slowest step.*

Maybe Questions 1 and 2 were multiple-choice.

To illustrate this process, consider the two-step mechanism for the low-temperature reaction between CO and NO_2.

$$NO_2(g) + NO_2(g) \longrightarrow NO_3(g) + NO(g) \qquad \text{(slow)}$$

$$CO(g) + NO_3(g) \longrightarrow CO_2(g) + NO_2(g) \qquad \text{(fast)}$$

$$CO(g) + NO_2(g) \longrightarrow CO_2(g) + NO(g)$$

Applying the above rules in order,

$$\text{rate of overall reaction} = \text{rate of 1st step} = k[NO_2]^2$$

This analysis explains why the rate expression for the two-step mechanism is different from that for the direct, one-step reaction.

Elimination of Intermediates

Sometimes the rate expression obtained by the process just described involves a reactive intermediate, that is, a species produced in one step of the mechanism and consumed in a later step. Ordinarily, concentrations of such species are too small to be determined experimentally. Hence they must be eliminated from the rate expression if it is to be compared with experiment. The final rate expression usually includes only those species that appear in the balanced equation for the reaction. Sometimes, the concentration of a catalyst is included, but never that of a reactive intermediate.

To illustrate this situation, consider the reaction between nitric oxide and chlorine, which is believed to proceed by a two-step mechanism:

$$\text{Step 1:} \qquad NO(g) + Cl_2(g) \underset{k_{-1}}{\overset{k_1}{\rightleftharpoons}} NOCl_2(g) \qquad \text{(fast)}$$

$$\text{Step 2:} \qquad NOCl_2(g) + NO(g) \overset{k_2}{\longrightarrow} 2NOCl(g) \qquad \text{(slow)}$$

$$2NO(g) + Cl_2(g) \longrightarrow 2NOCl(g)$$

The first step occurs rapidly and reversibly; a dynamic equilibrium is set up in which the rates of forward and reverse reactions are equal. Because the second step is slow and rate-determining, it follows that

$$\text{rate of overall reaction} = \text{rate of Step 2} = k_2[NOCl_2] \times [NO]$$

The rate expression just written is unsatisfactory in that it cannot be checked against experiment. The species $NOCl_2$ is a reactive intermediate whose concentration is too small to be measured accurately, if at all. To eliminate the $[NOCl_2]$ term from the rate expression, recall that the rates of forward and reverse reactions in Step 1 are equal, which means that

$$k_1[NO] \times [Cl_2] = k_{-1}[NOCl_2]$$

Solving for $[NOCl_2]$ and substituting in this rate expression,

$$[NOCl_2] = \frac{k_1[NO] \times [Cl_2]}{k_{-1}}$$

$$\text{rate of overall reaction} = k_2[NOCl_2] \times [NO] = \frac{k_2 k_1[NO]^2 \times [Cl_2]}{k_{-1}}$$

If the concentration of a species can't be measured, it better not appear in the rate expression.

The quotient k_2k_1/k_{-1} is the experimentally observed rate constant for the reaction, which is found to be second-order in NO and first-order in Cl_2, as predicted by this mechanism.

EXAMPLE 11.7   The decomposition of ozone, O_3, to diatomic oxygen, O_2, is believed to occur by a two-step mechanism:

$$\text{Step 1:} \qquad O_3(g) \underset{k_{-1}}{\overset{k_1}{\rightleftharpoons}} O_2(g) + O(g) \qquad \text{(fast)}$$

$$\text{Step 2:} \quad O_3(g) + O(g) \xrightarrow{k_2} 2O_2(g) \qquad \text{(slow)}$$

$$2O_3(g) \longrightarrow 3O_2(g)$$

Obtain the rate expression corresponding to this mechanism.

Strategy   Write the rate expression for the rate-determining second step. This will involve the unstable intermediate, O. To get rid of [O], use the fact that reactants and products are in equilibrium in Step 1, so forward and reverse reactions occur at the same rate.

Solution

(1) Overall rate = rate Step 2 = $k_2[O_3] \times [O]$

(2) $k_1[O_3] = k_{-1}[O_2] \times [O]$

$$[O] = \frac{k_1[O_3]}{k_{-1}[O_2]}$$

(3) Overall rate = $\dfrac{k_1k_2[O_3]^2}{k_{-1}[O_2]} = \dfrac{k[O_3]^2}{[O_2]}$

The quotient k_1k_2/k_{-1} is the observed rate constant, k.

Reality Check   Notice that the concentration of O_2, a product in the reaction, appears in the denominator. The rate is inversely proportional to the concentration of molecular oxygen, a feature that you would never have predicted from the balanced equation for the reaction. As the concentration of O_2 builds up, the rate slows down.

It is important to point out one of the limitations of mechanism studies. Usually more than one mechanism is compatible with the same experimentally obtained rate expression. To make a choice between alternative mechanisms, other evidence must be considered. A classic example of this situation is the reaction between hydrogen and iodine

$$H_2(g) + I_2(g) \longrightarrow 2HI(g)$$

for which the observed rate expression is

$$\text{rate} = k[H_2] \times [I_2]$$

For many years, it was assumed that the H_2-I_2 reaction occurs in a single step, a collision between an H_2 molecule and an I_2 molecule. That would, of course, be compatible with the rate expression above. However, there is now evidence to indicate that a quite different and more complex mechanism is involved (see Problem 61).

The Ozone Story

In recent years, a minor component of the atmosphere, ozone, has received a great deal of attention. Ozone, molecular formula O_3, is a pale blue gas with a characteristic odor that can be detected after lightning activity, in the vicinity of electric motors, or near a subway train.

Depending on its location in the atmosphere, ozone can be a villain or a beleaguered hero. In the lower atmosphere (the *troposphere*), ozone is bad news; it is a major component of photochemical smog, formed by the reaction sequence

$$NO_2(g) \longrightarrow NO(g) + O(g)$$
$$O_2(g) + O(g) \longrightarrow O_3(g)$$
$$NO_2(g) + O_2(g) \longrightarrow NO(g) + O_3(g)$$

Ozone is an extremely powerful oxidizing agent, which explains its toxicity to animals and humans. At partial pressures as low as 10^{-7} atm, it can cut in half the rate of photosynthesis by plants.

In the upper atmosphere (the *stratosphere*), the situation is quite different. There the partial pressure of ozone goes through a maximum of about 10^{-5} atm at an altitude of 30 km. From 95 to 99% of sunlight in the wavelength range 200 to 300 nm is absorbed by ozone in this region, commonly referred to as the "ozone layer." If this ultraviolet radiation were to reach the surface of the Earth, it could have several adverse effects. A decrease in ozone concentration of 5% could increase the incidence of skin cancer by 20%. Ultraviolet radiation is also a factor in diseases of the eye, including cataract formation.

Ozone molecules in the stratosphere can decompose by the reaction

$$O_3(g) + O(g) \longrightarrow 2O_2(g)$$

This reaction takes place rather slowly by direct collision between an O_3 molecule and an O atom. It can occur more rapidly by a two-step process in which a chlorine atom acts as a catalyst:

$$O_3(g) + Cl(g) \longrightarrow O_2(g) + ClO(g) \qquad k = 5.2 \times 10^9 \text{ L/mol·s at 220 K}$$
$$ClO(g) + O(g) \longrightarrow Cl(g) + O_2(g) \qquad k = 2.6 \times 10^{10} \text{ L/mol·s at 220 K}$$
$$O_3(g) + O(g) \longrightarrow 2O_2(g)$$

A single chlorine atom can bring about the decomposition of tens of thousands of ozone molecules. Bromine atoms can substitute for chlorine; indeed the rate constant for the Br-catalyzed reaction is larger than that for the reaction just cited.

The chlorine atoms that catalyze the decomposition of ozone come from chlorofluorocarbons (CFCs) used in many refrigerators and air conditioners. A major culprit is CF_2Cl_2, Freon, which forms Cl atoms when exposed to ultraviolet radiation at 200 nm:

$$CF_2Cl_2(g) \longrightarrow CF_2Cl(g) + Cl(g)$$

The two scientists who first suggested (in 1974) that CFCs could deplete the ozone layer, F. Sherwood Rowland and Mario Molina, won the 1995 Nobel Prize in chemistry, along with Paul Crutzen, who first suggested that oxides of nitrogen in the atmosphere could catalyze the decomposition of ozone.

The catalyzed decomposition of ozone is known to be responsible for the ozone hole (Figure A) that develops in Antarctica each year in September and October, at the end of winter in the Southern Hemisphere. No ozone is generated during the long, dark Antarctic winter. Meanwhile, a heterogeneous reaction occurring on clouds of ice crystals at $-85°C$ produces species such as Cl_2, Br_2, and $HOCl$. When the Sun reappears in September, these molecules decompose photochemically to form Cl or Br atoms.

In 1999, the ozone hole *decreased* in size.

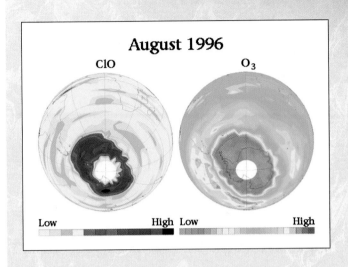

**Chlorine monoxide (ClO) and the Antarctic ozone (O₃)
hole in August 1996.** Chlorine monoxide is the major
ozone-destroying form of chlorine. Note the high levels of
ClO (*left*) where there are low levels of ozone (*right*).
(*Courtesy of NASA*)

Ozone depletion is by no means restricted to the Southern Hemisphere. In the extremely cold winter of 1994–1995, a similar "ozone hole" was found in the Arctic. Beyond that, the concentration of ozone in the atmosphere over parts of Siberia dropped by 40%.

In 1987 an international treaty was signed in Montreal to cut back on the use of CFCs. Production of Freon in the United States ended in 1996. It has been replaced in automobile air conditioners by a related compound with no chlorine atoms, $C_2H_2F_4$:

CHAPTER HIGHLIGHTS

Key Concepts

1. Determine the rate expression (reaction order) from—
   ■ initial rate data.
   (Examples 11.1, 11.2; Problems 19–28, 82)
   ■ concentration-time data, using Table 11.2.
   (Example 11.5; Problems 29, 30, 72, 73)
   ■ reaction mechanism.
   (Example 11.7; Problems 61–64)
2. Relate concentration to time for a first-order reaction.
   (Examples 11.3, 11.4; Problems 31–42, 65, 79)
3. Use the Arrhenius equation to relate rate constant to temperature.
   (Example 11.6; Problems 49–58, 67, 68)

Key Equations

Zero-order reaction	rate $= k$	$[A] = [A]_o - kt$
First-order reaction	rate $= k[A]$	$\ln [A]_o/[A] = kt$ $t_{1/2} = 0.693/k$
Second-order reaction	rate $= k[A]^2$	$1/[A] - 1/[A]_o = kt$
Arrhenius equation	$\ln \dfrac{k_2}{k_1} = \dfrac{E_a}{R}\left[\dfrac{1}{T_1} - \dfrac{1}{T_2}\right]$	

Key Terms

activation energy
catalyst
elementary step
half-life
initial rate
integrated rate equation

rate constant
rate-determining step
rate expression
reaction mechanism

reaction order
—first-order
—overall order
—second-order
—zero-order
reaction rate

Summary Problem

Hydrogen peroxide decomposes to water and oxygen according to the following reaction:

$$H_2O_2(aq) \longrightarrow H_2O + \tfrac{1}{2}O_2(g)$$

Its rate of decomposition is measured by titrating samples of the solution with potassium permanganate ($KMnO_4$) at certain intervals.

(a) Initial rate determinations at 25°C for the decomposition give the following data:

$[H_2O_2]$	Initial rate (mol/L·min)
0.200	1.03×10^{-4}
0.300	1.55×10^{-4}
0.500	2.59×10^{-4}

What is the order of the reaction? Write the rate equation for the decomposition. Calculate the rate constant and the half-life for the reaction at 25°C.

(b) Hydrogen peroxide is sold commercially as a 30.0% solution. If the solution is kept at room temperature (25°C), how long will it take for the solution to become 10.0% H_2O_2?

(c) It has been determined that at 40°C, the rate constant for the reaction is 1.93×10^{-3} min^{-1}. Calculate the activation energy for the decomposition of H_2O_2.

(d) Manufacturers recommend that solutions of hydrogen peroxide be kept in a refrigerator at 4°C. How much longer will it take a 30.0% solution to decompose to 10.0% if the solution is kept in a refrigerator instead of at room temperature?

(e) The rate constant for the catalyzed decomposition of H_2O_2 at 25°C is 2.95×10^8 min^{-1}. What is the half-life of the catalyzed reaction? What is the rate for the catalyzed decomposition of 0.500 M H_2O_2? How much faster is the rate of a catalyzed reaction than that of an uncatalyzed reaction at the same H_2O_2 concentration at 25°C?

(f) Hydrogen peroxide in basic solution oxidizes iodide ions to iodine. The proposed mechanism for this reaction is

$$H_2O_2(aq) + I^-(aq) \longrightarrow HOI(aq) + OH^-(aq) \quad \text{(slow)}$$

$$HOI(aq) + I^-(aq) \rightleftharpoons I_2(aq) + OH^-(aq) \quad \text{(fast)}$$

Write the overall redox reaction. Write a rate law consistent with this proposed mechanism.

Answers

(a) 1st order; rate $= k[H_2O_2]$; 5.17×10^{-4} min^{-1}; 22.3 h
(b) 35.4 h
(c) 68 kJ/mol
(d) 2.5×10^2 h
(e) 2.35×10^{-9} min; 1.48×10^8 mol/L·min; 5.71×10^{11} times faster or about 0.6 trillion times faster
(f) $H_2O_2(aq) + 2I^-(aq) \rightarrow I_2(aq) + 2OH^-(aq)$; rate $= k[H_2O_2] \times [I^-]$

Questions & Problems

Problem numbers in blue indicate that the answer is available in Appendix 6 at the back of the book.

WEB indicates that the solution is posted at **http://www.harcourtcollege.com/chem/general/masterton4/student/**

Meaning of Reaction Rate

1. Express the rate of the reaction

$$2N_2(g) + 5O_2(g) \longrightarrow 2N_2O_5(g)$$

in terms of
   (a) $\Delta[N_2O_5]$ **(b)** $\Delta[O_2]$

2. Express the rate of the reaction

$$2HCl(g) + O_2(g) \longrightarrow 2HOCl(g)$$

in terms of
   (a) $\Delta[HOCl]$ **(b)** $\Delta[O_2]$

3. For the reaction

$$5Br^-(aq) + BrO_3^-(aq) + 6H^+(aq) \longrightarrow 3Br_2(aq) + 3H_2O$$

it was found that at a particular instant bromine was being formed at the rate of 0.039 mol/L·s. At that instant, at what rate is

   (a) water being formed?
   (b) bromide ion being oxidized?
   (c) H^+ being consumed?

4. Consider the combustion of ethane:

$$C_2H_6(g) + 7O_2(g) \longrightarrow 4CO_2(g) + 6H_2O(g)$$

If the ethane is burning at the rate of 0.20 mol/L·s, at what rates are CO_2 and H_2O being produced?

5. Nitrosyl chloride (NOCl) decomposes to nitrogen oxide and chlorine gases.

   (a) Write a balanced equation using smallest whole-number coefficients for the decomposition.
   (b) Write an expression for the reaction rate in terms of $\Delta[NOCl]$.
   (c) The concentration of NOCl drops from 0.580 M to 0.238 M in 8.00 min. Calculate the average rate of reaction over this time interval.

6. Ammonia is produced by the reaction between nitrogen and hydrogen gases.

   (a) Write a balanced equation using smallest whole-number coefficients for the reaction.
   (b) Write an expression for the rate of reaction in terms of $\Delta[NH_3]$.
   (c) The concentration of ammonia increases from 0.257 M to 0.815 M in 15.0 min. Calculate the average rate of reaction over this time interval.

7. Experimental data are listed for the hypothetical reaction

$$A + B \longrightarrow C + D$$

Time (s)	0	1	2	3	4	5
[A]	0.200	0.166	0.139	0.117	0.100	0.089

   (a) Plot these data as in Figure 11.2.
   (b) Draw a tangent to the curve to find the instantaneous rate at 3 min.
   (c) Find the average rate over the 1- to 4-min interval.
   (d) Compare the instantaneous rate at 3 min with the average rate over the three-minute interval.

WEB 8. Experimental data are listed for the hypothetical reaction

$$A \longrightarrow B + C$$

Time	0	2	4	5	8	10
[B]	0	0.100	0.130	0.150	0.165	0.175

   (a) Plot these data as in Figure 11.2.

(b) Draw a tangent to the curve to find the instantaneous rate at 6 min.

(c) Find the average rate over the 2- to 8-min interval.

(d) Compare the instantaneous rate at 6 min with the average rate over the six-minute interval.

Rate Expressions

9. A reaction has two reactants A and B. What is the order with respect to each reactant and the overall order of the reaction described by each of the following rate expressions?

(a) rate = $k_1[A]^3$ **(b)** rate = $k_2[A] \times [B]$

(c) rate = $k_3[A] \times [B]^2$ **(d)** rate = $k_4[B]$

10. A reaction has two reactants X and Y. What is the order with respect to each reactant and the overall order of the reaction described by the following rate expressions?

(a) rate = $k_1[X]^2 \times [Y]$ **(b)** rate = $k_2[X]$

(c) rate = $k_3[X]^2 \times [Y]^2$ **(d)** rate = k_4

11. What will the units of the rate constants in Question 9 be if the rate is expressed in mol/L·min?

12. What will the units of the rate constants in Question 10 be if the rate is expressed in mol/L·s?

13. Complete the following table for the reaction

$$A(g) + B(g) \longrightarrow products$$

that is first-order in both A and B

	[A]	[B]	k(L/mol·min)	Rate (mol/L·min)
(a)	0.100	0.400	1.29	____
(b)	____	0.015	0.57	0.017
(c)	0.500	____	0.631	1.29
(d)	0.215	0.885	____	0.419

WEB 14. Complete the following table for the first-order reaction

$$A(g) \longrightarrow products$$

	[A]	k (h^{-1})	Rate (mol/L·h)
(a)	0.115	0.417	____
(b)	0.025	____	1.67
(c)	____	2.0×10^{-3}	0.0392

15. The decomposition of ammonia on tungsten at 1100°C is zero-order with a rate constant of 2.5×10^{-4} mol/L·min.

(a) Write the rate expression.

(b) Calculate the rate when $[NH_3] = 0.075\ M$.

(c) At what concentration of ammonia is the rate equal to the rate constant?

16. The decomposition of nitrogen dioxide is a second-order reaction. At 550 K, a 0.250 M sample decomposes at the rate of 1.17 mol/L·min.

(a) Write the rate expression.

(b) What is the rate constant at 550 K?

(c) What is the rate of decomposition when $[NO_2] = 0.800\ M$?

17. The reaction

$$NO(g) + \tfrac{1}{2}Br_2(g) \longrightarrow NOBr(g)$$

is second-order in nitrogen oxide and first-order in bromine. The rate of the reaction is 1.6×10^{-8} mol/L·min when the nitrogen oxide concentration is 0.020 M and the bromine concentration is 0.030 M.

(a) What is the value of k?

(b) At what concentration of bromine is the rate 3.5×10^{-7} mol/L·min and $[NO] = 0.043\ M$?

(c) At what concentration of nitrogen oxide is the rate 2.0×10^{-6} mol/L·min and the bromine concentration one fourth of the nitrogen oxide concentration?

18. The reaction

$$ICl(g) + \tfrac{1}{2}H_2(g) \longrightarrow \tfrac{1}{2}I_2(g) + HCl(g)$$

is first-order in both reactants. The rate of the reaction is 4.89×10^{-5} mol/L·s when the ICl concentration is 0.100 M and that of the hydrogen gas is 0.030 M.

(a) What is the value of k?

(b) At what concentration of hydrogen is the rate 5.00×10^{-4} mol/L·s and $[ICl] = 0.233\ M$?

(c) At what concentration of iodine chloride is the rate 0.0934 mol/L·s if the hydrogen concentration is three times that of ICl?

Determination of Reaction Order

19. For a reaction involving a single reactant, the following data are obtained:

Rate (mol/L·min)	0.020	0.016	0.013	0.010
[A]	0.100	0.090	0.080	0.070

(a) Determine the order of the reaction.

(b) Write the rate expression for the reaction.

(c) Calculate k for the experiment above.

20. For a reaction involving the decomposition of A, the following data are obtained:

Rate (mol/L·min)	0.020	0.020	0.019	0.021
[A]	0.100	0.090	0.080	0.070

(a) Determine the order of the reaction.

(b) Write the rate expression for the decomposition of A.

(c) Calculate k for the experiment.

21. When boron trifluoride reacts with ammonia, the following reaction occurs:

$$BF_3(g) + NH_3(g) \longrightarrow BF_3NH_3(g)$$

The following data are obtained at a particular temperature:

Expt.	[BF$_3$]	[NH$_3$]	Initial Rate (mol/L·s)
1	0.100	0.100	0.0341
2	0.200	0.233	0.159
3	0.200	0.0750	0.0512
4	0.300	0.100	0.102

(a) What is the order of the reaction with respect to BF$_3$, NH$_3$, and overall?
(b) Write the rate expression for the reaction.
(c) Calculate k for the reaction.
(d) When [BF$_3$] = 0.553 M and [NH$_3$] = 0.300 M, what is the rate of the reaction at the temperature of the experiment?

WEB **22.** The peroxysulfate ion reacts with the iodide ion in aqueous solution according to the following equation:

$$S_2O_8^{2-}(aq) + 3I^-(aq) \longrightarrow 2SO_4^{2-}(aq) + I_3^-(aq)$$

The following data are obtained at a certain temperature:

Expt.	[S$_2$O$_8^{2-}$]	[I$^-$]	Initial Rate (mol/L·min)
1	0.0200	0.0155	1.15 × 10^{-4}
2	0.0250	0.0200	1.85 × 10^{-4}
3	0.0300	0.0200	2.22 × 10^{-4}
4	0.0300	0.0275	3.06 × 10^{-4}

(a) What is the order of the reaction with respect to [S$_2$O$_8^{2-}$], [I$^-$], and overall?
(b) Write the rate expression for the reaction.
(c) Calculate k for the reaction.
(d) When [S$_2$O$_8^{2-}$] = 0.105 M and [I$^-$] = 0.0875 M, what is the rate of the reaction at the temperature of the experiment?

23. The rate of the reaction

$$HgCl_2(aq) + \tfrac{1}{2}C_2O_4^{2-}(aq) \longrightarrow Cl^-(aq) + CO_2(g) + \tfrac{1}{2}Hg_2Cl_2(s)$$

is followed by measuring the number of moles of Hg$_2$Cl$_2$ that precipitate per liter per second. The following data are obtained:

[HgCl$_2$]	[C$_2$O$_4^{2-}$]	Initial rate (mol/L·min)
0.020	0.020	6.24 × 10^{-8}
0.048	0.020	1.50 × 10^{-7}
0.020	0.033	1.70 × 10^{-7}
0.075	0.033	6.37 × 10^{-7}

(a) What is the order of the reaction with respect to mercury(II) chloride, oxalate ion, and overall?
(b) Write the rate expression for the reaction.
(c) Calculate k for the reaction.
(d) When [HgCl$_2$] = 0.085 M and [C$_2$O$_4^{2-}$] = 0.068 M, what is the rate of the reaction?

24. When a base is added to an aqueous solution of chlorine dioxide gas, the following reaction occurs.

$$2ClO_2(aq) + 2\ OH^-(aq) \longrightarrow ClO_3^-(aq) + ClO_2^-(aq) + H_2O$$

The reaction rate can be followed by monitoring the hydroxide concentration. The following data are obtained:

[ClO$_2$]	[OH$^-$]	Initial Rate (mol/L·s)
0.010	0.030	6.00 × 10^{-4}
0.010	0.075	1.50 × 10^{-3}
0.055	0.030	1.82 × 10^{-2}
0.055	0.062	3.75 × 10^{-2}

(a) What is the order of the reaction with respect to chlorine dioxide, hydroxide ion, and overall?
(b) Write the rate expression for the reaction.
(c) Calculate k for the reaction.
(d) When [ClO$_2$] = 0.25 M and [OH$^-$] = 0.036 M, what is the rate of the reaction?

25. The equation for the iodination of acetone in acidic solution is

$$CH_3COCH_3(aq) + I_2(aq) \longrightarrow CH_3COCH_2I(aq) + H^+(aq) + I^-(aq)$$

The rate of the reaction is found to be dependent not only on the concentration of the reactants but also on the hydrogen ion concentration. Hence the rate expression of this reaction is

$$\text{rate} = k[CH_3COCH_3]^m[I_2]^n[H^+]^p$$

The rate is obtained by following the disappearance of iodine using starch as an indicator. The following data are obtained:

[CH$_3$COCH$_3$]	[H$^+$]	[I$_2$]	Initial Rate (mol/L·s)
0.80	0.20	0.001	4.2 × 10^{-6}
1.6	0.20	0.001	8.2 × 10^{-6}
0.80	0.40	0.001	8.7 × 10^{-6}
0.80	0.20	0.0005	4.3 × 10^{-6}

(a) What is the order of the reaction with respect to each reactant?
(b) Write the rate expression for the reaction.
(c) Calculate k.
(d) What is the rate of the reaction when [H$^+$] = 0.933 M and [CH$_3$COCH$_3$] = 3[H$^+$] = 10[I$^-$]?

26. The equation for the reaction between iodide and bromate ions in acidic solution is

$$6I^-(aq) + BrO_3^-(aq) + 6H^+(aq) \longrightarrow 3I_2(aq) + Br^-(aq) + 3H_2O$$

The rate of the reaction is followed by measuring the appearance of I$_2$. The following data are obtained:

[I$^-$]	[BrO$_3^-$]	[H$^+$]	Initial Rate (mol/L·s)
0.0020	0.0080	0.020	8.89 × 10^{-5}
0.0040	0.0080	0.020	1.78 × 10^{-4}
0.0020	0.0160	0.020	1.78 × 10^{-4}
0.0020	0.0080	0.040	3.56 × 10^{-4}
0.0015	0.0040	0.030	7.51 × 10^{-5}

(a) What is the order of the reaction with respect to each reactant?

(b) Write the rate expression for the reaction.

(c) Calculate k.

(d) What is the hydrogen-ion concentration when the rate is 5.00×10^{-4} mol/L·s and [I$^-$] = $\frac{1}{2}$[BrO$_3^-$] = 0.0075 M?

27. In solution at a constant H$^+$ concentration, iodide ions react with hydrogen peroxide to produce iodine.

$$H^+(aq) + I^-(aq) + \tfrac{1}{2}H_2O_2(aq) \longrightarrow \tfrac{1}{2}I_2(aq) + H_2O$$

The reaction rate can be followed by monitoring the appearance of I$_2$. The following data were obtained

[I$^-$]	[H$_2$O$_2$]	Initial Rate (mol/L·min)
0.015	0.030	0.0022
0.035	0.030	0.0052
0.055	0.030	0.0082
0.035	0.050	0.0087

(a) Write the rate expression for the reaction.

(b) Calculate k.

(c) What is the rate of the reaction when 25.0 mL of a 0.100 M solution of KI is added to 25.0 mL of a 10.0% by mass solution of H$_2$O$_2$(d = 1.00 g/mL)? Assume volumes are additive.

28. Consider the reaction

$$Cr(H_2O)_6^{3+}(aq) + SCN^-(aq) \longrightarrow Cr(H_2O)_5SCN^{2+}(aq) + H_2O$$

The following data were obtained:

[Cr(H$_2$O)$_6^{3+}$]	[SCN$^-$]	Initial Rate (mol/L·min)
0.025	0.060	6.5 × 10^{-4}
0.025	0.077	8.4 × 10^{-4}
0.042	0.077	1.4 × 10^{-3}
0.042	0.100	1.8 × 10^{-3}

(a) Write the rate expression for the reaction.

(b) Calculate k.

(c) What is the rate of the reaction when 15 mg of KSCN is added to 1.50 L of a solution 0.0500 M in Cr(H$_2$O)$_6^{3+}$?

29. Nitrosyl bromide (NOBr) decomposes to nitrogen oxide and bromine. Use the following data to determine the order of the decomposition reaction of nitrosyl bromide.

Time (s)	0	6	12	18	24
[NOBr]	0.0286	0.0253	0.0229	0.0208	0.0190

30. In dilute acidic solution, sucrose (C$_{12}$H$_{22}$O$_{11}$) decomposes to glucose and fructose, both with molecular formula C$_6$H$_{12}$O$_6$. The following data are obtained for the decomposition of sucrose.

Time (min)	[C$_{12}$H$_{22}$O$_{11}$]
0	0.368
20	0.333
60	0.287
120	0.235
160	0.208

Write the rate expression for the reaction.

First-Order Reactions

31. At high temperatures, dinitrogen oxide gas decomposes to nitrogen and oxygen gases, according to the following equation:

$$2N_2O(g) \longrightarrow 2N_2(g) + O_2(g)$$

The following data are obtained in an experiment:

Time (h)	[N$_2$O]
0	0.100
0.5	0.0934
1	0.0873
1.5	0.0815
2	0.0762
3	0.0665
6	0.0442

(a) By plotting the data, show that the reaction is first-order.

(b) From the graph, determine k.

(c) Using k, find the time it takes to decrease the concentration to 0.0100 M.

(d) Calculate the rate of the reaction when [N$_2$O] = 0.0500 M.

WEB 32. The following data apply to the first-order decomposition of ethane:

$$C_2H_6(g) \longrightarrow products$$

Time (s)	[C$_2$H$_6$]
0	0.01000
200	0.00916
400	0.00839
600	0.00768
800	0.00703

(a) By plotting the data, show that the reaction is first-order.

(b) From the graph, determine k.

(c) Using k, find the time it takes to decrease the concentration to $0.00500\ M$.

(d) Calculate the rate of the reaction when $[C_2H_6] = 0.00400\ M$.

33. In the first-order decomposition of acetone at 500°C,

$$CH_3-CO-CH_3(g) \longrightarrow products$$

it is found that the concentration is $0.0300\ M$ after 200 min and $0.0200\ M$ after 400 min. Calculate the following.

(a) the rate constant

(b) the half-life

(c) the initial concentration

34. The first-order rate constant for the decomposition of a certain hormone in water at 25°C is $3.42 \times 10^{-4}\ day^{-1}$.

(a) If a $0.0200\ M$ solution of the hormone is stored at 25°C for two months, what will its concentration be at the end of that period?

(b) How long will it take for the concentration of the solution to drop from $0.0200\ M$ to $0.00350\ M$?

(c) What is the half-life of the hormone?

35. The decomposition of dimethyl ether (CH_3OCH_3) to methane, carbon monoxide, and hydrogen gases is found to be first-order. At 500°C, a 150.0-mg sample of dimethyl ether is reduced to 43.2 mg after three quarters of an hour. Calculate

(a) the rate constant.

(b) the half-life at 500°C.

(c) how long it will take to decompose 95% of the dimethyl ether.

36. The first-order rate constant for the decomposition of a certain drug at 25°C is $0.215\ month^{-1}$.

(a) If 10.0 g of the drug is stored at 25°C for one year, how many grams of the drug will remain at the end of the year?

(b) What is the half-life of the drug?

(c) How long will it take to decompose 65% of the drug?

37. The decomposition of dinitrogen pentaoxide to nitrogen dioxide and oxygen is first-order. Its rate constant is $1.91\ h^{-1}$.

(a) What is its half-life?

(b) What fraction of dinitrogen pentaoxide is decomposed after 28 min?

(c) How long will it take to decompose 39% of the N_2O_5?

(d) How fast is a $0.2500\ M$ solution decomposing?

38. The decomposition of sulfuryl chloride, SO_2Cl_2, to sulfur dioxide and chlorine gases is a first-order reaction. It is found that it takes 98 min to decompose a $0.300\ M$ SO_2Cl_2 to $0.274\ M$.

(a) What is the rate constant for the reaction?

(b) What is its half-life?

(c) What is the rate of decomposition when $[SO_2Cl_2] = 0.750\ M$?

(d) How long will it take to decompose SO_2Cl_2 so that 33% remains?

39. Iodine-131 is used to treat tumors in the thyroid. Its first-order half-life is 8.1 days. If a patient is given a sample containing 5.00 mg of I-131, how long will it take for 25% of the isotope to remain in her system?

40. Copper-64 is one of the metals used to study brain activity. Its decay constant is $0.0546\ h^{-1}$. If a solution containing 5.00 mg of Cu-64 is used, how many milligrams of Cu-64 remain after eight hours?

41. Argon-41 is used to measure the rate of gas flow. It has a decay constant of $6.3 \times 10^{-3}\ min^{-1}$.

(a) What is its half-life?

(b) How long will it take before only 1.00% of the original amount of Ar-41 is left?

42. A sample of sodium-24 chloride contains 0.050 mg of Na-24 to study the sodium balance of an animal. After 24.9 h, 0.016 mg of Na-24 is left. What is the half-life of Na-24?

Zero- and Second-Order Reactions

43. For the zero-order decomposition of HI on a gold surface

$$HI(g) \xrightarrow{Au} \tfrac{1}{2}H_2(g) + \tfrac{1}{2}I_2(g)$$

it takes 16.0 s for the pressure of HI to drop from 1.00 atm to 0.200 atm.

(a) What is the rate constant for the reaction?

(b) How long will it take for the pressure to drop from 0.150 atm to 0.0432 atm?

(c) What is the half-life of HI at a pressure of 0.500 atm?

44. For the zero-order decomposition of ammonia on tungsten

$$NH_3(g) \xrightarrow{W} \tfrac{1}{2}N_2(g) + \tfrac{3}{2}H_2(g)$$

the rate constant is $2.08 \times 10^{-4}\ mol/L \cdot s$.

(a) What is the half-life of a $0.250\ M$ solution of ammonia?

(b) How long will it take for the concentration of ammonia to drop from $1.25\ M$ to $0.388\ M$?

45. The decomposition of hydrogen iodide

$$HI(g) \longrightarrow \tfrac{1}{2}H_2(g) + \tfrac{1}{2}I_2(g)$$

is second-order. Its half-life is 85 s when the initial concentration is $0.15\ M$.

(a) What is k for the reaction?

(b) How long will it take to go from $0.300\ M$ to $0.100\ M$?

46. The decomposition of nitrogen dioxide

$$NO_2(g) \longrightarrow NO(g) + \tfrac{1}{2}O_2(g)$$

is second-order. It takes 125 s for the concentration of NO_2 to go from $0.800\ M$ to $0.0104\ M$.

(a) What is k for the reaction?

(b) What is the half-life of the reaction when the initial concentration of NO_2 is $0.500\ M$?

47. The decomposition of nitrosyl chloride

$$NOCl(g) \longrightarrow NO(g) + \tfrac{1}{2}Cl_2(g)$$

is a second-order reaction. If it takes 0.20 min to decompose

15% of a 0.300 M solution of nitrosyl chloride, what is k for the reaction?

48. The rate constant for the second-order reaction

$$NOBr(g) \longrightarrow NO(g) + \tfrac{1}{2}Br_2(g)$$

is 48 L/mol · min at a certain temperature. How long will it take to decompose 90.0% of a 0.0200 M solution of nitrosyl bromide?

Activation Energy, Reaction Rate, and Temperature

49. The following data were obtained for the gas-phase decomposition of acetaldehyde:

k (L/mol · s)	0.0105	0.101	0.60	2.92
t(K)	700	750	800	850

Plot these data (ln k versus $1/T$) and find the activation energy for the reaction.

50. The following data were obtained for the reaction:

$$SiH_4(g) \longrightarrow Si(s) + 2H_2(g)$$

k (s^{-1})	0.048	2.3	49	590
t(°C)	500	600	700	800

Plot these data (ln k versus $1/T$) and find the activation energy for the reaction.

51. Cold-blooded animals decrease their body temperature in cold weather to match that of their environment. The activation energy of a certain reaction in a cold-blooded animal is 65 kJ/mol. By what percentage is the rate of the reaction decreased if the body temperature of the animal drops from 35°C to 22°C?

52. The uncoiling of deoxyribonucleic acid (DNA) is a first-order reaction. Its activation energy is 420 kJ. At 37°C, the rate constant is 4.90×10^{-4} min^{-1}.

(a) What is the half-life of the uncoiling at 37°C (normal body temperature)?

(b) What is the half-life of the uncoiling if the organism has a temperature of 40°C ($\approx$ 104°F)?

(c) By what factor does the rate of uncoiling increase (per °C) over this temperature interval?

53. The activation energy for the reaction involved in the souring of raw milk is 75 kJ. Milk will sour in about eight hours at 21°C (70°F = room temperature). How long will raw milk last in a refrigerator maintained at 5°C? Assume the rate constant to be inversely related to souring time.

54. The chirping rate of a cricket X, in chirps per minute near room temperature, is

$$X = 7.2t - 32$$

where t is the temperature in °C.

(a) Calculate the chirping rates at 25°C and 35°C.

(b) Use your answers in (a) to estimate the activation energy for the chirping.

(c) What is the percentage increase for a 10°C rise in temperature?

55. For the reaction

$$HI(g) + CH_3I(g) \longrightarrow CH_4(g) + I_2(g)$$

the rate constant is 0.28 L/mol · s at 300°C and 0.0039 L/mol · s at 227°C.

(a) What is the activation energy of the reaction?

(b) What is k at 400°C?

56. For the reaction

$$C_2H_4(g) + H_2(g) \longrightarrow C_2H_6(g)$$

the activation energy is 181 kJ/mol. The rate constant at 500°C is 0.025 L/mol · s.

(a) At what temperature is the rate constant one-third its value at 500°C?

(b) What is the rate constant at 825°C?

57. The decomposition of ethyl bromide is a first-order reaction. Its activation energy is 226 kJ/mol. At 720 K, its half-life is 78 s. What is its half-life at 815 K?

58. The decomposition of ethyl chloride is a first-order reaction. Its activation energy is 248 kJ/mol. At 470°C, its half-life is 40.8 min. What is its half-life at 425°C?

Reaction Mechanisms

59. Write the rate expression for each of the following elementary steps:

(a) NO + $O_3 \rightarrow NO_2 + O_2$

(b) $2NO_2 \rightarrow 2NO + O_2$

(c) K + HCl $\rightarrow$ KCl + H

60. Write the rate expression for each of the following elementary steps:

(a) NO_3 + CO $\rightarrow NO_2 + CO_2$

(b) $I_2 \rightarrow 2I$

(c) NO + $O_2 \rightarrow NO_3$

61. For the reaction between hydrogen and iodine,

$$H_2(g) + I_2(g) \longrightarrow 2HI(g)$$

the experimental rate expression is rate = $k[H_2] \times [I_2]$. Show that this expression is consistent with the mechanism:

$$I_2(g) \rightleftharpoons 2I(g) \qquad \text{(fast)}$$

$$H_2(g) + I(g) + I(g) \longrightarrow 2HI(g) \qquad \text{(slow)}$$

62. For the reaction

$$2H_2(g) + 2NO(g) \longrightarrow N_2(g) + 2H_2O(g)$$

the experimental rate expression is rate = $k[NO]^2 \times [H_2]$. The following mechanism is proposed:

$$2NO \rightleftharpoons N_2O_2 \qquad \text{(fast)}$$

$$N_2O_2 + H_2 \longrightarrow H_2O + N_2O \qquad \text{(slow)}$$

$$N_2O + H_2 \longrightarrow N_2 + H_2O \qquad \text{(fast)}$$

Is this mechanism consistent with the rate expression?

63. Two mechanisms are proposed for the reaction

$$2NO(g) + O_2(g) \longrightarrow 2NO_2(g)$$

Mechanism 1: $NO + O_2 \rightleftharpoons NO_3$ (fast)

$NO_3 + NO \longrightarrow 2NO_2$ (slow)

Mechanism 2: $NO + NO \rightleftharpoons N_2O_2$ (fast)

$N_2O_2 + O_2 \longrightarrow 2NO_2$ (slow)

Show that each of these mechanisms is consistent with the observed rate law: rate $= k[NO]^2 \times [O_2]$.

WEB 64. At low temperatures, the rate law for the reaction

$$CO(g) + NO_2(g) \longrightarrow CO_2(g) + NO(g)$$

is as follows: rate = constant $\times [NO_2]^2$. Which of the following mechanisms is consistent with the rate law?

(a) $CO + NO_2 \rightarrow CO_2 + NO$
(b) $2NO_2 \rightleftharpoons N_2O_4$ (fast)
$N_2O_4 + 2CO \rightarrow 2CO_2 + 2NO$ (slow)
(c) $2NO_2 \rightarrow NO_3 + NO$ (slow)
$NO_3 + CO \rightarrow NO_2 + CO_2$ (fast)
(d) $2NO_2 \rightarrow 2NO + O_2$ (slow)
$O_2 + 2CO \rightarrow 2CO_2$ (fast)

Unclassified

65. Consider the decomposition of dinitrogen pentaoxide in CCl_4:

$$N_2O_5(g) \longrightarrow N_2O_4(g) + \tfrac{1}{2}O_2(g)$$

The rate constant for this first-order reaction is 0.037/min. A sample of N_2O_5 weighing 43.0 g is dissolved in CCl_4 and allowed to decompose.

(a) How long will it take to reduce the amount of N_2O_5 to 15.0 g?
(b) What total volume of O_2 at 27°C and 1.00 atm is produced if the reaction is stopped after 35 min?

66. The decomposition of nitrosyl bromide

$$NOBr(g) \longrightarrow NO(g) + \tfrac{1}{2}Br_2(g)$$

is a second-order reaction. Its rate constant is 0.80 L/mol·s. The heats of formation are 82.2 kJ/mol for NOBr, 90.2 kJ/mol for NO, and 30.9 kJ/mol for Br_2. If the initial concentration of NOBr is 0.250 M, how much heat is evolved or absorbed per liter of solution in one minute?

67. A reaction has an activation energy of 85 kJ. What percentage increase would one get for k for every 5°C increase in temperature around room temperature, 25°C?

68. How much faster would a reaction proceed at 45°C than at 27°C if the activation energy of the reaction is 133 kJ/mol?

69. For the first-order thermal decomposition of ozone

$$O_3(g) \longrightarrow O_2(g) + O(g)$$

$k = 3 \times 10^{-26}$ s^{-1} at 25°C. What is the half-life for this reaction in years? Comment on the likelihood that this reaction contributes to the depletion of the ozone layer.

70. Derive the integrated rate law, $[A] = [A]_o - kt$, for a zero-order reaction. (*Hint:* Start with the relation $-\Delta[A] = k \,\Delta t$.)

Conceptual Problems

71. The greatest increase in the reaction rate for the reaction between A and C, where rate $= k[A]^{\frac{1}{2}}\{C\}$, is caused by

(a) doubling [A]   (b) halving [C]
(c) halving [A]   (d) doubling [A] and [C]

72. Consider the decomposition of A represented by circles, where each circle represents a molecule of A.

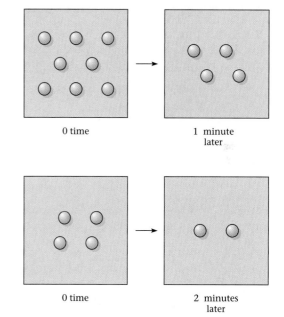

0 time   1 minute later

0 time   2 minutes later

Is the reaction zero-order, first-order, or second-order?

73. Consider the decomposition of B represented by squares, where each square represents a molecule of B.

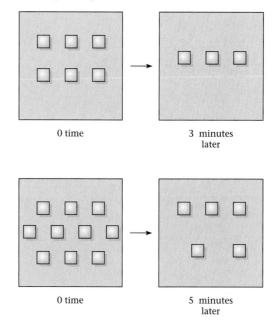

0 time   3 minutes later

0 time   5 minutes later

Is the reaction zero-order, first-order, or second-order?

74. Three first-order reactions have the following activation energies:

Reaction	A	B	C
E_a (kJ)	75	136	292

(a) Which reaction is the fastest?
(b) Which reaction has the largest half-life?
(c) Which reaction has the largest rate?

75. For a certain reaction, E_a is 135 kJ and $\Delta H = 45$ kJ. In the presence of a catalyst, the activation energy is 39% of that for the uncatalyzed reaction. Draw a diagram similar to Figure 11.11 but instead of showing two activated complexes (two humps) show only one activated complex (i.e., only one hump) for the reaction. What is the activation energy of the uncatalyzed reverse reaction?

76. The following reaction is second-order in A and first-order in B.

$$A + B \longrightarrow \text{products}$$

(a) Write the rate expression.
(b) Consider the following one-liter vessel in which each square represents a mole of A and each circle represents a mole of B.

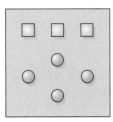

What is the rate of the reaction in terms of k?
(c) Assuming the same rate and k as (b), fill the similar one-liter vessel shown in the figure with an appropriate number of circles (representing B).

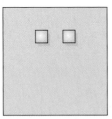

77. Consider the first-order decomposition reaction

$$A \longrightarrow B + C$$

where A (circles) decomposes to B (triangles) and C (squares). Given the following boxes representing the reaction at $t = 2$ minutes, fill in the boxes with products at the indicated time. Estimate the half-life of the reaction.

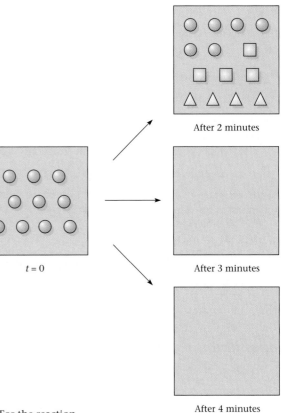

After 2 minutes

$t = 0$

After 3 minutes

After 4 minutes

78. For the reaction

$$A + B \longrightarrow C$$

the rate expression is rate $= k[A][B]$
(a) Given three test tubes, with different concentrations of A and B which test tube has the smallest rate?
  (1) 0.10 M A; 0.10 M B
  (2) 0.15 M A; 0.15 M B
  (3) 0.06 M A; 1.0 M B
(b) If the temperature is increased, describe (using the words *increases*, *decreases*, or *remains the same*) what happens to the rate, the value of k, and E_a.

79. The following experiments are performed for the first-order reaction:

$$A \longrightarrow \text{products}$$

Fill in the blanks. If the answer cannot be calculated with the given information, write NC on the blank(s) provided.

	Expt. 1	Expt. 2	Expt. 3	Expt. 4
[A]	0.100	0.200	0.100	0.100
Catalyst	No	No	Yes	No
Temperature	25°C	25°C	25°C	30°C
k (min^{-1})	0.5	_____	1	0.6
Rate (mol/L·min)	_____	_____	_____	_____
E_a (kJ)	32	_____	_____	_____

Challenge Problems

80. The gas-phase reaction between hydrogen and iodine

$$H_2(g) + I_2(g) \rightleftharpoons 2HI(g)$$

proceeds with a rate constant for the forward reaction at 700°C of 138 L/mol·s and an activation energy of 165 kJ/mol.

(a) Calculate the activation energy of the reverse reaction given that ΔH_f° for HI is 26.48 kJ/mol and ΔH_f° for $I_2(g)$ is 62.44 kJ/mol.

(b) Calculate the rate constant for the reverse reaction at 700°C. (Assume A in the equation $k = Ae^{-E_a/RT}$ is the same for both forward and reverse reactions.)

(c) Calculate the rate of the reverse reaction if the concentration of HI is 0.200 M. The reverse reaction is second-order in HI.

81. For a first-order reaction $aA \rightarrow$ products, where $a \neq 1$, the rate is $-\Delta[A]/a\,\Delta t$, or in derivative notation, $-\dfrac{1}{a}\dfrac{d[A]}{dt}$. Derive the integrated rate law for the first-order decomposition of a moles of reactant.

82. The following data apply to the reaction

$$A(g) + 3B(g) + 2C(g) \longrightarrow products$$

[A]	[B]	[C]	Rate
0.20	0.40	0.10	X
0.40	0.40	0.20	$8X$
0.20	0.20	0.20	X
0.40	0.40	0.10	$4X$

Determine the rate law for the reaction.

83. Using calculus, derive the equation for

(a) the concentration-time relation for a second-order reaction (see Table 11.2).

(b) the concentration-time relation for a third-order reaction, A $\rightarrow$ product.

84. In a first-order reaction, suppose that a quantity X of a reactant is added at regular intervals of time, Δt. At first the amount of reactant in the system builds up; eventually, however, it levels off at a saturation value given by the expression

$$\text{saturation value} = \frac{X}{1 - 10^{-a}}, \qquad \text{where } a = 0.30\,\frac{\Delta t}{t_{1/2}}$$

This analysis applies to prescription drugs, where you take a certain amount each day. Suppose you take 0.100 g of a drug three times a day and that the half-life for elimination is 2.0 days. Using this equation, calculate the mass of the drug in the body at saturation. Suppose further that side effects show up when 0.500 g of the drug accumulates in the body. As a pharmacist, what is the maximum dosage you could assign to a patient for an 8-h period without causing side effects?

12 GASEOUS CHEMICAL EQUILIBRIUM

Equilibrium between liquid
bromine and its gas.
(Charles D. Winters)

CHAPTER OUTLINE

12.1   The N$_2$O$_4$–NO$_2$ Equilibrium System

12.2   The Equilibrium Constant Expression

12.3   Determination of K

12.4   Applications of the Equilibrium Constant

12.5   Effect of Changes in Conditions on an Equilibrium System

A mong the topics covered in Chapter 9 was the equilibrium between liquid and gaseous water:

$$H_2O(l) \rightleftharpoons H_2O(g)$$

The state of this equilibrium system at a given temperature can be described in a simple way by citing the equilibrium pressure of water vapor: 0.034 atm at 25°C, 1.00 atm at 100°C, and so on.

Chemical reactions involving gases carried out in closed containers resemble in many ways the $H_2O(l)$–$H_2O(g)$ system. The reactions are reversible; reactants are not completely consumed. Instead, an equilibrium mixture containing both products and reactants is obtained. At equilibrium, forward and reverse reactions take place at the same rate. As a result, the amounts of all species at equilibrium remain constant with time.

Ordinarily, where gaseous chemical equilibria are involved, more than one gaseous substance is present. We might, for example, have a system,

$$aA(g) + bB(g) \rightleftharpoons cC(g) + dD(g)$$

in which the equilibrium mixture contains two gaseous products (C and D) and two gaseous reactants (A and B). To describe the position of this equilibrium, we must cite the ***partial pressure*** of each species, that is, P_C, P_D, P_A, and P_B.

Recall from Chapter 5 that the partial pressure of gas, i, in a mixture is given by the expression

$$P_i = n_i RT/V$$

It follows that, for an equilibrium mixture confined in a closed container of volume V at temperature T, *the partial pressure, P_i, of each species is directly proportional to the number of moles, n_i, of that species.*

It turns out that there is a relatively simple relationship between the partial pressures of different gases present at equilibrium in a reaction system. This relationship is expressed in terms of a quantity called the *equilibrium constant,* symbol K.* In this chapter, you will learn how to—

■ write the expression for K corresponding to any chemical equilibrium (Section 12.2).

> They can often be reversed by changing the temperature or pressure.

---

*Thermodynamically, the quantity that should appear for each species in the expression for K is the *activity* rather than the partial pressure. Activity is defined as the ratio of the equilibrium partial pressure to the standard pressure, 1 atm. This means that the activity and hence the equilibrium constant K are dimensionless, that is, a pure number without units.

Don't confuse the *equilibrium* constant *K* with the *rate* constant, *k*.

- calculate the value for *K* from experimental data for the equilibrium system (Section 12.3).

- use the value of *K* to predict the extent to which a reaction will take place (Section 12.4).

- use *K*, along with Le Châtelier's principle, to predict the result of disturbing an equilibrium system (Section 12.5).

12.1   THE N_2O_4–NO_2 EQUILIBRIUM SYSTEM

Consider what happens when a sample of N_2O_4, a colorless gas, is placed in a closed, evacuated container at 100°C. Instantly, a reddish-brown color develops. This color is due to nitrogen dioxide, NO_2, formed by decomposition of part of the N_2O_4 (Figure 12.1):

$$N_2O_4(g) \longrightarrow 2NO_2(g)$$

At first, this is the only reaction taking place. As soon as some NO_2 is formed, however, the reverse reaction can occur:

$$2NO_2(g) \longrightarrow N_2O_4(g)$$

As time passes, the partial pressure of N_2O_4 drops, and the forward reaction slows down. Conversely, the partial pressure of NO_2 increases, so the rate of the

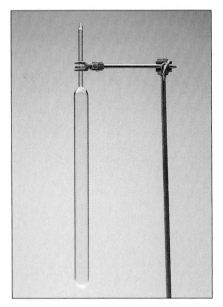
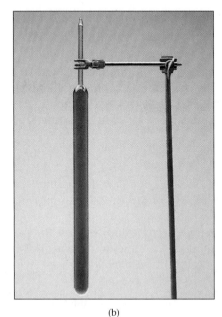
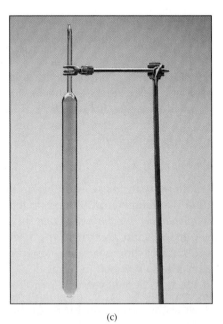

(a)   (b)   (c)

FIGURE 12.1

Gases of the N_2O_4–NO_2 system. (a) Pure dinitrogen tetroxide, N_2O_4, a colorless gas. (b) Pure nitrogen dioxide, NO_2, a brownish-red gas. (c) An equilibrium mixture of the two gases in which the partial pressure of NO_2 is directly related to the depth of the brown color. *(Marna G. Clarke)*

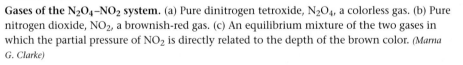

TABLE 12.1	Establishment of Equilibrium in the System $N_2O_4(g) \rightleftharpoons 2NO_2(g)$ (at 100°C)					
Time (s)	0	20	40	60	80	100
$P_{N_2O_4}$ (atm)	1.00	0.60	0.35	**0.22***	**0.22**	**0.22**
P_{NO_2} (atm)	0.00	0.80	1.30	**1.56**	**1.56**	**1.56**

*Boldface numbers are equilibrium pressures.

reverse reaction increases. Soon, these rates become equal. A dynamic equilibrium has been established.

$$N_2O_4(g) \rightleftharpoons 2NO_2(g)$$

At equilibrium, appreciable amounts of both gases are present. As time passes, their amounts or partial pressures remain constant, as long as the volume of the container and the temperature remain unchanged.

The characteristics just described are typical of all systems at equilibrium. First, *the forward and reverse reactions are taking place at the same rate*. This explains why *the concentrations of species present remain constant with time*. Moreover, these concentrations are independent of the direction from which equilibrium is approached.

The approach to equilibrium in the N_2O_4–NO_2 system is illustrated by the data in Table 12.1 and by Figure 12.2. Originally, only N_2O_4 is present; its pressure is 1.00 atm. Because there is no NO_2 around, its original pressure is zero. As equilibrium is approached, the overall reaction is

$$N_2O_4(g) \longrightarrow 2NO_2(g)$$

The amount and hence the partial pressure of N_2O_4 drop, rapidly at first, then more slowly. The partial pressure of NO_2 increases. Finally, both partial pressures level off and become constant. At equilibrium, at 100°C:

$$P_{N_2O_4} = 0.22 \text{ atm} \qquad P_{NO_2} = 1.56 \text{ atm}$$

These pressures don't change, because forward and reverse reactions occur at the same rate.

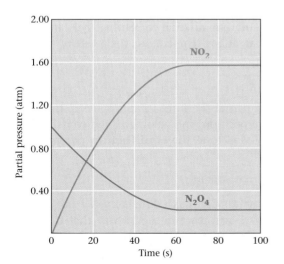

FIGURE 12.2
Approach to equilibrium in the N_2O_4–NO_2 system. The partial pressure of N_2O_4 starts off at 1.00 atm, drops sharply at first, and finally levels off at the equilibrium value of 0.22 atm. Meanwhile, the partial pressure of NO_2 rises from zero to its equilibrium value, 1.56 atm. **OHT**

TABLE 12.2 Equilibrium Measurements in the N_2O_4–NO_2 System at 100°C		Original Pressure (atm)	Equilibrium Pressure (atm)
Expt. 1	N_2O_4	1.00	0.22
	NO_2	0.00	1.56
Expt. 2	N_2O_4	0.00	0.07
	NO_2	1.00	0.86
Expt. 3	N_2O_4	1.00	0.42
	NO_2	1.00	2.16

There are many ways to approach equilibrium in the N_2O_4–NO_2 system. Table 12.2 gives data for three experiments in which the original conditions are quite different. Experiment 1 is that just described, starting with pure N_2O_4. Experiment 2 starts with pure NO_2 at a partial pressure of 1.00 atm. As equilibrium is approached, some of the NO_2 reacts to form N_2O_4:

$$2NO_2(g) \longrightarrow N_2O_4(g)$$

Finally, in Experiment 3, both N_2O_4 and NO_2 are present originally, each at a partial pressure of 1.00 atm.

Looking at the data in Table 12.2, you might wonder whether these three experiments have anything in common. Specifically, is there any relationship between the equilibrium partial pressures of NO_2 and N_2O_4 that is valid for all the experiments? It turns out that there is, although it is not an obvious one. The value of the quotient $(P_{NO_2})^2/P_{N_2O_4}$ is the same, about 11, in each case:

Expt. 1 $\quad \dfrac{(P_{NO_2})^2}{P_{N_2O_4}} = \dfrac{(1.56)^2}{0.22} = 11$

Expt. 2 $\quad \dfrac{(P_{NO_2})^2}{P_{N_2O_4}} = \dfrac{(0.86)^2}{0.07} = 11$

Expt. 3 $\quad \dfrac{(P_{NO_2})^2}{P_{N_2O_4}} = \dfrac{(2.16)^2}{0.42} = 11$

Note that this relationship holds only at equilibrium; initial pressures can have any values.

This relationship holds for any equilibrium mixture containing N_2O_4 and NO_2 at 100°C. More generally, it is found that, at any temperature, the quantity

$$\dfrac{(P_{NO_2})^2}{P_{N_2O_4}}$$

where P_{NO_2} and $P_{N_2O_4}$ are equilibrium partial pressures in atmospheres, is a constant, independent of the original composition, the volume of the container, or the total pressure. This constant is referred to as the **equilibrium constant K** for the system

$$N_2O_4(g) \rightleftharpoons 2NO_2(g)$$

See the *Saunders Interactive General Chemistry CD-ROM*, Screen 16.4, The Equilibrium Constant.

At 100°C, K for this system is 11; at 150°C, it has a different value, about 110. Any mixture of NO_2 and N_2O_4 at 100°C will react in such a way that the ratio $(P_{NO_2})^2/P_{N_2O_4}$ becomes equal to 11. At 150°C, reaction occurs until this ratio becomes 110.

12.2   THE EQUILIBRIUM CONSTANT EXPRESSION

For every gaseous chemical system, an **equilibrium constant expression** can be written stating the condition that must be attained at equilibrium. For the general system involving only gases,

$$a\,A(g) + b\,B(g) \rightleftharpoons c\,C(g) + d\,D(g)$$

where A, B, C, and D represent different substances and a, b, c, and d are their coefficients in the balanced equation

$$K = \frac{(P_C)^c \times (P_D)^d}{(P_A)^a \times (P_B)^b}$$

where P_C, P_D, P_A, and P_B are the partial pressures of the four gases at equilibrium.* *These partial pressures must be expressed in atmospheres.* Notice that in the expression for K—

- the equilibrium partial pressures of *products* (right side of equation) appear in the *numerator*.
- the equilibrium partial pressures of *reactants* (left side of equation) appear in the *denominator*.
- each partial pressure is raised to a *power* equal to its *coefficient* in the balanced equation.

See Screens 16.5, The Meaning of the Equilibrium Constant, & 16.6, Writing Equilibrium Exercises.

EXAMPLE 12.1   Ammonia can be made from hydrogen and nitrogen

$$N_2(g) + 3H_2(g) \rightleftharpoons 2NH_3(g)$$

Write the equilibrium constant expression for this system.

Strategy NH_3 is a product, so P_{NH_3} appears in the numerator; the partial pressures of N_2 and H_2 are in the denominator. The exponents are the coefficients in the balanced equation: 2 for NH_3, 1 for N_2, 3 for H_2.

Solution NH_3 is a product; H_2 and N_2 are reactants. The expression is

$$K = \frac{(P_{NH_3})^2}{P_{N_2} \times (P_{H_2})^3}$$

---

*This equilibrium constant is often given the symbol K_p to emphasize that it involves partial pressures. Other equilibrium expressions for gases are sometimes used, including K_c:

$$K_c = \frac{[C]^c \times [D]^d}{[A]^a \times [B]^b}$$

where the brackets represent equilibrium concentrations in moles per liter. The two constants, K_c and K_p, are simply related (see Problem 65):

$$K_p = K_c(RT)^{\Delta n_g}$$

where Δn_g is the change in the number of moles of gas in the equation, R is the gas law constant, 0.0821 L·atm/mol·K, and T is the Kelvin temperature. Throughout this chapter, we deal only with K_p, referring to it simply as K.

Changing the Chemical Equation

K is meaningless unless accompanied by a chemical equation.

It is important to realize that *the expression for K depends on the form of the chemical equation written to describe the equilibrium system.* To illustrate what this statement means, consider the N_2O_4–NO_2 system:

$$N_2O_4(g) \rightleftharpoons 2NO_2(g) \qquad K = \frac{(P_{NO_2})^2}{P_{N_2O_4}}$$

Many other equations could be written for this system, for example,

$$\tfrac{1}{2}N_2O_4(g) \rightleftharpoons NO_2(g)$$

In this case, the expression for the equilibrium constant would be

$$K' = \frac{P_{NO_2}}{(P_{N_2O_4})^{1/2}}$$

Comparing the expressions for *K* and *K'*, it is clear that *K'* is the square root of *K*. This illustrates a general rule, sometimes referred to as the **coefficient rule,** which states,

 See Screen 16.7, Manipulating Equilibrium Expressions.

If the coefficients in a balanced equation are multiplied by a factor n, the equilibrium constant is raised to the nth power:

$$K' = K^n$$

In this particular case, $n = \tfrac{1}{2}$ because the coefficients were divided by 2. Hence *K'* is the square root of *K*. If $K = 11$, then $K' = (11)^{1/2} = 3.3$.

Another equation that might be written to describe the N_2O_4–NO_2 system is

$$2NO_2(g) \rightleftharpoons N_2O_4(g)$$

for which the equilibrium constant expression is

$$K'' = \frac{P_{N_2O_4}}{(P_{NO_2})^2}$$

This chemical equation is simply the reverse of that written originally; N_2O_4 and NO_2 switch sides of the equation. Notice that *K''* is the reciprocal of *K*; the numerator and denominator have been inverted. This illustrates the **reciprocal rule:**

The equilibrium constants for forward and reverse reactions are the reciprocals of each other,

$$K'' = 1/K$$

If, for example, $K = 11$, then $K'' = 1/11 = 0.091$.

Adding Chemical Equations

A property of *K* that you will find very useful in this and succeeding chapters is expressed by the **rule of multiple equilibria,** which states

We will use this relationship frequently in later chapters.

If a reaction can be expressed as the sum of two or more reactions, K for the overall reaction is the product of the equilibrium constants of the individual reactions.

That is, if

$$\text{Reaction 3} = \text{Reaction 1} + \text{Reaction 2}$$

then

$$K \text{ (Reaction 3)} = K \text{ (Reaction 1)} \times K \text{ (Reaction 2)}$$

To illustrate the application of this rule, consider the following reactions at 700°C:

$$SO_2(g) + \tfrac{1}{2}O_2(g) \rightleftharpoons SO_3(g) \qquad\qquad K = 2.2$$

$$NO_2(g) \rightleftharpoons NO(g) + \tfrac{1}{2}O_2(g) \qquad K = 4.0$$

Adding these equations eliminates $\tfrac{1}{2}O_2$; the result is

$$SO_2(g) + NO_2(g) \rightleftharpoons SO_3(g) + NO(g)$$

For this overall reaction, $K = 2.2 \times 4.0 = 8.8$.

The validity of this rule can be demonstrated by writing the expression for K for the individual reactions and multiplying:

$$\frac{P_{SO_3}}{P_{SO_2} \times (P_{O_2})^{1/2}} \times \frac{P_{NO} \times (P_{O_2})^{1/2}}{P_{NO_2}} = \frac{P_{SO_3} \times P_{NO}}{P_{SO_2} \times P_{NO_2}}$$

The product expression, at the right, is clearly the equilibrium constant for the overall reaction, as it should be according to the rule of multiple equilibria.

Heterogeneous Equilibria

In the reactions described so far, all the reactants and products have been gaseous; the equilibrium systems are *homogeneous*. In certain reactions, one or more of the substances involved is a pure liquid or solid; the others are gases. Such a system is *heterogeneous*, because more than one phase is present. Examples include

$$CO_2(g) + H_2(g) \rightleftharpoons CO(g) + H_2O(l)$$

$$I_2(s) \rightleftharpoons I_2(g)$$

Experimentally, it is found that for such systems—

■ *the position of the equilibrium is independent of the amount of solid or liquid, as long as some is present.*
■ *terms for pure liquids or solids need not appear in the expression for K.*

To see why this is the case, consider the data shown in Table 12.3 for the system

$$CO_2(g) + H_2(g) \rightleftharpoons CO(g) + H_2O(l)$$

TABLE 12.3   Equilibrium Constant Expressions for
$CO_2(g) + H_2(g) \rightleftharpoons CO(g) + H_2O(l)$

	Expt. 1	Expt. 2	Expt. 3	Expt. 4
Mass $H_2O(l)$	8 g	6 g	4 g	2 g
P_{H_2O} (atm)	3×10^{-2}	3×10^{-2}	3×10^{-2}	3×10^{-2}
K_I	9×10^{-6}	9×10^{-6}	9×10^{-6}	9×10^{-6}
K_{II}	3×10^{-4}	3×10^{-4}	3×10^{-4}	3×10^{-4}

$$K_I = \frac{P_{CO} \times P_{H_2O}}{P_{CO_2} \times P_{H_2}} \qquad\qquad K_{II} = \frac{P_{CO}}{P_{CO_2} \times P_{H_2}}$$

at 25°C. In Experiments 1 through 4, there is liquid water present, so $P_{H_2O} = 0.03$ atm, the equilibrium vapor pressure of water at 25°C. Notice that—

1. In all cases, the quantity

$$K_I = \frac{P_{CO} \times P_{H_2O}}{P_{CO_2} \times P_{H_2}}$$

is constant at 9×10^{-6}.

2. In all cases (Experiments 1 through 4), the quantity

$$K_{II} = \frac{P_{CO}}{P_{CO_2} \times P_{H_2}}$$

is constant at 3×10^{-4}. This must be true because

$$K_{II} = K_I/P_{H_2O}$$

and P_{H_2O} is constant at 3×10^{-2} atm. It follows that K_{II} is a valid equilibrium constant for the system

$$CO_2(g) + H_2(g) \rightleftharpoons CO(g) + H_2O(l)$$

Because the expression for K_{II} is simpler than that for K_I, it is the equilibrium constant of choice for the heterogeneous system.

A similar argument can be applied to the equilibrium involved in the sublimation of iodine:

$$I_2(s) \rightleftharpoons I_2(g)$$

At a given temperature, the pressure of iodine vapor is constant, independent of the amount of solid iodine or any other factor. The equilibrium constant expression is

$$K = P_{I_2}$$

Applying the reciprocal rule (page 356), we can deduce the equilibrium constant expression for the reverse reaction

$$I_2(g) \rightleftharpoons I_2(s) \qquad K = 1/P_{I_2}$$

A heterogeneous equilibrium system: solid I_2–gaseous I_2.

(Charles D. Winters)

EXAMPLE 12.2   Write the expression for K for

(a) the reduction of black solid copper(II) oxide (1 mol) with hydrogen to form copper metal and steam.
(b) the reaction of 1 mol of steam with red-hot coke (carbon) to form a mixture of hydrogen and carbon monoxide, called water gas.

Strategy First write the chemical equation for the equilibrium system. Then write the expression for K, leaving out pure liquids and solids. Remember that gases are represented by their partial pressures.

Solution

(a) The reaction is $CuO(s) + H_2(g) \rightleftharpoons Cu(s) + H_2O(g)$

The expression for K is $K = \dfrac{P_{H_2O}}{P_{H_2}}$

If the product were $H_2O(l)$, $K = 1/P_{H_2}$.

(b) The reaction is $C(s) + H_2O(g) \rightleftharpoons H_2(g) + CO(g)$

The expression for K is $K = \dfrac{P_{H_2} \times P_{CO}}{P_{H_2O}}$

In this and succeeding chapters, a wide variety of different types of equilibria will be covered. They may involve gases, pure liquids or solids, and species in aqueous solution. It will always be true that in the expression for the equilibrium constant—

■ *gases enter as their partial pressures in atmospheres.*
■ *pure liquids or solids do not appear; neither does the solvent for a reaction in dilute solution.*
■ *species (ions or molecules) in water solution enter as their molar concentrations.*

To express *K* for any system, follow these rules.

Thus for the equilibrium attained when zinc metal reacts with acid:

$$Zn(s) + 2H^+(aq) \rightleftharpoons Zn^{2+}(aq) + H_2(g)$$

$$K = \dfrac{P_{H_2} \times [Zn^{2+}]}{[H^+]^2}$$

We will have more to say in later chapters about equilibria involving species in aqueous solution.

12.3   DETERMINATION OF K

Numerical values of equilibrium constants can be calculated if the partial pressures of products and reactants at equilibrium are known. Sometimes you will be given equilibrium partial pressures directly (Example 12.3). At other times you will be given the original partial pressures and the equilibrium partial pressure of one species (Example 12.4). In that case, the calculation of *K* is a bit more difficult, because you have to calculate the equilibrium partial pressures of all the species.

See Screen 16.8, Determining an Equilibrium Constant.

EXAMPLE 12.3   Ammonium chloride is sometimes used as a flux in soldering because it decomposes on heating:

$$NH_4Cl(s) \rightleftharpoons NH_3(g) + HCl(g)$$

The HCl formed removes oxide films from metals to be soldered. In a certain equilibrium system at 400°C, 12.0 g of NH$_4$Cl is present; the partial pressures of NH$_3$ and HCl are 3.0 atm and 5.0 atm, respectively. Calculate *K* at 400°C.

Strategy   First write the expression for *K*. Then substitute equilibrium partial pressures into that expression and solve.

Solution $K = (P_{NH_3}) \times (P_{HCl}) = (3.0) \times (5.0) = $ 15

EXAMPLE 12.4   Consider the equilibrium system

$$2HI(g) \rightleftharpoons H_2(g) + I_2(g)$$

Originally, a system contains only HI at a pressure of 1.00 atm at 520°C. The equilibrium partial pressure of H_2 is found to be 0.10 atm. Calculate
(a) P_{I_2} at equilibrium   (b) P_{HI} at equilibrium   (c) K

Strategy   To solve this problem, you must—

▪ carefully distinguish between *original* and *equilibrium* partial pressures.
▪ learn how to relate the equilibrium partial pressures of H_2, I_2, and HI, using the coefficients of the balanced equation.

Solution

(a) Note from the balanced equation that one mole of I_2 is formed for every mole of H_2. Putting it another way, as reaction proceeds from left to right, the increase in the number of moles of I_2 is the same as that for H_2:

$$\Delta n_{I_2} = \Delta n_{H_2}$$

But because $P = nRT/V$, it follows that for an equilibrium system at constant T and V, pressure is directly proportional to amount in moles, and the above relationship must hold for partial pressures as well:

$$\Delta P_{I_2} = \Delta P_{H_2}$$

Because both partial pressures start at zero, the equilibrium partial pressures of H_2 and I_2 must be equal. The equilibrium partial pressure of I_2, like that of H_2, is 0.10 atm.
    (b) Because two moles of HI are required to form one mole of H_2

$$\Delta n_{HI} = -2\,\Delta n_{H_2}$$

where the minus sign takes account of the fact that when the amount of hydrogen increases, the amount of HI decreases. But because P is directly proportional to n,

$$\Delta P_{HI} = -2\,\Delta P_{H_2} = -2(0.10\ \text{atm}) = -0.20\ \text{atm}$$

In other words, the partial pressure of HI decreases by 0.20 atm. Because it starts at 1.00 atm, its equilibrium value must be 1.00 atm − 0.20 atm = 0.80 atm.
    You will find it helpful, here and elsewhere, to summarize this reasoning in a table set up as follows:

If $\Delta P_{H_2} = x$, then $\Delta P_{I_2} = x$ and $\Delta P_{HI} = -2x$.

	2HI(g) $\rightleftharpoons$	H₂(g) +	I₂(g)
P_o (atm)	1.00	0.00	0.00
ΔP (atm)	−0.20	+0.10	+0.10
P_{eq} (atm)	0.80	0.10	0.10

(P_o = initial pressure; P_{eq} = equilibrium pressure)

Note that the numbers in the center row (ΔP) are related through the coefficients of the balanced equation (2HI, 1H_2, 1I_2).

(c) $K = \dfrac{P_{H_2} \times P_{I_2}}{(P_{HI})^2} = \dfrac{0.10 \times 0.10}{(0.80)^2} = $ 0.016

Reality Check   You might expect that K would be rather small (0.016). Recall that the equilibrium partial pressure of H_2 (0.10 atm) is only 1/10 of the original pressure of HI (1.00 atm).

Example 12.4 illustrates a principle that you will find very useful in solving equilibrium problems throughout this (and later) chapters. *Changes in partial pressures of reactants and products as a system approaches equilibrium, like changes in molar amounts, are related to one another through the coefficients of the balanced equation.*

12.4   APPLICATIONS OF THE EQUILIBRIUM CONSTANT

Sometimes, knowing only the magnitude of the equilibrium constant, it is possible to decide on the feasibility of a reaction. Consider, for example, a possible method for "fixing" atmospheric nitrogen—converting it to a compound—by reaction with oxygen:

$$N_2(g) + O_2(g) \rightleftharpoons 2NO(g)$$

$$K = \frac{(P_{NO})^2}{P_{N_2} \times P_{O_2}} = 1 \times 10^{-30} \text{ at } 25°C$$

Because K is so small, the partial pressure of NO in equilibrium with N_2 and O_2, and hence the amount of NO, must be extremely small, approaching zero. Clearly, this would not be a suitable way to fix nitrogen, at least at 25°C.

An alternate approach to nitrogen fixation involves reacting it with hydrogen:

$$N_2(g) + 3H_2(g) \rightleftharpoons 2NH_3(g)$$

$$K = \frac{(P_{NH_3})^2}{P_{N_2} \times (P_{H_2})^3} = 6 \times 10^5 \text{ at } 25°C$$

In this case, the equilibrium system must contain mostly ammonia. A mixture of N_2 and H_2 should be almost completely converted to NH_3 at equilibrium.

In general, if K is a very small number, the equilibrium mixture will contain mostly unreacted starting materials; for all practical purposes, the forward reaction does not go. Conversely, a large K implies a reaction that, at least in principle, is feasible; products should be formed in high yield. Frequently, K has an intermediate value, in which case you must make quantitative calculations concerning the *direction* or *extent* of reaction.

> The concept of equilibrium is most useful when K is neither very large nor very small.

Direction of Reaction; the Reaction Quotient (Q)

Consider the general gas-phase reaction

$$aA(g) + bB(g) \rightleftharpoons cC(g) + dD(g)$$

for which

$$K = \frac{(P_C)^c \times (P_D)^d}{(P_A)^a \times (P_B)^b}$$

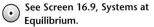

See Screen 16.9, Systems at Equilibrium.

As we have seen, for a given system at a particular temperature, the value of K is fixed.

In contrast, the *beginning* pressure ratio, Q,

$$Q = \frac{(P_C)^c \times (P_D)^d}{(P_A)^a \times (P_B)^b} \quad \text{(color is used to distinguish actual from equilibrium pressures)}$$

can have any value at all. If you start with pure reactants (A and B), P_C and P_D are zero, and the value of Q is zero:

$$Q = \frac{0 \times 0}{(P_A)^a \times (P_B)^b} = 0$$

Conversely, starting with only C and D, the value of Q is infinite:

$$Q = \frac{(P_C)^c \times (P_D)^d}{0 \times 0} \longrightarrow \infty$$

If all four substances are present, Q can have any value between zero and infinity.

The form of the expression for Q, known as the **reaction quotient,** is the same as that for the equilibrium constant, K. The difference is that the partial pressures that appear in Q are those that apply at a particular moment, not necessarily when the system is at equilibrium. By comparing the numerical value of Q with that of K, it is possible to decide in which direction the system will move to achieve equilibrium.

1. If $Q < K$ the reaction proceeds from left to right:

$$aA(g) + bB(g) \longrightarrow cC(g) + dD(g)$$

In this way, the partial pressures of products increase, while those of reactants decrease. As this happens, the reaction quotient Q increases and eventually at equilibrium becomes equal to K.

2. If $Q > K$ the partial pressures of products are "too high" and those of the reactants "too low" to meet the equilibrium condition. Reaction proceeds in the reverse direction:

$$aA(g) + bB(g) \longleftarrow cC(g) + dD(g)$$

increasing the partial pressures of A and B while reducing those of C and D. This lowers Q to its equilibrium value, K.

3. If perchance $Q = K$, the system is already at equilibrium, so the amounts of products and reactants remain unchanged.

> As equilibrium is approached, Q approaches K.

> $Q < K; \rightarrow$

> $Q > K; \leftarrow$

EXAMPLE 12.5 Consider the following system at 100°C:

$$N_2O_4(g) \rightleftharpoons 2NO_2(g) \qquad K = 11$$

Predict the direction in which reaction will occur to reach equilibrium starting with 0.20 mol of N_2O_4 and 0.20 mol of NO_2 in a 4.0-L container.

Strategy First, calculate the partial pressures of N_2O_4 and NO_2, using the ideal gas law, as applied to mixtures: $P_i = n_i RT/V$. Then, calculate Q. Finally, compare Q and K to predict the direction of reaction.

Solution

1. $P_{N_2O_4} = P_{NO_2} = \dfrac{nRT}{V} = \dfrac{(0.20 \text{ mol})(0.0821 \text{ L·atm/mol·K})(373 \text{ K})}{4.0 \text{ L}} = 1.5 \text{ atm}$

2. $Q = \dfrac{(P_{NO_2})^2}{P_{N_2O_4}} = \dfrac{(1.5)^2}{1.5} = 1.5$

3. Because $Q < 11$, reaction must proceed in the forward direction.

Reality Check When reaction occurs in the forward direction, the partial pressure of NO_2 increases, that of N_2O_4 decreases, and Q becomes equal to K.

Extent of Reaction; Equilibrium Partial Pressures

The equilibrium constant for a chemical system can be used to calculate the partial pressures of the species present at equilibrium. In the simplest case, one equilibrium pressure can be calculated, knowing all the others. Consider, for example, the system

$$N_2(g) + O_2(g) \rightleftharpoons 2NO(g)$$

$$K = \frac{(P_{NO})^2}{P_{N_2} \times P_{O_2}} = 1 \times 10^{-30} \text{ at } 25°C$$

In air, the partial pressures of N_2 and O_2 are about 0.78 and 0.21 atm, respectively. The expression for K can be used to calculate the equilibrium partial pressure of NO under these conditions:

$$(P_{NO})^2 = P_{N_2} \times P_{O_2} \times K$$

$$= (0.78)(0.21)(1 \times 10^{-30}) = 1.6 \times 10^{-31}$$

$$P_{NO} = (1.6 \times 10^{-31})^{1/2} = 4 \times 10^{-16} \text{ atm}$$

This is an extremely small pressure, as you might have guessed from the small magnitude of the equilibrium constant.

More commonly, K is used to determine the equilibrium partial pressures of all species, reactants, and products, knowing their original pressures. To do this, you follow what amounts to a four-step path.

1. *Using the balanced equation for the reaction, write the expression for K.*
2. *Express the equilibrium partial pressures of all species in terms of a single unknown, x.* To do this, apply the principle mentioned earlier: *The changes in partial pressures of reactants and products are related through the coefficients of the balanced equation.* To keep track of these values, make an equilibrium table, like the one illustrated in Example 12.6.
3. *Substitute the equilibrium terms into the expression for K. This gives an algebraic equation that must be solved for x.*
4. *Having found x, refer back to the table and calculate the equilibrium partial pressures of all species.*

Automobile exhaust. The reaction of nitrogen with oxygen at high temperatures forms some NO. *(Charles D. Winters)*

⌐ That's just as well; if N_2 and O_2 in the atmosphere were converted to NO, we'd be in big trouble.

⊙ See Screen 16.10, Estimating Equilibrium Concentrations.

EXAMPLE 12.6 For the system

$$CO_2(g) + H_2(g) \rightleftharpoons CO(g) + H_2O(g)$$

K is 0.64 at 900 K. Originally, only CO_2 and H_2 are present, each at a partial pressure of 1.00 atm. What are the equilibrium partial pressures of all species?

Strategy Follow the four-step procedure described above. Be particularly careful about step (2), the most critical one in the process.

Solution

(1) $K = \dfrac{P_{CO} \times P_{H_2O}}{P_{CO_2} \times P_{H_2}}$

(2) You must express all equilibrium partial pressures in terms of a single unknown, x. A reasonable choice for x is

x = change in partial pressure of CO as the system goes to equilibrium

$= \Delta P_{CO}$

All the coefficients in the balanced equation are the same, 1. It follows that all the partial pressures change by the same amount, x.

$$\Delta P_{H_2O} = \Delta P_{CO} = x$$

$$\Delta P_{H_2} = \Delta P_{CO_2} = -x$$

(The minus sign arises because H_2 and CO_2 are consumed, whereas CO and H_2O are formed.) The equilibrium table becomes

> You must distinguish carefully between initial pressures (P_O) and equilibrium pressures.

	$CO_2(g)$	+ $H_2(g)$	$\rightleftharpoons$ CO(g)	+ $H_2O(g)$
P_o (atm)	1.00	1.00	0	0
ΔP (atm)	$-x$	$-x$	$+x$	$+x$
P_{eq} (atm)	$1.00 - x$	$1.00 - x$	x	x

(3) Substituting into the expression for K,

$$0.64 = \frac{(x)(x)}{(1.00 - x)(1.00 - x)}$$

Taking the square root of both sides,

$$(0.64)^{1/2} = 0.80 = \frac{x}{1.00 - x}$$

Solving for x:

> If you should get a negative value for P_{eq}, something is wrong. Check your math and your reasoning.

$$x = 0.80 - 0.80x \qquad 1.80x = 0.80 \qquad x = 0.44$$

(4) Referring back to the equilibrium table,

$$P_{CO} = P_{H_2O} = x = \boxed{0.44 \text{ atm}}$$

$$P_{CO_2} = P_{H_2} = 1.00 - x = \boxed{0.56 \text{ atm}}$$

Reality Check To check the validity of the calculation, note that

$$\frac{P_{CO} \times P_{H_2O}}{P_{CO_2} \times P_{H_2}} = \frac{(0.44)^2}{(0.56)^2} = 0.64 = K$$

The arithmetic involved in equilibrium calculations can be relatively simple, as it was in Example 12.6. Sometimes it is more complex (Example 12.7). The reasoning involved, however, is the same. It is always helpful to set up an equilibrium table to summarize the analysis of the problem.

EXAMPLE 12.7 Consider the system

$$N_2O_4(g) \rightleftharpoons 2NO_2(g) \qquad K = 11 \text{ at } 100°C$$

Starting with pure N_2O_4 at a pressure of 1.00 atm, what will be the equilibrium partial pressures of NO_2 and N_2O_4?

Strategy Again, follow the four-step path. The second step is a bit trickier here because the coefficients in the balanced equation are not equal to one another. The third step is more difficult algebraically, as you will find.

Solution

(1) $K = \dfrac{(P_{NO_2})^2}{P_{N_2O_4}}$

(2) Some of the N_2O_4 will decompose; let x be the decrease in the partial pressure of N_2O_4, i.e., $\Delta P_{N_2O_4} = -x$. The balanced equation shows that one N_2O_4 yields two NO_2; it follows that the change in partial pressure of NO_2 must be $2x$. That is, $\Delta P_{NO_2} = 2x$. The equilibrium table becomes

$$N_2O_4(g) \rightleftharpoons 2NO_2(g)$$

P_o (atm)	1.00	0.00
ΔP (atm)	$-x$	$+2x$
P_{eq} (atm)	$1.00 - x$	$2x$

(3) $K = \dfrac{(2x)^2}{(1.00 - x)} = \dfrac{4x^2}{(1.00 - x)} = 11$

This time, you cannot solve for x as simply as in Example 12.6; the denominator on the left side is not a perfect square. Use the general method of solving a quadratic equation. This involves rearranging to the form

$$ax^2 + bx + c = 0$$

and applying the quadratic formula

$$x = \frac{-b \pm \sqrt{b^2 - 4ac}}{2a}$$

Putting the equilibrium constant expression in the desired form,

$$4x^2 + 11x - 11 = 0$$

So $a = 4, b = 11, c = -11$

$$x = \frac{-11 \pm \sqrt{121 + 16(11)}}{8} = \frac{-11 \pm \sqrt{297}}{8}$$

Noting that the square root of 297 is 17.2,

$$x = \frac{-11 \pm 17.2}{8} = \frac{6.2}{8} \quad \text{or} \quad \frac{-28.2}{8} = 0.78 \quad \text{or} \quad -3.52$$

Of the two answers, only 0.78 is plausible; a value of -3.52 would imply a negative partial pressure of NO_2, which is ridiculous.

(4) $P_{NO_2} = 2(0.78 \text{ atm}) = \boxed{1.56 \text{ atm}}$

$P_{N_2O_4} = 1.00 \text{ atm} - 0.78 \text{ atm} = \boxed{0.22 \text{ atm}}$

Reality Check   In other words, 22% of the N_2O_4 originally present is left when equilibrium is established; 78% is converted to NO_2.

For the most part, we'll avoid tedious calculations of this type.

You may be curious about what would have happened in Example 12.7 with a different choice of variable. Suppose, for example, x had been chosen to be the *increase* in the partial pressure of NO_2. In that case, the partial pressure of N_2O_4 at equilibrium would have decreased by $\frac{1}{2}x$, so

$$P_{NO_2} = x \qquad P_{N_2O_4} = 1.00 - \frac{1}{2}x$$

This leads to the algebraic equation

$$\frac{x^2}{(1.00 - \frac{1}{2}x)} = 11$$

Solving this equation, you should find that $x = 1.56$, so $P_{NO_2} = 1.56$ atm, $P_{N_2O_4} = 1.00$ atm $- \frac{1}{2}(1.56$ atm$) = 0.22$ atm. These are, of course, the same values obtained with different choice of unknown in Example 12.7. In general, it doesn't matter what choice of unknown you make, provided you are consistent in relating equilibrium pressures.

We should point out that the calculations involved in Examples 12.6 and 12.7 assume ideal gas behavior. At the conditions specified (1 atm, relatively high temperatures), this assumption is a good one. However, many industrial gas-phase reactions are carried out at very high pressures. In that case, intermolecular forces become important, and calculated yields based on ideal gas behavior may be seriously in error.

12.5 EFFECT OF CHANGES IN CONDITIONS ON AN EQUILIBRIUM SYSTEM

See Screen 16.11, Disturbing an Equilibrium (Le Châtelier's Principle).

Once a system has attained equilibrium, it is possible to change the ratio of products to reactants by changing the external conditions. We will consider three ways in which a chemical equilibrium can be disturbed:

1. Adding or removing a gaseous reactant or product.
2. Compressing or expanding the system.
3. Changing the temperature.

You can deduce the direction in which an equilibrium will shift when one of these changes is made by applying the following principle:

If a system at equilibrium is disturbed by a change in concentration, pressure, or temperature, the system will, if possible, shift to partially counteract the change.

Equilibrium systems, like people, resist change.

This law was first stated, in a more complex form, in 1884 by Henri Le Châtelier (1850–1936), a French chemist who had studied a variety of industrial equilibria, including those involved in the production of iron in the blast furnace. The statement is commonly referred to as **Le Châtelier's principle.**

Adding or Removing a Gaseous Species

See Screen 16.13, Disturbing an Equilibrium (Addition or Removal of a Reagent).

According to Le Châtelier's principle, *if a chemical system at equilibrium is disturbed by adding a gaseous species (reactant or product), the reaction will proceed in such a direction as to consume part of the added species. Conversely, if a gaseous species is removed, the system shifts to restore part of that species.* In that way, the change in concentration brought about by adding or removing a gaseous species is partially counteracted when equilibrium is restored.

To apply this general rule, consider the reaction

$$N_2O_4(g) \rightleftharpoons 2NO_2(g)$$

Suppose this system has reached equilibrium at a certain temperature. This equilibrium could be disturbed by—

■ *adding N_2O_4.* Reaction will occur in the forward direction (left to right). In this way, part of the N_2O_4 will be consumed.

(a)   (b)   (c)

Le Châtelier's principle in cobalt ion equilibrium. When Cl^- ions are added to a solution containing the $Co(H_2O)_6^{2+}$ ion, the equilibrium

$$[Co(H_2O)_6]^{2+}(aq) + 4Cl^-(aq) \rightleftharpoons [CoCl_4]^{2-}(aq) + 6H_2O$$

pink   blue

shifts to the right and the color changes from pink to blue. *(Charles D. Winters)*

■ *adding* NO_2, which causes the reverse reaction (right to left) to occur, using up part of the NO_2 added.

■ *removing* N_2O_4. Reaction occurs in the reverse direction to restore part of the N_2O_4.

■ *removing* NO_2, which causes the forward reaction to occur, restoring part of the NO_2 removed.

   It is possible to use K to calculate the extent to which reaction occurs when an equilibrium is disturbed by adding or removing a product or reactant. To show how this is done, consider the effect of adding hydrogen iodide to the $HI-H_2-I_2$ system (Example 12.8).

$$2HI(g) \rightleftharpoons H_2(g) + I_2(g)$$

EXAMPLE 12.8   In Example 12.4 you found that the $HI - H_2 - I_2$ system is in equilibrium at 520°C when $P_{HI} = 0.80$ atm and $P_{H_2} = P_{I_2} = 0.10$ atm. Suppose enough HI is added to raise its pressure temporarily to 1.00 atm. When equilibrium is restored, what are P_{HI}, P_{H_2}, and P_{I_2}?

Strategy   This problem is entirely analogous to Examples 12.6 and 12.7; the only real difference is that in this case the original pressures are those that prevail immediately after equilibrium is disturbed. Note that

$$Q = \frac{(0.10) \times (0.10)}{(1.00)^2} = 0.010 < K = 0.016$$

so some HI must dissociate to establish equilibrium.

Solution

(1) Recall from Example 12.4 that

$$K = \frac{P_{H_2} \times P_{I_2}}{(P_{HI})^2} = 0.016 \text{ at } 520°C$$

(2) The partial pressure of HI must decrease; those of H_2 and I_2 must increase. Let x be the increase in partial pressure of H_2. Then

$$\Delta P_{I_2} = \Delta P_{H_2} = x \qquad \Delta P_{HI} = -2x$$

The equilibrium table becomes

	2HI(g) ⇌	H₂(g)	+ I₂(g)
P_o (atm)	1.00	0.10	0.10
ΔP (atm)	$-2x$	$+x$	$+x$
P_{eq} (atm)	$1.00 - 2x$	$0.10 + x$	$0.10 + x$

(3) Substituting into the expression for K,

$$0.016 = \frac{(0.10 + x)^2}{(1.00 - 2x)^2}$$

To solve this equation, take the square root of both sides:

$$0.13 = \frac{0.10 + x}{1.00 - 2x}$$

from which you should find that $x = 0.024$.

(4) $P_{H_2} = P_{I_2} = 0.10 + x =$ 0.12 atm

$P_{HI} = 1.00 - 2x =$ 0.95 atm

> The system does not return to its original equilibrium state.

Reality Check Note that the equilibrium partial pressure of HI is intermediate between its value before equilibrium was established (0.80 atm) and that immediately afterward (1.00 atm). This is exactly what Le Châtelier's principle predicts; part of the added HI is consumed to re-establish equilibrium.

We should emphasize that adding a pure liquid or solid has no effect on a system at equilibrium. The rule is a simple one: *For a species to shift the position of an equilibrium, it must appear in the expression for K.*

Compression or Expansion

> See Screen 16.14, Disturbing an Equilibrium (Volume Changes).

To understand how a change in pressure can change the position of an equilibrium, consider again the N_2O_4–NO_2 system:

$$N_2O_4(g) \rightleftharpoons 2NO_2(g)$$

Suppose the system is compressed by pushing down the piston shown in Figure 12.3. The immediate effect is to increase the gas pressure because the same number of molecules are crowded into a smaller volume ($P = nRT/V$). According to Le Châtelier's principle, the system will shift to partially counteract this change. There is a simple way in which the gas pressure can be reduced. Some of the NO_2 molecules combine with each other to form N_2O_4 (cylinder at right of Figure 12.3). That is, reaction occurs in the reverse direction:

> When V decreases, NO_2 is converted to N_2O_4.

$$N_2O_4(g) \longleftarrow 2NO_2(g)$$

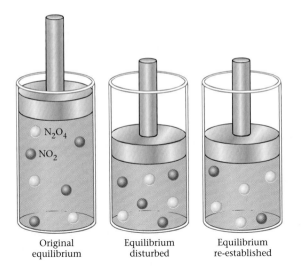

FIGURE 12.3
Effect of compression on the $N_2O_4(g) \rightleftharpoons 2NO_2(g)$ system at equilibrium.
The immediate effect (middle cylinder) is to crowd the same number of moles of gas into a smaller volume and so increase the total pressure. This is partially compensated for by the conversion of some of the NO_2 to N_2O_4, thereby reducing the total number of moles of gas.

Original
equilibrium

Equilibrium
disturbed

Equilibrium
re-established

This reduces the number of moles, n, and hence the pressure, P.

It is possible (but not easy) to calculate from K the extent to which NO_2 is converted to N_2O_4 when the system is compressed. The results of such calculations are given in Table 12.4. As the pressure is increased from 1.0 to 10.0 atm, more and more of the NO_2 is converted to N_2O_4. Notice that the total number of moles of gas decreases steadily as a result of this conversion.

The analysis we have just gone through for the N_2O_4–NO_2 system can be applied to any equilibrium system involving gases—

- *When the system is compressed, thereby increasing the total pressure, reaction takes place in the direction that decreases the total number of moles of gas.*
- *When the system is expanded, thereby decreasing the total pressure, reaction takes place in the direction that increases the total number of moles of gas.*

The application of this principle to several different systems is shown in Table 12.5 (page 370). In System 2, the number of moles of gas decreases from $\frac{3}{2}$ to 1 as the reaction goes to the right. Hence increasing the pressure causes the forward reaction to occur; a decrease in pressure has the reverse effect. Notice that it is the change in the number of moles of *gas* that determines which way the equilibrium shifts (System 4). When there is no change in the number of moles of gas (System 5), a change in pressure has no effect on the position of the equilibrium.

TABLE 12.4   Effect of Compression on the Equilibrium System $N_2O_4(g) \rightleftharpoons 2NO_2(g)$; $K = 11$ at 100°C OHT

P_{tot} (atm)	n_{NO_2}	$n_{N_2O_4}$	n_{tot}
1.0	0.92	0.08	1.00
2.0	0.82	0.13	0.95
5.0	0.64	0.22	0.86
10.0	0.50	0.29	0.79

TABLE 12.5 **Effect of Pressure on the Position of Gaseous Equilibria** OHT

System	Δn_{gas}*	P_{tot} Increases	P_{tot} Decreases
1. $N_2O_4(g) \rightleftharpoons 2NO_2(g)$	$+1$	$\longleftarrow$	$\longrightarrow$
2. $SO_2(g) + \frac{1}{2}O_2(g) \rightleftharpoons SO_3(g)$	$-\frac{1}{2}$	$\longrightarrow$	$\longleftarrow$
3. $N_2(g) + 3H_2(g) \rightleftharpoons 2NH_3(g)$	-2	$\longrightarrow$	$\longleftarrow$
4. $C(s) + H_2O(g) \rightleftharpoons CO(g) + H_2(g)$	$+1$	$\longleftarrow$	$\longrightarrow$
5. $N_2(g) + O_2(g) \rightleftharpoons 2NO(g)$	0	0	0

*Δn_{gas} is the change in the number of moles of gas as the forward reaction occurs.

We should emphasize that in applying this principle it is important to realize that an "increase in pressure" means that the system is compressed; that is, the volume is decreased. Similarly, a "decrease in pressure" corresponds to expanding the system by increasing its volume. There are other ways in which pressure can be changed. One way is to add an unreactive gas such as helium at constant volume. This increases the total number of moles and hence the total pressure. It has no effect, however, on the position of the equilibrium, because it does not change the partial pressures of any of the gases taking part in the reaction. Remember that the partial pressure of a gas A is given by the expression

$$P_A = n_A RT/V \qquad (n_A = \text{no. of moles of A})$$

Adding a different gas at constant volume does not change any of the quantities on the right side of this equation, so P_A stays the same.

EXAMPLE 12.9 The pressure on each of the following systems is decreased from 5 atm to 1 atm. Which way does the equilibrium shift?

(a) $2CO_2(g) \rightleftharpoons 2CO(g) + O_2(g)$
(b) $H_2(g) + I_2(g) \rightleftharpoons 2HI(g)$
(c) $H_2(g) + I_2(s) \rightleftharpoons 2HI(g)$

Strategy Using the coefficients of the balanced equation, determine what happens to the number of moles of gas. Then apply Le Châtelier's principle.

Solution

(a) $\rightarrow$ ($\Delta n_{gas} = +1$). The immediate effect of the expansion is to decrease the pressure. That is partially counteracted by reaction in the forward direction, thereby increasing the number of moles and hence the pressure.
(b) No effect. Same number of moles of gas on both sides.
(c) $\rightarrow$ (1 mol gas $\rightarrow$ 2 mol gas). Note that it is the number of moles of gas that is important; solids or liquids don't count.

> If the reaction involves no change in the number of moles of gas, pressure has no effect on the equilibrium position.

Change in Temperature

Le Châtelier's principle can be used to predict the effect of a change in temperature on the position of an equilibrium. In general, **an increase in temperature causes the endothermic reaction to occur.** This absorbs heat and so tends to reduce the temperature of the system, partially compensating for the original temperature increase.

 See Screen 16.12, Disturbing an Equilibrium (Temperature Changes).

Applying this principle to the N_2O_4–NO_2 system,

$$N_2O_4(g) \rightleftharpoons 2NO_2(g) \qquad \Delta H° = +57.2 \text{ kJ}$$

it follows that raising the temperature causes the forward reaction to occur, because that reaction absorbs heat. This is confirmed by experiment (Figure 12.4). At higher temperatures, more NO_2 is produced, and the reddish-brown color of that gas becomes more intense.

For the synthesis of ammonia,

$$N_2(g) + 3H_2(g) \rightleftharpoons 2NH_3(g) \qquad \Delta H° = -92.2 \text{ kJ}$$

an increase in temperature shifts the equilibrium to the left; some ammonia decomposes to the elements. This reflects the fact that the reverse reaction is endothermic:

$$2NH_3(g) \longrightarrow N_2(g) + 3H_2(g) \qquad \Delta H° = +92.2 \text{ kJ}$$

As pointed out earlier, the equilibrium constant of a system changes with temperature. The form of the equation relating K to T is a familiar one, similar to the Clausius-Clapeyron equation (Chapter 9) and the Arrhenius equation (Chapter 11). This one is called the **van't Hoff equation,** honoring Jacobus van't Hoff (1852–1911), who was the first to use the equilibrium constant, K. Coincidentally, van't Hoff was a good friend of Arrhenius. The equation is

$$\boxed{\ln \frac{K_2}{K_1} = \frac{\Delta H°}{R}\left[\frac{1}{T_1} - \frac{1}{T_2}\right]}$$

where K_2 and K_1 are the equilibrium constants at T_2 and T_1, respectively, $\Delta H°$ is the standard enthalpy change for the forward reaction, and R is the gas constant, 8.31 J/mol·K.

To illustrate how this equation is used, let us apply it to calculate the equilibrium constant at 100°C for the system

$$N_2(g) + 3H_2(g) \rightleftharpoons 2NH_3(g) \qquad \Delta H° = -92.2 \text{ kJ}$$

given that $K = 6 \times 10^5$ at 25°C. The relation is

$$\ln \frac{K \text{ at } 100°C}{6 \times 10^5} = \frac{-92{,}200 \text{ J/mol}}{8.31 \text{ J/mol·K}}\left[\frac{1}{298 \text{ K}} - \frac{1}{373 \text{ K}}\right] = -7.5$$

Taking inverse logarithms,

$$\frac{K \text{ at } 100°C}{6 \times 10^5} = 6 \times 10^{-4}$$

$$K \text{ at } 100°C = 4 \times 10^2$$

Notice that the equilibrium constant becomes smaller as the temperature increases. In general—

- if the forward reaction is exothermic, as is the case here ($\Delta H° = -92.2$ kJ), K decreases as T increases
- if the forward reaction is endothermic as in the decomposition of N_2O_4,

$$N_2O_4(g) \rightleftharpoons 2NO_2(g) \qquad \Delta H° = +57.2 \text{ kJ}$$

K increases as T increases. For this reaction K is 0.11 at 25°C and 11 at 100°C.

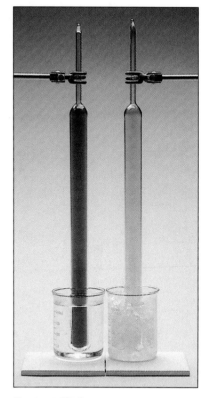

FIGURE 12.4
Effect of temperature on the N_2O_4–NO_2 **system at equilibrium.** At 0°C *(tube at right)*, N_2O_4, which is colorless, predominates. At 50°C *(tube at left)*, some of the N_2O_4 has disassociated to give the deep brown color of NO_2. *(Marna G. Clarke)*

EXAMPLE 12.10 Consider the system $I_2(g) \rightleftharpoons 2I(g)$; $\Delta H° = +151$ kJ. Suppose the system is at equilibrium at 1000°C. In which direction will reaction occur if
(a) I atoms are added?
(b) the system is compressed?
(c) the temperature is increased?

Strategy Apply Le Châtelier's principle to predict the effect of (a) adding a product, (b) increasing the pressure, or (c) increasing the temperature.

Solution

(a) $2I(g) \rightarrow I_2(g)$

(b) $2I(g) \rightarrow I_2(g)$ This decreases the number of moles of gas.

(c) $I_2(g) \rightarrow 2I(g)$ This reaction absorbs heat.

We should emphasize that of the three changes in conditions described in this section—

▪ adding or removing a gaseous species
▪ compressing or expanding the system
▪ changing the temperature

the only one that changes the value of the equilibrium constant is a change in temperature. In the other two cases, K remains constant.

CHEMISTRY

Beyond the Classroom

An Industrial Application of Gaseous Equilibrium

A wide variety of materials, both pure substances and mixtures, are made by processes that involve one or more gas-phase reactions. Among these is one of the most important industrial chemicals, ammonia.

The process used to make this chemical applies many of the principles of chemical equilibrium, discussed in this chapter, and chemical kinetics (Chapter 11).

Haber Process for Ammonia

Combined ("fixed") nitrogen in the form of protein is essential to all forms of life. There is more than enough elementary nitrogen in the air, about 4×10^{18} kg, to meet all our needs. The problem is to convert the element to compounds that can be used by plants to make proteins. At room temperature and atmospheric pressure, N_2 does not react with any nonmetal. However, in 1908 in Germany, Fritz Haber (Figure A) showed that nitrogen does react with hydrogen at high temperatures and pressures to form ammonia in good yield:

$$N_2(g) + 3H_2(g) \rightleftharpoons 2NH_3(g) \qquad \Delta H° = -92.2 \text{ kJ}$$

The Haber process, represented by this equation, is now the main source of fixed nitrogen. Its feasibility depends on choosing conditions under which nitrogen and hydrogen react rapidly to give a high yield of ammonia. At 25°C and atmospheric pressure, the position of the equilibrium favors the formation of $NH_3(K = 6 \times 10^5)$. Unfortunately, however, the rate of reaction is virtually zero. Equilibrium is reached more rapidly by raising the temperature. However, because the synthesis of ammonia is exothermic, high temperatures reduce K and hence the yield of ammonia. High pressures, on the other hand, have a favorable effect on both the rate of the reaction and the position of the equilibrium (Table A).

TABLE A   Effect of Temperature and Pressure on the Yield of Ammonia in the Haber Process ($P_{H_2} = 3P_{N_2}$)

°C	K	Mole Percent NH$_3$ in Equilibrium Mixture				
		10 atm	50 atm	100 atm	300 atm	1000 atm
200	0.30	51	74	82	90	98
300	3.7×10^{-3}	15	39	52	71	93
400	1.5×10^{-4}	4	15	25	47	80
500	1.4×10^{-5}	1	6	11	26	57
600	2.1×10^{-6}	0.5	2	5	14	31

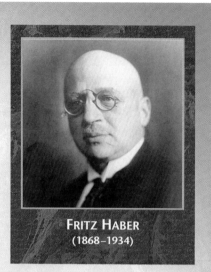

FRITZ HABER
(1868–1934)

FIGURE A

Fritz Haber. *(Oesper Collection in the History of Chemistry/University of Cincinnati)*

The values of the equilibrium constant K listed in Table A are those obtained from data at low pressures, where the gases behave ideally. At higher pressures the mole percent of ammonia observed is generally larger than the calculated value. For example, at 400°C and 300 atm, the observed mole percent of NH$_3$ is 47; the calculated value is only 41.

The goal of Haber's research was to find a catalyst to synthesize ammonia at a reasonable rate without going to very high temperatures. These days two different catalysts are used. One consists of a mixture of iron, potassium oxide, K$_2$O, and aluminum oxide, Al$_2$O$_3$. The other, which uses finely divided ruthenium, Ru, metal on a graphite surface, is less susceptible to poisoning by impurities. Reaction takes place at 450°C and a pressure of 200 to 600 atm. The ammonia formed is passed through a cooling chamber. Ammonia boils at −33°C, so it is condensed out as a liquid, separating it from unreacted nitrogen and hydrogen. The yield is typically less than 50%, so the reactants are recycled to produce more ammonia.

Fritz Haber

Haber had a distinguished scientific career. He received, among other honors, the 1918 Nobel Prize in chemistry. For more than 20 years, he was director of the prestigious Kaiser Wilhelm Institute for Physical Chemistry at Dahlem, Germany, a suburb of Berlin. He was forced to resign that position when Adolf Hitler came to power in 1933, presumably because of Haber's Jewish ancestry. A contributing factor may have been his comment after listening to a radio broadcast of one of Hitler's hysterical harangues. When asked what he thought of it, Haber replied, "Give me a gun so I can shoot him."

Haber, like all of us, suffered disappointment and tragedy in his life. In the years after World War I, he devised a scheme to pay Germany's war debts by extracting gold from seawater. The project, which occupied Haber for several years, was a total failure because the concentration of gold turned out to be only about one thousandth of the literature value.

During World War I, Haber was in charge of the German poison gas program. In April of 1915, the Germans used chlorine for the first time on the Western front, causing 5000 fatalities. Haber's wife Clara was aghast; she pleaded with her husband to forsake poison gas. When he adamantly refused to do so, she committed suicide.

On a lighter note, there is an amusing anecdote told by the author Morris Goran (*The Story of Fritz Haber*, 1967). It seems that in a PhD examination, Haber asked the candidate how iodine was prepared. The student made a wild guess that it was obtained from a tree (it isn't). Haber played along with him, asking where the iodine tree grew, what it looked like and, finally, when it bloomed. "In the fall," was the answer. Haber smiled, put his arm around the student, and said, "Well, my friend, I'll see you again when the iodine blossoms appear once more."

CHAPTER HIGHLIGHTS

Key Concepts

1. Relate the expression for K to the corresponding chemical equation.
   (Examples 12.1, 12.2; Problems 5–12)
2. Calculate K, knowing—
   ■ appropriate K's for other reactions.
   (Problems 13–18)
   ■ all the equilibrium partial pressures.
   (Example 12.3; Problems 19–22, 54)
   ■ all the original and one equilibrium partial pressure.
   (Example 12.4; Problems 23, 24)
3. Use the value of K to determine—
   ■ the direction of reaction.
   (Example 12.5; Problems 25–28)
   ■ equilibrium partial pressures of all species.
   (Examples 12.6, 12.7; Problems 31–42, 64)
4. Apply Le Châtelier's principle to predict the direction in which an equilibrium shifts when conditions are changed.
   (Examples 12.8–12.10; Problems 43–52, 59, 61)

Key Equations

Equilibrium constant expression	$K = \dfrac{(P_C)^c \times (P_D)^d}{(P_A)^a \times (P_B)^b}$
Coefficient rule	$K' = K^n$
Reciprocal rule	$K'' = 1/K$
Multiple equilibrium	$K_3 = K_1 \times K_2$
van't Hoff equation	$\ln \dfrac{K_2}{K_1} = \dfrac{\Delta H^\circ}{R}\left[\dfrac{1}{T_1} - \dfrac{1}{T_2}\right]$

Key Terms

■ endothermic
  equilibrium constant K
■ exothermic

Le Châtelier's principle
■ mole

■ partial pressure
  reaction quotient

Summary Problem

Consider phosgene, $COCl_2(g)$. It is an important intermediate in the manufacture of some plastics. It can decompose into carbon monoxide and chlorine gas. Its ΔH_f° is -220.1 kJ/mol.

(a) Write the chemical equation for the decomposition of phosgene using simplest whole-number coefficients.

(b) Write the equilibrium expression for the decomposition.
(c) At 500°C, after the reaction has reached equilibrium, the partial pressures of CO, Cl_2, and $COCl_2$ are 0.413 atm, 0.237 atm, and 0.217 atm, respectively. Calculate K at 500°C.
(d) Initially, a 5.00-L flask contains $COCl_2$, Cl_2, and CO at partial pressures of 0.689 atm, 0.250 atm, and 0.333 atm, respectively. After equilibrium is established at 500°C, the partial pressure of $COCl_2$ is 0.501 atm. Calculate the equilibrium partial pressures of chlorine and carbon monoxide gases.
(e) Initially a reaction vessel contains Cl_2, CO, and $COCl_2$ at 500°C, where both Cl_2 and CO gases have partial pressures of 0.750 atm. The partial pressure of phosgene is 0.500 atm. In what direction will the reaction proceed at 500°C? What is the partial pressure of each gas when equilibrium is established? Suppose enough $COCl_2$ is added to raise its pressure temporarily to 2.00 atm. In which direction will equilibrium shift? When equilibrium is restored, what is the partial pressure of each gas?
(f) In which direction will this system shift if at equilibrium
    (1) the system is expanded?
    (2) helium gas is added?
    (3) the temperature is increased?
(g) What is the equilibrium constant at 375°C?

Answers

(a) $COCl_2(g) \rightarrow Cl_2(g) + CO(g)$
(b) $K = \dfrac{P_{CO} \times P_{Cl_2}}{P_{COCl_2}}$
(c) 0.451

(d) P_{Cl_2} = 0.438 atm; P_{CO} = 0.521 atm
(e) $\leftarrow$; $P_{CO} = P_{Cl_2}$ = 0.558 atm; P_{COCl_2} = 0.692 atm;
    $\rightarrow$; $P_{CO} = P_{Cl_2}$ = 0.872 atm, P_{COCl_2} = 1.686 atm
(f) $\rightarrow$; no change; $\rightarrow$
(g) 0.0168

Problem numbers in blue indicate that the answer is available in Appendix 6 at the back of the book.
WEB indicates that the solution is posted at **http://www.harcourtcollege.com/chem/general/masterton4/student/**

Establishment of Equilibrium

1. The following data are for the system

$$A(g) \rightleftharpoons 2B(g)$$

Time (s)	0	30	45	60	75	90
P_A (atm)	0.500	0.390	0.360	0.340	0.325	0.325
P_B (atm)	0.000	0.220	0.280	0.320	0.350	0.350

(a) How long does it take the system to reach equilibrium?
(b) How does the rate of the forward reaction compare with the rate of the reverse reaction after 45 s? After 90 s?

2. The following data are for the system

$$A(g) \rightleftharpoons 2B(g)$$

Time (s)	0	20	40	60	80	100
P_A (atm)	1.00	0.83	0.72	0.65	0.62	0.62
P_B (atm)	0.00	0.34	0.56	0.70	0.76	0.76

(a) How long does it take the system to reach equilibrium?
(b) How does the rate of the forward reaction compare with the rate of the reverse reaction after 30 s? After 90 s?
3. Complete the table below for the reaction:

$$3A(g) + 2B(g) \rightleftharpoons C(g)$$

Time (s)	0	10	20	30	40	50	60
P_A (atm)	2.450	2.000	___	___	1.100	___	0.950
P_B (atm)	1.500	___	___	0.750	___	___	___
P_C (atm)	0.000	___	0.275	___	___	0.500	___

WEB 4. Complete the table below for the reaction:

$$3A(g) + B(g) \rightleftharpoons 2C(g)$$

Time (min)	0	1	2	3	4	5	6
P_A (atm)	1.000	0.778				0.325	
P_B (atm)	0.400		0.260		0.185		0.175
P_C (atm)	0.000			0.390			

Expressions for K

5. Write equilibrium constant expressions (K) for the following reactions:

(a) $IF(g) \rightleftharpoons \frac{1}{2}F_2(g) + \frac{1}{2}I_2(g)$

(b) $2C_5H_6(g) \rightleftharpoons C_{10}H_{12}(g)$

(c) $P_4O_{10}(g) + 6PCl_5(g) \rightleftharpoons 10POCl_3(g)$

6. Write equilibrium constant expressions (K) for the following reactions:

(a) $CS_2(g) + 4H_2(g) \rightleftharpoons CH_4(g) + 2H_2S(g)$

(b) $2H_2O_2(g) \rightleftharpoons 2H_2O(g) + O_2(g)$

(c) $C_3H_8(g) + 5O_2(g) \rightleftharpoons 3CO_2(g) + 4H_2O(g)$

7. Write equilibrium constant expressions (K) for the following reactions:

(a) $Ni(s) + 4CO(g) \rightleftharpoons Ni(CO)_4(g)$

(b) $HF(aq) \rightleftharpoons H^+(aq) + F^-(aq)$

(c) $Cl_2(g) + 2Br^-(aq) \rightleftharpoons Br_2(l) + 2Cl^-(aq)$

8. Write equilibrium constant expressions (K) for the following reactions:

(a) $2NO_3^-(aq) + 8H^+(aq) + 3Cu(s) \rightleftharpoons$
$\qquad 2NO(g) + 3Cu^{2+}(aq) + 4H_2O(l)$

(b) $2PbS(s) + 3O_2(g) \rightleftharpoons 2PbO(s) + 2SO_2(g)$

(c) $Ca^{2+}(aq) + CO_3^{2-}(aq) \rightleftharpoons CaCO_3(s)$

9. Given the following descriptions of reversible reactions, write a balanced equation (simplest whole-number coefficients) and the equilibrium constant expression (K) for each.

(a) Nitrogen gas reacts with solid sodium carbonate and solid carbon to produce carbon monoxide gas and solid sodium cyanide.

(b) Solid magnesium nitride reacts with water vapor to form magnesium hydroxide solid and ammonia gas.

(c) Ammonium ion in aqueous solution reacts with a strong base at 25°C, giving aqueous ammonia and water.

WEB 10. Given the following descriptions of reversible reactions, write a balanced net ionic equation (simplest whole-number coefficients) and the equilibrium constant expression (K) for each.

(a) Liquid acetone (C_3H_6O) is in equilibrium with its vapor.

(b) Hydrogen gas reduces nitrogen dioxide gas to form ammonia and steam.

(c) Hydrogen sulfide gas (H_2S) bubbled into an aqueous solution of lead(II) ions produces lead sulfide precipitate and hydrogen ions.

11. Write a chemical equation for an equilibrium system that would lead to the following expressions (a–e) for K.

(a) $K = \dfrac{P_{CH_3OH}}{(P_{CO})(P_{H_2})^2}$

(b) $K = \dfrac{(P_{IF})^2}{(P_{I_2})(P_{F_2})}$

(c) $K = \dfrac{P_{O_2}}{P_{CO_2}}$

(d) $K = \dfrac{[I^-]^2 P_{Cl_2}}{[Cl^-]^2}$

(e) $K = \dfrac{(P_{SO_2})^2 (P_{O_2})}{(P_{SO_3})^2}$

12. Write a chemical equation for an equilibrium system that would lead to the following expressions (a–e) for K.

(a) $K = \dfrac{(P_{C_3H_8})(P_{O_2})^5}{(P_{CO_2})^3 (P_{H_2O})^4}$

(b) $K = \dfrac{[Fe^{2+}]^2 P_{Cl_2}}{[Fe^{3+}]^2 [Cl^-]^2}$

(c) $K = \dfrac{(P_{N_2}) (P_{H_2})^3}{(P_{NH_3})^2}$

(d) $K = \dfrac{(P_{H_2O})^2 (P_{Cl_2})^2}{(P_{HCl})^4 (P_{O_2})}$

(e) $K = \dfrac{(P_{H_2O_2})^2}{(P_{H_2O})^2 (P_{O_2})}$

Calculation of K

13. At 627°C, $K = 0.76$ for the reaction

$$2SO_2(g) + O_2(g) \rightleftharpoons 2SO_3(g)$$

Calculate K at 627°C for
(a) the synthesis of one mole of sulfur trioxide gas.
(b) the decomposition of two moles of SO_3.

14. At 25°C, $K = 2.2 \times 10^{-3}$ for the reaction

$$ICl(g) \rightleftharpoons \frac{1}{2}I_2(g) + \frac{1}{2}Cl_2(g)$$

Calculate K at 25°C for
(a) the decomposition of ICl into one mole of iodine and chlorine.
(b) the formation of two moles of $ICl(g)$.

15. Given the following reactions and their equilibrium constants,

$$H_2(g) + S(s) \rightleftharpoons H_2S(g) \qquad K = 1.0 \times 10^{-3}$$

$$S(s) + O_2(g) \rightleftharpoons SO_2(g) \qquad K = 5.0 \times 10^6$$

Calculate K for the reaction

$$H_2(g) + SO_2(g) \rightleftharpoons H_2S(g) + O_2(g)$$

16. Given the following reactions and their equilibrium constants,

$$SnO_2(s) + 2H_2(g) \rightleftharpoons Sn(s) + 2H_2O(g) \qquad K = 21$$

$$CO(g) + H_2O(g) \rightleftharpoons CO_2(g) + H_2(g) \qquad K = 0.034$$

Calculate K for the reaction

$$SnO_2(s) + 2CO(g) \rightleftharpoons Sn(s) + 2CO_2(g)$$

17. Given the following data at a certain temperature,

$$2N_2(g) + O_2(g) \rightleftharpoons 2N_2O(g) \qquad K = 1.2 \times 10^{-35}$$

$$N_2O_4(g) \rightleftharpoons 2NO_2(g) \qquad K = 4.6 \times 10^{-3}$$

$$\tfrac{1}{2}N_2(g) + O_2(g) \rightleftharpoons NO_2(g) \qquad K = 4.1 \times 10^{-9}$$

Calculate K for the reaction between one mole of dinitrogen oxide gas and oxygen gas to give dinitrogen tetroxide gas.

18. Given the following data at 25°C,

$$2NO(g) \rightleftharpoons N_2(g) + O_2(g) \qquad K = 1 \times 10^{-30}$$

$$2NO(g) + Br_2(g) \rightleftharpoons 2NOBr(g) \qquad K = 8 \times 10^{1}$$

Calculate K for the formation of one mole of NOBr from its elements in the gaseous state at 25°C.

19. Calculate K for the formation of methyl alcohol at 100°C:

$$CO(g) + 2H_2(g) \rightleftharpoons CH_3OH(g)$$

given that at equilibrium, the partial pressures of the gases are $P_{CO} = 0.814$ atm, $P_{H_2} = 0.274$ atm, and $P_{CH_3OH} = 0.0512$ atm.

WEB 20. When carbon monoxide reacts with hydrogen gas, methane and steam are formed.

$$CO(g) + 3H_2(g) \rightleftharpoons CH_4(g) + H_2O(g)$$

At 1127°C, analysis at equilibrium shows that $P_{CO} = 0.921$ atm, $P_{H_2} = 1.21$ atm, $P_{CH_4} = 0.0391$ atm, and $P_{H_2O} = 0.0124$ atm. What is the equilibrium constant, K, for the reaction at 1127°C?

21. Ammonium carbamate solid ($NH_4CO_2NH_2$) decomposes at 313 K into ammonia and carbon dioxide gases. At equilibrium, analysis shows that there are 0.0451 atm of CO_2, 0.0961 atm of ammonia, and 0.159 g of ammonium carbamate.

   (a) Write a balanced equation for the decomposition of one mole of $NH_4CO_2NH_2$.

   (b) Calculate K at 313 K.

22. At 1123 K, methane and hydrogen sulfide gases react to form carbon disulfide and hydrogen gases. At equilibrium the concentrations of methane, hydrogen sulfide, carbon disulfide, and hydrogen gas are 0.00142 M, 6.14×10^{-4} M, 0.00266 M, and 0.00943 M, respectively.

   (a) Write a balanced equation for the formation of one mole of carbon disulfide gas.

   (b) Calculate K for the reaction at 1123 K.

23. Consider the decomposition of nitrosyl chloride, NOCl(g):

$$2NOCl(g) \rightleftharpoons 2NO(g) + Cl_2(g)$$

At 220°C in a sealed flask, originally, $P_{NOCl} = 1.000$ atm, $P_{NO} = 0.009$ atm, $P_{Cl_2} = 0.467$ atm. When equilibrium is established at

220°C, it is found that the partial pressure of NOCl has dropped by 11.6%. Calculate K for the decomposition at 220°C.

24. When nitrogen oxide gas and bromine gas with partial pressures of 1.577 atm and 0.427 atm, respectively, are sealed in a one-liter flask at 77°C, the following equilibrium is established:

$$2NO(g) + Br_2(g) \rightleftharpoons 2NOBr(g)$$

Given that the equilibrium partial pressure of NOBr at 77°C is 0.624 atm, calculate K for the reaction at 77°C.

K; Direction of Reaction

25. For the system

$$PCl_5(g) \rightleftharpoons PCl_3(g) + Cl_2(g)$$

K is 26 at 300°C. In a 5.0-L flask, a gaseous mixture consists of all three gases with partial pressures as follows: $P_{PCl_5} = 0.012$ atm, $P_{Cl_2} = 0.45$ atm, $P_{PCl_3} = 0.90$ atm.

   (a) Is the mixture at equilibrium? Explain.

   (b) If it is not at equilibrium, which way will the system shift to establish equilibrium?

26. A gaseous reaction mixture contains 0.30 atm SO_2, 0.16 atm Cl_2, and 0.50 atm SO_2Cl_2 in a 2.0-L container. $K = 0.011$ for the equilibrium system

$$SO_2Cl_2(g) \rightleftharpoons SO_2(g) + Cl_2(g)$$

   (a) Is the system at equilibrium? Explain.

   (b) If it is not at equilibrium, in which direction will the system move to reach equilibrium?

27. The reversible reaction between hydrogen chloride gas and one mole of oxygen gas produces steam and chlorine gas:

$$4HCl(g) + O_2(g) \rightleftharpoons 2Cl_2(g) + 2H_2O(g) \qquad K = 0.79$$

Predict the direction in which the system will move to reach equilibrium if one starts with

   (a) $P_{H_2O} = P_{HCl} = P_{O_2} = 0.20$ atm

   (b) $P_{HCl} = 0.30$ atm, $\quad P_{H_2O} = 0.35$ atm, $\quad P_{Cl_2} = 0.2$ atm, $P_{O_2} = 0.15$ atm

28. For the reaction

$$2NO_2(g) \rightleftharpoons 2NO(g) + O_2(g)$$

K at a certain temperature is 0.50. Predict the direction in which the system will move to reach equilibrium if one starts with

   (a) $P_{O_2} = P_{NO} = P_{NO_2} = 0.10$ atm

   (b) $P_{NO_2} = 0.0848$ atm, $P_{O_2} = 0.0116$ atm

   (c) $P_{NO_2} = 0.20$ atm, $P_{O_2} = 0.010$ atm, $P_{NO} = 0.040$ atm

K; Equilibrium Pressures

29. Consider the reaction

$$2NO(g) + Cl_2(g) \rightleftharpoons 2NOCl(g)$$

At a certain temperature, $K = 1.8$. Calculate P_{NO} if $P_{Cl_2} = 0.393$ atm and $P_{NOCl} = 0.217$ atm at equilibrium.

30. For the reaction

$$NO_2(g) \rightleftharpoons \tfrac{1}{2}O_2(g) + NO(g)$$

K is 0.0132 at 500 K. Calculate P_{O_2} if P_{NO_2} is 0.817 atm and P_{NO} is 0.281 atm at equilibrium.

31. At a certain temperature, K is 0.040 for the decomposition of two moles of bromine chloride gas (BrCl) to its elements. An equilibrium mixture at this temperature contains bromine and chlorine gases at equal partial pressures of 0.0493 atm. What is the equilibrium partial pressure of bromine chloride?

32. At a certain temperature, K is 1.3×10^5 for the reaction

$$2H_2(g) + S_2(g) \rightleftharpoons 2H_2S(g)$$

What is the equilibrium pressure of hydrogen sulfide if those of hydrogen and sulfur gases are 0.103 atm and 0.417 atm, respectively?

33. For the reaction

$$N_2(g) + 2H_2O(g) \rightleftharpoons 2NO(g) + 2H_2(g)$$

K is 1.54×10^{-3}. When equilibrium is established, the partial pressure of nitrogen is 0.168 atm, and that of NO is 0.225 atm. The total pressure of the system at equilibrium is 1.87 atm. What are the equilibrium partial pressures of hydrogen and steam?

34. Nitrogen dioxide can decompose to nitrogen oxide and oxygen.

$$2NO_2(g) \rightleftharpoons 2NO(g) + O_2(g)$$

K is 0.87 at a certain temperature. A 5.0-L flask at equilibrium is determined to have a total pressure of 1.25 atm and oxygen to have a partial pressure of 0.515 atm. Calculate P_{NO} and P_{NO_2} at equilibrium.

35. At 800 K, hydrogen iodide gas can be formed from hydrogen and iodine gases.

$$H_2(g) + I_2(g) \rightleftharpoons 2HI(g)$$

K is 59.2. What are the partial pressures of all the gases at equilibrium when the initial partial pressures of hydrogen gas and iodine gas are 0.450 atm?

36. Hydrogen cyanide, a highly toxic gas, can be prepared by reducing cyanogen gas, C_2N_2, with hydrogen gas.

$$C_2N_2(g) + H_2(g) \rightleftharpoons 2HCN(g)$$

At a certain temperature, the equilibrium constant is 47. What are the partial pressures of all gases at equilibrium if the initial partial pressures of the reactants are 0.500 atm?

37. At 460°C, the reaction

$$SO_2(g) + NO_2(g) \rightleftharpoons NO(g) + SO_3(g)$$

has $K = 84.7$. All gases are at an initial pressure of 1.25 atm.

(a) Calculate the partial pressure of each gas at equilibrium.

(b) Compare the total pressure initially with the total pressure at equilibrium. Would that relation be true of all gaseous systems?

38. The reaction

$$CO(g) + H_2O(g) \rightleftharpoons H_2(g) + CO_2(g)$$

has an equilibrium constant of 1.30 at 650°C. Carbon monoxide and steam both have initial partial pressures of 0.485 atm, while hydrogen and carbon dioxide start with partial pressures of 0.159 atm.

(a) Calculate the partial pressure of each gas at equilibrium.

(b) Compare the total pressure initially with the total pressure at equilibrium. Would that relation be true of all gaseous systems?

39. Solid ammonium carbamate, $NH_4CO_2NH_2$, decomposes at 25°C to ammonia and carbon dioxide.

$$NH_4CO_2NH_2(s) \rightleftharpoons 2NH_3(g) + CO_2(g)$$

The equilibrium constant for the decomposition at 25°C is 2.3×10^{-4}. At 25°C, 7.50 grams of $NH_4CO_2NH_2$ are sealed in a 10.0-L flask and allowed to decompose.

(a) What is the total pressure in the flask when equilibrium is established?

(b) What percentage of $NH_4CO_2NH_2$ decomposed?

(c) Can you state from the data calculated that the decomposition took place slowly?

WEB **40.** Solid ammonium iodide decomposes to ammonia and hydrogen iodide gases at sufficiently high temperatures.

$$NH_4I(s) \rightleftharpoons NH_3(g) + HI(g)$$

The equilibrium constant for the decomposition at 673 K is 0.215. Fifteen grams of ammonium iodide is sealed in a 5.0-L flask and heated to 673 K.

(a) What is the total pressure in the flask at equilibrium?

(b) How much ammonium iodide decomposes?

41. Phosphorus pentachloride gas can be produced by reacting phosphorus trichloride and chlorine gases at a certain temperature.

$$PCl_3(g) + Cl_2(g) \rightleftharpoons PCl_5(g)$$

At that temperature, $K = 0.42$. What is the partial pressure of chlorine gas at equilibrium if initial partial pressures are $P_{Cl_2} = 0.500$ atm, $P_{PCl_3} = 1.00$ atm?

42. The decomposition at 25°C of dinitrogen tetroxide

$$N_2O_4(g) \rightleftharpoons 2NO_2(g)$$

has an equilibrium constant of 0.144. What are the partial pressures at equilibrium of the two gases if a sealed flask initially contains only N_2O_4 at a pressure of 0.863 atm?

Le Châtelier's Principle

43. Consider the system

$$4NH_3(g) + 3O_2(g) \rightleftharpoons 2N_2(g) + 6H_2O(l) \qquad \Delta H = -1530.4 \text{ kJ}$$

(a) How will the amount of ammonia at equilibrium be affected by

(1) removing $O_2(g)$?

(2) adding $N_2(g)$?

(3) adding water?

(4) expanding the container at constant pressure?

(5) increasing the temperature?

(b) Which of the above factors will increase the value of K? Which will decrease it?

44. Consider the system

$$SO_3(g) \rightleftharpoons SO_2(g) + \tfrac{1}{2}O_2(g) \qquad \Delta H = 98.9 \text{ kJ}$$

(a) Predict whether the forward or reverse reaction will occur when the equilibrium is disturbed by

(1) adding oxygen gas.

(2) compressing the system at constant temperature.

(3) adding argon gas.

(4) removing $SO_2(g)$.

(5) decreasing the temperature.

(b) Which of the above factors will increase the value of K? Which will decrease it?

45. Predict the direction in which each of the following equilibria will shift if the pressure on the system is increased by compression.

(a) $H_2O(g) + C(s) \rightleftharpoons CO(g) + H_2(g)$

(b) $SbCl_5(g) \rightleftharpoons SbCl_3(g) + Cl_2(g)$

(c) $CO(g) + H_2O(g) \rightleftharpoons CO_2(g) + H_2(g)$

46. Predict the direction in which each of the following equilibria will shift if the pressure on the system is decreased by expansion.

(a) $Ni(s) + 4CO(g) \rightleftharpoons Ni(CO)_4(g)$

(b) $ClF_5(g) \rightleftharpoons ClF_3(g) + F_2(g)$

(c) $HBr(g) \rightleftharpoons \tfrac{1}{2}H_2(g) + \tfrac{1}{2}Br_2(g)$

47. For the reaction

$$CO_2(g) + H_2(g) \rightleftharpoons CO(g) + H_2O(g)$$

Equilibrium is established at a certain temperature when the partial pressures of CO, H_2O, CO_2, and H_2 are (in atm) 0.27, 0.41, 0.16, 0.68, respectively.

(a) Calculate K.

(b) If enough steam condenses to water to decrease its partial pressure to 0.30 atm, in which direction will the reaction proceed? What is the partial pressure of steam when equilibrium is re-established?

48. Water gas, a commercial fuel, is made by reacting hot coke with steam.

$$C(s) + H_2O(g) \rightleftharpoons CO(g) + H_2(g)$$

When equilibrium is established at 900 K, the partial pressures of CO, H_2, and H_2O are 0.22 atm, 0.63 atm, and 0.31 atm, respectively.

(a) Calculate K at 900 K.

(b) Enough steam is added to raise its partial pressure temporarily to 0.50 atm. What are the equilibrium partial pressures of all gases after equilibrium is re-established?

49. For the system

$$2SO_3(g) \rightleftharpoons 2SO_2(g) + O_2(g)$$

$K = 1.32$ at 627°C. What is the equilibrium constant at 555°C?

WEB 50. For the system

$$H_2(g) + I_2(g) \rightleftharpoons 2HI(g) \qquad \Delta H = -9.4 \text{ kJ}$$

$K = 62.5$ at 800 K. What is the equilibrium constant at 333°C?

51. For a certain reaction, $\Delta H° $ is +33 kJ. What is the ratio of the equilibrium constant at 400 K to that at 200 K?

52. What is the value of $\Delta H°$ for a reaction if K at 50°C is 40% of K at 37°C?

Unclassified

53. Isopropyl alcohol is the main ingredient in rubbing alcohol. It can decompose into acetone (the main ingredient in nail polish remover) and hydrogen gas according to the following reaction:

$$C_3H_7OH(g) \rightleftharpoons C_2H_6CO(g) + H_2(g)$$

At 180°C, the equilibrium constant for the decomposition is 0.45. If 20.0 mL ($d = 0.785$ g/mL) of isopropyl alcohol is placed in a 5.00-L vessel and heated to 180°C, what percentage remains undissociated at equilibrium?

54. Consider the equilibrium

$$C(s) + CO_2(g) \rightleftharpoons 2CO(g)$$

When this system is at equilibrium at 700°C in a 2.0-L container, there are 0.10 mol CO, 0.20 mol CO_2, and 0.40 mol C present. When the system is cooled to 600°C, an additional 0.040 mol C(s) forms. Calculate K at 700°C and again at 600°C.

55. At 1800 K, oxygen dissociates into gaseous atoms:

$$O_2(g) \rightleftharpoons 2O(g)$$

K for the system is 1.7×10^{-8}. If one mole of oxygen molecules is placed in a 5.0-L flask and heated to 1800 K, what percentage by mass of the oxygen dissociates? How many O atoms are there in the flask?

Conceptual Problems

56. The following data are for the system

$$A(g) \rightleftharpoons 2B(g)$$

Time (s)	0	20	40	60	80	100
P_A (atm)	1.00	0.83	0.72	0.65	0.62	0.62
P_B (atm)	0.00	0.34	0.56	0.70	0.76	0.76

Prepare a graph of P_A and P_B versus time and use it to answer the following questions:

(a) Estimate P_A and P_B after 30 s.

(b) Estimate P_A after 150 s.

(c) Estimate P_B when $P_A = 0.700$ atm.

57. The following data apply to the *unbalanced* equation:

$$A(g) \rightleftharpoons B(g)$$

Time (s)	0	50	100	150	200	250
P_A (atm)	2.00	1.25	0.95	0.80	0.71	0.68
P_B (atm)	0.10	0.60	0.80	0.90	0.96	0.98

(a) Based on the data, balance the equation (simplest whole-number coefficients).
(b) Has the system reached equilibrium? Explain.

58. For the reaction

$$C(s) + CO_2(g) \rightleftharpoons 2CO(g)$$

$K = 168$ at 1273 K. If one starts with 0.3 atm of CO_2 and 12.0 g of C at 1273 K, will the equilibrium mixture contain
(a) mostly CO_2?
(b) mostly CO?
(c) roughly equal amounts of CO_2 and CO?
(d) only C?

59. The system

$$3Z(g) + Q(g) \rightleftharpoons 2R(g)$$

is at equilibrium when the partial pressure of Q is 0.44 atm. Sufficient R is added to increase the partial pressure of Q temporarily to 1.5 atm. When equilibrium is re-established, the partial pressure of Q could be which of the following?
(a) 1.5 atm   **(b)** 1.2 atm   **(c)** 0.80 atm
(d) 0.44 atm   **(e)** 0.40 atm

60. The graph below is similar to that of Figure 12.2.

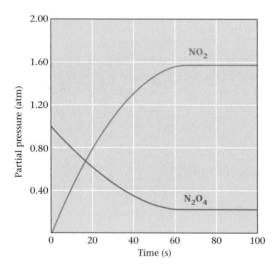

If after 100 s have elapsed the partial pressure of N_2O_4 is increased to 1.0 atm, what will the graph for N_2O_4 look like beyond 100 s?

61. The figures below represent the following reaction at equilibrium at different temperatures.

$$A_2(g) + 3B_2 \rightleftharpoons 2AB_3$$

where squares represent atom A and circles represent atom B. Is the reaction exothermic?

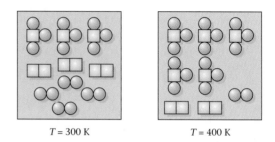

$T = 300$ K $T = 400$ K

62. Consider the statement, "The equilibrium constant for a mixture of hydrogen, nitrogen, and ammonia is 3.41." What information is missing from this statement?

63. Consider the statement, "The equilibrium constant for a reaction at 400 K is 792. It must be a very fast reaction." What is wrong with the statement?

Challenge Problems

64. Ammonia can decompose into its constituent elements according to the reaction

$$2NH_3(g) \rightleftharpoons N_2(g) + 3H_2(g)$$

The equilibrium constant for the decomposition at a certain temperature is 2.5. Calculate the partial pressures of all the gases at equilibrium if ammonia with a pressure of 1.00 atm is sealed in a 3.0-L flask.

65. Derive the relationship

$$K = K_c \times (RT)^{\Delta n_g}$$

where K_c is the equilibrium constant using molarities and Δn_g is the change in the number of moles of gas in the reaction (see page 355). (*Hint:* Recall that $P_A = n_A RT/V$ and $n_A/V = [A]$.)

66. Hydrogen iodide gas decomposes to hydrogen gas and iodine gas:

$$2HI(g) \rightleftharpoons H_2(g) + I_2(g)$$

To determine the equilibrium constant of the system, identical one-liter glass bulbs are filled with 3.20 g of HI and maintained at a certain temperature. Each bulb is periodically opened and analyzed for iodine formation by titration with sodium thiosulfate, $Na_2S_2O_3$.

$$I_2(aq) + 2S_2O_3{}^{2-}(aq) \longrightarrow S_4O_6{}^{2-}(aq) + 2I^-(aq)$$

It is determined that when equilibrium is reached, 37.0 mL of 0.200 M $Na_2S_2O_3$ is required to titrate the iodine. What is K at the temperature of the experiment?

67. For the system

$$SO_3(g) \rightleftharpoons SO_2(g) + \tfrac{1}{2}O_2(g)$$

at 1000 K, $K = 0.45$. Sulfur trioxide, originally at 1.00-atm pressure, partially dissociates to SO_2 and O_2 at 1000 K. What is its partial pressure at equilibrium?

68. At a certain temperature, the reaction

$$Xe(g) + 2F_2(g) \rightleftharpoons XeF_4(g)$$

gives a 50.0% yield of XeF_4, starting with Xe (P_{Xe} = 0.20 atm) and F_2 (P_{F_2} = 0.40 atm). Calculate K at this temperature. What must the initial pressure of F_2 be to convert 75.0% of the xenon to XeF_4?

69. A student studies the equilibrium

$$I_2(g) \rightleftharpoons 2I(g)$$

at a high temperature. She finds that the total pressure at equi-

librium is 40% greater than it was originally, when only I_2 was present. What is K for this reaction at that temperature?

70. Benzaldehyde, a flavoring agent, is obtained by the dehydrogenation of benzyl alcohol.

$$C_6H_5CH_2OH(g) \rightleftharpoons C_6H_5CHO(g) + H_2(g)$$

K for the reaction at 250°C is 0.56. If 1.50 g of benzyl alcohol is placed in a 2.0-L flask and heated to 250°C,

(a) what is the partial pressure of the benzaldehyde when equilibrium is established?

(b) how many grams of benzyl alcohol remain at equilibrium?

*There is nothing in the Universe but alkali and acid,
From which Nature composes all things.*

—Otto Tachenius (1671)

13 ACIDS AND BASES

The color of the hydrangeas
indicates the pH of the soil
where they have been grown.
(Charles D. Winters)

CHAPTER OUTLINE

13.1 BRØNSTED-LOWRY ACID-BASE MODEL

13.2 THE ION PRODUCT OF WATER

13.3 pH AND pOH

13.4 WEAK ACIDS AND THEIR EQUILIBRIUM CONSTANTS

13.5 WEAK BASES AND THEIR EQUILIBRIUM CONSTANTS

13.6 ACID-BASE PROPERTIES OF SALT SOLUTIONS

Among the solution reactions considered in Chapter 4 were those between acids and bases. In this chapter, we take a closer look at the properties of acidic and basic water solutions. In particular, we examine—

■ the ionization of water and the equilibrium between hydrated H^+ ions (H_3O^+ ions) and OH^- ions in water solution (Section 13.2).

■ the quantities pH and pOH, used to describe the acidity or basicity of water solutions (Section 13.3).

■ the types of species that act as weak acids and weak bases and the equilibria that apply in their water solutions (Sections 13.4, 13.5).

■ the acid-base properties of salt solutions (Section 13.6).

13.1 BRØNSTED-LOWRY ACID-BASE MODEL

In Chapter 4 we considered an acid to be a substance that produces an excess of H^+ ions in water. A base was similarly defined to be a substance that forms excess OH^- ions in water solution. This approach, first proposed by Svante Arrhenius in 1884, is a very practical one, but it has one disadvantage. It greatly restricts the number of reactions that can be considered to be of the acid-base type.

The model of acids and bases used in this chapter is a somewhat more general one developed independently by Johannes Brønsted (1879–1947) in Denmark and Thomas Lowry (1874–1936) in England in 1923. The **Brønsted-Lowry** model focuses on the nature of acids and bases and the reactions that take place between them. Specifically, it considers that—

■ *an **acid** is a proton (H^+ ion) donor.*
■ *a **base** is a proton (H^+ ion) acceptor.*
■ *in an acid-base reaction, a proton is transferred from an acid to a base.*

A Brønsted-Lowry acid-base reaction might be represented as

$$HB(aq) + A^-(aq) \rightleftharpoons HA(aq) + B^-(aq)$$

The species shown in red, HB and HA, acts as Brønsted-Lowry acids in the forward and reverse reactions, respectively; A^- and B^- (blue) act as Brønsted-Lowry bases. (We will use this color coding consistently throughout the chapter in writing Brønsted-Lowry acid-base equations.)

Sometimes called the Lowry-Brønsted model, at least in England.

See the *Saunders Interactive General Chemistry* CD-ROM, Screen 17.2, Brønsted Acids and Bases.

In connection with the Brønsted-Lowry model, there are some terms that are used frequently.

1. The species formed when a proton is removed from an acid is referred to as the **conjugate base** of that acid; B^- is the conjugate base of HB. The species formed when a proton is added to a base is called the **conjugate acid** of that base; HA is the conjugate acid of A^-. Thus we have

Conjugate Acid	Conjugate Base
HF	F^-
HSO_4^-	SO_4^{2-}
NH_4^+	NH_3

2. A species that can either accept or donate a proton is referred to as **amphiprotic.** An example is the H_2O molecule, which can gain a proton to form the hydronium ion, H_3O^+, or lose a proton, leaving the hydroxide ion, OH^-.

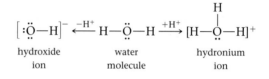

| hydroxide ion | water molecule | hydronium ion |

EXAMPLE 13.1

(a) What is the conjugate base of HNO_3? The conjugate acid of CN^-?
(b) The HCO_3^- ion, like the H_2O molecule, is amphiprotic. What is its conjugate base? Its conjugate acid?

Strategy   To form a conjugate base, remove H^+; the effect is to lower the number of hydrogen atoms by one and lower the charge by one unit. Conversely, a conjugate acid is formed by adding H^+; this adds a H atom and increases the charge by one unit.

Solution

(a) NO_3^-; HCN   (b) CO_3^{2-}; H_2CO_3

13.2   THE ION PRODUCT OF WATER

The acidic and basic properties of aqueous solutions are dependent on an equilibrium that involves the solvent, water. The reaction involved can be regarded as a Brønsted-Lowry acid-base reaction in which the H_2O molecule shows its amphiprotic nature:

$$H_2O + H_2O \rightleftharpoons H_3O^+(aq) + OH^-(aq)$$

Alternatively, and somewhat more simply, the reaction can be regarded as the ionization of a single H_2O molecule:

$$H_2O \rightleftharpoons H^+(aq) + OH^-(aq)$$

Recall (page 359, Chapter 12) that in the equilibrium constant expression for reactions in solution—

See Screen 17.3, The Acid-Base Properties of Water.

- solutes enter as their molarity, [ ].
- the solvent, H_2O in this case, does not appear: Its concentration is essentially the same in all dilute solutions.

Hence for the **ionization of water,** the equilibrium constant expression is

$$K_W = [H^+] \times [OH^-]$$

where K_W, referred to as the **ion product constant of water,** has a very small value. At 25°C,

$$K_W = 1.0 \times 10^{-14}$$

The concentrations of H^+ ($[H^+] = [H_3O^+]$) ions and OH^- ions in *pure water* at 25°C are readily calculated from K_W. Notice from the equation for ionization that these two ions are formed in equal numbers. Hence in pure H_2O,

$$[H^+] = [OH^-]$$

$$K_W = [H^+] \times [OH^-] = [H^+]^2 = 1.0 \times 10^{-14}$$

$$[H^+] = 1.0 \times 10^{-7}\ M = [OH^-]$$

An aqueous solution in which $[H^+]$ *equals* $[OH^-]$ is called a **neutral solution.** It has a $[H^+]$ of $1.0 \times 10^{-7}\ M$ at 25°C.

In most water solutions, the concentrations of H^+ and OH^- are not equal. The equation $[H^+] \times [OH^-] = 1.0 \times 10^{-14}$ indicates that these two quantities are inversely proportional to one another (Figure 13.1). When $[H^+]$ is very high, $[OH^-]$ is very low, and vice versa.

> There are very few H^+ and OH^- ions in pure water.

EXAMPLE 13.2   In a certain tap-water sample, $[H^+] = 3.0 \times 10^{-7}\ M$. What is the concentration of OH^-?

Strategy   Simply substitute into the expression for K_W and solve for $[OH^-]$.

Solution

$$[OH^-] = \frac{K_W}{[H^+]} = \frac{1.0 \times 10^{-14}}{3.0 \times 10^{-7}} = 3.3 \times 10^{-8}\ M$$

Reality Check   Because $[H^+]$ is greater than $10^{-7}\ M$, $[OH^-]$ must be less than $10^{-7}\ M$; it is.

An aqueous solution in which $[H^+]$ is greater than $[OH^-]$ is termed acidic. An aqueous solution in which $[OH^-]$ is greater than $[H^+]$ is basic (alkaline). Therefore,

if $[H^+] > 1.0 \times 10^{-7}\ M$, $[OH^-] < 1.0 \times 10^{-7}\ M$,   solution is acidic

if $[OH^-] > 1.0 \times 10^{-7}\ M$, $[H^+] < 1.0 \times 10^{-7}\ M$,   solution is basic

13.3   PH AND POH

As just pointed out, the acidity or basicity of a solution can be described in terms of its H^+ concentration. In 1909, Søren Sørensen, a biochemist working at the Carlsberg Brewery in Copenhagen, proposed an alternative method of specify-

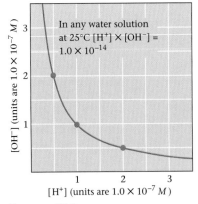

FIGURE 13.1
The $[H^+]$-$[OH^-]$ relationship. A graph of $[OH^-]$ versus $[H^+]$ looks very much like a graph of gas volume versus pressure. In both cases, the two variables are inversely proportional to one another. When $[H^+]$ gets larger, $[OH^-]$ gets smaller. **OHT**

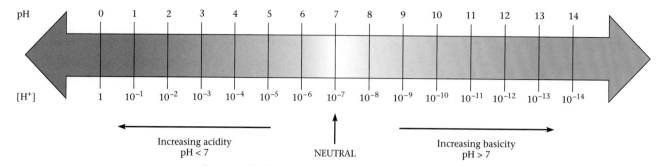

pH

$$0 \quad 1 \quad 2 \quad 3 \quad 4 \quad 5 \quad 6 \quad 7 \quad 8 \quad 9 \quad 10 \quad 11 \quad 12 \quad 13 \quad 14$$

$[H^+]$

$$1 \quad 10^{-1} \quad 10^{-2} \quad 10^{-3} \quad 10^{-4} \quad 10^{-5} \quad 10^{-6} \quad 10^{-7} \quad 10^{-8} \quad 10^{-9} \quad 10^{-10} \quad 10^{-11} \quad 10^{-12} \quad 10^{-13} \quad 10^{-14}$$

Increasing acidity
pH < 7

NEUTRAL

Increasing basicity
pH > 7

FIGURE 13.2

pH-[H^+] relationship. Acidity is inversely related to pH; the higher the H^+ ion concentration, the lower the pH. In neutral solution, $[H^+] = [OH^-] = 1.0 \times 10^{-7}$ M; pH = 7.00 at 25°C. **OHT**

> It is simpler to use numbers (pH = 4) than exponents ($[H^+] = 10^{-4}$ M) to describe acidity.

ing the acidity of a solution. He defined a term called **pH** (for "power of the hydrogen ion"):

$$pH = -\log_{10}[H^+] = -\log_{10}[H_3O^+]$$

or

$$[H^+] = [H_3O^+] = 10^{-pH}$$

See Screen 17.4, The pH Scale.

Figure 13.2 shows the relationship between pH and $[H^+]$. Notice that, as the defining equation implies, pH increases by one unit when the concentration of H^+ decreases by a power of 10. Moreover, *the higher the pH, the less acidic the solution.* Most aqueous solutions have hydrogen ion concentrations between 1 and 10^{-14} M and hence have a pH between 0 and 14.

The pH, like $[H^+]$ or $[OH^-]$, can be used to differentiate acidic, neutral, and basic solutions. At 25°C,

if pH < 7.0,   solution is acidic

if pH = 7.0,   solution is neutral

if pH > 7.0,   solution is basic

Table 13.1 shows the pH of some common solutions.

A similar approach is used for the hydroxide ion concentration. The **pOH** of a solution is defined as

$$pOH = -\log_{10}[OH^-]$$

Because $[H^+] \times [OH^-] = 1.0 \times 10^{-14}$ at 25°C, it follows that at this temperature

$$pH + pOH = 14.00$$

Thus a solution that has a pH of 6.20 must have a pOH of 7.80, and vice versa.

Sea water, a slightly basic solution. When these children emerge from the water, they may notice that their skin feels slippery, a result of the basic pH of the water (pH 8.3). *(Art Brewer/Tony Stone Images)*

TABLE 13.1	pH of Some Common Materials		
Lemon juice	2.2–2.4	Urine, human	4.8–8.4
Wine	2.8–3.8	Cow's milk	6.3–6.6
Vinegar	3.0	Saliva, human	6.5–7.5
Tomato juice	4.0	Drinking water	5.5–8.0
Beer	4–5	Blood, human	7.3–7.5
Cheese	4.8–6.4	Sea water	8.3

EXAMPLE 13.3 Calculate, at 25°C,

(a) the pH and pOH of a lemon juice solution in which $[H^+]$ is 5.0×10^{-3} M.
(b) the $[H^+]$ and $[OH^-]$ of human blood at pH 7.40.

Strategy These calculations can be carried out by substitution into the basic relations

$$pH = -\log_{10}[H^+] \qquad [H^+] \times [OH^-] = 1.0 \times 10^{-14} \qquad pH + pOH = 14.00$$

Use a calculator to find logarithms and inverse logarithms, following the rules for significant figures discussed in Appendix 3.

Solution

(a) You should find on your calculator that

$$\log_{10}(5.0 \times 10^{-3}) = -2.30$$

Hence pH $= -(-2.30) = $ **2.30** pOH $= 14.00 - $ pH $= $ **11.70**

(b) Because the pH is 7.40, $[H^+] = 10^{-7.40}$. To find $[H^+]$, enter -7.40 on your calculator. Then either—

- punch the 10^x key, if you have one, or
- punch the INV and then the LOG key

Either way, you should find that $[H^+] = $ **4.0×10^{-8} M.**
Knowing $[H^+]$, the concentration of OH^- is calculated as in Example 13.2:

$$[OH^-] = \frac{1.0 \times 10^{-14}}{4.0 \times 10^{-8}} = 2.5 \times 10^{-7} \text{ M}$$

Reality Check Because the pH of blood is 7.40, it must be slightly basic, which means that $[OH^-] > [H^+]$.

Knowing $[H^+]$, $[OH^-]$, pH, or pOH, you can calculate the other three quantities.

Remember to enter the minus sign first.

The number just calculated for the concentration of H^+ in blood, 4.0×10^{-8} M, is very small. You may wonder what difference it makes whether $[H^+]$ is 4.0×10^{-8} M, 4.0×10^{-7} M, or some other such tiny quantity. In practice, it makes a great deal of difference because a large number of biological processes involve H^+ as a reactant, so the rates of these processes depend on its concentration. If $[H^+]$ increases from 4.0×10^{-8} M to 4.0×10^{-7} M, the rate of a first-order reaction involving H^+ increases by a factor of 10. Indeed, if $[H^+]$ in blood increases by a much smaller amount, from 4.0×10^{-8} M to 5.0×10^{-8} M (pH 7.40 → 7.30), a condition called *acidosis* develops. The nervous system is depressed; fainting and even coma can result.

pH of Strong Acids and Strong Bases

As pointed out in Chapter 4, the following acids are strong

HCl	HBr	HI
$HClO_4$	HNO_3	H_2SO_4*

See Screen 17.5, Strong Acids and Bases.

*The *first* ionization of H_2SO_4 is complete:

$$H_2SO_4(aq) \longrightarrow H^+(aq) + HSO_4^-(aq)$$

FIGURE 13.3

Ionization of a strong acid. When HCl is added to water, there is a proton transfer from HCl to an H_2O molecule, forming a Cl^- ion and an H_3O^+ ion. In the reaction, HCl acts as a Brønsted-Lowry acid, H_2O as a Brønsted-Lowry base.

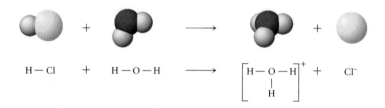

in the sense that they are completely ionized in water. The reaction of the strong acid HCl with water can be represented by the equation (Figure 13.3)

$$HCl(aq) + H_2O \longrightarrow H_3O^+(aq) + Cl^-(aq)$$

Because this reaction goes to completion, it follows that a 0.10 M solution of HCl is 0.10 M in both H_3O^+ (i.e., H^+) and Cl^- ions.

The strong bases

LiOH	NaOH	KOH
$Ca(OH)_2$	$Sr(OH)_2$	$Ba(OH)_2$

are completely ionized in dilute water solution. A 0.10 M solution of NaOH is 0.10 M in both Na^+ and OH^- ions.

> 1 M HCl has a pH of 0, 1 M NaOH a pH of 14.

The fact that strong acids and bases are completely ionized in water makes it relatively easy to calculate the pH and pOH of their solutions (Example 13.4).

EXAMPLE 13.4   What is the pH of

(a) 0.025 M HNO_3?
(b) a solution prepared by dissolving 10.0 g of $Ba(OH)_2$ per liter?

Strategy   To calculate pH, you can either use the defining equation (a) or the relation between pH and pOH (b). Note that both HNO_3 and $Ba(OH)_2$ are completely ionized in water.

Solution

(a) Because HNO_3 is a strong acid, $[H^+]$ = 0.025 M

$$pH = -\log_{10}(0.025) = \boxed{1.60}$$

> The pH of a 10^{-10} M solution of HNO_3 is not 10. Can you see why?

(b) $[Ba(OH)_2] = \dfrac{10.0 \text{ g}}{1 \text{ L}} \times \dfrac{1 \text{ mol}}{171.3 \text{ g}}$ = 0.0584 mol/L

Because 1 mol of $Ba(OH)_2$ yields 2 mol of OH^- ions,

$$[OH^-] = 2 \times 0.0584 \text{ mol/L} = 0.117 \text{ mol/L}$$

$$pOH = -\log_{10}(0.117) = 0.932 \qquad pH = 14.00 - 0.93 = \boxed{13.07}$$

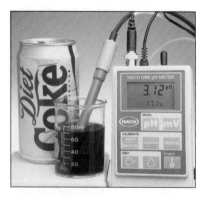

FIGURE 13.4

A pH meter with a digital readout. With a pH of 3.12, cola drinks are quite acidic. *(Charles D. Winters)*

Measuring pH

The pH of a solution can be measured by an instrument called a pH meter. A pH meter translates the H^+ ion concentration of a solution into an electrical signal that is converted into either a digital display or a deflection on a meter that reads pH directly (Figure 13.4). Later, in Chapter 18, we will consider the principle on which the pH meter works.

FIGURE 13.5
pH as shown by universal indicator. Universal indicator is deep red in strongly acidic solution *(upper left)*. It changes to yellow and green at pH 6 to 8, and then to deep violet in strongly basic solution *(lower right)*. *(Charles D. Winters)*

FIGURE 13.6
Influence of acidic or basic soil. Hydrangeas grown in strongly acidic soil (below pH 5) are blue. When they are grown in neutral or basic soil, the flowers are rosy pink. In weakly acidic soil (pH between 5 and 7) the flowers would be mauve. *(Charles D. Winters)*

A less accurate but more colorful way to measure pH uses a universal indicator, which is a mixture of acid-base indicators that shows changes in color at different pH values (Figure 13.5). A similar principle is used with pH paper. Strips of this paper are coated with a mixture of pH-sensitive dyes; these strips are widely used to test the pH of biological fluids, groundwater, and foods. Depending on the indicators used, a test strip can measure pH over a wide or narrow range.

Home gardeners frequently measure and adjust soil pH in an effort to improve the yield and quality of grass, vegetables, and flowers. Interestingly enough, the colors of many flowers depend on pH; among these are dahlias, delphiniums, and, in particular, hydrangeas (Figure 13.6). The first research in this area was carried out by Robert Boyle of gas-law fame, who published a paper on the relation between flower color and acidity in 1664.

13.4   WEAK ACIDS AND THEIR EQUILIBRIUM CONSTANTS

A wide variety of solutes behave as weak acids; that is, they react reversibly with water to form H_3O^+ ions. Using HB to represent a weak acid, its Brønsted-Lowry reaction with water is

$$HB(aq) + H_2O \longrightarrow H_3O^+(aq) + B^-(aq)$$

Typically, this reaction occurs to a very small extent; usually, fewer than 1% of the HB molecules are converted to ions.

Most weak acids fall into one of two categories:

1. ***Molecules containing an ionizable hydrogen atom.*** This type of weak acid was discussed in Chapter 4. There are literally thousands of molecular weak acids,

 See Screen 17.6, Weak Acids and Bases.

FIGURE 13.7

A weakly acidic cation. In water so-lution, the Al^{3+} ion is bonded to six water molecules in the $Al(H_2O)_6^{3+}$ ion *(left)*. In the $Al(H_2O)_5OH^{2+}$ ion *(right)*, one of the H_2O molecules has been re-placed by an OH^- ion.

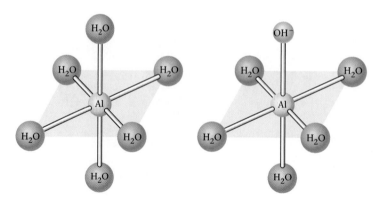

You can assume that all acids other than HCl, HBr, HI, HNO_3, $HClO_4$, and H_2SO_4 are weak.

most of them organic in nature. Among the molecular inorganic weak acids is nitrous acid:

$$HNO_2(aq) + H_2O \rightleftharpoons H_3O^+(aq) + NO_2^-(aq)$$

2. Cations. The ammonium ion, NH_4^+, behaves as a weak acid in water; a 0.10 M solution of NH_4Cl has a pH of about 5. The process by which the NH_4^+ ion lowers the pH of water can be represented by the (Brønsted-Lowry) equation:

$$NH_4^+(aq) + H_2O \rightleftharpoons H_3O^+(aq) + NH_3(aq)$$

Comparing the equation just written with that cited above for HNO_2, you can see that they are very similar. In both cases, a weak acid (HNO_2, NH_4^+) is converted to its conjugate base (NO_2^-, NH_3). The fact that the NH_4^+ ion has a +1 charge, whereas the HNO_2 molecule is neutral, is really irrelevant here.

To get red hydrangeas, add lime.

You may be surprised to learn that many metal cations act as weak acids in water solution. A 0.10 M solution of $Al_2(SO_4)_3$ has a pH close to 3; you can change the color of hydrangeas from red to blue by adding aluminum salts to soil. At first glance it is not at all obvious how a cation such as Al^{3+} can make a water solution acidic. However, the aluminum cation in water solution is really a hydrated species, $Al(H_2O)_6^{3+}$, in which six water molecules are bonded to the central Al^{3+} ion. This species can transfer a proton to a solvent water molecule to form an H_3O^+ ion:

$$Al(H_2O)_6^{3+}(aq) + H_2O \rightleftharpoons H_3O^+(aq) + Al(H_2O)_5(OH)^{2+}(aq)$$

Figure 13.7 shows the structures of the $Al(H_2O)_6^{3+}$ cation and its conjugate base, $Al(H_2O)_5(OH)^{2+}$.

A very similar equation can be written to explain why solutions of zinc salts are acidic. Here it appears that the hydrated cation in water solution contains four H_2O molecules bonded to a central Zn^{2+} ion. The Brønsted-Lowry equation is

All transition metal cations behave this way.

$$Zn(H_2O)_4^{2+}(aq) + H_2O \rightleftharpoons H_3O^+(aq) + Zn(H_2O)_3(OH)^+(aq)$$

EXAMPLE 13.5   Using the Brønsted-Lowry model, write equations to show why the following species behave as weak acids in water.

(a) acetic acid, $HC_2H_3O_2$, a weak acid found in vinegar

(b) the hydrated iron(II) cation, $Fe(H_2O)_6^{2+}$

Strategy   In each case, a proton is transferred to a solvent water molecule, which acts as a Brønsted base. The products are an H_3O^+ ion and the conjugate base of the weak acid.

Solution

(a) $HC_2H_3O_2(aq) + H_2O \rightleftharpoons H_3O^+(aq) + C_2H_3O_2^-(aq)$

(b) $Fe(H_2O)_6^{2+}(aq) + H_2O \rightleftharpoons H_3O^+(aq) + Fe(H_2O)_5(OH)^+(aq)$

The Equilibrium Constant for a Weak Acid

In discussing the equilibrium involved when a weak acid is added to water, it is convenient to represent the proton transfer

$$HB(aq) + H_2O \rightleftharpoons H_3O^+(aq) + B^-(aq)$$

as a simple ionization

$$HB(aq) \rightleftharpoons H^+(aq) + B^-(aq)$$

in which case the expression for the equilibrium constant becomes

$$K_a = \frac{[H^+] \times [B^-]}{[HB]}$$

The equilibrium constant K_a is called, logically enough, the **acid equilibrium constant** of the weak acid HB. The K_a values of some weak acids (in order of decreasing strength) are listed in Table 13.2. The weaker the acid, the smaller

See Screen 17.7, Determining K_a and K_b Values.

TABLE 13.2 Equilibrium Constants for Weak Acids and Their Conjugate Bases OHT

	Acid	K_a	Base	K_b
Sulfurous acid	H_2SO_3	1.7×10^{-2}	HSO_3^-	5.9×10^{-13}
Hydrogen sulfate ion	HSO_4^-	1.0×10^{-2}	SO_4^{2-}	1.0×10^{-12}
Phosphoric acid	H_3PO_4	7.1×10^{-3}	$H_2PO_4^-$	1.4×10^{-12}
Hexaaquairon(III) ion	$Fe(H_2O)_6^{3+}$	6.7×10^{-3}	$Fe(H_2O)_5OH^{2+}$	1.5×10^{-12}
Hydrofluoric acid	HF	6.9×10^{-4}	F^-	1.4×10^{-11}
Nitrous acid	HNO_2	6.0×10^{-4}	NO_2^-	1.7×10^{-11}
Formic acid	$HCHO_2$	1.9×10^{-4}	CHO_2^-	5.3×10^{-11}
Lactic acid	$HC_3H_5O_3$	1.4×10^{-4}	$C_3H_5O_3$	7.1×10^{-11}
Benzoic acid	$HC_7H_5O_2$	6.6×10^{-5}	$C_7H_5O_2^-$	1.5×10^{-10}
Acetic acid	$HC_2H_3O_2$	1.8×10^{-5}	$C_2H_3O_2^-$	5.6×10^{-10}
Hexaaquaaluminum(III) ion	$Al(H_2O)_6^{3+}$	1.2×10^{-5}	$Al(H_2O)_5OH^{2+}$	8.3×10^{-10}
Carbonic acid	H_2CO_3	4.4×10^{-7}	HCO_3^-	2.3×10^{-8}
Dihydrogen phosphate ion	$H_2PO_4^-$	6.2×10^{-8}	HPO_4^{2-}	1.6×10^{-7}
Hydrogen sulfite ion	HSO_3^-	6.0×10^{-8}	SO_3^{2-}	1.7×10^{-7}
Hypochlorous acid	HClO	2.8×10^{-8}	ClO^-	3.6×10^{-7}
Hydrocyanic acid	HCN	5.8×10^{-10}	CN^-	1.7×10^{-5}
Ammonium ion	NH_4^+	5.6×10^{-10}	NH_3	1.8×10^{-5}
Tetraaquazinc(II) ion	$Zn(H_2O)_4^{2+}$	3.3×10^{-10}	$Zn(H_2O)_3OH^+$	3.0×10^{-5}
Hydrogen carbonate ion	HCO_3^-	4.7×10^{-11}	CO_3^{2-}	2.1×10^{-4}
Hydrogen phosphate ion	HPO_4^{2-}	4.5×10^{-13}	PO_4^{3-}	2.2×10^{-2}

$$HB(aq) \rightleftharpoons H^+(aq) + B^-(aq) \qquad K_a = \frac{[H^+] \times [B^-]}{[HB]}$$

$$B^-(aq) + H_2O \rightleftharpoons HB(aq) + OH^-(aq) \qquad K_b = \frac{[HB] \times [OH^-]}{[B^-]}$$

the value of K_a. For example, HCN ($K_a = 5.8 \times 10^{-10}$) is a weaker acid than HNO$_2$, for which $K_a = 6.0 \times 10^{-4}$.

We sometimes refer to the **pK_a** value of a weak acid

$$pK_a = -\log_{10}K_a$$

HNO$_2$	$K_a = 6.0 \times 10^{-4}$	$pK_a = 3.22$
HCN	$K_a = 5.8 \times 10^{-10}$	$pK_a = 9.24$

There are several ways to determine K_a of a weak acid. A simple approach involves measuring [H$^+$] or pH in a solution prepared by dissolving a known amount of the weak acid to form a given volume of solution.

EXAMPLE 13.6   Aspirin is a weak organic acid whose molecular formula may be written as HC$_9$H$_7$O$_4$. A water solution of aspirin is prepared by dissolving 3.60 g per liter. The pH of this solution is found to be 2.60. Calculate K_a for aspirin.

Strategy   The approach used is very similar to that used in Chapter 12, except that partial pressures are replaced by concentrations in moles per liter. Note that you can readily calculate—

■ the original concentration of HC$_9$H$_7$O$_4$ ($\mathcal{M} = 180.15$ g/mol)

$$[HC_9H_7O_4]_0 = \frac{3.60 \text{ g}}{1 \text{ L}} \times \frac{1 \text{ mol}}{180.15 \text{ g}} = 2.00 \times 10^{-2} \text{ M}$$

■ the equilibrium concentration of H$^+$

$$[H^+]_{eq} = 10^{-2.60} = 2.5 \times 10^{-3} \text{ M}$$

Your task is to calculate the acid equilibrium constant:

$$HC_9H_7O_4(aq) \rightleftharpoons H^+(aq) + C_9H_7O_4^-(aq) \qquad K_a = \frac{[H^+] \times [C_9H_7O_4^-]}{[HC_9H_7O_4]}$$

Solution   From the chemical equation for the ionization of the weak acid, it should be clear that *1 mol* of C$_9$H$_7$O$_4^-$ is *produced*, and *1 mol* of HC$_9$H$_7$O$_4$ is *consumed* for every mole of H$^+$ produced. It follows that

$$\Delta[C_9H_7O_4^-] = \Delta[H^+] \qquad \Delta[HC_9H_7O_4] = -\Delta[H^+]$$

Originally, there is essentially no H$^+$ (ignoring the slight ionization of water). The same holds for the anion C$_9$H$_7$O$_4^-$; the only species present originally is the weak acid, HC$_9$H$_7$O$_4$, at a concentration of 0.0200 M.

Putting this information together in the form of a table:

	HC$_9$H$_7$O$_4(aq)$ $\rightleftharpoons$	H$^+(aq)$ +	C$_9$H$_7$O$_4^-(aq)$
[ ]$_0$	0.0200	0.0000	0.0000
Δ[ ]	−0.0025	+0.0025	+0.0025
[ ]$_{eq}$	0.0175	0.0025	0.0025

(Numbers in color are those given or implied in the statement of the problem; the other numbers are deduced using the ionization equation printed above the table. The symbols [ ]$_0$ and [ ]$_{eq}$ refer to original and equilibrium concentrations, respectively.)

All the information needed to calculate K_a is now available,

$$K_a = \frac{(2.5 \times 10^{-3})^2}{0.0175} = 3.6 \times 10^{-4}$$

It is always true that $\Delta[H^+] = \Delta[B^-] = -\Delta[HB]$.

Reality Check   Aspirin is a relatively *strong* weak acid; it would be located near the top of Table 13.2.

In discussing the ionization of a weak acid,

$$HB(aq) \rightleftharpoons H^+(aq) + B^-(aq)$$

we often refer to the **percent ionization:**

$$\% \text{ ionization} = \frac{[H^+]_{eq}}{[HB]_o} \times 100\%$$

For the aspirin solution referred to in Example 13.6,

$$\% \text{ ionization} = \frac{2.5 \times 10^{-3}}{2.0 \times 10^{-2}} \times 100\% = 12\%$$

As you might expect, percent ionization at a given concentration is directly related to K_a. Ibuprofen ($K_a = 2.5 \times 10^{-5}$), a weaker acid than aspirin, ($K_a = 3.6 \times 10^{-4}$) should be only 3.6% ionized at 0.020 M compared with 12% for aspirin.

Percent ionization also depends on the concentration of weak acid, increasing as the acid is diluted (Figure 13.8).

Calculation of [H⁺] in a Water Solution of a Weak Acid

Given the ionization constant of a weak acid and its original concentration, the H^+ concentration in solution is readily calculated. The approach used is the inverse of that followed in Example 13.6, where K_a was calculated knowing $[H^+]$; here, K_a is known and $[H^+]$ must be calculated.

See Screen 17.8, Estimating the pH of Weak Acid Solutions.

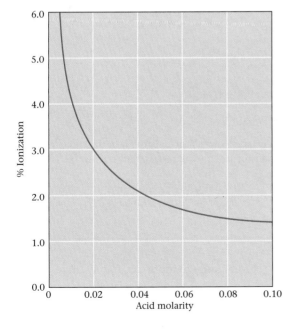

FIGURE 13.8
Percent ionization of acetic acid ($K_a = 1.8 \times 10^{-5}$). Like any weak acid, the percent ionization of acetic acid is inversely related to its molar concentration. **OHT**

EXAMPLE 13.7   Nicotinic acid, $HC_6H_4O_2N$ ($K_a = 1.4 \times 10^{-5}$), is another name for niacin, an important member of the vitamin B group. Determine $[H^+]$ in a solution prepared by dissolving 0.10 mol of nicotinic acid, HNic, in water to form one liter of solution.

Strategy   To start with, set up an equilibrium table similar to the one in the previous example. To accomplish this, note that—

■ the original concentrations of HNic, H^+, and Nic^- are 0.10 M, 0.00 M, and 0.00 M, respectively, ignoring, tentatively at least, the H^+ ions from the ionization of water.
■ the changes in concentration are related by the coefficients of the balanced equation, all of which are 1:

$$\Delta[Nic^-] = \Delta[H^+] \qquad \Delta[HNic] = -\Delta[H^+]$$

Letting $\Delta[H^+] = x$, it follows that $\Delta[Nic^-] = x$; $\Delta[HNic] = -x$. This information should enable you to express the equilibrium concentration of all species in terms of x. The rest is algebra; substitute into the expression for K_a and solve for $x = [H^+]$.

Solution   Setting up the table:

$HNic(aq) \rightleftharpoons H^+(aq) + Nic^-(aq)$

	0.10	0.00	0.00
$[\ ]_0$	0.10	0.00	0.00
$\Delta[\ ]$	$-x$	$+x$	$+x$
$[\ ]_{eq}$	$0.10 - x$	x	x

Substituting into the expression for K_a:

$$K_a = \frac{(x)(x)}{0.10 - x} = 1.4 \times 10^{-5}$$

This is a quadratic equation. It could be rearranged to the form $ax^2 + bx + c = 0$ and solved for x, using the quadratic formula. Such a procedure is time-consuming and, in this case, unnecessary. Nicotinic acid is a weak acid, only slightly ionized in water. The equilibrium concentration of HNic, $0.10 - x$, is probably only very slightly less than its original concentration, 0.10 M. So let's make the approximation $0.10 - x \approx 0.10$. This simplifies the equation written above:

$$\frac{x^2}{0.10} = 1.4 \times 10^{-5}$$

$$x^2 = 1.4 \times 10^{-6}$$

Taking square roots:

$$x = 1.2 \times 10^{-3} \, M = [H^+]$$

Reality Check   Note that the concentration of H^+, 0.0012 M, is—

■ much smaller than the original concentration of weak acid, 0.10 M. In this case, then, the approximation $0.10 - x \approx 0.10$ is justified. This will usually, but not always, be the case (see Example 13.8).
■ much larger than $[H^+]$ in pure water, 1×10^{-7} M, justifying the assumption that the ionization of water can be neglected. *This will always be the case, provided $[H^+]$ from the weak acid is $\geqq 10^{-6}$ M.*

In general, the value of K_a is seldom known more accurately than $\pm 5\%$. Hence in the expression

$$K_a = \frac{x^2}{a - x}$$

Equilibrium tables always help; be sure to include the equation.

where $x = [H^+]$ and $a = $ *original* concentration of weak acid, you can neglect the x in the denominator if doing so does not introduce an error of more than 5%. In other words,

$$\text{if } \frac{x}{a} \le 0.05$$

This is known as the 5% rule.

then

$$a - x \approx a$$

In most of the problems you will work, the approximation $a - x \approx a$ is valid, and you can solve for $[H^+]$ quite simply, as in Example 13.7, where $x = 0.012a$. Sometimes, though, you will find that the calculated $[H^+]$ is greater than 5% of the original concentration of weak acid. In that case, you can solve for x either by using the **quadratic formula** or the method of **successive approximations.**

Some household weak acids. Vinegar contains acetic acid. Fruit juices and sodas contain citric acid. Baking powder contains the Al^{3+} ion. *(Charles D. Winters)*

EXAMPLE 13.8   Calculate $[H^+]$ in a 0.100 M solution of nitrous acid, HNO_2, for which $K_a = 6.0 \times 10^{-4}$.

Strategy   The setup is identical with that in Example 13.7. However, you will find, on solving for x, that $x > 0.050a$, so the approximation $a - x \approx a$ fails. The simplest way to proceed is to use the calculated value of x to obtain a better estimate of $[HNO_2]$, then solve again for $[H^+]$. An alternative is to use the quadratic formula. (This is a particularly shrewd choice if you have a calculator that can be programmed to solve quadratic equations.)

Solution   Proceeding as in Example 13.7, you arrive at the equation

$$K_a = \frac{x^2}{0.100 - x} = 6.0 \times 10^{-4}$$

Making the same approximation as before, $0.100 - x \approx 0.100$,

$$x^2 = 0.100 \times 6.0 \times 10^{-4} = 6.0 \times 10^{-5}$$

$$x = 7.7 \times 10^{-3} \approx [H^+]$$

To check the validity of the approximation, note that

$$\frac{x}{a} = \frac{7.7 \times 10^{-3}}{1.0 \times 10^{-1}} = 0.077 > 0.050$$

So the approximation is not valid.

To obtain a better value for x, you have a choice of two approaches.

(1) *The method of successive approximations.* In the first approximation, you ignored the x in the denominator of the equation. This time, you use the approximate value of x just calculated, 0.0077, to find a more exact value for the concentration of HNO_2:

$$[HNO_2] = 0.100 - 0.0077 = 0.092 \, M$$

Substituting in the expression for K_a,

$$K_a = \frac{x^2}{0.092} = 6.0 \times 10^{-4} \qquad x^2 = 5.5 \times 10^{-5}$$

$$x = \boxed{7.4 \times 10^{-3} \, M} \approx [H^+]$$

WLM likes this method; CNH leans toward the use of the quadratic formula.

This value is closer to the true $[H^+]$, because 0.092 M is a better approximation for $[HNO_2]$ than was 0.100 M. If you're still not satisfied, you can go one step further. Using 7.4×10^{-3} for x instead of 7.7×10^{-3}, you can recalculate $[HNO_2]$ and solve again for x. If you do, you will find that your answer does not change. In other words, you have gone about as far as you can go.

(2) The quadratic formula. This gives an exact solution for x but is more time-consuming. Rewrite the equation

$$\frac{x^2}{0.100 - x} = 6.0 \times 10^{-4}$$

in the form $ax^2 + bx + c = 0$. Doing this:

$$x^2 + (6.0 \times 10^{-4}\, x) - (6.0 \times 10^{-5}) = 0$$

thus $a = 1; b = 6.0 \times 10^{-4}; c = -6.0 \times 10^{-5}$. Applying the quadratic formula,

$$x = \frac{-b \pm \sqrt{b^2 - 4ac}}{2a}$$

$$= \frac{-6.0 \times 10^{-4} \pm \sqrt{(6.0 \times 10^{-4})^2 + (24.0 \times 10^{-5})}}{2}$$

If you carry out the arithmetic properly, you should get two answers for x:

$$x = \boxed{7.4 \times 10^{-3}\, M} \quad \text{and} \quad -8.0 \times 10^{-3}\, M$$

The second answer is physically ridiculous; the concentration of H^+ cannot be a negative quantity. The first answer is the same one obtained by the method of successive approximations.

Polyprotic Weak Acids

Certain weak acids are **polyprotic;** they contain more than one ionizable hydrogen atom. Such acids ionize in steps, with a separate equilibrium constant for each step. Oxalic acid, a weak organic acid sometimes used to remove bloodstains, is *diprotic:*

$$H_2C_2O_4(aq) \rightleftharpoons H^+(aq) + HC_2O_4^-(aq) \qquad K_{a1} = 5.9 \times 10^{-2}$$

$$HC_2O_4^-(aq) \rightleftharpoons H^+(aq) + C_2O_4^{2-}(aq) \qquad K_{a2} = 5.2 \times 10^{-5}$$

Phosphoric acid, a common ingredient of cola drinks, is *triprotic:*

$$H_3PO_4(aq) \rightleftharpoons H^+(aq) + H_2PO_4^-(aq) \qquad K_{a1} = 7.1 \times 10^{-3}$$

$$H_2PO_4^-(aq) \rightleftharpoons H^+(aq) + HPO_4^{2-}(aq) \qquad K_{a2} = 6.2 \times 10^{-8}$$

$$HPO_4^{2-}(aq) \rightleftharpoons H^+(aq) + PO_4^{3-}(aq) \qquad K_{a3} = 4.5 \times 10^{-13}$$

The behavior of these acids is typical of all polyprotic acids in that—

- *the anion formed in one step* (e.g., $HC_2O_4^-$, $H_2PO_4^-$) *produces another H^+ ion in the next step.*
- *the acid equilibrium constant becomes smaller with each successive step.*

$$K_{a1} > K_{a2} > K_{a3}$$

Putting it another way, *the acids in successive steps become progressively weaker.* This is reasonable; it should be more difficult to remove a positively charged proton, H^+, from a negatively charged species like $H_2PO_4^-$ than from a neutral molecule like H_3PO_4.

Ordinarily, successive values of K_a for polyprotic acids decrease by a factor of at least 100 (Table 13.3). In that case, essentially all the H^+ ions in the solution come from the first step. This makes it relatively easy to calculate the pH of a solution of a polyprotic acid.

TABLE 13.3 Equilibrium Constants for Some Weak Polyprotic Acids at 25°C

Acid	Formula	K_{a1}	K_{a2}	K_{a3}
Carbonic acid*	H_2CO_3	4.4×10^{-7}	4.7×10^{-11}	
Oxalic acid	$H_2C_2O_4$	5.9×10^{-2}	5.2×10^{-5}	
Phosphoric acid	H_3PO_4	7.1×10^{-3}	6.2×10^{-8}	4.5×10^{-13}
Sulfurous acid	H_2SO_3	1.7×10^{-2}	6.0×10^{-8}	

*Carbonic acid is a water solution of carbon dioxide:

$$CO_2(g) + H_2O \rightleftharpoons H_2CO_3(aq)$$

The ionization constants listed are calculated assuming that all the carbon dioxide that dissolves is in the form of H_2CO_3.

EXAMPLE 13.9   The distilled water you use in the laboratory is slightly acidic because of dissolved CO_2, which reacts to form carbonic acid, H_2CO_3. Calculate the pH of a 0.0010 M solution of H_2CO_3.

Strategy   In principle, there are two different sources of H^+ ions from H_2CO_3:

$$H_2CO_3(aq) \rightleftharpoons H^+(aq) + HCO_3^-(aq) \qquad K_{a1} = 4.4 \times 10^{-7}$$

$$HCO_3^-(aq) \rightleftharpoons H^+(aq) + CO_3^{2-}(aq) \qquad K_{a2} = 4.7 \times 10^{-11}$$

In practice, essentially all the H^+ ions come from the first reaction, because K_{a1} is so much larger than K_{a2}. In other words, H_2CO_3 can be treated as if it were a weak monoprotic acid.

Solution   For the first reaction, $H_2CO_3(aq) \rightleftharpoons H^+(aq) + HCO_3^-(aq)$

$$K_{a1} = \frac{x^2}{0.0010 - x} = 4.4 \times 10^{-7} \qquad (x = [H^+] = [HCO_3^-])$$

Making the approximation $0.0010 - x \approx 0.0010$ and solving gives

$$x = (4.4 \times 10^{-10})^{1/2} = 2.1 \times 10^{-5}$$

$$\boxed{pH = -\log_{10}(2.1 \times 10^{-5}) = 4.68}$$

Note that the solution is indeed acidic, with a pH considerably less than 7.

Reality Check   To check the assumption that essentially all of the H^+ ions come from the first reaction, let x be the concentration of H^+ produced by the second reaction. Then we have

$$HCO_3^-(aq) \rightleftharpoons H^+(aq) + CO_3^{2-}(aq) \qquad K_a = 4.7 \times 10^{-11}$$
$$\quad (2.1 \times 10^{-5} - x) \qquad (2.1 \times 10^{-5} + x) \qquad x$$

$$\frac{(2.1 \times 10^{-5} + x)\,(x)}{(2.1 \times 10^{-5} - x)} = 4.7 \times 10^{-11}$$

Solving this equation by the quadratic formula, we find that $x = 4.7 \times 10^{-11}$. Because $2.1 \times 10^{-5} \gg 4.7 \times 10^{-11}$, it follows that virtually all of the H^+ ions come from the first reaction.

13.5 WEAK BASES AND THEIR EQUILIBRIUM CONSTANTS

See Screen 17.6, Weak Acids and Bases.

Like the weak acids, there are a large number of solutes that act as weak bases. It is convenient to classify weak bases into two groups, molecules and anions.

1. Molecules. As pointed out in Chapter 4, there are many molecular weak bases, including the organic compounds known as amines. The simplest weak base is ammonia, whose reversible, Brønsted-Lowry reaction with water is represented by the equation

$$NH_3(aq) + H_2O \rightleftharpoons NH_4^+(aq) + OH^-(aq)$$

This reaction proceeds to only a very small extent. In 0.10 M NH_3, the concentrations of NH_4^+ and OH^- are only about 0.0013 M.

2. Anions. An anion derived from a weak acid is itself a weak base. A typical example is the fluoride ion, F^-, which is the conjugate base of the weak acid HF. The reaction of the F^- ion with water is

$$F^-(aq) + H_2O \rightleftharpoons HF(aq) + OH^-(aq)$$

Only a few HF molecules ionize, so $[OH^-] \gg [H^+]$.

The OH^- ions formed make the solution basic. A 0.10 M solution of sodium fluoride, NaF, has a pH of about 8.1.

Notice the similarity between the equation just written and that for the NH_3 molecule cited above—

■ both weak bases (NH_3, F^-) accept protons to form the conjugate acid (NH_4^+, HF).
■ water acts as a Brønsted-Lowry acid in each case, donating a proton to form the OH^- ion.

EXAMPLE 13.10   Write an equation to explain why each of the following species produces a basic water solution.

(a) NO_2^- (b) CO_3^{2-} (c) HCO_3^-

Strategy   In each case, the weak base reacts reversibly with a water molecule, picking up a proton from it. Two species are formed. One is an OH^- ion, which makes the solution basic. The other product is the conjugate acid of the weak base.

Solution

(a) $NO_2^-(aq) + H_2O \rightleftharpoons HNO_2(aq) + OH^-(aq)$
(b) $CO_3^{2-}(aq) + H_2O \rightleftharpoons HCO_3^-(aq) + OH^-(aq)$
(c) $HCO_3^-(aq) + H_2O \rightleftharpoons H_2CO_3(aq) + OH^-(aq)$

The Equilibrium Constant for a Weak Base

See Screen 17.7, Determining K_a and K_b Values.

For the weak base ammonia

$$NH_3(aq) + H_2O \rightleftharpoons NH_4^+(aq) + OH^-(aq)$$

the **base equilibrium constant** is written in the usual way, omitting the term for solvent water.

$$K_b = \frac{[NH_4^+] \times [OH^-]}{[NH_3]}$$

For an anion B^- that acts as a weak base,

$$B^-(aq) + H_2O \rightleftharpoons HB(aq) + OH^-(aq)$$

a very similar expression can be written:

$$K_b = \frac{[HB] \times [OH^-]}{[B^-]}$$

Values of K_b are listed at the right of Table 13.2 (page 391). The larger the value of K_b, the stronger the base. Because K_b for NH_3 (1.8×10^{-5}) is larger than K_b for $C_2H_3O_2^-$ (5.6×10^{-10}), ammonia is a stronger base than the acetate ion.

The quantity **pK_b** is defined as

$$pK_b = -\log_{10}K_b$$

The pK_b of ammonia ($K_b = 1.8 \times 10^{-5}$) is 4.74.

Calculation of [OH⁻] in a Water Solution of a Weak Base

We showed in Section 13.4 how K_a of a weak acid can be used to calculate $[H^+]$ in a solution of that acid. In a very similar way, K_b can be used to find $[OH^-]$ in a solution of a weak base.

See Screen 17.9, Estimating the pH of Weak Base Solutions.

EXAMPLE 13.11   Calculate the pH of 0.10 M NaF (K_b F^- = 1.4×10^{-11}).

Strategy   The procedure is entirely analogous to that followed in Example 13.7. Take $[OH^-] = x$ and set up an equilibrium table, ignoring the OH^- ions present in pure water. Solve for x, making the usual approximation. Then calculate pOH and finally pH. You should obtain a pH greater than 7; a solution of sodium fluoride had better be basic!

Solution   The equilibrium table is

$$F^-(aq) + H_2O \rightleftharpoons HF(aq) + OH^-(aq)$$

[ ]₀	0.10	0.00	0.00
Δ[ ]	$-x$	$+x$	$+x$
[ ]eq	$0.10 - x$	x	x

Substituting into the equilibrium constant expression:

$$K_b = \frac{x^2}{0.10 - x} = 1.4 \times 10^{-11}$$

Assuming $0.10 - x \approx 0.10$ and solving for x,

$$x^2 = 1.4 \times 10^{-12} \qquad x = 1.2 \times 10^{-6}$$

Clearly $x \ll 0.10$, so the approximation is justified.

$$[OH^-] = 1.2 \times 10^{-6}\,M$$

$$pOH = -\log_{10}(1.2 \times 10^{-6}) = 5.92$$

$$pH = 14.00 - 5.92 = \boxed{8.08}$$

HF is among the strongest of weak acids.

Reality Check   The pH is not much above 7, which is what you would expect. The F^- ion is a very weak base.

Relation Between K_a and K_b

Equilibrium constants of weak bases can be measured in the laboratory by procedures very much like those used for weak acids. In practice, though, it is simpler to take advantage of a simple mathematical relationship between K_b for a weak base and K_a for its conjugate acid. This relationship can be derived by adding together the equations for the ionization of the weak acid HB and the reaction of the weak base B^- with water:

$$
\begin{array}{lll}
(1) & HB(aq) \rightleftharpoons H^+(aq) + B^-(aq) & K_I = K_a \text{ of HB} \\
(2) & B^-(aq) + H_2O(aq) \rightleftharpoons HB(aq) + OH^-(aq) & K_{II} = K_b \text{ of } B^- \\
\hline
(3) & H_2O \rightleftharpoons H^+(aq) + OH^-(aq) & K_{III} = K_w
\end{array}
$$

Because Equation (1) + Equation (2) = Equation (3), we have, according to the rule of multiple equilibria (Chapter 12),

$$K_I \times K_{II} = K_{III}$$

or

$$\boxed{(K_a \text{ of HB}) \times (K_b \text{ of } B^-) = K_w = 1.0 \times 10^{-14}}$$

Checking Table 13.2 (page 391), you can see that this relation holds in each case. For example,

$$K_a \, HNO_2 \times K_b \, NO_2^- = (6.0 \times 10^{-4})(1.7 \times 10^{-11})$$
$$= 10 \times 10^{-15} = 1.0 \times 10^{-14}$$

EXAMPLE 13.12 Taking K_b of methylamine, CH_3NH_2, to be 4.2×10^{-4}, calculate K_a of the $CH_3NH_3^+$ ion.

Strategy Substitute into the relation

$$K_a \times K_b = 1.0 \times 10^{-14}$$

Solution

$$K_a = \frac{1.0 \times 10^{-14}}{K_b} = \frac{1.0 \times 10^{-14}}{4.2 \times 10^{-4}} = 2.4 \times 10^{-11}$$

Reality Check Methylamine is a relatively strong weak base, so the $CH_3NH_3^+$ ion is a very weak acid.

From the general relation between K_a of a weak acid and K_b of its conjugate weak base, it should be clear that these two quantities are inversely related to each other. Table 13.4 shows this effect and some other interesting features. In particular,

1. Brønsted-Lowry acids (left column) can be divided into three categories:
(a) strong acids ($HClO_4$, . . .), which are stronger proton donors than the H_3O^+ ion.
(b) weak acids (HF, . . .), which are weaker proton donors than the H_3O^+ ion, but stronger than the H_2O molecule.

Common products containing strong acids and bases. Muriatic acid (sold in hardware stores) is hydrochloric acid, used to clean tile surfaces. Battery acid *(center)* is sulfuric acid. Oven cleaners like that shown contain sodium hydroxide. *(Charles D. Winters)*

TABLE 13.4 Relative Strengths of Brønsted-Lowry Acids and Bases OHT

K_a	Conjugate Acid	Conjugate Base	K_b
Very large	$HClO_4$	ClO_4^-	Very small
Very large	HCl	Cl^-	Very small
Very large	HNO_3	NO_3^-	Very small
	H_3O^+	H_2O	
6.9×10^{-4}	HF	F^-	1.4×10^{-11}
1.8×10^{-5}	$HC_2H_3O_2$	$C_2H_3O_2^-$	5.6×10^{-10}
1.4×10^{-5}	$Al(H_2O)_6^{3+}$	$Al(H_2O)_5(OH)^{2+}$	7.1×10^{-10}
4.4×10^{-7}	H_2CO_3	HCO_3^-	2.3×10^{-8}
2.8×10^{-8}	$HClO$	ClO^-	3.6×10^{-7}
5.6×10^{-10}	NH_4^+	NH_3	1.8×10^{-5}
4.7×10^{-11}	HCO_3^-	CO_3^{2-}	2.1×10^{-4}
	H_2O	OH^-	
Very small	C_2H_5OH	$C_2H_5O^-$	Very large
Very small	OH^-	O^{2-}	Very large
Very small	H_2	H^-	Very large

Species shown in deep color are un-stable because they react with H_2O molecules.

The —OH group in C_2H_5OH does not act as an OH^- ion.

(c) species such as C_2H_5OH, which are weaker proton donors than the H_2O molecule and hence do not form acidic water solutions.

2. Brønsted-Lowry bases (right column) can be divided similarly:

(a) strong bases (H^-, . . .), which are stronger proton acceptors than the OH^- ion.

(b) weak bases (F^-, . . .), which are weaker proton acceptors than the OH^- ion, but stronger than the H_2O molecule.

(c) the anions of strong acids (ClO_4^-, . . .), which are weaker proton acceptors than the H_2O molecule and hence do not form basic water solutions.

We should point out that, just as the three strong acids at the top of Table 13.4 are completely converted to H_3O^+ ions in aqueous solution:

$$HClO_4(aq) + H_2O \longrightarrow H_3O^+(aq) + ClO_4^-(aq)$$

$$HCl(aq) + H_2O \longrightarrow H_3O^+(aq) + Cl^-(aq)$$

$$HNO_3(aq) + H_2O \longrightarrow H_3O^+(aq) + NO_3^-(aq)$$

the three strong bases at the bottom of the table are completely converted to OH^- ions:

$$H^-(aq) + H_2O \longrightarrow OH^-(aq) + H_2(g)$$

$$O^{2-}(aq) + H_2O \longrightarrow OH^-(aq) + OH^-(aq)$$

$$C_2H_5O^-(aq) + H_2O \longrightarrow OH^-(aq) + C_2H_5OH(aq)$$

In other words, the species H^-, O^{2-}, and $C_2H_5O^-$ do not exist in water solution.

Hydride ion–water reaction. As water is dropped onto solid calcium hydride, the hydride ion (H^-) reacts immediately and vigorously to form $H_2(g)$ (which is ignited by the heat of reaction) and OH^-. *(Charles D. Winters)*

See Screen 17.10, Acid-Base
Properties of Salts.

Strongest one wins!

Learn the spectator ions, not the
players.

13.6 ACID-BASE PROPERTIES OF SALT SOLUTIONS

*A **salt** is an ionic solid containing a cation other than H^+ and an anion other
than OH^- or O^{2-}.* When a salt such as $NaCl$, K_2CO_3, or $Al(NO_3)_3$ dissolves in water, the cation and anion separate from one another.

$$NaCl(s) \longrightarrow Na^+(aq) + Cl^-(aq)$$

$$K_2CO_3(s) \longrightarrow 2K^+(aq) + CO_3^{2-}(aq)$$

$$Al(NO_3)_3(s) \longrightarrow Al^{3+}(aq) + 3NO_3^-(aq)$$

To predict whether a given salt solution will be acidic, basic, or neutral, you
consider three factors in turn.

1. *Decide what effect, if any, the cation has on the pH of water.*
2. *Decide what effect, if any, the anion has on the pH of water.*
3. *Combine the two effects to decide on the behavior of the salt.*

Cations: Weak Acids or Spectator Ions?

As we pointed out in Section 13.4, certain cations act as weak acids in water solution because of reactions such as

$$NH_4^+(aq) + H_2O \rightleftharpoons H_3O^+(aq) + NH_3(aq)$$

$$Zn(H_2O)_4^{2+}(aq) + H_2O \rightleftharpoons H_3O^+(aq) + Zn(H_2O)_3(OH)^+(aq)$$

Essentially all transition metal ions behave like Zn^{2+}, forming a weakly acidic solution. Among the main-group cations, Al^{3+} and, to a lesser extent, Mg^{2+}, act as
weak acids. In contrast the cations in Group 1 show little or no tendency to react
with water.

If we say that *to classify an ion as acidic or basic in water solution, it must change
the pH by more than 0.5 unit in 0.1 M solution,* then the cations derived from strong
bases:

- the alkali metal cations (Li^+, Na^+, K^+ . . .)
- the heavier alkaline earth cations (Ca^{2+}, Sr^{2+}, Ba^{2+})

are spectator ions as far as pH is concerned (Table 13.5).

Sodium chloride solution. When the
NaCl is completely dissolved, the solution will contain only Na^+, Cl^-, and
H_2O, and it will be neither acidic nor
basic. *(Charles D. Winters)*

TABLE 13.5 Acid-Base Properties of Ions* in Water Solution

	Spectator		Basic		Acidic	
Anion	Cl^- Br^- I^-	NO_3^- ClO_4^- SO_4^{2-}	$C_2H_3O_2^-$ F^- Many others	CO_3^{2-} PO_4^{3-}		
Cation	Li^+ Na^+ K^+	Ca^{2+} Sr^{2+} Ba^{2+}			NH_4^+ Mg^{2+} Transition metal ions	Al^{3+}

*For the acid-base properties of amphiprotic anions such as HCO_3^- or $H_2PO_4^-$, see the discussion at the
end of this section.

Anions: Weak Bases or Spectator Ions?

As pointed out in Section 13.5, anions that are the conjugate bases of weak acids act themselves as weak bases in water. They accept a proton from a water molecule, leaving an OH^- ion that makes the solution basic. The reactions of the fluoride and carbonate ions are typical:

$$F^-(aq) + H_2O \rightleftharpoons HF(aq) + OH^-(aq)$$
$$CO_3^{2-}(aq) + H_2O \rightleftharpoons HCO_3^-(aq) + OH^-(aq)$$

Most anions behave this way, except those derived from six strong acids (Cl^-, Br^-, I^-, NO_3^-, ClO_4^-, SO_4^{2-}), which show little or no tendency to react with water to form OH^- ions. Like the cations listed in the left column of Table 13.5, they act as spectator ions as far as pH is concerned.

Salts: Acidic, Basic, or Neutral?

If you know how the cation and anion of a salt affect the pH of water, it is a relatively simple matter to decide what the net effect will be (Example 13.13).

EXAMPLE 13.13   Consider water solutions of these four salts:

$$NH_4I, \quad Zn(NO_3)_3, \quad KClO_4, \quad Na_3PO_4$$

Which of these solutions are acidic? Basic? Neutral?

Strategy   First, decide what ions are present in the solution. Then classify each cation and anion as acidic, basic, or neutral, using Table 13.5. Finally, consider the combined effects of the two ions in each salt.

Solution   To implement this strategy, it is convenient to prepare a table.

Salt	Cation	Anion	Solution of Salt
NH_4I	NH_4^+ (acidic)	I^- (spectator)	acidic
$Zn(NO_3)_2$	Zn^{2+} (acidic)	NO_3^- (spectator)	acidic
$KClO_4$	K^+ (spectator)	ClO_4^- (spectator)	neutral
Na_3PO_4	Na^+ (spectator)	PO_4^{3-} (basic)	basic

These predictions are confirmed by experiment (Figure 13.9).

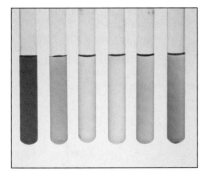

FIGURE 13.9
Universal indicator. The indicator is red in strong acid (far left) and purple in strong base (far right). In between these extremes, the dilute salt solutions are (left to right) weakly acidic NH_4I (pH about 4) and $Zn(NO_3)_2$ (pH about 5); neutral $KClO_4$; and basic Na_3PO_4 (pH about 12). *(Charles D. Winters)*

A weak acid–weak base reaction. The fizz is produced by CO_2 from the reaction of HCO_3^- as a weak base with citric acid, a weak organic acid. *(Charles D. Winters)*

Most ions of this type, like HCO_3^-, are basic.

The procedure described in Example 13.13 does not cover one combination—an acidic cation (e.g., NH_4^+) and a basic anion (e.g., F^-). In that case K_a of the cation must be compared with K_b of the anion. In general,

1. *If $K_a > K_b$, the salt is acidic.* An example is NH_4F ($K_aNH_4^+ = 5.6 \times 10^{-10}$; $K_bF^- = 1.4 \times 10^{-11}$), where the pH is about 6.2.
2. *If $K_b > K_a$, the salt is basic.* An example is NH_4ClO ($K_aNH_4^+ = 5.6 \times 10^{-10}$, $K_bClO^- = 3.6 \times 10^{-7}$), where the pH is 8.4.

A similar argument can be used to decide whether an amphiprotic anion such as HCO_3^- or $H_2PO_4^-$ is acidic or basic. In principle, these ions can act as either weak acids or weak bases:

$$HCO_3^-(aq) + H_2O \rightleftharpoons H_3O^+(aq) + CO_3^{2-}(aq) \qquad K_a = 4.7 \times 10^{-11}$$
$$HCO_3^-(aq) + H_2O \rightleftharpoons H_2CO_3(aq) + OH^-(aq) \qquad K_b = 2.3 \times 10^{-8}$$

Because $K_b > K_a$, you would predict, correctly, that the HCO_3^- ion produces a basic water solution. Sodium hydrogen carbonate, $NaHCO_3$ ("bicarbonate of soda"), is used as an antacid because its water solution is basic, with a pH of about 8.3. The reverse situation applies with the $H_2PO_4^-$ ion ($K_a = 6.2 \times 10^{-8}$, $K_b = 1.4 \times 10^{-12}$). Because $K_a > K_b$, the $H_2PO_4^-$ ion gives an acidic solution. The compound NaH_2PO_4 is weakly acidic, with a solution pH of about 4.7.

CHEMISTRY
Beyond the Classroom

Organic Acids

As pointed out earlier in this chapter, most weak acids are organic in nature; that is, they contain carbon and hydrogen atoms. Table A lists some of the organic acids found in foods. All of these compounds contain the *carboxyl* group:

$$-\underset{\underset{O}{\|}}{C}-O-H$$

The general equation for their reversible dissociation in water is

$$RCOOH(aq) \rightleftharpoons RCOO^-(aq) + H^+(aq)$$

Several different organic acids are formed in human metabolism, among them lactic acid. If you exercise strenuously, glucose may be converted to lactic acid in your muscles, producing tiredness and a painful sensation. Resting after vigorous exercise allows enough oxygen to reach the tissues and converts glucose to carbon dioxide and water.

Certain drugs, both prescription and over-the-counter, contain organic acids. Two of the most popular products of this type are the analgesics aspirin and ibuprofen (Advil, Nuprin, etc.).

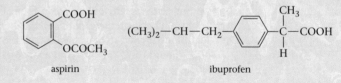

aspirin   ibuprofen

TABLE A   Some Naturally Occurring Organic Acids

Name		Source
Acetic acid	CH_3-COOH	Vinegar
Citric acid	$\underset{\displaystyle \underset{COOH}{|}}{HOOC-CH_2-\overset{\displaystyle \overset{OH}{|}}{C}-CH_2-COOH}$	Citrus fruits
Lactic acid	$CH_3-\underset{\displaystyle \underset{OH}{|}}{CH}-COOH$	Sour milk
Malic acid	$HOOC-CH_2-\underset{\displaystyle \underset{OH}{|}}{CH}-COOH$	Apples, watermelons, grape juice, wine
Oxalic acid	$HOOC-COOH$	Rhubarb, spinach, tomatoes
Quinic acid	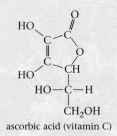	Cranberries
Tartaric acid	$HOOC-\underset{\displaystyle \underset{OH}{|}}{CH}-\underset{\displaystyle \underset{OH}{|}}{CH}-COOH$	Grape juice, wine

Because these compounds are acidic, they can cause stomach irritation unless taken with food or water. Beyond that, aspirin inhibits blood clotting, which explains why it is often prescribed to reduce the likelihood of a stroke or heart attack. Indeed, it is now recommended for a person in the throes of a heart attack.

To the general public, probably the best known organic acid is ascorbic acid (vitamin C):

Foods that contain organic acids. Table A identifies the acids in each of these foods. *(Charles D. Winters)*

ascorbic acid (vitamin C)

Vitamin C plays a variety of roles in human nutrition, some of which have been discovered only recently. For one thing, it is involved in the formation of collagen, the principal protein in connective tissue. Beyond that, vitamin C is required for the metabolism of several amino acids and for the absorption of iron. Moreover, because vitamin C is easily oxidized, it serves to protect other vitamins (A and E) from oxidation. The recommended daily allowance of vitamin C is 60 mg/day.

CHAPTER HIGHLIGHTS

Key Concepts

1. Using the Brønsted-Lowry model, classify a species as an acid or base and explain by a net ionic equation.
   (Examples 13.1, 13.5, 13.10; Problems 1–6, 25, 26, 49, 50)
2. Given $[H^+]$, $[OH^-]$, pH, or pOH, calculate the other three quantities.
   (Examples 13.2–13.4; Problems 7–22, 83)
3. Given the pH and original concentration of a weak acid solution, calculate K_a.
   (Example 13.6; Problems 33–36)
4. Given K_a of a weak acid and its original concentration, calculate $[H^+]$.
   (Examples 13.7–13.9; Problems 37–44, 47, 48, 68)
5. Given K_b of a weak base and its original concentration, calculate $[OH^-]$.
   (Example 13.11; Problems 55–58, 70)
6. Given either K_a or K_b for a conjugate acid-base pair, calculate the other equilibrium constant.
   (Example 13.12; Problems 53–56)
7. Predict whether a salt solution is acidic, basic, or neutral; if it is acidic or basic, write an equation to explain.
   (Example 13.13; Problems 59–62, 69)

Key Equations

Ionization of water $K_w = [H^+] \times [OH^-] = 1.0 \times 10^{-14}$ at 25°C

Expressions for K_a, K_b $K_a = \dfrac{[H^+] \times [B^-]}{[HB]}$

$$K_b = \dfrac{[OH^-] \times [HB]}{[B^-]}$$

$$K_a \times K_b = K_w$$

Quadratic formula $ax^2 + bx + c = 0$ $x = \dfrac{-b \pm \sqrt{b^2 - 4ac}}{2a}$

Key Terms

- acid
  - —Brønsted-Lowry
  - —conjugate
  - —equilibrium constant
  - —polyprotic
  - —strong
  - —weak
  amphiprotic

- base
  - —Brønsted-Lowry
  - —conjugate
  - —equilibrium constant
  - —strong
  - —weak
- molarity
  neutral solution
  pH

pK_a
pK_b
pOH
percent ionization
salt
successive approximations
water
—ion product constant

Consider aqueous solutions of nitric acid, sodium hydroxide, ammonia, and hydrogen fluoride.

(a) When aqueous solutions of nitric acid and ammonia are combined, ammonium and nitrate ions are produced. When aqueous solutions of hydrogen fluoride and sodium hydroxide are combined, a solution containing sodium and fluoride ions is obtained. Classify each reactant as strong or weak, acid or base.

(b) Using the Brønsted-Lowry model, write equations to explain the acidity or basicity of HNO_3, NH_3, and HF.

(c) Calculate the pH of 0.45 M solutions of HNO_3, NaOH, NH_4NO_3, NaF, NH_3, and HF.

(d) What is the conjugate base of NH_4^+? What is the value of K_b for the conjugate base? What is the conjugate acid of F^-? What is the value of K_a for the conjugate acid?

(e) Classify the salts NH_4NO_3 and NaF as acidic, basic, or neutral. Write net ionic equations to explain your answers.

(f) Would ammonium fluoride, NH_4F, be acidic or basic?

Answers

(a) HNO_3, strong acid; NaOH, strong base; NH_3, weak base; HF, weak acid

(b) $HNO_3(aq) + H_2O \rightarrow H_3O^+(aq) + NO_3^-(aq)$
$NH_3(aq) + H_2O \rightleftharpoons NH_4^+(aq) + OH^-(aq)$
$HF(aq) + H_2O \rightleftharpoons H_3O^+(aq) + F^-(aq)$

(c) HNO_3, pH = 0.35; NaOH, pH = 13.65; HF, pH = 1.75; NH_3, pH = 11.45; NH_4NO_3, pH = 4.80; NaF, pH = 8.40

(d) Conjugate base of NH_4^+ is NH_3. K_b for $NH_3 = 1.8 \times 10^{-5}$
Conjugate acid of F^- is HF. K_a for HF $= 6.9 \times 10^{-4}$

(e) NH_4NO_3 is acidic; $NH_4^+(aq) + H_2O \rightleftharpoons NH_3(aq) + H_3O^+(aq)$
NaF is basic; $F^-(aq) + H_2O \rightleftharpoons HF(aq) + OH^-(aq)$

(f) Ammonium fluoride is acidic.

Problem numbers in blue indicate that the answer is available in Appendix 6 at the back of the book.
WEB indicates that the solution is posted at **http://www.harcourtcollege.com/chem/general/masterton4/student/**

Brønsted-Lowry Acid-Base Model

1. For each of the following reactions, indicate the Brønsted-Lowry acids and bases. What are the conjugate acid/base pairs?

   (a) $CN^-(aq) + H_2O \rightleftharpoons HCN(aq) + OH^-(aq)$
   (b) $HCO_3^-(aq) + H_3O^+(aq) \rightleftharpoons H_2CO_3(aq) + H_2O$
   (c) $HC_2H_3O_2(aq) + HS^-(aq) \rightleftharpoons C_2H_3O_2^-(aq) + H_2S(aq)$

2. Follow the directions for Question 1 for the following reactions.

   (a) $H_3O^+(aq) + CN^-(aq) \rightleftharpoons HCN(aq) + H_2O$
   (b) $HNO_2(aq) + OH^-(aq) \rightleftharpoons NO_2^-(aq) + H_2O$
   (c) $HCHO_2(aq) + H_2O \rightleftharpoons CHO_2^-(aq) + H_3O^+(aq)$

3. According to the Brønsted-Lowry theory, which of the following would you expect to act as an acid? Which as a base?

   (a) $C_2H_5NH_3^+$ (b) HClO   (c) CN^-

4. According to the Brønsted-Lowry theory, which of the following would you expect to act as an acid? Which as a base?

   (a) CHO_2^- (b) NH_4^+ (c) HSO_3^-

5. Give the formula of the conjugate acid of
   (a) H_2O (b) HCO_3^- (c) CH_3NH_2
   (d) S^{2-} (e) $Fe(H_2O)_5(OH)^+$

6. Give the formula of the conjugate base of
   (a) $HC_2H_3O_2$ (b) $Zn(H_2O)_4^{2+}$ (c) $HBrO_2$
   (d) $CH_3NH_3^+$ (e) H_2S

$[H^+]$, $[OH^-]$, pH, and pOH

7. Find the pH and pOH of solutions with the following $[H^+]$. Classify each as acidic or basic.
   (a) 1.0 M (b) 1.7 × 10^{-4} M
   (c) 6.8 × 10^{-8} M (d) 9.3 × 10^{-11} M

WEB 8. Find the pH and pOH of solutions with the following $[H^+]$. Classify each as acidic or basic.
   (a) 6.0 M (b) 0.33 M
   (c) 4.6 × 10^{-8} M (d) 7.2 × 10^{-14} M

9. Calculate [H^+] and [OH^-] in solutions with the following pH.

(a) 4.0   (b) 8.52   (c) 0.00   (d) 12.60

10. Calculate [H^+] and [OH^-] in solutions with the following pH.

(a) 9.0   (b) 3.20   (c) −1.05   (d) 7.46

11. Solution A has [OH^-] = 3.2×10^{-4} M. Solution B has [H^+] = 6.9×10^{-9} M. Which solution is more basic? Which has the lower pH?

12. Solution 1 has pH 4.3. Solution 2 has [OH^-] = 3.4×10^{-7} M. Which solution is more acidic? Which has the higher pOH?

13. Solution A has a pH of 12.32. Solution B has [H^+] three times as large as that of solution A. Solution C has a pH half that of solution A.

(a) What is [H^+] for all three solutions?

(b) What is the pH of solutions B and C?

(c) Classify each solution as acidic, basic, or neutral.

14. Solution X has a pH of 4.35. Solution Y has [OH^-] ten times as large as solution X. Solution Z has a pH 4.0 units higher than that of solution X.

(a) Calculate the ratio of [H^+] in solutions X and Y and solutions X and Z.

(b) What is the pH of solutions Y and Z?

(c) Classify each solution as acidic, basic, or neutral.

15. Unpolluted rain water has a pH of about 5.5. Acid rain has been shown to have a pH as low as 3.0. Calculate the [H^+] ratio of acid rain to unpolluted rain.

16. Milk of magnesia has a pH of 10.5.

(a) Calculate [H^+].

(b) Calculate the ratio of the H^+ concentration of gastric juice, pH 1.5, to that of milk of magnesia.

17. Find [H^+] and the pH of the following solutions.

(a) 0.350 g of HBr dissolved in enough water to make 250.0 mL

(b) One hundred milliliters of 0.188 M HCl diluted with enough water to make 1.00 L of solution; how does a tenfold dilution affect the pH?

18. Find [H^+] and the pH of the following solutions.

(a) Two liters (d = 1.00 g/mL) of a 25.0% by mass solution of $HClO_4$. What is the pH of one liter of the same solution?

(b) Twelve grams of $HClO_4$ dissolved in enough water to make 2.00 L of solution. What is the pH of 1.00 L of solution in which the same mass of $HClO_4$ is dissolved?

19. What is the pH of a solution obtained by adding 145 mL of 0.575 M HCl to 493 mL of a HNO_3 solution with a pH of 1.39? Assume that volumes are additive.

20. What is the pH of a solution obtained by adding 5.00 g of HI to 295 mL of a 0.786 M solution of HNO_3? Assume that the HI addition does not change the volume of the resulting solution.

21. Find [OH^-], [H^+], and the pH and pOH of the following solutions.

(a) 0.27 M $Sr(OH)_2$

(b) a solution made by dissolving 13.6 g of KOH in enough water to make 2.50 L of solution

22. Find [OH^-], [H^+], and the pH and pOH of the following solutions.

(a) 45.0 mL of 0.0921 M $Ba(OH)_2$ diluted with enough water to make 350.0 mL of solution

(b) A solution made by dissolving 4.68 g of NaOH in enough water to make 635 mL of solution

23. What is the pH of a solution obtained by adding 2.00 g of $Ca(OH)_2$ and 14.6 g of NaOH to enough water to make 1.20 L of solution?

24. What is the pH of a solution obtained by adding 75.0 mL of 0.366 M CsOH to 250.0 mL of $Ba(OH)_2$ solution with a pH of 11.65? Assume that volumes are additive.

Ionization Expressions, Weak Acids

25. Using the Brønsted-Lowry model, write equations to show why the following species behave as weak acids in water.

(a) $Zn(H_2O)_3OH^+$ (b) HSO_4^- (c) HNO_2
(d) $Fe(H_2O)_6^{2+}$ (e) $HC_2H_3O_2$ (f) $H_2PO_4^-$

26. Follow the directions of Question 25 for the following species.

(a) $Ni(H_2O)_5OH^+$ (b) $Al(H_2O)_6^{3+}$ (c) H_2S
(d) HPO_4^{2-} (e) $HClO_2$ (f) $Cr(H_2O)_5(OH)^+$

27. Write the ionization equation and the K_a expression for each of the following acids.

(a) PH_4^+ (b) HS^- (c) $HBrO_2$

WEB **28.** Write the ionization equation and the K_a expression for each of the following acids.

(a) HSO_3^- (b) HPO_4^{2-} (c) HNO_2

29. Calculate K_a for the weak acids that have the following pK_a values.

(a) 4.6   (b) 9.3   (c) 11.7

30. Calculate pK_a for the weak acids that have the following K_a values.

(a) 6.3×10^{-4} (b) 1.7×10^{-11} (c) 1.9×10^{-8}

31. Consider these acids

Acid	A	B	C	D
K_a	1.6×10^{-3}	9×10^{-4}	2×10^{-6}	3×10^{-4}

(a) Arrange the acids in order of increasing acid strength from weakest to strongest.

(b) Which acid has the smallest pK_a value?

32. Consider these acids

Acid	A	B	C	D
pK_a	3.7	9.2	7.4	1.6

(a) Arrange the acids in order of decreasing acid strength from strongest to weakest.

(b) Which acid has the largest K_a value?

Equilibrium Calculations, Weak Acids

33. Caproic acid, $HC_6H_{11}O_2$, is found in coconut oil and is used in making artificial flavors. A solution is made by dissolving 0.450 mol of caproic acid in enough water to make 2.0 L of solution. The solution has $[H^+] = 1.7 \times 10^{-3}$ M. What is K_a for caproic acid?

34. Para-amino benzoic acid (PABA), $HC_7H_6NO_2$, is used in some sunscreen agents. A solution is made by dissolving 0.263 mol of PABA in enough water to make 750.0 mL of solution. The solution has $[H^+] = 2.6 \times 10^{-3}$ M. What is K_a for PABA?

35. Phenol, once known as carbolic acid, HC_6H_5O, is a weak acid used as an antiseptic and in the manufacture of plastics. A solution of phenol is made by dissolving 2.68 g in enough water to make 600.0 mL of solution. The resulting solution has pH 5.64. What is K_a for phenol?

36. Benzoic acid, $HC_7H_5O_2$, is present in many berries. A solution is prepared by dissolving 13.7 g in enough water to make 250.0 mL of solution. The solution has pH 2.27. What is K_a for benzoic acid?

37. Penicillin ($M = 356$ g/mol), an antibiotic often used to treat bacterial infections, is a weak acid. Its K_a is 1.7×10^{-3}. Calculate $[H^+]$ in solutions prepared by adding enough water to the following to make 725 mL.
   (a) 0.187 mol   (b) 127 g

38. Butyric acid, $HC_4H_7O_2$, is responsible for the odor of rancid butter and cheese. Its K_a is 1.51×10^{-5}. Calculate $[H^+]$ in solutions prepared by adding enough water to the following to make 1.30 L.
   (a) 0.279 mol   (b) 13.5 g

39. Uric acid, $HC_5H_3O_3N_4$, can accumulate in the joints. This accumulation causes severe pain and the condition is called *gout*. Its K_a is 5.1×10^{-6}. For a 0.894 M solution of uric acid, calculate
   (a) $[H^+]$ (b) $[OH^-]$
   (c) pH   (d) % ionization

40. Barbituric acid ($K_a = 1.1 \times 10^{-4}$) is used in the manufacture of some sedatives. For a 0.673 M solution of barbituric acid, calculate
   (a) $[H^+]$ (b) $[OH^-]$
   (c) pH   (d) % ionization

41. Picric acid, $HC_6H_2O_7N_3$, is an intensely yellow acid used in the manufacture of explosives and dyes. Its K_a is 0.16. A picric acid solution is prepared by dissolving 13.2 g of picric acid in enough water to make 550.0 mL of solution. For this solution, calculate
   (a) pH   (b) percent ionization

WEB 42. Trichloroacetic acid ($K_a = 2.0 \times 10^{-1}$) is sometimes used in the treatment of warts. Calculate the pH and percent ionization of a 4.0 M solution of trichloroacetic acid.

43. Using the K_a values listed in Table 13.2, calculate the pH of a 0.39 M solution of ammonium chloride.

44. When aluminum chloride is dissolved in water, $Al(H_2O)_6^{3+}$ and Cl^- ions are obtained. What is the pH of a 1.75 M solution of $AlCl_3$?

Polyprotic Acids

Use the K_a values listed in Table 13.3 for polyprotic acids.

45. Write the overall chemical equation and calculate K for the complete ionization of oxalic acid, $H_2C_2O_4$.

46. Write the overall chemical equation and calculate K for the complete ionization of H_3PO_4.

47. Calculate the pH of 0.25 M solution of sulfurous acid, H_2SO_3. Estimate $[HSO_3^-]$ and $[SO_3^{2-}]$.

48. Calculate the pH of 0.63 M H_2CO_3. Estimate $[HCO_3^-]$ and $[CO_3^{2-}]$.

Ionization Expressions; Weak Bases

49. Using the Brønsted-Lowry model, write an equation to show why each of the following species produces a basic aqueous solution.
   (a) $(CH_3)_3N$ (b) PO_4^{3-} (c) HPO_4^{2-}
   (d) $H_2PO_4^-$ (e) HS^- (f) $C_2H_5NH_2$

50. Follow the directions of Question 49 for the following species.
   (a) NH_3 (b) NO_2^- (c) $C_6H_5NH_2$
   (d) CO_3^{2-} (e) F^- (f) HCO_3^-

51. Using the equilibrium constants listed in Table 13.2, arrange the following 0.1 M aqueous solutions in order of increasing pH (from lowest to highest).
   (a) $NaNO_2$ (b) HCl   (c) NaF
   (d) $Zn(H_2O)_3(OH)(NO_3)$

52. Using the equilibrium constants listed in Table 13.2, arrange the following 0.1 M aqueous solutions in order of decreasing pH (from highest to lowest).
   (a) KOH   (b) NaCN   (c) HCO_3^- (d) $Ba(OH)_2$

Equilibrium Calculations; Weak Bases

53. Find the value of K_b for the conjugate base of the following organic acids.
   (a) gallic acid present in tea; $K_a = 3.9 \times 10^{-5}$
   (b) sorbic acid, an additive in breads to inhibit mold formation; $K_a = 1.7 \times 10^{-5}$

WEB 54. Find the value of K_a for the conjugate acid of the following bases.
   (a) morphine, an opiate, $K_b = 7.4 \times 10^{-7}$
   (b) ephedrine, used as a decongestant; $K_b = 1.4 \times 10^{-4}$

55. Consider sodium acrylate, $NaC_3H_3O_2$. K_a for acrylic acid (its conjugate acid) is 5.5×10^{-5}.
   (a) Write a balanced net ionic equation for the reaction that makes aqueous solutions of sodium acrylate basic.
   (b) Calculate K_b for the reaction in (a).
   (c) Find the pH of a solution prepared by dissolving 1.61 g of $NaC_3H_3O_2$ in enough water to make 835 mL of solution.

56. Codeine (Cod), a powerful and addictive painkiller, is a weak base.
   (a) Write a reaction to show its basic nature in water. Represent the codeine molecule as Cod.
   (b) The K_a for its conjugate acid is 1.2×10^{-8}. What is K_b for the reaction written in (a)?
   (c) What is the pH of a 0.0020 M solution of codeine?

57. The pH of a household ammonia cleaning solution is 11.68. How many grams of ammonia are needed in a 1.25-L solution to give the same pH?

58. A solution of sodium cyanide (NaCN) has a pH of 12.10. How many grams of NaCN are in 425 mL of solution with the same pH?

Salt Solutions

59. State whether 1 M solutions of the following salts in water are acidic, basic, or neutral.
  (a) $Zn(NO_3)_2$ (b) $SrCl_2$ (c) NH_4CN
  (d) KF   (e) $NaNO_3$

60. State whether 1 M solutions of the following salts in water would be acidic, basic, or neutral.
  (a) $Sr(NO_3)_2$ (b) Li_2SO_3 (c) $NaHCO_3$
  (d) NH_4NO_2 (e) Na_3PO_4

61. Write net ionic equations to explain the acidity or basicity of the various salts listed in Question 59.

62. Write net ionic equations to explain the acidity or basicity of the various salts listed in Question 60.

63. Arrange the following 0.1 M aqueous solutions in order of decreasing pH (highest to lowest).

$$Ba(NO_3)_2, \quad HNO_3, \quad NH_4NO_3, \quad Al(NO_3)_3, \quad NaF$$

64. Arrange the following 0.1 M aqueous solutions in order of increasing pH (lowest to highest).

$$KOH, \quad KF, \quad KCl, \quad ZnCl_2, \quad HCl$$

65. Write formulas for four salts that
  (a) contain Co^{3+} and are acidic.
  (b) contain Br^- and are neutral.
  (c) contain Sr^{2+} and are basic.
  (d) contain Li^+ and are neutral.

66. Write formulas for four salts that
  (a) contain K^+ and are basic.
  (b) contain K^+ and are neutral.
  (c) contain ClO_4^- and are neutral.
  (d) contain ClO_4^- and are acidic.

Unclassified

67. Give two examples of
  (a) an acid stronger than acetic acid, $HC_2H_3O_2$.
  (b) a base weaker than KOH.
  (c) a salt that dissolves in water to yield a solution with a pH > 7.
  (d) a salt that dissolves in water to give a pH = 7.

68. There are 324 mg of acetylsalicylic acid ($\mathcal{M}$ = 180.15 g/mol) per aspirin tablet. If two tablets are dissolved in water to give two ounces ($\frac{1}{16}$ quart) of solution, estimate the pH. K_a of acetylsalicylic acid is 3.6×10^{-4}.

69. Using data in Table 13.2, classify solutions of the following salts as acidic, basic, or neutral.
  (a) $NH_4C_2H_3O_2$ (b) $NH_4H_2PO_4$
  (c) $Al(NO_2)_3$ (d) NH_4F

70. A student is asked to bubble enough ammonia gas through water to make 4.00 L of an aqueous ammonia solution with a pH of 11.55. What volume of ammonia gas at 25°C and 1.00 atm pressure is necessary?

71. Consider the process

$$H_2O \rightleftharpoons H^+(aq) + OH^-(aq) \qquad \Delta H° = 55.8 \text{ kJ}$$

  (a) Will the pH of pure water at body temperature (37°C) be 7.0?
  (b) If not, calculate the pH of pure water at 37°C.

72. Is a saline (NaCl) solution at 80°C acidic, basic, or neutral?

73. Household bleach is prepared by dissolving chlorine in water.

$$Cl_2(g) + H_2O \rightleftharpoons H^+(aq) + Cl^-(aq) + HOCl(aq)$$

K_a for HOCl is 3.2×10^{-8}. How much chlorine must be dissolved in one liter of water so that the pH of the solution is 1.19?

Conceptual Problems

74. Which of the following is/are true about a 0.10 M solution of a strong acid, HY?
  (a) $[Y^-] = 0.10 \, M$ (b) $[HY] = 0.10 \, M$
  (c) $[H^+] = 0.10 \, M$ (d) pH = 1.0
  (e) $[H^+] + [Y^-] = 0.20 \, M$

75. Which of the following is/are true regarding a 0.10 M solution of a weak base, B^-?
  (a) $[HB] = 0.10 \, M$ (b) $[OH^-] \approx [HB]$

  (c) $[B^-] > [HB]$ (d) $[H^+] = \dfrac{1.0 \times 10^{-14}}{0.10}$

  (e) pH = 13.0

76. Each figure below represents an acid solution in equilibrium. Squares represent H^+ ions, and circles represent the anion. Water molecules are not shown. Which figure represents a strong acid? Which a weak acid?

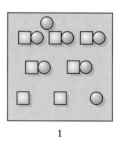

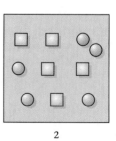

1   2

77. Each box represents an acid solution at equilibrium. Squares represent H^+ ions. Circles represent anions. (Although the anions have different identities in each figure, they are all represented as circles.) Water molecules are not shown. Assume that all solutions have the same volume.

(a) Which figure represents the strongest acid?

(b) Which figure represents the acid with the smallest K_a?

(c) Which figure represents the acid with the lowest pH?

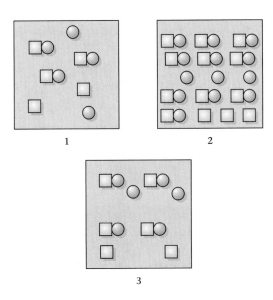

1

2

3

78. If □○ represents the weak acid HA(□ = H^+, ○ = A^-) in Figure a below, fill in Figure b with □○, □, and/or ○ to represent HA as being 10% ionized. (Water molecules are omitted.)

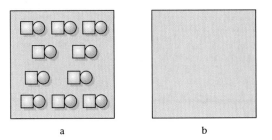

a

b

79. You are asked to determine whether an unknown white solid is acidic or basic. You also need to say whether the acid or base is weak or strong. You are given the molar mass of the solid and told that it is soluble in water. Describe an experiment that you can perform to obtain the desired characteristics of the white solid.

Challenge Problems

80. Using the Tables in Appendix 1, calculate ΔH for the reaction of the following.

(a) 1.00 L of 0.100 M NaOH with 1.00 L of 0.100 M HCl

(b) 1.00 L of 0.100 M NaOH with 1.00 L of 0.100 M HF, taking the heat of formation of HF(aq) to be -320.1 kJ/mol

81. Show by calculation that when the concentration of a weak acid decreases by a factor of 10, its percent ionization increases by a factor of $10^{1/2}$.

82. What is the freezing point of vinegar, which is an aqueous solution of 5.00% acetic acid, $HC_2H_3O_2$, by mass ($d = 1.006$ g/cm^3)?

83. The solubility of $Ca(OH)_2$ at 25°C is 0.153 g/100 g H_2O. Assuming that the density of a saturated solution is 1.00 g/mL, calculate the maximum pH one can obtain when $Ca(OH)_2$ is dissolved in water.

14 EQUILIBRIA IN ACID–BASE SOLUTIONS

Crystals of citric acid ($H_3C_5H_5O_7$) under polarized light. *(Jan Hinsch/Science Photo Library/Photo Researchers, Inc.)*

CHAPTER OUTLINE

14.1 BUFFERS

14.2 ACID-BASE INDICATORS

14.3 ACID-BASE TITRATIONS

In Chapter 13 we dealt with the equilibrium established when a single solute, either a weak acid or a weak base, is added to water. This chapter focuses on the equilibrium established when two different solutes are mixed in water solution. These solutes may be—

■ *a weak acid, HB, and its conjugate base, B⁻.* Solutions called *buffers* contain roughly equal amounts of these two species. The equilibria involved in buffer solutions are considered in Section 14.1.

■ *an acid and a base used in an acid-base titration.* This type of reaction was discussed in Chapter 4. Section 14.3 examines the equilibria involved, the way pH changes during the titration, and the choice of indicator for the titration.

We will also consider the equilibrium involved when an acid-base indicator is used to estimate pH (Section 14.2).

14.1 BUFFERS

Any solution containing appreciable amounts of both a weak acid and its conjugate base—

■ *is highly resistant to changes in pH brought about by addition of strong acid or strong base.*
■ *has a pH close to the pK_a of the weak acid.*

A solution showing these properties is called a **buffer,** because it cushions the "shock" (i.e., the drastic change in pH) that occurs when a strong acid or strong base is added to water.

To prepare a buffer, we can mix solutions of a weak acid HB and the sodium salt of that acid NaB, which consists of Na^+ and B^- ions. This mixture can react with either strong base

$$HB(aq) + OH^-(aq) \longrightarrow B^-(aq) + H_2O$$

or strong acid

$$B^-(aq) + H^+(aq) \longrightarrow HB(aq)$$

These reactions have very large equilibrium constants, as we will see in Section 14.3, and so go virtually to completion. As a result, the added H^+ or OH^- ions are consumed and do not directly affect the pH. This is the principle of buffer action, which explains why a buffered solution is much more resistant to a change in pH than one that is unbuffered (Figure 14.1, page 414).

See the *Saunders Interactive General Chemistry CD-ROM*, Screen 18.8, Buffer Solutions.

By definition, a buffer is any solution that resists a change in pH.

FIGURE 14.1

Comparison of pH changes in water and a buffer solution. The three tubes at the left show the effect of adding a few drops of strong acid or strong base to water. The pH changes drastically, giving a pronounced color change with universal indicator. This experiment is repeated in the three tubes at the right, using a buffer of pH 7 instead of water. This time the pH changes only very slightly, and there is no change in color of the indicator. *(Marna G. Clarke)*

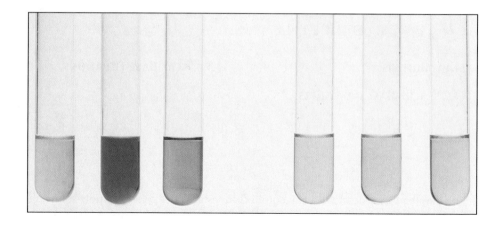

Buffers are widely used to maintain nearly constant pH in a variety of commercial products and laboratory procedures (Figure 14.2). For these applications and others, it is essential to be able to determine—

- the pH of a buffer system made by mixing a weak acid with its conjugate base.
- the appropriate buffer system to maintain a desired pH.
- the (small) change in pH that occurs when a strong acid or base is added to a buffer.
- the capacity of a buffer to absorb H^+ or OH^- ions.

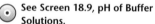

See Screen 18.9, pH of Buffer Solutions.

Determination of [H⁺] in a Buffer System

The concentration of H^+ ion in a buffer can be calculated if you know the concentrations of the weak acid HB and its conjugate base B^-. These three quantities are related through the acid equilibrium constant of HB:

$$HB(aq) \rightleftharpoons H^+(aq) + B^-(aq) \qquad K_a = \frac{[H^+] \times [B^-]}{[HB]}$$

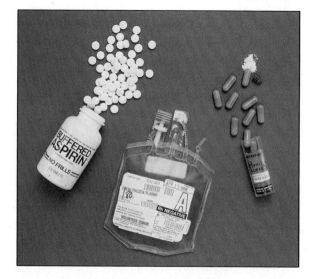

FIGURE 14.2

Some applications of buffers. Many products, including aspirin and blood plasma, are buffered. Buffer tablets are also available in the laboratory *(far right)* to make up a solution to a specified pH. *(Marna G. Clarke)*

Solving for $[H^+]$,

$$\boxed{[H^+] = K_a \times \frac{[HB]}{[B^-]}} \qquad (1)^*$$

This is a completely general equation, applicable to all buffer systems. The calculation of $[H^+]$, and hence pH, can be simplified if you keep two points in mind.

1. *You can always assume that equilibrium is established without appreciably changing the original concentrations of either HB or B^-.* That is,

$$[HB] = [HB]_o$$

$$[B^-] = [B^-]_o$$

To explain why these relations hold, consider, for example, what happens when the weak acid HB is added to water:

$$HB(aq) \rightleftharpoons H^+(aq) + B^-(aq)$$

We saw in Chapter 13 that under these conditions, it is usually a good approximation to take $[HB] \approx [HB]_o$. The approximation is even more valid when a considerable amount of B^- is added, as is the case with a buffer. By Le Châtelier's principle, the reverse reaction occurs, the ionization of HB is repressed, and $[HB] = [HB]_o$.

2. *Because the two species HB and B^- are present in the same solution, the ratio of their concentrations is also their mole ratio.* That is,

$$\frac{[HB]}{[B^-]} = \frac{\text{no. moles HB}/V}{\text{no. moles B}^-/V} = \frac{n_{HB}}{n_{B^-}}$$

where $n =$ amount in moles and $V =$ volume of solution. Hence Equation 1 can be rewritten as

$$[H^+] = K_a \times \frac{n_{HB}}{n_{B^-}} \qquad (2)$$

Frequently, Equation 2 is easier to work with than Equation 1.

> B^- comes from adding a salt of the weak acid, such as NaB.

EXAMPLE 14.1   Lactic acid, $C_3H_6O_3$, is a weak organic acid present in both sour milk and buttermilk. It is also a product of carbohydrate metabolism and is found in the blood after vigorous muscular activity. A buffer is prepared by dissolving 1.00 mol of lactic acid, HLac ($K_a = 1.4 \times 10^{-4}$) and 1.00 mol of sodium lactate, NaLac, in enough water to form 550 mL of solution. Calculate $[H^+]$ and the pH of the buffer.

Strategy   You are given K_a and n_{HLac} in the statement of the problem. Because sodium lactate, like all ionic solids, is completely ionized in water, $n_{Lac^-} = n_{NaLac} = 1.00$ mol. Thus you have all the quantities required to calculate $[H^+]$ by Equation 2.

A runner in pain. Exertion beyond the point at which the body can clear lactic acid from the muscles causes the pain. *(Gerard Vandystadt/Photo Researchers, Inc.)*

---

*By taking the logarithms of both sides of Equation 1 and multiplying through by −1, you can show (Problem 65) that, for any buffer system,

$$pH = pK_a + \log_{10} \frac{[B^-]}{[HB]}$$

This relation, known as the ***Henderson-Hasselbalch*** equation, is often used in biology and biochemistry to calculate the pH of buffers.

Solution

$$[H^+] = K_a \times \frac{n_{HLac}}{n_{Lac^-}}$$

$$= 1.4 \times 10^{-4} \times \frac{1.00}{1.00} = \boxed{1.4 \times 10^{-4}}$$

$$pH = -\log_{10}(1.4 \times 10^{-4}) = \boxed{3.85}$$

Reality Check When equal amounts of a weak acid and its conjugate base are present, as is true in this case, the pH of the buffer is equal to pK_a.

Notice in Example 14.1 that the total volume of solution is irrelevant. All that is required to solve for $[H^+]$ is the number of moles of lactic acid and sodium lactate. The pH of the buffer is 3.85 whether the volume of solution is 550 mL, 1 L, or even 10 L. This explains why *diluting a buffer with water does not change the pH.*

Choosing a Buffer System

See Screen 18.10, Preparing Buffer Solutions.

Suppose you want to make up a buffer in the laboratory with a specified pH (e.g., 4.0, 7.0, 10.0, . . .). Looking at the equation

$$[H^+] = K_a \times \frac{[HB]}{[B^-]} = K_a \times \frac{n_{HB}}{n_{B^-}}$$

it is evident that the pH of the buffer depends on two factors:

1. The acid equilibrium constant of the weak acid, K_a. The value of K_a has the greatest influence on buffer pH. Because HB and B^- are likely to be present in nearly equal amounts,

$$[H^+] \approx K_a \qquad pH \approx pK_a$$

Thus to make up a buffer with a pH close to 7, you should start with a conjugate weak acid–weak base pair in which the acid equilibrium constant of the weak acid is about 10^{-7}.

2. The ratio of the concentrations or amounts of HB and B^-. Small variations in pH can be achieved by adjusting this ratio. To obtain a slightly more

TABLE 14.1 Buffer Systems at Different pH Values

	Buffer System		
Desired pH	Weak Acid	Weak Base	K_a(Weak Acid)	pK_a
---	---	---	---	---
4	Lactic acid (HLac)	Lactate ion (Lac$^-$)	1.4×10^{-4}	3.85
5	Acetic acid (HC$_2$H$_3$O$_2$)	Acetate ion (C$_2$H$_3$O$_2^-$)	1.8×10^{-5}	4.74
6	Carbonic acid (H$_2$CO$_3$)	Hydrogen carbonate ion (HCO$_3^-$)	4.4×10^{-7}	6.36
7	Dihydrogen phosphate ion (H$_2$PO$_4^-$)	Hydrogen phosphate ion (HPO$_4^{2-}$)	6.2×10^{-8}	7.21
8	Hypochlorous acid (HClO)	Hypochlorite ion (ClO$^-$)	2.8×10^{-8}	7.55
9	Ammonium ion (NH$_4^+$)	Ammonia (NH$_3$)	5.6×10^{-10}	9.25
10	Hydrogen carbonate ion (HCO$_3^-$)	Carbonate ion (CO$_3^{2-}$)	4.7×10^{-11}	10.32

acidic buffer, add more weak acid, HB; addition of more weak base, B⁻, will make the buffer a bit more basic.

EXAMPLE 14.2   To prepare a buffer with a pH of 9.00, which of the systems in Table 14.1 would you use? What should be the ratio of concentrations, $[HB]/[B^-]$?

Strategy   Choose the buffer in which pK_a is closest to 9. Clearly, this is the NH_4^+–NH_3 buffer. Then use the equation $[H^+] = K_a \times [NH_4^+]/[NH_3]$ to calculate the necessary ratio. Note that since the pH is 9.00, $[H^+] = 1.0 \times 10^{-9}\ M$.

Solution   For the NH_4^+–NH_3 buffer

$$[H^+] = K_a \times \frac{[NH_4^+]}{[NH_3]}$$

Solving for the ratio,

$$\frac{[NH_4^+]}{[NH_3]} = \frac{[H^+]}{K_a} = \frac{1.0 \times 10^{-9}}{5.6 \times 10^{-10}} = \boxed{1.8}$$

You might add 1.8 mol of NH_4Cl to 1.0 mol of NH_3, or 9 mol of NH_4Cl to 5 mol of NH_3 or . . . .

Most often, buffers are prepared, as implied in Example 14.2, by mixing a weak acid and its conjugate base. However, a somewhat different approach can be followed: partial neutralization of a weak acid or a weak base gives a buffer. To illustrate, suppose that 0.10 mol of a strong acid, perhaps HCl, is added to 0.20 mol of NH_3. The following reaction occurs:

$$H^+(aq) + NH_3(aq) \longrightarrow NH_4^+(aq)$$

The limiting reactant, H^+, is consumed. Hence we have

	n_{H^+}	n_{NH_3}	$n_{NH_4^+}$
Original	0.10	0.20	0.00
Change	−0.10	−0.10	+0.10
Final	0.00	0.10	0.10

The solution formed contains appreciable amounts of both NH_4^+ and NH_3; it is a buffer. We will have more to say in Section 14.3 about this way of making a buffer.

Note that complete neutralization would not produce a buffer; only NH_4^+ ions would be present.

Effect of Added H⁺ or OH⁻ on Buffer Systems

The pH of a buffer does change slightly on addition of moderate amounts of a strong acid or strong base. Addition of H^+ ions converts an equal amount of weak base B⁻ to its conjugate acid HB:

$$H^+(aq) + B^-(aq) \longrightarrow HB(aq)$$

By the same token, addition of OH⁻ ions converts an equal amount of weak acid to its conjugate base B⁻:

$$HB(aq) + OH^-(aq) \longrightarrow B^-(aq) + H_2O$$

See Screen 18.11, Adding Reagents to Buffer Solutions.

Effect of a buffer. (a) Both solutions contain an acid-base indicator. The solution at the left is buffered at a pH of about 7. The solution at the right is mildly acidic (pH ≈ 5). (b) Five milliliters of 0.10 M HCl has been added to both solutions. The unchanged color of the indicator shows there is little pH change in the buffered solution. The pH meter shows the large change in pH of the unbuffered solution. *(Charles D. Winters)*

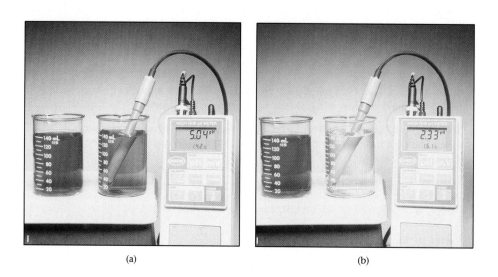

(a)   (b)

In either case, the ratio n_{HB}/n_{B^-} changes. This in turn changes the H^+ ion concentration and pH of the buffer. The effect ordinarily is small, as illustrated in Example 14.3.

EXAMPLE 14.3   Consider the buffer described in Example 14.1, where $n_{HLac} = n_{Lac^-} = 1.00$ mol ($K_a HLac = 1.4 \times 10^{-4}$). You will recall that in this buffer the pH is 3.85. Calculate the pH after addition of

(a) 0.10 mol of HCl   (b) 0.10 mol of NaOH

Strategy   First (1), write the chemical equation for the reaction that occurs when strong acid or base is added. Then (2), apply the principles of stoichiometry to find the numbers of moles of weak acid and weak base remaining after reaction. Finally (3), apply the relation $[H^+] = K_a \times n_{HB}/n_{B^-}$ to find $[H^+]$ and then the pH.

> To work this problem, you have to combine principles of stoichiometry and equilibrium.

Solution

(a) (1) The H^+ ions from the strong acid HCl convert Lac^- ions to HLac molecules:

$$H^+(aq) + Lac^-(aq) \longrightarrow HLac(aq)$$

(2) Because the reacting mole ratios are 1:1, the addition of 0.10 mol of H^+ produces 0.10 mol of HLac and consumes 0.10 mol of Lac^-. Originally, there was 1.00 mol of both HLac and Lac^-. Hence,

$$n_{HLac} = (1.00 + 0.10) \text{ mol} = 1.10 \text{ mol}$$

$$n_{Lac^-} = (1.00 - 0.10) \text{ mol} = 0.90 \text{ mol}$$

(3) $[H^+] = 1.4 \times 10^{-4} \times \dfrac{n_{HLac}}{n_{Lac^-}} = 1.4 \times 10^{-4} \times \dfrac{1.10}{0.90} = 1.7 \times 10^{-4} M$

$$pH = -\log_{10}[H^+] = \boxed{3.77}$$

(b) (1) The OH^- ions from the strong base NaOH convert HLac molecules to Lac^- ions:

$$HLac(aq) + OH^-(aq) \longrightarrow Lac^-(aq) + H_2O$$

(2) The reaction produces 0.10 mol of Lac^- and consumes 0.10 mol of HLac:

$$n_{HLac} = (1.00 - 0.10) \text{ mol} = 0.90 \text{ mol}$$

$$n_{Lac^-} = (1.00 + 0.10) \text{ mol} = 1.10 \text{ mol}$$

$$(3) \quad [H^+] = 1.4 \times 10^{-4} \times \frac{n_{HLac}}{n_{Lac^-}} = 1.4 \times 10^{-4} \times \frac{0.90}{1.10} = 1.1 \times 10^{-4} \, M$$

$$pH = -\log_{10}[H^+] = \boxed{3.94}$$

Reality Check   Notice that in both (a) and (b) the change in pH is quite small, less than 0.1 unit.

(a) pH 3.85 → 3.77   (b) pH 3.85 → 3.94

This is what is supposed to happen with buffer solutions. Addition of strong acid lowers the pH slightly; addition of strong bases raises it slightly.

Example 14.3 shows, among other things, how effectively a buffer "soaks up" H^+ or OH^- ions. That can be important. Suppose you are carrying out a reaction whose rate is first-order in H^+. If the pH increases from 5 to 7, perhaps by the absorption of traces of ammonia from the air, the rate will decrease by a factor of 100. A reaction that should have been complete in three hours will still be going on when you come back ten days later. Small wonder that chemists frequently work with buffered solutions to avoid disasters of that type.

Buffer Capacity

Figure 14.3 shows graphically what happens to a buffer system when OH^- or H^+ ions are added. We start at point A with the buffer referred to in Example 14.1, which contains 1.00 mol of both lactic acid (HLac) and its conjugate base (Lac$^-$). When OH^- ions are added, we move to the right of A; lactic acid molecules are consumed by the reaction

$$HLac(aq) + OH^-(aq) \longrightarrow Lac^-(aq) + H_2O$$

If H^+ ions are added instead, we move to the left of A; in this case lactate ions are used up.

$$Lac^-(aq) + H^+(aq) \longrightarrow HLac(aq)$$

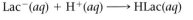

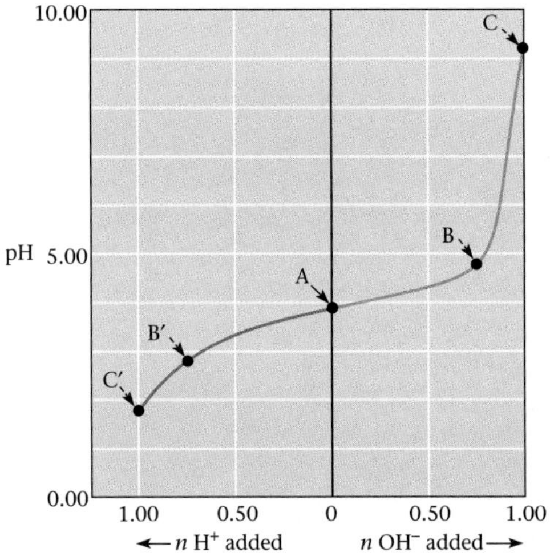

FIGURE 14.3

Variation of pH of a buffer solution as strong acid (red) or strong base (blue) is added. With the addition of strong acid, the pH changes along the curve to the left of A. With the addition of strong base, the pH changes along the curve to the right of A. **OHT**

Notice from Figure 14.3 that near point A the slope of the curve is quite small. Addition of either OH^- or H^+ ions produces a relatively small change in pH. The situation changes as more and more HLac molecules or Lac$^-$ ions are consumed. When the number of moles of these species is reduced by 80% (points B, B′), the pH changes more rapidly and the buffer becomes much less effective. When all of the HLac or Lac$^-$ is consumed (points C and C′), the system no longer acts as a buffer and the pH climbs or falls precipitously.

The principle just described is often stated in terms of **buffer capacity,** the amount of OH^- or H^+ ions that can be absorbed without an appreciable change in pH. The capacity of the buffer is directly related to the number of moles of weak acid (HLac) and weak base (Lac$^-$). When these species are consumed, the capacity of the buffer drops, slowly at first and then more rapidly. When HLac is all gone, the system has no capacity to absorb OH^- ions; if there is no Lac$^-$ left, it cannot absorb H^+ ions.

Biological buffers frequently have a greater capacity to absorb H^+ ions than OH^- ions. Blood is buffered at a pH of 7.40, $[H^+] = 4.0 \times 10^{-8} \, M$, in part by a mixture of the weak acid H_2CO_3 (a water solution of CO_2) and its conjugate base, HCO_3^-. The acid equilibrium constant of H_2CO_3 is 4.4×10^{-7}:

$$\frac{[H^+] \times [HCO_3^-]}{[H_2CO_3]} = 4.4 \times 10^{-7}$$

To maintain $[H^+]$ at $4.0 \times 10^{-8} \, M$, the concentration of HCO_3^- ion has to be much higher than that of H_2CO_3.

$$\frac{[HCO_3^-]}{[H_2CO_3]} = \frac{4.4 \times 10^{-7}}{[H^+]} = \frac{4.4 \times 10^{-7}}{4.0 \times 10^{-8}} = 11$$

This means that the H_2CO_3-HCO_3^- buffer in blood has a much higher capacity for absorbing H^+ ions

$$HCO_3^-(aq) + H^+(aq) \longrightarrow H_2CO_3(aq)$$

than for OH^- ions

$$H_2CO_3(aq) + OH^-(aq) \longrightarrow HCO_3^-(aq) + H_2O$$

This is indeed fortunate because life processes produce many more H^+ ions than OH^- ions.

14.2   ACID-BASE INDICATORS

As pointed out in Chapter 4, an *acid-base indicator* is useful in determining the *equivalence point* of an acid-base titration. This is the point at which reaction is complete; equivalent quantities of acid and base have reacted. If the indicator is chosen properly, the point at which it changes color (its **end point**) coincides with the equivalence point. To understand how and why an indicator changes color, we need to understand the equilibrium principle involved.

An acid-base indicator is derived from a weak acid HIn:

$$HIn(aq) \rightleftharpoons H^+(aq) + In^-(aq) \qquad K_a = \frac{[H^+] \times [In^-]}{[HIn]}$$

where the weak acid HIn and its conjugate base In$^-$ have different colors (Figure

> It's called an end point because that's when you stop the titration.

> For litmus, HIn is red; In$^-$ is blue.

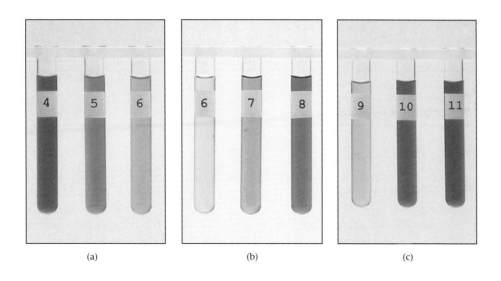

(a)   (b)   (c)

FIGURE 14.4

Acid-base indicators. Methyl red (a) goes from red at low pH to orange at about pH 5 to yellow at high pH. Bromthymol blue (b) is yellow at low pH, blue at high pH, and green at about pH 7. Phenolphthalein (c) goes from colorless to pink at about pH 9. *(Marna G. Clarke)*

14.4). The color that you see when a drop of indicator solution is added in an acid-base titration depends on the ratio

$$\frac{[HIn]}{[In^-]}$$

Three cases can be distinguished:

1. If $\dfrac{[HIn]}{[In^-]} \geq 10$

the principal species is HIn; you see the "acid" color, that of the HIn molecule.

2. If $\dfrac{[HIn]}{[In^-]} \leq 0.1$

the principal species is In^-; you see the "base" color, that of the In^- ion.

3. If $\dfrac{[HIn]}{[In^-]} \approx 1$

the color observed is intermediate between those of the two species, HIn and In^-.

The expression for the ionization of the indicator molecule, HIn,

$$K_a = \frac{[H^+] \times [In^-]}{[HIn]}$$

can be rearranged to give

$$\frac{[HIn]}{[In^-]} = \frac{[H^+]}{K_a}$$

From this expression, it follows that the ratio $[HIn]/[In^-]$ and hence the color of an indicator depend on two factors.

1. *$[H^+]$ or the pH of the solution.* At high $[H^+]$ (low pH), you see the color of the HIn molecule; at low $[H^+]$ (high pH), the color of the In^- ion dominates.

2. *K_a of the indicator.* Because K_a varies from one indicator to another, different indicators change colors at different pHs. A color change occurs when

$$[H^+] \approx K_a \qquad pH \approx pK_a$$

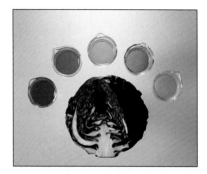

Red cabbage juice, a natural acid-base indicator. The picture shows *(left to right)* its colors at pH 1, 4, 7, 10, and 13. *(Charles Steele)*

TABLE 14.2 Colors and End Points of Indicators

	Color [HIn]	Color [In$^-$]	K_a	pH at End Point
Methyl red	Red	Yellow	1×10^{-5}	5
Bromthymol blue	Yellow	Blue	1×10^{-7}	7
Phenolphthalein	Colorless	Pink	1×10^{-9}	9

Because only a drop or two of indicator is used, it does not affect the pH of the solution.

Table 14.2 shows the characteristics of three indicators: methyl red, bromthymol blue, and phenolphthalein. These indicators change colors at pH 5, 7, and 9, respectively.

In practice it is usually possible to detect an indicator color change over a range of about two pH units. Consider, for example, what happens with bromthymol blue as the pH increases—

- below pH 6 (e.g., . . . 3, 4, 5) the color is pure yellow.
- between pH 6 and 7, the indicator starts to take on a greenish tinge; we might describe its color as yellow-green.
- at pH 7, the color is a 50-50 mixture of yellow and blue so it looks green.
- between pH 7 and 8, a blue color becomes visible.
- above pH 8 (or 9, 10, 11, . . .) the color is pure blue.

EXAMPLE 14.4 Consider bromthymol blue ($K_a = 1 \times 10^{-7}$). At pH 7.5,

(a) calculate the ratio [In$^-$]/[HIn].
(b) what is the color of the indicator at this point?

Strategy   In (a), use the general relation:

$$K_a \text{ of HIn} = \frac{[H^+] \times [In^-]}{[HIn]}$$

to find the required ratio. In (b), use Table 14.2 to deduce the color of the indicator.

Solution

(a) [In$^-$]/[HIn] = K_a/[H$^+$]
At pH 7.5, [H$^+$] = 3×10^{-8} M, so we have

$$[In^-]/[HIn] = (1 \times 10^{-7})/(3 \times 10^{-8}) = 3 \qquad \text{or} \qquad [In^-] = 3[HIn]$$

(b) At pH 7.5 we are halfway between 7.0 (green) and the beginning of the blue range at pH 8.0. The color is blue-green. This is reasonable; the blue color of In$^-$ should predominate over the yellow color of HIn because there are three In$^-$ ions for every HIn molecule.

14.3   ACID–BASE TITRATIONS

The discussion of acid-base titrations in Chapter 4 focused on stoichiometry. Here, the emphasis is on the equilibrium principles that apply to the acid-base reactions involved. It is convenient to distinguish between titrations involving—

- a strong acid (e.g., HCl) and a strong base (e.g., NaOH, Ba(OH)$_2$).

■ a weak acid (e.g., $HC_2H_3O_2$) and a strong base (e.g., NaOH).
■ a strong acid (e.g., HCl) and a weak base (e.g., NH_3).

See Screen 18.2, Acid-Base Reactions.

Strong Acid – Strong Base

As pointed out in Chapter 13, strong acids ionize completely in water to form H_3O^+ ions; strong bases dissolve in water to form OH^- ions. The neutralization reaction that takes place when *any* strong acid reacts with *any* strong base can be represented by a net ionic equation of the Brønsted-Lowry type:

$$H_3O^+(aq) + OH^-(aq) \longrightarrow 2H_2O$$

or, more simply, substituting H^+ ions for H_3O^+ ions, as

$$H^+(aq) + OH^-(aq) \longrightarrow H_2O$$

See Screen 18.3, Acid-Base Reactions (Strong Acids + Strong Bases).

The equation just written is the reverse of that for the ionization of water, so the equilibrium constant can be calculated by using the reciprocal rule (Chapter 12).

$$K = 1/K_W = 1/(1.0 \times 10^{-14}) = 1.0 \times 10^{14}$$

The enormous value of K means that for all practical purposes this reaction goes to completion, consuming the limiting reactant, H^+ or OH^-.

⌐ K must be large for a titration to work.

   Consider now what happens when HCl, a typical strong acid, is titrated with NaOH. Figure 14.5 shows how the pH changes during the titration. Two features of this curve are of particular importance:

1. At the equivalence point, when all the HCl has been neutralized by NaOH, a solution of NaCl, a neutral salt, is present. The pH at the equivalence point is 7.
2. Near the equivalence point, the pH rises very rapidly. Indeed, the pH may increase by as much as six units (from 4 to 10) when half a drop, ≈ 0.02 mL, of NaOH is added (Example 14.5).

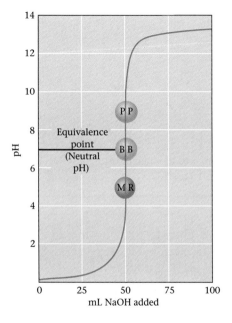

FIGURE 14.5

A strong acid-strong base titration. The curve represents the titration of 50.00 mL of 1.000 *M* HCl with 1.000 *M* NaOH. The solution at the equivalence point is neutral (pH = 7). The pH rises so rapidly near the equivalence point that any of the three indicators, methyl red (MR, end point pH = 5), bromthymol blue (BB, end point pH = 7), or phenolphthalein (PP, end point pH = 9) can be used. **OHT**

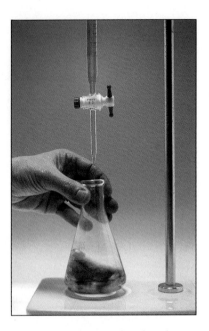

Titrating a solution of HCl with NaOH using phenolphthalein as the indicator. *(Charles D. Winters)*

We assume here that volumes are additive.

pH changes very rapidly near the end point.

EXAMPLE 14.5 When 50.00 mL of 1.000 *M* HCl is titrated with 1.000 *M* NaOH, the pH increases. Find the pH of the solution after the following volumes of 1.000 *M* NaOH have been added.

(a) 49.99 mL   (b) 50.01 mL

Strategy This is essentially a stoichiometry problem of the type discussed in Chapter 4. The reaction is $H^+(aq) + OH^-(aq) \rightarrow H_2O$. The number of moles of H^+ in 50.00 mL of 1.000 *M* HCl is

$$n_{H^+} = 50.00 \times 10^{-3} \text{ L} \times \frac{1.000 \text{ mol } H^+}{1 \text{ L}} = 50.00 \times 10^{-3} \text{ mol of } H^+$$

In each part of the problem, you start (1) by calculating the number of moles of OH^- added. Then (2), find the number of moles of H^+ or OH^- in excess. Finally (3), calculate $[H^+]$ and the pH.

Solution

(a) (1) $n_{OH^-} = 49.99 \times 10^{-3} \text{ L} \times \dfrac{1.000 \text{ mol } OH^-}{1 \text{ L}} = 49.99 \times 10^{-3} \text{ mol of } OH^-$

(2) Because H^+ and OH^- react in a 1:1 mole ratio,

n_{H^+} reacted $= 49.99 \times 10^{-3}$ mol

n_{H^+} excess $= 50.00 \times 10^{-3} \text{ mol} - 49.99 \times 10^{-3} \text{ mol}$

$$= 0.01 \times 10^{-3} \text{ mol} = 1 \times 10^{-5} \text{ mol}$$

(3) The total volume is almost exactly 100 mL:

$$[H^+] = \frac{1 \times 10^{-5} \text{ mol } H^+}{1 \times 10^{-1} \text{ L}} = 1 \times 10^{-4} \text{ M}$$

$$pH = -\log_{10}[H^+] = \boxed{4.0}$$

(b) (1) $n_{OH^-} = 50.01 \times 10^{-3} \text{ L} \times \dfrac{1.000 \text{ mol } OH^-}{1 \text{ L}} = 50.01 \times 10^{-3} \text{ mol of } OH^-$

(2) $n_{H^+ \text{ reacted}} = 50.00 \times 10^{-3}$ mol

$n_{OH^- \text{ excess}} = 50.01 \times 10^{-3} \text{ mol} - 50.00 \times 10^{-3} \text{ mol}$

$$= 0.01 \times 10^{-3} \text{ mol} = 1 \times 10^{-5} \text{ mol}$$

(3) $[OH^-] = \dfrac{1 \times 10^{-5} \text{ mol } OH^-}{1 \times 10^{-1} \text{ L}} = 1 \times 10^{-4} \text{ M}$

$$pOH = -\log_{10}[OH^-] = 4.0$$

$$pH = 14.00 - pOH = \boxed{10.0}$$

Reality Check Note that the pH changes from 4 to 10 when 0.02 mL of titrant is added at the end point. That is what is predicted by Figure 14.5.

From Figure 14.5 and Example 14.5, we conclude that any indicator that changes color between pH 4 and 10 should be satisfactory for a strong acid–strong base titration. Bromthymol blue (BB: end point pH = 7) would work very well, but so would methyl red (MR: end point pH = 5) or phenolphthalein (PP: end point pH = 9).

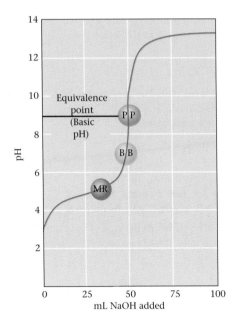

FIGURE 14.6

A weak acid–strong base titration. The curve represents the titration of 50.00 mL of 1.000 M acetic acid, a weak acid, $HC_2H_3O_2$, with 1.000 M NaOH. The solution at the equivalence point is basic (pH = 9.22). Phenolphthalein is a suitable indicator. Methyl red would change color much too early, when only about 33 mL of NaOH had been added. Bromthymol blue would change color slightly too quickly. **OHT**

Weak Acid–Strong Base

A typical weak acid–strong base titration is that of acetic acid with sodium hydroxide. The net ionic equation for the reaction is

$$HC_2H_3O_2(aq) + OH^-(aq) \longrightarrow C_2H_3O_2^-(aq) + H_2O$$

Notice that this reaction is the reverse of the reaction of the weak base $C_2H_3O_2^-$ (the acetate ion) with water (Chapter 13). It follows from the reciprocal rule that for this reaction,

$$K = \frac{1}{(K_b\ C_2H_3O_2^-)} = \frac{1}{(5.6 \times 10^{-10})} = 1.8 \times 10^9$$

Here again, K is a very large number; the reaction of acetic acid with a strong base goes essentially to completion.

Figure 14.6 shows how the pH changes as fifty milliliters of one molar acetic acid is titrated with one molar sodium hydroxide. Notice that—

- the pH starts off above 2; we are dealing with a weak acid, $HC_2H_3O_2$.
- there is a region, centered at the halfway point of the titration, where pH changes very slowly. The solution contains equal amounts of unreacted $HC_2H_3O_2$ and its conjugate base, $C_2H_3O_2^-$, formed by reaction with NaOH. As pointed out earlier, such a system acts as a buffer.
- at the equivalence point, there is a solution of sodium acetate, $NaC_2H_3O_2$. The acetate ion is a weak base, so the pH at the equivalence point must be greater than 7.

See Screen 18.5, Acid-Base Reactions (Weak Acids + Strong Bases).

For a titration to work, the reaction must go to completion, which requires $K > 10^4$ (approximately).

See Screen 18.12, Titration Curves.

EXAMPLE 14.6   50.00 mL of 1.000 M acetic acid, $HC_2H_3O_2$, is titrated with 1.000 M NaOH. Find the pH of the solution after the following volumes of 1.000 M NaOH have been added.

(a) 0.00 mL   (b) 25.00 mL   (c) 50.00 mL

Strategy In (a), before any base is added, we have a solution of 1.000 M $HC_2H_3O_2$ ($K_a = 1.8 \times 10^{-5}$). Its pH is calculated like that of any weak acid HB. Recall Example 13.7 (page 394), where we found that

$$[H^+]^2 \approx K_a \times [HB]_o$$

In (b), we are dealing with a buffer system; the pH is calculated as described in Section 14.1. In (c), we have a solution of a weak base, $C_2H_3O_2^-$ ($K_b = 5.6 \times 10^{-10}$). Recall Example 13.11 (page 399), where we found that for a weak base, B^-:

$$[OH^-]^2 \approx K_b \times [B^-]_o$$

Solution

(a) $[H^+]^2 \approx K_a HC_2H_3O_2 \times [HC_2H_3O_2] = 1.8 \times 10^{-5} \times 1.000 = 1.8 \times 10^{-5}$
$[H^+] = (1.8 \times 10^{-5})^{1/2} = 4.2 \times 10^{-3}\ M;$ pH = 2.38

> This is referred to as "half-neutralization."

(b) When 25.00 mL of NaOH has been added, we are halfway to the equivalence point (50.00 mL). It follows that half of the $HC_2H_3O_2$ molecules have been converted to $C_2H_3O_2^-$ ions, so

$$n_{HC_2H_3O_2} = n_{C_2H_3O_2^-}$$

Under these conditions, as pointed out in Section 14.1,

$$pH = pK_a \text{ of } HC_2H_3O_2 = 4.74$$

(c) $[OH^-]^2 \approx K_b C_2H_3O_2^- \times [C_2H_3O_2^-]$

To find $[C_2H_3O_2^-]$, note that all of the acetic acid molecules originally present have been converted to acetate ions. The volume of the solution is twice as large as it was originally, 100.00 mL instead of 50.00 mL. It follows that the concentration of acetate ion ($n_{C_2H_3O_2^-}$)/V solution) must be one half of the original concentration of $HC_2H_3O_2$. In other words,

$$[C_2H_3O_2^-] = 1.000\ M/2 = 0.5000\ M$$

Hence,

$$[OH^-]^2 = 5.6 \times 10^{-10} \times 0.5000 = 2.8 \times 10^{-10}$$
$$[OH^-] = (2.8 \times 10^{-10})^{1/2} = 1.7 \times 10^{-5}\ M$$
$$pOH = 4.78;\ pH = 9.22$$

> pH changes relatively slowly near the end point.

From Figure 14.6 and Example 14.6, it should be clear that the indicator used in this titration must change color at about pH 9. Phenolphthalein (end point pH = 9) is satisfactory. Methyl red (end point pH = 5) is not suitable. If we used methyl red, we would stop the titration much too early, when reaction is only about 65% complete. This situation is typical of *weak acid–strong base titrations*. For such a titration we choose an indicator that *changes color above pH 7*.

Strong Acid–Weak Base

The reaction between solutions of hydrochloric acid and ammonia can be represented by the Brønsted-Lowry equation:

$$H_3O^+(aq) + NH_3(aq) \longrightarrow NH_4^+(aq) + H_2O$$

Again, we simplify the equation by replacing H_3O^+ with H^+:

$$H^+(aq) + NH_3(aq) \longrightarrow NH_4^+(aq)$$

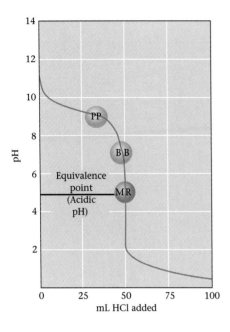

FIGURE 14.7

A strong acid–weak base titration. The curve represents the titration of 50.00 mL of 1.000 M ammonia, a weak base, with 1.000 M HCl. The solution at the equivalence point is acidic because of the NH_4^+ ion. Methyl red is a suitable indicator; phenolphthalein would change color much too early. **OHT**

To find the equilibrium constant, note that this equation is the reverse of that for the acid dissociation of the NH_4^+ ion. Hence

$$K = \frac{1}{K_a NH_4^+} = \frac{1}{5.6 \times 10^{-10}} = 1.8 \times 10^9$$

Because K is so large, the reaction goes virtually to completion.

Figure 14.7 shows how pH changes when fifty milliliters of one molar NH_3 is titrated with one molar HCl. In many ways, this curve is the inverse of that shown in Figure 14.6 for the weak acid–strong base case. In particular—

- the original pH is that of 1.000 M NH_3, a weak base ($K_b = 1.8 \times 10^{-5}$). As you would expect, it lies between 7 and 14:

$$[OH^-]^2 \approx K_b \times [NH_3]_o = (1.8 \times 10^{-5})(1.000)$$

$$[OH^-] = 4.2 \times 10^{-3}\ M;\ pOH = 2.38$$

$$pH = 14.00 - 2.38 = 11.62$$

- addition of 25 mL of HCl gives a buffer containing equal amounts of NH_3 and NH_4^+ ($K_a = 5.6 \times 10^{-10}$). Hence

$$pH = pK_a\ of\ NH_4^+ = 9.25$$

As strong acid is added, the pH drops.

- at the equivalence point, there is a 0.5000 M solution of NH_4^+, a weak acid. As you would expect, the pH is on the acid side.

$$[H^+]^2 \approx 0.5000 \times 5.6 \times 10^{-10}$$

$$[H^+] = 1.7 \times 10^{-5}\ M$$

$$pH = 4.77$$

From Figure 14.7, it should be clear that the only suitable indicator listed is methyl red. The other two indicators would change color too early, before the equivalence point. In general, *for the titration of a weak base with a strong acid, the indicator should change color on the acid side of pH 7.*

TABLE 14.3 Characteristics of Acid-Base Titrations [OHT]

Strong Acid–Strong Base

Example	Equation	K	Species at equivalence point	pH at equivalence point	Indicator*
NaOH—HCl	$H^+(aq) + OH^-(aq) \longrightarrow H_2O$	$K = 1/K_w$ 1.0×10^{14}	Na^+, Cl^-	7.00	MR, BB, PP
Ba(OH)$_2$—HNO$_3$	$H^+(aq) + OH^-(aq) \longrightarrow H_2O$	1.0×10^{14}	Ba^{2+}, NO_3^-	7.00	MR, BB, PP

Weak Acid–Strong Base

Example	Equation	K	Species at equivalence point	pH at equivalence point	Indicator*
HC$_2$H$_3$O$_2$—NaOH	$HC_2H_3O_2(aq) + OH^-(aq) \longrightarrow$ $C_2H_3O_2^-(aq) + H_2O$	$K = 1/K_b$ 1.8×10^9	Na^+, $C_2H_3O_2^-$	9.22†	PP
HF—KOH	$HF(aq) + OH^-(aq) \longrightarrow$ $F^-(aq) + H_2O$	6.9×10^{10}	K^+, F^-	9.42†	PP

Strong Acid–Weak Base

Example	Equation	K	Species at equivalence point	pH at equivalence point	Indicator*
NH$_3$—HCl	$NH_3(aq) + H^+(aq) \longrightarrow NH_4^+(aq)$	$K = 1/K_a$ 1.8×10^9	NH_4^+, Cl^-	4.78†	MR
ClO$^-$—HCl	$ClO^-(aq) + H^+(aq) \longrightarrow HClO(aq)$	3.6×10^7	HClO, Cl^-	3.93†	MR

*MR = methyl red (end point pH = 5); BB = bromthymol blue (end point pH = 7); PP = phenolphthalein (end point pH = 9).

†When 1 M acid is titrated with 1 M base.

Summary

Table 14.3 summarizes our discussion of acid-base titrations. Notice that for these three types of titrations—

> H$^+$ and OH$^-$ ions are the reacting species in strong acids and strong bases, respectively.

- **the equations (second column) written to describe the reactions are quite different.** Strong acids and bases are represented by H$^+$ and OH$^-$ ions, respectively; weak acids and weak bases are represented by their chemical formulas.
- **the equilibrium constants (K) for all these reactions are very large, indicating that the reactions go essentially to completion.**
- **the pH at the equivalence point is determined by what species are present.** When only "spectator ions" are present, as in a strong acid–strong base titration, the pH at the equivalence point is 7. Basic anions (F$^-$, C$_2$H$_3$O$_2^-$), formed during a weak acid–strong base titration, make the pH at the equivalence point greater than 7. Conversely, acidic species (NH$_4^+$, HClO) formed during a strong acid–weak base titration make the pH at the equivalence point less than 7.

EXAMPLE 14.7

Consider the titration of nitrous acid, HNO$_2$, with barium hydroxide.

(a) Write a balanced net ionic equation for the reaction.
(b) Calculate K for the reaction.
(c) Is the solution at the equivalence point acidic, basic, or neutral?
(d) What would be an appropriate indicator for the titration?

Strategy Once you realize that this is a weak acid–strong base titration, the problem unravels; follow the rules cited in Table 14.3.

Solution

(a) $HNO_2(aq) + OH^-(aq) \rightarrow NO_2^-(aq) + H_2O$

(b) $K = 1/K_b\ NO_2^-$. From Table 13.2 (page 391), $K_b\ NO_2^- = 1.7 \times 10^{-11}$.

$K = 1/(1.7 \times 10^{-11}) = 5.9 \times 10^{10}$

(c) At the equivalence point we have a solution of $NaNO_2$, which is basic due to the NO_2^- ion.

(d) phenolphthalein

Acid Rain

CHEMISTRY

Beyond the Classroom

N atural rainfall has a pH of about 5.5. It is slightly acidic because of dissolved carbon dioxide, which reacts with water to form the weak acid H_2CO_3. In contrast, the average pH of rainfall in the eastern United States and southeastern Canada is about 4.4, corresponding to an H^+ ion concentration more than ten times the normal value. In an extreme case, rain with a pH of 1.5 was recorded in Wheeling, West Virginia (distilled vinegar has a pH of 2.4).

Acid rain results from the presence of two strong acids in polluted air: H_2SO_4 and, to a lesser extent, HNO_3. Nitric acid comes largely from a three-step process. The first step involves the formation of NO from the elements by high-temperature combustion, as in an automobile engine. This is then oxidized to NO_2, which in turn undergoes the following reaction:

$$2NO_2(g) + H_2O \longrightarrow HNO_3(aq) + HNO_2(aq)$$

There are two main sources of SO_2, the precursor of sulfuric acid. Coal used in power plants contains up to 4% sulfur, mostly in the form of minerals such as pyrite, FeS_2. Combustion forms sulfur dioxide:

$$4FeS_2(s) + 11O_2(g) \longrightarrow 2Fe_2O_3(s) + 8SO_2(g)$$

Large quantities of sulfur dioxide are also produced in the metallurgical processes used to treat sulfide ores. A single nickel smelter in Sudbury, Ontario, produces 1% of the global supply of SO_2 by the reaction

$$2NiS(s) + 3O_2(g) \longrightarrow 2NiO(s) + 2SO_2(g)$$

Sulfur dioxide formed by reactions such as these is slowly converted to SO_3 in the atmosphere

$$SO_2(g) + \tfrac{1}{2}O_2(g) \longrightarrow SO_3(g)$$

Reaction of SO_3 with water

$$SO_3(g) + H_2O \longrightarrow H_2SO_4(aq)$$

produces sulfuric acid as tiny droplets high in the atmosphere. These may be carried by prevailing winds as far as 1500 km. The acid rain that falls in the Adirondacks of New York or the Green Mountains of Vermont comes from sulfur dioxide produced by power plants in Ohio and Illinois (Figure A).

The effects of acid rain are particularly severe in areas where the bedrock is granite or other materials incapable of neutralizing H^+ ions. As the concentration of acid builds up in a lake, aquatic life, from algae to brook trout, dies. The end product is a crystal-clear, totally sterile lake.

(continued)

FIGURE A

Air pollutants and acid rain.
Sulfur dioxide and nitrogen oxides
(NO_x) can move long distances on
the wind before conversion to acid
rain.

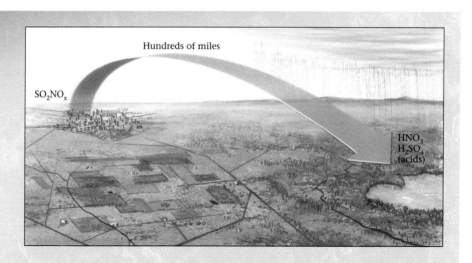

FIGURE B

**Acid rain–damaged trees in the
Great Smoky Mountains.** *(Frederica
Georgia/Photo Researchers, Inc.)*

Acid rain adversely affects trees as well (Figure B). It appears that the damage is
largely due to the leaching of metal cations from the soil. In particular, H^+ ions in acid
rain can react with insoluble aluminum compounds in the soil, bringing Al^{3+} ions into
solution. The following reaction is typical:

$$Al(OH)_3(s) + 3H^+(aq) \longrightarrow Al^{3+}(aq) + 3H_2O$$

Al^{3+} ions in solution can have a direct toxic effect on the roots of trees. Perhaps more important, Al^{3+} can replace Ca^{2+}, an ion that is essential to the growth of all plants.

Strong acids in the atmosphere can also attack building materials such as limestone
or marble (calcium carbonate)

$$CaCO_3(s) + 2H^+(aq) \longrightarrow Ca^{2+}(aq) + CO_2(g) + H_2O$$

The relatively soluble calcium compounds formed this way [$CaSO_4$, $Ca(NO_3)_2$] are gradually washed away (Figure C). This process is responsible for the deterioration of the Greek
ruins on the Acropolis in Athens. These structures suffered more damage in the twentieth
century than in the preceding 2000 years.

To reduce acid rain, sulfur dioxide emissions from power plants and smelters must be
brought under control. One way to do this is to spray a water solution of calcium hydroxide directly into the smokestack. This "scrubbing" operation brings about the reaction

$$Ca^{2+}(aq) + 2\,OH^-(aq) + SO_2(g) + \tfrac{1}{2}O_2(g) \longrightarrow CaSO_4(s) + H_2O$$

One problem with this approach is that it produces large quantities of solid calcium sulfate as a waste product. A quite different method is used at the Sudbury smelter referred
to earlier. There the SO_2 is separated from the stack gas and converted to sulfuric acid by
the contact process. More than 10^6 tons of H_2SO_4, worth \$100 million, are produced
each year.

FIGURE C

Acid rain damage on a 16th-century sculpture in Cracow, Poland. Acid rain is a serious
problem in Poland, where rain with the highly
acidic pH of 1.7 has been recorded. Since 1973
the United Nations has listed the historic building where this sculpture sits as being in urgent
need of repair. *(Simon Frasier/Science Photo
Library/Photo Researchers, Inc.)*

CHAPTER HIGHLIGHTS

Key Concepts

1. Calculate the pH of a buffer as it is originally prepared.
   (Example 14.1; Problems 13, 14, 19, 20, 23–30)
2. Choose a buffer to get a specified pH.
   (Example 14.2; Problems 15–18, 21, 22)
3. Calculate the pH of a buffer after H^+ or OH^- ions are added.
   (Example 14.3; Problems 25–30, 37, 38)
4. Determine the color of an acid-base indicator at a given pH.
   (Example 14.4; Problems 41, 42)
5. Calculate the pH during an acid-base titration.
   (Examples 14.5, 14.6; Problems 43–48, 63
6. Choose the proper indicator for an acid-base titration.
   (Example 14.7; Problems 39, 40, 64)
7. Calculate K for an acid-base reaction.
   (Example 14.7; Problems 5–8)

Key Equations

The key equation, used over and over again in this chapter, in one form or another, is

$$K_a = \frac{[H^+] \times [B^-]}{[HB]} = [H^+] \times \frac{n_{B^-}}{n_{HB}}$$

Key Terms

- acid
- —conjugate
- —equilibrium constant
- —strong
- —weak
- acid-base indicator
- base
- —conjugate
- —equilibrium constant
- —strong
- —weak

buffer
—capacity
end point

- equivalence point
- pH
- titration

Summary Problem

Consider formic acid, $HCHO_2$, ($K_a = 1.9 \times 10^{-4}$) and its conjugate base, CHO_2^- ($K_b = 5.3 \times 10^{-11}$).

(a) Write net ionic equations and calculate K for the reactions in aqueous solution between the following.

(1) $HCHO_2$ and NaOH
(2) CHO_2^- and HCl
(3) $HCHO_2$ and NaCN ($K_b = 1.7 \times 10^{-5}$)

(b) A buffer is prepared by dissolving 25.0 g of sodium formate, $NaCHO_2$, in 1.30 L of a solution of $0.660\ M\ HCHO_2$. Assume that the volume remains the same after dissolving $NaCHO_2$.

    (1) Calculate the pH of the buffer.
    (2) Calculate the pH of the buffer after 0.100 mol of HNO_3 is added.
    (3) Calculate the pH of the buffer after 10.0 g of KOH is added.
    (4) Calculate the pH of the buffer after 75.0 mL of $0.517\ M\ HNO_3$ is added.
    (5) What is the buffer capacity?

(c) Thirty-five milliliters of $0.739\ M\ HCHO_2$ is titrated with $0.431\ M$ KOH.

    (1) What is the pH of the $HCHO_2$ solution before titration?
    (2) How many milliliters of KOH are required to reach the equivalence point?
    (3) What is the pH of the solution halfway to neutralization?
    (4) What is the pH of the solution at the equivalence point?

Answers

(a) **(1)** $HCHO_2(aq) + OH^-(aq) \rightleftharpoons H_2O + CHO_2^-(aq)$ $K = 1.9 \times 10^{10}$
    (2) $CHO_2^-(aq) + H^+(aq) \rightleftharpoons HCHO_2(aq)$ $K = 5.3 \times 10^3$
    (3) $HCHO_2(aq) + CN^-(aq) \rightleftharpoons HCN(aq) + CHO_2^-(aq)$ $K = 3.2 \times 10^5$

(b) **(1)** 3.35   **(2)** 3.16   **(3)** 3.63   **(4)** 3.29
    (5) The buffer can theoretically absorb a maximum of 0.37 mol of H^+ and 0.86 mol of OH^-.

(c) **(1)** 1.93   **(2)** 60.0 mL   **(3)** 3.72   **(4)** 8.58

Questions & Problems

Problem numbers in blue indicate that the answer is available in Appendix 6 at the back of the book.

WEB indicates that the solution is posted at **http://www.harcourtcollege.com/chem/general/masterton4/student/**

Equilibrium constants required to solve these problems can be found in the tables in Chapter 13 or in Appendix 1.

Acid–Base Reactions

1. Write a net ionic equation for the reaction between aqueous solutions of
    (a) sodium acetate ($NaC_2H_3O_2$) and nitric acid.
    (b) hydrobromic acid and strontium hydroxide.
    (c) hypochlorous acid and sodium cyanide.
    (d) sodium hydroxide and nitrous acid.

WEB 2. Write a net ionic equation for the reaction between aqueous solutions of
    (a) ammonia and hydrofluoric acid.
    (b) perchloric acid and rubidium hydroxide.
    (c) sodium sulfite and hydriodic acid.
    (d) nitric acid and calcium hydroxide.

3. Write a balanced net ionic equation for the reaction of each of the following aqueous solutions with OH^- ions.
    (a) $Fe(H_2O)_6^{3+}$ **(b)** sodium hydrogen carbonate
    (c) ammonium chloride

4. Write a balanced net ionic equation for the reaction of each of the following aqueous solutions with H^+ ions.
    (a) sodium acetate   **(b)** cesium hydroxide
    (c) sodium hydrogen carbonate

5. Calculate K for the reactions in Question 1.
6. Calculate K for the reactions in Question 2.
7. Calculate K for the reactions in Question 3.
8. Calculate K for the reactions in Question 4.

Buffers

9. Calculate $[H^+]$ and pH in a solution in which $[NH_4^+]$ is $0.500\ M$ and $[NH_3]$ is
    (a) $0.50\ M$ **(b)** $0.20\ M$ **(c)** $0.10\ M$ **(d)** $0.010\ M$

10. Calculate $[OH^-]$ and pH in a solution in which $[HClO]$ is $0.25\ M$ and $[ClO^-]$ is
    (a) $0.50\ M$ **(b)** $0.30\ M$ **(c)** $0.15\ M$ **(d)** $0.075\ M$

11. The buffer capacity indicates how much OH^- or H^+ ions a buffer can react with. What is the buffer capacity of the buffers in Problem 9?

12. The buffer capacity indicates how much OH^- or H^+ ions a buffer can react with. What is the buffer capacity of the buffers in Problem 10?

13. A buffer is prepared by dissolving 0.0250 mol of sodium nitrite, $NaNO_2$, in 250.0 mL of $0.0410\ M$ nitrous acid, HNO_2. Assume no volume change after HNO_2 is dissolved. Calculate the pH of this buffer.

14. A buffer is prepared by dissolving 0.037 mol of potassium fluoride in 135 mL of $0.0237\ M$ hydrofluoric acid. Assume no volume change after KF is dissolved. Calculate the pH of this buffer.

15. Consider the weak acids in Table 13.2. Which acid-base pair would be best for a buffer at a pH of
  (a) 3.5   (b) 7.3   (c) 10.0

16. Follow the instructions of Question 15 for a pH of
  (a) 2.0   (b) 4.0   (c) 8.5

17. To make a buffer with pH = 3.0 from $HCHO_2$ and CHO_2^-,
  (a) what must the $[HCHO_2]/[CHO_2^-]$ ratio be?
  (b) how many moles of $HCHO_2$ must be added to a liter of 0.139 M $NaCHO_2$ to give this pH?
  (c) how many grams of $NaCHO_2$ must be added to 350.0 mL of 0.159 M $HCHO_2$ to give this pH?
  (d) What volume of 0.236 M $HCHO_2$ must be added to 1.00 L of a 0.500 M solution of $NaCHO_2$ to give this pH? (Assume that volumes are additive.)

WEB 18. A sodium hydrogen carbonate–sodium carbonate buffer is to be prepared with a pH of 9.40.
  (a) What must the $[HCO_3^-]/[CO_3^{2-}]$ ratio be?
  (b) How many moles of sodium hydrogen carbonate must be added to a liter of 0.225 M Na_2CO_3 to give this pH?
  (c) How many grams of sodium carbonate must be added to 475 mL of 0.336 M $NaHCO_3$ to give this pH? (Assume no volume change.)
  (d) What volume of 0.200 M $NaHCO_3$ must be added to 735 mL of a 0.139 M solution of Na_2CO_3 to give this pH? (Assume that volumes are additive.)

19. A buffer solution is prepared by adding 15.00 g of sodium acetate ($NaC_2H_3O_2$) and 12.50 g of acetic acid to enough water to make 500 mL (three significant figures) of solution.
  (a) What is the pH of the buffer?
  (b) The buffer is diluted by adding enough water to make 1.50 L of solution. What is the pH of the diluted buffer?

20. A buffer solution is prepared by adding 5.50 g of ammonium chloride and 0.0188 mol of ammonia to enough water to make 155 mL of solution.
  (a) What is the pH of the buffer?
  (b) If enough water is added to double the volume, what is the pH of the solution?

21. A bottle with 0.500 L of aqueous ammonia is labeled "13.2% by mass ammonia, d = 0.9430 g/mL." How many grams of ammonium chloride should be added to produce a buffer with a pH of 9.45?

22. A student is asked to make an acetic acid–acetate buffer with a pH of 4.10. The only materials on hand are 300.0 mL of white vinegar (5.00% by mass acetic acid, d = 1.006 g/mL) and potassium acetate. How should the student proceed?

23. An aqueous solution of 0.043 M weak acid, HX, has a pH of 3.92. What is the pH of the solution if 0.021 mol of KX is dissolved in one liter of the weak acid?

24. A solution with a pH of 8.73 is prepared by adding water to 0.614 mol of NaX to make 2.50 L of solution. What is the pH of the solution after 0.219 mol of HX is added?

25. A buffer is made up of 0.300 L each of 0.500 M KH_2PO_4 and 0.317 M K_2HPO_4. Assuming that volumes are additive, calculate

  (a) the pH of the buffer.
  (b) the pH of the buffer after the addition of 0.0500 mol of HCl to 0.600 L of buffer.
  (c) the pH of the buffer after the addition of 0.0500 mol of NaOH to 0.600 L of buffer.

26. A buffer is made up of 355 mL each of 0.200 M $NaHCO_3$ and 0.134 M Na_2CO_3. Assuming that volumes are additive, calculate
  (a) the pH of the buffer.
  (b) the pH of the buffer after the addition of 0.0300 mol of HCl to 0.710 L of buffer.
  (c) the pH of the buffer after the addition of 0.0300 mol of KOH to 0.710 L of buffer.

27. Enough water is added to the buffer in Question 25 to make the total volume 10.0 L. Calculate
  (a) the pH of the buffer.
  (b) the pH of the buffer after the addition of 0.0500 mol of HCl to 0.600 L of diluted buffer.
  (c) the pH of the buffer after the addition of 0.0500 mol of NaOH to 0.600 L of diluted buffer.
  (d) Compare your answers to Question 25(a)–(c) to your answers to (a)–(c) in this problem.
  (e) Comment on the effect of dilution on the pH of a buffer and on its buffer capacity.

28. Enough water is added to the buffer in Question 26 to make the total volume 10.0 L. Calculate
  (a) the pH of the buffer.
  (b) the pH of the buffer after the addition of 0.0300 mol of HCl to 0.710 L of diluted buffer.
  (c) the pH of the buffer after the addition of 0.0030 mol of NaOH to 0.710 L of diluted buffer.
  (d) Compare your answers to Question 26(a)–(c) to your answers to (a)–(c) in this problem.
  (e) Comment on the effect of dilution on the pH of a buffer and on its buffer capacity.

29. A buffer is prepared in which the ratio $[HCO_3^-]/[CO_3^{2-}]$ is 4.0.
  (a) What is the pH of this buffer?
  (b) Enough strong acid is added to convert 25% of CO_3^{2-} to HCO_3^-. What is the pH of the resulting solution?
  (c) Enough strong base is added to make the pH 10.83. What is the ratio of $[CO_3^{2-}]$ to $[HCO_3^-]$ at this point?

30. A buffer is prepared using the conjugate acid-base pair HNO_2/NO_2^- where the ratio $[HNO_2]/[NO_2^-]$ is 2.50.
  (a) What is the pH of this buffer?
  (b) Enough strong base is added to convert 33% of HNO_2 to NO_2^-. What is the pH of the resulting solution?
  (c) Enough strong base is added to make the pH 3.13. What is the ratio of $[HNO_2]/[NO_2^-]$ at this point?

31. Which of the following would form a buffer if added to 650.0 mL of 0.40 M $Sr(OH)_2$?
  (a) 1.00 mol of HF   (b) 0.75 mol of HF
  (c) 0.30 mol of HF   (d) 0.30 mol of NaF
  (e) 0.30 mol of HCl

Explain your reasoning in each case.

32. Which of the following would form a buffer if added to 250.0 mL of 0.150 M SnF$_2$?

(a) 0.100 mol of HCl   (b) 0.060 mol of HCl
(c) 0.040 mol of HCl   (d) 0.040 mol of NaOH
(e) 0.040 mol of HF

Explain your reasoning in each case.

33. Calculate the pH of a solution prepared by mixing 1.73 g of propionic acid (HC$_3$H$_5$O$_2$) and 0.53 g of NaOH in water (K_a propionic acid $= 1.4 \times 10^{-5}$).

34. Calculate the pH of a solution prepared by mixing 5.60 g of Na$_3$PO$_4$ and 75.0 mL of 0.225 M HCl.

35. Calculate the pH of a solution prepared by mixing 100.0 mL of 1.20 M ethanolamine, C$_2$H$_5$ONH$_2$, with 50.0 mL of 1.0 M HCl. K_a for C$_2$H$_5$ONH$_3{}^+$ is 3.6×10^{-10}.

36. Calculate the pH of a solution prepared by mixing 2.00 g of butyric acid (HC$_4$H$_7$O$_2$) with 0.50 g of NaOH in water (K_a butyric acid $= 1.5 \times 10^{-5}$).

37. Blood is buffered mainly by the HCO$_3{}^-$-H$_2$CO$_3$ buffer system. The normal pH of blood is 7.40.

(a) Calculate the [H$_2$CO$_3$]/[HCO$_3{}^-$] ratio.
(b) What does the pH become if 15% of the HCO$_3{}^-$ ions are converted to H$_2$CO$_3$?
(c) What does the pH become if 15% of the H$_2$CO$_3$ molecules are converted to HCO$_3{}^-$ ions?

38. There is a buffer system (H$_2$PO$_4{}^-$-HPO$_4{}^{2-}$) in blood that helps keep the blood pH at about 7.40. (K_aH$_2$PO$_4{}^-$ $= 6.2 \times 10^{-8}$).

(a) Calculate the [H$_2$PO$_4{}^-$]/[HPO$_4{}^{2-}$] ratio at the normal pH of blood.
(b) What percentage of the HPO$_4{}^{2-}$ ions are converted to H$_2$PO$_4{}^-$ when the pH goes down to 6.80?
(c) What percentage of H$_2$PO$_4{}^-$ ions are converted to HPO$_4{}^{2-}$ when the pH goes up to 7.80?

Titrations and Indicators

39. Given three acid-base indicators: methyl orange (end point at pH 4), bromthymol blue (end point at pH 7), and phenolphthalein (end point at pH 9), which would you select for the following acid-base titrations?

(a) perchloric acid with an aqueous solution of ammonia
(b) nitrous acid with lithium hydroxide
(c) hydrobromic acid with strontium hydroxide
(d) sodium fluoride with nitric acid

40. Given the acid-base indicators in Question 39, select a suitable indicator for the following titrations:

(a) acetic acid with sodium hydroxide
(b) sulfurous acid with sodium hydroxide
(c) sulfuric acid with potassium hydroxide
(d) sodium acetate with hydrochloric acid

41. Alizarin yellow has a pH range of 10.1–12.1. What is its K_a? It is yellow at pH 10 and red at pH 12. What is the color of a solution with pH 10.9 that has a few drops of alizarin yellow?

WEB **42.** Methyl yellow is an indicator with $K_a = 3.56 \times 10^{-4}$. It is red in acid solution and yellow in alkaline solution. What is its pH range? What would its color be at its pK_a?

43. A 0.1375 M solution of potassium hydroxide is used to titrate 35.00 mL of 0.257 M hydrobromic acid. (Assume that volumes are additive).

(a) Write a balanced net ionic equation for the reaction that takes place during titration.
(b) What are the species present at the equivalence point?
(c) What volume of potassium hydroxide is required to reach the equivalence point?
(d) What is the pH of the solution before any KOH is added?
(e) What is the pH of the solution halfway to the equivalence point?
(f) What is the pH of the solution at the equivalence point?

44. A 0.4000 M solution of nitric acid is used to titrate 50.00 mL of 0.237 M barium hydroxide. (Assume that volumes are additive.)

(a) Write a balanced net ionic equation for the reaction that takes place during titration.
(b) What are the species present at the equivalence point?
(c) What volume of nitric acid is required to reach the equivalence point?
(d) What is the pH of the solution before any HNO$_3$ is added?
(e) What is the pH of the solution halfway to the equivalence point?
(f) What is the pH of the solution at the equivalence point?

45. Consider the titration of benzoic acid (HBen) with sodium hydroxide. In an experiment, 25.00 mL of 0.175 M benzoic acid is titrated with 0.469 M KOH.

(a) Write a balanced net ionic equation for the reaction that takes place during titration.
(b) What are the species present at the equivalence point?
(c) What volume of potassium hydroxide is required to reach the equivalence point?
(d) What is the pH of the solution before any KOH is added?
(e) What is the pH of the solution halfway to the equivalence point?
(f) What is the pH of the solution at the equivalence point?

46. Consider the titration of sodium lactate (NaLac) with hydrochloric acid. In the titration, 50.0 mL of 0.224 M sodium lactate is titrated with 0.1035 M HCl.

(a) Write a balanced net ionic equation for the reaction that takes place during titration.
(b) What are the species present at the equivalence point?
(c) What volume of hydrochloric acid is required to reach the equivalence point?
(d) What is the pH of the solution before any HCl is added?
(e) What is the pH of the solution halfway to the equivalence point?
(f) What is the pH of the solution at the equivalence point?

47. At 25°C and 1.00 atm pressure, one liter of ammonia is bubbled into 725 mL of water. Assume that all the ammonia dissolves and the volume of the solution is the volume of the water. A 50.0-mL portion of the prepared solution is titrated with 0.2193 M HNO$_3$. Calculate the pH of the solution

   (a) before titration.
   (b) halfway to the equivalence point.
   (c) at the equivalence point.

48. Consider a 10.0% (by mass) solution of hypochlorous acid. Assume the density of the solution to be 1.00 g/mL. A 30.0-mL sample of the solution is titrated with 0.419 M KOH. Calculate the pH of the solution

   (a) before titration.
   (b) halfway to the equivalence point.
   (c) at the equivalence point.

Unclassified

49. For an aqueous solution of acetic acid to be called "distilled white vinegar" it must contain 5.0% acetic acid by mass. A solution with a density of 1.05 g/mL has a pH of 2.95. Can the solution be called "distilled white vinegar"?

50. Consider an unknown monoprotic acid, HX. One experiment titrates a 50.0-mL aqueous solution containing 2.500 g of the unknown acid. It requires 34.2 mL of 0.652 M NaOH to reach the equivalence point. A second experiment uses an identical 50.0-mL solution of the unknown acid. This time 17.1 mL of 0.652 M NaOH is added. The pH after the NaOH addition is 3.17.

   (a) What is the molar mass of HX?
   (b) What is K_a for HX?
   (c) What is K_b for X$^-$?

51. Consider an aqueous solution of HF. The molar heat of formation for aqueous HF is -320.1 kJ/mol.

   (a) What is the pH of a 0.100 M solution of HF at 100°C?
   (b) Compare with the pH of a 0.100 M solution of HF at 25°C.

52. A solution of an unknown weak acid at 25°C has an osmotic pressure of 0.878 atm and a pH of 6.76. What is K_b for its conjugate base? (Assume that, in the equation for π [Chapter 10], $i \approx 1$.)

Conceptual Problems

53. Each symbol in the box below represents a mole of a component in one liter of a buffer solution; $\bigcirc$ represents the anion (X$^-$), $\square\bigcirc$ = the weak acid, (HX), $\square$ = H$^+$, and $\triangle$ = OH$^-$. Water

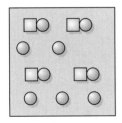

molecules and the few H$^+$ and OH$^-$ from the dissociation of HX and X$^-$ are not shown.

   (a) Fill in a similar box (representing one liter of the same solution) after two moles of H$^+$ (2$\square$) have been added. Indicate whether the resulting solution is an acid, base, or buffer.
   (b) Follow the directions of part (a) for the resulting solution after two moles of OH$^-$ (2$\triangle$) have been added.
   (c) Follow the directions of part (a) for the resulting solution after five moles of OH$^-$ (5$\triangle$) have been added.

Hint: Write the equation for the reaction before you draw the results.

54. The box contains ten moles of a weak acid, $\square\bigcirc$, in a liter of solution. Using the same symbols as in Question 53 ($\bigcirc$ = anion, $\triangle$ = OH$^-$), show what happens on

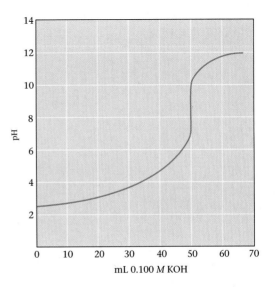

   (a) the addition of two moles of OH$^-$ (2$\triangle$)
   (b) the addition of five moles of OH$^-$ (5$\triangle$)
   (c) the addition of ten moles of OH$^-$ (10$\triangle$)
   (d) the addition of 12 moles of OH$^-$ (12$\triangle$)

Which addition, (a)–(d), represents neutralization halfway to the equivalence point?

55. The following is the titration curve for the titration of 50.00 mL of a 0.100 M acid with 0.100 M KOH.

(a) Is the acid strong or weak?

(b) If the acid is weak, what is its K_a?

(c) Estimate the pH at the equivalence point.

56. Carbonic acid, H_2CO_3, is a solution of CO_2 in water. Use Le Châtelier's principle to explain the build-up of CO_2 in the blood when H^+ ions are added.

57. Explain why

(a) the pH decreases when lactic acid is added to a sodium lactate solution.

(b) the pH of 0.1 M NH_3 is less than 13.0.

(c) a buffer resists changes in pH caused by the addition of H^+ or OH^-.

(d) a solution with a low pH is not necessarily a strong acid solution.

58. Indicate whether each of the following statements is true or false. If the statement is false, restate it to make it true.

(a) The formate ion (CHO_2^-) concentration in 0.10 M $HCHO_2$ is the same as in 0.10 M $NaCHO_2$.

(b) A buffer can be destroyed by adding too much strong acid.

(c) A buffer can be made up by any combination of weak acid and weak base.

(d) Because K_a for HCO_3^- is 4.7×10^{-11}, K_b for HCO_3^- is 2.1×10^{-4}.

59. The solutions in three test tubes labeled A, B, and C all have the same pH. The test tubes are known to contain 1.0×10^{-3} M HCl, 6.0×10^{-3} M $HCHO_2$, and 4×10^{-2} M $C_6H_5NH_3^+$. Describe a procedure for identifying the solutions.

Challenge Problems

60. The species called *glacial acetic acid* is 98% acetic acid by mass ($d = 1.0542$ g/mL). What volume of glacial acetic acid must be added to 100.0 mL of 1.25 M NaOH to give a buffer with a pH of 4.20?

61. Explain why it is not possible to prepare a buffer with a pH of 6.50 by mixing NH_3 and NH_4Cl.

62. Four grams of a monoprotic weak acid are dissolved in water to make 250.0 mL of solution with a pH of 2.56. The solution is divided into two equal parts, A and B. Solution A is titrated with strong base to its equivalence point. Solution B is added to solution A after solution A is neutralized. The pH of the resulting solution is 4.26. What is the molar mass of the acid?

63. Fifty cm^3 of 1.000 M nitrous acid is titrated with 0.850 M NaOH. What is the pH of the solution

(a) before any NaOH is added?

(b) at half-neutralization?

(c) at the equivalence point?

(d) when 0.10 mL less than the volume of NaOH to reach the equivalence point is added?

(e) when 0.10 mL more than the volume of NaOH to reach the equivalence point is added?

(f) Use your data to construct a plot similar to that shown in Figure 14.6 (pH versus volume NaOH added).

64. In a titration of 50.00 mL of 1.00 M $HC_2H_3O_2$ with 1.00 M NaOH, a student used bromcresol green as an indicator ($K_a = 1.0 \times 10^{-5}$). About how many milliliters of NaOH would it take to reach the end point with this indicator? Is there a better indicator that she could have used for this titration?

65. Starting with the relation

$$[H^+] = K_a \frac{[HB]}{[B^-]}$$

derive the Henderson-Hasselbalch equation

$$pH = pK_a + \log_{10} \frac{[B^-]}{[HB]}$$

66. What is the pH of a 0.1500 M H_2SO_4 solution if

(a) the ionization of HSO_4^- is ignored?

(b) the ionization of HSO_4^- is taken into account? (K_a for HSO_4^- is 1.1×10^{-2}.)

67. Two students were asked to determine the K_b of an unknown base. They were given a bottle with a solution in it. The bottle was labeled "aqueous solution of a monoprotic strong acid." They were also given a pH meter, a buret, and an appropriate indicator. They reported the following data:

volume of acid required for neutralization = 21.0 mL

pH after 7.00 mL of strong acid added = 8.95

Use the students' data to determine the K_b of the unknown base.

COMPLEX IONS

15

Chlorophyll is a chelate with
magnesium as its central metal
atom. It is responsible for the
green portions of the leaf.
(George Semple)

CHAPTER OUTLINE

15.1 COMPOSITION OF COMPLEX IONS

15.2 GEOMETRY OF COMPLEX IONS

15.3 ELECTRONIC STRUCTURE OF COMPLEX IONS

15.4 FORMATION CONSTANTS OF COMPLEX IONS

In previous chapters we have referred from time to time to compounds of the transition metals. Many of these have relatively simple formulas such as $CuSO_4$, $CrCl_3$, and $Fe(NO_3)_3$. These compounds are ionic: The transition metal is present as a simple cation (Cu^{2+}, Cr^{3+}, Fe^{3+}). In that sense, they resemble the ionic compounds formed by the main-group metals, such as $CaSO_4$ and $Al(NO_3)_3$.

It has been known for more than a century, however, that transition metals also form a variety of ionic compounds with more complex formulas such as

$$[Cu(NH_3)_4]SO_4 \qquad [Cr(NH_3)_6]Cl_3 \qquad K_3[Fe(CN)_6]$$

In these so-called *coordination compounds,* the transition metal is present as a complex ion, enclosed within the brackets. In the three compounds listed above, the following complex ions are present:

$$Cu(NH_3)_4{}^{2+} \qquad Cr(NH_3)_6{}^{3+} \qquad Fe(CN)_6{}^{3-}$$

The charges of these complex ions are balanced by those of simple anions or cations (e.g., $SO_4{}^{2-}$, $3Cl^-$, $3K^+$).

This chapter is devoted to complex ions and the important role they play in inorganic chemistry. We consider in turn—

- the composition of complex ions* (Section 15.1).

- the geometry of complex ions (Section 15.2).

- the electronic structure of the central metal ion (or atom) in a complex (Section 15.3).

- the equilibrium constant for the formation of a complex ion (Section 15.4).

> Cu^{2+} is a simple cation; $Cu(NH_3)_4{}^{2+}$ is a complex ion.

15.1 COMPOSITION OF COMPLEX IONS

When ammonia is added to an aqueous solution of a copper(II) salt, a deep, almost opaque, blue color develops (Figure 15.1). This color is due to the formation of the $Cu(NH_3)_4{}^{2+}$ ion, in which four NH_3 molecules are bonded to a central Cu^{2+} ion. The formation of this species can be represented by the equation

FIGURE 15.1
Ions of copper complexes. When ammonia is added to Cu^{2+}, the $[Cu(NH_3)_4]^{2+}$ complex ion forms. The ammonia-containing ion is an intense deep blue, almost violet. The hydrated copper complex ion $[Cu(H_2O)_4]^{2+}$ is light blue. *(Charles D. Winters)*

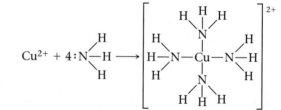

*The system used to name complex ions is described in Appendix 4.

The nitrogen atom of each NH_3 molecule contributes a pair of unshared electrons to form a covalent bond with the Cu^{2+} ion. This bond and others like it, where both electrons are contributed by the same atom, are referred to as *coordinate covalent bonds*.

The $Cu(NH_3)_4^{2+}$ ion is commonly referred to as a **complex ion,** a charged species in which a **central metal cation** is bonded to molecules and/or anions referred to collectively as **ligands.** The number of atoms bonded to the central metal cation is referred to as its **coordination number.** In the $Cu(NH_3)_4^{2+}$ complex ion—

- the central metal cation is Cu^{2+}.
- the ligands are NH_3 molecules.
- the coordination number is 4.

Complex ions are commonly formed by transition metals, particularly those toward the right of a transition series ($_{24}Cr \rightarrow {}_{30}Zn$ in the first transition series). Nontransition metals, including Al, Sn, and Pb, form a more limited number of stable complex ions.

Cations of these metals invariably exist in aqueous solution as complex ions. Consider, for example, the zinc(II) cation. In a water solution of $Zn(NO_3)_2$, the $Zn(H_2O)_4^{2+}$ ion is present. Treatment with ammonia converts this to $Zn(NH_3)_4^{2+}$; addition of sodium hydroxide forms $Zn(OH)_4^{2-}$.

When a complex ion is formed from a simple cation, the electron pairs required for bond formation come solely from the ligands. Reactions such as these, in which one species donates an electron pair to another, are referred to as *Lewis acid-base reactions*. In particular—

- a **Lewis base** is a species that donates a pair of electrons. Ligands such as the H_2O molecule, the NH_3 molecule, or the OH^- ion act as Lewis bases in complex ion formation.
- a **Lewis acid** is a species that accepts a pair of electrons. Cations such as Cu^{2+} and Zn^{2+} act as Lewis acids when they form complex ions. Thus we have

$$Cu^{2+}(aq) + 4NH_3(aq) \longrightarrow Cu(NH_3)_4^{2+}(aq)$$
Lewis acid   Lewis base

$$Zn^{2+}(aq) + 4OH^-(aq) \longrightarrow Zn(OH)_4^{2-}(aq)$$
Lewis acid   Lewis base

Notice that—

- species that act as Lewis bases (e.g., H_2O, NH_3, OH^-) are also Brønsted-Lowry bases in the sense that they can accept a proton.
- species that act as Lewis acids (e.g., Cu^{2+}, Zn^{2+}) need not be Brønsted-Lowry acids (proton donors). In that sense the Lewis model broadens the concept of an acid.

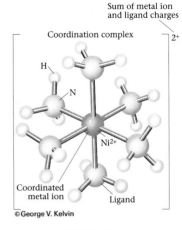

Sum of metal ion and ligand charges

Coordination complex

Coordinated metal ion

Ligand

©George V. Kelvin

$[Ni(NH_3)_6]^{2+}$

The $[Ni(NH_3)_6]^{2+}$ complex ion. In this complex ion, the ligands are NH_3 molecules. As in all complex ions, the ligands form coordinate covalent bonds between their atoms with unshared electron pairs and the central metal ion. The central Ni^{2+} has a coordination number of 6.

The Lewis acid-base model is the most general of the three we have considered.

See the *Saunders Interactive General Chemistry CD-ROM*, Screens 17.11, Lewis Acids and Bases, & 17.12, Cationic Lewis Acids.

Charges of Complexes

The charge of a complex is readily determined by applying a simple principle:

charge of complex = oxid. no. central metal + charges of ligands

The application of this principle is shown in Table 15.1 (page 440), where we list the formulas of several complexes formed by platinum(II), which shows a coordi-

TABLE 15.1 **Complexes of Pt^{2+} with NH$_3$ and Cl$^-$**

Complex	Oxid. No. of Pt	Ligands	Total Charge of Ligands	Charge of Complex
Pt(NH$_3$)$_4$$^{2+}$	+2	4NH$_3$	0	+2
Pt(NH$_3$)$_3$Cl$^+$	+2	3NH$_3$, 1Cl$^-$	−1	+1
Pt(NH$_3$)$_2$Cl$_2$	+2	2NH$_3$, 2Cl$^-$	−2	0
Pt(NH$_3$)Cl$_3$$^-$	+2	1NH$_3$, 3Cl$^-$	−3	−1
PtCl$_4$$^{2-}$	+2	4Cl$^-$	−4	−2

nation number of 4. Notice that one of the species, Pt(NH$_3$)$_2$Cl$_2$, is a neutral complex rather than a complex ion; the charges of the two Cl$^-$ ions just cancel that of the central Pt^{2+} ion.

> Note that this complex contains two Cl$^-$ ions, not a Cl$_2$ molecule.

EXAMPLE 15.1 Consider the Co(H$_2$O)$_4$Cl$_2$$^+$ ion.

(a) What is the oxidation number of cobalt?
(b) What is the formula of the coordination compound containing this cation and the Br$^-$ anion? The SO$_4$$^{2-}$ anion?

Strategy To find the oxidation number, apply the relation

charge complex ion = oxid. no. central atom + charges of ligands

You know the charge of the complex and those of the ligands. To find the formulas of the coordination compounds, apply the principle of electrical neutrality.

Solution

(a) +1 = oxid. no. Co + 2(−1);   oxid. no. Co = +3
(b) [Co(H$_2$O)$_4$Cl$_2$]Br   [Co(H$_2$O)$_4$Cl$_2$]$_2$SO$_4$ (compare with NaBr and Na$_2$SO$_4$)

The brackets enclose the formula of the complex ion.

Ligands; Chelating Agents

> Unshared pair = lone pair.

In principle, any molecule or anion with an unshared pair of electrons can act as a Lewis base. In other words, it can donate a lone pair to a metal cation to form a coordinate covalent bond. In practice, a ligand usually contains an atom of one of the more electronegative elements (C, N, O, S, F, Cl, Br, I). Several hundred different ligands are known. Those most commonly encountered in general chemistry are NH$_3$ and H$_2$O molecules and CN$^-$, Cl$^-$, and OH$^-$ ions.

$$:N{-}H \quad :O: \quad [:C{\equiv}N:]^- \quad [:\ddot{C}l:]^- \quad [:\ddot{O}{-}H]^- $$

Some ligands have more than one atom with an unshared pair of electrons and hence can form more than one bond with a central metal atom. Ligands of this type are referred to as *chelating agents;* the complexes formed are referred to as **chelates** (from the Greek *chela,* crab's claw). Two of the most common chelating

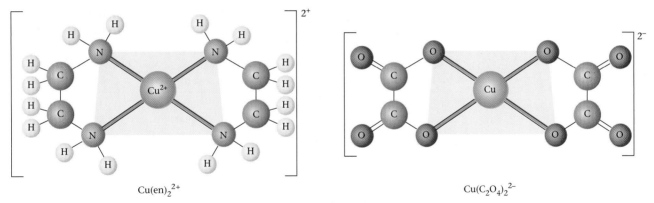

$Cu(en)_2^{2+}$ $Cu(C_2O_4)_2^{2-}$

FIGURE 15.2

Chelates formed by Cu^{2+} with the ethylenediamine molecule (*en*) and the oxalate ion $(C_2O_4^{2-})$.

agents are the oxalate anion (abbreviated *ox*) and the ethylenediamine molecule (abbreviated *en*), whose Lewis structures are

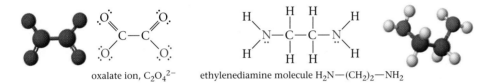

oxalate ion, $C_2O_4^{2-}$ ethylenediamine molecule $H_2N-(CH_2)_2-NH_2$

(The atoms which form bonds with the central metal are shown in color.)

Figure 15.2 shows the structure of the chelates formed by copper(II) with these ligands. Notice that in both of these complex ions, the coordination number of copper(II) is 4. The central cation is bonded to four atoms, two from each ligand.

For a ligand to act as a chelating agent, it must have at least two lone pairs of electrons. Moreover, these electron pairs must be far enough removed from one another to give a chelate ring with a stable geometry. In the chelates shown in Figure 15.2, the ring is five-membered, containing a Cu^{2+} ion and four nonmetal atoms. Most chelates contain five- or six-membered rings. Smaller or larger rings are less stable. For example, even though the hydrazine molecule

$$H-\overset{\cdot\cdot}{N}-\overset{\cdot\cdot}{N}-H$$
$$\quad\;\; | \quad\;\; |$$
$$\quad\;\; H \quad H$$

has two lone pairs, it does not form chelates. To do so, it would have to form a three-membered ring with 60° bond angles.

Coordination Number

As shown in Table 15.2 (page 442), the most common coordination number is 6. A coordination number of 4 is less common. A value of 2 is restricted largely to Cu^+, Ag^+, and Au^+.

Odd coordination numbers are rare.

TABLE 15.2 Coordination Number and Geometry of Complex Ions*

Metal Ion	Coordination Number	Geometry	Example
Ag^+, Au^+, Cu^+	2	Linear	$Ag(NH_3)_2^+$
Cu^{2+}, Ni^{2+}, **Pd^{2+}**, **Pt^{2+}**	4	Square planar	$Pt(NH_3)_4^{2+}$
Al^{3+}, Au^+, **Cd^{2+}**, Co^{2+}, Cu^+, Ni^{2+}, **Zn^{2+}**	4	Tetrahedral	$Zn(NH_3)_4^{2+}$
Al^{3+}, Co^{2+}, **Co^{3+}**, **Cr^{3+}**, Cu^{2+}, **Fe^{2+}**, **Fe^{3+}**, Ni^{2+}, **Pt^{4+}**	6	Octahedral	$Co(NH_3)_6^{3+}$

*Cations that commonly show only one coordination number are in bold type.

A few cations show only one coordination number in their complexes. Thus Co^{3+} always shows a coordination number of 6, as in

$$Co(NH_3)_6^{3+} \qquad Co(NH_3)_4Cl_2^+ \qquad Co(en)_3^{3+}$$

Other cations, such as Al^{3+} and Ni^{2+}, have variable coordination numbers, depending on the nature of the ligand.

EXAMPLE 15.2 Write formulas (including charges) of all the complex ions formed by cobalt(III) with NH_3 and/or ethylenediamine(en) molecules as ligands.

Strategy As indicated in Table 15.2, the coordination number of Co^{3+} is 6. This means that Co^{3+} forms six bonds with ligands, *including two with each en molecule.* The complex ions can contain 0 to 3 *en* molecules. Both ligands are neutral, so each complex ion has a charge of +3.

Solution There are four in all:

$$Co(NH_3)_6^{3+} \qquad Co(NH_3)_4(en)^{3+} \qquad Co(NH_3)_2(en)_2^{3+} \qquad Co(en)_3^{3+}$$

Reality Check Notice that, moving from left to right, one *en* replaces two NH_3 molecules. This makes sense; an *en* molecule has two lone pairs; NH_3 has only one.

15.2 GEOMETRY OF COMPLEX IONS

The physical and chemical properties of complex ions and of the coordination compounds they form depend on the spatial orientation of ligands around the central metal atom. Here we consider the geometries associated with the coordination numbers 2, 4, and 6. With that background, we then examine the phenomenon of **geometric isomerism,** where two or more complex ions have the same chemical formula but different properties because of their different geometries.

Coordination Number = 2

Complex ions in which the central metal forms only two bonds to ligands are *linear;* that is, the two bonds are directed at a 180° angle. The structures of $CuCl_2^-$,

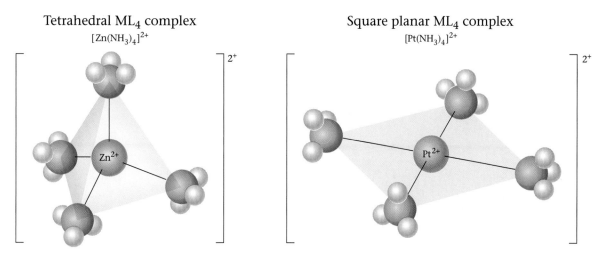

FIGURE 15.3

Geometry of four-coordinate complexes. Complexes in which the central metal has a co-ordination number of 4 may be tetrahedral or square-planar. **OHT**

$Ag(NH_3)_2{}^+$, and $Au(CN)_2{}^-$ may be represented as

$$(Cl-Cu-Cl)^- \qquad \begin{bmatrix} H & & H \\ H-N-Ag-N-H \\ H & & H \end{bmatrix}^+ \qquad (N\equiv C-Au-C\equiv N)^-$$

Coordination Number = 4

Four-coordinate metal complexes may have either of two different geometries (Figure 15.3). The four bonds from the central metal may be directed toward the corners of a regular *tetrahedron*. This is what we would expect from VSEPR model (recall Chapter 7). Two common **tetrahedral** complexes are $Zn(NH_3)_4{}^{2+}$ and $CoCl_4{}^{2-}$.

Square planar complexes, in which the four bonds are directed toward the corners of a square, are more common. Certain complexes of copper(II) and nickel(II) show this geometry; it is characteristic of the complexes of Pd^{2+} and Pt^{2+}, including $Pt(NH_3)_4{}^{2+}$.

Coordination Number = 6

We saw in Chapter 7 that octahedral geometry is characteristic of many molecules (e.g., SF_6) in which a central atom is surrounded by six other atoms. (Remember; an *octa*hedron has eight sides, which is irrelevant here; it has *six* corners, which is important.) All complex ions in which the coordination number is 6 are **octahedral.** The metal ion (or atom) is at the center of the *octahedron;* the six ligands are at the corners. This geometry is shown in Figure 15.4 (page 444).

It is important to realize that—

■ the six ligands can be considered to be equidistant from the central metal.
■ an octahedral complex can be regarded as a derivative of a square planar complex. The two extra ligands are located above and below the square, on a line perpendicular to the square at its center.

The structure is often shown simply as ⊿.

FIGURE 15.4
An octahedral, six-coordinate complex. The drawing at the left shows six ligands (represented by spheres) at the corners of an octahedron with a metal atom at the center. A simpler way to represent an octahedral complex is shown at the right.   **OHT**

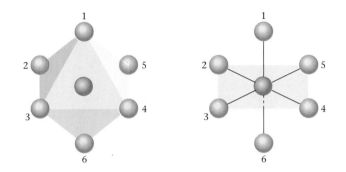

Geometric Isomerism

Two or more species with different physical and chemical properties but the same formula are said to be **isomers** of one another. Complex ions can show many different kinds of isomerism, only one of which we will consider. *Geometric isomers* are ones that differ only in the spatial orientation of ligands around the central metal atom. Geometric isomerism is found in square planar and octahedral complexes. It cannot occur in tetrahedral complexes where all four positions are equivalent.

1. *Square planar.* There are two compounds with the formula $Pt(NH_3)_2Cl_2$, differing in water solubility, melting point, chemical behavior, and biological activity.* Their structures are

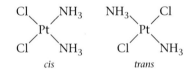

The complex in which the two chloride ions are at adjacent corners of the square, as close to one another as possible, is called the ***cis*** isomer. It is made by reacting ammonia with the $PtCl_4^{2-}$ ion. In the ***trans*** isomer, made by reacting $Pt(NH_3)_4^{2+}$ with HCl, the two chloride ions are at opposite corners of the square, as far apart as possible.

Geometric isomerism can occur with any square complex of the type Mabcd, Ma_2bc, or Ma_2b_2, where M refers to the central metal and a, b, c, and d are different ligands. Conversely, geometric isomerism cannot occur with a square complex of the type Ma_4 or Ma_3b. Thus there are two different square complexes with the formula $Pt(NH_3)_2ClBr$ but only one with the formula $Pt(NH_3)_3Cl^+$.

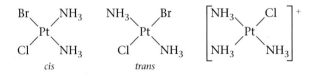

2. *Octahedral.* To understand how geometric isomerism can arise in octahedral complexes, refer back to Figure 15.4. Notice that for any given position of a

cis = close together, *trans* = far apart.

*The *cis* isomer ("cisplatin") is an effective anticancer drug. This reflects the ability of the two Cl atoms to interact with the nitrogen atoms of DNA, a molecule responsible for cell reproduction. The *trans* isomer is ineffective in chemotherapy, presumably because the Cl atoms are too far apart to react with a DNA molecule.

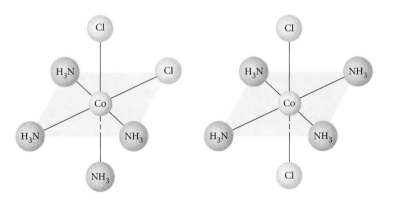

FIGURE 15.5
Cis and *trans* isomers of a complex ion. **OHT**

ligand, four other positions are at the same distance from that ligand, and a fifth is at a greater distance.

Taking position 1 in Figure 15.4 as a point of reference, you can see that groups at 2, 3, 4, and 5 are equidistant from 1; 6 is farther away. In other words, positions 1 and 2, 1 and 3, 1 and 4, 1 and 5 are *cis* to one another; positions 1 and 6 are *trans*. Hence a complex ion like $Co(NH_3)_4Cl_2^+$ (Figure 15.5) can exist in two different isomeric forms. In the *cis* isomer, the two Cl^- ions are at adjacent corners of the octahedron, as close together as possible. In the *trans* isomer they are at opposite corners, as far away from one another as possible.

EXAMPLE 15.3   How many geometric isomers are possible for the neutral complex $[Co(NH_3)_3Cl_3]$?

Strategy   It's best to approach a problem of this type systematically. You might start (1) by putting two NH_3 molecules *trans* to one another and see how many different structures can be obtained by placing the third NH_3 molecule at different positions. Then (2) start over with two NH_3 molecules *cis* to one another, and follow the same procedure.

Solution

(1) Starting with two NH_3 molecules *trans* to one another (Figure 15.6a), it clearly doesn't matter where you put the third NH_3 molecule. All the available vacancies are equivalent in the sense that they are *cis* to the two NH_3 molecules already in place. Choosing one of

FIGURE 15.6
Isomers of $Co(NH_3)_3Cl_3$ (Example 15.3). You may be tempted to write down additional structures, but you will find that they are equivalent to one of the two isomers shown.

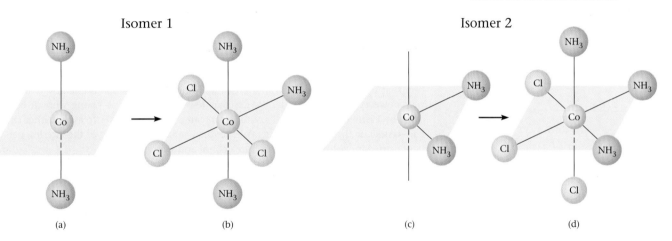

Isomer 1   Isomer 2

(a)   (b)   (c)   (d)

these positions arbitrarily for NH_3 gives the first isomer (Figure 15.6b); the Cl^- ions occupy the three remaining positions.

(2) Start again with the two NH_3 molecules *cis* to one another (Figure 15.6c). If you place the third NH_3 molecule at either of the other corners of the square, you will simply reproduce the first isomer. (Remember that the symmetry of a regular octahedron requires that the distance across the diagonal of the square be the same as that from "top" to "bottom.") The only way you can get a "new" isomer is to place the third NH_3 molecule above or below the square so that all three NH_3 molecules are *cis* to one another. Putting that molecule arbitrarily at the top gives the second isomer (Figure 15.6d). That's all there are.

> Unless you're careful, you'll come up with too many isomers.

Geometric isomerism can also occur in chelated octahedral complexes (Figure 15.7). Notice that an *ethylenediamine molecule,* here and indeed in all complexes, **can only bridge** **cis** **positions.** It is not long enough to connect to *trans* positions.

CHEMISTRY ▪ The Human Side

Alfred Werner
(1866–1919)

The basic ideas concerning the structure and geometry of complex ions presented in this chapter were developed by one of the most gifted individuals in the history of inorganic chemistry, Alfred Werner. His theory of coordination chemistry was published in 1893 when Werner was 26 years old. In his paper Werner made the revolutionary suggestion that metal ions such as Co^{3+} could show two different kinds of valences. For the compound $Co(NH_3)_6Cl_3$, Werner postulated a central Co^{3+} ion joined by "primary valences" (ionic bonds) to three Cl^- ions and by "secondary valences" (coordinate covalent bonds) to six NH_3 molecules. Moreover, he made the inspired guess that the six secondary valences were directed toward the corners of a regular octahedron.

Werner spent the next 20 years obtaining experimental evidence to prove his theory. (At the University of Zürich, there remain several *thousand* samples of coordination compounds prepared by Werner and his students.) He was able to show, for example, that the electrical conductivities in water solution decreased in the order $[Co(NH_3)_6]Cl_3 > [Co(NH_3)_5Cl]Cl_2 > [Co(NH_3)_4Cl_2]Cl$ in much the same way as with simple salts, for example, $ScCl_3 > CaCl_2 > NaCl$. Another property he studied was isomerism. In 1907 he was able to isolate a second geometric isomer of $[Co(NH_3)_4Cl_2]Cl$, in complete accord with his theory. Six years later, Werner won the Nobel Prize in chemistry.

By all accounts, Werner was a superb lecturer. Sometimes as many as 300 students crowded into a hall with a capacity of 150 to hear him speak. So great was his reputation that students in theology and law came to hear him talk about chemistry. There was, however, a darker side to Werner that few students saw. A young woman badgered by Werner during an oral examination came to see him later to ask whether she had passed; he threw a chair at her. Werner died at 52 of hardening of the arteries, perhaps caused in part by his addiction to alcohol and strong black cigars.

(Photo credit: E. F. Smith Memorial Collection, Van Pelt Library, University of Pennsylvania)

FIGURE 15.7
Cis and *trans* isomers of
$[Co(en)_2Cl_2]^+$. The ethylenediamine
molecule is represented by —en—. It
has two bonds per molecule. The *cis*
isomer is reddish-purple and the *trans*
is dark green. *(Charles D. Winters)*

15.3   ELECTRONIC STRUCTURE OF COMPLEX IONS

Over the years, several different models have been developed to treat electronic
structure and bonding in complex ions. One of the simplest and most useful is the
crystal field model, originally proposed by Hans Bethe and J. H. van Vleck. Here
the magnetic properties and brilliant colors (Figure 15.8) of coordination compounds are explained in terms of the effect of ligands on the energies of electronic
levels in transition metal cations. We will apply this model to octahedral complexes. Before doing so, it may be helpful to review the electronic structure of uncomplexed transition metal cations, originally covered in Chapter 6.

> "Crystal field" isn't a very descrip-
> tive term, but it's part of the jargon.

Transition Metal Cations

Recall (page 165) that in a simple transition metal cation—

- there are no outer s electrons. Electrons beyond the preceding noble gas are located in an inner d sublevel (3d for the first transition series).
- electrons are distributed among the five d orbitals in accordance with Hund's
  rule, giving the maximum number of unpaired electrons.

To illustrate these rules, consider the Fe^{2+} ion. Because the atomic number of
iron is 26, this +2 ion must contain $26 - 2 = 24e^-$. Of these electrons, the first 18
have the argon structure; the remaining six are located in the 3d sublevel. The abbreviated electron configuration is

$$Fe^{2+} \qquad [Ar]3d^6$$

FIGURE 15.8
**Brightly colored coordination com-
pounds.** Most coordination compounds are brilliantly colored, a property that can be explained by the
crystal field model. *(Marna G. Clarke)*

These six electrons are spread over all five orbitals; the orbital diagram is

3d

Fe^{2+} [Ar]   ($\uparrow\downarrow$)($\uparrow$)($\uparrow$)($\uparrow$)($\uparrow$)

The Fe^{2+} ion is *paramagnetic,* with four unpaired electrons.

EXAMPLE 15.4   For the Cr^{3+} ion, derive the electron configuration, orbital diagram, and number of unpaired electrons.

Strategy   First (1), find the total number of electrons (Z Cr = 24). Then (2), find the electron configuration; the first 18 electrons form the argon core, and the remaining electrons enter the 3d sublevel. Finally (3), apply Hund's rule to obtain the orbital diagram.

Solution

(1) no. of electrons = 24 − 3 = 21
(2) electron configuration [Ar]$3d^3$

3d

(3) orbital diagram (beyond Ar) ($\uparrow$)($\uparrow$)($\uparrow$)( )( )   3 unpaired electrons

---

The shapes of the five d orbitals are shown in Figure 15.9. These orbitals are given the symbols

d_{z^2} $d_{x^2-y^2}$ d_{xy} d_{yz} d_{xz}

In the uncomplexed transition metal cation, all of these orbitals have the same energy.

Octahedral Complexes

As six ligands approach a central metal ion to form an octahedral complex, they change the energies of electrons in the d orbitals. The effect (Figure 15.10) is to split the five d orbitals into two groups of different energy.

1. A higher-energy pair, the $d_{x^2-y^2}$ and d_{z^2} orbitals
2. A lower-energy trio, the d_{xy}, d_{yz}, and d_{xz} orbitals

The difference in energy between the two groups is called the **crystal field splitting energy** and given the symbol Δ_o (the subscript "o" stands for "octahedral").

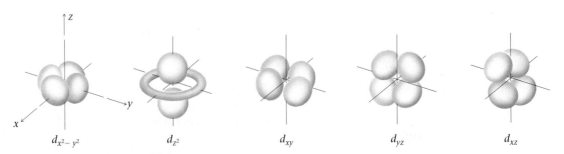

$d_{x^2-y^2}$ d_{z^2} d_{xy} d_{yz} d_{xz}

FIGURE 15.9
The five d orbitals and their relation to ligands on the *x, y,* and *z* axes. **OHT**

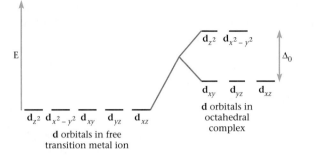

FIGURE 15.10

Change in the energy levels of d orbitals in complex formation. In a free transition metal ion, all five d orbitals have the same energy. When the ion forms an octahedral complex, two orbitals (d_{z^2}, $d_{x^2-y^2}$) have a higher energy than the other three (d_{xy}, d_{yz}, d_{xz}). The difference in energy between the two d orbital levels in the octahedral complex is known as the *crystal field splitting energy* and has the symbol Δ_o.

To see why this splitting occurs, consider what happens when six ligands (e.g., H_2O, CN^-, NH_3) approach a central metal cation along the x-, y-, and z-axes (Figure 15.9). The unshared electron pairs on these ligands repel the electrons in the d orbitals of the cation. The repulsion is greatest for the $d_{x^2-y^2}$ and d_{z^2} orbitals, which have their maximum electron density directly along the x-, y-, and z-axes, respectively. Electrons in the other three orbitals (d_{xy}, d_{yz}, d_{xz}) are less affected because their densities are concentrated between the axes rather than along them.

> This gives the d_{xy}, d_{yz}, and d_{xz} orbitals a lower energy.

Depending on the magnitude of Δ_o, a cation may form either of two different kinds of complexes. This is illustrated for the Fe^{2+} ion in the two diagrams shown in Figure 15.11.

1. In the $Fe(CN)_6^{4-}$ ion shown at the left, Δ_o is so large that the six 3d electrons of the Fe^{2+} ion pair up in the lower-energy orbitals. In this way, the Fe^{2+} ion achieves its most stable (lowest energy) electronic structure. The complex is diamagnetic, with no unpaired electrons.

2. In the $Fe(H_2O)_6^{2+}$ ion shown at the right, Δ_o is relatively small, and the six 3d electrons of the Fe^{2+} ion spread out over all the orbitals as expected by Hund's rule. In other words, the electron distribution in the complex is the same as in the Fe^{2+} ion itself (where Δ_o is 0!). The $Fe(H_2O)_6^{2+}$ ion is paramagnetic, with four unpaired electrons.

> When Δ_o is small, the electron distribution is the same as in the simple cation; if Δ_o is large, Hund's rule is not strictly followed.

More generally, it is possible to distinguish between—

- **low-spin complexes,** containing the minimum number of unpaired electrons, formed when Δ_o is large; electrons are concentrated in the lower-energy orbitals.

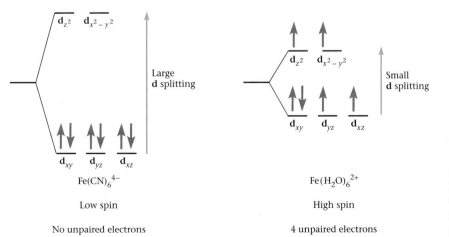

FIGURE 15.11

Electron distribution in low-spin and high-spin complexes of Fe^{2+}. Depending on the magnitude of Δ_o, either of two different complexes may be formed by Fe^{2+}. **OHT**

- **high-spin complexes,** containing the maximum number of unpaired electrons, formed when Δ_o is small; electrons are spread out among the d orbitals, exactly as in the uncomplexed transition metal cation.

For a given cation—

- the high-spin complex always contains more unpaired electrons than the low-spin complex and so is more strongly paramagnetic.
- the value of Δ_o is determined by the nature of the ligand. So-called "strong-field" ligands (e.g., CN^-), which interact strongly with d-orbital electrons, produce a large Δ_o. Conversely, "weak-field" ligands (e.g., H_2O) interact weakly with d-orbital electrons, producing a small Δ_o.

Remember:
strong-field ligands→large Δ_o→
low spin
weak-field ligands→small Δ_o→
high spin

$[Co(H_2O)_6]^{2+}$, a high-spin complex ion. *(Photo by Charles D. Winters; model by Susan M. Young)*

EXAMPLE 15.5   Derive the structure of the Co^{2+} ion in the low-spin and high-spin octahedral complexes.

Strategy   First, (1) find out how many 3d electrons there are in Co^{2+}. Now (2), consider how those electrons will be distributed if Hund's rule is followed; that corresponds to the high-spin complex. Finally (3), sprinkle as many electrons as possible into the lower three orbitals; this is the low-spin complex.

Solution

(1) no. of electrons = 27 − 2 = 25; the electron configuration is $[Ar]3d^7$.

(2)   (↑)(↑)
    (↑↓)(↑↓)(↑)   high-spin   3 unpaired electrons

(3)   (↑)( )
    (↑↓)(↑↓)(↑↓)   low-spin   1 unpaired electron

Reality Check   Both types of complexes are known. The $Co(CN)_6^{4-}$ ion is of the low-spin type; $Co(H_2O)_6^{2+}$ is a high-spin complex.

This model of the electronic structure of complex ions explains why high-spin and low-spin complexes occur only with ions that have four to seven electrons (d^4, d^5, d^6, d^7). With three or fewer electrons, only one distribution is possible; the same is true with eight or more electrons.

( )( )   (↑)(↑)
(↑)(↑)(↑)   (↑↓)(↑↓)(↑↓)
3 electrons   8 electrons

Color

Most coordination compounds are brightly colored except for those of cations whose d sublevels are completely empty (such as Sc^{3+}) or completely filled (such as Zn^{2+}). These colors are readily explained, at least qualitatively, by the model we have just described. The energy difference between two sets of d orbitals in a complex is ordinarily equal to that of a photon in the visible region. Hence by absorbing visible light, an electron may be able to move from the lower energy set of d orbitals to the higher one. This removes some of the component wavelengths of white light, so that the light reflected or transmitted by the complex is colored. The relation between absorbed and transmitted light can be deduced from the color wheel shown in Figure 15.12. For example, a complex that absorbs blue-green light appears orange, and vice versa.

The color we see is what is left over after certain wavelengths are absorbed.

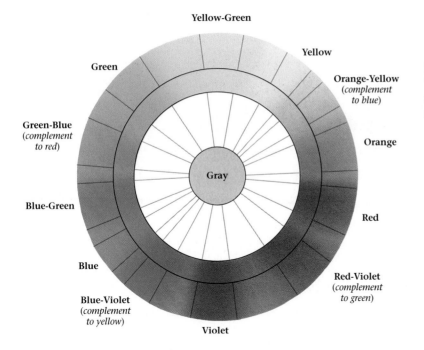

FIGURE 15.12
The color wheel. The color of a complex is complementary to (180° across from) the color of the light absorbed. **OHT**

From the color (absorption spectrum) of a complex ion, it is sometimes possible to deduce the value of Δ_o, the crystal field splitting energy. The situation is particularly simple in $_{22}Ti^{3+}$, where there is only one 3d electron. Consider, for example, the $Ti(H_2O)_6^{3+}$ ion, which has an intense purple color. This ion absorbs at 510 nm, in the green region. The purple (red-violet) color of a solution of $Ti(H_2O)_6^{3+}$ is what is left over when the green component is subtracted from the visible spectrum. Using the equation relating energy to wavelength (Chapter 6), it is possible to calculate the energy of a photon with a wavelength of 510 nm:

$$E = \frac{hc}{\lambda} = \frac{(6.626 \times 10^{-34}\ J\cdot s)(2.998 \times 10^8\ m/s)}{510 \times 10^{-9}\ m} = 3.90 \times 10^{-19}\ J$$

The energy in kilojoules per mole is

$$E = 3.90 \times 10^{-19}\ J \times \frac{1\ kJ}{10^3\ J} \times \frac{6.022 \times 10^{23}}{1\ mol} = 2.35 \times 10^2\ kJ/mol$$

This is the energy absorbed in raising the 3d electron from a lower to a higher orbital. In other words, in the $Ti(H_2O)_6^{3+}$ ion, the two sets of d orbitals are separated by this amount of energy. The splitting energy, Δ_o, is 235 kJ/mol.

The smaller the value of Δ_o, the longer the wavelength of the light absorbed. This effect is shown in Table 15.3 and Figure 15.13 (page 452). When weak-field

TABLE 15.3	**Colors of Complex Ions of Co^{3+}**		
Complex	**Color Observed**	**Color Absorbed**	**Approximate Wavelength (nm) Absorbed**
$Co(NH_3)_6^{3+}$	Yellow	Violet	430
$Co(NH_3)_5NCS^{2+}$	Orange	Blue-green	470
$Co(NH_3)_5H_2O^{3+}$	Red	Green-blue	500
$Co(NH_3)_5Cl^{2+}$	Purple	Yellow-green	530
$trans\text{-}Co(NH_3)_4Cl_2^+$	Green	Red	680

FIGURE 15.13

A spectrochemical series. The five cations listed in Table 15.3 vary in color from yellow $[Co(NH_3)_6]Cl_3$ to green [*trans*-$Co(NH_3)_4Cl_2$]Cl. *(Ray Boyington/S. Ruven Smith)*

This series is based on experiment, not theory.

ligands such as NCS^-, H_2O, or Cl^- are substituted for NH_3, the light absorbed shifts to longer wavelengths (lower energies). On the basis of observations like this, ligands can be arranged in order of decreasing tendency to split the d orbitals. A short version of such a *spectrochemical series* is

$$CN^- > NO_2^- > en > NH_3 > NCS^- > H_2O > F^- > Cl^-$$

strong field   decreasing Δ_o weak field

15.4   FORMATION CONSTANTS OF COMPLEX IONS

The equilibrium constant for the formation of a complex ion is called a **formation constant** (or stability constant) and given the symbol K_f. A typical example is

$$Cu^{2+}(aq) + 4NH_3(aq) \rightleftharpoons Cu(NH_3)_4^{2+}(aq) \qquad K_f = \frac{[Cu(NH_3)_4^{2+}]}{[Cu^{2+}] \times [NH_3]^4}$$

Table 15.4 lists formation constants of complex ions. In each case, K_f applies to the formation of the complex by a reaction of the type just cited. Notice that for each complex ion listed, K_f is a large number, 10^5 or greater. This means that equilibrium considerations strongly favor complex formation. Consider, for example, the system

$$Ag^+(aq) + 2NH_3(aq) \rightleftharpoons Ag(NH_3)_2^+(aq)$$

$$K_f = \frac{[Ag(NH_3)_2^+]}{[Ag^+] \times [NH_3]^2} = 1.7 \times 10^7$$

The large K_f value means that the forward reaction goes virtually to completion. Addition of ammonia to a solution of $AgNO_3$ will convert nearly all the Ag^+ ions to $Ag(NH_3)_2^+$.

The stabilities of different complexes of the same cation are directly related to their formation constants; the larger the K_f value, the more stable the complex. For the reaction

$$Ag^+(aq) + 2S_2O_3^{2-}(aq) \rightleftharpoons Ag(S_2O_3)_2^{3-}(aq)$$

$$K_f = \frac{[Ag(S_2O_3)_2^{3-}]}{[Ag^+] \times [S_2O_3^{2-}]^2} = 1 \times 10^{13}$$

This K_f is larger than that for $Ag(NH_3)_2^+$ ($K_f = 1.7 \times 10^7$). Hence the $Ag(S_2O_3)_2^{3-}$ complex ion is more stable than $Ag(NH_3)_2^+$. The reaction

$$Ag(NH_3)_2^+(aq) + 2S_2O_3^{2-}(aq) \longrightarrow Ag(S_2O_3)_2^{3-}(aq) + 2NH_3(aq)$$

TABLE 15.4   Formation Constants of Complex Ions [OHT]

Complex Ion	K_f	Complex Ion	K_f
$AgCl_2^-$	1.8×10^5	$CuCl_2^-$	1×10^5
$Ag(CN)_2^-$	2×10^{20}	$Cu(NH_3)_4^{2+}$	2×10^{12}
$Ag(NH_3)_2^+$	1.7×10^7	$FeSCN^{2+}$	9.2×10^2
$Ag(S_2O_3)_2^{3-}$	1×10^{13}	$Fe(CN)_6^{3-}$	4×10^{52}
$Al(OH)_4^-$	1×10^{33}	$Fe(CN)_6^{4-}$	4×10^{45}
$Cd(CN)_4^{2-}$	2×10^{18}	$Hg(CN)_4^{2-}$	2×10^{41}
$Cd(NH_3)_4^{2+}$	2.8×10^7	$Ni(NH_3)_6^{2+}$	9×10^8
$Cd(OH)_4^{2-}$	1.2×10^9	$PtCl_4^{2-}$	1×10^{16}
$Co(NH_3)_6^{2+}$	1×10^5	$trans\text{-}Pt(NH_3)_2Cl_2$	3×10^{28}
$Co(NH_3)_6^{3+}$	1×10^{23}	$cis\text{-}Pt(NH_3)_2Cl_2$	3×10^{29}
$Co(NH_3)_5Cl^{2+}$	2×10^{28}	$Zn(CN)_4^{2-}$	6×10^{16}
$Co(NH_3)_5NO_2^{2+}$	1×10^{24}	$Zn(NH_3)_4^{2+}$	3.6×10^8
		$Zn(OH)_4^{2-}$	3×10^{14}

is spontaneous with a large equilibrium constant

$$K = \frac{K_f Ag(S_2O_3)_2^{3-}}{K_f Ag(NH_3)_2^+} = \frac{1 \times 10^{13}}{1.7 \times 10^7} = 6 \times 10^5$$

EXAMPLE 15.6   Determine the ratios

(a) $[Ag(NH_3)_2^+]/[Ag^+]$ in 0.10 M NH_3
(b) $[Ag(S_2O_3)_2^{3-}]/[Ag^+]$ in 0.10 M $S_2O_3^{2-}$

Strategy   In each case, write down the expression for K_f and solve for the desired ratio.

Solution

(a) $K_f = 1.7 \times 10^7 = \dfrac{[Ag(NH_3)_2^+]}{[Ag^+] \times [NH_3]^2}$

$\dfrac{[Ag(NH_3)_2^+]}{[Ag^+]} = K_f \times [NH_3]^2 = 1.7 \times 10^7 (0.10)^2 = \boxed{1.7 \times 10^5}$

This means that in 0.10 M NH_3 there are 170,000 $Ag(NH_3)_2^+$ complex ions for every Ag^+ ion.
(b) Proceeding as in (a),

$\dfrac{[Ag(S_2O_3)_2^{3-}]}{[Ag^+]} = K_f \times [S_2O_3^{2-}]^2 = 1 \times 10^{13}(0.10)^2 = \boxed{1 \times 10^{11}}$

Reality Check   Note that this ratio for $Ag(S_2O_3)_2^{3-}$ is much higher than for the comparable $Ag(NH_3)_2^+$ complex. This is really what we mean when we say that the $Ag(S_2O_3)_2^{3-}$ complex is "more stable" than $Ag(NH_3)_2^+$.

⌐ Essentially all of the Ag^+ ions are complexed.

If you've ever developed your own photographic film, you've probably made use of the high stability of the $Ag(S_2O_3)_2^{3-}$ complex ion. In the fixing process, any unreacted silver halide (e.g., AgBr) remaining on the developed film is removed by washing with a solution of sodium thiosulfate, $Na_2S_2O_3$, commonly called "hypo." The following reaction occurs.

$$AgBr(s) + 2S_2O_3^{2-}(aq) \longrightarrow Ag(S_2O_3)_2^{3-}(aq) + Br^-(aq)$$

Chelates: Natural and Synthetic

Chelating agents, capable of bonding to metal atoms at more than one position, are abundant in nature. Certain species of soybeans synthesize and secrete organic chelating agents that extract iron from insoluble compounds in the soil, thereby making it available to the plant. Mosses and lichens growing on boulders use a similar process to obtain the metal ions they need for growth.

Many important natural products are chelates in which a central metal atom is bonded into a large organic molecule. In chlorophyll, the green coloring matter of plants, the central atom is magnesium; in the essential vitamin B_{12}, it is cobalt. The structure of heme, the pigment responsible for the red color of blood, is shown in Figure A. There is an Fe^{2+} ion at the center of the complex surrounded by four nitrogen atoms at the corners of a square. A fifth coordination position around the iron is occupied by an organic molecule globin, which in combination with heme, gives the protein referred to as hemoglobin.

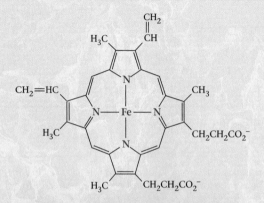

FIGURE A
Structure of heme. In the hemoglobin molecule, the Fe^{2+} ion is at the center of an octahedron, surrounded by four nitrogen atoms, a globin molecule, and a water molecule.

Ribbon molecule of hemoglobin. Hemoglobin picks up oxygen molecules where the structure is shown in red. Iron atoms are shown as white spheres.

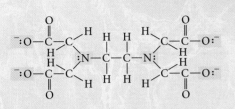

The sixth coordination position of Fe^{2+} in heme is occupied by a water molecule that can be replaced reversibly by oxygen to give a derivative known as oxyhemoglobin, which has the bright red color characteristic of arterial blood.

$$\text{hemoglobin} + O_2 \rightleftharpoons \text{oxyhemoglobin} + H_2O$$

The position of this equilibrium is sensitive to the pressure of oxygen. In the lungs, where the blood is saturated with air (P_{O_2} = 0.2 atm), the hemoglobin is almost completely converted to oxyhemoglobin. In the tissues, the partial pressure of oxygen drops, and the oxyhemoglobin breaks down to release O_2 essential for metabolism. In this way, hemoglobin acts as an oxygen carrier, absorbing oxygen in the lungs and liberating it to the tissues.

Unfortunately, hemoglobin forms a complex with carbon monoxide that is considerably more stable than oxyhemoglobin. The equilibrium constant for the reaction

$$\text{hemoglobin} \cdot O_2(aq) + CO(g) \rightleftharpoons \text{hemoglobin} \cdot CO(aq) + O_2(g)$$

is about 200 at body temperature. Consequently, the carbon monoxide complex is formed preferentially in the lungs even at CO concentrations as low as one part per thousand. When this happens, the flow of oxygen to the tissues is cut off, resulting eventually in muscular paralysis and death.

Many chelating agents have been synthesized in the laboratory to take advantage of their ability to tie up *(sequester)* metal cations. Perhaps the best-known synthetic chelating agent is the ethylenediaminetetraacetate ion (EDTA) shown in Figure B. EDTA can bond at as many as six positions, thereby occupying all the octahedral sites surrounding a central metal cation.

EDTA forms 1:1 complexes with a large number of cations, including those of some of the main-group metals. The complex formed by calcium with EDTA is used to treat lead poisoning. When a solution containing the Ca-EDTA complex is given by injection, the calcium is displaced by lead. The more stable Pb-EDTA complex is eliminated in the urine. EDTA has also been used to remove radioactive isotopes of metals, notably plutonium, from body tissues.

You may have noticed Ca-EDTA on the list of ingredients of many prepared foods, ranging from beer to mayonnaise. EDTA acts as a scavenger to pick up traces of metal ions that catalyze the chemical reactions responsible for flavor deterioration, loss of color, or rancidity. Typically, Ca-EDTA is added at a level of 30 to 800 ppm.

Household products containing EDTA. *(Charles D. Winters)*

CHAPTER HIGHLIGHTS

Key Concepts

1. Relate the composition of a complex ion to its charge, the coordination number, and the oxidation number of the central metal ion.
   (Examples 15.1, 15.2; Problems 1–8)
2. Sketch the geometry of complex ions and identify geometric isomers.
   (Example 15.3; Problems 13–20)
3. Give the electron configuration and/or orbital diagram of a transition metal ion.
   (Example 15.4; Problems 21–24)
4. Distinguish between high-spin and low-spin complexes.
   (Example 15.5; Problems 25–32)
5. Determine the ratio of the concentration of complex ion to that of metal cation, given K_f for the complex.
   (Example 15.6; Problems 37–40, 48)

Key Terms

central metal cation	formation constant	ligand
chelate	geometric isomerism	low-spin complex
cis-	high-spin complex	octahedral
complex ion	isomer	square planar
coordination number	Lewis	tetrahedral
crystal field	—acid	*trans-*
—splitting energy	—base	∎ transition metal

Summary Problem

Consider the complex compound, $Pt(NH_3)_2(CN)_4$.
(a) Identify the ligands and their charges.
(b) What is the oxidation number of platinum?
(c) Suppose the ammonia molecules are replaced by OH^- ions. Write the formula of the resulting complex.
(d) What is the coordination number of platinum in the compound?
(e) Describe the geometry of the compound.
(f) How many possible geometric isomers are there?
(g) Write the abbreviated electron configuration for platinum(IV) ion.
(h) Cyanide ligands are ligands with the strongest crystal field (large Δ_o). Show the electron distribution for platinum in this compound. Is this a high-spin or a low-spin complex?
(i) Is the compound paramagnetic or diamagnetic?
(j) If there were only four ligands, all of them chloride ions, the complex compound would be $PtCl_4$. Write the expression for K for the formation of this complex.
(k) Calculate $[Cl^-]$ in a solution where $[Pt^{4+}] = [PtCl_4] = 0.20\ M$, taking $K_f = 1 \times 10^{16}$.

Answers

(a) $NH_3 = 0$; $CN^- = -1$

(b) $+4$

(c) $[Pt(OH)_2(CN)_4]^{2-}$

(d) 6

(e) octahedral

(f) two

(g) $[Xe]\ 4f^{14}\ 5d^6$

(h) $\underline{\uparrow\downarrow}\ \ \overline{\underline{\uparrow\downarrow}}\ \underline{\uparrow\downarrow}$ low-spin

(i) diamagnetic

(j) $K_f = \dfrac{[PtCl_4]}{[Pt^{4+}][Cl^-]^4}$

(k) $1 \times 10^{-4}\ M$

Questions & Problems

Problem numbers in blue indicate that the answer is available in Appendix 6 at the back of the book.

WEB indicates that the solution is posted at **http://www.harcourtcollege.com/chem/general/masterton4/student/**

Composition of Complex Ions and Coordination Compounds

1. Consider the complex ion $[Co(en)_2(SCN)Cl]^+$.
   (a) Identify the ligands and their charges.
   (b) What is the oxidation number of cobalt?
   (c) What is the formula for the sulfide salt of this ion?

WEB **2.** Consider the complex ion $[Ni(H_2O)_2Cl_2(OH)_2]^{2-}$.
   (a) Identify the ligands and their charges.
   (b) What is the oxidation number of nickel?
   (c) What is the formula for the sodium salt of this ion?

3. Chromium(III) forms many complexes, among them those with the following ligands. Give the formula and charge of each chromium complex ion described below.
   (a) two oxalate ions ($C_2O_4^{2-}$) and two water molecules
   (b) five ammonia molecules and one sulfate ion
   (c) one ethylenediamine molecule, two ammonia molecules, and two iodide ions

4. Platinum(II) forms many complexes, among them those with the following ligands. Give the formula and charge of each complex.
   (a) two ammonia molecules and one oxalate ion ($C_2O_4^{2-}$)
   (b) two ammonia molecules, one thiocyanate ion (SCN^-), and one bromide ion
   (c) one ethylenediamine molecule and two nitrite ions

5. What is the coordination number of the metal in the following complexes?
   (a) $[Co(NH_3)_5(SO_4)]^+$ **(b)** $[Ni(CN)_4]^{2-}$
   (c) $[Mo(Cl)_6]^{3-}$ **(d)** $[Zn(C_2O_4)_2]^{2-}$

6. What is the coordination number of the metal in the following complexes?
   (a) $[Co(en)_2(SCN)Cl]^+$ **(b)** $[Zn(en)(C_2O_4)]$
   (c) $[Ag(NH_3)Cl]$ **(d)** $[Cu(H_2O)_4]^{2+}$

7. Refer to Table 15.2 to predict the formula of the complex formed by
   (a) Pt^{4+} with NH_3 **(b)** Ag^+ with CN^-
   (c) Zn^{2+} with $C_2O_4^{2-}$ **(d)** Cd^{2+} with CN^-

8. Refer to Table 15.2 to predict the formula of the complex formed by
   (a) Ag^+ with H_2O **(b)** Pt^{2+} with H_2O
   (c) Pd^{2+} with Br^- **(d)** Fe^{3+} with $C_2O_4^{2-}$

9. What is the mass percent of nitrogen in the $Co(en)_3^{3+}$ complex ion?

10. What is the mass percent of chromium in the chloride salt of $[Cr(H_2O)_5(OH)]^{2+}$?

11. Vitamin B_{12} is a coordination compound with cobalt as its central atom. It contains 4.4% cobalt by mass and has a molar mass of 1.3×10^3 g/mol. How many cobalt atoms are there in a molecule of vitamin B_{12}?

12. There are four iron atoms in each hemoglobin molecule. The mass percent of iron in a hemoglobin molecule is 0.35%. Estimate the molar mass of hemoglobin.

Geometry of Complex Ions

13. Sketch the geometry of
   (a) $Ag(CN)_2^-$
   (b) $Co(H_2O)_2(Cl)_2$ (tetrahedral)
   (c) cis-$[Ni(H_2O)_2(Cl)_2]^+$
   (d) cis-$[Pt(en)_2(Br_2)]^{2+}$
   (e) $trans$-$[Cr(H_2O)_4(Cl)_2]^+$

14. Sketch the geometry of
   (a) cis-$[Cu(H_2O)_2(Br)_4]^{2-}$
   (b) $Zn(en)(Cl)_2$ (tetrahedral)
   (c) $trans$-$[Ni(NH_3)_2(en)_2]^{2+}$
   (d) $[Cu(C_2O_4)_3]^{4-}$
   (e) $trans$-$Ni(H_2O)_2(Cl)_2$

15. The acetylacetonate ion ($acac^-$)

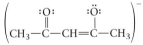

forms complexes with many metal ions. Sketch the geometry of $Fe(acac)_3$.

WEB 16. The compound 1,2-diaminocyclohexane

(abbreviated "dech") is a ligand in the promising anticancer complex *cis*-Pd(H_2O)$_2$(dech)$^{2+}$. Sketch the geometry of this complex.

17. Which of the following octahedral complexes show geometric isomerism? If geometric isomers are possible, draw their structures.
    (a) [Co(*en*)Cl$_4$]$^-$
    (b) [Ni(C_2O_4)$_2$ClBr]$^{4-}$
    (c) [Cd(NH_3)$_2$Cl$_4$]$^{2-}$

18. Follow the directions of Question 17 for the following.
    (a) [Cr(NH_3)$_2$(SCN)$_4$]$^{2-}$
    (b) [Co(NH_3)$_3$(NO_2)$_3$]$^-$
    (c) [Mn(H_2O)$_2$(NH_3)$_3$OH]$^+$

19. Draw all the structural formulas for the octahedral complexes of Cr^{3+} with only *en* and/or H_2O as ligands.

20. Draw all the structural formulas for the octahedral complexes of [Fe(NH_3)$_3$Cl$_2$Br].

Electronic Structure of Metal Ions

21. Give the electronic configuration for
    (a) Fe^{3+} (b) V^{2+} (c) Zn^{2+}
    (d) Cu^+ (e) Mn^{4+}

WEB 22. Give the electronic configuration for
    (a) Ti^{3+} (b) Cr^{2+} (c) Ru^{4+}
    (d) Pd^{2+} (e) Mo^{3+}

23. Write an abbreviated orbital diagram and determine the number of unpaired electrons in each species in Question 21.

24. Write an abbreviated orbital diagram and determine the number of unpaired electrons in each species in Question 22.

Electron Distributions and Crystal-Field Energy

25. Give the electron distribution in low-spin and high-spin complexes of
    (a) Co^{3+} (b) Mn^{2+}

26. Follow the directions of Question 25 for
    (a) Fe^{2+} (b) Zn^{2+}

27. Explain why Mn^{3+} forms high-spin and low-spin octahedral complexes but Mn^{4+} does not.

28. For complexes of V^{3+}, only one distribution of electrons is possible. Explain.

29. [Cr(CN)$_6$]$^{4-}$ is less paramagnetic than [Cr(H_2O)$_6$]$^{2+}$. Explain.

30. Why is [Co(NH_3)$_6$]$^{3+}$ diamagnetic while [CoF$_6$]$^{3-}$ is paramagnetic?

31. Give the number of unpaired electrons in octahedral complexes with weak-field ligands for
    (a) Mn^{3+} (b) Co^{3+} (c) Rh^{3+}
    (d) Ti^{2+} (e) Mo^{2+}

32. For the species in Question 31, indicate the number of unpaired electrons with strong-field ligands.

33. MnF$_6$$^{2-}$ has a crystal field splitting energy, Δ_o, of 2.60×10^2 kJ/mol. What is the wavelength responsible for this energy?

34. Ti(NH_3)$_6$$^{3+}$ has a d-orbital electron transition at 399 nm. Find Δ_o at this wavelength.

35. The wavelength of maximum absorption of Cu(NH_3)$_4$$^{2+}$ is 580 nm (orange-yellow). What color is a solution of Cu(NH_3)$_4$$^{2+}$?

36. A solution of Fe(CN)$_6$$^{3-}$ appears red. Using Table 1.4 (page 19), estimate the wavelength of maximum absorption.

Formation Constants of Complex Ions

37. At what concentration of $S_2O_3^{2-}$ is 75% of the Ag^+ in a solution converted to Ag(S_2O_3)$_2$$^{3-}$?

WEB 38. Use the data in Table 15.4 to calculate the ratio [Al^{3+}]/[Al(OH)$_4$$^-$] at pH 2.0, 7.0, and 11.0.

39. At what concentration of cyanide ion is
    (a) [Cd^{2+}] = $10^{-8} \times$ [Cd(CN)$_4$$^{2-}$]?
    (b) [Fe^{2+}] = $10^{-20} \times$ [Fe(CN)$_6$$^{4-}$]?

40. At what concentration of ammonia is
    (a) [Cd^{2+}] = [Cd(NH_3)$_4$$^{2+}$]?
    (b) [Co^{2+}] = [Co(NH_3)$_6$$^{2+}$]?

Unclassified

41. A chemist synthesizes two coordination compounds. One compound decomposes at 280°C, the other at 240°C. Analysis of the compounds gives the same mass percent data: 52.6% Pt, 7.6% N, 1.63% H, and 38.2% Cl. Both compounds contain a +4 central ion.
    (a) What is the simplest formula of the compounds?
    (b) Draw structural formulas for the complexes present.

42. Analysis of a coordination compound gives the following results: 22.0% Co, 31.4% N, 6.78% H, and 39.8% Cl. One mole of the compound dissociates in water to form four moles of ions.
    (a) What is the simplest formula of the compound?
    (b) Write an equation for its dissociation in water.

43. For the system

$$\text{hemoglobin} \cdot O_2(aq) + CO(g) \rightleftharpoons \text{hemoglobin} \cdot CO(aq) + O_2(g)$$

$K = 2.0 \times 10^2$. What must be the ratio of P_{CO}/P_{O_2} if 12.0% of the hemoglobin in the bloodstream is converted to the CO complex?

44. Oxyhemoglobin is red, and hemoglobin is blue. Use Le Châtelier's principle to explain why venous blood is blue and arterial blood is bright red.

45. What are the concentrations of Cu^{2+}, NH_3, and Cu(NH_3)$_4$$^{2+}$ at equilibrium when 18.8 g of Cu(NO_3)$_2$ is added to 1.0 L of a 0.400 M solution of aqueous ammonia? Assume that the reaction goes to completion and forms Cu(NH_3)$_4$$^{2+}$.

Conceptual Problems

46. Explain why
    (a) oxalic acid removes rust stains.
    (b) there are no geometric isomers of tetrahedral complexes.
    (c) cations such as Co^{2+} act as Lewis acids.
    (d) $C_2O_4^{2-}$ is a chelating agent.
    (e) NH_3 can be a ligand but NH_4^+ is not.

47. Indicate whether each of the following is true or false. If the statement is false, correct it.
    (a) The coordination number of iron(III) in $Fe(NH_3)_4(en)^{3+}$ is 5.
    (b) $Ni(CN)_6^{4-}$ is expected to absorb at a longer wavelength than $Ni(NH_3)_6^{2+}$.
    (c) Complexes of Cr^{3+} are brightly colored, and those of Zn^{2+} are colorless.
    (d) Ions with seven or more d electrons cannot form both high- and low-spin octahedral complexes.

48. Consider three complexes of Ag^+ and their formation constants, K_f.

Complex ion	K_f
$Ag(NH_3)_2^+$	1.6×10^7
$Ag(CN)_2^-$	5.6×10^{18}
$AgBr_2^-$	1.3×10^7

Which statements are true?
    (a) $Ag(NH_3)_2^+$ is more stable than $Ag(CN)_2^-$.
    (b) Adding a strong acid (HNO_3) to a solution that is 0.010 M in $Ag(NH_3)_2^+$ will tend to dissociate the complex ion into Ag^+ and NH_4^+.
    (c) Adding a strong acid (HNO_3) to a solution that is 0.010 M in $AgBr_2^-$ will tend to dissociate the complex ion into Ag^+ and Br^-.
    (d) To dissolve AgI, one can add either NaCN or HCN as a source of the cyanide-complexing ligand. Fewer moles of NaCN would be required.
    (e) Solution A is 0.10 M in Br^- and contains the complex ion $AgBr_2^-$. Solution B is 0.10 M in CN^- and contains the complex ion $Ag(CN)_2^-$. Solution B will have more particles of complex ion per particle of Ag^+ than solution A.

Challenge Problems

49. A child eats 10.0 g of paint containing 5.0% Pb. How many grams of the sodium salt of EDTA, $Na_4(EDTA)$, should he receive to bring the lead into solution as $Pb \cdot EDTA$?

50. A certain coordination compound has the simplest formula $PtN_2H_6Cl_2$. It has a molar mass of about 600 g/mol and contains both a complex cation and a complex anion. What is its structure?

51. Two coordination compounds decompose at different temperatures but have the same mass percent analysis data: 20.25% Cu, 15.29% C, 7.07% H, 26.86% N, 10.23% S, and 20.39% O. Each contains Cu^{2+}.
    (a) Determine the simplest formula of the compounds.
    (b) Draw the structural formulas of the complex ion in each case.

52. In the $Ti(H_2O)_6^{3+}$ ion, the splitting between the d levels, Δ_o, is 55 kcal/mol. What is the color of this ion? Assume that the color results from a transition between upper and lower d levels.

*S*urrounded by beakers, by strange coils,
By ovens and flasks with twisted necks,
The chemist, fathoming the whims of attractions,
Artfully imposes on them their precise meetings.

—Sully-Prudhomme
The Naked World (translated by William Dock)

16 PRECIPITATION EQUILIBRIA

The minerals shown contain insoluble sulfides of iron, arsenic,
and lead (clockwise from left).
(Charles D. Winters)

CHAPTER OUTLINE

16.1 PRECIPITATE FORMATION; SOLUBILITY PRODUCT CONSTANT (K_{sp})

16.2 DISSOLVING PRECIPITATES

16.3 QUALITATIVE ANALYSIS

In this chapter, we consider two types of equilibria, both in water solution—

■ that between a precipitate and its ions (Section 16.1), for example,

$$AgCl(s) \rightleftharpoons Ag^+(aq) + Cl^-(aq)$$

■ that between a precipitate and a species used to dissolve it (Section 16.2), for example,

$$AgCl(s) + 2NH_3(aq) \rightleftharpoons Ag(NH_3)_2^+(aq) + Cl^-(aq)$$

Equilibria such as these have applications in fields as diverse as geology, medicine, and agriculture. In chemistry you are most likely to meet up with precipitation equilibria in the laboratory when you carry out experiments in qualitative analysis (Section 16.3).

FIGURE 16.1
Precipitation of strontium chromate. A solution prepared by mixing solutions of $Sr(NO_3)_2$ and K_2CrO_4 is in equilibrium with yellow $SrCrO_4(s)$. In such a solution $[Sr^{2+}] \times [CrO_4^{2-}] = 3.6 \times 10^{-5}$. *(Charles D. Winters)*

See the *Saunders Interactive General Chemistry CD-ROM*, Screen 19.2, **Precipitation Reactions.**

16.1 PRECIPITATE FORMATION; SOLUBILITY PRODUCT CONSTANT (K_{sp})

As we saw in Chapter 4, a precipitate forms when a cation from one solution combines with an anion from another solution to form an insoluble ionic solid. We also considered how to predict whether such a reaction would occur and, if so, how to represent it by a net ionic equation.

Precipitation reactions, like all reactions, reach a position of equilibrium. Suppose, for example, solutions of $Sr(NO_3)_2$ and K_2CrO_4 are mixed. In this case, Sr^{2+} ions combine with CrO_4^{2-} ions to form a yellow precipitate of strontium chromate, $SrCrO_4$ (Figure 16.1). Very quickly, an equilibrium is established between the solid and the corresponding ions in solution:

$$SrCrO_4(s) \rightleftharpoons Sr^{2+}(aq) + CrO_4^{2-}(aq)$$

K_{sp} Expression

The equilibrium constant expression for the dissolving of $SrCrO_4$ can be written following the rules cited in Chapters 12 and 13. In particular, the solid does not appear in the expression; the concentration of each ion is raised to a power equal to its coefficient in the chemical equation.

$$K_{sp} = [Sr^{2+}] \times [CrO_4^{2-}]$$

See Screen 19.4, **Solubility Product Constant.**

Although the solid doesn't appear in K_{sp}, it must be present for equilibrium.

The symbol K_{sp} represents a particular type of equilibrium constant known as the **solubility product constant.** Like all equilibrium constants, K_{sp} has a fixed value for a given system at a particular temperature. At 25°C, K_{sp} for $SrCrO_4$ is about 3.6×10^{-5}; that is,

$$[Sr^{2+}] \times [CrO_4^{2-}] = 3.6 \times 10^{-5}$$

This relation says that the product of the two ion concentrations at equilibrium must be 3.6×10^{-5}, regardless of how equilibrium is established.

EXAMPLE 16.1   Write expressions for K_{sp} for

(a) $PbCl_2$ (b) Ag_2CrO_4

Strategy   Start by writing the chemical equation for the solution process (solid on the left, ions in solution on the right). Then write the expression for K_{sp}, noting that

—solids do not appear.
—the concentration of each ion is raised to a power equal to its coefficient in the chemical equation.

Solution

(a) $PbCl_2(s) \rightleftharpoons Pb^{2+}(aq) + 2Cl^-(aq)$

$\quad K_{sp} = [Pb^{2+}] \times [Cl^-]^2$

(b) $Ag_2CrO_4(s) \rightleftharpoons 2Ag^+(aq) + CrO_4^{2-}(aq)$

$\quad K_{sp} = [Ag^+]^2 \times [CrO_4^{2-}]$

K_{sp} and the Equilibrium Concentrations of Ions

The relation

See Screen 19.5, Determining K_{sp}.

$$K_{sp} \, SrCrO_4 = [Sr^{2+}] \times [CrO_4^{2-}] = 3.6 \times 10^{-5}$$

can be used to calculate the equilibrium concentration of one ion if you know that of the other. Suppose, for example, the concentration of CrO_4^{2-} in a certain solution in equilibrium with $SrCrO_4$ is known to be 2.0×10^{-3} M. It follows that

$$[Sr^{2+}] = \frac{K_{sp} \, SrCrO_4}{[CrO_4^{2-}]} = \frac{3.6 \times 10^{-5}}{2.0 \times 10^{-3}} = 1.8 \times 10^{-2} \, M$$

If in another case, $[Sr^{2+}] = 1.0 \times 10^{-4}$ M,

$$[CrO_4^{2-}] = \frac{K_{sp} \, SrCrO_4}{[Sr^{2+}]} = \frac{3.6 \times 10^{-5}}{1.0 \times 10^{-4}} = 3.6 \times 10^{-1} \, M$$

Example 16.2 illustrates the same kind of calculation for a different electrolyte; the math is a bit more difficult, but the principle is the same.

EXAMPLE 16.2   Calcium phosphate, $Ca_3(PO_4)_2$, is a water-insoluble mineral, large quantities of which are used to make commercial fertilizers. Taking its K_{sp} value from Table 16.1, calculate

(a) the concentration of PO_4^{3-} in equilibrium with the solid if $[Ca^{2+}] = 1 \times 10^{-9}\ M$.
(b) the concentration of Ca^{2+} in equilibrium with the solid if $[PO_4^{3-}] = 1 \times 10^{-5}\ M$.

Strategy   The first step is to write down the K_{sp} expression:

$$Ca_3(PO_4)_2(s) \rightleftharpoons 3Ca^{2+}(aq) + 2PO_4^{3-}(aq)$$

$$K_{sp} = [Ca^{2+}]^3 \times [PO_4^{3-}]^2 = 1 \times 10^{-33}$$

Now substitute the concentration of one ion and solve for that of the other.

Solution

(a) $[PO_4^{3-}]^2 = \dfrac{1 \times 10^{-33}}{[Ca^{2+}]^3} = \dfrac{1 \times 10^{-33}}{(1 \times 10^{-9})^3} = 1 \times 10^{-6}$ $[PO_4^{3-}] = \boxed{1 \times 10^{-3}\ M}$

(b) $[Ca^{2+}]^3 = \dfrac{1 \times 10^{-33}}{[PO_4^{3-}]^2} = \dfrac{1 \times 10^{-33}}{(1 \times 10^{-5})^2} = 1 \times 10^{-23}$ $[Ca^{2+}] = \boxed{2 \times 10^{-8}\ M}$

(To find a cube root on your calculator, use the $\boxed{y^x}$ key, where $x = 1/3 = 0.333333.\ \dots$)   You can also use the $\boxed{\sqrt[x]{y}}$ key, where $x = 3$.

Reality Check   To check your answer, use the calculated concentration to determine K_{sp}. For example, in part (b), $[Ca^{2+}]^3 \times [PO_4^{3-}]^2 = (2 \times 10^{-8})^3 \times (1 \times 10^{-5})^2 = 8 \times 10^{-34}$, which is very close to the value listed in Table 16.1, 1×10^{-33}.

TABLE 16.1   Solubility Product Constants at 25°C OHT

		K_{sp}			K_{sp}
Acetates	$AgC_2H_3O_2$	1.9×10^{-3}	Hydroxides	$Al(OH)_3$	2×10^{-31}
				$Fe(OH)_2$	5×10^{-17}
Bromides	$AgBr$	5×10^{-13}		$Fe(OH)_3$	3×10^{-39}
	Hg_2Br_2	6×10^{-23}		$Mg(OH)_2$	6×10^{-12}
	$PbBr_2$	6.6×10^{-6}		$Tl(OH)_3$	2×10^{-44}
Carbonates	Ag_2CO_3	8×10^{-12}		$Zn(OH)_2$	4×10^{-17}
	$BaCO_3$	2.6×10^{-9}		$Ca(OH)_2$	4.0×10^{-6}
	$CaCO_3$	4.9×10^{-9}			
	$MgCO_3$	6.8×10^{-6}			
	$SrCO_3$	5.6×10^{-10}	Iodides	AgI	1×10^{-16}
	$PbCO_3$	1×10^{-13}		Hg_2I_2	5×10^{-29}
				PbI_2	8.4×10^{-9}
Chlorides	$AgCl$	1.8×10^{-10}			
	Hg_2Cl_2	1×10^{-18}			
	$PbCl_2$	1.7×10^{-5}	Phosphates	Ag_3PO_4	1×10^{-16}
				$AlPO_4$	1×10^{-20}
Chromates	Ag_2CrO_4	1×10^{-12}		$Ca_3(PO_4)_2$	1×10^{-33}
	$BaCrO_4$	1.2×10^{-10}		$Mg_3(PO_4)_2$	1×10^{-24}
	$PbCrO_4$	2×10^{-14}			
	$SrCrO_4$	3.6×10^{-5}			
Fluorides	BaF_2	1.8×10^{-7}	Sulfates	$BaSO_4$	1.1×10^{-10}
	CaF_2	1.5×10^{-10}		$CaSO_4$	7.1×10^{-5}
	MgF_2	7×10^{-11}		$PbSO_4$	1.8×10^{-8}
	PbF_2	7.1×10^{-7}		$SrSO_4$	3.4×10^{-7}

K_{sp} and Precipitate Formation

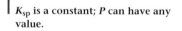

See Screen 19.7, Can a Precipitation Reaction Occur?

K_{sp} values can be used to make a general prediction concerning precipitate formation. To do this, a quantity called the **ion product, P,** is compared with the solubility product constant, K_{sp}. The ion product, P, is entirely analogous to the reaction quotient, Q, discussed in Chapter 12. The form of the expression for P is exactly the same as that for the equilibrium constant, K_{sp}. The difference is that the concentrations used to calculate P are the initial values that apply at a particular moment. Those that appear in K_{sp} are equilibrium concentrations. Putting it another way, the value of P is expected to change as a precipitation reaction proceeds, approaching K_{sp} and eventually becoming equal to it.

Three cases can be distinguished (Figure 16.2):

K_{sp} is a constant; P can have any value.

1. If $P > K_{sp}$, the solution contains a higher concentration of ions than it can hold at equilibrium. *A precipitate forms,* decreasing the concentrations until the ion product becomes equal to K_{sp} and equilibrium is established.
2. If $P < K_{sp}$, the solution contains a lower concentration of ions than is required for equilibrium with the solid. The solution is unsaturated. *No precipitate forms;* equilibrium is not established.
3. If $P = K_{sp}$, the solution is just saturated with ions and is at the point of precipitation.

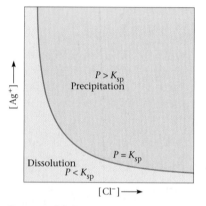

FIGURE 16.2

Equilibrium curve for silver chloride. Silver chloride(s) is in contact with Ag^+ and Cl^- ions in aqueous solution. The product P of the concentration of the ions $[Ag^+] \times [Cl^-]$ is equal to K_{sp} (curved line) when equilibrium exists. If $P > K_{sp}$, $AgCl(s)$ tends to precipitate out until equilibrium is reached. If $P < K_{sp}$, additional solid dissolves. **OHT**

EXAMPLE 16.3
Sodium chromate is added to a solution in which the original concentration of Sr^{2+} is 1.0×10^{-3} M. Assuming $[Sr^{2+}]$ stays constant,

(a) will a precipitate of $SrCrO_4$ ($K_{sp} = 3.6 \times 10^{-5}$) form when the concentration of $CrO_4^{2-} = 3.0 \times 10^{-2}$ M?

(b) will a precipitate form when the concentration of CrO_4^{2-} is 5.0×10^{-2} M?

Strategy First calculate P, using the concentration quoted. Then compare with K_{sp} and follow the rules just cited.

Solution

(a) $P = (1.0 \times 10^{-3}) \times (3.0 \times 10^{-2}) = 3.0 \times 10^{-5}$

Because P is less than K_{sp}, 3.6×10^{-5}, no precipitate forms.

(b) $P = (1.0 \times 10^{-3}) \times (5.0 \times 10^{-2}) = 5.0 \times 10^{-5}$ $P > K_{sp}$

A precipitate forms, reducing the concentrations of Sr^{2+} and CrO_4^{2-} until the ion product becomes equal to 3.6×10^{-5}.

The two solutions dilute each other.

The calculations in Example 16.3 were simplified by assuming that the concentration of one ion (Sr^{2+}) stayed constant while the other ion (CrO_4^{2-}) was added. That assumption will not hold if precipitation comes about through the mixing of two solutions. In that case, the concentrations of the ions present will decrease as volume increases. The concentrations used to calculate the ion product P must be those obtained after mixing, that is, in the total volume.

EXAMPLE 16.4
A student mixes 0.200 L of 0.0060 M $Sr(NO_3)_2$ solution with 0.100 L of 0.015 M K_2CrO_4 solution to give a total volume of 0.300 L. Will a precipitate of $SrCrO_4$ ($K_{sp} = 3.6 \times 10^{-5}$) form under these conditions?

Strategy   This is similar to Example 16.3 with one twist: You must first calculate the concentration of each ion after mixing, taking the total volume into account. To do this, it's convenient to find the number of moles of each ion and then divide by the volume after mixing, 0.300 L.

Solution

$$n_{Sr^{2+}} = 0.200 \text{ L} \times 0.0060 \text{ mol/L} = 1.2 \times 10^{-3} \text{ mol}$$

$$[Sr^{2+}] = 1.2 \times 10^{-3} \text{ mol}/0.300 \text{ L} = 4.0 \times 10^{-3} \text{ } M$$

$$n_{CrO_4^{2-}} = 0.100 \text{ L} \times 0.015 \text{ mol/L} = 1.5 \times 10^{-3} \text{ mol}$$

$$[CrO_4^{2-}] = 1.5 \times 10^{-3} \text{ mol}/0.300 \text{ L} = 5.0 \times 10^{-3} \text{ } M$$

$$P = (4.0 \times 10^{-3})(5.0 \times 10^{-3}) = 2.0 \times 10^{-5}$$

Because P is less than K_{sp} (3.6×10^{-5}), no precipitate forms.

Reality Check   A different way to solve this problem is to calculate the concentrations of the ions by using the relation $(V_{final} \times M_{final}) = (V_{initial} \times M_{initial})$. If you do this, you get the same answer, which is reassuring.

The development we have just gone through is the basis of an experimental method of separating ions in solution. Suppose, for example, that we slowly add SO_4^{2-} ions to a solution that is 0.10 M in both Ba^{2+} and Ca^{2+}. Barium sulfate, which is much less soluble than calcium sulfate, will start to precipitate when the concentration of SO_4^{2-} is very small, about $1.1 \times 10^{-9} M$.

$$P = (1.0 \times 10^{-1}) \times (1.1 \times 10^{-9}) = 1.1 \times 10^{-10} = K_{sp}BaSO_4$$

> The less soluble solid precipitates first; the more soluble solid stays in solution.

At this point, calcium ions will stay in solution, because

$$P = 1.1 \times 10^{-10} < K_{sp}CaSO_4 (7.1 \times 10^{-5})$$

As more sulfate ions are added, $BaSO_4$ will continue to precipitate. In contrast, $CaSO_4$ will not precipitate until the concentration of SO_4^{2-} reaches $7.1 \times 10^{-4} M$.

$$[SO_4^{2-}] = \frac{K_{sp}CaSO_4}{[Ca^{2+}]} = \frac{7.1 \times 10^{-5}}{1.0 \times 10^{-1}} = 7.1 \times 10^{-4} \text{ } M$$

By this time, essentially all the Ba^{2+} ions (about 99.9998%) will have been removed from solution as $BaSO_4$.

The process just described is referred to as *fractional precipitation*. It is particularly useful in separating cations from one another in qualitative analysis (Section 16.3). The calculations involved are further illustrated in Example 16.10, page 473.

K_{sp} and Water Solubility

At least in principle, the water solubility, s, of an ionic compound in moles per liter can be calculated from its solubility product constant, K_{sp}. The approach followed is illustrated in Example 16.5.

> The relationship between K_{sp} and s depends on the type of solid.

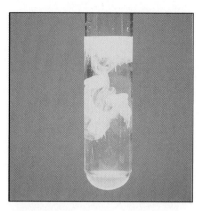

(a)

(b)

Barium sulfate crystals. (a) Precipitation of barium sulfate from a mixture of barium chloride and sulfuric acid. (b) A crystal of the mineral barite, which is mostly barium sulfate. *(Charles D. Winters)*

EXAMPLE 16.5 Calculate the water solubility of

(a) $BaSO_4$ ($K_{sp} = 1.1 \times 10^{-10}$) in moles per liter   (b) BaF_2 ($K_{sp} = 1.8 \times 10^{-7}$) in grams per liter

Strategy To relate molar solubility, s, to K_{sp}, start by writing the equation for the dissolving of the solid in water:

$$BaSO_4(s) \longrightarrow Ba^{2+}(aq) + SO_4{}^{2-}(aq)$$

$$BaF_2(s) \longrightarrow Ba^{2+}(aq) + 2F^-(aq)$$

Now express the equilibrium concentration of each ion in terms of s and substitute into the expression for K_{sp}. Knowing the numerical value of K_{sp}, solve for s. In (b), an additional conversion (moles to grams) is required.

Solution

(a) For every mole of $BaSO_4$ that dissolves, 1 mol of Ba^{2+} and 1 mol of $SO_4{}^{2-}$ enter the solution. So the concentrations of Ba^{2+} and $SO_4{}^{2-}$ are equal to each other and to the molar solubility.

$$[Ba^{2+}] = s \qquad [SO_4{}^{2-}] = s$$

$$K_{sp} = [Ba^{2+}] \times [SO_4{}^{2-}] = s \times s = s^2$$

$$s = (K_{sp})^{1/2} = (1.1 \times 10^{-10})^{1/2} = \boxed{1.0 \times 10^{-5} \text{ mol/L}}$$

(b) For every mole of BaF_2 that dissolves, 1 mol of Ba^{2+} and 2 mol of F^- form.

$$[Ba^{2+}] = s \qquad [F^-] = 2s$$

$$K_{sp} = [Ba^{2+}] \times [F^-]^2 = s \times (2s)^2 = 4s^3$$

$$s = (K_{sp}/4)^{1/3} = (1.8 \times 10^{-7}/4)^{1/3} = 3.6 \times 10^{-3} \text{ mol/L}$$

To find the solubility in grams per liter, note that the molar mass of BaF_2 is 175.3 g/mol. So

$$\text{solubility} = \frac{3.6 \times 10^{-3} \text{ mol } BaF_2}{1 \text{ L}} \times \frac{175.3 \text{ g } BaF_2}{1 \text{ mol } BaF_2} = \boxed{0.63 \text{ g } BaF_2/\text{L}}$$

Frequently we find that the experimentally determined solubility of an ionic solid is larger than that predicted from K_{sp}. Consider, for example, $PbCl_2$, where the solubility calculated from the relation

$$4s^3 = K_{sp} = 1.7 \times 10^{-5}$$

Sulfide minerals and K_{sp}. Sulfides are among the least soluble ionic compounds. Their K_{sp} values are often smaller than 10^{-25}. For this reason, many sulfides are found as minerals, for example *(clockwise from the left)*, iron pyrite (FeS_2), yellow orpiment (As_2S_3), and black galena (PbS). *(Charles D. Winters)*

is 0.016 mol/L. The measured solubility is considerably larger, 0.036 mol/L. The explanation for this is that some of the lead in $PbCl_2$ goes into solution in the form of species other than Pb^{2+}. In particular, we can detect the presence of the ions $Pb(OH)^+$ and $PbCl^+$.

K_{sp} and the Common Ion Effect

See Screen 19.8, The Common Ion Effect.

How would you expect the solubility of barium sulfate in water

$$BaSO_4(s) \rightleftharpoons Ba^{2+}(aq) + SO_4^{2-}(aq)$$

to compare with that in a 0.10 M solution of Na_2SO_4, which contains the same (*common*) anion, SO_4^{2-}? A moment's reflection should convince you that the solubility in 0.10 M Na_2SO_4 must be *less* than that in pure water. Recall (Example 16.5) that when $BaSO_4$ is dissolved in pure water, $[SO_4^{2-}]$ is 1.0×10^{-5} M. Increasing the concentration of SO_4^{2-} to 0.10 M should, by Le Châtelier's principle, drive the above equilibrium to the *left*, repressing the solubility of barium sulfate. This is, indeed, the case (Example 16.6).

A "common" ion comes from two sources such as $BaSO_4$ and Na_2SO_4.

EXAMPLE 16.6   Taking K_{sp} of $BaSO_4$ to be 1.1×10^{-10}, estimate its solubility (moles per liter) in 0.10 M Na_2SO_4 solution.

Strategy   Let s = solubility = no. of moles of $BaSO_4$ that dissolve per liter. Relate $[Ba^{2+}]$ and $[SO_4^{2-}]$ to s; notice that in this case *there are two different sources of SO_4^{2-}*. Substitute into the expression for K_{sp} and solve for s.

Solution   Every mole of $BaSO_4$ that dissolves forms 1 mol of Ba^{2+} and 1 mol of SO_4^{2-}. Because there was no Ba^{2+} present originally, its concentration becomes s:

$$[Ba^{2+}] = s$$

The concentration of SO_4^{2-} was originally 0.10 M; it increases by s moles per liter. So

$$[SO_4^{2-}] = 0.10 + s$$

Setting up this reasoning in tabular form:

	$[Ba^{2+}]$	$[SO_4^{2-}]$
Original	0	0.10
Change	+s	+s
Equilibrium	s	0.10 + s

It follows that

$$K_{sp} = 1.1 \times 10^{-10} = [Ba^{2+}] \times [SO_4^{2-}] = s(0.10 + s)$$

This is a quadratic equation; as usual, we look for ways to avoid solving it by brute force. Note that because $BaSO_4$ is very insoluble, a reasonable approximation would seem to be

$$s + 0.10 \approx 0.10$$

With that assumption,

$$s(0.10) = 1.1 \times 10^{-10}; \qquad s = \boxed{1.1 \times 10^{-9} \text{ mol/L}}$$

The solubility is indeed much less than 0.10 M, so the approximation is justified.

Reality Check   The solubility in 0.10 M Na_2SO_4 is much less than that in pure water ($1.1 \times 10^{-9} \ll 1.0 \times 10^{-5}$), which is exactly what we predicted.

Common ion effect. The tube at the left contains a saturated solution of silver acetate ($AgC_2H_3O_2$). Originally the tube at the right also contained a saturated solution of silver acetate. With the addition of a solution of silver nitrate ($AgNO_3$), the solubility equilibrium of the silver acetate is shifted by the common ion Ag^+ and additional silver acetate precipitates. *(Charles D. Winters)*

FIGURE 16.3
Sodium chloride and the common ion effect. The flask contains a saturated solution of NaCl. Addition of either sodium hydroxide (a), which contains the Na^+ common ion, or hydrochloric acid (b), which contains the Cl^- common ion, causes NaCl to precipitate from the solution. *(Marna G. Clarke)*

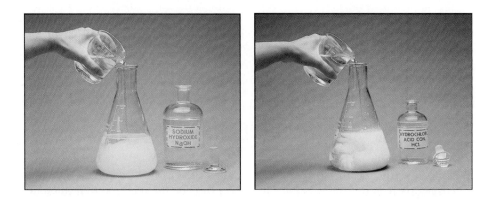

The effect illustrated in Example 16.6 is a general one. *An ionic solid is less soluble in a solution containing a **common ion** than it is in water* (Figure 16.3).

16.2 DISSOLVING PRECIPITATES

Many different methods can be used to bring water-insoluble ionic solids into solution. Most commonly, this is done by adding a reagent to react with either the anion or the cation. The two most useful reagents for this purpose are—

- a strong acid, H^+, used to react with basic anions.
- a complexing agent, most often NH_3 or OH^-, added to react with metal cations.

See Screen 19.11, Solubility and pH.

Strong Acid

Water-insoluble metal hydroxides can be brought into solution with a strong acid such as HCl. The reaction with zinc hydroxide is typical:

$$Zn(OH)_2(s) + 2H^+(aq) \longrightarrow Zn^{2+}(aq) + 2H_2O$$

We can imagine that this reaction occurs in two (reversible) steps:

dissolving $Zn(OH)_2$ in water: $Zn(OH)_2(s) \rightleftharpoons Zn^{2+}(aq) + 2OH^-(aq)$

neutralizing OH^- ions by H^+: $\dfrac{2H^+(aq) + 2OH^-(aq) \rightleftharpoons 2H_2O}{Zn(OH)_2(s) + 2H^+(aq) \rightleftharpoons Zn^{2+}(aq) + 2H_2O}$

The equilibrium constant for the neutralization (Example 16.7) is so large that the overall reaction goes essentially to completion.

$BaSO_4$ cannot be dissolved by acid. Explain.

EXAMPLE 16.7   Consider the following reaction:

$$Zn(OH)_2(s) + 2H^+(aq) \rightleftharpoons Zn^{2+}(aq) + 2H_2O$$

(a) Determine K for this system, applying the rule of multiple equilibria (page 356) to the two-step process referred to above.
(b) Using K, calculate the molar solubility, s, of $Zn(OH)_2$ in acid at pH 5.0.

Strategy   Start by splitting the overall equation into two simpler equations, one for the dissolving of $Zn(OH)_2$ in water, the other for the neutralization of OH^- by H^+ ions. Find the equilibrium constants K_1 and K_2 for the two reactions. Then apply the relation $K = K_1 \times K_2$ to find the equilibrium constant for the overall reaction.

Solution

(a) The first reaction is

$$Zn(OH)_2(s) \rightleftharpoons Zn^{2+}(aq) + 2OH^-(aq)$$

Hence $K_1 = K_{sp}\, Zn(OH)_2$. The second reaction,

$$2H^+(aq) + 2OH^-(aq) \rightleftharpoons 2H_2O$$

is the *reverse* of that for the ionization of water ($H_2O \rightleftharpoons H^+(aq) + OH^-(aq)$) with all the coefficients multiplied by *two*. Applying the reciprocal rule and the coefficient rule (Chapter 12) in succession: $K_2 = 1/K_W{}^2$.

Hence

$$K = K_1 \times K_2 = \frac{K_{sp}Zn(OH)_2}{K_W{}^2} = \frac{4 \times 10^{-17}}{(1 \times 10^{-14})^2} = \boxed{4 \times 10^{11}}$$

The fact that the equilibrium constant is huge implies that $Zn(OH)_2$ is extremely soluble in strong acid.

(b) For the reaction $Zn(OH)_2(s) + 2H^+(aq) \rightleftharpoons Zn^{2+}(aq) + 2H_2O$:

$$K = \frac{[Zn^{2+}]}{[H^+]^2}$$

For every mole of $Zn(OH)_2$ that dissolves, a mole of Zn^{2+} is formed, in other words, $[Zn^{2+}] = s$. Hence

$$s = [H^+]^2 \times K$$

At pH 5.0,

$$[H^+] = 1 \times 10^{-5}\ M;$$

$$s = (1 \times 10^{-5})^2(4 \times 10^{11}) = \boxed{4 \times 10^1\ mol/L}$$

Reality Check   This extremely high concentration, 40 *M*, means that $Zn(OH)_2$ is completely soluble in acid at pH 5 (or any lower pH).

Strong acid can also be used to dissolve many water-insoluble salts in which the anion is a weak base. In particular, H^+ ions will dissolve—

■ *virtually all carbonates* ($CO_3{}^{2-}$). The product is the weak acid H_2CO_3, which then decomposes into CO_2 and H_2O. The equation for the reaction of H^+ ions with $ZnCO_3$ is typical

$$ZnCO_3(s) + 2H^+(aq) \longrightarrow Zn^{2+}(aq) + H_2CO_3(aq)$$

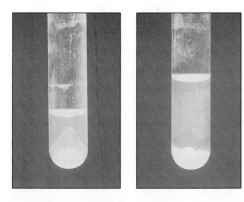

Dissolving a precipitate in acid. The tube at the left contains white silver chloride (AgCl) at the bottom and yellow silver phosphate (Ag_3PO_4) on top of the chloride. Adding a strong acid dissolves the silver phosphate, which has a basic anion but leaves the silver chloride undissolved. *(Charles D. Winters)*

■ *many sulfides* (S^{2-}). The driving force behind this reaction is the formation of the weak acid H_2S, much of which evolves as a gas. The reaction in the case of zinc sulfide is

$$ZnS(s) + 2H^+(aq) \longrightarrow Zn^{2+}(aq) + H_2S(aq)$$

EXAMPLE 16.8 Write balanced net ionic equations to explain why each of the following precipitates dissolves in strong acid.

(a) $Cr(OH)_3$ (b) Ag_2CO_3 (c) MnS

Strategy Compare the equations given above for dissolving $Zn(OH)_2$, $ZnCO_3$, and ZnS. In each case, the cation is brought into solution; the other product is H_2O, H_2CO_3, or H_2S.

Solution

(a) $Cr(OH)_3(s) + 3H^+(aq) \rightarrow Cr^{3+}(aq) + 3H_2O$

(b) $Ag_2CO_3(s) + 2H^+(aq) \rightarrow 2Ag^+(aq) + H_2CO_3(aq)$

(c) $MnS(s) + 2H^+(aq) \rightarrow Mn^{2+}(aq) + H_2S(aq)$

⊙ **See Screen 19.12, Complex Ion Formation and Solubility.**

Complex Formation

Ammonia and sodium hydroxide are commonly used to dissolve precipitates containing a cation that forms a stable complex with NH_3 or OH^- (Table 16.2). The reactions with zinc hydroxide are typical:

$$Zn(OH)_2(s) + 4NH_3(aq) \longrightarrow Zn(NH_3)_4^{2+}(aq) + 2\,OH^-(aq)$$

$$Zn(OH)_2(s) + 2OH^-(aq) \longrightarrow Zn(OH)_4^{2-}(aq)$$

The equilibrium constant for the solubility reaction is readily calculated. Consider, for example, the reaction by which zinc hydroxide dissolves in ammonia. Again, imagine that the reaction occurs in two steps:

⌐ **$Mg(OH)_2$ cannot be dissolved by NH_3 or NaOH. Explain.**

$$Zn(OH)_2(s) \rightleftharpoons Zn^{2+}(aq) + 2\,OH^-(aq) \qquad K_{sp}\,Zn(OH)_2$$

$$\underline{Zn^{2+}(aq) + 4NH_3(aq) \rightleftharpoons Zn(NH_3)_4^{2+}(aq) \qquad K_f\,Zn(NH_3)_4^{2+}}$$

$$Zn(OH)_2(s) + 4NH_3(aq) \rightleftharpoons Zn(NH_3)_4^{2+}(aq) + 2\,OH^-(aq)$$

Applying the rule of multiple equilibria,

$$K = K_{sp}\,Zn(OH)_2 \times K_f\,Zn(NH_3)_4^{2+} = (4 \times 10^{-17})(3.6 \times 10^8) = 1 \times 10^{-8}$$

In general, for any reaction of this type

$$K = K_{sp} \times K_f$$

where K_{sp} is the solubility product constant of the solid and K_f is the formation constant of the complex.

TABLE 16.2   Complexes of Cations with NH_3 and OH^-

Cation	NH_3 Complex	OH^- Complex
Ag^+	$Ag(NH_3)_2^+$	—
Cu^{2+}	$Cu(NH_3)_4^{2+}$ (blue)	—
Cd^{2+}	$Cd(NH_3)_4^{2+}$	—
Sn^{4+}	—	$Sn(OH)_6^{2-}$
Sb^{3+}	—	$Sb(OH)_4^-$
Al^{3+}	—	$Al(OH)_4^-$
Ni^{2+}	$Ni(NH_3)_6^{2+}$ (blue)	—
Zn^{2+}	$Zn(NH_3)_4^{2+}$	$Zn(OH)_4^{2-}$

EXAMPLE 16.9   Consider the reaction by which silver chloride dissolves in ammonia:

$$AgCl(s) + 2NH_3(aq) \rightleftharpoons Ag(NH_3)_2^+(aq) + Cl^-(aq)$$

(a) Taking K_{sp} AgCl $= 1.8 \times 10^{-10}$ and K_f Ag$(NH_3)_2^+ = 1.7 \times 10^7$, calculate K for this reaction.
(b) Calculate the number of moles of AgCl that dissolve in one liter of 6.0 M NH_3.

Strategy   Apply the rule of multiple equilibria to find K. Then work with the expression for K to find the molar solubility of AgCl.

Solution

(a) $K = K_{sp} \times K_f = (1.8 \times 10^{-10})(1.7 \times 10^7) = $ **3.1×10^{-3}**
(b) $AgCl(s) + 2NH_3(aq) \rightleftharpoons Ag(NH_3)_2^+(aq) + Cl^-(aq)$

$$K = \frac{[Ag(NH_3)_2^+] \times [Cl^-]}{[NH_3]^2} = 3.1 \times 10^{-3}$$

For every mole of AgCl that dissolves, two moles of NH_3 are consumed; one mole of Ag$(NH_3)_2^+$ and one mole of Cl^- are formed. We can set up an equilibrium table

	$[NH_3]$	$[Ag(NH_3)_2^+]$	$[Cl^-]$
Original	6.0	0	0
Change	$-2x$	$+x$	$+x$
Equilibrium	$6.0 - 2x$	x	x

where x is the number of moles per liter that dissolves.

We then have the equation $3.1 \times 10^{-3} = \dfrac{x^2}{(6.0 - 2x)^2}$

Taking the square root of both sides $0.056 = \dfrac{x}{6.0 - 2x}$

Solving, **$x = 0.30$**

Dissolving a precipitate by complex formation. At the left silver chloride (AgCl) is precipitated by adding a sodium chloride (NaCl) solution to a silver nitrate (AgNO₃) solution. At the right, as an ammonia solution is added, the AgCl precipitate is dissolving to form the complex ion Ag(NH₃)₂⁺. *(Charles D. Winters)*

AgCl(*s*)   [Ag(NH₃)₂]⁺(*aq*)

⊙ **See Screen 19.9, Using Solubility.**

⌐ Hg₂²⁺ is a polyatomic cation; there is a covalent bond between the mercury atoms.

16.3 QUALITATIVE ANALYSIS

In the general chemistry laboratory, you will most likely carry out at least one experiment dealing with the qualitative analysis of metal cations. The objective is to separate and identify the cations present in an "unknown" solution. A scheme of analysis for 22 different cations is shown in Table 16.3. As you can see, the general approach is to use precipitation reactions to divide the ions into four different groups. The ions within a group are then brought into solution, separated from one another, and identified.

Group I contains the only three common cations that form insoluble chlorides: Ag^+, Pb^{2+}, and Hg_2^{2+}. Addition of HCl precipitates AgCl, PbCl₂, and Hg₂Cl₂, all of which are white solids. To separate the cations, lead chloride is brought into solution in hot water

$$PbCl_2(s) \longrightarrow Pb^{2+}(aq) + 2Cl^-(aq)$$

and AgCl is dissolved in ammonia through complex ion formation

$$AgCl(s) - 2NH_3(aq) \longrightarrow Ag(NH_3)_2^+(aq) + Cl^-(aq)$$

Group II consists of six different cations, all of which form very insoluble sulfides with characteristic colors (Figure 16.4). These compounds are precipitated by adding hydrogen sulfide, a toxic, foul-smelling gas, at a pH of 0.5. At this rather high H⁺ ion concentration, the equilibrium

$$H_2S(aq) \rightleftharpoons 2H^+(aq) + S^{2-}(aq)$$

TABLE 16.3	**Cation Groups of Qualitative Analysis**	
Group	**Cations**	**Precipitating Reagent/Conditions**
I	Ag^+, Pb^{2+}, Hg_2^{2+}	6 *M* HCl
II	Cu^{2+}, Bi^{3+}, Hg^{2+}, Cd^{2+}, Sn^{4+}, Sb^{3+}	0.1 *M* H₂S at a pH of 0.5
III	Al^{3+}, Cr^{3+}, Co^{2+}, Fe^{2+}, Mn^{2+}, Ni^{2+}, Zn^{2+}	0.1 *M* H₂S at a pH of 9
IV	Ba^{2+}, Ca^{2+}, Mg^{2+}; Na^+, K^+, NH_4^+	0.2 *M* (NH₄)₂CO₃ at a pH of 9.5. No precipitates with Na⁺, K⁺, NH₄⁺; separate tests for identification

lies far to the left. The concentration of S^{2-} is extremely low but sufficient to precipitate the very insoluble Group II sulfides such as CuS and Bi_2S_3.

$$Cu^{2+}(aq) + H_2S(aq) \longrightarrow CuS(s) + 2H^+(aq)$$

$$2Bi^{3+}(aq) + 3H_2S(aq) \longrightarrow Bi_2S_3(s) + 6H^+(aq)$$

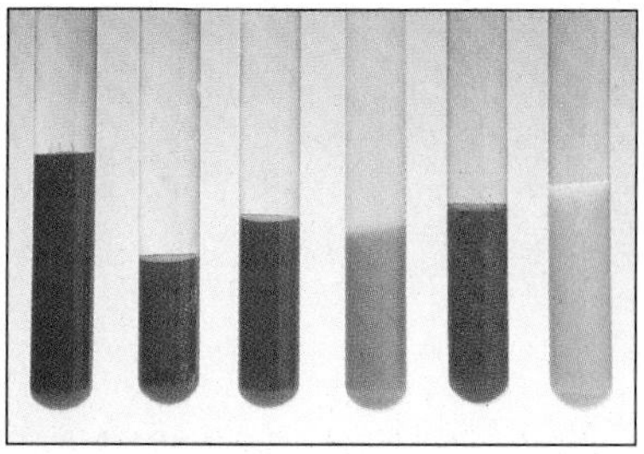

Group III cations form sulfides that are considerably more soluble than those of Group II. Consequently, they do not precipitate at pH 0.5, allowing for their separation from Group II. However, at pH 9, where the concentration of S^{2-} is considerably higher, five Group III cations precipitate as sulfides; Al^{3+} and Cr^{3+} come down as hydroxides in this basic solution (Figure 16.5):

$$Al^{3+}(aq) + 3OH^-(aq) \longrightarrow Al(OH)_3(s)$$

$$Cr^{3+}(aq) + 3OH^-(aq) \longrightarrow Cr(OH)_3(s)$$

FIGURE 16.4

Group II sulfides. *(From left to right)* CuS, Bi_2S_3, and HgS are black; CdS is orange-yellow; Sb_2S_3 is brilliant red-orange; and SnS_2 is yellow. *(Charles D. Winters)*

EXAMPLE 16.10   For the general solution reaction of a metal sulfide, MS:

$$MS(s) + 2H^+(aq) \rightleftharpoons M^{2+}(aq) + H_2S(aq)$$

$K = 1 \times 10^{-1}$ for NiS; $K = 1 \times 10^{-16}$ for CuS. Calculate the molar solubility of NiS and CuS in 0.3 M H^+ and 0.1 M H_2S, the conditions under which Group II is precipitated.

Strategy   For every mole of MS that dissolves, one mole of M^{2+} is formed. Hence the molar solubility s is equal to $[M^{2+}]$. To calculate s from the equilibrium constant, set up the expression for K.

Solution

$$K = \frac{[M^{2+}] \times [H_2S]}{[H^+]^2} = s \times \frac{(0.1)}{(0.3)^2} = s \times 1$$

In other words, for this particular system, $s = K$. Hence for NiS, $s = 1 \times 10^{-1}\ M$; for CuS, $s = 1 \times 10^{-16}\ M$.

Reality Check   Clearly, NiS is appreciably soluble under these conditions, but CuS is not. This explains why CuS precipitates in Group II while Ni^{2+} stays in solution and does not precipitate until $[H^+]$ is lowered in Group III.

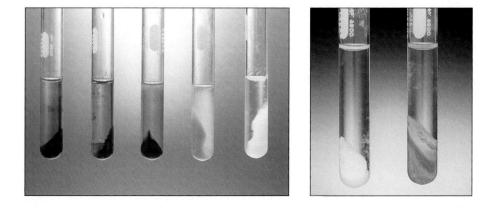

FIGURE 16.5

Group III sulfides and hydroxides. Five Group III cations precipitate as sulfides. These are NiS, CoS, and FeS (all of which are black), MnS (light pink), and ZnS (white). Two cations precipitate as hydroxides, $Al(OH)_3$ (white) and $Cr(OH)_3$ (gray-green). *(Charles D. Winters)*

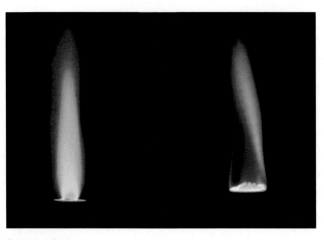

FIGURE 16.6
Flame tests for Na+ (yellow) and K+ (violet). A drop of solution is picked up on a platinum loop and immersed in the flame. The test for K+ is best done with a filter that hides the strong Na+ color. *(Ray Boyington/S. Ruven Smith)*

Group IV cations have soluble chlorides and sulfides. However, the alkaline earth cations in this group (Mg^{2+}, Ca^{2+}, Ba^{2+}) can be precipitated as carbonates, thereby separating them from the other three cations in this group. The reaction with Mg^{2+} is typical:

$$Mg^{2+}(aq) + CO_3{}^{2-}(aq) \longrightarrow MgCO_3(s)$$

The alkali metal cations (Na^+, K^+) are identified by flame tests (Figure 16.6). The $NH_4{}^+$ ion is heated with strong base to form NH_3, the only gas that turns red litmus blue:

$$NH_4{}^+(aq) + OH^-(aq) \longrightarrow NH_3(g) + H_2O$$

> The fewer insoluble compounds a cation forms, the closer it is to the end of the "qual" scheme.

EXAMPLE 16.11   Write balanced net ionic equations for two reactions by which $Al(OH)_3$, precipitated in Group III, can be brought into solution.

Strategy   Like any metal hydroxide, $Al(OH)_3$ dissolves in strong acid. Another way to dissolve it is suggested by Table 16.2, which lists $Al(OH)_4{}^-$ as a stable complex ion.

Solution

$$Al(OH)_3(s) + 3H^+(aq) \longrightarrow Al^{3+}(aq) + 3H_2O$$

$$Al(OH)_3(s) + OH^-(aq) \longrightarrow Al(OH)_4{}^-(aq)$$

Hydroxides like $Al(OH)_3$ that react with either strong acid (H^+) or strong base (OH^-) are described as being *amphoteric*.

Fluoridation of Drinking Water

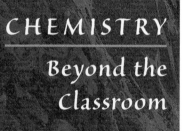

Many years ago, it was found that children in areas where the drinking water contained relatively high concentrations of F^- ions from natural sources showed a remarkably low incidence of tooth decay. Data of this sort led to the fluoridation of public water supplies, carried out first in Grand Rapids, Michigan, in 1945. Today about half the communities in the United States and Canada add metal fluorides to drinking water as needed to raise $[F^-]$ to 0.7 to 1.2 parts per million. Since 1945 the average number of cavities in schoolage children has dropped from about seven to about three. How much of this decrease is due to fluoridation is hard to say. Improvements in nutrition and dental hygiene have certainly contributed as well.

On average, an adult takes in about 2.4 mg/day of F^- ions from drinking fluoridated water. This compares with perhaps 0.4 mg/day from dietary sources; tea and seafoods are rich in fluoride. Another source these days is dental products. Most toothpastes contain small amounts of a metal fluoride, typically sodium fluoride (NaF) or tin(II) fluoride, SnF_2 (Figure A). Mouthwashes are usually fluoridated, as are the gels dentists use to coat teeth.

The protective action of F^- ions is explained at least in part by a chemical reaction with tooth enamel composed of hydroxyapatite, whose formula may be written as $Ca(OH)_2 \cdot 3Ca_3(PO_4)_2$. The OH^- ions in this compound are readily replaced by F^-, which has the same charge as OH^- and is nearly the same size. The product is fluorapatite, $CaF_2 \cdot 3Ca_3(PO_4)_2$. Just as CaF_2 ($K_{sp} = 1.5 \times 10^{-10}$) is much less soluble than $Ca(OH)_2$ ($K_{sp} = 4.0 \times 10^{-6}$), so fluorapatite is less soluble than hydroxyapatite. Beyond that, fluorapatite is less soluble in acid, because the F^- ion is a much weaker base than $OH-$:

$$F^-(aq) + H^+(aq) \rightleftharpoons HF(aq) \qquad K = 1.4 \times 10^3$$

$$OH^-(aq) + H^+(aq) \rightleftharpoons H_2O \qquad K = 1.0 \times 10^{14}$$

Tooth decay starts when bacteria in the mouth convert sugar or other carbohydrates into organic acids. These acids attack the enamel of the tooth, exposing the interior to decay.

In some ground water, $[F^-]$ is well above that produced by fluoridation. To understand why this happens, consider the solubility of calcium fluoride, CaF_2, a common mineral in many areas (Figure B).

$$s = (K_{sp}/4)^{1/3} = (1.5 \times 10^{-10}/4)^{1/3} = 3.3 \times 10^{-4} \text{ mol/L}$$

This corresponds to a fluoride concentration of 0.013 g/L or 13 ppm:

$$[F^-] = 2s = 6.6 \times 10^{-4} \frac{\text{mol}}{\text{L}} \times 19.0 \frac{\text{g}}{\text{mol}} = 0.013 \frac{\text{g}}{\text{L}} = 13 \text{ ppm}$$

Fortunately, F^- ion concentrations are rarely, if ever, as high as 13 ppm in North America. In general—

- if $[F^-] > 10$ ppm, longtime exposure can lead to a crippling bone disease, skeletal fluorosis, which is common in parts of India, China, and Africa.
- if $[F^-] > 5$ ppm, children's teeth can develop with mottled enamel. Chalky white patches form, along with yellowish stains.

Taking this into account, the Environmental Protection Agency has set an upper limit of 4 ppm for F^- in drinking water. A concentration of 1 ppm is believed to be ideal.

FIGURE A
Dental products containing F^- ions. *(Charles D. Winters)*

FIGURE B
A fluorite crystal (CaF_2).
(Charles D. Winters)

CHAPTER HIGHLIGHTS

Key Concepts

1. Set up the expression for K_{sp} for any ionic solid.
   (Example 16.1; Problems 1–4)
2. Use the value of K_{sp} to—
   ■ calculate the concentration of one ion, knowing that of the other.
   (Example 16.2; Problems 5–8)
   ■ determine whether a precipitate will form.
   (Examples 16.3, 16.4; Problems 9–14, 29, 30, 44)
   ■ calculate the solubility in water or a solution containing a common ion.
   (Examples 16.5, 16.6; Problems 15–20)
   ■ determine which ion will precipitate first from a solution.
   (Example 16.10; Problems 41, 42)
3. Calculate K for—
   ■ the dissolving of a metal hydroxide in strong acid.
   (Example 16.7; Problem 28)
   ■ the dissolving of a precipitate in a complexing agent.
   (Example 16.9; Problems 27, 31–36)
4. Write balanced net ionic equations to explain why a precipitate dissolves in—
   ■ strong acid.
   (Examples 16.8, 16.11; Problems 21, 22)
   ■ NH_3 or OH^-.
   (Example 16.11; Problems 23–26)

Key Terms

common ion
■ complex ion
■ equilibrium rules
  —coefficient
  —multiple
  —reciprocal

■ formation constant, K_f
  ion product, P
■ net ionic equation

■ pH
■ precipitate
  solubility product constant, K_{sp}

Summary Problem

Consider zinc hydroxide. It can be formed when solutions of zinc nitrate and sodium hydroxide are mixed. Its K_{sp} is 4×10^{-17}.
(a) Write the K_{sp} expression for zinc hydroxide.
(b) How many moles of NaOH must be added to 575 mL of 0.0300 M $Zn(NO_3)_2$ to just start the precipitation of zinc hydroxide?
(c) Will a precipitate form if 10.00 mL of 0.0200 M NaOH is added to 250.0 mL of 0.098 M $Zn(NO_3)_2$?
(d) How many milligrams of zinc hydroxide are necessary to prepare 700.0 mL of a saturated $Zn(OH)_2$ solution?

(e) How many milligrams of zinc hydroxide can be dissolved in 2.00 L of a 0.083 M $ZnCl_2$ solution?

(f) Which would precipitate first, $Zn(OH)_2$ or $Al(OH)_3$, on adding 0.075 M NaOH to a 0.500-L solution containing 0.0100 mol of Zn^{2+} and 0.0100 mol of Al^{3+}? (K_{sp} $Al(OH)_3$ = 2×10^{-31})

(g) Write the equation for dissolving $Zn(OH)_2$ with a strong acid. Calculate K for this reaction.

(h) Write the equation for dissolving $Zn(OH)_2$ with a strong base. Calculate K for this reaction.

(i) Write the equation for dissolving $Zn(OH)_2$ with an aqueous solution of ammonia. Assume that the ammonia instead of the hydroxide complex is formed. Calculate K for this reaction.

(j) How many milligrams of $Zn(OH)_2$ can be dissolved by complex formation in 2.50 L of 0.166 M NaOH?

Answers

(a) $K_{sp} = [Zn^{2+}][OH^-]^2$
(b) 2×10^{-8}
(c) yes
(d) 0.2 mg
(e) 0.002 mg
(f) $Al(OH)_3$
(g) $Zn(OH)_2(s) + 2H^+(aq) \rightleftharpoons Zn^{2+}(aq) + 2H_2O$ $K = 4 \times 10^{11}$
(h) $Zn(OH)_2(s) + 2\,OH^-(aq) \rightleftharpoons Zn(OH)_4{}^{2-}(aq)$ $K = 0.01$
(i) $Zn(OH)_2(s) + 4NH_3(aq) \rightleftharpoons Zn(NH_3)_4{}^{2+}(aq) + 2\,OH^-(aq)$ $K = 1 \times 10^{-8}$
(j) 7×10^1 mg

Questions & Problems

Problem numbers in blue indicate that the answer is available in Appendix 6 at the back of the book.

WEB indicates that the solution is posted at **http://www.harcourtcollege.com/chem/general/masterton4/student/**

Expression for K_{sp}

1. Write the equilibrium equation and the K_{sp} expression for each of the following.
  (a) AgCl   **(b)** $Al_2(CO_3)_3$ **(c)** MnS_2 **(d)** $Mg(OH)_2$

2. Write the equilibrium equation and the K_{sp} expression for each of the following.
  (a) Co_2S_3 **(b)** $PbCl_2$ **(c)** $Zn_2P_2O_7$ **(d)** $Sc(OH)_3$

3. Write the equilibrium equations on which the following K_{sp} expressions are based.
  (a) $[Hg_2{}^{2+}][Cl^-]^2$ **(b)** $[Pb^{2+}][CrO_4{}^{2-}]$
  (c) $[Mn^{4+}][O^{2-}]^2$ **(d)** $[Al^{3+}]^2[S^{2-}]^3$

4. Write the equilibrium equations on which the following K_{sp} expressions are based.
  (a) $[Ca^{2+}][CO_3{}^{2-}]$ **(b)** $[Co^{3+}][OH^-]^3$
  (c) $[Ag^+]^2[S^{2-}]$ **(d)** $[Pb^{2+}][Cl^-]^2$

K_{sp} and Precipitation

5. Complete the following table for concentrations in equilibrium with calcium phosphate.

	[Ca^{2+}]	[PO$_4{}^{3-}$]
(a)	_____	4×10^{-7}
(b)	9×10^{-2}	_____
(c)	$3\,[PO_4{}^{3-}]$	
(d)	_____	$7\,[Ca^{2+}]$

WEB 6. Complete the following table for concentrations in equilibrium with silver carbonate.

	[Ag$^+$]	[CO$_3{}^{2-}$]
(a)	_____	7×10^{-4}
(b)	0.003	_____
(c)		$\frac{1}{2}\,[Ag^+]$
(d)	$12\,[CO_3{}^{2-}]$	_____

7. Calculate the concentration of each of the following ions in equilibrium with 0.0034 M Ba^{2+}.
  (a) F^- **(b)** $SO_4{}^{2-}$ **(c)** $CrO_4{}^{2-}$

8. Calculate the concentration of each of the following ions in equilibrium with 0.019 M Br^-.

(a) Pb^{2+} (b) Hg_2^{2+} (c) Ag^+

9. Iron(III) chloride is added to a solution of 0.075 M sodium hydroxide.

(a) At what concentration of Fe^{3+} does a precipitate start to form?

(b) Enough iron(III) chloride is added to make $[Fe^{3+}]$ = 0.0030 M. What is $[OH^-]$? What percentage of the original hydroxide ion has precipitated?

10. Calcium nitrate is added to 0.15 M sodium phosphate.

(a) At what concentration of Ca^{2+} does a precipitate first start to form?

(b) What is $[Ca^{2+}]$, when only 15% of the original PO_4^{3-} remains?

11. Water from a well is found to contain 3.0 mg of calcium ion per liter. If 0.50 mg of sodium sulfate is added to one liter of the well water without changing its volume, will a precipitate form? What should $[SO_4^{2-}]$ be to just start precipitation?

12. Before lead in paint was discontinued, lead chromate was a common pigment in yellow paint. A 1.0-L solution is prepared by mixing 0.50 mg of lead nitrate with 0.020 mg of potassium chromate. Will a precipitate form? What should $[Pb^{2+}]$ be to just start precipitation?

13. A solution is prepared by mixing 35.00 mL of a 0.061 M solution of zinc nitrate with 20.0 mL of KOH with a pH of 9.00. Assume that volumes are additive.

(a) Will precipitation occur?

(b) Calculate $[Zn^{2+}]$, $[NO_3^-]$, $[K^+]$, and the pH after equilibrium is established.

14. A solution is prepared by mixing 13.00 mL of 0.0021 M aqueous $Hg_2(NO_3)_2$ with 25.0 mL of 0.015 M HCl. Assume that volumes are additive.

(a) Will precipitation occur?

(b) Calculate $[Hg_2^{2+}]$, $[Cl^-]$, and $[NO_3^-]$ after equilibrium is established.

Solubility

15. At 25°C, 54 mg of lead iodate must be added to 100.0 g of water to prepare a saturated solution. Assuming that the volume of the solution is 100.0 mL, what is the K_{sp} for lead iodate?

16. At 25°C, 1.5 mg of zinc cyanide ($Zn(CN)_2$) must be dissolved in 100.0 mL to prepare a saturated solution. What is the K_{sp} for $Zn(CN)_2$?

17. Calculate the solubility (in grams per liter) of silver chloride in the following.

(a) pure water

(b) 0.025 M $BaCl_2$

(c) 0.17 M $AgNO_3$

18. Calculate the solubility (in grams per liter) of magnesium hydroxide in the following.

(a) pure water

(b) 0.041 M $Ba(OH)_2$

(c) 0.0050 M $MgCl_2$

19. K_{sp} for silver acetate ($AgC_2H_3O_2$) at 80°C is estimated to be

2×10^{-2}. Ten grams of silver acetate is added to 1.0 L of water at 25°C.

(a) Will all the silver acetate dissolve at 25°C?

(b) If the solution (assume the volume to be 1.0 L) is heated to 80°C, will all the silver acetate dissolve?

20. One gram of $PbCl_2$ is dissolved in 1.0 L of hot water. When the solution is cooled to 25°C, will some of the $PbCl_2$ crystallize out? If so, how much?

Dissolving Precipitates

21. Write net ionic equations for the reactions of each of the following with strong acid.

(a) $CaCO_3$ (b) NiS (c) $Al(OH)_3$ (d) $Sb(OH)_4^-$

(e) $AgCl$

22. Write net ionic equations for the reaction of H^+ with

(a) Cu_2S (b) Hg_2Cl_2 (c) $SrCO_3$

(d) $Cu(NH_3)_4^{2+}$ (e) $Ca(OH)_2$

23. Write a net ionic equation for the reaction with ammonia by which

(a) silver chloride dissolves.

(b) aluminum ion forms a precipitate.

(c) copper(II) forms a complex ion.

WEB **24.** Write a net ionic equation for the reaction with ammonia by which

(a) $Cu(OH)_2$ dissolves.

(b) Cd^{2+} forms a complex ion.

(c) Pb^{2+} forms a precipitate.

25. Write a net ionic equation for the reaction with OH^- by which

(a) Ni^{2+} forms a precipitate.

(b) Sn^{4+} forms a complex ion.

(c) $Al(OH)_3$ dissolves.

26. Write a net ionic equation for the reaction with OH^- by which

(a) Sb^{3+} forms a precipitate

(b) antimony(III) hydroxide dissolves when more OH^- is added.

(c) Sb^{3+} forms a complex ion.

Solution Equilibria

27. Calculate K for the reaction where silver bromide is dissolved by 6 M aqueous ammonia.

28. Calculate K for the reaction where aluminum hydroxide is dissolved by 6 M HCl.

29. Consider the reaction

$$AgCl(s) + I^-(aq) \rightleftharpoons AgI(s) + Cl^-(aq)$$

(a) Calculate K for the reaction.

(b) Will AgI precipitate if NaI is added to a saturated AgCl solution?

30. Consider the reaction

$$Zn(OH)_2(s) + 2CN^-(aq) \rightleftharpoons Zn(CN)_2(s) + 2OH^-(aq)$$

(a) Calculate K for the reaction. (K_{sp} $Zn(CN)_2$ = 8.0×10^{-12})

(b) Will $Zn(CN)_2$ precipitate if NaCN is added to a saturated $Zn(OH)_2$ solution?

31. Aluminum hydroxide reacts with an excess of hydroxide ions to form the complex ion $Al(OH)_4^-$.

(a) Write an equation for this reaction.

(b) Calculate K.

(c) Determine the solubility of $Al(OH)_3$ (in mol/L) at pH 12.0.

32. Consider the reaction

$$Cu(OH)_2(s) + 4NH_3(aq) \rightleftharpoons Cu(NH_3)_4^{2+}(aq) + 2OH^-(aq)$$

(a) Calculate K given that for $Cu(OH)_2$ $K_{sp} = 2 \times 10^{-19}$ and for $Cu(NH_3)_4^{2+}$ $K_f = 2 \times 10^{12}$.

(b) Determine the solubility of $Cu(OH)_2$ (in mol/L) in $4.5\ M\ NH_3$.

33. Calculate the molar solubility of silver bromide in $1.5\ M$ aqueous NH_3.

WEB 34. Calculate the molar solubility of lead chloride in a solution $2.0\ M$ in OH^-. The K_f for $Pb(OH)_3^-$ is 3.8×10^{14}.

35. For the reaction

$$Fe(OH)_2 + 6CN^-(aq) \rightleftharpoons Fe(CN)_6^{4-}(aq) + 2OH^-(aq)$$

(a) Calculate K.

(b) Calculate $[CN^-]$ at equilibrium when 10.0 g of $Fe(OH)_2$ is dissolved in 1.00 L of solution.

36. For the reaction

$$CdC_2O_4(s) + 4NH_3(aq) \rightleftharpoons Cd(NH_3)_4^{2+}(aq) + C_2O_4^{2-}(aq)$$

(a) Calculate K. (K_{sp} for CdC_2O_4 is 1.5×10^{-8}.)

(b) Calculate $[NH_3]$ at equilibrium when 2.00 g of CdC_2O_4 is dissolved in 1.00 L of solution.

Qualitative Analysis

37. Complete the following table.

Species	Test/Reagent	Response to Test/Reagent
(a) _____	Flame test	Yellow
(b) $AgCl(s)$	_____	Dissolves
(c) $MgCO_3(s)$	H^+	_____
(d) Cu^{2+}	H_2S at pH 0.5	_____

38. Complete the following table for cations.

Cation	Analytical Group	Precipitating Agent	Precipitate Formed
(a) Ag^+	_____	_____	_____
(b) Bi^{3+}	_____	_____	_____
(c) Co^{2+}	_____	_____	_____
(d) Mg^{2+}	_____	_____	_____

39. A solution contains several unknown metal ions. All tests done prove negative except that a precipitate forms on addition of H_2S in basic solution. What group of cations is present?

40. A solution is known to contain several unknown metal ions. Tests for K^+, Na^+, and NH_4^+ are negative. When the solution is treated with dilute HCl, a precipitate forms. The precipitate is separated and the filtrate is acidified to a pH of 0.5. H_2S is added to the filtrate. No precipitate forms. The solution is then treated with base to a pH of 9. H_2S is again added. No precipitate forms. Addition of ammonium carbonate gives a precipitate. What possible cations are present?

41. Consider the following reactions:

$$HgS(s) + 2H^+(aq) \rightleftharpoons Hg^{2+}(aq) + H_2S(aq) \qquad K = 1 \times 10^{-32}$$

$$CoS(s) + 2H^+(aq) \rightleftharpoons Co^{2+}(aq) + H_2S(aq) \qquad K = 3 \times 10^{-6}$$

Under the conditions pH = 0.50 and $[H_2S] = 0.10\ M$, which sulfide will precipitate first?

42. Consider the following reactions:

$$CdS(s) + 2H^+(aq) \rightleftharpoons Cd^{2+}(aq) + H_2S(aq) \qquad K = 1 \times 10^{-9}$$

$$MnS(s) + 2H^+(aq) \rightleftharpoons Mn^{2+}(aq) + H_2S(aq) \qquad K = 5 \times 10^6$$

Under the conditions pH = 0.50 and $[H_2S] = 0.10\ M$, which sulfide will precipitate first?

Unclassified

43. Predict what effect each of the following has on the position of the equilibrium

$$PbCl_2(s) \rightleftharpoons Pb^{2+}(aq) + 2Cl^-(aq) \qquad \Delta H = 23.4\ kJ$$

(a) addition of $1\ M\ Pb(NO_3)_2$ solution

(b) increase in temperature

(c) addition of Ag^+ forming AgCl

(d) addition of $1\ M$ hydrochloric acid

44. A town adds 2.0 ppm of F^- ion to fluoridate its water supply (fluoridation of water reduces the incidence of dental caries). If the concentration of Ca^{2+} in the water is $3.5 \times 10^{-4}\ M$, will a precipitate of CaF_2 form when the water is fluoridated?

45. When 25.0 mL of $0.500\ M$ iron(II) sulfate is combined with 35.0 mL of $0.332\ M$ barium hydroxide, two different precipitates are formed.

(a) Write a net ionic equation for the reaction that takes place.

(b) Estimate the mass of the precipitates formed.

(c) What are the equilibrium concentrations of the ions in solution?

46. Consider a 1.50-L aqueous solution of $3.75\ M\ NH_3$, where 17.5 g of NH_4Cl is dissolved. To this solution, 5.00 g of $MgCl_2$ is added.

(a) What is $[OH^-]$ before $MgCl_2$ is added?

(b) Will a precipitate form?

(c) What is $[Mg^{2+}]$ after equilibrium is established?

Conceptual Problems

47. Shown on page 480 is a representation of the ionic solid MX, where M cations are represented by squares and X anions are represented by circles. Fill in the box after the arrow to represent what happens to the solid after it has been completely dissolved in water. For simplicity, do not represent the water molecules.

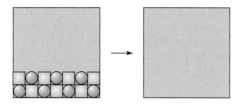

48. The box below represents one liter of a saturated solution of the species $\square^\circ$, where squares represent the cation and circles represent the anion. Water molecules, though present, are not shown.

Complete the next three figures by filling one-liter boxes to the right of the arrow, showing the state of the ions after water is added to form saturated solutions. The species represented to the left of the arrow is the solid form of the ions represented above. Do not show the water molecules.

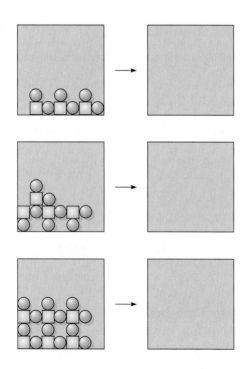

49. Using the same saturation data and species representation described in Question 48, complete the picture below.

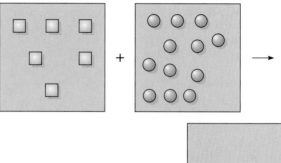

50. Marble used for statuary is almost pure $CaCO_3$. When marble is exposed to H^+ ions, the following reaction occurs.

$$CaCO_3(s) + H^+(aq) \rightleftharpoons HCO_3^-(aq) + Ca^{2+}(aq)$$

Why is acid rain the primary culprit in the disintegration of marble statuary exposed to the elements? Assume a pH of 3.00 for acid rain and a pH of 5.4 for that of normal rainwater.

51. Consider the insoluble salts JQ, K_2R, L_2S_3, MT_2, and NU_3. They are formed from the metal ions, J^+, K^+, L^{3+}, M^{2+}, N^{3+}, and the nonmetal ions Q^-, R^{2-}, S^{2-}, T^-, and U^-. All the salts have the same K_{sp}, 1×10^{-10}, at 25°C.

(a) Which salt has the highest molar solubility?

(b) Does the salt with the highest molar solubility have the highest solubility in g salt/100 g water?

(c) Can the solubility of each salt in g/100 g water be determined from the information given? If yes, calculate the solubility of each salt in g/100 g water. If no, why not?

52. A plot of the solubility of a certain compound (g/100 g H_2O) against temperature (°C) is a straight line with a positive slope. Is dissolving that compound an exothermic process?

Challenge Problems

53. Ammonium chloride solutions are slightly acidic, so they are better solvents than water for insoluble hydroxides such as $Mg(OH)_2$. Find the solubility of $Mg(OH)_2$ in moles per liter in $0.2\ M$ NH_4Cl and compare with the solubility in water. *Hint:* Find K for the reaction

$$Mg(OH)_2(s) + 2NH_4^+(aq) \longrightarrow Mg^{2+}(aq) + 2NH_3(aq) + 2H_2O$$

54. What is the solubility of CaF_2 in a buffer solution containing $0.30\ M$ $HCHO_2$ and $0.20\ M$ $NaCHO_2$? *Hint:* Consider the equation

$$CaF_2(s) + 2H^+(aq) \longrightarrow Ca^{2+}(aq) + 2HF(aq)$$

and solve the equilibrium problem.

55. What is the I^- concentration just as AgCl begins to precipitate when 1.0 M $AgNO_3$ is slowly added to a solution containing 0.020 M Cl^- and 0.020 M I^-?

56. The concentrations of various cations in sea water, in moles per liter, are

Ion	Na^+	Mg^{2+}	Ca^{2+}	Al^{3+}	Fe^{3+}
Molarity (M)	0.46	0.056	0.01	4×10^{-7}	2×10^{-7}

(a) At what $[OH^-]$ does $Mg(OH)_2$ start to precipitate?

(b) At this concentration, will any of the other ions precipitate?

(c) If enough OH^- is added to precipitate 50% of the Mg^{2+}, what percentage of each of the other ions will precipitate?

(d) Under the conditions in (c), what mass of precipitate will be obtained from one liter of sea water?

57. Consider the equilibrium

$$Zn(NH_3)_4{}^{2+}(aq) + 4OH^-(aq) \rightleftharpoons Zn(OH)_4{}^{2-}(aq) + 4NH_3(aq)$$

(a) Calculate K for this reaction.

(b) What is the ratio $[Zn(NH_3)_4{}^{2+}]/[Zn(OH)_4{}^{2-}]$ in a solution 1.0 M in NH_3?

58. Use the equilibrium constants in Appendix 1 to calculate K for the reaction

$$Ag(NH_3)_2{}^+(aq) + 2H^+(aq) + Cl^-(aq) \rightleftharpoons AgCl(s) + 2NH_4{}^+(aq)$$

17 SPONTANEITY OF REACTION

Solid ammonium nitrate
(NH_4NO_3) decomposing into
N_2O and H_2O gases results in a
large increase in entropy.

(Charles D. Winters)

CHAPTER OUTLINE

17.1 SPONTANEOUS PROCESSES

17.2 ENTROPY, S

17.3 FREE ENERGY, G

17.4 STANDARD FREE ENERGY CHANGE, ΔG°

17.5 EFFECT OF TEMPERATURE, PRESSURE, AND CONCENTRATION ON REACTION SPONTANEITY

17.6 THE FREE ENERGY CHANGE AND THE EQUILIBRIUM CONSTANT

17.7 ADDITIVITY OF FREE ENERGY CHANGES; COUPLED REACTIONS

The goal of this chapter is to answer a basic question: Will a given reaction occur "by itself" at a particular temperature and pressure, without the exertion of any outside force? In that sense, is the reaction *spontaneous*? This is a critical question in just about every area of science and technology. A synthetic organic chemist looking for a source of acetylene, C_2H_2, would like to know whether this compound can be made by heating the elements together. A metallurgist trying to produce titanium metal from TiO_2 would like to know what reaction to use; would hydrogen, carbon, or aluminum be feasible reducing agents?

To develop a general criterion for spontaneity, we will apply the principles of *thermodynamics,* the science that deals with heat and energy effects. Three different thermodynamic functions are of value.

1. ΔH, the change in enthalpy (Chapter 8); a *negative* value of ΔH tends to make a reaction spontaneous.
2. ΔS, the change in entropy (Section 17.2); a *positive* value of ΔS tends to make a reaction spontaneous.
3. ΔG, the change in free energy (Sections 17.3, 17.4); a reaction at constant temperature and pressure will be *spontaneous* if ΔG is *negative,* no ifs, ands, or buts.

Besides serving as a general criterion for spontaneity, the free energy change can be used to—

■ determine the effect of temperature, pressure, and concentration on reaction spontaneity (Section 17.5).

■ calculate the equilibrium constant for a reaction (Section 17.6).

■ determine whether coupled reactions will be spontaneous (Section 17.7).

17.1 SPONTANEOUS PROCESSES

All of us are familiar with certain **spontaneous processes.** For example—

■ an ice cube melts when added to a glass of water at room temperature

$$H_2O(s) \longrightarrow H_2O(l)$$

See the *Saunders Interactive General Chemistry CD-ROM,* Screen 20.2, **Reaction Spontaneity.**

■ a mixture of hydrogen and oxygen burns if ignited by a spark

$$2H_2(g) + O_2(g) \longrightarrow 2H_2O(l)$$

■ an iron (steel) tool exposed to moist air rusts

$$2Fe(s) + \frac{3}{2}O_2(g) + 3H_2O(l) \longrightarrow 2Fe(OH)_3(s)$$

In other words, these three reactions are spontaneous at 25°C and 1 atm.

The word "spontaneous" does not imply anything about how rapidly a reaction occurs. Some spontaneous reactions, notably the rusting of iron, are quite slow. Often a reaction that is spontaneous does not occur without some sort of stimulus to get the reaction started. A mixture of hydrogen and oxygen shows no sign of reaction in the absence of a spark or match. Once started, though, a spontaneous reaction continues by itself without further input of energy from the outside.

If a reaction is spontaneous under a given set of conditions, the reverse reaction must be nonspontaneous. For example, water does not spontaneously decompose to the elements by the reverse of the reaction referred to above.

$$2H_2O(l) \longrightarrow 2H_2(g) + O_2(g) \qquad \text{nonspontaneous}$$

However, it is often possible to bring about a nonspontaneous reaction by supplying energy in the form of work. Electrolysis can be used to decompose water to the elements. To do this, electrical energy must be furnished, perhaps from a storage battery.

> A spark is OK, but a continuous input of energy isn't.

The Energy Factor

Many spontaneous processes proceed with a decrease of energy. Boulders roll downhill, not uphill. A storage battery discharges when you leave your car's headlights on. Extrapolating to chemical reactions, one might guess that spontaneous reactions would be exothermic ($\Delta H < 0$). A century ago, P. M. Berthelot in Paris and Julius Thomsen in Copenhagen proposed this as a general principle, applicable to all reactions.

It turns out that almost all exothermic chemical reactions are indeed spontaneous at 25°C and 1 atm. Consider, for example, the formation of water from the elements and the rusting of iron:

$$2H_2(g) + O_2(g) \longrightarrow 2H_2O(l) \qquad \Delta H = -571.6 \text{ kJ}$$

$$2Fe(s) + \frac{3}{2}O_2(g) + 3H_2O(l) \longrightarrow 2Fe(OH)_3(s) \qquad \Delta H = -788.6 \text{ kJ}$$

For both of these spontaneous reactions, ΔH is a negative quantity.

On the other hand, this simple rule fails for many familiar phase changes. An example is the melting of ice. This takes place spontaneously at 1 atm above 0°C, even though it is endothermic:

$$H_2O(s) \longrightarrow H_2O(l) \qquad \Delta H = +6.0 \text{ kJ}$$

There is still another basic objection to using the sign of ΔH as a general criterion for spontaneity. Endothermic reactions that are nonspontaneous at room temperature often become spontaneous when the temperature is raised. Consider, for example, the decomposition of limestone:

$$CaCO_3(s) \longrightarrow CaO(s) + CO_2(g) \qquad \Delta H = +178.3 \text{ kJ}$$

At 25°C and 1 atm, this reaction does not occur. Witness the existence of the white cliffs of Dover and other limestone deposits over eons of time. However, if the temperature is raised to about 1100 K, limestone decomposes to give off carbon dioxide gas at 1 atm. In other words, this endothermic reaction becomes spontaneous at high temperatures. This is true despite the fact that ΔH remains about 178 kJ, nearly independent of temperature.

The Randomness Factor

Clearly, the direction of a spontaneous change is not always determined by the tendency for a system to go to a state of lower energy. There is another natural tendency that must be taken into account to predict the direction of spontaneity. *Nature tends to move spontaneously from a state of lower probability to one of higher probability.* Or, as G. N. Lewis put it,

> Each system which is left to its own will, over time, change toward a condition of maximum probability.

To illustrate what these statements mean, consider a pastime far removed from chemistry: tossing dice. If you've ever shot craps (and maybe even if you haven't), you know that when a pair of dice is thrown, a 7 is much more likely to come up than a 12. Figure 17.1 shows why this is the case. There are six different ways to throw a 7 and only one way to throw a 12. Over time, dice will come up 7 six times as often as 12. A 7 is a state of "high" probability; a 12 is a state of "low" probability.

Now let's consider a process a bit closer to chemistry (Figure 17.2). Two different gases, let us say H_2 and N_2, are originally contained in different glass bulbs, separated by a stopcock. When the stopcock is opened, the two different kinds of molecules distribute themselves evenly between the two bulbs. Eventually, half of the H_2 molecules will end up in the left bulb and half in the right; the same holds for the N_2 molecules. Each gas achieves its own most probable distribution, independent of the presence of the other gas.

We could explain the results of this experiment the way we did before; the final distribution is clearly much more probable than the initial distribution. There is, however, another useful way of looking at this process. The system has gone from a highly ordered state (all the H_2 molecules on the left, all the N_2 molecules

FIGURE 17.1
Certain "states" are more probable than others. For example, when you toss a pair of dice, a 7 is much more likely to come up than a 12.

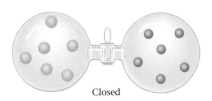

Closed

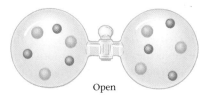

Open

FIGURE 17.2
Spontaneous change. Different kinds of gas molecules mix spontaneously, going from a more ordered to a more random state.   **OHT**

FIGURE 17.3
Probability. Some states are much more probable than others. If you shake red and black marbles with each other, the random distribution at the left is much more probable than the highly ordered distribution at the right. *(Charles Steele)*

See Screen 20.4, Entropy.

on the right) to a more disordered, or random, state in which the molecules are distributed evenly between the two bulbs. The same situation holds when marbles rather than molecules are mixed (Figure 17.3). *In general, nature tends to move spontaneously from more ordered to more random states.*

This statement is quite easy for parents or students to understand. Your room tends to get messy because an ordered room has few options for objects to be moved around (socks on the floor is not an option for an orderly room). The comedian Bill Cosby insists that with an army of 80 two-year-olds he could take over any country in the world, because they have a remarkable ability for disorganization.

EXAMPLE 17.1   Choose the state that is more random and thus more probable:

(a) a chessboard before the first move or while a game is in process.
(b) a solved jigsaw puzzle or the unassembled pieces in a box.
(c) Humpty Dumpty after the fall or Humpty Dumpty together again.

Solution

(a) The chess game in progress is more random. There are many more ways the pieces on the board could be arranged when the game is going on.
(b) A solved jigsaw puzzle is quite orderly. A box has all the pieces at random.
(c) Humpty Dumpty after the fall is more probable. Putting him together again is non-spontaneous; all the king's horses and all the king's men couldn't do it.

17.2 ENTROPY, S

The randomness factor can be treated quantitatively in terms of a function called **entropy,** symbol S. In general, the more probable a state or the more random the distribution of molecules, the greater the entropy.

Entropy, like enthalpy (Chapter 8), is a state property. That is, the entropy depends only on the state of a system, not on its history. The entropy change is determined by the entropies of the final and initial states, not on the path followed from one state to another.

$$\Delta S = S_{\text{final}} - S_{\text{initial}}$$

Several factors influence the amount of entropy that a system has in a particular state. In general—

■ *a liquid has a higher entropy than the solid from which it is formed.* In a solid, the atoms, molecules, or ions are fixed in position; in the liquid, these particles are free to move past one another. In that sense, the liquid structure is more random, the solid more ordered.
■ *a gas has a higher entropy than the liquid from which it is formed.* When vaporization occurs, the particles acquire greater freedom to move about. They are distributed throughout the entire container instead of being restricted to a small volume.
■ *increasing the temperature of a substance increases its entropy.* Raising the temperature increases the kinetic energy of the molecules (or atoms or ions) and hence their freedom of motion. In the solid, the molecules vibrate with a greater amplitude at higher temperatures. In a liquid or a gas, they move about more rapidly.

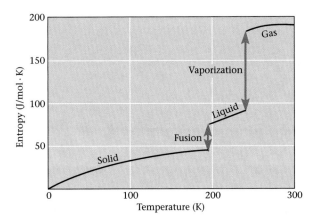

FIGURE 17.4

Molar entropy of ammonia as a function of temperature. Note the large increases in entropy upon fusion (melting) and vaporization.

These effects are shown in Figure 17.4, where the entropy of ammonia, NH_3, is plotted versus temperature. Note that the entropy of solid ammonia at 0 K is zero. This reflects the fact that molecules are completely ordered in the solid state at this temperature; there is no randomness whatsoever. More generally, the **third law of thermodynamics** tells us that *a completely ordered pure crystalline solid has an entropy of zero at 0 K.*

Notice from the figure that the effect of temperature on entropy is due almost entirely to phase changes. The slope of the curve is small in regions where only one phase is present. In contrast, there is a large jump in entropy when the solid melts and an even larger one when the liquid vaporizes. This behavior is typical of all substances; melting and vaporization are accompanied by relatively large increases in entropy.

> We'll get to the second law shortly.

EXAMPLE 17.2   Predict whether ΔS is positive or negative for each of the following processes:

> What is the sign of ΔS for:
> egg → omelet?

(a) taking Dry Ice from a freezer where its temperature is $-80°C$ and allowing it to warm to room temperature.
(b) dissolving bromine in hexane.
(c) condensing gaseous bromine to liquid bromine.

Strategy   Consider the relative disorder of final and initial states; remember that entropy increases in the order solid < liquid < gas.

Solution

(a) This process involves an increase in temperature and a phase change from solid to gas: $\Delta S > 0.$
(b) The solution is more random than the original liquids: $\Delta S > 0.$
(c) This is a phase change from gas to liquid, from disorder to order, $\Delta S < 0.$

Standard Molar Entropies

The entropy of a substance, unlike its enthalpy, can be evaluated directly. The details of how this is done are beyond the level of this text, but Figure 17.4 shows the results for one substance, ammonia. From such a plot you can read off the **standard molar entropy** at 1 atm pressure and any given temperature, most often 25°C. This quantity is given the symbol $S°$ and has the units of joules per mole per kelvin ($J/mol \cdot K$). From Figure 17.4, it appears that

$$S° \ NH_3(g) \text{ at } 25°C \approx 192 \ J/mol \cdot K$$

Standard molar entropies of elements, compounds, and aqueous ions are listed in Table 17.1. Notice that—

TABLE 17.1 Standard Entropies at 25°C ($J/mol \cdot K$) of Elements and Compounds at 1 atm, Aqueous Ions at 1 M [OHT]

Elements

Ag(s)	42.6	$Cl_2(g)$	223.0	$I_2(s)$	116.1	$O_2(g)$	205.0
Al(s)	28.3	Cr(s)	23.8	K(s)	64.2	Pb(s)	64.8
Ba(s)	62.8	Cu(s)	33.2	Mg(s)	32.7	$P_4(s)$	164.4
$Br_2(l)$	152.2	$F_2(g)$	202.7	Mn(s)	32.0	S(s)	31.8
C(s)	5.7	Fe(s)	27.3	$N_2(g)$	191.5	Si(s)	18.8
Ca(s)	41.4	$H_2(g)$	130.6	Na(s)	51.2	Sn(s)	51.6
Cd(s)	51.8	Hg(l)	76.0	Ni(s)	29.9	Zn(s)	41.6

Compounds

AgBr(s)	107.1	$CaCl_2(s)$	104.6	$H_2O(g)$	188.7	$NH_4NO_3(s)$	151.1
AgCl(s)	96.2	$CaCO_3(s)$	92.9	$H_2O(l)$	69.9	NO(g)	210.7
AgI(s)	115.5	CaO(s)	39.8	$H_2O_2(l)$	109.6	$NO_2(g)$	240.0
$AgNO_3(s)$	140.9	$Ca(OH)_2(s)$	83.4	$H_2S(g)$	205.7	$N_2O_4(g)$	304.2
$Ag_2O(s)$	121.3	$CaSO_4(s)$	106.7	$H_2SO_4(l)$	156.9	NaCl(s)	72.1
$Al_2O_3(s)$	50.9	$CdCl_2(s)$	115.3	HgO(s)	70.3	NaF(s)	51.5
$BaCl_2(s)$	123.7	CdO(s)	54.8	KBr(s)	95.9	NaOH(s)	64.5
$BaCO_3(s)$	112.1	$Cr_2O_3(s)$	81.2	KCl(s)	82.6	NiO(s)	38.0
BaO(s)	70.4	CuO(s)	42.6	$KClO_3(s)$	143.1	$PbBr_2(s)$	161.5
$BaSO_4(s)$	132.2	$Cu_2O(s)$	93.1	$KClO_4(s)$	151.0	$PbCl_2(s)$	136.0
$CCl_4(l)$	216.4	CuS(s)	66.5	$KNO_3(s)$	133.0	PbO(s)	66.5
$CHCl_3(l)$	201.7	$Cu_2S(s)$	120.9	$MgCl_2(s)$	89.6	$PbO_2(s)$	68.6
$CH_4(g)$	186.2	$CuSO_4(s)$	107.6	$MgCO_3(s)$	65.7	$PCl_3(g)$	311.7
$C_2H_2(g)$	200.8	$Fe(OH)_3(s)$	106.7	MgO(s)	26.9	$PCl_5(g)$	364.5
$C_2H_4(g)$	219.5	$Fe_2O_3(s)$	87.4	$Mg(OH)_2(s)$	63.2	$SiO_2(s)$	41.8
$C_2H_6(g)$	229.5	$Fe_3O_4(s)$	146.4	$MgSO_4(s)$	91.6	$SnO_2(s)$	52.3
$C_3H_8(g)$	269.9	HBr(g)	198.6	MnO(s)	59.7	$SO_2(g)$	248.1
$CH_3OH(l)$	126.8	HCl(g)	186.8	$MnO_2(s)$	53.0	$SO_3(g)$	256.7
$C_2H_5OH(l)$	160.7	HF(g)	173.7	$NH_3(g)$	192.3	$ZnI_2(s)$	161.1
CO(g)	197.6	HI(g)	206.5	$N_2H_4(l)$	121.2	ZnO(s)	43.6
$CO_2(g)$	213.6	$HNO_3(l)$	155.6	$NH_4Cl(s)$	94.6	ZnS(s)	57.7

Cations

$Ag^+(aq)$	72.7	$Hg^{2+}(aq)$	−32.2
$Al^{3+}(aq)$	−321.7	$K^+(aq)$	102.5
$Ba^{2+}(aq)$	9.6	$Mg^{2+}(aq)$	−138.1
$Ca^{2+}(aq)$	−53.1	$Mn^{2+}(aq)$	−73.6
$Cd^{2+}(aq)$	−73.2	$Na^+(aq)$	59.0
$Cu^+(aq)$	40.6	$NH_4^+(aq)$	113.4
$Cu^{2+}(aq)$	−99.6	$Ni^{2+}(aq)$	−128.9
$Fe^{2+}(aq)$	−137.7	$Pb^{2+}(aq)$	10.5
$Fe^{3+}(aq)$	−315.9	$Sn^{2+}(aq)$	−17.4
$H^+(aq)$	0.0	$Zn^{2+}(aq)$	−112.1

Anions

$Br^-(aq)$	82.4	$HPO_4^{2-}(aq)$	−33.5
$CO_3^{2-}(aq)$	−56.9	$HSO_4^-(aq)$	131.8
$Cl^-(aq)$	56.5	$I^-(aq)$	111.3
$ClO_3^-(aq)$	162.3	$MnO_4^-(aq)$	191.2
$ClO_4^-(aq)$	182.0	$NO_2^-(aq)$	123.0
$CrO_4^{2-}(aq)$	50.2	$NO_3^-(aq)$	146.4
$Cr_2O_7^{2-}(aq)$	261.9	$OH^-(aq)$	−10.8
$F^-(aq)$	−13.8	$PO_4^{3-}(aq)$	−222
$HCO_3^-(aq)$	91.2	$S^{2-}(aq)$	−14.6
$H_2PO_4^-(aq)$	90.4	$SO_4^{2-}(aq)$	20.1

- *elements have nonzero standard entropies.* This means that in calculating the standard entropy change for a reaction, $\Delta S°$, elements as well as compounds must be taken into account.
- *standard molar entropies of pure substances* (elements and compounds) *are always positive quantities* ($S° > 0$).
- *aqueous ions may have negative S° values.* This is a consequence of the arbitrary way in which ionic entropies are defined, taking

$$S° \text{ H}^+(aq) = 0$$

The fluoride ion has a standard entropy 13.8 units less than that of H^+; hence $S° \text{ F}^-(aq) = -13.8 \text{ J/mol} \cdot \text{K}$.

You will also notice that gases, as a group, have higher entropies than liquids or solids. Moreover, among substances of similar structure and physical state, entropy usually increases with molar mass. Compare, for example, the hydrocarbons

$$CH_4(g) \qquad S° = 186.2 \text{ J/mol} \cdot \text{K}$$
$$C_2H_6(g) \qquad S° = 229.5 \text{ J/mol} \cdot \text{K}$$
$$C_3H_8(g) \qquad S° = 269.9 \text{ J/mol} \cdot \text{K}$$

As the molecule becomes more complex, there are more ways for the atoms to move about with respect to one another, leading to a higher entropy.

Decomposition of ammonium nitrate. The value of ΔS for this reaction,

$$NH_4NO_3(s) \longrightarrow N_2O(g) + 2H_2O\ (g)$$

is positive. That can be predicted since a solid decomposes to gases. *(Charles D. Winters)*

$\Delta S°$ for Reactions

Table 17.1 can be used to calculate the **standard entropy change,** $\Delta S°$, for reactions, using the relation

$$\boxed{\Delta S° = \sum S°_{\text{products}} - \sum S°_{\text{reactants}}}$$

In taking these sums, the standard molar entropies are multiplied by the number of moles specified in the balanced chemical equation.

To show how this relation is used, consider the reaction

$$CaCO_3(s) \longrightarrow CaO(s) + CO_2(g)$$

$$\Delta S° = S°CaO(s) + S°CO_2(g) - S°CaCO_3(s)$$

$$\Delta S° = 1 \text{ mol}\left(39.8 \ \frac{J}{\text{mol} \cdot \text{K}}\right) + 1 \text{ mol}\left(213.6 \ \frac{J}{\text{mol} \cdot \text{K}}\right) - 1 \text{ mol}\left(92.9 \ \frac{J}{\text{mol} \cdot \text{K}}\right)$$

$$= 39.8 \text{ J/K} + 213.6 \text{ J/K} - 92.9 \text{ J/K} = +160.5 \text{ J/K}$$

You will notice that $\Delta S°$ for the decomposition of calcium carbonate is a positive quantity. This is reasonable because the gas formed, CO_2, has a much higher molar entropy than either of the solids, CaO or $CaCO_3$. As a matter of fact, it is almost always found that *a reaction that results in an increase in the number of moles of gas is accompanied by an increase in entropy. Conversely, if the number of moles of gas decreases, $\Delta S°$ is a negative quantity.* Consider, for example, the reaction

$$2H_2(g) + O_2(g) \longrightarrow 2H_2O(l)$$

$$\Delta S° = 2S° \text{ H}_2O(l) - 2S° \text{ H}_2(g) - S° \text{ O}_2(g)$$

$$= 139.8 \text{ J/K} - 261.2 \text{ J/K} - 205.0 \text{ J/K} = -326.4 \text{ J/K}$$

⊙ **See Screen 20.5, Calculating ΔS for a Chemical Reaction.**

⌐ The units of $S°$ are J/mol·K: those of $\Delta S°$ are J/K. Why the difference?

⌐ If there is no change to the number of moles of gas, ΔS is usually small.

EXAMPLE 17.3 Calculate $\Delta S°$ for the process by which calcium hydroxide dissolves in water: $Ca(OH)_2(s) \longrightarrow Ca^{2+}(aq) + 2OH^-(aq)$.

Strategy Apply the relation

$$\Delta S° = \sum S°_{products} - \sum S°_{reactants}$$

Use data in Table 17.1; take account of the coefficients in the balanced equation.

Solution

$$\Delta S° = S° \, Ca^{2+}(aq) + 2S° \, OH^-(aq) - S° \, Ca(OH)_2(s)$$
$$= -53.1 \text{ J/K} + 2(-10.8 \text{ J/K}) - 83.4 \text{ J/K} = \boxed{-158.1 \text{ J/K}}$$

The Second Law of Thermodynamics

See Screen 20.6, The Second Law of Thermodynamics.

The relationship between entropy change and spontaneity can be expressed through a basic principle of nature known as the second law of thermodynamics. One way to state this law is to say that, *in a spontaneous process, there is a net increase in entropy, taking into account both system and surroundings.* That is,

$$\Delta S_{universe} = (\Delta S_{system} + \Delta S_{surroundings}) > 0 \qquad \text{spontaneous process}$$

In this sense, the universe is running down.

(Recall from Chapter 8 that the system is that portion of the universe on which attention is focused; the surroundings include everything else.)

Notice that the second law refers to the total entropy change, involving both system and surroundings. For many spontaneous processes, the entropy change for the system is a *negative* quantity. Consider, for example, the rusting of iron, a spontaneous process:

$$2Fe(s) + \tfrac{3}{2}O_2(g) + 3H_2O(l) \longrightarrow 2Fe(OH)_3(s)$$

$\Delta S°$ for this system at 25°C and 1 atm can be calculated from a table of standard entropies; it is found to be -358.4 J/K. The negative sign of $\Delta S°$ is entirely consistent with the second law. All the law requires is that the entropy change of the surroundings be greater than 358.4 J/K, so that $\Delta S_{universe} > 0$.

In principle, the second law can be used to determine whether a reaction is spontaneous. To do that, however, requires calculating the entropy change for the surroundings, which is not easy. We follow a conceptually simpler approach (Section 17.3), which deals only with the thermodynamic properties of chemical *systems*.

17.3 FREE ENERGY, G

See Screen 20.7, Gibbs Free Energy.

As pointed out earlier, there are two thermodynamic quantities that affect reaction spontaneity. One of these is the enthalpy, H; the other is the entropy, S. The problem is to put these two quantities together in such a way as to arrive at a single function whose sign will determine whether a reaction is spontaneous. This problem was first solved more than a century ago by J. Willard Gibbs, who introduced a new quantity, now called the **Gibbs free energy** and given the symbol G. Gibbs showed that for a reaction taking place at constant pressure and temperature, ΔG represents that portion of the total energy change that is available (i.e., "free") to do useful work. If, for example, ΔG for a reaction is -270 kJ, it is possible to obtain 270 kJ of useful work from the reaction. Conversely, if ΔG is $+270$ kJ, at least that much energy in the form of work must be supplied to make the reaction take place.

The basic definition of the Gibbs free energy is

$$G = H - TS$$

where T is the absolute (Kelvin) temperature. The free energy of a substance, like its enthalpy and entropy, is a state property; its value is determined only by the state of a system, not by how it got there. Putting it another way, ΔG for a reaction depends only on the nature of products and reactants and the conditions (temperature, pressure, and concentration). It does *not* depend on the path by which the reaction is carried out.

The sign of the free energy change can be used to determine the spontaneity of a reaction carried out at constant temperature and pressure.

1. **If ΔG is negative, the reaction is spontaneous.**
2. **If ΔG is positive, the reaction will not take place spontaneously.** Instead, the reverse reaction will be spontaneous.
3. **If ΔG is 0, the system is at equilibrium;** there is no tendency for reaction to occur in either direction.

Putting it another way, ΔG is a measure of the driving force of a reaction. *Reactions, at constant pressure and temperature, go in such a direction as to decrease the free energy of the system.* This means that the direction in which a reaction takes place is determined by the relative free energies of products and reactants. If the products at the specified conditions of temperature, pressure, and concentration have a lower free energy than the reactants ($G_{products} < G_{reactants}$), the forward reaction will occur (Figure 17.5). If the reverse is true ($G_{reactants} < G_{products}$), the reverse reaction is spontaneous. Finally, if $G_{products} = G_{reactants}$, there is no driving force to make the reaction go in either direction.

Relation Among ΔG, ΔH, and ΔS

From the defining equation for free energy, it follows that at constant temperature

$$\Delta G = \Delta H - T\Delta S$$

where ΔG, ΔH, and ΔS are the changes in free energy, enthalpy, and entropy for a reaction. This relation, known as the *Gibbs-Helmholtz equation,* is perhaps the

> A spontaneous process is capable of producing useful work.

> A system has its minimum free energy at equilibrium.

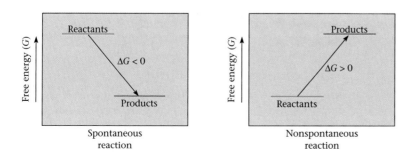

FIGURE 17.5

Sign of free energy and spontaneity. For a spontaneous reaction, the free energy of the products is less than that of the reactants: $\Delta G° < 0$. For a nonspontaneous reaction, the reverse is true, $\Delta G° > 0$. **OHT**

most important equation in chemical thermodynamics. As you can see, there are two factors that tend to make ΔG negative and hence lead to a spontaneous reaction:

1. ***A negative value of*** ΔH***.*** Exothermic reactions ($\Delta H < 0$) tend to be spontaneous, inasmuch as they contribute to a negative value of ΔG. On the molecular level, this means that there will be a tendency to form "strong" bonds at the expense of "weak" ones.
2. ***A positive value of*** ΔS***.*** If the entropy change is positive ($\Delta S > 0$), the term $-T\Delta S$ will make a negative contribution to ΔG. Hence there will be a tendency for a reaction to be spontaneous if the products are less ordered than the reactants.

In many physical changes, the entropy increase is the major driving force. This situation applies when two liquids with similar intermolecular forces, such as benzene (C_6H_6) and toluene (C_7H_8), are mixed. There is no change in enthalpy but the entropy increases because the molecules of benzene and toluene are mixed randomly in solution.*

In certain reactions, ΔS is nearly zero, and ΔH is the only important component of the driving force for spontaneity. An example is the synthesis of hydrogen fluoride from the elements:

$$\tfrac{1}{2}H_2(g) + \tfrac{1}{2}F_2(g) \longrightarrow HF(g)$$

For this reaction, ΔH is a large negative number, -271.1 kJ, showing that the bonds in HF are stronger than those in the H_2 and F_2 molecules. As you might expect for a gaseous reaction in which there is no change in the number of moles, ΔS is very small, about 0.0070 kJ/K. The free energy change, ΔG, at 1 atm is -273.2 kJ at 25°C, almost identical to ΔH. Even at very high temperatures, the difference between ΔG and ΔH is small, amounting to only about 14 kJ at 2000 K.

17.4   STANDARD FREE ENERGY CHANGE, $\Delta G°$

Although the Gibbs-Helmholtz equation is valid under all conditions we will apply it only under **standard conditions,** where—

"Standard conditions" has a quite different meaning from "STP."

■ gases are at one atmosphere partial pressure.
■ ions or molecules in solution are at one molar concentration.

In other words, we will use the equation in the form

$$\boxed{\Delta G° = \Delta H° - T\Delta S°}$$

where $\Delta G°$ is the **standard free energy change** (1 atm, 1 M); $\Delta H°$ is the standard enthalpy change, which can be calculated from heats of formation, $\Delta H_f°$ (listed in Table 8.3, Chapter 8); and $\Delta S°$ is the standard entropy change (Table 17.1).

You will recall that the sign of ΔG can be correlated with the spontaneity of reaction. We can do the same thing with $\Delta G°$ provided we restrict our attention to standard conditions (1 atm, 1 M).

---

* The formation of a *water* solution is often accompanied by a *decrease* in entropy because hydrogen bonding or hydration effects lead to a highly ordered solution structure. Recall from Example 17.3 that when one mole of $Ca(OH)_2$ dissolves in water, $\Delta S° = -158.1$ J/K.

CHEMISTRY ▪ The Human Side

Two theoreticians working in the latter half of the nineteenth century changed the very nature of chemistry by deriving the mathematical laws that govern the behavior of matter undergoing physical or chemical change. One of these was James Clerk Maxwell, whose contributions to kinetic theory were discussed in Chapter 5. The other was J. Willard Gibbs, Professor of Mathematical Physics at Yale from 1871 until his death in 1903.

In 1876 Gibbs published the first portion of a remarkable paper in the *Transactions of the Connecticut Academy of Sciences* titled, "On the Equilibrium of Heterogeneous Substances." When the paper was completed in 1878 (it was 323 pages long), the foundation was laid for the science of chemical thermodynamics. For the first time, the concept of free energy appeared. Included as well were the basic principles of chemical equilibrium (Chapter 12), phase equilibrium (Chapter 9), and the relations governing energy changes in electrical cells (Chapter 18).

If Gibbs had never published another paper, this single contribution would have placed him among the greatest theoreticians in the history of science. Generations of experimental scientists have established their reputations by demonstrating in the laboratory the validity of the relationships that Gibbs derived at his desk. Many of these relationships were rediscovered by others; an example is the Gibbs-Helmholtz equation developed in 1882 by Helmholtz, a prestigious German physiologist and physicist who was completely unaware of Gibbs's work.

J. Willard Gibbs is often cited as an example of the "prophet without honor in his own country." His colleagues in New Haven and elsewhere in the United States seem not to have realized the significance of his work until late in his life. During his first ten years as a professor at Yale he received no salary. In 1920, when he was first proposed for the Hall of Fame of Distinguished Americans at New York University, he received 9 votes out of a possible 100. Not until 1950 was he elected to that body. Even today the name of J. Willard Gibbs is generally unknown among educated Americans outside of those interested in the natural sciences.

Admittedly, Gibbs himself was largely responsible for the fact that for many years his work did not attract the attention it deserved. He made little effort to publicize it; the *Transactions of the Connecticut Academy of Sciences* was hardly the leading scientific journal of its day. Gibbs was one of those rare individuals who seem to have no inner need for recognition by contemporaries. His satisfaction came from solving a problem in his mind; having done so, he was ready to proceed to other problems. His papers are not easy to read; he seldom cites examples to illustrate his abstract reasoning. Frequently, the implications of the laws that he derives are left for the readers to grasp on their own. One of his colleagues at Yale confessed many years later that none of the members of the Connecticut Academy of Sciences understood his paper on thermodynamics; as he put it, "We knew Gibbs and took his contributions on faith."

(Burndy Library/AIP Emilio Segre Visual Archives)

J. Willard Gibbs
(1839–1903)

⌐They started paying him when he got an offer from Johns Hopkins.

1. If $\Delta G°$ is negative, the reaction is spontaneous at standard conditions. For example, the following reaction is spontaneous at 25°C:

$$CaO(s) + CO_2(g, 1 \text{ atm}) \longrightarrow CaCO_3(s) \qquad \Delta G° \text{ at } 25°C = -130.4 \text{ kJ}$$

2. If $\Delta G°$ is positive, the reaction is nonspontaneous at standard conditions.

Precipitation of silver chloride. A spontaneous reaction for which $\Delta G°$ is negative. *(Charles D. Winters)*

The reaction

$$AgCl(s) \longrightarrow Ag^+(aq, 1\ M) + Cl^-(aq, 1\ M) \qquad \Delta G° \text{ at } 25°C = +55.7 \text{ kJ}$$

is nonspontaneous at 25°C. The reverse reaction is spontaneous; when solutions of $AgNO_3$ and HCl are mixed in such a way that $[Ag^+] = [Cl^-] = 1\ M$, silver chloride precipitates.

3. If $\Delta G°$ is 0, the system is at equilibrium at standard conditions; there is no tendency for the reaction to occur in either direction. An example is the vaporization of water at 100°C and 1 atm:

$$H_2O(l) \longrightarrow H_2O(g, 1\ atm) \qquad \Delta G° = 0$$

under these conditions (i.e., at the normal boiling point), the molar free energies of liquid and gaseous water are identical. Hence $\Delta G° = 0$ and the system is at equilibrium.

Calculation of $\Delta G°$ at 25°C; Free Energies of Formation

To illustrate the use of the Gibbs-Helmholtz equation

$$\Delta G° = \Delta H° - T\Delta S°$$

we apply it first to find $\Delta G°$ for a reaction at 25°C (Example 17.4). To do this, the units of $\Delta H°$ and $\Delta S°$ must be consistent. Using $\Delta H°$ in kilojoules, it is necessary to convert $\Delta S°$ from joules per kelvin to *kilojoules per kelvin*.

EXAMPLE 17.4 For the reaction $CaSO_4(s) \longrightarrow Ca^{2+}(aq) + SO_4^{2-}(aq)$, calculate

(a) $\Delta H°$ (b) $\Delta S°$ (c) $\Delta G°$ at 25°C

Strategy $\Delta H°$ is calculated from Table 8.3, Chapter 8, using the relation

$$\Delta H° = \sum \Delta H°_{f\ products} - \sum \Delta H°_{f\ reactants}$$

$\Delta S°$ is obtained from Table 17.1 using the relation

$$\Delta S° = \sum S°_{products} - \sum S°_{reactants}$$

Convert $\Delta S°$ to kJ/K using the relation $1\ J = 10^{-3}$ kJ. Finally, calculate $\Delta G°$ from the Gibbs-Helmholtz equation with $T = 298$ K.

Solution

(a) $\Delta H° = \Delta H°_f\ Ca^{2+}(aq) + \Delta H°_f\ SO_4^{2-}(aq) - \Delta H°_f\ CaSO_4(s)$
$= -542.8 \text{ kJ} - 909.3 \text{ kJ} - (-1434.1 \text{ kJ}) = \boxed{-18.0 \text{ kJ}}$

(b) $\Delta S° = S°\ Ca^{2+}(aq) + S°\ SO_4^{2-}(aq) - S°\ CaSO_4(s)$
$= -53.1 \text{ J/K} + 20.1 \text{ J/K} - 106.7 \text{ J/K} = -139.7 \text{ J/K}$

$= -139.7\ \dfrac{J}{K} \times \dfrac{10^{-3}\ kJ}{1\ J} = \boxed{-0.1397\ \dfrac{kJ}{K}}$

(c) $\Delta G° = \Delta H° - T\Delta S°$
$= -18.0 \text{ kJ} - 298 \text{ K} \times (-0.1397 \text{ kJ/K}) = \boxed{+23.6 \text{ kJ}}$

Reality Check $\Delta G°$ is positive, so this reaction at standard conditions

$$CaSO_4(s) \longrightarrow Ca^{2+}(aq, 1\ M) + SO_4^{2-}(aq, 1\ M)$$

should not be spontaneous. In other words, calcium sulfate should not dissolve in water to give a 1 M solution. This is indeed the case. The solubility of $CaSO_4$ at 25°C is considerably less than 1 mol/L.

This conversion is necessary to find $\Delta G°$ in kilojoules.

The Gibbs-Helmholtz equation can be used to calculate the **standard free energy of formation** of a compound. This quantity, ΔG_f°, is analogous to the enthalpy of formation, ΔH_f°. It is defined as the free energy change per mole when a compound is formed from the elements in their stable states at 1 atm.

EXAMPLE 17.5   Calculate the standard free energy of formation, ΔG_f°, at 25°C for $CH_4(g)$, using data in Tables 8.3 and 17.1.

Strategy   By definition, ΔG_f° for CH_4 is ΔG° for the reaction $C(s) + 2H_2(g) \longrightarrow CH_4(g)$. The calculations are entirely analogous to those in Example 17.4.

Solution

$$\Delta H^\circ = \Delta H_f^\circ \, CH_4(g) = -74.8 \text{ kJ}$$

$$\Delta S^\circ = S^\circ \, CH_4(g) - S^\circ \, C(s) - 2S^\circ \, H_2(g)$$

$$= 186.2 \text{ J/K} - 5.7 \text{ J/K} - 2(130.6 \text{ J/K}) = -80.7 \text{ J/K} = -0.0807 \text{ kJ/K}$$

At 25°C, that is, 298 K,

$$\Delta G^\circ = \Delta H^\circ - 298 \, \Delta S^\circ = -74.8 \text{ kJ} - 298 \text{ K}(-0.0807 \text{ kJ/K}) = -50.7 \text{ kJ}$$

Hence the free energy of formation of $CH_4(g)$ at 25°C is -50.7 kJ/mol.

Tables of standard free energies of formation at 25°C of compounds and ions in solution are given in Appendix 1 (along with standard heats of formation and standard entropies). Notice that, for most compounds, ΔG_f° is a negative quantity, which means that the compound can be formed spontaneously from the elements. This is true for water:

$$H_2(g) + \tfrac{1}{2}O_2(g) \longrightarrow H_2O(l) \qquad \Delta G_f^\circ \, H_2O(l) = -237.2 \text{ kJ/mol}$$

and, at least in principle, for methane:

$$C(s) + 2H_2(g) \longrightarrow CH_4(g) \qquad \Delta G_f^\circ \, CH_4(g) = -50.7 \text{ kJ/mol}$$

A few compounds, including acetylene, have positive free energies of formation ($\Delta G_f^\circ \, C_2H_2(g) = +209.2$ kJ/mol). These compounds cannot be made from the elements at ordinary temperatures and pressures; indeed, they are potentially unstable with respect to the elements. In the case of acetylene, the reaction

$$C_2H_2(g) \longrightarrow 2C(s) + H_2(g) \qquad \Delta G^\circ \text{ at } 25°C = -209.2 \text{ kJ}$$

occurs with explosive violence unless special precautions are taken.

Values of ΔG_f° can be used to calculate free energy changes for reactions. The relationship is entirely analogous to that given for enthalpies in Chapter 8:

$$\boxed{G^\circ_{\text{reaction}} = \Sigma \Delta G_f^\circ \text{ products} - \Sigma \Delta G_f^\circ \text{ reactants}}$$

If you calculate ΔG° in this way, you should keep in mind an important limitation. $\Delta G^\circ_{\text{reaction}}$ *is valid only at the temperature at which ΔG_f° data are tabulated, in this case 25°C.* ΔG° varies considerably with temperature, so this approach is not even approximately valid at other temperatures.

ΔG_f° for an element, like ΔH_f°, is zero.

To find ΔG° at temperatures other than 25°C, use the Gibbs-Helmholtz equation.

Calculation of $\Delta G°$ at Other Temperatures

To a good degree of approximation, the temperature variation of $\Delta H°$ and $\Delta S°$ can be neglected.* This means that to apply the Gibbs-Helmholtz equation

$$\Delta G° = \Delta H° - T\Delta S°$$

at temperatures other than 25°C, you need only change the value of T. The quantities $\Delta H°$ and $\Delta S°$ can be calculated in the usual way from tables of standard enthalpies and entropies.

EXAMPLE 17.6   Calculate $\Delta G°$ for the reaction $Cu(s) + H_2O(g) \longrightarrow CuO(s) + H_2(g)$ at 500 K.

Strategy   Calculate $\Delta H°$ using Table 8.3 and $\Delta S°$ using Table 17.1. Then apply the Gibbs-Helmholtz equation with $T = 500$ K.

Solution   From Table 8.3

$$\Delta H° = \Delta H_f° \, CuO(s) - \Delta H_f° \, H_2O(g)$$

$$= -157.3 \text{ kJ} + 241.8 \text{ kJ} = +84.5 \text{ kJ}$$

From Table 17.1

$$\Delta S° = S° \, CuO(s) + S° \, H_2(g) - S° \, Cu(s) - S° \, H_2O(g)$$

$$= 42.6 \text{ J/K} + 130.6 \text{ J/K} - 33.2 \text{ J/K} - 188.7 \text{ J/K}$$

$$= -48.7 \text{ J/K} = -0.0487 \text{ kJ/K}$$

Hence since $\Delta G° = \Delta H° - T\Delta S°$:

$$\Delta G° = +84.5 \text{ kJ} - 500 \text{ K} \times (-0.0487 \text{ kJ/K}) = \boxed{+108.9 \text{ kJ}}$$

Reality Check $\Delta G°$ is positive, so the reaction is not spontaneous. Instead, the reverse reaction occurs at 1 atm and 500 K (227°C). Under these conditions, CuO is reduced to copper in a stream of H_2 gas.

From Example 17.6 and the preceding discussion, it should be clear that $\Delta G°$, unlike $\Delta H°$ and $\Delta S°$, is strongly dependent on temperature. This comes about, of course, because of the T in the Gibbs-Helmholtz equation:

$$\Delta G° = \Delta H° - T\Delta S°$$

Comparing this equation with that of a straight line,

$$y = b + mx$$

it is clear that a plot of $\Delta G°$ versus T should be linear, with a slope of $-\Delta S°$ and a y-intercept (at 0 K) of $\Delta H°$ (Figure 17.6).

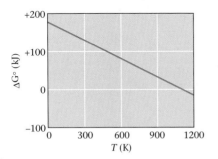

FIGURE 17.6
Variation of $\Delta G°$ with temperature for the decomposition of calcium carbonate. For the reaction

$$CaCO_3(s) \longrightarrow CaO(s) + CO_2(g)$$

$\Delta G°$ decreases from +178.3 kJ at 0 K (the value of $\Delta H°$) to −14.3 kJ at 1200 K. The slope of the straight line is $-\Delta S° = -0.1605$ kJ/K.

---

* As far as the Gibbs-Helmholtz equation is concerned, there is another reason for ignoring the temperature dependence of ΔH and ΔS. These two quantities always change in the same direction as the temperature changes (that is, if ΔH becomes more positive, so does ΔS). Hence the two effects tend to cancel each other.

17.5   EFFECT OF TEMPERATURE, PRESSURE, AND CONCENTRATION ON REACTION SPONTANEITY

A change in reaction conditions can, and often does, change the direction in which a reaction occurs spontaneously. The Gibbs-Helmholtz equation in the form

$$\Delta G° = \Delta H° - T\Delta S°$$

is readily applied to deduce the effect of temperature on reaction spontaneity. To consider the effect of pressure or concentration, we need an analogous expression for the dependence of ΔG on these factors.

Temperature

When the temperature of a reaction system is increased, the sign of $\Delta G°$, and hence the direction in which the reaction proceeds spontaneously, may or may not change. Whether it does or does not depends on the signs of $\Delta H°$ and $\Delta S°$. The four possible situations, deduced from the Gibbs-Helmholtz equation, are summarized in Figure 17.7.

See Screen 20.8, Free Energy and Temperature.

If $\Delta H°$ and $\Delta S°$ have opposite signs (Figure 17.7, I and III), it is impossible to reverse the direction of spontaneity by a change in temperature alone. The two terms $\Delta H°$ and $-T\Delta S°$ reinforce one another. Hence $\Delta G°$ has the same sign at all temperatures. Reactions of this type are rather uncommon. One such reaction is that discussed in Example 17.6:

$$Cu(s) + H_2O(g) \longrightarrow CuO(s) + H_2(g)$$

Here $\Delta H°$ is $+84.5$ kJ and $\Delta S°$ is -0.0487 kJ/K. Hence

$$\Delta G° = \Delta H° - T\Delta S°$$
$$= +84.5 \text{ kJ} + T(0.0487 \text{ kJ/K})$$

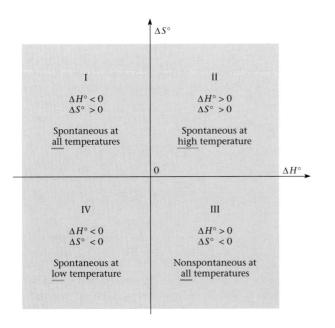

FIGURE 17.7

Effect of temperature on spontaneity. The spontaneity of a reaction is dependent on temperature when $\Delta H°$ and $\Delta S°$ have the same sign (II and IV), which is the most common situation. Spontaneity does not vary with temperature when $\Delta H°$ and $\Delta S°$ have different signs (I and III). **OHT**

Clearly $\Delta G°$ is positive at all temperatures. The reaction cannot take place spontaneously at 1 atm regardless of temperature.

It is more common to find that $\Delta H°$ and $\Delta S°$ have the same sign (Figure 17.7, II and IV). When this happens, the enthalpy and entropy factors oppose each other. $\Delta G°$ changes sign as temperature increases, and the direction of spontaneity reverses. At low temperatures, $\Delta H°$ predominates, and the exothermic reaction, which may be either the forward or the reverse reaction, occurs. As the temperature rises, the quantity $T\Delta S°$ increases in magnitude and eventually exceeds $\Delta H°$. At high temperatures, the reaction that leads to an increase in entropy occurs. In most cases, 25°C is a "low" temperature, at least at a pressure of 1 atm. This explains why exothermic reactions are usually spontaneous at room temperature and atmospheric pressure.

An example of a reaction for which $\Delta H°$ and $\Delta S°$ have the same sign is the decomposition of calcium carbonate:

$$CaCO_3(s) \longrightarrow CaO(s) + CO_2(g)$$

Here $\Delta H° = +178.3$ kJ and $\Delta S° = +160.5$ J/K. Hence

$$\Delta G° = +178.3 \text{ kJ} - T(0.1605 \text{ kJ/K})$$

For this reaction at "low" temperatures, such as 298 K,

$$\Delta G° = +178.3 \text{ kJ} - 47.8 \text{ kJ} = +130.5 \text{ kJ}$$

the $\Delta H°$ term predominates, $\Delta G°$ is positive, and the reaction is nonspontaneous at 1 atm. Conversely, at "high" temperatures, such as 2000 K,

$$\Delta G° = +178.3 \text{ kJ} - 321.0 \text{ kJ} = -142.7 \text{ kJ}$$

the $T\Delta S°$ term predominates, $\Delta G°$ is negative, and the reaction is spontaneous at 1 atm.

> Copper metal doesn't react with water vapor, period.

> CaO is made commercially by heating $CaCO_3$.

EXAMPLE 17.7

At what temperature does $\Delta G°$ become zero for the reaction

$$CaCO_3(s) \longrightarrow CaO(s) + CO_2(g)$$

Strategy This is another application of the Gibbs-Helmholtz equation. Setting $\Delta G° = 0$ gives $\Delta H° = T\Delta S°$; solving for T gives $T = \Delta H°/\Delta S°$. Since you know $\Delta H°$ and $\Delta S°$, T is readily calculated.

Solution From the preceding discussion.

$$\Delta H° = +178.3 \text{ kJ} \qquad \Delta S° = 0.1605 \text{ kJ/K}$$

Hence

$$T = \frac{\Delta H°}{\Delta S°} = \frac{178.3 \text{ kJ}}{0.1605 \text{ kJ/K}} = \boxed{1110 \text{ K}} \text{ (that is, } 1.110 \times 10^3 \text{ K)}$$

Reality Check The equation

$$T = \Delta H°/\Delta S°$$

offers a general method of calculating the temperature at which a system is at equilibrium at 1 atm. For example, it can be used to calculate the normal boiling point or freezing point of a substance.

Some forms of calcium carbonate. Limestone, seashells, and the mineral calcite do not decompose under everyday conditions. *(Charles D. Winters)*

The development presented in Example 17.7 is an important one from a practical standpoint. It tells us the temperature at which the direction of spontaneity changes. In this case, one can deduce that calcium carbonate must be heated to about 1110 K ($\approx 840°C$) to decompose it to calcium oxide and carbon dioxide at 1 atm. In another case, for the reaction

$$Fe_2O_3(s) + 3H_2(g) \longrightarrow 2Fe(s) + 3H_2O(g)$$

it turns out that $\Delta G°$, which is positive at low temperatures, becomes zero at about 400°C. This means that to reduce hematite ore, Fe_2O_3, to iron in a stream of hydrogen, the system must be heated to 400°C.

Pressure and Concentration

All of the free energy calculations to this point have involved the standard free energy change, $\Delta G°$. It is possible, however, to write a general relation for the free energy change, ΔG, valid under any conditions:

$$\Delta G = \Delta G° + RT \ln Q$$

The quantity Q that appears in this equation is the "reaction quotient" referred to in Chapter 12. It has the same mathematical form as the equilibrium constant, K; the difference is that the terms that appear in Q are arbitrary, initial pressures or concentrations rather than equilibrium values.

To use this general expression for ΔG, you need to express T in degrees K, R in kilojoules per kelvin (8.31×10^{-3} kJ/K). As far as Q is concerned, the general rules for expressing the equilibrium constant are the same as those for expressing the equilibrium constant—

- gases enter as their partial pressures in atmospheres.
- species in aqueous solution enter as their molar concentrations.
- pure liquids or solids do not appear; neither does the solvent in a dilute solution.

P and T affect ΔG but not ΔH; P affects ΔS.

EXAMPLE 17.8   Consider the reaction of zinc with strong acid:

$$Zn(s) + 2H^+(aq) \longrightarrow Zn^{2+}(aq) + H_2(g)$$

(a) Write the expression for Q.
(b) Calculate $\Delta G°$ at 25°C.
(c) Find ΔG at 25°C when $P_{H_2} = 750$ mm Hg, $[Zn^{2+}] = 0.10$ M, $[H^+] = 1.0 \times 10^{-4}$ M.

Strategy   In (a), follow the rules cited above for Q. In (b), calculate $\Delta G°$ using the Gibbs-Helmholtz equation, as in Examples 17.4–17.7. Finally (c), determine ΔG using the relation $\Delta G = \Delta G° + RT \ln Q$. Note that the pressure of H_2 must be expressed in atmospheres.

Solution

(a) $Q = \dfrac{[Zn^{2+}] \times P_{H_2}}{[H^+]^2}$

(b) $\Delta H° = \Delta H_f° \, Zn^{2+} = -153.9$ kJ

$\Delta S° = S° \, Zn^{2+} + S° \, H_2 - S° \, Zn = -23.1$ J/K = -0.0231 kJ/K

$\Delta G° = \Delta H° - T\,\Delta S° = -153.9$ kJ + 298 K $\times$ (0.0231 kJ/K) = $\boxed{-147.0 \text{ kJ}}$

(c) $\Delta G = \Delta G° + RT \ln Q$

$= -147.0$ kJ + 298 K $(8.31 \times 10^{-3}$ kJ/K$) \left[\ln \dfrac{(0.10)(750/760)}{(1.0 \times 10^{-4})^2}\right]$

$= -147.0$ kJ + 39.9 kJ = $\boxed{-107.1 \text{ kJ}}$

FIGURE 17.8
A saturated solution of strontium chromate. The solution of $SrCrO_4$ contains 0.006 mol/L of Sr^{2+} and CrO_4^{2-}. For the reaction at 25°C: $SrCrO_4(s) \rightleftharpoons Sr^{2+}(aq, 0.006\ M) + CrO_4^{2-}$ (aq, 0.006 M), $\Delta G° = 0$.
(Charles D. Winters)

Remember, when $\Delta G = 0$ the system is at equilibrium.

Reality Check   The free energy change in (c) is less negative than at standard conditions ($\Delta G = -107.1\text{kJ}$ versus $\Delta G° = -147.0$ kJ) largely because the concentration of H^+, at $1.0 \times 10^{-4}\ M$, is so low.

As Example 17.8 implies, changes in pressure and/or concentration can have a considerable effect on ΔG. Sometimes, when $\Delta G°$ is close to zero, changing the pressure from 1 atm to some other value can change the direction of spontaneity. Consider, for example, the reaction at 300°C:

$$NH_4Cl(s) \longrightarrow NH_3(g) + HCl(g) \qquad \Delta G° = +13.0 \text{ kJ}$$

The positive sign of $\Delta G°$ implies that ammonium chloride will not decompose at 300°C to give NH_3 and HCl, both at 1 atm pressure. However, notice what happens if $P_{NH_3} = P_{HCl} = 0.10$ atm, in which case $Q = (0.10)^2$:

$$\Delta G = \Delta G° + RT \ln(0.10)^2$$

$$= +13.0 \text{ kJ} + (8.31 \times 10^{-3})(573) \text{ kJ} \times \ln 0.010$$

$$= +13.0 \text{ kJ} - 21.9 \text{ kJ} = -8.9 \text{ kJ}$$

This means that NH_3 and HCl *can* be formed, each at 0.10 atm pressure, by heating ammonium chloride to 300°C.

Another example of how a change in concentration can change the direction of spontaneity involves the reaction

$$SrCrO_4(s) \longrightarrow Sr^{2+}(aq) + CrO_4^{2-}(aq) \qquad \Delta G° = +25.3 \text{ kJ at 25°C}$$

Because $\Delta G°$ is a positive quantity, $SrCrO_4$ does not dissolve spontaneously to form a 1 M solution at 25°C (Figure 17.8). Suppose, however, the concentrations of Sr^{2+} and CrO_4^{2-} are reduced to 0.0010 M:

$$\Delta G = \Delta G° + RT \ln [Sr^{2+}] \times [CrO_4^{2-}]$$

$$= +23.6 \text{ kJ} + (8.31 \times 10^{-3})(298) \text{ kJ} \times [\ln (1.0 \times 10^{-3})^2]$$

$$= +25.3 \text{ kJ} - 34.2 \text{ kJ} = -8.9 \text{ kJ}$$

Strontium chromate should, and does, dissolve spontaneously to form a 0.0010-M water solution. As you might expect, its solubility at 25°C lies between 1 M and 0.0010 M, at about 0.0060 M.

$$\Delta G \text{ at } 0.0060\ M = +25.3 \text{ kJ} + RT \ln (0.0060)^2 = 0$$

17.6   THE FREE ENERGY CHANGE AND THE EQUILIBRIUM CONSTANT

Throughout this chapter we have stressed the relation between the free energy change and reaction spontaneity. For a reaction to be spontaneous at standard conditions (1 atm, 1 M), $\Delta G°$ must be negative. Another indicator of reaction spontaneity is the equilibrium constant, K; if K is greater than 1, the reaction is spontaneous at standard conditions. As you might suppose, the two quantities $\Delta G°$ and K are related. The nature of that relationship is readily deduced by starting with the general equation

$$\Delta G = \Delta G° + RT \ln Q$$

Suppose now that reaction takes place until equilibrium is established, at which point $Q = K$ and $\Delta G = 0$.

$$0 = \Delta G° + RT \ln K$$

or

$$\boxed{\Delta G° = -RT \ln K}$$

See Screen 20.9, Thermodynamics and the Equilibrium Constant.

The equation just written is generally applicable to any system. The equilibrium constant may be the K referred to in our discussion of gaseous equilibrium (Chapter 12), or any of the solution equilibrium constants (K_w, K_a, K_b, K_{sp}, . . .) discussed in subsequent chapters. Notice that $\Delta G°$ is the *standard* free energy change (gases at *1 atm*, species in solution at *1 M*). That is why, in the expression for K, gases enter as their partial pressures in *atmospheres* and ions or molecules in solution as their *molarities*.

EXAMPLE 17.9   Using $\Delta G_f°$ tables in Appendix 1, calculate the solubility product constant, K_{sp}, for AgBr at 25°C.

Strategy   Write out the chemical equation for the solution reaction $AgBr(s) \rightleftharpoons Ag^+(aq) + Br^-(aq)$. Now apply the equation

$$\Delta G° = \Sigma \Delta G_f° \text{ products} - \Sigma \Delta G_f° \text{ reactants}$$

to calculate $\Delta G°$. Then use the equation $\Delta G° = -RT \ln K_{sp}$ to find K_{sp}.

How would you find K_{sp} at 100°C?

Solution

(1) $\Delta G° = \Delta G_f° \, Ag^+ + \Delta G_f° \, Br^- - \Delta G_f° \, AgBr$
     $= 77.1 \text{ kJ} - 104.0 \text{ kJ} + 96.9 \text{ kJ} = 70.0 \text{ kJ}$

(2) $\ln K_{sp} = \dfrac{-\Delta G°}{RT} = \dfrac{-70.0 \text{ kJ}}{(8.31 \times 10^{-3} \text{ kJ/K})(298 \text{ K})} = -28.3$

     $K_{sp} = \boxed{5 \times 10^{-13}}$

Reality Check   This is the value listed in Chapter 16 for K_{sp} of AgBr. Solubility products are usually determined this way.

The relationship $\Delta G° = -RT \ln K$ allows us to relate the standard free energy change to the extent of reaction. Consider, for example, the simple equilibrium system

$$A(g) \rightleftharpoons B(g)$$

at 25°C. Figure 17.9 shows how the extent of reaction varies with the value of $\Delta G°$ for this system. Notice that—

- if $\Delta G°$ is greater than about +20 kJ, the equilibrium constant is so small that virtually no A is converted to B. We would say that, for all practical purposes, the reaction does not occur.
- if $\Delta G°$ is less than about −20 kJ, the equilibrium constant is so large that virtually all of A is converted to B. In essence, the reaction goes to completion.
- only if $\Delta G°$ lies between +20 kJ and −20 kJ will the equilibrium mixture contain appreciable amounts of both A and B. In particular, if $\Delta G° = 0$, $K = 1$ and the equilibrium system contains equal amounts of A and B.

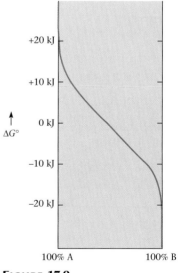

FIGURE 17.9
Dependence of extent of reaction on the value of $\Delta G°$ for the general system $A(g) \rightleftharpoons B(g)$.   **OHT**

17.7 ADDITIVITY OF FREE ENERGY CHANGES; COUPLED REACTIONS

Free energy changes for reactions, like enthalpy or entropy changes, are additive. That is,

$$\text{if Reaction 3} = \text{Reaction 1} + \text{Reaction 2}$$
$$\text{then } \Delta G_3 = \Delta G_1 + \Delta G_2$$

This relation can be regarded as the free energy equivalent of Hess's law (Chapter 8). To illustrate its application, consider the synthesis of $CuCl_2$ from the elements:

$$Cu(s) + \tfrac{1}{2}Cl_2(g) \longrightarrow CuCl(s) \qquad \Delta G° \text{ at } 25°C = -119.9 \text{ kJ}$$
$$\underline{CuCl(s) + \tfrac{1}{2}Cl_2(g) \longrightarrow CuCl_2(s) \qquad \Delta G° \text{ at } 25°C = -55.8 \text{ kJ}}$$
$$Cu(s) + Cl_2(g) \longrightarrow CuCl_2(s)$$

For the overall reaction, $\Delta G°$ at $25°C = -119.9 \text{ kJ} - 55.8 \text{ kJ} = -175.7 \text{ kJ}$.

Because free energy changes are additive, it is often possible to bring about a nonspontaneous reaction by coupling it with a reaction for which $\Delta G°$ is a large negative number. As an example, consider the preparation of iron metal from hematite ore. The reaction

$$Fe_2O_3(s) \longrightarrow 2Fe(s) + \tfrac{3}{2}O_2(g) \qquad \Delta G° \text{ at } 25°C = +742.2 \text{ kJ}$$

is clearly nonspontaneous; even at temperatures as high as 2000°C, $\Delta G°$ is a positive quantity. Suppose, though, that this reaction is "coupled" with the spontaneous oxidation of carbon monoxide:

$$CO(g) + \tfrac{1}{2}O_2(g) \longrightarrow CO_2(g) \qquad \Delta G° \text{ at } 25°C = -257.1 \text{ kJ}$$

The overall reaction is spontaneous:

$$Fe_2O_3(s) \longrightarrow 2Fe(s) + \tfrac{3}{2}O_2(g)$$
$$\underline{3CO(g) + \tfrac{3}{2}O_2(g) \longrightarrow 3CO_2(g)}$$
$$Fe_2O_3(s) + 3CO(g) \longrightarrow 2Fe(s) + 3CO_2(g)$$

For the coupled reaction at 25°C,

$$\Delta G° = +742.2 \text{ kJ} + 3(-257.1 \text{ kJ}) = -29.1 \text{ kJ}$$

The negative sign of $\Delta G°$ implies that although Fe_2O_3 does not spontaneously decompose, it can be converted to iron by reaction with carbon monoxide. This is in fact the reaction used in the blast furnace when iron ore consisting mainly of Fe_2O_3 is reduced to iron (Chapter 20).

Coupled reactions are common in human metabolism. Spontaneous processes, such as the oxidation of glucose,

$$C_6H_{12}O_6(aq) + 6O_2(g) \longrightarrow 6CO_2(g) + 6H_2O \qquad \Delta G° = -2870 \text{ kJ at } 25°C$$

ordinarily do not serve directly as a source of energy. Instead these reactions are used to bring about a nonspontaneous reaction:

$$ADP(aq) + HPO_4^{2-}(aq) + 2H^+(aq) \longrightarrow ATP(aq) + H_2O$$
$$\Delta G° = +31 \text{ kJ at } 25°C$$

⌐ **Many industrial processes involve coupled reactions.**

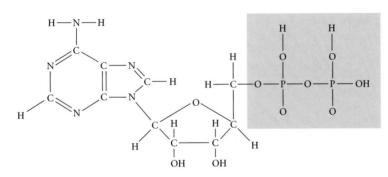

adenosine diphosphate (ADP)

FIGURE 17.10
Structures of ADP and ATP.

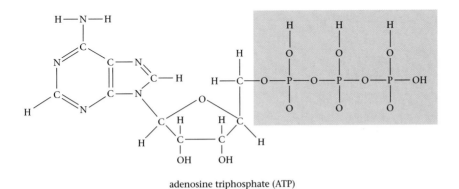

adenosine triphosphate (ATP)

ADP (adenosine diphosphate) and ATP (adenosine triphosphate) are complex organic molecules (Figure 17.10) that, in essence, differ only by the presence of an extra phosphate group in ATP. In the coupled reaction with glucose, about 38 mol of ATP are synthesized for every mole of glucose consumed. This gives an overall free energy change for the coupled reaction of

$$-2870 \text{ kJ} + 38(+31 \text{ kJ}) \approx -1700 \text{ kJ}$$

In a very real sense, your body "stores" energy available from the metabolism of foods in the form of ATP. This molecule in turn supplies the energy required for all sorts of biochemical reactions taking place in the body. It does this by reverting to ADP, that is, by reversing the above reaction. The amount of ATP consumed is amazingly large; a competitive sprinter may hydrolyze as much as 500 g (about 1 lb) of ATP per minute.

⌐ Conversion of ATP to ADP gives you energy in a hurry.

EXAMPLE 17.10   The lactic acid ($C_3H_6O_3(aq)$, $\Delta G_f^\circ = -559$ kJ) produced in muscle cells by vigorous exercise eventually is absorbed into the bloodstream, where it is metabolized back to glucose ($\Delta G_f^\circ = -919$ kJ) in the liver. The reaction is

$$2C_3H_6O_3(aq) \longrightarrow C_6H_{12}O_6(aq)$$

(a) Calculate ΔG° at 25°C for this reaction, using free energies of formation.
(b) If the hydrolysis of ATP to ADP is coupled with this reaction, how many moles of ATP must react to make the process spontaneous?

Strategy Use the relation

$$\Delta G° = \sum \Delta G°_{f \text{ products}} - \sum \Delta G°_{f \text{ reactants}}$$

to find $\Delta G°$. To solve part (b), note that 31 kJ of energy is produced per mole of ATP consumed.

Solution

(a) $\Delta G° = \Delta G°_f C_6H_{12}O_6(aq) - 2 \Delta G°_f C_3H_6O_3(aq)$

$= -919$ kJ $+ 2(559$ kJ$) = \boxed{+199 \text{ kJ}}$

(b) 199 kJ $\times \dfrac{1 \text{ mol ATP}}{31 \text{ kJ}} = \boxed{6.4 \text{ mol ATP}}$

(a)

(b)

FIGURE A
Glacier melting in Alaska.
(a) Hiker on a ridge above the Muir Glacier in Glacier Bay National Park, August 26, 1978. (b) Hiker on the same ridge, June 27, 1993.
(Tom Bean)

Global Warming

The thermodynamics of our planet is governed in large part by the interaction of the Earth with its atmosphere. If it were not for the presence of two relatively minor components of the atmosphere, water vapor and carbon dioxide, the Earth would be essentially barren of life. Its mean temperature would be about $-15°C$, roughly 35°C lower than it is today.

Water vapor and carbon dioxide, unlike N_2, O_2, and Ar, strongly absorb infrared radiation (heat) given off by the Earth. In this way, CO_2 and H_2O act as an insulating blanket to prevent heat from escaping into outer space. This is often referred to as the **greenhouse effect.**

Of the two gases, water vapor absorbs more infrared radiation than carbon dioxide because its concentration is higher. This property of water vapor accounts for the fact that the temperature drops less on nights when there is a heavy cloud cover. In desert regions, where there is very little water vapor, large variations between day and night temperatures are common.

Although the concentration of water vapor in the atmosphere varies greatly with location, it remains relatively constant over time. In contrast, the concentration of carbon dioxide has increased by more than 20% over the past century, as a result of human activities. Increased combustion of fossil fuels is mainly responsible. Every gram of fossil fuel burned releases about three grams of carbon dioxide into the atmosphere. Part of this CO_2 is used by plants in photosynthesis or is absorbed by the oceans, but at least half of it remains.

It has been estimated that unless preventive action is taken, increasing CO_2 levels could raise the Earth's temperature by 3°C by the year 2100. That seems like a small amount until you realize that, during the last ice age, the temperature was only about 5°C lower than it is today. Looking at it another way, a temperature increase of 3°C could drastically change the world as we know it by—

- raising sea level by as much as one meter, caused in part by the melting of glaciers (Figure A). This could flood many coastal areas, including most of the state of Florida.
- bringing about more frequent and more intense summer heat waves and, perhaps, other natural disasters including floods and hurricanes.
- changing the pattern of precipitation worldwide; rainfall would most likely increase in temperate regions (including the United States and Canada) and decrease in the tropics and subtropics.

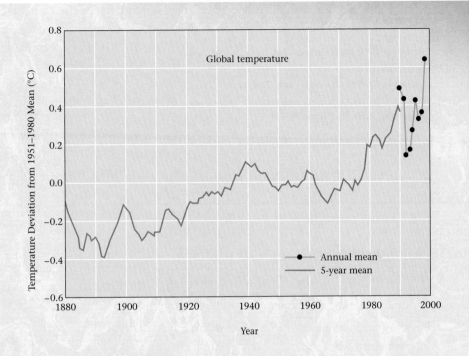

FIGURE B
Global mean temperature based on readings from worldwide meteorological stations. In this plot, the mean temperature for 1951–1980 (which was about 14°C) is taken as zero. The solid red line is a plot of the 5-year means since 1880. The blue line is the annual mean temperature for 1990 to 1998, which illustrates that 1998 had the highest annual mean temperature (about 14.7°C) recorded up to that year. *(NASA Goddard Space Institute for Space Studies)*

It has even been suggested that, illogical as it sounds, global warming could bring about a new ice age by shutting down the Gulf Stream that warms much of Europe.

Average global temperatures have increased by about 0.5°C in the past century (Figure B). The year 1998 was the hottest since at least 1880, when temperature records were first kept. Indeed, scientists at the National Oceanic and Atmospheric Administration suggest that global temperatures in 1998 may have been the highest in 1200 years. Nobody can be sure that this warming trend is a direct consequence of increasing concentrations of CO_2. However, there was sufficient concern to bring about a conference at Kyoto, Japan, in December of 1997. There the United States, Canada, and other industrialized countries agreed to reduce the emission of carbon dioxide and other greenhouse gases by an average of 6% by the year 2012.

CHAPTER HIGHLIGHTS

Key Concepts

1. Deduce the sign of ΔS based on randomness.
   (Examples 17.1, 17.2; Problems 5–10)
2. Calculate $\Delta S°$ from a table of standard entropies.
   (Example 17.3; Problems 11–16, 27, 31, 32)
3. Calculate $\Delta G°$ from—
   ▪ the Gibbs-Helmholtz equation.
   (Examples 17.4, 17.6; Problems 17–20, 27, 28)
   ▪ free energies of formation.
   (Example 17.5; Problems 21–26)

4. Calculate the temperature at which $\Delta G° = 0$.
   (Example 17.7; Problems 35–38, 43–48)
5. Calculate ΔG from $\Delta G°$ and Q.
   (Example 17.8; Problems 49–54)
6. Relate $\Delta G°$ to K.
   (Example 17.9; Problems 61–68, 70)
7. Calculate $\Delta G°$ for coupled reactions.
   (Example 17.10; Problems 55–60, 71)

Key Equations

Entropy change	$\Delta S° = \sum S°_{\text{products}} - \sum S°_{\text{reactants}}$
Gibbs-Helmholtz equation	$\Delta G = \Delta H - T\Delta S$
Free energy change	$\Delta G° = \sum \Delta G°_{f\ \text{products}} - \sum \Delta G°_{f\ \text{reactants}}$
	$\Delta G° = -RT \ln K$
	$\Delta G = \Delta G° + RT \ln Q$

Key Terms

- atmosphere
- endothermic
- enthalpy, H
  entropy, S

- exothermic
  free energy (Gibbs), G
- molarity
- reaction quotient, Q
  spontaneous process

standard
—conditions
—entropy change, $\Delta S°$
—free energy change, $\Delta G°$
—free energy of formation, $\Delta G°_f$
- —enthalpy of formation, $\Delta H°_f$
—molar entropy, $S°$

Summary Problem

Consider formic acid, HCOOH. It is given off by many insects when they sting their victims. Its name comes from *formica,* the Latin name for ant. Extensively used by the textile industry, it can be obtained by the exposure of methyl alcohol (an alcohol present in gasohol) to air.

$$CH_3OH(aq) + O_2(g) \longrightarrow HCOOH(aq) + H_2O(l)$$

The following data may be useful:

$$CH_3OH(aq): \Delta H°_f = -246 \text{ kJ/mol}; \ S° = 133.1 \text{ J/mol·K}$$

$$HCOOH(aq): \Delta H°_f = -425 \text{ kJ/mol}; \ S° = 163 \text{ J/mol·K}$$

(a) Calculate $\Delta H°$ and $\Delta S°$ for this process.
(b) Is the reaction spontaneous at 25°C? At 100°C?
(c) The heat of vaporization for formic acid is 23.1 kJ/mol. Its normal boiling point is 101°C. Calculate $\Delta S°$ for the reaction

$$HCOOH(l) \longrightarrow HCOOH(g)$$

(d) What is the standard molar entropy for $HCOOH(g)$ taking $S°$ for $HCOOH(l)$ to be 129 J/mol·K?
(e) Calculate ΔG at 25°C for the formation of formic acid from methyl alcohol (equation given above) when $[HCOOH] = 0.250\ M$, $P_{O_2} = 0.75$ atm, and $[CH_3OH] = 0.0100\ M$.
(f) Calculate $\Delta G°$ for the ionization of formic acid at 25°C. ($K_a = 1.9 \times 10^{-4}$)

Answers

(a) $\Delta H° = -465$ kJ; $\Delta S° = -105$ J/K
(b) $\Delta G° = -434$ kJ; yes; yes
(c) 61.8 J/K at 25°C
(d) 191 J/mol·K
(e) -425 kJ
(f) 21.2 kJ

Questions & Problems

Problem numbers in blue indicate that the answer is available in Appendix 6 at the back of the book.
WEB indicates that the solution is posted at **http://www.harcourtcollege.com/chem/general/masterton4/student/**

Spontaneity

1. Which of the following processes are spontaneous?
   (a) a snowman melting in the sun
   (b) building a sand castle
   (c) a drop of ink dispersing in water
   (d) outlining your chemistry notes

2. Which of the following processes are spontaneous?
   (a) building a house of cards
   (b) cleaning up your desk
   (c) folding clothes from a laundry basket
   (d) wind scattering leaves in a pile

3. Based on your experience, predict which of the following reactions are spontaneous.
   (a) $CO_2(s) \rightarrow CO_2(g)$ at 25°C
   (b) $NaCl(s) \rightarrow NaCl(l)$ at 25°C
   (c) $2NaCl(s) \rightarrow 2Na(s) + Cl_2(g)$
   (d) $CO_2(g) \rightarrow C(s) + O_2(g)$

4. Based on your experience, predict which of the following reactions are spontaneous.
   (a) $Zn(s) + 2H^+(aq) \rightarrow Zn^{2+}(aq) + H_2(g)$
   (b) $CaCO_3(s) + 2H_2O(l) \rightarrow Ca(OH)_2(s) + H_2CO_3(aq)$
   (c) $CH_4(g) + O_2(g) \rightarrow CO_2(g) + 2H_2O(g)$
   (d) $Ag^+(aq) + Cl^-(aq) \rightarrow AgCl(s)$

Entropy, $\Delta S°$

5. Predict the sign of ΔS for the following.
   (a) ice cream melting

(b) boiling water
(c) dissolving instant coffee in hot water
(d) sugar, $C_{12}H_{22}O_{11}$, decomposing to carbon and steam

6. Predict the sign of ΔS for the following
   (a) a lake freezing
   (b) precipitating lead chloride
   (c) a candle burning
   (d) weeding a garden

7. Predict the sign of $\Delta S°$ for each of the following reactions.
   (a) $2Fe_2O_3(s) + 3C(s) \rightarrow 4Fe(s) + 3CO_2(g)$
   (b) $2C(s) + O_2(g) \rightarrow 2CO(g)$
   (c) $H_2(g) + O_2(g) \rightarrow H_2O_2(l)$
   (d) $N_2(g) + 2H_2(g) \rightarrow N_2H_4(g)$

8. Predict the sign of $\Delta S°$ for each of the following reactions.
   (a) $Mg(s) + 2H_2O(l) \rightarrow Mg(OH)_2(s) + H_2(g)$
   (b) $2K(s) + Cl_2(g) \rightarrow 2KCl(s)$
   (c) $2NH_3(g) \rightarrow N_2(g) + 3H_2(g)$
   (d) $2CH_3OH(l) + 3O_2(g) \rightarrow 2CO_2(g) + 4H_2O(g)$

9. Predict the sign of $\Delta S°$ for each of the following reactions.
   (a) $H_2(g) + Ni^{2+}(aq) \rightarrow 2H^+(aq) + Ni(s)$
   (b) $Cu(s) + 2H^+(aq) \rightarrow H_2(g) + Cu^{2+}(aq)$
   (c) $N_2O_4(g) \rightarrow 2NO_2(g)$

WEB 10. Predict the sign of $\Delta S°$ for each of the following reactions.
   (a) $O_3(g) \rightarrow O_2(g) + O(g)$
   (b) $PCl_3(g) + Cl_2(g) \rightarrow PCl_5(g)$
   (c) $CuSO_4(s) + 5H_2O(l) \rightarrow CuSO_4 \cdot 5H_2O(s)$

11. Use Table 17.1 to calculate $\Delta S°$ for each of the following reactions.

(a) $CO(g) + 2H_2(g) \rightarrow CH_3OH(l)$
(b) $N_2(g) + O_2(g) \rightarrow 2NO(g)$
(c) $BaCO_3(s) \rightarrow BaO(s) + CO_2(g)$
(d) $2NaCl(s) + F_2(g) \rightarrow 2NaF(s) + Cl_2(g)$

12. Use Table 17.1 to calculate $\Delta S°$ for each of the following reactions.
(a) $4NH_3(g) + 7O_2(g) \rightarrow 4NO_2(g) + 6H_2O(g)$
(b) $2H_2O_2(l) + N_2H_4(l) \rightarrow N_2(g) + 4H_2O(g)$
(c) $C(s) + O_2(g) \rightarrow CO_2(g)$
(d) $CH_4(g) + 3Cl_2(g) \rightarrow CHCl_3(l) + 3HCl(g)$

13. Use Table 17.1 to calculate $\Delta S°$ for each of the following reactions.
(a) $2Na(s) + 2H_2O(l) \rightarrow 2Na^+(aq) + 2OH^-(aq) + H_2(g)$
(b) $Zn(s) + 2Ag^+(aq) \rightarrow 2Ag(s) + Zn^{2+}(aq)$
(c) $2NO_3^-(aq) + 8H^+(aq) + 3Cu(s) \rightarrow 3Cu^{2+}(aq) +$
$2NO(g) + 4H_2O(l)$

14. Use Table 17.1 to calculate $\Delta S°$ for each of the following reactions.
(a) $Zn(s) + 2H^+(aq) \rightarrow Zn^{2+}(aq) + H_2(g)$
(b) $H^+(aq) + OH^-(aq) \rightarrow H_2O(l)$
(c) $NH_3(g) + H_2O(l) \rightarrow NH_4^+(aq) + OH^-(aq)$

15. Use Table 17.1 to calculate $\Delta S°$ for each of the following reactions.
(a) $2HNO_3(l) + 3H_2S(g) \rightarrow 4H_2O(l) + 2NO(g) + 3S(s)$
(b) $PCl_5(g) + 4H_2O \rightarrow 5Cl^-(aq) + H_2PO_4^-(aq) + 6H^+(aq)$
(c) $MnO_4^-(aq) + 3Fe^{2+}(aq) + 4H^+(aq) \rightarrow 3Fe^{3+}(aq) +$
$MnO_2(s) + 2H_2O(l)$

16. Use Table 17.1 to calculate $\Delta S°$ for each of the following reactions.
(a) $2H_2S(s) + 3O_2(g) \rightarrow 2H_2O(g) + 2SO_2(g)$
(b) $Ag(s) + 2H^+(aq) + NO_3^-(aq) \rightarrow Ag^+(aq) + H_2O(l) +$
$NO_2(g)$
(c) $SO_4^{2-}(aq) + 4H^+(aq) + Ni(s) \rightarrow Ni^{2+}(aq) + SO_2(g) +$
$2H_2O(l)$

$\Delta G°$ and the Gibbs-Helmholtz Equation

17. Calculate $\Delta G°$ at 45°C for reactions for which
(a) $\Delta H° = 215$ kJ; $\Delta S° = 32.5$ J/K
(b) $\Delta H° = 638$ kJ; $\Delta S° = -215.2$ J/K
(c) $\Delta H° = -7.34$ kJ; $\Delta S° = -0.337$ kJ/K

18. Calculate $\Delta G°$ at 45°C for reactions for which
(a) $\Delta H° = -79.6$ kJ; $\Delta S° = 433.1$ J/K
(b) $\Delta H° = 837.4$ kJ; $\Delta S° = 173.8$ J/K
(c) $\Delta H° = -34.9$ kJ; $\Delta S° = -0.039$ kJ/K

19. Calculate $\Delta G°$ at 355 K for each of the reactions in Question 11. State whether the reactions are spontaneous.

20. Calculate $\Delta G°$ at 415 K for each of the reactions in Question 12. State whether the reactions are spontaneous.

21. From the values for $\Delta G_f°$ given in Appendix 1, calculate $\Delta G°$ at 25°C for each of the reactions in Question 13.

22. Follow the directions of Problem 21 for each of the reactions in Question 14.

23. Use standard entropies and heats of formation to calculate $\Delta G_f°$ at 25°C for
(a) tin(IV) oxide(s)

(b) silver nitrate(s)
(c) hydrogen peroxide(l)

WEB 24. Follow the directions of Question 23 for the following compounds.
(a) calcium carbonate
(b) magnesium oxide
(c) phosphorus pentachloride

25. A student warned his friends not to swim in a river close to an electric plant. He claimed that the ozone produced by the plant turned the river water to hydrogen peroxide, which would bleach hair. The reaction is

$$O_3(g) + H_2O(l) \longrightarrow H_2O_2(aq) + O_2(g)$$

Assuming that the river water is at 25°C and all species are at standard concentrations, show by calculation whether his claim is plausible. Take $\Delta G_f°$ $O_3(g)$ at 25°C to be $+163.2$ kJ/mol and $\Delta G_f°$ $H_2O_2(aq) = -134$ kJ/mol.

26. It has been proposed that wood alcohol, CH_3OH, a relatively inexpensive fuel to produce, be decomposed to produce methane. Methane is a natural gas commonly used for heating homes. Is the decomposition of wood alcohol to methane and oxygen thermodynamically feasible at 25°C and 1 atm?

27. Sodium carbonate, also called "washing soda," can be made by heating sodium hydrogen carbonate:

$$2NaHCO_3(s) \longrightarrow Na_2CO_3(s) + CO_2(g) + H_2O(l)$$

$$\Delta H° = +135.6 \text{ kJ}; \Delta G° = +34.6 \text{ kJ at } 25°C$$

(a) Calculate $\Delta S°$ for this reaction. Is the sign reasonable?
(b) Calculate $\Delta G°$ at 0 K; at 1000 K.

28. The reaction between magnesium metal and liquid water produces solid $Mg(OH)_2$ and hydrogen gas. Calculate $\Delta G°$ for the formation of one mole of $Mg(OH)_2$ at 25°C and at 15°C.

29. The alcohol in most liqueurs is ethanol, C_2H_5OH. It is produced by the fermentation of the glucose in fruit or grain.

$$C_6H_{12}O_6(aq) \longrightarrow 2C_2H_5OH(l) + 2CO_2(g)$$

$$\Delta H° = -82.4 \text{ kJ}; \Delta G° = -219.8 \text{ kJ at } 25°C$$

(a) Calculate $\Delta S°$. Is the sign reasonable?
(b) Calculate $S°$ for $C_6H_{12}O_6(aq)$.
(c) Calculate $\Delta H_f°$ for $C_6H_{12}O_6(aq)$.

30. Oxygen can be made in the laboratory by reacting sodium peroxide and water.

$$2Na_2O_2(s) + 2H_2O(l) \longrightarrow 4NaOH(s) + O_2(g)$$

$$\Delta H° = -109.0 \text{ kJ}; \Delta G° = -148.4 \text{ kJ at } 25°C$$

(a) Calculate $\Delta S°$. Is the sign reasonable?
(b) Calculate $S°$ for $Na_2O_2(s)$.
(c) Calculate $\Delta H_f°$ for $Na_2O_2(s)$.

31. When permanganate ions in aqueous solution react with cobalt metal in strong acid, the equation for the reaction that takes place is

$$2MnO_4^-(aq) + 16H^+(aq) + 5Co(s) \longrightarrow$$
$$2Mn^{2+}(aq) + 5Co^{2+}(aq) + 8H_2O(l)$$

$\Delta H° = -2024.6$ kJ; $\Delta G°$ at 25°C $= -1750.9$ kJ

(a) Calculate $\Delta S°$ for the reaction at 25°C.
(b) Calculate $S°$ for Co^{2+} given $S°$ for Co is 30.04 J/mol·K.

32. Phosgene, $COCl_2$, can be formed by the reaction of chloroform, $CHCl_3(l)$, with oxygen:

$$2CHCl_3(l) + O_2(g) \longrightarrow 2COCl_2(g) + 2HCl(g)$$

$$\Delta H° = -353.2 \text{ kJ}; \Delta G° = -452.4 \text{ kJ at 25°C}$$

(a) Calculate $\Delta S°$ for the reaction. Is the sign reasonable?
(b) Calculate $S°$ for phosgene.
(c) Calculate $\Delta H_f°$ for phosgene.

Temperature Dependence of Spontaneity

33. Discuss the effect of temperature change on the spontaneity of the following reactions at 1 atm.

(a) $2PbO(s) + 2SO_2(g) \rightarrow 2PbS(s) + 3O_2(g)$

$$\Delta H° = +830.8 \text{ kJ}; \Delta S° = +168 \text{ J/K}$$

(b) $2As(s) + 3F_2(g) \rightarrow 2AsF_3(l)$

$$\Delta H° = -1643 \text{ kJ}; \Delta S° = -0.316 \text{ kJ/K}$$

(c) $CO(g) \rightarrow C(s) + \frac{1}{2}O_2(g)$

$$\Delta H° = 110.5 \text{ kJ}; \Delta S° = -89.4 \text{ J/K}$$

34. Discuss the effect of temperature change on the spontaneity of the following reactions at 1 atm.

(a) $Al_2O_3(s) + 2Fe(s) \rightarrow 2Al(s) + Fe_2O_3(s)$

$$\Delta H° = +851.5 \text{ kJ}; \Delta S° = +38.5 \text{ J/K}$$

(b) $N_2H_4(l) \rightarrow N_2(g) + 2H_2(g)$

$$\Delta H° = -50.6 \text{ kJ}; \Delta S° = 0.3315 \text{ kJ/K}$$

(c) $SO_3(g) \rightarrow SO_2(g) + \frac{1}{2}O_2(g)$

$$\Delta H° = 98.9 \text{ kJ}; \Delta S° = +0.0939 \text{ kJ/K}$$

35. At what temperature does $\Delta G°$ become zero for each of the reactions in Problem 33? Explain the significance of your answers.

36. Over what temperature range are the reactions in Problem 34 spontaneous?

37. For the reaction

$$SO_2(g) + 2H_2S(g) \longrightarrow 3S(s) + 2H_2O(g)$$

calculate the temperature at which $\Delta G° = 0$.

WEB 38. For the reaction

$$NH_4Cl(s) \longrightarrow NH_3(g) + HCl(g)$$

calculate the temperature at which $\Delta G° = 0$.

39. For the decomposition of Ag_2O,

$$2Ag_2O(s) \longrightarrow 4Ag(s) + O_2(g)$$

(a) obtain an expression for $\Delta G°$ as a function of temperature. Prepare a table of $\Delta G°$ values at 100-K intervals between 100 K and 500 K.

(b) calculate the temperature at which $\Delta G°$ becomes zero.

40. Earlier civilizations smelted iron from ore by heating it with charcoal from a wood fire:

$$2Fe_2O_3(s) + 3C(s) \longrightarrow 4Fe(s) + 3CO_2(g)$$

(a) Obtain an expression for $\Delta G°$ as a function of temperature. Prepare a table of $\Delta G°$ values at 100-K intervals between 100 K and 500 K.
(b) Calculate the lowest temperature at which the smelting could be carried out.

41. Two possible ways of producing iron from iron ore are
(a) $Fe_2O_3(s) + \frac{3}{2}C(s) \rightarrow 2Fe(s) + \frac{3}{2}CO_2(g)$
(b) $Fe_2O_3(s) + 3H_2(g) \rightarrow 2Fe(s) + 3H_2O(g)$
Which of these reactions proceeds spontaneously at the lower temperature?

42. It is desired to produce tin from its ore, cassiterite, SnO_2, at as low a temperature as possible. The ore could be
(a) decomposed by heating, producing tin and oxygen.
(b) heated with hydrogen gas, producing tin and water vapor.
(c) heated with carbon, producing tin and carbon dioxide.
Based solely on thermodynamic principles, which method would you recommend? Show calculations.

43. Organ pipes in unheated churches develop "tin disease," in which white tin is converted to gray tin. Given

white Sn: $\Delta H_f° = 0.00$ kJ/mol; $S° = 51.55$ J/mol·K

gray Sn: $\Delta H_f° = -2.09$ kJ/mol; $S° = 44.14$ J/mol·K

calculate the equilibrium temperature for the transition.

44. Red phosphorus is formed by heating white phosphorus. Calculate the temperature at which the two forms are at equilibrium, given

white P: $\Delta H_f° = 0.00$ kJ/mol; $S° = 41.09$ J/mol·K

red P: $\Delta H_f° = -17.6$ kJ/mol; $S° = 22.80$ J/mol·K

45. Sulfur has about 20 different allotropes. The most common are rhombic sulfur (the stable form at 25°C and 1 atm) and monoclinic sulfur. They differ in their crystal structures. Given

$S(s, \text{monoclinic})$: $\Delta H_f° = 0.30$ kJ/mol, $S° = 0.0326$ kJ/mol·K

at what temperature are the two forms in equilibrium?

46. Pencil "lead" is almost pure graphite. Graphite is the stable elemental form of carbon at 25°C and 1 atm. Diamond is an allotrope of graphite. Given

diamond: $\Delta H_f° = 1.9$ kJ/mol; $S° = 2.4$ J/mol·K

at what temperature are the two forms in equilibrium at 1 atm?

$$C \text{ (graphite)} \Longrightarrow C \text{ (diamond)}$$

47. Given the following data for mercury

Hg(l): $S° = 76.0$ J/mol·K

Hg(g): $S° = 175.0$ J/mol·K $\Delta H_f° = 61.32$ kJ/mol

estimate the normal boiling point ($\Delta G° = 0$) of Hg.

48. Given the following data for iodine

$$I_2(s): S° = 116.1 \text{ J/mol·K}$$

$$I_2(g): S° = 260.6 \text{ J/mol·K} \qquad \Delta H_f° = 62.4 \text{ kJ/mol}$$

estimate the temperature at which iodine sublimes at 1 atm.

$$I_2(s) \rightleftharpoons I_2(g)$$

Effect of Concentration/Pressure on Spontaneity

49. Show by calculation, using Appendix 1, whether the reaction

$$CaCO_3(s) \rightleftharpoons Ca^{2+}(aq) + CO_3{}^{2-}(aq)$$

is spontaneous at 25°C
   (a) when $[Ca^{2+}] = [CO_3{}^{2-}] = 1.0 \text{ M}$.
   (b) when $[Ca^{2+}] = [CO_3{}^{2-}] = 1.0 \times 10^{-10}$ M.

WEB 50. Show by calculation whether the reaction

$$HF(aq) \rightleftharpoons H^+(aq) + F^-(aq) \qquad \Delta G° = +18.0 \text{ kJ}$$

is spontaneous at 25°C
   (a) when $[H^+] = [F^-] = [HF] = 1.0$ M.
   (b) when $[H^+] = [F^-] = 1.0 \times 10^{-3}$ M; $[HF] = 1.0$ M.

51. For the reaction

$$O_2(g) + 4H^+(aq) + 4Fe^{2+}(aq) \longrightarrow 2H_2O(l) + 4Fe^{3+}(aq)$$

   (a) calculate $\Delta G°$ at 25°C.
   (b) calculate ΔG at 25°C when $[Fe^{2+}] = [Fe^{3+}] = 0.250$ M, $P_{O_2} = 0.755$ atm, and the pH of the solution is 3.12.

52. For the reaction

$$2H_2O(l) + 2Cl^-(aq) \longrightarrow H_2(g) + Cl_2(g) + 2\,OH^-(aq)$$

   (a) calculate $\Delta G°$ at 25°C.
   (b) calculate ΔG at 25°C when $P_{H_2} = P_{Cl_2} = 0.250$ atm, $[Cl^-] = 0.335$ M, and the pH of the solution is 11.98.

53. Consider the reaction

$$2SO_2(g) + O_2(g) \longrightarrow 2SO_3(g)$$

   (a) Calculate $\Delta G°$ at 25°C.
   (b) If the partial pressures of SO_2 and SO_3 are kept at 0.400 atm, what partial pressure should O_2 have so that the reaction just becomes nonspontaneous (i.e., $\Delta G = +1.0$ kJ)?

54. Consider the reaction

$$AgCl(s) \longrightarrow Ag^+(aq) + Cl^-(aq)$$

   (a) Calculate $\Delta G°$ at 25°C.
   (b) What should the concentrations of Ag^+ and Cl^- be so that $\Delta G = -1.0$ kJ (just spontaneous)? Take $[Ag^+] = [Cl^-]$.
   (c) The K_{sp} for AgCl is 1.8×10^{-10}. Is the answer to (b) reasonable? Explain.

Additivity of Coupled Reactions

55. Given that, at 25°C,

$$2Cu(s) + \tfrac{1}{2}O_2(g) \longrightarrow Cu_2O(s) \qquad \Delta G° = -146.0 \text{ kJ}$$

$$Cu_2O(s) + \tfrac{1}{2}O_2(g) \longrightarrow 2CuO(s) \qquad \Delta G° = -113.4 \text{ kJ}$$

calculate $\Delta G_f°$ CuO(s) at 25°C.

56. Given that, at 25°C,

$$Fe(s) + Cl_2(g) \longrightarrow FeCl_2(s) \qquad \Delta G° = -302.3 \text{ kJ}$$

$$Fe(s) + \tfrac{3}{2}Cl_2(g) \longrightarrow FeCl_3(s) \qquad \Delta G° = -334.0 \text{ kJ}$$

calculate $\Delta G°$ at 25°C for the reaction

$$2FeCl_2(s) + Cl_2(g) \longrightarrow 2FeCl_3(s)$$

57. Theoretically, one can obtain zinc from an ore containing zinc sulfide, ZnS, by the reaction

$$ZnS(s) \longrightarrow Zn(s) + S(s)$$

   (a) Show by calculation that this reaction is not feasible at 25°C.
   (b) Show that by coupling the above reaction with the reaction

$$S(s) + O_2(g) \longrightarrow SO_2(g)$$

the overall reaction, where Zn is obtained by roasting in oxygen, is feasible at 25°C.

58. Natural gas, which is mostly methane, CH_4, is a resource that the United States has in abundance. In principle, ethane can be obtained from methane by the reaction

$$2CH_4(g) \longrightarrow C_2H_6(g) + H_2(g)$$

   (a) Calculate $\Delta G°$ at 25°C for the reaction. Comment on the feasibility of this reaction at 25°C.
   (b) Couple the reaction above with the formation of steam from the elements:

$$H_2(g) + \tfrac{1}{2}O_2(g) \longrightarrow H_2O(g) \qquad \Delta G° = -228.6 \text{ kJ}$$

What is the equation for the overall reaction? Comment on the feasibility of the overall reaction.

59. How many moles of ATP must be converted to ADP by the reaction

$$ATP(aq) + H_2O \longrightarrow ADP(aq) + HPO_4{}^{2-}(aq) + 2H^+(aq)$$
$$\Delta G° = -31 \text{ kJ}$$

to bring about a nonspontaneous biochemical reaction in which $\Delta G° = +372$ kJ?

60. Consider the following reactions at 25°C:

$$C_6H_{12}O_6(aq) + 6O_2(g) \longrightarrow 6CO_2(g) + 6H_2O$$
$$\Delta G° = -2870 \text{ kJ}$$

$$ADP(aq) + HPO_4{}^{2-}(aq) + 2H^+(aq) \longrightarrow ATP(aq) + H_2O$$
$$\Delta G° = 31 \text{ kJ}$$

Write an equation for a coupled reaction between glucose, $C_6H_{12}O_6$, and ADP in which $\Delta G° = -390$ kJ.

Free Energy and Equilibrium

61. Consider the reaction

$$CaCO_3(s) \rightleftharpoons CaO(s) + CO_2(g)$$

Use the appropriate tables to calculate
  (a) $\Delta G°$ at 375°C   (b) K at 375°C

WEB **62.** Consider the reaction

$$H_2O(l) \rightleftharpoons H^+(aq) + OH^-(aq)$$

Use the appropriate tables to calculate
  (a) $\Delta G°$ at 25°C   (b) K_w at 25°C

63. Consider the reaction

$$N_2O(g) + NO_2(g) \longrightarrow 3NO(g) \qquad K = 4.4 \times 10^{-19}$$

  (a) Calculate $\Delta G°$ for the reaction at 25°C.
  (b) Calculate $\Delta G_f°$ for N_2O.

64. Consider the following reaction at 25°C:

$$Cl_2(g) \rightleftharpoons 2Cl(g) \qquad K = 1.0 \times 10^{-37}$$

  (a) Calculate $\Delta G°$ for the reaction at 25°C.
  (b) Calculate $\Delta G_f°$ for $Cl(g)$.

65. For the reaction

$$NH_4Cl(s) \rightleftharpoons NH_3(g) + HCl(g)$$

  (a) calculate K at 25°C.
  (b) calculate $\Delta G°$ at 100°C.

66. For the reaction

$$O_3(g) \longrightarrow O(g) + O_2(g)$$

$\Delta G°$ at 298 K is −68.6 kJ and is −46.3 kJ at 1200 K. Calculate K at both temperatures.

67. Use the values for $\Delta G_f°$ in Appendix 1 to calculate K_{sp} for barium sulfate at 25°C. Compare with the value given in Chapter 16.

68. Given that $\Delta H_f°$ for HF(aq) is −320.1 kJ/mol and $S°$ for HF(aq) is 88.7 J/mol·K, find K_a for HF at 25°C.

Unclassified

69. At 1200 K, an equilibrium mixture of CO and CO_2 gases contains 98.31 mol percent CO and some solid carbon. The total pressure of the mixture is 1.00 atm. For the system

$$CO_2(g) + C(s) \rightleftharpoons 2CO(g)$$

calculate
  (a) P_{CO} and P_{CO_2} (b) K   (c) $\Delta G°$ at 1200 K

70. At 25°C, a 0.13 M solution of a weak acid, HB, has a pH of 3.71. What is $\Delta G°$ for

$$H^+(aq) + B^-(aq) \rightleftharpoons HB(aq)$$

71. Some bacteria use light energy to convert carbon dioxide and water to glucose and oxygen:

$$6CO_2(g) + 6H_2O(l) \longrightarrow C_6H_{12}O_6(aq) + 6O_2(g)$$

$$\Delta G° = 2870 \text{ kJ at 25°C}$$

Other bacteria, those that do not have light available to them, couple the reaction

$$H_2S(g) + \tfrac{1}{2}O_2(g) \longrightarrow H_2O(l) + S(s)$$

to the glucose synthesis above. Coupling the two reactions, the overall reaction is

$$24H_2S(g) + 6CO_2(g) + 6O_2(g) \longrightarrow$$
$$C_6H_{12}O_6(aq) + 18H_2O(l) + 24S(s)$$

Show that the reaction is spontaneous at 25°C.

72. It has been proposed that if ammonia, methane, and oxygen gas are combined at 25°C in their standard states, glycine, the simplest of all amino acids, can be formed.

$$2CH_4(g) + NH_3(g) + \tfrac{5}{2}O_2(g) \longrightarrow NH_2CH_2COOH(s) + 3H_2O(l)$$

Given $\Delta G_f°$ for glycine = −368.57 kJ/mol,
  (a) will this reaction proceed spontaneously at 25°C?
  (b) does the equilibrium constant favor the formation of products?
  (c) do your calculations for (a) and (b) indicate that glycine is formed as soon as the three gases are combined?

73. A student is asked to prepare a 0.030 M aqueous solution of $PbCl_2$.
  (a) Is this possible at 25°C? (*Hint:* Is dissolving 0.030 mol of $PbCl_2$ at 25°C possible?)
  (b) If the student used water at 100°C, would this be possible?

Conceptual Problems

74. Determine whether each of the following statements is true or false. If false, modify it to make it true.
  (a) An exothermic reaction is spontaneous.
  (b) When $\Delta G°$ is positive, the reaction cannot occur under any conditions.
  (c) $\Delta S°$ is positive for a reaction in which there is an increase in the number of moles.
  (d) If $\Delta H°$ and $\Delta S°$ are both negative, $\Delta G°$ will be negative.

75. Which of the following quantities can be taken to be independent of temperature? Independent of pressure?
  (a) ΔH for a reaction   (b) ΔS for a reaction
  (c) ΔG for a reaction   (d) S for a substance

76. Fill in the blanks:
  (a) $\Delta H°$ and $\Delta G°$ become equal at _____ K.
  (b) $\Delta G°$ and ΔG are equal when Q = _____.
  (c) $S°$ for steam is _____ than $S°$ for water.

77. Fill in the blanks:
  (a) At equilibrium, ΔG is _____.
  (b) For $C_6H_6(l) \rightleftharpoons C_6H_6(g)$, $\Delta H°$ is _____ (+, −, 0).
  (c) When a pure solid melts, the temperature at which liquid and solid are in equilibrium and $\Delta G° = 0$ is called _____.

78. In your own words, explain why
    (a) $\Delta S°$ is negative for a reaction in which the number of moles of gas decreases.
    (b) we take $\Delta S°$ to be independent of T, even though entropy increases with T.
    (c) a solid has lower entropy than its corresponding liquid.

79. Consider the graph below.

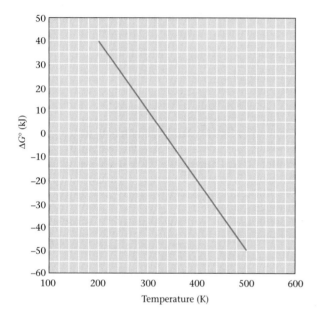

    (a) Describe the relationship between the spontaneity of the process and temperature.
    (b) Is the reaction exothermic?
    (c) Is $\Delta S° > 0$?
    (d) At what temperature is the reaction at standard conditions likely to be at equilibrium?
    (e) What is K for the reaction at 27°C?

80. The relationship between $\Delta G°$ and T is linear. Draw the graph of a reaction with the properties described below. You need only label the point at which $\Delta G° = 0$
    ■ the reaction is exothermic.
    ■ $\Delta n_g < 0$ (Δn_g = moles of gas products − moles of gas reactants).
    ■ at 300 K, the system is at equilibrium and $K = 1$.

Challenge Problems

81. The normal boiling point for ethyl alcohol is 78.4°C. $S°$ for $C_2H_5OH(g)$ is 282.7 J/mol·K. At what temperature is the vapor pressure of ethyl alcohol 357 mm Hg?

82. $\Delta H_f°$ for iodine gas is 62.4 kJ/mol, and $S°$ is 260.7 J/mol·K. Calculate the equilibrium partial pressures of $I_2(g)$, $H_2(g)$, and $HI(g)$ for the system

$$2HI(g) \rightleftharpoons H_2(g) + I_2(g)$$

at 500°C if the initial partial pressures are all 0.200 atm.

83. The heat of fusion of ice is 333 J/g. For the process

$$H_2O(s) \longrightarrow H_2O(l)$$

determine
    (a) $\Delta H°$ **(b)** $\Delta G°$ at 0°C   **(c)** $\Delta S°$
    (d) $\Delta G°$ at −20°C   **(e)** $\Delta G°$ at 20°C

84. The overall reaction that occurs when sugar is metabolized is

$$C_{12}H_{22}O_{11}(s) + 12O_2(g) \longrightarrow 12CO_2(g) + 11H_2O(l)$$

For this reaction, $\Delta H°$ is −5650 kJ and $\Delta G°$ is −5790 kJ at 25°C.
    (a) If 25% of the free energy change is actually converted to useful work, how many kilojoules of work are obtained when one gram of sugar is metabolized at body temperature, 37°C?
    (b) How many grams of sugar would a 120-lb woman have to eat to get the energy to climb the Jungfrau in the Alps, which is 4158 m high? ($w = 9.79 \times 10^{-3}$ mh, where w = work in kilojoules, m is body mass in kilograms, and h is height in meters.)

85. Hydrogen has been suggested as the fuel of the future. One way to store it is to convert it to a compound that can be heated to release the hydrogen. One such compound is calcium hydride, CaH_2. This compound has a heat of formation of −186.2 kJ/mol and a standard entropy of 42.0 J/mol·K. What is the minimum temperature to which calcium hydride would have to be heated to produce hydrogen at one atmosphere pressure?

86. When a copper wire is exposed to air at room temperature, it becomes coated with a black oxide, CuO. If the wire is heated above a certain temperature, the black oxide is converted to a red oxide, Cu_2O. At a still higher temperature, the oxide coating disappears. Explain these observations in terms of the thermodynamics of the reactions

$$2CuO(s) \longrightarrow Cu_2O(s) + \tfrac{1}{2}O_2(g)$$

$$Cu_2O(s) \longrightarrow 2Cu(s) + \tfrac{1}{2}O_2(g)$$

and estimate the temperatures at which the changes occur.

ELECTROCHEMISTRY

18

Copper strips are oxidized when exposed to moist air. The same reaction can occur in an electro–chemical cell. *(Paul Silverman/Fundamental Photographs)*

CHAPTER OUTLINE

18.1   VOLTAIC CELLS

18.2   STANDARD VOLTAGES

18.3   RELATIONS BETWEEN $E°$, $\Delta G°$,
         AND K

18.4   EFFECT OF CONCENTRATION
         ON VOLTAGE

18.5   ELECTROLYTIC CELLS

18.6   COMMERCIAL CELLS

E lectrochemistry is the study of the interconversion of electrical and chemical energy. This conversion takes place in an electrochemical cell that may be a(n)—

- **voltaic cell** (Section 18.1), in which a spontaneous reaction generates electrical energy.

- **electrolytic cell** (Section 18.5), in which electrical energy is used to bring about a nonspontaneous reaction.

All of the reactions considered in this chapter are of the oxidation-reduction type. You will recall from Chapter 4 that such a reaction can be split into two **half-reactions.** In one half-reaction, referred to as *reduction,* electrons are consumed; in the other, called *oxidation,* electrons are produced. There can be no net change in the number of electrons; the number of electrons consumed in reduction must be exactly equal to the number produced in the oxidation half-reaction.

In an electrochemical cell, these two half-reactions occur at two different **electrodes,** which most often consist of metal plates or wires. Reduction occurs at the **cathode;** a typical half-reaction might be

$$\text{cathode:}\quad Cu^{2+}(aq) + 2e^- \longrightarrow Cu(s)$$

Oxidation takes place at the **anode,** where a species such as zinc metal produces electrons:

$$\text{anode:}\quad Zn(s) \longrightarrow Zn^{2+}(aq) + 2e^-$$

It is always true that in an electrochemical cell, *anions* move to the *anode; cations* move to the *cathode.*

One of the most important characteristics of a cell is its *voltage,* which is a measure of reaction spontaneity. Cell voltages depend on the nature of the half-reactions occurring at the electrodes (Section 18.2) and on the concentrations of species involved (Section 18.4). From the voltage measured at standard concentrations, it is possible to calculate the standard free energy change and the equilibrium constant (Section 18.3) of the reaction involved.

The principles discussed in this chapter have a host of practical applications. Whenever you start your car, turn on a flashlight, or take a logarithm on a calculator, you are making use of a voltaic cell. Many of our most important elements, including hydrogen and chlorine, are made in electrolytic cells. These applications, among others, are discussed in Section 18.6.

⊙ See the *Saunders Interactive General Chemistry CD-ROM,* Screen 21.2, Redox Reactions.

Cathode = reduction; anode = oxidation. The consonants go together, as do the vowels.

18.1 VOLTAIC CELLS

In principle at least, any spontaneous redox reaction can serve as a source of energy in a **voltaic cell.** The cell must be designed in such a way that oxidation occurs at one electrode (anode) with reduction at the other electrode (cathode). The electrons produced at the anode must be transferred to the cathode, where they are consumed. To do this, the electrons move through an external circuit, where they do electrical work.

To understand how a voltaic cell operates, let us start with some simple cells that are readily made in the general chemistry laboratory.

The Zn-Cu^{2+} Cell (Zn | Zn^{2+} || Cu^{2+} | Cu)

When a piece of zinc is added to a water solution containing Cu^{2+} ions, the following redox reaction takes place:

$$Zn(s) + Cu^{2+}(aq) \longrightarrow Zn^{2+}(aq) + Cu(s)$$

In this reaction, copper metal plates out on the surface of the zinc. The blue color of the aqueous Cu^{2+} ion fades as it is replaced by the colorless aqueous Zn^{2+} ion (Figure 18.1). Clearly, this redox reaction is spontaneous; it involves electron transfer from a Zn atom to a Cu^{2+} ion.

To design a voltaic cell using the Zn-Cu^{2+} reaction as a source of electrical energy, the electron transfer must occur indirectly; that is, the electrons given off by zinc atoms must be made to pass through an external electric circuit before they reduce Cu^{2+} ions to copper atoms. One way to do this is shown in Figure 18.2. The voltaic cell consists of two **half-cells—**

■ a Zn anode dipping into a solution containing Zn^{2+} ions, shown in the beaker at the far right.
■ a Cu cathode dipping into a solution containing Cu^{2+} ions (blue), shown in the beaker at the center of Figure 18.2.

The "external circuit" consists of a voltmeter with leads (red and black) to the anode and cathode.

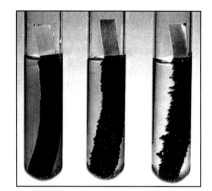

FIGURE 18.1
Zn-Cu^{2+} spontaneous redox reaction. The blue of the Cu^{2+} ion in solution fades *(from left to right)* as the ion is reduced to metallic Cu that deposits on the zinc strip in the solution. *(Charles D. Winters)*

⌐ The Zn electrode must not come in contact with Cu^{2+} ions. Why?

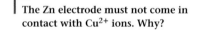 See Screen 21.4, Electrochemical Cells.

FIGURE 18.2
A Zn-Cu^{2+} voltaic cell. In this voltaic cell, a voltmeter *(left)* is connected to a half-cell consisting of a Cu cathode in a solution of blue Cu^{2+} ions and a half-cell consisting of a Zn anode in a solution of colorless Zn^{2+} ions. The following spontaneous reaction takes place in this cell: $Zn(s) + Cu^{2+}(aq) \rightarrow Zn^{2+}(aq) + Cu(s)$. The salt bridge allows ions to pass from one solution to the other to complete the circuit and prevents direct contact between Zn atoms and the Cu^{2+} ions. *(Charles D. Winters)*

See Screen 21.5, Electrochemical Cells and Potentials.

Let us trace the flow of electric current through the cell.

1. At the zinc *anode* (connected to the red wire shown in Figure 18.2), electrons are produced by the oxidation half-reaction

$$Zn(s) \longrightarrow Zn^{2+}(aq) + 2e^-$$

This electrode, which "pumps" electrons into the external circuit, is ordinarily marked as the negative pole of the cell.

2. Electrons generated at the anode move through the external circuit (right to left in Figure 18.2) to the copper *cathode,* connected through the black wire to the voltmeter. At the cathode, the electrons are consumed, reducing Cu^{2+} ions present in the solution around the electrode:

$$Cu^{2+}(aq) + 2e^- \longrightarrow Cu(s)$$

The copper electrode, which "pulls" electrons from the external circuit, is considered to be the positive pole of the cell.

3. As the above half-reactions proceed, a surplus of positive ions (Zn^{2+}) tends to build up around the zinc electrode. The region around the copper electrode tends to become deficient in positive ions as Cu^{2+} ions are consumed. To maintain electrical neutrality, cations must move toward the copper cathode or, alternatively, anions must move toward the zinc anode. In practice, both migrations occur.

In the cell shown in Figure 18.2, movement of ions occurs through a **salt bridge** connecting the two beakers. The salt bridge shown is an inverted glass U-tube, plugged with glass wool at each end. The tube is filled with a solution of a salt that takes no part in the electrode reactions; potassium nitrate, KNO_3, is frequently used. As current is drawn from the cell, K^+ ions move from the salt bridge into the cathode half-cell. At the same time, NO_3^- ions move into the anode half-cell. In this way, electrical neutrality is maintained without Cu^{2+} ions coming in contact with the Zn electrode, which would short-circuit the cell.

The cell shown in Figure 18.2 is often abbreviated as

$$Zn \mid Zn^{2+} \| Cu^{2+} \mid Cu$$

In this notation—

Anode ‖ cathode.

- the **anode** reaction **(oxidation)** is shown at the left. Zn atoms are oxidized to Zn^{2+} ions.
- the salt bridge (or other means of separating the half cells) is indicated by the symbol ‖.
- the **cathode** reaction **(reduction)** is shown at the right. Cu^{2+} ions are reduced to Cu atoms.
- a single vertical line indicates a phase boundary, such as that between a solid electrode and an aqueous solution.

Notice that the anode half-reaction comes first in the cell notation, just as the letter *a* comes before *c*.

Other Salt Bridge Cells

Cells similar to that shown in Figure 18.2 can be set up for many different spontaneous redox reactions. Consider, for example, the reaction

$$Ni(s) + Cu^{2+}(aq) \longrightarrow Ni^{2+}(aq) + Cu(s)$$

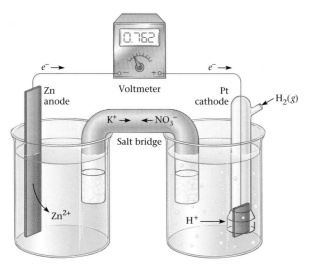

FIGURE 18.3

A Zn-H₂ voltaic cell. Hydrogen is bubbled over a platinum electrode surrounded by a solution containing H^+ ions. The spontaneous cell reaction is

$$Zn(s) + 2H^+(aq) \longrightarrow Zn^{2+}(aq) + H_2(g)$$

The platinum electrode does not participate in the reaction. **OHT**

This reaction, like that between Zn and Cu^{2+}, can serve as a source of electrical energy in a voltaic cell. The cell is similar to that shown in Figure 18.2 except that, in the anode compartment, a nickel electrode is surrounded by a solution of a nickel(II) salt, such as $NiCl_2$ or $NiSO_4$. The cell notation is $Ni \mid Ni^{2+} \parallel Cu^{2+} \mid Cu$.

Another spontaneous redox reaction that can serve as a source of electrical energy is that between zinc metal and H^+ ions:

$$Zn(s) + 2H^+(aq) \longrightarrow Zn^{2+}(aq) + H_2(g)$$

A voltaic cell using this reaction is similar to the Zn-Cu^{2+} cell; the $Zn \mid Zn^{2+}$ half-cell and the salt bridge are the same. Because no metal is involved in the cathode half-reaction, an *inert* electrode that conducts an electric current is used. Frequently, the cathode is made out of platinum (Figure 18.3). Hydrogen gas is bubbled over the cathode, which is surrounded by H^+ ions from a solution of HCl. The half-reactions occurring in the cell are

> Nichrome or graphite can also be used.

anode:	$Zn(s) \longrightarrow Zn^{2+}(aq) + 2e^-$	(oxidation)
cathode:	$2H^+(aq) + 2e^- \longrightarrow H_2(g)$	(reduction)

The cell notation is $Zn \mid Zn^{2+} \parallel H^+ \mid H_2 \mid Pt$. The symbol Pt is used to indicate the presence of an inert platinum electrode.

EXAMPLE 18.1   When chlorine gas is bubbled through a water solution of NaBr, a spontaneous redox reaction occurs:

$$Cl_2(g) + 2Br^-(aq) \longrightarrow 2Cl^-(aq) + Br_2(l)$$

This reaction can serve as a source of electrical energy in the voltaic cell shown in Figure 18.4 (page 518). In this cell,

(a) what is the cathode reaction? The anode reaction?
(b) which way do electrons move in the external circuit?
(c) which way do anions move within the cell? Cations?

Strategy   Split the reaction into two half-reactions. Remember that oxidation occurs at the anode, reduction at the cathode. Anions move to the anode, cations to the cathode. Electrons are produced at the anode and transferred through the external circuit to the cathode, where they are consumed.

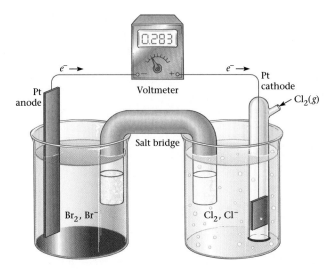

FIGURE 18.4

A Cl_2-Br^- voltaic cell. In this cell the spontaneous cell reaction is

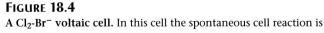

$$Cl_2(g) + 2Br^-(aq) \longrightarrow 2Cl^-(aq) + Br_2(l)$$

Both electrodes are platinum, with the anode at the left and the cathode at the right. **OHT**

Solution

(a) cathode: $Cl_2(g) + 2e^- \longrightarrow 2Cl^-(aq)$ (reduction)

   anode: $2Br^-(aq) \longrightarrow Br_2(l) + 2e^-$ (oxidation)

(b) From anode to cathode (left to right in Figure 18.4).

(c) Anions move to the anode (right to left); cations move to the cathode (left to right).

To summarize our discussion of the structure of voltaic cells, note that—

- *a voltaic cell consists of two half-cells.* They are joined by an external electric circuit through which electrons move and a salt bridge through which ions move.
- *each half-cell consists of an electrode dipping into a water solution.* If a metal participates in the cell reaction, either as a product or as a reactant, it is ordinarily used as the electrode; otherwise, an inert electrode such as platinum is used.
- *in one half-cell, oxidation occurs at the anode; in the other, reduction takes place at the cathode.* The overall cell reaction is the sum of the half-reactions taking place at the anode and cathode.

18.2 STANDARD VOLTAGES

See Screen 21.6, Standard Potentials.

The driving force behind the spontaneous reaction in a voltaic cell is measured by the cell voltage, which is an *intensive* property, independent of the number of electrons passing through the cell. Cell voltage depends on the nature of the redox reaction and the concentrations of the species involved; for the moment, we'll concentrate on the first of these factors.

The standard voltage for a given cell is that measured when the current flow is essentially zero, *all ions and molecules in solution are at a concentration of 1 M, and all gases are at a pressure of 1 atm.* To illustrate, consider the Zn-H$^+$ cell referred to earlier. Let us suppose that the half-cells are set up in such a way that the concentrations of Zn^{2+} and H$^+$ are both 1 M and the pressure of H$_2(g)$ is 1 atm. Under these conditions, the cell voltage at very low current flow is +0.762 V. This quantity is referred to as the **standard voltage** and is given the symbol $E°$.

$$Zn(s) + 2H^+(aq, 1 M) \longrightarrow Zn^{2+}(aq, 1 M) + H_2(g, 1 atm) \qquad E° = +0.762 \text{ V}$$

$E°_{red}$ and $E°_{ox}$

Any redox reaction can be split into two half-reactions, an oxidation and a reduction. It is possible to associate standard voltages $E°_{ox}$ (standard oxidation voltage) and $E°_{red}$ (standard reduction voltage) with the oxidation and reduction half-reactions. The standard voltage for the overall reaction, $E°$, is the sum of these two quantities

$$\boxed{E° = E°_{red} + E°_{ox}}$$

To illustrate, consider the reaction between Zn and H$^+$ ions, for which the standard voltage is +0.762 V.

$$+0.762 \text{ V} = E°_{red}(H^+ \longrightarrow H_2) + E°_{ox}(Zn \longrightarrow Zn^{2+})$$

There is no way to measure the standard voltage for a half-reaction; only $E°$ can be measured directly. To obtain values for $E°_{ox}$ and $E°_{red}$, the value zero is arbitrarily assigned to the standard voltage for reduction of H$^+$ ions to H$_2$ gas:

> You need two half-cells to measure a voltage.

$$2H^+(aq, 1 M) + 2e^- \longrightarrow H_2(g, 1 atm) \qquad E°_{red}(H^+ \longrightarrow H_2) = 0.000 \text{ V}$$

Using this convention, it follows that the standard voltage for the oxidation of zinc must be +0.762 V; that is,

$$Zn(s) \longrightarrow Zn^{2+}(aq, 1 M) + 2e^- \qquad E°_{ox}(Zn \longrightarrow Zn^{2+}) = +0.762 \text{ V}$$

As soon as one half-reaction voltage is established, others can be calculated. For example, the standard voltage for the Zn-Cu^{2+} cell shown in Figure 18.2 is found to be +1.101 V. Knowing that $E°_{ox}$ for zinc is +0.762 V, it follows that

$$E°_{red}(Cu^{2+} \longrightarrow Cu) = E° - E°_{ox}(Zn \longrightarrow Zn^{2+})$$
$$= 1.101 \text{ V} - 0.762 \text{ V} = +0.339 \text{ V}$$

Standard half-cell voltages are ordinarily obtained from a list of **standard potentials** such as those in Table 18.1 (pages 520 and 521). The potentials listed are the standard voltages for reduction half-reactions, that is,

$$\text{standard potential} = E°_{red}$$

For example, the standard potentials listed in the table for Zn$^{2+} \longrightarrow$ Zn and Cu$^{2+} \longrightarrow$ Cu are -0.762 V and $+0.339$ V, respectively; it follows that at 25°C

$$Zn^{2+}(aq, 1 M) + 2e^- \longrightarrow Zn(s) \qquad E°_{red} = -0.762 \text{ V}$$
$$Cu^{2+}(aq, 1 M) + 2e^- \longrightarrow Cu(s) \qquad E°_{red} = +0.339 \text{ V}$$

TABLE 18.1 Standard Potentials in Water Solution at 25°C OHT

Acidic Solution, $[H^+] = 1\ M$

			$E°_{red}$ (V)
$Li^+(aq) + e^-$	$\longrightarrow$	$Li(s)$	-3.040
$K^+(aq) + e^-$	$\longrightarrow$	$K(s)$	-2.936
$Ba^{2+}(aq) + 2e^-$	$\longrightarrow$	$Ba(s)$	-2.906
$Ca^{2+}(aq) + 2e^-$	$\longrightarrow$	$Ca(s)$	-2.869
$Na^+(aq) + e^-$	$\longrightarrow$	$Na(s)$	-2.714
$Mg^{2+}(aq) + 2e^-$	$\longrightarrow$	$Mg(s)$	-2.357
$Al^{3+}(aq) + 3e^-$	$\longrightarrow$	$Al(s)$	-1.68
$Mn^{2+}(aq) + 2e^-$	$\longrightarrow$	$Mn(s)$	-1.182
$Zn^{2+}(aq) + 2e^-$	$\longrightarrow$	$Zn(s)$	-0.762
$Cr^{3+}(aq) + 3e^-$	$\longrightarrow$	$Cr(s)$	-0.744
$Fe^{2+}(aq) + 2e^-$	$\longrightarrow$	$Fe(s)$	-0.409
$Cr^{3+}(aq) + e^-$	$\longrightarrow$	$Cr^{2+}(aq)$	-0.408
$Cd^{2+}(aq) + 2e^-$	$\longrightarrow$	$Cd(s)$	-0.402
$PbSO_4(s) + 2e^-$	$\longrightarrow$	$Pb(s) + SO_4^{2-}(aq)$	-0.356
$Tl^+(aq) + e^-$	$\longrightarrow$	$Tl(s)$	-0.336
$Co^{2+}(aq) + 2e^-$	$\longrightarrow$	$Co(s)$	-0.282
$Ni^{2+}(aq) + 2e^-$	$\longrightarrow$	$Ni(s)$	-0.236
$AgI(s) + e^-$	$\longrightarrow$	$Ag(s) + I^-(aq)$	-0.152
$Sn^{2+}(aq) + 2e^-$	$\longrightarrow$	$Sn(s)$	-0.141
$Pb^{2+}(aq) + 2e^-$	$\longrightarrow$	$Pb(s)$	-0.127
$2H^+(aq) + 2e^-$	$\longrightarrow$	$H_2(g)$	0.000
$AgBr(s) + e^-$	$\longrightarrow$	$Ag(s) + Br^-(aq)$	0.073
$S(s) + 2H^+(aq) + 2e^-$	$\longrightarrow$	$H_2S(aq)$	0.144
$Sn^{4+}(aq) + 2e^-$	$\longrightarrow$	$Sn^{2+}(aq)$	0.154
$SO_4^{2-}(aq) + 4H^+(aq) + 2e^-$	$\longrightarrow$	$SO_2(g) + 2H_2O$	0.155
$Cu^{2+}(aq) + e^-$	$\longrightarrow$	$Cu^+(aq)$	0.161
$Cu^{2+}(aq) + 2e^-$	$\longrightarrow$	$Cu(s)$	0.339
$Cu^+(aq) + e^-$	$\longrightarrow$	$Cu(s)$	0.518
$I_2(s) + 2e^-$	$\longrightarrow$	$2I^-(aq)$	0.534
$Fe^{3+}(aq) + e^-$	$\longrightarrow$	$Fe^{2+}(aq)$	0.769
$Hg_2^{2+}(aq) + 2e^-$	$\longrightarrow$	$2Hg(l)$	0.796
$Ag^+(aq) + e^-$	$\longrightarrow$	$Ag(s)$	0.799
$2Hg^{2+}(aq) + 2e^-$	$\longrightarrow$	$Hg_2^{2+}(aq)$	0.908
$NO_3^-(aq) + 4H^+(aq) + 3e^-$	$\longrightarrow$	$NO(g) + 2H_2O$	0.964
$AuCl_4^-(aq) + 3e^-$	$\longrightarrow$	$Au(s) + 4Cl^-(aq)$	1.001
$Br_2(l) + 2e^-$	$\longrightarrow$	$2Br^-(aq)$	1.077
$O_2(g) + 4H^+(aq) + 4e^-$	$\longrightarrow$	$2H_2O$	1.229
$MnO_2(s) + 4H^+(aq) + 2e^-$	$\longrightarrow$	$Mn^{2+}(aq) + 2H_2O$	1.229
$Cr_2O_7^{2-}(aq) + 14H^+(aq) + 6e^-$	$\longrightarrow$	$2Cr^{3+}(aq) + 7H_2O$	1.33
$Cl_2(g) + 2e^-$	$\longrightarrow$	$2Cl^-(aq)$	1.360
$ClO_3^-(aq) + 6H^+(aq) + 5e^-$	$\longrightarrow$	$\frac{1}{2}Cl_2(g) + 3H_2O$	1.458
$Au^{3+}(aq) + 3e^-$	$\longrightarrow$	$Au(s)$	1.498
$MnO_4^-(aq) + 8H^+(aq) + 5e^-$	$\longrightarrow$	$Mn^{2+}(aq) + 4H_2O$	1.512
$PbO_2(s) + SO_4^{2-}(aq) + 4H^+(aq) + 2e^-$	$\longrightarrow$	$PbSO_4(s) + 2H_2O$	1.687
$H_2O_2(aq) + 2H^+(aq) + 2e^-$	$\longrightarrow$	$2H_2O$	1.763
$Co^{3+}(aq) + e^-$	$\longrightarrow$	$Co^{2+}(aq)$	1.953
$F_2(g) + 2e^-$	$\longrightarrow$	$2F^-(aq)$	2.889

continued

O = strongest oxidizing agent;
R = strongest reducing agent.

TABLE 18.1   *continued*

Basic Solution, [OH⁻] = 1 M

		$E°_{red}$ (V)
$Fe(OH)_2(s) + 2e^-$	$\longrightarrow Fe(s) + 2OH^-(aq)$	-0.891
$2H_2O + 2e^-$	$\longrightarrow H_2(g) + 2OH^-(aq)$	-0.828
$Fe(OH)_3(s) + e^-$	$\longrightarrow Fe(OH)_2(s) + OH^-(aq)$	-0.547
$S(s) + 2e^-$	$\longrightarrow S^{2-}(aq)$	-0.445
$NO_3^-(aq) + 2H_2O + 3e^-$	$\longrightarrow NO(g) + 4OH^-(aq)$	-0.140
$NO_3^-(aq) + H_2O + 2e^-$	$\longrightarrow NO_2^-(aq) + 2OH^-(aq)$	0.004
$ClO_4^-(aq) + H_2O + 2e^-$	$\longrightarrow ClO_3^-(aq) + 2OH^-(aq)$	0.398
$O_2(g) + 2H_2O + 4e^-$	$\longrightarrow 4OH^-(aq)$	0.401
$ClO_3^-(aq) + 3H_2O + 6e^-$	$\longrightarrow Cl^-(aq) + 6OH^-(aq)$	0.614
$ClO^-(aq) + H_2O + 2e^-$	$\longrightarrow Cl^-(aq) + 2OH^-(aq)$	0.890

To obtain the standard voltage for an oxidation half-reaction, all you have to do is ***change the sign of the standard potential listed in Table 18.1.*** For example, knowing that

$$Zn^{2+}(aq) + 2e^- \longrightarrow Zn(s) \qquad E°_{red} = -0.762 \text{ V}$$

it follows that

$$Zn(s) \longrightarrow Zn^{2+}(aq) + 2e^- \qquad E°_{ox} = +0.762 \text{ V}$$

Again, we have

$$Cu^{2+}(aq) + 2e^- \longrightarrow Cu(s) \qquad E°_{red} = +0.339 \text{ V}$$

so:

$$Cu(s) \longrightarrow Cu^{2+}(aq) + 2e^- \qquad E°_{ox} = -0.339 \text{ V}$$

Another way to express this principle is to say that ***standard voltages for forward and reverse half-reactions are equal in magnitude but opposite in sign.***

Strength of Oxidizing and Reducing Agents

As pointed out in Chapter 4, an oxidizing agent is a species that can gain electrons; it is a reactant in a reduction half-reaction. Table 18.1 lists reduction half-reactions from left to right; it follows that ***oxidizing agents are located in the left column of the table.*** All of the species listed in that column (Li^+, . . . , F_2) are, in principle, oxidizing agents. A "strong" oxidizing agent is one that has a strong attraction for electrons and hence *can readily oxidize other species.* In contrast, a "weak" oxidizing agent does not gain electrons readily. It is capable of reacting only with those species that are very easily oxidized.

The strength of an oxidizing agent is directly related to the standard voltage for its reduction, $E°_{red}$. ***The more positive $E°_{red}$ is, the stronger the oxidizing agent.*** Looking at Table 18.1, you can see that ***oxidizing strength increases moving down the left column.*** The Li^+ ion, at the top of that column, is a very weak oxidizing agent with a large negative reduction voltage ($E°_{red} = -3.040$ V). In practice, cations of the Group 1 metals (Li^+, Na^+, K^+, . . .) and the Group 2 metals (Mg^{2+}, Ca^{2+}, . . .) never act as oxidizing agents in water solution.

Further down the left column of Table 18.1, the H^+ ion ($E^\circ_{red} = 0.000$ V) is a reasonably strong oxidizing agent. It is capable of oxidizing many metals, including magnesium and zinc:

$$Mg(s) + 2H^+(aq) \longrightarrow Mg^{2+}(aq) + H_2(g) \qquad E^\circ = +2.357 \text{ V}$$

$$Zn(s) + 2H^+(aq) \longrightarrow Zn^{2+}(aq) + H_2(g) \qquad E^\circ = +0.762 \text{ V}$$

The strongest oxidizing agents are those at the bottom of the left column. Species such as $Cr_2O_7^{2-}$ ($E^\circ_{red} = +1.33$ V) and Cl_2 ($E^\circ_{red} = +1.360$ V) are commonly used as oxidizing agents in the laboratory. The fluorine molecule, at the bottom of the left column, should be the strongest of all oxidizing agents. In practice, fluorine is seldom used as an oxidizing agent in water solution. The F_2 molecule takes electrons away from just about anything, including water, often with explosive violence.

The arguments we have just gone through can be applied, in reverse, to **reducing agents, listed in the column to the right of Table 18.1** (Li, . . . , F^-). In principle, at least, all of them can supply electrons to another species in a redox reaction. Their strength as reducing agents is directly related to their E°_{ox} values. **The more positive E°_{ox} is, the stronger the reducing agent.** Looking at the values, remembering that $E^\circ_{ox} = -E^\circ_{red}$,

$$Li(s) \longrightarrow Li^+(aq) + e^- \qquad E^\circ_{ox} = +3.040 \text{ V}$$

$$H_2(g) \longrightarrow 2H^+(aq) + 2e^- \qquad E^\circ_{ox} = \ \ 0.000 \text{ V}$$

$$2F^-(aq) \longrightarrow F_2(g) + 2e^- \qquad E^\circ_{ox} = -2.889 \text{ V}$$

It should be clear that **reducing strength decreases moving down the table.** Actually, the five metals at the top of the right column, (Li, K, Ba, Ca, Na) cannot be used as reducing agents in water solution because they react directly with water, reducing it to hydrogen gas:

$$Li(s) + H_2O \longrightarrow Li^+(aq) + OH^-(aq) + \tfrac{1}{2}H_2(g)$$

$$Ba(s) + 2H_2O \longrightarrow Ba^{2+}(aq) + 2OH^-(aq) + H_2(g)$$

Lithium as a reducing agent.
Lithium, the strongest reducing agent listed in Table 18.1, is shown reacting with water. The heat of reaction is igniting the hydrogen produced. (*Charles D. Winters*)

EXAMPLE 18.2   Consider the following species in acidic solution: $Cr_2O_7^{2-}$, NO_3^-, Br^-, Mg, and Sn^{2+}. Using Table 18.1,

(a) classify each of these as an oxidizing and/or reducing agent.
(b) arrange the oxidizing agents in order of increasing strength.
(c) do the same with the reducing agents.

Strategy   Remember that oxidizing agents are located in the left column of Table 18.1, reducing agents in the right column. Large positive values of E°_{red} and E°_{ox} are associated with strong oxidizing and reducing agents, respectively.

Solution

(a) Oxidizing agents:

Sn^{2+} ($E^\circ_{red} = -0.141$ V), NO_3^- $E^\circ_{red} = +0.964$ V), $Cr_2O_7^{2-}$ ($E^\circ_{red} = +1.33$ V)

Reducing agents:

Mg $E^\circ_{ox} = +2.357$ V), Sn^{2+} ($E^\circ_{ox} = -0.154$ V), Br^- ($E^\circ_{ox} = -1.077$ V)

Note that the Sn^{2+} ion can act as either an oxidizing agent, when it is reduced to Sn, or a reducing agent, when it is oxidized to Sn^{4+}.

(b) Comparing values of E°_{red}:

$$Sn^{2+} < NO_3^- < Cr_2O_7^{2-}$$

This ranking correlates with the positions of these species in the left column of Table 18.1; $Cr_2O_7^{2-}$ is near the bottom, Sn^{2+} closest to the top.

(c) Comparing values of E°_{ox}:

$$Br^- < Sn^{2+} < Mg$$

Reality Check The dichromate ion, $Cr_2O_7^{2-}$, is a powerful oxidizing agent in acidic solution. Magnesium metal is frequently used as a reducing agent for cations such as H^+ or Cu^{2+}.

Calculation of E° from E°_{red} and E°_{ox}

As pointed out earlier, the standard voltage for a redox reaction is the sum of the standard voltages of the two half-reactions, reduction and oxidation; that is,

$$E^\circ = E^\circ_{red} + E^\circ_{ox}$$

This simple relation makes it possible, using Table 18.1, to calculate the standard voltages for more than 3000 different redox reactions.

EXAMPLE 18.3   Using Table 18.1, calculate E° for a voltaic cell in which the reaction is

$$2Fe^{3+}(aq) + 2I^-(aq) \longrightarrow 2Fe^{2+}(aq) + I_2(s)$$

Strategy   Split the reaction into two half-reactions, a reduction and an oxidation. Using Table 18.1, find the appropriate values for E°_{red} and E°_{ox}. Add these to obtain E°.

Solution   The oxidation number of iron drops from $+3$ to $+2$, so the reduction half-reaction is

$$2Fe^{3+}(aq) + 2e^- \longrightarrow 2Fe^{2+}(aq) \qquad E^\circ_{red} = +0.769 \text{ V}$$

The oxidation number of iodine increases from -1 to 0, so the oxidation half-reaction is

$$2I^-(aq) \longrightarrow I_2(s) + 2e^- \qquad E^\circ_{ox} = -0.534 \text{ V}$$

(Notice that the sign of the standard potential is changed, because this is an oxidation.)

Just as the half-equations are added to obtain the overall equation, the half-reaction voltages are added to obtain the overall voltage.

$2Fe^{3+}(aq) + 2e^- \longrightarrow 2Fe^{2+}(aq)$	$E^\circ_{red} = +0.769$ V
$2I^-(aq) \longrightarrow I_2(s) + 2e^-$	$E^\circ_{ox} = -0.534$ V
$2Fe^{3+}(aq) + 2I^-(aq) \longrightarrow 2Fe^{2+}(aq) + I_2(s)$	$E^\circ = +0.235$ V

Two general points concerning cell voltages are illustrated by Example 18.3.

1. The calculated voltage, E°, is always a positive quantity for a reaction taking place in a voltaic cell.

It would be nice if you could double the voltage by rewriting the equation, but you can't.

2. The quantities $E°$, $E°_{ox}$, and $E°_{red}$ are independent of how the equation for the cell reaction is written. You *never* multiply the voltage by the coefficients of the balanced equation.

Spontaneity of Redox Reactions

To determine whether a given redox reaction is spontaneous, apply a simple principle:

If the calculated voltage for a redox reaction is a positive quantity, the reaction will be spontaneous. If the calculated voltage is negative, the reaction is not spontaneous.

Ordinarily, this principle is applied at standard concentrations (1 atm for gases, 1 M for species in aqueous solution). Hence it is the sign of $E°$ that serves as the criterion for spontaneity. To show how this works, consider the problem of oxidizing nickel metal to Ni^{2+} ions. This cannot be accomplished by using 1 M Zn^{2+} ions:

$$Ni(s) + Zn^{2+}(aq, 1\ M) \longrightarrow Ni^{2+}(aq, 1\ M) + Zn(s)$$
$$E° = E°_{ox}\ Ni + E°_{red}\ Zn^{2+} = +0.236\ V - 0.762\ V = -0.526\ V$$

Since you add to obtain $E°$ it makes no difference which you write first, $E°_{red}$ or $E°_{ox}$.

Sure enough, if you immerse a bar of nickel in a solution of 1 M $ZnSO_4$, nothing happens. Suppose, however, you add the nickel bar to a solution of 1 M $CuSO_4$ (Figure 18.5; left beaker). As time passes, the nickel is oxidized and the Cu^{2+} ions reduced (beaker at right of Figure 18.5) through the spontaneous redox reaction

$$Ni(s) + Cu^{2+}(aq, 1\ M) \longrightarrow Ni^{2+}(aq, 1\ M) + Cu(s)$$
$$E° = E°_{ox}\ Ni + E°_{red}\ Cu^{2+} = +0.236\ V + 0.339\ V = +0.575\ V$$

EXAMPLE 18.4   Using standard potentials listed in Table 18.1, decide whether, at standard concentrations,

(a) Fe(s) will be oxidized to Fe^{2+} by treatment with hydrochloric acid (HCl).
(b) Cu(s) will be oxidized to Cu^{2+} by treatment with hydrochloric acid.
(c) Cu(s) will be oxidized to Cu^{2+} by treatment with nitric acid (HNO_3).

Strategy   The oxidation half-reaction is given in each case; you must decide on the nature of the reduction half-reaction. Once that is done, look up the appropriate standard potentials and combine them to find out whether $E°$ is positive or negative.

Solution

(a) The oxidation half-reaction is

$$Fe(s) \longrightarrow Fe^{2+}(aq) + 2e^- \qquad E°_{ox} = +0.409\ V$$

Hydrochloric acid consists of H^+ and Cl^- ions. Of these two ions, only H^+ is listed in the left column of Table 18.1; the Cl^- ion cannot be reduced. The reduction half-reaction must be

$$2H^+(aq) + 2e^- \longrightarrow H_2(g) \qquad E°_{red} = 0.000\ V$$

Because the calculated voltage is positive,

$$E° = +0.409\ V + 0.000\ V = +0.409\ V$$

The following redox reaction, obtained by summing the half-reactions,

$$Fe(s) + 2H^+(aq) \longrightarrow Fe^{2+}(aq) + H_2(g)$$

FIGURE 18.5
Oxidation of Ni by Cu^{2+}. Nickel metal reacts spontaneously with Cu^{2+} ions, producing Cu metal and Ni^{2+} ions. Copper plates out on the surface of the nickel, and the blue color of Cu^{2+} is replaced by the green color of Ni^{2+}. *(Marna G. Clarke)*

should and does occur (Figure 18.6).

(b) Proceeding in the same way,

$$Cu(s) \longrightarrow Cu^{2+}(aq) + 2e^- \qquad E°_{ox} = -0.339 \text{ V}$$

$$\underline{2H^+(aq) + 2e^- \longrightarrow H_2(g)524 \qquad E°_{red} = 0.000 \text{ V}}$$

$$Cu(s) + 2H^+(aq) \longrightarrow Cu^{2+}(aq) + H_2(g) \qquad E° = -0.339 \text{ V}$$

No reaction occurs when copper is added to 1 M hydrochloric acid.

(c) There are two possible oxidizing agents in this case, the H^+ ion and the NO_3^- ion. Looking at Table 18.1, you should find that $E°_{red}$ for the NO_3^- ion is $+0.964$ V. It follows that $E°$ is positive:

$$E° = E°_{ox} \text{ for Cu} + E°_{red} \text{ for } NO_3^- = -0.339 \text{ V} + 0.964 \text{ V} = +0.625 \text{ V}$$

This means that copper is oxidized (to Cu^{2+}) by nitric acid, which is reduced to NO (or NO_2); see Figure 18.7.

FIGURE 18.6
Oxidation of Fe by H^+. Finely divided iron in the form of steel wool reacts with hydrochloric acid to evolve hydrogen: $Fe(s) + 2H^+(aq) \rightarrow Fe^{2+}(aq) + H_2(g)$. The hydrogen gas is creating foam as it is produced along the strands of the steel wool. *(Charles D. Winters)*

18.3 RELATIONS BETWEEN $E°$, $\Delta G°$, AND K

As pointed out previously, the value of the standard cell voltage, $E°$, is a measure of the spontaneity of a cell reaction. In Chapter 17, we showed that the standard free energy change, $\Delta G°$, is a general criterion for reaction spontaneity. As you might suppose, these two quantities are simply related to one another and to the equilibrium constant, K, for the cell reaction.

$E°$ and $\Delta G°$

It can be shown by a thermodynamic argument that we will not go through here that these two quantities are related by a simple equation

$$\boxed{\Delta G° = -nFE°}$$

where—

- $\Delta G°$ is the standard free energy change (gases at 1 atm, species in solution at 1 M) for the reaction.

Spontaneous reaction: $E° > 0$, $\Delta G° < 0$, $K > 1$.

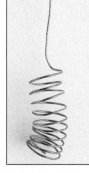 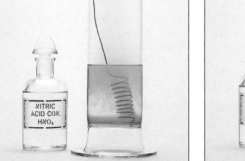 

FIGURE 18.7
Oxidation of Cu by HNO_3. Copper metal is comparatively inactive but it reacts with concentrated nitric acid. The brown fumes are $NO_2(g)$, a reduction product of HNO_3. The copper is oxidized to Cu^{2+} ions, which impart their color to the solution. *(Marna G. Clarke)*

$F = 9.648 \times 10^4$ C/mol
$= 9.648 \times 10^4$ J/mol·V.

- $E°$ is the standard voltage for the cell reaction.
- n is the number of moles of electrons transferred in the reaction.
- F, called the **Faraday constant,** has the value: 9.648×10^4 J/mol·V (J = joule, V = volt).

Notice from this equation that $\Delta G°$ and $E°$ have opposite signs. This is reasonable; a spontaneous reaction is one for which $\Delta G°$ is *negative* but $E°$ is *positive*.

EXAMPLE 18.5   Calculate $\Delta G°$ for the reaction

$$Cl_2(g) + 2Br^-(aq) \longrightarrow 2Cl^-(aq) + Br_2(l)$$

using data in Table 18.1.

Strategy   (1) Split the reaction into two half-reactions; find $E°_{ox}$ and $E°_{red}$ from Table 18.1 and add to obtain $E°$. Then (2) use the relation $\Delta G° = -nFE°$.

Solution

(1) The balanced half-equations for reduction and oxidation are

$Cl_2(g) + 2e^- \longrightarrow 2Cl^-(aq)$	$E°_{red} = +1.360$ V
$2Br^-(aq) \longrightarrow Br_2(l) + 2e^-$	$E°_{ox} = -1.077$ V
$Cl_2(g) + 2Br^-(aq) \longrightarrow 2Cl^-(aq) + Br_2(l)$	$E° = +0.283$ V

(2) Note from the half-equations that $n = 2$:

$$\Delta G° = -2 \text{ mol} (9.648 \times 10^4 \text{ J/mol·V})(0.283 \text{ V}) = -5.46 \times 10^4 \text{ J} = \boxed{-54.6 \text{ kJ}}$$

Reality Check   Notice that $E°$ is *positive*, and $\Delta G°$ is *negative*, indicating a spontaneous reaction. This reaction can serve as a basis for a voltaic cell (Figure 18.4) or as a way of testing for Br^- ions in solution (Figure 18.8).

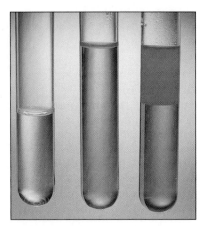

FIGURE 18.8
Oxidation of Br⁻ by Cl₂. When a water solution saturated with $Cl_2(g)$ is added to a solution containing Br^- ions *(left)*, a redox reaction occurs. The Br_2 formed gives the water solution a light orange color *(center)*. Extraction with a small amount of organic solvent gives the characteristic reddish-orange color of Br_2 *(right)*.
(Charles D. Winters)

The approach used in Example 18.5 to find the number of moles of electrons transferred, n, is generally useful. What you do is to break down the equation for the cell reaction into two half-equations, oxidation and reduction. The quantity n is the number of electrons appearing in either half-equation.

$E°$ and K

Redox reactions, like all reactions, eventually reach a state of equilibrium. It is possible to calculate the equilibrium constant for a redox reaction from the standard voltage. To do that, we start with the relation obtained in Chapter 17:

$$\boxed{\Delta G° = -RT \ln K}$$

We have just seen that $\Delta G° = -nFE°$. It follows that

$$RT \ln K = nFE°$$

or

$$E° = \frac{RT}{nF} \ln K$$

The quantity RT/F is readily evaluated at 25°C, the temperature at which standard potentials are tabulated.

$$\frac{RT}{F} = \frac{8.31 \text{ J/mol·K} \times 298 \text{ K}}{9.648 \times 10^4 \text{ J/mol·V}} = 0.0257 \text{ V}$$

Hence
$$E° = \frac{(0.0257\ V)}{n} \ln K \qquad (\text{at } 25°C)$$

Notice that if the standard voltage is positive, $\ln K$ is also positive, and K is greater than one. Conversely, if the standard voltage is negative, $\ln K$ is also negative, and K is less than 1.

EXAMPLE 18.6   For the reaction

$$3Ag(s) + NO_3^-(aq) + 4H^+(aq) \longrightarrow 3Ag^+(aq) + NO(g) + 2H_2O$$

calculate the equilibrium constant K, using data in Table 18.1.

Strategy   (1) Use Table 18.1 to find $E°$. (2) Split the equation into two half-equations to find n. Then (3) use the relation $nE° = (0.0257\ V) \ln K$ to find K.

Solution

(1) $E° = E°_{\text{red}}\ NO_3^- + E°_{\text{ox}}\ Ag = +0.964\ V - 0.799\ V = +0.165\ V$

(2) Ag is oxidized to Ag^+; NO_3^- is reduced to NO. The half-equations are

$$3Ag(s) \longrightarrow 3Ag^+(aq) + 3e^-$$

$$NO_3^-(aq) + 4H^+(aq) + 3e^- \longrightarrow NO(g) + 2H_2O$$

Clearly, $n = 3$.

(3) $\ln K = \dfrac{nE°}{(0.0257\ V)} = \dfrac{3(+0.165\ V)}{(0.0257\ V)} = 19.3; K = \boxed{2 \times 10^8}$

Reality Check   Because $E° > 0$, K is relatively large (compare Table 18.2).

Table 18.2 lists values of K and $\Delta G°$ corresponding to various values of $E°$ with $n = 2$. Notice that—

- if $E°$ is greater than about $+0.10\ V$, K is very large, and the reaction goes virtually to completion.
- if $E°$ is smaller than about $-0.10\ V$, K is very small, so the reaction does not proceed to any appreciable extent.

Only if the standard voltage falls within a rather narrow range, say $+0.10$ to $-0.10\ V$, will the value of K (and that of $\Delta G°$) be such that the reaction will produce an equilibrium mixture containing appreciable amounts of both reactants and products.

> Most redox reactions either go to completion or not at all.

TABLE 18.2   Relation Between $E°$, K, and $\Delta G°$ ($n = 2$) OHT

$E°$ (V)	K	$\Delta G°$ (kJ)	$E°$ (V)	K	$\Delta G°$ (kJ)
+2.00	4×10^{67}	−400	−2.00	3×10^{-68}	+400
+1.00	6×10^{33}	−200	−1.00	2×10^{-34}	+200
+0.50	8×10^{16}	−100	−0.50	1×10^{-17}	+100
+0.25	3×10^8	−50	−0.25	4×10^{-9}	+50
+0.10	2×10^3	−20	−0.10	0.0004	+20
+0.05	50	−10	−0.05	0.02	+10
0.00	1	0			

18.4   EFFECT OF CONCENTRATION ON VOLTAGE

To this point, we have dealt only with "standard" voltages, that is, voltages when all gases are at 1 atm pressure and all species in aqueous solution are at a concentration of 1 M. When the concentration of a rectant or product changes, the voltage changes as well. Qualitatively, the direction in which the voltage shifts is readily predicted when you realize that cell voltage is directly related to reaction spontaneity.

1. Voltage will *increase* if the concentration of a reactant is increased or that of a product is decreased. Either of these changes increases the driving force behind the redox reaction, making it more spontaneous.

2. Voltage will *decrease* if the concentration of a reactant is decreased or that of a product is increased. Either of these changes makes the redox reaction less spontaneous.

When a voltaic cell operates, supplying electrical energy, the concentration of reactants decreases and that of the products increases. As time passes, the voltage drops steadily. Eventually it becomes zero, and we say that the cell is "dead." At that point, the redox reaction taking place within the cell is at equilibrium, and there is no driving force to produce a voltage.

⌐ That's what happens when you leave your car lights on.

Nernst Equation

To obtain a quantitative relation between cell voltage and concentration, it is convenient to start with the general expression for the free energy change discussed in Chapter 17:

$$\Delta G = \Delta G^\circ + RT \ln Q$$

where Q is the reaction quotient in which products appear in the numerator, reactants in the denominator. But $\Delta G^\circ = -nFE^\circ$ and $\Delta G = -nFE$, so we have

$$-nFE = -nFE^\circ + RT \ln Q$$

Solving for the cell voltage E,

$$\boxed{E = E^\circ - \frac{RT}{nF} \ln Q}$$

◉ See Screen 21.7, Electrochemical Cells at Nonstandard Conditions.

This relationship is known as the **Nernst equation,** after Walther Nernst, a brilliant and egocentric colleague of Arrhenius, who first proposed it in 1888. Recalling that, at 25°C, the quantity RT/F is 0.0257 V,

$$E = E^\circ - \frac{(0.0257\ \text{V})}{n} \ln Q$$

In this equation, E is the cell voltage, E° is the standard voltage, n is the number of moles of electrons exchanged in the reaction, and Q is the reaction quotient. Notice that—

■ if $Q > 1$, which means that the concentrations of products are high relative to those of reactants, $\ln Q$ is positive and $E < E^\circ$.

■ if $Q < 1$ (concentrations of products low relative to reactants), $\ln Q$ is negative and $E > E^\circ$.

■ if $Q = 1$, as is the case under standard conditions, $\ln Q = 0$ and $E = E^\circ$.

Remember that gases enter Q as their partial pressures in atmospheres. Species in water solution enter as their molar concentrations. Pure liquids and solids do not appear in the expression for Q. For example,

$$aA(s) + bB(aq) \longrightarrow cC(aq) + dD(g) \qquad Q = \frac{[C]^c \times (P_D)^d}{[B]^b}$$

EXAMPLE 18.7   Consider a voltaic cell in which the following reaction occurs:

$$O_2(g) + 4H^+(aq) + 4Br^-(aq) \longrightarrow 2H_2O + 2Br_2(l)$$

Calculate the cell voltage, E, when O_2 is at 1.0 atm pressure, $[H^+] = [Br^-] = 0.10\ M$.

Strategy   First (1) set up the Nernst equation, following the rules for Q listed above. Then (2) calculate $E°$, using the standard potentials in Table 18.1. Finally (3), using the Nernst equation, calculate E.

Solution

(1) $Q = \dfrac{1}{(P_{O_2}) \times [H^+]^4 \times [Br^-]^4}$

To find n, break the reaction down into two half-reactions:

$$O_2(g) + 4H^+(aq) + 4e^- \longrightarrow 2H_2O$$

$$4Br^-(aq) \longrightarrow 2Br_2(l) + 4e^-$$

Clearly, $n = 4$. The Nernst equation must then be

$$E = E° - \frac{(0.0257\ \text{V})}{4} \ln \frac{1}{(P_{O_2}) \times [H^+]^4 \times [Br^-]^4}$$

(2) $E° = E°_{red}\ O_2 + E°_{ox}\ Br^- = +1.229\ \text{V} - 1.077\ \text{V} = +0.152\ \text{V}$

(3) $E = +0.152\ \text{V} - \dfrac{(0.0257\ \text{V})}{4} \ln \dfrac{1}{1.0(0.10)^4(0.10)^4}$

$\qquad = +0.152\ \text{V} - \dfrac{(0.0257\ \text{V})}{4} \ln (1.0 \times 10^8)$

$\qquad = +0.152\ \text{V} - \dfrac{(0.0257\ \text{V})(18.4)}{4} = \boxed{+0.034\ \text{V}}$

Reality Check E is less than $E°$ because the concentrations of reactants, H^+ and Br^-, are less than 1 M.

The Nernst equation can also be used to determine the effect of changes in concentration on the voltage of an individual half-cell, $E°_{red}$ or $E°_{ox}$. Consider, for example, the half-reaction

$$MnO_4^-(aq) + 8H^+(aq) + 5e^- \longrightarrow Mn^{2+}(aq) + 4H_2O \qquad E°_{red} = +1.512\ \text{V}$$

The Nernst equation takes the form

$$E_{red} = +1.512\ \text{V} - \frac{(0.0257\ \text{V})}{5} \ln \frac{[Mn^{2+}]}{[MnO_4^-] \times [H^+]^8}$$

where E_{red} is the observed reduction voltage corresponding to any given concentrations of Mn^{2+}, MnO_4^-, and H^+.

Use of the Nernst Equation to Determine Ion Concentrations

In chemistry, the most important use of the Nernst equation lies in the experimental determination of the concentration of ions in solution. Suppose you measure the cell voltage E and know the concentration of all but one species in the two half-cells. It should then be possible to calculate the concentration of that species by using the Nernst equation (Example 18.8).

> This approach is particularly useful for species at very low concentrations.

EXAMPLE 18.8   Consider a voltaic cell in which the reaction is

$$Zn(s) + 2H^+(aq) \longrightarrow Zn^{2+}(aq) + H_2(g)$$

It is found that the cell voltage is $+0.460$ V when $[Zn^{2+}] = 1.0$ M, $P_{H_2} = 1.0$ atm. What must be the concentration of H^+ in the H_2-H^+ half-cell?

Strategy   This example is handled exactly like Example 18.7, except that in the last step you solve for $[H^+]$ instead of E.

Solution

(1) Setting up the Nernst equation with $n = 2$,

$$E = E° - \frac{(0.0257 \text{ V})}{2} \ln \frac{[Zn^{2+}] \times (P_{H_2})}{[H^+]^2}$$

(2) $E° = E°_{ox}$ Zn $+ E°_{red}$ H$^+$ $= +0.762$ V
(3) All that remains is to substitute for $E, E°_{tot}, [Zn^{2+}], P_{H_2}$, and solve for $[H^+]$:

$$+0.460 \text{ V} = +0.762 \text{ V} - \frac{0.0257 \text{ V}}{2} \ln \frac{1 \times 1}{[H^+]^2}$$

Solving

$$\ln \frac{1}{[H^+]^2} = \frac{2(+0.460 \text{ V} - 0.762 \text{ V})}{-0.0257 \text{ V}} = 23.5$$

$$\frac{1}{[H^+]^2} = 1.6 \times 10^{10} \qquad [H^+] = \boxed{8 \times 10^{-6} M}$$

Reality Check   Notice that the concentration of H^+, a reactant, is much less than 1 M. That explains why the cell voltage, $+0.460$ V, is less than the standard voltage, $+0.762$ V.

As Example 18.8 implies, the $Zn|Zn^{2+}\|H^+|H_2|Pt$ cell can be used to measure the concentration of H^+ or pH of a solution. Indeed, cells of this type are commonly used to measure pH; Figure 18.9 shows a schematic diagram of a cell used with a pH meter. The pH meter, referred to in Chapter 13, is actually a high-resistance voltmeter calibrated to read pH rather than voltage. The cell connected to the pH meter consists of two half-cells. One of these is a reference half-cell of known voltage. The other half-cell contains a solution of known pH separated by a thin, fragile *glass electrode* from a solution whose pH is to be determined. The voltage of this cell is a linear function of the pH of the solution in the beaker.

Specific ion electrodes, similar in design to the glass electrode, have been developed to analyze for a variety of cations and anions. One of the first to be used extensively was a fluoride ion electrode that is sensitive to F^- at concentrations as low as 0.1 part per million and hence is ideal for monitoring fluoridated water supplies. An electrode that is specific for Cl^- ions is used to diagnose cystic fibrosis.

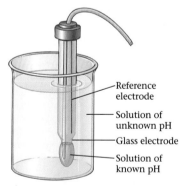

Reference electrode

Solution of unknown pH

Glass electrode

Solution of known pH

FIGURE 18.9

Electrode for a pH meter. The pH of a solution can be determined with the aid of a "glass electrode." The voltage between the glass electrode and the reference electrode is directly related to pH. The leads from the electrode are connected to a pH meter. **OHT**

Attached directly to the skin, it detects the abnormally high concentrations of sodium chloride in sweat that are a characteristic symptom of this disorder. Diagnoses that used to require an hour or more can now be carried out in a few minutes; as a result, large numbers of children can be screened rapidly and routinely.

The general approach illustrated by Example 18.8 is widely used to determine equilibrium constants for solution reactions. The pH meter in particular can be used to determine acid or base equilibrium constants by measuring the pH of solutions containing known concentrations of weak acids or bases. Specific ion electrodes are readily adapted to the determination of solubility product constants. For example, a chloride ion electrode can be used to find $[Cl^-]$ in equilibrium with $AgCl(s)$ and a known $[Ag^+]$. From that information, K_{sp} of AgCl can be calculated.

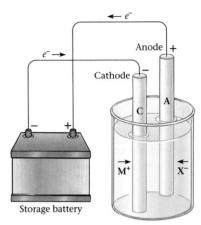

FIGURE 18.10
Diagram of an electrolytic cell.
Electrons enter the cathode from an external source. Cations move to the cathode, where they are reduced and anions move to the anode, where they are oxidized. **OHT**

18.5 ELECTROLYTIC CELLS

In an **electrolytic cell,** a nonspontaneous redox reaction is made to occur by pumping electrical energy into the system. A generalized diagram for such a cell is shown in Figure 18.10. The storage battery at the left provides a source of direct electric current. From the terminals of the battery, two wires lead to the electrolytic cell. This consists of two electrodes, A and C, dipping into a solution containing ions M^+ and X^-.

The battery acts as an electron pump, pushing electrons into the *cathode,* C, and removing them from the *anode,* A. To maintain electrical neutrality, some process within the cell must consume electrons at C and liberate them at A. This process is an oxidation-reduction reaction; when carried out in an electrolytic cell, it is called **electrolysis.** At the cathode, an ion or molecule undergoes reduction by accepting electrons. At the anode, electrons are produced by the oxidation of an ion or molecule.

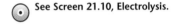

See Screen 21.10, Electrolysis.

⌐Oxidation occurs at the anode, reduction at the cathode, in both voltaic and electrolytic cells.

Quantitative Relationships

There is a simple relationship between the amount of electricity passed through an electrolytic cell and the amounts of substances produced by oxidation or reduction at the electrodes. From the balanced half-equations

$$Ag^+(aq) + e^- \longrightarrow Ag(s)$$

$$Cu^{2+}(aq) + 2e^- \longrightarrow Cu(s)$$

$$Au^{3+}(aq) + 3e^- \longrightarrow Au(s)$$

you can deduce that

$$1 \text{ mol of } e^- \longrightarrow 1 \text{ mol of Ag (107.9 g of Ag)}$$

$$2 \text{ mol of } e^- \longrightarrow 1 \text{ mol of Cu (63.55 g of Cu)}$$

$$3 \text{ mol of } e^- \longrightarrow 1 \text{ mol of Au (197.0 g of Au)}$$

Relations of this type, obtained from balanced half-equations, can be used in many practical calculations involving electrolytic cells. Frequently, the relations between electrical units provided in Table 18.3 (page 532) will be required as well.

See Screen 21.11, Coulometry.

TABLE 18.3	Electrical Units		
Quantity	**Unit**	**Defining Relation**	**Conversion Factors**
Charge	**coulomb (C)**	$1 \text{ C} = 1 \text{ A} \cdot \text{s} = 1 \text{ J/V}$	$1 \text{ mol } e^- = 9.648 \times 10^4 \text{ C}$
Current	**ampere (A)**	$1 \text{ A} = 1 \text{ C/s}$	
Potential	**volt (V)**	$1 \text{ V} = 1 \text{ J/C}$	
Power	watt (W)	$1 \text{ W} = 1 \text{ J/s}$	
Energy	joule (J)	$1 \text{ J} = 1 \text{ V} \cdot \text{C}$	$1 \text{ kWh} = 3.600 \times 10^6 \text{ J}$

EXAMPLE 18.9 Chromium metal can be electroplated from an acidic solution of CrO_3.

(a) How many grams of chromium will be plated by 1.00×10^4 C?
(b) How long will it take to plate one gram of chromium using a current of 6.00 A?

Strategy The first thing to do is to write the half-equation for the reduction. Following the rules cited in Chapter 4, you should arrive at

$$CrO_3(aq) + 6H^+(aq) + 6e^- \longrightarrow Cr(s) + 3H_2O$$

From this equation, it should be clear that 6 mol of $e^- = 1$ mol of Cr (52.0 g of Cr). That relationship, along with the relations $1 \text{ mol } e^- = 9.648 \times 10^4$ C and $1 \text{ A} = 1$ C/s, can be used to calculate the required quantities. Follow a conversion-factor approach.

Solution

(a) $1.00 \times 10^4 \text{ C} \times \dfrac{1 \text{ mol } e^-}{9.648 \times 10^4 \text{ C}} \times \dfrac{52.0 \text{ g Cr}}{6 \text{ mol } e^-} = \boxed{0.898 \text{ g of Cr}}$

(b) The most straightforward approach is to calculate the number of coulombs first and then the time.

(1) $1.00 \text{ g Cr} \times \dfrac{6 \text{ mol } e^-}{52.0 \text{ g Cr}} \times \dfrac{9.648 \times 10^4 \text{ C}}{1 \text{ mol } e^-} = 1.11 \times 10^4 \text{ C}$

(2) Because 1 A = 1 C/s,

$$\text{time (s)} = \frac{\text{charge (C)}}{\text{current (A)}} = \frac{1.11 \times 10^4 \text{ C}}{6.00 \text{ C/s}} = \boxed{1.85 \times 10^3 \text{ s}} \qquad \text{(about half an hour)}$$

EXAMPLE 18.10 Consider the electroplating of chromium, referred to in Example 18.9. If the applied voltage is 4.5 V, calculate the amount of electrical energy absorbed in plating 1.00 g of Cr, first in joules and then in kilowatt-hours.

Strategy Recall that, in Example 18.9b, you calculated the number of coulombs required to plate one gram of Cr. Multiplying that by the number of volts will give the energy in joules, which can then be converted to kilowatt-hours.

Solution

$$\text{energy (J)} = \text{voltage (V)} \times \text{charge (C)} = 4.5 \text{ V} \times 1.1 \times 10^4 \text{ C} = \boxed{5.0 \times 10^4 \text{ J}}$$

$$\text{energy in kWh} = 5.0 \times 10^4 \text{ J} \times \frac{1 \text{ kWh}}{3.6 \times 10^6 \text{ J}} = \boxed{1.4 \times 10^{-2} \text{ kWh}}$$

Reality Check At 3¢ per kilowatt-hour, that's about 0.04¢ per gram of Cr or 40¢ per kilogram. That sounds about right.

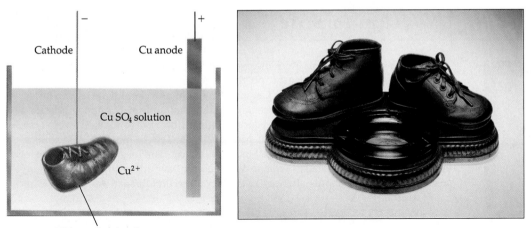

FIGURE 18.11

Electroplating. Copper metal can be plated onto a baby's shoe by electrolysis. The shoe's surface is coated with graphite to make it conduct electricity. *(Charles D. Winters)*

In working Examples 18.9 and 18.10, we have in effect assumed that the electrolyses were 100% efficient in converting electrical energy into chemical energy. In practice, this is almost never the case. Some electrical energy is wasted in side reactions at the electrodes and in the form of heat. This means that the actual yield of products is less than the theoretical yield.

Nobody is 100% efficient.

Cell Reactions (Water Solution)

As is always the case, a reduction half-reaction occurs at the cathode of an electrolytic cell. This half-reaction may be—

- *the reduction of a cation to the corresponding metal.* This commonly occurs with transition metal cations, which are relatively easy to reduce. Examples include

$$Ag^+(aq) + e^- \longrightarrow Ag(s) \qquad E^\circ_{red} = +0.799 \text{ V}$$
$$Cu^{2+}(aq) + 2e^- \longrightarrow Cu(s) \qquad E^\circ_{red} = +0.339 \text{ V}$$

This type of half-reaction is characteristic of electroplating processes, in which a metal object serves as the cathode (Figure 18.11).

- *the reduction of a water molecule to hydrogen gas*

$$2H_2O + 2e^- \longrightarrow H_2(g) + 2OH^-(aq) \qquad E^\circ_{red} = -0.828 \text{ V}$$

This half-reaction commonly occurs when the cation in solution is very difficult to reduce. For example, electrolysis of a solution containing K^+ ions ($E^\circ_{red} = -2.936$ V) or Na^+ ions ($E^\circ_{red} = -2.714$ V) yields hydrogen gas at the cathode (Table 18.4, page 535).

At the anode of an electrolytic cell, the half-reaction may be—

- *the oxidation of an anion to the corresponding nonmetal*

$$2I^-(aq) \longrightarrow I_2(s) + 2e^- \qquad E^\circ_{ox} = -0.534 \text{ V}$$

CHEMISTRY ▪ The Human Side

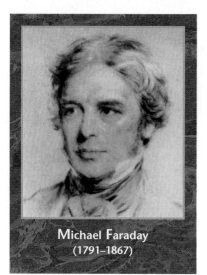

Michael Faraday
(1791–1867)

The laws of electrolysis were discovered by Michael Faraday, perhaps the most talented experimental scientist of the nineteenth century. Faraday lived his entire life in what is now greater London. The son of a blacksmith, he had no formal education beyond the rudiments of reading, writing, and arithmetic. Apprenticed to a bookbinder at the age of 13, Faraday educated himself by reading virtually every book that came into the shop. One that particularly impressed him was a textbook, *Conversations in Chemistry,* written by Mrs. Jane Marcet. Anxious to escape a life of drudgery as a tradesman, Faraday wrote to Sir Humphry Davy at the Royal Institution, requesting employment. Shortly afterward, a vacancy arose, and Faraday was hired as a laboratory assistant.

Davy quickly recognized Faraday's talents and as time passed allowed him to work more and more independently. In his years with Davy, Faraday published papers covering almost every field of chemistry. They included studies on the condensation of gases (he was the first to liquefy ammonia), the reaction of silver compounds with ammonia, and the isolation of several organic compounds, the most important of which was benzene. In 1825, Faraday began a series of lectures at the Royal Institution that were brilliantly successful. That same year he succeeded Davy as director of the laboratory. As Faraday's reputation grew, it was said that "Humphrey Davy's greatest discovery was Michael Faraday." Perhaps it was witticisms of this sort that led to an estrangement between master and protégé. Late in his life, Davy opposed Faraday's nomination as a Fellow of the Royal Society and is reputed to have cast the only vote against him.

To Michael Faraday, science was an obsession; one of his biographers described him as a "work maniac." An observer (Faraday had no students) said of him,

> . . . if he had to cross the laboratory for anything, he did not walk, he ran; the quickness of his perception was equalled by the calm rapidity of his movements.

In 1839, he suffered a nervous breakdown, the result of overwork. For much of the rest of his life, Faraday was in poor health. He gradually gave up more and more of his social engagements but continued to do research at the same pace as before.

Faraday developed the laws of electrolysis between 1831 and 1834. In mid-December of 1833, he began a quantitative study of the electrolysis of several metal cations, including Sn^{2+}, Pb^{2+}, and Zn^{2+}. Despite taking a whole day off for Christmas, he managed to complete these experiments, write up the results of three years' work, and get his paper published in the *Philosophic Transactions of the Royal Society* on January 9, 1834. In this paper, Faraday introduced the basic vocabulary of electrochemistry, using for the first time the terms "anode," "cathode," "ion," "electrolyte," and "electrolysis."

(Photo credit: Oesper Collection in the History of Chemistry, University of Cincinnati)

Electrolysis of a water solution of KI gives a saturated solution of iodine at the anode.

▪ *the oxidation of a water molecule to oxygen gas*

$$2H_2O \longrightarrow O_2(g) + 4H^+(aq) + 4e^- \qquad E^\circ_{ox} = -1.229 \text{ V}$$

This half-reaction occurs when the anion cannot be oxidized. Examples include nitrate and sulfate anions, where the nonmetal present is already in its highest oxidation state (+5 for N, +6 for S).

TABLE 18.4	Electrolysis of Water Solutions	
Solution	**Cathode Product**	**Anode Product**
$CuBr_2(aq)$	$Cu(s)$	$Br_2(l)$
$AgNO_3(aq)$	$Ag(s)$	$O_2(g)$
$KI(aq)$	$H_2(g)$	$I_2(s)$
$Na_2SO_4(aq)$	$H_2(g)$	$O_2(g)$

18.6   COMMERCIAL CELLS

To a chemist, electrochemical cells are of interest primarily for the information they yield concerning the spontaneity of redox reactions, the strengths of oxidizing and reducing agents, and the concentrations of trace species in solution. The viewpoint of an engineer is somewhat different; applications of electrolytic cells in electroplating and electrosynthesis are of particular importance. To the layman, electrochemistry is important primarily because of commercial voltaic cells, which supply the electrical energy for instruments ranging in size from pacemakers to automobiles.

Electrolysis of Aqueous NaCl

From a commercial standpoint, the most important electrolysis carried out in water solution is that of sodium chloride (Figure 18.12, page 536). At the anode, Cl^- ions are oxidized to chlorine gas:

$$\text{anode:} \qquad 2Cl^-(aq) \longrightarrow Cl_2(g) + 2e^-$$

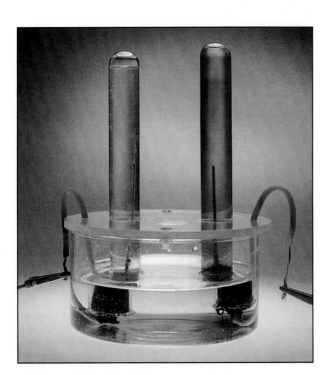

Electrolysis of potassium iodide (KI) solution. The electrolysis of aqueous KI is similar to that of aqueous NaCl. The cathode reaction *(left)* is the reduction of water to $H_2(g)$ and OH^-, as shown by the pink color of phenolphthalein indicator in the water. The anode reaction *(right)* is the oxidation of $I^-(aq)$ to $I_2(aq)$, as shown by the brown color of the solution. *(Charles D. Winters)*

FIGURE 18.12
Electrolysis of aqueous NaCl (brine). The anode and cathode half-cells are separated by a membrane that is permeable to ions but not to water. Thus the flow of ions maintains charge balance. The products, NaOH(*aq*), Cl$_2$, and H$_2$, are all valuable industrial chemicals.

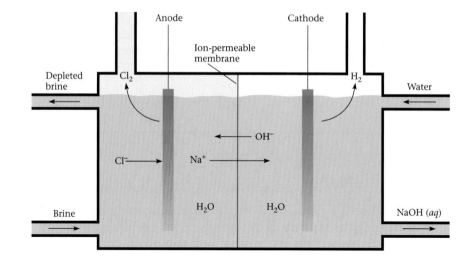

At the cathode, the half-reaction involves H$_2$O molecules, which are easier to reduce ($E^\circ_{red} = -0.828$ V) than Na$^+$ ions ($E^\circ_{red} = -2.714$ V).

$$\text{cathode:} \quad 2H_2O + 2e^- \longrightarrow H_2(g) + 2OH^-(aq)$$

The overall cell reaction is obtained by summing the half-reactions:

$$2Cl^-(aq) + 2H_2O \longrightarrow Cl_2(g) + H_2(g) + 2OH^-(aq)$$

Chlorine gas bubbles out of solution at the anode. At the cathode, hydrogen gas is formed, and the solution around the electrode becomes strongly basic.

The products of this electrolysis have a variety of uses. Chlorine is used to purify drinking water; large quantities of it are consumed in making plastics such as polyvinyl chloride (PVC). Hydrogen, prepared in this and many other industrial processes, is used chiefly in the synthesis of ammonia (Chapter 12). Sodium hydroxide (lye), obtained on evaporation of the electrolyte, is used in processing pulp and paper, in the purification of aluminum ore, in the manufacture of glass and textiles, and for many other purposes.

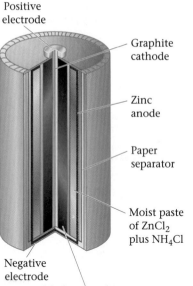

FIGURE 18.13
Zn-MnO$_2$ dry cell. This cell produces 1.5 V and will deliver a current of about half an ampere for six hours. **OHT**

⊙ **See Screen 21.8, Batteries.**

Primary (Nonrechargeable) Voltaic Cells

The construction of the ordinary dry cell (Leclanché cell) used in flashlights is shown in Figure 18.13. The zinc wall of the cell is the anode. The graphite rod through the center of the cell is the cathode. The space between the electrodes is filled with a moist paste. This contains MnO$_2$, ZnCl$_2$, and NH$_4$Cl. When the cell operates, the half-reaction at the anode is

$$Zn(s) \longrightarrow Zn^{2+}(aq) + 2e^-$$

At the cathode, manganese dioxide is reduced to species in which Mn is in the +3 oxidation state, such as Mn$_2$O$_3$:

$$2MnO_2(s) + 2NH_4^+(aq) + 2e^- \longrightarrow Mn_2O_3(s) + 2NH_3(aq) + H_2O$$

The overall reaction occurring in this voltaic cell is

$$Zn(s) + 2MnO_2(s) + 2NH_4^+(aq) \longrightarrow Zn^{2+}(aq) + Mn_2O_3(s) + 2NH_3(aq) + H_2O$$

If too large a current is drawn from a Leclanché cell, the ammonia forms a gaseous insulating layer around the carbon cathode. When this happens, the voltage drops sharply and then returns slowly to its normal value of 1.5 V. This problem can be avoided by using an alkaline dry cell, in which the paste between the electrodes contains KOH rather than NH_4Cl. In this case the overall cell reaction is simply

$$Zn(s) + 2MnO_2(s) \longrightarrow ZnO(s) + Mn_2O_3(s)$$

No gas is produced. The alkaline dry cell, although more expensive than the Leclanché cell, has a longer shelf life and provides more current.

Another important primary battery is the mercury cell. It usually comes in very small sizes and is used in hearing aids, watches, cameras, and some calculators. The anode of this cell is a zinc-mercury amalgam; the reacting species is zinc. The cathode is a plate made up of mercury(II) oxide, HgO. The electrolyte is a paste containing HgO and sodium or potassium hydroxide. The electrode reactions are

anode: $Zn(s) + 2OH^-(aq) \longrightarrow Zn(OH)_2(s) + 2e^-$

cathode: $\dfrac{HgO(s) + H_2O + 2e^- \longrightarrow Hg(l) + 2OH^-(aq)}{Zn(s) + HgO(s) + H_2O \longrightarrow Zn(OH)_2(s) + Hg(l)}$

Notice that the overall reaction does not involve any ions in solution, so there are no concentration changes when current is drawn. As a result, the battery maintains a constant voltage of about 1.3 V throughout its life.

Storage (Rechargeable) Voltaic Cells

A storage cell, unlike an ordinary dry cell, can be recharged repeatedly. This can be accomplished because the products of the reaction are deposited directly on the electrodes. By passing a current through a storage cell, it is possible to reverse the electrode reactions and restore the cell to its original condition.

The rechargeable 12-V lead storage battery used in automobiles consists of six voltaic cells of the type shown in Figure 18.14. A group of lead plates, the grills of which are filled with spongy gray lead, forms the anode of the cell. The multiple cathode consists of another group of plates of similar design filled with lead(IV) oxide, PbO_2. These two sets of plates alternate through the cell. They are immersed in a water solution of sulfuric acid, H_2SO_4, which acts as the electrolyte.

In August 1999, scientists in Israel announced the development of a new alkaline "superbattery" in which MnO_2 is replaced by K_2FeO_4.

A Leclanché dry cell cut in half. Compare this with the diagram in Figure 18.13. *(Charles D. Winters)*

A 12-V storage battery can deliver 300 A for a minute or so.

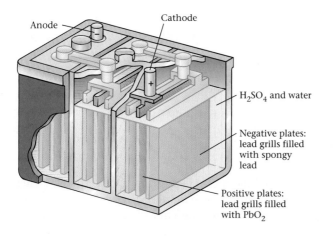

Anode

Cathode

H_2SO_4 and water

Negative plates: lead grills filled with spongy lead

Positive plates: lead grills filled with PbO_2

FIGURE 18.14

Lead storage battery. Three advantages of the lead storage battery are its ability to deliver large amounts of energy for a short time, the ease of recharging, and a nearly constant voltage from full charge to discharge. A disadvantage is its high mass-to-energy ratio. **OHT**

When a lead storage battery is supplying current, the lead in the anode grids is oxidized to Pb^{2+} ions. These immediately react with SO_4^{2-} ions in the electrolyte, precipitating $PbSO_4$ (lead sulfate) on the plates. At the cathode, lead dioxide is reduced to Pb^{2+} ions, which also precipitate as $PbSO_4$:

anode: $$Pb(s) + SO_4^{2-}(aq) \longrightarrow PbSO_4(s) + 2e^-$$

cathode: $$\frac{PbO_2(s) + 4H^+(aq) + SO_4^{2-}(aq) + 2e^- \longrightarrow PbSO_4(s) + 2H_2O}{Pb(s) + PbO_2(s) + 4H^+(aq) + 2SO_4^{2-}(aq) \longrightarrow 2PbSO_4(s) + 2H_2O}$$

Deposits of lead sulfate slowly build up on the plates, partially covering and replacing the lead dioxide. As the cell discharges, the concentration of sulfuric acid decreases. For every mole of lead reacting, two moles of H_2SO_4 ($4H^+$, $2SO_4^{2-}$) are replaced by two moles of water. The state of charge of a storage battery can be checked by measuring the density of the electrolyte. When the battery is fully charged, the density is in the range of 1.25 to 1.30 g/cm^3. A density below 1.20 g/cm^3 indicates a low sulfuric acid concentration and hence a partially discharged cell.

A lead storage battery can be recharged and thus restored to its original condition. To do this, a direct current is passed through the cell in the reverse direction. While a storage battery is being recharged, it acts as an electrolytic cell. The overall cell reaction is the reverse of that occurring when the battery discharges:

$$2PbSO_4(s) + 2H_2O \longrightarrow Pb(s) + PbO_2(s) + 4H^+(aq) + 2SO_4^{2-}(aq)$$

The electrical energy required to bring about this nonspontaneous reaction in an automobile is furnished by an alternator equipped with a rectifier to convert alternating to direct current.

As you may have found from experience, lead storage batteries do not endure forever, particularly if they are allowed to stand for some time in a discharged state. Also, repeated quick-charging can cause Pb, PbO_2, and $PbSO_4$ to flake off the electrodes. This collects as a sludge at the bottom of the battery, often short-circuiting one or more cells. Discharged batteries are also susceptible to freezing, when sulfuric acid concentration is low. If freezing occurs, the electrodes may warp and come in contact with one another.

One of the advantages of the lead storage battery is that its voltage stays constant at 2 V per cell over a wide range of sulfuric acid concentrations. Only when the battery is nearly completely discharged does the voltage drop. It is also true that the cell voltage is virtually independent of temperature. You have trouble starting your car on a cold morning because the conductivity of the electrolyte drops off sharply with temperature; the voltage is still 2 V per cell at $-40°C$.

Another type of rechargeable voltaic cell is the Nicad storage battery, used for small appliances, tools, and calculators. The anode in this cell is made of cadmium metal, and the cathode contains nickel(IV) oxide, NiO_2. The electrolyte is a concentrated solution of potassium hydroxide. The discharge reactions are

anode: $$Cd(s) + 2OH^-(aq) \longrightarrow Cd(OH)_2(s) + 2e^-$$

cathode: $$\frac{NiO_2(s) + 2H_2O + 2e^- \longrightarrow Ni(OH)_2(s) + 2OH^-(aq)}{Cd(s) + NiO_2(s) + 2H_2O \longrightarrow Cd(OH)_2(s) + Ni(OH)_2(s)}$$

The insoluble hydroxides of cadmium and nickel deposit on the electrodes. Hence the half-reactions are readily reversed during recharging. Nicad batteries are more expensive than lead storage batteries for a given amount of electrical energy delivered but also have a longer life.

Corrosion of Metals

Most metals corrode when exposed to the atmosphere, reacting with oxygen, water vapor, or carbon dioxide. Gold and platinum are among the few metals that retain their shiny appearance indefinitely when exposed to air; these metals are very difficult to oxidize (E_{ox}° Pt = −1.320 V, Au = −1.498 V).

Aluminum (E_{ox}° = +1.68 V) reacts readily with oxygen of the air:

$$4Al(s) + 3O_2(g) \longrightarrow 2Al_2O_3(s)$$

However, the Al_2O_3 coating, which is only about 10^{-8} m thick, adheres tightly to the surface of the metal. This prevents further corrosion and explains why aluminum cookware does not disintegrate on exposure to air.

Copper in moist air (Figure A) slowly acquires a dull green coating. The green material is a 1:1 mole mixture of $Cu(OH)_2$ and $CuCO_3$:

$$2Cu(s) + H_2O(g) + CO_2(g) + O_2(g) \longrightarrow Cu(OH)_2(s) + CuCO_3(s)$$

Several other elements, including zinc and lead, react similarly. The products, $Zn(OH)_2 \cdot ZnCO_3$ and $Pb(OH)_2 \cdot PbCO_3$, are white and adhere tightly to the metal, preventing further corrosion. In the case of lead, the protective coating dissolves in acetic acid, primarily because Pb^{2+} forms a very stable complex with the acetate ion. It has been suggested that the ancient Romans suffered from lead poisoning because they stored wine (containing some acetic acid) in pottery vessels glazed with lead compounds.

From an economic standpoint, the most important corrosion reaction is that involving iron and steel. About 20% of all the iron produced each year goes to replace products whose usefulness has been destroyed by rust. When a piece of iron is exposed to water containing dissolved oxygen, the half-reactions of oxidation

$$Fe(s) \longrightarrow Fe^{2+}(aq) + 2e^-$$

and reduction

$$\tfrac{1}{2}O_2(g) + H_2O + 2e^- \longrightarrow 2OH^-(aq)$$

occur at different locations. The surface of a piece of corroding iron consists of a series of tiny voltaic cells. At *anodic areas,* iron is oxidized to Fe^{2+} ions; at *cathodic areas,* elementary oxygen is reduced to OH^- ions. Electrons are transferred through the iron, which acts like the external conductor of an ordinary voltaic cell. The electrical circuit is completed by the flow of ions through the water solution or film covering the iron.

Many characteristics of corrosion are most readily explained in terms of an electrochemical mechanism. A perfectly dry metal surface is not attacked by oxygen; iron exposed to dry air does not corrode. This seems plausible if corrosion occurs through a voltaic cell, which requires a water solution through which ions can move to complete the circuit. The fact that corrosion occurs more readily in seawater than in fresh water has a similar explanation. The dissolved salts in seawater supply the ions necessary for the conduction of current.

The existence of discrete cathodic and anodic areas on a piece of corroding iron requires that adjacent surface areas differ from each other chemically. This can happen if there are differences in oxygen concentration along the metal surface, as when a drop of water adheres to the surface of a piece of iron exposed to the air (Figure B, page 540). The metal around the edges of the drop is in contact with water containing a high concentration of dissolved oxygen. The water touching the metal beneath the center of the drop is depleted in oxygen, because it is cut off from contact with air. As a result, the area around the edge of the drop, where the oxygen concentration is high, becomes cathodic; oxygen

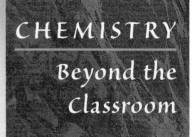

CHEMISTRY
Beyond the Classroom

⊙ **See Screen 21.9, Corrosion.**

FIGURE A
The Statue of Liberty. The statue was erected in 1885. It is constructed of sheets of copper that have corroded over the years. *(Andy Levin/Photo Researchers, Inc.)*

(continued)

molecules are reduced there. Directly beneath the drop is an anodic area where the iron is oxidized. A particle of dirt on the surface of an iron object can act in much the same way as a drop of water to cut off the supply of oxygen to the area beneath it and thereby establish anodic and cathodic areas. This explains why garden tools left covered with soil are particularly susceptible to rusting.

FIGURE B
Corrosion of iron under a drop of water. The Fe^{2+} ions migrate toward the edge of the drop, where they precipitate as $Fe(OH)_2$, which later forms $Fe(OH)_3$. **OHT**

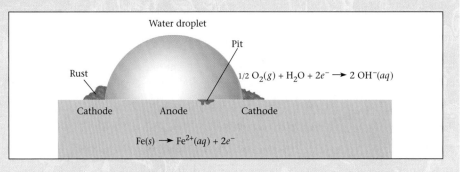

Water droplet

Pit

Rust

$1/2\ O_2(g) + H_2O + 2e^- \longrightarrow 2\ OH^-(aq)$

Cathode   Anode   Cathode

$Fe(s) \longrightarrow Fe^{2+}(aq) + 2e^-$

CHAPTER HIGHLIGHTS

Key Concepts

1. Draw a diagram for a voltaic cell, labeling electrodes and direction of current flow.
   (Example 18.1; Problems 3–6)
2. Use standard potentials (Table 18.1) to—
   ■ compare the relative strengths of different oxidizing agents; different reducing agents.
   (Example 18.2; Problems 7–14)
   ■ calculate $E°$ and/or reaction spontaneity.
   (Examples 18.3, 18.4; Problems 15–20, 23–32)
3. Relate $E°$ to $\Delta G°$ and K.
   (Examples 18.5, 18.6; Problems 33–42)
4. Use the Nernst equation to relate voltage to concentration.
   (Examples 18.7, 18.8; Problems 43–56, 66, 76, 77)
5. Relate mass of product to amount of electricity (coulombs) or amount of energy (joules) used in electrolysis reactions.
   (Examples 18.9, 18.10; Problems 57–65)

Key Equations

Standard voltage	$E° = E°_{ox} + E°_{red}$
$E°,\ \Delta G°,\ K$	$E° = \dfrac{\Delta G°}{-nF} = \dfrac{RT \ln K}{nF}$
Nernst equation	$E = E° - \dfrac{RT}{nF} \ln Q$

ampere	electrolytic cell	■ reducing agent
anode	Faraday constant	■ reduction
cathode	■ $\Delta G°$	salt bridge
coulomb	half-cell	standard potential
$E°, E°_{ox}, E°_{red}$	■ half-reaction	volt
electrode	■ oxidation	voltaic cell
electrolysis	■ oxidizing agent	

Summary Problem

A voltaic cell consists of two half-cells. One of the half-cells contains a platinum electrode surrounded by permanganate and manganese(II) ions. The other half-cell contains a platinum electrode surrounded by chlorate ions and chlorine gas. Assume that the cell reaction, which produces a positive voltage, involves both permanganate and chlorate ions in acidic solution. Take $t = 25°C$.

(a) Write the anode half-equation, the cathode half-equation, and the overall equation for the cell.

(b) Write the cell description in abbreviated notation.

(c) Calculate $E°$ for the cell.

(d) For the redox reaction in (a), calculate K and $\Delta G°$.

(e) Calculate the pH of the cell when the cell voltage is 0.041 V, all ionic species except H^+ are at 0.150 M, and $P_{Cl_2} = 0.915$ atm.

An electrolytic cell contains a solution of manganese(II) nitrate. Assume that manganese plates out at one electrode and oxygen gas is evolved at the other electrode.

(f) Write the anode half-equation, the cathode half-equation, and the overall equation for the electrolysis.

(g) How long will it take to deposit 7.25 g of manganese metal, using a current of 10.0 A?

(h) A current of 8.50 A is passed through the cell for one hour and 15 minutes. Before electrolysis, the cell has 435 mL of 0.893 M $Mn(NO_3)_2$. What is $[Mn^{2+}]$ after electrolysis? What is the pH of the solution neglecting the H^+ originally present? What is the volume of oxygen given off at 735 mm Hg and 25°C? Assume 100% efficiency and no change in volume during electrolysis.

Answers

a. anode: $Cl_2(g) + 6H_2O \rightarrow 2ClO_3^-(aq) + 12H^+(aq) + 10e^-$
  cathode: $MnO_4^-(aq) + 8H^+(aq) + 5e^- \rightarrow Mn^{2+}(aq) + 4H_2O$
  overall: $Cl_2(g) + 2MnO_4^-(aq) + 4H^+(aq) \rightarrow 2ClO_3^-(aq) + 2Mn^{2+}(aq) + 2H_2O$

b. $(Pt)|Cl_2|ClO_3^-\|MnO_4^-|Mn^{2+}|(Pt)$

c. 0.054 V

d. $\Delta G° = -52$ kJ; $K = 1.4 \times 10^9$

e. 0.95

f. anode: $2H_2O \rightarrow O_2(g) + 4H^+(aq) + 4e^-$
  cathode: $Mn^{2+}(aq) + 2e^- \rightarrow Mn(s)$
  overall: $2Mn^{2+}(aq) + 2H_2O \rightarrow 2Mn(s) + O_2(g) + 4H^+(aq)$

g. 42.4 min

h. 0.44 M; 0.041; 2.5 L

Questions & Problems

Problem numbers in blue indicate that the answer is available in Appendix 6 at the back of the book.

WEB indicates that the solution is posted at **http://www.harcourtcollege.com/chem/general/masterton4/student/**

Voltaic Cells

1. Write a balanced chemical equation for the overall cell reaction represented as
   (a) $Zn \mid Zn^{2+} \parallel Cr^{3+} \mid Cr$
   (b) $Sn \mid Sn^{2+} \parallel O_2 \mid H_2O \mid Pt$
   (c) $Al \mid Al^{3+} \parallel I_2 \mid I^- \mid Pt$

WEB 2. Write a balanced chemical equation for the overall cell reaction represented as
   (a) $Pt \mid H_2 \mid H^+ \parallel Fe^{3+} \mid Fe^{2+} \mid Pt$
   (b) $Cd \mid Cd^{2+} \parallel Ni^{2+} \mid Ni$
   (c) $Pt \mid Cl^- \mid Cl_2 \parallel MnO_4^- \mid Mn^{2+} \mid Pt$

3. Draw a diagram for a salt-bridge cell for each of the following reactions. Label the anode and cathode, and indicate the direction of current flow throughout the circuit.
   (a) $Zn(s) + Cd^{2+}(aq) \rightarrow Zn^{2+}(aq) + Cd(s)$
   (b) $2AuCl_4^-(aq) + 3Cu(s) \rightarrow$
                   $2Au(s) + 8Cl^-(aq) + 3Cu^{2+}(aq)$
   (c) $Fe(s) + Cu(OH)_2(s) \rightarrow Cu(s) + Fe(OH)_2(s)$

4. Follow the directions for Question 3 for the following reactions:
   (a) $Sn(s) + 2Ag^+(aq) \rightarrow Sn^{2+}(aq) + 2Ag(s)$
   (b) $H_2(g) + Hg_2Cl_2(s) \rightarrow 2H^+(aq) + 2Cl^-(aq) + 2Hg(l)$
   (c) $Pb(s) + PbO_2(s) + 4H^+(aq) + 2SO_4^{2-}(aq) \rightarrow$
                          $2PbSO_4(s) + 2H_2O$

5. Consider a salt-bridge cell in which the anode is a platinum rod immersed in an aqueous solution of sodium iodide containing solid iodine crystals. The cathode is another platinum rod immersed in an aqueous solution of sodium bromide with bromine liquid. Sketch a diagram of the cell, indicating the flow of the current throughout. Write the half-equations for the electrode reactions, the overall equation, and the abbreviated notation for the cell.

6. Follow the directions for Question 5 for a salt-bridge cell in which the anode is a manganese rod immersed in an aqueous solution of manganese(II) sulfate. The cathode is a chromium strip immersed in an aqueous solution of chromium(III) sulfate.

Strength of Oxidizing and Reducing Species

7. Which species in each pair is the better reducing agent?
   (a) Ag or Au
   (b) Cl^- or I^-
   (c) Mn^{2+} or Fe^{2+}
   (d) H_2 in acidic solution or H_2 in basic solution

8. Which species in each pair is the better oxidizing agent?
   (a) Br_2 or H_2O_2
   (b) SO_4^{2-} or MnO_4^-
   (c) $AgBr(s)$ or Ag^+
   (d) O_2 in acidic solution or O_2 in basic solution

9. Using Table 18.1, arrange the following oxidizing agents in order of increasing strength.

   Al^{3+} $AgBr$ F_2 ClO_3^- (acidic) Ni^{2+}

10. Using Table 18.1, arrange the following reducing agents in order of increasing strength.

   Br^- Zn Co $PbSO_4$ H_2S (acidic)

11. Consider the following species

   Cr^{3+} $Hg(l)$ H_2 (acidic) Sn^{2+} Br_2 (acidic)

Classify each species as oxidizing agent, reducing agent, or both. Arrange the oxidizing agents in order of increasing strength. Do the same for the reducing agents.

12. Follow the directions for Question 11 for the following species:

   Cu^+ Zn Ni^{2+} Fe^{2+} H^+ (acidic)

13. Use Table 18.1 to select
   (a) an oxidizing agent that converts I^- to I_2 but not Cl^- to Cl_2.
   (b) a reducing agent that converts Co^{2+} to Co but not Zn^{2+} to Zn.
   (c) an oxidizing agent that converts Cl^- to Cl_2 but not F^- to F_2.

14. Use Table 18.1 to select
   (a) a reducing agent that converts Pb^{2+} to Pb but not Tl^+ to Tl.
   (b) an oxidizing agent that converts Fe to Fe^{2+} but not Co to Co^{2+}.
   (c) a reducing agent that converts Au^{3+} to Au but not $AuCl_4^-$ to Au.

Calculation of E°

15. Calculate $E°$ for the following voltaic cells:
   (a) $MnO_2(s) + 4H^+(aq) + 2I^-(aq) \rightarrow$
                     $Mn^{2+}(aq) + 2H_2O + I_2(s)$
   (b) $H_2(g) + 2OH^-(aq) + S(s) \rightarrow 2H_2O + S^{2-}(aq)$
   (c) an $Ag-Ag^+$ half-cell and an $Au-AuCl_4^-$ half-cell

16. Calculate $E°$ for the following voltaic cells:
   (a) $Pb(s) + 2Ag^+(aq) \rightarrow Pb^{2+}(aq) + 2Ag(s)$
   (b) $O_2(g) + 4Fe^{2+}(aq) + 4H^+(aq) \rightarrow 2H_2O + 4Fe^{3+}(aq)$
   (c) a $Cd-Cd^{2+}$ half-cell and a $Zn-Zn^{2+}$ half-cell

17. Using Table 18.1, calculate $E°$ for the reaction between
   (a) iron and water to produce iron(II) hydroxide and hydrogen gas.
   (b) iron and iron(III) ions to give iron(II) ions.
   (c) iron(II) hydroxide and oxygen in basic solution.

WEB 18. Using Table 18.1, calculate $E°$ for the reaction between
   (a) chromium(II) ions and tin(IV) ions to produce chromium(III) ions and tin(II) ions.
   (b) manganese(II) ions and hydrogen peroxide to produce solid manganese dioxide (MnO_2).

(c) iron and oxygen in base

19. Calculate $E°$ for the following cells:
(a) $Pb \mid Pb^{2+} \parallel Sn^{4+} \mid Sn^{2+} \mid Pt$
(b) $Cu \mid Cu^{2+} \parallel NO_3^- \mid NO \mid Pt$
(c) $Mn \mid Mn^{2+} \parallel Ni^{2+} \mid Ni$

20. Calculate $E°$ for the following cells:
(a) $Al \mid Al^{3+} \parallel NO_3^-, H^+ \mid NO \mid Pt$
(b) C (graphite) $\mid Cr^{2+} \mid Cr^{3+} \parallel O_2 \mid H^+, H_2O \mid$ C (graphite)
(c) $Cu \mid Cu^{2+} \parallel I_2 \mid I^- \mid Pt$

21. Suppose $E°_{red}$ for $H^+ \rightarrow H_2$ were taken to be 0.300 V instead of 0.000 V. What would be
(a) $E°_{ox}$ for $H_2 \rightarrow H^+$?
(b) $E°_{red}$ for $Br_2 \rightarrow Br^-$?
(c) $E°$ for the cell in 19(c)? Compare your answer to that obtained in 19(c).

22. Suppose $E°_{red}$ for $Ag^+ \rightarrow Ag$ were set equal to zero instead of that of $H^+ \rightarrow H_2$. What would be
(a) $E°_{red}$ for $H^+ \rightarrow H_2$?
(b) $E°_{ox}$ for $Ca \rightarrow Ca^{2+}$?
(c) $E°$ for the cell in 20(c)? Compare your answer to that obtained in 20(c).

Spontaneity and E°

23. Which of the following reactions is (are) spontaneous at standard conditions?
(a) $2NO_3^-(aq) + 8H^+(aq) + 6Cl^- \rightarrow$
$$2NO(g) + 4H_2O + 3Cl_2(g)$$
(b) $O_2(g) + 4H^+(aq) + 4Cl^-(aq) \rightarrow 2H_2O + 2Cl_2(g)$
(c) $3Fe(s) + 2AuCl_4^-(aq) \rightarrow 2Au(s) + 8Cl^-(aq) + 3Fe^{2+}(aq)$

24. Which of the following reactions is (are) spontaneous at standard conditions?
(a) $Zn(s) + 2Fe^{3+}(aq) \rightarrow Zn^{2+}(aq) + 2Fe^{2+}(aq)$
(b) $Cu(s) + 2H^+(aq) \rightarrow Cu^{2+}(aq) + H_2(g)$
(c) $2Br^-(aq) + I_2(s) \rightarrow Br_2(l) + 2I^-(aq)$

25. Using Table 18.1, calculate $E°$ and decide whether the following ions will reduce chlorate ions to chlorine gas in acidic solution at standard conditions.
(a) I^- (b) F^- (c) Cu^+

26. Using Table 18.1, calculate $E°$ and decide whether the following ions will oxidize chloride ions to chlorine gas at standard conditions.
(a) MnO_4^- (b) $Cr_2O_7^{2-}$ (c) NO_3^-

27. Write the equation for the reaction, if any, that occurs when each of the following experiments is performed under standard conditions.
(a) Sulfur is added to mercury.
(b) Manganese dioxide in acidic solution is added to liquid mercury.
(c) Aluminum metal is added to a solution of potassium ions.

28. Write the equation for the reaction, if any, that occurs when each of the following experiments is performed under standard conditions.
(a) Crystals of iodine are added to an aqueous solution of potassium bromide.

(b) Liquid bromine is added to an aqueous solution of sodium chloride.
(c) A chromium wire is dipped into a solution of nickel(II) chloride.

29. Which of the following species will react with 1 M HNO_3?
(a) I^- (b) Fe (c) Ag (d) Pb

30. Which of the following species will be oxidized by 1 M HCl?
(a) Au (b) Mg (c) Cu (d) F^-

31. Predict what reaction, if any, will occur when liquid bromine is added to an acidic aqueous solution of each of the following species at standard conditions.
(a) $Ca(NO_3)_2$ (b) FeI_2 (c) AgF

32. Using Table 18.1, predict what reactions, if any, will occur when the following are mixed under standard conditions in acidic solution.
(a) Co^{2+}, Fe^{2+}, Co (b) Fe^{2+}, Fe^{3+}, Ag^+
(c) $Hg(l), ClO_3^-, H^+$

E°, ΔG°, and K

33. Consider a cell reaction at 25°C where $n = 4$. Fill in the following table.

	$\Delta G°$	$E°$	K
(a)	_____	_____	1.6×10^{-3}
(b)	_____	0.117 V	_____
(c)	-5.8 kJ	_____	_____

34. Consider a cell reaction at 25°C where $n = 2$. Fill in the following table.

	$\Delta G°$	$E°$	K
(a)	19 kJ	_____	_____
(b)	_____	0.035 V	_____
(c)	_____	_____	0.095

35. For a certain cell, $\Delta G° = 25.0$ kJ. Calculate $E°$ if n is
(a) 1 (b) 2 (c) 4
Comment on the effect that the number of electrons exchanged has on the voltage of a cell.

36. For a certain cell, $E° = 1.20$ V. Calculate $\Delta G°$ if n is
(a) 1 (b) 2 (c) 3
Comment on the effect that the number of electrons exchanged has on the spontaneity of a reaction.

37. Calculate $E°$, $\Delta G°$, and K at 25°C for the reaction

$$Cl^-(aq) + 3ClO_4^-(aq) \longrightarrow 4ClO_3^-(aq)$$

WEB **38.** Calculate $E°$, $\Delta G°$, and K at 25°C for the reaction

$$2Fe(OH)_3(s) + NO_2^-(aq) \longrightarrow NO_3^-(aq) + H_2O + 2Fe(OH)_2(s)$$

39. Calculate $\Delta G°$ at 25°C for each of the reactions referred to in Question 15. Assume smallest whole-number coefficients.

40. Calculate $\Delta G°$ at 25°C for each of the reactions referred to in Question 16. Assume smallest whole-number coefficients.

41. Calculate K at 25°C for each of the reactions referred to in Question 17. Assume smallest whole-number coefficients.

42. Calculate K at 25°C for each of the reactions referred to in Question 18. Assume smallest whole-number coefficients.

Nernst Equation

43. Consider a voltaic cell in which the following reaction takes place.

$$Br_2(l) + 2I^-(aq) \longrightarrow 2Br^-(aq) + I_2(s)$$

(a) Calculate $E°$.
(b) Write the Nernst equation for the cell.
(c) Calculate E when the concentration of iodide ion is a third of the concentration of the bromide ion.

44. Consider a voltaic cell in which the following reaction takes place.

$$2Cr(s) + 6H^+(aq) \longrightarrow 2Cr^{3+}(aq) + 3H_2(g)$$

(a) Calculate $E°$.
(b) Write the Nernst equation for the cell.
(c) Calculate E under the following conditions: $[H^+] = 0.00931\ M$, $[Cr^{3+}] = 1.21\ M$, $P_{H_2} = 0.929$ atm.

45. Consider a voltaic cell in which the following reaction takes place.

$$2NO_3^-(aq) + 3H_2(g) \longrightarrow 2NO(g) + 2OH^-(aq) + 2H_2O$$

(a) Calculate $E°$.
(b) Write the Nernst equation for the cell.
(c) Calculate E under the following conditions: $[NO_3^-] = 0.0315\ M$, $P_{NO} = 0.922$ atm, $P_{H_2} = 0.437$ atm, pH = 11.50.

46. Consider a voltaic cell in which the following reaction takes place.

$$2Fe^{2+}(aq) + H_2O_2(aq) + 2H^+(aq) \longrightarrow 2Fe^{3+}(aq) + 2H_2O$$

(a) Calculate $E°$.
(b) Write the Nernst equation for the cell.
(c) Calculate E under the following conditions: $[Fe^{2+}] = 0.00813\ M$, $[H_2O_2] = 0.914\ M$, $Fe^{3+} = 0.199\ M$, pH = 2.88.

47. Calculate voltages of the following cells at 25°C and under the following conditions.
(a) Fe | Fe^{2+} (0.010 M) ‖ Cu^{2+} (0.10 M) | Cu
(b) Pt | Sn^{2+} (0.10 M) | Sn^{4+} (0.010 M) ‖ Co^{2+} (0.10 M) | Co

48. Calculate voltages of the following cells at 25°C and under the following conditions.
(a) Zn | Zn^{2+} (0.50 M) ‖ Cd^{2+} (0.020 M) | Cd
(b) Cu | Cu^{2+} (0.0010 M) ‖ H$^+$ (0.010 M) | H$_2$ (1.00 atm) | Pt

49. Consider the reaction

$$2Fe^{3+}(aq) + 2I^-(aq) \longrightarrow 2\ Fe^{2+}(aq) + I_2(s)$$

At what concentration of I$^-$ is the voltage zero, if all other species are at standard conditions?

50. Consider the reaction

$$2Au(s) + 8Cl^-(aq) + 3Br_2(l) \longrightarrow 6Br^-(aq) + +2AuCl_4^-(aq)$$

Calculate [Cl$^-$] when the voltage is zero and all the other ionic species have a concentration of 0.200 M.

51. Complete the following cell notation.

$$Zn\ |\ Zn^{2+}\ (1.00\ M)\ \|\ H^+(\ \ )\ |\ H_2(1.0\ atm)\ |\ Pt \qquad E = +0.40\ V$$

52. Complete the following cell notation

$$Ag,\ Br^-\ (3.73\ M)\ |\ AgBr\ \|\ H^+(\ \ )\ |\ H_2\ (1.0\ atm)\ |\ Pt$$
$$E = -0.030\ V$$

53. The reaction

$$2Cr^{3+}(aq) + 3Cl_2\ (g) + 7H_2O \longrightarrow$$
$$Cr_2O_7^{2-}(aq) + 6Cl^-(aq) + 14H^+\ (aq)$$

is spontaneous at standard conditions. If the pH is adjusted to 0.90 and all the other species are kept at standard conditions, is the reaction still spontaneous?

54. The reaction

$$2H_2O(l) + 2Br_2(l) \longrightarrow O_2(g) + 4H^+(aq) + 4Br^-(aq)$$

is nonspontaneous at standard conditions. If the pH is adjusted to 4.50 and all the other species are kept at standard conditions, is the reaction still nonspontaneous?

55. Consider a cell in which the reaction is

$$Pb(s) + 2H^+(aq) \longrightarrow Pb^{2+}(aq) + H_2(g)$$

(a) Calculate $E°$ for this cell.
(b) Chloride ions are added to the Pb | Pb^{2+} half-cell to precipitate PbCl$_2$. The voltage is measured to be +0.210 V. Taking $[H^+] = 1.0\ M$ and $P_{H_2} = 1.0$ atm, calculate [Pb^{2+}].
(c) Taking [Cl$^-$] in (b) to be 0.10 M, calculate K_{sp} of PbCl$_2$.

56. Consider a cell in which the reaction is

$$2Ag(s) + Cu^{2+}(aq) \longrightarrow 2Ag^+(aq) + Cu(s)$$

(a) Calculate $E°$ for this cell.
(b) Chloride ions are added to the Ag | Ag$^+$ half-cell to precipitate AgCl. The measured voltage is +0.060 V. Taking $[Cu^{2+}] = 1.0\ M$, calculate [Ag$^+$].
(c) Taking [Cl$^-$] in (b) to be 0.10 M, calculate K_{sp} of AgCl.

Electrolytic Cells

57. The electrolysis of an aqueous solution of NaCl has the overall equation

$$2H_2O + 2Cl^-(aq) \longrightarrow H_2(g) + Cl_2(g) + 2OH^-(aq)$$

During the electrolysis, 0.228 mol of electrons passes through the cell.
(a) How many electrons does this represent?
(b) How many coulombs does this represent?
(c) Assuming 100% yield, what masses of H$_2$ and Cl$_2$ are produced?

58. An electrolytic cell produces aluminum from Al$_2$O$_3$ at the rate of ten kilograms a day. Assuming a yield of 100%,
(a) how many moles of electrons must pass through the cell in one day?
(b) how many amperes are passing through the cell?

(c) how many moles of oxygen (O_2) are being produced simultaneously?

59. A baby's spoon with an area of 6.25 cm² is plated with silver from $AgNO_3$ using a current of 2.00 A for two hours and 25 minutes.

(a) If the current efficiency is 82.0%, how many grams of silver are plated?

(b) What is the thickness of the silver plate formed (d = 10.5 g/cm³)?

60. A metallurgist wants to goldplate a thin sheet with the following dimensions: 1.5 in × 8.5 in × 0.0012 in. The gold plating must be 0.0020 in thick.

(a) How many grams of gold (d = 19.3 g/cm³) are required?

(b) How long will it take to plate the sheet from AuCN using a current of 7.00 A? (Assume 100% efficiency.)

61. Aluminum can be produced by the electrolysis of Al_2O_3 at a voltage of 2.5 V.

(a) How many joules of electrical energy are required to form ten pounds of aluminum?

(b) What is the cost of the electrical energy in (a) at the rate of 7.5¢ per kilowatt-hour?

WEB 62. A lead storage battery delivers a current of 5.00 A for 95 minutes at a voltage of 12.0 V.

(a) How many grams of lead are converted to $PbSO_4$?

(b) How much electrical energy is produced in kilowatt-hours?

Unclassified

63. Consider the electrolysis of $CuCl_2$ to form Cu(s) and $Cl_2(g)$. Calculate the minimum voltage required to carry out this reaction at standard conditions. If a voltage of 1.50 V is actually used, how many kilojoules of electrical energy are consumed in producing 2.00 g of Cu?

64. Hydrogen gas is produced when water is electrolyzed.

$$2H_2O(g) \longrightarrow 2H_2(g) + O_2(g)$$

A balloonist wants to fill a balloon with hydrogen gas. How long must a current of 12.0 A be used in the electrolysis of water to fill the balloon to a volume of 10.00 L and a pressure of 0.924 atm at 22°C?

65. An electrolysis experiment is performed to determine the value of the Faraday constant (number of coulombs per mole of electrons). In this experiment, 28.8 g of gold is plated out from a AuCN solution by running an electrolytic cell for two hours with a current of 2.00 A. What is the experimental value obtained for the Faraday constant?

66. Consider the following reaction at 25°C.

$$O_2(g) + 4H^+(aq) + 4Br^-(aq) \longrightarrow 2H_2O + 2Br_2(l)$$

If [H^+] is adjusted by adding a buffer that is 0.100 M in sodium acetate and 0.100 M in acetic acid, the pressure of oxygen gas is 1.00 atm, and the bromide concentration is 0.100 M, what is the calculated cell voltage? (K_a acetic acid = 1.8×10^{-5}.)

67. Given the standard reduction potential for $Zn(OH)_4^{2-}$:

$$Zn(OH)_4^{2-}(aq) + 2e^- \longrightarrow Zn(s) + 4OH^-(aq) \qquad E_{red}^\circ = -1.19 \text{ V}$$

Calculate the formation constant (K_f) for the reaction

$$Zn^{2+}(aq) + 4OH^-(aq) \rightleftharpoons Zn(OH)_4^{2-}(aq)$$

68. Consider the following reaction carried out at 1000°C.

$$CO(g) + \tfrac{1}{2}O_2(g) \longrightarrow CO_2(g)$$

Assuming that all gases are at 1.00 atm, calculate the voltage produced at the given conditions. (Use Appendix 1 and assume that ΔH° and ΔS° do not change with an increase in temperature.)

69. Atomic masses can be determined by electrolysis. In one hour, a current of 0.600 A deposits 2.42 g of a certain metal, M, which is present in solution as M^+ ions. What is the atomic mass of the metal?

Conceptual Problems

70. Explain why

(a) a salt bridge is used in a voltaic cell.

(b) in a salt bridge with KNO_3, the K^+ ions move from the salt bridge to the cathode.

(c) a lead storage battery won't work if the level of H_2SO_4 is low.

71. Choose the figure below that best represents the results after the electrolysis of water. (Circles represent hydrogen atoms and squares represent oxygen atoms.)

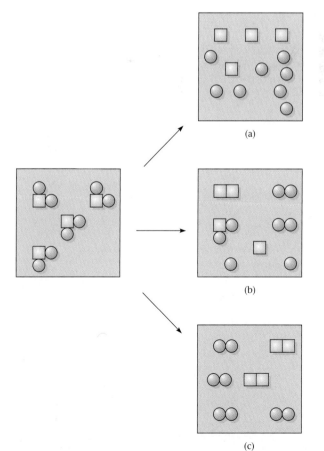

72. Which of the changes below will increase the voltage of the following cell?

$$Co \,|\, Co^{2+} \,(0.010 \, M) \,\|\, H^+ \,(0.010 \, M) \,|\, H_2 \,(0.500 \, atm) \,|\, Pt$$

(a) Increase the volume of $CoCl_2$ solution from 100 mL to 300 mL.
(b) Increase $[H^+]$ from 0.010 M to 0.500 M.
(c) Increase the pressure of H_2 from 0.500 atm to 1 atm.
(d) Increase the mass of the Co electrode from 15 g to 25 g.
(e) Increase $[Co^{2+}]$ from 0.010 M to 0.500 M.

73. For the cell

$$Zn \,|\, Zn^{2+} \,\|\, Cu^{2+} \,|\, Cu$$

$E°$ is 1.10 V. A student prepared the same cell in the lab at standard conditions. Her experimental $E°$ was 1.0 V. A possible explanation for the difference is that

(a) a larger volume of Zn^{2+} than Cu^{2+} was used.
(b) the zinc electrode had twice the mass of the copper electrode.
(c) $[Zn^{2+}]$ was smaller than 1 M.
(d) $[Cu^{2+}]$ was smaller than 1 M.
(e) the copper electrode had twice the surface area of the zinc electrode.

74. The standard potential for the reduction of AgSCN is 0.0895 V.

$$AgSCN(s) + e^- \longrightarrow Ag(s) + SCN^-(aq)$$

Find another electrode potential to use together with the above value and calculate K_{sp} for AgSCN.

75. Consider the following standard reduction potentials:

$$Tl^+(aq) + e^- \longrightarrow Tl(s) \qquad E°_{red} = -0.34 \text{ V}$$

$$Tl^{3+}(aq) + 3e^- \longrightarrow Tl(s) \qquad E°_{red} = 0.74 \text{ V}$$

$$Tl^{3+}(aq) + 2e^- \longrightarrow Tl^+(aq) \qquad E°_{red} = 1.28 \text{ V}$$

and the following abbreviated cell notations:

$$\text{(1) } Tl \,|\, Tl^+ \,\|\, Tl^{3+} \,|\, Tl^+ \,|\, Pt$$

$$\text{(2) } Tl \,|\, Tl^{3+} \,\|\, Tl^{3+} \,|\, Tl^+ \,|\, Pt$$

$$\text{(3) } Tl \,|\, Tl^+ \,\|\, Tl^{3+} \,|\, Tl$$

(a) Write the overall equation for each cell.
(b) Calculate $E°$ for each cell.

(c) Calculate $\Delta G°$ for each overall equation.
(d) Comment on whether $\Delta G°$ and/or $E°$ are state properties.
(*Hint:* A state property is path-independent.)

Challenge Problems

76. In a fully charged lead storage battery, the electrolyte consists of 38% sulfuric acid by mass. The solution has a density of 1.286 g/cm^3. Calculate E for the cell. (Assume all the H^+ ions come from the first dissociation of H_2SO_4, which is complete; $K_a \, HSO_4^- = 1.0 \times 10^{-2}$.)

77. Consider a voltaic cell in which the following reaction occurs.

$$Zn(s) + Sn^{2+}(aq) \longrightarrow Zn^{2+}(aq) + Sn(s)$$

(a) Calculate $E°$ for the cell.
(b) When the cell operates, what happens to the concentration of Zn^{2+}? The concentration of Sn^{2+}?
(c) When the cell voltage drops to zero, what is the ratio of the concentration of Zn^{2+} to that of Sn^{2+}?
(d) If the concentration of both cations is 1.0 M originally, what are the concentrations when the voltage drops to zero?

78. In biological systems, acetate ion is converted to ethyl alcohol in a two-step process:

$$CH_3COO^-(aq) + 3H^+(aq) + 2e^- \longrightarrow CH_3CHO(aq) + H_2O$$
$$E°' = -0.581 \text{ V}$$

$$CH_3CHO(aq) + 2H^+(aq) + 2e^- \longrightarrow C_2H_5OH(aq)$$
$$E°' = -0.197 \text{ V}$$

($E°'$ is the standard reduction voltage at 25°C and pH of 7.00.)

(a) Calculate $\Delta G°'$ for each step and for the overall conversion.
(b) Calculate $E°'$ for the overall conversion.

79. Consider the cell

$$Pt \,|\, H_2 \,|\, H^+ \,\|\, H^+ \,|\, H_2 \,|\, Pt$$

In the anode half-cell, hydrogen gas at 1.0 atm is bubbled over a platinum electrode dipping into a solution that has a pH of 7.0. The other half-cell is identical to the first except that the solution around the platinum electrode has a pH of 0.0. What is the cell voltage?

The soul, perhaps, is a gust of gas
And wrong is a form of right —
But we know that Energy equals Mass
By the square of the Speed of Light.

—Morris Bishop
$E = MC^2$

NUCLEAR REACTIONS

19

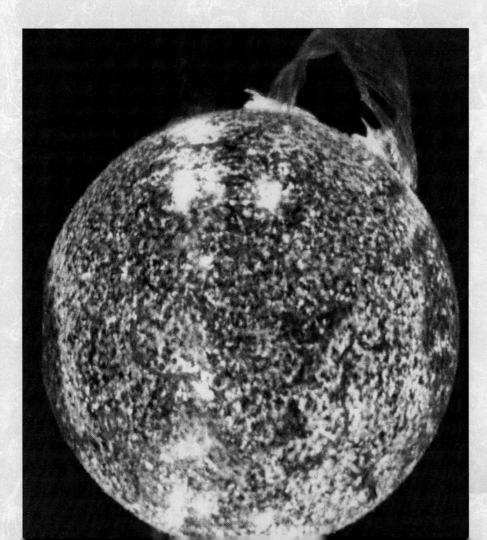

The Sun's energy is the result of nuclear fusion involving the isotopes of hydrogen. *(NASA)*

CHAPTER OUTLINE

19.1   RADIOACTIVITY

19.2   RATE OF RADIOACTIVE DECAY

19.3   MASS-ENERGY RELATIONS

19.4   NUCLEAR FISSION

19.5   NUCLEAR FUSION

The "ordinary chemical reactions" discussed to this point involve changes in the outer electronic structures of atoms or molecules. In contrast, nuclear reactions result from changes taking place within atomic nuclei. You will recall (Chapter 2) that atomic nuclei are represented by symbols such as

$$^{12}_{6}C \qquad ^{14}_{6}C$$

The atomic number Z (number of protons in the nucleus) is shown as a left subscript. The mass number A (number of protons + number of neutrons in the nucleus) appears as a left superscript. Nuclei with the same number of protons but different numbers of neutrons are called *isotopes*. The symbols written above represent two isotopes of the element carbon ($Z = 6$). One isotope has six neutrons in the nucleus and hence has a mass number of $6 + 6 = 12$. The heavier isotope has eight neutrons and hence a mass number of 14.

In this chapter we will take a closer look at some of the more relevant topics in nuclear chemistry. Many of these topics will be ones that you have heard about on television or read about in newspapers or on the Internet. Specifically, we will consider—

■ *radioactivity,* both natural and induced (Section 19.1).

■ *rate of nuclear reactions* as related to the "age" of organic materials (Section 19.2).

■ *mass-energy relations,* as expressed by the Einstein equation: $\Delta E = \Delta mc^2$ (Section 19.3).

■ *nuclear fission,* a major but controversial source of energy today (Section 19.4).

■ *nuclear fusion,* a potentially limitless source of energy for the future (Section 19.5).

The reactions that we discuss in this chapter will be represented by **nuclear equations.** An equation of this type uses nuclear symbols such as those written above; in other respects it resembles an ordinary chemical equation. A nuclear equation must be balanced with respect to nuclear charge (atomic number) and nuclear mass (mass number). To see what that means, consider an equation that we will have a lot more to say about later in this chapter:

$$^{14}_{7}N + ^{1}_{0}n \longrightarrow ^{14}_{6}C + ^{1}_{1}H$$

The reactants are an N-14 nucleus and a neutron; the products are a C-14 nucleus and a H-1 nucleus. The atomic numbers add to 7 on both sides:

$$\text{reactants: } 7 + 0 = 7 \qquad \text{products} = 6 + 1 = 7$$

whereas the mass numbers add to 15:

$$\text{reactants: } 14 + 1 = 15 \qquad \text{products} = 14 + 1 = 15$$

19.1 RADIOACTIVITY

A radioactive nucleus spontaneously decomposes ("decays") with the evolution of energy. As pointed out in Chapter 2, a few such nuclei occur in nature, accounting for *natural radioactivity*. Many more can be made ("induced") in the laboratory by bombarding stable nuclei with high-energy particles.

Mode of Decay

Naturally occurring radioactive nuclei commonly decompose by—

- **alpha particle emission,** in which an ordinary helium nucleus, ^4_2He, is given off. Uranium-238 behaves this way:

$$^{238}_{92}\text{U} \longrightarrow {}^4_2\text{He} + {}^{234}_{90}\text{Th}$$

Notice that when alpha emission occurs, the atomic number decreases by two units; the mass number decreases by four units.

- **beta particle emission,** which produces an electron, given the symbol $^0_{-1}e$. An example of beta emission is

$$^{234}_{90}\text{Th} \longrightarrow {}^0_{-1}e + {}^{234}_{91}\text{Pa}$$

Notice that the product nucleus, $^{234}_{91}\text{Pa}$, has the same mass number as the reactant, $^{234}_{90}\text{Th}$, but its atomic number is one unit larger.

- **gamma radiation emission,** which consists of high-energy photons. Because gamma emission changes neither the mass number nor the atomic number, it is ordinarily omitted in writing nuclear equations.

Radioactive nuclei produced "artificially" in the laboratory can show α-, β-, or γ-emission. They can also decompose by—

- **positron emission.** A **positron** is identical to an electron except that it has a charge of +1 rather than −1. A positron has the symbol ^0_1e.

$$^{40}_{19}\text{K} \longrightarrow {}^0_1e + {}^{40}_{18}\text{Ar}$$

- **K-electron capture,** in which an electron in the innermost energy level ($n = 1$) "falls" into the nucleus.

$$^{82}_{37}\text{Rb} + {}^0_{-1}e \longrightarrow {}^{82}_{36}\text{Kr}$$

Notice that the result of K-electron capture is the same as positron emission; mass number remains unchanged whereas atomic number decreases by one unit. Electron capture is more common with heavy nuclei, presumably because the $n = 1$ level is closer to the nucleus.

Promethium is by far the rarest of the "rare earths."

EXAMPLE 19.1 Promethium ($Z = 61$) is essentially nonexistent in nature; all of its isotopes are radioactive. Write balanced nuclear equations for the decomposition of

(a) Pm-142 by positron emission; by K-electron capture.
(b) Pm-147 by beta emission.

Strategy Use the principle that atomic and mass numbers must balance to find the symbol of the product nucleus.

Solution

(a) $^{142}_{61}\text{Pm} \rightarrow\, ^{0}_{1}e +\, ^{142}_{60}\text{Nd}$

$^{142}_{61}\text{Pm} +\, ^{0}_{-1}e \rightarrow\, ^{142}_{60}\text{Nd}$

(b) $^{147}_{61}\text{Pm} \rightarrow\, ^{0}_{-1}e +\, ^{147}_{62}\text{Sm}$

Bombardment Reactions

More than 1500 radioactive isotopes have been prepared in the laboratory. The number of such isotopes per element ranges from 1 (hydrogen and boron) to 34 (indium). They are all prepared by bombardment reactions in which a stable nucleus is converted to one that is radioactive, which in turn decays to stable products. The bombarding particle may be—

- a *neutron,* usually of rather low energy, produced in a fission reactor (Section 19.4). A typical reaction of this type is

$$^{27}_{13}\text{Al} +\, ^{1}_{0}n \longrightarrow\, ^{28}_{13}\text{Al}$$

The product nucleus, Al-28, is radioactive, decaying by beta emission:

$$^{28}_{13}\text{Al} \longrightarrow\, ^{28}_{14}\text{Si} +\, ^{0}_{-1}e$$

- a *charged particle* (electron, positron, α-particle, . . .), which can be accelerated to very high velocities in electric and/or magnetic fields. In this way, the charged particle acquires enough energy to bring about a nuclear reaction, despite electrostatic repulsion with the components of the atom.

The first radioactive isotopes to be made in the laboratory were prepared in 1934 by Irene Curie and her husband, Frederic Joliot. They achieved this by bombarding certain stable isotopes with high-energy alpha particles. One reaction was

$$^{27}_{13}\text{Al} +\, ^{4}_{2}\text{He} \longrightarrow\, ^{30}_{15}\text{P} +\, ^{1}_{0}n$$

The product, phosphorus-30, is radioactive, decaying by positron emission:

$$^{30}_{15}\text{P} \longrightarrow\, ^{30}_{14}\text{Si} +\, ^{0}_{1}e$$

An interesting application of bombardment reactions is the preparation of the so-called transuranium elements. During the past 50 years, a series of elements with atomic numbers greater than that of uranium have been synthesized. Much of this work was done by a group at the University of California at Berkeley, under the direction of Glenn Seaborg and later Albert Ghiorso. A Russian group at Dubna near Moscow, led by G. N. Flerov, made significant contributions. The heavy elements (107 through 112) were prepared in the late twentieth century by a German group at Darmstadt under the direction of Peter Armbruster. Most of the nuclei produced have half-lives of a few microseconds. Quite recently (January 1999) the Russian group at Dubna reported the formation of a single, long-lived atom of element 114 ($t_{1/2} = 30$ s) by the reaction

$$^{48}_{20}\text{Ca} +\, ^{244}_{94}\text{Pu} \longrightarrow\, ^{289}_{114}? +\, 3\,^{1}_{0}n$$

TABLE 19.1 Synthesis of Transuranium Elements	
Neutron Bombardment	
Neptunium, plutonium	$^{238}_{92}U + ^{1}_{0}n \rightarrow ^{239}_{93}Np + ^{0}_{-1}e$
	$^{239}_{93}Np \rightarrow ^{239}_{94}Pu + ^{0}_{-1}e$
Americium	$^{239}_{94}Pu + 2\,^{1}_{0}n \rightarrow ^{241}_{95}Am + ^{0}_{-1}e$
Positive Ion Bombardment	
Curium	$^{239}_{94}Pu + ^{4}_{2}He \rightarrow ^{242}_{96}Cm + ^{1}_{0}n$
Californium	$^{242}_{96}Cm + ^{4}_{2}He \rightarrow ^{245}_{98}Cf + ^{1}_{0}n$
Rutherfordium	$^{249}_{98}Cf + ^{12}_{6}C \rightarrow ^{257}_{104}Rf + 4\,^{1}_{0}n$
Dubnium	$^{249}_{98}Cf + ^{15}_{7}N \rightarrow ^{260}_{105}Db + 4\,^{1}_{0}n$
Seaborgium	$^{249}_{98}Cf + ^{18}_{8}O \rightarrow ^{263}_{106}Sg + 4\,^{1}_{0}n$

The names of these elements reflect the names of their discoverers and other nuclear scientists.

Three months later, the Berkeley group reported the synthesis of isotopes of elements 116 and 118.

Some of the reactions used to prepare transuranium elements are listed in Table 19.1. Neutron bombardment is effective for the lower members of the series (elements 93 through 95), but the yield of product decreases sharply with increasing atomic number. To form the heavier transuranium elements, it is necessary to bombard appropriate targets with high-energy positive ions. By using relatively heavy bombarding particles such as carbon-12, one can achieve a considerable increase in atomic number.

Applications

A large number of radioactive nuclei have been used both in industry and in many areas of basic and applied research. A few of these are discussed below.

Medicine

Radioactive isotopes are commonly used in cancer therapy, usually to eliminate any malignant cells left after surgery. Cobalt-60 is most often used; γ-rays from

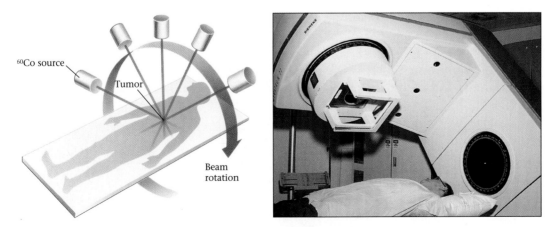

Cobalt-60 cancer therapy. Gamma rays from the rotating radiation source are concentrated at the location of the diseased tissue. *(Beverly March, Courtesy of Long Island Jewish Hospital)*

TABLE 19.2 Diagnostic Uses of Radioactive Isotopes

Isotope	Use
$^{11}_{6}$C	Brain scan (PET); see text
$^{24}_{11}$Na	Circulatory disorders
$^{32}_{15}$P	Detection of eye tumors
$^{59}_{26}$Fe	Anemia
$^{67}_{31}$Ga	Scan for lung tumors, abscesses
$^{75}_{34}$Se	Pancreas scan
$^{99}_{43}$Tc	Imaging of brain, liver, kidneys, bone marrow
$^{133}_{54}$Xe	Lung imaging
$^{201}_{81}$Tl	Heart disorders

this source are focused at small areas where cancer is suspected. Certain types of cancer can be treated internally with radioactive isotopes. If a patient suffering from cancer of the thyroid drinks a solution of NaI containing radioactive iodide ions (^{131}I or ^{123}I), the iodine moves preferentially to the thyroid gland. There the radiation destroys malignant cells without affecting the rest of the body.

More commonly, radioactive nuclei are used for diagnosis (Table 19.2). Positron emission tomography (PET) is a technique used to study brain disorders. The patient is given a dose of glucose ($C_6H_{12}O_6$) containing a small amount of carbon-11, a positron emitter. The brain is then scanned to detect positron emission from the radioactive "labeled" glucose. In this way, differences in glucose uptake and metabolism in patients with normal and abnormal brains are established. For example, PET scans have determined that the brain of a schizophrenic metabolizes only about 20% as much glucose as that of most people.

Chemistry

Radioactive nuclei are used extensively in chemical analysis. One technique of particular importance is *neutron activation analysis*. This procedure depends on the phenomenon of induced radioactivity. A sample is bombarded by neutrons, bringing about such reactions as

$$^{84}_{38}\text{Sr} + ^{1}_{0}n \longrightarrow ^{85}_{38}\text{Sr}$$

Ordinarily the element retains its chemical identity, but the isotope formed is radioactive, decaying by gamma emission. The magnitude of the energy change and hence the wavelength of the gamma ray vary from one element to another and so can serve for the qualitative analysis of the sample. The intensity of the radiation depends on the amount of the element present in the sample; this permits quantitative analysis of the sample. Neutron activation analysis can be used to analyze for 50 different elements in amounts as small as one picogram (10^{-12} g).

One application of neutron activation analysis is in the field of archaeology. By measuring the amount of strontium in the bones of prehistoric humans, it is possible to get some idea of their diet. Plants contain considerably more strontium than animals do, so a high strontium content suggests a largely vegetarian diet. Strontium analyses of bones taken from ancient farming communities consistently show a difference by sex; women have higher strontium levels than men. Apparently, in those days, women did most of the farming; men spent a lot of time away from home hunting and eating their kill.

FIGURE 19.1

Smoke detector. Most smoke detectors use a tiny amount of a radioactive isotope to produce a current flow that drops off sharply in the presence of smoke particles, emitting an alarm in the process. *(Marna G. Clarke)*

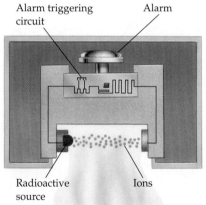

CHEMISTRY ▪ The Human Side

The history of radiochemistry is in no small measure the story of two remarkable women, Marie and Irene Curie, and their husbands, Pierre Curie and Frederic Joliot. Marie Curie was born Maria Sklodowska in 1867 in Warsaw, Poland, then a part of the Russian empire. In 1891 she emigrated to Paris to study at the Sorbonne, where she met and married a French physicist, Pierre Curie. The Curies were associates of Henri Becquerel, the man who discovered that uranium salts are radioactive. They showed that thorium, like uranium, is radioactive and that the amount of radiation emitted is directly proportional to the amount of uranium or thorium in the sample.

In 1898, Marie and Pierre Curie isolated two new radioactive elements, which they named radium and polonium. To obtain a few milligrams of these elements, they started with several tons of pitchblende ore and carried out a long series of tedious separations. Their work was done in a poorly equipped, unheated shed where the temperature reached 6°C (43°F) in winter. Four years later, in 1902, Marie determined the atomic mass of radium to within 0.5%, working with a tiny sample.

In 1903, the Curies received the Nobel Prize in physics (with Becquerel) for the discovery of radioactivity. Three years later, Pierre Curie died at the age of 46, the victim of a tragic accident. He stepped from behind a carriage in a busy Paris street and was run down by a horse-driven truck. That same year, Marie became the first woman instructor at the Sorbonne. In 1911, she won the Nobel Prize in chemistry for the discovery of radium and polonium, thereby becoming the first person to win two Nobel Prizes.

When Europe exploded into war in 1914, scientists largely abandoned their studies to go to the front. Marie Curie, with her daughter Irene, then 17 years old, organized medical units equipped with X-ray machinery. These were used to locate foreign metallic objects in wounded soldiers. Many of the wounds were to the head; French soldiers came out of the trenches without head protection because their government had decided that helmets looked too German. In November of 1918, the Curies celebrated the end of World War I; France was victorious, and Marie's beloved Poland was free again.

In 1921, Irene Curie began research at the Radium Institute. Five years later she married Frederic Joliot, a brilliant young physicist who was also an assistant at the Institute. In 1931, they began a research program in nuclear chemistry that led to several important discoveries and at least one near miss. The Joliot-Curies were the first to demonstrate induced radioactivity. They also discovered the positron, a particle that scientists had been seeking for many years. They narrowly missed finding another, more fundamental particle, the neutron. That honor went to James Chadwick in England. In 1935, Irene Curie and Frederic Joliot received the Nobel Prize in physics. The award came too late for Irene's mother, who had died of leukemia in 1934. Twenty-two years later, Irene Curie-Joliot died of the same disease. Both women acquired leukemia through prolonged exposure to radiation.

(Photo credit: Mütter Museum, Philadelphia College of Physicians)

MARIE AND PIERRE CURIE WITH DAUGHTER IRENE, AT THEIR HOME NEAR PARIS.

Chadwick, a student of Rutherford, discovered the neutron in 1932.

Commercial Applications

Most smoke alarms (Figure 19.1) use a radioactive species, typically americium-241. A tiny amount of this isotope is placed in a small ionization chamber; decay of Am-241 ionizes air molecules within the chamber. Under the influence of a

FIGURE 19.2
Food preservation. Strawberries irradiated with gamma rays from radioactive isotopes to keep them fresh. *(Nordion International)*

potential applied by a battery, these ions move across the chamber, producing an electric current. If smoke particles get into the chamber, the flow of ions is impeded and the current drops. This is detected by electronic circuitry, and an alarm sounds. The alarm also goes off if the battery voltage drops, indicating that it needs to be replaced.

Another potential application of radioactive species is in food preservation (Figure 19.2). It is well known that gamma rays can kill insects, larvae, and parasites such as trichina that cause trichinosis in pork. Radiation can also inhibit sprouting of onions and potatoes. Perhaps most important from a commercial standpoint, it can extend the shelf lives of many foods for weeks or even months. Many chemicals used to preserve foods have later been shown to have adverse health effects, so irradiation is an attractive alternative. Finally, irradiation can destroy microorganisms such as *E. coli,* which explains its use in treating beef.

19.2 RATE OF RADIOACTIVE DECAY

As pointed out in Chapter 11, radioactive decay is a first-order process. This means that the following equations apply:

$$\text{rate} = kX \tag{1}$$

$$\boxed{\ln \frac{X_0}{X} = kt} \tag{2}$$

$$k = \frac{0.693}{t_{1/2}} \tag{3}$$

where k is the first-order rate constant, $t_{1/2}$ is the half-life, X is the amount of radioactive species present at time t, and X_0 is the amount at $t = 0$.

Because of the way in which rate of decay is measured (Figure 19.3), it is often described by the **activity (A)** of the sample, which expresses the number of atoms decaying in unit time. Equation (1) can be written

$$A = kN \tag{4}$$

where A is the activity, k the first-order rate constant, and N the number of radioactive nuclei present.

Activity can be expressed in terms of the number of atoms decaying per second, or becquerels (Bq).

$$1 \text{ Bq} = 1 \text{ atom/s}$$

Alternatively, activity may be cited in disintegrations per minute or, perhaps most commonly, in **curies** (Ci)

$$1 \text{ Ci} = 3.700 \times 10^{10} \text{ atom/s}$$

FIGURE 19.3
A liquid scintillation counter. This instrument is used to detect radiation and measure disintegrations per minute quickly and accurately.
(Courtesy of Beckman Instruments)

EXAMPLE 19.2 The half-life of radium-226 is 1.60×10^3 y $= 5.05 \times 10^{10}$ s.

(a) Calculate k in s^{-1}.
(b) What is the activity in curies of a 1.00-g sample of Ra-226?
(c) What is the mass in grams of a sample of Ra-226 that has an activity of 1.00×10^9 atom/s?

Strategy Use Equation (3) to find k. In (b) and (c), use Equation (4). To relate N to mass in grams, note that Avogadro's number (6.022×10^{23}) of atoms of Ra-226 weigh 226 g.

Solution

(a) $k = 0.693/(5.05 \times 10^{10} \text{ s}) = \boxed{1.37 \times 10^{-11}/s}$

(b) First find N for the 1.00-g sample:

$$N = 1.00 \text{ g} \times \frac{6.022 \times 10^{23} \text{ atoms}}{226 \text{ g}} = 2.66 \times 10^{21} \text{ atoms}$$

Now apply Equation (4):

$$A = (1.37 \times 10^{-11}/s) \times 2.66 \times 10^{21} \text{ atoms} \times \frac{1 \text{ Ci}}{3.700 \times 10^{10} \text{ atom/s}} = \boxed{0.985 \text{ Ci}}$$

(c) $N = A/k = \dfrac{1.00 \times 10^9 \text{ atoms/s}}{1.37 \times 10^{-11}/s} = 7.30 \times 10^{19}$ atoms

$$\text{mass Ra-226} = 7.30 \times 10^{19} \text{ atoms} \times \frac{226 \text{ g}}{6.022 \times 10^{23} \text{ atoms}} = \boxed{0.0274 \text{ g}}$$

Reality Check Note the answer to part (b). The curie was supposed to be the activity of a one-gram sample of radium, the element discovered by the Curies; it isn't quite.

Age of Organic Material

During the 1950s, Professor W. F. Libby of the University of Chicago and others worked out a method for determining the age of organic material. It is based on the decay rate of carbon-14. The method can be applied to objects from a few hundred up to 50,000 years old. It has been used to determine the authenticity of canvases of Renaissance painters and to check the ages of relics left by prehistoric cave dwellers.

Carbon-14 is produced in the atmosphere by the interaction of neutrons from cosmic radiation with ordinary nitrogen atoms:

$$^{14}_{7}\text{N} + ^{1}_{0}n \longrightarrow ^{14}_{6}\text{C} + ^{1}_{1}\text{H}$$

The carbon-14 formed by this nuclear reaction is eventually incorporated into the carbon dioxide of the air. A steady-state concentration, amounting to about one

The Iceman. In 1991 hikers discovered the Iceman in a glacier near the Italian-Austrian border. Remarkably, because of exposure to cold winds before he was buried by the ice, his body had been dehydrated. Radiocarbon dating showed the Iceman to be about 5300 years old. *(Sygma)*

atom of carbon-14 for every 10^{12} atoms of carbon-12, is established in atmospheric CO_2. More specifically, the concentration of C-14 is such that a sample containing one gram of carbon has an activity of 13.6 atoms/min*. A living plant, taking in carbon dioxide, has this same activity, as do plant-eating animals or human beings.

When a plant or animal dies, the intake of radioactive carbon stops. Consequently, the radioactive decay of carbon-14

$$^{14}_{6}C \longrightarrow ^{14}_{7}N + ^{0}_{-1}e \qquad t_{1/2} = 5730 \text{ y}$$

takes over, and the C-14 activity drops. The activity of a sample is directly proportional to the amount of C-14, so Equation (2) can be rewritten as

> Since activity is directly proportional to amount, $x_0/x = A_0/A$.

$$\ln \frac{A_0}{A} = kt$$

where A_0 is the original activity, assumed to be 13.6 atoms/min, and A is the measured activity today; t is the age of the sample.

EXAMPLE 19.3 A tiny piece of paper taken from the Dead Sea Scrolls, believed to date back to the first century A.D., was found to have an activity per gram of carbon of 10.8 atoms/min. Taking A_0 to be 13.6 atoms/min, estimate the age of the scrolls.

Strategy First (1), calculate the rate constant, knowing that the half-life is 5730 y. Then (2) find t, using the equation $\ln (A_0/A) = kt$.

Solution

(1) $k = 0.693/5730 \text{ y} = 1.21 \times 10^{-4}/\text{y}$

(2) $\ln \dfrac{13.6}{10.8} = 0.231 = 1.21 \times 10^{-4}/\text{y} \times t$

$$t = 0.231/(1.21 \times 10^{-4}/\text{y}) = \boxed{1.91 \times 10^3 \text{ y}}$$

The Shroud of Turin. This piece of linen first appeared in about 1350 A.D. It bears a faint image (enhanced here) of a man who appears to have been crucified, possibly Jesus Christ. *(Santi Visali/The IMAGE Bank)*

As you can imagine, it is not easy to determine accurately activities of the order of 10 atoms decaying per minute, about one "event" every six seconds. Elaborate precautions have to be taken to exclude background radiation. Moreover, relatively large samples must be used to increase the counting rate. Recently a technique has been developed whereby C-14 atoms can be counted very accurately in a specially designed mass spectrometer. This method was used to date the Shroud of Turin, using six samples with a total mass of about 0.1 g. Analysis by an international team of scientists in 1988 showed that the flax used to make the linen of which the Shroud is composed grew in the fourteenth century A.D. That means this remarkable burial garment could not have been used for the body of Christ.

---

*Actually, this was the activity in 1950, prior to nuclear testing, which raised the C-14 content considerably. Moreover, it is now known that the C-14 content has varied significantly over the past several thousand years; in very accurate work, a correction must be made for this effect.

19.3 MASS-ENERGY RELATIONS

The energy change accompanying a nuclear reaction can be calculated from the relation

$$\Delta E = c^2 \, \Delta m$$

where

Δm = change in mass = mass products − mass reactants

ΔE = energy products − energy reactants

c = speed of light

In a spontaneous nuclear reaction, the products weigh less than the reactants (Δm negative). In this case, the energy of the products is less than that of the reactants (ΔE negative) and energy is evolved to the surroundings.

Such a reaction is exothermic; ΔH, like ΔE, is negative.

In an "ordinary chemical reaction," Δm is immeasurably small. In a nuclear reaction, on the other hand, Δm is appreciable, amounting to 0.002% or more of the mass of reactants. The change in mass can readily be calculated from a table of nuclear masses (Table 19.3, page 558).

To obtain a more useful form of the equation $\Delta E = c^2 \, \Delta m$, we substitute for c its value in meters per second.

$$c = 3.00 \times 10^8 \text{ m/s}$$

Thus

$$\Delta E = 9.00 \times 10^{16} \, \frac{\text{m}^2}{\text{s}^2} \times \Delta m$$

But

$$1 \text{ J} = 1 \text{ kg} \cdot \frac{\text{m}^2}{\text{s}^2}; \qquad 1 \frac{\text{m}^2}{\text{s}^2} = 1 \frac{\text{J}}{\text{kg}}$$

So

$$\Delta E = 9.00 \times 10^{16} \, \frac{\text{J}}{\text{kg}} \times \Delta m$$

This equation can be used to calculate the energy change in *joules*, if you know Δm in *kilograms*. Ordinarily, Δm is expressed in *grams*; ΔE is calculated in *kilojoules*. The relationship between ΔE and Δm in these units can be found by using conversion factors:

$$\Delta E = 9.00 \times 10^{16} \, \frac{\text{J}}{\text{kg}} \times \frac{1 \text{ kg}}{10^3 \text{ g}} \times \frac{1 \text{ kJ}}{10^3 \text{ J}} \times \Delta m$$

$$\boxed{\Delta E = 9.00 \times 10^{10} \, \frac{\text{kJ}}{\text{g}} \times \Delta m}$$

This is the most useful form of the mass-energy relation.

EXAMPLE 19.4 For the radioactive decay of radium, $^{226}_{88}\text{Ra} \rightarrow {}^{222}_{86}\text{Rn} + {}^{4}_{2}\text{He}$, calculate ΔE in kilojoules when 1.02 g of radium decays.

Strategy (1) Using Table 19.3, calculate Δm when one mole of radium decays. Then (2) find Δm when 1.02 g of radium decays (1 mol Ra = 226.0 g Ra). Finally, (3) find ΔE using the relation:

$$\Delta E = 9.00 \times 10^{10} \, \frac{\text{kJ}}{\text{g}} \times \Delta m$$

TABLE 19.3   Nuclear Masses on the ^{12}C Scale* OHT

	At. No.	Mass No.	Mass (amu)		At. No.	Mass No.	Mass (amu)
e	0	0	0.00055				
n	0	1	1.00867	Br	35	79	78.8992
H	1	1	1.00728		35	81	80.8971
	1	2	2.01355		35	87	86.9028
	1	3	3.01550	Rb	37	89	88.8913
He	2	3	3.01493	Sr	38	90	89.8869
	2	4	4.00150	Mo	42	99	98.8846
Li	3	6	6.01348	Ru	44	106	105.8832
	3	7	7.01436	Ag	47	109	108.8790
Be	4	9	9.00999	Cd	48	109	108.8786
	4	10	10.01134		48	115	114.8791
B	5	10	10.01019	Sn	50	120	119.8748
	5	11	11.00656	Ce	58	144	143.8817
C	6	11	11.00814		58	146	145.8868
	6	12	11.99671	Pr	59	144	143.8809
	6	13	13.00006	Sm	62	152	151.8857
	6	14	13.99995	Eu	63	157	156.8908
O	8	16	15.99052	Er	68	168	167.8951
	8	17	16.99474	Hf	72	179	178.9065
	8	18	17.99477	W	74	186	185.9138
F	9	18	17.99601	Os	76	192	191.9197
	9	19	18.99346	Au	79	196	195.9231
Na	11	23	22.98373	Hg	80	196	195.9219
Mg	12	24	23.97845	Pb	82	206	205.9295
	12	25	24.97925		82	207	206.9309
	12	26	25.97600		82	208	207.9316
Al	13	26	25.97977	Po	84	210	209.9368
	13	27	26.97439		84	218	217.9628
	13	28	27.97477	Rn	86	222	221.9703
Si	14	28	27.96924	Ra	88	226	225.9771
S	16	32	31.96329	Th	90	230	229.9837
Cl	17	35	34.95952	Pa	91	234	233.9934
	17	37	36.95657	U	92	233	232.9890
Ar	18	40	39.95250		92	235	234.9934
K	19	39	38.95328		92	238	238.0003
	19	40	39.95358		92	239	239.0038
Ca	20	40	39.95162	Np	93	239	239.0019
Ti	22	48	47.93588	Pu	94	239	239.0006
Cr	24	52	51.92734		94	241	241.0051
Fe	26	56	55.92066	Am	95	241	241.0045
Co	27	59	58.91837	Cm	96	242	242.0061
Ni	28	59	58.91897	Bk	97	245	245.0129
Zn	30	64	63.91268	Cf	98	248	248.0186
	30	72	71.91128	Es	99	251	251.0255
Ge	32	76	75.90380	Fm	100	252	252.0278
As	33	79	78.90288		100	254	254.0331

*Note that these are *nuclear masses*. The masses of the corresponding atoms can be calculated by adding the masses of each extranuclear electron (0.000549). For example, for an *atom* of $^4_2$He we have 4.00150 + 2(0.000549) = 4.00260. Similarly, for an atom of $^{12}_6$C, 11.99671 + 6(0.000549) = 12.00000.

Solution

(1) Using Table 19.3 for one mole of Ra-226,

$$\Delta m = \text{mass 1 mol He-4} + \text{mass 1 mol Rn-222} - \text{mass 1 mol Ra-226}$$

$$= 4.0015 \text{ g} + 221.9703 \text{ g} - 225.9771 \text{ g}$$

$$= -0.0053 \text{ g/mol Ra}$$

(2) When 1.02 g of radium decays,

$$\Delta m = \frac{-0.0053 \text{ g}}{\text{mol Ra}} \times \frac{1 \text{ mol Ra}}{226.0 \text{ g Ra}} \times 1.02 \text{ g Ra} = -2.4 \times 10^{-5} \text{ g}$$

(3) $\Delta E = 9.00 \times 10^{10} \dfrac{\text{kJ}}{\text{g}} \times (-2.4 \times 10^{-5} \text{ g}) = \boxed{-2.2 \times 10^6 \text{ kJ}}$

Notice that the change in mass doesn't show up until you get to the third decimal place.

Reality Check In ordinary chemical reactions, the energy change is of the order of 50 kJ/g or less. The energy change in this nuclear reaction is greater by a factor of

$$2.2 \times 10^6/50 = 4 \times 10^4$$

In other words, about 40,000 times as much energy is evolved.

Nuclear Binding Energy

It is always true that *a nucleus weighs less than the individual protons and neutrons of which it is composed.* Consider, for example, the ^6_3Li nucleus, which contains three protons and three neutrons. According to Table 19.3, one mole of Li-6 nuclei weighs 6.01348 g. In contrast, the total mass of three moles of neutrons and three moles of protons is

$$3(1.00867 \text{ g}) + 3(1.00728 \text{ g}) = 6.04785 \text{ g}$$

Clearly, one mole of Li-6 weighs *less* than the corresponding protons and neutrons. For the process

$$^6_3\text{Li} \longrightarrow 3\,^1_1\text{H} + 3\,^1_0 n$$

$\Delta m = 6.04785 \text{ g} - 6.01348 \text{ g} = 0.03437 \text{ g/mol Li}$.

The quantity just calculated is referred to as the *mass defect*. The corresponding energy difference is

$$\Delta E = 9.00 \times 10^{10} \frac{\text{kJ}}{\text{g}} \times 0.03437 \frac{\text{g}}{\text{mol Li}} = 3.09 \times 10^9 \frac{\text{kJ}}{\text{mol Li}}$$

It takes a lot of energy to blow a nucleus apart.

This energy is referred to as the **binding energy.** It follows that 3.09×10^9 kJ of energy would have to be absorbed to decompose one mole of Li-6 nuclei into protons and neutrons.

$$^6_3\text{Li} \longrightarrow 3\,^1_0 n + 3\,^1_1\text{H} \qquad \Delta E = 3.09 \times 10^9 \frac{\text{kJ}}{\text{mol Li}}$$

By the same token, 3.09×10^9 kJ of energy would be evolved when one mole of Li-6 is formed from protons and neutrons.

EXAMPLE 19.5   Calculate the binding energy of Be-9 in kilojoules per mole.

Strategy   Use Table 19.3 to find Δm for the decomposition of Be-9 into neutrons and protons. Then calculate $\Delta E = 9.00 \times 10^{10}$ kJ/g $\times \Delta m$.

Solution

(1) $^{9}_{4}\text{Be} \rightarrow 5\,^{1}_{0}n + 4\,^{1}_{1}\text{H}$

(2) $\Delta m = 5(1.00867 \text{ g}) + 4(1.00728 \text{ g}) - 9.00999 \text{ g} = 0.06248 \text{ g/mol Be}$

$$\Delta E = 9.00 \times 10^{10} \frac{\text{kJ}}{\text{g}} \times 0.06248 \frac{\text{g}}{\text{mol Be}} = \boxed{5.62 \times 10^9 \text{ kJ/mol Be}}$$

The binding energy of a nucleus is, in a sense, a measure of its stability. The greater the binding energy, the more difficult it would be to decompose the nucleus into protons and neutrons. As you might expect, the binding energy as calculated above increases steadily as the nucleus gets heavier, containing more protons and neutrons. A better measure of the relative stabilities of different nuclei is the **binding energy per mole of nuclear particles (nucleons)**. This quantity is calculated by dividing the binding energy per mole of nuclei by the number of particles per nucleus. Thus

Li-6: $\quad 3.09 \times 10^9 \dfrac{\text{kJ}}{\text{mol Li-6}} \times \dfrac{1 \text{ mol Li-6}}{6 \text{ mol nucleons}} = 5.15 \times 10^8 \dfrac{\text{kJ}}{\text{mol}}$

Be-9: $\quad 5.62 \times 10^9 \dfrac{\text{kJ}}{\text{mol Be-9}} \times \dfrac{1 \text{ mol Be-9}}{9 \text{ mol nucleons}} = 6.24 \times 10^8 \dfrac{\text{kJ}}{\text{mol}}$

Figure 19.4 shows a plot of this quantity, binding energy/mole of nucleons, versus mass number. Notice that the curve has a broad maximum in the vicinity of mass numbers 50 to 80. Consider what would happen if a heavy nucleus such as $^{235}_{92}\text{U}$

FIGURE 19.4

Binding energy per nucleon, a measure of nuclear stability. The binding energy has its maximum value for nuclei of intermediate mass, falling off for very heavy or very light nuclei. The form of this curve accounts for the fact that both fission and fusion give off large amounts of energy.   **OHT**

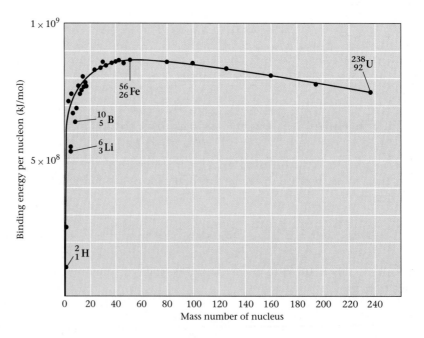

were to split into smaller nuclei with mass numbers near the maximum. This process, referred to as **nuclear fission,** should result in an evolution of energy. The same result is obtained if very light nuclei such as ^2_1H combine with one another to form a heavier nucleus. Indeed, this process, called **nuclear fusion,** should evolve even more energy, because the binding energy per nucleon increases very sharply at the beginning of the curve.

Both very heavy and very light nuclei are relatively unstable; they decompose with the evolution of energy.

19.4 NUCLEAR FISSION

The process of nuclear fission was discovered more than half a century ago in 1938 by Lise Meitner and Otto Hahn in Germany. With the outbreak of World War II a year later, interest focused on the enormous amount of energy released in the process. At Los Alamos, in the mountains of New Mexico, a group of scientists led by J. Robert Oppenheimer worked feverishly to produce the fission, or "atomic" bomb. Many of the members of this group were exiles from Nazi Germany. They were spurred on by the fear that Hitler would obtain the bomb first. Their work led to the explosion of the first atomic bomb in the New Mexico desert at 5:30 A.M. on July 16, 1945. Less than a month later (August 6, 1945), the world learned of this new weapon when another bomb was exploded over Hiroshima. This bomb killed 70,000 people and completely devastated an area of 10 square kilometers. Three days later, Nagasaki and its inhabitants met a similar fate. On August 14, Japan surrendered, and World War II was over.

A nuclear fission explosion. Such a dramatic and destructive release of energy had never been seen before the development of the "atomic bomb" during World War II. *(FPG International)*

The Fission Process ($^{235}_{92}U$)

Several isotopes of the heavy elements undergo fission if bombarded by neutrons of high enough energy. In practice, attention has centered on two particular isotopes, $^{235}_{92}U$ and $^{239}_{94}Pu$. Both of these can be split into fragments by relatively low-energy neutrons.

Our discussion concentrates on the uranium-235 isotope. It makes up only about 0.7% of naturally occurring uranium. The more abundant isotope, uranium-238, does not undergo fission. The first process used to separate these isotopes, and until recently the only one available, was that of gaseous effusion (Chapter 5). The volatile compound uranium hexafluoride, UF_6, which sublimes at 56°C, is used for this purpose.

When a uranium-235 atom undergoes fission, it splits into two unequal fragments and a number of neutrons and beta particles. The fission process is complicated by the fact that different uranium-235 atoms split up in many different ways. For example, while one atom of $^{235}_{92}U$ is splitting to give isotopes of rubidium ($Z = 37$) and cesium ($Z = 55$), another may break up to give isotopes of bromine ($Z = 35$) and lanthanum ($Z = 57$), while still another atom yields isotopes of zinc ($Z = 30$) and samarium ($Z = 62$):

$$^1_0n + {}^{235}_{92}U \longrightarrow \begin{cases} {}^{90}_{37}Rb + {}^{144}_{55}Cs + 2\,{}^1_0n \\ {}^{87}_{35}Br + {}^{146}_{57}La + 3\,{}^1_0n \\ {}^{72}_{30}Zn + {}^{160}_{62}Sm + 4\,{}^1_0n \end{cases}$$

More than 200 isotopes of 35 different elements have been identified among the fission products of uranium-235.

The stable neutron-to-proton ratio near the middle of the periodic table, where the fission products are located, is considerably smaller (~1.2) than that of uranium-235 (1.5). Hence the immediate products of the fission process contain too many neutrons for stability; they decompose by beta emission. In the case of rubidium-90, three steps are required to reach a stable nucleus:

$$\ce{^{90}_{37}Rb} \longrightarrow \ce{^{90}_{38}Sr} + \ce{^{0}_{-1}e} \qquad t_{1/2} = 2.8 \text{ min}$$

$$\ce{^{90}_{38}Sr} \longrightarrow \ce{^{90}_{39}Y} + \ce{^{0}_{-1}e} \qquad t_{1/2} = 29 \text{ y}$$

$$\ce{^{90}_{39}Y} \longrightarrow \ce{^{90}_{40}Zr} + \ce{^{0}_{-1}e} \qquad t_{1/2} = 64 \text{ h}$$

The radiation hazard associated with fallout from nuclear weapons testing arises from radioactive isotopes such as these. One of the most dangerous is strontium-90. In the form of strontium carbonate, $SrCO_3$, it is incorporated into the bones of animals and human beings, where it remains for a lifetime.

Notice from the fission equations written above that two to four neutrons are produced by fission for every one consumed. Once a few atoms of uranium-235 split, the neutrons produced can bring about the fission of many more uranium-235 atoms. This creates the possibility of a *chain reaction,* whose rate increases exponentially with time. This is precisely what happens in the atomic bomb. The energy evolved in successive fissions escalates to give a tremendous explosion within a few seconds.

For nuclear fission to result in a chain reaction, the sample must be large enough so that most of the neutrons are captured internally. If the sample is too small, most of the neutrons escape, breaking the chain. The *critical mass* of uranium-235 required to maintain a chain reaction in a bomb appears to be about 1 to 10 kg. In the bomb dropped on Hiroshima, the critical mass was achieved by using a conventional explosive to fire one piece of uranium-235 into another.

The Hiroshima bomb was equivalent to 20,000 tons of TNT.

The evolution of energy in nuclear fission is directly related to the decrease in mass that takes place. About 80,000,000 kJ of energy is given off for every gram of $\ce{^{235}_{92}U}$ that reacts. This is about 40 times as great as the energy change for simple nuclear reactions such as radioactive decay. The heat of combustion of coal is only about 30 kJ/g; the energy given off when TNT explodes is still smaller, about 2.8 kJ/g. Putting it another way, the fission of one gram of $\ce{^{235}_{92}U}$ produces as much energy as the combustion of 2700 kg of coal or the explosion of 30 metric tons (3×10^4 kg) of TNT.

Nuclear Reactors

About 18% of the electrical energy produced in the United States comes from nuclear reactors using the fission of U-235. A so-called light water reactor (LWR) is shown schematically in Figure 19.5. The *fuel rods* are cylinders that contain fissionable material, uranium dioxide (UO_2) pellets, in a zirconium alloy tube. The uranium in these reactors is "enriched" so that it contains about 3% U-235, the fissionable isotope. The *control rods* are cylinders composed of substances, such as boron and cadmium, that absorb neutrons. Increased absorption of neutrons slows down the chain reaction. By varying the depth of the control rods within the fuel-rod assembly, the speed of the chain reaction can be controlled. Water at a pressure of 140 atm is passed through the reactor to absorb the heat given off by fission. The water, coming out of the reactor core at 320°C, circulates through a

Uranium fuel rods ready for use in a nuclear reactor. *(Courtesy of Westinghouse Electric Co., Commercial Nuclear Fuel Division)*

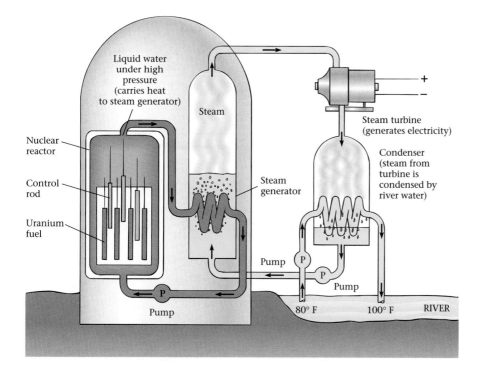

FIGURE 19.5

A pressurized water nuclear reactor. The control rods are made of a material such as cadmium or boron, which absorb neutrons effectively. The fuel rods contain uranium oxide enriched in U-235. **OHT**

closed loop containing a heat exchanger that produces steam at 270°C. This steam is used to drive a turbogenerator that produces electrical energy.

In the light water reactor, the circulating water serves another purpose in addition to heat transfer. It acts to slow down or *moderate* the neutrons given off by fission. This is necessary if the chain reaction is to continue; fast neutrons are not readily absorbed by U-235. Reactors in Canada use "heavy water," D_2O, which has an important advantage over H_2O. Its moderating properties are such that naturally occurring uranium can be used as a fuel; enrichment in U-235 is not necessary.

Thirty years ago it was generally supposed that nuclear fission would replace fossil fuels (oil, natural gas, coal) as an energy source. That hasn't happened for at least three reasons.

1. *Nuclear energy is more expensive,* at least in the United States. Electrical energy produced in a nuclear reactor costs about 7¢ per kilowatt-hour, as compared to only 3¢ per kilowatt-hour for fossil fuel plants.

2. *Nuclear accidents* at Three Mile Island, Pennsylvania, in 1979 and Chernobyl, Russia, in 1986 have had a devastating effect on public opinion in the United States and Europe. At Three Mile Island, only about 50 curies of radiation were released to the environment; there were no casualties. The explosion at Chernobyl was a very different story. About 100 million curies were released, leading to at least 31 fatalities. Moreover, 135,000 people had to be permanently evacuated from the Chernobyl area.

3. *Disposal of radioactive wastes* from nuclear reactors has proved to be a serious political problem. The "NIMBY" syndrome (*not in my back yard*) applies here. There are plans to deposit low-level nuclear waste in a salt bed half a mile underground in Carlsbad, New Mexico. High-level waste is supposed to be buried in Yucca Mountain, Nevada.

On September 30, 1999, a fuel-processing accident at a nuclear reactor at Tokaimura, Japan, severely injured three workers.

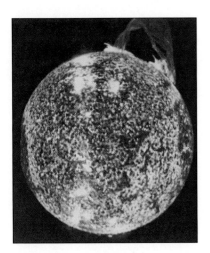

The Sun. Most of its energy is created by the fusion of hydrogen. *(NASA)*

19.5 NUCLEAR FUSION

Recall (Figure 19.4) that very light nuclei, such as those of hydrogen, are unstable with respect to fusion into heavier isotopes. Indeed, the energy available from nuclear fusion is considerably greater than that given off in the fission of an equal mass of a heavy element (Example 19.6).

EXAMPLE 19.6   Calculate ΔE, in kilojoules per gram of reactants, in

(a) a fusion reaction, $^2_1H + {}^2_1H \rightarrow {}^4_2He$.

(b) a fission reaction, $^{235}_{92}U \rightarrow {}^{90}_{38}Sr + {}^{144}_{58}Ce + {}^1_0n + 4\ {}^{\ \ 0}_{-1}e$.

Strategy   This problem is entirely analogous to Example 19.4. First, find Δm for the equation as written, and then find Δm for one gram of reactant. Finally, calculate ΔE:
$\Delta E = 9.00 \times 10^{10}$ kJ/g $\times \Delta m$.

Solution

(a) (1) Using Table 19.3,

$$\Delta m = 4.00150\ g - 2(2.01355\ g) = -0.02560\ g$$

(2) The total mass of reactant, ^2_1H, is 4.027 g:

$$\Delta m = \frac{-0.02560\ g}{4.027\ g\ \text{reactant}} = -6.357 \times 10^{-3}\ g/g\ \text{reactant}$$

(3) $\Delta E = 9.00 \times 10^{10}\ \dfrac{kJ}{g} \times \dfrac{(-6.357 \times 10^{-3}\ g)}{g\ \text{reactant}} = \boxed{\dfrac{-5.72 \times 10^8\ kJ}{g\ \text{reactant}}}$

(b) (1) $\Delta m = 89.8869\ g + 143.8817\ g + 1.0087\ g + 4(0.00055\ g) - 234.9934\ g$
    $= -0.2139\ g$

(2) The total mass of reactant, U-235, is 235.0 g:

$$\Delta m = \frac{-0.2139\ g}{235.0\ g\ \text{reactant}} = 9.102 \times 10^{-4}\ g/g\ \text{reactant}$$

(3) $\Delta E = 9.00 \times 10^{10}\ \dfrac{kJ}{g} \times \left(\dfrac{-9.102 \times 10^{-4}\ g}{g\ \text{reactant}}\right) = \boxed{\dfrac{-8.19 \times 10^7\ kJ}{g\ \text{reactant}}}$

Reality Check   Comparing the answers to (a) and (b), it appears that the fusion reaction produces about seven times as much energy per gram of reactant (57.2×10^7 versus 8.19×10^7 kJ) as does the fission reaction. This factor varies from about 3 to 10, depending on the particular reactions chosen to represent the fusion and fission processes.

As an energy source, nuclear fusion possesses several additional advantages over nuclear fission. In particular, light isotopes suitable for fusion are far more abundant than the heavy isotopes required for fission. You can calculate, for example (Problem 65), that the fusion of only 2×10^{-9} % of the deuterium (^2_1H) in sea water would meet the total annual energy requirements of the world.

Unfortunately, fusion processes, unlike neutron-induced fission, have very high activation energies. To overcome the electrostatic repulsion between two deuterium nuclei and cause them to react, they have to be accelerated to velocities of about 10^6 m/s, about 10,000 times greater than ordinary molecular velocities at room temperature. The corresponding temperature for fusion, as calculated from kinetic theory, is of the order of $10^{9\circ}$C. In the hydrogen bomb, temperatures of this magnitude were achieved by using a fission reaction to trigger nuclear fusion.

Nuclear fusion has been the "fuel of the future" for at least 40 years.

If fusion reactions are to be used to generate electricity, it will be necessary to develop equipment in which very high temperatures can be maintained long enough to allow fusion to occur and give off energy. In any conventional container, the reactant nuclei would quickly lose their high kinetic energies by collisions with the walls.

One fusion reaction currently under study is a two-step process involving deuterium and lithium as the basic starting materials:

$$\frac{2}{1}H + \frac{3}{1}H \longrightarrow \frac{4}{2}He + \frac{1}{0}n$$

$$\underline{\frac{6}{3}Li + \frac{1}{0}n \longrightarrow \frac{4}{2}He + \frac{3}{1}H}$$

$$\frac{2}{1}H + \frac{6}{3}Li \longrightarrow 2\ \frac{4}{2}He$$

This process is attractive because it has a lower activation energy than other fusion reactions.

One possible way to achieve nuclear fusion is to use magnetic fields to confine the reactant nuclei and prevent them from touching the walls of the container, where they would quickly slow down below the velocity required for fusion. Using 400-ton magnets, it is possible to sustain the reaction for a fraction of a second. To achieve a net evolution of energy, this time must be extended to about one second. A practical fusion reactor would have to produce 20 times as much energy as it consumes. It is likely to take at least 20 years to achieve that goal with magnetic confinement.

Another approach to nuclear fusion is shown in Figure 19.6. Tiny glass pellets (about 0.1 mm in diameter) filled with frozen deuterium and tritium serve as a target. The pellets are illuminated by a powerful laser beam, which delivers 10^{12} kilowatts of power in one nanosecond (10^{-9} s). The reaction is the same as with magnetic confinement; unfortunately, at this point energy breakeven seems many years away.

In April, 1997, the Tokomac fusion reactor at Princeton shut down when government funding was withdrawn.

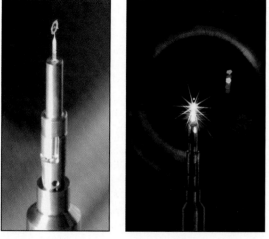

(a)   (b)

FIGURE 19.6

Laser fusion. (a) A mixture of deuterium and tritium is sealed inside the tiny capsule (1 mm in diameter) at the tip of the laser target. (b) Exposure of the capsule to the energy of a powerful laser beam produces a 0.5-picosecond burst of energy from the fusion of the deuterium and tritium. *(a, Courtesy of the University of California, Lawrence Livermore National Laboratory, and the U.S. Department of Energy; b, Lawrence Livermore National Laboratory)*

CHEMISTRY Beyond the Classroom

Biological Effects of Radiation

The harmful effects of radiation result from its high energy, sufficient to form unstable free radicals (species containing unpaired electrons) such as

$$H\cdot \qquad H—\overset{..}{\underset{..}{O}}\cdot \qquad :\overset{..}{\underset{..}{O}}—\overset{..}{\underset{..}{O}}\cdot$$

These free radicals can react with and in that sense destroy organic molecules essential to life.

The extent of damage from radiation depends mainly on two factors. These are the amount of radiation absorbed and the type of radiation. The former is commonly expressed in *rads* (radiation *a*bsorbed *d*ose). A rad corresponds to the absorption of 10^{-2} J of energy per kilogram of tissue:

$$1 \text{ rad} = 10^{-2} \text{ J/kg}$$

The biological effect of radiation is expressed in *rems* (radiation *e*quivalent for *m*an). The number of rems is found by multiplying the number of rads by a "damage" factor, *n*:

$$\text{no. of rems} = n(\text{no. of rads})$$

where *n* is 1 for gamma and beta radiation, 5 for low-energy neutrons, and 10 to 20 for high-energy neutrons and alpha particles. Table A lists some of the effects to be expected when a person is exposed to a single dose of radiation at various levels.

TABLE A Effect of Exposure to a Single Dose of Radiation

Dose (rems)	Probable Effect
0 to 25	No observable effect
25 to 50	Small decrease in white blood cell count
50 to 100	Lesions, marked decrease in white blood cells
100 to 200	Nausea, vomiting, loss of hair
200 to 500	Hemorrhaging, ulcers, possible death
500+	Fatal

Small doses of radiation repeated over long periods of time can have very serious consequences. Many of the early workers in the field of radioactivity developed cancer in this way. Cases are known in which cancers developed as long as 40 years after initial exposure. Studies have shown an abnormally large number of cases of leukemia among the survivors of the atomic bombs dropped on Hiroshima and Nagasaki. Radiation can also have genetic effects; that is, it can produce mutations in plants and animals by bringing about changes in genes. The children born to survivors of Nagasaki and Hiroshima show an abnormally high number of congenital defects.

The average exposure to radiation of people living in the United States is about 360 mrem (0.36 rem) per year. Notice (Figure A) that 82% of the radiation comes from natural sources; the greatest single source by far is radon (55%). The level of exposure to radon depends on location. In 1985 a man named Stanley Watras, who happened to work at a nuclear power plant, found that he was setting off the radiation monitors when he went to work in the morning. It turned out that the house he lived in had a radon level 2000 times the national average.

$^{222}_{86}$Rn, a radioactive isotope of radon, is a decay product of naturally occurring uranium-238. Because it is gaseous and chemically inert, radon seeps through cracks in

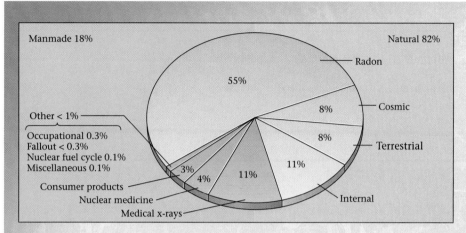

FIGURE A

Sources of average radiation exposure of the U.S. population. Notice that only 18% is from manmade sources and, of that, less than one percent is from nuclear reactor fuel and the fallout from nuclear weapons that have been detonated. *(Source: National Council on Radiation Protection and Measurement, Report #93, "Ionizing Radiation Exposure of the Population of the United States," 1987)*

FIGURE B

A commercially available home-test kit for radon. *(Charles D. Winters)*

concrete and masonry from the ground into houses. There its concentration builds up, particularly if the house is tightly insulated. Inhalation of radon-222 can cause health problems because its decay products, including Po-218 and Po-214, are intensely radioactive and readily absorbed in lung tissue. The Environmental Protection Agency (EPA) estimates that radon inhalation causes between 5,000 and 20,000 of the 130,000 deaths from lung cancer annually in the United States. The EPA recommends that special ventilation devices be used to remove radon from basements if tests show (Figure B) that the radiation level exceeds 4×10^{-12} Ci/L.

CHAPTER HIGHLIGHTS

Key Concepts

1. Write balanced nuclear equations.
   (**Example 19.1; Problems 1–12**)
2. Relate activity to rate constant and number of atoms.
   (**Examples 19.2, 19.3; Problems 13–24, 41, 43, 62**)
3. Relate rate constant to half-life.
   (**Examples 19.2, 19.3; Problems 25–30, 42**)
4. Relate Δm to ΔE.
   (**Examples 19.4, 19.6; Problems 31, 32, 37–40, 44, 52**)
5. Calculate binding energies.
   (**Example 19.5; Problems 33–36**)

Key Equations

Rate of decay $\ln X_0/X = kt$ $k = 0.693/t_{1/2}$ $A = kN$

Mass-energy $\Delta E = 9.00 \times 10^{10} \dfrac{kJ}{g} \times \Delta m$

Key Terms

activity
alpha particle
■ atomic number
beta particle
binding energy
curie (Ci)

gamma radiation
■ half-life
■ isotope
K-electron capture
■ mass number

nuclear equation
nuclear fission
nuclear fusion
positron
■ radioactivity

Summary Problem

Consider the isotopes of gallium.
(a) Write the nuclear symbol for Ga-67, which is used medically to scan for various tumors. How many protons are there in the nucleus? How many neutrons?
(b) Write the equation for the decomposition of Ga-67 by K-capture.
(c) Write nuclear equations to represent the bombardment of Ga-64 by a neutron and the decay of the product nucleus by beta emission.
(d) Calculate ΔE in kJ when 1.00 g of Ga-71 (nuclear mass = 70.9476 amu) is bombarded by an α-particle producing As-72 (nuclear mass = 71.9493 amu) and neutrons.
(e) What are the mass defect and binding energy of Ga-71?
(f) The half-life of Ga-67 is 77.9 hours. What is the rate constant for Ga-67? What is the activity (in curies) of a 5.00×10^2 mg sample of Ga-67?
(g) How long will it take Ga-67 to decay and lose 95.0% of its radioactivity?

Answers

(a) $^{67}_{31}\text{Ga}$; 31 protons, 36 neutrons
(b) $^{67}_{31}\text{Ga} + ^{0}_{-1}e \rightarrow ^{67}_{30}\text{Zn}$
(c) $^{64}_{31}\text{Ga} + ^{1}_{0}n \rightarrow ^{65}_{31}\text{Ga} \rightarrow ^{0}_{-1}e + ^{65}_{32}\text{Ge}$
(d) 3.32×10^7 kJ
(e) mass defect = 0.6249 g; binding energy = 5.62×10^{10} kJ
(f) $k = 8.90 \times 10^{-3}$ h^{-1}; $A = 3.00 \times 10^5$ Ci
(g) 337 h $\approx$ 14 days

Questions & Problems

Problem numbers in blue indicate that the answer is available in Appendix 6 at the back of the book.
WEB indicates that the solution is posted at **http://www.harcourtcollege.com/chem/general/masterton4/student/**

Nuclear Equations

1. Lead-210 is used to prepare eyes for corneal transplants. Its decay product is bismuth-210. Identify the emission from lead-210.

2. Smoke detectors contain a small amount of americium-241. Its decay product is neptunium-237. Identify the emission from americium-241.
3. Write balanced nuclear equations for

(a) the alpha emission resulting in the formation of Pa-233.

(b) the loss of a positron by Y-85.

(c) the fusion of two C-12 nuclei to give sodium-23 and another particle.

(d) the fission of Pu-239 to give tin-130, another nucleus and an excess of two neutrons.

4. Write balanced nuclear equations for

(a) the loss of an alpha particle by Th-230.

(b) the loss of a beta particle by lead-210.

(c) the fission of U-235 to give Ba-140, another nucleus and an excess of two neutrons.

(d) the K-capture of Ar-37.

5. Thorium-231 is the product of alpha emission and is radioactive, emitting beta radiation. Determine

(a) the parent nucleus of ^{231}Th.

(b) the product of ^{231}Th decay.

WEB 6. Rubidium-87, a beta emitter, is the product of positron emission. Determine

(a) the product of ^{87}Rb decay.

(b) the parent nucleus from which ^{87}Rb is formed.

7. Consider the "new" isotope $^{282}_{115}$X. Compare the product nuclides after beta emission and positron emission.

8. Follow the directions for Question 7 but compare the product nuclides after K-capture and positron emission.

9. Write balanced nuclear equations for the bombardment of

(a) U-238 with a nucleus with neutrons to produce Fm-249 and five neutrons.

(b) Al-26 with an alpha particle to produce P-30.

(c) Cu-63 with a nucleus with neutrons producing Zn-63 and a neutron.

(d) Al-27 with deuterium ($^2_1$H) to produce an alpha particle and another nucleus.

10. Write balanced nuclear equations for the bombardment of

(a) Fe-54 with an alpha particle to produce another nucleus and two protons.

(b) Mo-96 with deuterium ($^2_1$H) to produce a neutron and another nucleus.

(c) Ar-40 with an unknown particle to produce K-43 and a proton.

(d) a nucleus with a neutron to produce a proton and P-31.

11. Balance the following equations by filling in the blanks.

(a) $^{238}_{92}U + ^1_1H \rightarrow ^{238}_{93}Np +$ _____

(b) $^{241}_{95}Am + ^4_2He \rightarrow 2\,^1_0n +$ _____

(c) $^4_2He +$ _____ $\rightarrow ^1_0n + ^{12}_6C$

(d) _____ $+ ^{27}_{13}Al \rightarrow ^{24}_{11}Si + ^4_2He$

12. Balance the following nuclear equations by filling in the blanks.

(a) $^{121}_{51}Sb + ^4_2He \rightarrow ^1_1H +$ _____

(b) $^{238}_{92}U + ^1_0n \rightarrow ^{0}_{-1}e +$ _____

(c) $^{14}_7N +$ _____ $\rightarrow ^1_1H + ^{17}_8O$

(d) _____ $+ ^4_2He \rightarrow ^{27}_{14}Si + ^1_0n$

Rate of Nuclear Decay

13. How many atoms decay in one hour and ten minutes in a 15.0-mCi (millicuries) source?

14. Food can be preserved by radiation with gamma rays. A γ-ray source has an activity of 2793 Ci. How many disintegrations are there per minute?

15. A Geiger counter counts 0.070% of all particles emitted by a sample. What is the activity that registers 19.4×10^3 counts in one minute?

16. A scintillation counter registers emitted radiation caused by the disintegration of nuclides. If each atom of nuclide emits one count, what is the activity of a sample that registers 3.00×10^4 disintegrations in five minutes?

17. Yttrium-87 has a rate constant of 2.6×10^{-6} s^{-1}. What is the activity of a 5.00-mg sample?

WEB 18. Krypton-87 has a rate constant of 1.5×10^{-4} s^{-1}. What is the activity of a 2.00-mg sample?

19. Np-237 (at. mass = 237.0 amu) has a half-life of 2.20×10^6 years. This isotope decays by alpha emission.

(a) Write a balanced nuclear reaction for the decay of neptunium-237.

(b) What is the activity (in millicuries) of a 0.500-g sample of the isotope?

20. Lead-210 has a half-life of 20.4 years. This isotope decays by beta particle emission.

(a) Write a balanced nuclear equation for the decay of lead-210.

(b) What is the activity (in millicuries) of a 0.500-g sample of the isotope (at. mass = 210.0 amu)?

21. Bromine-82 has a half-life of 36 hours. A sample containing Br-82 was found to have an activity of 1.2×10^5 disintegrations/min. How many grams of Br-82 were present in the sample? Assume that there were no other radioactive nuclides in the sample.

22. Consider the information about lead-210 given in Problem 20. A counter registers 1.3×10^4 disintegrations in five minutes. How many grams of Pb-210 are there?

23. Fluorine-18 has a decay constant of 6.31×10^{-3} min^{-1}. How many counts will one get on a Geiger counter in one minute from 1.00 mg of fluorine-18? Assume the sensitivity of the counter is such that it intercepts 0.50% of the emitted radiation.

24. Chlorine-36 decays by beta emission. It has a decay constant of 2.3×10^{-6} y^{-1}. How many β-particles are emitted in one minute from a 1.00-mg sample of Cl-36? How many curies does this represent?

25. A sample of a beam from the tomb of an ancient Egyptian pharaoh was analyzed in 2000 and gave 6.0 counts per minute (cpm) in a scintillation counter. A sample of freshly cut wood containing the same amount of carbon gave 15.3 cpm. In what year, approximately, did the king die? ($t_{1/2}$ C-14 = 5730 y)

26. An oil painting attributed to Rembrandt (1606–1669) is checked by ^{14}C dating. The ^{14}C content ($t_{1/2}$ C-14 = 5730 y) of the canvas is 0.975 times that of a living plant. Could the painting have been by Rembrandt?

27. A sample of a wooden artifact gives 5.0 disintegrations/min/g of carbon-14. The half-life of carbon-14 is 5730 years and the activity of C-14 in wood just cut down from a tree is 15.3 disintegrations/min/g. How old is the wooden artifact?

28. The remains of an ancient cave were unearthed. Analysis from charcoal in the cave gave 12.0 disintegrations/min/g of carbon-14. The half-life of C-14 is 5730 years. Analysis of a tree cut down when the cave was unearthed showed 22.1 disintegrations/min/g of carbon-14. How old are the remains in the cave?

29. The radioactive isotope tritium, ^3_1H, is produced in nature in much the same way as $^{14}_6C$. Its half-life is 12.3 years. Estimate the ^3_1H ratio of the tritium of water in the area to the tritium in a bottle of wine claimed to be 25 years old.

30. Consider the information given about tritium in Problem 29. Estimate the age of a sample of Scotch whiskey that has a tritium content three-fifths that of the water in the area where the whiskey was produced.

Mass Changes

31. Bk-245 decays by alpha emission.
(a) Calculate Δm in grams when one mole of Bk-245 decays.
(b) Calculate ΔE in kilojoules when one gram of ^{245}Bk decays.

32. Th-230 decays by alpha emission.
(a) Calculate Δm in grams when one mole of Th-230 decays.
(b) Calculate ΔE in kilojoules when one gram of ^{230}Th decays.

33. For carbon-14, calculate
(a) the mass defect.
(b) the binding energy.

WEB **34.** For beryllium-10, calculate
(a) the mass defect.
(b) the binding energy.

35. Which has the larger binding energy, fluorine-19 or oxygen-17?

36. Which has the larger binding energy, Mg-26 or Al-26?

37. Compare the energies given off per gram of reactant in the two fusion processes considered to occur during the formation of a star.
(a) $^2_1H + ^1_1H \rightarrow ^3_2He$
(b) $2^3_2He \rightarrow ^4_2He + 2^1_1H$

38. Some of the Sun's energy comes from the reaction

$$4^1_1H \longrightarrow ^4_2He + 2^0_1e$$

Calculate the energy change in this reaction per gram of hydrogen.

39. Consider the fission reaction in which U-235 is bombarded by neutrons. The products of the bombardment are rubidium-89, cerium-144, beta particles, and more neutrons.
(a) Write a balanced nuclear equation for the bombardment.
(b) Calculate ΔE when one gram of U-235 undergoes fission.
(c) The detonation of TNT, an explosive, evolves 2.76 kJ/g. How many kilograms of TNT are required to produce the same amount of energy as one milligram of U-235?

40. The decomposition of ammonium nitrate, an explosive, evolves 37.0 kJ/mol. Use the reaction given in Problem 39 to calculate the mass of ammonium nitrate (in kilograms) required to produce the same amount of energy as that produced when one milligram of U-235 undergoes fission.

Unclassified

41. How many disintegrations per second occur in a basement that is 40 × 40 × 10 feet if the radiation level from radon is the allowed 4×10^{-12} Ci/L?

42. Iodine-131 is used in the treatment of tumors in the thyroid gland. Its half-life is 8.1 days. Suppose that, due to a shipment delay, the I-131 in a hospital's pharmacy is 2.0 days old.
(a) What percentage of the I-131 has disintegrated?
(b) A patient is scheduled to receive 15.0 mg of I-131. What dosage (in mg) should the hospital pharmacist recommend for this patient if the 2-day-old bottle of I-131 is used?

43. Smoke detectors contain small amounts of americium-241. Am-241 decays by emitting α-particles and has a decay constant of 1.51×10^{-3} y^{-1}. If a smoke detector gives off ten disintegrations per second, how many grams of Am-241 are present in the detector?

44. An explosion used five tons (1 ton = 2000 lb) of ammonium nitrate ($\Delta E = -37.0$ kJ/mol).
(a) How much energy was released by the explosion?
(b) How many grams of TNT ($\Delta E = -2.76$ kJ/g) are needed to release the energy calculated in (a)?
(c) How many grams of U-235 are needed to obtain the same amount of energy calculated in (a)? (See the equation in Problem 39.)

45. The amount of oxygen dissolved in a sample of water can be determined by using thallium metal containing a small amount of the isotope Tl-204. When excess thallium is added to oxygen-containing water, the following reaction occurs.

$$2Tl(s) + \tfrac{1}{2}O_2(g) + H_2O \longrightarrow 2Tl^+(aq) + 2OH^-(aq)$$

After reaction, the activity of a 25.0-mL water sample is 745 counts per minute (cpm), caused by the presence of Tl$^+$-204 ions. The activity of Tl-204 is 5.53×10^5 cpm per gram of thallium metal. Assuming that O_2 is the limiting reactant in the above equation, calculate its concentration in moles per liter.

46. A 35-mL sample of 0.050 M AgNO$_3$ is mixed with 35 mL of 0.050 M NaI labeled with I-131. The following reaction occurs.

$$Ag^+(aq) + I^-(aq) \longrightarrow AgI(s)$$

The filtrate is found to have an activity of 2.50×10^3 counts per minute per millimeter. The 0.050 M NaI solution had an activity of 1.25×10^{10} counts per minute per milliliter. Calculate K_{sp} for AgI.

47. A 100.0-g sample of water containing tritium, ^3_1H, emits 2.89×10^3 beta particles per second. Tritium has a half-life of 12.3 years. What percentage of all the hydrogen atoms in the water sample is tritium?

48. Use the half-life of tritium given in Problem 47 to calculate the activity in curies of 1.00 mL of $^3_1H_2$ at STP.

49. One of the causes of the explosion at Chernobyl may have been the reaction between zirconium, which coated the fuel rods, and steam.

$$Zr(s) + 2H_2O(g) \longrightarrow ZrO_2(s) + 2H_2(g)$$

If half a metric ton of zirconium reacted, what pressure was exerted by the hydrogen gas produced at 55°C in the containment chamber, which had a volume of 2.0×10^4 L?

50. To measure the volume of the blood in an animal's circulatory system, the following experiment was performed. A 5.0-mL sample of an aqueous solution containing 1.7×10^5 counts per minute (cpm) of tritium was injected into the bloodstream. After an adequate period of time to allow for the complete circulation of the tritium, a 5.0-mL sample of blood was withdrawn and found to have 1.3×10^3 cpm on the scintillation counter. Assuming that only a negligible amount of tritium has decayed during the experiment, what is the volume of the animal's circulatory system?

51. A chelate of Cr^{3+} and $C_2O_4^{2-}$ is made by a reaction that involves Na_2CrO_4 and oxalic acid, $H_2C_2O_4$, a reducing agent. The sodium chromate has an activity of 765 counts per minute per gram, from Cr-51. The oxalic acid has an activity of 512 counts per minute per gram. It is labeled with C-14. Because Cr-51 and C-14 emit different particles during decay, their activities can be counted independently. A sample of the chelate was found to have a Cr-51 count of 314 cpm and a C-14 count of 235 cpm. How many oxalate ions are bound to one Cr^{3+} ion?

52. Consider the fission reaction

$$^1_0n + {}^{235}_{92}U \longrightarrow {}^{89}_{37}Rb + {}^{144}_{58}Ce + 3\,{}^{0}_{-1}e + 3\,{}^1_0n$$

How many liters of octane, C_8H_{18}, the primary component of gasoline, must be burned to $CO_2(g)$ and $H_2O(g)$ to produce as much energy as the fission of one gram of U-235 fuel? Octane has a density of 0.703 g/mL; its heat of formation is -249.9 kJ/mol.

53. Radium-226 decays by alpha emission to radon-222. Suppose that 25.0% of the energy given off by one gram of radium is converted to electrical energy. What is the minimum mass of lithium that would be needed for the voltaic cell $Li|Li^+\|Cu^{2+}|Cu$, at standard conditions, to produce the same amount of electrical work ($\Delta G°$)?

54. Polonium-210 decays to Pb-206 by alpha emission. Its half-life is 138 days. What volume of helium at 25°C and 1.20 atm would be obtained from a 25.00-g sample of Po-210 left to decay for 75 hours?

Conceptual Problems

55. Explain how
  (a) alpha and beta radiation are separated by an electric field.
  (b) radioactive C-11 can be used as a tracer to study brain disorders.

(c) a self-sustaining chain reaction occurs in nuclear fission.

56. Classify the following statements as true or false. If false, correct the statement to make it true.
  (a) The mass number increases in beta emission.
  (b) A radioactive species with a large rate constant, k, decays very slowly.
  (c) Fusion gives off less energy per gram of fuel than fission.

57. A positron and an electron have the same mass but opposite charges. When a positron meets an electron, annihilation occurs; that is, both masses are converted entirely into energy in the form of a γ-ray. Calculate
  (a) the energy emitted by the annihilation of a mole of positrons with electrons.
  (b) the energy of one gamma ray.

58. The principle behind the home smoke detector is described on page 553. Americium-241 is present in such detectors. It has a decay constant of 1.51×10^{-3} y^{-1}. You are urged to check the battery in the detector at least once a year. You are, however, never encouraged to check how much Am-241 remains undecayed. Explain why.

59. Suppose the $^{14}C/^{12}C$ ratio in plants a thousand years ago was 10% higher than it is today. What effect, if any, would this have on the calculated age of an artifact found by the C-14 method to be a thousand years old?

60. The following data are obtained for a radioactive isotope. Plot the data and determine the half life of the isotope.

Time (h)	0.00	0.50	1.00	1.50	2.00	2.50
Activity (disintegrations/h)	14,472	13,095	11,731	10,615	9,605	8,504

61. The cleavage of ATP (adenosine triphosphate) to ADP (adenosine diphosphate) and H_3PO_4 may be written as follows:

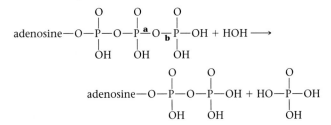

It is interesting to determine which bond (the P—O bond marked **a** or the O—P bond marked **b**) is cleaved by hydrolysis (reaction with water).
  (a) Outline an experiment (using radioactivity) that can be used to determine where the cleavage results.
  (b) Describe the results that would lead you to conclude that cleavage results at **a**.
  (c) Describe the results that would lead you to conclude that cleavage results at **b**.
Results show that the cleavage occurs at **b**.

Challenge Problems

62. An activity of 20 picocuries (20×10^{-12} Ci) of radon-222 per liter of air in a house constitutes a health hazard to anyone living there. The half-life of radon-222 is 3.82 days. Calculate the concentration of radon in air (moles per liter) that corresponds to a 20-picocurie activity level.

63. Plutonium-239 decays by the reaction

$$^{239}_{94}\text{Pu} \longrightarrow ^{235}_{92}\text{U} + ^{4}_{2}\text{He}$$

Its rate constant is 5.5×10^{-11}/min. In a one-gram sample of Pu-239,

(a) how many grams decompose in 45 minutes?

(b) how much energy in kilojoules is given off in 45 minutes?

(c) what radiation dosage in rems (page 566) is received by a 75-kg man exposed to a gram of Pu-239 for 45 minutes?

64. It is possible to estimate the activation energy for fusion by calculating the energy required to bring two deuterons close enough to one another to form an alpha particle. This energy can be obtained by using Coulomb's law in the form

$E = 8.99 \times 10^9\ q_1q_2/r$, where q_1 and q_2 are the charges of the deuterons (1.60×10^{-19} C), r is the radius of the He nucleus, about 2×10^{-15} m, and E is the energy in joules.

(a) Estimate E in joules per alpha particle.

(b) Using the equation $E = mv^2/2$, estimate the velocity (meters per second) each deuteron must have if a collision between the two of them is to supply the activation energy for fusion (m is the mass of the deuteron in kilograms).

65. Consider the reaction

$$2\,^{2}_{1}\text{H} \longrightarrow ^{4}_{2}\text{He}$$

(a) Calculate ΔE in kilojoules per gram of deuterium fused.

(b) How much energy is potentially available from the fusion of all the deuterium in sea water? The percentage of deuterium in water is about 0.0017%. The total mass of water in the oceans is 1.3×10^{24} g.

(c) What fraction of the deuterium in the oceans would have to be consumed to supply the annual energy requirements of the world (2.3×10^{17} kJ)?

The fields of Nature long prepared and fallow
the silent cyclic chemistry
The slow and steady ages plodding, the unoccupied surface
ripening, the rich ores forming beneath;

—Walt Whitman
Song of the Redwood Tree

CHEMISTRY OF THE METALS

20

An open-pit copper mine in Utah. *(David R. Frazier)*

CHAPTER OUTLINE

20.1   METALLURGY

20.2   REACTIONS OF THE ALKALI AND ALKALINE EARTH METALS

20.3   REDOX CHEMISTRY OF THE TRANSITION METALS

As you can see from the periodic table on the inside cover of this book, the overwhelming majority of elements, about 88%, are metals (shown in blue). In discussing the descriptive chemistry of the metals, we concentrate on—

■ the *main-group metals in Groups 1 and 2* at the far left of the periodic table. These are commonly referred to as the **alkali metals** (Group 1) and **alkaline earth metals** (Group 2). The group names reflect the strongly basic nature of the oxides (K_2O, CaO, . . .) and hydroxides (KOH, $Ca(OH)_2$, . . .) of these elements.

■ *the transition metals,* located in the center of the periodic table. There are three series of transition metals, each consisting of ten elements, located in the fourth, fifth, and sixth periods. We focus on a few of the more important transition metals (Figure 20.1), particularly those toward the right of the first series.

Section 20.1 deals with the processes by which these metals are obtained from their principal ores. Section 20.2 describes the reactions of the alkali and alkaline earth metals, particularly those with hydrogen, oxygen, and water. Section 20.3 considers the redox chemistry of the transition metals, their cations (e.g., Fe^{2+}, Fe^{3+}), and their oxoanions (e.g., CrO_4^{2-}).

20.1   METALLURGY

An ore is a natural source from which a metal can be extracted profitably.

The processes by which metals are extracted from their **ores** fall within the science of **metallurgy.** As you might expect, the chemical reactions involved depend on the type of ore (Figure 20.2). We consider some typical processes used to obtain metals from chloride, oxide, sulfide, or "native" ores.

FIGURE 20.1

Metals and the periodic table. The periodic table groups discussed in this chapter are Groups 1 and 2, the alkali and alkaline earth metals (shaded in blue), and the transition metals (shaded in yellow). Symbols are shown for the more common metals. **OHT**

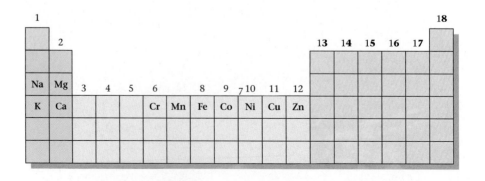

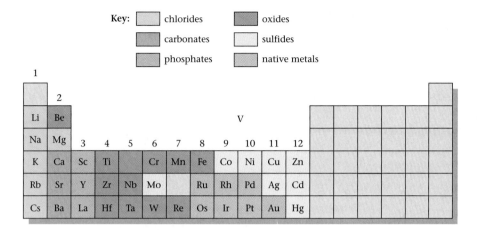

FIGURE 20.2
Principal ores of the Group 1, Group 2, and transition metals.

Chloride Ores: Na from NaCl

Sodium metal is obtained by the electrolysis of molten sodium chloride (Figure 20.3). The electrode reactions are quite simple:

$$\text{cathode:} \quad 2Na^+(l) + 2e^- \longrightarrow 2Na(l)$$
$$\underline{\text{anode:} \quad\quad 2Cl^-(l) \longrightarrow Cl_2(g) + 2e^-}$$
$$2NaCl(l) \longrightarrow 2Na(l) + Cl_2(g)$$

The cell is operated at about 600°C to keep the electrolyte molten; calcium chloride is added to lower the melting point. About 14 kJ of electrical energy is required to produce one gram of sodium, which is drawn off as a liquid (mp of Na = 98°C). The chlorine gas produced at the anode is a valuable by-product.

$Na^+(l)$ and $Cl^-(l)$ are the ions present in molten NaCl.

It's cheaper to electrolyze NaCl (*aq*), but you don't get sodium metal that way.

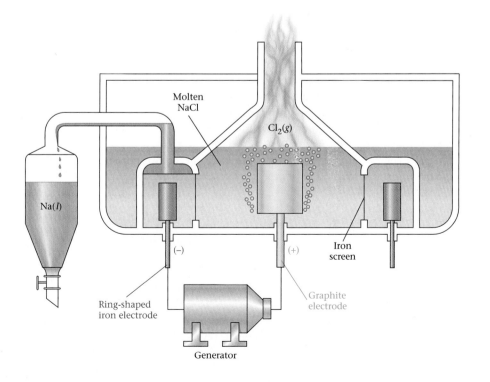

FIGURE 20.3
Electrolysis of molten sodium chloride. Calcium chloride ($CaCl_2$) is added to lower the melting point. The iron screen prevents sodium and chlorine from coming into contact with each other. **OHT**

EXAMPLE 20.1   Taking $\Delta H°$ and $\Delta S°$ for the reaction

$$2NaCl(l) \longrightarrow 2Na(l) + Cl_2(g)$$

to be +820 kJ and +0.180 kJ/K, respectively, calculate

(a) $\Delta G°$ at the electrolysis temperature, 600°C.
(b) the voltage required to carry out the electrolysis.

Strategy   To find $\Delta G°$, apply the Gibbs-Helmholtz equation, $\Delta G° = \Delta H° - T\,\Delta S°$ (Chapter 17). To find $E°$, use the relation $\Delta G° = -nFE°$ (Chapter 18).

Solution

(a) $\Delta G° = +820 \text{ kJ} - 873 \text{ K}(0.180 \text{ kJ/K}) = \boxed{+663 \text{ kJ}}$

(b) $E° = \dfrac{-\Delta G°}{nF} = \dfrac{-6.63 \times 10^5 \text{ J}}{(2 \text{ mol})(9.648 \times 10^4 \text{ J/mol·V})} = -3.44 \text{ V}$

At least $\boxed{3.44 \text{ V}}$ must be applied to carry out the electrolysis.

Reality Check   Notice that the value of $\Delta G°$ calculated in (a), +663 kJ, is for two moles of Na, 45.98 g. The value of the free energy change per gram is

$$663 \text{ kJ}/45.98 \text{ g} = 14.4 \text{ kJ/g}$$

This is consistent with the statement in the text that "about 14 kJ of electrical energy is required to produce one gram of sodium. . . . "

Oxide Ores: Al from Al_2O_3, Fe from Fe_2O_3

Oxides of very reactive metals such as calcium or aluminum are reduced by electrolysis. In the case of aluminum, bauxite ore, Al_2O_3, is used.

$$2Al_2O_3(l) \longrightarrow 4Al(l) + 3O_2(g)$$

Cryolite, Na_3AlF_6, is added to Al_2O_3 to produce a mixture melting at about 1000°C. (A mixture of AlF_3, NaF, and CaF_2 may be substituted for cryolite.) The cell is heated electrically to keep the mixture molten so that ions can move through it, carrying the electric current. About 30 kJ of electrical energy is consumed per gram of aluminum formed. The high energy requirement explains in large part the value of recycling aluminum cans.

The process for obtaining aluminum from bauxite was worked out in 1886 by Charles Hall, a chemistry student at Oberlin College. The problem that Hall faced was to find a way to electrolyze Al_2O_3 at a temperature below its melting point of 2000°C. His general approach was to look for ionic compounds in which Al_2O_3 would dissolve at a reasonable temperature. After several unsuccessful attempts, Hall found that cryolite was the ideal "solvent." Curiously enough, the same electrolytic process was worked out by Paul Heroult in France, also in 1886.

With less active metals, a chemical reducing agent can be used to reduce a metal cation to the element. The most common reducing agent in metallurgical processes is carbon, in the form of coke or, more exactly, carbon monoxide formed from the coke.

The most important metallurgical process involving carbon is the reduction of hematite ore, which consists largely of iron(III) oxide, Fe_2O_3, mixed with silicon dioxide, SiO_2. Reduction occurs in a blast furnace (Figure 20.4a) typically 30 m high and 10 m in diameter. The furnace is lined with refractory brick, capable of withstanding temperatures that may go as high as 1800°C. The solid charge, ad-

Chemistry majors can be very productive.

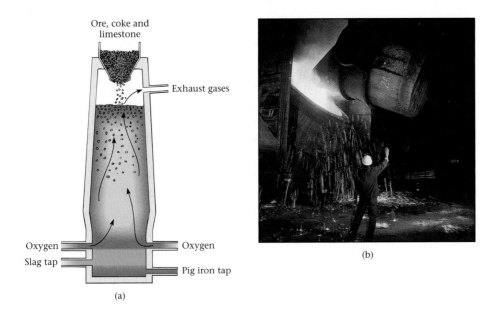

Ore, coke and limestone

Exhaust gases

Oxygen

Oxygen

Slag tap

Pig iron tap

(a)

(b)

FIGURE 20.4

Iron and steel production. (a) A blast furnace for the production of pig iron. (b) A basic oxygen furnace, where the carbon content of pig iron is lowered by heating it with oxygen to form steel. *(b, Courtesy of Bethlehem Steel Corporation)* **OHT**

mitted at the top of the furnace, consists of iron ore, coke, and limestone ($CaCO_3$). To get the process started, a blast of compressed air or pure O_2 at 500°C is blown into the furnace through nozzles near the bottom. Several different reactions occur, of which three are most important:

1. *Conversion of carbon to carbon monoxide.* In the lower part of the furnace, coke burns to form carbon dioxide, CO_2. As the CO_2 rises through the solid mixture, it reacts further with the coke to form carbon monoxide, CO. The overall reaction is

$$2C(s) + O_2(g) \longrightarrow 2CO(g) \qquad \Delta H = -221 \text{ kJ}$$

The heat given off by this reaction maintains a high temperature within the furnace.

2. *Reduction of Fe^{3+} ions to Fe.* The carbon monoxide reacts with the iron(III) oxide in the ore:

$$Fe_2O_3(s) + 3CO(g) \longrightarrow 2Fe(l) + 3CO_2(g)$$

Molten iron, formed at a temperature of 1600°C, collects at the bottom of the furnace. Four or five times a day, it is drawn off. The daily production of iron from a single blast furnace is about 1000 metric tons. This is enough to make 500 Cadillacs or 1000 Hyundais.

3. *Formation of slag.* The limestone added to the furnace decomposes at about 800°C:

$$CaCO_3(s) \longrightarrow CaO(s) + CO_2(g)$$

The calcium oxide formed reacts with impurities in the iron ore to form a glassy material called *slag*. The main reaction is with SiO_2 to form calcium silicate, $CaSiO_3$:

$$CaO(s) + SiO_2(s) \longrightarrow CaSiO_3(l)$$

The slag, which is less dense than molten iron, forms a layer on the surface of the metal. This makes it possible to draw off the slag through an opening in the furnace above that used to remove the iron. The slag is used to make cement and as a base in road construction.

The product that comes out of the blast furnace, called "pig iron," is highly impure. On the average, it contains about 4% carbon along with lesser amounts of silicon, manganese, and phosphorus. To make steel from pig iron, the carbon content must be lowered below 2%. Most of the steel produced in the world today is made by the basic oxygen process (Figure 20.4b). The "converter" is filled with a mixture of about 70% molten iron from the blast furnace, 25% scrap iron or steel, and 5% limestone. Pure oxygen under a pressure of about 10 atm is blown through the molten metal. The reaction

$$C(s) + O_2(g) \longrightarrow CO_2(g)$$

occurs rapidly; at the same time, impurities such as silicon are converted to oxides, which react with limestone to form a slag. When the carbon content drops to the desired level, the supply of oxygen is cut off. At this stage, the steel is ready to be poured. The whole process takes from 30 min to 1 h and yields about 200 metric tons of steel in a single "blow."

> Where do you suppose the phrase "pig iron" came from?

> Fe_3O_4 is really $FeO \cdot Fe_2O_3$.

EXAMPLE 20.2 Write balanced equations for the reduction of each of the following oxide ores by carbon monoxide:

(a) ZnO (b) MnO_2 (c) Fe_3O_4

Strategy In each case, the products are solid metal and $CO_2(g)$. The equations are most simply balanced by the process described in Chapter 3.

Solution

(a) $ZnO(s) + CO(g) \rightarrow Zn(s) + CO_2(g)$
(b) $MnO_2(s) + 2CO(g) \rightarrow Mn(s) + 2CO_2(g)$
(c) $Fe_3O_4(s) + 4CO(g) \rightarrow 3Fe(s) + 4CO_2(g)$

Sulfide Ores: Cu from Cu_2S

Sulfide ores, after preliminary treatment, most often undergo *roasting*, that is, heating with air or pure oxygen. With a relatively reactive transition metal such as zinc, the product is the oxide

$$2ZnS(s) + 3O_2(g) \longrightarrow 2ZnO(s) + 2SO_2(g)$$

which can then be reduced to the metal with carbon. With sulfides of less reactive metals such as copper or mercury, the free metal is formed directly on roasting. The reaction with cinnabar, the sulfide ore of mercury, is

$$HgS(s) + O_2(g) \longrightarrow Hg(g) + SO_2(g)$$

Among the several ores of copper, one of the most important is chalcocite, which contains copper(I) sulfide, Cu_2S, in highly impure form. Rocky material typically lowers the fraction of copper in the ore to 1% or less. The Cu_2S is concentrated by a process called *flotation* (Figure 20.5), which raises the fraction of copper to 20%–40%. The concentrated ore is then converted to the metal by blowing air through it at a high temperature, typically above 1000°C. (Pure O_2 is often used instead of air.) The overall reaction that occurs is a simple one:

> Because Cu can be produced by simple "roasting," it was one of the first metals known.

$$Cu_2S(s) + O_2(g) \longrightarrow 2Cu(s) + SO_2(g)$$

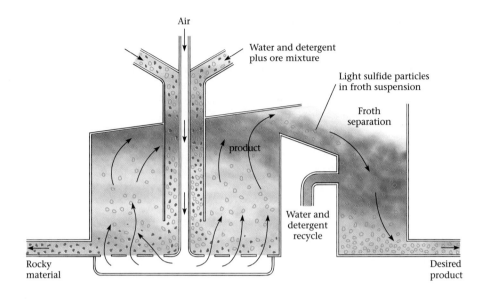

Air

Water and detergent
plus ore mixture

Light sulfide particles
in froth suspension

Froth
separation

product

Water and
detergent
recycle

Rocky
material

Desired
product

FIGURE 20.5
Flotation process for concentrating a sulfide ore. Low-grade sulfide ores, including Cu_2S, are often concentrated by flotation. The finely divided sulfide particles are trapped in soap bubbles; the rocky waste sinks to the bottom and is discarded. **OHT**

The solid produced is called "blister copper." It has an irregular appearance due to air bubbles that enter the copper while it is still molten. Blister copper is impure, containing small amounts of several other metals.

Copper is purified by electrolysis. The anode, which may weigh as much as 300 kg, is made of blister copper. The electrolyte is 0.5 to 1.0 M $CuSO_4$, adjusted to a pH of about 0 with sulfuric acid. The cathode is a piece of pure copper, weighing perhaps 150 kg. The half-reactions are

$$\text{oxidation:} \qquad Cu(s, \text{impure}) \longrightarrow Cu^{2+}(aq) + 2e^-$$

$$\text{reduction:} \qquad Cu^{2+}(aq) + 2e^- \longrightarrow Cu(s, \text{pure})$$

An open-pit copper mine. This mine in Bingham Canyon, Utah, is 4 km in diameter and 0.8 km deep. It is the largest human-created hole on Earth. *(Agricultural Stabilization and Conservation Service, USDA)*

The overall reaction, obtained by adding these two half-reactions, is

$$Cu(s, \text{impure}) \longrightarrow Cu(s, \text{pure})$$

Thus the net effect of electrolysis is to transfer copper metal from the impure blister copper used as one electrode to the pure copper sheet used as the other electrode. Electrolytic copper is 99.95% pure.

EXAMPLE 20.3   A major ore of bismuth is bismuth(III) sulfide, Bi_2S_3. By roasting in air, it is converted to the corresponding oxide; sulfur dioxide is a by-product. What volume of SO_2, at 25°C and 1.00 atm, is formed from one metric ton (10^6 g) of ore containing 1.25% Bi_2S_3?

Strategy   First (1), write a balanced equation for the reaction, which is very similar to that for ZnS, except that Zn^{2+} is replaced by Bi^{3+}. (2) Using the balanced equation, calculate the number of moles of SO_2. Finally (3), use the ideal gas law to calculate the volume of SO_2.

Solution

(1) The product is $Bi_2O_3(s)$; the equation is

$$2Bi_2S_3(s) + 9O_2(g) \longrightarrow 2Bi_2O_3(s) + 6SO_2(g)$$

(2) $n_{SO_2} = 0.0125 \times 10^6 \text{ g } Bi_2S_3 \times \dfrac{1 \text{ mol } Bi_2S_3}{514.2 \text{ g } Bi_2S_3} \times \dfrac{6 \text{ mol } SO_2}{2 \text{ mol } Bi_2S_3} = 72.9 \text{ mol } SO_2$

(3) $V_{SO_2} = \dfrac{nRT}{P} = \dfrac{(72.9 \text{ mol})(0.0821 \text{ L} \cdot \text{atm/mol} \cdot \text{K})(298 \text{ K})}{1.00 \text{ atm}} = \boxed{1.78 \times 10^3 \text{ L}}$

Reality Check   That is 1.78 cubic meters. The roasting of sulfide ores is a major source of the gaseous pollutant SO_2, which contributes to acid rain (Chapter 14).

Native Metals: Au

A few very unreactive metals, notably silver and gold, are found in nature in elemental form, mixed with large amounts of rocky material. For countless centuries, people have extracted gold by taking advantage of its high density (19.3 g/mL). In ancient times, gold-bearing sands were washed over sheepskins, which retained the gold; this is believed to be the source of the "Golden Fleece" of Greek mythology. The forty-niners in California obtained gold by swirling gold-bearing sands with water in a pan. Less dense impurities were washed away, leaving gold nuggets or flakes at the bottom of the pan.

Nowadays, the gold content of ores is much too low for these simple mechanical separation methods to be effective. Instead, the ore is treated with very dilute (0.01 M) sodium cyanide solution, through which air is blown. The following redox reaction takes place:

$$4Au(s) + 8CN^-(aq) + O_2(g) + 2H_2O \longrightarrow 4Au(CN)_2^-(aq) + 4 OH^-(aq)$$

The oxidizing agent is O_2, which takes gold to the +1 state. The cyanide ion acts as a complexing ligand, forming the stable $Au(CN)_2^-$ ion. Metallic gold is recovered from solution by adding zinc; the gold in the complex ion is reduced to the metal.

$$Zn(s) + 2Au(CN)_2^-(aq) \longrightarrow Zn(CN)_4^{2-}(aq) + 2Au(s)$$

Modern gold jewelry. The gold was probably obtained by the cyanide process. *(Tiffany & Co.)*

20.2 REACTIONS OF THE ALKALI AND ALKALINE EARTH METALS

The metals in Groups 1 and 2 are among the most reactive of all elements (Table 20.1). Their low ionization energies and high E_{ox}° values explain why they are so readily oxidized to cations, +1 cations for Group 1, +2 for Group 2. The alkali metals and the heavier alkaline earth metals (Ca, Sr, Ba) are commonly stored under dry mineral oil or kerosene to prevent them from reacting with oxygen or water vapor in the air. Magnesium is less reactive; it is commonly available in the form of ribbon or powder. Beryllium, as one would expect from its position in the periodic table, is the least metallic element in these two groups. It is also the least reactive toward water, oxygen, or other nonmetals.

⌐ **Not under water!**

EXAMPLE 20.4   Write balanced equations for the reaction of

(a) sodium with hydrogen.   (b) barium with oxygen.

Strategy   The formulas of the products can be deduced from Table 20.1. Note that barium forms two products with oxygen, BaO and BaO_2.

Solution

(a) $2Na(s) + H_2(g) \rightarrow 2NaH(s)$

(b) $2Ba(s) + O_2(g) \rightarrow 2BaO(s)$

  $Ba(s) + O_2(g) \rightarrow BaO_2(s)$

⌐ BaO_2 **is barium peroxide.**

TABLE 20.1   Reactions of the Alkali and Alkaline Earth Metals OHT

Reactant	Product	Comments
	Alkali Metals (M)	
$H_2(g)$	$MH(s)$	On heating in hydrogen gas
$X_2(g)$	$MX(s)$	X = F, Cl, Br, I
$N_2(g)$	$M_3N(s)$	Only Li reacts; product contains N^{3-} ion
$S(s)$	$M_2S(s)$	On heating
$O_2(g)$	$M_2O(s)$	Li; product is an oxide (O^{2-} ion)
	$M_2O_2(s)$	Na; product is a peroxide (O_2^{2-} ion)
	$MO_2(s)$	K, Rb, Cs; product is a superoxide (O_2^- ion)
$H_2O(l)$	$H_2(g)$, M^+, OH^-	Violent reaction with Na, K
	Alkaline Earth Metals (M)	
$H_2(g)$	$MH_2(s)$	All except Be; heating required
$X_2(g)$	$MX_2(s)$	Any halogen
$N_2(g)$	$M_3N_2(s)$	All except Be; heating required
$S(s)$	$MS(s)$	On heating
$O_2(g)$	$MO(s)$	All; product contains O^{2-} ion
	$MO_2(s)$	Ba; product contains O_2^{2-} ion
$H_2O(l)$	$H_2(g)$, M^{2+}, OH^-	Ca, Sr, Ba

Reaction of calcium hydride (CaH$_2$) with water.

(Charles D. Winters)

Reaction with Hydrogen

The compounds formed by the reaction of hydrogen with the alkali and alkaline earth metals contain H$^-$ ions; for example, sodium hydride consists of Na$^+$ and H$^-$ ions. These white crystalline solids are often referred to as saline hydrides because of their physical resemblance to NaCl. Chemically, they behave quite differently from sodium chloride; for example, they react with water to produce hydrogen gas. Typical reactions are

$$NaH(s) + H_2O(l) \longrightarrow H_2(g) + Na^+(aq) + OH^-(aq)$$

$$CaH_2(s) + 2H_2O(l) \longrightarrow 2H_2(g) + Ca^{2+}(aq) + 2OH^-(aq)$$

In this way, saline hydrides can serve as compact, portable sources of hydrogen gas for inflating life rafts and balloons.

Reaction with Water

The alkali metals react vigorously with water (Figure 20.6) to evolve hydrogen and form a water solution of the alkali hydroxide. The reaction of sodium is typical:

$$2Na(s) + 2H_2O(l) \longrightarrow 2Na^+(aq) + 2\,OH^-(aq) + H_2(g) \qquad \Delta H° = -368.6 \text{ kJ}$$

The heat evolved in the reaction frequently causes the hydrogen to ignite.

Among the Group 2 metals, Ca, Sr, and Ba react with water in much the same way as the alkali metals. The reaction with calcium is

$$Ca(s) + 2H_2O(l) \longrightarrow Ca^{2+}(aq) + 2\,OH^-(aq) + H_2(g)$$

Beryllium does not react with water at all. Magnesium reacts very slowly with boiling water but reacts more readily with steam at high temperatures:

$$Mg(s) + H_2O(g) \longrightarrow MgO(s) + H_2(g)$$

This reaction, like that of sodium with water, produces enough heat to ignite the hydrogen. Firefighters who try to put out a magnesium fire by spraying water on it have discovered this reaction, often with tragic results. The best way to extinguish burning magnesium is to dump dry sand on it.

FIGURE 20.6
Reaction of sodium with water. When a *very small* piece of sodium is added to water, it reacts violently with the water. The OH$^-$ ions formed in the reaction turn the acid-base indicator phenolphthalein from colorless to pink. *(Marna G. Clarke)*

Reaction with Oxygen

Note from Table 20.1 that several different products are possible when an alkali or alkaline earth metal reacts with oxygen. The product may be a normal oxide (O^{2-} ion), a **peroxide** (O_2^{2-} ion), or **superoxide** (O_2^{-} ion).

$$[:\ddot{O}:]^{2-} \qquad [:\ddot{O}-\ddot{O}:]^{2-} \qquad [:\ddot{O}-\ddot{O}\cdot]^{-}$$

oxide ion   peroxide ion   superoxide ion

> The superoxide ion has an unpaired electron; KO_2 is paramagnetic.

Lithium is the only Group 1 metal that forms the normal oxide in good yield by direct reaction with oxygen. The other Group 1 oxides (Na_2O, K_2O, Rb_2O, Cs_2O) must be prepared by other means. In contrast, the Group 2 metals usually react with oxygen to give the normal oxide. Beryllium and magnesium must be heated strongly to give BeO and MgO (Figure 20.7). Calcium and strontium react more readily to give CaO and SrO. Barium, the most reactive of the Group 2 metals, catches fire when exposed to moist air. The product is a mixture of the normal oxide BaO (Ba^{2+}, O^{2-} ions) and the peroxide BaO_2 (Ba^{2+}, O_2^{2-} ions). The higher the temperature, the greater the fraction of BaO in the mixture.

The oxides of these metals react with water to form hydroxides:

$$Li_2O(s) + H_2O(l) \longrightarrow 2LiOH(s)$$

$$CaO(s) + H_2O(l) \longrightarrow Ca(OH)_2(s)$$

The oxides of the Group 1 and Group 2 metals are sometimes referred to as "basic anhydrides" (bases without water) because of this reaction. The reaction with CaO is referred to as the "slaking" of lime; it gives off 65 kJ of heat per mole of $Ca(OH)_2$ formed. A similar reaction with MgO takes place slowly to form $Mg(OH)_2$, the antacid commonly referred to as "milk of magnesia."

When sodium burns in air, the principal product is yellowish sodium peroxide, Na_2O_2:

$$2Na(s) + O_2(g) \longrightarrow Na_2O_2(s)$$

Addition of sodium peroxide to water gives hydrogen peroxide, H_2O_2:

$$Na_2O_2(s) + 2H_2O(l) \longrightarrow 2Na^+(aq) + 2OH^-(aq) + H_2O_2(aq)$$

Through this reaction, sodium peroxide finds use as a bleaching agent in the pulp and paper industry.

The heavier alkali metals (K, Rb, Cs) form the superoxide when they burn in air. For example,

$$K(s) + O_2(g) \longrightarrow KO_2(s)$$

Potassium superoxide is used in self-contained breathing devices for firefighters and miners. It reacts with the moisture in exhaled air to generate oxygen:

$$4KO_2(s) + 2H_2O(g) \longrightarrow 3O_2(g) + 4KOH(s)$$

The carbon dioxide in the exhaled air is removed by reaction with the KOH formed:

$$KOH(s) + CO_2(g) \longrightarrow KHCO_3(s)$$

A person using a mask charged with KO_2 can rebreathe the same air for an extended period of time. This allows that person to enter an area where there are poisonous gases or oxygen-deficient air.

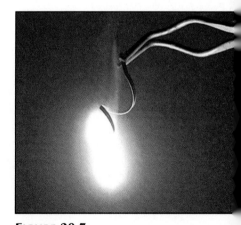

FIGURE 20.7
Burning magnesium. A piece of magnesium ribbon bursts into flame when heated in the air; the product of the reaction is MgO(s). *(Charles D. Winters)*

EXAMPLE 20.5 Consider the compounds strontium hydride, radium peroxide, and cesium superoxide.

(a) Give the formulas of these compounds.
(b) Write equations for the formation of these compounds from the elements.
(c) Write equations for the reactions of strontium hydride and radium peroxide with water.

Strategy The formulas can be deduced from the charges of the ions (Sr^{2+}, Ra^{2+}, Cs^+; H^-, O_2^{2-}, O_2^-). If you know the formulas, the equations are readily written. In part (c), note that—

- hydrides on reaction with water give $H_2(g)$ and a solution of the metal hydroxide.
- peroxides on reaction with water give $H_2O_2(aq)$ and a solution of the metal hydroxide.

Solution

(a) SrH_2, RaO_2, CsO_2
(b) $Sr(s) + H_2(g) \rightarrow SrH_2(s)$
$Ra(s) + O_2(g) \rightarrow RaO_2(s)$
$Cs(s) + O_2(g) \rightarrow CsO_2(s)$
(c) $SrH_2(s) + 2H_2O(l) \rightarrow 2H_2(g) + Sr^{2+}(aq) + 2\,OH^-(aq)$
$RaO_2(s) + 2H_2O(l) \rightarrow H_2O_2(aq) + Ra^{2+}(aq) + 2\,OH^-(aq)$

⌐ RaO_2 is a peroxide; CsO_2 is a superoxide. Explain.

20.3 REDOX CHEMISTRY OF THE TRANSITION METALS

The transition metals, unlike those in Groups 1 and 2, typically show several different oxidation numbers in their compounds. This tends to make their redox chemistry more complex (and more colorful). Only in the lower oxidation states (+1, +2, +3) are the transition metals present as cations (e.g., Ag^+, Zn^{2+}, Fe^{3+}). In higher oxidation states (+4 to +7) a transition metal is covalently bonded to a nonmetal atom, most often oxygen.

Reactions of the Transition Metals with Oxygen

Table 20.2 lists the formulas of the oxides formed when the more common transition metals react with oxygen. With one exception (Cu_2O), the transition metal is present as a +2 and/or a +3 ion. In Co_3O_4 and in other oxides of general formula M_3O_4, there are two different types of cations: +2 and +3. To be specific, there are twice as many +3 as +2 cations; we might show the composition of Co_3O_4 as

$$1Co^{2+}:2Co^{3+}:4O^{2-}$$

We should point out that many oxides of the transition metals beyond those listed in Table 20.2 can be prepared indirectly. For example, although silver does not react directly with oxygen, silver(I) oxide, Ag_2O, can be made by treating a solution of a silver salt with strong base.

$$2Ag^+(aq) + 2\,OH^-(aq) \longrightarrow Ag_2O(s) + H_2O$$

In another case, cobalt(II) oxide can be prepared by heating the carbonate in the absence of air.

$$CoCO_3(s) \longrightarrow CoO(s) + CO_2(g)$$

Many oxides of the transition metals can be *nonstoichiometric;* the atom ratios of the elements differ slightly from whole-number values. Stoichiometric nickel(II)

TABLE 20.2 Products of Reactions of the Transition Metals with Oxygen

Metal	Product
Cr	Cr_2O_3
Mn	Mn_3O_4
Fe	Fe_2O_3, Fe_3O_4
Co	Co_3O_4
Ni	NiO
Cu	Cu_2O, CuO
Zn	ZnO
Ag	—
Cd	CdO
Au	—
Hg	HgO

oxide, in which the atom ratio is exactly 1:1 (i.e., $Ni_{1.00}O_{1.00}$), is a green, noncon-ducting solid. If this compound is heated to 1200°C in pure oxygen, it turns black and undergoes a change in composition to something close to $Ni_{0.97}O_{1.00}$. This nonstoichiometric solid was one of the early semiconductors; it has the structure shown in Figure 20.8. A few Ni^{2+} ions are missing from the crystal lattice; to maintain electrical neutrality, two Ni^{2+} ions are converted to Ni^{3+} ions for each missing Ni^{2+}.

Nonstoichiometry is relatively common among "mixed" metal oxides, in which more than one metal is present. In 1986 it was discovered that certain com-pounds of this type showed the phenomenon of *superconductivity;* on cooling to about 100 K, their electrical resistance drops to zero (Figure 20.9). A typical for-mula here is $YBa_2Cu_3O_x$, where x varies from 6.5 to 7.2, depending on the method of preparation of the solid.

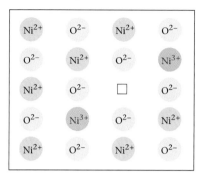

FIGURE 20.8

Nonstoichiometric nickel(II) oxide. For each missing Ni^{2+} ion, two others are converted to Ni^{3+} ions to main-tain charge balance.

Reaction of Transition Metals with Acids

Any metal with a positive standard oxidation voltage, E°_{ox}, can be oxidized by the H^+ ions present in a 1 M solution of a strong acid. All the transition metals in the left column of Table 20.3 (page 586) react spontaneously with dilute solutions of such strong acids as HCl, HBr, and H_2SO_4. The products are hydrogen gas and a cation of the transition metal. A typical reaction is that of nickel:

$$Ni(s) + 2H^+(aq) \longrightarrow Ni^{2+}(aq) + H_2(g)$$

The end result of this reaction is shown in Figure 20.10 (page 586). With metals that can form more than one cation, such as iron, the reaction product with H^+ in the absence of air is ordinarily the cation of lower charge, for example, Fe^{2+}:

$$Fe(s) + 2H^+(aq) \longrightarrow Fe^{2+}(aq) + H_2(g)$$

Metals with negative values of E°_{ox}, listed at the right of Table 20.3, are too in-active to react with hydrochloric acid. The H^+ ion is not a strong enough oxidiz-ing agent to convert a metal such as copper ($E^\circ_{ox} = -0.339$ V) to a cation.

FIGURE 20.9

A superconductor. A pellet of superconducting material, previously cooled to 77 K with liq-uid nitrogen, floats above a magnet. *(Courtesy of Edmund Scientific)*

TABLE 20.3		Ease of Oxidation of Transition Metals					
Metal		**Cation**	E°_{ox} **(V)**	**Metal**		**Cation**	E°_{ox} **(V)**
Mn	$\longrightarrow$	Mn^{2+}	+1.182	Cu	$\longrightarrow$	Cu^{2+}	−0.339
Cr	$\longrightarrow$	Cr^{2+}	+0.912	Ag	$\longrightarrow$	Ag^{+}	−0.799
Zn	$\longrightarrow$	Zn^{2+}	+0.762	Hg	$\longrightarrow$	Hg^{2+}	−0.852
Fe	$\longrightarrow$	Fe^{2+}	+0.409	Au	$\longrightarrow$	Au^{3+}	−1.498
Cd	$\longrightarrow$	Cd^{2+}	+0.402				
Co	$\longrightarrow$	Co^{2+}	+0.282				
Ni	$\longrightarrow$	Ni^{2+}	+0.236				

However, copper can be oxidized by nitric acid. The oxidizing agent is the nitrate ion, NO_3^{-}, which may be reduced to NO_2 or NO:

$$3Cu(s) + 8H^{+}(aq) + 2NO_3^{-}(aq) \longrightarrow 3Cu^{2+}(aq) + 2NO(g) + 4H_2O$$

$$E^{\circ} = E^{\circ}_{ox}\ Cu + E^{\circ}_{red}\ NO_3^{-} = -0.339V + 0.964V = +0.625V$$

EXAMPLE 20.6   Write balanced equations for the reactions, if any, at standard conditions of

(a) chromium with hydrochloric acid.   (b) silver with nitric acid.

Strategy   Use standard potentials to decide whether reaction will occur. Note that with HCl only the H^{+} ion can be reduced (to H_2); in HNO_3, the NO_3^{-} ion can be reduced to NO ($E^{\circ}_{red} = +0.964$ V).

Solution

(a) $Cr(s) \longrightarrow Cr^{2+}(aq) + 2e^{-}$ $E^{\circ}_{ox} = +0.912$ V

$2H^{+}(aq) + 2e^{-} \longrightarrow H_2(g)$ $E^{\circ}_{red} = 0.000$ V

The reaction is

$$Cr(s) + 2H^{+}(aq) \longrightarrow Cr^{2+}(aq) + H_2(g) \qquad E^{\circ} = +0.912\ V$$

On exposure to air, Cr^{2+} is oxidized to Cr^{3+}.

If chromium metal is added to hydrochloric acid in the absence of air, it slowly reacts, forming blue Cr^{2+} and bubbles of hydrogen gas.

(b) $Ag(s) \longrightarrow Ag^{+}(aq)$ $E^{\circ}_{ox} = -0.799$ V

$NO_3^{-}(aq) \longrightarrow NO(g)$ $E^{\circ}_{red} = +0.964$ V

FIGURE 20.10
Oxidation of nickel. Nickel reacts slowly with hydrochloric acid to form $H_2(g)$ and Ni^{2+} ions in solution. Evaporation of the solution formed gives green crystals of $NiCl_2 \cdot 6H_2O$.
(Charles D. Winters)

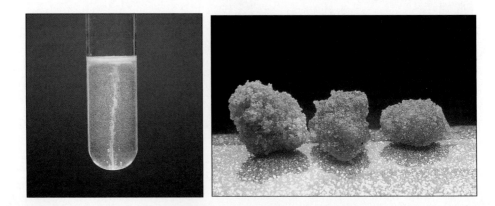

Because $E°$ is a positive quantity, $+0.165$ V, a redox reaction should occur. The balanced half-equations are

oxidation: $Ag(s) \longrightarrow Ag^+(aq) + e^-$

reduction: $NO_3^-(aq) + 4H^+(aq) + 3e^- \longrightarrow NO(g) + 2H_2O$

To obtain the final balanced equation, multiply the first half-equation by 3 and add to the second. The result is

$$3Ag(s) + NO_3^-(aq) + 4H^+(aq) \longrightarrow 3Ag^+(aq) + NO(g) + 2H_2O$$

Although gold is not oxidized by nitric acid, it can be brought into solution in *aqua regia*, a 3:1 mixture by volume of 12 M HCl and 16 M HNO$_3$:

$$Au(s) + 4H^+(aq) + 4Cl^-(aq) + NO_3^-(aq) \longrightarrow AuCl_4^-(aq) + NO(g) + 2H_2O$$

The nitrate ion of the nitric acid acts as the oxidizing agent. The function of the hydrochloric acid is to furnish Cl$^-$ ions to form the very stable complex ion $AuCl_4^-$.

> "Aqua regia" = royal water.

Equilibria Between Different Cations of a Transition Metal

Several transition metals form more than one cation. For example, chromium forms Cr^{2+} and Cr^{3+}; copper forms Cu^+ and Cu^{2+}. Table 20.4 lists values of $E°_{red}$ for several such systems. Using the data in this table and in Table 18.1 (page 520), it is possible to decide on the relative stabilities of different transition metal cations in water solution.

Cations for which $E°_{red}$ is a large, positive number are readily reduced and hence tend to be unstable in water solution. A case in point is the Mn^{3+} ion, which reacts spontaneously with water:

$$2Mn^{3+}(aq) + H_2O(l) \longrightarrow 2Mn^{2+}(aq) + \tfrac{1}{2}O_2(g) + 2H^+(aq)$$

$$E° = E°_{red}\ Mn^{3+} + E°_{ox}\ H_2O = +1.559\ V - 1.229\ V = +0.330\ V$$

As a result of this reaction, Mn^{3+} cations are never found in water solution. Manganese(III) occurs only in insoluble oxides and hydroxides such as Mn_2O_3 and $MnO(OH)$.

> Co^{3+} is also rare, except in complexes.

The cations in the center column of Table 20.4 (Cr^{2+}, Mn^{2+}, . . .) are in an intermediate oxidation state. They can either be oxidized to a cation of higher charge ($Cr^{2+} \rightarrow Cr^{3+}$) or reduced to the metal ($Cr^{2+} \rightarrow Cr$). With certain cations

TABLE 20.4   Ease of Reduction $E°_{red}$ (V) of Transition Metal Cations

Chromium	Cr^{3+}	$\xrightarrow{-0.408\ V}$	Cr^{2+}	$\xrightarrow{-0.912\ V}$	Cr
Manganese	Mn^{3+}	$\xrightarrow{+1.559\ V}$	Mn^{2+}	$\xrightarrow{-1.182\ V}$	Mn
Iron	Fe^{3+}	$\xrightarrow{+0.769\ V}$	Fe^{2+}	$\xrightarrow{-0.409\ V}$	Fe
Cobalt	Co^{3+}	$\xrightarrow{+1.953\ V}$	Co^{2+}	$\xrightarrow{-0.282\ V}$	Co
Copper	Cu^{2+}	$\xrightarrow{+0.161\ V}$	Cu^+	$\xrightarrow{+0.518\ V}$	Cu
Gold	Au^{3+}	$\xrightarrow{+1.400\ V}$	Au^+	$\xrightarrow{+1.695\ V}$	Au
Mercury	Hg^{2+}	$\xrightarrow{+0.908\ V}$	Hg_2^{2+}	$\xrightarrow{+0.796\ V}$	Hg

of this type, these two half-reactions occur simultaneously. Consider, for example, the Cu^+ ion. In water solution, copper(I) salts **disproportionate,** undergoing reduction (to copper metal) and oxidation (to Cu^{2+}) at the same time:

$$2Cu^+(aq) \longrightarrow Cu(s) + Cu^{2+}(aq); \ E° = +0.518 \ V - 0.161 \ V = +0.357 \ V$$

As a result, the only stable copper(I) species are insoluble compounds such as CuCN or complex ions such as $Cu(CN)_2^-$.

The Cu^+ ion is one of the few species in the center column of Table 20.4 that disproportionates in water, undergoing simultaneous reduction to the metal and oxidation to a cation of higher charge. However, cations of this type may be unstable for a quite different reason. Water ordinarily contains dissolved air; the O_2 in air may oxidize the cation. When a blue solution of chromium(II) salt is exposed to air, the color quickly changes to violet or green as the Cr^{3+} ion is formed by the reaction

$$2Cr^{2+}(aq) \longrightarrow 2Cr^{3+}(aq) + 2e^- \qquad E°_{ox} = +0.408 \ V$$

$$\frac{1}{2}O_2(g) + 2H^+(aq) + 2e^- \longrightarrow H_2O \qquad E°_{red} = +1.229 \ V$$

$$\overline{2Cr^{2+}(aq) + \tfrac{1}{2}O_2(g) + 2H^+(aq) \longrightarrow 2Cr^{3+}(aq) + H_2O \qquad E° = +1.637 \ V}$$

As a result of this reaction, chromium(II) salts are difficult to prepare and even more difficult to store.

The Fe^{2+} ion ($E°_{ox} = -0.769 \ V$) is much more stable toward oxidation than Cr^{2+}. However, iron(II) salts in water solution are slowly converted to iron(III) by dissolved oxygen. In acidic solution, the reaction is

$$2Fe^{2+}(aq) + \tfrac{1}{2}O_2(g) + 2H^+(aq) \longrightarrow 2Fe^{3+}(aq) + H_2O$$

$$E° = E°_{ox} \ Fe^{2+} + E°_{red} \ O_2 = -0.769 \ V + 1.229 \ V = +0.460 \ V$$

A similar reaction takes place in basic solution. Iron(II) hydroxide is pure white when first precipitated, but in the presence of air it turns first green and then brown as it is oxidized by O_2:

$$2Fe(OH)_2(s) + \tfrac{1}{2}O_2(g) + H_2O(l) \longrightarrow 2Fe(OH)_3(s)$$

EXAMPLE 20.7 Using Table 20.4, find

(a) three different cations, in addition to Mn^{3+}, that react with H_2O to form $O_2(g)$ ($E°_{ox} \ H_2O = -1.229 \ V$).
(b) another cation, in addition to Cu^+, that disproportionates in water.
(c) two other cations, in addition to Cr^{2+} and Fe^{2+}, that are oxidized by $O_2(g)$ dissolved in water ($E°_{red} \ O_2 = +1.229 \ V$)

Strategy In each case, look for a reaction in which $E°$ is a positive quantity. In (a), $E°_{red}$ for the cation must exceed $+1.229 \ V$. In (b), $E°_{red} + E°_{ox}$ for the cation must be a positive quantity. In (c), the cation must have an $E°_{ox}$ value no smaller than $-1.229 \ V$.

Solution

(a) Co^{3+} $E°_{red} = +1.953 \ V$), Au^{3+} ($E°_{red} = +1.400 \ V$) and Au^+ ($E°_{red} = +1.695 \ V$)
(b) Au^+ ($E°_{red} + E°_{ox} = +1.695 \ V - 1.400 \ V = 0.295 \ V$)
(c) Cu^+ ($E°_{ox} = -0.161 \ V$) and Hg_2^{2+} ($E°_{ox} = -0.908 \ V$)

Reality Check All of these cations, at least in principle, are unstable in water solutions exposed to air.

Oxoanions of the Transition Metals (CrO_4^{2-}, $Cr_2O_7^{2-}$, MnO_4^-)

Chromium in the +6 state forms two different oxoanions, the yellow chromate ion, CrO_4^{2-}, and the red dichromate ion, $Cr_2O_7^{2-}$ (Figure 20.11). The chromate ion is stable in basic or neutral solution; in acid, it is converted to the dichromate ion:

$$2CrO_4^{2-}(aq) + 2H^+(aq) \rightleftharpoons \underset{\text{red}}{Cr_2O_7^{2-}(aq)} + H_2O(l) \qquad K = 3 \times 10^{14}$$
$$\underset{\text{yellow}}{}$$

The dichromate ion in acidic solution is a powerful oxidizing agent,

$$Cr_2O_7^{2-}(aq) + 14H^+(aq) + 6e^- \longrightarrow 2Cr^{3+}(aq) + 7H_2O(l) \qquad E^\circ_{red} = +1.33 \text{ V}$$

As you might expect from the half-equation for its reduction, the oxidizing strength of the dichromate ion decreases as the concentration of H^+ decreases (increasing pH).

The $Cr_2O_7^{2-}$ ion can act as an oxidizing agent in the solid state as well as in water solution. In particular, it can oxidize the NH_4^+ ion to molecular nitrogen. When a pile of ammonium dichromate is ignited, a spectacular reaction occurs (Figure 20.12).

$$\underset{\text{red}}{(NH_4)_2Cr_2O_7(s)} \longrightarrow N_2(g) + 4H_2O(g) + \underset{\text{green}}{Cr_2O_3(s)}$$

The ammonium dichromate resembles a tiny volcano as it burns, emitting hot gases, sparks, and a voluminous green dust of chromium(III) oxide.

Chromates and dichromates are gradually disappearing from chemistry teaching laboratories because of concern about their toxicity. Long-term exposure of industrial workers to dust containing chromates has, in a few cases, been implicated in lung cancer. More commonly, repeated contact with chromate salts leads to skin disorders; a few people are extremely allergic to CrO_4^{2-} and $Cr_2O_7^{2-}$ ions, breaking into a rash on first exposure.

The permanganate ion, MnO_4^-, has an intense purple color easily visible even in very dilute solution.* Crystals of solid potassium permanganate, $KMnO_4$, have a deep purple, almost black color. This compound is used to treat such diverse

FIGURE 20.11

Oxidation states of chromium. The first two containers (green and violet) contain +3 chromium. The $Cr(H_2O)_4Cl_2^+$ ion is green; $Cr(H_2O)_6^{3+}$ is violet. The two bottles at the right contain +6 chromium; the CrO_4^{2-} ion is yellow, and the CrO_7^{2-} ion is red. *(Charles D. Winters)*

⌐ This makes a great lecture demonstration.

FIGURE 20.12

Ammonium dichromate volcano. The compound ammonium dichromate, $(NH_4)_2Cr_2O_7$, has a reddish-orange color owing to the presence of the CrO_7^{2-} ion. When ignited, it decomposes to give finely divided Cr_2O_3 (which is green), nitrogen gas, and water vapor. *(Charles D. Winters)*

*The purple color of old bottles exposed to the sun for a long time is due to MnO_4^- ions. These are formed when ultraviolet light oxidizes manganese compounds in the glass.

ailments as "athlete's foot" and rattlesnake bites. These applications depend on the fact that the MnO_4^- ion is a very powerful oxidizing agent. This is especially true in acidic solution, where MnO_4^- is reduced to Mn^{2+}:

$$MnO_4^-(aq) + 8H^+(aq) + 5e^- \longrightarrow Mn^{2+}(aq) + 4H_2O \qquad E^\circ_{red} = +1.512 \text{ V}$$

In basic solution, MnO_4^- is reduced to MnO_2, with a considerably smaller value of E°_{red}:

$$MnO_4^-(aq) + 2H_2O + 3e^- \longrightarrow MnO_2(s) + 4\,OH^-(aq) \qquad E^\circ_{red} = +0.596 \text{ V}$$

However, even in basic solution, MnO_4^- can oxidize water:

$$4MnO_4^-(aq) + 2H_2O \longrightarrow 4MnO_2(s) + 3O_2(g) + 4\,OH^-(aq)$$

$$E^\circ = E^\circ_{red}\, MnO_4^- + E^\circ_{ox}\, H_2O = +0.596 \text{ V} - 0.401 \text{ V} = +0.195 \text{ V}$$

This reaction accounts for the fact that laboratory solutions of $KMnO_4$ slowly decompose, producing a brownish solid (MnO_2) and gas bubbles (O_2).

CHEMISTRY Beyond the Classroom

Essential Metals in Nutrition

Four of the main-group cations are essential in human nutrition (Table A). Of these, the most important is Ca^{2+}. About 90% of the calcium in the body is found in bones and teeth, largely in the form of hydroxyapatite, $Ca(OH)_2 \cdot 3Ca_3(PO_4)_2$. Calcium ions in bones and teeth exchange readily with those in the blood; about 0.6 g of Ca^{2+} enters and leaves your bones every day. In a normal adult this exchange is in balance, but in elderly people, particularly women, there is sometimes a net loss of bone calcium, leading to the disease known as osteoporosis.

Although sodium is an essential metal, it can cause hypertension. About 50% of people with high blood pressure (10% of the general population) are sensitive to high concentrations of Na^+ ions. For that reason, it is recommended that the daily consumption of Na^+ be limited to 2.4 g. The simplest way to cut back on Na^+ is to read the label carefully before buying processed foods, which account for about 75% of our sodium intake. For example, a single serving of packaged scalloped potatoes can contain as much as 0.5 g of Na^+ ion as opposed to only 0.006 g for a large baked potato.

Of the ten trace elements known to be essential to human nutrition, seven are transition metals. For the most part, transition metals in biochemical compounds are present as complex ions, chelated by organic ligands. You will recall (Chapter 15) that hemoglobin has such a structure with Fe^{2+} as the central ion of the complex. The Co^{3+} ion occupies a similar position in vitamin B_{12}, octahedrally coordinated to organic molecules somewhat similar to those in hemoglobin.

As you can judge from Table A, transition metal cations are frequently found in enzymes. The Zn^{2+} ion alone is known to be a component of at least 70 different enzymes. One of these, referred to as "alcohol dehydrogenase," is concentrated in the liver, where it acts to break down alcohols. Another zinc-containing enzyme is involved in the normal functioning of oil glands in the skin, which accounts for the use of Zn^{2+} compounds in the treatment of acne.

Although Zn^{2+} is essential to human nutrition, compounds of the two elements below zinc in the periodic table, Cd and Hg, are extremely toxic. This reflects the fact that Cd^+ and Hg^{2+}, in contrast to Zn^{2+}, form very stable complexes with ligands containing sulfur atoms. As a result, these two cations react with and thereby deactivate enzymes containing —SH groups.

Foods high in iron.
(Charles D. Winters)

TABLE A   Essential Metal Ions

Major Species (Main-Group Cations)

Ion	Amount in Body	Daily Requirement	Function in Body	Rich Sources*
Na^+	63 g	~2 g	Principal cation *outside* cell fluid	Table salt, many processed and preserved foods
K^+	150 g	~2 g	Principal cation *within* cell fluid	Fruits, nuts, fish, instant coffee, wheat bran
Mg^{2+}	21 g	0.35 g	Activates enzymes for body processes	Chocolate, nuts, instant coffee, wheat bran
Ca^{2+}	1160 g	0.8 g	Bone and tooth formation	Milk, cheese, broccoli, canned salmon (with bones)

Trace Species (Transition Metal Cations)

Ion	Amount in Body	Daily Requirement	Function in Body	Rich Sources*
Fe^{2+}, Fe^{3+}	5000 mg	15 mg	Component of hemoglobin, myoglobin	Liver, meat, clams, spinach
Zn^{2+}	3000 mg	15 mg	Component of many enzymes, hormones	Oysters, crab, meat, nuts
Cu^{2+}, Cu^+	100 mg	2 mg	Iron metabolism, component of enzymes	Lobster, cherries, chocolate
Mn^{2+}	15 mg	3 mg	Metabolism of carbohydrates, lipids	Beet greens, nuts, blueberries
Mo(IV,V,VI)	10 mg	0.3 mg	Fe, N metabolism; component of enzymes	Legumes, ice cream
Cr^{3+}	5 mg	0.2 mg	Glucose metabolism; affects action of insulin	Corn, clams, nuts
Co^{2+}, Co^{3+}	1 mg	0.005 mg	Component of vitamin B_{12}	Liver, shellfish

*Other, more mundane, sources can be found in any nutrition textbook.

CHAPTER HIGHLIGHTS

Key Concepts

1. Calculate $\Delta G°$ from thermodynamic data.
   (Example 20.1; Problems 5, 6, 35, 44)
2. Write balanced equations to represent—
   ■ metallurgical processes.
   (Examples 20.2, 20.3; Problems 1–4, 7, 8, 38)
   ■ reactions of Group 1 and Group 2 metals.
   (Examples 20.4, 20.5; Problems 15, 16)
   ■ redox reactions of transition metals.
   (Example 20.6; Problems 19–24)
3. Determine $E°$ and reaction spontaneity from standard potentials.
   (Example 20.7; Problems 25–30, 39)

Key Terms

- alkali metal
- alkaline earth metal
- disproportionation
- $E°_{ox}$, $E°_{red}$
- electrolysis

- main-group metal
- metallurgy
- ore
- oxidation
- oxoanion

- peroxide
- reduction
- superoxide
- transition metal

Summary Problem

Consider the alkaline earth metal strontium and the transition metal manganese.

(a) Write a balanced equation for the reaction of strontium with oxygen; with water.

(b) Give the formulas of strontium hydride, strontium nitride, and strontium sulfide.

(c) Electrolysis of strontium chloride gives strontium metal and chlorine gas. What mass of the chloride must be electrolyzed to form one kilogram of strontium? One liter of $Cl_2(g)$ at STP?

(d) Give the formulas of three manganese compounds containing manganese in different oxidation states.

(e) Write a balanced equation for the reaction of manganese metal with hydrochloric acid; Mn^{3+} with water; Mn with O_2.

(f) Write a balanced equation for the reduction of the principal ore of manganese, pyrolusite, MnO_2, with CO.

(g) For the reaction

$$MnO(s) + H_2(g) \rightarrow Mn(s) + H_2O(g), \quad \Delta H° = +143 \text{ kJ}, \quad \Delta S° = +0.0297 \text{ kJ/K}$$

would it be feasible to reduce MnO to the metal by heating with hydrogen?

(h) Given that $E°_{red}$ $Mn^{3+} \rightarrow Mn^{2+} = +1.559$ V, $E°_{red}$ $Mn^{2+} \rightarrow Mn = -1.182$ V, and $E°_{red}$ $O_2(g) \rightarrow H_2O = +1.229$ V, show by calculation whether Mn^{2+} in water solution will disproportionate; whether it will be oxidized to Mn^{3+} by dissolved oxygen; whether it will be reduced by hydrogen gas to the metal.

(i) Given

$$MnO_4^-(aq) + 2H_2O + 3e^- \longrightarrow MnO_2(s) + 4\,OH^-(aq) \qquad E°_{red} = +0.59 \text{ V}$$

write a balanced equation for the reaction of MnO_4^- in basic solution with Fe^{2+} to form $Fe(OH)_3$. Calculate E_{red} for the MnO_4^- ion at pH 9.0, taking $[MnO_4^-] = 1$ M.

Answers

(a) $2Sr(s) + O_2(g) \rightarrow 2SrO(s)$
$Sr(s) + 2H_2O \rightarrow Sr^{2+}(aq) + 2\,OH^-(aq) + H_2(g)$

(b) SrH_2; Sr_3N_2; SrS

(c) 1.809 kg; 7.07 g

(d) $MnCl_2$, Mn_2O_3; $KMnO_4$

(e) $Mn(s) + 2H^+(aq) \rightarrow Mn^{2+}(aq) + H_2(g)$
$2Mn^{3+}(aq) + H_2O \rightarrow 2Mn^{2+}(aq) + \frac{1}{2}O_2(g) + 2H^+(aq)$
$3Mn(s) + 2O_2(g) \rightarrow Mn_3O_4(s)$

(f) $MnO_2(s) + 2CO(g) \rightarrow Mn(s) + 2CO_2(g)$

(g) no; $T_{calc} \approx 4800$ K

(h) no; in all cases, $E°$ is negative

(i) $MnO_4^-(aq) + 3Fe^{2+}(aq) + 5\,OH^-(aq) + 2H_2O \rightarrow$
$MnO_2(s) + 3Fe(OH)_3(s) \qquad\qquad E_{red} = +0.98$ V

Problem numbers in blue indicate that the answer is available in Appendix 6 at the back of the book.
WEB indicates that the solution is posted at **http://www.hartcourtcollege.com/chem/general/masterton4/student/**

Metallurgy

1. Write a balanced equation to represent the electrolysis of molten sodium chloride. What volume of Cl_2 at STP is formed at the anode when 1.00 g of sodium is formed at the cathode?

2. Write a balanced equation to represent the electrolysis of aluminum oxide. If 2.00 L of O_2 at 25°C and 751 mm Hg is formed at the anode, what mass of Al is formed at the cathode?

3. Write a balanced equation to represent
  (a) the roasting of nickel(II) sulfide to form nickel(II) oxide.
  (b) the reduction of nickel(II) oxide by carbon monoxide.

4. Write a balanced equation to represent the roasting of copper(I) sulfide to form "blister copper."

5. Show by calculation whether the reaction in Problem 3(b) is spontaneous at 25°C and 1 atm.

6. Calculate $\Delta G°$ at 200°C for the reaction in Problem 4.

7. Write a balanced equation for the reaction that occurs when
  (a) finely divided gold is treated with $CN^-(aq)$ in the presence of $O_2(g)$.
  (b) finely divided zinc metal is added to the solution formed in (a).

WEB 8. Write a balanced equation for the reaction that occurs when
  (a) iron(III) oxide is reduced with carbon monoxide.
  (b) the excess carbon in pig iron is removed by the basic oxygen process.

9. How many cubic feet of air (assume 21% by volume of oxygen in air) at 25°C and 1.00 atm are required to react with coke to form the CO needed to convert one metric ton of hematite ore (92% Fe_2O_3) to iron?

10. Zinc is produced by electrolytic refining. The electrolytic process, which is similar to that for copper, can be represented by the two half-reactions

$$Zn(impure, s) \longrightarrow Zn^{2+} + 2e^-$$

$$Zn^{2+} + 2e^- \longrightarrow Zn(pure, s)$$

For this process, a voltage of 3.0 V is used. How many kilowatt hours are needed to produce one metric ton of pure zinc?

11. When 2.876 g of a certain metal sulfide is roasted in air, 2.368 g of the metal oxide is formed. If the metal has an oxidation number of +2, what is its molar mass?

12. Chalcopyrite, $CuFeS_2$, is an important source of copper. A typical chalcopyrite ore contains about 0.75% Cu. What volume of sulfur dioxide at 25°C and 1.00 atm pressure is produced when one boxcar load (4.00×10^3 ft³) of chalcopyrite ore ($d = 2.6$ g/cm³) is roasted? Assume all the sulfur in the ore is converted to SO_2 and no other source of sulfur is present.

Reactions of Alkali Metals and Alkaline Earth Metals

13. Give the formula and name of the compound formed by strontium with
  (a) nitrogen   (b) bromine
  (c) water   (d) oxygen

14. Give the formula and name of the compound formed by potassium with
  (a) nitrogen   (b) iodine   (c) water
  (d) hydrogen   (e) sulfur

15. Write a balanced equation and give the names of the products for the reaction of
  (a) magnesium with chlorine.
  (b) barium peroxide with water.
  (c) lithium with sulfur.
  (d) sodium with water.

WEB 16. Write a balanced equation and give the names of the products for the reaction of
  (a) sodium peroxide and water.
  (b) calcium and oxygen.
  (c) rubidium and oxygen.
  (d) strontium hydride and water.

17. To inflate a life raft with hydrogen to a volume of 25.0 L at 25°C and 1.10 atm, what mass of calcium hydride must react with water?

18. What mass of KO_2 is required to remove 90.0% of the CO_2 from a sample of 1.00 L of exhaled air (37°C, 1.00 atm) containing 5.00 mole percent CO_2?

Redox Chemistry of Transition Metals

19. Write a balanced equation to show
  (a) the reaction of chromate ion with strong acid.
  (b) the oxidation of water to oxygen gas by permanganate ion in basic solution.
  (c) the reduction half-reaction of chromate ion to chromium(III) hydroxide in basic solution.

20. Write a balanced equation to show
  (a) the formation of gas bubbles when cobalt reacts with hydrochloric acid.
  (b) the reaction of copper with nitric acid.
  (c) the reduction half-reaction of dichromate ion to Cr^{3+} in acid solution.

21. Write a balanced redox equation for the reaction of mercury with aqua regia, assuming the products include $HgCl_4^{2-}$ and $NO_2(g)$.

22. Write a balanced redox equation for the reaction of cadmium with aqua regia, assuming the products include $CdCl_4^{2-}$ and $NO(g)$.

23. Balance the following redox equations.

(a) $Cu(s) + NO_3^-(aq) \rightarrow Cu^{2+}(aq) + NO_2(g)$ (acidic)

(b) $Cr(OH)_3(s) + ClO^-(aq) \rightarrow$
$$CrO_4^{2-}(aq) + Cl^-(aq) \text{ (basic)}$$

24. Balance the following redox equations.

(a) $Fe(s) + NO_3^-(aq) \rightarrow Fe^{3+}(aq) + NO_2(g)$ (acidic)

(b) $Cr(OH)_3(s) + O_2(g) \rightarrow CrO_4^{2-}(aq)$ (basic)

25. Show by calculation which of the following metals will react with hydrochloric acid (standard concentrations).

    (a) Cd   (b) Cr   (c) Co   (d) Ag   (e) Au

26. Show by calculation which of the metals in Problem 25 will react with nitric acid to form NO (standard concentrations).

27. Of the cations listed in Table 20.4, show by calculation which one (besides Cu^+) will disproportionate at standard conditions.

WEB **28.** Using Table 18.1 (Chapter 18) calculate $E°$ for

(a) $2Co^{3+}(aq) + H_2O \rightarrow 2Co^{2+}(aq) + \frac{1}{2}O_2(g) + 2H^+(aq)$

(b) $2Cr^{2+}(s) + I_2(s) \rightarrow 2Cr^{3+}(aq) + 2I^-(aq)$

29. Using Table 20.4, calculate, for the disproportionation of Fe^{2+}:

(a) the equilibrium constant, K.

(b) the concentration of Fe^{3+} in equilibrium with $0.10\ M\ Fe^{2+}$:

30. Using Table 20.4, calculate, for the disproportionation of Au^+:

(a) K.

(b) the concentration of Au^+ in equilibrium with $0.10\ M\ Au^{3+}$.

Unclassified

31. A sample of sodium liberates 2.73 L hydrogen at 752 mm Hg and 22°C when it is added to a large amount of water. How much sodium is used?

32. A self-contained breathing apparatus contains 248 g of potassium superoxide. A firefighter exhales 116 L of air at 37°C and 748 mm Hg. The volume percent of water in exhaled air is 6.2. What mass of potassium superoxide is left after the water in the exhaled air reacts with it?

33. Taking $K_{sp}\ PbCl_2 = 1.7 \times 10^{-5}$ and assuming $[Cl^-] = 0.20\ M$, calculate the concentration of Pb^{2+} at equilibrium.

34. The equilibrium constant for the reaction

$$2CrO_4^{2-}(aq) + 2H^+(aq) \rightleftharpoons Cr_2O_7^{2-}(aq) + H_2O$$

is 3×10^{14}. What must the pH be so that the concentrations of chromate and dichromate ion are both $0.10\ M$?

35. Using data in Appendix 1, estimate the temperature at which Fe_2O_3 can be reduced to iron, using hydrogen gas as a reducing agent (assume $H_2O(g)$ is the other product).

36. A 0.500-g sample of zinc-copper alloy was treated with dilute hydrochloric acid. The hydrogen gas evolved was collected by water displacement at 27°C and a total pressure of 755 mm Hg. The volume of the water displaced by the gas is 105.7 mL. What is the percent composition, by mass, of the alloy? (Vapor

pressure of H_2O at 27°C is 26.74 mm Hg). Assume only the zinc reacts.

37. One type of stainless steel contains 22% nickel by mass. How much nickel sulfide ore, NiS, is required to produce one metric ton of stainless steel?

38. Silver is obtained in much the same manner as gold, using NaCN solution and O_2. Describe with appropriate equations the extraction of silver from argentite ore, Ag_2S. (The products are SO_2 and $Ag(CN)_2^-$, which is reduced with zinc.)

39. Iron(II) can be oxidized to iron(III) by permanganate ion in acidic solution. The permanganate ion is reduced to manganese(II) ion.

(a) Write the oxidation half-reaction, the reduction half-reaction, and the overall redox equation.

(b) Calculate $E°$ for the reaction.

(c) Calculate the percent Fe in an ore if a 0.3500-g sample is dissolved and the Fe^{2+} formed requires for titration 55.63 mL of a 0.0200 M solution of $KMnO_4$.

40. Of the cations listed in the center column of Table 20.4, which one is the

(a) strongest reducing agent?

(b) strongest oxidizing agent?

(c) weakest reducing agent?

(d) weakest oxidizing agent?

Challenge Problems

41. A sample of 20.00 g of barium reacts with oxygen to form 22.38 g of a mixture of barium oxide and barium peroxide. Determine the composition of the mixture.

42. Rust, which you can take to be $Fe(OH)_3$, can be dissolved by treating it with oxalic acid. An acid-base reaction occurs, and a complex ion is formed.

(a) Write a balanced equation for the reaction.

(b) What volume of 0.10 M $H_2C_2O_4$ would be required to remove a rust stain weighing 1.0 g?

43. A 0.500-g sample of steel is analyzed for manganese. The sample is dissolved in acid and the manganese is oxidized to permanganate ion. A measured excess of Fe^{2+} is added to reduce MnO_4^- to Mn^{2+}. The excess Fe^{2+} is determined by titration with $K_2Cr_2O_7$. If 75.00 mL of 0.125 M $FeSO_4$ is added and the excess requires 13.50 mL of 0.100 M $K_2Cr_2O_7$ to oxidize Fe^{2+}, calculate the percent by mass of Mn in the sample.

44. Calculate the temperature in °C at which the equilibrium constant (K) for the following reaction is 1.00.

$$MnO_2(s) \longrightarrow Mn(s) + O_2(g)$$

45. A solution of potassium dichromate is made basic with sodium hydroxide; the color changes from red to yellow. Addition of silver nitrate to the yellow solution gives a precipitate. This precipitate dissolves in concentrated ammonia but reforms when nitric acid is added. Write balanced net ionic equations for all the reactions in this sequence.

For what can so fire us,
Enrapture, inspire us,
As Oxygen? What so delicious to quaff?
It is so stimulating,
And so titillating,
E'en grey-beards turn freshy, dance, caper, and laugh.

—John Shield
Oxygen Gas

CHEMISTRY OF THE NONMETALS

21

The electrical charge that creates this spectacular display also changes some of the oxygen in the air to ozone. *(Paul & Linda Marie Ambrose/FPG International)*

CHAPTER OUTLINE

21.1 THE ELEMENTS AND THEIR PREPARATION

21.2 HYDROGEN COMPOUNDS OF NONMETALS

21.3 OXYGEN COMPOUNDS OF NONMETALS

21.4 OXOACIDS AND OXOANIONS

Approximately 18 elements are classified as nonmetals; they lie above and to the right of the "stairway" that runs diagonally across the periodic table (Figure 21.1). As the word "nonmetal" implies, these elements do not show metallic properties; in the solid state they are brittle as opposed to ductile, insulators rather than conductors. Most of the nonmetals, particularly those in Groups 15 to 17 of the periodic table, are molecular in nature (e.g., N_2, O_2, F_2). The noble gases (Group 18) consist of individual atoms attracted to each other by weak dispersion forces. Carbon in Group 14 has a network covalent structure.

As indicated in Figure 21.1, this chapter concentrates on the more common and/or more reactive nonmetals, namely—

- nitrogen and phosphorus in Group 15.

- oxygen and sulfur in Group 16.

- the halogens (F, Cl, Br, I) in Group 17.

We will consider

- the properties of these elements and methods of preparing them (Section 21.1).

- their hydrogen compounds (Section 21.2).

- their oxides (Section 21.3).

- their oxoacids and oxoanions (Section 21.4).

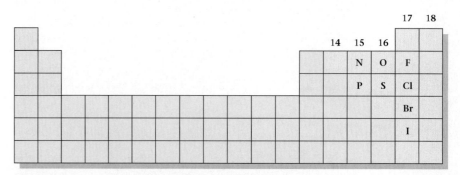

FIGURE 21.1

Nonmetals and the periodic table. The location of all nonmetals in the periodic table is shown in yellow. Symbols are given for the nonmetallic elements discussed in this chapter. **OHT**

TABLE 21.1 **Properties of Nonmetallic Elements** OHT								
	Nitrogen	Phosphorus	Oxygen	Sulfur	Fluorine	Chlorine	Bromine	Iodine
Outer electron configuration	$2s^22p^3$	$3s^23p^3$	$2s^22p^4$	$3s^23p^4$	$2s^22p^5$	$3s^23p^5$	$4s^24p^5$	$5s^25p^5$
Molecular formula	N_2	P_4	O_2	S_8	F_2	Cl_2	Br_2	I_2
Molar mass (g/mol)	28	124	32	257	38	71	160	254
State (25°C, 1 atm)	gas	solid	gas	solid	gas	gas	liquid	solid
Melting point (°C)	-210	44	-218	119	-220	-101	-7	114
Boiling point (°C)	-196	280	-183	444	-188	-34	59	184
Bond enthalpy* (kJ/mol)	941	200	498	226	153	243	193	151
E°_{red}	—	—	—	—	$+2.889$ V	$+1.360$ V	$+1.077$ V	$+0.534$ V

*In the element (triple bond in N_2, double bond in O_2).

21.1   THE ELEMENTS AND THEIR PREPARATION

Table 21.1 lists some of the properties of the eight nonmetals considered in this chapter. Notice that all of these elements are molecular; those of low molar mass (N_2, O_2, F_2, Cl_2) are gases at room temperature and atmospheric pressure (Figure 21.2). Stronger dispersion forces cause the nonmetals of higher molar mass to be either liquids (Br_2) or solids (I_2, P_4, S_8).

Chemical Reactivity

Of the eight nonmetals listed in Table 21.1, **nitrogen** is by far the least reactive. Its inertness is due to the strength of the triple bond holding the N_2 molecule together (B.E. $N\equiv N = 941$ kJ/mol). This same factor explains why virtually all chemical explosives are compounds of nitrogen (e.g., nitroglycerin, trinitrotoluene, ammonium nitrate, lead azide). These compounds detonate exothermically to form molecular nitrogen. The reactions with ammonium nitrate and lead azide are

$$2NH_4NO_3(s) \longrightarrow 4H_2O(l) + O_2(g) + 2N_2(g) \qquad \Delta H^\circ = -412 \text{ kJ}$$

$$Pb(N_3)_2(s) \longrightarrow Pb(s) + 3N_2(g) \qquad \Delta H^\circ = -476 \text{ kJ}$$

FIGURE 21.2
Chlorine (Cl_2), bromine (Br_2), and iodine (I_2). The bulbs contain *(left to right)* gaseous chlorine and the vapors in equilibrium with liquid bromine and solid iodine. *(Marna G. Clarke)*

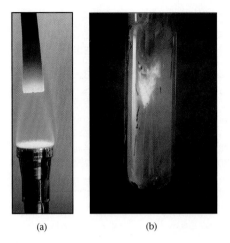

(a)   (b)

FIGURE 21.3
Reaction of chlorine with copper. When a heated piece of copper (a) is plunged into a cylinder containing chlorine gas, the copper reacts vigorously, giving off sparks (b). The equation for the reaction is

$$Cu(s) + Cl_2(g) \longrightarrow CuCl_2(s)$$

(Charles D. Winters)

Fluorine is the most electronegative element.

Fluorine is the most reactive of all elements, in part because of the weakness of the F—F bond (B.E. F—F = 153 kJ/mol), but mostly because it is such a powerful oxidizing agent (E°_{red} = +2.889 V). Fluorine combines with every element in the periodic table except He, Ne, and Ar. With a few metals, it forms a surface film of metal fluoride, which adheres tightly enough to prevent further reaction. This is the case with nickel, where the product is NiF_2. Fluorine gas is ordinarily stored in containers made of a nickel alloy, such as stainless steel (Fe, Cr, Ni) or Monel (Ni, Cu). Fluorine also reacts with many compounds including water, which is oxidized to a mixture of O_2, O_3, H_2O_2, and OF_2.

A species disproportionates when it is oxidized and reduced at the same time.

Chlorine is somewhat less reactive than fluorine. Although it reacts with nearly all metals (Figure 21.3), heating is often required. This reflects the relatively strong bond in the Cl_2 molecule (B.E. Cl—Cl = 243 kJ/mol). Chlorine disproportionates in water, forming Cl^- ions (oxid. no. Cl = −1) and HClO molecules (oxid. no. Cl = +1).

$$Cl_2(g) + H_2O \rightleftharpoons Cl^-(aq) + H^+(aq) + HClO(aq)$$

The hypochlorous acid, HClO, formed by this reaction is a powerful oxidizing agent (E°_{red} = +1.630 V); it kills bacteria, apparently by destroying certain enzymes essential to their metabolism. The taste and odor that we associate with "chlorinated water" are actually due to compounds such as CH_3NHCl, produced by the action of hypochlorous acid on bacteria.

Chlorinated water. The oxidizing power of chlorine-containing chemicals keeps swimming pool water free of disease-causing microorganisms.
(VCG/FPG International)

EXAMPLE 21.1   For the reaction $Cl_2(g) + H_2O \rightleftharpoons Cl^-(aq) + H^+(aq) + HClO(aq)$,

(a) write the expression for the equilibrium constant K.
(b) given that $K = 2.7 \times 10^{-5}$, calculate the concentration of HClO in equilibrium with $Cl_2(g)$ at 1.0 atm.

Strategy   To answer (a), note that, in the expression for K, gases enter as their partial pressures in atmospheres, species in aqueous solution as their molarities; water does not appear, because it is the solvent. To answer (b), note that H^+ ions, Cl^- ions, and HClO molecules are formed in equimolar amounts.

Solution

(a) $K = \dfrac{[HClO] \times [H^+] \times [Cl^-]}{P_{Cl_2}}$

(b) Let x = [HClO]. Since H^+, Cl^-, and HClO all have coefficients of 1 in the balanced equation, $[H^+] = [Cl^-] = [HClO] = x$. Substituting in the expression for K:

$$K = \frac{[HClO] \times [H^+] \times [Cl^-]}{P_{Cl_2}} \qquad 2.7 \times 10^{-5} = \frac{x^3}{1.0}$$

Solving

$$x = (2.7 \times 10^{-5})^{1/3} = \boxed{0.030\ M}$$

Reality Check   In other words, the concentration of hypochlorous acid in a solution formed by bubbling chlorine gas through water should be about 0.03 mol/L.

The oxidizing power of the halogens makes them hazardous to work with. Fluorine is the most dangerous, but it is very unlikely that you will ever come across it in a teaching laboratory. You are most likely to encounter chlorine as its saturated water solution, called "chlorine water." Remember that the pressure of chlorine gas over this solution (if it is freshly prepared) is 1 atm and that chlorine was used as a poison gas in World War I. Use small quantities of chlorine water and don't breathe the vapors. Bromine, although not as strong an oxidizing agent as chlorine, can cause severe burns if it comes in contact with your skin, particularly if it gets under your fingernails.

Of the four halogens, iodine is the weakest oxidizing agent. "Tincture of iodine," a 10% solution of I_2 in alcohol, is sometimes used as an antiseptic. Hospitals most often use a product called "povidone-iodine," a quite powerful iodine-containing antiseptic and disinfectant, which can be diluted with water to the desired strength. These applications of molecular iodine should not delude you into thinking that the solid is harmless. On the contrary, if $I_2(s)$ is allowed to remain in contact with your skin, it can cause painful burns that are slow to heal.

Occurrence and Preparation

Of the eight nonmetals considered here, three (nitrogen, oxygen, and sulfur) occur in nature in elemental form. ***Nitrogen*** and ***oxygen*** are obtained from air, where their mole fractions are 0.7808 and 0.2095, respectively. When liquid air at $-200°C$ (73 K) is allowed to warm, the first substance that boils off is nitrogen (bp N_2 = 77 K). After most of the nitrogen has been removed, further warming gives oxygen (bp O_2 = 90 K). About 2×10^{10} kg of O_2 and lesser amounts of N_2 are produced annually in the United States from liquid air.

At the close of the Civil War in 1865, oil prospectors in Louisiana discovered (to their disgust) elemental ***sulfur*** in the caprock of vast salt domes up to 20 km^2 in area. The sulfur lies 60 to 600 m below the surface of the Earth. The process used to mine sulfur is named after its inventor, Herman Frasch, an American chemical engineer (born in Germany). A diagram of the Frasch process is shown in Figure 21.4 (page 600). The sulfur is heated to its melting point (119°C) by pumping superheated water at 165°C down one of three concentric pipes. Compressed air is used to bring the sulfur to the surface. The air and sulfur form a frothy mixture that rises through the middle pipe. On cooling, the sulfur solidifies, filling huge vats that may be 0.5 km long. The sulfur obtained in this way has a purity approaching 99.9%.

The ***halogens*** are far too reactive to occur in nature as the free elements. Instead, they are found as halide anions—

All the noble gases except He are obtained from liquid air.

FIGURE 21.4

Frasch process for mining sulfur. Superheated water at 165°C is sent down through the outer pipe to form a pool of molten sulfur (mp = 119°C) at the base. Compressed air, pumped down the inner pipe, brings the sulfur to the surface. Sulfur deposits are often 100 m or more beneath the Earth's surface, covered with quicksand and rock. **OHT**

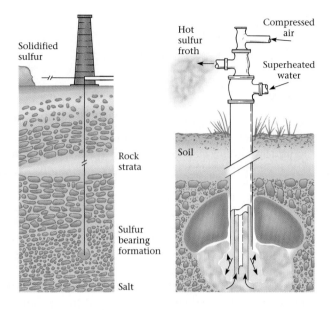

Br$_2$ and I$_2$ can also be obtained from sea water.

- F$^-$ in the mineral calcium fluoride, CaF$_2$ (fluorite).
- Cl$^-$ in huge underground deposits of sodium chloride, NaCl (rock salt), underlying parts of Oklahoma, Texas, and Kansas.
- Br$^-$(aq) and I$^-$(aq) in brine wells in Arkansas (conc. Br$^-$ = 0.05 M) and Michigan (conc. I$^-$ = 0.001 M), respectively.

The fluoride and chloride ions are very difficult to oxidize ($E°_{ox}$ F$^-$ = −2.889 V; $E°_{ox}$ Cl$^-$ = −1.360 V). Hence the elements fluorine and chlorine are ordinarily prepared by electrolytic oxidation, using a high voltage. As pointed out in Chapter 18, chlorine is prepared by the electrolysis of aqueous sodium chloride:

$$2Cl^-(aq) + 2H_2O \longrightarrow Cl_2(g) + H_2(g) + 2\,OH^-(aq)$$

The process used to prepare fluorine was developed by Henri Moissan in Paris more than a century ago; it won him one of the early (1906) Nobel Prizes in chemistry. The electrolyte is a mixture of HF and KF in a 2:1 mole ratio. At 100°C, fluorine is generated by the decomposition of hydrogen fluoride:

$$2HF(l) \longrightarrow H_2(g) + F_2(g)$$

Potassium fluoride furnishes the ions required to carry the electric current.

Bromide and iodide ions are easier to oxidize ($E°_{ox}$ Br$^-$ = −1.077 V; $E°_{ox}$ I$^-$ = −0.534 V), so bromine and iodine can be prepared by chemical oxidation. Commonly, the oxidizing agent is chlorine gas ($E°_{red}$ = +1.360 V):

$$Cl_2(g) + 2Br^-(aq) \longrightarrow 2Cl^-(aq) + Br_2(l)$$

$$Cl_2(g) + 2I^-(aq) \longrightarrow 2Cl^-(aq) + I_2(s)$$

Allotropy

Several nonmetals show the phenomenon of **allotropy;** they exist in two or more different structural forms in the same physical state. For example, graphite, diamond, and buckminsterfullerene, referred to in Chapter 1, are allotropic forms of

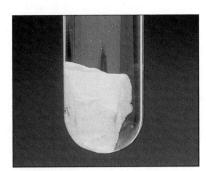

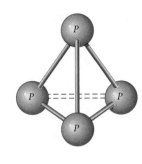

FIGURE 21.5
White phosphorus. *(Charles D. Winters)*

the element carbon. One of the simplest forms of allotropy is shown by ***oxygen,*** which can exist in the gas phase as either O_2 or O_3 (ozone). Commercially, ozone is prepared by passing O_2 gas through a high-voltage (10^4 V) electric discharge:

$$3O_2(g) \longrightarrow 2O_3(g)$$

At atmospheric pressure, the reverse reaction

$$2O_3(g) \longrightarrow 3O_2(g) \qquad \Delta H^\circ = -285.4 \text{ kJ} \qquad \Delta S^\circ = +137.5 \text{ J/K}$$

is thermodynamically spontaneous at all temperatures. Kinetically, however, ozone stays around long enough to find some application as a substitute for chlorine in disinfecting municipal water supplies.

Phosphorus forms several allotropes in the solid state, of which white and red phosphorus are the most common.

1. *White phosphorus* (Figure 21.5) consists of P_4 molecules. It is a soft, waxy substance with a low melting point (44°C) and boiling point (280°C). Like most molecular substances, white phosphorus is readily soluble in such nonpolar solvents as CCl_4. The chemical reactivity of white phosphorus is so great that it is stored under water to protect it from O_2. A piece of P_4 exposed to air in a dark room glows because of the light given off on oxidation (Figure 21.6). White

Ozone and lightning. The electrical discharge that causes lightning produces ozone from oxygen in the air. *(Paul and Linda Marie Ambrose/FPG International)*

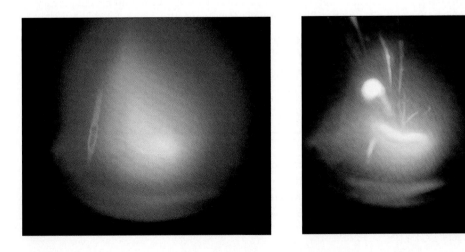

FIGURE 21.6
Reaction of phosphorus with oxygen. When white phosphorus is exposed to oxygen gas, it first glows (phosphorescence) and then bursts into flame. The reaction is

$$P_4(s) + 5O_2(g) \longrightarrow P_4O_{10}(s)$$

(Charles D. Winters)

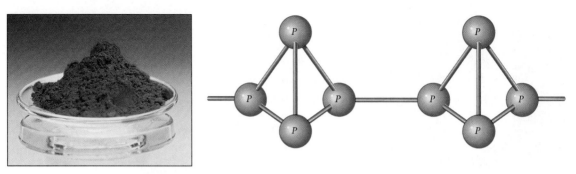

FIGURE 21.7
Red phosphorus. *(Charles D. Winters)*

phosphorus is extremely toxic. As little as 0.1 g taken internally can be fatal. Direct contact with the skin produces painful burns.

2. *Red phosphorus* (Figure 21.7) is the form usually found in the laboratory. This allotrope has properties quite different from those of white phosphorus. It is much higher-melting (mp = 590°C at 43 atm) and insoluble in common solvents. The low volatility of red phosphorus makes it much less toxic than the white form. It is also less reactive and must be heated to 250°C to burn in air. These properties are consistent with the structure of red phosphorus, which is known to be network covalent (Figure 21.7). This allotrope can be made by heating white phosphorus in the absence of air to about 300°C.

Before the undesirable properties of white phosphorus were known, it was used in matches. Today, two different kinds of matches are available, neither of which contains white phosphorus. The heads of "strike-anywhere" matches contain a mixture of a sulfide of phosphorus, P_4S_3, potassium chlorate, $KClO_3$, and powdered glass. When struck against a rough surface, the mixture ignites. Safety matches contain sulfur and potassium chlorate; the special surface against which they are struck contains red phosphorus and powdered glass. Friction sets off a reaction between red phosphorus and $KClO_3$.

FIGURE 21.8
Naturally occurring crystals of monoclinic sulfur *(above)* **and rhombic sulfur** *(below).* *(a, Gregory G. Dimijian/Photo Researchers, Inc.; b, Kip Peticolas/Fundamental Photographs, Inc.)*

EXAMPLE 21.2   For the allotropic conversion P(white) → P(red), $\Delta H°$ is −17.6 kJ and $\Delta S°$ is −18.3 J/K. At what temperature are the two allotropes in equilibrium at 1 atm?

Strategy   A system is at equilibrium at 1 atm when $\Delta G° = 0$. To find the temperature at which that happens, use the Gibbs-Helmholtz equation: $\Delta G° = \Delta H° - T \Delta S°$.

Solution   Setting $\Delta G° = 0$, it follows that $\Delta H° = T \Delta S°$. Hence

$$T = \frac{\Delta H°}{\Delta S°} = \frac{-17.6 \text{ kJ}}{-0.0183 \text{ kJ/K}} = 962 \ K \approx \boxed{690°C}$$

Reality Check   At any temperature below about 700°C, red phosphorus is the stable allotrope.

In the solid state, ***sulfur*** can have more than 20 different allotropic forms. You will no doubt be relieved to learn that we will talk about only two of these, rhombic and monoclinic sulfur (Figure 21.8). Both consist of ring molecules of formula S_8. They differ only in the way that molecules are packed in the solid, reflected in different crystal structures.

At room temperature, rhombic sulfur is the stable allotrope. However, because the process

$$S_8(\text{rhombic}) \longrightarrow S_8(\text{monoclinic}) \qquad \Delta H° = +2.6 \text{ kJ}$$

is endothermic, the monoclinic form is stable at high temperatures. The two forms are in equilibrium at 96°C; if rhombic sulfur is heated to that temperature, it slowly converts to monoclinic sulfur. More commonly, the monoclinic allotrope is prepared by freezing liquid sulfur at the melting point (119°C) and then cooling quickly to room temperature. Typically, at 25°C, monoclinic crystals stay around for a day or more before converting to the rhombic form.

The free-flowing, pale yellow liquid formed when sulfur melts contains S_8 molecules. However, on heating to 160°C, a striking change occurs. The liquid becomes so viscous that it cannot be poured readily. At the same time its color changes to a deep reddish-brown. These effects reflect a change in molecular structure. The S_8 rings break apart and then link to one another to form long chains such as

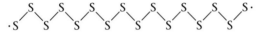

Liquid sulfur between 160 and 250°C contains a high proportion of such chains. They vary in length from eight to perhaps a million atoms. The chains become tangled, producing a highly viscous liquid. The deep color is due to the absorption of light by the unpaired electrons at the ends of the chains.

If liquid sulfur at 200°C is quickly poured into water, a rubbery mass results. This is referred to as "plastic sulfur." It consists of long-chain molecules that did not have time to rearrange to the S_8 molecules stable at room temperature. Within a few hours, the plastic sulfur loses its elasticity as it converts to rhombic crystals.

Under these conditions, sulfur is a polymer.

Making plastic sulfur.
(Charles D. Winters)

21.2   HYDROGEN COMPOUNDS OF NONMETALS

Table 21.2 lists some of the more important hydrogen compounds of the nonmetals. (Those of carbon are discussed in Chapter 22.) The physical states listed are those observed at 25°C and 1 atm. The remainder of this section is devoted to a discussion of the chemical properties of the compounds shown in boldface in the table.

TABLE 21.2   Hydrogen Compounds of the Nonmetals*

Group 15	Group 16	Group 17
Ammonia, NH$_3$(g)	Water, H$_2$O(l)	**Hydrogen fluoride, HF(g)**
Hydrazine, N$_2$H$_4$(l)	**Hydrogen peroxide, H$_2$O$_2$(l)**	
Hydrazoic acid, HN$_3$(l)		
Phosphine, PH$_3$(g)	**Hydrogen sulfide, H$_2$S(g)**	**Hydrogen chloride, HCl(g)**
Diphosphine, P$_2$H$_4$(l)		
		Hydrogen bromide, HBr(g)
		Hydrogen iodide, HI(g)

*Compounds discussed in this section are shown in bold type.

Ammonia, NH_3

Ammonia is one of the most important industrial chemicals; more than ten million tons of NH_3 are produced annually in the United States. You will recall (Chapter 12) that it is made by the Haber process

$$N_2(g) + 3H_2(g) \rightleftharpoons 2NH_3(g) \qquad 450°C, 200–600 \text{ atm, solid catalyst}$$

Ammonia is used to make fertilizers and a host of different nitrogen compounds, notably nitric acid, HNO_3.

The NH_3 molecule acts as a *Brønsted-Lowry base* in water, accepting a proton from a water molecule:

$$NH_3(aq) + H_2O \rightleftharpoons NH_4^+(aq) + OH^-(aq) \qquad K_b = 1.8 \times 10^{-5}$$

Ammonia can also act as a *Lewis base* when it reacts with a metal cation to form a complex ion

> The NH_3 molecule donates an electron pair to Ag^+.

$$2NH_3(aq) + Ag^+(aq) \longrightarrow Ag(NH_3)_2^+(aq)$$

or with an electron-deficient compound such as boron trifluoride:

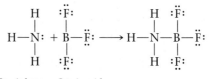

Lewis base   Lewis acid

Ammonia is often used to precipitate insoluble metal hydroxides such as $Al(OH)_3$. The OH^- ions formed when ammonia reacts with water precipitate the cation from solution as the hydroxide. The overall equation for the reaction is

$$Al^{3+}(aq) + 3NH_3(aq) + 3H_2O \longrightarrow Al(OH)_3(s) + 3NH_4^+(aq)$$

Nitrogen cannot have an oxidation number lower than -3, which means that when NH_3 takes part in a redox reaction, it always acts as a *reducing agent*. Ammonia may be oxidized to elementary nitrogen or to a compound of nitrogen. An important redox reaction of ammonia is that with hypochlorite ion:

$$2NH_3(aq) + ClO^-(aq) \longrightarrow N_2H_4(aq) + Cl^-(aq) + H_2O$$

Hydrazine, N_2H_4, is made commercially by this process. Certain by-products of this reaction, notably NH_2Cl and $NHCl_2$, are both toxic and explosive, so solutions of household bleach and ammonia should never be mixed with one another.

Hydrogen Sulfide, H_2S

In water solution, hydrogen sulfide acts as a *Brønsted-Lowry acid;* it can donate a proton to a water molecule:

$$H_2S(aq) + H_2O \rightleftharpoons HS^-(aq) + H_3O^+(aq) \qquad K_a = 1.0 \times 10^{-7}$$

Like ammonia, hydrogen sulfide can act as a precipitating agent toward metal cations (recall the discussion of qualitative analysis in Chapter 16). The reaction with Cd^{2+} is typical:

$$Cd^{2+}(aq) + H_2S(aq) \longrightarrow CdS(s) + 2H^+(aq)$$

Like ammonia, hydrogen sulfide (oxid. no. S = −2) can act only as a reducing agent when it takes part in redox reactions. Most often the H_2S is oxidized to elementary sulfur, as in the reaction

$$2H_2S(aq) + O_2(g) \longrightarrow 2S(s) + 2H_2O$$

The sulfur formed is often very finely dispersed, which explains why aqueous solutions of hydrogen sulfide in contact with air have a milky appearance.

If you've worked with H_2S in the laboratory, you won't soon forget its rotten-egg odor. In a sense, it's fortunate that hydrogen sulfide has such a distinctive odor. This gas is highly toxic, as poisonous as HCN. At a concentration of 10 parts per million, H_2S can cause headaches and nausea; at 100 ppm it can be fatal.

EXAMPLE 21.3   When a solution containing Cu^{2+} is treated with hydrogen sulfide, a black precipitate forms. When another portion of the solution is treated with ammonia, a blue precipitate forms. This precipitate dissolves in excess ammonia to form a deep blue solution containing the $Cu(NH_3)_4^{2+}$ ion. Write balanced net ionic equations to explain these observations.

Strategy   Before you can write the equations, you must identify the products. The black precipitate is CuS. The blue precipitate must be copper(II) hydroxide, $Cu(OH)_2$. The identity of the final product is given: $Cu(NH_3)_4^{2+}$. The equations are readily written if you know the formulas of the products.

Solution

$$Cu^{2+}(aq) + H_2S(aq) \longrightarrow CuS(s) + 2H^+(aq)$$

$$Cu^{2+}(aq) + 2NH_3(aq) + 2H_2O \longrightarrow Cu(OH)_2(s) + 2NH_4^+(aq)$$

$$Cu(OH)_2(s) + 4NH_3(aq) \longrightarrow Cu(NH_3)_4^{2+}(aq) + 2OH^-(aq)$$

Hydrogen Peroxide

In hydrogen peroxide, oxygen has an oxidation number of −1, intermediate between the extremes for the element, 0 and −2. This means that H_2O_2 can act as either an oxidizing agent, in which case it is reduced to H_2O, or as a reducing agent, when it is oxidized to O_2. In practice, hydrogen peroxide is an extremely strong oxidizing agent:

$$H_2O_2(aq) + 2H^+(aq) + 2e^- \longrightarrow 2H_2O \qquad E^\circ_{red} = +1.763 \text{ V}$$

but a very weak reducing agent:

$$H_2O_2(aq) \longrightarrow O_2(g) + 2H^+(aq) + 2e^- \qquad E^\circ_{ox} = -0.695 \text{ V}$$

Hydrogen peroxide tends to decompose in water, which explains why its solutions soon lose their oxidizing power. The reaction involved is *disproportionation*, combining the two half-reactions referred to above:

$$H_2O_2(aq) + 2H^+(aq) + 2e^- \longrightarrow 2H_2O \qquad\qquad E^\circ_{red} = +1.763 \text{ V}$$

$$\underline{H_2O_2(aq) \longrightarrow O_2(g) + 2H^+(aq) + 2e^- \qquad E^\circ_{ox} = -0.695 \text{ V}}$$

$$2H_2O_2(aq) \longrightarrow O_2(g) + 2H_2O \qquad\qquad E^\circ = +1.068 \text{ V}$$

Disproportionation of H_2O_2, catalyzed by MnO_2. *(Charles D. Winters)*

⌐ Half of the H_2O_2 molecules are reduced to H_2O; half are oxidized to O_2.

This reaction is catalyzed by a wide variety of materials, including I^- ions, MnO_2, metal surfaces (Pt, Ag), and even by traces of OH^- ions dissolved from glass.

You are most likely to come across hydrogen peroxide as its water solution. Solutions available in a drugstore for medicinal use contain 3 to 6 percent by mass of H_2O_2. Industrially, concentrations as high as 86% H_2O_2 are available. All water solutions of hydrogen peroxide contain stabilizers to prevent disproportionation from taking place during storage.

EXAMPLE 21.4   Using Table 18.1 (page 520), decide whether hydrogen peroxide will react with (oxidize or reduce) the following ions in acid solution at standard concentrations.

(a) I^- (b) Sn^{2+} (c) Co^{2+}

Strategy   For a redox reaction to occur spontaneously, either

$$E^\circ_{ox} X + E^\circ_{red} H_2O_2 > 0$$

or

$$E^\circ_{red} X + E^\circ_{ox} H_2O_2 > 0$$

where X is one of the ions listed and $E^\circ_{red} H_2O_2 = +1.763$ V; $E^\circ_{ox} H_2O_2 = -0.695$ V.

Solution

(a) $E^\circ_{red} H_2O_2 + E^\circ_{ox} I^- = +1.763$ V $- 0.534$ V $= +1.229$ V

> Hydrogen peroxide oxidizes I^- ions to I_2.

(b) $E^\circ_{red} H_2O_2 + E^\circ_{ox} Sn^{2+} = +1.763$ V $- 0.154$ V $= +1.609$ V
$E^\circ_{ox} H_2O_2 + E^\circ_{red} Sn^{2+} = -0.695$ V $- 0.141$ V $= -0.836$ V

> Hydrogen peroxide oxidizes Sn^{2+} to Sn^{4+}.

(c) Co^{2+} can be reduced to Co ($E^\circ_{red} = -0.282$ V) or oxidized to Co^{3+} ($E^\circ_{ox} = -1.953$ V). Neither gives a positive combination with H_2O_2; no reaction.

Hydrogen Fluoride and Hydrogen Chloride

The most common hydrogen halides are HF (U.S. production $= 3 \times 10^8$ kg/yr) and HCl (3×10^9 kg/yr). They are most familiar as water solutions, referred to as hydrofluoric acid and hydrochloric acid, respectively. Recall (Chapter 13) that hydrofluoric acid is weak, incompletely dissociated in water, whereas HCl is a strong acid.

$$HF(aq) \rightleftharpoons H^+(aq) + F^-(aq) \qquad K_a = 6.9 \times 10^{-4}$$

$$HCl(aq) \longrightarrow H^+(aq) + Cl^-(aq) \qquad K_a \longrightarrow \infty$$

Hydrofluoric and hydrochloric acids undergo very similar reactions with bases such as OH^- or CO_3^{2-} ions. The equations for these reactions look somewhat different because of the difference in acid strength. Thus for the reaction of hydrochloric acid with a solution of sodium hydroxide, the equation is simply

$$H^+(aq) + OH^-(aq) \longrightarrow H_2O$$

With hydrofluoric acid, the equation is

$$HF(aq) + OH^-(aq) \longrightarrow H_2O + F^-(aq)$$

HF appears in this equation because hydrofluoric acid is weak, containing many more HF molecules than H^+ ions. Similarly, for the reaction of these two acids with a solution of sodium carbonate:

hydrochloric acid: $\quad 2H^+(aq) + CO_3{}^{2-}(aq) \longrightarrow CO_2(g) + H_2O$

hydrofluoric acid: $\quad 2HF(aq) + CO_3{}^{2-}(aq) \longrightarrow CO_2(g) + H_2O + 2F^-(aq)$

Concentrated hydrofluoric acid reacts with glass, which can be considered to be a mixture of SiO_2 and ionic silicates such as calcium silicate, $CaSiO_3$:

$$SiO_2(s) + 4HF(aq) \longrightarrow SiF_4(g) + 2H_2O$$

$$CaSiO_3(s) + 6HF(aq) \longrightarrow SiF_4(g) + CaF_2(s) + 3H_2O$$

As you might guess, HF solutions are never stored in glass bottles; plastic is used instead. Hydrofluoric acid is sometimes used to etch glass. The glass object is first covered with a thin protective coating of wax or plastic. Then the coating is removed from the area to be etched and the glass is exposed to the HF solution. Thermometer stems and burets can be etched or light bulbs frosted in this way.

Hydrogen fluoride is a very unpleasant chemical to work with. If spilled on the skin, it removes Ca^{2+} ions from the tissues, forming insoluble CaF_2. A white patch forms that is agonizingly painful to the touch. To make matters worse, HF is a local anesthetic, so a person may be unaware of what's happening until it's too late.

Rest assured that you will never have occasion to use hydrofluoric acid in the general chemistry laboratory. Neither are you likely to come in contact with HBr or HI, both of which are relatively expensive. Hydrochloric acid, on the other hand, is a workhorse chemical of the teaching laboratory. It is commonly available as "dilute HCl" (6 M) or "concentrated HCl" (12 M). You will use it as a source of H^+ ions for such purposes as—

■ dissolving insoluble carbonates or hydroxides.

$$Ag_2CO_3(s) + 2H^+(aq) \longrightarrow 2Ag^+(aq) + CO_2(g) + H_2O$$

$$Zn(OH)_2(s) + 2H^+(aq) \longrightarrow Zn^{2+}(aq) + 2H_2O$$

■ converting a weak base such as NH_3 to its conjugate acid.

$$NH_3(aq) + H^+(aq) \longrightarrow NH_4{}^+(aq)$$

■ generating $H_2(g)$ by reaction with a metal.

$$Zn(s) + 2H^+(aq) \longrightarrow Zn^{2+}(aq) + H_2(g)$$

> Hydrochloric acid contains H^+ and Cl^- ions; hydrofluoric acid contains mostly HF molecules.

> How do you suppose HF(aq) was stored B.P. (before plastics)?

EXAMPLE 21.5   Write balanced net ionic equations to explain why

(a) aluminum hydroxide dissolves in hydrochloric acid.
(b) carbon dioxide gas is evolved when calcium carbonate is treated with hydrochloric acid.
(c) carbon dioxide gas is evolved when calcium carbonate is treated with hydrofluoric acid.

Strategy   In each case, decide on the nature of the products. You can reason by analogy with the equations cited in the text.

Solution

(a) $Al(OH)_3(s) + 3H^+(aq) \rightarrow Al^{3+}(aq) + 3H_2O$

(b) $CaCO_3(s) + 2H^+(aq) \rightarrow Ca^{2+}(aq) + CO_2(g) + H_2O$

(c) $CaCO_3(s) + 2HF(aq) \rightarrow Ca^{2+}(aq) + CO_2(g) + H_2O + 2F^-(aq)$

A test for limestone. Geologists use a few drops of hydrochloric acid to identify limestone, which is calcium carbonate. (*Charles D. Winters*)

TABLE 21.3 Nonmetal Oxides*		
Group 15	Group 16	Group 17
$N_2O_5(s)$, $N_2O_4(g)$, $NO_2(g)$		$OF_2(g)$, $O_2F_2(g)$
$N_2O_3(d)$, $NO(g)$, $N_2O(g)$		
$P_4O_{10}(s)$, $P_4O_6(s)$	$SO_3(l)$, $SO_2(g)$	$Cl_2O_7(l)$, $Cl_2O_6(l)$
		$ClO_2(g)$, $Cl_2O(g)$
		$BrO_2(d)$, $Br_2O(d)$
		$I_2O_5(s)$, $I_4O_9(s)$, $I_2O_4(s)$

*The states listed are those observed at 25°C and 1 atm. Compounds that decompose below 25°C are listed as (*d*). Oxides shown in boldface are discussed in the text.

21.3 OXYGEN COMPOUNDS OF NONMETALS

Table 21.3 lists some of the more familiar nonmetal oxides. Curiously enough, only 5 of the 21 compounds shown are thermodynamically stable at 25°C and 1 atm (P_4O_{10}, P_4O_6, SO_3, SO_2, I_2O_5). The others, including all of the oxides of nitrogen and chlorine, have positive free energies of formation at 25°C and 1 atm. For example, $\Delta G_f^\circ NO_2(g) = +51.3$ kJ; $\Delta G_f^\circ ClO_2(g) = +120.6$ kJ. Kinetically, these compounds stay around long enough to have an extensive chemistry. Nitrogen dioxide, a reddish-brown gas, is a major factor in the formation of photochemical smog (Figure 21.9). Chlorine dioxide, a yellow gas, is widely used as an industrial bleach and water purifier, even though it tends to explode at partial pressures higher than 50 mm Hg.

> All such compounds are potentially unstable.

Molecular Structures of Nonmetal Oxides

The Lewis structures of the oxides of nitrogen are shown in Figure 21.10. Two of these species, NO and NO_2, are paramagnetic, with one unpaired electron. When nitrogen dioxide is cooled, it dimerizes; the unpaired electrons combine to form a single bond between the two nitrogen atoms:

> Dimerization occurs when two identical molecules combine.

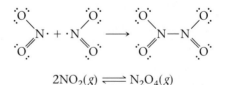

$$2NO_2(g) \rightleftharpoons N_2O_4(g)$$

A similar reaction occurs when an equimolar mixture of NO and NO_2 is cooled. Two odd electrons, one from each molecule, pair off to form an N—N bond:

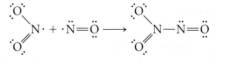

$$NO_2(g) + NO(g) \rightleftharpoons N_2O_3(g)$$

At −20°C, dinitrogen trioxide separates from the mixture as a blue liquid.

FIGURE 21.9
Smog over New York City. From a distance, you can clearly see the layer of smog containing reddish-brown nitrogen dioxide. *(Peter Arnold/Peter Arnold, Inc.)*

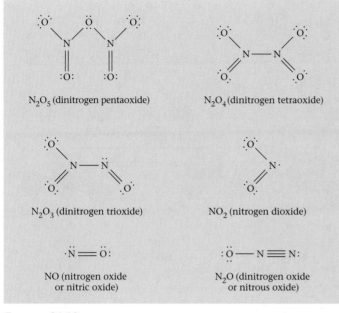

FIGURE 21.10
Lewis structures of the oxides of nitrogen. Many other resonance forms are possible.

Nitric oxide (NO) is a minor but villainous component of the atmosphere. It is involved in the formation of both smog (Chapter 11) and acid rain (Chapter 14). You may be surprised to learn that small amounts of NO are also produced in the human body, where it has a generally beneficial effect. In particular, it has the ability to dilate blood vessels, lowering blood pressure and reducing the likelihood of strokes or heart attacks. Beyond that, NO is effective in treating what television commercials refer to as "erectile dysfunction"; it increases blood flow to the penis.

Perhaps the best-known oxide of nitrogen is N_2O, commonly called nitrous oxide or "laughing gas." Nitrous oxide is frequently used as an anesthetic, particularly in dentistry. It is also the propellant gas used in whipped cream containers; N_2O is nontoxic, virtually tasteless, and quite soluble in vegetable oils. The N_2O molecule, like all those in Figure 21.10, can be represented as a resonance hybrid.

EXAMPLE 21.6   Consider the N_2O molecule shown in Figure 21.10.

(a) Draw another resonance form of N_2O.
(b) What is the bond angle in N_2O?
(c) Is the N_2O molecule polar or nonpolar?

Strategy   Recall the discussion in Chapter 7, Sections 7.1 to 7.3, where the principles of resonance, molecular geometry, and polarity were considered.

Solution

(a) :Ö=N=N̈:   or   :O≡N—N̈:
(b) In any of the resonance forms, the central nitrogen atom, insofar as geometry is concerned, would behave as if it were surrounded by two electron pairs. The bond angle is 180°; the molecule is linear, like BeF_2.

(c) Polar (unsymmetrical).

FIGURE 21.11
Ball-and-stick models of the oxides
of phosphorus.

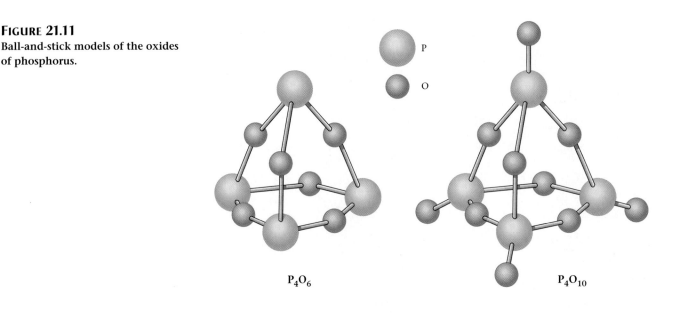

P_4O_6 P_4O_{10}

The structures of SO_2 and SO_3 were referred to in Chapter 7. These molecules are often cited as examples of resonance; sulfur trioxide, for example, has three equivalent resonance structures:

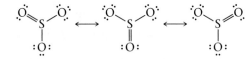

As you might expect, the SO_3 molecule is nonpolar, with 120° bond angles.

Of the two oxides of phosphorus, P_4O_6 and P_4O_{10}, the latter is the more stable; it is formed when white phosphorus burns in air:

$$P_4(s) + 5O_2(g) \longrightarrow P_4O_{10}(s)$$

Note from Figure 21.11 that in both oxides, as in P_4 itself, the four phosphorus atoms are at the corners of a tetrahedron. The P_4O_6 molecule can be visualized as derived from P_4 by inserting an oxygen atom between each pair of phosphorus atoms. In P_4O_{10}, an extra oxygen atom is bonded to each phosphorus.

Reactions of Nonmetal Oxides with Water

Many nonmetal oxides are acidic in the sense that they react with water to form acids. Looking at the reaction

$$SO_3(g) + H_2O(l) \longrightarrow H_2SO_4(l)$$

you can see that sulfur trioxide is an **acidic oxide.** Notice that in this reaction, the nonmetal does not change oxidation number; sulfur is in the +6 state in both SO_3 and H_2SO_4. Other acidic oxides include N_2O_5 and N_2O_3:

What is the acid derived from CO_2?
Ans. H_2CO_3.

+5 nitrogen: $N_2O_5(s) + H_2O(l) \longrightarrow 2HNO_3(l)$

+3 nitrogen: $N_2O_3(g) + H_2O(l) \longrightarrow 2HNO_2(aq)$

The products are nitric acid, HNO_3, and an aqueous solution of nitrous acid, HNO_2.

One of the most important reactions of this type involves the +5 oxide of phosphorus, P_4O_{10}. Here the product is phosphoric acid, H_3PO_4:

$$P_4O_{10}(s) + 6H_2O(l) \longrightarrow 4H_3PO_4(s)$$

This reaction is used to prepare high-purity phosphoric acid and salts of that acid for use in food products. Phosphoric acid, H_3PO_4, is added in small amounts to soft drinks to give them a tart taste. It is present to the extent of about 0.05 mass percent in colas, 0.01 mass percent in root beers.

Household products containing phosphoric acid or its salts. *(Charles D. Winters)*

EXAMPLE 21.7   Give the formula of the oxide that reacts with water to form

(a) H_3PO_3 (b) $HClO$ (c) H_2SO_3

Strategy   Referring to Table 21.3, find an oxide in which the central nonmetal atom has the same oxidation number as in the acid.

Solution

(a) +3 P: H_3PO_3 and P_4O_6

(b) +1 Cl: $HClO$ and Cl_2O

(c) +4 S: H_2SO_3 and SO_2

21.4   OXOACIDS AND OXOANIONS

Table 21.4 lists some of the more important oxoacids of the nonmetals. In all these compounds, the ionizable hydrogen atoms are bonded to oxygen, not to the central nonmetal atom. Dissociation of one or more protons from the oxoacid gives the corresponding oxoanion (Figure 21.12).

In this section we discuss the principles that allow you to predict the relative acid strengths of oxoacids such as those listed in Table 21.4 (page 612). Then we consider their strengths as oxidizing and/or reducing agents. Finally, we take a closer look at the chemistry of three important oxoacids: HNO_3, H_2SO_4, and H_3PO_4.

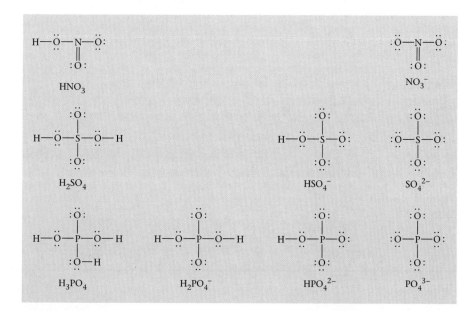

FIGURE 21.12
Lewis structures of the oxoacids HNO_3, H_2SO_4, H_3PO_4, and the oxoanions derived from them.   **OHT**

TABLE 21.4	Oxoacids of the Nonmetals	
Group 15	Group 16	Group 17
HNO_3, HNO_2* H_3PO_4, H_3PO_3	H_2SO_4, H_2SO_3*	$HClO_4$, $HClO_3$*, $HClO_2$*, $HClO$* $HBrO_4$*, $HBrO_3$*, $HBrO$* HIO_4, H_5IO_6, HIO_3, HIO*

*These compounds cannot be isolated from water solution.

Acid Strength

The acid equilibrium constants of the oxoacids of the halogens are listed in Table 21.5. Notice that the value of K_a increases with—

- *increasing oxidation number of the central atom* ($HClO < HClO_2 < HClO_3 < HClO_4$)
- *increasing electronegativity of the central atom* ($HIO < HBrO < HClO$)

These trends are general ones, observed with other oxoacids of the nonmetals. Recall, for example, that nitric acid, HNO_3 (oxid. no. N = +5), is a strong acid, completely ionized in water. In contrast, nitrous acid, HNO_2 (oxid. no. N = +3) is a weak acid ($K_a = 6.0 \times 10^{-4}$). The electronegativity effect shows up with the strengths of the oxoacids of sulfur and selenium:

$$K_{a1} \, H_2SO_3 = 1.7 \times 10^{-2} \qquad K_{a1} \, H_2SeO_3 = 2.7 \times 10^{-3} \text{ (E.N. S = 2.6, Se = 2.5)}$$

Trends in acid strength can be explained in terms of molecular structure. In an oxoacid molecule, the hydrogen atom that dissociates is bonded to oxygen, which in turn is bonded to a nonmetal atom, X. The ionization in water of an oxoacid H—O—X can be represented as

$$H{-}O{-}X(aq) \rightleftharpoons H^+(aq) + XO^-(aq)$$

For a proton, with its +1 charge, to separate from the molecule, the electron density around the oxygen should be as low as possible. This will weaken the O—H bond and favor ionization. The electron density around the oxygen atom is decreased when—

- *X is a highly electronegative atom such as Cl.* This draws electrons away from the oxygen atom and makes hypochlorous acid stronger than hypoiodous acid.
- *Additional, strongly electronegative oxygen atoms are bonded to X.* These tend to draw electrons away from the oxygen atom bonded to H. Thus we would pre-

TABLE 21.5	Equilibrium Constants of Oxoacids of the Halogens OHT				
Oxid. State		K_a		K_a		K_a
+7	$HClO_4$	$\sim 10^7$	$HBrO_4$	$\sim 10^6$	HIO_4*	1.4×10^1
+5	$HClO_3$	$\sim 10^3$	$HBrO_3$	3.0	HIO_3	1.6×10^{-1}
+3	$HClO_2$	1.0×10^{-2}	—	—	—	—
+1	$HClO$	2.8×10^{-8}	$HBrO$	2.6×10^{-9}	HIO	2.4×10^{-11}

*Estimated; in water solution the stable species is H_5IO_6, whose first K_a value is 5×10^{-4}.

dict that the ease of dissociation of a proton, and hence K_a, should increase in the following order, from left to right:

$$X-O-H < O-X-O-H < O-X-O-H < O-X-O-H$$

$$\overset{O}{\underset{O}{|}}$$

oxid. no. X = +1   +3   +5   +7

EXAMPLE 21.8   Consider sulfurous acid, H_2SO_3.

(a) Show its Lewis structure and that of the HSO_3^- and SO_3^{2-} ions.
(b) How would its acid strength compare with that of H_2SO_4? H_2TeO_3?

Strategy   The structure can be obtained by removing an oxygen atom from H_2SO_4 (Figure 21.12). Relative acid strengths can be predicted on the basis of the electronegativity and oxidation number of the central nonmetal atom, following the rules cited above.

Solution

(a)   H—Ö—S̈—Ö—H   (H—Ö—S̈—Ö:)⁻   (:Ö—S̈—Ö:)²⁻

           :Ö: :Ö: :Ö:

    sulfurous acid   hydrogen sulfite ion   sulfite ion

(b) $H_2SO_3 < H_2SO_4$ (oxid. no. S = +4, +6)

     $H_2SO_3 > H_2TeO_3$ (S more electronegative than Te)

Reality Check   Both H_2SO_3 and H_2TeO_3 are weak acids; H_2SO_4 is a strong acid.

Oxidizing and Reducing Strength

Many of the reactions of oxoacids and oxoanions involve oxidation and reduction. There are certain general principles that apply, regardless of the particular species involved.

    1. *A species in which a nonmetal is in its highest oxidation state can act only as an oxidizing agent, never as a reducing agent.* Consider, for example, the ClO_4^- ion, in which chlorine is in its highest oxidation state, +7. In any redox reaction in which this ion takes part, chlorine must be reduced to a lower oxidation state. When that happens, the ClO_4^- ions act as an oxidizing agent, taking electrons away from something else. The same argument applies to—

> Many oxoanions are very powerful oxidizing agents.

- the SO_4^{2-} ion (highest oxid. no. S = +6).
- the NO_3^- ion (highest oxid. no. N = +5).

Note that, in general, *the highest oxidation number of a nonmetal is given by the second digit of its group number* (**17** for Cl, **16** for S, **15** for N).
    2. *A species in which a nonmetal is in an intermediate oxidation state can act as either an oxidizing agent or a reducing agent.* Consider, for example, the ClO_3^- ion (oxid. no. Cl = +5). It can be oxidized to the perchlorate ion, in which case ClO_3^- acts as a reducing agent:

$$ClO_3^-(aq) + H_2O \longrightarrow ClO_4^-(aq) + 2H^+(aq) + 2e^- \qquad E^\circ_{ox} = -1.226 \text{ V}$$

Chlorate ion as an oxidizing agent. The effect of molten potassium perchlorate on the gummy bear is shown in the test tube. The spontaneous reaction is

$$C_6H_{12}O_6(s) + 4KClO_3(l) \longrightarrow 6CO_2(g) + 6H_2O(g) + 4KCl(s)$$
sugar

(Charles D. Winters)

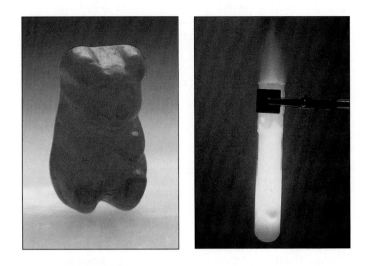

Alternatively, the ClO_3^- ion can be reduced, perhaps to a Cl^- ion. When that occurs, ClO_3^- acts as an oxidizing agent:

$$ClO_3^-(aq) + 6H^+(aq) + 6e^- \longrightarrow Cl^-(aq) + 3H_2O \qquad E°_{red} = +1.442 \text{ V}$$

Notice that the ClO_3^- ion is a much stronger oxidizing agent ($E°_{red} = +1.442$ V) than reducing agent ($E°_{ox} = -1.226$ V). This is generally true of oxoanions and oxoacids in an intermediate oxidation state, at least in acidic solution. Compare, for example,

HClO $E°_{red}$ (to Cl_2) = +1.630 V; $E°_{ox}$ (to $HClO_2$) = −1.157 V

HNO$_2$ $E°_{red}$ (to NO) = +1.036 V; $E°_{ox}$ (to NO_2) = −1.056 V

3. Sometimes, with a species such as ClO_3^-, oxidation and reduction occur together, resulting in disproportionation:

$$4ClO_3^-(aq) \longrightarrow 3ClO_4^-(aq) + Cl^-(aq) \qquad E° = +0.216 \text{ V}$$

In general, a *species in an intermediate oxidation state is expected to disproportionate if the sum $E°_{ox} + E°_{red}$ is a positive number.*

4. The oxidizing strength of an oxoacid or oxoanion is greatest at high [H⁺] (low pH). Conversely, its reducing strength is greatest at low [H⁺] (high pH).
This principle has a simple explanation. Looking back at the half-equations written above, you can see that

■ when ClO_3^- acts as an oxidizing agent, the H^+ ion is a reactant, so increasing its concentration makes the process more spontaneous.

■ when ClO_3^- acts as a reducing agent, the H^+ ion is a product; to make the process more spontaneous, [H⁺] should be lowered.

EXAMPLE 21.9   Calculate E_{red} and E_{ox} for the ClO_3^- ion in neutral solution, at pH 7.00, assuming all other species are at standard concentration ($E°_{red} = +1.442$ V; $E°_{ox} = -1.226$ V). Will the ClO_3^- ion disproportionate at pH 7.00?

Strategy   First (1) set up the Nernst equation for the reduction half-reaction and calculate E_{red}. Then (2) repeat the calculation for the oxidation half-reaction, finding E_{ox}. Finally (3), add $E_{red} + E_{ox}$; if the sum is positive, disproportionation should occur.

Solution

(1) $ClO_3^-(aq) + 6H^+(aq) + 6e^- \rightarrow Cl^-(aq) + 3H_2O$

$$E_{red} = +1.442 \text{ V} - \frac{0.0257}{6} \ln \frac{1}{[H^+]^6}$$

$$= +1.442 \text{ V} - \frac{0.0257}{6} \ln \frac{1}{(1.0 \times 10^{-7})^6}$$

$$= +1.442 \text{ V} - \frac{0.0257}{6} (96.7) = \boxed{+1.028 \text{ V}}$$

As expected, decreasing the concentration of H^+ makes the half-reaction less spontaneous and hence makes E_{red} less positive.

(2) $ClO_3^-(aq) + H_2O \rightarrow ClO_4^-(aq) + 2H^+(aq) + 2e^-$

$$E_{ox} = -1.226 \text{ V} - \frac{0.0257}{2} \ln[H^+]^2$$

$$= -1.226 \text{ V} - \frac{0.0257}{2} \ln(1.0 \times 10^{-7})^2$$

$$= -1.226 \text{ V} - \frac{0.0257}{2} (-32.2) = \boxed{-0.812 \text{ V}}$$

(3) $E = +1.028 \text{ V} - 0.812 \text{ V} = +0.216 \text{ V};$ disproportionation should occur.

Reality Check

Notice that E is the same as $E°$, +0.216 V. You could have predicted that (and saved a lot of work!), because the overall equation for disproportionation

$$4ClO_3^-(aq) \longrightarrow 3ClO_4^-(aq) + Cl^-(aq)$$

does not involve H^+ or OH^- ions.

Nitric Acid, HNO_3

Commercially, nitric acid is made by a three-step process developed by the German physical chemist Wilhelm Ostwald (1853–1932). The starting material is ammonia, which is burned in an excess of air at 900°C, using a platinum-rhodium catalyst:

$$4NH_3(g) + 5O_2(g) \longrightarrow 4NO(g) + 6H_2O(g)$$

The gaseous mixture formed is cooled and mixed with more air to convert NO to NO_2:

$$2NO(g) + O_2(g) \longrightarrow 2NO_2(g)$$

Finally, nitrogen dioxide is bubbled through water to produce nitric acid:

$$3NO_2(g) + H_2O(l) \longrightarrow NO(g) + 2HNO_3(aq)$$

Nitric acid is a strong acid, completely ionized to H^+ and NO_3^- ions in dilute water solution:

$$HNO_3(aq) \longrightarrow H^+(aq) + NO_3^-(aq)$$

Many of the reactions of nitric acid are those associated with all strong acids. For example, dilute (6 M) nitric acid can be used to dissolve aluminum hydroxide

$$Al(OH)_3(s) + 3H^+(aq) \longrightarrow Al^{3+}(aq) + 3H_2O$$

or to generate carbon dioxide gas from calcium carbonate:

$$CaCO_3(s) + 2H^+(aq) \longrightarrow Ca^{2+}(aq) + CO_2(g) + H_2O$$

Referring back to Example 21.5, you will find that these equations are identical with those written for the reactions of hydrochloric acid with $Al(OH)_3$ and $CaCO_3$. It is the H^+ ion that reacts in either case: Cl^- and NO_3^- ions take no part in the reactions and hence do not appear in the equation.

Concentrated (16 M) nitric acid is a strong oxidizing agent; the nitrate ion is reduced to nitrogen dioxide. This happens when 16 M HNO_3 reacts with copper metal (Figure 21.13):

$$Cu(s) + 4H^+(aq) + 2NO_3^-(aq) \longrightarrow Cu^{2+}(aq) + 2NO_2(g) + 2H_2O$$

> You can always smell NO_2 over 16 M HNO_3.

Dilute nitric acid (6 M) gives a wide variety of reduction products, depending on the nature of the reducing agent. With inactive metals such as copper ($E^\circ_{ox} = -0.339$ V), the major product is usually NO (oxid. no. N = +2):

$$3Cu(s) + 2NO_3^-(aq) + 8H^+(aq) \longrightarrow 3Cu^{2+}(aq) + 2NO(g) + 4H_2O$$

With very dilute acid and a strong reducing agent such as zinc ($E^\circ_{ox} = +0.762$ V) reduction may go all the way to the NH_4^+ ion (oxid. no. N = −3):

$$4Zn(s) + NO_3^-(aq) + 10H^+(aq) \longrightarrow 4Zn^{2+}(aq) + NH_4^+(aq) + 3H_2O$$

As you would expect, the oxidizing strength of the NO_3^- ion drops off sharply as pH increases; the reduction voltage (to NO) is +0.964 V at pH 0, +0.412 V at pH 7, and −0.140 V at pH 14. The nitrate ion is a very weak oxidizing agent in basic solution.

FIGURE 21.13
A copper penny dissolving in nitric acid. Copper metal is comparatively inactive, but it reacts with concentrated nitric acid. The brown fumes are $NO_2(g)$, a reduction product of HNO_3. The copper is oxidized to Cu^{2+} ions, which impart their color to the solution. (The penny is an old one made of solid copper. Newer pennies have a coating of copper over a zinc core.) *(Charles D. Winters)*

EXAMPLE 21.10 Write a balanced net ionic equation for the reaction of nitric acid with insoluble copper(II) sulfide; the products include Cu^{2+}, S(s), and $NO_2(g)$.

Strategy Follow the general procedure for writing and balancing redox equations in Chapter 4 and reviewed in Chapter 18. Note that because nitric acid is strong, it should be represented as H^+ and NO_3^- ions.

Solution

(1) The "skeleton" half-equations are

oxidation: $CuS(s) \longrightarrow S(s)$

reduction: $NO_3^-(aq) \longrightarrow NO_2(g)$

(2) The balanced half-equations are

oxidation: $CuS(s) \longrightarrow Cu^{2+}(aq) + S(s) + 2e^-$

reduction: $NO_3^-(aq) + 2H^+(aq) + e^- \longrightarrow NO_2(g) + H_2O$

(3) Multiplying the reduction half-equation by 2 and adding to the oxidation half-equation, you should obtain, after simplification,

$$CuS(s) + 2NO_3^-(aq) + 4H^+(aq) \longrightarrow Cu^{2+}(aq) + S(s) + 2NO_2(g) + 2H_2O$$

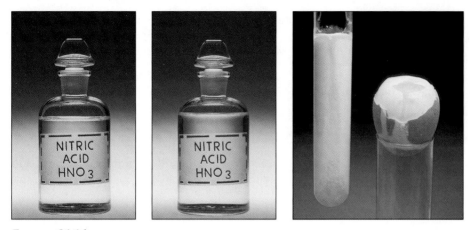

FIGURE 21.14
Nitric acid. A water solution of nitric acid slowly turns yellow because of the $NO_2(g)$ formed by decomposition. Nitric acid also reacts with proteins [casein in milk (test tube) and albumen in eggs] to give a characteristic yellow color. *(Charles D. Winters)*

Concentrated nitric acid (16 *M*) is colorless when pure. In sunlight, it turns yellow (Figure 21.14) because it decomposes to $NO_2(g)$:

$$4HNO_3(aq) \longrightarrow 4NO_2(g) + 2H_2O + O_2(g)$$

The yellow color that appears on your skin if it comes in contact with nitric acid has quite a different explanation. Nitric acid reacts with proteins to give a yellow material called xanthoprotein.

Sulfuric Acid, H_2SO_4

Sulfuric acid is made commercially by a three-step *contact process*. First, elemental sulfur is burned in air to form sulfur dioxide:

$$S(s) + O_2(g) \longrightarrow SO_2(g)$$

The sulfur dioxide is converted to sulfur trioxide by bringing it into contact with oxygen on the surface of a solid catalyst:

$$SO_2(g) + \tfrac{1}{2}O_2(g) \longrightarrow SO_3(g)$$

The catalyst used is divanadium pentoxide, V_2O_5, at a temperature of 450°C to 600°C. Sulfur trioxide is an acidic oxide that reacts with water to form sulfuric acid.

$$SO_3(g) + H_2O(l) \longrightarrow H_2SO_4(l)$$

Sulfuric acid is a strong acid, completely ionized to H^+ and HSO_4^- ions in dilute water solution. The HSO_4^- ion, in contrast, is only partially ionized.

$$H_2SO_4(aq) \longrightarrow H^+(aq) + HSO_4^-(aq) \qquad K_a \longrightarrow \infty$$

$$HSO_4^-(aq) \rightleftharpoons H^+(aq) + SO_4^{2-}(aq) \qquad K_a = 1.0 \times 10^{-2}$$

FIGURE 21.15

Sugar plus sulfuric acid. When sulfuric acid is added to sugar (sucrose), an exothermic reaction occurs. The elements of water are removed from the sugar, $C_{12}H_{22}O_{11}$, leaving a mass of black carbon. *(Charles D. Winters)*

Most of the H^+ ions come from the first ionization; in $0.10 M$ H_2SO_4, $[H^+] = 0.11 M$.

Sulfuric acid is a relatively weak oxidizing agent; most often it is reduced to sulfur dioxide:

$$SO_4^{2-}(aq) + 4H^+(aq) + 2e^- \longrightarrow SO_2(g) + 2H_2O \qquad E°_{red} = +0.155 \text{ V}$$

This is the case when copper metal is oxidized by hot concentrated sulfuric acid:

$$Cu(s) + 4H^+(aq) + SO_4^{2-}(aq) \longrightarrow Cu^{2+}(aq) + 2H_2O + SO_2(g)$$

When dilute sulfuric acid ($3 M$) reacts with metals, it is ordinarily the H^+ ion rather than the SO_4^{2-} ion that is reduced. For example, zinc reacts with dilute sulfuric acid to form hydrogen gas:

$$Zn(s) + 2H^+(aq) \longrightarrow Zn^{2+}(aq) + H_2(g)$$

Concentrated sulfuric acid, in addition to being an acid and an oxidizing agent, is also a dehydrating agent. Small amounts of water can be removed from organic liquids such as gasoline by extraction with sulfuric acid. Sometimes it is even possible to remove the elements of water from a compound by treating it with $18 M$ H_2SO_4. This happens with table sugar, $C_{12}H_{22}O_{11}$; the product is a black char that is mostly carbon (Figure 21.15).

$$C_{12}H_{22}O_{11}(s) \longrightarrow 12C(s) + 11H_2O$$

When concentrated sulfuric acid dissolves in water, a great deal of heat is given off, nearly 100 kJ per mole of H_2SO_4. Sometimes enough heat is evolved to bring the solution to the boiling point. To prevent this and to avoid splattering, the acid should be added slowly to water, with constant stirring. If it comes in contact with the skin, concentrated sulfuric acid can cause painful chemical burns. Never add water to concentrated H_2SO_4!

Phosphoric Acid, H_3PO_4

In water solution, $+5$ phosphorus can be present as—

- the H_3PO_4 molecule ($K_a = 7.1 \times 10^{-3}$), which is a (relatively strong) weak acid. A $0.10 M$ solution of H_3PO_4 has a pH of 1.6.
- the $H_2PO_4^-$ ion ($K_a = 6.2 \times 10^{-8}$; $K_b = 1.4 \times 10^{-12}$) is a (very) weak acid. A $0.10 M$ solution of NaH_2PO_4 has a pH of 5.
- the HPO_4^{2-} ion ($K_a = 4.5 \times 10^{-13}$; $K_b = 1.6 \times 10^{-7}$) is a (very) weak base. A $0.10 M$ solution of Na_2HPO_4 has a pH of 10.
- the PO_4^{3-} ion ($K_b = 2.2 \times 10^{-2}$) is a (relatively strong) weak base. A $0.10 M$ solution of Na_3PO_4 has a pH of 12.7.

Of the three compounds NaH_2PO_4, Na_2HPO_4, and Na_3PO_4, two are used as cleaning agents: NaH_2PO_4 in acid-type cleaners, Na_3PO_4 in strongly basic cleaners. The principal use of Na_2HPO_4 is in the manufacture of cheese; J. L. Kraft discovered 75 years ago that this compound is an excellent emulsifying agent. To this day no one quite understands why this happens.

A great many "phosphates" are used in commercial fertilizers. Perhaps the most important of these is calcium dihydrogen phosphate, $Ca(H_2PO_4)_2$. In relatively pure form, this compound is known as "triple superphosphate of lime." A 1:2 mol mixture of $Ca(H_2PO_4)_2$ and gypsum, $CaSO_4 \cdot 2H_2O$, is commonly referred to as "superphosphate of lime."

Arsenic and Selenium

Arsenic and selenium, which fall directly below phosphorus and sulfur in the periodic table, are of interest for a variety of reasons. Arsenic is a true metalloid. A metallic allotrope, called gray arsenic, has an electrical conductivity approaching that of lead. Another allotrope, yellow arsenic, is distinctly nonmetallic; it has the molecular formula As_4, analogous to white phosphorus, P_4. Selenium is properly classified as a nonmetal, although one of its allotropes has a somewhat metallic appearance and is a semiconductor. Another allotrope of selenium has the molecular formula Se_8, analogous to sulfur.

The principal use of elemental arsenic is in its alloys with lead. The "lead" storage battery contains a trace of arsenic along with 3% antimony. Lead shot, formed by allowing drops of molten metal to fall through air, contains from 0.5 to 2.0% arsenic. The presence of arsenic raises the surface tension of the liquid and hence makes the shot more nearly spherical.

Selenium has long been used as an additive in glassmaking. Particles of colloidally dispersed selenium give the ruby-red color seen in stained glass windows and traffic lights (Figure A). Very small amounts of a sulfide of selenium are used in certain dandruff shampoos (e.g., Selsun Blue).

The principal use of selenium today takes advantage of its *photoconductivity;* the electrical conductivity of selenium increases by a factor of 1000 when it is exposed to light. Modern photocopiers use a plate in which a thin film of photoconductive material, usually selenium, is deposited on an aluminum base. A high potential is placed on the plate, which is then exposed to a light and dark image pattern, obtained by illuminating the article to be copied. In the light areas, the electrostatic charge on the plate is reduced as a result of a photoconductive discharge; the dark areas retain their original charge. An image is formed on the plate when toner particles (black or colored) are attracted to the high-charge areas. The image is then transferred to a piece of plain paper by charging it sufficiently to pull the toner particles off the plate.

FIGURE A
Stained glass colors. Addition of a selenium compound produces red glass, and addition of a cobalt compound produces blue glass. *(David Michael Davis/FPG International)*

Physiological Properties

The "arsenic poison" referred to in true-crime dramas is actually the oxide of arsenic, As_2O_3, rather than the element itself. Less than 0.1 g of this white, slightly soluble powder can be fatal. The classic symptoms of acute arsenic poisoning involve various unpleasant gastrointestinal disturbances, severe abdominal pain, and burning of the mouth and throat.

In the modern forensic chemistry laboratory (Figure B) arsenic is detected by analysis of hair samples, where the element tends to concentrate in chronic arsenic poisoning. A single strand of hair is sufficient to establish the presence or absence of the element. The technique most commonly used is neutron activation analysis, described in Chapter 19. If the concentration found is greater than about 0.0003%, poisoning is indicated; normal arsenic levels are much lower than this.

This technique was applied in the early 1960s to a lock of hair taken from Napoleon Bonaparte (1769–1821) on St. Helena. Arsenic levels of up to 50 times normal suggested he may have been a victim of poisoning, perhaps on orders from the French royal family.

It has long been known that high concentrations of selenium are toxic. The cattle disease known picturesquely as "blind staggers" arises from grazing on grass growing in soil with a high selenium content. It is now known, however, that selenium is an essential element in human nutrition. A selenium-containing enzyme, glutathione peroxidase, in combination with vitamin E, destroys harmful free radicals in the body.

There is considerable evidence to suggest that selenium compounds are anticarcinogens. For one thing, tests with laboratory animals show that the incidence and size of malignant tumors are reduced when a solution containing Na_2SeO_3 is injected at the part per million level. Beyond that evidence, statistical studies show an inverse correlation between selenium levels in the soil and the incidence of certain types of cancer.

FIGURE B
A forensic toxicology laboratory. The young woman is working on a research project in the laboratory at Hartford Hospital. *(Photo courtesy of Professor S. Ruven Smith)*

CHAPTER HIGHLIGHTS

Key Concepts

1. Carry out equilibrium calculations for solution reactions.
   (Example 21.1; Problems 47–52)
2. Apply the Gibbs-Helmholtz equation.
   (Example 21.2; Problems 53–58)
3. Write balanced net ionic equations for solution reactions.
   (Examples 21.3, 21.5, 21.10; Problems 15–26, 69)
4. Draw Lewis structures for compounds of the nonmetals.
   (Examples 21.6, 21.8; Problems 31–38)
5. Relate oxoacids to the corresponding oxides and compare their acid strengths.
   (Examples 21.7, 21.8; Problems 7, 8, 67)
6. Carry out electrochemical calculations involving $E°$, the Nernst equation, and/or electrolysis.
   (Examples 21.4, 21.9; Problems 59–66)

Key Terms

- acid
  acidic oxide
  allotrope
- base
- disproportionation

- $E°_{ox}$, $E°_{red}$
- electronegativity
- Lewis structure
- nonmetal
- oxidation number

- oxidizing agent
- oxoacid
- oxoanion
- reducing agent
- resonance

Summary Problem

Consider bromine, whose chemistry is quite similar to that of chlorine.

(a) Write equations for the reaction of bromine with I^- ions; for the reaction of Br_2 with water (disproportionation).

(b) Write equations for the preparation of bromine by the electrolysis of aqueous NaBr, by the reaction of chlorine with bromide ions in aqueous solution.

(c) Write equations for the reaction of hydrobromic acid with OH^- ions; with CO_3^{2-} ions; with ammonia.

(d) Consider the species Br^-, Br_2, BrO^-, BrO_4^-. In a redox reaction, which of these species can act only as an oxidizing agent? Only as a reducing agent? Which can act as either oxidizing or reducing agents?

(e) When aqueous sodium bromide is heated with a concentrated solution of sulfuric acid, the products include liquid bromine and sulfur dioxide. Write a balanced equation for the redox reaction involved.

(f) Write Lewis structures for the following species: Br^-, BrO^-, BrO_4^-. What is the bond angle in the BrO_4^- ion? For which of these ions is the conjugate acid the weakest?

(g) Give the formula for the acidic oxide of HBrO. Knowing that K_a of HBrO is 2.6×10^{-9}, calculate the ratio [HBrO]/[BrO$^-$] at pH 10.00.

(h) For the reduction of HBrO to Br^-, what is the change in voltage when the pH increases by one unit?

Answers

(a) $Br_2(l) + 2I^-(aq) \rightarrow 2Br^-(aq) + I_2(s)$
$Br_2(l) + H_2O \rightarrow HBrO(aq) + H^+(aq) + Br^-(aq)$

(b) $2Br^-(aq) + 2H_2O \rightarrow Br_2(l) + H_2(g) + 2\,OH^-(aq)$
$2Br^-(aq) + Cl_2(aq) \rightarrow Br_2(l) + 2Cl^-(aq)$

(c) $H^+(aq) + OH^-(aq) \rightarrow H_2O$; $2H^+(aq) + CO_3{}^{2-}(aq) \rightarrow CO_2(g) + H_2O$
$H^+(aq) + NH_3(aq) \rightarrow NH_4{}^+(aq)$

(d) $BrO_4{}^-$; Br^-; Br_2; BrO^-

(e) $2Br^-(aq) + SO_4{}^{2-}(aq) + 4H^+(aq) \rightarrow Br_2(l) + SO_2(g) + 2H_2O$

(f) $[:\ddot{B}r:]^-$ $[:\ddot{B}r-\ddot{O}:]^-$ $[:\ddot{O}-\overset{\displaystyle :\ddot{O}:}{\underset{\displaystyle :\ddot{O}:}{\overset{|}{\underset{|}{Br}}}}-\ddot{O}:]^-$ $109.5°$; BrO^-

(g) Br_2O; 0.038 **(h)** -0.0296 V

Questions & Problems

Problem numbers in blue indicate that the answer is available in Appendix 6 at the back of the book.
WEB indicates that the solution is posted at **http://www.harcourtcollege.com/chem/general/masterton4/student/**

Formulas, Equations, and Reactions

1. Name the following species.
(a) HIO_4 (b) $BrO_2{}^-$ (c) HIO (d) $NaClO_3$

2. Name the following compounds.
(a) $HBrO_3$ (b) KIO (c) $NaClO_2$ (d) $NaBrO_4$

3. Write the formula for each of the following compounds.
(a) chloric acid   (b) periodic acid
(c) hypobromous acid   (d) hydriodic acid

4. Write the formula for each of the following compounds.
(a) potassium bromite   (b) calcium bromide
(c) sodium periodate   (d) magnesium hypochlorite

5. Write the formula of a compound of each of the following elements that *cannot* act as an oxidizing agent.
(a) N   (b) S   (c) Cl

6. Write the formula of an oxoanion of each of the following elements that *cannot* act as a reducing agent.
(a) N   (b) S   (c) Cl

7. Give the formula for the acidic oxide of
(a) HNO_3 (b) HNO_2 (c) H_2SO_4

8. Write the formula of the acid formed when each of these acidic oxides reacts with water.
(a) SO_2 (b) Cl_2O (c) P_4O_6

9. Write the formulas of the following compounds.
(a) ammonia   (b) laughing gas
(c) hydrogen peroxide   (d) sulfur trioxide

10. Write the formula for the following compounds.
(a) sodium azide   (b) sulfurous acid
(c) hydrazine   (d) sodium dihydrogen phosphate

11. Write the formula of a compound of hydrogen with
(a) nitrogen, which is a gas at 25°C and 1 atm.
(b) phosphorus, which is a liquid at 25°C and 1 atm.
(c) oxygen, which contains an O—O bond.

12. Write the formula of a compound of hydrogen with

(a) sulfur.
(b) nitrogen, which is a liquid at 25°C and 1 atm.
(c) phosphorus, which is a poisonous gas at 25°C and 1 atm.

13. Give the formula of
(a) an anion in which S has an oxidation number of -2.
(b) two anions in which S has an oxidation number of $+4$.
(c) two different acids of sulfur.

14. Give the formula of a compound of nitrogen that is
(a) a weak base.   (b) a strong acid.
(c) a weak acid.   (d) capable of oxidizing copper.

15. Write a balanced net ionic equation for
(a) the electrolytic decomposition of hydrogen fluoride.
(b) the oxidation of iodide ion to iodine by hydrogen peroxide in acidic solution. Hydrogen peroxide is reduced to water.

16. Write a balanced net ionic equation for
(a) the oxidation of iodide to iodine by sulfate ion in acidic solution. Sulfur dioxide gas is also produced.
(b) The preparation of iodine from an iodide salt and chlorine gas.

17. Write a balanced net ionic equation for the disproportionation reaction
(a) of iodine to give iodate and iodide ions in basic solution.
(b) of chlorine gas to chloride and perchlorate ions in basic solution.

WEB 18. Write a balanced net ionic equation for the disproportionation reaction of
(a) hypochlorous acid to chlorine gas and chlorous acid in acidic solution.
(b) chlorate ion to perchlorate and chlorite ions.

19. Complete and balance the following equations. If no reaction occurs, write NR.

(a) $Cl_2(g) + I^-(aq) \rightarrow$ (b) $F_2(g) + Br^-(aq) \rightarrow$
(c) $I_2(s) + Cl^-(aq) \rightarrow$ (d) $Br_2(l) + I^-(aq) \rightarrow$

20. Complete and balance the following equations. If no reaction occurs, write NR.

(a) $Cl_2(g) + Br^-(aq) \rightarrow$ (b) $I_2(s) + Cl^-(aq) \rightarrow$
(c) $I_2(s) + Br^-(aq) \rightarrow$ (d) $Br_2(l) + Cl^-(aq) \rightarrow$

21. Write a balanced equation for the preparation of

(a) F_2 from HF   (b) Br_2 from NaBr
(c) NH_4^+ from NH_3

22. Write a balanced equation for the preparation of

(a) N_2 from $Pb(N_3)_2$ (b) O_2 from O_3
(c) S from H_2S

23. Write a balanced equation for the reaction of ammonia with

(a) Cu^{2+} (b) H^+ (c) Al^{3+}

24. Write a balanced equation for the reaction of hydrogen sulfide with

(a) Cd^{2+} (b) OH^- (c) $O_2(g)$

25. Write a balanced net ionic equation for the reaction of nitric acid with

(a) a solution of $Ca(OH)_2$.
(b) $Ag(s)$; assume the nitrate ion is reduced to $NO_2(g)$
(c) $Cd(s)$; assume the nitrate ion is reduced to $N_2(g)$.

26. Write a balanced net ionic equation for the reaction of sulfuric acid with

(a) $CaCO_3(s)$.
(b) a solution of NaOH.
(c) Cu; assume the SO_4^{2-} ion is reduced to SO_2.

Allotropy

27. Which of the following elements show allotropy?

(a) F   (b) O   (c) S   (d) Cl

28. Which of the following elements show allotropy?

(a) Br   (b) N   (c) P   (d) I

29. Describe the structural difference between

(a) ozone and ordinary oxygen.
(b) white phosphorus and red phosphorus.
(c) liquid sulfur at 120° and 180°.

30. Describe how

(a) ozone can be prepared from O_2.
(b) monoclinic sulfur can be prepared from rhombic sulfur.
(c) red phosphorus can be made from white phosphorus.

Molecular Structure

31. Give the Lewis structure of

(a) NO_2 (b) NO   (c) SO_2 (d) SO_3

WEB **32.** Give the Lewis structure of

(a) Cl_2O (b) N_2O (c) P_4 (d) N_2

33. Which of the molecules in Question 31 are polar?

34. Which of the molecules in Question 32 are polar?

35. Give the Lewis structure of

(a) HNO_3 (b) H_2SO_4 (c) H_3PO_4

36. Give the Lewis structures of the conjugate bases of the species in Question 35.

37. Give the Lewis structure of

(a) the strongest oxoacid of bromine.
(b) a hydrogen compound of nitrogen in which there is an
—N—N— bond.
(c) an acid added to cola drinks.

38. Give the Lewis structure of

(a) an oxide of nitrogen in the +5 state.
(b) the strongest oxoacid of nitrogen.
(c) a tetrahedral oxoanion of sulfur.

Stoichiometry

39. The average concentration of bromine (as bromide) in sea water is 65 ppm. Calculate

(a) the volume of sea water ($d = 64.0$ lb/ft^3) in cubic feet required to produce one kilogram of liquid bromine.
(b) the volume of chlorine gas in liters, measured at 20°C and 762 mm Hg, required to react with this volume of sea water.

40. A 425-gal tank is filled with water containing 175 g of sodium iodide. How many liters of chlorine gas at 758 mm Hg and 25°C will be required to oxidize all the iodide to iodine?

41. Iodine can be prepared by allowing an aqueous solution of hydrogen iodide to react with manganese dioxide, MnO_2. The reaction is

$$2I^-(aq) + 4H^+(aq) + MnO_2(s) \longrightarrow Mn^{2+}(aq) + 2H_2O + I_2(s)$$

If an excess of hydrogen iodide is added to 0.200 g MnO_2, how many grams of iodine are obtained, assuming 100% yield?

42. When a solution of hydrogen bromide is prepared, 1.283 L of HBr gas at 25°C and 0.974 atm is bubbled into 250.0 mL of water. Assuming all the HBr dissolves with no volume change, what is the molarity of the hydrobromic acid solution produced?

43. When ammonium nitrate explodes, nitrogen, steam, and oxygen gas are produced. If the explosion is carried out by heating one kilogram of ammonium nitrate sealed in a rigid bomb with a volume of one liter, what is the total pressure produced by the gases before the bomb ruptures? Assume that the reaction goes to completion and that the final temperature is 500°C. (3 significant figures)

44. Sulfur dioxide can be removed from the smokestack emissions of power plants by reacting it with hydrogen sulfide, producing sulfur and water. What volume of hydrogen sulfide at 27°C and 755 mm Hg is required to remove the sulfur dioxide produced by a power plant that burns one metric ton of coal containing 5.0% sulfur by mass? How many grams of sulfur are produced by the reaction of H_2S with SO_2?

45. A 1.500-g sample containing sodium nitrate was heated to form $NaNO_2$ and O_2. The oxygen evolved was collected over water at 23°C and 752 mm Hg; its volume was 125.0 mL. Calculate the percentage of $NaNO_3$ in the sample. The vapor pressure of water at 23°C is 21.07 mm Hg.

46. Chlorine can remove the foul smell of H_2S in water. The reaction is

$$H_2S(aq) + Cl_2(aq) \longrightarrow 2H^+(aq) + 2Cl^-(aq) + S(s)$$

If the contaminated water has 5.0 ppm hydrogen sulfide by mass, what volume of chlorine gas at STP is required to remove all the H_2S from 1.00×10^3 gallons of water ($d = 1.00$ g/mL)? What is the pH of the solution after treatment with chlorine?

Equilibria

47. The equilibrium constant at 25°C for the reaction

$$Br_2(l) + H_2O \rightleftharpoons H^+(aq) + Br^-(aq) + HBrO(aq)$$

is 1.2×10^{-9}. This is the system present in a bottle of "bromine water." Assuming that HBrO does not ionize appreciably, what is the pH of the bromine water?

48. Calculate the pH and the equilibrium concentration of HClO in a 0.10 M solution of hypochlorous acid. K_a HClO = 2.8×10^{-8}.

49. At equilibrium, a gas mixture has a partial pressure of 0.7324 atm for HBr and 2.80×10^{-3} atm for both hydrogen and bromine gases. What is K for the formation of two moles of HBr from H_2 and Br_2?

WEB 50. Given

$$HF(aq) \rightleftharpoons H^+(aq) + F^-(aq) \qquad K_a = 6.9 \times 10^{-4}$$

$$HF(aq) + F^-(aq) \rightleftharpoons HF_2^-(aq) \qquad K = 2.7$$

calculate K for the reaction

$$2HF(aq) \rightleftharpoons H^+(aq) + HF_2^-(aq)$$

51. What is the concentration of fluoride ion in a water solution saturated with BaF_2, $K_{sp} = 1.8 \times 10^{-7}$?

52. Calculate the solubility in grams per 100 mL of BaF_2 in a 0.10 M $BaCl_2$ solution.

Thermodynamics

53. Determine whether the following redox reaction is spontaneous at 25°C and 1 atm:

$$2KIO_3(s) + Cl_2(g) \longrightarrow 2KClO_3(s) + I_2(s)$$

Use data in Appendix 1 and the following information: ΔH_f° $\Delta KIO_3(s) = -501.4$ kJ/mol, S° $KIO_3(s) = 151.5$ J/mol·K. What is the lowest temperature at which the reaction is spontaneous?

54. Follow the directions for Problem 53 for the reaction

$$2KBrO_3(s) + Cl_2(g) \longrightarrow 2KClO_3(s) + Br_2(l)$$

The following thermodynamic data may be useful:

$$\Delta H_f^\circ \text{ } KBrO_3 = -360.2 \text{ kJ/mol}; \text{ } S^\circ \text{ } KBrO_3 = 149.2 \text{ J/mol·K}$$

55. Consider the equilibrium system

$$HF(aq) \rightleftharpoons H^+(aq) + F^-(aq)$$

Given ΔH_f° HF(aq) = -320.1 kJ/mol,

$$\Delta H_f^\circ \text{ } F^-(aq) = -332.6 \text{ kJ/mol; } S^\circ \text{ } F^-(aq) = -13.8 \text{ J/mol·K;}$$

$$K_a \text{ HF} = 6.9 \times 10^{-4} \text{ at 25°C}$$

calculate S° for HF(aq).

56. Applying the Tables in Appendix 1 to

$$4HCl(g) + O_2(g) \longrightarrow 2Cl_2(g) + 2H_2O(l)$$

determine

(a) whether the reaction is spontaneous at 25°C and 1 atm.

(b) K for the reaction at 25°C.

57. Consider the reaction

$$4NH_3(g) + 5O_2(g) \longrightarrow 4NO(g) + 6H_2O(g)$$

(a) Calculate ΔH° for this reaction. Is it exothermic or endothermic?

(b) Would you expect ΔS° to be positive or negative? Calculate ΔS°.

(c) Is the reaction spontaneous at 25°C and 1 atm?

(d) At what temperature, if any, is the reaction at equilibrium at 1 atm pressure?

58. Data are given in Appendix 1 for white phosphorus, $P_4(s)$. $P_4(g)$ has the following thermodynamic values: $\Delta H_f^\circ = 58.9$ kJ/mol, $S^\circ = 280.0$ J/K·mol. What is the temperature at which white phosphorus sublimes at 1 atm pressure?

Electrochemistry

59. In the electrolysis of a KI solution, using 5.00 V, how much electrical energy in kilojoules is consumed when one mole of I_2 is formed?

WEB 60. If an electrolytic cell producing fluorine uses a current of 7.00×10^3 A (at 10.0 V), how many grams of fluorine gas can be produced in two days (assuming that the cell operates continuously at 95% efficiency)?

61. Sodium hypochlorite is produced by the electrolysis of cold sodium chloride solution. How long must a cell operate to produce 1.500×10^3 L of 5.00% NaClO by mass if the cell current is 2.00×10^3 A? Assume that the density of the solution is 1.00 g/cm³.

62. Sodium perchlorate is produced by the electrolysis of sodium chlorate. If a current of 1.50×10^3 A passes through an electrolytic cell, how many kilograms of sodium perchlorate are produced in an eight-hour run?

63. Taking E_{ox}° $H_2O_2 = -0.695$ V, determine which of the following species will be reduced by hydrogen peroxide (use Table 18.1 to find E_{red}° values).

(a) $Cr_2O_7^{2-}$ **(b)** Fe^{2+} **(c)** I_2 **(d)** Br_2

64. Taking E_{red}° $H_2O_2 = +1.763$ V, determine which of the following species will be oxidized by hydrogen peroxide (use Table 18.1 to find E_{ox}° values).

(a) Co^{2+} **(b)** Cl^- **(c)** Fe^{2+} **(d)** Sn^{2+}

65. Consider the reduction of nitrate ion in acidic solution to nitrogen oxide ($E_{red}^\circ = 0.964$ V) by sulfur dioxide that is oxidized to sulfate ion ($E_{red}^\circ = 0.155$ V). Calculate the voltage of a cell involving this reaction in which all the gases have pressures of 1.00 atm, all the ionic species (except H^+) are at 0.100 M, and the pH is 4.30.

66. For the reaction in Problem 65 if gas pressures are at 1.00 atm and ionic species are at 0.100 M (except H^+), at what pH will the voltage be 1.000 V?

Unclassified

67. Choose the strongest acid from each group.
(a) $HClO$, $HBrO$, HIO (b) HIO, HIO_3, HIO_4
(c) HIO, $HBrO_2$, $HBrO_4$

68. What intermolecular forces are present in the following?
(a) Cl_2 (b) HBr (c) HF
(d) $HClO_4$ (e) MgI_2

69. Write a balanced equation for the reaction of hydrofluoric acid with SiO_2. What volume of 2.0 M HF is required to react with one gram of silicon dioxide?

70. State the oxidation number of N in
(a) NO_2^- (b) NO_2 (c) HNO_3 (d) NH_4^+

71. The density of sulfur vapor at one atmosphere pressure and 973 K is 0.8012 g/L. What is the molecular formula of the vapor?

72. Give the formula of a substance discussed in this chapter that is used
(a) to disinfect water.   (b) in safety matches.
(c) to prepare hydrazine.   (d) to etch glass.

73. Why does concentrated nitric acid often have a yellow color even though pure HNO_3 is colorless?

74. Explain why
(a) acid strength increases as the oxidation number of the central nonmetal atom increases.
(b) nitrogen dioxide is paramagnetic.
(c) the oxidizing strength of an oxoanion is inversely related to pH.
(d) sugar turns black when treated with concentrated sulfuric acid.

Challenge Problems

75. Suppose you wish to calculate the mass of sulfuric acid that can be obtained from an underground deposit of sulfur 1.00 km^2 in area. What additional information do you need to make this calculation?

76. The reaction

$$4HF(aq) + SiO_2(aq) \longrightarrow SiF_4(aq) + 2H_2O$$

can be used to release gold that is distributed in certain quartz (SiO_2) veins of hydrothermal origin. If the quartz contains $1.0 \times 10^{-3}\%$ Au by weight and the gold has a market value of $425 per troy ounce, would the process be economically feasible if commercial HF (50% by weight, $d = 1.17$ g/cm^3) costs 75¢ a liter? (1 troy ounce = 31.1 g.)

77. The amount of sodium hypochlorite in a bleach solution can be determined by using a given volume of bleach to oxidize excess iodide ion to iodine; ClO^- is reduced to Cl^-. The amount of iodine produced by the redox reaction is determined by titration with sodium thiosulfate, $Na_2S_2O_3$; I_2 is reduced to I^-. The sodium thiosulfate is oxidized to sodium tetrathionate, $Na_2S_4O_6$. In this analysis, potassium iodide was added in excess to 5.00 mL of bleach ($d = 1.00$ g/cm^3). If 25.00 mL of 0.0700 M $Na_2S_2O_3$ was required to reduce all the iodine produced by the bleach back to iodide, what is the mass percent of NaClO in the bleach?

78. What is the minimum amount of sodium azide, NaN_3, that can be added to an automobile airbag to give a volume of 20.0 L of $N_2(g)$ on inflation? Make any reasonable assumptions required to obtain an answer, but state what these assumptions are.

ORGANIC CHEMISTRY

22

Crystals of insulin under polarized light. *(Phillip A. Harrington/
Fran Heyl Associates)*

CHAPTER OUTLINE

22.1   SATURATED HYDROCARBONS: ALKANES

22.2   UNSATURATED HYDROCARBONS: ALKENES AND ALKYNES

22.3   AROMATIC HYDROCARBONS AND THEIR DERIVATIVES

22.4   OXYGEN COMPOUNDS; FUNCTIONAL GROUPS

22.5   ISOMERISM IN ORGANIC COMPOUNDS

Organic chemistry deals with the compounds of carbon, of which there are literally millions. More than 90% of all known compounds contain carbon atoms. There is a simple explanation for this remarkable fact. Carbon atoms bond to one another to a far greater extent than do atoms of any other element. Carbon atoms may link together to form chains or rings.

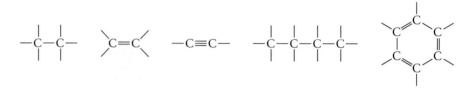

The bonds may be single (one electron pair), double (two electron pairs), or triple (three electron pairs).

There are a wide variety of different organic compounds that have quite different structures and properties. However, all these substances have certain features in common:

1. *Organic compounds are ordinarily molecular rather than ionic.* Most of the compounds we discuss consist of small, discrete molecules. Many of them are gases or liquids at room temperature.
2. *Each carbon atom forms a total of four covalent bonds.* This is illustrated by the structures written above. A particular carbon atom may form four single bonds, two single bonds and a double bond, two double bonds, or one single bond and a triple bond. One way or another, though, the bonds add up to four.
3. *Carbon atoms may be bonded to each other or to other nonmetal atoms, most often hydrogen, a halogen, oxygen, or nitrogen.* In most organic compounds—

- a hydrogen or halogen atom (F, Cl, Br, I) forms one covalent bond, —H, —X
- an oxygen atom forms two covalent bonds, —O— or =O
- a nitrogen atom forms three covalent bonds, —N—, =N—, or ≡N

In this chapter, we consider—

- the simplest type of organic compound, called a **hydrocarbon,** which contains only two kinds of atoms, hydrogen and carbon. Hydrocarbons can be classified as alkanes (Section 22.1), alkenes and alkynes (Section 22.2), and aromatics (Section 22.3).

> Carbon always follows the octet rule.

- organic compounds containing oxygen atoms in addition to carbon and hydrogen (Section 22.4).

- the phenomenon of isomerism, which is very common among organic compounds. This topic is introduced in Section 22.1 but discussed more generally in Section 22.5.

22.1   SATURATED HYDROCARBONS: ALKANES

One large and structurally simple class of hydrocarbons includes those substances in which all the carbon-carbon bonds are single bonds. These are called **saturated hydrocarbons,** or **alkanes.** In the alkanes the carbon atoms are bonded to each other in chains, which may be long or short, straight or branched.

The ratio of hydrogen to carbon atoms is a maximum in alkanes, hence the term "saturated" hydrocarbon. The general formula of an alkane containing n carbon atoms is

$$C_nH_{2n+2}$$

The simplest alkanes are those for which $n = 1$ (CH_4), $n = 2$ (C_2H_6), or $n = 3$ (C_3H_8):

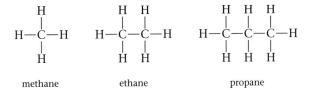

methane   ethane   propane

Around the carbon atoms in these molecules and indeed in any saturated hydrocarbon, there are four single bonds involving sp^3 hybrid orbitals. As would be expected from the VSEPR model, these bonds are directed toward the corners of a regular tetrahedron. The bond angles are approximately 109.5°, the tetrahedral angle. This means that in propane (C_3H_8) and in the higher alkanes, the carbon atoms are arranged in a "zigzag" pattern (Figure 22.1).

Two different alkanes are known with the molecular formula C_4H_{10}. In one of these, called butane, the four carbon atoms are linked in a "straight chain." In the other, called 2-methylpropane,* there is a "branched chain." The longest chain in the molecule contains three carbon atoms; there is a CH_3 branch from the central

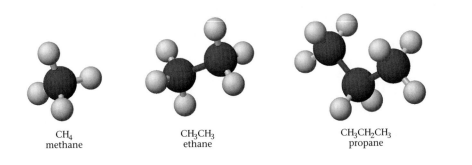

CH₄
methane

CH₃CH₃
ethane

CH₃CH₂CH₃
propane

FIGURE 22.1
The three simplest alkanes. The bond angles in methane, ethane, and propane are all close to 109.5°, the tetrahedral angle. **OHT**

See the *Saunders Interactive General Chemistry CD-ROM,* Screen 3.6, Alkanes: A Class of Compounds.

See Screen 11.2, Carbon-Carbon Bonds.

*The nomenclature of alkanes is discussed in Appendix 4.

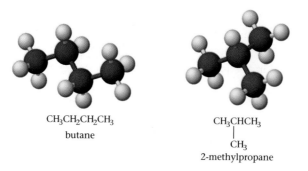

CH₃CH₂CH₂CH₃
butane

CH₃CHCH₃
|
CH₃
2-methylpropane

carbon atom. The geometries of these molecules are shown in Figure 22.2. The
structures are

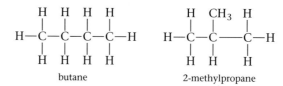

butane   2-methylpropane

In structural isomers, the atoms are
bonded in different patterns; the
skeletons are different.

Compounds having the same molecular formula but different molecular structures are called **structural isomers.** Butane and 2-methylpropane are referred to
as structural isomers of C_4H_{10}. They are two distinct compounds with their own
characteristic physical and chemical properties.

EXAMPLE 22.1   Draw structures for the isomers of C_5H_{12}.

Strategy   Start with the straight-chain structure, stringing all five carbon atoms one after the other. Then work with structures containing four carbon atoms in a chain with one
branch; find all the nonequivalent structures of this type. Continue this process using a
three-carbon chain, which is the shortest one that can be drawn for C_5H_{12}.

Solution   For simplicity, we show only the carbon atoms; there is an H atom attached
to each bond extending from a C atom.

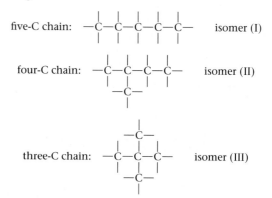

five-C chain: isomer (I)

four-C chain: isomer (II)

three-C chain: isomer (III)

Reality Check   Working with only pencil and paper, you might be tempted to draw
other structures, such as

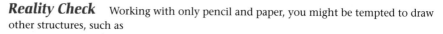

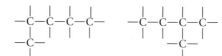

If you want a really tedious job,
try drawing the 366,319 structural
isomers of $C_{20}H_{42}$.

However, a few moments of reflection (or access to a molecular model kit) should convince you that these are in fact equivalent to structures written previously. In particular, the first one, like isomer I, has a five-carbon chain in which no carbon atom is attached to more than two other carbons. The second structure, like II, has a four-carbon chain with one carbon atom bonded to three other carbons. Structures I, II, and III represent the three possible isomers of C_5H_{12}; there are no others.

Structural isomers have different properties. For example, the three isomers of C_5H_{12} listed in Example 22.1 boil at different temperatures.

Structure	Normal Boiling Point
(I)	36°C
(II)	28°C
(III)	10°C

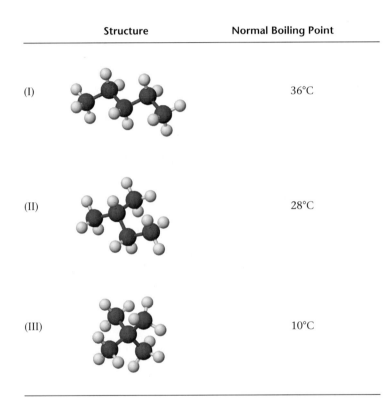

In general, long "skinny" isomers like I tend to have higher boiling points than highly branched isomers like III. Dispersion forces become weaker as chain-branching increases. This principle is used in the seasonal blending of gasoline. A greater percentage of more volatile, highly branched alkanes is included in winter, making combustion occur more readily at lower temperatures.

Sources and Uses of Alkanes

Natural gas, transmitted around the United States and Canada by pipeline, consists largely of methane (80 to 90%), with smaller amounts of C_2H_6, C_3H_8, and C_4H_{10}. Cylinders of "bottled gas," used with campstoves, barbecue grills, and the like, contain liquid propane (C_3H_8) and butane (C_4H_{10}) (Figure 22.3). The pressure remains constant as long as any liquid is present, then drops abruptly to zero, indicating that it's time for a recharge.

FIGURE 22.3
Bottled gas. The barbecue grill is fueled by a mixture of liquid propane (C_3H_8) and liquid butane (C_4H_{10}).
(David Frazier/Photolibrary)

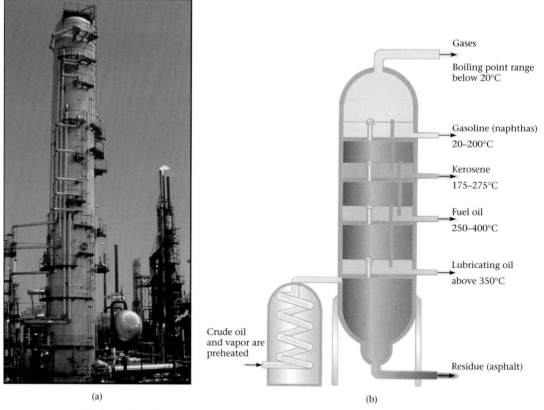

(a)   (b)

FIGURE 22.4

Petroleum distillation. (a) A distillation tower at a petroleum refinery. (b) A diagram show-ing the boiling points of the petroleum fractions separated by distillation. *(a, courtesy of Ashland Oil Company)* **OHT**

The higher alkanes are most often obtained from petroleum, a dark brown, viscous liquid dispersed through porous rock deposits. Distillation of petroleum gives a series of fractions of different boiling points (Figure 22.4). The most impor-tant of these is gasoline; distillation of a liter of petroleum gives about 250 mL of "straight-run" gasoline. It is possible to double the yield of gasoline by con-verting higher- or lower-boiling fractions to hydrocarbons in the gasoline range (C_5 to C_{12}).

Gasoline-air mixtures tend to ignite prematurely or "knock" rather than burn smoothly. The *octane number* of a gasoline is a measure of its resistance to knock. It is determined by comparing the knocking characteristics of a gasoline sample to those of isooctane and heptane:

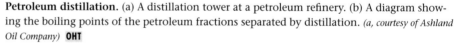

Isooctane, which is highly branched, burns smoothly with little knocking and is assigned an octane number of 100. Heptane, being unbranched, knocks badly. It is given an octane number of zero. Gasoline with the same knocking properties as a mixture of 90% isooctane and 10% heptane is rated as "90 octane."

To obtain premium gasoline of octane number above 80, it is necessary to use additives of one type or another. Until the mid-1970s, the principal antiknock agent was tetraethyllead, $(C_2H_5)_4Pb$. Its use was phased out because it poisons catalytic converters and contaminates the environment with lead compounds. An additive that is widely used today is methyl-*t*-butyl ether (MTBE):

$$CH_3-\underset{\underset{CH_3}{|}}{\overset{\overset{CH_3}{|}}{C}}-O-CH_3$$

This compound was first added to gasoline in small quantities in the 1970s. In the early 1990s, larger amounts of MTBE, up to 15% by volume, were added to gasoline sold in urban areas to ensure complete combustion, thereby lowering carbon monoxide levels. Recently, however, MTBE, which is very soluble in water, has been found occasionally in ground water at concentrations high enough to give an offensive taste and odor.

22.2 UNSATURATED HYDROCARBONS: ALKENES AND ALKYNES

In an **unsaturated hydrocarbon,** at least one of the carbon-carbon bonds in the molecule is a multiple bond. As a result, there are fewer hydrogen atoms in an unsaturated hydrocarbon than there are in a saturated one with the same number of carbons. Unsaturated hydrocarbons are of two major types—

See Screen 11.3, Hydrocarbons.

■ **alkenes,** in which there is one carbon-carbon double bond in the molecule:

$$\overset{\diagdown}{\underset{\diagup}{C}}=\overset{\diagup}{\underset{\diagdown}{C}}$$

Replacing a single bond with a double bond eliminates two hydrogen atoms. Hence the general formula of an alkene is

$$C_nH_{2n}$$

as compared to C_nH_{2n+2} for an alkane.

■ **alkynes,** in which there is one carbon-carbon triple bond in the molecule:

$$-C\equiv C-$$

Again, two hydrogen atoms are "lost" when a double bond is converted to a triple bond. Hence the general formula of an alkyne is

$$C_nH_{2n-2}$$

Alkenes

The simplest alkene is ethylene, C_2H_4

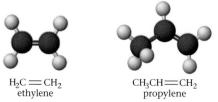

FIGURE 22.5
The two simplest alkenes. Ethylene is planar, as is the —CH=CH₂ region of propylene.

$H_2C=CH_2$
ethylene

$CH_3CH=CH_2$
propylene

Ethylene is produced in larger amounts than any other organic compound; about 1.7×10^7 metric tons of C_2H_4 are manufactured in the United States each year. Ethylene is used primarily as a starting material for the preparation of other organic compounds, including ethyl alcohol, ethylene glycol, and, in particular, the polymer known as polyethylene (page 272). Smaller amounts of ethylene are used to ripen fruit, picked while still unripe to avoid spoilage.

⌐ That's how we get red tomatoes that taste like sawdust.

You may recall that we discussed the bonding in ethylene in Chapter 7. The double bond in ethylene and other alkenes consists of a sigma bond and a pi bond. The ethylene molecule is planar. There is no rotation about the double bond, because that would require "breaking" the pi bond. The bond angles in ethylene are approximately 120°, corresponding to sp² hybridization about each carbon atom. The geometries of ethylene and the next member of the alkene series, C_3H_6, are shown in Figure 22.5.

Ethylene and other alkenes are considerably more reactive than the corresponding alkanes. In particular, they undergo addition reactions, in which a small molecule such as H_2, HBr, or Br_2 adds to the double bond. The reaction with bromine is typical (Figure 22.6):

⊙ See Screen 11.4, Hydrocarbons and Addition Reactions.

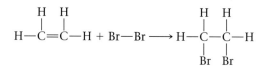

Alkynes

The most common alkyne by far is the first member of the series, acetylene. Recall from Chapter 7 that the C_2H_2 molecule is linear, with 180° bond angles. The triple bond consists of a sigma bond and two pi bonds; each carbon atom is sp-hybridized. The geometries of acetylene and the next member of the series, C_3H_4, are shown in Figure 22.7.

Acetylene can be made in the laboratory by allowing water to drop on calcium carbide.

$$CaC_2(s) + H_2O(l) \longrightarrow C_2H_2(g) + CaO(s)$$

FIGURE 22.6
Test for an alkene. At the left is a solution of bromine in carbon tetrachloride. Addition of a few drops of an alkene causes the red color to disappear as the alkene reacts with the bromine. *(Charles D. Winters)*

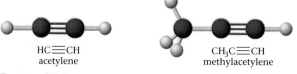

$HC\equiv CH$
acetylene

$CH_3C\equiv CH$
methylacetylene

FIGURE 22.7
The two simplest alkynes. Both molecules contain four atoms in a straight line. **OHT**

The gas produced in this way has a garlic-like odor due to traces of phosphine, PH_3, formed from Ca_3P_2 in the impure calcium carbide. Commercially, acetylene is made by heating methane to about 1500°C in the absence of air:

$$2CH_4(g) \longrightarrow C_2H_2(g) + 3H_2(g)$$

Thermodynamically, acetylene is unstable with respect to decomposition to the elements:

$$C_2H_2(g) \longrightarrow 2C(s) + H_2(g) \qquad \Delta G° = -209.2 \text{ kJ at } 25°C$$

At high pressures, this reaction can occur explosively. For that reason, cylinders of acetylene do not contain the pure gas. Instead the cylinder is packed with an inert, porous material that holds a solution of acetylene gas in acetone.

You are probably most familiar with acetylene as a gaseous fuel used in welding and cutting metals (Figure 22.8). When mixed with pure oxygen in a torch, acetylene burns at temperatures above 2000°C. The heat comes from the reaction

$$C_2H_2(g) + \tfrac{5}{2}O_2(g) \longrightarrow 2CO_2(g) + H_2O(l) \qquad \Delta H = -1300 \text{ kJ}$$

The reaction gives off a brilliant white light, which served as a source of illumination in the headlights of early automobiles.

Calcium carbide *(left)* and its reaction with water *(right)*. *(Charles D. Winters)*

Also miners' headlamps.

EXAMPLE 22.2   What is the molecular formula of

(a) the alkane, alkene, and alkyne containing six carbon atoms?
(b) the alkane containing ten hydrogen atoms?

Strategy   Apply the general formulas: C_nH_{2n+2} (alkane), C_nH_{2n} (alkene), C_nH_{2n-2} (alkyne).

Solution

(a) In each case, $n = 6$. Hence

  alkane: C_6H_{14} ($2 \times 6 + 2 = 14$ H atoms)

  alkene: C_6H_{12} ($2 \times 6 = 12$ H atoms)

  alkyne: C_6H_{10} ($2 \times 6 - 2 = 10$ H atoms)

(b) Here, $2n + 2 = 10$, so $n = (10 - 2)/2 = 4$; the formula is C_4H_{10}.

FIGURE 22.8
Welding with an oxyacetylene torch. *(Charles D. Winters)*

22.3   AROMATIC HYDROCARBONS AND THEIR DERIVATIVES

Aromatic hydrocarbons, sometimes referred to as *arenes*, can be considered to be derived from benzene, C_6H_6. Benzene is a transparent, volatile liquid (bp = 80°C) that was discovered by Michael Faraday in 1825. Its formula, C_6H_6, suggests a high degree of unsaturation, yet its properties are quite different from those of alkenes or alkynes.

As pointed out in Chapter 7, the atomic orbital (valence bond) model regards benzene as a resonance hybrid of the two structures

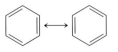

There are three pi bonds in benzene.

This model is consistent with many of the properties of benzene. The molecule is a planar hexagon with bond angles of 120°. The hybridization of each carbon is sp². However, this structure is misleading in one respect. Chemically, benzene does not behave as if there were double bonds present. In particular, benzene does not undergo addition reactions like the one shown in Figure 22.6.

A more satisfactory model of the electron distribution in benzene, based on molecular orbital theory (Appendix 5), assumes that—

- each carbon atom forms three sigma bonds, one to a hydrogen atom and two to adjacent carbon atoms.
- the three electron pairs remaining are spread symmetrically over the entire molecule to form "delocalized" pi bonds.

This model is often represented by the structure

where it is understood that—

- there is a carbon atom at each corner of the hexagon.
- there is an H atom bonded to each carbon atom.
- the circle in the center of the molecule represents the six delocalized electrons.

At one time, benzene was widely used as a solvent, both commercially and in research and teaching laboratories. Its use for that purpose has largely been abandoned because of its toxicity. Chronic exposure to benzene vapor leads to various blood disorders and, in extreme cases, aplastic anemia and leukemia. It appears that the culprits are oxidation products of the aromatic ring, formed in an attempt to solubilize benzene and thus eliminate it from the body.

Derivatives of Benzene

Among the more common monosubstituted benzenes are the following:

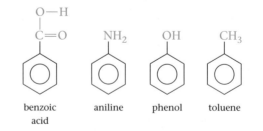

benzoic acid   aniline   phenol   toluene

Phenol was the first commercial antiseptic; its introduction into hospitals in the 1870s led to a dramatic decrease in deaths from postoperative infections. Its use for this purpose has long since been abandoned because phenol burns exposed tissue, but many modern antiseptics are phenol derivatives. Toluene has largely replaced benzene as a solvent because it is much less toxic. Oxidation of toluene in the body gives benzoic acid, which is readily eliminated and has none of the toxic properties of the oxidation products of benzene. Indeed, benzoic acid or its sodium salt (Na^+, $C_6H_5COO^-$ ions) is widely used as a preservative in foods and beverages, including fruit juices and soft drinks.

Condensed Ring Structures

In another type of aromatic hydrocarbon, two or more benzene rings are fused together. Naphthalene is the simplest compound of this type. Fusion of three benzene rings gives two different isomers, anthracene and phenanthrene.

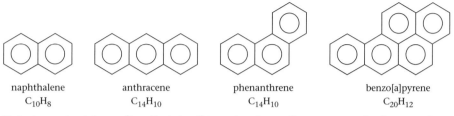

| naphthalene | anthracene | phenanthrene | benzo[a]pyrene |
| $C_{10}H_8$ | $C_{14}H_{10}$ | $C_{14}H_{10}$ | $C_{20}H_{12}$ |

It is important to realize that in these structures there are no hydrogen atoms bonded to the carbons at the juncture of two rings. The eight hydrogen atoms in naphthalene are located as shown below.

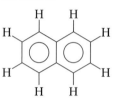

Certain compounds of this type are potent carcinogens. One of the most dangerous is benzo[a]pyrene, which has been detected in cigarette smoke. It is believed to be a cause of lung cancer, to which smokers are susceptible.

22.4 OXYGEN COMPOUNDS; FUNCTIONAL GROUPS

In many oxygen-containing organic compounds, an oxygen atom is part of a **functional group,** which takes the place of one of the hydrogen atoms of a hydrocarbon. Table 22.1 lists some of the more common functional groups of this type. In this section we will discuss alcohols, carboxylic acids, and esters.

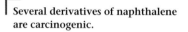

Several derivatives of naphthalene are carcinogenic.

Cigarette smoke. Benzo[a]pyrene present in cigarette smoke can cause lung cancer not only in smokers but also in nonsmokers who inhale the cigarette smoke regularly. *(Charles D. Winters)*

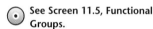

See Screen 11.5, Functional Groups.

TABLE 22.1 Oxygen-Containing Functional Groups [OHT]

Group	Class	Example	Name
—OH	Alcohol	$CH_3CH_2\,OH$	Ethyl alcohol
—O—	Ether	CH_3—O—CH_3	Dimethyl ether
O ‖ —C—H	Aldehyde	O ‖ CH_3 —C—H	Acetaldehyde
O ‖ —C—	Ketone	O ‖ CH_3 —C— CH_3	Acetone
O ‖ —C—OH	Carboxylic acid	O ‖ CH_3—C—OH	Acetic acid
O ‖ —C—O—	Ester	O ‖ CH_3 —C—O— CH_3	Methyl acetate

You may recall from Chapter 15 that electron transitions from one energy level to another often result in the absorption of radiation in the visible region (wavelength 400–700 nm). That effect accounts for the bright colors of many coordination compounds. In contrast most organic compounds are not colored because they do not absorb visible light.

However, as we saw in Chapter 1, organic compounds typically absorb infrared radiation ($\lambda > 700$ nm). Absorption can come about through the stretching and bending of covalent bonds, which ordinarily requires considerably less energy than electron transitions. The wavelengths of absorption bands in the infrared are characteristic of the type of bond involved (Table 22.2). For example, compounds containing a C=O bond show a characteristic absorption wavelength in the region between 5700 and 6000 nm.

From the measured absorption spectrum of an organic compound, it is possible, using data such as those given in Table 22.2, to make deductions about its structure. The approach followed is illustrated in Example 22.3.

EXAMPLE 22.3   A certain organic compound with an oxygen-containing functional group shows infrared absorption bands centered at 3920 nm, 5780 nm, and 9560 nm. Of the functional groups listed in Table 22.1, which would you expect it to contain?

Strategy   Use Table 22.2 to deduce, as best you can, what bonds are present. Then use Table 22.1 to find out what functional group contains these bonds.

Solution   The band at 3920 nm corresponds to an O—H bond, the one at 5780 nm to a C=O bond, and the one at 9560 nm to a C—O bond. This indicates the presence of the

$$-\overset{\displaystyle \|}{\underset{\displaystyle O}{C}}-O-H$$

group, corresponding to a carboxylic acid.

Alcohols

An **alcohol** is derived from a hydrocarbon by replacing a hydrogen atom by an —OH group. The first two members of the series are methyl alcohol and ethyl alcohol.

$$CH_3OH \qquad CH_3CH_2OH$$

methyl alcohol   ethyl alcohol

TABLE 22.2   Characteristic Infrared Absorption Bands

Bond	(nm)	Bond	(nm)
C—H	3300–3500	C—O	7700–10,000
C=C	5900–6200	C=O	5700–6000
C≡C	4400–4800	O—H (alcohol)	2700–2800
		O—H (acid)	3000–4200

FIGURE 22.9

Making wine. Wine *(far right)* is produced from the glucose in grape juice *(left)* by fermentation. The reaction is $C_6H_{12}O_6(aq) \rightarrow 2C_2H_5OH(aq) + 2CO_2(g)$. The purpose of the bubble chamber in the fermentation jug *(center)* is to allow the carbon dioxide to escape but prevent oxygen from entering and oxidizing ethanol to acetic acid. *(Courtesy of Professor James M. Bobbit, Twenty Mile Vineyard)*

About 3×10^9 kg of methyl alcohol are produced annually in the United States from *synthesis gas,* a mixture of carbon monoxide and hydrogen:

$$CO(g) + 2H_2(g) \xrightarrow[\text{250 atm, 350°C}]{\text{ZnO, Cr}_2\text{O}_3} CH_3OH(g)$$

Methyl alcohol is also formed as a by-product when charcoal is made by heating wood in the absence of air. For this reason, it is sometimes called wood alcohol. Methyl alcohol is used in jet fuels and as a solvent, gasoline additive, and starting material for several industrial syntheses. It is a deadly poison; ingestion of as little as 25 mL can be fatal. The antidote in this case is a solution of sodium hydrogen carbonate, $NaHCO_3$.

"Gasohol" contains 10% methyl or ethyl alcohol.

Ethyl alcohol, the most common alcohol, can be made by the fermentation of sugar (Figure 22.9) or grain—hence its common name "grain alcohol." It is the active ingredient of alcoholic beverages, in which it is present at various concentrations (4 to 8% in beer, 12 to 15% in wine, and 40% or more in distilled spirits). The "proof" of an alcoholic beverage is twice the volume percentage of ethyl alcohol; an 86-proof bourbon whiskey contains 43% ethyl alcohol. The flavor of alcoholic beverages is due to impurities; ethyl alcohol itself is tasteless and colorless.

Industrial ethyl alcohol is made by the reaction of ethylene with water, using sulfuric acid as a catalyst.

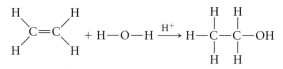

Frequently, ethyl alcohol that is not destined for human consumption is "denatured" by adding small quantities of methyl alcohol or benzene. This avoids the high federal tax on beverage alcohol; denatured alcohol is poisonous.

Ethylene glycol, an antifreeze. The antifreeze easily mixes with water because, like water, it contains —OH groups. *(Charles D. Winters)*

Certain alcohols contain two or more —OH groups per molecule. Perhaps the most familiar compounds of this type are ethylene glycol and glycerol:

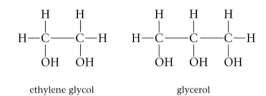

ethylene glycol   glycerol

Ethylene glycol is widely used as an antifreeze. Glycerol is formed as a by-product in making soaps and detergents. It is a viscous, sweet-tasting liquid, used in making drugs, antibiotics, plastics, and explosives (nitroglycerin).

Carboxylic Acids

A **carboxylic acid** is derived from a hydrocarbon by replacing a hydrogen atom by a carboxyl group:

$$-\underset{\underset{O}{\|}}{C}-OH$$

often abbreviated —COOH. The most familiar acid of this type is acetic acid:

$$CH_3-\underset{\underset{O}{\|}}{C}-OH$$

often written CH_3COOH. (In Chapter 13, we represented acetic acid by the formula $HC_2H_3O_2$ to emphasize that the molecule contains only one ionizable hydrogen atom.) Acetic acid is the active ingredient of vinegar, responsible for its sour taste. "White" vinegar is made by adding pure acetic acid to water, forming a 5% solution. "Brown" or "apple cider" vinegar is made from apple juice; the ethanol formed by fermentation is oxidized to acetic acid:

$$C_2H_5OH(aq) + O_2(g) \longrightarrow CH_3COOH(aq) + H_2O$$

> Homemade wine often contains some vinegar.

The most important chemical property of carboxylic acids is implied by their name; they act as weak acids in water solution.

$$RCOOH(aq) \rightleftharpoons H^+(aq) + RCOO^-(aq)$$

> R represents any hydrocarbon group, such as CH_3— or CH_3CH_2—.

Carboxylic acids vary considerably in strength. Among the strongest is trichloroacetic acid ($K_a = 0.20$); a 0.10 M solution of Cl_3C—COOH is about 73% ionized. Trichloroacetic acid is an ingredient of over-the-counter preparations used to treat canker sores and remove warts.

Treatment of a carboxylic acid with the strong base NaOH forms the sodium salt of the acid. With acetic acid, the acid-base reaction is

$$CH_3-\underset{\underset{O}{\|}}{C}-OH(aq) + OH^-(aq) \longrightarrow CH_3-\underset{\underset{O}{\|}}{C}-O^-(aq) + H_2O$$

Evaporation gives the salt sodium acetate, which contains Na^+ and CH_3COO^- ions. Many of the salts of carboxylic acids have important uses. Sodium and cal-

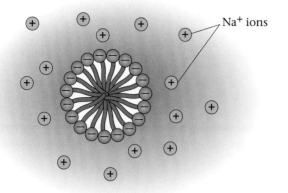

A soap micelle. In the center of the micelle, greasy dirt is dissolved by the hydrocarbon chains of the soap's carboxylate ions. The ionic COO⁻ ions encircle the micelle at the interface with water.

cium propionate (Na^+ or Ca^{2+} ions, $CH_3CH_2COO^-$ ions) are added to bread, cake, and cheese to inhibit the growth of mold. *Soaps* are sodium salts of long-chain carboxylic acids such as stearic acid:

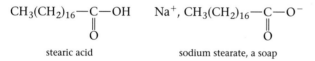

$$CH_3(CH_2)_{16}-\overset{\displaystyle O}{\underset{\displaystyle \|}{C}}-OH \qquad Na^+, CH_3(CH_2)_{16}-\overset{\displaystyle O}{\underset{\displaystyle \|}{C}}-O^-$$

stearic acid   sodium stearate, a soap

The cleaning action of soap reflects the nature of the long-chain carboxylate anion. The long hydrocarbon group is a good solvent for greases and oils, and the ionic COO⁻ group gives water solubility.

Esters

The reaction between an alcohol and a carboxylic acid forms an **ester,** containing the functional group

$$-O-\overset{\displaystyle O}{\underset{\displaystyle \|}{C}}-$$

often abbreviated —COO—. The reaction between methyl alcohol and acetic acid is typical

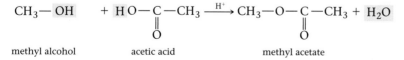

$$CH_3-OH \quad + \quad HO-\overset{\displaystyle O}{\underset{\displaystyle \|}{C}}-CH_3 \xrightarrow{H^+} CH_3-O-\overset{\displaystyle O}{\underset{\displaystyle \|}{C}}-CH_3 + H_2O$$

methyl alcohol   acetic acid   methyl acetate

The common name of an ester consists of two words. The first word (methyl, ethyl, . . .) is derived from that of the alcohol; the second word (formate, acetate, . . .) is the name of the acid with the -ic suffix replaced by -ate. Thus ethyl formate (Table 22.3, page 640) is made from ethyl alcohol and formic acid

$$CH_3CH_2OH \qquad HCOOH$$

ethyl alcohol   formic acid

TABLE 22.3   **Properties of Esters** OHT

Ester	Structure	Odor, Flavor
Ethyl formate	$CH_3CH_2-O-C-H$ (with $\parallel$ O below C)	Rum
Isobutyl formate	$(CH_3)_2-CHCH_2-O-C-H$ (with $\parallel$ O below C)	Raspberry
Methyl butyrate	$CH_3-O-C-(CH_2)_2CH_3$ (with $\parallel$ O below C)	Apple
Ethyl butyrate	$CH_3CH_2-O-C-(CH_2)_2CH_3$ (with $\parallel$ O below C)	Pineapple
Isopentyl acetate	$(CH_3)_2-CH-(CH_2)_2-O-C-CH_3$ (with $\parallel$ O below C)	Banana
Octyl acetate	$CH_3-(CH_2)_7-O-C-CH_3$ (with $\parallel$ O below C)	Orange
Pentyl propionate	$CH_3-(CH_2)_4-O-C-CH_2CH_3$ (with $\parallel$ O below C)	Apricot

Most esters have pleasant odors.

EXAMPLE 22.4   Show the structure of

(a) the three-carbon alcohol with an —OH group at the end of the chain.
(b) the three-carbon carboxylic acid.
(c) the ester formed when these two compounds react.

Strategy   To work (a) and (b), start by drawing the parent carbon chain. Put the functional group in the specified position; fill out the structural formula with hydrogen atoms. In (c), remember that an —OH group comes from the acid, an H atom from the alcohol.

Solution

(a) $CH_3CH_2CH_2-OH$

(b) CH_3CH_2-C-OH (with $\parallel$ O below C)

(c) $CH_3CH_2CH_2-O-C-CH_2CH_3$ (with $\parallel$ O below C)

22.5   ISOMERISM IN ORGANIC COMPOUNDS

Isomers are distinctly different compounds, with different properties, that have the same molecular formula. In Section 22.1, we considered structural isomers of alkanes. You will recall that butane and 2-methylpropane have the same molecular formula, C_4H_{10}, but different structural formulas. In these, as in all structural isomers, the order in which the atoms are bonded to each other differs.

Structural isomerism is common among all types of organic compounds. For example, there are

1. Three structural isomers of the alkene C_4H_8:

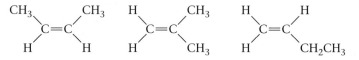

2. Three structural isomers with the molecular formula C_3H_8O. Two of these are alcohols, the third is an ether.

$$CH_3CH_2 - \overset{\overset{\displaystyle H}{|}}{\underset{\underset{\displaystyle H}{|}}{C}} - OH \qquad CH_3 - \overset{\overset{\displaystyle H}{|}}{\underset{\underset{\displaystyle OH}{|}}{C}} - CH_3 \qquad CH_3 - O - CH_2CH_3$$

EXAMPLE 22.5   Consider the molecule $C_3H_6Cl_2$, which is derived from propane, C_3H_8, by substituting two Cl atoms for H atoms. Draw the structural isomers of $C_3H_6Cl_2$.

Strategy   All of these compounds have the same carbon skeleton, —C—C—C—. To find the isomers, notice that the two carbon atoms at the ends of the chain are equivalent. The carbon atom at the center of the chain is different because it is bonded to two carbon atoms rather than one.

Solution

There are four isomers. In one, both Cl atoms are bonded to a carbon atom at the end of the chain. In another, both Cl atoms are bonded to the central carbon atom:

$$CH_3 - CH_2 - \overset{\overset{\displaystyle Cl}{|}}{\underset{\underset{\displaystyle Cl}{|}}{C}} - H \qquad CH_3 - \overset{\overset{\displaystyle Cl}{|}}{\underset{\underset{\displaystyle Cl}{|}}{C}} - CH_3$$

In the other two isomers, the Cl atoms are bonded to different carbons:

$$H - \overset{\overset{\displaystyle H}{|}}{\underset{\underset{\displaystyle Cl}{|}}{C}} - CH_2 - \overset{\overset{\displaystyle H}{|}}{\underset{\underset{\displaystyle Cl}{|}}{C}} - H \qquad CH_3 - \overset{\overset{\displaystyle H}{|}}{\underset{\underset{\displaystyle Cl}{|}}{C}} - \overset{\overset{\displaystyle H}{|}}{\underset{\underset{\displaystyle Cl}{|}}{C}} - H$$

Geometric Isomers

As we saw earlier, there are three structural isomers of the alkene C_4H_8. You may be surprised to learn that there are actually *four* different alkenes with this molecular formula. The "extra" compound arises because of a phenomenon called *geometric isomerism*. There are two different geometric isomers of the structure shown at the left at the top of this page.

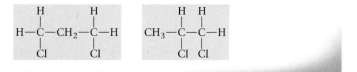

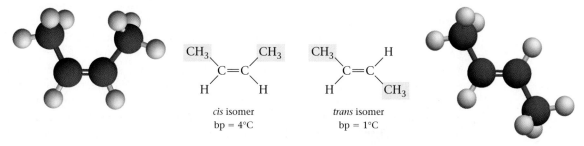

cis isomer
bp = 4°C

trans isomer
bp = 1°C

In the *cis* isomer, the two CH_3 groups (or the two H atoms) are as close to one another as possible. In the *trans* isomer, the two identical groups are farther apart. The two forms exist because there is no free rotation about the carbon—carbon double bond. The situation is analogous to that with *cis-trans* isomers of square planar complexes (Chapter 15). In both cases, the difference in geometry is responsible for isomerism; the atoms are bonded to each other in the same way.

Geometric or *cis-trans* isomerism is common among alkenes. It occurs when both of the double-bonded carbon atoms are joined to two different atoms or groups. The other two structural isomers of C_4H_8 shown on page 641 do not show *cis-trans* isomerism. In both cases the carbon atom at the left is joined to two identical hydrogen atoms.

EXAMPLE 22.6 Draw all the isomers of the molecule $C_2H_2Cl_2$ in which two of the H atoms of ethylene are replaced by Cl atoms.

Strategy Start with ethylene and see how many isomers, structural or geometric, you can make by substituting two chlorine atoms for hydrogen.

Solution The isomers of $C_2H_2Cl_2$ are:

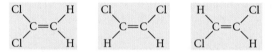

Reality Check Notice that the structural isomer shown on the left does not show *cis-trans* isomerism because two Cl atoms are bonded to the same carbon.

Cis and *trans* isomers differ from one another in their physical and, to a lesser extent, their chemical properties. They may also differ in their physiological behavior. For example, the compound *cis*-9-tricosene

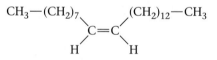

is a sex attractant secreted by the female housefly. The *trans* isomer is totally ineffective in this area.

Optical Isomers

Optical isomers arise when at least one carbon atom in a molecule is bonded to four different atoms or groups. Consider, for example, the methane derivative CHClBrI. As you can see from Figure 22.10, there are two different forms of this molecule, which are mirror images of one another. The mirror images are not superimposable; that is, you cannot place one molecule over the other so that identical groups are touching. In this sense, the two isomers resemble right- and left-hand gloves.

There is another way to convince yourself that the two structures shown in Figure 22.10 are different isomers. Imagine that you are located directly above the hydrogen atom, looking down at the rest of the molecule. In the structure at the left, the atoms Cl, Br, I in that order are arranged in a clockwise pattern. With the structure at the right, in order to move from Cl to Br to I, you have to move counterclockwise.

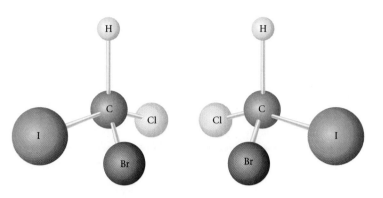

FIGURE 22.10
Optical isomers. These isomers of CHClBrI are mirror images of each other. **OHT**

A carbon atom with four different atoms or groups attached to it is referred to as a **chiral center.** A molecule containing such a carbon atom shows optical isomerism. It exists in two different forms that are nonsuperimposable mirror images. These forms are referred to as **enantiomers.** Molecules may contain more than one chiral center, in which case there can be more than two enantiomers.

EXAMPLE 22.7   In the following structural formulas, locate each chiral carbon atom.

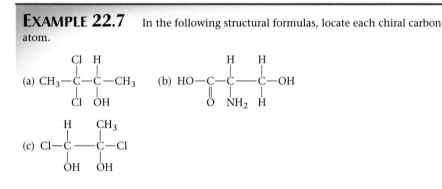

Strategy   Look for carbon atoms that have four different atoms or groups attached.

Solution   In (a), the third carbon atom from the left is chiral; it is bonded to H, CH_3, OH, and CH_3-CCl_2. In (b), the second carbon atom is chiral. In (c), both carbons are chiral.

The term "optical isomerism" comes from the effect that enantiomers have on plane-polarized light, such as that produced by a Polaroid lens (Figure 22.11, page 644). When this light is passed through a solution containing a single enantiomer, the plane is rotated from its original position. One isomer rotates it to the right (clockwise) and the other to the left (counterclockwise). If both isomers are present in equal amounts, we obtain what is known as a **racemic mixture.** In this case, the two rotations offset each other and there is no effect on plane-polarized light.

Enantiomers have identical physical properties. For example, the two isomers of lactic acid shown below have the same melting point, 52°C, and density, 1.25 g/mL.

$$
\begin{array}{ccc}
& COOH & COOH \\
& | & | \\
(I) \quad HO-C-H & (II) \quad H-C-OH \\
& | & | \\
& CH_3 & CH_3 \\
\end{array}
$$

enantiomers of lactic acid

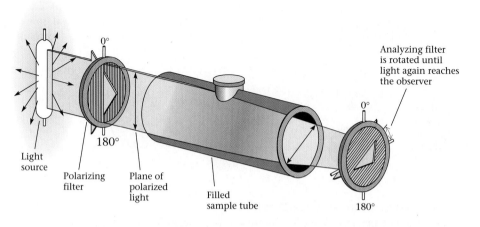

On the other hand, because enantiomers differ in their three-dimensional structure, they often interact with biochemical molecules in different ways. As a result, they may show quite different physiological properties. Consider, for example, amphetamine, often used illicitly as an "upper" or "pep pill." Amphetamine consists of two enantiomers:

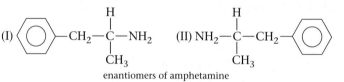

enantiomers of amphetamine

Enantiomer I, called Dexedrine, is by far the stronger stimulant. It is from two to four times as active as Benzedrine, the racemic mixture of the two isomers.

In the 1990s, a large number of so-called chiral drugs containing a single enantiomer came on the market. One of the first of these was naproxen (trade name Aleve).

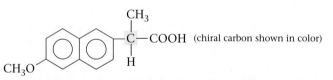

The naproxen that you buy over the counter at your local pharmacy consists of a single enantiomer, which is an effective analgesic (pain killer). The other enantiomer is not a pain killer; even worse, it can cause liver damage.

Of the 100 best-selling drugs on the market today, about half consist of a single enantiomer. Chiral drugs are used to treat all kinds of illnesses, from arthritis to asthma, from heartburn to heart disease. In general, these drugs are prepared by either of two different methods. One possibility is to resolve a racemic mixture into its components. This was first done by Louis Pasteur more than 150 years ago. He used a magnifying glass and a pair of tweezers to separate crystals of enantiomers. Today resolution can be accomplished much more quickly by using modern separation techniques such as liquid chromatography (Chapter 1).

A different approach to making chiral drugs is *asymmetric synthesis.* An optically inactive precursor is converted to the drug by a reaction that uses a special catalyst, usually an enzyme (Chapter 11). If all goes well, the product is a single enantiomer with the desired physiological effect.

Cholesterol

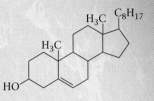

An organic compound that we hear a great deal about nowadays is cholesterol, whose structure is shown in Figure A. Your body contains about 140 g of cholesterol; it is synthesized in the liver at the rate of 2 to 3 g/day. Cholesterol is essential to life for two reasons. It is a major component of all cell membranes and it serves as the starting material for the synthesis of sex hormones, adrenal hormones, bile acids, and vitamin D.

Cholesterol, which is water-insoluble, is transported through the blood in the form of soluble protein complexes known as lipoproteins. There are two types of complexes: low-density (LDL), which contain mostly cholesterol, and high-density (HDL), which contain relatively little cholesterol. Commonly, these complexes are referred to as "LDL cholesterol" and "HDL cholesterol," respectively.

LDL or "bad" cholesterol builds up as a plaque-like deposit on the interior walls of arteries. This process used to be called hardening of the arteries; today it is referred to as *atherosclerosis*. It can lead to cardiovascular diseases, including strokes and heart attacks. In contrast, HDL or "good" cholesterol retards or even reduces arterial deposits.

The relative amounts of LDL and HDL cholesterol in your bloodstream depend, at least in part, on your diet. In particular, they depend on the total amount and the type of fat that you eat. Fats (triglycerides) are esters of glycerol with long-chain carboxylic acids. The general structure of a fat can be represented as

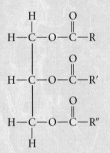

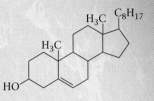

FIGURE A
Cholesterol.

See Screen 11.7, Fats and Oils.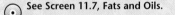

where R, R′, and R″ are hydrocarbon residues typically containing 14 to 22 carbon atoms. Depending on the nature of these residues, we can distinguish among—

- *saturated fats,* in which R, R′, and R″ contain no multiple bonds. These are "bad" in the sense that they lead to high levels of LDL relative to HDL cholesterol.
- *monounsaturated fats,* in which each residue contains one double bond. These are much better nutritionally unless, God forbid, the hydrogen atoms on opposite sides of the double bonds are *trans* to one another. So-called *trans* fats are bad news.
- *polyunsaturated fats,* in which the hydrocarbon residues contain more than one double bond, are good guys, particularly if a double bond is located on the third or sixth carbon atom from the —CH₃ end of the residue. These so-called omega fats are the best of all.

Figure B (page 646) shows the composition of various fats (solids) and oils (liquids). Now you know why sunflower seeds are better for you than coconuts.

(continued)

FIGURE B
The composition of some common fats and oils. Fatty acids provide the R, R', and R″ hydrocarbon residues in the general structure of a fat or oil shown in the text. Fats and oils all contain a mixture of saturated, monounsaturated, and polyunsaturated fatty acids.

*Cholesterol Content (mg/Tbsp): Lard 12; Beef Tallow 14; Butterfat 33. No cholesterol in any vegetable-based oil.

CHAPTER HIGHLIGHTS

Key Concepts

1. Draw—
   ▨ structural isomers.
   (Examples 22.1, 22.5; Problems 11–20)
   ▨ geometric isomers.
   (Example 22.6; Problems 21–24)
   ▨ optical isomers, containing chiral carbon atoms.
   (Example 22.7; Problems 25–28)

2. Distinguish between alkanes, alkenes, and alkynes.
   (Example 22.2; Problems 1–4)

3. Given the IR spectrum of an oxygen-containing organic compound, state the functional group(s) present.
   (Example 22.3; Problems 9, 10)

4. Write structural formulas for alcohols, carboxylic acids, and esters.
   (Example 22.4; Problems 5–8)

alcohol	enantiomer	hydrocarbon
alkane	ester	optical isomerism
alkene	functional group	racemic mixture
alkyne	■ geometric isomer	saturated hydrocarbon
aromatic hydrocarbon	■ —cis	structural isomer
carboxylic acid	■ —trans	unsaturated hydrocarbon
chiral center		

Summary Problem

Consider the hydrocarbon that has the structure

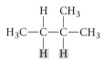

(a) Classify the hydrocarbon as an alkane, alkene, or alkyne.
(b) Draw the structure of all hydrocarbons that are structurally isomeric with it.
(c) Draw the structure obtained by removing the two hydrogen atoms shown in red.
(d) Classify the hydrocarbon obtained in (c) as an alkane, alkene, or alkyne.
(e) Does the hydrocarbon referred to in (c) have structural isomers? Geometric isomers?
(f) Does the hydrocarbon referred to in (c) have a chiral carbon? Which one?
(g) Draw the structures of the two alcohols formed by replacing an H atom shown in red with an —OH group.
(h) Are the two alcohols in (g) structural isomers? Geometric isomers?
(i) Give the molecular formula of the ester formed when one of the alcohols in (g) reacts with formic acid, HCOOH.

Answers
(a) alkane

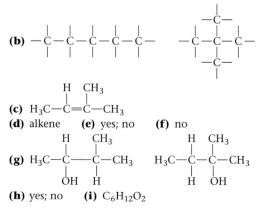

(h) yes; no   **(i)** $C_6H_{12}O_2$

Questions & Problems

Problem numbers in blue indicate that the answer is available in Appendix 6 at the back of the book.
WEB indicates that the solution is posted at **http://www.harcourtcollege.com/chem/general/masterton4/student/**

Hydrocarbons

1. Classify each of the following hydrocarbons as alkanes, alkenes, or alkynes.
(a) C_6H_{14} (b) C_3H_4 (c) C_9H_{18}

2. Classify each of the following hydrocarbons as alkanes, alkenes, or alkynes.
(a) $C_{12}H_{24}$ (b) C_7H_{12} (c) $C_{13}H_{28}$

3. Write the formula for
(a) an alkene with two carbon atoms.
(b) an alkane with 22 hydrogen atoms.
(c) an alkyne with three carbon atoms.

WEB **4.** Write the formula for
(a) an alkyne with 16 hydrogen atoms.
(b) an alkene with 44 hydrogen atoms.
(c) an alkane with ten carbon atoms.

Functional Groups

5. Classify each of the following as a carboxylic acid, ester, and/or alcohol.
(a) $HO-CH_2-CH_2-CH_2-OH$

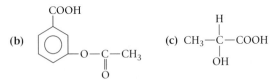

(b)

(c) $CH_3-\overset{\overset{\displaystyle H}{|}}{\underset{\underset{\displaystyle OH}{|}}{C}}-COOH$

6. Classify each of the following as a carboxylic acid, ester, and/or alcohol.
(a) $CH_3-(CH_2)_3-OH$
(b) $CH_3-CH_2-\overset{\overset{\displaystyle}{}}{\underset{\underset{\displaystyle O}{\|}}{C}}-O-CH_2-CH_3$
(c) $CH_3-CH_2-O-\overset{\overset{\displaystyle}{}}{\underset{\underset{\displaystyle O}{\|}}{C}}-(CH_2)_6-COOH$

7. Give the structure of
(a) a four-carbon straight-chain alcohol with an —OH group not at the end of the chain.
(b) a five-carbon straight-chain carboxylic acid
(c) the ester formed when these two compounds react.

8. Give the structure of
(a) a three-carbon alcohol with the —OH group on the center carbon.
(b) a four-carbon branched-chain carboxylic acid
(c) the ester formed when these two compounds react.

9. A certain organic compound with an oxygen-containing functional group shows infrared absorption bands centered at 8760 nm, 9377 nm, and 5893 nm. Of the functional groups listed in Table 22.1, which would you expect it to contain?

10. A certain organic compound with an oxygen-containing functional group shows infrared absorption bands centered at 5732 nm and 3305 nm. Of the functional groups listed in Table 22.1, which would you expect it to contain?

Isomerism

11. Draw the structural isomers of the alkane C_6H_{14}.

12. Draw the structural isomers of the alkene C_4H_8.

13. Draw the structural isomers of C_4H_9Cl in which one hydrogen atom of a C_4H_{10} molecule has been replaced by chlorine.

WEB **14.** Draw the structural isomers of $C_3H_6Cl_2$ in which two of the hydrogen atoms of C_3H_8 have been replaced by chlorine atoms.

15. There are three compounds with the formula C_6H_4ClBr in which two of the hydrogen atoms of the benzene molecule have been replaced by halogen atoms. Draw structures for these compounds.

16. There are three compounds with the formula $C_6H_3Cl_3$ in which three of the hydrogen atoms of the benzene molecule have been replaced by chlorine atoms. Draw structures for these compounds.

17. Write structures for all the structural isomers of double-bonded compounds with the molecular formula C_5H_{10}.

18. Write structures for all the structural isomers of compounds with the molecular formula C_4H_6ClBr in which Cl and Br are bonded to a double-bonded carbon.

19. Draw structures for all the alcohols with molecular formula $C_5H_{12}O$.

20. Draw structures for all the saturated carboxylic acids with four carbon atoms per molecule.

21. Of the compounds in Problem 17, which ones show geometric isomerism? Draw the *cis*- and *trans*- isomers.

22. Of the compounds in Problem 18, which ones show geometric isomerism? Draw the *cis*- and *trans*- isomers.

23. Maleic acid and fumaric acid are the *cis*- and *trans*- isomers, respectively, of $C_2H_2(COOH)_2$, a dicarboxylic acid. Draw and label their structures.

24. For which of the following is geometric isomerism possible?
(a) $(CH_3)_2C=CCl_2$ (b) $CH_3ClC=CCH_3Cl$
(c) $CH_3BrC=CCH_3Cl$

25. Which of the following can show optical isomerism?
(a) 2-bromo-2-chlorobutane
(b) 2-methylpropane
(c) 2,2-dimethyl-l-butanol
(d) 2,2,4-trimethylpentane

26. Which of the following compounds can show optical isomerism?

(a) dichloromethane
(b) 1,2-dichloroethane
(c) bromochlorofluoromethane
(d) 1-bromoethanol

27. Locate the chiral carbon(s), if any, in the following molecules.

(a) $HO-\overset{\overset{\displaystyle H}{|}}{\underset{\underset{\displaystyle O}{\|}}{C}}-\overset{\overset{\displaystyle H}{|}}{\underset{\underset{\displaystyle OH}{|}}{C}}-\overset{\overset{\displaystyle H}{|}}{\underset{\underset{\displaystyle OH}{|}}{C}}-\overset{}{\underset{\underset{\displaystyle O}{\|}}{C}}-H$

(b) $CH_3-\overset{}{\underset{\underset{\displaystyle O}{\|}}{C}}-\overset{}{\underset{\underset{\displaystyle O}{\|}}{C}}-OH$

(c) $CH_3-CH_2-\overset{\overset{\displaystyle H}{|}}{\underset{\underset{\displaystyle NH_2}{|}}{C}}-COOH$

WEB 28. Locate the chiral carbon(s), if any, in the following molecules.

(a) $CH_3-\overset{\overset{\displaystyle H}{|}}{\underset{\underset{\displaystyle OH}{|}}{C}}-\overset{\overset{\displaystyle H}{|}}{\underset{\underset{\displaystyle OH}{|}}{C}}-H$ (b) $H-C=\overset{}{\underset{\underset{\displaystyle H}{|}}{C}}-CH_2-OH$
where the C=C has H below left and H below right

(c) $CH_3-\overset{\overset{\displaystyle Cl}{|}}{\underset{\underset{\displaystyle Cl}{|}}{C}}-\overset{\overset{\displaystyle F}{|}}{\underset{\underset{\displaystyle H}{|}}{C}}-Cl$

Unclassified

29. Which of the following are expected to show bond angles of 109.5°? 120°? 180°?

(a) H_3C-CH_3 (b) $H_3C-\overset{\overset{\displaystyle H}{|}}{C}=CH_2$

(c) $H_3C-C\equiv C-CH_3$

30. What does the "circle" represent in the structural formula of benzene?

31. State whether the following statements are true. If the statements are not true, rewrite them to make them true.

(a) Straight-chain hydrocarbons can have any number of hydrogen atoms (i.e., both odd and even).
(b) Alkanes can show geometric isomerism.
(c) Alkenes can have chiral carbon atoms.
(d) Cholesterol is a condensed-ring hydrocarbon. (See Figure A, page 645.)

32. Write a reaction that shows how each of the following compounds are prepared.

(a) methanol   (b) acetic acid   (c) ethyl acetate

33. Calculate [H^+] and the pH of a $0.10\,M$ solution of chloroacetic acid ($K_a = 1.5 \times 10^{-3}$); use successive approximations.

34. Explain what is meant by the following.

(a) a saturated fat
(b) a soap
(c) the "proof" of an alcoholic beverage
(d) denatured alcohol

Challenge Problems

35. The structure of cholesterol is given in the text. (See Figure A.) What is its molecular formula?

36. Draw structures for all the alcohols with molecular formula $C_6H_{14}O$.

37. What mass of propane must be burned to bring one quart of water in a saucepan at 25°C to the boiling point? Use data in Appendix 1 and elsewhere and make any reasonable assumptions.

Units, Constants, and Reference Data

SI Units

As pointed out in Chapter 1, most quantities are commonly expressed in a variety of different units. For example, temperature can be reported in °F, °C, or K. Volume is sometimes quoted in liters or cubic centimeters, at other times in quarts, cubic feet, and so on. In 1960, the General Conference of Weights and Measures attempted to end this proliferation of units. They recommended adopting a self-consistent set of units based on the metric system. In the so-called International System of Units (SI) each quantity is represented by a single unit or decimal multiple thereof, using the prefixes listed in Table 1.2, page 8. For example, the fundamental SI unit of length is the meter; the length of an object can be expressed in meters, kilometers (1 km = 10^3 m), or nanometers (1 nm = 10^{-9} m), but *not* in inches, angstroms, or other non-SI units.

Table 1 lists the SI base and derived units for quantities commonly referred to in general chemistry. Perhaps the least familiar of these units to the beginning chemistry student are the ones used to represent force, pressure, and energy.

The *newton* is defined as the force required to impart an acceleration of one meter per second squared to a mass of one kilogram. Recall that Newton's second law can be stated as

$$\text{force} = \text{mass} \times \text{acceleration}$$

The *pascal* is defined as the pressure exerted by a force of one newton acting on an area of one square meter (recall that pressure = force/area).

The *joule* is defined as the work done when a force of one newton (kg·m/s²) acts through a distance of one meter (recall that work = force × distance).

Over the past 40 years, the International System of Units has met with a decidedly mixed reception, at least in the United States. On the one hand, scientists have adopted the joule as the base unit of energy, replacing the calorie. On the

TABLE 1　SI Units

Base Units			Derived Units		
Quantity	Unit	Symbol	Quantity	Unit	Symbol
length	meter	m	volume	cubic meter	m³
mass	kilogram	kg	force	newton	N
time	second	s	pressure	pascal	Pa
temperature	kelvin	K	energy	joule	J
amount of substance	mole	mol	electric charge	coulomb	C
electric current	ampere	A	electric potential	volt	V

other hand, volumes are seldom if ever expressed in m^3. The cubic meter is simply too large a unit; the medium-sized test tube used in the general chemistry laboratory has a volume of about $0.00001\ m^3$. The density of water is commonly expressed as $1.00\ g/cm^3$, not as

$$\frac{1.00\ g}{cm^3} \times \frac{1\ kg}{10^3\ g} \times \frac{(10^2\ cm)^3}{1\ m^3} = 1.00 \times 10^3\ kg/m^3$$

(A cubic meter of water weighs about one metric ton!)

FUNDAMENTAL CONSTANTS

Acceleration of gravity (standard)	$9.8066\ m/s^2$
Atomic mass unit (amu)	$1.6605 \times 10^{-24}\ g$
Avogadro constant	$6.0221 \times 10^{23}/mol$
Electronic charge	$1.6022 \times 10^{-19}\ C$
Electronic mass	$9.1094 \times 10^{-28}\ g$
Faraday constant	$9.6485 \times 10^4\ C/mol$
Gas constant	$0.082058\ L \cdot atm/mol \cdot K$
	$8.3145\ J/mol \cdot K$
Planck constant	$6.6261 \times 10^{-34}\ J \cdot s$
Rydberg constant	$2.1799 \times 10^{-18}\ J$
Velocity of light	$2.9979 \times 10^8\ m/s$
π	3.1416
e	2.7183
$\ln x$	$2.3026\ \log_{10} x$

VAPOR PRESSURE OF WATER (mm Hg)

T(°C)	vp	T(°C)	vp	T(°C)	vp	T(°C)	vp
0	4.58	21	18.65	35	42.2	92	567.0
5	6.54	22	19.83	40	55.3	94	610.9
10	9.21	23	21.07	45	71.9	96	657.6
12	10.52	24	22.38	50	92.5	98	707.3
14	11.99	25	23.76	55	118.0	100	760.0
16	13.63	26	25.21	60	149.4	102	815.9
17	14.53	27	26.74	65	187.5	104	875.1
18	15.48	28	28.35	70	233.7	106	937.9
19	16.48	29	30.04	80	355.1	108	1004.4
20	17.54	30	31.82	90	525.8	110	1074.6

THERMODYNAMIC DATA

	ΔH_f° (kJ/mol)	S° (kJ/K·mol)	ΔG_f° (kJ/mol) at 25°C		ΔH_f° (kJ/mol)	S° (kJ/K·mol)	ΔG_f° (kJ/mol) at 25°C
Ag(s)	0.0	+0.0426	0.0	Cd(s)	0.0	+0.0518	0.0
Ag$^+$(aq)	+105.6	+0.0727	+77.1	Cd^{2+}(aq)	−75.9	−0.0732	−77.6
AgBr(s)	−100.4	+0.1071	−96.9	CdCl$_2$(s)	−391.5	+0.1153	−344.0
AgCl(s)	−127.1	+0.0962	−109.8	CdO(s)	−258.2	+0.0548	−228.4
AgI(s)	−61.8	+0.1155	−66.2	Cl$_2$(g)	0.0	+0.2230	0.0
AgNO$_3$(s)	−124.4	+0.1409	−33.4	Cl$^-$(aq)	−167.2	+0.0565	−131.2
Ag$_2$O(s)	−31.0	+0.1213	−11.2	ClO$_3^-$(aq)	−104.0	+0.1623	−8.0
Al(s)	0.0	+0.0283	0.0	ClO$_4^-$(aq)	−129.3	+0.1820	−8.5
Al^{3+}(aq)	−531.0	−0.3217	−485.0	Cr(s)	0.0	+0.0238	0.0
Al$_2$O$_3$(s)	−1675.7	+0.0509	−1582.3	CrO$_4^{2-}$(aq)	−881.2	+0.0502	−727.8
Ba(s)	0.0	+0.0628	0.0	Cr$_2$O$_3$(s)	−1139.7	+0.0812	−1058.1
Ba^{2+}(aq)	−537.6	+0.0096	−560.8	Cr$_2$O$_7^{2-}$(aq)	−1490.3	+0.2619	−1301.1
BaCl$_2$(s)	−858.6	+0.1237	−810.4	Cu(s)	0.0	+0.0332	0.0
BaCO$_3$(s)	−1216.3	+0.1121	−1137.6	Cu$^+$(aq)	+71.7	+0.0406	+50.0
BaO(s)	−553.5	+0.0704	−525.1	Cu^{2+}(aq)	+64.8	−0.0996	+65.5
BaSO$_4$(s)	−1473.2	+0.1322	−1362.3	CuO(s)	−157.3	+0.0426	−129.7
Br$_2$(l)	0.0	+0.1522	0.0	Cu$_2$O(s)	−168.6	+0.0931	−146.0
Br$^-$(aq)	−121.6	+0.0824	−104.0	CuS(s)	−53.1	+0.0665	−53.6
C(s)	0.0	+0.0057	0.0	Cu$_2$S(s)	−79.5	+0.1209	−86.2
CCl$_4$(l)	−135.4	+0.2164	−65.3	CuSO$_4$(s)	−771.4	+0.1076	−661.9
CHCl$_3$(l)	−134.5	+0.2017	−73.7	F$_2$(g)	0.0	+0.2027	0.0
CH$_4$(g)	−74.8	+0.1862	−50.7	F$^-$(aq)	−332.6	−0.0138	−278.8
C$_2$H$_2$(g)	+226.7	+0.2008	+209.2	Fe(s)	0.0	+0.0273	0.0
C$_2$H$_4$(g)	+52.3	+0.2195	+68.1	Fe^{2+}(aq)	−89.1	−0.1377	−78.9
C$_2$H$_6$(g)	−84.7	+0.2295	−32.9	Fe^{3+}(aq)	−48.5	−0.3159	−4.7
C$_3$H$_8$(g)	−103.8	+0.2699	−23.5	Fe(OH)$_3$(s)	−823.0	+0.1067	−696.6
CH$_3$OH(l)	−238.7	+0.1268	−166.3	Fe$_2$O$_3$(s)	−824.2	+0.0874	−742.2
C$_2$H$_5$OH(l)	−277.7	+0.1607	−174.9	Fe$_3$O$_4$(s)	−1118.4	+0.1464	−1015.5
CO(g)	−110.5	+0.1976	−137.2	H$_2$(g)	0.0	+0.1306	0.0
CO$_2$(g)	−393.5	+0.2136	−394.4	H$^+$(aq)	0.0	0.0000	0.0
CO$_3^{2-}$(aq)	−677.1	−0.0569	−527.8	HBr(g)	−36.4	+0.1986	−53.4
Ca(s)	0.0	+0.0414	0.0	HCl(g)	−92.3	+0.1868	−95.3
Ca^{2+}(aq)	−542.8	−0.0531	−553.6	HCO$_3^-$(aq)	−692.0	+0.0912	−586.8
CaCl$_2$(s)	−795.8	+0.1046	−748.1	HF(g)	−271.1	+0.1737	−273.2
CaCO$_3$(s)	−1206.9	+0.0929	−1128.8	HI(g)	+26.5	+0.2065	+1.7
CaO(s)	−635.1	+0.0398	−604.0	HNO$_3$(l)	−174.1	+0.1556	−80.8
Ca(OH)$_2$(s)	−986.1	+0.0834	−898.5	H$_2$O(g)	−241.8	+0.1887	−228.6
CaSO$_4$(s)	−1434.1	+0.1067	−1321.8	H$_2$O(l)	−285.8	+0.0699	−237.2

continued

THERMODYNAMIC DATA (continued)

	ΔH_f° (kJ/mol)	S° (kJ/K·mol)	ΔG_f° (kJ/mol) at 25°C		ΔH_f° (kJ/mol)	S° (kJ/K·mol)	ΔG_f° (kJ/mol) at 25°C
$H_2O_2(l)$	−187.8	+0.1096	−120.4	$NO_2^-(aq)$	−104.6	+0.1230	−32.2
$H_2PO_4^-(aq)$	−1296.3	+0.0904	−1130.3	$NO_3^-(aq)$	−205.0	+0.1464	−108.7
$HPO_4^{2-}(aq)$	−1292.1	−0.0335	−1089.2	$N_2O_4(g)$	+9.2	+0.3042	+97.9
$H_2S(g)$	−20.6	+0.2057	−33.6	$Na(s)$	0.0	+0.0512	0.0
$H_2SO_4(l)$	−814.0	+0.1569	−690.1	$Na^+(aq)$	−240.1	+0.0590	−261.9
$HSO_4^-(aq)$	−887.3	+0.1318	−755.9	$NaCl(s)$	−411.2	+0.0721	−384.2
$Hg(l)$	0.0	+0.0760	0.0	$NaF(s)$	−573.6	+0.0515	−543.5
$Hg^{2+}(aq)$	+171.1	−0.0322	+164.4	$NaOH(s)$	−425.6	+0.0645	−379.5
$HgO(s)$	−90.8	+0.0703	−58.6	$Ni(s)$	0.0	+0.0299	0.0
$I_2(s)$	0.0	+0.1161	0.0	$Ni^{2+}(aq)$	−54.0	−0.1289	−45.6
$I^-(aq)$	−55.2	+0.1113	−51.6	$NiO(s)$	−239.7	+0.0380	−211.7
$K(s)$	0.0	+0.0642	0.0	$O_2(g)$	0.0	+0.2050	0.0
$K^+(aq)$	−252.4	+0.1025	−283.3	$OH^-(aq)$	−230.0	−0.0108	−157.2
$KBr(s)$	−393.8	+0.0959	−380.7	$P_4(s)$	0.0	+0.1644	0.0
$KCl(s)$	−436.7	+0.0826	−409.1	$PCl_3(g)$	−287.0	+0.3117	−267.8
$KClO_3(s)$	−397.7	+0.1431	−296.3	$PCl_5(g)$	−374.9	+0.3645	−305.0
$KClO_4(s)$	−432.8	+0.1510	−303.2	$PO_4^{3-}(aq)$	−1277.4	−0.222	−1018.7
$KNO_3(s)$	−494.6	+0.1330	−394.9	$Pb(s)$	0.0	+0.0648	0.0
$Mg(s)$	0.0	+0.0327	0.0	$Pb^{2+}(aq)$	−1.7	+0.0105	−24.4
$Mg^{2+}(aq)$	−466.8	−0.1381	−454.8	$PbBr_2(s)$	−278.7	+0.1615	−261.9
$MgCl_2(s)$	−641.3	+0.0896	−591.8	$PbCl_2(s)$	−359.4	+0.1360	−314.1
$MgCO_3(s)$	−1095.8	+0.0657	−1012.1	$PbO(s)$	−219.0	+0.0665	−188.9
$MgO(s)$	−601.7	+0.0269	−569.4	$PbO_2(s)$	−277.4	+0.0686	−217.4
$Mg(OH)_2(s)$	−924.5	+0.0632	−833.6	$S(s)$	0.0	+0.0318	0.0
$MgSO_4(s)$	−1284.9	+0.0916	−1170.7	$S^{2-}(aq)$	+33.1	−0.0146	+85.8
$Mn(s)$	0.0	+0.0320	0.0	$SO_2(g)$	−296.8	+0.2481	−300.2
$Mn^{2+}(aq)$	−220.8	−0.0736	−228.1	$SO_3(g)$	−395.7	+0.2567	−371.1
$MnO(s)$	−385.2	+0.0597	−362.9	$SO_4^{2-}(aq)$	−909.3	+0.0201	−744.5
$MnO_2(s)$	−520.0	+0.0530	−465.2	$Si(s)$	0.0	+0.0188	0.0
$MnO_4^-(aq)$	−541.4	+0.1912	−447.2	$SiO_2(s)$	−910.9	+0.0418	−856.7
$N_2(g)$	0.0	+0.1915	0.0	$Sn(s)$	0.0	+0.0516	0.0
$NH_3(g)$	−46.1	+0.1923	−16.5	$Sn^{2+}(aq)$	−8.8	−0.0174	−27.2
$NH_4^+(aq)$	−132.5	+0.1134	−79.3	$SnO_2(s)$	−580.7	+0.0523	−519.6
$NH_4Cl(s)$	−314.4	+0.0946	−203.0	$Zn(s)$	0.0	+0.0416	0.0
$NH_4NO_3(s)$	−365.6	+0.1511	−184.0	$Zn^{2+}(aq)$	−153.9	−0.1121	−147.1
$N_2H_4(l)$	+50.6	+0.1212	+149.2	$ZnI_2(s)$	−208.0	+0.1611	−209.0
$NO(g)$	+90.2	+0.2107	+86.6	$ZnO(s)$	−348.3	+0.0436	−318.3
$NO_2(g)$	+33.2	+0.2400	+51.3	$ZnS(s)$	−206.0	+0.0577	−201.3

EQUILIBRIUM CONSTANTS OF WEAK ACIDS (K_a)

H_3AsO_4	5.7×10^{-3}	HNO_2	6.0×10^{-4}	$N_2H_5^+$	1.0×10^{-8}
$H_2AsO_4^-$	1.8×10^{-7}	H_3PO_4	7.1×10^{-3}	$Al(H_2O)_6^{3+}$	1.2×10^{-5}
$HAsO_4^{2-}$	2.5×10^{-12}	$H_2PO_4^-$	6.2×10^{-8}	$Ag(H_2O)_2^+$	1.2×10^{-12}
$HBrO$	2.6×10^{-9}	HPO_4^{2-}	4.5×10^{-13}	$Ca(H_2O)_6^{2+}$	2.2×10^{-13}
$HCHO_2$	1.9×10^{-4}	H_2S	1.0×10^{-7}	$Cd(H_2O)_4^{2+}$	4.0×10^{-10}
$HC_2H_3O_2$	1.8×10^{-5}	H_2SO_3	1.7×10^{-2}	$Fe(H_2O)_6^{3+}$	6.7×10^{-3}
HCN	5.8×10^{-10}	HSO_3^-	6.0×10^{-8}	$Fe(H_2O)_6^{2+}$	1.7×10^{-7}
H_2CO_3	4.4×10^{-7}	HSO_4^-	1.0×10^{-2}	$Mg(H_2O)_6^{2+}$	3.7×10^{-12}
HCO_3^-	4.7×10^{-11}	H_2Se	1.5×10^{-4}	$Mn(H_2O)_6^{2+}$	2.8×10^{-11}
$HClO_2$	1.0×10^{-2}	H_2SeO_3	2.7×10^{-3}	$Ni(H_2O)_6^{2+}$	2.2×10^{-10}
$HClO$	2.8×10^{-8}	$HSeO_3^-$	5.0×10^{-8}	$Pb(H_2O)_6^{2+}$	6.7×10^{-7}
HF	6.9×10^{-4}	$CH_3NH_3^+$	2.4×10^{-11}	$Sc(H_2O)_6^{3+}$	1.1×10^{-4}
HIO	2.4×10^{-11}	NH_4^+	5.6×10^{-10}	$Zn(H_2O)_4^{2+}$	3.3×10^{-10}
HN_3	2.4×10^{-5}				

EQUILIBRIUM CONSTANTS OF WEAK BASES (K_b)

AsO_4^{3-}	4.0×10^{-3}	N_3^-	4.2×10^{-10}	$HSeO_3^-$	3.7×10^{-12}
$HAsO_4^{2-}$	5.6×10^{-8}	NH_3	1.8×10^{-5}	$AlOH^{2+}$	8.3×10^{-10}
$H_2AsO_4^-$	1.8×10^{-12}	N_2H_4	1.0×10^{-6}	$AgOH$	8.3×10^{-3}
BrO^-	3.8×10^{-6}	NO_2^-	1.7×10^{-11}	$CaOH^+$	4.5×10^{-2}
CH_3NH_2	4.2×10^{-4}	PO_4^{3-}	2.2×10^{-2}	$CdOH^+$	2.5×10^{-5}
CHO_2^-	5.3×10^{-11}	HPO_4^{2-}	1.6×10^{-7}	$FeOH^{2+}$	1.5×10^{-12}
$C_2H_3O_2^-$	5.6×10^{-10}	$H_2PO_4^-$	1.4×10^{-12}	$FeOH^+$	5.9×10^{-8}
CN^-	1.7×10^{-5}	HS^-	1.0×10^{-7}	$MgOH^+$	2.7×10^{-3}
CO_3^{2-}	2.1×10^{-4}	SO_3^{2-}	1.7×10^{-7}	$MnOH^+$	3.6×10^{-4}
HCO_3^-	2.3×10^{-8}	HSO_3^-	5.9×10^{-13}	$NiOH^+$	4.5×10^{-5}
ClO_2^-	1.0×10^{-12}	SO_4^{2-}	1.0×10^{-12}	$PbOH^+$	1.5×10^{-8}
ClO^-	3.6×10^{-7}	HSe^-	6.7×10^{-11}	$ScOH^{2+}$	9.1×10^{-11}
F^-	1.4×10^{-11}	SeO_3^{2-}	2.0×10^{-7}	$ZnOH^+$	3.0×10^{-5}
IO^-	4.2×10^{-4}				

FORMATION CONSTANTS OF COMPLEX IONS (K_f)

$AgBr_2^-$	2×10^7	$Cu(CN)_2^-$	1×10^{24}	$PdBr_4^{2-}$	6×10^{13}
$AgCl_2^-$	1.8×10^5	$Cu(NH_3)_4^{2+}$	2×10^{12}	$PtBr_4^{2-}$	6×10^{17}
$Ag(CN)_2^-$	2×10^{20}	$FeSCN^{2+}$	9.2×10^2	$PtCl_4^{2-}$	1×10^{16}
AgI_2^-	5×10^{10}	$Fe(CN)_6^{3-}$	4×10^{52}	$Pt(CN)_4^{2-}$	1×10^{41}
$Ag(NH_3)_2^+$	1.7×10^7	$Fe(CN)_6^{4-}$	4×10^{45}	$c\text{-}Pt(NH_3)_2Cl_2$	3×10^{29}
$Al(OH)_4^-$	1×10^{33}	$HgBr_4^{2-}$	1×10^{21}	$t\text{-}Pt(NH_3)_2Cl_2$	3×10^{28}
$Cd(CN)_4^{2-}$	2×10^{18}	$HgCl_4^{2-}$	1×10^{15}	$c\text{-}Pt(NH_3)_2(H_2O)_2^{2+}$	4×10^{23}
CdI_4^{2-}	4.0×10^5	$Hg(CN)_4^{2-}$	2×10^{41}	$c\text{-}Pt(NH_3)_2I_2$	2×10^{33}
$Cd(NH_3)_4^{2+}$	2.8×10^7	HgI_4^{2-}	1×10^{30}	$t\text{-}Pt(NH_3)_2I_2$	5×10^{32}
$Cd(OH)_4^{2-}$	1.2×10^9	$Hg(NH_3)_4^{2+}$	2×10^{19}	$c\text{-}Pt(NH_3)_2(OH)_2$	1×10^{39}
$Co(NH_3)_6^{2+}$	1×10^5	$Ni(CN)_4^{2-}$	1×10^{30}	$Pt(NH_3)_4^{2+}$	2×10^{35}
$Co(NH_3)_6^{3+}$	1×10^{23}	$Ni(NH_3)_6^{2+}$	9×10^8	$Zn(CN)_4^{2-}$	6×10^{16}
$Co(NH_3)_5Cl^{2+}$	2×10^{28}	PbI_4^{2-}	1.7×10^4	$Zn(NH_3)_4^{2+}$	3.6×10^8
$Co(NH_3)_5NO_2^{2+}$	1×10^{24}	$Pb(OH)_3^-$	8×10^{13}	$Zn(OH)_4^{2-}$	3×10^{14}

SOLUBILITY PRODUCT CONSTANTS (K_{sp})

$AgBr$	5×10^{-13}	$Co(OH)_2$	2×10^{-16}	$NiCO_3$	1.4×10^{-7}
$AgC_2H_3O_2$	1.9×10^{-3}	$Co_3(PO_4)_2$	1×10^{-35}	$Ni_3(PO_4)_2$	1×10^{-32}
Ag_2CO_3	8×10^{-12}	$CuCl$	1.7×10^{-7}	NiS	1×10^{-21}
$AgCl$	1.8×10^{-10}	$CuBr$	6.3×10^{-9}	$PbBr_2$	6.6×10^{-6}
Ag_2CrO_4	1×10^{-12}	CuI	1.2×10^{-12}	$PbCO_3$	1×10^{-13}
AgI	1×10^{-16}	$Cu_3(PO_4)_2$	1×10^{-37}	$PbCl_2$	1.7×10^{-5}
Ag_3PO_4	1×10^{-16}	CuS	1×10^{-36}	$PbCrO_4$	2×10^{-14}
Ag_2S	1×10^{-49}	Cu_2S	1×10^{-48}	PbF_2	7.1×10^{-7}
$AgSCN$	1.0×10^{-12}	$FeCO_3$	3.1×10^{-11}	PbI_2	8.4×10^{-9}
AlF_3	1×10^{-18}	$Fe(OH)_2$	5×10^{-17}	$Pb(OH)_2$	1×10^{-20}
$Al(OH)_3$	2×10^{-31}	$Fe(OH)_3$	3×10^{-39}	PbS	1×10^{-28}
$AlPO_4$	1×10^{-20}	FeS	2×10^{-19}	$Pb(SCN)_2$	2.1×10^{-5}
$BaCO_3$	2.6×10^{-9}	GaF_3	2×10^{-16}	$PbSO_4$	1.8×10^{-8}
$BaCrO_4$	1.2×10^{-10}	$Ga(OH)_3$	1×10^{-35}	$Sc(OH)_3$	2×10^{-31}
BaF_2	1.8×10^{-7}	$GaPO_4$	1×10^{-21}	SnS	3×10^{-28}
$BaSO_4$	1.1×10^{-10}	Hg_2Br_2	6×10^{-23}	$SrCO_3$	5.6×10^{-10}
Bi_2S_3	1×10^{-99}	Hg_2Cl_2	1×10^{-18}	$SrCrO_4$	3.6×10^{-5}
$CaCO_3$	4.9×10^{-9}	Hg_2I_2	5×10^{-29}	SrF_2	4.3×10^{-9}
CaF_2	1.5×10^{-10}	HgS	1×10^{-52}	$SrSO_4$	3.4×10^{-7}
$Ca(OH)_2$	4.0×10^{-6}	Li_2CO_3	8.2×10^{-4}	$TlCl$	1.9×10^{-4}
$Ca_3(PO_4)_2$	1×10^{-33}	$MgCO_3$	6.8×10^{-6}	$TlBr$	3.7×10^{-6}
$CaSO_4$	7.1×10^{-5}	MgF_2	7×10^{-11}	TlI	5.6×10^{-8}
$CdCO_3$	6×10^{-12}	$Mg(OH)_2$	6×10^{-12}	$Tl(OH)_3$	2×10^{-44}
$Cd(OH)_2$	5×10^{-15}	$Mg_3(PO_4)_2$	1×10^{-24}	Tl_2S	1×10^{-20}
$Cd_3(PO_4)_2$	1×10^{-33}	$Mn(OH)_2$	2×10^{-13}	$ZnCO_3$	1.1×10^{-10}
CdS	1×10^{-29}	MnS	5×10^{-14}	$Zn(OH)_2$	4×10^{-17}

PROPERTIES OF THE ELEMENTS

Element	At. No.	mp(°C)	bp(°C)	E.N.	Ion. Ener. (kJ/mol)	At. Rad. (nm)	Ion. Rad (nm)
H	1	−259	−253	2.2	1312	0.037	(−1)0.208
He	2	−272	−269		2372	0.05	
Li	3	186	1326	1.0	520	0.152	(+1)0.060
Be	4	1283	2970	1.6	900	0.111	(+2)0.031
B	5	2300	2550	2.0	801	0.088	
C	6	3570	subl.	2.5	1086	0.077	
N	7	−210	−196	3.0	1402	0.070	
O	8	−218	−183	3.5	1314	0.066	(−2)0.140
F	9	−220	−188	4.0	1681	0.064	(−1)0.136
Ne	10	−249	−246		2081	0.070	
Na	11	98	889	0.9	496	0.186	(+1)0.095
Mg	12	650	1120	1.3	738	0.160	(+2)0.065
Al	13	660	2327	1.6	578	0.143	(+3)0.050
Si	14	1414	2355	1.9	786	0.117	
P	15	44	280	2.2	1012	0.110	
S	16	119	444	2.6	1000	0.104	(−2)0.184
Cl	17	−101	−34	3.2	1251	0.099	(−1)0.181
Ar	18	−189	−186		1520	0.094	
K	19	64	774	0.8	419	0.231	(+1)0.133
Ca	20	845	1420	1.0	590	0.197	(+2)0.099
Sc	21	1541	2831	1.4	631	0.160	(+3)0.081
Ti	22	1660	3287	1.5	658	0.146	
V	23	1890	3380	1.6	650	0.131	
Cr	24	1857	2672	1.6	653	0.125	(+3)0.064
Mn	25	1244	1962	1.5	717	0.129	(+2)0.080
Fe	26	1535	2750	1.8	759	0.126	(+2)0.075
Co	27	1495	2870	1.9	758	0.125	(+2)0.072
Ni	28	1453	2732	1.9	737	0.124	(+2)0.069
Cu	29	1083	2567	1.9	746	0.128	(+1)0.096
Zn	30	420	907	1.6	906	0.133	(+2)0.074
Ga	31	30	2403	1.6	579	0.122	(+3)0.062
Ge	32	937	2830	2.0	762	0.122	
As	33	814	subl.	2.2	944	0.121	
Se	34	217	685	2.5	941	0.117	(−2)0.198
Br	35	−7	59	3.0	1140	0.114	(−1)0.195
Kr	36	−157	−152	3.3	1351	0.109	
Rb	37	39	688	0.8	403	0.244	(+1)0.148
Sr	38	770	1380	0.9	550	0.215	(+2)0.113
Y	39	1509	2930	1.2	616	0.180	(+3)0.093
Zr	40	1852	3580	1.4	660	0.157	
Nb	41	2468	5127	1.6	664	0.143	

continued

Element	At. No.	mp(°C)	bp(°C)	E.N.	Ion. Ener. (kJ/mol)	At. Rad. (nm)	Ion. Rad (nm)
Mo	42	2610	5560	1.8	685	0.136	
Tc	43	2200	4700	1.9	702	0.136	
Ru	44	2430	3700	2.2	711	0.133	
Rh	45	1966	3700	2.2	720	0.134	
Pd	46	1550	3170	2.2	805	0.138	
Ag	47	961	2210	1.9	731	0.144	(+1)0.126
Cd	48	321	767	1.7	868	0.149	(+2)0.097
In	49	157	2000	1.7	558	0.162	(+3)0.081
Sn	50	232	2270	1.9	709	0.140	
Sb	51	631	1380	2.0	832	0.141	
Te	52	450	990	2.1	869	0.137	(−2)0.221
I	53	114	184	2.7	1009	0.133	(−1)0.216
Xe	54	−112	−107	3.0	1170	0.130	
Cs	55	28	690	0.8	376	0.262	(+1)0.169
Ba	56	725	1640	0.9	503	0.217	(+2)0.135
La	57	920	3469	1.1	538	0.187	(+3)0.115
Ce	58	795	3468	1.1	528	0.182	(+3)0.101
Pr	59	935	3127	1.1	523	0.182	(+3)0.100
Nd	60	1024	3027	1.1	530	0.182	(+3)0.099
Pm	61	1027	2727	1.1	536	0.181	
Sm	62	1072	1900	1.1	543	0.180	
Eu	63	826	1439	1.1	547	0.204	(+2)0.097
Gd	64	1312	3000	1.1	592	0.179	(+3)0.096
Tb	65	1356	2800	1.1	564	0.177	(+3)0.095
Dy	66	1407	2600	1.1	572	0.177	(+3)0.094
Ho	67	1461	2600	1.1	581	0.176	(+3)0.093
Er	68	1497	2900	1.1	589	0.175	(+3)0.092
Tm	69	1356	2800	1.1	597	0.174	(+3)0.091
Yb	70	824	1427	1.1	603	0.193	(+3)0.089
Lu	71	1652	3327	1.1	524	0.174	(+3)0.089
Hf	72	2225	5200	1.3	654	0.157	
Ta	73	2980	5425	1.5	761	0.143	
W	74	3410	5930	1.7	770	0.137	
Re	75	3180	5885	1.9	760	0.137	
Os	76	2727	4100	2.2	840	0.134	
Ir	77	2448	4500	2.2	880	0.135	
Pt	78	1769	4530	2.2	870	0.138	
Au	79	1063	2966	2.4	890	0.144	(+1)0.137
Hg	80	−39	357	1.9	1007	0.155	(+2)0.110
Tl	81	304	1457	1.8	589	0.171	(+3)0.095
Pb	82	328	1750	1.9	716	0.175	
Bi	83	271	1560	1.9	703	0.146	
Po	84	254	962	2.0	812	0.165	
At	85	302	334	2.2			
Rn	86	−71	−62		1037	0.14	
Fr	87	27	677	0.7			
Ra	88	700	1140	0.9	509	0.220	
Ac	89	1050	3200	1.1	490	0.20	
Th	90	1750	4790	1.3	590	0.180	
Pa	91	1600	4200	1.4	570		
U	92	1132	3818	1.4	590	0.14	

EXPONENTS AND LOGARITHMS

The mathematics you will use in general chemistry is relatively simple. You will, however, be expected to—

- make calculations involving exponential numbers, such as 6.022×10^{23} or 1.6×10^{-10}.
- work with logarithms or inverse logarithms, particularly in problems involving pH:

$$pH = -\log_{10} (\text{conc. } H^+)$$

This appendix reviews each of these topics.

EXPONENTIAL NOTATION

Chemists deal frequently with very large or very small numbers. In one gram of the element carbon there are

$$50,140,000,000,000,000,000,000 \text{ atoms of carbon}$$

At the opposite extreme, the mass of a single atom is

$$0.00000000000000000000001994 \text{ g}$$

Numbers such as these are very awkward to work with. For example, neither of the numbers just written could be entered directly on a calculator. Operations involving very large or very small numbers can be simplified by using **exponential (scientific)** notation. To express a number in exponential notation, write it in the form

$$C \times 10^n$$

where C is a number between 1 and 10 (for example, 1, 2.62, 5.8) and n is a positive or negative integer such as 1, -1, -3. To find n, count the number of places that the decimal point must be moved to give the coefficient, C. If the decimal point must be moved to the *left, n* is a *positive* integer; if it must be moved to the *right, n* is a *negative* integer. Thus

$$26.23 = 2.623 \times 10^1 \qquad \text{(decimal point moved 1 place to left)}$$
$$5609 = 5.609 \times 10^3 \qquad \text{(decimal point moved 3 places to left)}$$
$$0.0918 = 9.18 \times 10^{-2} \qquad \text{(decimal point moved 2 places to right)}$$

Multiplication and Division

A major advantage of exponential notation is that it simplifies the processes of multiplication and division. To *multiply, add exponents:*

$$10^1 \times 10^2 = 10^{1+2} = 10^3 \qquad 10^6 \times 10^{-4} = 10^{6+(-4)} = 10^2$$

To *divide, subtract* exponents:

$$10^3/10^2 = 10^{3-2} = 10^1 \qquad 10^{-3}/10^6 = 10^{-3-6} = 10^{-9}$$

It often happens that multiplication or division yields an answer that is not in standard exponential notation. For example,

$$(5.0 \times 10^4) \times (6.0 \times 10^3) = (5.0 \times 6.0) \times 10^4 \times 10^3 = 30 \times 10^7$$

The product is not in standard exponential notation because the coefficient, 30, does not lie between 1 and 10. To correct this situation, rewrite the coefficient as 3.0×10^1 and then add exponents:

$$30 \times 10^7 = (3.0 \times 10^1) \times 10^7 = 3.0 \times 10^8$$

In another case,

$$0.526 \times 10^3 = (5.26 \times 10^{-1}) \times 10^3 = 5.26 \times 10^2$$

Exponential Notation on the Calculator

On all scientific calculators it is possible to enter numbers in exponential notation. The method used depends on the brand of calculator. Most often, it involves using a key labeled $\boxed{\text{EXP}}$, $\boxed{\text{EE}}$, or $\boxed{\text{EEX}}$. Check your instruction manual for the procedure to be followed. To make sure you understand it, try entering the following numbers:

$$2.4 \times 10^6 \qquad 3.16 \times 10^{-8} \qquad 6.2 \times 10^{-16}$$

Multiplication, division, and many other operations can be carried out directly on your calculator. Try the following exercises, using your calculator:

(a) $(6.0 \times 10^2) \times (4.2 \times 10^{-4}) = ?$
(b) $\dfrac{6.0 \times 10^2}{4.2 \times 10^{-4}} = ?$
(c) $(2.50 \times 10^{-9})^{1/2} = ?$
(d) $3.6 \times 10^{-4} + 4 \times 10^{-5} = ?$

The answers, expressed in exponential notation and following the rules of significant figures, are as follows: (a) 2.5×10^{-1} (b) 1.4×10^6 (c) 5.00×10^{-5}
(d) 4.0×10^{-4}.

LOGARITHMS AND INVERSE LOGARITHMS

The logarithm of a number n to the base m is defined as the power to which m must be raised to give the number n. Thus

$$\text{if } m^x = n, \text{ then } \log_m n = x$$

In other words, a *logarithm* is an *exponent,* which may be a whole number (e.g., 1, -1) but more often is not (e.g., 1.500, -0.301).

In general chemistry, you will encounter two kinds of logarithms.

1. *Common logarithms,* where the base is 10, ordinarily denoted as $\log_{10}$. If $10^x = n$, then $\log_{10} n = x$. Examples include the following:

$$\log_{10} 100 = 2.00 \qquad (10^2 = 100)$$

$$\log_{10} 1 = 0.000 \qquad (10^0 = 1)$$

$$\log_{10} 0.001 = -3.000 \qquad (10^{-3} = 0.001)$$

2. *Natural logarithms,* where the base is the quantity $e = 2.718. \ldots$ Many of the equations used in general chemistry are expressed most simply in terms of natural logarithms, denoted as ln. If $e^x = n$, then $\ln n = x$.

$$\ln 100 = 4.606 \qquad (\text{i.e., } 100 = e^{4.606})$$

$$\ln 1 = 0 \qquad (\text{i.e., } 1 = e^0)$$

$$\ln 0.001 = -6.908 \qquad (\text{i.e., } 0.001 = e^{-6.908})$$

Notice that

(a) $\log_{10} 1 = \ln 1 = 0$. The logarithm of 1 to any base is zero, because any number raised to the zero power is 1. That is

$$10^0 = e^0 = 2^0 = \cdots = n^0 = 1$$

(b) **Numbers larger than 1 have a positive logarithm; numbers smaller than 1 have a negative logarithm. For example,**

$$\log_{10} 100 = 2 \qquad \ln 100 = 4.606$$

$$\log_{10} 10 = 1 \qquad \ln 10 = 2.303$$

$$\log_{10} 0.1 = -1 \qquad \ln 0.1 = -2.303$$

$$\log_{10} 0.01 = -2 \qquad \ln 0.01 = -4.606$$

(c) The common and natural logarithms of a number are related by the expression

$$\ln n = 2.303 \log_{10} n$$

That is, the natural logarithm of a number is $2.303 \ldots$ times its base-10 logarithm.

(d) Finally, if $x \leq 0$, both $\log x$ and $\ln x$ are undefined.

An **inverse** logarithm (antilogarithm) is the number corresponding to a given logarithm. In general,

$$\text{if } m^x = n, \text{ then inverse } \log_m x = n$$

For example,

$$10^2 = 100 \qquad \text{inverse } \log_{10} 2 = 100$$

$$10^{-3} = 0.001 \qquad \text{inverse } \log_{10} (-3) = 0.001$$

In other words, the numbers whose base-10 logarithms are 2 and -3 are 100 and 0.001, respectively. The same reasoning applies to natural logarithms.

$$e^{4.606} = 100 \qquad \text{inverse } \ln 4.606 = 100$$

$$e^{-6.908} = 0.001 \qquad \text{inverse } \ln -6.908 = 0.001$$

The numbers whose natural logarithms are 4.606 and -6.908 are 100 and 0.001, respectively. Notice that if the inverse logarithm is positive, the corresponding number is larger than 1; if it is negative, the number is smaller than 1.

Finding Logarithms and Inverse Logarithms on a Calculator

To obtain a base-10 logarithm using a calculator, all you need do is enter the number and press the $\boxed{\text{LOG}}$ key. This way you should find that

$$\log_{10} 2.00 = 0.301 \ . \ . \ .$$

$$\log_{10} 0.526 = -0.279 \ . \ . \ .$$

Similarly, to find a natural logarithm, you enter the number and press the $\boxed{\text{LN } X}$ key.

$$\ln 2.00 = 0.693 \ . \ . \ .$$

$$\ln 0.526 = -0.642 \ . \ . \ .$$

To find the logarithm of an exponential number, you enter the number in exponential form and take the logarithm in the usual way. This way you should find that

$$\log_{10} 2.00 \times 10^3 = 3.301 \ . \ . \ . \qquad \ln 2.00 \times 10^3 = 7.601 \ . \ . \ .$$

$$\log_{10} 5.3 \times 10^{-12} = -11.28 \ . \ . \ . \qquad \ln 5.3 \times 10^{-12} = -25.96 \ . \ . \ .$$

The base-10 logarithm of an exponential number can be found in a somewhat different way by applying the relation

$$\log_{10} (C \times 10^n) = n + \log_{10} C$$

Thus

$$\log_{10} 2.00 \times 10^3 = 3 + \log_{10} 2.00 = 3 + 0.301 = 3.301$$

$$\log_{10} 5.3 \times 10^{-12} = -12 + \log_{10} 5.3 = -12 + 0.72 = -11.28$$

The method used to find inverse logarithms depends on the type of calculator. On certain calculators, you enter the number and then press, in succession, the $\boxed{\text{INV}}$ and either $\boxed{\text{LOG}}$ or $\boxed{\text{LN } X}$ keys. With other calculators, you press the $\boxed{10^x}$ or $\boxed{e^x}$ key. Either way, you should find that

$$\text{inverse } \log_{10} 1.632 = 42.8 \ . \ . \ . \qquad \text{inverse } \ln 1.632 = 5.11 \ . \ . \ .$$

$$\text{inverse } \log_{10} - 8.82 = 1.5 \times 10^{-9} \qquad \text{inverse } \ln -8.82 = 1.5 \times 10^{-4}$$

Significant Figures in Logarithms and Inverse Logarithms

For base-10 logarithms, the rules governing significant figures are as follows:

1. *In taking the logarithm of a number, retain after the decimal point in the log as many digits as there are significant figures in the number.* (This part of the logarithm is often referred to as the *mantissa;* digits that precede the decimal point comprise the *characteristic* of the logarithm.) To illustrate this rule, consider the following:

$$\log_{10} 2.00 = 0.301 \qquad \log_{10} (2.00 \times 10^3) = 3.301$$

$$\log_{10} 2.0 = .30 \qquad \log_{10} (2.0 \times 10^1) = 1.30$$

$$\log_{10} 2 = 0.3 \qquad \log_{10} (2 \times 10^{-3}) = 0.3 - 3 = -2.7$$

2. In taking the inverse logarithm of a number, retain as many significant figures as there are after the decimal point in the number. Thus

$$\text{inverse } \log_{10} 0.301 = 2.00 \qquad \text{inverse } \log_{10} 3.301 = 2.00 \times 10^3$$

$$\text{inverse } \log_{10} 0.30 = 2.0 \qquad \text{inverse } \log_{10} 1.30 = 2.0 \times 10^1$$

$$\text{inverse } \log_{10} 0.3 = 2 \qquad \text{inverse } \log_{10} (-2.7) = 2 \times 10^{-3}$$

These rules take into account the fact that, as mentioned earlier,

$$\log_{10} (C \times 10^n) = n + \log_{10} C$$

The digits that appear before (to the left of) the decimal point specify the value of n, that is, the power of 10 involved in the expression. In that sense, they are not experimentally significant. In contrast, the digits that appear after (to the right of) the decimal point specify the value of the logarithm of C; the number of such digits reflects the uncertainty in C. Thus,

$$\log_{10} 209 = 2.320 \qquad \text{(3 sig. fig.)}$$

$$\log_{10} 209.0 = 2.3201 \qquad \text{(4 sig. fig.)}$$

The rules for significant figures involving natural logarithms and inverse logarithms are somewhat more complex than those for base-10 logs. However, for simplicity we will assume that the rules listed above apply here as well. Thus

$$\ln 209 = 5.342 \qquad \text{(3 sig. fig.)}$$

$$\ln 209.0 = 5.3423 \qquad \text{(4 sig. fig.)}$$

Operations Involving Logarithms

Because logarithms are exponents, the rules governing the use of exponents apply as well. The rules that follow are valid for all types of logarithms, regardless of the base. We illustrate the rules with natural logarithms; that is where you are most likely to use them in working with this text.

Multiplication: $\ln(xy) = \ln x + \ln y$

Example: $\ln(2.50 \times 1.25) = \ln 2.50 + \ln 1.25 = 0.916 + 0.223 = 1.139$

Division: $\ln(x/y) = \ln x - \ln y$

Example: $\ln(2.50/1.25) = 0.916 - 0.223 = 0.693$

Raising to a Power: $\ln(x^n) = n \ln x$

Example: $\ln(2.00)^4 = 4 \ln 2.00 = 4(0.693) = 2.772$

Extracting a Root: $\ln(x^{1/n}) = \dfrac{1}{n} \ln x$

Example: $\ln(2.00)^{1/3} = \dfrac{\ln 2.00}{3} = \dfrac{0.693}{3} = 0.231$

Taking a Reciprocal: $\ln(1/x) = -\ln x$

Example: $\ln(1/2.00) = -\ln 2.00 = -0.693$

NOMENCLATURE

COMPLEX IONS (CHAPTER 15)

To name a complex ion, it is necessary to show—

- the number and identity of each ligand attached to the central metal ion.
- the identity and oxidation number of the central metal ion.
- whether the complex is a cation or an anion.

To accomplish this, a simple set of rules is followed.

1. The names of anionic ligands are obtained by substituting the suffix *-o* for the normal ending. Examples include

Cl^-	chlor*o*	SO_4^{2-}	sulfat*o*
OH^-	hydrox*o*	CO_3^{2-}	carbonat*o*

Ordinarily, the names of molecular ligands are unchanged. Two important exceptions are

$$H_2O \quad aqua \quad NH_3 \quad ammine$$

2. The number of ligands of a particular type is ordinarily indicated by the Greek prefixes *di, tri, tetra, penta, hexa:*

$Cu(H_2O)_4^{2+}$	*tetra*ammineaquacopper(II)
$Cr(NH_3)_6^{3+}$	*hexa*amminechromium(III)

If the name of the ligand is itself complex (e.g., ethylenediamine), the number of such ligands is indicated by the prefixes *bis, tris,* . . . The name of the ligand is enclosed in parentheses:

$$Cr(en)_3^{3+} \quad \textit{tris}(\text{ethylenediamine})\text{chromium(III)}$$

3. If more than one type of ligand is present, they are named in alphabetical order (without regard for prefixes):

$Cu(NH_3)_2(H_2O)_2^{2+}$	diammine*aqua*copper(II)
$Cr(NH_3)_5Cl^{2+}$	penta*amminechloro*chromium(III)

4. As you can deduce from the preceding examples, the oxidation number of the central metal ion is indicated by a Roman numeral written at the end of the name.

5. If the complex is an anion, the suffix *-ate* is inserted between the name of the metal and the oxidation number:

$$Zn(OH)_4^{2-} \quad \text{tetrahydroxozinc}\textit{ate}\text{(II)}$$

Coordination compounds are named in much the same way as simple ionic compounds. The cation is named first, followed by the anion. Examples include

$[Cr(NH_3)_4Cl_2]NO_3$ tetraamminedichlorochromium(III) nitrate

$K_2[PtCl_6]$ potassium hexachloroplatinate(IV)

$[Co(NH_3)_2(en)_2]Br_3$ diamminebis(ethylenediamine)cobalt(III) bromide

ALKANES (CHAPTER 22)

More than 50 years ago, the International Union of Pure and Applied Chemistry devised a system that could be used to name alkanes along with other types of organic compounds. For straight-chain alkanes such as

$$CH_3{-}CH_2{-}CH_3 \qquad CH_3{-}CH_2{-}CH_2{-}CH_3$$

propane   butane

the IUPAC name consists of a single word. These names, for up to eight carbon atoms, are listed in Table 1.

With alkanes containing a **branched chain,** such as

$$\begin{array}{c} H \\ | \\ CH_3{-}C{-}CH_3 \\ | \\ CH_3 \end{array}$$

2-methylpropane

the name is more complex. A branched-chain alkane such as 2-methylpropane can be considered to be derived from a straight-chain alkane by replacing one or more hydrogen atoms by alkyl groups. The name consists of two parts—

■ *a suffix that identifies the parent straight-chain alkane.* To find the suffix, count the number of carbon atoms in the longest continuous chain. For a three-

TABLE 1	Nomenclature of Alkanes		
Straight-Chain Alkanes		**Alkyl Groups**	
methane	CH_4	methyl	$CH_3{-}$
ethane	CH_3CH_3	ethyl	$CH_3{-}CH_2{-}$
propane	$CH_3CH_2CH_3$	propyl	$CH_3{-}CH_2{-}CH_2{-}$
butane	$CH_3(CH_2)_2CH_3$		
pentane	$CH_3(CH_2)_3CH_3$	isopropyl	$CH_3{-}\overset{\displaystyle H}{\underset{\displaystyle CH_3}{C}}{-}$
hexane	$CH_3(CH_2)_4CH_3$		
heptane	$CH_3(CH_2)_5CH_3$	butyl	$CH_3{-}CH_2{-}CH_2{-}CH_2{-}$
octane	$CH_3(CH_2)_6CH_3$		

carbon chain, the suffix is *propane;* for a four-carbon chain it is *butane,* and so on.

- *a prefix that identifies the branching alkyl group (Table 1) and indicates by a number the carbon atom where branching occurs.* In 2-methyl-propane, referred to above, the methyl group is located at the second carbon from the end of the chain:

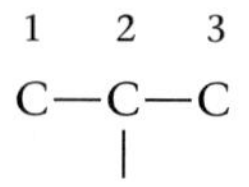

$$
\begin{array}{ccc}
1 & 2 & 3 \\
C & \!\!-C-\!\! & C \\
& | &
\end{array}
$$

Following this system, the IUPAC names of the isomers of pentane are

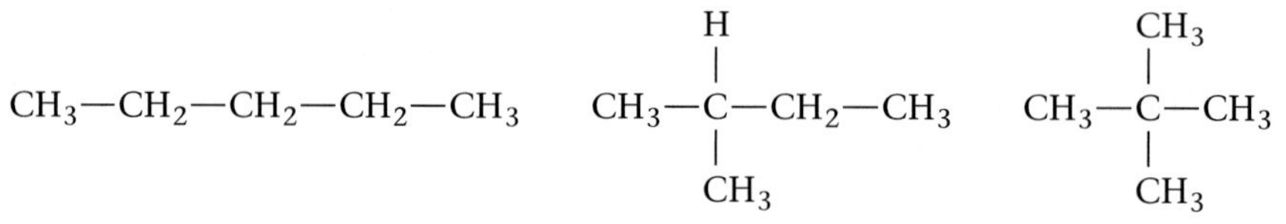

$$CH_3-CH_2-CH_2-CH_2-CH_3 \qquad CH_3-\underset{\underset{CH_3}{|}}{\overset{\overset{H}{|}}{C}}-CH_2-CH_3 \qquad CH_3-\underset{\underset{CH_3}{|}}{\overset{\overset{CH_3}{|}}{C}}-CH_3$$

pentane   2-methylbutane   2,2-dimethylpropane

Notice that—

- *if the same alkyl group is at two branches, the prefix "di-" is used (2,2-dimethylpropane).* If there were three methyl branches, we would write trimethyl, and so on.
- *the number in the name is made as small as possible.* Thus we refer to 2-methylbutane, numbering the chain from the left, rather than from the right.

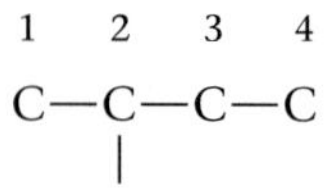

$$
\begin{array}{cccc}
1 & 2 & 3 & 4 \\
C-\!\! & C-\!\! & C-\!\! & C \\
& | & &
\end{array}
$$

MOLECULAR ORBITALS

In Chapter 7, we used valence bond theory to explain bonding in molecules. It accounts, at least qualitatively, for the stability of the covalent bond in terms of the overlap of atomic orbitals. By invoking hybridization, valence bond theory can account for the molecular geometries predicted by electron-pair repulsion. Where Lewis structures are inadequate, as in SO_2, the concept of resonance allows us to explain the observed properties.

A major weakness of valence bond theory has been its inability to predict the magnetic properties of molecules. We mentioned this problem in Chapter 7 with regard to the O_2 molecule, which is paramagnetic, even though it has an even number (12) of valence electrons. The octet rule, or valence bond theory, would predict that all the electrons in O_2 should be paired, which would make it diamagnetic.

This discrepancy between experiment and theory (and many others) can be explained in terms of an alternative model of covalent bonding, the molecular orbital (MO) approach. Molecular orbital theory treats bonds in terms of orbitals characteristic of the molecule as a whole. To apply this approach, we carry out three basic operations.

1. The atomic orbitals of atoms are combined to give a new set of molecular orbitals characteristic of the molecule as a whole. *The number of molecular orbitals formed is equal to the number of atomic orbitals combined.* When two H atoms combine to form H_2, two s orbitals, one from each atom, yield two molecular orbitals. In another case, six p orbitals, three from each atom, give a total of six molecular orbitals.

2. The molecular orbitals are arranged in order of increasing energy. The relative energies of these orbitals are ordinarily deduced from experiment. Spectra and magnetic properties of molecules are used.

3. The *valence electrons* in a molecule are distributed among the available molecular orbitals. The process followed is much like that used with electrons in atoms. In particular, we find the following:

(a) *Each molecular orbital can hold a maximum of two electrons.* When an orbital is filled the two electrons have opposed spins, in accordance with the Pauli principle (Chapter 7).

(b) *Electrons go into the lowest energy molecular orbital available.* A higher orbital starts to fill only when each orbital below it has its quota of two electrons.

(c) *Hund's rule is obeyed.* When two orbitals of equal energy are available to two electrons, one electron goes into each, giving two half-filled orbitals.

DIATOMIC MOLECULES OF THE ELEMENTS

To illustrate molecular orbital theory, we apply it to the diatomic molecules of the elements in the first two periods of the periodic table.

Hydrogen and Helium (Combination of 1s Orbitals)

Molecular orbital (MO) theory predicts that two 1s orbitals will combine to give two molecular orbitals. One of these has an energy lower than that of the atomic orbitals from which it is formed (Figure 1). Placing electrons in this orbital gives a species that is more stable than the isolated atoms. For that reason the lower molecular orbital in Figure 1 is called a **bonding orbital.** The other molecular orbital has a higher energy than the corresponding atomic orbitals. Electrons entering it are in an unstable, higher energy state. It is referred to as an **antibonding orbital.**

The electron density in these molecular orbitals is shown at the right of Figure 1. Notice that the bonding orbital has a higher density between the nuclei. This accounts for its stability. In the antibonding orbital, the chance of finding the electron between the nuclei is very small. The electron density is concentrated at the far ends of the "molecule." This means that the nuclei are less shielded from each other than they are in the isolated atoms.

The electron density in both molecular orbitals is symmetrical about the axis between the two nuclei. This means that both of these are sigma orbitals. In MO notation, the 1s bonding orbital is designated as σ_{1s}. The antibonding orbital is given the symbol σ_{1s}^*. An asterisk designates an antibonding orbital.

In the H_2 molecule, there are two 1s electrons. They fill the σ_{1s} orbital, giving a single bond. In the He_2 molecule, there would be four electrons, two from each atom. These would fill the bonding and antibonding orbitals. As a result, the number of bonds (the *bond order*) in He_2 is zero. The general relation is

$$\text{no. of bonds} = \text{bond order} = \frac{B - AB}{2}$$

where B is the number of electrons in bonding orbitals and AB is the number of electrons in antibonding orbitals. In H_2, B = 2 and AB = 0, so we have one bond. In He_2, B = AB = 2, so the number of bonds is zero. The He_2 molecule should not and does not exist.

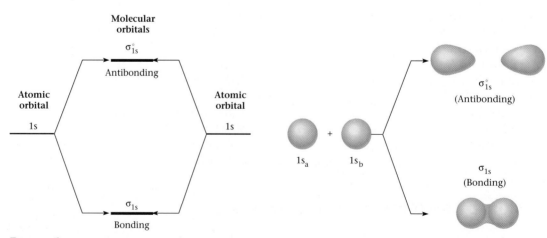

FIGURE 1
Molecular orbital formation. Two molecular orbitals are formed by combining two 1s atomic orbitals.

Second Period Elements (Combination of 2s and 2p Orbitals)

Among the diatomic molecules of the second period elements are three familiar ones, N_2, O_2, and F_2. The molecules Li_2, B_2, and C_2 are less common but have been observed and studied in the gas phase. In contrast, the molecules Be_2 and Ne_2 are either highly unstable or nonexistent. Let us see what molecular orbital theory predicts about the structure and stability of these molecules. We start by considering how the atomic orbitals containing the valence electrons (2s and 2p) are used to form molecular orbitals.

Combining two 2s atomic orbitals, one from each atom, gives two molecular orbitals. These are very similar to the ones shown on page A.18. They are designated as σ_{2s} (sigma, bonding, 2s) and σ_{2s}^* (sigma, antibonding, 2s).

Consider now what happens to the 2p orbitals. In an isolated atom, there are three such orbitals, oriented at right angles to each other. We call these atomic orbitals, p_x, p_y, and p_z (top of Figure 2). When two p_x atomic orbitals, one from

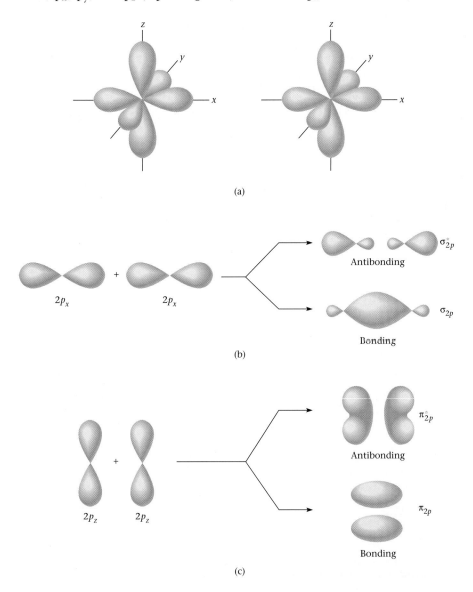

(a)

(b)

(c)

FIGURE 2

(See text, page A.20.) When p orbitals from two different atoms (a) overlap, there are two quite different possibilities. If they overlap head-to-head (b), two sigma molecular orbitals are produced. If, on the other hand, they overlap side-to-side (c), two pi molecular orbitals result.

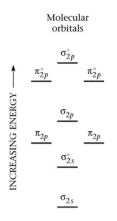

Molecular
orbitals

INCREASING ENERGY $\longrightarrow$

σ_{2p}°

π_{2p}° π_{2p}°

σ_{2p}

π_{2p} π_{2p}

σ_{2s}°

σ_{2s}

FIGURE 3
Relative energies, so far as filling order
is concerned, for the molecular orbitals formed by combining 2s and 2p
atomic orbitals.

each atom, overlap head-to-head, they form two sigma orbitals. These are a bond-
ing orbital, σ_{2p}, and an antibonding orbital, σ_{2p}^*. The situation is quite different
when the p_z orbitals overlap. Because they are oriented parallel to one another,
they overlap side-to-side (Figure 2c). The two molecular orbitals formed in this
case are pi orbitals; one is a bonding orbital, π_{2p}, the other a nonbonding orbital,
π_{2p}^*. In an entirely similar way, the p_y orbitals of the two atoms interact to form
another pair of pi molecular orbitals, π_{2p} and π_{2p}^* (these orbitals are not shown in
Figure 2).

The relative energies of the molecular orbitals available for occupancy by the
valence electrons of the second period elements are shown in Figure 3. This order
applies at least through N_2.*

To obtain the MO structure of the diatomic molecules of the elements in the
second period, we fill the available molecular orbitals in order of increasing energy. The results are shown in Table 1. Note the agreement between MO theory
and the properties of these molecules. In particular, the number of unpaired electrons predicted agrees with experiment. There is also a general correlation between
the predicted bond order:

$$\text{bond order} = \frac{B - AB}{2}$$

and the bond energy. We would expect a double bond (C_2 or O_2) to be stronger
than a single bond (Li_2, B_2, and F_2). A triple bond, as in N_2, should be still
stronger.

A major triumph of MO theory is its ability to explain the properties of O_2. It
explains how the molecule can have a double bond and, at the same time, have
two unpaired electrons.

EXAMPLE   Using MO theory, predict the electronic structure, bond order, and number of unpaired electrons in the peroxide ion, O_2^{2-}.

Strategy   First, find the number of valence electrons. Then construct an orbital diagram, filling the available molecular orbitals (Figure 3) in order of increasing energy.

Solution   Recall that oxygen is in Group **16** of the periodic table.

$$\text{no. valence } e^- = 2(6) + 2 = 14$$

The orbital diagram is

	σ_{2s}	σ_{2s}^*	π_{2p}	π_{2p}	σ_{2p}	π_{2p}^*	π_{2p}^*	σ_{2p}^*
O_2^{2-}	($\uparrow\downarrow$)	($\uparrow\downarrow$)	($\uparrow\downarrow$)	($\uparrow\downarrow$)	($\uparrow\downarrow$)	($\uparrow\downarrow$)	($\uparrow\downarrow$)	( )

There are eight electrons in bonding orbitals, six in antibonding orbitals.

$$\text{bond order} = \frac{8 - 6}{2} = 1$$

There are no unpaired electrons. These conclusions are in agreement with the Lewis structure of the peroxide ion: $(:\ddot{O}-\ddot{O}:)^{2-}$.

---

*It appears that beyond N_2 the σ_{2p} orbital lies below the two π_{2p} orbitals. This does not affect the filling
order, because in O_2 all of these orbitals are filled.

TABLE 1   Predicted and Observed Properties of Diatomic Molecules of Second Period Elements

	Occupancy of Orbitals							
	σ_{2s}	σ_{2s}^*	π_{2p}	π_{2p}	σ_{2p}	π_{2p}^*	π_{2p}^*	σ_{2p}^*
Li_2	(↑↓)	( )	( )	( )	( )	( )	( )	( )
Be_2	(↑↓)	(↑↓)	( )	( )	( )	( )	( )	( )
B_2	(↑↓)	(↑↓)	(↑)	(↑)	( )	( )	( )	( )
C_2	(↑↓)	(↑↓)	(↑↓)	(↑↓)	( )	( )	( )	( )
N_2	(↑↓)	(↑↓)	(↑↓)	(↑↓)	(↑↓)	( )	( )	( )
O_2	(↑↓)	(↑↓)	(↑↓)	(↑↓)	(↑↓)	(↑)	(↑)	( )
F_2	(↑↓)	(↑↓)	(↑↓)	(↑↓)	(↑↓)	(↑↓)	(↑↓)	( )
Ne_2	(↑↓)	(↑↓)	(↑↓)	(↑↓)	(↑↓)	(↑↓)	(↑↓)	(↑↓)

	Predicted Properties		Observed Properties	
	Number of Unpaired e^-	Bond order	Number of Unpaired e^-	Bond Energy (kJ/mol)
Li_2	0	1	0	105
Be_2	0	0	0	Unstable
B_2	2	1	2	289
C_2	0	2	0	628
N_2	0	3	0	941
O_2	2	2	2	494
F_2	0	1	0	153
Ne_2	0	0	0	Nonexistent

POLYATOMIC MOLECULES; DELOCALIZED π Electrons

The bonding in molecules containing more than two atoms can also be described in terms of molecular orbitals. We will not attempt to do this; the energy level structure is considerably more complex than the one we considered. However, one point is worth mentioning. In polyatomic species, *a pi molecular orbital can be spread over the entire molecule* rather than being concentrated between two atoms.

This principle can be applied to species such as the nitrate ion, whose Lewis structure is:

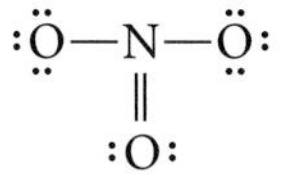

$$:\ddot{O}—N—\ddot{O}:$$
$$\|$$
$$:O:$$

Valence bond theory (Chapter 7) explains the fact that the three N—O bonds are identical by invoking the idea of resonance, with three contributing structures. MO theory, on the other hand, considers that the skeleton of the nitrate ion is established by the three sigma bonds while the electron pair in the pi orbital is *delocalized,* shared by all of the atoms in the molecule. According to MO theory, a similar interpretation applies with all of the "resonance hybrids" described in Chapter 7, including SO_2, SO_3, and CO_3^{2-}.

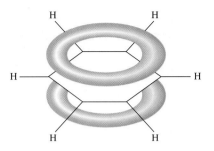

FIGURE 4
In benzene, three electron pairs are not localized on particular carbon atoms. Instead, they are spread out over two lobes of the shape shown, one above the plane of the benzene ring and the other below it.

Another species in which delocalized pi orbitals play an important role is benzene, C_6H_6. There are 30 valence electrons in the molecule, 24 of which are required to form the sigma bond framework:

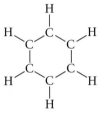

The remaining six electrons are located in three π orbitals, which according to MO theory extend over the entire molecule. Figure 4 is one way of representing this structure; more commonly it is shown simply as

METALS; BAND THEORY

In Chapter 9, we considered a simple picture of metallic bonding, the electron–sea model. The molecular orbital approach leads to a refinement of this model known as *band theory*. Here, a crystal of a metal is considered to be one huge molecule. Valence electrons of the metal are fed into delocalized molecular orbitals, formed in the usual way from atomic orbitals. A huge number of these MOs are grouped into an energy band; the energy separation between adjacent MOs is extremely small.

For purposes of illustration, consider a lithium crystal weighing one gram, which contains roughly 10^{23} atoms. Each Li atom has a half-filled 2s atomic orbital (elect. conf. Li = $1s^2 2s^1$). When these atomic orbitals combine, they form an equal number, 10^{23}, of molecular orbitals. These orbitals are spread over an energy band covering about 100 kJ/mol. It follows that the spacing between adjacent MOs is of the order of

$$\frac{100 \text{ kJ/mol}}{10^{23}} = 10^{-21} \text{ kJ/mol}$$

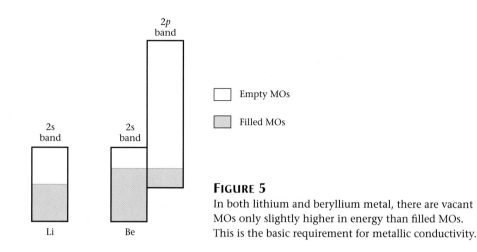

FIGURE 5
In both lithium and beryllium metal, there are vacant
MOs only slightly higher in energy than filled MOs.
This is the basic requirement for metallic conductivity.

Because each lithium atom has one valence electron and each molecular orbital can hold two electrons, it follows that the lower half of the valence band
(shown in color in Figure 5) is filled with electrons. The upper half of the band is
empty. Electrons near the top of the filled MOs can readily jump to empty MOs
only an infinitesimal distance above them. This is what happens when an electrical field is applied to the crystal; the movement of electrons through delocalized
MOs accounts for the electrical conductivity of lithium metal.

The situation in beryllium metal is more complex. We might expect all of the
2s molecular orbitals to be filled because beryllium has the electron configuration
$1s^2 2s^2$. However, in a crystal of beryllium, the 2p MO band overlaps the 2s (Figure
5). This means that, once again, there are vacant MOs that differ only infinitesi-
mally in energy from filled MOs below them. This is indeed the basic requirement
for electron conductivity; it is characteristic of all metals, including lithium and
beryllium.

Materials in which there is a substantial difference in energy between occupied and vacant MOs are poor electron conductors. Diamond, where the gap between the filled valence band and the empty conduction band is 500 kJ/mol, is an
insulator. Silicon and germanium, where the gaps are 100 kJ/mol and 60 kJ/mol respectively, are semiconductors.

Answers to Even-Numbered and Challenge Questions & Problems

Chapter 1

2. (a) element   (b) compound
    (c) mixture   (d) compound
4. (a) solution   (b) solution
    (c) heterogeneous mixture
6. (a) chromatography   (b) filter
8. (a) Na   (b) N   (c) Ni   (d) Pb
10. (a) silicon   (b) sulfur   (c) iron   (d) zinc
12. (a) mass—g   (b) volume—mL   (c) temperature—°C
14. (a) <   (b) =   (c) >
16. 177°C; 4.50×10^3 K
18. 35°C; 95°F
20. (a) 6   (b) 4   (c) 4   (d) 3   (e) ambiguous
22. 23.6 cm^3
24. (a) 0.528 g/cm^3 (b) 2.71 mi/h
    (c) 1481 g   (d) 82 g/cm^3
26. (a) 7.49 g   (b) 298.69 cm
    (c) 1×10^1 lb   (d) 12.0 oz
28. (a) 4.0206×10^3 mL   (b) 1.006 g
    (c) 1.001×10^2°C
30. a and b
32. (a) 2.23×10^{-2} L   (b) 1.36 in^3 (c) 0.0236 qt
34. (a) 1.15078 mi   (b) 1852 m   (c) 25 mph
36. 4.36×10^4 ft^2
38. 7.9%
40. 3.97×10^{14} L
42. as a source for silver
44. 1.49 g/mL
46. 1.6 g/mL
48. 0.039 mm
50. 252 g
52. (a) 18.0 g KCl   (b) 108 g H_2O (c) no; yes
54. (a) physical   (b) physical
    (c) physical   (d) chemical
56. 0.08818 cm^3 Pb   763 cm^3 O   Oxygen is a gas; lead is a solid.
58. 0.376 in^3
60. 2.35 cm
62. (a) Chemical properties show the behavior of the species in a reaction; physical properties are intrinsic qualities.
    (b) Distillation vaporizes the liquid; filtration removes the solid.

(c) The solute is a component of the solution.

64.
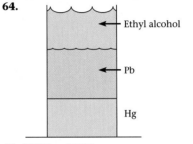

66. 320°F = 160°C
67. 1.2 km^2
68. 21.9 cm
69. 8.1×10^{-3} g Pb/year

Chapter 2

2. In a chemical reaction, mass is neither created nor destroyed.
4. (a) Conservation of Mass   (b) none
    (c) Law of Constant Composition
6. Rutherford; see pages 31, 32
8. $^{80}_{34}$Se
10. (a) $^{12}_{6}$C, $^{13}_{6}$C
    (b) same number of protons and electrons; different number of neutrons
12. (a) isobars: Ca-41, K-41, and Ar-41
       isotopes: Ca-40, Ca-41
    (b) same number of protons
    (c) same mass number
14. (a) 34   (b) 41   (c) 34   (d) 41 n, 34 p$^+$, 36e$^-$
16.

$^{64}_{30}$Zn	0	30	34	30
$^{28}_{14}$Si^{4+}	+4	14	14	10
$^{32}_{11}$S^{2-}	−2	16	16	18

18. (a) 14 p$^+$, 14 e$^-$ (b) 21 p$^+$, 22 e$^-$
    (c) 35 p$^+$, 34 e$^-$ (d) 70 p$^+$, 70 e$^-$
20. (a) sulfur   (b) scandium   (c) selenium
    (d) silicon   (e) strontium
22. (a) nonmetal   (b) metal   (c) nonmetal
    (d) metalloid   (e) metal
24. (a) 5   (b) 3   (c) 2
26. (a) 13   (b) 2   (c) 17 and 18

28. (a) C_2H_7N **(b)** C_3H_8O
30. (a) H_2O **(b)** NH_3 **(c)** N_2H_4 **(d)** SF_6 **(e)** PCl_5
32. (a) diselenium dichloride   **(b)** dinitrogen tetroxide
(c) phosphine   **(d)** iodine heptafluoride
(e) silicon carbide
34. KCl, K_2S, $CaCl_2$, CaS
36. (a) $Co(C_2H_3O_2)_2$ **(b)** BaO **(c)** Al_2S_3
(d) $KMnO_4$ **(e)** $NaHCO_3$
38. (a) potassium nitrate   **(b)** ammonium sulfate
(c) magnesium phosphate   **(d)** iron(III) chloride
(e) chromium(VII) oxide
40. (a) hydrochloric acid   **(b)** chloric acid
(c) iron(III) sulfite   **(d)** barium nitrite
(e) sodium hypochlorite
42. $Na_2Cr_2O_7$, bromine triiodide, $Cu(ClO)_2$, disulfur dichloride, K_3N
44. (a) N_2O_3—dinitrogen trioxide
(b) $Ca_3(PO_4)_2$—calcium phosphate
(c) $ZnSO_3$—zinc sulfite
46. (a) molecule: 2 p^+, 2 e^-, 0 n; anion: 1 p^+, 2 e^-, 0 n; cation: 1 p^+, 0 e^-, 0 n
(b) MgH_2—magnesium hydride
(c) acid
48. (a) usually true   **(b)** always true   **(c)** usually true
50.

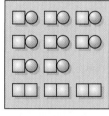

52.

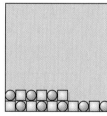

54. $^{234}_{90}Th$
56. (c)
57. (a) Ratio of C in ethane to C in ethylene/g H is 3:2.
(b) CH_3, CH_2; C_2H_6, C_2H_4
58. 3.71 g/cm³; lots of space between atoms
59. 1.4963×10^{-23} g
60. (a) 2.5×10^{24} molecules
(b) 2.3×10^{-20}
(c) $\approx 2.8 \times 10^2$ molecules

Chapter 3

2. (a) 3.959   **(b)** 1.994   **(c)** 0.1996
4. 16.00 amu
6. 11.0 amu

8. 9.77%; 1st isotope: 4.3%; 2nd isotope: 83.5% (The number of significant figures may vary according to how you solved the problem.)
10. (a) two—HCl-35 and HCl-37   **(b)** 36 and 38
(c)

12. (a) 6.022×10^{23} g   **(b)** 4.015×10^{13} cookies
14. (a) 3.054×10^{-22} g   **(b)** 3.275×10^9 atoms
16. (a) 0.35744 mol   **(b)** 2.152×10^{23} atoms
(c) 9.039×10^{24}
18. (a) two hundred   **(b)** two hundred
(c) 1.204×10^{25} **(d)** 3.255×10^{23}
20. (a) 342.30 g/mol   **(b)** 44.02 g/mol
(c) 286.44 g/mol
22. (a) 0.2895 mol   **(b)** 6.582×10^{-4} mol
(c) 0.00733 mol
24. (a) 245 g   **(b)** 4.90×10^2 g   **(c)** 298 g
26. (a) 0.5600 mol; 3.372×10^{23} molecules; 1.012×10^{24} N atoms
(b) 210.2 g; 5.573×10^{23} molecules; 1.672×10^{24} N atoms
(c) 4.68×10^6 g; 2.06×10^4 mol; 3.72×10^{28} N atoms
(d) 9.4 g; 0.042 mol; 2.5×10^{22} molecules
28. 30.93% Al; 45.86% O; 2.889% H; 20.32% Cl
30. 162.5 mg
32. (a) 41.1%   **(b)** 0.0233 g
34. 38.40% C; 1.50% H; 52.3% Cl; 7.80% O
36. $AsCl_5$
38. (a) $C_5H_8O_4NNa$ **(b)** $ZrSiO_4$ **(c)** C_5H_7N
40. $C_8H_8O_3$
42. $C_{12}H_{11}NO_2$
44. C_3H_8N; $C_6H_{16}N_2$
46. 51.18%; 3.825 g
48. (a) $TiO_2(s) + 2Cl_2(g) + 2C(s) \rightarrow TiCl_4(l) + 2CO(g)$
(b) $3Br_2(l) + I_2(s) \rightarrow 2IBr_3(g)$
(c) $C_2H_8N_2(s) + 2N_2O_4(g) \rightarrow 3N_2(g) + 2CO_2(g) + 4H_2O(g)$
50. (a) $2K(s) + S(s) \rightarrow K_2S(s)$
(b) $Mg(s) + S(s) \rightarrow MgS(s)$
(c) $2Al(s) + 3S(s) \rightarrow Al_2S_3(s)$
(d) $Ca(s) + S(s) \rightarrow CaS(s)$
(e) $Fe(s) + S(s) \rightarrow FeS(s)$
52. (a) $2F_2(g) + H_2O(l) \rightarrow OF_2(g) + 2HF(g)$

(b) $7O_2(g) + 4NH_3(g) \rightarrow 4NO_2(g) + 6H_2O(l)$
(c) $Au_2S_3(s) + 3H_2(g) \rightarrow 2Au(s) + 3H_2S(g)$
(d) $2NaHCO_3(s) \rightarrow Na_2CO_3(s) + H_2O(l) + CO_2(g)$
(e) $SO_2(g) + 4HF(l) \rightarrow SF_4(g) + 2H_2O(l)$

54. (a) 6.35 mol   **(b)** 1.36 mol
(c) 1.337 mol   **(d)** 0.220 mol
56. (a) 882.2 g   **(b)** 16.7 g
(c) 2.368 g   **(d)** 38.60 g
58. (a) $SiO_2(s) + 2C(s) \rightarrow Si(s) + 2CO(g)$
(b) 0.4528 mol
(c) 89.72 g
60. (a) $8.8 \times 10^2\, g$ **(b)** $2.4 \times 10^2\, L$
62. $9.2 \times 10^4\, L$
64. (a) $2Al(s) + 3S(s) \rightarrow Al_2S_3(s)$
(b) Al   **(c)** 0.590 mol   **(d)** 0.48 mol
66. (a) $3Si(s) + 2N_2(g) \rightarrow Si_3N_4(s)$
(b) $1.1 \times 10^2\, kg$
68. (a) 193 g   **(b)** 63.9%   **(c)** 118 g
70. $2.76 \times 10^3\, g$; 526 mL
72. (a)

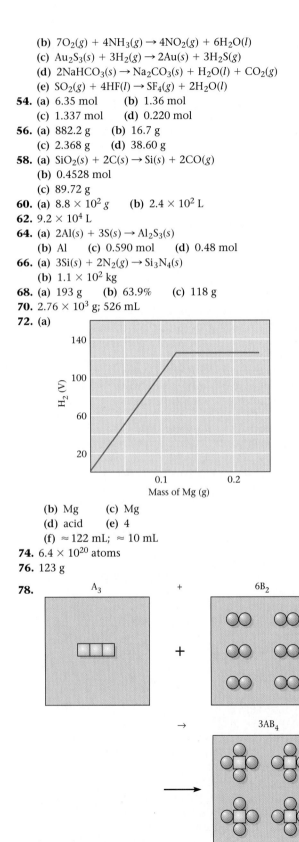

(b) Mg   **(c)** Mg
(d) acid   **(e)** 4
(f) ≈ 122 mL; ≈ 10 mL
74. 6.4×10^{20} atoms
76. 123 g
78.

80. (a) $X + 3Y \rightarrow XY_3$
(b) 5 mol X; 10 mol Y   **(c)** 3 mol XY_3; 2 mol X; 1 mol Y
82. (a) false—theoretical yield is 4.0 mol
(b) false—theoretical yield is 2.9×10^2 g
(c) true
(d) false—need only to know the amount of limiting reactant
(e) true
(f) false—total mass of reactants equal to total mass of products
(g) false—2 mol HF consumed/mol CCl_4 used
(h) true
84. 12.5 g
85. 893 g/mol
86. 6.01×10^{23} atoms
87. 3.657 g CaO; 2.972 g Ca_3N_2
88. 34.7%
89. (a) V_2O_3; V_2O_5 **(b)** 2.271 g
90. 28%

Chapter 4

2. (a) Add 7.54 g to enough water to make 0.400 L of solution.
(b) Add 56.0 g to enough water to make 1.75 L of solution.
4. (a) 0.0205 mol   **(b)** 1.78 L   **(c)** 152 g   **(d)** 0.180 M
6. (a) 1.2396 mol   **(b)** 2.0660 mol
(c) 1.6528 mol   **(d)** 1.2396 mol
8. (a) Na_2SO_4—soluble   **(b)** $Fe(NO_3)_3$—soluble
(c) AgCl—insoluble   **(d)** $Cr(OH)_3$—insoluble
10. (a) sodium phosphate   **(b)** sodium carbonate
(c) sodium hydroxide
12. (a) $Fe^{3+}(aq) + 3OH^-(aq) \rightarrow Fe(OH)_3(s)$
(b) $Ag^+(aq) + Cl^-(aq) + Cu^{2+}(aq) + 2OH^-(aq) \rightarrow$ $AgCl(s) + Cu(OH)_2(s)$
14. (a) no reaction
(b) $2Ag^+(aq) + CO_3{}^{2-}(aq) \rightarrow Ag_2CO_3(s)$
(c) $2Co^{3+}(aq) + 3CO_3{}^{2-}(aq) \rightarrow Co_2(CO_3)_3(s)$
(d) $3Ba^{2+}(aq) + 2PO_4{}^{3-}(aq) \rightarrow Ba_3(PO_4)_2(s)$
(e) no reaction
16. (a) $3Ba^{2+}(aq) + 2PO_4{}^{3-}(aq) \rightarrow Ba_3(PO_4)_2(s)$
(b) $Zn^{2+}(aq) + 2OH^-(aq) \rightarrow Zn(OH)_2(s)$
(c) no reaction
(d) $Co^{3+}(aq) + PO_4{}^{3-}(aq) \rightarrow CoPO_4(s)$
18. (a) 4.85 mL   **(b)** 22.2 mL   **(c)** 36.1 mL
20. (a) 10.4 mL   **(b)** 0.179 g
22. (a) H_2SO_3 **(b)** $HClO_2$ **(c)** H^+
(d) H^+ **(e)** $HCHO_2$
24. (a) OH^- **(b)** $C_6H_5NH_2$ **(c)** $(CH_3)_2NH$ **(d)** OH^-
26. (a) strong acid   **(b)** strong base
(c) weak acid   **(d)** weak base
28. (a) $HC_2H_3O_2(aq) + OH^-(aq) \rightarrow H_2O + C_2H_3O_2{}^-(aq)$

(b) $(C_2H_5)_2NH(aq) + H^+(aq) \rightarrow (C_2H_5)_2NH_2^+(aq)$
(c) $HCN(aq) + OH^-(aq) \rightarrow H_2O + CN^-(aq)$

30. (a) $H^+(aq) + C_2H_5NH_2(aq) \rightarrow C_2H_5NH_3^+(aq)$
   (b) correct
   (c) $HC_2H_3O_2(aq) + OH^-(aq) \rightarrow C_2H_3O_2^-(aq) + H_2O$
   (d) correct
   (e) correct

32. 0.126 M

34. (a) 8.14 mL   **(b)** 282 mL   **(c)** 84.4 mL

36. 19 g

38. 71.6%

40. one

42. (a) N_2O_3: O = -2; N = $+3$
   (b) SO_2: O = -2; S = $+4$
   (c) $Cr_2O_7^{2-}$: O = -2; Cr = $+6$
   (d) OCl^-: O = -2; Cl = $+1$

44. (a) P_2O_5: P = $+5$; O = -2
   (b) NH_3: N = -3; H = $+1$
   (c) CO_3^{2-}: O = -2; C = $+4$
   (d) $S_2O_3^{2-}$: S = $+2$; O = -2
   (e) N_2H_4: H = $+1$; N = -2

46. (a) reduction   **(b)** reduction
   (c) oxidation   **(d)** oxidation

48. (a) $TiO_2(s) + 2H_2O + e^- \rightarrow Ti^{3+}(aq) + 4OH^-(aq)$
   (b) $Zn^{2+}(aq) + 2e^- \rightarrow Zn(s)$
   (c) $2NH_4^+(aq) \rightarrow N_2(g) + 8H^+(aq) + 6e^-$
   (d) $CH_3OH(aq) \rightarrow CH_2O(aq) + 2e^- + 2H^+(aq)$

50. (a) oxidation: $Mn^{2+}(aq) + 4H_2O \rightarrow MnO_4^-(aq) + 5e^- + 8H^+(aq)$
   (b) reduction: $CrO_4^{2-}(aq) + 3e^- + 4H_2O \rightarrow Cr^{3+}(aq) + 8OH^-(aq)$
   (c) reduction: $PbO_2(s) + 2e^- + 2H_2O \rightarrow Pb^{2+}(aq) + 4OH^-(aq)$
   (d) reduction: $ClO_2^-(aq) + 2e^- + 2H^+(aq) \rightarrow ClO^-(aq) + H_2O$

52. (a) reduction: $H_2O_2(aq) + e^- \rightarrow H_2O$; H_2O_2—reduced, oxidizing agent

oxidation: $Ni^{2+}(aq) \rightarrow Ni^{3+}(aq) + e^-$; Ni^{2+}—oxidized, reducing agent
(b) reduction: $Cr_2O_7^{2-}(aq) + 3e^- \rightarrow Cr^{3+}(aq)$; CrO_7^{2-}— reduced, oxidizing agent
   oxidation: $Sn^{2+}(aq) \rightarrow Sn^{4+}(aq) + 2e^-$; Sn^{2+}— oxidized, reducing agent

54. (a) $H_2O_2(aq) + 2Ni^{2+}(aq) + 2H^+(aq) \rightarrow 2Ni^{3+}(aq) + 2H_2O$
   (b) $Cr_2O_7^{2-}(aq) + 3Sn^{2+}(aq) + 14H^+(aq) \rightarrow 2Cr^{3+}(aq) + 3Sn^{4+}(aq) + 7H_2O$

56. (a) $8Ni^{2+}(aq) + IO_4^-(aq) + 8H^+(aq) \rightarrow 8Ni^{3+}(aq) + I^-(aq) + 4H_2O$
   (b) $O_2(g) + 4Br^-(aq) + 4H^+(aq) \rightarrow 2Br_2(l) + 2H_2O$
   (c) $3Ca(s) + Cr_2O_7^{2-}(aq) + 14H^+(aq) \rightarrow 3Ca^{2+}(aq) + 2Cr^{3+}(aq) + 7H_2O$
   (d) $IO_3^-(aq) + 3Mn^{2+}(aq) + 3H_2O \rightarrow I^-(aq) + 3MnO_2(s) + 6H^+(aq)$

58. (a) $2Ni(OH)_2(s) + N_2H_4(aq) \rightarrow 2Ni(s) + N_2(g) + 4H_2O$
   (b) $3Fe(OH)_3(s) + Cr^{3+}(aq) + 5OH^-(aq) \rightarrow 3Fe(OH)_2(s) + CrO_4^{2-}(aq) + 4H_2O$
   (c) $2MnO_4^-(aq) + 3BrO_3^-(aq) + H_2O \rightarrow 2MnO_2(s) + 3BrO_4^-(aq) + 2OH^-(aq)$
   (d) $2H_2O_2(aq) + IO_4^-(aq) \rightarrow 2O_2(g) + IO_2^-(aq) + 2H_2O$

60. (a) $Zn(s) + 2H^+(aq) \rightarrow H_2(g) + Zn^{2+}(aq)$
   (b) $2NO_3^-(aq) + 3CuS(s) + 8H^+(aq) \rightarrow 3Cu^{2+}(aq) + 3S(s) + 2NO(g) + 4H_2O$
   (c) $4Sb^{3+}(aq) + IO_4^-(aq) + 8H^+(aq) \rightarrow 4Sb^{5+}(aq) + I^-(aq) + 4H_2O$

62. (a) $2MnO_4^-(aq) + 2H^+(aq) + 3H_2C_2O_4(aq) \rightarrow 2MnO_2(s) + 4H_2O + 6CO_2(g)$
   (b) 0.657 M   **(c)** 0.522 g

64. (a) $2H^+(aq) + NO_3^-(aq) + Ag(s) \rightarrow Ag^+(aq) + NO_2(g) + H_2O$
   (b) 27.5 g

66. 13.3%

68. 5.04%

70. 0.379%

72. 7.21% NH_3; no

74. (a)

(b)

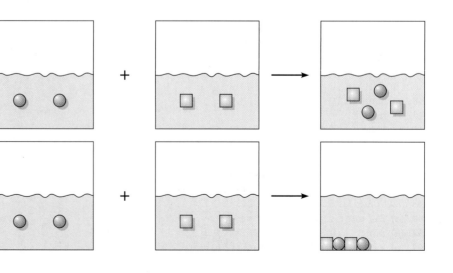

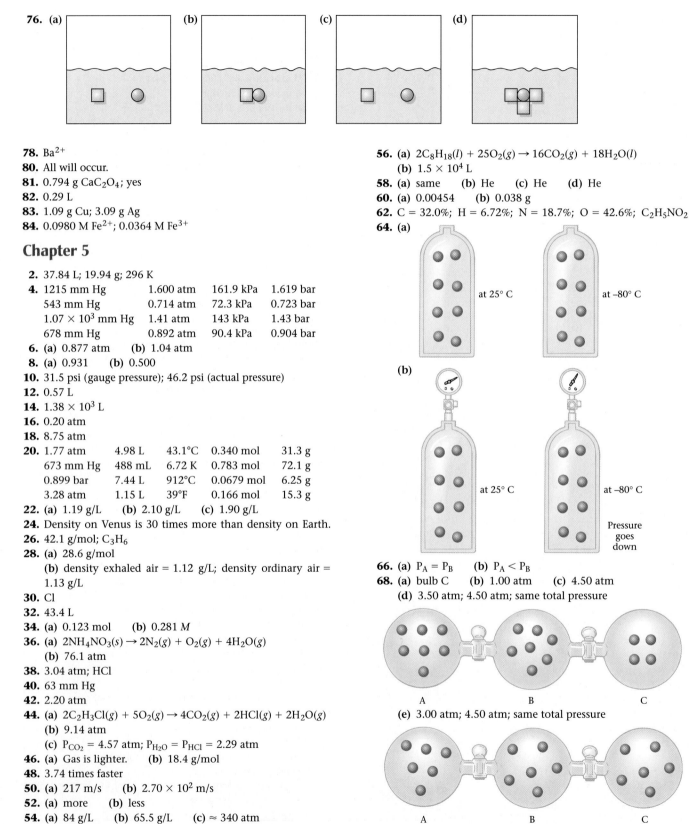

76. (a) (b) (c) (d)

78. Ba^{2+}
80. All will occur.
81. 0.794 g CaC_2O_4; yes
82. 0.29 L
83. 1.09 g Cu; 3.09 g Ag
84. 0.0980 M Fe^{2+}; 0.0364 M Fe^{3+}

Chapter 5

2. 37.84 L; 19.94 g; 296 K
4.

1215 mm Hg	1.600 atm	161.9 kPa	1.619 bar
543 mm Hg	0.714 atm	72.3 kPa	0.723 bar
1.07×10^3 mm Hg	1.41 atm	143 kPa	1.43 bar
678 mm Hg	0.892 atm	90.4 kPa	0.904 bar

6. (a) 0.877 atm (b) 1.04 atm
8. (a) 0.931 (b) 0.500
10. 31.5 psi (gauge pressure); 46.2 psi (actual pressure)
12. 0.57 L
14. 1.38×10^3 L
16. 0.20 atm
18. 8.75 atm
20.

1.77 atm	4.98 L	43.1°C	0.340 mol	31.3 g
673 mm Hg	488 mL	6.72 K	0.783 mol	72.1 g
0.899 bar	7.44 L	912°C	0.0679 mol	6.25 g
3.28 atm	1.15 L	39°F	0.166 mol	15.3 g

22. (a) 1.19 g/L (b) 2.10 g/L (c) 1.90 g/L
24. Density on Venus is 30 times more than density on Earth.
26. 42.1 g/mol; C_3H_6
28. (a) 28.6 g/mol
   (b) density exhaled air = 1.12 g/L; density ordinary air = 1.13 g/L
30. Cl
32. 43.4 L
34. (a) 0.123 mol (b) 0.281 M
36. (a) $2NH_4NO_3(s) \rightarrow 2N_2(g) + O_2(g) + 4H_2O(g)$
   (b) 76.1 atm
38. 3.04 atm; HCl
40. 63 mm Hg
42. 2.20 atm
44. (a) $2C_2H_3Cl(g) + 5O_2(g) \rightarrow 4CO_2(g) + 2HCl(g) + 2H_2O(g)$
   (b) 9.14 atm
   (c) $P_{CO_2} = 4.57$ atm; $P_{H_2O} = P_{HCl} = 2.29$ atm
46. (a) Gas is lighter. (b) 18.4 g/mol
48. 3.74 times faster
50. (a) 217 m/s (b) 2.70×10^2 m/s
52. (a) more (b) less
54. (a) 84 g/L (b) 65.5 g/L (c) ≈ 340 atm

56. (a) $2C_8H_{18}(l) + 25O_2(g) \rightarrow 16CO_2(g) + 18H_2O(l)$
   (b) 1.5×10^4 L
58. (a) same (b) He (c) He (d) He
60. (a) 0.00454 (b) 0.038 g
62. C = 32.0%; H = 6.72%; N = 18.7%; O = 42.6%; $C_2H_5NO_2$
64. (a)

at 25° C   at −80° C

(b)

at 25° C   at −80° C

Pressure goes down

66. (a) $P_A = P_B$ (b) $P_A < P_B$
68. (a) bulb C (b) 1.00 atm (c) 4.50 atm
   (d) 3.50 atm; 4.50 atm; same total pressure

A   B   C

   (e) 3.00 atm; 4.50 atm; same total pressure

A   B   C

70. Ideal gas pressures: at 40°C = 25.0 mm Hg; at 70°C = 27.3 mm Hg; at 100°C = 29.7 mm Hg

Vapor pressures: at 40°C = 55.3 mm Hg; at 70°C = 233.7 mm Hg; at 100°C = 760 mm Hg

Vapor pressure implies the presence of some liquid together with the vapor. As T increases, more liquid becomes gas, increasing n and thus P.

71. 5.74 g/mol; at T = 15°C, He and H_2 are in outer space. Ar is present in Earth's atmosphere.

72. 3.0 ft from NH_3 end

73. 0.0456 L-atm/mol-°R

74. 78.7%

75. 6.62 m

76. 0.897 atm

77. $\dfrac{V_A}{V} = \dfrac{n_A}{n}$; but $\dfrac{n_A}{n} \neq \dfrac{mass_A}{mass_{TOT}}$

Chapter 6

2. (a) 7.09×10^{14} s^{-1} (b) 4.70×10^{-19} J
   (c) 283 kJ/mol

4. (a) infrared   (b) 2.001×10^{14} s^{-1} (c) 1.326×10^{-19} J

6. 1.9×10^{20} photons

8. microwave: 0.0239 kJ/mol; sun: $\approx 1.2 \times 10^3$ kJ/mol

10. (a) 6.169×10^{14} s^{-1} (b) visible   (c) no

12.

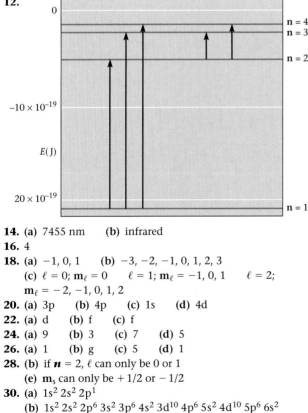

14. (a) 7455 nm   (b) infrared

16. 4

18. (a) $-1, 0, 1$ (b) $-3, -2, -1, 0, 1, 2, 3$
   (c) $\ell = 0$; $\mathbf{m}_\ell = 0$ $\ell = 1$; $\mathbf{m}_\ell = -1, 0, 1$ $\ell = 2$; $\mathbf{m}_\ell = -2, -1, 0, 1, 2$

20. (a) 3p   (b) 4p   (c) 1s   (d) 4d

22. (a) d   (b) f   (c) f

24. (a) 9   (b) 3   (c) 7   (d) 5

26. (a) 1   (b) g   (c) 5   (d) 1

28. (b) if $\mathbf{n}$ = 2, ℓ can only be 0 or 1
   (e) $\mathbf{m}_s$ can only be $+1/2$ or $-1/2$

30. (a) $1s^2 2s^2 2p^1$
   (b) $1s^2 2s^2 2p^6 3s^2 3p^6 4s^2 3d^{10} 4p^6 5s^2 4d^{10} 5p^6 6s^2$

(c) $1s^2 2s^2$
(d) $1s^2 2s^2 2p^6 3s^2 3p^6 4s^2 3d^{10} 4p^6 5s^2 4d^{10} 5p^6 6s^2 4f^{14} 5d^{10} 6p^3$
(e) $1s^2 2s^2 2p^6 3s^2 3p^6 4s^2 3d^{10} 4p^5$

32. (a) [$_{54}$Xe] $6s^2 4f^{14} 5d^{10}$ (b) [$_{10}$Ne] $3s^2 3p^1$
   (c) [$_{18}$Ar] $4s^2 3d^{10} 4p^3$ (d) [$_{54}$Xe] $6s^2 4f^{14} 5d^4$
   (e) [$_{54}$Xe] $6s^2 4f^{14} 5d^{10} 6p^5$

34. (a) B   (b) Nd   (c) Zn   (d) Mg

36. (a) 6/12   (b) 12/25   (c) 18/42

38. (a) excited   (b) excited   (c) ground
   (d) impossible   (e) excited   (f) impossible

40.

	1s	2s	2p	3s	3p
(a) Li:	(↑↓)	(↑)			
(b) P:	(↑↓)	(↑↓)	(↑↓)(↑↓)(↑↓)	(↑↓)	(↑)(↑)(↑)

	1s	2s	2p	3s	3p
(c) F:	(↑↓)	(↑↓)	(↑↓)(↑↓)(↑)		
(d) Fe:	(↑↓)	(↑↓)	(↑↓)(↑↓)(↑↓)	(↑↓)	(↑↓)(↑↓)(↑↓)

   4s   **3d**
   (↑↓) (↑↓)(↑)(↑)(↑)(↑)

42. (a) Mg   (b) P   (c) O

44. (a) I, At   (b) Kr   (c) Si, Ge, As, Sb, Te   (d) Li, B, F

46. (a) 3   (b) 1   (c) 6

48. (a) Ca   (b) K, Ga   (c) none   (d) none

50. (a) P: $1s^2 2s^2 2p^6 3s^2 3p^3$ P^{3-}: $1s^2 2s^2 2p^6 3s^2 3p^6$
   (b) Ca: $1s^2 2s^2 2p^6 3s^2 3p^6 4s^2$ Ca^{2+}: $1s^2 2s^2 2p^6 3s^2 3p^6$
   (c) Ti: $1s^2 2s^2 2p^6 3s^2 3p^6 4s^2 3d^2$
                              Ti^{4+}: $1s^2 2s^2 2p^6 3s^2 3p^6 3d^3$
   (d) Mn^{2+} $1s^2 2s^2 2p^6 3s^2 3p^6 3d^5$
                              Fe^{3+}: $1s^2 2s^2 2p^6 3s^2 3p^6 3d^3$

52. (a) none   (b) none   (c) none   (d) 5

54. (a) Cl < S < Mg   (b) Mg < S < Cl
   (c) Mg < S < Cl

56. (a) Sb   (b) Cs   (c) Cs

58. (a) N   (b) Ba^{2+} (c) Se   (d) Co^{3+}

60. (a) $Co^{3+} < Co^{2+} < Co$ (b) $Cl < Cl^- < Br^-$

62. 1.8×10^{19}

64. (a) Cl—chlorine
   (b) theoretically: Pb—lead; actually: Sn—tin
   (c) Mn—manganese   (d) Li—lithium
   (e) Kr—krypton

66. (a) (2) and (4)   (b) (1) and (3)   (c) (1)
   (d) (2)   (e) (1)

68. (a) no. of orbitals = $2\mathbf{n} + 1$
   (b) number of orbitals = $2\ell + 1$ (c) none

70. (a) Two electrons cannot have the same set of quantum numbers.
   (b) Fill orbitals to get maximum number of unpaired electrons.
   (c) wavelength
   (d) $n = 1, 2, 3, 4, \ldots$

72. (a) true   (b) false; inversely proportional to $\mathbf{n}^2$
   (c) false; as soon as 4p is full

74. (a) The extra electron adds to the repulsion between outer electrons, making the anion larger than its parent atom.
   (b) Sc loses 3 electrons to become like Ar.

(c) The elements become more metallic.

76. $+2954$ kJ

77. $\Delta E = 2.180 \times 10^{-18}\left(\dfrac{1}{4} - \dfrac{1}{n^2}\right)$

$$\lambda = \frac{hc}{\Delta E} = \frac{(6.626 \times 10^{-34})(2.998 \times 10^8) \times 10^9}{2.180 \times 10^{-18}\left(\dfrac{1}{4} - \dfrac{1}{n^2}\right)}$$

$$= \frac{91.12}{\left(\dfrac{1}{4} - \dfrac{1}{n^2}\right)} = \frac{364.5\,n^2}{n^2 - 4}$$

78. $1s^4\,1p^4$

79. (a) s sublevel: $m_\ell = 0 = 3\ e^-$; p sublevel: $m_\ell = -1, 0, 1 = 9\ e^-$ d sublevel: $m_\ell = -2, -1, 0, 1, 2 = 15\ e^-$
  (b) $n = 3$, $\ell = 0, 1, 2$; total electrons $= 27$
  (c) $1s^3\,2s^3\,2p^2$ $1s^3\,2s^3\,2p^9\,3s^2$

80. (a) 3.42×10^{-19} J   (b) 581 nm

Chapter 7

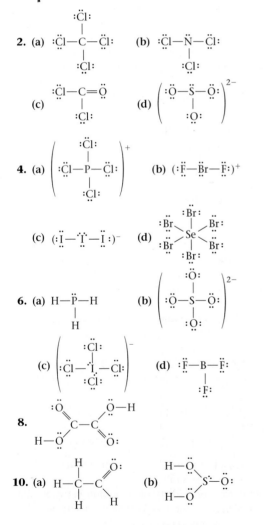

2. (a)   (b)

  (c)   (d)

4. (a)   (b)

  (c)   (d)

6. (a) H—P—H   (b)

  (c)   (d)

8.

10. (a)   (b)

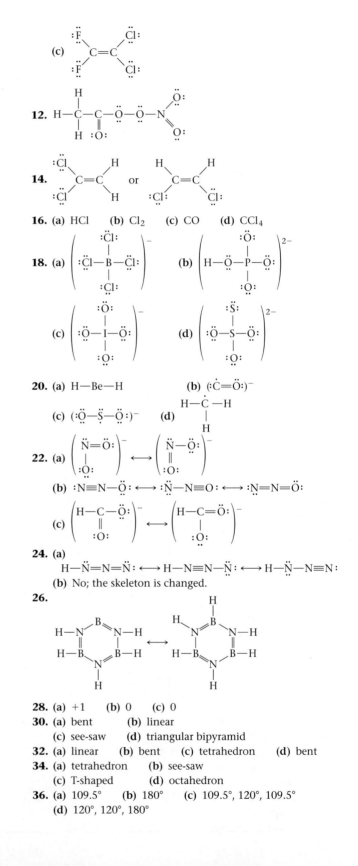

(c)

12. H—C—C—O—O—N

14.   or

16. (a) HCl   (b) Cl_2 (c) CO   (d) CCl_4

18. (a)   (b)

  (c)   (d)

20. (a) H—Be—H   (b) $(\ddot{C}=\ddot{O}\!:)^-$
  (c) $(\!:\!\ddot{O}-\ddot{S}-\ddot{O}\!:)^-$ (d)

22. (a)

  (b) $:N\equiv N-\ddot{O}\!: \longleftrightarrow :\ddot{N}-N\equiv O\!: \longleftrightarrow :\ddot{N}=N=\ddot{O}\!:$

  (c)

24. (a)
  $H-\ddot{N}=N=\ddot{N}\!: \longleftrightarrow H-N\equiv N-\ddot{\ddot{N}}\!: \longleftrightarrow H-\ddot{\ddot{N}}-N\equiv N\!:$
  (b) No; the skeleton is changed.

26.

28. (a) $+1$ (b) 0   (c) 0
30. (a) bent   (b) linear
  (c) see-saw   (d) triangular bipyramid
32. (a) linear   (b) bent   (c) tetrahedron   (d) bent
34. (a) tetrahedron   (b) see-saw
  (c) T-shaped   (d) octahedron
36. (a) 109.5°   (b) 180°   (c) 109.5°, 120°, 109.5°
  (d) 120°, 120°, 180°

38. (a)

$$H-\underset{\underset{H}{|}}{\overset{\overset{H}{|}}{C}}-\underset{\overset{||}{O}}{\overset{||}{C}}-\ddot{\overset{..}{O}}-\ddot{\overset{..}{O}}-\underset{\overset{||}{O}}{\overset{||}{C}}-\underset{\underset{H}{|}}{\overset{\overset{H}{|}}{C}}-H$$

(b) around CH_3-109.5°; around C—O-120°;
around O—O-109.5°

40. (1) 120°   (2) 109.5°   (3) 109.5°

42. a

44. a, b, d

46. First structure is polar; dipoles do not cancel.

48. (a) sp^2 (b) sp   (c) sp^3d (d) sp^3d

50. (a) sp   (b) sp^2 (c) sp^3 (d) sp^2

52. (a) sp^3 (b) sp^3d (c) sp^3d (d) sp^3d^2

54. (a) 6 e^- pairs; sp^3d^2 (b) 6 e^- pairs; sp^3d^2
(c) 5 e^- pairs; sp^3d

56. All B atoms and N atoms are sp^2.

58. (a) sp^3 (b) sp^2 (c) sp   (d) sp^2

60. (a) sp^2 (b) sp^2 (c) sp^3

62. (a) 12 σ, 3 π

64. 4 σ (b) 3 σ, 1 π (c) 2 σ, 2 π (d) 4 σ, 1 π

66. Formal charges: Be = -2; Cl = $+1$; Cl = $+1$
Structure with single bonds has formal charges of zero on all atoms.

68. (b) $:\ddot{F}-\underset{\underset{:\ddot{F}:}{|}}{\overset{..}{S}}-\ddot{F}:$ (d) $:\ddot{F}-\underset{\underset{:\ddot{F}::\ddot{F}:}{|}}{\overset{\overset{:\ddot{F}::\ddot{F}:}{}}{S}}-\ddot{F}:$

70.

AX_2E_2	2	2	bent	sp^3	polar
AX_3	3	0	triangular planar	sp^2	nonpolar
AX_4E_2	4	2	square planar	sp^3d^2	nonpolar
AX_5	5	0	triangular bipyramid	sp^3d	nonpolar

72. C

74. PH_3 and H_2S—unshared pairs on central atom

76. (a) NH_3 (b) CN^-
(c) CH_2O (d) CH_2O
(e) XeF_2

78. $x = 3$; T-shaped; polar; sp^3d; 90°, 180°; 3 σ bonds

79. $\underset{\underset{H}{\diagdown}}{\overset{\overset{H}{\diagdown}}{N}}-\underset{\overset{H}{\diagup}}{\overset{\overset{H}{}}{N}}$ bent; 109.5°; polar

80. 6; octahedron; sp^3d^2

81. (a) $\left(:\ddot{O}-\underset{\underset{:\ddot{O}:}{|}}{\overset{\overset{:\ddot{O}:}{|}}{S}}-\ddot{O}:\right)^{2-}$ $\left(:\ddot{O}=\underset{\underset{:\ddot{O}:}{|}}{\overset{\overset{:\ddot{O}:}{|}}{S}}=\ddot{O}:\right)^{2-}$

(b) tetrahedral for both   (c) sp^3; sp^3
(d) 1st structure: O = -1; S = 2
2nd structure: S = 0; O = -1; O = -1, O = 0

82. $:\ddot{Cl}-\underset{\underset{:\ddot{Cl}:}{|}}{\overset{\overset{:\ddot{O}:}{|}}{P}}-\ddot{Cl}:$ Formal charges: P = $+1$; Cl = 0; O = -1

$:\ddot{Cl}-\underset{\underset{:\ddot{Cl}:}{|}}{\overset{\overset{:O:}{||}}{P}}-\ddot{Cl}:$ Formal charges: P = 0; Cl = 0; O = 0

Chapter 8

2. 1.55°C

4. 71.5 kJ

6. (a) $KBr(s) \rightarrow K^+(aq) + Br^-(aq)$
(b) endothermic   (c) 223 J   (d) 19.8 kJ

8. (a) 89.3 kJ   (b) -89.3 kJ   (c) -3.57×10^3 kJ

10. 1.12×10^4 J/°C

12. 21.8°C

14. 3.03×10^3 kJ

16. (a) $CaO(s) + 3C(s) \rightarrow CO(g) + CaC_2(s)$ $\Delta H = 464.8$ kJ
(b) endothermic
(c)

(d) 7.25 kJ   (e) 1.550 g

18. (a) -55.8 kJ   (b) -3.10 kJ

20. (a) $Sr(s) + C(g) + \frac{3}{2}O_2(g) \rightarrow SrCO_3(s)$
$\Delta H = -1.220 \times 10^3$ kJ
(b) -332 kJ

22. 0.0367 g

24. condensing water

26. -48.9 kJ

28. -306.9 kJ

30. $+11.3$ kJ

32. (a) $N_2(g) + 2O_2(g) \rightarrow N_2O_4(g)$ $\Delta H° = 9.2$ kJ
(b) $Ca(s) + S(s) + 2O_2(g) \rightarrow CaSO_4(s)$ $\Delta H° = -1434.1$ kJ
(c) $Ag(s) + \frac{1}{2}Cl_2(g) \rightarrow AgCl(s)$ $\Delta H° = -127.1$ kJ
(d) $\frac{1}{2}H_2(g) + \frac{1}{2}I_2(s) \rightarrow HI(g)$ $\Delta H° = 26.5$ kJ

34. (a) -1675.7 kJ/mol   (b) -205.4 kJ

36. 9.38 kJ evolved

38. (a) -446.0 kJ   (b) -1085.8 kJ   (c) -18.4 kJ

40. (a) -125.7 kJ   (b) 163.9 kJ

42. (a) $CaCO_3(s) + 2NH_3(g) \rightarrow CaCN_2(s) + 3H_2O(\ell)$
$\Delta H° = 90.1$ kJ
(b) -351.6 kJ/mol

44. -1140.0 kJ/mol

46. 51.8 kJ

48. 1.20×10^2 L·atm

50. (a) 31 J   **(b)** −11 J
52. (a) 40.7 kJ   **(b)** 3.1 kJ   **(c)** −37.6 kJ
54. (a) $C_3H_8(g) + 5O_2(g) \rightarrow 3CO_2(g) + 4H_2O(\ell)$
$$\Delta H° = -2219.9 \text{ kJ}$$
   (b) −2212.4 kJ
56. (a) $q_{halogenbulb} = 8.1 \times 10^5$ J; $q_{fluorescentbulb} = 2.2 \times 10^5$ J
   (b) halogen bulb: 15°C; fluorescent bulb: 4.1°C
58. 29 min
60. 7.7 h
62. 42.1°C
64. 1.16×10^5 mol
66. (a) $Hg(s) \rightarrow Hg(\ell)$ $\Delta H° = 2.33$ kJ
   (b) $Br_2(\ell) \rightarrow Br_2(g)$ $\Delta H° = 29.6$ kJ
   (c) $C_6H_6(\ell) \rightarrow C_6H_6(g)$ $\Delta H° = -9.84$ kJ
   (d) $Hg(g) \rightarrow Hg(\ell)$ $\Delta H° = -59.4$ kJ
   (e) $C_{10}H_8(s) \rightarrow C_{10}H_8(g)$ $\Delta H° = 62.6$ kJ
68. Piston goes down.
70. a, b, c
71. (a) 1.69×10^5 J   **(b)** 505 g
72. 22%
73. (a) −851.5 kJ   **(b)** 6.6×10^{3}°C   **(c)** yes
74. 76.0%

Chapter 9

2. (a) at 35°C: P = 237 mm Hg   at 45°C: P = 245 mm Hg
   (b) at 35°C: 237 mm Hg > P_{eq} at 45°C: 245 mm Hg < P_{eq}
   (c) at 35°C: P = 203 mm Hg   at 45°C: P = 245 mm Hg
   (d) at 35°C: liquid and gas   at 45°C: gas only
4. (a) 6.26 mg   **(b)** 0.15 mm Hg   **(c)** 0.466 mm Hg
6. (a) yes   **(b)** 24.9 L   **(c)** 32.2 mm Hg
8. (a) 29.2 kJ/mol   **(b)** 61°C
10. 97°C
12. 20°C
14. (a) solid   **(b)** liquid   **(c)** vapor
16. (a) solid   **(b)** liquid   **(c)** liquid
18. (a)

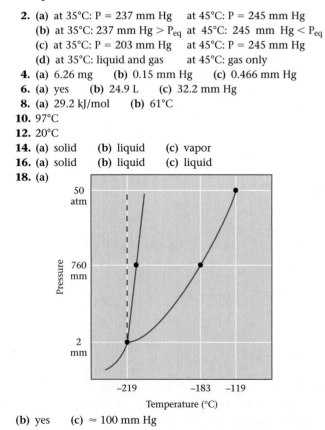

(b) yes   **(c)** ≈ 100 mm Hg

20. (a)

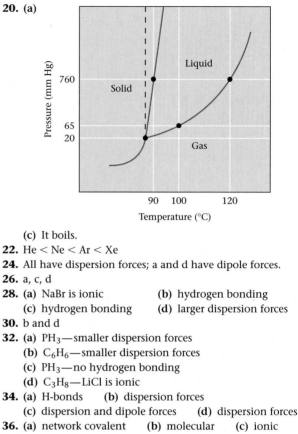

(c) It boils.
22. He < Ne < Ar < Xe
24. All have dispersion forces; a and d have dipole forces.
26. a, c, d
28. (a) NaBr is ionic   **(b)** hydrogen bonding
   (c) hydrogen bonding   **(d)** larger dispersion forces
30. b and d
32. (a) PH_3—smaller dispersion forces
   (b) C_6H_6—smaller dispersion forces
   (c) PH_3—no hydrogen bonding
   (d) C_3H_8—LiCl is ionic
34. (a) H-bonds   **(b)** dispersion forces
   (c) dispersion and dipole forces   **(d)** dispersion forces
36. (a) network covalent   **(b)** molecular   **(c)** ionic
38. (a) metallic, network covalent
   (b) network covalent, metallic   **(c)** metallic
40. (a) metallic   **(b)** molecular   **(c)** network covalent
   (d) ionic   **(e)** molecular
42. (a) $C_6H_{12}O_6$ **(b)** Na_2CO_3
   (c) diamond   **(d)** steel
44. (a) C atoms   **(b)** Si and C atoms
   (c) ions   **(d)** C_2H_2 molecules
46. body-centered
48. 0.219 nm
50. (a) 0.151 nm   **(b)** no; no
52. (a) 0.700 nm   **(b)** 0.404 nm
54. (a) 4   **(b)** 8 for Cl^- at the corners; 2 for Cl^- at the face
56. network covalent
58. 0.055 mL
60. b and d
62. (a) The covalent bond is the force within molecules; hydrogen bonds are forces between molecules.
   (b) Normal boiling point is the temperature where vapor pressure is 760 mm Hg. Boiling point is the temperature where vapor pressure is pressure above the liquid.
   (c) Triple point is the point where all three phases are in equilibrium. Critical point is the last point at which liquid and vapor can be at equilibrium.
   (d) The vapor pressure curve is part of a phase diagram.
   (e) Increasing temperature increases the vapor pressure.

Volume has no effect on vapor pressure as long as liquid is present.

64. 6.05×10^{23} atoms/mol

65. (a) liquid and vapor   **(b)** 26.7 mm Hg   **(c)** 3.4 atm

66. 41%

67. 80 atm; 0.60°C; see *Scientific American,* February 2000

68. $\dfrac{r_{cation}}{r_{anion}} = 0.414$

69. $P_{C_3H_8}$ is the vapor pressure of propane, which decreases exponentially with T. P_{N_2} is gas pressure, which decreases linearly with T.

Chapter 10

2. (a) 35.8%   **(b)** 64.3%   **(c)** 0.421

4. 0.838 M

6. 1.21×10^{-9} mol

8. (a) 334.9 g   **(b)** 80.18 mL   **(c)** 0.0494 M

10.

	m	Mass % solv	ppm solute	$X_{solvent}$
(a)	6.17	45.50	5.450×10^5	0.900
(b)	0.006543	99.8731	1269	0.9999
(c)	0.873	85.5	1.45×10^5	0.985
(d)	0.2560	95.26	4.74×10^4	0.9954

12. (a) Dissolve 128 g $Ba(OH)_2$ in 1.00 L of solution.
(b) Dilute 125 mL to 1.00 L of solution.

14. (a) $CoCl_2$: 0.278 M; Co^{2+}: 0.278 M; Cl^-: 0.556 M
(b) 45.2 g

16. 14.6 M; 57.8 m; 0.510

18.

	Density (g/mL)	M	m	Mass % solute
(a)	1.06	0.886	0.939	11.0
(b)	1.15	2.27	2.66	26.0
(c)	1.23	2.71	3.11	29.1

20. (a) H_2O_2—H-bonds with H_2O **(b)** NaOH—ionic
(c) HCl—ionic in water
(d) methyl alcohol—H-bonds with H_2O

22. −13.0 kJ   **(b)** no

24. (a) 8.42×10^{-7} mol/L·mm Hg   **(b)** 0.0174 g
(c) 0.00282 g

26. (a) 0.0022 g   **(b)** 0.0019 g   **(c)** 14%

28. (a) 468 mm Hg   **(b)** 554 mm Hg
(c) 6.30×10^2 mm Hg

30. Dissolve 89 g in 1.00 L of solution.

32. 34.1 mm Hg

34. 6.0×10^3 g/mol

36. (a) 14 g; 0.95 g   **(b)** 14.1 g; 0.96 g

38. −22.4°C; no

40. 5.1°C/m

42. $C_6H_4N_2O_4$

44. simplest formula; C_5H_7; molecular formula: $C_{40}H_{56}$

46. 0.32 M

48. 6.50×10^4 g/mol

50. freezing: $CH_3OH > KMnO_4 > Ni(NO_3)_2 > Al_2(SO_4)_3$
boiling: $Al_2(SO_4)_3 > Ni(NO_3)_2 > KMnO_4 > CH_3OH$

52. 0.086 m

54. (a) 2.002 M **(b)** 2.17 m **(c)** 196 atm   **(d)** −16.1°C

56. (a) 4.8°C/m   **(b)** 1.3×10^2 g/mol

58. (a) $BaSO_4$ **(b)** 10.0 g   **(c)** 3.27 M

60.

6 M Na$^+$
3 M S^{2-}

62. (a) ≈200 g   **(b)** ≈475 g

64. (a) greater   **(b)** less   **(c)** greater   **(d)** less

66. (b)

68. (a) Seawater contains ions.
(b) Supersaturated solution is disturbed, solute crystallizes out.
(c) They need to be isotonic so cells do not shrink or burst.
(d) More oxygen is available (for fish).
(e) Pressure decreases, and carbon dioxide is less soluble.

70. (a) Check electrical conductivity.
(b) Solubility of gas decreases with increasing temperature.
(c) Number of moles of water is large so X_{solute} is small.
(d) Colligative property; the presence of a solute decreases vapor pressure.

72. 1.1 g/mL

73. Add 1.03×10^3 g of water.

74. $m = \dfrac{n \text{ solute}}{\text{kg solvent}}$; in 1 L of solution n solute $= M$

$\text{kg solvent} = \dfrac{\text{mass solution (g)} - \text{mass solute (g)}}{1000}$

$= \dfrac{(1000 \times d) - M(\mathcal{M})}{1000} = d - \dfrac{M(\mathcal{M})}{1000}$

$m = \dfrac{M}{d - \dfrac{M(\mathcal{M})}{1000}}$

In dilute solution, $m \rightarrow M/d$; and for water, $d = 1.00$ g/mL

75. 48%

76. 0.0018 g/cm^3—intoxicated

77. (a) 2.08 M **(b)** 1.872 mol   **(c)** 47.4 L

78. $V_{gas} = \dfrac{n_{gas} \times RT}{P_{gas}}$; $n_{gas} = k \times P_{gas}$; $V_{gas} = kRT$
V_{gas} depends only on temperature.

Chapter 11

2. (a) rate $= \dfrac{\Delta[\text{HOCl}]}{2\Delta t}$ **(b)** rate $= \dfrac{-\Delta[\text{O}_2]}{\Delta t}$

4. For CO_2: 0.80 mol/L·s   For H_2O: 1.20 mol/L·s

6. (a) $N_2(g) + 3H_2(g) \rightarrow 2NH_3(g)$

(b) rate $= \dfrac{\Delta[\text{NH}_3]}{2\Delta t}$ **(c)** 0.0186 mol/L·min

8. (c) 0.011 mol/L·min **(d)** inst. rate < ave. rate

10. (a) 2, 1, 3 **(b)** 1, 0, 1 **(c)** 2, 2, 4 **(d)** 0, 0, 0

12. (a) $L^2/mol^2 \cdot s$ **(b)** s^{-1} **(c)** $L^3/mol^3 \cdot s$ **(d)** mol/L·s

14. (a) 0.0480 mol/L·h **(b)** 67 h^{-1} **(c)** $2.0 \times 10^1 \ M$

16. (a) rate = $k[NO_2]^2$ **(b)** 18.8 L/mol·min
(c) 12.0 mol/L·min

18. (a) 0.016 L/mol·s **(b)** 0.13 M **(c)** 1.4 M

20. (a) 0 **(b)** rate = k **(c)** 0.020 mol/L·min

22. (a) 1, 1, 2 **(b)** rate = $k[S_2O_8^{2-}]^2[I^-]$
(c) 0.371 L/mol·min **(d)** 0.00341 mol/L·min

24. (a) 1 for OH^-, 2 for ClO_2, 3 overall
(b) rate = $k[OH^-][ClO_2]^2$
(c) $2.0 \times 10^2 \ L^2/mol^2 \cdot s$ **(d)** 0.45 mol/L·s

26. (a) 1st order in $[I^-]$, 2nd order in $[H^+]$, 1st order in $[BrO_3^-]$
(b) rate = $k[I^-][H^+]^2[BrO_3^-]$
(c) $1.4 \times 10^4 \ L^3/mol^3 \cdot s$ **(d)** 0.018 M

28. (a) rate = $k[Cr(H_2O)_6^{3-}][SCN^-]$ **(b)** 0.43 L/mol·min
(c) 2.2×10^{-6} mol/L·min

30. rate = $k[C_{12}H_{22}O_{11}]^2$

32. (a) linear plot obtained for $\ln[C_2H_6]$ vs. t
(b) $4 \times 10^{-4} \ s^{-1}$
(c) 2×10^3 s **(d)** 2×10^{-6} mol/L·s

34. (a) 0.0196 M **(b)** 14.0 years **(c)** 5.55 years

36. (a) 0.758 g **(b)** 3.22 months **(c)** 4.9 months

38. (a) $9.3 \times 10^{-4} \ min^{-1}$ **(b)** 7.5×10^2 min
(c) 6.9×10^{-4} mol/L·min **(d)** 2.0×10^1 h

40. 3.23 mg

42. 15 h

44. (a) 10.0 min **(b)** 1.2 h

46. (a) 0.759 L/mol·s **(b)** 2.64 s

48. 9.4 min

50. $E_a = 2.2 \times 10^2$ kJ/mol

52. (a) 1.41×10^3 min **(b)** 3×10^2 min
(c) $k_2/k_1 \approx 5$ for 3°C; ≈ 2 for each 1°C

54. (a) At 25°C: 148 chirps/min; at 35°C: 220 chirps/min
(b) 3.0×10^1 kJ **(c)** 49%

56. (a) 471°C **(b)** 1.0×10^2 L/mol·s

58. 543 min

60. (a) rate = $k[NO_3][CO]$ **(b)** rate = $k[I_2]$
(c) rate = $k[NO][O_2]$

62. yes

64. c and d

66. 5.4 kJ

68. 20 times as great

70. $-\Delta[A] = k\Delta t$; $\Delta[A] = [A] - [A]_o$ and $\Delta t = t - 0$
$-([A] - [A]_o) = k(t - 0)$; $[A] = [A]_o - kt$

72. 2nd order

74. (a) A **(b)** C **(c)** A

76. (a) rate = $k[A]^2[B]$ **(b)** rate = $36k$
(c) Add 9 circles to vessel.

78. (a) (1)
(b) Rate increases; k increases; E_a remains the same.

80. (a) 174 kJ **(b)** 46 L/mol·s **(c)** 1.8 mol/L·s

81. $\dfrac{-d[A]}{a \, dt} = k[A]$; $\dfrac{-d[A]}{[A]} = k \, a \, dt$; $\ln \dfrac{[A]_o}{[A]} = a \, k \, t$

82. rate = $k[A]^2[B][C]$

83. (a) $\dfrac{-d[A]}{dt} = k[A]^2$; $\dfrac{-d[A]}{[A]^2} = k \, dt$; $\dfrac{1}{[A]} - \dfrac{1}{[A_o]} = k \, t$
(b) $\dfrac{-d[A]}{dt} = k[A]^3$; $\dfrac{-d[A]}{[A]^3} = k \, dt$; $\dfrac{1}{2[A]^2} - \dfrac{1}{2[A_o]^2} = k \, t$;
$\dfrac{1}{[A]^2} - \dfrac{1}{[A_o]^2} = 2 \, k \, t$

84. 0.90 g; 54 mg no more than three times a day

Chapter 12

2. (a) ≈ 80 s **(b)** faster; about the same

4.

Time (min)	0	1	2	3
P_A (atm)	1.000	0.778	0.580	0.415
P_B (atm)	0.400	0.326	0.260	0.205
P_C (atm)	0.000	0.148	0.280	0.390
Time (min)	4	5	6	
P_A (atm)	0.355	0.325	0.325	
P_B (atm)	0.185	0.175	0.175	
P_C (atm)	0.430	0.450	0.450	

6. (a) $K = \dfrac{(P_{CH_4})(P_{H_2S})^2}{(P_{CS_2})(P_{H_2})^4}$ **(b)** $K = \dfrac{(P_{H_2O})^2(P_{O_2})}{(P_{H_2O_2})^2}$

(c) $K = \dfrac{(P_{CO_2})^3(P_{H_2O})^4}{(P_{C_3H_8})(P_{O_2})^5}$

8. (a) $K = \dfrac{[Cu^{2+}]^3 (P_{NO})^2}{[H^+]^8[NO_3^-]^2}$ **(b)** $K = \dfrac{(P_{SO_2})^2}{(P_{O_2})^3}$

(c) $K = \dfrac{1}{[Ca^{2+}][CO_3^{2-}]}$

10. (a) $C_3H_6O(\ell) \rightleftharpoons C_3H_6O(g)$ $K = P_{C_3H_6O}$
(b) $7H_2(g) + 2NO_2(g) \rightleftharpoons 2NH_3(g) + 4H_2O(g)$
$K = \dfrac{(P_{NH_3})^2 (P_{H_2O})^4}{(P_{H_2})^7 (P_{NO_2})^2}$
(c) $H_2S(g) + Pb^{2+}(aq) \rightleftharpoons PbS(s) + 2H^+(aq)$
$K = \dfrac{[H^+]^2}{(P_{H_2S}) [Pb^{2+}]}$

12. (a) $3CO_2(g) + 4H_2O(g) \rightleftharpoons C_3H_8(g) + 5O_2(g)$
(b) $2Fe^{3+}(aq) + 2Cl^-(aq) \rightleftharpoons 2Fe^{2+}(aq) + Cl_2(g)$
(c) $2NH_3(g) \rightleftharpoons N_2(g) + 3H_2(g)$
(d) $4HCl(g) + O_2(g) \rightleftharpoons 2H_2O(g) + 2Cl_2(g)$
(e) $2H_2O(g) + 2O_2(g) \rightleftharpoons 2H_2O_2(g)$

14. (a) 4.8×10^{-6} **(b)** 2.1×10^5

16. 0.024

18. 9×10^{15}

20. 2.97×10^{-4}

22. (a) $CH_4(g) + 2H_2S(g) \rightleftharpoons CS_2(s) + 4H_2(g)$ **(b)** 334

24. 3.73

26. (a) no; $Q \neq K$ **(b)** ←

28. (a) → **(b)** → **(c)** →

30. 0.00147 atm

32. 24 atm

34. $P_{NO} = 0.42$ atm; $P_{NO_2} = 0.32$ atm

36. $P_{H_2} = P_{C_2N_2} = 0.11$ atm; $P_{HCN} = 0.77$ atm
38. (a) $P_{H_2} = P_{CO_2} = 0.343$ atm; $P_{CO} = P_{H_2O} = 0.301$ atm
　　(b) $P_{init} = P_{final} = 1.288$ atm; only if $\Delta n_g = 0$
40. (a) 0.928 atm　　(b) 6.1 g
42. $P_{NO_2} = 0.318$ atm　　$P_{N_2O_4} = 0.704$ atm
44. (a) (1) ←　　(2) ←　　(3) no change　　(4) →　　(5) ←
　　(b) Decreasing T decreases K.
46. (a) ←　　(b) →　　(c) no change
48. (a) 0.45
　　(b) $P_{CO} = 0.28$ atm　　$P_{H_2} = 0.69$ atm　　$P_{H_2O} = 0.44$ atm
50. 98
52. −59 kJ
54. $K_{700°C} = 2.0$　　$K_{600°C} = 0.06$
56. (a) $P_A = 0.75$ atm　　$P_B = 0.50$ atm　　(b) 0.62 atm
　　(c) 0.60 atm
58. (b)
60. The curve will resemble the original curve but would level off sooner at higher N_2O_4 concentration.
62. temperature and equation for the reaction
64. $P_{N_2} = 0.33$ atm; $P_{H_2} = 0.99$ atm; $P_{NH_3} = 0.34$ atm

65.
$$K = \frac{(P_C)^c(P_D)^d}{(P_A)^a(P_B)^b} = \frac{([C]RT)^c([D]RT)^d}{([A]RT)^a([B]RT)^b} = \frac{[C]^c[D]^d}{[A]^a[B]^b}(RT)^{(c+d)-(a+b)}$$
$$\Delta n_g = (c+d) - (a+b) \qquad K = K_c(RT)^{\Delta n_g}$$

66. 0.0442
67. 0.52 atm
68. 25; 0.65 atm
69. 1.1
70. 0.22 atm; 0.4 g

Chapter 13

	Acids	Bases	Pairs
2. (a)	H_3O^+, HCN	CN^-, H_2O	H_3O^+/H_2O and HCN/CN^-
(b)	HNO_2, H_2O	OH^-, NO_2^-	HNO_2/NO_2^- and H_2O/OH^-
(c)	$HCHO_2$, H_3O^+	H_2O, CHO_2^-	$HCHO_2/CHO_2^-$ and H_3O^+/H_2O

4. as acid: HSO_3^-, NH_4^+　　as base: CHO_2^-, HSO_3^-
6. (a) $C_2H_3O_2^-$　　(b) $Zn(H_2O)_3OH^+$
　　(c) BrO_2^-　　(d) CH_3NH_2　　(e) HS^-

8.

	pH	pOH	Classification
(a)	−0.78	14.78	acidic
(b)	0.48	13.52	acidic
(c)	7.34	6.66	basic
(d)	13.14	0.86	basic

10.

	$[H^+]$	$[OH^-]$		$[H^+]$	$[OH^-]$
(a)	1×10^{-9}	1×10^{-5}	(b)	6.3×10^{-4}	1.6×10^{-11}
(c)	11	8.9×10^{-16}	(d)	3.5×10^{-8}	2.9×10^{-7}

12. (1) more acidic; (1) has a higher pOH
14. (a) $pH_X:pH_Y = 10:1$;　　$pH_X:pH_Z = 1 \times 10^4:1$
　　(b) $pH_Y = 5.35$; $pH_Z = 8.35$
　　(c) X and Y—acidic; Z—basic
16. (a) 3.2×10^{-11}　　(b) $1 \times 10^9:1$
18. (a) −0.396　　(b) 0.922
20. 0.0367
22. (a) $[OH^-] = 0.0237$ M; $[H^+] = 4.2 \times 10^{-13}$ M;
　　pH = 12.37; pOH = 1.63
　　(b) $[OH^-] = 0.184$ M; $[H^+] = 5.4 \times 10^{-14}$ M;
　　pH = 13.26; pOH = 0.74
24. 12.94
26. (a) $Ni(H_2O)_5OH^+(aq) + H_2O \rightleftharpoons$
　　　　　　　　　　　　$Ni(H_2O)_4(OH)_2(aq) + H_3O^+(aq)$
　　(b) $Al(H_2O)_6^{3+}(aq) + H_2O \rightleftharpoons$
　　　　　　　　　　　　$Al(H_2O)_5OH^{2+}(aq) + H_3O^+(aq)$
　　(c) $H_2S(aq) + H_2O \rightleftharpoons H_3O^+(aq) + HS^-(aq)$
　　(d) $HPO_4^{2-}(aq) + H_2O \rightleftharpoons PO_4^{3-} + H_3O^+(aq)$
　　(e) $HClO_2(aq) + H_2O \rightleftharpoons H_3O^+(aq) + ClO_2^-(aq)$
　　(f) $Cr(H_2O)_5OH^+(aq) + H_2O \rightleftharpoons$
　　　　　　　　　　　　$Cr(H_2O)_4(OH)_2(aq) + H_3O^+(aq)$
28. (a) $HSO_3^-(aq) \rightleftharpoons H^+(aq) + SO_3^{2-}(aq)$　　$K_a = \dfrac{[H^+][SO_3^{2-}]}{[HSO_3^-]}$

　　(b) $HPO_4^{2-}(aq) \rightleftharpoons H^+(aq) + PO_4^{3-}(aq)$　　$K_a = \dfrac{[H^+][PO_4^{3-}]}{[HPO_4^{2-}]}$

　　(c) $HNO_2(aq) \rightleftharpoons NO_2^-(aq) + H^+(aq)$　　$K_a = \dfrac{[H^+][NO_2^-]}{[HNO_2]}$

30. (a) 3.20　　(b) 10.77　　(c) 7.72
32. (a) D > A > C > B　　(b) D
34. 1.9×10^{-5}
36. 6.6×10^{-5}
38. (a) 0.00180 M　　(b) 0.00133 M
40. (a) 0.0086 M　　(b) 1.2×10^{-12} M
　　(c) 2.07　　(d) 1.3%
42. 0.10; 20%
44. 2.34
46. $H_3PO_4(aq) \rightleftharpoons 3H^+(aq) + PO_4^{3-}(aq)$　　$K = 2.0 \times 10^{-22}$
48. pH = 3.28; $[HCO_3^-] = 5.3 \times 10^{-4}$ M;
　　$[CO_3^{2-}] = 4.7 \times 10^{-11}$ M
50. (a) $NH_3(aq) + H_2O \rightleftharpoons NH_4^+(aq) + OH^-(aq)$
　　(b) $NO_2^-(aq) + H_2O \rightleftharpoons HNO_2(aq) + OH^-(aq)$
　　(c) $C_6H_5NH_2(aq) + H_2O \rightleftharpoons C_6H_5NH_3^+(aq) + OH^-(aq)$
　　(d) $CO_3^{2-}(aq) + H_2O \rightleftharpoons HCO_3^-(aq) + OH^-(aq)$
　　(e) $F^-(aq) + H_2O \rightleftharpoons HF(aq) + OH^-(aq)$
　　(f) $HCO_3^-(aq) + H_2O \rightleftharpoons H_2CO_3(aq) + OH^-(aq)$
52. $Ba(OH)_2$ > KOH > NaCN > $NaHCO_3$
54. (a) 1.4×10^{-8}　　(b) 7.1×10^{-11}
56. (a) $Cod(aq) + H_2O \rightleftharpoons CodH^+(aq) + OH^-(aq)$
　　(b) $K_b = 8.3 \times 10^{-7}$　　(c) 9.61
58. 2.1×10^2 g
60. (a) neutral　　(b) basic　　(c) basic
　　(d) acidic　　(e) basic
62. (a) $H^+(aq) + OH^-(aq) \rightleftharpoons H_2O$
　　(b) $SO_3^{2-}(aq) + H_2O \rightleftharpoons HSO_3^-(aq) + OH^-(aq)$
　　(c) $HCO_3^-(aq) + H_2O \rightleftharpoons OH^-(aq) + H_2CO_3(aq)$
　　　　　　　　　　　　　　$K_b = 2.3 \times 10^{-8}$
　　　　$HCO_3^-(aq) \rightleftharpoons H^+(aq) + CO_3^{2-}(aq)$　　$K_a = 4.7 \times 10^{-11}$

(d) $NH_4^+(aq) \rightleftharpoons NH_3(aq) + H^+(aq)$ $K_a = 5.6 \times 10^{-10}$
$NO_2^-(aq) + H_2O \rightleftharpoons HNO_2(aq) + OH^-(aq)$
$K_b = 1.7 \times 10^{-11}$
(e) $PO_4^{3-}(aq) + H_2O \rightleftharpoons HPO_4^{2-}(aq) + OH^-(aq)$

64. $HCl < ZnCl_2 < KCl < KF < KOH$

66. **(a)** KF, KNO_2, K_2SO_3, KCN **(b)** KCl, KNO_3, $KClO_4$, KBr
(c) $NaClO_4$, $KClO_4$, $Ba(ClO_4)_2$, $LiClO_4$
(d) $Al(ClO_4)_3$, $Co(ClO_4)_2$, $Fe(ClO_4)_2$, $Zn(ClO_4)_2$

68. 2.35

70. 68 L

72. neutral

74. a, c, d, e

76. (1) weak acid; (2) strong acid

78.

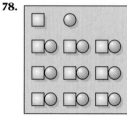

80. **(a)** -5.58 kJ **(b)** -6.83 kJ

81. % ionization $= \dfrac{[H^+]}{[HA]_o} \times 100$; $[H^+]^2 \approx K_a \times [HA]_o$;

$[H^+] \approx K_a^{1/2} \times [HA]_o^{1/2}$;

% ionization $= \dfrac{K_a^{1/2}}{[HA]_o^{1/2}} \times 100$;

% ionization is inversely proportional to $[HA]_o^{1/2}$

82. $-1.64°C$

83. 12.62

Chapter 14

2. **(a)** $NH_3(aq) + HF(aq) \rightleftharpoons NH_4^+(aq) + F^-(aq)$
(b) $H^+(aq) + OH^-(aq) \rightleftharpoons H_2O$
(c) $SO_3^{2-}(aq) + H^+(aq) \rightleftharpoons HSO_3^-(aq)$
(d) $H^+(aq) + OH^-(aq) \rightleftharpoons H_2O$

4. **(a)** $C_2H_3O_2^-(aq) + H^+(aq) \rightleftharpoons HC_2H_3O_2(aq)$
(b) $H^+(aq) + OH^-(aq) \rightleftharpoons H_2O$
(c) $HCO_3^-(aq) + H^+(aq) \rightleftharpoons H_2CO_3(aq)$

6. **(a)** 1.2×10^6 **(b)** 1.0×10^{14}
(c) 1.7×10^7 **(d)** 1.0×10^{14}

8. **(a)** 5.6×10^4 **(b)** 1.0×10^{14} **(c)** 2.3×10^6

10. **(a)** 7.85; 7.1×10^{-7} **(b)** 7.63; 4.3×10^{-7}
(c) 7.33; 2.1×10^{-7} **(d)** 7.03; 1.1×10^{-7}

12. Theoretically, one liter of the buffer can absorb:
(a) 0.50 mol acid; 0.250 mol base
(b) 0.30 mol acid and 0.250 mol base
(c) 0.15 mol acid; 0.250 mol base
(d) 0.075 mol acid; 0.250 mol base

14. 4.22

16. **(a)** HSO_4^-/SO_4^{2-} **(b)** $HC_3H_5O_3/C_3H_5/O_3^-$
(c) HCN/CN^-

18. **(a)** 8.5 **(b)** 1.9 mol **(c)** 2.0 g **(d)** 4.3 L

20. **(a)** 8.51 **(b)** pH remains the same.

22. Add 5.6 g of $KC_2H_3O_2$ to the vinegar.

24. 4.52

26. **(a)** 10.15 **(b)** 9.57 **(c)** 10.60

28. **(a)** 10.15 **(b)** 1.43 **(c)** 12.55
(d) There are large fluctuations in pH when base and acid are added.
(e) Buffer capacity is diminished by dilution.

30. **(a)** 2.82 **(b)** 3.26 **(c)** 1.2

32. b, c, e

34. 12.36

36. 4.91

38. **(a)** 0.64 **(b)** 54% **(c)** 47%

40. **(a)** phenolphthalein **(b)** phenolphthalein
(c) all three indicators **(d)** methyl orange

42. **(a)** 2.45–4.45 **(b)** orange

44. **(a)** $H^+(aq) + OH^-(aq) \rightleftharpoons H_2O$ **(b)** NO_3^-, Ba^{2+}, H_2O
(c) 59.2 mL **(d)** 13.68 **(e)** 13.17 **(f)** 7.00

46. **(a)** $H^+(aq) + Lac^-(aq) \rightleftharpoons HLac$
(b) Lac^-, Na^+, $HLac$, Cl^-, H^+, H_2O
(c) 108 mL **(d)** 8.60 **(e)** 3.85 **(f)** 2.50

48. **(a)** 3.64 **(b)** 7.55 **(c)** 10.55

50. **(a)** 112 g/mol **(b)** 6.8×10^{-4} **(c)** 1.5×10^{-11}

52. 1.2×10^{-2}

54. **(a)**

56. $H^+(aq) + HCO_3^-(aq) \rightleftharpoons H_2CO_3(aq) \rightleftharpoons CO_2(g) + H_2O$
An increase in H^+ ions shifts the equilibrium to the right, forming CO_2.

58. **(a)** False; $[CHO_2^-]$ is 0.1 M only in 0.1 M $NaCHO_2$.
(b) True
(c) False; a buffer can be made up by combining a weak acid and its conjugate base in approximately equal concentrations.
(d) False; K_b for HCO_3^- is 2.3×10^{-8}; K_b for CO_3^{2-} is 2.1×10^{-4}.

60. 33 mL

61. $\dfrac{[NH_4^+]}{[NH_3]} = 5.7 \times 10^2$

62. 1.1×10^2 g/mol

63. **(a)** 1.61 **(b)** 3.22 **(c)** 8.44 **(d)** 5.92 **(e)** 10.89

64. ≈ 30 mL; phenolphthalein

65. $-\log[H^+] = -\log K_a - \log\left(\dfrac{[HB]}{[B^-]}\right)$;

$pH = pK_a - \log\dfrac{[HB]}{[B^-]} = pK_a + \log\dfrac{[B^-]}{[HB]}$

66. (a) 0.8239   (b) 0.80

67. 4.6×10^{-6}

Chapter 15

2. (a) $H_2O = 0$; $Cl = -1$; $OH^- = -1$ (b) 2+
   (c) $Na_2[Ni(H_2O)_2Cl_2(OH)_2]$

4. (a) $Pt(NH_3)_2(C_2O_4)$ (b) $Pt(NH_3)_2(SCN)Br$
   (c) $Pt(en)(NO_2)_2$

6. (a) 6   (b) 4   (c) 2   (d) 4

8. (a) $[Ag(H_2O)_2]^+$ (b) $[Pt(H_2O)_4]^{2+}$
   (c) $[Pd(Br)_4]^{2-}$ (d) $[Fe(C_2O_4)_3]^{3-}$

10. 22.61%

12. 6.4×10^4 g/mol

14.

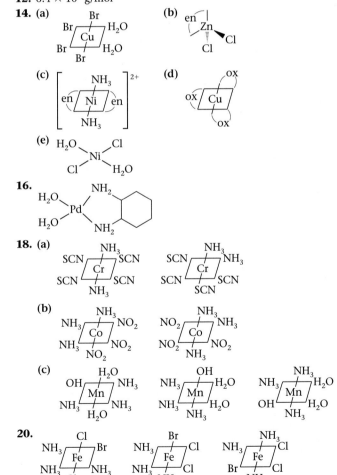

16.

18. (a)

   (b)

   (c)

20.

22. (a) $1s^2\,2s^2\,2p^6\,3s^2\,3p^6\,3d^1$
   (b) $1s^2\,2s^2\,2p^6\,3s^2\,3p^6\,3d^4$
   (c) $1s^2\,2s^2\,2p^6\,3s^2\,3p^6\,4s^2\,3d^{10}\,4p^6\,4d^4$

   (d) $1s^2\,2s^2\,2p^6\,3s^2\,3p^6\,4s^2\,3d^{10}\,4p^6\,4d^8$
   (e) $1s^2\,2s^2\,2p^6\,3s^2\,3p^6\,4s^2\,3d^{10}\,4p^6\,4d^3$

24. (a) [Ar] (↑)( )( )( )( ); 1
   (b) [Ar] (↑)(↑)(↑)(↑)( ); 4
   (c) [Kr] (↑)(↑)(↑)(↑)( ); 4
   (d) [Kr] (↑↓)(↑↓)(↑↓)(↑)(↑); 2
   (e) [Kr] (↑)(↑)(↑)( )( ); 3

26. (a) high spin   low spin

   (b) high spin   low spin

28. V^{3+} has an outer electron configuration of $3d^2$. For outer configurations of $3d^0$ to $3d^3$ only one distribution is possible.

30. NH_3 has a large Δ_o; low spin; no unpaired electrons.
   F^- has a small Δ_o; high spin; unpaired electrons present

32. (a) 2   (b) 0   (c) 0   (d) 2   (e) 2

34. 3.00×10^2 kJ/mol

36. ≈ 500 nm; green

38. 1×10^{15} 1×10^{-5} 1×10^{-21}

40. (a) 0.014 M   (b) 0.1 M

42. (a) $CoN_6H_{18}Cl_3$
   (b) $Co(NH_3)_6Cl_3(s) \rightleftharpoons Co(NH_3)_6^{3+}(aq) + 3Cl^-(aq)$

44. Lungs saturated with oxygen so the shift is to oxyhemoglobin formation. Tissues low in oxygen, so in venal blood, the shift is to hemoglobin formation.

46. (a) It chelates with Fe to form a stable complex.
   (b) All positions are equivalent.
   (c) They are electron-pair acceptors from ligands.
   (d) 2 bonds/molecule
   (e) Electron pairs are available in NH_3 but not in NH_4^+.

48. b, d, e

49. 0.92 g

50. $[Pt(NH_3)_4][PtCl_4]$ or $[Pt(NH_3)_3Cl][Pt(NH_3)Cl_3]$

51. (a) $CuC_4H_{22}N_6SO_4$
   (b)

52. red-violet

Chapter 16

2. (a) $Co_2S_3(s) \rightleftharpoons 2Co^{3+}(aq) + 3S^{2-}(aq)$ $K_{sp} = [Co^{3+}]^2[S^{2-}]^3$
   (b) $PbCl_2(s) \rightleftharpoons 2Cl^-(aq) + Pb^{2+}(aq)$ $K_{sp} = [Cl^-]^2[Pb^{2+}]$
   (c) $Zn_2P_2O_7(s) \rightleftharpoons 2Zn^{2+}(aq) + P_2O_7^{4-}(aq)$
   $K_{sp} = [Zn^{2+}]^2[P_2O_7^{4-}]$
   (d) $Sc(OH)_3(s) \rightleftharpoons Sc^{3+}(aq) + 3\,OH^-(aq)$
   $K_{sp} = [Sc^{3+}][OH^-]^3$

4. (a) $CaCO_3(s) \rightleftharpoons Ca^{2+}(aq) + CO_3^{2-}(aq)$
   (b) $Co(OH)_3(s) \rightleftharpoons Co^{3+}(aq) + 3\,OH^-(aq)$
   (c) $Ag_2S(s) \rightleftharpoons 2Ag^+(aq) + S^{2-}(aq)$
   (d) $PbCl_2(s) \rightleftharpoons Pb^{2+}(aq) + 2Cl^-(aq)$

6. (a) 1×10^{-4} M   (b) 9×10^{-7} M
   (c) 3×10^{-4} M   (d) 4×10^{-5} M

8. (a) 1.8×10^{-2} M (b) 2×10^{-19} M (c) 3×10^{-11} M
10. (a) 4×10^{-11} M (b) 1×10^{-10} M
12. yes; 2×10^{-7} M
14. (a) yes
    (b) $[Hg_2^{2+}] = 1 \times 10^{-14}$ M; $[Cl^-] = 8.4 \times 10^{-3}$ M;
    $[NO_3^-] = 1.4 \times 10^{-3}$ M
16. 8.3×10^{-12}
18. (a) 6×10^{-3} g/L   (b) 5×10^{-8} g/L   (c) 1×10^{-3} g/L
20. no
22. (a) $Cu_2S(s) + 2H^+(aq) \rightarrow 2Cu^+(aq) + H_2S(aq)$
    (b) no reaction
    (c) $SrCO_3(s) + 2H^+(aq) \rightarrow Sr^{2+}(aq) + CO_2(g) + H_2O$
    (d) $Cu(NH_3)_4^{2+}(aq) + 4H^+(aq) \rightarrow Cu^{2+}(aq) + 4NH_4^+(aq)$
    (e) $Ca(OH)_2(s) + 2H^+(aq) \rightarrow Ca^{2+}(aq) + 2H_2O$
24. (a) $Cu(OH)_2(s) + 4NH_3(aq) \rightarrow Cu(NH_3)_4^{2+}(aq) + 2OH^-(aq)$
    (b) $Cd^{2+}(aq) + 4NH_3(aq) \rightarrow Cd(NH_3)_4^{2+}(aq)$
    (c) $Pb^{2+}(aq) + 2NH_3(aq) + 2H_2O \rightarrow$
    $Pb(OH)_2(s) + 2NH_4^+(aq)$
26. (a) $Sb^{3+}(aq) + 3\,OH^-(aq) \rightarrow Sb(OH)_3(s)$
    (b) $Sb(OH)_3(s) + OH^-(aq) \rightarrow Sb(OH)_4^-(aq)$
    (c) $Sb^{3+}(aq) + 4OH^-(aq) \rightarrow Sb(OH)_4^-(aq)$
28. 2×10^{11}
30. (a) 5×10^{-6} (b) no
32. (a) 4×10^{-7} (b) 0.03 mol/L
34. 2.4×10^3 M
36. (a) 0.42   (b) 0.12 M
38.

	Group	Ppt. agent	Ppt. formed
(a)	1	Cl^-	AgCl
(b)	2	S^{2-} (pH – 0.5)	Bi_2S_3
(c)	3	S^{2-} (pH – 9)	CoS
(d)	4	CO_3^{2-}	$MgCO_3$

40. Ag^+, Pb^{2+}, Hg_2^{2+}, Ba^{2+}, Mg^{2+}, Ca^{2+}
42. Cd^{2+}
44. no
46. (a) 3.1×10^{-4} M (b) yes   (c) 6×10^{-5} M

48.

(1)   (2)

(3)

50. Acid rain has more H^+ available to dissolve the $CaCO_3$.
52. No—increasing temperature favors an endothermic process.
53. in water: 1×10^{-4} mol/L   in NH_4Cl: 0.06 mol/L
54. 1.9×10^{-4} mol/L
55. 1×10^{-8} M
56. (a) 1×10^{-5} M (b) $Al(OH)_3$, $Fe(OH)_3$
    (c) virtually all   (d) 1.6 g
57. (a) 8×10^5 (b) 4×10^3
58. 1.0×10^{21}

Chapter 17

2. d
4. a, c, d
6. (a) −   (b) −   (c) +   (d) −
8. (a) +   (b) −   (c) +   (d) +
10. (a) +   (b) −   (c) −
12. (a) −112.0 J/K   (b) 605.9 J/K
    (c) 2.9 J/K   (d) −93.1 J/K
14. (a) −23.1 J/K   (b) 80.7 J/K   (c) −159.6 J/K
16. (a) −152.8 J/K   (b) 193.6 J/K   (c) 209.0 J/K
18. (a) −217 kJ   (b) 782.1 kJ   (c) −22 kJ
20. (a) −1087.1 kJ; spontaneous
    (b) −894 kJ; spontaneous
    (c) −394.7 kJ; spontaneous
    (d) −298.0 kJ; spontaneous
22. (a) −147.1 kJ   (b) −80.0 kJ   (c) 17.2 kJ
24. (a) −1128.9 kJ/mol   (b) −569.4 kJ/mol
    (c) −305.1 kJ/mol
26. $\Delta G° = 115.6$ kJ; no
28. at 25°C: −359.2 kJ; at 15°C: −359.0 kJ
30. (a) 0.132 kJ/K; yes   (b) 0.0956 kJ/mol-K
    (c) −510.9 kJ/mol
32. (a) 0.333 kJ/K   (b) 284 J/mol-K   (c) −218.8 kJ/mol
34. (a) T-dependent; spontaneous at high T
    (b) T-independent; spontaneous at any T
    (c) T-dependent; spontaneous at high T
36. (a) spontaneous above 2.2×10^4 K
    (b) T-independent; spontaneous at any T
    (c) spontaneous above 1.05×10^3 K
38. 346°C
40. (a)

$\Delta G°$	412.1	356.3	300.5	244.7	188.9
T	100	200	300	400	500

    $\Delta G = 467.9 - 0.5581\,T$
    (b) 838 K
42. b
44. 962 K
46. not at equilibrium at any temperature when P = 1 atm
48. 432 K
50. (a) $\Delta G = 18.0$ kJ; not spontaneous
    (b) $\Delta G = -16.2$ kJ; spontaneous
52. (a) 422.4 kJ   (b) 397.9 kJ
54. (a) 55.7 kJ   (b) 1.08×10^{-5} M (c) yes—close to K_{sp}
56. −63.4 kJ
58. (a) + 68.5 kJ—not feasible   (b) −160.1 kJ—feasible

60. (a) $C_6H_{12}O_6(s) + 6O_2(g) + 80\,ADP + 80\,HPO_4^{2-}(aq) +$
$160H^+(aq) \rightarrow 80\,ATP + 86H_2O + 6CO_2(g)$

62. (a) 80.0 kJ   **(b)** 9.3×10^{-15}

64. (a) 211 kJ   **(b)** 105 kJ/mol

66. (a) 1.1×10^{12} at 298 K   **(b)** 1.0×10^2 at 1200 K

68. 6.9×10^{-4}

70. -37.3 kJ

72. (a) yes   **(b)** yes   **(c)** no

74. (a) False— . . . is often spontaneous.
(b) False— . . . under standard conditions
(c) False— . . . moles of gas
(d) False—If $\Delta H° < 0$, and $\Delta S° > 0$, . . .

76. (a) 0 K   **(b)** 1   **(c)** larger

78. (a) Entropy decreases because $S°_{gas} > S°_{solid}$ or $S°_{liq}$
(b) $\Delta S°$ is the difference between entropy of products and reactants.
(c) A solid has fewer options for placement of particles, i.e., more orderly.

80.

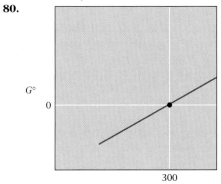

$G°$

0

300

Temperature

81. 61°C

82. $P_{H_2} = P_{I_2} = 0.07$ atm $P_{HI} = 0.46$ atm

83. (a) 6.00 kJ   **(b)** 0   **(c)** 0.0220 kJ/K
(d) 0.43 kJ   **(e)** -0.45 kJ

84. (a) 4.2 kJ   **(b)** $\approx 5.2 \times 10^2$ g

85. 1430 K

86. (1) $\Delta G° = 146 - 0.1104\ T$; becomes spontaneous above 1050°C
(2) $\Delta G° = 168.6 - 0.0758\ T$; becomes spontaneous above 1950°C

Chapter 18

2. (a) $H_2(g) + 2Fe^{3+}(aq) \rightarrow 2H^+(aq) + 2Fe^{2+}(aq)$
(b) $Cd(s) + Ni^{2+}(aq) \rightarrow Cd^{2+}(aq) + Ni(s)$
(c) $10Cl^-(aq) + 16H^+(aq) + 2MnO_4^-(aq) \rightarrow$
$5Cl_2(g) + 2Mn^{2+}(aq) + 8H_2O$

4. (a) Sn anode, Ag cathode; e^- move from Sn to Ag. Anions move to Sn and cations to Ag.
(b) Pt both anode and cathode. Anode has H_2 gas and H^+; cathode has chloride ions, Hg and Hg_2Cl_2. e^- move from anode to cathode. Anions move to the anode; cations move to the cathode.
(c) Pb anode and PbO_2 cathode. e^- move from Pb to PbO_2.

Anions move to Pb; cations to PbO_2.

6. anode: $Mn(s) \rightarrow Mn^{2+}(aq) + 2e^-$
cathode: $Cr^{3+}(aq) + 3e^- \rightarrow Cr(s)$
overall: $3Mn(s) + 2Cr^{3+}(aq) \rightarrow 3Mn^{2+}(aq) + 2Cr(s)$
cell notation: $Mn\,|\,Mn^{2+}\,\|\,Cr^{3+}\,|\,Cr$

8. (a) H_2O_2 **(b)** MnO_4^- **(c)** Ag^+ **(d)** O_2 (acidic)

10. (a) $PbSO_4 < Br^- < H_2S < Co < Zn$

12. oxidizing agents: $Fe^{2+} < Ni^{2+} < H^+ < Cu^+$
reducing agents: $Fe^{2+} < Cu^+ < Zn$
both: Fe^{2+}, Cu^+

14. (a) Co through Sn   **(b)** Cr^{3+} through Tl^+
(c) Br^- through Cl_2

16. (a) 0.926 V   **(b)** 0.460 V   **(c)** 0.360 V

18. (a) 0.562 V   **(b)** 0.534 V   **(c)** 1.292 V

20. (a) 2.64 V   **(b)** 1.637 V   **(c)** 0.195 V

22. (a) -0.799 V   **(b)** 3.668 V   **(c)** 0.195 V; same

24. Reaction (a)

26. (a) 0.152 V; yes   **(b)** -0.030 V; no   **(c)** -0.396 V; no

28. (a) no reaction   **(b)** no reaction
(c) $2Cr(s) + 3Ni^{2+}(aq) \rightarrow 2Cr^{3+}(aq) + 3Ni(s)$

30. b

32. (a) no reaction
(b) $Fe^{2+}(aq) + Ag^+(aq) \rightarrow Fe^{3+}(aq) + Ag(s)$
(c) $10\,Hg(\ell) + 12H^+(aq) + 2ClO_3^-(aq) \rightarrow$
$5Hg_2^{2+}(aq) + Cl_2(g) + 6H_2O$

34.

	$\Delta G°$	$E°$	K
(a)	19 kJ	-0.098 V	5×10^{-4}
(b)	-6.8 kJ	0.035 V	15
(c)	5.8 kJ	-0.030 V	0.095

36. (a) -116 kJ   **(b)** -232 kJ   **(c)** -347 kJ
There is no effect on spontaneity, since the sign of $\Delta G°$ does not change, nor does the position of equilibrium.

38. $E° = -0.551$ V $\Delta G° = +106$ kJ $K = 2.4 \times 10^{-19}$

40. (a) -179 kJ   **(b)** -178 kJ   **(c)** -69.5 kJ

42. (a) 1×10^{19} **(b)** 1×10^{18} **(c)** 2×10^{87}

44. (a) 0.744 V   **(b)** $E = 0.744 - \dfrac{0.0257}{6}\ln\dfrac{[Cr^{3+}]^2\,(P_{H_2})^3}{[H^+]^6}$
(c) 0.623 V

46. (a) 0.994 V
(b) $E = 0.994 - \dfrac{0.0257}{2}\ln\dfrac{[Fe^{3+}]^2}{[Fe^{2+}]^2\,[H_2O_2][H^+]^2}$
(c) 0.740 V

48. (a) 0.319   **(b)** -0.369 V

50. 0.022 M

52. $H^+ = 1.4$ M

54. $E = 0.114$ V; no

56. (a) -0.460 V   **(b)** 1.6×10^{-9} **(c)** 1.6×10^{-10}

58. (a) 1112 mol   **(b)** 1242 amps   **(c)** 278.0 mol

60. (a) 16 g   **(b)** 19 min

62. (a) 31 g   **(b)** 0.095 kWhr

64. 1.70 hr

66. -0.188 V

68. 0.896 V

70. (a) It allows current to flow without bringing reactants into direct contact.

(b) K^+ is a cation and cations move to the cathode.

(c) H_2SO_4 participates in the cell reactions.

72. b

74. 1×10^{-12}

76. 2.007 V

77. (a) 0.621 V   **(b)** increases; decreases   **(c)** 1×10^{21}
(d) $[Zn^{2+}] = 2.0\ M;\ [Sn^{2+}] = 2 \times 10^{-21}\ M$

78. (a) 1.12×10^5 J; 0.38×10^5 J; 1.50×10^5 J
(b) -0.389 V

79. 0.414 V

Chapter 19

2. α-particle

4. (a) $^{230}_{90}\text{Th} \rightarrow ^4_2\text{He} + ^{226}_{88}\text{Ra}$
(b) $^{210}_{82}\text{Pb} \rightarrow ^0_{-1}e + ^{210}_{83}\text{Bi}$
(c) $^{235}_{92}\text{U} + ^1_0 n \rightarrow ^{140}_{56}\text{Ba} + 3^1_0 n + ^{93}_{36}\text{Kr}$
(d) $^{37}_{18}\text{Ar} + ^0_{-1}e \rightarrow ^{37}_{17}\text{Cl}$

6. (a) $^{87}_{38}\text{Sr}$ **(b)** $^{87}_{38}\text{Sr}$

8. The product ($^{282}_{114}\text{Z}$) is the same.

10. (a) $^{54}_{26}\text{Fe} + ^4_2\text{He} \rightarrow 2^1_1\text{H} + ^{56}_{26}\text{Fe}$
(b) $^{96}_{42}\text{Mo} + ^2_1\text{H} \rightarrow ^1_0 n + ^{97}_{43}\text{Tc}$
(c) $^{40}_{18}\text{Ar} + ^4_2\text{He} \rightarrow ^{43}_{19}\text{K} + ^1_1\text{H}$
(d) $^{31}_{16}\text{S} + ^1_0 n \rightarrow ^1_1\text{H} + ^{31}_{15}\text{P}$

12. (a) $^{124}_{52}\text{Te}$ **(b)** $^{239}_{93}\text{Np}$ **(c)** ^4_2He **(d)** $^{24}_{12}\text{Mg}$

14. 6.200×10^{15} disintegrations

16. 2.70×10^{-9} Ci

18. 5.6×10^4 Ci

20. (a) $^{210}_{82}\text{Pb} \rightarrow ^0_{-1}e + ^{210}_{83}\text{Bi}$
(b) 4.18×10^4 mCi

22. 1.4×10^{-11} g

24. 7.3×10^7 β-particles/minute; 3.3×10^{-5} Ci

26. no

28. 5.05×10^3 years

30. 9.07 years

32. (a) -0.0051 g   **(b)** -2.0×10^6 kJ/g

34. (a) 0.06980 g   **(b)** 6.28×10^9 kJ/mol

36. Mg-26

38. -5.92×10^8 kJ/g H

40. 167 kg

42. (a) 16%   **(b)** 18 mg

44. (a) 2.10×10^6 kJ   **(b)** 7.61×10^5 g   **(c)** 2.72×10^{-2} g

46. 1.0×10^{-16}

48. 2.60 Ci

50. 6.5×10^2 mL

52. 2.47×10^3 mL

54. 38 mL

56. (a) atomic number (not mass number) increases
(b) decays very rapidly
(c) more energy

58. 99.8% remains after one year. Decay is slow.

60. $t_{\frac{1}{2}} = 3.3$ hours

62. 5.8×10^{-19} mol/L

63. (a) 2.5×10^{-9} g   **(b)** 5.5×10^{-3} kJ   **(c)** 73 rems

64. (a) 1×10^{-13} J   **(b)** 6×10^6 m/s

65. (a) -5.72×10^8 kJ/g   **(b)** -1.3×10^{28} kJ
(c) 1.8×10^{-11}

Chapter 20

2. $2Al_2O_3(l) \rightarrow 4Al(l) + 3\ O_2(g)$; 2.91 g

4. $Cu_2S(s) + O_2(g) \rightarrow 2Cu(s) + SO_2(g)$

6. -211.9 kJ

8. (a) $Fe_2O_3(s) + 3CO(g) \rightarrow 2Fe(l) + 3CO_2(g)$
(b) $C(s) + O_2(g) \rightarrow CO_2(g)$

10. 2.5×10^3 kWh

12. 1.7×10^6 L

14. (a) potassium nitride, K_3N
(b) potassium iodide, KI
(c) potassium hydroxide, KOH
(d) potassium hydride, KH
(e) potassium sulfide, K_2S

16. (a) $Na_2O_2(s) + 2H_2O \rightarrow 2Na^+(aq) + 2OH^-(aq) + H_2O_2(aq)$
sodium and hydroxide ions, hydrogen peroxide
(b) $2Ca(s) + O_2(g) \rightarrow 2CaO(s)$; calcium oxide
(c) $Rb(s) + O_2(g) \rightarrow RbO_2(s)$; rubidium superoxide
(d) $SrH_2(s) + 2H_2O \rightarrow Sr^{2+}(aq) + 2OH^-(aq) + 2H_2(g)$
strontium and hydroxide ions, hydrogen gas

18. 0.126 g

20. (a) $Co(s) + 2H^+(aq) \rightarrow Co^{2+}(aq) + H_2(g)$
(b) $3Cu(s) + 2NO_3^-(aq) + 8H^+(aq) \rightarrow$
$$3Cu^{2+}(aq) + 2NO(g) + 4H_2O$$
(c) $Cr_2O_7^{2-}(aq) + 6e^- + 14H^+(aq) \rightarrow 2Cr^{3+}(aq) + 7H_2O$

22. $3Cd(s) + 12Cl^-(aq) + 2NO_3^-(aq) + 8H^+(aq) \rightarrow$
$$3CdCl_4^{2-}(aq) + 2NO(g) + 4H_2O$$

24. (a) $Fe(s) + 3NO_3^-(aq) + 6H^+(aq) \rightarrow$
$$Fe^{3+}(aq) + 3NO_2(g) + 3H_2O$$
(b) $4Cr(OH)_3(s) + 3\ O_2(g) + 8OH^-(aq) \rightarrow$
$$4CrO_4^{2-}(aq) + 10H_2O$$

26. (a) Cd ($E° = 1.366$ V)   **(b)** Cr ($E° = 1.708$ V)
(c) Co ($E° = 1.246$ V)   **(d)** Ag ($E° = 0.165$ V)

28. (a) 0.724 V   **(b)** 0.942 V

30. (a) 9×10^9 **(b)** 2×10^{-4} M

32. 208 g

34. 6.7

36. 53.8% Zn, 46.2% Cu

38. $2Ag_2S(s) + 8CN^-(aq) + 3\ O_2(g) + 2H_2O \rightarrow$
$$4Ag(CN)_2^-(aq) + 2SO_2(g) + 4OH^-(aq)$$
$2Ag(CN)_2^-(aq) + Zn(s) \rightarrow Zn(CN)_4^{2-}(aq) + 2Ag(s)$

40. (a) Cr^{2+} **(b)** Au^+ **(c)** Co^{2+} **(d)** Mn^{2+}

41. 2% BaO_2

42. (a) $Fe(OH)_3(s) + 3H_2C_2O_4(aq) \rightarrow$
$$Fe(C_2O_4)_3^{3-}(aq) + 3H_2O + 3H^+(aq)$$
(b) 0.28 L

43. 2.80%

44. 2.83×10^3 K

45. $Cr_2O_7^{2-}(aq) + 2OH^-(aq) \rightarrow 2CrO_4^{2-}(aq) + H_2O$
$2Ag^+(aq) + CrO_4^{2-}(aq) \rightarrow Ag_2CrO_4(s)$

$$Ag_2CrO_4(s) + 4NH_3(aq) \rightarrow 2Ag(NH_3)_2^+(aq) + CrO_4^{2-}(aq)$$
$$2Ag(NH_3)_2^+(aq) + 4H^+(aq) + CrO_4^{2-}(aq) \rightarrow$$
$$Ag_2CrO_4(s) + 4NH_4^+(aq)$$

Chapter 21

2. (a) bromic acid   **(b)** potassium hypoiodite
(c) sodium chlorite   **(d)** sodium perbromate
4. (a) $KBrO_2$ **(b)** $CaBr_2$
(c) $NaIO_4$ **(d)** $Mg(ClO)_2$
6. (a) NO_3^- **(b)** SO_4^{2-} **(c)** ClO_4^-
8. (a) H_2SO_3 **(b)** $HClO$ **(c)** H_3PO_3
10. (a) NaN_3 **(b)** H_2SO_3
(c) N_2H_4 **(d)** NaH_2PO_4
12. (a) H_2S **(b)** N_2H_4 **(c)** PH_3
14. (a) NH_3, N_2H_4 **(b)** HNO_3
(c) HNO_2 **(d)** HNO_3
16. (a) $2I^-(aq) + SO_4^{2-}(aq) + 4H^+(aq) \rightarrow$
$$I_2(s) + SO_2(g) + 2H_2O$$
(b) $2I^-(aq) + Cl_2(g) \rightarrow I_2(s) + 2Cl^-(aq)$
18. (a) $3HClO(aq) \rightarrow Cl_2(g) + HClO_2(g) + H_2O$
(b) $2ClO_3^-(aq) \rightarrow ClO_4^-(aq) + ClO_2^-(aq)$
20. (a) $Cl_2(g) + 2Br^-(aq) \rightarrow 2Cl^-(aq) + Br_2(l)$
(b) NR   **(c)** NR   **(d)** NR
22. (a) $Pb(N_3)_2(s) \rightarrow 3N_2(g) + Pb(s)$
(b) $2\,O_3(g) \rightarrow 3\,O_2(g)$
(c) $2H_2S(g) + O_2(g) \rightarrow 2S(s) + 2H_2O$
24. (a) $Cd^{2+}(aq) + H_2S(aq) \rightarrow CdS(s) + 2H^+(aq)$
(b) $H_2S(aq) + OH^-(aq) \rightarrow H_2O + HS^-(aq)$
(c) $2H_2S(aq) + O_2(g) \rightarrow 2H_2O + 2S(s)$
26. (a) $2H^+(aq) + CaCO_3(s) \rightarrow CO_2(g) + H_2O + Ca^{2+}(aq)$
(b) $H^+(aq) + OH^-(aq) \rightarrow H_2O$
(c) $Cu(s) + 4H^+(aq) + SO_4^{2-}(aq) \rightarrow$
$$Cu^{2+}(aq) + 2H_2O + SO_2(g)$$
28. P
30. (a) Pass O_2 through 10^4 V of electric discharge.
(b) Freeze liquid sulfur at 119°C and cool quickly to room temperature.
(c) Heat white phosphorus in the absence of air to about 300°C.

32. (a) $:\ddot{C}l-\ddot{O}-\ddot{C}l:$ **(b)** $:\ddot{O}-N\equiv N:$

(c) :P$\overset{\overset{\textstyle P}{\diagup\diagdown}}{\underset{\underset{\textstyle P}{\diagdown\diagup}}{\text{---}|\text{---}}}$P: **(d)** $:N\equiv N:$

34. a and b

36. (a) $\left(:\ddot{O}-N=\ddot{O}:\atop \underset{:O:}{|} \right)^{-}$ **(b)** $\left(H-\ddot{O}-\underset{\underset{:O:}{|}}{\overset{\overset{:\ddot{O}:}{|}}{S}}-\ddot{O}: \right)^{-}$

(c) $\left(H-\ddot{O}-\underset{\underset{:\underset{\cdot\cdot}{O}:}{|}}{\overset{\overset{:\ddot{O}:}{|}}{P}}-\ddot{O}-H \right)^{-}$

38. (a) $:\ddot{O}-\underset{\underset{:O:}{\|}}{N}-\ddot{O}-\underset{\underset{:O:}{\|}}{N}-\ddot{O}:$ **(b)** $H-\ddot{O}-\underset{\underset{:\underset{\cdot\cdot}{O}:}{|}}{N}=\ddot{O}$

(c) $\left(:\ddot{O}-\underset{\underset{:\underset{\cdot\cdot}{O}:}{|}}{\overset{\overset{:\ddot{O}:}{|}}{S}}-\ddot{O}: \right)^{2-}$

40. 14.3 L
42. 0.204 M
44. 7.7×10^4 L; 1.5×10^5 g
46. 12 L; 3.53
48. 4.28, 0.10 M
50. 1.9×10^{-3}
52. 0.012 g/100 mL
54. yes; 0 K
56. (a) yes   **(b)** 2×10^{16}
58. 510 K
60. 2.26×10^5 g
62. 27.4 kg
64. b, c, d
66. 4.6
68. (a) dispersion
(b) dispersion, dipole
(c) dispersion, H— bonds
(d) dispersion, H— bonds
(e) no intermolecular forces; not a molecule
70. (a) +3   **(b)** +4   **(c)** +5   **(d)** −3
72. (a) $HClO$ **(b)** S, $KClO_3$
(c) NH_3, NaClO   **(d)** HF
74. (a) See text page 612.
(b) has an unpaired electron
(c) H^+ is a reactant.
(d) C is a product.
75. density of sulfur; depth of deposit, purity of S
76. no
77. 1.30%
78. Assume reaction is: $NaN_3(s) \rightarrow Na(s) + \frac{3}{2}N_2(g)$
Assume 25°C, 1 atm pressure
mass of NaN_3 is 35 g

Chapter 22

2. (a) alkene   **(b)** alkyne   **(c)** alkane
4. (a) C_9H_{16} **(b)** $C_{22}H_{44}$ **(c)** $C_{10}H_{22}$
6. (a) alcohol   **(b)** ester   **(c)** ester, acid

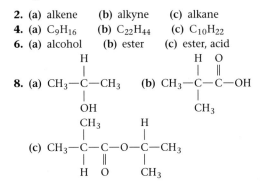

10. C=O (ketone or aldehyde)

12.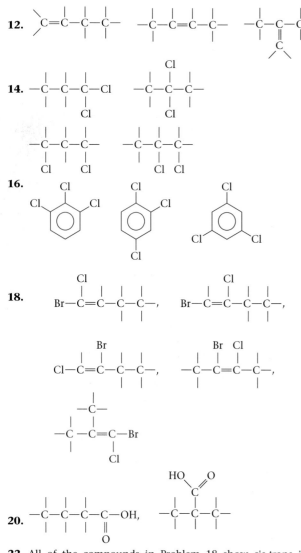

14.

16.

18.

20.

22. All of the compounds in Problem 18 show cis-trans isomerism except

where there are two —CH₃ groups attached to the same carbon

24. b, c

26. c and d

28. (a)—center carbon   **(b)** none
   (c)—carbon atom at right

30. three pairs of electrons spread around the ring

32. (a) $CO(g) + 2H_2(g) \rightarrow CH_3OH(g)$
   (b) $C_2H_5OH(aq) + O_2(g) \rightarrow CH_3COOH(aq) + H_2O$
   (c) $CH_3CH_2OH(aq) + CH_3COOH(aq) \rightarrow$
$$CH_3CH_2{-}\underset{\underset{O}{\|}}{C}{-}CH_3(aq)$$

34. (a) no multiple bonds
   (b) sodium salt of a long-chain carboxylic acid
   (c) twice the volume of alcohol present
   (d) ethanol with additives (mainly methanol) that make it unpalatable, sometimes even poisonous

35. $C_{27}H_{46}O$

36.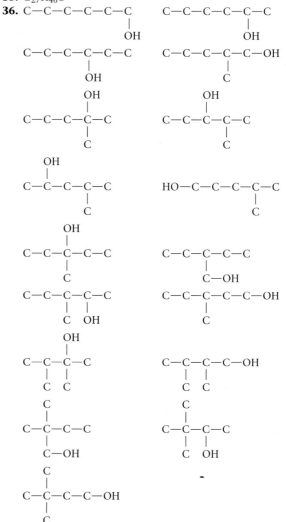

37. approximately 6 g (Assume $H_2O(l)$ is a product; neglect heat capacity of pan.)

Note: Boldface terms are defined in the context used in the text. Italic page numbers indicate figures; t indicates table; q indicates an end-of-chapter question.

A

A (mass number), 33
Abbreviated electron configuration
Brief notation in which only those electrons beyond the preceding noble gas are shown. The abbreviated electron configuration of the Fe atom is [Ar] $4s^2 3d^6$, 158
Absolute temperature, 10, 116
Absolute zero, 10
Absorption spectrum, 18, *19*
Abundance (elements), 4t
Abundance (isotopic) Mole percent of an isotope in a natural sample of an element, 57, 58
  of boron, 77q
  of chlorine, 58, *58*
  of chromium, 77q
  of neon, 77q
  of oxygen, 76q
  of strontium, 76q
  of sulfur, 58
  use in calculating atomic mass, 58
Acetaldehyde, 315, 635t
Acetaminophen, 78q
Acetate ion, 43t, 391t, 402t
Acetic acid *389*, 391t, *393*, 405t
  glacial, 436q
  reaction with strong base, 425, 428t
  titration with strong base, 95, *95*, 425, 428t
Acetic anhydride, 80q
Acetol, 301
Acetone, 122, 635t
Acetylacetonate ion, 457q
Acetylene (ethyne), 199, 632, 633, *633*
  bonding in, 183, 632
  decomposition of, 633
  hybridization of, 206
  pi and sigma bonds, 207, *208*, 632
  preparation of, 632, 633, *633*
  welding, *633*
Acetylperoxide, 212q
Acetylsalicylic acid, 61
Acid A substance that dissolves in water to produce a solution in which $[H^+]$ is greater than $10^{-7}\,M$, 91–97, 382–411
  Arrhenius, 92
  Brønsted-Lowry, 383, 384
  carboxylic, 404, 405

conjugate, 385
Lewis, 439
organic, 404, 405, 638
polyprotic, 396, 397, 397t
reacting species for, 94t
strong, 92, *92*, 387
  reaction with strong base, 93, 423, *423*, 424, 428t
  reactions with weak bases, 94, 94t, 95, 426, *427*, 428t
weak, 93, *389–397*
  reactions with strong bases, 94, 94t, 425, *425*, 426
Acid-base indicator, 95, 96, *389*, 420–422, *421*
Acid-base reactions, 93–95
  writing of equation for, 94t, 428t
Acid-base titration Procedure used to determine the concentration of an acid or base. The volume of a solution of an acid (or base) of known molarity required to react with a known volume of base (or acid) is measured, 95, *95*, 96, 422–429
  strong acid–strong base, 423, *423*, 424, 428t
  strong acid–weak base, 426, *427*, 424, 428t
  weak acid–strong base, 425, *425*, 426, 428t
Acid equilibrium constant, K_a The equilibrium constant for the ionization of a weak acid, 391, 391t, A.5
Acid rain, 2, 387, 429, 430
Acid strength, 92, 401t, 611–613
  hydrohalic acids, 606, 607
  oxoacids, 611–618
Acidic ion Ion that forms H^+ ions in water. The ammonium ion is acidic because of the reaction $NH_4^+(aq) \rightleftharpoons H^+(aq) + NH_3(aq)$, 402, 402t, 404
Acidic oxide Nonmetal oxide which reacts with water to form an acid, 610
Acidic solution An aqueous solution with a pH less than 7; *91*, 385, 386, 481t, 520t, 521t
Acidosis, 387
Acne, 590
Acre, 25q
Acropolis, 430

Actinides Elements 89 (Ac) through 102 (No) in the periodic table, 160, 161
Activated complex A species formed by collision of energetic particles, that can react to form products, 326
Activation energy The minimum energy that must be possessed by a pair of molecules if collision is to result in reaction, 323–331
  Arrhenius equation, 328–331
  catalyzed reaction, *331, 335*
  diagrams, *326, 331, 335*
Activity, nuclear Rate of radioactive decay; number of atoms decaying per unit time, 554–556
Activity, thermodynamic, 351
Actual yield, 73
Addition, uncertainties in, 13
Addition polymer Polymer produced by a monomer, usually a derivative of ethylene, adding to itself; no other product is formed, 272
Addition reaction, 632, *632*
Adenosine diphosphate (ADP), *503*
Adenosine triphosphate (ATP), *503*
Adrenaline, 305q
Age
  organic matter, 555, *555*, 556
Air
  composition, 599
  fractional distillation, 599
  liquefaction, 599
Air pollution, 338, 429, 430
  carbon monoxide, 455
  nitrogen oxides, 429
  sulfur oxides, 429
  sulfuric acid, 429, 430
Airbags, 135
Alchemy, 16
Alcohol A compound containing an OH group attached to a hydrocarbon chain. The simplest example is CH_3OH, 40, 636–638
  in water, solubilities of, 287, 287t
Alcohol dehydrogenase, 590
Alcoholic beverages
  ethyl alcohol in, 637
  intoxication from, 48, 49
  proof, 637
Aldehyde, 635t
Alizarin yellow, 434q

Alkali metal A metal in Group 1 of the periodic table, 36, 581–584
  chemical properties, 581t
  electron configurations of, 158t, 160
  oxidation number of, 99
  reaction with halogens, 581t
  reaction with hydrogen, 581t, 582
  reaction with nitrogen, 581t
  reaction with oxygen, 581t, 583, 584
  reaction with sulfur, 581t
  reaction with water, *506*, 582, *582*
Alkaline, 385
Alkaline dry cell, 537
Alkaline earth metal A metal in Group 2 of the periodic table, 36, 581–584
  chemical properties, 581t
  electron configurations of, 158t, 160
  oxidation number of, 99
  reaction with halogens, 581t
  reaction with hydrogen, 581t, 582
  reaction with nitrogen, 581t
  reaction with oxygen, 581t, 583, 584
  reaction with sulfur, 581t
  reaction with water, 582
Alkaloid, 93, *93*, 94
Alkane Hydrocarbon containing only single carbon-carbon bonds. The simplest example is methane, CH_4, 627–631
  isomerism, 628, 629
  nomenclature, A.15, A.16
Alkene A hydrocarbon containing one carbon-carbon double bond. The simplest alkene is C_2H_4, 631, 632
Alkyl group A hydrocarbon group derived from the alkane by loss of a hydrogen atom. An example is the methyl group, CH_3—, 638
Alkyne A hydrocarbon containing one carbon-carbon triple bond. Example: $HC\equiv CH$, 631, 632
Allotrope One of two or more forms of an element in the same physical state. Graphite and diamond are allotropes of carbon; O_2 and O_3 are allotropes of oxygen, 20, 600–603, 619
  arsenic, 619
  carbon, 20, 21, 263
  oxygen, 601
  phosphorus, 509, 601, 602
  selenium, 619
  sulfur, 602, 603
  tin, 509q
Alpha particle A helium nucleus; He^{2+} ion, 31
Alpha radiation, *35*, 549
Alpha scattering, *32*
Alum, 79q
Aluminum

alloys of, 142q
and iron oxide, 230, *230*
corrosion of, in air, 539
crystal structure, 277q
preparation of, 4, 5, 576, 577
reaction with oxygen, 539
Aluminum hydroxide, *473*, 474
Aluminum(III) ion, 41
  acidity of, 390, 391t, *395*
  complexes of, 442t, 470t
  hydration of, 390, *390*
  qualitative analysis, 472t, 473
  toxicity, 430
Aluminum oxide, *41*, 576
Amalgam, 266q
Americium-241, 551, 553, 554
Amethyst, *264*
Amine Organic compound in which there is an —N— functional group.

An example is methylamine, CH_3NH_2, 40, 93–95, 105
  primary, 105t
  reaction with strong acid, 94, 105
  secondary, 105t
  tertiary, 105t
Ammonia
  base 93, 604
  bond angle, 195
  Brønsted-Lowry base, 398, 604
  buffer with ammonium ion, 416t
  complexes of, 470t, 604
  entropy, 487, *487*, 488
  formation constants, complexes, 453t
  gas diffusion of, hydrogen chloride and, *130*
  hydrogen bonding, 260t
  Lewis base, 439, 604
  ligand in complex ions, 440, 604
  molecule, geometry of, *39*
  nitric acid from, 615
  precipitating agent, 604
  preparation, Haber, 372, 373
  properties, 604
  reaction with acid, 94, 426, 427, *427*
  reaction with hypochlorite ion, 604
  reaction with metal cations, 440, 604
  reaction with water, 93, 398, 604
  reducing agent, 604
  titration with HCl, 427, *427*
  uses of, 69, 604, 615
Ammonium carbamate, 378q
Ammonium chloride, 500
Ammonium dichromate, 589, *589*
Ammonium iodide, 403
Ammonium ion, 41, 43t
  acid-base properties, 402t
  buffer with ammonia, 416t
  qualitative analysis, 474
  reaction with hydroxide ion, 474

Ammonium nitrate, *482, 489*
Ampere Rate of flow of electric current such that one coulomb passes a given point in one second, 532t
Amphetamine, 644
Amphiprotic species Capable of acting as either a Brønsted-Lowry acid or base, 384
Amphoteric species Metal hydroxide that dissolves in both strong acid (H^+) and strong base (OH^-), 474
Amplitude Height of a standing wave, *145*, 151
Amu. *See* atomic mass unit
Analysis, chemical
  qualitative, 472–474
  simplest formula from, 64–66
Angstrom (A), 14t
Anion Species carrying a negative charge. Examples include Cl^- and OH^-, 40
  basic, 402t, 403
  monatomic, 40, 41
  naming, 44
  polyatomic, 43, 43t
  pronunciation, 40
  size relative to cation, 167, *168*
  spectator, 89
  voltaic cell, 514
Anode Electrode at which oxidation occurs and toward which anions move, 514, 531
Answers to problems, A.24–A.42
Antacid, *96*, 97, 98t
Anthracene, 635
Antibonding orbital(s), A.18
Antifreeze solution(s), "permanent," 294, 638
Antilogarithm, A.11
Antimony, *71*
  reaction with iodine, 70, *71*
Antimony(III)
  complexes with hydroxide, 470t
  qualitative analysis, 472t, *473*
  sulfide, *473*
Antimony triiodide, *55, 71*
Approximations, successive, 395
Aqua regia, 111q, 587
Archaeology, neutron activation analysis in, 552
Arene, 633
Argon, 208
Armbuster, Peter, 550
Aromatic hydrocarbon A hydrocarbon containing one or more benzene rings, 633–635
Arrhenius, Svante, 92, 97, *97*
Arrhenius acid Species that, upon addition to water, increases $[H^+]$, 92
Arrhenius base Species that, upon addition to water, increases $[OH^-]$, 92

Arrhenius equation Equation that expresses the temperature dependence of the rate constant: $\ln k_2/k_1 = E_a(1/T_1 - 1/T_2)/R$, 328–330

Arsenic, 619

Arsenic(III) oxide, 619

Asbestos, 264

Aspartame, 333

Aspirin, 392
  buffered, 414
  ionization constant, 392
  structure, 213q

Assay, *285*

Asymmetric synthesis, 644

-ate suffix, 47

Atherosclerosis, 645

Atmosphere, standard Unit of pressure equal to 101.325 kPa; equivalent to the pressure exerted by a column of mercury 760 mm high, 116

Atmospheric pressure, 116, *116*

Atom Smallest particle of an element that retains the chemical properties of that element, 29–34
  central, in Lewis structures, 183
  components of, 31–35
  decaying per second, 555
  electron configuration of, 156-162
  electronegativity of, 170, 171, *171*
  formal charge, 187, 188
  individual, masses of, 59, 60
  ionization energy of, 169, *169*, 170
  Lewis structure of, 182t
  light and, 144–148
  masses of, individual, 59, 60
  noble-gas structure of, 181
  orbital diagram of, 162–164
  in periodic table, properties of, 161–171
  properties of, 29
  quantum mechanical, 150, 151
  radius of, 166, *166*, *167*, 168
  spectrum of, 147, *147*, 148
  terminal, in Lewis structures, 183

Atom ratio, compound, 63

Atomic bomb, 561, *561*

Atomic mass Average mass of an atom compared to that of another element, based upon the C-12 scale, 56–60
  of chlorine, 58
  determination by electrolysis, 545q
  isotopic abundances, 57–59
  measurement of, *57*
  of sulfur, 58

Atomic mass unit (amu) The unit used to express atomic masses; 1/12 of the mass of a carbon-12 atom, 56

Atomic number Number of protons in the nucleus of an atom, 33

Atomic orbital An electron cloud with an energy state characterized by given values of **n**, ℓ, and **m**$_\ell$, has a capacity of two electrons of opposite spin, 151, 153, *153*, 154, 203–208
  hybridization of. *See* hybrid orbital order of filling of, *157*

Atomic orbital model. *See* valence bond model

Atomic oxygen, 172

Atomic radius One half of the distance of closest approach between two nuclei in the ordinary form of an element, 166, *166*, *167*, 168
  experimental determination of, *166*
  relation of, to ionization energy, 169, 170
  trends of, in periodic table, 166

Atomic spectrum Diagram showing the wavelength at which light is emitted by excited electrons in an atom, 147, *147*, 148
  of hydrogen, 147, 148, 148t, *149*, *149*
  of mercury, *147*
  of neon, 209
  of sodium, 147, *147*

Atomic theory Dalton's theory of the atomic nature of matter, 29
  modern postulates of, 29

Atomic weight. *See* atomic mass

ATP. *See* adenosine triphosphate

Attractive forces, 257–262
  dipole, 258, 259, *259*
  dispersion, 257–258
  of gases, 134
  of hydrogen bonds, 259–262

Aufbau (building up) principle Rule stating that sublevels are filled in order of increasing energy, 156–158

Aurora borealis, 171, *171*,*172*, *172*

Average speed, of gas molecules, 128–130

Average translational energy, 129

Avogadro, Amadeo, 125, *125*

Avogadro's law, 119, 125

Avogadro's number (N_A) 6.022×10^{23}; the number of units in a mole, 59, 60
  calculation from cell dimensions, 278

B

Baking soda, *60*, *62*

Balance, analytical, 9, *9*

Balanced equation An equation for a chemical reaction that has the same number of atoms of each element on both sides, 67
  acid solution, 101
  basic solution, 101
  redox, 100–102
  trial and error method, 67, 68

Balloon, 118
  buoyant force on, 142q
  gas density and, 123, *123*

Balmer, Johann, 148

Balmer series, 148t, *149*
  equation, 177q

Band spectra, 172, *172*

Band theory, A.22

Bar Unit of pressure = 10^5 Pa = 1.013 atm, 117

Barbituric acid, 409q

Barite, *466*

Barium
  chemical properties, 581t
  reaction of, with oxygen, 583

Barium ion
  qualitative analysis, 472t, 474
  reaction with sulfate, *466*

Barium peroxide, 583

Barium sulfate, *466*

Barometer
  mercury, 116, *116*

Bartlett, Neil, 209

Base Compound that dissolves in water to give a solution in which [OH^-] is greater than 10^{-7} M, 92, 382–411
  Arrhenius, 92
  Brønsted-Lowry, 383, 384
  conjugate, 385
  Lewis, 439
  organic, 93
  reacting species for, 94t
  strong, 92t, 387
    reactions with strong acids, 423, *423*, 424, 428t
    reactions with weak acids, 94, 94t, 425, 426, 428t
  weak, 93
    reactions with strong acids, 94, 94t, 425, *425*, 426

Base equilibrium constant, K_b The equilibrium constant for the ionization of a weak base, 391t, 398, A.5

Basic ion An anion that forms OH^- ions in water; the CO_3^{2-} ion is basic because of the reaction $CO_3^{2-}(aq) + H_2O \rightleftharpoons OH^-(aq) + HCO_3^-(aq)$, 402, 402t, 404

Basic oxygen process, *577*, 578

Basic solution An aqueous solution with a pH greater than 7 (at 25°C), *91*, 385, 386
  balancing redox equation in, 101
  standard potentials in, 520t

Basketball, 138q

Battery, 537, 538
  lead storage, 537, *537*, 538
  Nicad, 538

Bauxite, 576

Becquerel (unit), 555

Becquerel, Henri, 553

Beer, 293, 637

Belt of stability, 34, *35*

Bends, 139q, 291

Bent molecule Molecule containing three atoms in which the bond angle is less than 180°; examples include H_2O and SO_2, 195, 196

Benzaldehyde, 381q

Benzedrine, 644

Benzene
   boiling point, freezing point constants, 294t
   derivatives of, 633–635
   properties, 276q, 634
   resonance, 186
   toxicity, 634

Benzoic acid, 391t, 634

Benz[a]pyrene(3,4-), 635

Berthelot, P. M., 484

Beryllium
   chemical properties, 581t
   reactivity, 581

Beryllium fluoride, 193, *194*, 196t

Beryllium hydride, 196

Berzelius, Jons Jacob, 97, 125

Beta particle, 35

Beta radiation Electron emission from unstable nuclei, *35*

Bicarbonate of soda, 60, 62

Bimolecular step, 334

Binary molecular compound, 45, 46

Binding energy Energy equivalent of the mass defect; measure of nuclear stability, 559, 560

Bipyramid, triangular, 193, *194*

Bis, 553, A.14

Bismuth, 580

Bismuth(III) ion, 472t, 473

Bismuth subsalicylate, 788

Bismuth(III) sulfide, 473, *473*, 580

Bjerrum, Niels, 148

Blast furnace, *577*

Bleach, household, 111q

Blister copper, 578, 579

Blood alcohol, test for, 48, 49

Blood buffers, 420, 434q

Blood plasma, 414

Blood pressure, 590, 609

Body-centered cubic (BCC) A cubic unit cell with an atom at each corner and one at the center, 268, *268*, *269*

Bohr, Niels, 148

Bohr equation, H atom, 148

Bohr model Model of the hydrogen atom developed by Niels Bohr; it predicts that the electronic energy is $-R_H/n^2$, where R_H is the Rydberg constant and n is the principal quantum number, 148–150

Boiling point Temperature at which the vapor pressure of a liquid equals the applied pressure, leading to the formation of vapor bubbles, 17, 251

concentration and, 292–294
constant, molal, 294t
elevation, 292–294, *293*
hydrogen compounds of, 260, *260*
intermolecular forces and, 157–262
molar mass and, 258, 258t
molecular substances of, 257–262
nonpolar versus polar substances of, 258, *259*
normal, 251
pressure and, 251, 252
vapor bubbles, *251*

Boiling point constants, (k_b), 294t

Boiling point elevation (ΔT_b) Increase in the boiling point caused by addition of a nonvolatile solute, 292–294, *293*

Boltzmann, Ludwig, 128

Bomb calorimeter Device used to measure heat flow, in which a reaction is carried out within a sealed metal container, 221, 222, *222*

Bombardment reaction, 550, 551

Bond A linkage between atoms
   covalent, 38, 39, 179–214
   double, 183
   hydrogen, 159–262
   ionic, 41
   metallic, 265, 266
   multiple, 183
   nonpolar, 200
   pi, 207, 208
   polar, 200
   sigma, 207, 208
   single, 183
   triple, 183

Bond angle The angle between two covalent bonds, 192, 193, *194*, *198*

Bond energy. *See* bond enthalpy

Bond enthalpy The enthalpy change (kJ/mol) when a particular type of bond is broken in the gas state, 232–235, 234t

Bonding orbital, A.18

Borazine, 212q

Boric acid, 109q, 189

Boron, 77q, 278q

Boron trifluoride, 188t, 189

Bottled gas, *629*

Boyle, Robert, 118

Boyle's law Relation stating that at constant T and n, the pressure of a gas is inversely proportional to its volume, 118, *118*

Brackett series, 175q

Bragg, W. H., and W. L., 271

Branched chain alkane Saturated hydrocarbon in which not all the carbon atoms are in a single continuous chain, e.g., C—C—C, 627, 628
$\big|$
C

Brass, *6,*

Breath analyzer, 48, 49

Bromide ion
   in brines, 600
   oxidation of, 600
   reaction of, with chlorine, 518, 526, *526*

Bromine, 248
   atoms, 338
   burns, 599
   chemical reactivity, 599
   color of, *248, 597*
   oxides, 608t
   oxoacids, 611t
   preparation of, 518, 526
   properties of, 597t
   reaction of, with water, 623q
   toxicity of, 599

Bromthymol blue, 421, 422t

Brønsted, Johannes, 383

Brønsted-Lowry acid A species that donates a proton in an acid-base reaction; in the reaction HB(aq) + A^-(aq) $\rightleftharpoons$ HA(aq) + B^-(aq), HB is a Brønsted-Lowry acid, 383
   relative strengths, 401t

Brønsted-Lowry base A species that accepts a proton in an acid-base reaction; in the reaction HB(aq) + A^-(aq) $\rightleftharpoons$ HA(aq) + B^-(aq), A^- is a Brønsted-Lowry base, 383
   relative strengths, 401t

Buchner funnel, 8

Buckminsterfullerene, 21, *21*

Buckyball, 21

Buffer A system whose pH changes only slightly when strong acid or base is added. A buffer ordinarily consists of a weak acid and its conjugate base, present in roughly equal amounts, 413–420, *414*
   calculation of pH, 414–416
   choice of system, 416, 417
   effect of added acid or base, 417, *418*, 419

Buffer capacity Amount of strong acid or base that can be added to a buffer without causing a drastic change in pH, 419, 420

Buffer systems, different pH of, 416t

Buret, 8, *8*

Butane, 628, *628*

Butyric acid, 409q

C

Cabbage juice, 421

Cadmium, 538
   control rods, 562
   oxidation to cadmium(II), 586t
   reaction of, with oxygen, 584t

Cadmium(II) ion
   complexes, 442t, 470t

qualitative analysis, 472, *473*
toxicity, 590
Cadmium(II) sulfide, *473*
Caffeine, 105, *105*, 242q
Calcite, *498*
Calcium
analysis for, 113q
chemical properties, 581t
reaction of, with water, 582
Calcium carbide, 632, *633*
Calcium carbonate, *61*, 137, *498*
decomposition, thermodynamics of, 498
reaction of, with acids, 607
Calcium chloride
in highway ice removal, 299
melting point of, 42
reaction of, with sodium carbonate, 89
Calcium fluoride, 475, *475*
Calcium hydride, 401, 582
Calcium hydroxide, 92t, 490
formation from calcium, 582
formation from calcium hydride, 582
formation from calcium oxide, 583
reaction with sulfur dioxide, 430
Calcium ion
complex with EDTA, 455
in nutrition, 590
qualitative analysis of, 474
Calcium oxide, 583
Calcium phosphate, 462, 463, 618
Calcium propionate, 639
Calcium silicate. *See* slag
Calcium sulfate, 430, 494, 618
Calculators, electronic, A.10, A.12
Calculus, 321, 349q
Calorie, 218
Calorimeter Device used to measure
heat flow, 221, 222
bomb, 221, *221*, 222
coffee-cup, 220, 221
heat capacity of, 222
Calorimetry, 219–222
Calvin, Melvin, 191
Camphor, 294t, 305q
Cancer
following radioactive exposure, 553, 566
treatment with coordination com-
pounds, 444
treatment with radioactive isotope, 551,
551
Candy, *290*
Cannizzaro, Stanislao, 125
Capacity for electrons
orbital, 155
principal level, 156, 156t
sublevel, 156, 156t
Caproic acid, 408q
Carbohydrate, 238
Carbon
allotropy, 20, *20*, 21, *21*, 263
in metallurgy of iron, 577
molecular allotrope, 21, *21*

organic compounds of, 625–649
reaction of, with oxygen, 577, 578
Carbon-11, 552t
Carbon-12 scale Atomic mass scale
where the C-12 isotope is assigned a
mass of exactly 12 amu, 56, 57
Carbon-14, 555, 556
Carbon dioxide
atmospheric, 504
critical behavior of, *252*, 253
from carbonate ion, 224
geometry of molecule, 197
global warming, 504
melting point of, 24q
molar volume of gas, 133t
sublimation of, 255
supercritical, 253
Carbon monoxide
preparation of, 577
reaction with hemoglobin, 455
reaction with metal oxides, 576–578
reaction with nitrogen dioxide, 323, 324
reaction with oxygen, 227
toxicity of, 455
Carbon tetrachloride, 202, *246*
Carbonate ion, 43t
acid-base properties of, 391t, 402t
buffer with hydrogen carbonate ion, 420
precipitation, *88, 89*
reaction with acid, 232, *232*, 469, 470
reaction with water, 398
Carbonate ores, *575*
Carbonated beverages, bottling of, 291
Carbonic acid, 397, 398
buffer systems, 416
equilibrium constants, 398t
Carboxylic acid Organic compound
containing the —C—OH group,
$\overset{\parallel}{\underset{O}{}}$
404, 638
salts, 638, 639
Carcinogens
benzene, 634
benzo[a]pyrene, 635
naphthalene derivatives, 635
Carlsberg brewery, 385
β-Carotene, 306q
Carrier gas, chromatography, 7
Cassiterite, 509q
Catalase, 303
Catalysis Process in which a species,
called a catalyst, affects the reaction
rate without being consumed,
331–333
heterogeneous, 331–332
homogeneous, 332–333
Catalytic converter, *332*
Cataracts, 338
Cathode Electrode at which reduction
occurs and toward which cations
move, 514, 531

Cathode-ray tube, 31, *31*
Cation Ion with a positive charge, such
as Na^+, 40–43
acidic, 390, 391, 402t
main-group, 42t
monatomic, 41
naming, 44
polyatomic, 43t
post-transition metal, 42, 43
pronunciation, 40
qualitative analysis groups, 472–474
size of, relative to anion, 167, *168*
spectator, 89, 402t
transition metal, 42–44, *87*, 165, 166
Cation exchange, 264
Cavendish, Henry, 115
Cell notation, 516
Cell voltage Voltage associated with an
electrochemical cell, 518–531
concentration and, 528–531
Celsius, Anders, 9
Celsius degree Unit of temperature
based on there being 100° between
the freezing and boiling points of
water, 9, *9*
Cement, 578
Centi- Metric prefix indicating a multiple
of 10^{-2}, 8t
Centimeter, 8
Central atom Atom at the center of a
molecule, to which two or more
atoms are bonded. C is the central
atom in CH_4, 183, 196t
Central metal cation Monatomic
metal cation to which all the lig-
ands are bonded in a complex ion,
439
Cerium oxide, 161
Cesium, 581t
Cesium chloride, *270*
CFCs, 338
and uncatalyzed decomposition of
ozone, 338, 339
Chain reaction, 562
Chalcocite, 578
Chalcopyrite, 593q
Charcoal, 4
Charles, Jacques, 118
Charles's and Gay-Lussac's law
Relation stating that at constant P
and n, the volume of a gas is directly proportional to its absolute
temperature, 118, *118*, 120
Chelate, 440, 441, 454, 455
Chelating agent Complexing ligand
that forms more than one bond
with a central metal atom; the com-
plex formed is called a chelate, 440,
441
Chemical analysis
qualitative, 472–474
simplest formula from, 64–66

Chemical equation Expression that describes the nature and relative amounts of reactants and products in a reaction. *See* equation, net ionic, 67, 68

Chemical formulas, *see* Formula

Chemical kinetics Study of reaction rates, 309–349

Chemical property Property of a substance that is observed during a chemical reaction, 17

Chemical reaction A process in which one or more substances, called reactants, are converted to product(s), 17

Chemistry Beyond the Classroom
  Acid rain, 429, 430
  Airbags, 135
  Amines, 105, 106
  Arsenic, 619
  Aurora, 171, *171*, 172
  Biological effects of radiation, 566, 567
  Breath analysis for ethyl alcohol, 48, 49
  Chelates: natural and synthetic, 440, 441
  Cholesterol, 645, 646
  Corrosion of metals, 539, 540
  Energy balance in the human body, 238, 239
  Essential metals in nutrition, 590, 591
  Fluoridation of drinking water, 475
  Fullerenes, 21
  Global warming, 504, 505
  Hydrates, 74, 75
  Industrial applications of gaseous equilibria, 372, 373
  Maple syrup, 309
  Noble gases, 208, 209
  Organic acids, 404, 405
  Ozone story, 338, 339
  Polymers, 272, 273
  Selenium, 619
Chemistry: The Human Side
  Arrhenius, Svante, 97
  Avogadro, Amadeo, 125
  Curie, Marie and Irene, 553
  Dalton, John, 30
  Eyring, Henry, 327
  Faraday, Michael, 534
  Gibbs, J. Willard, 493
  Haber, Fritz, 373
  Hodgkin, Dorothy, 271
  Lavoisier, Antoine, 16
  Lewis, G. N., 191
  Seaborg, Glenn, 160
  Werner, Alfred, 446
Chemotherapy, 444
Chernobyl, 563

Chiral center Carbon atom bonded to four different groups, 643

Chiral drug, 644

Chlorate ion, 43t, 613–615
  disproportionation, 614, 615
Chlorides
  ores, *575*
  solubility of, 88
Chloride ion, 600
  ligand in complex ions, 440
Chlorination, 598
Chlorine, 5
  chemical reactivity of, 598, 599
  color of, *597*
  disproportionation of, 598
  hydrate, 74
  isotopic composition of, 58, *58*
  oxides, 338, 343q, 608t
  oxoacids, 611t, 612t
  oxoanions, 43t, 45t, 613–615
  preparation of, from sodium chloride, 535, 536, *536*, 600
  properties, 597t
  reaction with bromide ions, 518, 526, *526*, 600
  reaction with copper, *598*
  reaction with iodide ions, 600
  reaction with nitrogen oxide, 336
  reaction with water, 598, 599
  toxicity of, 599
  uses of, 536
Chlorine dioxide, 608
Chlorine monoxide, 338
Chlorine water, 598
Chlorine-bromide ion voltaic cell, *518*
Chlorite ion, 45t
Chloroform, 275q
Chlorophyll, 82q, *437*
Cholesterol, 645, 646
Chromate ion, 43t, 589, *589*
  conversion to dichromate, 589
  toxicity, 589
Chromatography, 6, 7
Chrome plating, 532
Chromium
  chemistry of, 589
  electron configuration of, 161
  isotopic composition of, 77q
  oxides of, 30
  plating, 532
  reaction with acids, 586
  reaction with oxygen, 584t
Chromium(III) hydroxide, *473*
Chromium(II) ion, 586, 586t, 587t, 588
Chromium(III) ion, 587t, 588, *589*
  complexes of, 442t
  in nutrition, 591t
  qualitative analysis of, 472, 473
Chromium(III) oxide, *30*
Chromium(VI) oxide, *30*
Cinnabar, 5

Cis isomer Geometric isomer in which two like groups are relative closely to another, e.g.,

            a   a
             \ /
              C
             / \
            b   b

  complex ions, 444, 445
  organic, 641, 642
Cis-platin, 78q, 444
Citric acid, 405t, 412
Clapeyron, B. P. E., 250
Clathrate, 74
Clausius, Rudolph, 128, 250

Clausius-Clapeyron equation An equation expressing the temperature dependence of vapor pressure: $\ln(P_2/P_1) = \Delta H_{vap}(1/T_1 - 1/T_2)/R$, 250, 251

Coal, 429
Cobalt
  isotopes of, 32
  oxidation to cobalt(II), 586t
  reaction of, with oxygen, 584t
Cobalt-60, 551
Cobalt(II) chloride, *74, 165*
Cobalt(II) ion
  complexes of, 442t, 450, *450*
  in nutrition, 591t
  oxidation of, to cobalt(III), 587t
  qualitative analysis, 472t, 473
  reduction to cobalt, 587t
Cobalt(II) oxide, 584
Cobalt(III) ion
  complexes, *367*, 442, 442t, 445, 447
  in nutrition, 571t
  reduction of, to cobalt(II), 587t
Cobalt(II) sulfide, *473*
Codeine, 409q

Coefficient A number preceding a formula in a chemical equation, 68
  fractional, 68
  in balanced equation, 67, 68

Coefficient rule Rule which states that when the coefficients of a chemical equation are multiplied by a number n, the equilibrium constant is raised to the nth power, 356

Coffee-cup calorimeter Calorimeter in which essentially all of the heat given off by a reaction is absorbed by a known amount of water, 220, 221

Coke, 577
Cold pack, *224*

Colligative property A physical property of a solution that primarily depends on the concentration rather than the kind of solute particle. Vapor pressure lowering is a colligative property; color is not, 291–300
  electrolytes, 298–300
  molar mass determination from, 297, 298
  nonelectrolytes, 291–298

Collision frequency, 325
Collision model, 323–326
  reaction rate, 323–326
  steric factor, 324, *324*
Color, 18, 19t
Color wheel, 451
Combining volumes, law of, 124
Combustion, 124
Common ion effect Decrease in solubility of an ionic solute in a solution containing a high concentration of one of the ions of the solute, 467, *467*, 468
Complex ion An ion containing a central metal atom bonded to two or more ligands. The species $Ag(NH_3)_2^+$ and $Zn(H_2O)_4^{2+}$ are complex ions, 437–459
  charge of, 440, 440t
  color of, *367*, 447, 450, 451, 451t, *452*
  composition of, 438–442
  coordination number of, 439, 442t
  electronic structure of, 447–452
  formation constants, 452, 453
  geometry of, 442–446
  high spin, 450
  isomerism of, 444–446
  low spin, 449
  naming of, A.3–A.4, A.14–A.15
Compound(s) Substance containing more than one element, 5
  coordination, 438
  decomposition of, to the elements, 5, 6
  enthalpy of formation of, 229t, A.3, A.4
  free energy of formation of, A.3, A.4
  ionic. *See* ionic compounds
  molecular, naming of, 45, 46
  nonstoichiometric, 584, 585, *585*
  standard entropy of, 488, A.3, A.4
Compression, effect on equilibrium, 368–370, 370t
Concentrated solution, 84, 85
Concentration A measure of the relative amounts of solute and solvent in a solution; may be expressed as molarity, mole fraction, etc., 84
  calculation of, from solubility product constant, 462, 463
  cell voltage and, 528–531
  conversions between, 284–286
  effect on equilibrium, 366–368
  effect on free energy change, 499, 500
  reaction rate and, 313–317
  time dependence, reactant, 317–322
  units of, 281–286
Concentration cell, 546q
Concentration quotient (Q), 361, 362, 499
Condensation, 248
Condensation polymer A polymer formed from monomer units by

splitting out a small unit, most often a water molecule, 273
Condensed ring structure, 635
Condensed structural formula Formula of a molecule which indicates the functional group present, e.g., CH_3COOH for acetic acid, 39
Conductivity The relative ease with which a sample transmits electricity. Because a much larger electrical current will flow through an aluminum rod at a given voltage than through a glass rod of the same shape, aluminum is a better electrical conductor than glass, 86
  ionic, 265
  metallic, 266
Coniine, 105
Conjugate acid The acid formed by adding a proton to a base; NH_4^+ is the conjugate acid of NH_3, 384
Conjugate base The base formed by removing a proton from an acid; NH_3 is the conjugate base of NH_4^+, 384
Conservation of energy, law of, 235
Conservation of mass, law of, 30
Constant(s), A.2
Constant composition, law of, 30
Constant pressure process, 236, 237
Constant volume process, 236, 237
Contact process, sulfuric acid, 617
Control rod, 562
Conversion(s)
  concentrations, 284–286
  electrical, 532t
  energy, 532
  English system, 14t
  mass relations in reactions, 67–73
  metric, 14t
  metric-English, 14t, 15
  mole-gram, 61, 62
  multiple, 15
  pressure, 117
  solution, molarity, 85
  thermochemical, 225, 226
Conversion factor A ratio, numerically equal to 1, by which a quantity can be converted from one set of units to another, 13–15
Coordinate covalent bond, 439
Coordination compound Compound in which there is a complex ion. Examples include $[Cu(NH_3)_4]Cl_2$ and $K_4[Fe(CN)_6]$, 438
  color of, *367*, 447, 450, 451, 451t, *452*
  electrical conductivities, 446
  naming of, A.15
Coordination number The number of bonds from the central metal to the ligands in a complex ion, 439

cations, 442t
  geometry, of complex ions, 442t
Copper, *60*
  blister, 579
  corrosion of, in air, 513
  electrical conductivity of, 266
  electrolytic, 579
  electron configuration of, 161
  electroplating, 533
  electrorefining of, 579
  flame color, *143*
  metallurgy, *573*, 578–580
  oxidation to copper(II), 586, 586t
  reaction with chlorine, *598*
  reaction with nitric acid, 524, 525, *525*, 616, *616*
  reaction with oxygen, 513, 584
  reaction with sulfuric acid, 618
  uses of, 4
Copper-64, 345q
Copper(I) ion
  complexes, 442t, 443
  disproportionation of, 588
  in nutrition, 591t
Copper(II) ion
  complexes of, 438, *438*, 442t
  and nutrition, 591t
  qualitative analysis, 472t, 473, *473*
  reaction with nickel, 516, 524, *524*
  reaction with zinc, 515, *515*, 516, 519
  reduction of, to copper(I), 587t
Copper(II) oxide, *358*
Copper(II) sulfate, 5, *165*, *282*
Copper(I) sulfide, 578
Copper(II) sulfide, *473*
Core electron An electron in an inner, complete level, 168
Corrosion, 539, 540
  aluminum, 539
  copper, 513, 539
  electrochemical mechanism of, 539
  gold, 539
  iron, *539*, 540
  lead, 539
  platinum, 539
  zinc, 539
Corundum, *41*
Cosby, Bill, 486
Cosmic radiation, 555
Coulomb A unit of quantity of electricity; 9.648×10^4 C = 1 mol e^-, 532, 532t
Coulomb's law, 265
Coupled reactions Two reactions that add to give a third, 502–504
Covalent bond A chemical link between two atoms produced by sharing electrons in the region between the atoms, 38, 39, 179–214
  energy of, 232–235, 234t
  molecules, 256–262
  nature of, 180

Covalent bond *(continued)*
network, 263, 264
nonpolar, 263
polar, 200
polarity of, electronegativity and, 200
stability, 180
Cricket, chirping rate, 346q
Critical mass, 562
Critical pressure The pressure at the critical temperature; the highest vapor pressure that a liquid can have, 252, 253
Critical temperature The highest temperature at which a substance can exhibit liquid-vapor equilibrium. Above that temperature, liquid cannot exist, 252, 253
Crutzen, Paul, 338
Cryolite, 576
Crystal structures, 267–270
Crystal field model Electrostatic model of the bonding in complex ions. The only effect of the ligands is to change the relative energies of the d orbitals of the central metal atom, 447–452
Crystal field splitting energy The difference in energy between the two sets of d orbitals in a complex ion, 448–452
Cubic cell, 268–270
body-centered, *268*, 268t
face-centered, *268*, 268t
simple, *268*, 268t
Cubic centimeter, 8
Cubic meter, 10
Cucumber pickles, *296*
Curie (unit) Radiation corresponding to the radioactive decay of 3.700×10^{10} atoms/s, 555
Curie, Irene, 550, 553, *553*
Curie, Marie, 553, *553*
Curie, Pierre, 553 *553*
Curl, Robert, 21
Cyanide ion, 391t
complexes of, 440, 449, 452
in gold metallurgy, 580
Cyanogen, 378q
Cylinder, graduated, 10
Cyclohexane, 294t,
Cyclotene, 301
Cystic fibrosis, 530

D

d orbital, 154
designation, 448
geometry of, *448*
splitting in octahedral field, 449, *449*
d sublevel, 152, 154
Dacron, 273
Dalton, John, 29, *30*, 125, 126, 171

Dalton's law A relation stating that the total pressure of a gas mixture is the sum of the partial pressures of its components, 126
Davy, Sir Humphrey, 534
Dead cell, 528
Dead Sea scrolls, 556
DeBroglie, Louis, 150
Debye, Peter, 300
Decay radioactive, 34, 554–556
Dech, 458q
Deci-, 8t
Deliquescence, 298
Denatured alcohol, 637
Density The ratio of the mass of a substance to its volume, 17, 18
calculation of, of gas, 122
Dental cavities, 475
Dental filling, composition of, 266
Deoxyribonucleic acid. *See* DNA
Deuterium, 33, *33*
Deviations, from ideal gas law, 133–135
Dexedrine, 644
Diagonal stairway, periodic table, *36*, 37, 596
Diamagnetic A term indicating that a substance does not contain unpaired electrons and so is not attracted into a magnetic field, 163, 449
Diamond, 20, 263
Diatomic elements, *40*
Dice, 485
Dichromate ion, 43t
conversion of, to chromate, 589
oxidizing strength, 589
reaction with ammonium ion, 589, *589*
toxicity of, 589
Dihydrogen phosphate ion, 43t, 618
acid-base properties of, 391t, 404
buffer systems, 416t, 434q
Dilute solution, 84, 85
Dilution effects
buffer, 416
concentration of solutes, 282, *282*
precipitation, 464, 465
Dimethyl ether, 635t
Dimethyl hydrazine, 241
Dimethylamine, 105
Dimethylhydrazine, 241
Dinitrogen oxide (nitrous oxide), 609
decomposition of, 331
molecular structure of, *609*
uses of, 609
Dinitrogen pentoxide
rate of decomposition, 317–319, *318*
reaction of, with water, 610
structure of, *609*
Dinitrogen tetroxide, 67, *189*
equilibrium with nitrogen dioxide, 352–354, *352*
molecular structure, *609*

Dinitrogen trioxide, *609*, 610
Diopside, 264
Diphosphine, 603t
Dipole Species in which there is a separation of charge, i.e., a positive pole at one point and a negative pole at a different point, 201
temporary, 257, 258
Dipole force An attractive force between polar molecules, 258, 259
Dipole moment, 201
Diprotic acid, 396, 397t
Direction of reaction, 361, 362, 366–372
free energy change, 491, 493
reaction quotient, 361, 362
Dispersion force An attractive force between molecules that arises because of the presence of temporary dipoles, 257, 258
Disproportionation A reaction in which a species undergoes oxidation and reduction simultaneously, 588
of chlorine, 596
of copper(I), 588
of hydrogen peroxide, 605
of oxoanions, 614
Dissolving precipitates, 468–472
Distillation, 6, *7*
Distribution, of molecular speeds, 132, *132*, 133
Division
exponential notation, A.10
logarithms, A.13
significant figures, 12
DNA, 444
Double bond Two shared electron pairs between two bonded atoms, 183
Lewis structures, 184, 185
Drain cleaners, 308q
Drugs
aspirin, 392
ibuprofen, 392
morphine, 105, 106
opium, *105*
Dry cell, 536, *536*, 537
Dry Ice. *See* carbon dioxide
Dynamic equilibrium. *See* equilibrium
Dubnium, 551t
Ductility, 266
Duet rule (H), 183

E

$E°$. *See* standard cell voltage
$E°_{ox}$. *See* standard oxidation voltage
$E°_{red}$. *See* standard reduction voltage
EDTA, 455
Effective collision, *324*
Effective nuclear charge Positive charge felt by the outermost electrons in an atom; approximately

equal to the atomic number minus the number of electrons in inner, complete levels, 168

Efflorescence Loss of water by a hydrate, 74

Effusion Movement of gas molecules through a pinhole or capillary, 131, *131*, 132

Einstein, Albert, 146

Einstein equation, 557

Elastic collision, 128

Electric charge, 532t

Electric current, 532t

Electric power, 532t

Electrical conductivity
  ionic solids, 265
  metals, 266
  molecular substances, 256
  network covalent, 263

Electrical neutrality The principle that, in any compound, the total positive charge must equal the total negative charge, 43

Electrical units, 532t

Electrochemistry, 513–546

Electrode Anode or cathode in an electrochemical cell, 514

Electrolysis Passage of a direct electric current through a liquid containing ions, producing chemical changes at the electrodes, 124, 531–536
  aluminum oxide, 576
  efficiency of, 533
  purification of copper, 579
  quantitative relations, 531, 532
  sodium chloride molten, 575, *575*, 576
  sodium chloride solution, 535, 536, *536*
  water, 6, 124

Electrolyte A substance that exists as ions in water solution, such as NaCl or HCl, 86
  colligative properties, 298–300
  strong, 86, *86*
  weak, 93

Electrolytic cell A cell in which the flow of electrical energy from an external source causes a redox reaction to occur, 531–536

Electromagnetic radiation, types of, 146

Electron The negatively charged component of an atom, found outside the nucleus, 31, 32t
  arrangements in atoms, 148–163
  arrangements in monatomic ions, 164–166
  charge, A.2
  direction of flow of, 516
  discovery of, 32
  excited, 149, 157
  kinetic energy, 150
  lone pair, 182

mass of, 558t, A.2

pairs, two to six, ideal geometries for, *192, 194*

properties of, 32t

quantum numbers, 151–154

shared covalent bond, 180

spin, 154, *154*

unpaired in orbital, 162, 163

unshared pair, 182

valence, 182, 182t

Electron cloud, 151, *151*

Electron configuration An expression giving the population of electrons in each sublevel. The electron configuration of the Li atom is $1s^2 2s^1$, 156–162, 164–166
  abbreviated, 158, 158t
  elements, 1–36, 156, 157, *157*
  from periodic table, 161, 162

Electron dot structure, 182

Electron pair acceptor, 439

Electron pair donor, 439

Electron pair repulsion The principle used to predict molecular geometry. Electron pairs around a central atom tend to orient themselves to be as far apart as possible, 192, *192*

Electron spin The property of an electron described by the quantum number $\mathbf{m}_s$, which can be $+1/2$ or $-1/2$, 154, *154*

Electron-sea model A model of metallic bonding in which cations act as fixed points in a mobile "sea" of electrons, 265

Electronegativity A property of an atom that increases with its tendency to attract electrons to a bond, 170, *191*
  bond polarity, 200
  effect of, on acid strength, 611–613

Electronic structure, 144–178

Electroplating, 532, *533*

Element A substance whose atoms are all chemically the same, containing a definite number of protons, 3–5, 114, 116, 118, 550, 551
  actinide, 161
  atoms of, 29
  enthalpy of formation of, 230
  essential, *38*, 590, 591
  lanthanide, 160, 161
  main-group, 36
  metals, 573–594
  molecular, *40*
  names and symbols of, 4t
  "new" (114, 116, 118), 550, 551
  nonmetals, 595–624
  periodic table, 35–38
  properties, A.7, A.8
  standard entropy of, 488t, A.3–A.4

toxic, *38*

transition, 36

Elementary step, 334

Elements of Chemistry (Lavoisier), 16

Emission spectra, *143, 147*, 147, 148

Empirical (simplest) formula, 64

Enantiomer Optical isomer, 643

End point The point during a titration at which the indicator changes color, 420–424, 422t

Endothermic A process in which heat is absorbed by a system; ΔH is positive for an endothermic reaction, 218, *218*, 223, *223*

Energy A property of a system which can be altered only by exchanging heat or work with the surroundings, 235–237
  balance, in human body, 238, 239
  binding, 559, 560
  input, 238
  internal, 235
  ionization. *See* ionization energy
  law of conservation of, 235
  metabolic, 238, 239
  nuclear, 557–565
  output, 238, 239
  units, relation between, 218

Energy change (ΔE), 235
  relation to enthalpy change, 239
  relation to mass change, 557–561

Energy level of atoms Energy corresponding to the principal quantum number, $\mathbf{n}$, 146–149

Energy values, foods, 238t

English-metric conversions, 14t, 15

Enthalpy (H) A property of a system that reflects its capacity to exchange heat (q) with the surroundings; defined so that $\Delta H = q$ for a constant pressure process, 222, 223
  bond, 232–235, 234t

Enthalpy change (ΔH) The difference in enthalpy between products and reactants, 222, 238
  endothermic, exothermic reactions, 223
  enthalpy of formation, calculation from, 229–232
  entropy change and, 497, *497*
  forward and reverse reactions, 226
  Hess's law and, 227
  phase change, 227t
  relation to amount, 225, 226
  relation to energy change, 237
  spontaneity and, 484, 485
  standard, 229, 230
  thermochemical equation, 223, 224

Enthalpy of formation ΔH when one mole of a species is formed from the elements
  calculation from $\Delta H°$, 228–232

Enthalpy of formation *(continued)*
of compounds, 229t, A.3–A.4
of elements, 230
of hydrogen ion, 229
of ions, 229t
relation to enthalpy change, 230
standard, 228
Entropy (S) A property of a system related to its degree of order; highly ordered systems have low entropy, 482–490
factors affecting, 486, 487
gases, liquids, solids, 486
molar mass and, 489
randomness and, 486
standard, 488t , A.3–A.4
temperature and, *487*
Entropy change (ΔS) The difference in entropy between products and reactants
enthalpy change and, 489, 490
phase change, 487
reaction, calculation of, 489, 490
solution formation, 490, 492
spontaneity and, 491, 492
system and surroundings, 490
units of, 494
Environment
acid rain, 387
CFCs, 121
corrosion of metals, 539, 540
fluoride in water, limits, 475
lead pollution, 27q
nitric acid, 429
nuclear reactors, 562, 563
ozone layer, 338, 339
radon, 566, 567
road salts, 299
smog, 338, 608
sulfur dioxide, 429, 430
sulfuric acid, 430
Environmental Protection Agency, 567
Enzyme, 332, 333
Epsom salts, 74t, 79q
Equation
Acid-base reactions, 93–95, *94,* 428t
Arrhenius, 328–330
Bohr, 148
chemical, writing and balancing of, 67, 68
Clausius-Clapeyron, 250, 251
dilution of solution, 282
dissolving ionic solids, 86, 87, 469, 470
Einstein, 557
Henderson-Hasselbalch, 415, 436q
Nernst, 528–530
net ionic, 89, 90
Planck, 146
precipitation reactions, 88–91
redox, balancing of, 98–104
thermochemical, 223–228

van't Hoff, 371
Equilateral triangle, 193
Equilibrium A state of dynamic balance in which rates of forward and reverse reactions are equal; the system does not change with time, 248, 351
acid-base, 413–436
chemical, in gas phase, 350–381
effect of changes upon, 366–371
liquid-vapor, 248–254
partial pressure, 351, 355
precipitation, 460–472
solid-liquid, 256
solid-vapor, 255
solute-solvent, 288–291
table, 360
Equilibrium constant, K A number characteristic of an equilibrium system at a particular temperature. For the system $A(g) + 2B(aq) \rightleftharpoons 2\ C(g)$, the expression for K is $(P_C)^2/(P_A)[B]^2$, 354
applications of, 361–366
determination of, 359–361
direction of reaction, 361, 362
expression, 355–359
extent of reaction, 363–366
heterogeneous equilibrium, 357–359
magnitude of, 361
from Nernst equation, 531
polyprotic acid, 396, 397, 397t
relation between K_a, K_b, 400
relation to chemical equation, 355, 356
relation to $E°$, 576, 577, *577*
relation to $\Delta G°$, 500, 501
relation to K_c, 355, 380q
temperature effect on, 371
water, 384, 385
weak acid, 391, 391t, 392, A.5
weak base, 391t, 398, 399, A.5
Equivalence point The point during a titration when reaction is complete; equivalent quantities of the two reactants have been used, 96, *423, 425, 427*
Erectile dysfunction, 609
Erlenmeyer flask, 8
Ester An organic compound containing the —C—O— functional
group, 639, 640
properties of, 640t
Ethane, 627, *627*
Ethanol. *See* ethyl alcohol
Ethanolamine, 434q
Ethene. *See* ethylene
Ether, 635t
Ethyl alcohol, 637

concentration in blood (BAC), 48, 49t
concentration in breath (BrA), 48
concentration in beverages, 637
infrared spectrum, *19*
preparation, 637
proof, 637
solubility, 287
Ethyl butyrate, 640t
Ethyl ether, 275q
Ethyl formate, 640t
Ethylamine, 95
Ethylene (ethene), 272, 631, 632
bonding in, 183, 632
ethyl alcohol from, 637
hybridization, 206
molecular geometry, 207
polymerization of, 272, 273
Ethylene glycol, 273, 287, 294, *638*
Ethylenediamine, 441
complexes of, 441, 446, *447*
Ethylenediaminetetraacetic acid. *See* EDTA
Europium oxide, 161
Evaporation, 248
Evergreens, effect of acid rain on, 430, *430*
Exact numbers, 13
Excess reactant, 71, 72
Excited state An electronic state that has a higher energy than the ground state, 149, 157
Exclusion principle The rule stating that no two electrons can have the identical set of four quantum numbers, 154, 155
Exothermic Describes a process in which heat is evolved by a system; ΔH is negative for an exothermic reaction, 218, 223, *223*
Expanded octet More than four electron pairs about a central atom, 190, 191
Expansion, effect on equilibrium, 368–370, 370t
Expansion work, *236*
Experimental yield The amount of product actually obtained in a reaction, 73
Exponential notation, A.9–A.10
Extensive property A property of a substance which depends upon the amount of sample; volume is an extensive property, 17
Extent of reaction, 363–366, *327*
Eyring, Henry, 327, *327*

F

f orbitals, 154, 155
f sublevels, 152, 154, 155
Face-centered cubic (FCC) A cubic unit cell with atoms at each corner and one at the center of each face, *268,* 268t

Fahrenheit, Daniel, 9
Fahrenheit degree (°F), 9, *9*
Families, in periodic table, 36
Faraday, Michael, 74, 534, *534*
Faraday constant The constant that gives the number of coulombs equivalent to one mole of electrons; 96480 C/mol e^-, 526, A.2
Fat, 645, 646
Feldspar, 6
Fermentation, 16, 637
Fertilizer, 327, 618
Filtration A process for separating a solid-liquid mixture by passing it through a barrier with fine pores, such as filter paper, 6
Fire fighting, 582
First ionization energy, *169*
First law of thermodynamics The statement that the change in energy, ΔE, of a system, is the sum of the heat flow into the system, q, and the work done on the system, w, 235–237
First-order reaction A reaction whose rate depends upon reactant concentration raised to the first power, 318–321
    concentration-time dependence, 320
    half-life expression, 321
    radioactivity, 554–556
Fission (nuclear) The splitting of a heavy nucleus by a neutron into two lighter nuclei, accompanied by the release of energy, 561–563
    critical mass, 562
    discovery, 561
    energy evolution, 561, 562, 564
    products, 561
    reactors, 562, 563
    waste products, 563
Five percent rule The empirical rule that the approximation $a - x \approx a$ is valid if $x \leq 0.05a$, 395
Fixation of nitrogen, 361, 372
Flame tests, *143*, 474, *474*
Flerov, G. N., 550
Florida, flooding of, 504
Flotation, 578, 579
Fluorapatite, 475
Fluoridation, 475
Fluoride ion, 391t, 402t
    complexes, 452
    concentration of, in water, 475
    dietary sources, 475
    in tooth decay, 475
    reaction with water, 308, 309
Fluorine
    chemical reactivity, 598
    oxides, 608t
    oxidizing strength of, 598

physical properties, 597t
preparation from hydrogen fluoride, 600
reaction with water, 598
toothpaste, 475, *475*
Fluorite, 475, *475*
Fluorosis, 475
Foam formation, prevention of, 301
Food(s), energy values of, 238t
Food irradiation, 554, *554*
Force The product of mass times acceleration, A.1
Forces, types of
    dipole, 258, 259
    dispersion, 257, 258
    intermolecular, 257–262, 267t
    intramolecular, 267t
Ford Taurus, 135
Forensic chemistry, 619
Formal charge The charge that an atom would have if the bonding electrons in a molecule were equally shared, 187, 188
Formate ion, 391t
Formation constant (K_f) Equilibrium constant for the formation of a complex ion from the corresponding cation and ligands, 452, 453, A.6
Formic acid, 391t, 639
Formula An expression used to describe the relative numbers of atoms of different elements present in a substance
    empirical. *See* simplest formula
    from chemical analysis, 64-66
    molecular. *See* molecular formula
    of ionic compounds, 43
    percent composition from, 62, 63
    structural, 39, 40
Fossil fuels, 563
Fractional precipitation Method of separating ions in solution by selectively precipitating one ion, leaving other(s) in solution, 465, 473
Franklin, Benjamin, 171
Frasch process The process used to extract native sulfur from underground deposits, 599, *600*
Free energy (G) A quantity, defined as $H - TS$, which decreases in a reaction that is spontaneous at constant T and P, 490
    relation to H, S, 491
Free energy change (ΔG) The difference in free energy between products and reactants, 491–504
    additivity, 502–504
    cell voltage and, 525, 526
    concentration dependence, 499, 500
    coupled reactions, 502–504
    equilibrium constant and, 500, 501

pressure dependence, 499, 500
reaction, calculation of, 494–496
relation of, to ΔH, ΔS, 492
spontaneity and, 491, 493
temperature dependence, 496–499
Free energy of formation ΔG for the formation of a species from the elements, 495, A.3, A.4
Free radical, 188, 310
Freeze drying, 255
Freezing point The temperature at which the solid and liquid phases of a substance are at equilibrium, 256
    effect of concentration, 293
    effect of pressure, 256
Freezing point constants, 294t
Freezing point lowering The decrease in the freezing point of a liquid caused by adding a solute, 293, 294
    calculation of molar mass, 297
Freons, 121
Frequency The number of complete wave cycles per unit time, 145
    calculation of, H spectrum, 149
Fuel cell, 236
Fuel rod, 562
Fuller, Buckminster, 21
Fullerene, 21
Fumaric acid, 648q
Functional group A small group of atoms in an organic molecule that gives the molecule its characteristic chemical properties, 635, 635t
Fusion, heat of, 227t
Fusion, nuclear A reaction between small atomic nuclei to form a larger one, releasing energy in the process, 564, 565
    activation energy, 564
    energy evolved, 564
    laser, 565
    magnetic, 565

G

G. See free energy
g orbital, 177q
Gadolinium oxide, 161
Galena, 243q, *460*, *466*
Gallium-67, 552t
Gamma radiation High-energy photons emitted by radioactive nuclei, 35, *35*, 549
Gas(es), 114–142
    chemical equilibrium, 350–381
    collected over water, 126, *126*, 127
    condensable, 253t
    effusion of, 131, *131*, 132
    entropy of, 486, 489
    equilibrium of, with liquid, 248–253
    expansion of, against constant pressure, 236, *236*

Gas(es) *(continued)*
final and initial state of, problems of, 120, 121
involved in reactions, volumes of, 124
kinetic theory of, 128–133
measurements on, 115–117
mixtures of, mole fractions of, 127, 128
partial pressures in, 125–128
molecules, speeds and energies, distribution of, 132, *132*, 133
noble, compounds of, 209
particle(s)
attractive forces between, 134
average speeds of, 130
volume, 134, 135
permanent, 253t
pressure, 116, 117
kinetic theory and, 128, 129
real vs. ideal 133–135
solubility of
effect of pressure on, 290, 291
effect of temperature on, 288–290
Gas constant, R The constant that appears in the ideal gas law; R = 0.0821 L·atm/mol·K = 8.31 J/mol·K, 119, 119t, A.2
evaluation of, 119
values of, in different units, 119t
Gas hydrates, 74
Gas law calculations
calculation of one variable, 121
final and initial states, 120, 121
molar mass and density, 122, 123
stoichiometry of reaction, 123, 124
Gaseous equilibrium, 350–381
Gas-liquid chromatography, 6, *7*
Gasohol, 637
Gasoline, 630, 631
Gauge pressure, 138q
Gay-Lussac, Joseph, 118, 124
General Electric, synthetic diamonds, 20
Geometric isomerism A type of isomerism that arises when two species have the same molecular formulas but different geometric structures, 444–447
Geometry
complex ions, 442–446
d orbitals, 448
molecular. *See* molecular geometry
p orbitals, *153*
unit cells, *268, 269*
Ghiorso, Albert, 550
Giauque, William, 191
Gibbs, J. Willard, 493, *493*
Gibbs free energy. *See* free energy
Gibbs-Helmholtz equation The relation ΔG = ΔH − TΔS, 491, 492
Gillespie, R. J., 192

Glass
coloring of, *60*, 619
etching of, 607
purple, 589
reaction with hydrogen fluoride, 607
Glass electrode, 530, *530*
Glauber's salt, 74t
Global warming, 2, 504, 505
carbon dioxide, 504
infrared absorption, 504
water vapor, 505
Glucose, 552
Glycerol, 638
Gold, *17*, 373, 580
metallurgy, 580
oxidation to gold(III), 586t
reaction with aqua regia, 587
resistance of, to corrosion, 539
Gold(I) ion
complexes, 442t, 443
disproportionation, 588
Gold(III) ion, 586t, 587, 587t
Golden fleece, 580
Gout, 74
Graduated cylinder, 10
Graham, Thomas, 131
Graham's law The relation stating that the rate of effusion of a gas is inversely proportional to the square root of its molar mass, 131
Granite, *6*
Graphite, 20, *20*, 263
equilibrium of, with diamond, 20
Greek prefixes, 46t
Greenhouse effect, 504
Ground state The lowest allowed energy state of a species, 149, 156
Group A vertical column of the periodic table, 36
Group 1 metals. *See* alkali metals
Group 2 metals. *See* alkaline earth metals
Groups I, II, III, IV cation groups in qualitative analysis, 472–474
Gypsum, 74t

H

H . *See* enthalpy
Haber, Clara, 373
Haber, Fritz, 372, *373*
Haber process An industrial process used to make ammonia from the elements, 372, 373
Hahn, Otto, 561
Half-cell Half of an electrochemical cell, at which oxidation or reduction occurs, 515, 516, 518
Half-equation An equation written to describe a half-reaction of oxidation or reduction, 100
balancing, 100–102, 517

Half-life The time required to convert half of the original amount of reactant to product
first-order, 319, 320
second-order, 322t
zero-order, 322t
Half-reaction, 98, 518
Halide ions, 600
Hall, Charles, 4, 576
Hall process, 4, 576
Halogen An element of Group 17, 36
colors of vapors, 597
preparation, 600
properties of, 597t
reaction with metals in Groups 1, 2, 581t
reactivity, 598, 599
standard potentials of, 520t, 521t
Hard water, 264
HDL, 645
Heat A form of energy that flows between two samples because of their difference in temperature, 216
Heat capacity The amount of heat required to raise the temperature one degree Celsius, 218
calorimeter, 222
Heat content. *See* enthalpy
Heat flow The amount of heat, *q*, flowing into a system (+) or out of it (−)
magnitude of, 218, 219
measurement of, 219–222
principles of, 216–219
relation to enthalpy, 222
sign of, 217, 218, *235*
Heat of formation. *See* enthalpy of formation
Heat of fusion ΔH for the conversion of one mole of a solid to a liquid at constant T and P, 227t
Heat of vaporization ΔH for the conversion of one mole of a liquid to a gas at constant T and P, 227t
Clausius-Clapeyron equation, 250, 251
Heavy water reactor, 563
Hectare, 25q
Helium, 208, 209
in balloons, 123
in chromatography, 7
solubility of, in blood, 291
Helmholtz, H., 493
Hematite, 576
Heme, 454
Hemoglobin, 454, *454, 455*
complex with carbon monoxide, 455
Henderson-Hasselbalch equation, 415, 436q
Henry's law, 290
Heptane, 630
Heroult, P.L.T., 5, 576
Hertz A unit of frequency; one cycle per second, 145

Hess, Germaine, 227

Hess's law A relation stating that the heat flow in a reaction that is the sum of two other reactions is equal to the sum of the heat flows in those two reactions, 227

Heterogeneous Non-uniform composition, 6

Heterogeneous catalysis, 331, 332

Heterogeneous equilibrium, 357–359

Hibernation, 346q

High spin complex A complex that, for a particular metal cation, has the largest number of unpaired electrons, 450

Hiroshima, 561

History of chemistry
 Arrhenius, 97
 Avogadro, 125
 Curies, 553
 Dalton, 30
 Eyring, 327
 Faraday, 534
 Gibbs, 493
 Haber, 372, 373
 Hodgkin, 271
 Lavoisier, 16
 Lewis, G. N., 191
 Mendeleev, 37
 Seaborg, 160
 Werner, 446

Hitler, Adolf, 373

Hodgkin, Dorothy, 271

Homogeneous Uniform in composition, 6

Homogeneous catalysis, 332

Homogeneous equilibrium, 357

Honey, 289

Hot air balloons, 123

Human body
 acidosis, 387
 ADP, ATP, 503
 antacids, 98t
 bends, 291
 biological effects of radiation on, 566, 567
 blood plasma, 414
 bones and teeth, 475
 calcium in urine, 113q
 carbon monoxide toxicity, 455
 cholesterol, 645, 646
 energy balance of, 238, 239
 enzymes, 332, 333
 hemoglobin, 454, 455
 isotonic solution, 296, 297
 metabolism, 502, 503
 pH of body fluids, 386t
 stomach acid, 113q

Humpty-Dumpty, 486

Hund, Friedrich, 162

Hund's rule A relation stating that, ordinarily, electrons will not pair in an orbital until all orbitals of equal energy contain one electron, 162, 163

Hybrid orbital An orbital made from a mixture of individual atomic orbitals. An sp^2 orbital is formed by mixing an s with two p orbitals, 203–207
 geometries, 205t

Hybridization Mixing of two or more orbitals or structures, 203–207

Hydrangeas, *389*

Hydrate A compound containing bound water such $BaCl_2 \cdot H_2O$, 74, 75

Hydrazine, 261, 441
 preparation of, 604

Hydrazoic acid, 603t

Hydride A compound of hydrogen, specifically one containing H^- ions, 99, 401

Hydriodic acid, 47, 92t

Hydrobromic acid, 47, 92t

Hydrocarbon An organic compound containing only carbon and hydrogen atoms, 627–635
 alkane (saturated), 627–631
 alkene, 631, 632
 alkyne, 632, 633
 aromatic, 633-635
 solubility of, 286, 287

Hydrochloric acid, 47, 92t, *93*
 Brønsted-Lowry acid, *388*
 concentration of, 85
 ionization of, in water, 92
 label, 285
 reaction with carbonates, 224
 reaction with hydroxides, 606, 607
 reaction with iron, *525*
 reaction with transition metals, 585–587
 titration with ammonia, 426, 427, 428t
 titration with sodium hydroxide, *423*, *424*, 428t

Hydrofluoric acid, 92
 etching glass, 607
 reaction with calcium silicate, 607
 reaction with carbonate ion, 607
 reaction with hydroxide ion, 606
 reaction with silicon dioxide, 607
 reaction with strong base, 606
 safety hazard, 607

Hydrogen
 compounds of, boiling points of, 260t
 with nonmetals, 603t
 formation by electrolysis, 484, 533, 535
 fuel, 512q
 in balloons, 123
 oxidation number of, 99
 reaction with Group 1, Group 2 metals, 581t, 582

reaction with iodine, 337, 346q
 reaction with nitrogen, 372, 373
 reaction with oxygen, 484

Hydrogen atom
 Bohr model, 148–150
 quantum mechanical model of, 150, 151
 spectrum of, *147*, 148, 148t, 149

Hydrogen bomb, 564

Hydrogen bond An attractive force between molecules found when a hydrogen atom is bonded to N, O, or F, 260
 effect on boiling point, 260t
 effect on solubility, 287
 strength of, 260
 water and ice, 261, *262*

Hydrogen carbonate ion, 43t
 acid-base properties, 404

Hydrogen chloride, 606
 Brønsted-Lowry acid, *388*

Hydrogen cyanide, 391t, 392

Hydrogen fluoride, 606, 607
 hydrogen bonding in, 260
 preparation from elements, 492
 preparation of fluorine from, 600

Hydrogen iodide, 322

Hydrogen ion
 acceptor, donor, 383
 balancing redox equations, 101, 102
 concentration of, in acid-base titrations, 423–427
 in buffers, 414–416
 in weak acids, 393–396
 hydroxide ion and, 385
 indicator color and, 420–422
 pH and, 386, 387
 reaction with precipitates, 468–470, *469*
 reaction with carbonate ion, 232, *232*, 469, 470
 reaction with chromate ion, 589
 reaction with hydroxide ion, 423, 469
 reaction with iron, *525*
 reaction with metals, *98*
 reaction with sulfides, 470, 473
 reaction with weak bases, 94, 94t
 reaction with zinc, *98*
 reaction with zinc hydroxide, 469
 standard enthalpy, 229
 standard entropy, 488
 standard potential, 519

Hydrogen molecule
 dispersion forces, 257, 258
 electron density, *180*
 Lewis structure, 181

Hydrogen peroxide, 124
 concentration, water solutions, 606
 disproportionation, 605
 oxidizing agent, 605

Hydrogen peroxide *(continued)*
rate of decomposition of, 332, 333, *333*
reducing agent, 605
uses, 590
Hydrogen phosphate ion, 43t, 391t, 618
Hydrogen sulfate ion, *611,* 617
Hydrogen sulfide
Brønsted acid, 604
dissociation equilibrium, 604
precipitating agent, 604
preparation of sulfur from, 605
properties, 605
reaction with cations, 604
reducing agent, 605
toxicity, 605
use in qualitative analysis, 472, 473
Hydrogen sulfite ion, 391t
Hydrohalic acid, 47
Hydrolysis of salts, 402–404
Hydronium ion The H_3O^+ ion characteristic of acidic water solutions,
384
Hydroxide
solubility in, 470t
solubility in acid, 468, 469
solubility in water, *88*
Hydroxide ion, 43t
balancing redox equations, 101
concentration in solution of weak bases,
399
effect on indicator color, 420–422
ligand in complex ions, 440, 470t
pOH and, 386
reaction with hydrogen ion, 93, 423
reaction with weak acids, 94, 425, 426,
428t
Hydroxyapatite, 475
Hypo- prefix, 45, 45t
Hypochlorite ion, 45t, 391t
reaction with ammonia, 604
Hypochlorous acid, 47, 391t
oxidizing strength, 598
preparation, 598, 599

I

i (Van't Hoff factor), 299
Ibuprofen, 393, *404*
-ic suffix, 47
Ice
density, 261
heavy (D_2O), *33*
heat of fusion, 227
melting point, pressure and, 256
structure of, 261, *262*
sublimation, 255
vapor pressure, 255
Ice beer, 293
Ice removers
calcium chloride, 299
potassium chloride, 299
sodium chloride, 299, *299*

Iced tea, 245q
Iceman, 555
-ide suffix, 44
Ideal bond angle, *194,* 196t
Ideal gas law A relation between pressure, volume, temperature, and
amount for any gas at moderate
pressures: $PV = nRT$, 118–123
calculations involving vaporization,
248, 249
volumes of gases in reactions, 123, 124
Ideal geometry The geometry a molecule would have if the effect of unshared pairs were neglected,
192–195
Indicator, acid-base, 96, 389, 420–422,
421, 422t
pH range, 422
Induced dipole, 258
Induced radioactivity, 549
Industrial equilibria, 372, 373
Inert electrode, 517
Infrared radiation Light having a
wavelength greater than about
700 nm, 19, 48, 145, *146,* 504
Initial rate, 315
Insoluble compounds, *88*
Insulin, 305
Intensive property A property of a sub-
stance which is independent of the
amount of sample; density is an in-
tensive property, 17
Intermediate, reactive, 336
Intermolecular force, 257–262, 266t
Internal energy. *See* Energy
International system of units (SI), 10, A.1
Intoxication, tests for, 48, 49
Intoxilizer 5000, 48
Intravenous feeding, 296
Inverse logarithms, A.11
Invisible ink, 74
Iodide ion, 600
brines, 600
catalysis of hydrogen peroxide decomposition, 332
oxidation, 600
Iodine, *357*
color of, *71, 597*
oxides of, 608t
oxoacids of, 611t
preparation, 600
properties of, 50, 597t
reaction with antimony, 70, *71*
reaction with hydrogen, 337, 346q
safety hazards, 599
subliming of, 255, *255*
tincture of, 599
Iodine-131, 50, 123
Iodine chloride, 258, *259*
Ion charged species, 40, 41
acid-base properties of, 402t

acidic vs. basic, 402–404
complex, 437–459
concentrations of, from Nernst equation, 530, 531
electron configuration of, 164–166
enthalpy of formation, 229t, A.3–A.4
free energy of formation of, A.3–A.4
monatomic, electron arrangements in,
164–166
noble-gas structures, 42, 164
polyatomic. *See* polyatomic ions
standard entropy of, 473t, A.3–A.4
transition metal, charges of, 42, 43, *87*
Ion pair, 300
Ion product (P) An expression which
has the same form as K_{sp} but involves arbitrary rather than equilibrium concentrations of ions, 464,
464
Ionic atmosphere, 300
Ionic bond The electrostatic force of
attraction between oppositely
charged ions in an ionic compound, 41
Ionic compound A compound made up
of cations and anions, 41, 42
conductivity of, 265
crystal structure of, *41,* 270, *270*
formulas of, 42–44
melting point of, 265
naming of, 45
properties of, 264, 265
solubility of, in water, *87,* 287, 288
structure of, 263
Ionic radius The radius assigned to a
monatomic ion, *167,* 168, 169
Ionizable hydrogen atom, 389
Ionization
of water, 384, 385
of weak acid, 389, 390
Ionization energy The energy that
must be absorbed to remove an
electron from a species, 169, *169,*
170
correlation of, with atomic radius, 169,
170
trends in periodic table, 169
Iron
corrosion, 539, 540
metallurgy, 576–578
oxidation to iron(II), 585
reaction with acid, *525,* 585
reaction with oxygen, 484, 584
steel from, 578
Iron-59, 552t
Iron(II) chloride, *165*
Iron(II) hydroxide, 588
Iron(III) hydroxide, *90,* 588
Iron(II) ion
complexes, 442t, 449
nutrition, 591t

preparation of, from iron, 585
qualitative analysis of, 472t, 473
reaction with oxygen, 539
reaction with permanganate ion, 103, *103*, 104
reduction to iron, 587
Iron(III) ion
complexes, 442t
nutrition, 591t
reduction to iron(II), 587
Iron(III) oxide, *83*
blast furnace, 576
reaction with aluminum, 230, *230*
reaction with carbon monoxide, 502, 577
Iron(II) sulfide, 473
Isobar, 51q
Isobutyl formate, 640t
Isoelectronic species, 164, *164*
Isomer A species with the same formula as another species but having different properties. Structural and geometric isomers are possible, 186
cis-trans, 444–447, 641, 642
complex ions, 444–447
octahedral complexes, 444–447
optical, 642–644
organic, 640–644
square planar complexes, 444
structural, 627–629, 641
Isooctane, 630
Isopentyl acetate, 640t
Isotonic solution, 296
Isotope An atom having the same number of protons as another atom but with different numbers of neutrons, 33, 34
abundance and masses of, 58
separation of, 132
-ite suffix, 45, 45t

J

Joliot, Frederic, 550, 553
Joule The base SI unit of energy, equal to the kinetic energy of a two-kilogram mass moving at a speed of one meter per second, 146, 218, 532t
Joule, James, 218

K

K. See equilibrium constant
K_a. *See* acid equilibrium constant
K_b. *See* base equilibrium constant
K_f. *See* formation constant
K_p. *See* K
K_{sp}. *See* solubility product constant
K_w. *See* water ion product constant
K-electron capture The natural radioactive process in which an inner electron ($\mathbf{n}$ = 1) enters the nucleus, converting a proton to a neutron, 549
Kelvin (K), 10

Kelvin, Lord, 10
Kelvin scale (K) The scale obtained by taking the lowest attainable temperature to be 0 K; the size of degree is the same as °C, 10, 116
Ketone, 635t
Kilo- The prefix indicating a multiple of 1000, 8t
Kilocalorie, 218
Kilogram, 9, 10
Kilojoule, 146, 218
Kilometer, 8
Kilowatt hour (kWh) Unit of energy; 1 kWh = 3600 kJ, 532t
Kinetic energy, 129
electron, 150
Kinetic theory A model of molecular motion used to explain many of the properties of gases, 128–133, *128*
postulates of, 128
Kinetics, chemical, 309–349
Knot, 25q
Kroto, Henry, 21
Krypton, 208
Krypton difluoride, 209
Kyoto conference, 505

L

Lactate ion, 391t, 415
Lactic acid, 391t, 405t, 415
Lanthanides Elements 57 (La) through 70 (Yb) in the periodic table, 160, 161
Laser fusion, 565, *565*
Lattice structures, *264, 268*
Laughing gas, 609
Lauryl alcohol, 306q
Lavoisier, Antoine, 16, *16*, 115
Law
Avogadro's, 119, 125
Boyle's, 118
Charles's and Gay-Lussac's, 118
Dalton's, 126
Graham's, 131
Henry's, 290
Hess's, 227
Raoult's, 292
Law of combining volumes, 124
Law of conservation of energy A natural law stating that energy can neither be created nor destroyed; it can only be converted from one form to another, 235
Law of conservation of mass, 30
Law of constant composition, 30
Law of multiple proportions A relation stating that when two elements A and B form two compounds, the relative amounts of B that combine with a fixed amount of A are in a ratio of small integers, 30

Law of partial pressures, 126
Law of thermodynamics
first, 235–237
second, 490
third, 487
Le Châtelier, Henri, 366
Le Châtelier's principle A relation stating that when a system at equilibrium is disturbed it responds in such a way as to partially counteract that change, 366–372
Lead, 4, 513, 619
Lead-210, 568q
Lead azide, 597
Lead(II) chloride, 462, 472
Lead(II) ion
EDTA complex, 455
qualitative analysis, 472, 472t
toxicity, 4
Lead(IV) oxide, 538
Lead pencil, 20
Lead storage battery, *537*, 538
Lead(II) sulfate, 53q, 538
Leclanché cell. *See* dry cell
LDL, 645
Lemon juice, 386, 386t
Length, 8
Length volume, and mass units, relations between, 14t
Lewis, G. N., 181, 191, *191*
Lewis acid A species that accepts an electron pair in an acid-base reaction, 439
Lewis base A species that donates an electron pair in an acid-base reaction, 439
Lewis structure An electronic structure of a molecule or ion in which electrons are shown by dashes or dots (electron pairs), 181–191
chelates, 441
hybridization and, 205
molecular geometry and, 195, 196
molecules with expanded octets, 190, 191
polarity, and, 202
resonance forms, 185–187
rules for writing, 183, 184, 190
Libby, W. F., 555
Lichen, 454
Ligand A molecule or anion bonded to the central metal in a complex ion, 440, 441
effect of, on d orbital energies, 449
strong field, 452
weak field, 452
Light, 144–148
frequency of, 145, *145*
infrared, 145, *146*
particle nature of, 145, 146
speed of, 145, A.2

Light *(continued)*
    types of, 145, *146*
    ultraviolet, 145, *146*
    visible, 145, *146*
    wavelength of, 145
Light sticks, 330
Light water reactor, 563
Limestone, *498*
Limiting reactant The least abundant reactant based on the equation for a reaction; dictates the maximum amount of product that can be formed, 72, 90
Limonite, 63
Line spectrum, 147, *147*
Linear complex, 442, 443
Linear molecule A triatomic molecule in which the bond angle is 180°; examples include BeF_2 and CO_2, 192, 193
Linear plot
    $\Delta G°$ vs. T, 496
    ln k vs. $1/T$, *329*, 329, 330
    ln K vs. $1/T$, 371
    ln P vs. $1/T$, 249, *250*
    reaction order, 322t
Liquefaction of gas, 253
Liquids, 246–278
    boiling point, 251, 252
    characteristics, 3
    critical behavior, 252, 253
    entropy, 486
    freezing point vs. pressure, 256, *256*
    surface tension, 247, *247*
    vapor pressure, 248–251
Liquid-vapor equilibrium, 248–254
Liter A unit of volume; 1 L = 1000 cm^3, 8
Lithium, *522*, 581, 583
Lithium chloride, *270*
Lithium oxide, 581
Litmus, 420
Logarithm, A.10–A.13
    common, A.11
    natural, A.11
    significant figures in, A.12–A.13
London forces, 257, 258
Lone pair, 182
Low spin complex A complex that, for a particular metal ion, has the smallest possible number of unpaired electrons, 438, 449
Lowry, Thomas, 383
Luster The characteristic shiny appearance of a metal surface, 266
Lyman series, 148t
Lysozyme, 305q

M

Magnesium
    chemical properties, 581t
    reaction with oxygen, *228*, 583
    reaction with steam, 582
Magnesium ion
    in nutrition, 591t
    qualitative analysis of, 472t, 474
Magnesium sulfate, 300t
Magnetic fusion, 565
Main-group cations, in nutrition, 590, 591t
Main-group element An element in one of the groups numbered 1 to 2 or **13** to **18** of the periodic table, 32
    electronegativity of, 170, 171t
    ionization energies of, *169*
    Lewis structures of, 182t
    periodic table and, 158–160, *159*
    sizes of atoms and ions of, *167*
Maleic acid, 648q
Malic acid, 405t
Malleability The ability to be shaped, as by pounding with a hammer; characteristic of metals, 266, *266*
Maltase, 333
Maltose, 333
Manganate ion, 101
Manganese, *101*, 266
    chemistry of, 589, 590
    oxidation to manganese(II), 586t
    reaction with oxygen, 584t
Manganese(II) chloride, *165*
Manganese(II) ion, *101*
    nutrition, 591t
    oxidation to manganese(III), 587t
    qualitative analysis, 472t, 473
Manganese(III) ion, 587
Manganese(III) oxide, 536, 587
Manganese(IV) oxide, *101*
Manganese(II) sulfide, *473*
Manometer, 116, 117, *117*
Maple sap, 301, *301*
    boiling point of, 301
    concentration of, by reverse osmosis, 301
    osmotic pressure of, 301
Maple syrup, 301, 302
Marbles, mixing of, 486
Marcet, Jane, 534
Marina, Mario, 338
Martini, 301q
Mass An extensive property reflecting the amount of matter in a sample, 9
    atomic. *See* atomic mass
    conservation of, law of, 30
    length and volume units, 14t
    molar. *See* molar mass
    nuclear, 558t
    of individual atoms, 60
Mass defect The difference between the mass of a nucleus and the sum of the masses of the neutrons and pro-

tons of which it is composed, 559
Mass number An integer equal to the sum of the number of protons and neutrons in an atomic nucleus, 33, 34
Mass percent 100 times the ratio of the mass of a component to the total mass of a sample, 283, 284
    conversion to other concentration units, 285, 285t
    of elements in compound, 62, 63
    of solute, 283, 284
Mass relations
    in electrolysis, 531, 532
    from equations, 69, 70
    in formulas, 62–66
    in reactions, 67–73
    in nuclear reactions, 557–565
    in solution reactions, 90, 91, 104
Mass spectrometer, 57, *57*
Mass units, length, and volume, relation between, 14t
Mass-energy relations, 557–561
Matches, 602
Mathematics, review of, A.9–A.13
Matter
    classification of, *3*
    definition of, 3
    types of, 3–7
Maxwell, James Clerk, 128, 132
Maxwell distribution A relation describing the way in which molecular speeds or energies are shared among gas molecules, 132, *132*
Measurement, 7–15
    uncertainties in, 10–13
Mechanism A sequence of steps that occurs during the course of a reaction, 333–337
Mega-, 8t
Megagram, 9
Meitner, Lise, 561
Melting point The temperature at which the solid and liquid phases of a substance are in equilibrium with each other, 17, 257
    of ionic solids, 265
    of metals, 267
    of molecular substances, 257
    of network covalent substances, 263
Membrane, semipermeable, 295
Mendeleev, Dmitri, 37
Mendeleevium, 161
Mercury, 4, *5*
    drinking water limits, 284
    metallurgy of, 578
    oxidation to mercury(II), 586t
    reaction of, with oxygen 584t
    spectrum of, 147
    uses, 4

vapor pressure, 278q
Mercury(I) chloride, 472
Mercury(I) ion, 43t
  qualitative analysis, 472t
  reduction to mercury, 587t
Mercury(II) ion
  qualitative analysis, 472t, 473, *473*
  reduction to mercury(I), 587t
  toxicity, 590
Mercury(II) oxide, 6, 17
Mercury(II) sulfide, *473*
Metabolic energy, 238, 239
Metal A substance having characteristic luster, malleability, and high electrical conductivity; 37, 573–594
  alkali, 36, 581–584
  alkaline earth, 36, 581–584
  band theory, A.22–A.23
  ductility, of, 266
  electrical conductivity of, 266
  luster of, 266
  main-group, 36
  malleability of, 266, *266*
  melting point of, 267
  nutrition, 590, 591
  periodic table, 37, 574
  properties of, 266
  qualitative analysis of, 472–474
  solubility of, 266
  thermal conductivity of, 266
  transition, 36, 160, 584–590
  unit cells of, 267–270, *268*
Metalloid An element such as Si that has properties intermediate between those of metals and nonmetals, 37
Metallurgy The science and processes of extracting metals from their ores, 574–580
  aluminum, 576
  bismuth, 580
  copper, 578–580
  gold, 580
  iron, 576–578, *578*
  mercury, 578
  native metals, 575t
  oxide ores, 575t
  sodium, 575, *575*
  sulfide ores, 575t
  zinc, 578
Meter, 8
Methane, 75
Methanol. *See* methyl alcohol
Methyl acetate, 639
Methyl acetylene, 632
Methyl alcohol, 37, 275q, 637
  preparation, 637
  toxicity, 637
Methyl butyrate, 640t
Methyl red, 421, 422t
Methyl salicylate, 78q
Methyl yellow, 434q

Methylamine, 8, 93, 94
Metric system A measuring system where all units of a particular type (e.g., volume) are related to one another by powers of ten, 8, 8t
Metric ton, 9
Metric-English conversion, 14t
Mica, 6
Micro-, 8t
Microwave, 146, 245q
Milli- The prefix on a metric unit indicating a multiple of 10^{-3}, 8t
Milligram, 9
Millikan, Robert, 81q
Milliliter, 8
Millimeter of mercury (mm Hg) The unit of pressure: 1 atm = 760 mm Hg, 116, *116*
Mixed oxides, 585, *585*
Mixture Two or more substances combined so that each substance retains its chemical identity, 6, 7
  heterogeneous, 6, *6*
  homogeneous, 6, *6*
  separation of, 7
Moissan, Henri, 600
Molal boiling point constant, 294t
Molal freezing point constant, 294t
Molality A concentration unit defined as the number of moles of solute per kilogram of solvent, 284
  boiling point elevation and, 293
  conversion to other concentration units, 285, 285t, 286
  freezing point depression and, 293
  relation of, to molarity, 286
Molar mass The mass of one mole of a substance, 60, 61
  determination of, gases, 122
  effect of, on boiling point, 258, 258t
  from gaseous effusion, 131, 132
  of nonelectrolytes, from colligative properties, 297, 298
Molar volume, real gases, 133–135
Molarity A concentration unit defined to be the number of moles of solute per liter of solution, 84, 85
  conversion of, to other concentration units, 285, 285t, 286
  ions in solution, 86, 87
  preparation from concentrated solution, 282, *282*
  relation of, to molality, 286
Mole A collection of 6.0122×10^{23} items. The mass in grams of one mole of a substance is numerically equal to its formula mass, 60–62, *61*
  electrons, redox reaction, 532t
  elements and compounds, *61*
  ions per mole solute, 86, 87
  ratio, buffers, 415

  relation to molality, 284
  relation to molarity, 84
  relation to mole fraction, 127, 283
  relation to volume of gas, 124
Mole fraction A concentration unit defined as the number of moles of a component divided by the total number of moles, 127, 128, 283
  conversion of, to other concentration units, 285t
  partial pressure and, 127, 128
  vapor pressure lowering and, 292
Mole-gram conversions, 61, 62
Molecular elements, 40, *40*
Molecular formula A formula in which the number of atoms of each type in a molecule is indicated as a subscript after the symbol of the atom, 39
  from simplest formula, 66
Molecular geometry The shape of a molecule, describing the relative positions of atoms, 192–200
  effect of unshared pairs on, 193–197
  of molecules with 2 to 6 electron pairs, 192–194, *194*
  of molecules with expanded octets, *194, 198*
  with 2, 3, or 4 electron pairs around the central atom, 196t
Molecular orbital, A.17–A.23
Molecular substance A substance built up of discrete molecules such as H_2 or CH_4, 39
  intermolecular forces, 257–262, 267t
  naming, 45, 46
  physical properties of, 156, 157, 167t
Molecule An aggregate of a few atoms that is the fundamental building block in all gases and many solids and liquids; held together by covalent bonds between the atoms, 38–40
  elements, *40*
  expanded octets, 190
  forces between, 257–262, 267t
  geometry of, 192–200
  weak acids, 92, 389, 390
  weak bases, 93, 398
Monatomic ion, 41
  charges of, 42, 43
  electron arrangements in, 164–166
  naming of, 44
  noble-gas structures, 42, 164
  radii of, *167*, 168, 169
Monoclinic sulfur, 602, *602*, 603
Monomer A small molecule that joins with other monomers to form a polymer, 272
Monoprotic acid An acid molecule containing one ionizable H atom
Morphine, 105, 106

Moss, 454

MTBE, 631

Multiple bond A double or triple bond, 183
  enthalpy, 268t
  geometry with, 197, 199, 200
  hybridization in, 206, 207
  pi and sigma bonds, 207, 208

Multiple equilibria, rule of A rule stating that if Equation 1 + Equation 2 = Equation 3, then $K_3 = K_1 \times K_2$, 356, 357

Multiple proportions, law of, 30, *30*

Multiplication
  exponential notation in, A.10
  logarithms in, A.13
  significant figures in, 12

Muscular activity, energy consumption and, 239

Mylar, 273

N

N_A (Avogadro's number), 59

Nagasaki, 561

Naming. *See* nomenclature

Nano- The prefix on a metric unit indicating a multiple of 10^{-9}, 8t

Nanometer, 8

Nanotube, 21, *21*

Naphthalene, 303, 635

Napoleon Bonaparte, 619

Naproxen, 644

Native metals, 575t, 580

Natural gas, 236

Natural logarithm, 250, A.11

Natural radioactivity, 34, 35

Nautical mile, 25q

Neon, 77q, 209

Neptunium, 551t

Nernst, Walter, 528

Nernst equation An equation relating cell voltage E to the standard voltage $E°$ and the concentrations of reactants and products: $E = E° - (0.0257/n)(\ln Q)$, 528–531

Net ionic equation A chemical equation for a reaction involving ions in which only those species that actually react are included, 89, 90
  acid-base reaction, 93–95, 94t
  dissolving ionic solid, 86, 469, 470
  formation of acidic solution, 92, 389, 390
  formation of basic solution, 93, 398
  precipitation reaction, 89, 90
  redox reaction, 100–104

Network covalent Having a structure in which all the atoms in a crystal are linked by a network of covalent bonds, 263, *263*, 264, 267t
  conductivity, 263

melting point, 263
  solubility, 263

Neutral complex, 440

Neutral salts, 403t

Neutral solution A water solution with a pH of 7 (at 25°C), 386, *386*

Neutralization A reaction between a strong acid and base to form a neutral solution, 93, 423, 424

Neutron A particle in an atomic nucleus with zero charge and a mass of approximately 1 amu, 12, 12t
  mass of, 558

Neutron activation analysis, 552

Neutron-to-proton ratio, *35*

Newton (N), A.1, A.1t

Newton, Sir Isaac, 6, 147

Niacin, 394

Nicad storage battery, 538

Nichrome, 517

Nickel
  alloys of, 598
  oxidation to nickel(II), 586t
  reaction with acid, 586
  reaction with copper(II), 516, 524, *524*
  reaction with fluorine, 598

Nickel(II) chloride, 586, *586*

Nickel(II) hydroxide, *88*, 538

Nickel(II) ion
  complexes of, 439, 442t
  preparation from nickel, *586*
  qualitative analysis, 472t, 473

Nickel(II) oxide, 585, *585*

Nickel (IV) oxide, 538

Nickel(II) sulfide, *473*

Nickel-copper(II) voltaic cell, 516, 517

Nicotinic acid. *See* niacin

Nitrate ion, 43t
  resonance forms of, 186

Nitrates, *88*

Nitric acid, 92t, 107, 615–617
  air pollution, 429
  decomposition in sunlight, 617
  oxidizing agent, 616
  preparation from ammonia, 615
  preparation from dinitrogen pentoxide, 610
  reaction with aluminum hydroxide, 616
  reaction with calcium carbonate, 616
  reaction with copper, 524, 525, *525*, 586, *616*
  reaction with copper(II) sulfide, 616
  reaction with protein, 617
  reaction with silver, 586, 587
  reaction with zinc, 616
  structure, *611*

Nitric oxide. *See* nitrogen oxide

Nitrite ion, 45t, 391t

Nitrogen
  hydrogen compounds, 603t
  inertness of, 597

oxides, 608, 608t, 609, *609*
  oxoanions of, 45t, *611*
  oxyacids, 611t
  preparation from liquid air, 599
  properties of, 597t
  reaction with Group 1, 2 metals, 581t
  reaction with hydrogen, 372, 373
  solubility of, in blood, 291

Nitrogen dioxide, *45*, *189*, *352*
  air pollution, 608, *608*
  equilibrium with dinitrogen tetroxide, 352–354, 371
  from nitric acid, 616
  reaction with carbon monoxide, 333, 334
  reaction with nitrogen oxide, 608
  reaction with water, 615
  structure, 188, 189

Nitrogen fixation, 372

Nitrogen oxide (nitric oxide), 46
  equilibrium with the elements, 361, *363*
  from nitric acid, 616
  reaction with chlorine, 336
  reaction with nitrogen dioxide, 608
  structure, 188, 189

Nitroglycerine, 139q

Nitrosyl chloride, 377q

Nitrous acid, 47
  acid strength of, 391t
  Brønsted-Lowry acid, 390
  preparation of, from dinitrogen trioxide, 610

Nitrous oxide. *See* dinitrogen oxide

Nobelium, 161

Noble gas An element in group **18** at the far right of the periodic table, 36, 37
  boiling point of, 258t

Noble-gas structure
  atoms, 181
  ions with, 42, 164, 165
  molecules, 183

Nomenclature
  acids, 47
  alkanes, A.15, A.16
  anions, 44
  binary molecular compounds, 45, 46
  cations, 44
  complex anions, A.14
  complex cations, A.14
  coordination compounds, A.15
  electrodes, 514
  esters, 639
  ionic compounds, 45
  oxoacids, 47
  oxoanions, 44, 45t
  polyatomic ions, 43t

Nonelectrolyte A substance such as CH_3OH that exists as molecules rather than ions in water solution
  boiling point elevation and freezing point depression, 292–294

molar masses of, determination of, from colligative properties, 297, 298
osmotic pressure, 294–297
vapor pressure lowering, 291, 292
Nonmetal One of the elements in the upper right part of the periodic table that does not show metallic properties, 37
allotropy, 600–603
expanded octets, 190
hydrogen compounds, 603–607, 603t
molecular compounds, 45, 46
molecular formulas, 40
noble-gas structures of, 183
oxoacids of, 47, 611–618, 611t
oxoanions of, 44, 45t, 611–618
oxygen compounds of, 608–611, 608t
in periodic table, 37, 596
preparation of, 599, 600
properties of, 597t
reactivity of, 597–599
Nonpolar bond A covalent bond in which the electrons are equally shared by two atoms, so there are no positive and negative ends, 200
Nonpolar molecule A molecule in which there is no separation of charge and hence no positive and negative poles, 200–203, 258t, 259t
Nonspontaneous reaction, 484
Nonstoichiometric solid, 585
Normal boiling point The boiling point at 1 atm pressure, 251
Northern lights, 171, *171*, 172
Novocaine, 105
Nuclear accidents, 563
Nuclear energy, 557–565
Nuclear equation, 548, 549
Nuclear fallout, 556, 566
Nuclear fission. *See* fission, nuclear
Nuclear fusion. *See* fusion, nuclear
Nuclear masses, 558t
Nuclear reactions, 547–572
Nuclear reactor, 562, 563
Nuclear stability, 34, 35
Nuclear symbol A symbol giving the atomic number and mass number of a nucleus. Example: $^{14}_{6}C$, 33
Nucleon A proton or neutron
Nucleus The small, dense, positively charged region at the center of the atom, 31, 32
Number
atomic, 33
Avogadro's, 59, 60, 125
coordination, 439, 442
exact, 13
mass, 33, 34
oxidation, 99–102
quantum, 151–154
spelled out, 13

Nutrition
acidosis, 387
aspartame, 333
calcium ion, 590, 591t
caloric value, 238
cholesterol, 645, 646
chromium(III), 590, 591t
cobalt (II, III), 590, 591t
copper(I, II), 590, 591t
EDTA in foods, 455
energy consumption, 238, 239
energy balance, 239
enzymes, 332, 333
essential trace elements, 38, *38*
fats, 645, 646
fluoride ion, 475
food irradiation, 554
hemoglobin, 454, *454*, 455
iron(II, III), 590, 590t
isotonic solution, 296
magnesium ion, 591t
main-group cations in, 590, 591t
manganese(II), 590, 591t
molybdenum(IV, V, VI), 591t
osteoporosis, 590
potassium ion, 590, 591t
selenium, 619
sodium, 590, 591t
vitamin C, 66, 212q
weight control, 238, 239
zinc(II), 590, 591t
Nyholm, R. S., 192

O

Octahedral Having the symmetry of a regular octahedron. In an octahedral species, a central atom is surrounded by six other atoms, one above, one below, and four at the corners of a square, 193, *194*
Octahedral complex, 443–445
crystal field model, 447–454
Octane, 124
Octane number, 630, 631
Octet rule The principle that bonded atoms (except H) tend to have a share in eight valence electrons, 183
exceptions to, 188–191
expanded octets, 190, 191
Octyl acetate, 640t
Odd electron species, 188, 189
Opium, *93*
Oppenheimer, J. Robert, 561
Opposed spins, 154
Optical isomer Isomer which rotates the plane of plane-polarized light, 643
Orbital Region of space in which there is a high probability of finding an electron within an atom, 151–154
d, 154, *448*
f, 154

hybrid. *See* hybrid orbital(s)
hybridization, 204, 205
p, 153, *153*
pi, 207, 208
s, 153, *153*, 154
sigma, 207, 208
Orbital diagram A sketch showing electron populations of atomic orbitals; electron spins are indicated by up and down arrows, 162, 163
Orbital quantum number, $\mathbf{m}_\ell$, 153, 154
Order of reaction An exponent to which the concentration of a reactant must be raised to give the observed dependence of rate upon concentration, 314–317
Ore A natural mineral deposit from which a metal can be extracted profitably, 574
Organic acids, 404, 405, 638, 639
Organic chemistry, 625–649
functional groups, 635–640
hydrocarbons, 627–635
isomerism, 640–644
Organic material, age of, 555, 556
Orpiment, *1, 460, 466*
Osmosis A process by which a solvent moves through a semipermeable membrane from a region where its vapor pressure is high to one where it is low, 294, 297
cucumber and, *296*
prune and, *296*
reverse, *296*
Osmotic pressure The pressure that must be applied to a solution to prevent osmosis, 296
molar mass from, 298
Osteoporosis, 590
Ostwald, Wilhelm, 615
Ostwald process, 615
-ous suffix, 47
Oven, self-cleaning, 161
Overall order, 316
Ox bladder, 16
Oxalate ion, 441, *441*
Oxalic acid, 396, 397t
Oxidation A half-reaction in which there is an increase in oxidation number, 98–100, 514
Oxidation number A number that is assigned to an element in a molecule or ion to reflect, qualitatively, its state of oxidation, 99–102
charge of complex ion, 440
effect of, on acid strength, 611–613
Groups 1, 2; 99
hydrogen, 99
oxygen, 99
rules for assigning, 99–100
transition metal ions, 577, 578

Oxidation-reduction reaction, 98–104
Oxide ion, 581t, 583
Oxide ores, *575*, *576–578*
Oxidizing agent A species that accepts
    electrons in a redox reaction, 99,
    521, 522
    strength, 521, 522
Oxyacetylene torch, *633*
Oxoacid An acid containing oxygen,
    such as HNO_3 or H_2SO_4, 47,
    611–618, 611t
    acid strength of, 611, 612
    chlorine, 47, 612
    Lewis structures of, 611
    naming of, 47
    nonmetal oxides and, 610, 611
    oxidizing, reducing strengths of,
        613–615
    weak, K_a values of, 391t, 397t
Oxoanion An anion containing oxygen,
    such as NO_3^-, or SO_4^{2-}, 611–618
    Lewis structures of, 611
    naming of, 47
    oxidizing and reducing strength of,
        613–615
    transition metals, 589, 590
Oxygen
    allotropy, 601
    compounds of nonmetals, 608–611
    discovery of, 6
    electronic structure, 189
    from liquid air, 599
    hydrogen compounds of, 608t
    isotopic composition of, 77q
    molar volume, gas, 133t
    molecular speed of, *132*
    molecular structure, 99
    paramagnetism, *189*
    physical properties, 597t
    preparation of, from sodium peroxide,
        508q
    reaction with chromium(II), 588
    reaction with copper(I) sulfide, 578
    reaction with hydrogen, 484
    reaction with iron, 584
    reaction with iron(II), 588, 589
    reaction with magnesium, *228*, *583*
    reaction with metals, 581t, 583–585
    reaction with metal cations, 588
    reaction with phosphorus, 601, *601*, 602
    reaction with sulfur, 350
    solubility of, in water, 290
    steel manufacture, 578
    transport by hemoglobin, 455
Oxyhemoglobin, 455
Ozone
    catalyzed decomposition of, 338
    hole, *339*
    layer, 2, 115, 338
    mechanism of decomposition of, 338
    preparation of, 601

smog formation, 338
toxicity of, 338
upper atmosphere, 338
water treatment, 601

P

p group element, *159*, 160
p orbital, *153*
p sublevel, 152–154
Palladium(II) ion, 442t
Paraaminobenzoic acid, 409q
Parallel spins, 154
Paramagnetic Showing magnetic prop-
    erties caused by the presence of un-
    paired electrons, 163
    complex ions, 449
Partial pressure The part of the total
    pressure in a gas mixture that can
    be attributed to a particular compo-
    nent. The partial pressure of a com-
    ponent may be found by multiply-
    ing the total pressure by the mole
    fraction, 125–128
    equilibrium expressions involving,
        351–381
    mole fraction and, 127, 128
    of water vapor, 126, 127
Parts per billion, 284
Parts per million For liquids and solids,
    the number of grams of solute per
    million grams of sample, 284
Pascal An SI unit of pressure; the pres-
    sure exerted by the force of 1 new-
    ton on an area of 1 square meter,
    117
Paschen series, 148t
Pasteur, Louis, 644
Pauli, Wolfgang, 154
Pauli exclusion principle, 154, 155
Pauling, Linus, 203
Penicillin, 409q
Pentyl propionate, 640t
Per-, prefix, 45, 47
Percent composition Mass percents of
    the elements in a compound, 62
    from formula, 62, 63
    simplest formula from, 64, 65
Percent ionization For a weak acid HB,
    % ionization $= 100 \times [H^+]/[HB]_o$,
    393, 393
Percent yield A quantity equal to 100 ×
    actual yield/theoretical yield, 73
Perchlorate ion, 43t, 613
Perchloric acid, 47, 611t, 612t
Period A horizontal row of elements in
    the periodic table, 36
Periodic function A physical or chemi-
    cal property of elements that varies
    periodically with atomic number,
    166
Periodic table An arrangement of the

elements in rows and columns ac-
    cording to atomic numbers such
    that elements with similar chemical
    properties fall in the same column,
    35–38, *36*, 159, 162
    atomic radius and, 166–168, *166*,
        167
    atoms with expanded octet in, 190
    development of, 37
    electron configurations and, 161, 162
    electronegativity and, 170, 171
    filling sublevels, 158-161
    ionic radius and, 168, 169
    ionization energy and, 169, *169*, 170
    Lewis structures and, 182t
    ores, 575
    trends in, 166–171
Permanent gas, 253t
Permanganate ion, 43t, *101*, 103, *103*,
    589, 590
    oxidizing strength, 590
    reaction with iron(II), *103*
    reaction with water, 590
Peroxide A compound containing the
    peroxide ion, O_2^{2-}, 581t, 583
PET scan, 552
Petroleum, 630, *630*
Pfund series, 175q
pH Defined as $-\log_{10}[H^+]$, 385–389
    acid-base titrations, 423–427, 428t
    buffer, 434–436
    common materials, 386t
    effect on color of flowers, 389, *389*
    effect on oxidizing, reducing strength,
        614
    measurement of, 388, *388*, 389
    pOH and, 386, 387
    relation to acidity, 386
    salt solutions, *403*
    strong acid, 367, 368
    strong base, 367, 368
    weak acid, 393–397
    weak base, 399
pH control, body, 420, 434q
pH meter, *388*, 530, *530*
pH paper, 389
Phase change, 255, 256
    enthalpy change, 227t
    entropy change, 486
Phase diagram A graph of pressure vs.
    temperature showing the conditions
    for equilibrium between phases,
    254, *254*, 256
Phenanthrene, 635
Phenol, 409q, 634
Phenolphthalein, *421*, 422t
Philosopher's stone, 16
Phosgene, 509q
Phosphate ion, 43t, 391t, 402, *403*, *611*,
    618
Phosphate ores, 575

Phosphine, 46, 603t
Phosphorescence, *601*
Phosphoric acid, 396, 618
    equilibrium constants of, 397t
    preparation of, from phosphorus pen-
        toxide, 611
Phosphorous acid, 611t
Phosphorus
    allotropy, 509q, 601, *601*, 602, *602*
    hydrides, 603t
    matches, 602
    oxides, 608t
    oxoacids, 611t
    physical properties, 597t
    reaction of, with oxygen, *601*
    red, *602*
    white, *601*
Phosphorus-32, 552t
Phosphorus pentachloride, 190
Phosphorus pentafluoride, *194, 198*
Phosphorus pentoxide (P_4O_{10}), 608t, *610*
Phosphorus trioxide (P_4O_6), 608t, *610*
Phosphoryl chloride, 214q
Photochemical smog, 338
Photoconductivity, 619
Photoelectric effect, 146, 178q
Photon An individual particle of radiant
    energy, 146
Physical property A property such as
    melting point or density which can
    be determined without changing
    the chemical identity of a sub-
    stance, 17
Pi bond A bond in which electrons are
    concentrated in orbitals located off
    the internuclear axis; in a double
    bond there is one pi bond, in a
    triple bond there are two, 207,
    208, A.21, A.22
Pico-, 8t
Picric acid, 409q
Pig iron, 578
Pipet, 8, *8*
pK_a Defined as $-\log_{10}K_a$, where K_a is the
    ionization constant of a weak acid,
    392
pK_b Defined as $-\log_{10}K_b$, where K_b is the
    ionization constant of a weak base,
    399
Planck, Max, 146
Planck's constant, 146, A.2
Plaster, 74
Plaster of Paris, 74
Plastic sulfur, *603*
Platinum
    catalyst, 332
    catalytic converter, 332
    electrode, 517, 518
    resistance of, to corrosion, 539
Platinum(II) complexes, 440t, 442t, *443*,
    444

Platinum(IV) complexes, 442t, 456
Plutonium, 551t, 561, 572q
pOH Defined as pOH = $-\log_{10}[OH^-]$,
    386
Polar bond A chemical bond that has
    positive and negative ends; charac-
    teristic of all bonds between unlike
    atoms, 200
Polar molecule A molecule in which
    there is a separation of charge and
    hence positive and negative poles,
    200–202
    boiling points, 259t
    dipole forces between, 258, 259
    partial charge, 201, 202
Polarity, of molecules, 200–202
Polarized light, 644, *644*
Polyatomic ion, 43, 44
    Lewis structures, 182–184
Polyester A large molecule made up of
    ester units, 273
Polyethylene, 272, 273
    branched, 272
    linear, 272, *273*
Polymer A huge molecule made up of
    many small units linked together
    chemically
    addition, 272, 273
    condensation, 273
    polyester, 273
Polyprotic acid An acid containing
    more than one ionizable H atom.
    Examples include H_2SO_4 and
    H_3PO_4, 396, 397, 397t
Porsche, 141q
Positron A particle with the same mass
    as the electron but the opposite
    charge, 549
Post-transition metal Lower members
    of periodic groups **13**, **14**, and **15**,
    such as Pb and Bi, 36
Potassium
    chemical properties, 581t
    flame test, *474*
Potassium chloride, 299
Potassium chromate, *85*
Potassium dichromate, *43, 266*
    breath analyzer, 48
Potassium fluoride, 601
Potassium hydroxide, 583
Potassium iodide, *535*
Potassium ion
    in nutrition, 590, 591t
    qualitative analysis, 472, *474*
Potassium nitrate, *43*
Potassium permanganate, *19, 43*
    properties of, 589, 590
Potassium superoxide, 583
Potential, standard, 519, 520t, 521t
Pound per square inch, 116
Povidone-iodine, 599

Powell, H. M., 192
Precipitate A solid that forms when two
    solutions are mixed, 88
    dissolving in acid, 468–470
    dissolving in complexing agent, *469*
    formation, *88*
    solubility product constants and,
        461–467
Precipitation diagram, 88, *88*
Precipitation equilibria, 461–467
Precipitation reaction Formation of
    an insoluble solid when two elec-
    trolyte solutions are mixed, 88–91
Prefixes
    bis-, A.14
    Greek, 46t
    hypo-, 47
    metric, 8t
    naming complex ions, A.14
    per-, 47
    tris-, A.14
Prescription drug, saturation value of,
    349q
Pressure Force per unit area, 116
    atmospheric, *116*
    conversions, 117
    critical, 253
    effect on boiling point, 251, 252
    effect on equilibrium, 368–370, 369t
    effect on equilibrium yield, 372t
    effect on melting point, 256
    effect on reaction rate, 372
    effect on gas solubility, 290, 291
    gas, kinetic theory and, 128, 129
    liquid column, 140q
    osmotic, 298
    relation between units, 117
    vapor. *See* vapor pressure
Pressure cooker, 252
Priestley, Joseph, 6
Primary amine, 105
Primary cell, 536, 537
Principal energy level The energy
    level designated by the quantum
    number **n**
    capacity for electrons, 152
Principal quantum number (**n**), 148, 149,
    152
Probability, electronic, 150
    thermodynamic, 485
Problems, answers to, A.24–A.42
Product A substance formed as a result
    of a reaction; appears on the right
    side of an equation, 67
Proof (alcohol), 637
Propane, 627
Properties
    chemical, 17
    colligative, 291–300
    extensive, 17
    intensive, 17

Properties *(continued)*
  physical, 17–19
  state, 217
Propylamine, 260, 274
Propylene, 632
Proton The nucleus of the hydrogen
    atom, the H^+ ion, 32, *32. See also*
    hydrogen ion
  acceptor-donor, 383
  mass of 558
Prunes, wrinkled, *296*
psi, 116
Pyramid, triangular, 195, 196t
Pyrite, *460, 466*

Q

q (heat flow), 218
Q (concentration quotient), 361, 362, 499,
    528
Quadratic equation, 365
Quadratic formula The formula used
    to obtain the two roots of a qua-
    dratic equation $ax^2 + bx + c = 0$.
    The formula is
    $x = (-b \pm \sqrt{b^2 - 4ac})/2a$, 365,
    396
Qualitative analysis The determina-
    tion of the nature of the species
    present in a sample; most often ap-
    plied to cations, 472–474
  Group I, 472, 472t
  Group II, 472t, 473
  Group III, 472t, 473
  Group IV, 472t, 474
Quantum mechanics, 150, 151
Quantum number A number used to
    describe energy levels available to
    electrons in atoms; there are four
    such numbers, 151–156
  first (**n**), 152
  fourth ($\mathbf{m}_s$), 154
  second (ℓ), 152, 153
  third ($\mathbf{m}_\ell$), 153, 154
Quart, 14t
Quartz, 264, *264*
Quinic acid, 405t

R

R (gas constant), 119, 119t
Racemic mixture Mixture containing
    equal numbers of two enantiomers
    of a substance, 643
Rad, 566
Radiation (nuclear), 34, 35
  alpha, 35, 549
  beta, 35, 549
  biological effect, 566
  in food preservation, 554
  gamma, 35, 549
Radio wave, 146

Radioactive isotope, 34
Radioactivity The ability possessed by
    some natural and synthetic isotopes
    to undergo nuclear transformation
    to other isotopes, 34–35
  biological effects, 566
  induced, 549
  natural, 549
  rate of decay, 554–556
Radium, 553, 557
Radius
  atomic, 166, *167,* 168
  ionic, *167,* 168, 169
Radon, 51q, 566
Ramsey, Sir William, 208
Randomness, 486
Rankine temperature, 142q
Raoult, François, 292
Raoult's law A relation between the va-
    por pressure (*P*) of a component of a
    solution and that of the pure com-
    ponent (*P°*) at the same tempera-
    ture; $P = XP°$, where *X* is the mole
    fraction, 292
Rare earth. *See* lanthanide
Rast method, 305q
Rate, of radioactive decay, 554–556
Rate constant The proportionality con-
    stant in the rate equation for a reac-
    tion
  determination, 314
  temperature effect, 329, 330
Rate-determining step The slowest
    step in a multistep mechanism, 334,
    335
Rate equation, 314
Rate expression A mathematical rela-
    tionship describing the dependence
    of reaction rate upon the concentra-
    tion (s) of reactant(s), 314
Rayleigh, Lord, 208
Reactant The starting material in a reac-
    tion; appears on the left side of an
    equation
  limiting, 67
Reacting species A species that takes
    part in a reaction. When hydrochlo-
    ric acid reacts with NaOH, the react-
    ing species is the H^+ ion, 94t
Reaction
  acid-base, 91–98
  in aqueous solution, 83–113
  nuclear, 547–572
  order of, 314–317
  oxidation-reduction, 98–104
  precipitation, 88–91
  redox, 98–104
  volumes of gases involved in, 123, 124
Reaction mechanism, 333–337
Reaction quotient (Q) An expression
    with the same form as *K* but involv-

ing arbitrary rather than equilib-
    rium partial pressures, 361, 362,
    499, 528
Reaction rate The ratio of the change
    in concentration of a species di-
    vided by the time interval over
    which the change occurs,
    309–349
  concentration effect on, 313–317
  defining equation, 311
  measurement, 312, 313
  temperature effect on, 328–330
Real (vs. ideal) gas, 133
Real gases, 133–135
Realgar, 1
Reciprocal rule The rule that states that
    the equilibrium constant for a reac-
    tion is the reciprocal of that for the
    same reaction in the reverse direc-
    tion, 356
Red phosphorus, 602, *602*
Redox reaction A reaction involving
    oxidation and reduction, 98–104,
    513–546
  spontaneity of, 524, 525
  standard voltage in, 523, 524
  transition metals, 584–590
Reducing agent A species that furnishes
    electrons to another in a redox reac-
    tion, 99
  strength of, 522
Reduction A half-reaction in which a
    species decreases in oxidation num-
    ber; 98, 100, 514
Refinery, *630*
Rem, 566
Rembrandt, 569q
Resonance A model used to rationalize
    the properties of a species for which
    a single Lewis structure is inade-
    quate; resonance forms differ only
    in the distribution of electrons,
    185–187
Reverse osmosis, *296*
Reversible reaction, 351
Rhombic sulfur, *602,* 603
Roasting A metallurgical process in
    which a sulfide ore is heated in air,
    forming either the free metal or the
    metal oxide, 578
Rock salt, 600
Rowland, F. Sherwood, 338
Rubidium, 581t
Ruby, *41*
Rule of multiple equilibria, 356, 357
Rust, *83, 99*
Ruthenium, 373
Rutherford, Ernest, 31, *31,* 32
Rutherford scattering, *32*
Rutherfordium, 551t
Rydberg constant (R_H), 149, A.2

S

s group elements, *159,* 160
s orbital, *152*
s sublevel, 152
SI unit, 10, A.1
sp hybrid, 204, 205t
sp^2 hybrid, 204, 205t
sp^3 hybrid, 204, 205t
sp^{3}d hybrid, 205, 205t
sp^{3}d^2 hybrid, 205, 205t
STP, 119, *119*
Safety matches, 602
Salicylic acid, 80q
Saline hydrides, 582
Salt An ionic solid containing any cation
    other than H$^+$ and any anion other
    than OH$^-$, or O^{2-}, 402
    pH of solution of, 402–404
Salt bridge A U-tube containing a salt
    solution; used to bridge the gap be-
    tween two halves of a voltaic cell,
    516
Sapphire, *41*
Saturated fat, 646
Saturated hydrocarbon An alkane; a
    hydrocarbon in which all the car-
    bon-carbon bonds are single,
    627–631
Saturated solution A solution in equi-
    librium with undissolved solute,
    288
Schrödinger, Erwin, 151
Schrödinger equation, 151
Scientific notation, A.9
Scintillation counter, *555*
Scotch whiskey, 570q
Scrubber, 430
Scuba diving, 291
Seaborg, Glenn, 160, *160,* 550
Seaborgium, 160, 551
Sea water, 386t
Second law of thermodynamics A
    basic law of nature, one form of
    which states that all spontaneous
    processes occur with an increase in
    entropy, 490
Second-order reaction A reaction
    whose rate depends on the second
    power of reactant concentration,
    321, 322
    characteristics, 322t
    concentration-time relation of, 321
Secondary amine, 105
See-saw molecule, *198*
Selenium, *37, 60*
    anticarcinogen, 619
    nutrition, 38, 619
    uses, 619
Selenium-75, 51q, 552t
Selenous acid, 612

Selsun blue, 619
Semiconductor, 37
Semipermeable membrane, 295
Sequester, 455
Sevin, 78q
Sex attractant, 642
Sex hormones, 645
Shielding electron, 168
Shroud of Turin, 556
Sidgwick, N. V., 192
Sigma bond A chemical bond in which
    electron density on the internuclear
    axis is high, as with all single
    bonds. In a multiple bond, one and
    only one of the electron pairs forms
    a sigma bond, 207, 208
Significant figure A meaningful digit
    in a measured quantity, 10–13
    counting of, 10–12
    logarithms, A.12–A.13
    multiplication and division, 12
Silicates, 263, 264
Silicon, 37, 76, 243q
Silicon dioxide, 263, *263,* 264
Silicon nitride, 80q
Silver, *37*
    electrical conductivity of, 266
    metallurgy, 580
    oxidation to silver(I), 586t
    plating, 545q
    reaction with nitric acid, 586, 587
Silver acetate, *467*
Silver chloride, *469, 472, 494*
Silver oxide, 584
Silver phosphate, *469*
Silver(I) ion
    complexes, 442t, 443
    qualitative analysis, 472, 472t
Simple cubic cell A unit cell in which
    there are atoms at each corner of a
    cube, 268, *268*
Simplest formula A formula of a com-
    pound which gives the simplest
    atom ratio of the elements present.
    The simplest formula of hydrogen
    peroxide is HO, 64
    determination of, 64, 65
    from percent composition, 65
    molecular formula from, 66
Single bond A pair of electrons shared
    between two bonded atoms
    enthalpy, 183, 234t
Skating, pressure on ice, 279q
Skeleton structure A structure of a
    species in which only sigma bonds
    are shown, 163, 164
Skin cancer, 338
Slag, 577
Slow step, 334, 335
Smalley, Richard, 21
Smog, 338

Smoke alarm, 552, 553
Soap, 639
Socrates, 105
Soda water, 6
Sodium, *5*
    chemical properties, 581t
    crystal structure of, 277q
    flame test, *474*
    preparation of, 575, *575*
    reaction with hydrogen, 582
    reaction with oxygen, 583
    spectrum of, 147, *147*
    water and, 582, *582*
Sodium-24, 552t
Sodium acetate, 280, *289*
Sodium azide, 135
Sodium benzoate, 81q
Sodium carbonate, *44*
    reaction with calcium chloride, 88, *89*
    reaction with hydrochloric acid, *224*
Sodium chloride, *5, 402*
    crystal structure of, *41*
    dissolving of, in water, 86
    electrolysis of, aqueous, 535, 536, *536*
        molten, 575, *575*
    in highway ice removal, 299, *299*
    melting point of, 42
    source of chlorine, *536, 575*
Sodium content, foods, 590, 591t
Sodium cyanide, 580
Sodium dihydrogen phosphate, 618
Sodium hydrogen carbonate, 62
Sodium phosphate, 618
Sodium hydroxide, 92t
    formation from sodium, 582
    formation from sodium chloride, 535,
        536
    formation from sodium hydride, 582
    uses of, 536
Sodium ion
    nutrition, 590, 591t
    qualitative analysis, 472t, 474
Sodium peroxide, 583
Sodium phosphate, 618
Sodium propionate, 638, 639
Sodium stearate, 639
Sodium ureate, 74, *75*
Softening, of water, 264
Solids, 246–278
    characteristics of, 3
    crystal structures of, 267–270
    entropy of, 486
    equilibrium of, with ions, 461–467
    expression for K, 357, 359
    ionic, 263t, 264, 265
    melting point vs. pressure, 256
    metallic, 263t, 265, 266
    network covalent, 263, 263t, 264
    nonstoichiometric, 585, *585*
    solubility of, effect of T on, 289, *289,*
        290

Solids *(continued)*
  sublimation of, 255, 256
  types of, 267t
Solubility The amount of a solute that
    dissolves in a given amount of sol-
    vent at a specified temperature, 18
  alcohols in water, 287, 287t
  calculation from K_{sp}, 465, 466
  common ion effect, 467, *467*, *468*
  effect of pressure upon, 290, 291
  effect of temperature upon, 288–290
  estimation of, from free energy change,
    500
  gases in liquids, 289–291
  hydroxide in acid, 468, 469
  ionic solids, *88*, 265, 465, 466
  metals, 266
  molecular substances, 257
  network covalent solids, 263
  principles of, 286–291
  silver chloride in ammonia, 471
  solids in liquids, 287–289
  sucrose in water, *289*
  vitamins in water, 288
Solubility product constant (K_{sp})
    The equilibrium constant for the re-
    action by which a slightly soluble
    ionic solid dissolves in water,
    461–468, 463t
  calculation of, from ΔG_f°, 501
  common ion effect, 467, *467*, *468*
  equilibrium concentrations of ions, 462,
    463
  expression for, 461, 462
  Nernst equation and, 531
  precipitate formation, 464, 465
  solubility in relation to, 465, 466
  values of, 463t, A.6
Solute The solution component present
    in smaller amount than the
    solvent, 6
  interaction of, with solvent, 286
Solution A homogeneous mixture con-
    taining a solvent and one or more
    solutes, 6
  acidic, 91, 385–386
  basic, 91, 385–386
  colligative properties of, 291–300
  concentrated, 84, 85
  concentration units of, 281–286
  dilute, 84, 85
  reactions in, 83–113
  saturated, 288
  stoichiometry, 90, 91, 104
  supersaturated, 289, *289*
Solvent A substance, usually a liquid,
    which is the major component of a
    solution, 6
  in expression for K, 359
  interaction of, with solute, 286–288
Sorbonne, 553

Sorensen, Soren, 385
Sour milk, 346q
Soybeans, 454
Specific gravity, 285
Specific heat The amount of heat re-
    quired to raise the temperature of
    one gram of a substance by 1°C,
    218, 219
  determination of, 219
Specific ion electrode, 530, 531
  chloride ion, 531
  fluoride ion, 530
  hydrogen ion, 530
Spectator ion An ion that, although
    present, takes no part in a reaction,
    89, 402t
Spectrochemical series, 451, *452*, 452t
Spectrum
  atomic 147, *147*
  emission, 147, *147*
  hydrogen, and Bohr model, 148, 148t,
    149
Speed
  of gas molecules, 129, 130
  of light, 145, 557, A.2
Spin
  parallel and opposed, 154
  quantum number, 154, 155
Splitting energy. *See* crystal field splitting
    energy
Spontaneity of a reaction, 482–512
  concentration and, 499–500
  pressure and, 499–500
  sign of ΔG, 491
  sign of ΔH, 497
  sign of ΔS, 497
  temperature and, 497, 498
Spontaneous process A process that
    can occur by itself without input of
    work from outside; $\Delta G < 0$ for a
    spontaneous process at constant T
    and P, 483, 484
  energy factor, 484, 485
  randomness factor, 485, 486
Spontaneous redox reaction, 524, 525
Square planar The geometry of a
    species in which four groups are lo-
    cated at the corners of a square, sur-
    rounding the central atom, *198*
Square planar complex, 443, 444
Square pyramid A pyramid that has a
    square as its base, 198
Stability, nuclear, 34, 35
Standard atmosphere, 116
Standard cell voltage (E°) The voltage
    of a cell in which all species are in
    their standard states (1 atm for
    gases, 1 M for ions in solution), 519
  calculation, 523, 524
  equilibrium constant and, 526, 527, 527t
  free energy and, 525, 526

reaction spontaneity and, 524, 525
Standard enthalpy change (ΔH°) The
    enthalpy change at 1 atm, 1 M, at a
    specified temperature, usually 25°C,
    229, 230
Standard enthalpy of formation (ΔH_f°),
    228, 229t, A.3, A.4
Standard entropy change (ΔS°) The
    entropy change at 1 atm, 1 M, at a
    specified temperature, usually 25°C,
    489
  calculation from S°, 489, 490
  sign of, prediction, 489
Standard free energy changes (ΔG°)
    ΔG when reactants and products are
    in their standard states,
    492–494
  additivity, 502, 503
  calculation from E°, 525, 526
  calculation from ΔG_f°, 494, 495
  calculation from ΔH°, ΔS°, 494, 496
  equilibrium constant and, 500, 501
  temperature dependence, 496
Standard free energy of formation (ΔG_f°),
    495, A.3–A.4
  determination of, 495
  use in calculating ΔG, 495
Standard molar entropy (S°) Entropy
    of a substance in its standard state
    (per mole)
  of compounds, 488t, A.3–A.4
  of elements, 488t, A.3–A.4
  of hydrogen ion, 489
  of ions in solution, 488t, A.3–A.4
  use of, in calculating ΔS°, 489, 490
Standard oxidation voltage (E_{ox}°)
    Voltage associated with an oxida-
    tion half-reaction when all gases are
    at 1 atm, all aqueous solutes at 1 M,
    519
  from standard potentials, 521
  relation to reducing strength, 522
Standard potential Identical with the
    standard reduction voltage, 519,
    520t, 521t
Standard reduction voltage (E_{red}°)
    The voltage associated with a reduc-
    tion half-reaction when all gases are
    at 1 atm and all aqueous solutions
    are at 1 M, 519, 520t, 521t
  of hydrogen ion, 519
  standard potential and, 519
  strength of oxidizing agent, 521, 522
**Standard temperature and pressure
    (STP)** Conditions of 0°C, 1 atm,
    119, *119*
State A condition of a system at fixed
    temperature, pressure, and chemical
    composition, 217
State property A property of a system
    whose value is fixed when its tem-

perature, pressure, and composition are specified, 217
Stearic acid, 639
Steel, *577, 578*
Stoichiometry Relationships between amounts (grams, moles) of reactants and products, 56–73, 90, 91, 104
Storage cell, 537, 538
Straight-chain alkane A saturated hydrocarbon in which all the carbon atoms are arranged in a single continuous chain, 627
Stratosphere, 338
Strong acid A species that is completely ionized to H^+ ions in dilute water solution, 92, 92t, 387
  dissolving precipitates, 468–470
  reaction with strong base, 93, 94t, 423, *423*, 424
  reaction with weak base, 94, 94t, 426, *427*, 428t
  titration, strong base, *423*, 424
    weak base, 426, 427, *427*
Strong base A species that is completely ionized to OH^- ions in dilute water solution, 92, 92t, 387
  reaction with strong acid, 93, 94t, 423, *423*, 424, 428t
  reaction with weak acid, 94, 94t, 425, *425*, 426
  titration, strong acid, *423*, 424
    weak acid, 425, *425, 426*
Strong electrolyte A compound that is completely ionized to ions in dilute water solution, 86, *86*
Strong field ligand, 452
Strontium, 242q, 581t
Strontium-90, 34, 562
Strontium chromate, *461, 462, 463*
Structural formula A formula showing the arrangement of bonded atoms in a molecule, 39
Structural isomers Two or more species having the same molecular formula but different structural formulas. Examples: C_2H_5OH and CH_3—O—CH_3, 628, 629, 641
Subatomic particles, properties of, 32t
Sublevel A subdivision of an energy level, designated as an s, p, d, or f sublevel
  capacity for electrons, 152
  number of, in principal level, 156t
  quantum no., 152, 153
  relative energies, *157*
  s, p, d, f, 152
Sublimation A change in state from solid to gas, 255
  Dry Ice, 255
  ice, 255
  iodine, 255, *255*

Substance, pure, 3
Subtraction, uncertainties in, 13
Successive approximations A technique used to solve quadratic or higher-order equations. A first, approximate answer is refined to obtain a more nearly exact solution, 395
Sucrose, 289, *289, 614, 618*
Suffixes
  -ate, 45
  -ic, 44
  -ide, 44
  -ite, 45
  -ous, 44
Sugar. *See* sucrose
Sulfate ion, 43t
  acid-base properties of, 402t
Sulfates, 45t
  solubility, 88
Sulfides
  group II, *473*
  group III, *473*
  ores, 578–580
  reaction with acid, 470
  solubility, 473
Sulfite ion, 45t, 391t
Sulfur, *28, 37*
  allotropic forms, *602, 603*
  extraction, Frasch process, 599, *600*
  liquid, properties of, 603
  molecular structure, 603
  monoclinic, 602, *602*
  oxides, 608t
  oxoacids, 611t
  oxoanions of, 45t
  physical properties, 597t
  plastic, *603*
  reaction with Group 1, Group 2, metals, 581t
  reaction with oxygen, 350
  safety matches, 602
Sulfur dioxide
  air pollution, 387
  conversion to sulfur trioxide, 387
  preparation from sulfur, 350, 617
  resonance forms, 185
Sulfur hexafluoride, *194, 198,* 205
Sulfur trioxide
  air pollution, 387
  preparation from sulfur dioxide, 387, 617
  reaction with water, 388
  resonance structures, 187
Sulfuric acid, 92, 92t, 617, 618
  air pollution, 387, 388, 429
  dehydrating agent, *618*
  lead storage battery, *537*, 538
  oxidizing agent, 618
  preparation, contact process, 617
  reaction with copper, 618

  reaction with limestone, 430
  reaction with sucrose, *618*
  reaction with water, 618
  structure, 612
Sulfurous acid, 47, 611t, 613
Sulfuryl chloride, 345q
Sun, 547
Superconductors, 585, *585*
Supercritical fluids, 253
Superoxide A compound containing the O_2^- ion, 583
Superphosphate of lime, 618
Supersaturated solution A solution containing more solute than allowed at equilibrium; unstable to addition of solute, 289, *289*
Surroundings Everything outside the system being studied, 216, 490
Symbol A one- or two-letter abbreviation for the name of an element, 4
  nuclear, 33
System The sample of matter under observation, 216
Système Internationale (SI), 10, A.1

T

T-shaped molecule, *198*
Talc, 264
Tartaric acid, 109q, 405q
Technetium-99, 552t
Temperature, 9, 10
  changes in, equilibrium system and, 370, 371
  critical, 252, 253, 253t
  effect on entropy, 487, *487*
  effect on free energy change, 497–499
  effect on reaction rate, 328–330
  effect on solubility, 288–290
  effect on vapor pressure, 249, *249*, 250
  effect on yield at equilibrium, 372
  phase diagram for pure substances, 254–256
  relation to *u*, for gases, 130
  relation to *V*, for gases, 118, *118*
  scales, conversions between, 9, 10
Terephthalic acid, 273
Terminal atom An atom at one end of a molecule in contrast to the central atom, 183
Termolecular step, 334
Tertiary amine, 105
Tetraethyl lead, 332, 631
Tetrahedral Having the symmetry of a regular tetrahedron. In a tetrahedron, a central atom is surrounded by four other atoms at the corners; the bond angles are 109.5°; 193
  complex, 443, *443*
Tetrathionate ion, 110q
Thallium-201, 552t
Thallium-204, 570q

Thatcher, Margaret, 271

Theoretical yield The amount of product obtained from the complete conversion of the limiting reactant, 70–72

Thermal conductivity The ability to conduct heat, 266

Thermite reaction, 230, *230*

Thermochemical equation A chemical equation in which the value of ΔH is specified, 223–228

Thermochemistry, 215–245
    rules of, 225–227

Thermodynamic data, A.3–A.4

Thermodynamics The study of heat and work effects in physical and chemical changes, 235–237, 482–512
    First Law, 235
    Second law, 490
    Third law, 487

Thermometer, 9, *9*

Thiocyanate ion, 452

Thiosulfate ion, 110q

Third law of thermodynamics A natural law that states that the entropy of a perfectly ordered, pure crystalline solid is 0 at 0 K, 487

Thomsen, J., 484

Thomson, J. J., 31, *31*

Three Mile Island, 563

Thymine, 212q

Thyroid gland, 552

Tin
    Allotropes, 509q

Tin(IV), 472t, 473

Tin(IV) sulfide, *473*

Tincture of iodine, 599

Titan rocket motor, *67*

Titanic, 6

Titanium(III) complexes, 459

Titration A process in which one reagent is added to another with which it reacts; an indicator is used to determine the point at which equivalent quantities of the two reagents have been added, 95
    acid-base, 95, *95, 96*
    redox, 104, *104*
    strong acid–strong base, 423, *423*, 424, *424*, 428t
    weak acid–strong base, 425, *425*, 426, 428t
    weak base–strong acid, 426, 427, *427*, 428t

Titration curve, 423, 425, 427

Toluene, 634

Ton, metric, 9

Tooth decay, 475

Toothpaste, 475

Torr, 116

Torricelli, Evangelista, 116

Trace elements, essential, 39, 591t

Trans isomer A geometric isomer in which two identical groups are as far apart as possible, e.g.,

    a   b
     \ /
      C
     / \
    b   a

    complex ion, 444, 445
    organic, 641, 642

Transition metal Any metal in the central groups (3–12) of the 4th, 5th and 6th periods of the periodic table, 36
    chemistry of, 584–590
    ease of oxidation, 566t
    electron configuration (atoms of), *159*, 160
    oxoanions of, 589, 590
    reaction with acids, 585–587
    reaction with oxygen, 584, 585

Transition metal ions, 43, *87*
    charges of, 87
    complex ions, 437–459
    electron configuration, 165, 166
    equilibrium between, 587, 588
    naming of, 44
    redox reactions, 584–590
    reduction, ease of, 587t

Transition state model, 326, 327

Translational energy, 129

Transmutation, 34

Triangle, equilateral, 193

Triangular bipyramid A solid with five vertices; may be regarded as two pyramids fused through a base that is an equilateral triangle, 193, 194

Triangular planar The geometry of an AX_3 molecule where A is at the center of the equilateral triangle formed by three X atoms, 193

Triangular pyramid The geometry of an AX_3E molecule in which atom A lies directly above the center of an equilateral triangle formed by three X atoms, 195, 196t

Trichloroacetic acid, 409q

Tricosene (*cis*-9-), 642

Trinitrotoluene (TNT), 570q

Triple bond Three electron pairs shared between two bonded atoms, 183

Triple point The temperature and pressure at which the solid, liquid, and vapor phase of a substance can co-exist in equilibrium, 254

Triprotic acid, 396, 397t

Tris, 553, A.14

Tritium, 565, 570q

Troposphere, 338

Turin, Shroud of, 556

Turquoise, 78q

U

Ultraviolet radiation Light with a wavelength less than 400 nm but greater than 10 nm, 19, 146, *146*

Uncertainty
    in addition and subtraction, 13
    in multiplication and division, 12
    in significant figures, 11, 12

Unimolecular step, 334

Unit(s),
    concentration, 281–286
    conversion of, 13–15
    electrical, 532t
    energy, 218
    entropy, 489, 494
    force, A.1
    frequency, 145
    length, 8
    mass, 9
    pressure, 116, 117
    radiation, 566
    radioactivity, 555
    SI, 10, A.1
    temperature, 9, 10
    volume, 8

Unit cell The smallest unit of a crystal that, repeated again and again, generates the whole crystal, 267
    number of atoms, 268, 268t, 269

Universal indicator, 389, *389*

Unpaired electron, 188, 189

Unsaturated fat, 645

Unsaturated hydrocarbon, 631–633

Unsaturated solution A solution that contains less solute than the equilibrium (saturated) value, 288

Unshared electron pair A pair of electrons that "belongs" to a single atom and is not involved in bonding
    effect on geometry, 182
    hybridization, 205
    space occupied, 195, 196

Uranium-235, 561–563

Uranium-238, 561

Uranium dioxide, 562

Uranium hexafluoride, 132

Uranium isotopes, separation of, 132

Urea, 132

Urey, Harold, 184

Uric acid, 409q

V

Valence bond model Model of the electronic structure of molecules in which electrons are assigned to orbitals, pure or hybridized, of individual atoms, 203

Valence electron An electron in the outermost principal energy level. For a main-group element, the number of valence electrons is given by the last digit of the group number, 182
counting, 183
group number in periodic table, 182t

Vanadium(V) oxide, 617
Vanillin, 301
Van't Hoff, Jacobus, 371
Van't Hoff equation, 371
Van't Hoff factor, 299
Vapor-liquid equilibrium, 248–253

Vapor pressure The pressure exerted by a vapor when it is in equilibrium with a liquid, 248–251
dependence of, on temperature, 249–251
independence of volume, 249, 250
movement of water, 295, *295*
of water, A.2

Vapor pressure lowering, 291, 292

Vaporization
heat of, 227t

Venus, atmosphere, 138q
Vinegar, 95, *95*, 303q
Visible spectrum, 19, 146, *146*
Vitamin B-6, *288*
Vitamin B-12, 454
Vitamin C, 66, 405
Vitamin D2, *288*

Vitriol
blue, 74t
green, 74t
red, 74t

Volt A unit of electric potential: 1 V = 1 J/C, 532

Voltage
concentration effect on, 528–530
standard, 518–525

Voltaic cell A device in which a spontaneous reaction produces electrical energy
calculation of voltage of, 515–518
cell notation, 516–517
commercial, 536–538
salt bridge cells, 516–518

Volume(s), 8
length, mass units, 14t
molar, of gas, 134, *134*, 135
of gas involved in reaction 123, 124
of gas particles, 134

Volumetric flask, *85*

VSEPR model Valence Shell Electron Pair Repulsion model, used to predict molecular geometry; states that electron pairs around a central atom

tend to be as far apart as possible, 192

W

w. *See* work
Warts, 409q
Washing soda, 74q

Water
boiling point, freezing point constants of, 294
bond angle, 195
Brønsted-Lowry acid, 384
Brønsted-Lowry base, 384
complexes with cations, 390, *390*, 439, 440
conductivity of, 86
electrolysis of, *124*, 484
heat of fusion of, 227t
heat of vaporization of, 227t
infrared radiation and, 504
ligand in complex ions, 440
liquid in gaseous equilibrium, 248, 249, 357, 357t, 359
molecular geometry of, *39*, 195
organic solutes in, 287
oxidation to oxygen, 534
phase diagram for, 254, *254*
polarity of, 201, *202*
reaction with alkali, alkaline earth metals, 582, *582*
reaction with chlorine, 598
reaction with lithium, *522*
reaction with magnesium, 582
reaction with metal hydrides, 582
reaction with metal oxides, 583
reaction with nonmetal oxides, 610, 611
reaction with permanganate ion, 590
reaction with peroxides, 583
reaction with superoxides, 583
reaction with weak bases, 398
reduction to hydrogen, 533
specific heat of, 219t
vapor pressure of, 126, 127, A.2

Water gas, 379q

Water ion product constant The product $[H^+] \times [OH^-] = 1.0 \times 10^{-14}$ at 25°C, 384, 385

Water softening, 264
Watras, Stanley, 566
Watt, 532t

Wave
amplitude of, 145, *145*
frequency, 145, *145*

Wave equation, 151
Wave function, 151

Wavelength A characteristic property of a wave related to its color and equal to the length of a full wave, 145, *145*

Weak acid An acid that is only partially dissociated to H^+ ions in water, 92

cations, 390
equilibrium constant, 391t, 397t, A.5
molecular, 389, 390
reaction with strong base, 94, 94t, 423, *423*, 424
titration, strong base, *95*, 425, *425*, 426

Weak base A base that is only partially dissociated to form OH^- ions in water, 93
anion, 398
equilibrium constant, 391t, A.5
molecular, 39
reaction with strong acid, 94, 94t, 425, *425*, 426
titration, strong acid, 426, 427, *427*

Weak electrolyte A species that, in water solution, forms an equilibrium mixture of molecules and ions, 93

Weak field ligand, 452
Weight, 9
Weight loss, exercise, 239
Werner, Alfred, 446

White phosphorus
chemical reactivity of, 601, *601*
equilibrium of, with red, 602
physical properties of, 597t
reaction of, with oxygen, 601, *601*
solubility of, 601
toxicity of, 602

Wine (Twenty Mile Vineyard), *637*
Wine cooler, 80q
Wintergreen, oil of, 78q
Wood alcohol. *See* methyl alcohol

Work Any form of energy except heat exchanged between system and surroundings; includes expansion work and electrical work, 236
electrical, 236, 532
expansion, 236

X

X-ray, *146*
X-ray diffraction, 271
Xenon, 209, 277q, A.8
Xenon-133, 552t
Xenon difluoride, 209
Xenon hexafluoride, 209
Xenon tetrafluoride, 209, *209*, 381q
Xenon trioxide, 209

Y

Yield
experimental, 73
percent, 73
theoretical, 70–72
limiting reactant and, 70–73
Yttrium oxide, 161

Z

Z (atomic number), 33
Zeolite, 264

Zero-order reaction A reaction whose
    rate is independent of reactant con-
    centration, 321, 322, 322t
Zeros, significant, 11
Zinc
    corrosion of, in air, 513
    dry cell, 536, *536*, 537
    gold metallurgy, 580
    oxidation to zinc(II), 586t

reaction with acid, 98, 517
reaction with nitric acid, 616
reaction with oxygen, 584t
sulfide ore, 578
Zinc-copper(II) voltaic cell, 515, *515*, 516
Zinc-hydrogen ion voltaic cell, 517, *517*
Zinc(II) carbonate, 469
Zinc(II) hydroxide, 468–470
Zinc(II) ion

Brønsted-Lowry acid, 390
complexes, 442t, *443*
hydration of, 390
nutrition, 591t
qualitative analysis, 472t, 473
treatment of acne, 590
Zinc(II) sulfide, 470, *473*
Zirconium, 571q